A Member of the International Code Family™

INTERNATIONAL CODE COUNCIL®

INTERNATIONAL RESIDENTIAL CODE®

FOR ONE- AND TWO-FAMILY DWELLINGS

COMMENTARY - Vol. II

2003

First Printing

Publication Date: September, 2004

ISBN # 1-58001-134-9

PRINTED IN THE U.S.A.

PREFACE

The principal purpose of the *Commentary* is to provide a basic volume of knowledge and facts relating to building construction as it pertains to the regulations set forth in the 2003 *International Residential Code*®. The person who is serious about effectively designing, constructing, and regulating buildings and structures will find the *Commentary* to be a reliable data source and reference for almost all components of the built environment.

As a follow-up to the *International Residential Code*, we offer a companion document, the International Residential Code *Commentary—Volume II.* This volume covers Chapters 12 through 43 and Appendices A through L of the 2003 *International Residential Code.* (Volume I covers Chapters 1 through 11.) The basic appeal of the *Commentary* is that it provides in a small package and at reasonable cost thorough coverage of many issues likely to be dealt with when using the *International Residential Code.* It then supplements that coverage with historical and technical background. Reference lists, information sources, and bibliographies are also included.

Throughout the creation of this volume, strenuous effort has been made to keep the vast quantity of material accessible and its method of presentation useful. With a comprehensive yet concise summary of each section, the *Commentary* provides a convenient reference for regulations applicable to the construction of buildings and structures. In the chapters that follow, discussions focus on the full meaning and implications of the code text. Guidelines suggest the most effective method of application and the consequences of not adhering to the code text. Illustrations are provided to aid understanding, but they do not necessarily illustrate the only methods of achieving code compliance.

The format of the *Commentary* includes the full text of each IRC section, table, and figure in the code, followed immediately by the commentary for that text. At the time of printing, the *Commentary* reflects the most up-to-date text of the 2003 *International Residential Code.* Each chapter's narrative includes a section on "General Comments" and "Purpose," and each section's narrative usually includes a discussion about why the requirement commands the conditions set forth. All code text is reproduced as it appears in the *International Residential Code*, and all commentary is indented below the code text, beginning with the symbol ❖. All code figures and tables are reproduced as they appear in the IRC. Commentary figures and tables are identified in the text by the word "Commentary" (as in "see Commentary Figure P2803.4"), and each has a full border.

The *Commentary* is to be used in conjunction with the *International Residential Code* and not as a substitute for the code. **The *Commentary* is advisory only;** the code official alone possesses the authority and responsibility for interpreting the code.

Comments and recommendations are encouraged, for through your input we can improve future editions. Please direct your comments to the Codes and Standards Development Department at the Chicago District Office.

TABLE OF CONTENTS

Part X—Appendices

Part V — Mechanical

Chapter 12: Mechanical Administration

General Comments

Chapter 12 provides regulations for the administration of the mechanical provisions of the *International Residential Code*®. Though this may be the smallest chapter in the IRC, it is very important in that it defines the application of the mechanical provisions to both existing and new construction. It also relates this chapter to the administrative provisions in Chapter 1.

Section M1201 addresses this set of mechanical regulations' relationship with Chapter 1 and the validity of the standards that are referred to. Section M1202 provides the applicability to existing mechanical systems. While the IRC® is mainly a code that deals with new systems, this section indicates that some existing situations may fall under the requirements in the chapter.

Purpose

A set of mechanical regulations is intended to be adopted as a legally enforceable document that can safeguard health, safety, property, and public welfare. Such regulations cannot be effective without adequate provisions for their administration and enforcement. The official charged with the administration and enforcement of mechanical regulations has a great responsibility, and with this responsibility goes authority. No matter how detailed the mechanical regulations may be, the building official must, to some extent, exercise judgement in determining code compliance. She or he has the responsibility to establish that the homes in which the citizens of the community reside are designed and constructed to be reasonably free from hazards associated with the presence and use of mechanical equipment, appliances, and systems.

SECTION M1201 GENERAL

M1201.1 Scope. The provisions of Chapters 12 through 24 shall regulate the design, installation, maintenance, alteration and inspection of mechanical systems that are permanently installed and utilized to provide control of environmental conditions within buildings. These chapters shall also regulate those mechanical systems, system components, equipment and appliances specifically addressed in this code.

❖ This section lists the chapters in the code that regulate mechanical systems. It indicates that the design, installation, and maintenance of mechanical equipment used to control the environmental conditions within the building are regulated by these chapters. It also states that other mechanical systems specifically addressed within these chapters are so regulated. Other provisions in this code reference the *International Mechanical Code*® and the *International Fuel Gas Code*®. This makes virtually all mechanical systems and equipment within a dwelling regulated in some form or another.

M1201.2 Application. In addition to the general administration requirements of Chapter 1, the administrative provisions of this chapter shall also apply to the mechanical requirements of Chapters 13 through 24.

❖ This section makes reference to the administrative requirements of Chapter 1 to include those administrative provisions and make them applicable to the mechanical chapters.

[EB] SECTION M1202 EXISTING MECHANICAL SYSTEMS

M1202.1 Additions, alterations or repairs. Additions, alterations, renovations or repairs to a mechanical system shall conform to that required for a new mechanical system without requiring the existing mechanical system to comply with all of the requirements of this code. Additions, alterations or repairs shall not cause an existing mechanical system to become unsafe, hazardous or overloaded. Minor additions, alterations or repairs to existing mechanical systems shall meet the provisions for new construction, unless such work is done in the same manner and arrangement as was in the existing system, is not hazardous, and is approved.

❖ Major alterations or additions to existing mechanical systems must comply with the provisions of this code. However, the code does not require existing systems to be upgraded to comply with the new requirements of this code unless the alteration or addition renders the existing system unsafe, overloaded, or hazardous.

Minor additions, alterations, or repairs to existing mechanical systems may be made following the same manner and arrangement as was found for the existing systems, as long as the work does not render the existing and new system unsafe, hazardous, or overloaded. The building official has to make a judgement call on the extent of the addition, alteration, or repair and determine whether the work is "minor," thereby allowing the use of the provisions of the old code, or that it is "major," thereby requiring the new work to be performed in compliance with the new code. Only when the new work makes the existing system overloaded, unsafe, or hazardous does the building official need to require the existing system to be brought up to the new provisions of the code.

M1202.2 Existing installations. Except as otherwise provided for in this code, a provision in this code shall not require the removal, alteration or abandonment of, nor prevent the continued utilization and maintenance of, an existing mechanical system lawfully in existence at the time of the adoption of this code.

❖ An existing mechanical system is generally "grandfathered" with code adoption, provided the system meets a minimum level of safety. Frequently the criteria for determining this level of safety are the regulations (or code) under which the existing building was originally constructed. If there are no previous code criteria that apply, the building official is to apply those provisions of the code that are reasonably applicable to existing buildings. Provisions dealing with hazard abatement in existing buildings and provisions dealing with maintenance, as contained in the property maintenance codes and fire prevention codes, dictate a specific level of safety.

M1202.3 Maintenance. Mechanical systems, both existing and new, and parts thereof shall be maintained in proper operating condition in accordance with the original design and in a safe and sanitary condition. Devices or safeguards that are required by this code shall be maintained in compliance with the code edition under which installed. The owner or the owner's designated agent shall be responsible for maintenance of the mechanical systems. To determine compliance with this provision, the building official shall have the authority to require a mechanical system to be reinspected.

❖ This section allows the continued use of an existing mechanical system if it was originally lawfully permitted and installed. Periodic adoption of new codes or new provisions that may affect an existing system does not require the system to be upgraded to comply with the new provisions. However, this is predicated on the system being maintained by the owner in proper operating condition, in accordance with the manufacturer's requirements. The building official has the authority, as provided for in this section, to require the inspection of a mechanical system to determine whether work was performed to maintain the equipment in a safe operating condition. If the system becomes hazardous or unsafe, the building official has the authority to require repair, modification, or even replacement to render the installation safe.

Bibliography

The following resource materials are referenced in this chapter or are relevant to the subject matter addressed in this chapter.

ICC EC-2003, *ICC Electrical Code.* Falls Church, VA: International Code Council, 2003.

IFGC-2003, *International Fuel Gas Code.* Falls Church, VA: International Code Council, 2003.

IMC-2003, *International Mechanical Code.* Falls Church, VA: International Code Council, 2003.

Legal Aspects of Code Administration. Country Club Hills, IL: Building Officials and Code Administrators International, 1984; Whittier, CA: International Conference of Building Officials, 1984; Birmingham, AL: Southern Building Code Congress International, 1984.

"Permit Fee Schedule." *The Code Official*. Country Club Hills, IL: Building Officials and Code Administrators International. Published semiannually in May/June and November/December.

Readings in Code Administration, Volume 1: History/Philosophy/Law. Country Club Hills, IL: Building Officials and Code Administrators International, 1974.

Chapter 13
General Mechanical System Requirements

General Comments

This chapter contains the provisions that apply to various types of mechanical appliances. The approval of appliances and their proper installation is the main theme. Section M1301 is the scope of the chapter and addresses its relationship with the *International Mechanical Code* and the *International Fuel Gas Code*. Section M1302 indicates that all mechanical appliances must be listed and labeled by an approved agency. Section M1303 addresses the information that is needed on the labels. Section M1304 discusses the proper design of appliances considering the appliances' type of fuel and the geographical location of the installed appliance. Section M1305 addresses access to installed appliances for servicing and potential replacement. Section M1306 contains the allowance for reduced clearances between appliances and combustible construction. Section M1307 contains the criteria for the safe installation of appliances. Section M1308 is a cross-reference to the proper sections in the building portion of the *International Residential Code* for the drilling and notching of structural members of the building.

Purpose

This chapter provides requirements for the safe and proper installation of mechanical equipment and appliances.

SECTION M1301 GENERAL

M1301.1 Scope. The provisions of this chapter shall govern the installation of mechanical systems not specifically covered in other chapters applicable to mechanical systems. Installations of mechanical appliances, equipment and systems not addressed by this code shall comply with the applicable provisions of the *International Mechanical Code* and the *International Fuel Gas Code.*

❖ This section provides general requirements for mechanical systems not specifically covered in other chapters of the *International Residential Code*. In addition, it refers to the *International Mechanical Code* and the *International Fuel Gas Code* for regulations governing equipment not addressed by the IRC.

M1301.1.1 Flood-resistant installation. In areas prone to flooding as established by Table R301.2(1), mechanical appliances, equipment and systems shall be located or installed in accordance with Section R323.1.5.

❖ The local jurisdiction must fill in Table R301.2(1) upon adoption of this code, including the FLOOD HAZARDS information. Mechanical appliances, equipment, and systems that are located in flood hazard areas must be installed above the design flood elevation or must be designed and installed to prevent the entrance of water into the components and to resist the forces of the flood waters on the components. See the commentary for Section R323.1.5.

SECTION M1302 APPROVAL

M1302.1 General. All appliances shall be listed and bear the label of an approved agency unless otherwise approved in accordance with Section R104.11.

❖ Mechanical appliances must be listed and labeled by an approved agency to show that they comply with applicable national standards. The code requires listing and labeling for appliances such as boilers, furnaces, space heaters, cooking appliances, and clothes dryers. The code also requires listing for system components. The label is the primary, if not the only, assurance to the installer, the inspector, and the end user that a similar appliance has been tested and evaluated by an approved agency and performed safely and efficiently when installed and operated in accordance with its listing.

The label is part of the information that the building official is to consider in the approval of appliances. The only exception to the labeling requirement occurs when a building official approves a specific appliance in accordance with the authority granted in Section R104.11.

The building official should exercise caution when considering the approval of unlisted and unlabeled appliances. Approval of unlabeled appliances must be based on documentation that demonstrates compliance with applicable standards. Where no product standards exist, documentation must be provided that the appliance is appropriate for the intended use and will provide the same level of performance as would be provided by a listed and labeled appliance. One fundamental principle of the code is the reliance on the listing and labeling process to assure appliance performance; approvals granted in accordance with Section R104.11 must be justified with supporting documentation. To the building official, the installer, and the end-user, very little is known about the performance of an appliance that is not tested and built to an appliance standard.

SECTION M1303 LABELING OF APPLIANCES

M1303.1 Label information. A permanent factory-applied nameplate(s) shall be affixed to appliances on which shall appear, in legible lettering, the manufacturer's name or trademark, the model number, serial number, and the seal or mark of the testing agency. A label shall also include the following

1. Electrical appliances. Electrical rating in volts, amperes and motor phase; identification of individual electrical components in volts, amperes or watts, and motor phase; and in Btu/h (W) output and required clearances.
2. Absorption units. Hourly rating in Btu/h (W), minimum hourly rating for units having step or automatic modulating controls, type of fuel, type of refrigerant, cooling capacity in Btu/h (W), and required clearances.
3. Fuel-burning units. Hourly rating in Btu/h (W), type of fuel approved for use with the appliance and required clearances.
4. Electric comfort heating appliances. Name and trademark of the manufacturer; the model number or equivalent; the electric rating in volts, ampacity and phase; Btu/h (W) output rating; individual marking for each electrical component in amperes or watts, volts and phase; required clearances from combustibles; and a seal indicating approval of the appliance by an approved agency.

❖ This section requires that the label be a permanent nameplate. In general, labels other than metal tags or plates usually consist of material that is similar in appearance to a decal, and the label, its adhesive, and the printed information are all durable and water resistant. Because of the important information on a label, the label must be permanent—not susceptible to damage and legible for the life of the appliance. The standards appliances are tested in accordance with usually specify the required label material, the method of attachment and the required label information. The code requires that the label be affixed permanently and prominently on the appliance or equipment and specifies the information that must appear on the label. The manufacturer may be required by the relevant standard or may voluntarily provide additional information on the label. See Commentary Figure M1303.1.

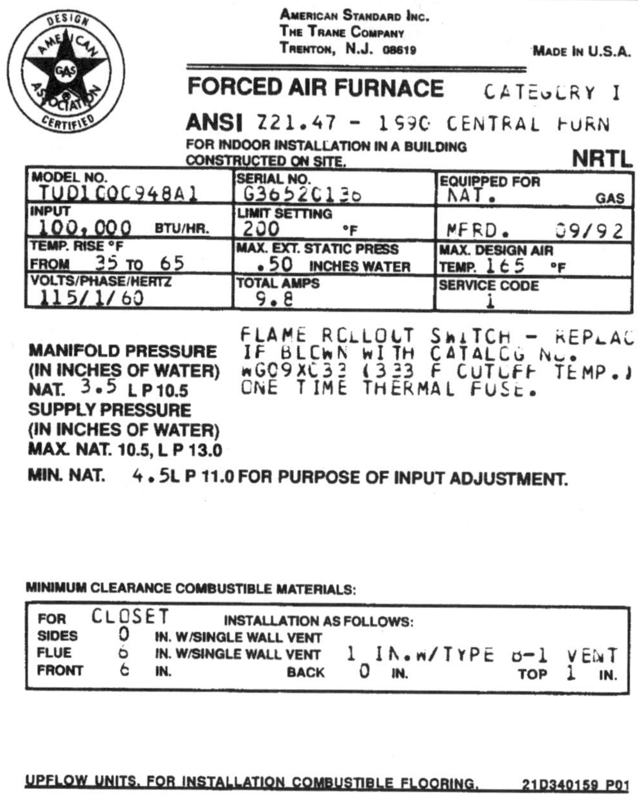

Figure M1303.1
TYPICAL LABEL FOR A CATEGORY I GAS-FIRED FURNACE
(Courtesy of The Trane Company, an American Standard Company)

SECTION M1304 TYPE OF FUEL

M1304.1 Fuel types. Fuel-fired appliances shall be designed for use with the type of fuel to which they will be connected and the altitude at which they are installed. Appliances that comprise parts of the building mechanical system shall not be converted for the usage of a different fuel, except where approved and converted in accordance with the manufacturer's instructions. The fuel input rate shall not be increased or decreased beyond the limit rating for the altitude at which the appliance is installed.

❖ An element of information used for the approval of appliances is the label, which assures that the appliance has been tested in accordance with a valid standard and performed acceptably when installed and operated in accordance with the appliance listing. Manufacturers usually design mechanical appliances to operate on a specific type of fuel. Thus, the fuel used in the appliance test must be the type of fuel specified by the manufacturer on the label. When an appliance is converted to a different type of fuel, the original label that appears on the appliance is no longer valid. Since the original approval of the appliance was based in part on the label, the appliance is no longer approved for use.

Field conversions will more likely allow for the safe operation of the appliance if, as required, the conversion is approved by the building official and done in accordance with the manufacturer's installation instructions. Fuel conversions that are not performed correctly can adversely affect the performance of burners, the venting of combustion gases, and the proper clearance to combustibles.

Before a fuel conversion is performed, the manufacturer must be contacted for installation instructions outlining the procedures to follow for proper operation of the appliance. In most cases, conversion kits from the manufacturer are available along with the installation instructions. Once a conversion has been completed, a supplemental label must be installed to update the information contained on the original label, thereby alerting any service personnel of the modifications that have been made.

All fuel-fired appliances are designed to operate with a maximum and minimum British thermal units per hour (Btu/h) input capacity. This capacity is field adjusted to suit the elevation because of the change in air density at different elevations. Alteration of Btu/h input beyond the allowable limits can result in hazardous overfiring or underfiring. Either condition can cause operation problems that include overheating, vent failure, corrosion, poor draft, and poor combustion.

SECTION M1305 APPLIANCE ACCESS

M1305.1 Appliance access for inspection service, repair and replacement. Appliances shall be accessible for inspection, service, repair and replacement without removing permanent construction. Thirty inches (762 mm) of working space shall be provided in front of the control side to service an appliance. Room heaters shall be permitted to be installed with at least an 18-inch (457 mm) working space. A platform shall not be required for room heaters.

❖ Because mechanical equipment and appliances require routine maintenance, repair, and possible replacement, access is required. Additionally, manufacturer's installation instructions usually contain access recommendations or requirements. As a result, the provisions stated herein supplement the manufacturer's installation instructions.

The provisions of this section specify that access must be provided to components that require observation, inspection, adjustment, servicing, repair, or replacement. Access is also necessary for conducting operating procedures such as startup or shutdown.

The code states that "accessible" means "access that requires the removal of an access panel of similar removable obstruction." An appliance or piece of equipment is not accessible if any portion of the structure's permanent finish materials, such as drywall, plaster, paneling, built-in furniture or cabinets, or any other similar permanently affixed building component, must be removed before access is achieved.

The intent is to provide access to all components such as controls, gauges, burners, filters, blowers, or motors that require observation, inspection, adjustment, servicing, repair, or replacement.

M1305.1.1 Central furnaces. Central furnaces within compartments or alcoves shall have a minimum working space clearance of 3 inches (76 mm) along the sides, back and top with a total width of the enclosing space being at least 12 inches (305 mm) wider than the furnace. Furnaces having a firebox open to the atmosphere shall have at least a 6-inch (152 mm) working space along the front combustion chamber side. Combustion air openings at the rear or side of the compartment shall comply with the requirements of Chapter 17.

Exception: This section shall not apply to replacement appliances installed in existing compartments and alcoves where the working space clearances are in accordance with the equipment or appliance manufacturer's installation instructions.

❖ Central furnaces installed in compartments or alcoves must have clearances from the enclosure so the furnace can be removed, maintained, or repaired as necessary. The minimum clearances specified in the code apply even though the manufacturer's instructions might permit a lesser clearance. Clearances provide access, ventilation, cooling of the appliance and equipment, and protection for surrounding combustibles. The front (fire box) clearance helps protect combustibles against flame roll-out and allows free movement of combustion air. The exception exempts replacement appliances and equipment installed in existing compartments or alcoves provided that the installation complies with the manufacturer's instructions.

M1305.1.2 Appliances in rooms. Appliances installed in a compartment, alcove, basement or similar space shall be accessed by an opening or door and an unobstructed passageway measuring not less than 24 inches (610 mm) wide and large enough to allow removal of the largest appliance in the space, provided that a level service space of not less than 30 inches (762 mm) deep and the height of the appliance, but not less than 30 inches (762 mm), is present at the front or service side of the appliance with the door open.

❖ This section specifies an access opening and passageway to afford service personnel reasonable access to appliances and to allow for the passage of system components. Quite often appliances such as furnaces, boilers, and water heaters are installed in spaces with little or no forethought about future access for maintenance or replacement.

M1305.1.3 Appliances in attics. Attics containing appliances requiring access shall be provided with an opening and a clear and unobstructed passageway large enough to allow removal of the largest appliance, but not less than 30 inches (762 mm) high and 22 inches (559 mm) wide and not more than 20 feet (6096 mm) in length when measured along the centerline of the passageway from the opening to the appliance. The passageway shall have continuous solid flooring in accordance with Chapter 5 not less than 24 inches (610 mm) wide. A level service space at least 30 inches (762 mm) deep and 30 inches (762 mm) wide shall be present along all sides of the appliance where access is required. The clear access opening dimensions shall be a minimum of 20 inches by 30 inches (508 mm by 762 mm), where such dimensions are large enough to allow removal of the largest appliance.

Exception: The passageway and level service space are not required where the appliance is capable of being serviced and removed through the required opening.

❖ There is not always sufficient room for mechanical equipment and appliances to be installed in spaces such as basements, alcoves, utility rooms, and furnace rooms. In an effort to save floor space or simplify an installation, people often install appliances and mechanical equipment on roofs, in attics, or in similar remote locations. Access to appliances and equipment could be difficult because of roof slope, stone roof ballast, or the lack of a walking surface, such as might occur in an attic or similar space with exposed ceiling joists. This section requires a suitable access opening, passageway, and workspace that will allow reasonably easy access without endangering the service person. See Commentary Figure M1305.1.3. The longer the attic passageway, the more the service person will be exposed to extreme temperatures and the risk of injury.

M1305.1.3.1 Electrical requirements. A lighting fixture controlled by a switch located at the required passageway opening and a receptacle outlet shall be provided at or near the appliance location in accordance with Chapter 38.

❖ A lighting outlet and receptacle outlet encourage and facilitate appliance maintenance. The receptacle will accommodate power tools, drop lights, and diagnostic instruments. Also, these provisions negate the need for extension cords, which can be hazardous to service personnel.

M1305.1.4 Appliances under floors. Underfloor spaces containing appliances requiring access shall be provided with an unobstructed passageway large enough to remove the largest appliance, but not less than 30 inches (762 mm) high and 22 inches (559 mm) wide, nor more than 20 feet (6096 mm) in length when measured along the centerline of the passageway from the opening to the appliance. A level service space at least 30 inches (762 mm) deep and 30 inches (762 mm) wide shall be present at the front or service side of the appliance. If the depth of the passageway or the service space exceeds 12 inches (305 mm) below the adjoining grade, the walls of the passageway shall be lined with concrete or masonry extending 4 inches (102 mm) above the adjoining grade in accordance with Chapter 4. The rough-framed access opening dimensions shall be a minimum of 22 inches by 30 inches (559 mm by 762 mm), where the dimensions are large enough to remove the largest appliance.

Exception: The passageway is not required where the level service space is present when the access is open, and the appliance is capable of being serviced and removed through the required opening.

❖ This section, which applies to crawl spaces, has concepts similar to those of Section M1305.1.3. The more difficult the access to appliances and equipment is, the less likely that the appliance or equipment will be inspected and serviced on a regular basis. Attic and crawl space installations suffer from the "out-of-sight, out-of-mind" syndrome.

M1305.1.4.1 Ground clearance. Appliances supported from the ground shall be level and firmly supported on a concrete slab or other approved material extending above the adjoining ground. Appliances suspended from the floor shall have a clearance of not less than 6 inches (152 mm) from the ground.

❖ This section's provision provides a buffer from the corrosive effects of an appliance's contact with the ground. If supported on the ground, the appliance is to rest on a material that will be a barrier between the appliance and the ground. Concrete is the material that is prescribed, but other approved materials could be used if they provide the same level of protection as the concrete. If the appliance is suspended from the floor assembly above the ground, then a minimum separation of 6 inches is called out.

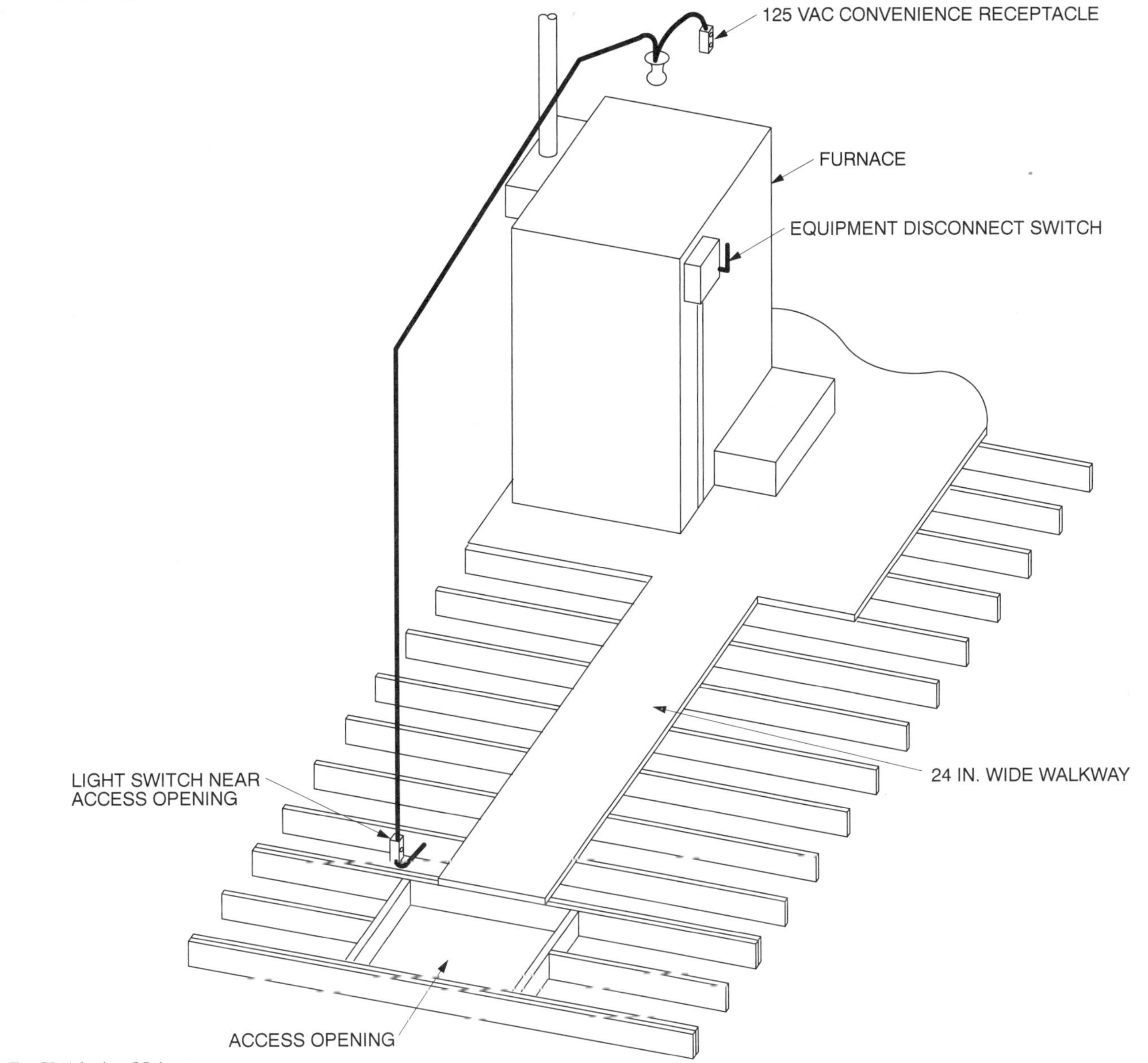

For SI: 1 inch = 25.4 mm.

Figure M1305.1.3
REMOTE LOCATION REQUIREMENTS (ATTIC INSTALLATION)

M1305.1.4.2 Excavations. Excavations for appliance installations shall extend to a depth of 6 inches (152 mm) below the appliance and 12 inches (305 mm) on all sides, except that the control side shall have a clearance of 30 inches (762 mm).

❖ This section allows for appliances to be located below the level of the ground surface in underfloor areas. Again, the concern is to provide a separation from the ground itself. Minimum clearances are provided in this section. The 30-inch-deep (762 mm) workspace is required for adequate access for servicing the appliance.

M1305.1.4.3 Electrical requirements. A lighting fixture controlled by a switch located at the required passageway opening and a receptacle outlet shall be provided at or near the appliance location in accordance with Chapter 38.

❖ This is the same requirement as found in Section M1305.1.3.1. Lighting and a power supply are required for the service of mechanical appliances.

SECTION M1306
CLEARANCES FROM COMBUSTIBLE CONSTRUCTION

M1306.1 Appliance clearance. Appliances shall be installed with the clearances from unprotected combustible materials as indicated on the appliance label and in the manufacturer's installation instructions.

❖ This section requires appliances to be installed with clearances from unprotected combustibles as indicated on the label for the listed appliance and the manufacturer's installation instructions. It does not include provisions or guidelines for the installations of unlisted appliances.

M1306.2 Clearance reduction. Reduction of clearances shall be in accordance with the appliance manufacturer's instructions and Table M1306.2. Forms of protection with ventilated air space shall conform to the following requirements:

1. Not less than 1-inch (25.4 mm) air space shall be provided between the protection and combustible wall surface.
2. Air circulation shall be provided by having edges of the wall protection open at least 1 inch (25.4 mm).
3. If the wall protection is mounted on a single flat wall away from corners, air circulation shall be provided by having the bottom and top edges, or the side and top edges open at least 1 inch (25.4 mm).
4. Wall protection covering two walls in a corner shall be open at the bottom and top edges at least 1 inch (25.4 mm).

❖ If it is not possible to install an appliance with the required clearances from combustibles, this section allows the clearances to be reduced in accordance with Table M1306.2, provided that the appliance manufacturer's instructions allow the reductions to the extent desired. The table includes several forms of protection, depending on the extent of reduction. The section requires that a 1-inch air space be maintained between the protection and the combustible wall surface to allow the air circulation needed to keep the temperature rise within acceptable limits.

M1306.2.1 Solid fuel appliances. Table M1306.2 shall not be used to reduce the clearance required for solid fuel appliances listed for installation with minimum clearances of 12 inches (305 mm) or less. For appliances listed for installation with minimum clearances greater than 12 inches (305 mm), Table M1306.2 shall not be used to reduce the clearance to less than 12 inches (305 mm).

❖ Because solid-fuel-burning appliances can produce radiant high-intensity heat and have wide variations in heat output, this section poses restrictions on the use of Table M1306.2 for types of solid-fuel-burning appliances that are listed for clearances of 12 inches or less. If the listed clearance is greater than 12 inches, Table M1306.2 is applicable, provided that the reduced clearance is not less than 12 inches. See Commentary Figure M1306.2.1.

SECTION M1307 APPLIANCE INSTALLATION

M1307.1 General. Installation of appliances shall conform to the conditions of their listing and label and the manufacturer's installation instructions. The manufacturer's operating and installation instructions shall remain attached to the appliance.

❖ The appliance label and the manufacturer's installation instructions are extensions of the code and must be enforced by the building official. Where the code provisions are more stringent, the code provisions must be used. The manufacturer's instructions must be available at the job site when the equipment is being inspected.

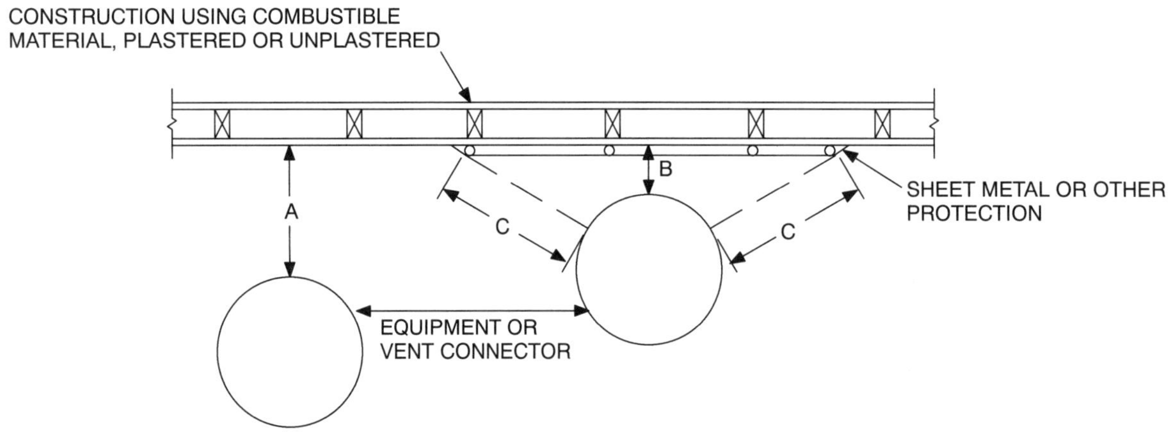

Note: "A" equals the required clearance with no protection. "B" equals the reduced clearance permitted in accordance with Table M1306.2. The protection applied to the construction using combustible material shall extend far enough in each direction to make "C" equal to "A."

FIGURE M1306.1 REDUCED CLEARANCE DIAGRAM

❖ Figure M1306.1 provides a graphic image of the reduced clearances allowed by the IRC. It works in conjunction with Table M1306.2.

TABLE M1306.2
REDUCTION OF CLEARANCES WITH SPECIFIED FORMS OF PROTECTION[a,b,c,d,e,f,g,h,i,j,k]

TYPE OF PROTECTION APPLIED TO AND COVERING ALL SURFACES OF COMBUSTIBLE MATERIAL WITHIN THE DISTANCE SPECIFIED AS THE REQUIRED CLEARANCE WITH NO PROTECTION [SEE FIGURES M1306.1 AND M1306.2]	WHERE THE REQUIRED CLEARANCE WITH NO PROTECTION FROM APPLIANCE, VENT CONNECTOR, OR SINGLE WALL METAL PIPE IS:									
	36 inches		18 inches		12 inches		9 inches		6 inches	
	Allowable clearances with specified protection (Inches)[b]									
	Use column 1 for clearances above an appliance or horizontal connector. Use column 2 for clearances from an appliance, vertical connector and single-wall metal pipe.									
	Above column 1	Sides and rear column 2	Above column 1	Sides and rear column 2	Above column 1	Sides and rear column 2	Above column 1	Sides and rear column 2	Above column 1	Sides and rear column 2
$3^1/_2$-in. thick masonry wall without ventilated air space	—	24	—	12	—	9	—	6	—	5
$^1/_2$-in. insulation board over 1-in. glass fiber or mineral wool batts	24	18	12	9	9	6	6	5	4	3
24 gage sheet metal over 1-in. glass fiber or mineral wool batts reinforced with wire on rear face with ventilated air space	18	12	9	6	6	4	5	3	3	3
$3^1/_2$-in. thick masonry wall with ventilated air space	—	12	—	6	—	6	—	6	—	6
24 gage sheet metal with ventilated air space	18	12	9	6	6	4	5	3	3	2
$^1/_2$-in. thick insulation board with ventilated air space	18	12	9	6	6	4	5	3	3	3
24 gage sheet metal with ventilated air space over 24 gage sheet metal with ventilated air space	18	12	9	6	6	4	5	3	3	3
1-in. glass fiber or mineral wool batts sandwiched between two sheets 24 gage sheet metal with ventilated air space	18	12	9	6	6	4	5	3	3	3

For SI: 1 inch = 25.4 mm, 1 pound per cubic foot = 16.019 kg/m^3, °C = [(°F)-32/1.8], 1 Btu per inch per square foot per hour per °F = 0.0144 (W/m^2 · K).

a. Reduction of clearances from combustible materials shall not interfere with combustion air, draft hood clearance and relief, and accessibility of servicing.

b. Clearances shall be measured from the surface of the heat producing appliance or equipment to the outer surface of the combustible material or combustible assembly.

c. Spacers and ties shall be of noncombustible material. No spacer or tie shall be used directly opposite appliance or connector.

d. Where all clearance reduction systems use a ventilated air space, adequate provision for air circulation shall be provided as described. (See Figures M1306.1 and M1306.2.)

e. There shall be at least 1 inch between clearance reduction systems and combustible walls and ceilings for reduction systems using ventilated air space.

f. If a wall protector is mounted on a single flat wall away from corners, adequate air circulation shall be permitted to be provided by leaving only the bottom and top edges or only the side and top edges open with at least a 1-inch air gap.

g. Mineral wool and glass fiber batts (blanket or board) shall have a minimum density of 8 pounds per cubic foot and a minimum melting point of 1,500°F.

h. Insulation material used as part of a clearance reduction system shall have a thermal conductivity of 1.0 Btu per inch per square foot per hour per °F or less. Insulation board shall be formed of noncombustible material.

i. There shall be at least 1 inch between the appliance and the protector. In no case shall the clearance between the appliance and the combustible surface be reduced below that allowed in Table M1306.2.

j. All clearances and thicknesses are minimum; larger clearances and thicknesses are acceptable.

k. Listed single-wall connectors shall be permitted to be installed in accordance with the terms of their listing and the manufacturer's instructions.

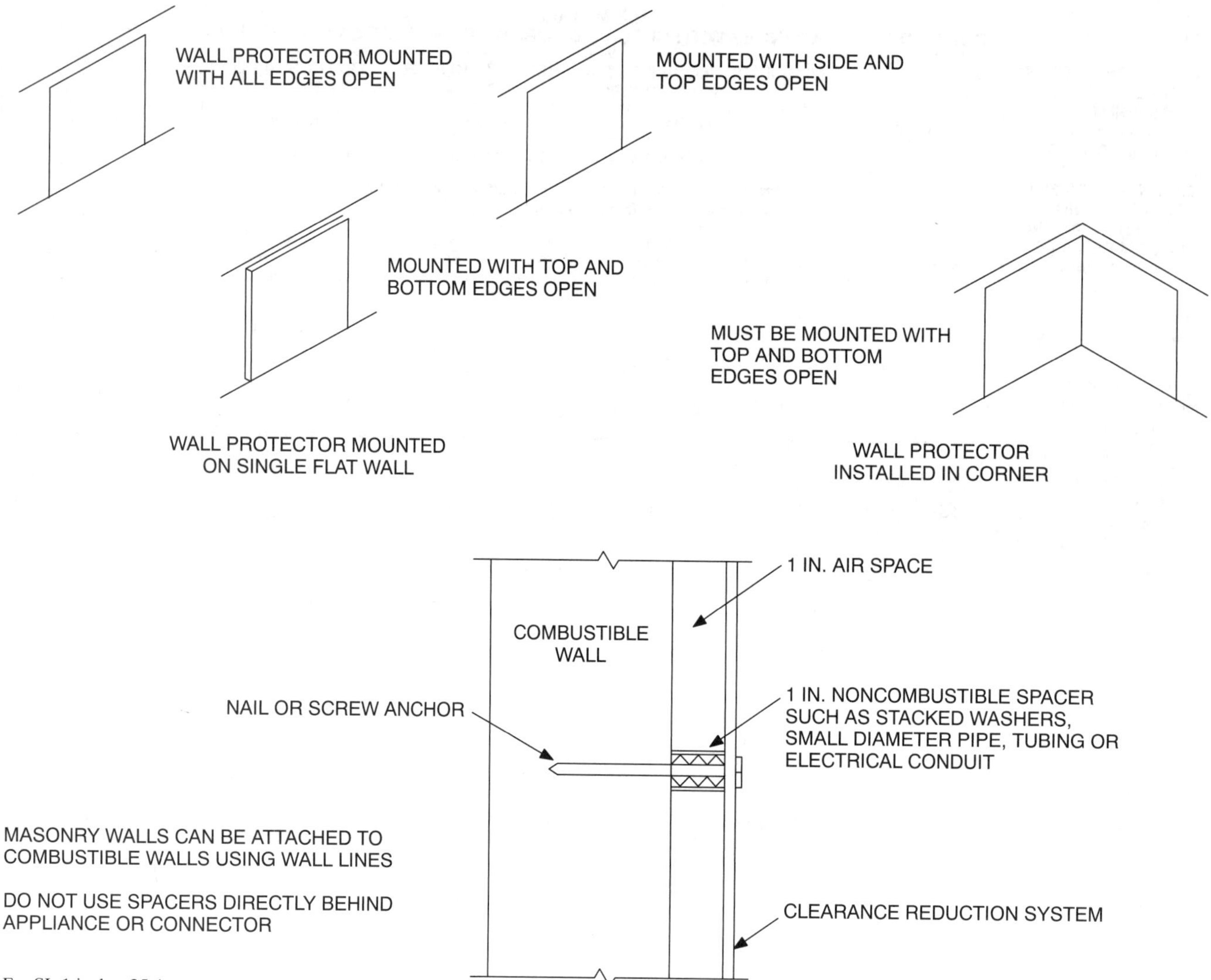

For SI: 1 inch = 25.4 mm.

FIGURE M1306.2
WALL PROTECTOR CLEARANCE REDUCTION SYSTEM

❖ See the commentary for Section M1306.2.

M1307.2 Anchorage of appliances. Appliances designed to be fixed in position shall be fastened or anchored in an approved manner. In Seismic Design Categories D_1 and D_2, water heaters shall be anchored or strapped to resist horizontal displacement due to earthquake motion. Strapping shall be at points within the upper one-third and lower one-third of the appliance's vertical dimensions. At the lower point, the strapping shall maintain a minimum distance of 4 inches (102 mm) above the controls.

❖ This section requires the anchorage of all fixed appliances to prevent movement or the possible tipping over of the appliance. In areas of high seismic risk, all water heaters must be secured directly to the building. Water heaters can weigh several hundred pounds, and if not properly secured, the water, gas, and electrical connections may be damaged or severed, creating a hazardous situation. Commentary Figure M1307.2 depicts the proper anchorage of a residential water heater.

M1307.3 Elevation of ignition source. Appliances having an ignition source shall be elevated such that the source of ignition is not less than 18 inches (457 mm) above the floor in garages. For the purpose of this section, rooms or spaces that are not part of the living space of a dwelling unit and that communicate with a private garage through openings shall be considered to be part of the garage.

❖ To reduce the hazard of fire or explosion from possible gasoline leaks from automobiles, appliances having an ignition source must be elevated to keep the ignition source a safe distance above garage floors. Commentary Figure M1307.3 shows a water heater installation that complies with these requirements. In the context of this section, a source of ignition could be a pilot flame, burner, burner igniter, or electrical component capable of producing a spark. The term "ignition

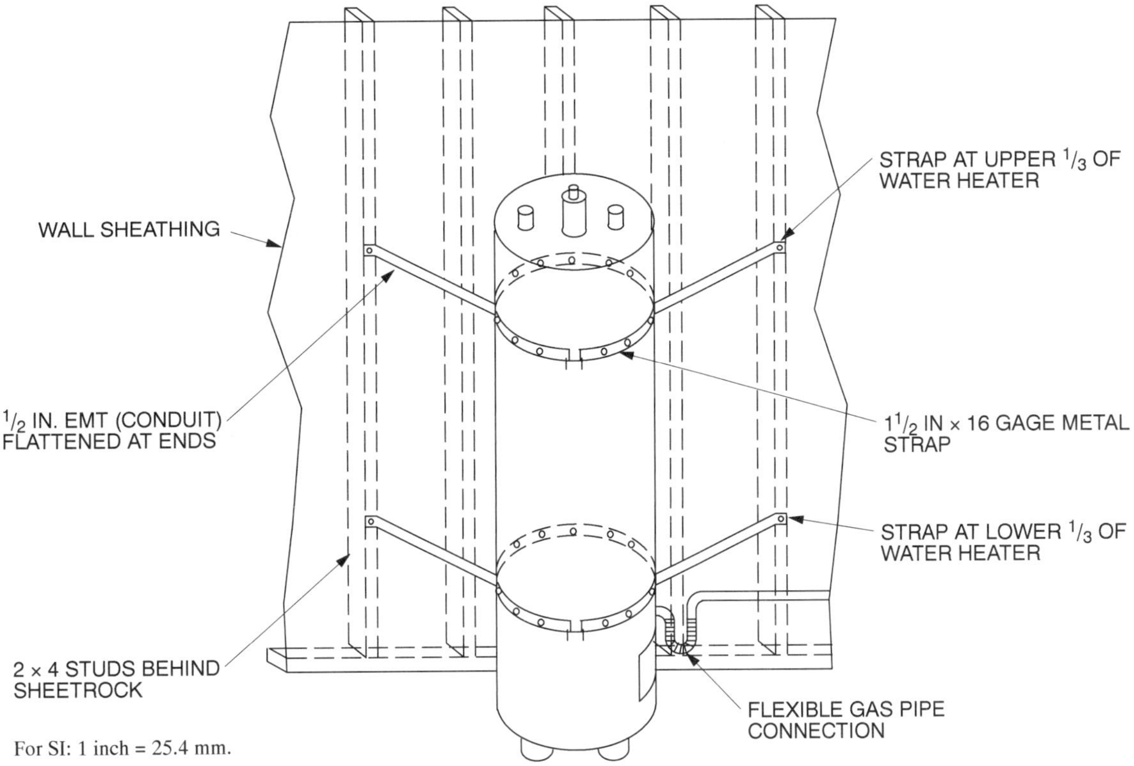

For SI: 1 inch = 25.4 mm.

NOTE: A MINIMUM 4-INCH CLEARANCE MUST BE MAINTAINED ABOVE THE WATER HEATER CONTROLS.

Figure M1307.2
ANCHORAGE OF WATER HEATER

source" is defined and can be interpreted as meaning an intentional source of ignition for any flammable vapors that may be present. See the definition of "ignition source."

An appliance installed in a closet or room that is accessible only from the garage must be considered as part of the garage for application of this section. Even though the room may be separated from the garage by walls and a door, there are no practical means of making the door vapor-tight nor is there any assurance the door will remain closed during normal use. An appliance room that is accessed only from the outdoors of from the living space would not be considered as part of the garage. Rooms such as utility rooms or laundry rooms that communicate both with the garage and the living space are considered as part of the living space and not part of the garage.

Even though appliances in the garage must be protected from impact by automobiles by the installation of substantial barriers, elevation of the appliance may not be sufficient to guard against damage from impact unless the building official determines that the construction of the platform is of substantial construction material capable of withstanding impact.

M1307.3.1 Protection from impact. Appliances located in a garage or carport shall be protected from impact by automobiles.

❖ Mechanical appliances installed in garages and carports must be protected from vehicular impact.

Although the code does not specify methods of protection, the most apparent method would be to locate the appliance where it could not reasonably be struck by a vehicle. A practical method of protection would be to place a formidable and permanent barrier between the motor vehicles and the appliance. This barrier could include such items as an effectively located vehicle wheel stop that is anchored in place, an elevated platform higher than the vehicle's bumpers, or one or more concrete-filled steel pipes. Final approval of the method of protection is left to the local building official.

M1307.4 Hydrogen generating and refueling operations. Ventilation shall be required in accordance with Section 1307.4.1, 1307.4.2 or 1307.4.3 in private garages which contain hydrogen generating appliances or refueling systems. Such spaces shall be used for the storage of not more than three hydrogen-fueled passenger motor vehicles and have a floor area not exceeding 850 square feet (79 m^2). The maximum rated output capacity of hydrogen generating appliances shall not exceed 4 SCFM of hydrogen for each 250 square feet

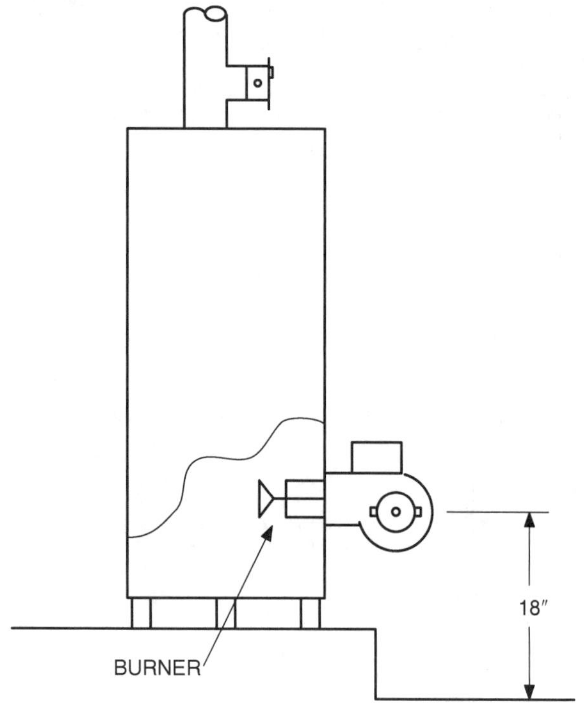

Figure M1307.3
OIL-FIRED WATER HEATER INSTALLATION IN A GARAGE

(23.2 m^2) of floor area in such spaces. For the purpose of this section, rooms or spaces that are not part of the living space of a dwelling unit and that communicate directly with a private garage through openings shall be considered to be part of the private garage.

❖ This section intends to minimize the potential for explosions by limiting the source of hydrogen gas and by requiring sufficient ventilation to dissipate any leakage.

M1307.4.1 Natural ventilation. Indoor locations intended for hydrogen generating or refueling operations shall communicate with the outdoors in accordance with Sections 1307.4.1.1 through 1307.4.1.2. The minimum dimension of air openings shall be not less than 3 inches (76 mm). Where ducts are used, they shall be of the same cross-sectional area as the free area of the openings to which they connect. In such locations, equipment and appliances having an ignition source shall be located such that the source of ignition is not less than 12 inches (228 mm) below the ceiling.

❖ This section parallels the intent of Section 305.3, but with respect to the ceiling instead of the floor. This section does not apply to spaces that house only vehicles and that do not contain hydrogen-generating or refueling operations. Because it is more buoyant, hydrogen will dissipate more quickly than natural gas and much more quickly than either propane or gasoline, both of which have vapors that are heavier than air and will linger at an accident site. However, hydrogen and natural gas can both accumulate in unventilated pockets at the top of indoor structures and could represent a risk in such situations.

Similarly, gasoline fumes can accumulate at the floor level in unventilated spaces, posing a different risk. Thus, ignition sources must be avoided at the top of any unventilated spaces for hydrogen gas. Also, hydrogen is odorless, colorless and burns with a flame that is not generally visible to the human eye. This means that it is unlikely that people will be able to detect unsafe conditions without appropriate instrumentation [similar to carbon monoxide (CO) accumulation in a structure].

M1307.4.1.1 Two openings. Two permanent openings, one located entirely within 12 inches (305 mm) of the ceiling of the garage, and one located entirely within 12 inches (305 mm) of the floor of the garage, shall be provided in the same exterior wall. The openings shall communicate directly, or by ducts, with the outdoors. Each opening shall directly communicate with the outdoors horizontally, and have a minimum free area of $^1/_2$ square foot per 1,000 cubic feet (1 m^2/610 m^3) of garage volume.

❖ The location requirement will prevent the openings from being more than 12 inches tall because the required openings must be entirely within the 12 inches of wall space measured down from the ceiling and up from the floor. The openings must be in the same wall to help create a gravity flow of gases driven by the natural buoyancy of hydrogen gas. The bottom opening is an air inlet and the top opening is an air outlet.

M1307.4.1.2 Louvers and grilles. In calculating free area required by Section 1307.4.1, the required size of openings shall be based on the net free area of each opening. If the free area through a design of louver or grille is known, it shall be used in calculating the size opening required to provide the free area specified. If the design and free area are not known, it shall be assumed that wood louvers will have a 25-percent free area and metal louvers and grilles will have a 75-percent free area. Louvers and grilles shall be fixed in the open position.

❖ This section recognizes that louvers and grilles are usually installed over air inlets and outlets to prevent rain, snow and animals from entering the building. When louvers or grilles are used, the solid portion of the louver or grille must be considered when determining the unobstructed (net clear) area of the opening.

Air openings are sized based on a free, unobstructed area for the passage of air into the space. Louvers or grilles placed over these openings reduce the area of the openings because of the area occupied by the solid portions of the grille or louver. The reduction in area must be considered because only the unobstructed area can be credited toward the required opening size.

The reduction in opening area caused by the presence of grilles or louvers will always require openings to be larger than determined from the sizing ratios of this chapter and larger than any duct of the minimum required size that might connect to these openings.

M1307.4.2 Mechanical ventilation. Indoor locations intended for hydrogen generating or refueling operations shall be ventilated in accordance with Section 502.16 of the *International Mechanical Code.*

❖ Section 502.16 of the *International Mechanical Code* (IMC) provides criteria for the design and operation of a mechanical ventilation system in repair garages for natural-gas-fueled and hydrogen-fueled vehicles. This section makes Section 502.16 of the IMC applicable to indoor spaces containing hydrogen-generating and/or refueling operations.

M1307.4.3 Specially engineered installations. As an alternative to the provisions of Sections 1307.4.1 and 1307.4.2, the necessary supply of air for, ventilation and dilution of flammable gases shall be provided by an approved engineered system.

❖ The code is not intended to inhibit innovative ideas or technological advances. A comprehensive regulatory document such as a fuel gas code cannot envision and then address all future innovations in the industry. As a result, a performance code must be applicable to and provide a basis for the approval of an increasing number of newly developed, innovative materials, systems and methods for which no code text or referenced standards yet exist. The fact that a material, product or method of construction is not addressed in the code is not an indication that prohibition of the material, product or method is intended. The code official is expected to apply sound technical judgement in accepting materials, systems or methods that, while not anticipated by the drafters of the current code text, can be demonstrated to offer equivalent performance. By virtue of its text, the code regulates new and innovative construction practices while addressing the relative safety of building occupants. The code official is responsible for determining whether a requested alternative provides a level of protection of the public health, safety and welfare as required by the code.

M1307.5 Electrical appliances. Electrical appliances shall be installed in accordance with Chapters 14, 15, 19, 20 and 33 through 42 of this code.

❖ Electrical appliances must be installed in accordance with the electrical provisions of Chapters 33 through 42 and the installation requirements for specific appliances in Chapters 14, 15, 19, and 20.

SECTION M1308 MECHANICAL SYSTEMS INSTALLATION

M1308.1 Drilling and notching. Wood-framed structural members shall be drilled, notched or altered in accordance with the provisions of Sections R502.8, R602.6, R602.6.1 and R802.7. Holes in cold-formed, steel-framed, load-bearing members shall only be permitted in accordance with Sections R505.2, R603.2 and R804.2. In accordance with the provisions of Sections R505.3.5, R603.3.4 and R804.3.5, cutting and notching of flanges and lips of cold-formed, steel-framed, load-bearing members shall not be permitted.

❖ During mechanical system installation, it is usually necessary to penetrate structural members. The building chapters of the *International Residential Code* regulate how these penetrations should be made so as not to weaken the structural members. The provisions of this section refer to those requirements, which are strict regarding the cutting and notching of steel members and prohibit the cutting and notching of flanges and lips of cold-formed, steel-framed, load-bearing members. See the commentary for Sections R505.3.5, R603.3.4 and R804.3.5.

M1308.2 Protection against physical damage. In concealed locations where piping, other than cast-iron or galvanized steel, is installed through holes or notches in studs, joists, rafters or similar members less than 1.5 inches (38 mm) from the nearest edge of the member, the pipe shall be protected by shield plates. Protective shield plates shall be a minimum of 0.062-inch-thick (1.6 mm) steel, shall cover the area of the pipe where the member is notched or bored, and shall extend a minimum of 2 inches (51 mm) above sole plates and below top plates.

❖ This section is intended to minimize the possibility of damage to refrigerant piping and other mechanical piping from nails, screws, or other fasteners. Because nails and screws sometimes miss the stud, rafter joist, or top or sole plate, the shield must protect the pipe through the full width of the member and must extend not less than 2 inches (51 mm) above or below the sole or top plates, respectively. Commentary Figure M1308.2 shows typical shield plates. Cast-iron and galvanized steel pipe have wall thicknesses greater than the required thickness of the shield plate, which makes them inherently resistant to nail and screw penetrations. The same requirement exists in Section P2603.2.1 for plumbing piping; see the commentary for Section P2603.2.1.

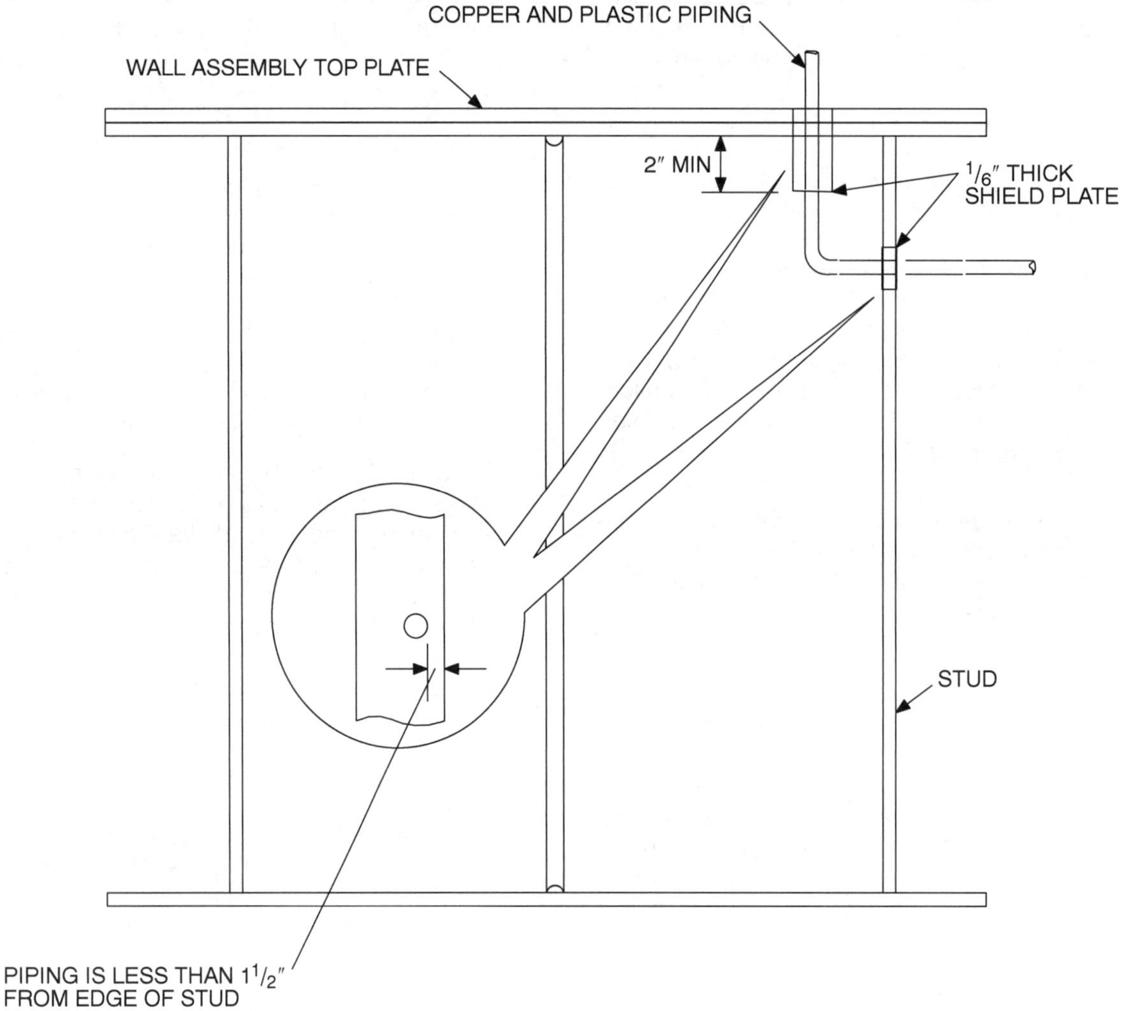

Figure M1308.2
PIPE PROTECTION

Bibliography

The following resource materials are referenced in this chapter or are relevant to the subject matter addressed in this chapter.

ASCE 7-02, *Minimum Design Loads for Buildings and Other Structures*. New York: American Society of Civil Engineers, 2002.

ASHRAE 90.1-2001, *Energy Efficient Design of New Buildings except Low-Rise Residential Buildings.* Atlanta: American Society of Heating, Refrigerating and Air-Conditioning Engineers, 2001.

ASHRAE-2000, *HVAC Systems and Equipment Handbook.* Atlanta: American Society of Heating, Refrigerating and Air-Conditioning Engineers, 2000.

ASHRAE-2001, *Fundamentals Handbook*. Atlanta: American Society of Heating, Refrigerating and Air-Conditioning Engineers, 2001.

ASTM C 411-97, *Test Method for Hot-Surface Performance of High-Temperature Thermal Insulation*. West Conshohocken, PA: ASTM International, 1997.

ASTM E 84-01, *Test Method for Surface Burning Characteristics of Building Materials.* West Conshohocken, PA: ASTM International, 2001.

ASTM E 119-00a, *Test Methods for Fire Tests of Building Construction and Materials*. West Conshohocken, PA: ASTM International, 2000.

ASTM E 136-99, *Test Method for Behavior of Materials in a Vertical Tube Furnace at 750°C.* West Conshohocken, PA: ASTM International, 1999.

MSS SP-69-96, *Pipe Hangers and Supports — Selection and Application*. Vienna, VA: Manufacturers Standardization Society of the Valve and Fittings Industry, 1996.

NFPA 70-02, *National Electrical Code*. Quincy, MA: National Fire Protection Association, 2002.

PEI RP-300-93, *Recommended Practices for Installation and Testing of Vapor Recovery Systems at Vehicle Fueling Sites.* Tulsa, OK: Petroleum Equipment Institute, 1993.

UL Subject 1618-86, *Outline of Investigation for Wall Protectors and Hearth Extensions*. Northbrook, IL: Underwriters Laboratories, 1986.

Chapter 14: Heating and Cooling Equipment

General Comments

This chapter is concerned with the installation and hazards associated with heating and cooling appliances. These types of appliances have risks either due to the potential for fire or due to the accidental release of refrigerants. Both situations are undesirable in dwellings that are covered by the *International Residential Code*.

The chapter is a collection of requirements for various heating and cooling appliances, dedicated to single topics by section. The common theme is that all of these types of appliances use energy in one form or another, and the improper installation of such appliances would present a hazard to the occupants of the dwellings.

Section M1401 consists of the general requirements that apply to all such appliances. Section M1402 contains the provisions for central furnaces. Section M1403 addresses heat pump equipment. Section M1404 refers the reader to Section M1411 for the criteria for refrigeration cooling equipment. Section M1405 contains the provisions for baseboard convectors. Section M1406 addresses radiant heating systems. Section M1407 consists of the criteria for duct heaters. Section M1408 contains the provisions for vented floor furnaces. Section M1409 contains provisions for vented wall furnaces. Section M1410 addresses vented room heaters. Section M1411 consists of the provisions for refrigeration cooling equipment. Section M1412 contains the provisions for absorption cooling equipment. Section M1413 addresses evaporative cooling equipment, and Section M1414 contains the provisions for fireplace stoves.

Purpose

This chapter provides, in a simple format, the requirements for heating and cooling appliances commonly found in dwellings covered by the *International Residential Code*. These regulations pertain to the safety of the occupants by establishing minimum requirements for the design, construction, installation and operation of these systems.

SECTION M1401 GENERAL

M1401.1 Installation. Heating and cooling equipment and appliances shall be installed in accordance with the manufacturer's installation instructions and the requirements of this code.

❖ As discussed in the commentary to Section M1307.1, the manufacturer's installation instructions are an enforceable extension of the code. The manufacturer's instructions must be available at the job site when the equipment and appliances are being inspected.

M1401.2 Access. Heating and cooling equipment shall be located with respect to building construction and other equipment to permit maintenance, servicing and replacement. Clearances shall be maintained to permit cleaning of heating and cooling surfaces; replacement of filters, blowers, motors, controls and vent connections; lubrication of moving parts; and adjustments.

❖ As does Section M1305.1, this section requires that heating and cooling equipment be located with clearances sufficient to allow maintenance, servicing, and replacement. It requires that clearances be sufficient to permit cleaning of the heating and cooling surfaces and replacement of the filters, blowers, motors, and control and vent connections.

M1401.3 Sizing. Heating and cooling equipment shall be sized based on building loads calculated in accordance with ACCA Manual J or other approved heating and cooling calculation methodologies.

❖ The Air Conditioning Contractors of America's Manual J, *Load Calculations for Residential Winter and Summer Air Conditioning*, contains a simplified method of calculating heating and cooling loads. It includes a room-by-room calculation method that allows the designer to determine the required capacity of the heating and cooling equipment. In addition, it provides for an estimate of the air-flow requirements for each of the areas in the house. The estimate can be used in sizing the duct system for the types of heating and cooling units that use air as the media for heat transfer. Other approved methods may be used with the building official's approval.

M1401.4 Exterior installations. Equipment installed outdoors shall be listed and labeled for outdoor installation. Supports and foundations shall prevent excessive vibration, settlement or movement of the equipment. Supports and foundations shall be level and conform to the manufacturer's installation instructions.

❖ This section points to the importance of verifying that equipment installed outdoors be listed and labeled accordingly. It requires the equipment be supported in accordance with manufacturers' installation instructions.

M1401.5 Flood hazard. In areas prone to flooding as established by Table R301.2(1), heating and cooling equipment and appliances shall be located or installed in accordance with Section R323.1.5.

❖ The local jurisdiction must fill in Table R301.2(1) upon adoption of this code, including the "Flood Hazards" information. Mechanical appliances, equipment, and systems that are located in flood hazard areas must be installed above the design flood elevation or must be designed and installed to prevent the entrance of water into the components and to resist the forces of the flood waters on the components. See the commentary for Section R323.1.5.

SECTION M1402 CENTRAL FURNACES

M1402.1 General. Oil-fired central furnaces shall conform to ANSI/UL 727. Electric furnaces shall conform to UL 1995.

❖ This section requires oil-fired central furnaces to meet the requirements in UL 727, and it requires electric furnaces to meet the requirements of UL 1995. Furnaces can come in many configurations including upflow, counterflow (downflow), horizontal flow, and indoor and outdoor units. (See Chapter 24 for gas-fired furnaces.)

M1402.2 Clearances. Clearances shall be provided in accordance with the listing and the manufacturer's installation instructions.

❖ The clearances indicated on the equipment label and in the manufacturers' installation instructions must always be followed, as indicated throughout these mechanical provisions.

M1402.3 Combustion air. Combustion air shall be supplied in accordance with Chapter 17. Combustion air openings shall be unobstructed for a distance of not less than 6 inches (152 mm) in front of the openings.

❖ This section refers to Chapter 17 for combustion-air requirements and specifically requires a minimum 6-inch (152 mm) unobstructed clearance in front of the fire box to ensure adequate distribution of combustion air to the appliance. Fuel gas-fired furnaces should meet the combustion-air requirements in Section G2407.

SECTION M1403 HEAT PUMP EQUIPMENT

M1403.1 Heat pumps. The minimum unobstructed total area of the outside and return air ducts or openings to a heat pump shall be not less than 6 square inches per 1,000 Btu/h (13 208 mm^2/kW) output rating or as indicated by the conditions of the listing of the heat pump. Electric heat pumps shall conform to UL 1995.

❖ Heat pumps are central heating units that use the condenser of a refrigeration circuit as the heat source, often backed up by auxiliary electric resistance heating elements. These types of units require a certain amount of flow of return air and/or outside air to ensure that heat will be removed from the condenser, which guarantees proper operation of the refrigeration cycle. This section requires a minimum duct opening of 6 square inches per 1,000 Btu/h output rating. It also specifies the standards to which the heat pump must conform.

M1403.2 Foundations and supports. Supports and foundations for the outdoor unit of a heat pump shall be raised at least 3 inches (76 mm) above the ground to permit free drainage of defrost water, and shall conform to the manufacturer's installation instructions.

❖ This section requires outdoor heat pump units to be on a support or foundation raised 3 inches (76 mm) above the surrounding ground. This is similar to other support criteria where the protection from corrosion and the deleterious effects of ponded water is a concern.

SECTION M1404 REFRIGERATION COOLING EQUIPMENT

M1404.1 Compliance. Refrigeration cooling equipment shall comply with Section M1411.

❖ This section is a cross-reference to the actual provisions found in Section M1411.

SECTION M1405 BASEBOARD CONVECTORS

M1405.1 General. Electric baseboard convectors shall be installed in accordance with the manufacturer's installation instructions and Chapters 33 through 42 of this code.

❖ Electric baseboard convectors must be installed in accordance with manufacturers' installation instructions and with the electrical requirements of the IRC. The latter are based on the electrical provisions of the *National Electrical Code*, published by the National Fire Protection Association. See Commentary Figure M1405.1 for an illustration of a typical electric baseboard convector.

SECTION M1406 RADIANT HEATING SYSTEMS

M1406.1 General. Electric radiant heating systems shall be installed in accordance with the manufacturer's installation instructions and Chapters 33 through 42 of this code.

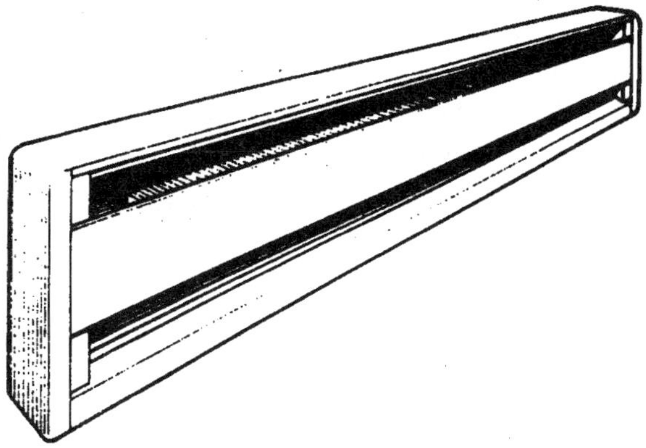

Figure M1405.1
TYPICAL ELECTRIC BASEBOARD CONVECTOR

❖ Radiant heating panels contain electrical elements that transmit heat when an electrical current passes through them. If the system is not installed properly, electricity and the heat emitted could create hazards for the occupants of the dwelling. This section refers to the manufacturers' installation instructions and the electrical provisions in Chapters 33 through 42 for safe installations.

M1406.2 Clearances. Clearances for radiant heating panels or elements to any wiring, outlet boxes and junction boxes used for installing electrical devices or mounting lighting fixtures shall comply with Chapters 33 through 42 of this code.

❖ This section provides certain minimum clearances so the heating elements are not damaged or the electrical circuit continuity is not interrupted, thus rendering portions or all the radiant heating systems inoperable. This section is also intended to protect the electrical systems, equipment, and components from damage from the heating elements.

M1406.3 Installation of radiant panels. Radiant panels installed on wood framing shall conform to the following requirements:

1. Heating panels shall be installed parallel to framing members and secured to the surface of framing members or mounted between framing members.
2. Panels shall be nailed or stapled only through the unheated portions provided for this purpose and shall not be fastened at any point closer than $^1/_4$ inch (6.4 mm) from an element.
3. Unless listed and labeled for field cutting, heating panels shall be installed as complete units.

❖ Radiant heat panels mounted on wood framing may act as an ignition source if improperly installed. Because these systems generate heat, they must be mounted to wood framing by fastening through the unheated portions of each panel. Metallic fasteners, such as nails or screws, are excellent conductors of heat as well as electricity; therefore, they must not be driven through any portion of a panel closer than $^1/_4$ inch (6.4 mm) to a heating element.

M1406.4 Installation in concrete or masonry. Radiant heating systems installed in concrete or masonry shall conform to the following requirements:

1. Radiant heating systems shall be identified as being suitable for the installation, and shall be secured in place, as specified in the manufacturer's installation instructions.
2. Radiant heating panels or radiant heating panel sets shall not be installed where they bridge expansion joints unless protected from expansion and contraction.

❖ Fire hazards created by electric radiant heating systems installed in masonry construction are minimal, but these installations require special mounting hardware and must be installed in accordance with the manufacturers' installation instructions. Radiant panels must be protected from expansion and contraction when mounted over masonry expansion joints to avoid structural damage.

M1406.5 Gypsum panels. Where radiant heating systems are used on gypsum assemblies, operating temperatures shall not exceed 125°F (52°C).

❖ The maximum allowable temperature of radiant heating systems installed in gypsum assemblies is 125°F (52°C). Temperatures that are higher can cause the gypsum to disintegrate by driving off the water of hydration from the gypsum's crystalline structure.

M1406.6 Finish surfaces. Finish materials installed over radiant heating panels or systems shall be installed in accordance with the manufacturer's installation instructions. Surfaces shall be secured so that nails or other fastenings do not pierce the radiant heating elements.

❖ Finish materials must be carefully installed over electric radiant heating systems so that nails or screws do not penetrate the heating elements. Fasteners that are accidentally driven through these elements could short out or destroy the elements or cause the surfaces to overheat and start a fire. The finish surfaces must be installed in accordance with the instructions supplied by the manufacturer of the radiant heating system.

SECTION M1407 DUCT HEATERS

M1407.1 General. Electric duct heaters shall be installed in accordance with the manufacturer's installation instructions and Chapters 33 through 42 of this code. Electric furnaces shall be tested in accordance with UL 1995.

❖ Electric duct heaters are designed to be installed in the airflow of an air duct and use electric resistance heating elements to heat the air supply. The code requires that their installation comply with the manufacturers' installation instructions and the electrical provisions in the IRC.

M1407.2 Installation. Electric duct heaters shall be installed so that they will not create a fire hazard. Class 1 ducts, duct coverings and linings shall be interrupted at each heater to provide the clearances specified in the manufacturer's installation instructions. Such interruptions are not required for duct heaters listed and labeled for zero clearance to combustible materials. Insulation installed in the immediate area of each heater shall be classified for the maximum temperature produced on the duct surface.

❖ These heaters are typically installed in ducts that may contain combustible linings, coverings, or insulating materials that may also be combustible. This section requires that these heaters be listed for zero clearance to combustible material or that the material is kept at the required distance from the heaters as specified by the manufacturer or the listing.

M1407.3 Installation with heat pumps and air conditioners. Duct heaters located within 4 feet (1219 mm) of a heat pump or air conditioner shall be listed and labeled for such installations. The heat pump or air conditioner shall additionally be listed and labeled for such duct heater installations.

❖ Heat pumps are typically sized so that they do not provide all the heat required in a building. Thus, they are usually supplemented with electrical resistance heating to meet the occasional periods of peak load requirements. Certain types of air conditioning systems are designed to reheat the air after it is cooled in order to provide different levels of cooling to rooms having varying cooling needs. The section requires that the heaters be listed for installation in close proximity to the heat pumps or air conditioners or that the heaters be kept a certain distance away from these units.

M1407.4 Access. Duct heaters shall be accessible for servicing, and clearance shall be maintained to permit adjustment, servicing and replacement of controls and heating elements.

❖ This section requires proper access to the appliance, which should be provided through the ceiling, floor, or wall panels. Section M1305 has additional criteria for access to the installations of appliances in specific locations.

M1407.5 Fan interlock. The fan circuit shall be provided with an interlock to prevent heater operation when the fan is not operating.

❖ A fan interlock is required to prevent the duct heater from operating when there is no air flow. Without air flow, the duct heater may overheat, possibly creating a fire hazard.

SECTION M1408 VENTED FLOOR FURNACES

M1408.1 General. Vented floor furnaces shall conform to UL 729 and be installed in accordance with their listing, the manufacturer's installation instructions and the requirements of this code.

❖ Floor furnaces are vented appliances that are installed in an opening in the floor. Such units supply heat to the room by gravity convection and direct radiation and typically serve as the sole source of space heating. Such furnaces are common in cottages, small homes, seasonally occupied structures, and rural homes. Because the floor grille can become very hot, extreme care must be exercised to prevent occupants, especially children, from contacting the grille by walking or falling on it. Also, care must be taken to avoid a fire hazard caused by placement of materials or furnishings on or near the furnace floor grille.

M1408.2 Clearances. Vented floor furnaces shall be installed in accordance with their listing and the manufacturer's installation instructions.

❖ Vented floor furnaces must be installed in accordance with the manufacturers' installation intructions, which are an enforceable extension of the code.

M1408.3 Location. Location of floor furnaces shall conform to the following requirements:

1. Floor registers of floor furnaces shall be installed not less than 6 inches (152 mm) from a wall.
2. Wall registers of floor furnaces shall be installed not less than 6 inches (152 mm) from the adjoining wall at inside corners.
3. The furnace register shall be located not less than 12 inches (305 mm) from doors in any position, draperies or similar combustible objects.
4. The furnace register shall be located at least 5 feet (1524 mm) below any projecting combustible materials.

5. The floor furnace burner assembly shall not project into an occupied under-floor area.
6. The floor furnace shall not be installed in concrete floor construction built on grade.
7. The floor furnace shall not be installed where a door can swing within 12 inches (305 mm) of the grill opening.

❖ Because of the heat generated in the combustion chamber that is emitted through the floor register, there are minimum clearances necessary from the face of the air registers to surrounding walls, overhead projections, and other combustible items, such as doors and draperies. See Commentary Figure M1408.3 for required clearances.

M1408.4 Access. An opening in the foundation not less than 18 inches by 24 inches (457 mm by 610 mm), or a trap door not less than 22 inches by 30 inches (559 mm by 762 mm) shall be provided for access to a floor furnace. The opening and passageway shall be large enough to allow replacement of any part of the equipment.

❖ Either a floor opening (trap door) or a foundation opening must be provided to allow for maintenance or replacement of the furnace. Without such access, appliances are not as likely to receive the necessary maintenance for efficient and safe operation.

M1408.5 Installation. Floor furnace installations shall conform to the following requirements:

1. Thermostats controlling floor furnaces shall be located in the room in which the register of the floor furnace is located.
2. Floor furnaces shall be supported independently of the furnace floor register.
3. Floor furnaces shall be installed not closer than 6 inches (152 mm) to the ground. Clearance may be reduced to 2 inches (51 mm), provided that the lower 6 inches (152 mm) of the furnace is sealed to prevent water entry.
4. Where excavation is required for a floor furnace installation, the excavation shall extend 30 inches (762 mm) beyond the control side of the floor furnace and 12 inches (305 mm) beyond the remaining sides. Excavations shall slope outward from the perimeter of the base of the excavation to the surrounding grade at an angle not exceeding 45 degrees (0.39 rad) from horizontal.
5. Floor furnaces shall not be supported from the ground.

❖ Floor furnaces must have supports independent of the floor register and the ground below. A 6-inch (152 mm) clearance must also be maintained from the ground to protect against moisture, rust, and corrosion of the fur-

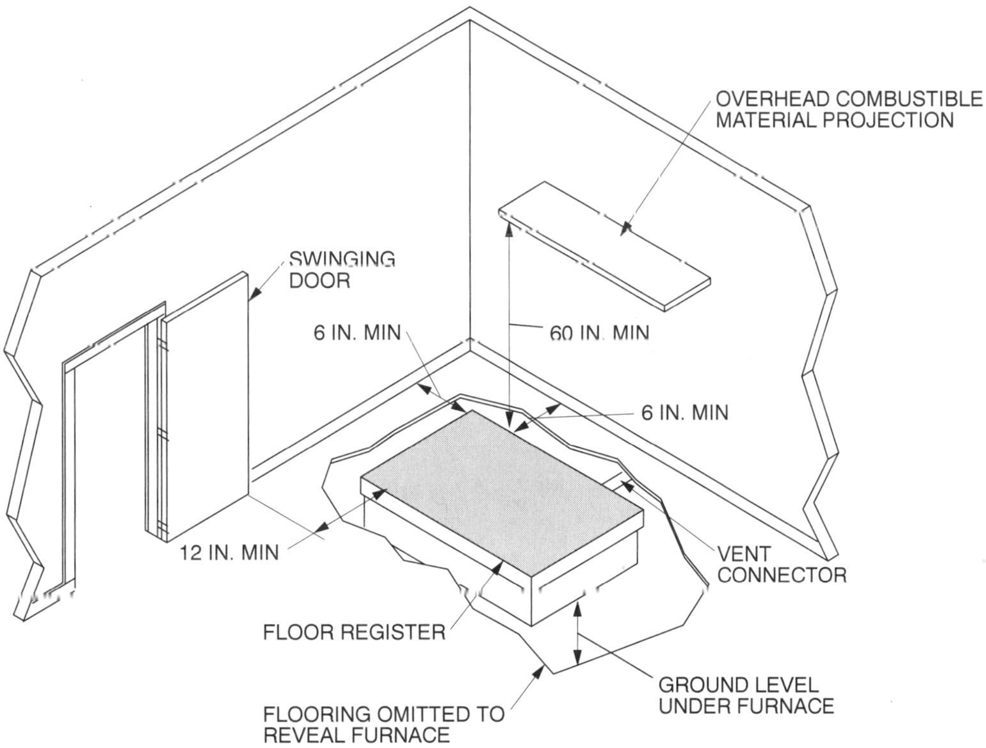

For SI: 1 inch = 25.4 mm.

Figure M1408.3
FLOOR FURNACE CLEARANCES

nace casing. This may be reduced to 2 inches (51 mm) if the unit is sealed against flooding. When it is necessary to excavate the ground to provide space for a floor furnace, a 30-inch clearance must be provided for access to the control side. A 12-inch (305 mm) clearance is required on all sides.

SECTION M1409 VENTED WALL FURNACES

M1409.1 General. Vented wall furnaces shall conform to UL 730 and be installed in accordance with their listing, the manufacturer's installation instructions and the requirements of this code.

❖ Vented wall furnaces are a type of room heater typically designed to be installed within a stud cavity in frame construction. Wall furnaces are typically used in cottages, room additions, and homes in mild climates. Some units are designed to serve a single room, and others are designed as through-the-wall units to serve adjacent rooms. Wall furnaces are ductless; however, some units are listed for use with a surface-mounted supply outlet extension. Wall furnaces can be either gravity or forced-air type.

M1409.2 Location. The location of vented wall furnaces shall conform to the following requirements:

1. Vented wall furnaces shall be located so as not to cause a fire hazard to walls, floors, combustible furnishings or doors. Vented wall furnaces installed between bathrooms and adjoining rooms shall not circulate air from bathrooms to other parts of the building.
2. Vented wall furnaces shall not be located where a door can swing within 12 inches (305 mm) of the furnace air inlet or outlet measured at right angles to the opening. Doorstops or door closers shall not be installed to obtain this clearance.

❖ Since the heat discharged from these units can ignite nearby wall or floor surfaces, trim, and doors, care must be exercised in the installation of these heaters to avoid these problems. Doors must be located to prevent swinging within 12 inches (305 mm) of the furnace to avoid interference with the flow of air from or to the heater, as well as to maintain adequate clearance from combustible door material. Through-the-wall units serving both a bathroom and an adjacent space must not recirculate air between the two spaces.

M1409.3 Installation. Vented wall furnace installations shall conform to the following requirements:

1. Required wall thicknesses shall be in accordance with the manufacturer's installation instructions.
2. Ducts shall not be attached to a wall furnace. Casing extensions or boots shall only be installed when listed as part of a listed and labeled appliance.
3. A manual shut off valve shall be installed ahead of all controls.

❖ Most wall furnaces are not designed to force air through ducts, especially the gravity types that do not use fans. Attachment of ducts would add resistance to the flow of air through the furnace, thereby causing abnormally high temperature rises across the heat exchanger and abnormally high temperatures on furnace surfaces. Some wall furnaces are listed and designed for use with supply duct extensions designed for wall mounting that are intended to improve heat distribution. Such duct extensions are factory-built and supplied only by the furnace manufacturer.

The fuel-line connection to every appliance must be provided with an individual shutoff valve to permit maintenance, repair, or replacement of the appliance or its components.

M1409.4 Access. Vented wall furnaces shall be provided with access for cleaning of heating surfaces; removal of burners; replacement of sections, motors, controls, filters and other working parts; and for adjustments and lubrication of parts requiring such attention. Panels, grilles and access doors that must be removed for normal servicing operations shall not be attached to the building construction.

❖ Proper access for servicing and replacement of vented wall furnaces is very important, as it is with the other mechanical appliances regulated by this code. See the commentary for Section M1408.4.

SECTION M1410 VENTED ROOM HEATERS

M1410.1 General. Vented room heaters shall be tested in accordance with UL 1482 or UL 896 and installed in accordance with their listing, the manufacturer's installation instructions and the requirements of this code.

❖ This section addresses vented space/room heaters, including direct-vent and vent-connected or chimney-connected appliances. As required by the appliance standard, such appliances are equipped with safety controls that will prevent fuel flow to the burners in the event of ignition system failure.

M1410.2 Floor mounting. Room heaters shall be installed on noncombustible floors or approved assemblies constructed of noncombustible materials that extend at least 18 inches (457 mm) beyond the appliance on all sides.

Exceptions:

1. Listed room heaters shall be installed on non-combustible floors, assemblies constructed of non-combustible materials or listed floor protectors with materials and dimensions in accordance with the appliance manufacturer's instructions.
2. Room heaters listed for installation on combustible floors without floor protection shall be installed in accordance with the appliance manufacturer's instructions.

❖ Room heaters must be installed on noncombustible floors or approved floor assemblies constructed of noncombustible materials that extend 18 inches (457 mm) beyond the appliance on all sides, or they must be listed for installation on combustible floors. They must also be installed following the manufacturers' installation instructions.

SECTION M1411
REFRIGERATION COOLING EQUIPMENT

M1411.1 Approved refrigerants. Refrigerants used in direct refrigerating systems shall conform to the applicable provisions of ANSI/ASHRAE 34.

❖ Classification of refrigerants is based on ASHRAE 34, which numbers and classifies refrigerants in accordance with their potential hazards. Refrigerants are identified by their number preceded by the letter R. Trademark names of manufacturers such as "Freon" are not used in the code.

Each refrigerant is classified into a safety group, which is based on two factors: flammability and toxicity. The safety group in which an individual refrigerant is classified relates to the potential hazard to the building occupants and is one of the factors used to determine the maximum quantities of refrigerants allowed by the code.

The safety classification designations of ASHRAE 34 are necessary to provide for regulations applicable to the newly developed and implemented ozone-friendly refrigerants intended to replace the ozone-depleting refrigerants such as R-11 and R-12 now used in many existing refrigeration systems.

M1411.2 Refrigeration coils in warm-air furnaces. Where a cooling coil is located in the supply plenum of a warm-air furnace, the furnace blower shall be rated at not less than 0.5-inch water column (124 Pa) static pressure unless the furnace is listed and labeled for use with a cooling coil. Cooling coils shall not be located upstream from heat exchangers unless listed and labeled for such use. Conversion of existing furnaces for use with cooling coils shall be permitted provided the furnace will operate within the temperature rise specified for the furnace.

❖ This section alerts the installer that if a cooling coil is not a part of a listed furnace or air conditioner system and is added to a warm air furnace, the following requirements must be met.

1. The furnace blower must be rated at $^1/_2$-inches (12.7 mm) of water-column minimum static pressure. Thus the furnace blower will be capable of pushing the larger cooled air quantities required for air conditioning as opposed to the smaller air quantities needed for heating.
2. Unless the furnace heat exchanger is corrosion resistant (usually stainless steel), the cooling coil unit must be installed downstream of the heat exchanger. As cooled air condenses the moisture in the air, condensation can corrode downstream metal parts.

M1411.3 Condensate disposal. Condensate from all cooling coils or evaporators shall be conveyed from the drain pan outlet to an approved place of disposal. Condensate shall not discharge into a street, alley or other areas so as to cause a nuisance.

❖ Appliances and equipment containing evaporators or cooling coils, including refrigeration, dehumidification, and comfort cooling equipment, can produce condensate from the water vapor in the atmosphere. A drainage system is necessary to dispose of the condensate and prevent damage to the structure.

M1411.3.1 Auxiliary and secondary drain systems. In addition to the requirements of Section M1411.3, a secondary drain or auxiliary drain pan shall be required for each cooling or evaporator coil where damage to any building components will occur as a result of overflow from the equipment drain pan or stoppage in the condensate drain piping. Drain piping shall be a minimum of $^3/_4$-inch (19.1 mm) nominal pipe size. One of the following methods shall be used:

1. An auxiliary drain pan with a separate drain shall be provided under the coils on which condensation will occur. The auxiliary pan drain shall discharge to a conspicuous point of disposal to alert occupants in the event of a stoppage of the primary drain. The pan shall have a minimum depth of 1.5 inches (38 mm), shall not be less than 3 inches (76 mm) larger than the unit or the coil dimensions in width and length and shall be constructed of corrosion resistant material. Metallic pans shall have a minimum thickness of not less than 0.0276-inch (0.7 mm) galvanized sheet metal. Nonmetallic pans shall have a minimum thickness of not less than 0.0625 inch (1.6 mm).
2. A separate overflow drain line shall be connected to the drain pan provided with the equipment. Such overflow drain shall discharge to a conspicuous point of disposal to alert occupants in the event of a stoppage of the primary drain. The overflow drain line shall connect to the drain pan at a higher level than the primary drain connection.
3. An auxiliary drain pan without a separate drain line shall be provided under the coils on which condensate will occur. Such pan shall be equipped with a water level detection device that will shut off the equipment served prior to overflow of the pan. The auxiliary drain pan shall be constructed in accordance with Item 1 of this section.

❖ An auxiliary (redundant) drain pan or a secondary drain is required for equipment locations where condensate overflow would cause damage to a building or its contents. The purpose of the auxiliary drain pan and secondary drain is to catch condensate spilling from the primary drain pan in the equipment. This back-up protects the building from structural and finish damage.

Condensate drains are notorious for clogging because of debris (lint, dust) from air-handling systems and the natural affinity to produce slime growths in drain pans and pipes. It is relatively common for condensate overflows to cause damage to buildings. This section provides three options for preventing such damage where the equipment is located in spaces such as attics, above suspended ceilings and furred spaces, and locations on upper stories. One of the three methods must be used.

The first method uses an auxiliary drain pan below the primary drain pan with an independent drain line that discharges to a location that is easily observable to notify the building occupants that a problem with the primary pan exists. The code prescribes the depth of the pan and specific material thicknesses to assure that the pan will be corrosion resistant and will have sufficient holding capacity and capture ability.

The second method uses an independent overflow drain line connected to the primary drain pan at a point higher than the primary drain line. Most evaporator coil pans are factory provided with an overflow drain tap that can be used for this purpose. As in the first method, the point of discharge must be easily observable.

The third method uses a water-level detection device, usually a float switch or electronic sensor, located in the auxiliary drain pan. These detection devices will shut down the equipment prior to the pan overflowing. There is no requirement for a separate drain line in this method.

M1411.3.2 Drain pipe materials and sizes. Components of the condensate disposal system shall be cast iron, galvanized steel, copper, polybutylene, polyethylene, ABS, CPVC or PVC pipe or tubing. All components shall be selected for the pressure and temperature rating of the installation. Condensate waste and drain line size shall be not less than $^3/_4$-inch (91mm) internal diameter and shall not decrease in size from the drain pan connection to the place of condensate disposal. Where the drain pipes from more than one unit are manifolded together for condensate drainage, the pipe or tubing shall be sized in accordance with an approved method. All horizontal sections of drain piping shall be installed in uniform alignment at a uniform slope.

❖ Condensate drains must be constructed of a material listed in this section, which are corrosion resistant. Such drains must be straight runs of pipe without sags or dips that would trap liquid. If coiled stock tubing is used, it would be difficult to maintain a uniform slope, and the low points (dips) in the tubing would trap water, causing the drain to be double trapped (air bound). Airbound drains will not allow flow. When drains are merged, the piping must be properly sized for the aggregate flow.

M1411.4 Insulation of refrigerant piping. Piping and fittings for refrigerant vapor (suction) lines shall be insulated with insulation having a thermal resistivity of at least R = 4.0 hr · ft^2 · °F/Btu and having external surface permeance not exceeding 0.05 perms [2.87 ng/(s · m^2 · Pa)] when tested in accordance with ASTM E 96.

❖ To improve the efficiency of the cooling cycle, the refrigerant vapor (suction) lines must be insulated.

SECTION M1412 ABSORPTION COOLING EQUIPMENT

M1412.1 Approval of equipment. Absorption systems shall be installed in accordance with the manufacturer's installation instructions.

❖ Absorption refrigeration systems employ two fluids (an absorbent and a refrigerant) and a heat source to remove heat through evaporation at a lower pressure and then reject the heat through condensation at a higher pressure. Typical absorption systems use ammonia as the refrigerant and water as the absorbent or lithium bromide as the absorbent and water as the refrigerant.

This section states that absorption systems must be installed in accordance with manufacturers' installation instructions, which are an enforceable extension of the code.

M1412.2 Condensate disposal. Condensate from the cooling coil shall be disposed of as provided in Section M1411.3.

❖ This section and Section M1411.3 provide for the proper and safe disposal of condensate.

M1412.3 Insulation of piping. Refrigerant piping, brine piping and fittings within a building shall be insulated to prevent condensation from forming on piping.

❖ Condensation will deteriorate piping and cause serious damage to other parts of buildings, so the piping must be insulated.

M1412.4 Pressure-relief protection. Absorption systems shall be protected by a pressure-relief device. Discharge from the pressure-relief device shall be located so as not to create a hazard to persons or property.

❖ In addition to requiring the installation of this type of equipment to be in compliance with the manufacturers' installation instructions, this section specifically requires the installation of pressure-relief devices that are to be discharged to a safe location.

SECTION M1413 EVAPORATIVE COOLING EQUIPMENT

M1413.1 General. Cooling equipment that utilizes evaporation of water for cooling shall be installed in accordance with the manufacturer's installation instructions. Evaporative coolers shall be installed on a level platform or base not less than 3 inches (76 mm) above the adjoining ground and secured to prevent displacement. Openings in exterior walls shall be flashed in accordance with Section R703.8.

❖ This section requires exterior equipment to be installed on a level platform at least 3 inches (76 mm) above the surrounding grade to protect the equipment from the effects of corrosion.

M1413.2 Protection of potable water. The potable water system shall be protected from backflow in accordance with the provisions in Section P2902.

❖ Since this type of equipment usually uses potable water and may have chemicals added to improve performance, this section requires backflow prevention devices installed on the potable water supply to protect it from contamination in compliance with the plumbing provisions of this code.

SECTION 1414 FIREPLACE STOVES

M1414.1 General. Fireplace stoves shall be listed, labeled and installed in accordance with the terms of the listing. Fireplace stoves shall be tested in accordance with UL 737.

❖ All units must be installed in accordance with the method by which they are tested. In this case, the units are tested in accordance with UL 737, and each set of manufacturer's instructions must indicate such. A fireplace stove is a free standing, chimney-connected, solid-fuel-burning heater designed to be operated with the fire chamber doors in either the open or closed position.

M1414.2 Hearth extensions. Hearth extensions for fireplace stoves shall be installed in accordance with the listing of the fireplace stove. The supporting structure for a hearth extension for a fireplace stove shall be at the same level as the supporting structure for the fireplace unit. The hearth extension shall be readily distinguishable from the surrounding floor area.

❖ Fireplace stoves must be installed in strict accordance with the manufacturers' installation instructions. These instructions will include specifications for the materials, dimensions, and installation of a hearth extension. The hearth extension must be of noncombustible materials and must be supported at the same level as the fireplace stove. The floor supporting the hearth extension can be of combustible construction. Since a fireplace stove is a free-standing heater, in most installations the fireplace stove will be installed on the hearth extension. The dimensions of the hearth extension will be dictated by the manufacturer's installation instructions and will include dimensions that the hearth extension must extend beyond the sides, back, and front of the fireplace stove.

A fireplace stove can be operated with the chamber doors in the open position, similar to a traditional masonry fireplace. While the hearth extension will protect the floor structure on all sides from radiant heat, in front of the stove the hearth extension will protect the floor structure from sparks, embers, coals, or burning solid fuel that might escape from the firebox. The nature of solid fuel can cause burning material to be expelled from the fire box, and large pieces of the fuel can fall from the grate or roll out of the firebox. In addition to the required hearth extensions, screens should be used as the initial and primary protection against sparks and embers from the firebox.

SECTION M1415 MASONRY HEATERS

M1415.1 General. Masonry heaters shall be constructed in accordance with Section R1006.

❖ Masonry heaters are a type of solid-fuel burning appliance and are covered in Section R1006, which requires installation in accordance with ASTM E 1602 or the terms of the listing. Such heaters are assembled on the job site from locally obtained raw materials or from factory components shipped to the job site. Masonry heaters are designed to be much more efficient than conventional fireplaces by incorporating immense thermal mass (masonry) and multiple pass flueways that provide greater heat exchange surface area and longer combustion gas retention time. Masonry heaters provide heat to the conditioned space by direct radiation of the heat energy stored in the immense thermal mass. Some factory-built kits are listed. Clearances to combustibles and convection cooling airflow are extremely important since these heaters store heat at high temperatures for long periods of time.

Bibliography

The following resource material is referenced in this chapter or is relevant to the subject matter addressed in this chapter.

ASHRAE 34-2001, *Designation and Classification of Refrigerants*. Atlanta: American Society of Heating, Refrigerating and Air-Conditioning Engineers, 2001.

ASTM E 1602-94, *Guide for Construction of Solid Fuel Burning Masonry Heaters*. West Conshohocken, PA: ASTM International, 1994

Manual J: Load Calculations for Residential Winter and Summer Air Conditioning. Arlington, VA: Air Conditioning Contractors of America, 1997.

NFPA 70–02, *National Electrical Code*. Quincy, MA: National Fire Protection Association, 2002.

UL 727–98, *Standard for Safety for Oil–Fired Central Furnaces*. Northbrook, IL: Underwriters Laboratories, 1998.

UL 737–96, *Fireplace Stoves*. Northbrook, IL: Underwriters Laboratories, 1996.

UL 1995–98, *Heating and Cooling Equipment*. Northbrook, IL: Underwriters Laboratories, 1998.

Chapter 15: Exhaust Systems

General Comments

As the title states, this chapter is a compilation of all exhaust-system-related code requirements applicable to those uses covered by the *International Residential Code*. The code regulates the materials used for constructing and installing such duct systems. Air brought into the building for ventilation, combustion, or makeup purposes is protected from contamination by the provisions found in this chapter.

Section M1501 pertains to the exhaust ducts from clothes dryers. Section M1502 has the criteria for exhaust ducts for range hoods. Section M1503 requires that microwave ovens be listed and labeled and properly installed. Section M1504 contains the regulations for exhausting domestic open-top broilers, and Section M1505 indicates that if the exhaust duct construction is not found in this chapter then the reader must use the criteria for ducts in Chapter 16.

The building official must be aware of the effect that exhaust systems can have on space conditioning and ventilation systems. For example, improperly designed exhaust systems can create negative pressure conditions, which can adversely affect the operation of fuel-burning appliances and the operation of other exhaust systems.

Purpose

The purpose of Chapter 15 is to set forth provisions associated with exhaust systems so hazards associated with exhaust systems and air contaminants may be avoided.

SECTION M1501 CLOTHES DRYERS EXHAUST

M1501.1 General. Dryer exhaust systems shall be independent of all other systems, shall convey the moisture to the outdoors and shall terminate on the outside of the building. Exhaust duct terminations shall be in accordance with the dryer manufacturer's installation instructions. Screens shall not be installed at the duct termination. Exhaust ducts shall not be connected with sheet-metal screws or fastening means which extend into the duct. Exhaust ducts shall be equipped with a backdraft damper. Exhaust ducts shall be constructed of minimum 0.016-inch-thick (0.406 mm) rigid metal ducts, having smooth interior surfaces with joints running in the direction of air flow. Flexible transition ducts used to connect the dryer to the exhaust duct system shall be limited to single lengths, not to exceed 8 feet (2438 mm) in length and shall be listed and labeled in accordance with UL 2158A. Transition ducts shall not be concealed within construction.

Exception: This section shall not apply to listed and labeled condensing (ductless) clothes dryers.

❖ Clothes dryer exhaust systems are a potential hazard because of the combustible lint and debris that can accumulate in the duct if the system is not designed and constructed properly. The combination of combustible lint and the elevated temperature of the exhaust can create a fire hazard in the system. The dryer exhaust system must be independent of other systems to prevent the hazard associated with one exhaust entering into or affecting other systems or areas in the building. The discharge must terminate outdoors because the high levels of moisture in the exhaust can cause condensation to form on exposed surfaces or in insulation material if discharged indoors. This can cause structural damage, deterioration of building material, and contribute to the growth of mold and fungus. The exhaust also contains highly combustible clothes fibers and, in the case of gas-fired units, products of combustion.

The exception specifically excludes condensing clothes dryers from the requirement to exhaust to the outdoors. Condensing dryers are electrically heated and recirculate the same air through the clothes drum. An air-to-air heat exchanger condenses the water vapor in the clothes drum air by using room air as the condensing medium. The condensate is collected in a reservoir that is either manually emptied or automatically emptied by an integral condensate pump. These dryers do not discharge heated air to the outdoors and have very little effect on the indoor temperature; thus, they are energy efficient.

M1501.2 Exhaust duct size. The diameter of the exhaust duct shall be as required by the clothes dryer's listing and the manufacturer's installation instructions.

❖ This section requires that the diameter of the exhaust duct must be that recommended by the dryer manufacturer. Ducts that are too small would restrict air movement and would cause the dryer to consume more energy because of the longer drying time. They could also cause appliance overheating. Ducts that are too large will cause a decrease in the flow velocity, causing lint and debris to drop out of the exhaust air stream.

M1501.3 Length limitation. The maximum length of a clothes dryer exhaust duct shall not exceed 25 feet (7620 mm) from the dryer location to the wall or roof termination. The maximum length of the duct shall be reduced 2.5 feet (762 mm) for each

45-degree (0.79 rad) bend and 5 feet (1524 mm) for each 90-degree (1.6 rad) bend. The maximum length of the exhaust duct does not include the transition duct.

Exceptions:

1. Where a clothes dryer booster fan is installed and listed and labeled for the application, the maximum length of the exhaust duct, including any transition duct, shall be permitted to be in accordance with the booster fan manufacturer's installation instructions. Where a clothes dryer booster fan is installed and not readily accessible from the room in which the dryer is located, a permanent identifying label shall be placed adjacent to where the exhaust duct enters the wall. The label shall bear the words "This dryer exhaust system is equipped with a remotely located booster fan."
2. Where the make and model of the clothes dryer to be installed is known and the manufacturer's installation instructions for such dryer are provided to the building official, the maximum length of the exhaust duct, including any transition duct, shall be permitted to be in accordance with the dryer manufacturer's installation instructions.

❖ The duct length, the number and angle of fittings, and the smoothness of the duct interior all contribute to the total friction loss of the air flow. When the friction loss is excessive, a slowdown of the air velocity occurs, allowing lint and debris to accumulate in the duct and thus creating a fire hazard. The manufacturers' instructions and the code limitations must be adhered concerning duct length, the number of elbows installed, and the type of termination fittings.

Exception 1 allows the duct length to exceed 25 feet (7620 mm) if a booster fan listed for this application is installed in the duct to maintain or increase the flow velocity. This exception allows the booster fan manufacturer to specify the duct length. If the booster fan is not located in the same room as the clothes dryer, a permanent label or sign is required to notify residents that a booster fan exists in the duct system.

The maximum exhaust duct length of 25 feet (7620 mm) is the normal installation requirement for domestic clothes dryers. The 25-foot (7620 mm) limit is based on the worst case scenario where the dryer is rated for a maximum duct length of 25 feet (7620 mm). Exception 2 allows longer exhaust duct lengths when the manufacturer's instructions specify the additional length. The make and model of the dryer must be provided to the code official, along with the installation instructions, to permit the code official to inspect the duct installation based on the manufacturer's instructions.

A point to consider is the situation that will exist when the original occupant moves out and the new occupant does not install a similar dryer that allows the longer duct length. If the duct is not marked in a conspicuous manner, the new resident will be unaware of the potentially hazardous situation that has been created by the dryer and exhaust mismatch.

SECTION M1502
RANGE HOODS

M1502.1 General. Range hoods shall discharge to the outdoors through a single-wall duct. The duct serving the hood shall have a smooth interior surface, shall be air tight and shall be equipped with a backdraft damper. Ducts serving range hoods shall not terminate in an attic or crawl space or areas inside the building.

Exception: Where installed in accordance with the manufacturer's installation instructions, and where mechanical or natural ventilation is otherwise provided, listed and labeled ductless range hoods shall not be required to discharge to the outdoors.

❖ A domestic kitchen exhaust system is one that serves appliances typically found in residential occupancies such as within dwelling units. When compared to commercial cooking operations, residential cooking operations are far less frequent, of shorter duration, have lower heat output, and produce fewer grease-laden vapors. However, airborne contaminant control may be even more important in residential cooking operations because of the lower or nonexistent ventilation rates typical of dwelling units.

The ducts must be sealed air tight to prevent leakage of air and grease into wall and ceiling cavities. Ducts should be sealed with a material that is suitable for long-term exposure to elevated temperatures. Backdraft dampers prevent the filtration of outdoor air when the exhaust system is not operating. The hood manufacturer's instructions may require that the ducts be installed with a minimum clearance to combustibles. See the commentary for Section M1307.1.

The exception to this section allows the use of ductless (recirculating) range hoods that have no means for discharge to the outdoors. Exhaust to the outdoors would be required if natural or mechanical ventilation was not provided.

M1502.2 Duct material. Single-wall ducts serving range hoods shall be constructed of galvanized steel, stainless steel or copper.

Exception: Ducts for domestic kitchen cooking appliances equipped with down draft exhaust systems shall be permitted to be constructed of schedule 40 PVC pipe provided that the installation complies with all of the following:

1. The duct shall be installed under a concrete slab poured on grade,
2. The underfloor trench in which the duct is installed shall be completely backfilled with sand or gravel,
3. The PVC duct shall extend not greater than 1 inch (25.4 mm) above the indoor concrete floor surface,
4. The PVC duct shall extend not greater than 1 inch (25.4 mm) above grade outside of the building, and

5. The PVC ducts shall be solvent cemented.

❖ Ducts used for dwelling unit kitchen exhaust must be constructed of noncombustible materials to reduce the potential fire hazard associated with the collection of grease or the absorption of grease into a more porous duct material. This section requires domestic kitchen exhaust ducts to be constructed of sheet steel, aluminum or copper.

The exception allows the use of plastic ducts to exhaust the downdraft type domestic kitchen cooking appliances when all five of the specified installation criteria are met. Plastic ducts are not allowed in aboveground applications because the plastic cannot contain fire within the confines of the duct. This exception allows plastic for the duct material because the duct must be below a slab on grade, and the trench for the duct must be completely backfilled with sand or gravel to (1) limit the amount of combustion air available to sustain a duct fire and (2) prevent a fire within the duct from spreading to the rest of the building. Additionally, the fumes and smoke from the burning plastic will be exhausted to the outdoors, not introduced into the occupied space. The plastic material also has the advantage of corrosion resistance. Metal ducts below a slab are susceptible to deterioration.

M1502.3 Kitchen exhaust rates. Where domestic kitchen cooking appliances are equipped with ducted range hoods or down-draft exhaust systems, the fans shall be sized in accordance with Section M1506.3.

❖ This section does not require domestic kitchen cooking appliances to be provided with a hood that exhausts to the outdoors. It only provides guidance for sizing the exhaust fans for hoods that do exhaust to the outdoors. Mechanical ventilation would only be required where natural ventilation is not provided. See Section R303.1.

SECTION M1503 INSTALLATION OF MICROWAVE OVENS

M1503.1 Installation of microwave oven over a cooking appliance. The installation of a listed and labeled cooking appliance or microwave oven over a listed and labeled cooking appliance shall conform to the terms of the upper appliance's listing and label and the manufacturer's installation instructions. The microwave oven shall conform to UL 923.

❖ A cooking appliance or microwave oven installed above cooking appliances must be listed for use above the lower appliance. Installation clearances must comply with the manufacturers' installation instructions for both the upper and lower appliances. UL 923 is the appropriate standard pertaining to microwave ovens.

SECTION M1504 OVERHEAD EXHAUST HOODS

M1504.1 General. Domestic open-top broiler units shall be provided with a metal exhaust hood, not less than 28 gage, with a clearance of not less than 0.25 inch (6.4 mm) between the hood and the underside of combustible material or cabinets. A clearance of at least 24 inches (610 mm) shall be maintained between the cooking surface and the combustible material or cabinet. The hood shall be at least as wide as the broiler unit and shall extend over the entire unit. Such exhaust hood shall discharge to the outdoors and shall be equipped with a backdraft damper or other means to control infiltration/exfiltration when not in operation. Broiler units incorporating an integral exhaust system, and listed and labeled for use without an exhaust hood, need not be provided with an exhaust hood.

❖ The surface material of the hood must not come in direct contact with combustible material. Therefore a $^{1}/_{4}$-inch (6.4 mm) clearance between the hood and the underside of combustible material or cabinets is required. Otherwise, a grease fire of hot gases from the broiler unit may heat the hood to temperatures that could cause ignition of combustible material mounted above the hood.

A clearance of at least 24 inches (610 mm) must be maintained between the cooking surface and overhead cabinets constructed of combustible materials. The hood must be at least as wide as the broiler unit and extend over the entire unit. These clearances help ensure that combustible material will not be ignited in the event of a fire on the broiler. The clearances also provide space for practical cooking operations.

SECTION M1505 EXHAUST DUCTS

M1505.1 Ducts. Where exhaust duct construction is not specified in this chapter, such construction shall comply with Chapter 16.

❖ This section provides a reference to Chapter 16 for any exhaust duct construction not specified in Chapter 15.

SECTION M1506 MECHANICAL VENTILATION

M1506.1 General. Where toilet rooms and bathrooms are mechanically ventilated, the ventilation equipment shall be installed in accordance with this section.

❖ Section R303.3 requires either natural or mechanical ventilation of toilet rooms and bathrooms. This section brings mechanical ventilation into the exhaust chapter, which is both appropriate and convenient.

M1506.2 Recirculation of air. Exhaust air from bathrooms and toilet rooms shall not be recirculated within a residence or to another dwelling unit.

❖ This section prohibits the recirculation of exhaust air from toilet rooms and bathrooms within a dwelling or to another dwelling. Recirculating (ductless) exhaust fans are not allowed if mechanical ventilation of a space is required, but such fans are allowed in naturally ventilated spaces.

Recirculation to other dwelling units is prohibited because odors, disease causing organisms, and excess moisture could be transported from one dwelling unit to another.

M1506.3 Ventilation rate. Ventilation systems shall be designed to have the capacity to exhaust the minimum air flow rate determined in accordance with Table M1506.3.

❖ The ventilation rates for kitchens, toilet rooms, and bathrooms are specified in Table M1506.3.

TABLE M1506.3
MINIMUM REQUIRED EXHAUST RATES FOR ONE- AND TWO-FAMILY DWELLINGS

AREA TO BE VENTILATED	VENTILATION RATES
Kitchens	100 cfm intermittent or 25 cfm continuous
Bathrooms—Toilet Rooms	Mechanical exhaust capacity of 50 cfm intermittent or 20 cfm continuous

For SI: 1 cubic foot per minute = 0.0004719 m^3/s.

❖ The exhaust rates specified in this table are identical to those in Table 403.3 of the *International Mechanical Code*. The continuous rates are lower than the intermittent rates, reflecting the fact that an exhaust fan that runs continuously will provide some degree of overventilation during periods in which contaminants are not being produced. An intermittent exhaust fan will run for short periods of time only, requiring a higher exhaust rate to effectively control odors, cooking effluent, and moisture.

Bibliography

The following resource material is referenced in this chapter or is relevant to the subject matter addressed in this chapter.

ASHRAE Fundamentals Handbook. Atlanta: American Society of Heating, Refrigerating and Air–Conditioning Engineers, Inc., 2001.

ASHRAE HVAC Systems and Equipment Handbook. Atlanta: American Society of Heating, Refrigerating and Air–Conditioning Engineers, Inc., 2000.

ASHRAE 62-01, *Ventilation for Acceptable Indoor Air Quality.* Atlanta: American Society of Heating, Refrigerating and Air-Conditioning Engineers, Inc., 2001.

ASTM E 84–01, *Test Method for Surface Burning Characteristics of Building Materials*. West Conshohocken, PA: ASTM International, 2001.

ASTM E 814–00, *Test Method for Fire Tests of Through–Penetration Fire Stops*. West Conshohocken, PA: ASTM International, 2000.

DOE 10 CFR, Part 20-95, *Standards for Protection against Radiation.* Washington, DC: United States Department of Energy, 1995.

NFPA 70–02, *National Electrical Code*. Quincy, MA: National Fire Protection Association, 2002.

SMACNA-95, *HVAC Duct Construction Standards—Metal and Flexible*. Chantilly, VA: Sheet Metal and Air-Conditioning Contractors National Association, 1995.

UL 181–96, *Factory–Made Air Ducts and Air Connectors*; Northbrook, IL: Underwriters Laboratories, 1996.

UL 923-95, *Microwave Cooking Appliances*. Northbrook, IL: Underwriters Laboratories, 1995.

Chapter 16: Duct Systems

General Comments

This chapter addresses duct systems by instituting requirements for the protection of the occupants of the building and the building itself. This chapter regulates the materials and methods of construction. Duct construction affects the performance of the entire air distribution system.

Section M1601 contains the material requirements that are generally applied to all duct construction and installation. Section M1602 is specific to those regulations for return air and Section M1603 contains regulations for supply air.

Purpose

Chapter 16 provides requirements for the installation of supply, return, and exhaust air systems. The chapter contains no information on the design of these systems from the standpoint of air movement, but it is concerned with the structural integrity of the systems and the overall impact of the systems on the fire-safety performance of the building. Design considerations such as duct sizing, maximum efficiency, cost effectiveness, occupant comfort, and convenience are the responsibility of the designer.

SECTION M1601 DUCT CONSTRUCTION

M1601.1 Duct design. Duct systems serving heating, cooling and ventilation equipment shall be fabricated in accordance with the provisions of this section and ACCA Manual D or other approved methods.

❖ Material and construction requirements for duct systems can be divided into two major categories: those for above-ground ducts and those for underground ducts. The demands placed on these categories are different; therefore, they are treated separately in the code and in the following commentary.

Adequately sized supply air and return air ductwork is essential for efficient and proper circulation of conditioned air. Ducts must be large enough to allow sufficient air flow through appliances to satisfy refrigerant coil air flow demands and to avoid excessive temperature rise. As a practical matter, restriction of the supply duct system is an infrequently encountered problem. However, an inadequately sized return system is often found in the field, particularly on heat pump installations. The code does not provide specific requirements for sizing duct systems, but instead relies on ACCA *Manual D* or other sizing criteria, as approved by the code official.

M1601.1.1 Above-ground duct systems. Above-ground duct systems shall conform to the following:

1. Equipment connected to duct systems shall be designed to limit discharge air temperature to a maximum of 250°F (121°C).
2. Factory-made air ducts shall be constructed of Class 0 or Class 1 materials as designated in Table M1601.1.1(1).
3. Fibrous duct construction shall conform to the SMACNA "Fibrous Glass Duct Construction Standards" or NAIMA "Fibrous Glass Duct Construction Standards."
4. Minimum thicknesses of metal duct material shall be as listed in Table M1601.1.1(2). Galvanized steel shall conform to ASTM A 525.
5. Gypsum products are permitted to be used to construct return air ducts or plenums, provided that the air temperature does not exceed 125°F (52°C) and exposed surfaces are not subject to condensation.
6. Duct systems shall be constructed of materials having a flame-spread index not greater than 200.
7. Stud wall cavities and the spaces between solid floor joists to be utilized as air plenums shall comply with the following conditions:
   - 7.1. Such cavities or spaces shall not be utilized as a plenum for supply air.
   - 7.2. Such cavities or spaces shall not be part of a required fire-resistance-rated assembly.
   - 7.3. Stud wall cavities shall not convey air from more than one floor level.
   - 7.4. Stud wall cavities and joist space plenums shall be isolated from adjacent concealed spaces by tight-fitting fire blocking in accordance with Section R602.8.

❖ Metallic ducts are usually constructed using galvanized sheet steel. Duct size is based on required airflow, system pressure, flow velocity, and pressure losses caused by friction. Duct material thickness is determined by duct size, static pressure of the system, distance between supports, and whether the duct is reinforced. Metallic ducts must be constructed with the minimum thicknesses specified in Table M1601.1.1(2), which bases the minimum required duct thickness on the geometry of the duct, the material used, and the major dimension of the duct (the diameter for round ducts and the widest side for rectangular ducts).

Nonmetallic ducts and duct materials must be tested and classified in accordance with the provisions of UL 181. Only Class 0 and Class 1 ducts can be used. Class 0 indicates flame-spread and smoke-developed indexes of zero; Class 1 indicates a flame spread index not greater than 25 and a smoke-developed index not greater than 50, when tested to ASTM E 84.

UL 181 requires that a nonmetallic duct be tested to determine its fire-performance characteristics, corrosion resistance, mold-growth resistance, humidity resistance, leakage resistance, temperature resistance, erosion resistance, and structural performance. Air ducts that conform to the requirements of UL 181 are identified by the manufacturer's or vendor's name, rated velocity, negative and positive pressure classification, and duct material class.

Fibrous ducts are constructed of a composite material of rigid (high-density) fiberglass board and a factory-applied facing, typically reinforced aluminum. The surface of the fibrous duct that is exposed to the airflow is sealed with a fiber-bonding adhesive that prevents erosion of the fiberglass material. The factory-applied exterior duct-board facing contributes to the strength and rigidity of the composite material, acts as a heat reflector, serves as a vapor barrier, and is an integral component of the joining method used to construct fibrous ducts. The material is available in board form for shop or field fabrication into rectangular sections or in ten-sided duct form, which approximates a circular cross section.

Fibrous ducts take advantage of the inherent insulating qualities of the glass fiber material. The air friction factors for fibrous ducts are greater than those for sheet metal because of the relatively rough surface finish of the former.

Construction of fibrous glass ducts must conform to the requirements of the SMACNA *Fibrous Glass Duct Construction Standards*, which provides details for the design and fabrication of air distribution systems using fibrous glass ducts. The SMACNA standards referenced in this section and the previous section are enforceable extensions of the code.

The maximum discharge temperature permitted by industry standards for warm air heating systems is 250°F (121°C). This section prohibits nonmetallic ducts from being used in applications in which the air temperature would exceed this maximum because the material has not been tested to withstand higher temperatures, and high temperatures will cause accelerated aging of the duct material.

Gypsum board is a composite material commonly used for the construction of air plenums and shafts. Gypsum board can reduce construction costs because it is a common component of building construction assemblies. By serving a dual purpose, gypsum board eliminates the need for independent duct construction. The use of gypsum board to form ducts and plenums is specifically regulated to prevent deterioration of the gypsum board material. Air temperatures that exceed 125°F (52°C) will, over time, dry both the paper facing and the gypsum of the gypsum board, leading to deterioration of the panel.

Gypsum board can also deteriorate when exposed to moisture, which will happen if the surface temperature of the gypsum board is lower than the airstream dew-point temperature, causing water to condense on the surface of the gypsum board. For these reasons, gypsum board cannot be used for air distribution systems using evaporative cooling equipment. It is further restricted to return-air system applications only, a maximum airstream temperature of 125°F (52°C), and an airstream dew-point temperature continuously below the temperature of the gypsum board surface. Evaporative cooling equipment, such as "swamp coolers," uses water as a refrigerant. The resulting addition of moisture to the airstream could cause deterioration of the gypsum board.

Common in residential construction is the stud-space and joist-space being used as a plenum. These spaces are limited to use for return air only because the negative pressures within the return-air plenum with respect to surrounding spaces will decrease the likelihood of spreading smoke to other spaces via the plenum. Also, the temperature and moisture content of heated, cooled, and conditioned supply air could cause a fire hazard or deterioration of the construction materials exposed in the spaces.

The space must not be a part of a fire-resistance-rated assembly because the ASTM E 119 test does not consider the impact of air movement within the assembly on the fire-resistance rating. This restriction is a concession to the convenience and cost-saving potential of this method of moving air. The use of stud spaces inherently means the interconnection of different floor levels by the concealed space. Because of the hazard of such an interconnection, the use of this type of plenum is limited to return air from one floor level only for each independent stud cavity. All cavities not used for air movement must be isolated from the plenum by fireblocking constructed and installed in accordance with this code.

Commentary Figure M1601.1.1(1) shows an example of an acceptable stud and joist space installation. The bottom plate of the wall is cut away for the plenum to function, and fireblocking is provided in the joist and stud space to limit communication of the plenum with other spaces. The stud cavity shown is being used to return air from one floor level only, conducting it to the space below, where the air-handling equipment is located. Commentary Figure M1601.1.1(2) shows an example of an unacceptable stud and joist space installation.

Whether viewed as a shaft or as a duct, a stud-cavity plenum penetrates floor assemblies and is, therefore, subject to the floor penetration protection requirements of Section R321.3.1 which may require floor penetration protection.

Since air is returned from more than one floor level through the same stud cavity, this section prohibits this type of installation because it creates a direct connection from one floor level to another by means of a concealed cavity. Such direct connections would act as a chase or chimney, allowing fire and smoke to spread quickly upward through the building. Thus, stud-cavity return-air plenums are subject to the same restrictions and requirements as floor openings and penetrations regulated by Section R321.3. Stud-space and joist-space plenums must be viewed as an exception to the fireblocking provisions of Section R602.8 because one or more fireblocks (wall plates) must be removed or relocated to construct the plenums. The code does not mention the type of materials allowed for "panning" the bottom of open joists to create joist-space plenums. Traditionally, sheet metal has been used; however, composite materials are also used. The building official must determine what materials are acceptable for joist panning.

TABLE M1601.1.1(1)
CLASSIFICATION OF FACTORY-MADE AIR DUCTS

DUCT CLASS	MAXIMUM FLAME-SPREAD RATING
0	0
1	25

❖ This table, referred to by Section M1601.1.1, contains the maximum flame-spread ratings for the two classes of factory-made duct construction.

M1601.1.2 Underground duct systems. Underground duct systems shall be constructed of approved concrete, clay, metal or plastic. The maximum duct temperature for plastic ducts shall not be greater than 150°F (66°C). Metal ducts shall be protected from corrosion in an approved manner or shall be completely encased in concrete not less than 2 inches (51 mm) in thickness. Nonmetallic ducts shall be installed in accordance with the manufacturer's installation instructions. Plastic pipe and fitting materials shall conform to cell classification 12454-B of ASTM D 1248 or ASTM D 1784 and external loading properties of ASTM D 2412. All ducts shall slope to an accessible point for drainage. Where encased in concrete, ducts shall be sealed and secured prior to pouring any concrete. Metallic ducts having an approved protective coating and nonmetallic ducts shall be installed in accordance with the manufacturer's installation instructions.

❖ Ducts installed underground must be able to resist the forces imposed on them by the materials that encase them, the forces created by floodwaters in and around them, and corrosion. The ASHRAE *Handbook of HVAC Systems and Equipment* recommends that all underground ducts and fittings be round to provide optimum structural performance. Unlike round ducts, square or rectangular ducts offer little resistance to deformation or collapse caused by the structural loads associated with burial.

Metal ducts must have either a protective coating to resist corrosion or be completely encased in concrete, a minimum of 2 inches (52 mm) thick all around. Concrete-encased ducts may eventually corrode; however, the air passageway will be maintained because of the remaining concrete enclosure. All nonmetallic ducts and metallic ducts with factory-applied protective coatings must be approved, and all such duct installations must be in accordance with the manufacturers' installation instructions. Application of any field-applied protective coating must be approved. Great care must be taken to protect underground ducts from damage prior to placing the concrete or installing the permanent structure above them. Plastic duct and fitting systems designed for underground applications are available and allow corrosion-resistant and waterproof installations.

Plastic ducts have the advantage of being corrosion resistant. The external coating properties of ASTM R 2412 assure that the pipe has the ability to resist deformation from the loads associated with direct burial. Plastic ducts rapidly lose strength as their temperature approaches their maximum service temperature. At temperatures above 150°F (65.5°C), PVC pipe is substantially weakened and deformation and/or collapse is possible.

Underground ducts must be sloped to drain to an accessible point in the event that water enters the duct through duct openings or from the subsoil. Water can cause corrosion, deterioration of duct materials, and duct blockages; therefore, sloping the duct to drain to a collection point will allow water to be removed. The code does not require ducts to be water tight. This section requires the duct to be sealed before the concrete is poured to prevent concrete from entering the duct.

M1601.2 Factory-made ducts. Factory-made air ducts or duct material shall be approved for the use intended, and shall be installed in accordance with the manufacturer's installation instructions. Each portion of a factory-made air duct system shall bear a listing and label indicating compliance with UL 181 and UL 181A or UL 181B.

❖ Several types of factory-made air ducts are available for a variety of applications; they are listed products that comply with UL 181. Air ducts are designed to convey air to or from a conditioned area. Labeling for compliance with UL 181 ensures that ducts have been fabricated from materials deemed satisfactory for the conditions of use.

TABLE M1601.1.1(2)
GAGES OF METAL DUCTS AND PLENUMS USED FOR HEATING OR COOLING

TYPE OF DUCT	SIZE (inches)	MINIMUM THICKNESS (inch)	EQUIVALENT GALVANIZED SHEET GAGE	APPROXIMATE ALUMINUM B & S GAGE
Round ducts and enclosed rectangular ducts	14 or less over 14	0.013 0.016	30 28	26 24
Exposed rectangular ducts	14 or less over 14	0.016 0.019	28 26	24 22

For SI: 1 inch = 25.4 mm.

❖ The metal thicknesses allowed by Table R603.3 have been used in dwelling unit construction without evidence of failure. Therefore, the lighter gages are justified in this limited application.

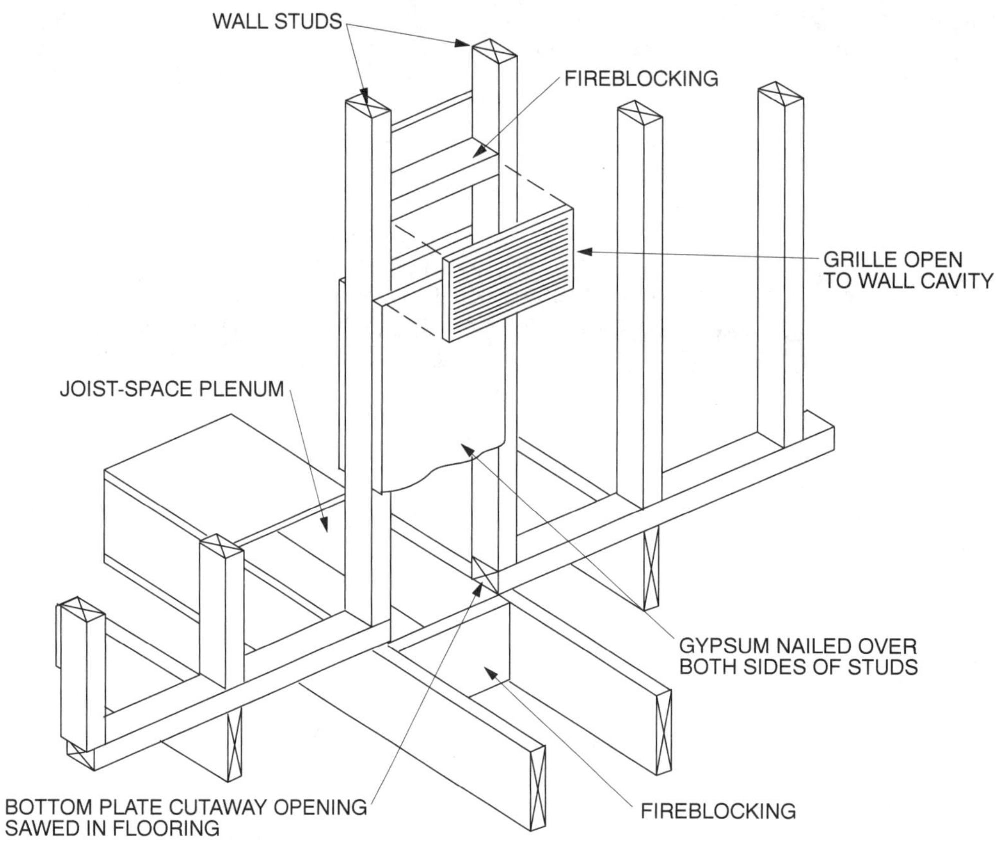

Figure M1601.1.1(1)
ACCEPTABLE STUD AND JOIST-SPACE INSTALLATION

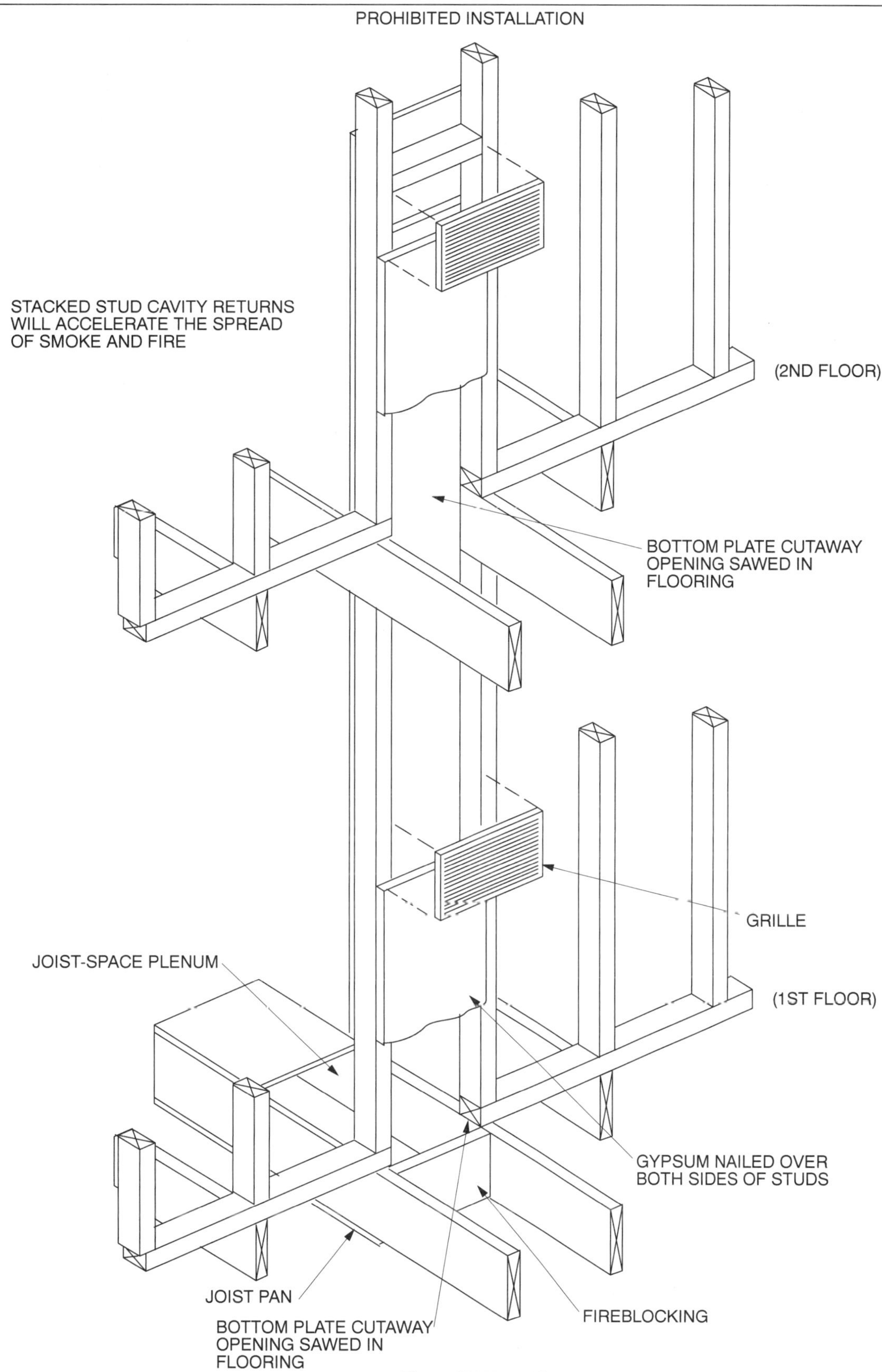

Figure M1601.1.1(2)

UNACCEPTABLE STUD AND JOIST-SPACE PLENUM INSTALLATION

M1601.2.1 Duct insulation materials. Duct insulation materials shall conform to the following requirements:

1. Duct coverings and linings shall have a flame-spread index not greater than 25, and a smoke-developed index not greater than 50 when tested in accordance with ASTM E 84. Specimens tested in accordance with ASTM E 84 shall be representative of the final composite material.
2. Duct coverings and linings shall not flame, glow, smolder or smoke when tested in accordance with ASTMC 411 at the temperature to which they are exposed in service. The test temperature shall not fall below 250°F (121°C).
3. External duct insulation and factory-insulated flexible ducts shall be legibly printed or identified at intervals not greater than 36 inches (914 mm) with the name of the manufacturer; the thermal resistance *R*-value at the specified installed thickness; and the flame spread and smoke-developed indexes of the composite materials. All duct insulation product *R*-values shall be based on insulation only, excluding air films, vapor retarders or other duct components, and shall be based on tested *C*-values at 75°F (24°C) mean temperature at the installed thickness, in accordance with recognized industry procedures. The installed thickness of duct insulation used to determine its *R*-values shall be determined as follows:
   - 3.1. For duct board, duct liner and factory-made rigid ducts not normally subjected to compression, the nominal insulation thickness shall be used.
   - 3.2. For duct wrap, the installed thickness shall be assumed to be 75 percent (25-percent compression) of nominal thickness.
   - 3.3. For factory-made flexible air ducts, the installed thickness shall be determined by dividing the difference between the actual outside diameter and nominal inside diameter by two.

❖ Since duct systems connect most rooms in a building, they can provide a path for fire and smoke to travel throughout that building. Duct coverings and linings are exposed to the surrounding environment and to the airstream in the duct. To reduce the possible spread of fire and smoke, duct coverings and linings must be tested to ASTM E 84. These materials are limited to a maximum flame spread index of 25 and a smoke-developed index of 50, which correspond to Class 1 material. In addition, duct coverings and linings must be rated for the air temperatures expected; this avoids degradation of these materials.

In order to verify that duct coverings and linings will not present a fire hazard, such materials must be tested in the form in which they will be installed at their rated temperatures but to not less than 250°F (121°C) in accordance with the procedures of ASTM C 411. This minimum temperature for testing represents the maximum temperature that industry standards will permit in the airstream of a warm-air heating appliance. It is important that the duct coverings and linings are tested in their composite form rather than having tests conducted on each component that comprises the product (i.e., the insulation, the facing, and the adhesive). Each component could pass the ASTM E 84 tests individually, but could fail when combined into the final assembly to be installed in the field.

To assist inspectors, the code requires duct insulation to have a label with the manufacturer's name, thermal resistance (*R*), and the flame/smoke indices.

A third-party agency must provide quality control inspections at the manufacturer's facility in accordance with the requirements for labeling. Testing performed by an independent agency must determine the insulating *R*-value, the flame spread index, and the smoke-developed index. The 36-inch (914 mm) label intervals are intended to increase the likelihood that every cut piece of insulation and flexible duct will have a label.

The thermal performance of duct insulation is dependent on its "installed" condition including the compressed condition, as is the case for duct wrap insulations. This section is intended to provide manufacturers, installers, and inspectors with specific guidance for meeting the intent of the code. For example, installed duct wrap is assumed to have a thickness of 75 percent of the nominal uninstalled thickness. The *R*-value on the product will account for the decreased thermal resistance caused by compression of the product.

M1601.2.2 Vibration isolators. Vibration isolators installed between mechanical equipment and metal ducts shall be fabricated from approved materials and shall not exceed 10 inches (254 mm) in length.

❖ Isolators must be built from materials that will withstand the temperatures and pressures of typical conditioned air passing through the duct. The 10-inch (254 mm) length limitation prevents installation of long lengths of nonrigid duct material, which could adversely affect the integrity of the duct system.

M1601.3 Installation. Duct installation shall comply with Sections M1601.3.1 through M1601.3.6.

❖ Due to increasing energy costs, ducts are being insulated more frequently to reduce losses. The following sections discuss the details of individual duct insulation installation requirements.

M1601.3.1 Joints and seams. Joints of duct systems shall be made substantially airtight by means of tapes, mastics, gasketing or other approved closure systems. Closure systems used with rigid fibrous glass ducts shall comply with UL 181A and shall be marked "181A-P" for pressure-sensitive tape, "181A-M" for mastic or "181A-H" for heat-sensitive tape. Closure systems used with flexible air ducts and flexible air connectors shall comply with UL 181B and shall be marked "181B-FX" for pressure-sensitive tape or "181B-M" for mastic. Duct connections to flanges of air distribution system equipment or sheet metal fittings shall be mechanically fastened. Crimp joints for round ducts shall have a contact lap of at least 1.5 inches (38 mm) and shall be mechanically fastened by means of at least three sheet metal screws or rivets equally spaced around the joint.

❖ Air leaking from ducts in unconditioned spaces must be minimized to prevent energy losses. Joints must be made airtight by approved methods, which include tapes, mastics, gasketing, and other approved closure systems. Tape alone cannot be substituted for mechanical fasteners, as these are needed to ensure a structurally sound duct system.

M1601.3.2 Support. Metal ducts shall be supported by 0.5-inch (12.7 mm) wide 18-gage metal straps or 12-gage galvanized wire at intervals not exceeding 10 feet (3048 mm) or other approved means. Nonmetallic ducts shall be supported in accordance with the manufacturer's installation instructions.

❖ These material requirements and the 10-foot (3048 mm) maximum spacing requirement prescribe structural support for metal ductwork to limit deflection and maintain alignment. Nonmetallic ducts must be supported in accordance with the manufacturer's installation instructions because these ducts are produced in various configurations.

M1601.3.3 Fireblocking. Duct installations shall be fireblocked in accordance with Section R602.8.

❖ This section states that all openings around ducts at ceiling and floor levels must be fireblocked in accordance with Section R602.8. The fireblocks retard the spread of fire to other areas in the dwelling.

M1601.3.4 Duct insulation. Duct insulation shall be installed in accordance with the following requirements:

1. A vapor retarder having a maximum permeance of 0.05 perm [(2.87 ng/(s · m^2 · Pa)] in accordance with ASTM E 96, or aluminum foil with a minimum thickness of 2 mils (0.051 mm), shall be installed on the exterior of insulation on cooling supply ducts that pass through nonconditioned spaces conducive to condensation.
2. Exterior duct systems shall be protected against the elements.
3. Duct coverings shall not penetrate a fireblocked wall or floor.

❖ Where a duct is installed outdoors or in an unconditioned area (such as an attic), it can be exposed to humidity and temperature differentials that can create condensation on the outside of the duct. A vapor retarder must be installed on the exterior of the duct to protect the duct and/or insulation from damage due to moisture. Duct insulation and coverings must not penetrate fireblocked assemblies. If the exterior insulation and vapor retarder burn in a fire, a pathway will open for the fire to spread through the duct penetration.

M1601.3.5 Factory-made air ducts. Factory-made air ducts shall not be installed in or on the ground, in tile or metal pipe, or within masonry or concrete.

❖ Factory-made air ducts must be listed and labeled for underground installations. Ducts not listed for in-ground installations may not have sufficient strength to withstand the loads applied by these types of installations. Therefore, they must not be installed in tile, metal pipe, concrete or masonry.

M 1601.3.6 Duct separation. Ducts shall be installed with at least 4 inches (102 mm) separation from earth except where they meet the requirements of Section M1601.1.2.

❖ A physical separation from earth is the best method of protecting metal ducts from the effects of corrosion.

M1601.3.7 Ducts located in garages. Ducts in garages shall comply with the requirements of Section R309.1.1.

❖ Ducts in garages and ducts penetrating separation walls or ceilings between garages and living spaces must be designed to prevent fire and smoke from easily entering the living spaces. See the commentary for Section R309.1.1.

M1601.3.8 Flood hazard areas. In areas prone to flooding as established by Table R301.2(1), duct systems shall be located or installed in accordance with Section R323.1.5.

❖ In dwellings located in flood hazard areas, ducts must be installed above the design flood elevation or must be capable of preventing water from entering the ducts and capable of withstanding the forces of buoyancy and moving water. See the commentary for Section R323.1.5.

M1601.4 Under-floor plenums. An under-floor space used as a supply plenum shall conform to the requirements of this section. Fuel gas lines and plumbing waste cleanouts shall not be located within the space.

❖ The code permits an under-floor crawl space to be used as a supply plenum if certain conditions are fulfilled. A counterflow furnace is normally used in the space above to discharge the conditioned air to an under-floor plenum. Floor registers are installed in openings cut through the floor for delivery of the conditioned air to the space above. A noncombustible receptacle must be installed under each floor register to catch matches, embers, or other sources of ignition that might fall through the floor registers. See Commentary Figure M1601.4 for an example of an under-floor plenum system.

M1601.4.1 General. The space shall be cleaned of loose combustible materials and scrap, and shall be tightly enclosed. The ground surface of the space shall be covered with a moisture barrier having a minimum thickness of 4 mils (0.102 mm).

❖ All loose combustible scrap must be removed from an under-floor space used as a plenum. The space must be substantially airtight since the furnace will place the plenum under a slight positive pressure in order to force conditioned air out of the registers into the rooms above. If there are leaks in the plenum walls, the conditioned air will escape to the outside and result in energy loss.

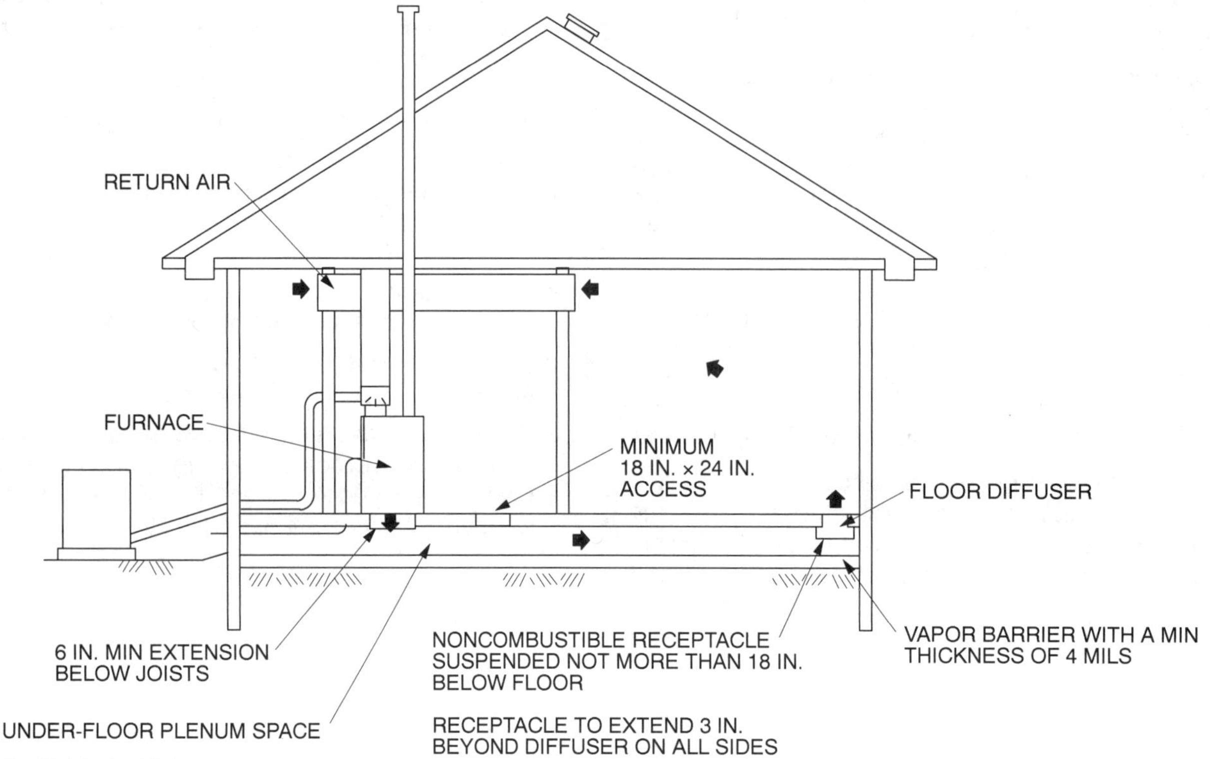

For SI: 1 inch = 25.4 mm.

Figure M1601.4
UNDER-FLOOR PLENUM SYSTEM

M1601.4.2 Materials. The under-floor space, including the sidewall insulation, shall be formed by materials having flame-spread ratings not greater than 200 when tested in accordance with ASTM E 84.

❖ The enclosing materials of an under-floor plenum, including the sidewall insulation, are limited to materials having a flame-spread index less than 200.

M1601.4.3 Furnace connections. A duct shall extend from the furnace supply outlet to not less than 6 inches (152 mm) below the combustible framing. This duct shall comply with the provisions of Section M1601.1. A noncombustible receptacle shall be installed below any floor opening into the plenum in accordance with the following requirements:

1. The receptacle shall be securely suspended from the floor members and shall not be more than 18 inches (457 mm) below the floor opening.
2. The area of the receptacle shall extend 3 inches (76 mm) beyond the opening on all sides.
3. The perimeter of the receptacle shall have a vertical lip at least 1 inch (25.4 mm) high at the open sides.

❖ A duct must be installed through the floor from the furnace outlet to at least 6 inches (153 mm) below combustible framing in the under-floor plenum. A noncombustible receptacle must be installed below the opening in the plenum to deflect the warm air from the furnace. The duct through the floor provides assurance that air is delivered to the plenum without leaking into the furnace enclosure. The receptacle deflects the supply air from the furnace to provide improved air distribution in the plenum. This receptacle must be suspended from the floor members and must not be more than 18 inches (457 mm) below the floor opening to ensure proper deflection of supply air. In addition, the receptacle must extend 3 inches (76 mm) beyond the opening on all sides and the perimeter and have a vertical lip at least 1 inch (25 mm) high on all sides to catch any sparks, embers, or other sources of ignition.

M1601.4.4 Access. Access to an under-floor plenum shall be provided through an opening in the floor with minimum dimensions of 18 inches by 24 inches (457 mm by 610 mm).

❖ An 18-inch by 24-inch (457 mm by 610 mm) opening in the floor is required to provide access to the under-floor plenum to permit maintenance and repairs in the crawl space.

M1601.4.5 Furnace controls. The furnace shall be equipped with an automatic control that will start the air-circulating fan when the air in the furnace bonnet reaches a temperature not greater than 150°F (66°C). The furnace shall additionally be equipped with an approved automatic control that limits the outlet air temperature to 200°F (93°C).

❖ Because exposed wood joists often form under-floor plenums and flooring, the outlet air temperature from the furnace must be limited to 200°F (93°C).

Furnaces must also be equipped with an automatic control that starts the air circulation fan when the air temperature in the furnace reaches 150°F (66°C). This requirement serves to inhibit the plenum air temperature from approaching 150°F (66°C) because of the volume of the under-floor space cools the heated air delivered by the furnace through mixing with the cooler plenum air.

SECTION M1602
RETURN AIR

M1602.1 Return air. Return air shall be taken from inside the dwelling. Dilution of return air with outdoor air shall not be prohibited.

❖ This section allows the mixture of outdoor air with that from inside the building. Outdoor air is commonly mixed with return air as a means of providing makeup air for exhaust systems. Duct sizing must be as required by ACCA *Manual D* or other acceptable sizing methods.

M1602.2 Prohibited sources. Outdoor and return air for a forced-air heating or cooling system shall not be taken from the following locations:

1. Closer than 10 feet (3048 mm) from an appliance vent outlet, a vent opening from a plumbing drainage system or the discharge outlet of an exhaust fan, unless the outlet is 3 feet (914 mm) above the outside air inlet.
2. Where there is the presence of flammable vapors; or where located less than 10 feet (3048 mm) above the surface of any abutting public way or driveway; or where located at grade level by a sidewalk, street, alley or driveway.
3. A room or space, the volume of which is less than 25 percent of the entire volume served by such system. Where connected by a permanent opening having an area sized in accordance with ACCA Manual D, adjoining rooms or spaces shall be considered as a single room or space for the purpose of determining the volume of such rooms or spaces.

   Exception: The minimum volume requirement shall not apply where the amount of return air taken from a room or space is less than or equal to the amount of supply air delivered to such room or space.
4. A closet, bathroom, toilet room, kitchen, garage, mechanical room, furnace room or other dwelling unit.
5. A room or space containing a fuel-burning appliance where such room or space serves as the sole source of return air.

   Exceptions:

   1. The fuel-burning appliance is a direct-vent appliance or an appliance not requiring a vent in accordance with Section M1801.1 or Chapter 24.
   2. The room or space complies with the following requirements:
      2.1. The return air shall be taken from a room or space having a volume exceeding 1 cubic foot for each 10 Btu/h (9.6 L/W) of combined input rating of all fuel-burning appliances therein.
      2.2. The volume of supply air discharged back into the same space shall be approximately equal to the volume of return air taken from the space.
      2.3. Return-air inlets shall not be located within 10 feet (3048 mm) of any appliance firebox or draft hood in the same room or space.
   3. Rooms or spaces containing solid-fuel burning appliances, provided that return-air inlets are located not less than 10 feet (3048 mm) from the firebox of such appliances.

❖ Recirculation of air contaminated with objectionable odors, fumes, or flammable vapors is prohibited because of potential health and safety hazards. Therefore, return air ducts must not serve kitchens, bathrooms, attached garages, or other dwelling units. Also, outdoor air intakes must be located carefully to preclude objectionable or contaminated air from being drawn into the air distribution system. Air intakes above the roof must be located a minimum of 10 feet laterally from a plumbing vent or fuel-burning appliance vent, unless the source of noxious fumes is more than 3 feet higher than the outdoor air intake. In that case, a lesser lateral separation is acceptable.

M1602.3 Inlet opening protection. Outdoor air inlets shall be covered with screens having not less than $^1/_4$-inch (6.4 mm) openings and not greater than $^1/_2$-inch (12.7 mm) openings.

❖ An outdoor air intake must be covered with a $^1/_4$-inch (6.4 mm) to $^1/_2$-inch (12.7 mm) mesh screen to prevent the intrusion of animals and insects.

Bibliography

The following resource material is referenced in this chapter or is relevant to the subject matter addressed in this chapter.

ACCA Manual D-95, *Residential Duct Systems.* Washington D.C.: Air Conditioning Contractors of America, 1995.

ASHRAE HVAC Systems and Equipment Handbook. Atlanta: American Society of Heating, Refrigerating and Air-Conditioning Engineers, 2000.

ASTM E 84-01, *Test Method for Surface-Burning Characteristics of Building Materials*. West Conshohocken, PA: ASTM International, 2001.

Fibrous Glass Duct Construction Standards. Chantilly, VA: Sheet Metal and Air-Conditioning Contractors' National Association, 1992.

HVAC Duct Construction Standards—Metal and Flexible. Chantilly, VA: Sheet Metal and Air Conditioning Contractors' National Association, 1995.

UL 181-96, *Factory-Made Air Ducts and Air Connectors*. Northbrook, IL: Underwriters Laboratories, 1996.

Chapter 17: Combustion Air

General Comments

Complete combustion of solid and liquid fuel is essential for the proper operation of appliances, control of harmful emissions, and for achieving maximum fuel efficiency. Combustion air provides, among other things, the oxygen necessary for the complete and efficient burning of fuel. If insufficient quantities of oxygen are supplied, the combustion process will be incomplete, creating dangerous by-products and wasting energy in the form of unburned fuel (hydrocarbons). The byproducts of incomplete combustion are poisonous, corrosive and combustible and can cause serious appliance or equipment malfunctions that pose fire or explosion hazards.

Combustion air also serves other purposes in addition to providing oxygen. It ventilates and cools appliances and the rooms or spaces that enclose them, and it plays an important role in producing and controlling draft in vents and chimneys. An insufficient combustion air supply could cause an appliance to overheat and discharge combustion by-products into the building.

A combustion air supply is also necessary to prevent oxygen depletion, which threatens the safety of the building occupants. Both building-occupant respiration and the combustion process of fuel-burning appliances consume the available oxygen in a room or space. The depletion of available oxygen promotes incomplete fuel combustion.

Adequate combustion air supply is extremely important, but it is one aspect of mechanical installation that is often overlooked, ignored, or compromised. Depending on the appliance type, location and building construction, supplying combustion air can be easy, or it can involve complex designs and extraordinary methods. In any case, the importance of a proper combustion air supply cannot be overemphasized.

The combustion air provisions in the *International Residential Code* are confined to three sections: Section M1701 is the general set of criteria that covers all cases, Section M1702 covers the condition where all air is obtained from within the building, and Section M1703 covers the condition where all air is from outside of the building. If fuel-gas appliances are used, then the provisions of Chapter 24 must be followed.

Purpose

The provisions in this chapter apply to the combustion air necessary for the complete combustion of fuel, dilution of flue gases, and ventilation of mechanical appliances and the space in which they are located. An inadequate combustion air supply to such appliances will result in an extremely hazardous condition to both life and property.

SECTION M1701 GENERAL

M1701.1 Air supply. Liquid and solid fuel-burning appliances shall be provided with a supply of air for fuel combustion, draft hood dilution and ventilation of the space in which the appliance is installed, in accordance with Section M1702 or Section M1703. The methods of providing combustion air in this chapter do not apply to fireplaces, fireplace stoves and direct-vent appliances.

❖ Combustion air includes the air necessary for complete fuel combustion, the air required for draft hood dilution, and the air necessary for ventilation of the appliance enclosure. A lack of combustion air results in incomplete fuel combustion, which causes soot production, increased carbon monoxide production, serious appliance malfunction, and the risk of fire or explosion. Lack of draft-hood dilution air can result in improper draft and appliance venting. Incomplete fuel combustion and improper draft and venting compound each other and greatly increase the risk of the release of carbon monoxide, thereby endangering the occupants of the space or building. Lack of the air necessary for ventilation of the appliance enclosure can result in excessive temperatures in that enclosure, thus introducing the risks of overheating the appliance and fire.

Potential sources of combustion air are (1) the indoor atmosphere in which the appliance is located, (2) outdoor openings and ducts, (3) a combination of the two and (4) any other approved arrangement for the dependable introduction of outdoor air into the building or directly to the fuel-burning appliance.

M1701.1.1 Buildings of unusually tight construction. In buildings of unusually tight construction, combustion air shall be obtained from outside the sealed thermal envelope. In buildings of ordinary tightness, insofar as infiltration is concerned, all or a portion of the combustion air for fuel-burning appliances may be obtained from infiltration when the room or space has a volume of 50 cubic feet per 1,000 Btu/h (4.83 L/W) input.

❖ The definition of unusually tight construction can be found in Section R202. Most buildings built today have features meeting these requirements listed except for one. The feature that usually determines whether a building falls under this category is the one stating that walls and ceilings comprising the building envelope must have a continuous water vapor retarder with a rating of 1 perm or less. The remaining provisions of this definition are typically required in most buildings today to comply with local or federal energy regulations. When a building is so tight that it does not allow sufficient infiltration, indoor air cannot be a source of combustion air. This section indicates that the interior of buildings of "ordinary tightness" can be used as a source of combustion air, provided that the interior volume is at least 50 cubic feet per 1,000 Btu/hr (4.83 L/w) of the total appliance input rating.

M1701.2 Exhaust and ventilation system. Air requirements for the operation of exhaust fans, kitchen ventilation systems, clothes dryers and fireplaces shall be considered in determining the adequacy of a space to provide combustion air.

❖ The provisions of this chapter rely on natural (gravity) movement of air. Combustion air can be adversely affected by other mechanical systems or appliances in the same room or space. Exhaust fans, kitchen hoods, clothes dryers, central vacuum systems and even fireplaces can create enough air-pressure imbalance to affect the delicate flow of combustion air to the appliance.

M1701.3 Volume dampers prohibited. Volume dampers shall not be installed in combustion air openings.

❖ To avoid the inadvertent closure of combustion air openings, this section states that volume dampers are not allowed. Although not mentioned in this section, the building official might allow automatic dampers if they are interlocked with the fuel source to prevent operation of the appliance when the damper is closed.

M1701.4 Prohibited sources. Combustion air ducts and openings shall not connect appliance enclosures with space in which the operation of a fan may adversely affect the flow of combustion air. Combustion air shall not be obtained from an area in which flammable vapors present a hazard. Fuel-fired appliances shall not obtain combustion air from any of the following rooms or spaces:

1. Sleeping rooms.
2. Bathrooms.
3. Toilet rooms.

Exception: The following appliances shall be permitted to obtain combustion air from sleeping rooms, bathrooms and toilet rooms:

1. Solid fuel-fired appliances provided that the room is not a confined space and the building is not of unusually tight construction.
2. Appliances installed in an enclosure in which all combustion air is taken from the outdoors and the enclosure is equipped with a solid weatherstripped door and self-closing device.

❖ Exhaust fans, as just one example, can produce negative pressures in a space, thereby reducing the availability of combustion air. Air that could have served as combustion air could be consumed instead as makeup air for an exhaust system. Openings to the outdoors and infiltration might not be able to provide the required combustion air while also providing makeup air for exhaust fans or appliances such as clothes dryers. This section applies to all other sections of this chapter.

M1701.5 Opening area. The free area of each opening shall be used for determining combustion air. Unless otherwise specified by the manufacturer or determined by actual measurement, the free area shall be considered 75 percent of the gross area for metal louvers and 25 percent of the gross area for wood louvers.

❖ This section provides a rule-of-thumb to calculate the free area of combustion-air openings fitted with louvers. Metal louvers obstruct only approximately 25 percent of the opening, while wood louvers obstruct approximately 75 percent of the opening. Prefabricated louvers may have the manufacturer's "free area" designation, which can be used in lieu of the above method.

M1701.6 Opening location. In areas prone to flooding as established by Table R301.2(1), openings shall be located at or above the design flood elevation established in Section R323.1.5.

❖ Combustion air inlet openings must be located above the design flood elevation in areas designated as flood hazard areas. This is to prevent the entrance of water into the combustion air opening and to prevent the opening from being blocked by rising water. See the commentary for Section R323.1.5.

SECTION M1702 ALL AIR FROM INSIDE THE BUILDING

M1702.1 Required volume. Where the volume of the space in which fuel-burning appliances are installed is greater than 50 cubic feet per 1,000 Btu/h (4.83 L/W) of aggregate input rating in buildings of ordinary tightness, insofar as infiltration is concerned, normal infiltration shall be regarded as adequate to provide combustion air. Rooms communicating directly with the space in which the appliances are installed through openings not furnished with doors shall be considered part of the required volume.

❖ If the building is of ordinary tightness (as opposed to unusually tight construction) and of sufficient volume [50 cu. ft./1,000 Btu/hr (4.83 L/w)], there will be no need to obtain combustion air from outside the building. The space volume can include rooms communicating directly with the space in which the appliance is installed through unobstructed openings not furnished with doors.

M1702.2 Confined space. Where the space in which the appliance is located does not meet the criterion specified in Section M1702.1, two permanent openings to adjacent spaces shall be provided so that the combined volume of all spaces meets the criterion. One opening shall be within 12 inches (305 mm) of the top and one within 12 inches (305 mm) of the bottom of the space, as illustrated in Figure M1702.2. Each opening shall have a free area equal to a minimum of 1 square inch per 1,000 Btu/h (2200 mm^2/kW) input rating of all appliances installed within the space, but not less than 100 square inches (64 415 mm^2).

❖ The provisions of Section M1702.1 require rather large room volumes to provide the necessary air volume. For example, if the building room or space has 10-foot (30 480 mm) ceilings, the room or space in which the fuel-burning appliances are located must have a floor area of 5 square feet (0.04645 m^2) for each 1,000 Btu/h (293 W) of total appliance input. In open buildings without interior partitions, compliance with Section M1702.1 might be easy; however, appliances are often installed in confined spaces or rooms that are not much larger than necessary to contain the appliances themselves. Appliances and equipment located in confined spaces not having the room air volume prescribed in Section M1702.1 [50 cubic feet per 1,000 Btu/h (4.83 L/w)] must obtain combustion air from the outdoors or adjacent interior spaces or both.

This section provides a method of coupling adjacent rooms or spaces so the combined total volume of the communicating rooms or spaces complies with the minimum volume criteria of Section M1702.1. When the air is obtained from adjacent interior spaces, at least two openings must be provided, one of which must be near the top of the room and the other near the bottom. See Commentary Figure M1702.2(1). Openings located in this configuration create a natural (gravity) airflow in the confined space by virtue of convection. Appliances warm the air in the room or space in which they are contained, and the lighter (warmer) air will rise and move out through the upper opening. Cooler (heavier) replacement air will enter the lower opening, thus creating a natural (convective) circulation. When such permanent openings are provided to connect adjacent interior spaces, the spaces communicating with each other can be considered as one for the purpose of figuring the volume of available inside air.

Opening protectors such as fire, smoke, and ceiling dampers could be required by the building code depending on the location of the combustion air openings. The operation (closure) of a damper can create a hazard by interfering with the combustion air supply.

M1702.3 Unusually tight construction. Where the space is of adequate volume in accordance with Section M1702.1 or Section M1702.2, but is within a building sealed so tightly that in-

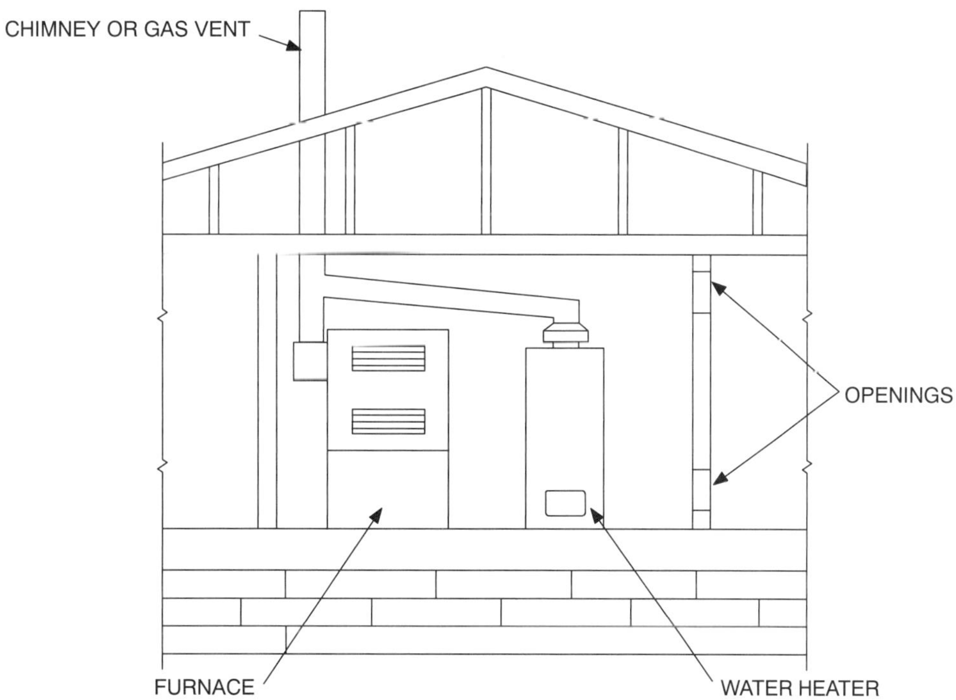

For SI: 1 square inch = 645.16 mm^2, 1 British thermal unit per hour = 0.2931 W.

Note: Each opening shall have a free area of not less than 1 square inch per 1,000 Btu/h of the total input rating of all appliances in the enclosure, but not less than 100 square inches.

FIGURE M1702.2
APPLIANCES LOCATED IN CONFINED SPACES—ALL AIR TAKEN FROM ADJACENT SPACES WITHIN THE BUILDING

❖ See the commentary for Section M1702.2.

filtration air is not adequate for combustion, combustion air shall be obtained from outdoors or from spaces freely communicating with the outdoors in accordance with Section M1703.

❖ Air taken directly from inside the building is an acceptable source of combustion air, provided that a sufficient volume of air is available for the appliances served and the building construction allows sufficient infiltration.

The provisions of this section rely on building-envelope infiltration as the only source of combustion air. The combustion air supply can be obtained from indoor air, provided that the building infiltration is sufficient to provide makeup air at the rate combustion air is taken from the space. Modern construction methods, energy-conservation efforts that have produced very tight buildings, and air depletion by equipment and appliances make it unlikely that the inside air alone will be adequate in many buildings. Sections M1702.1 or M1702.2 provide guidance for obtaining combustion air from the outdoors where the building is of unusually tight construction or the indoor volume does not comply with Section M1702.1

SECTION M1703
ALL AIR FROM OUTDOORS

M1703.1 Outdoor air. Where the space in which fuel-burning appliances are located does not meet the criterion for indoor air specified in Section M1702, outside combustion air shall be supplied as specified in Section M1703.2.

❖ This section relies on the traditional method of providing combustion air by means of two direct openings or ducts to the outdoors. A single-opening method that applies only to fuel gas-fired appliances can be found in the fuel gas chapter (Chapter 24). Openings to spaces that are sufficiently ventilated with outside air, such as attic or crawl spaces, are acceptable alternatives to a direct connection to the outside. Attic and crawl spaces can be acceptable sources of combustion air only where such spaces are provided with adequate ventilation openings directly to the outdoors. See Commentary Figures M1703.1(1) and M1703.1(2). Attic and crawl spaces ventilated by mechanical means are not an acceptable source of combustion air.

Combustion air ducts and openings that penetrate components of wall, floor, ceiling, and roof assemblies must be installed as required by the building code. Transfer openings and ducts that connect an attic space to a room or space below cause a breech in the continuity of the ceiling. This breech would allow fire to spread from the room or space below to the concealed attic space.

When designing combustion air installations, consideration should be given to the effect that openings to the outdoors can have on appliances, plumbing systems, and building occupants. For example, depending on the location, openings to the outdoors can:

- cause drafts that can blow out pilot lights or interfere with appliance ignition;

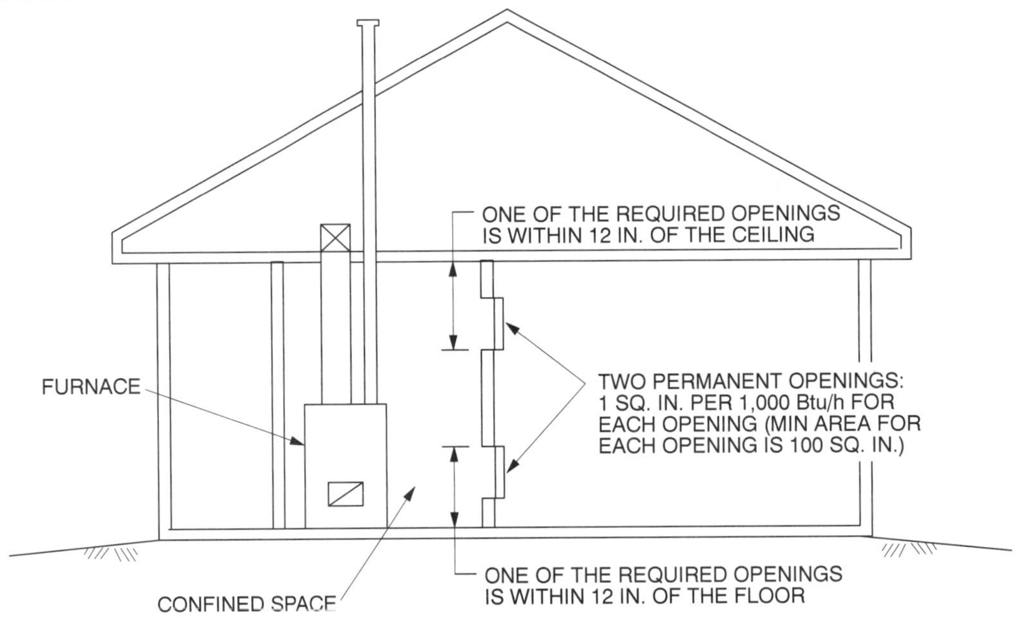

For SI: 1 inch = 25.4 mm, 1 square inch = 645 mm^2, 1 Btu/h = 0.2931 W.

Figure M1702.2(1)
INSIDE AIR—AIR FROM ADJACENT SPACES (LOCATION)

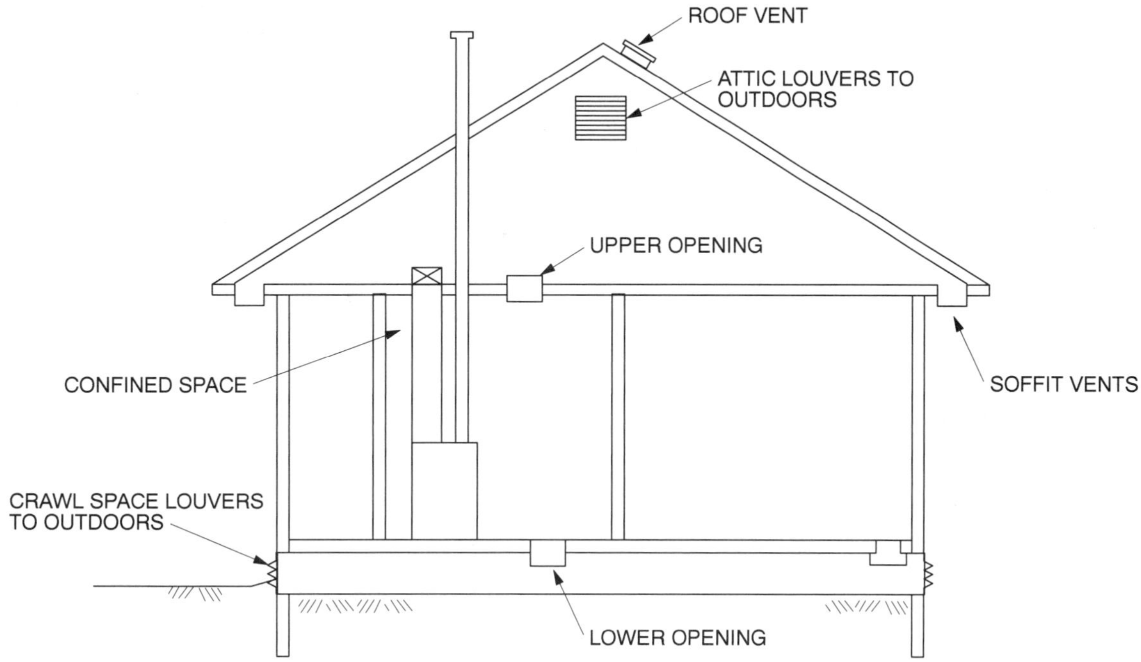

Figure M1703.1(1)
OUTDOOR AIR—AIR FROM VENTILATED ATTIC/CRAWL SPACE

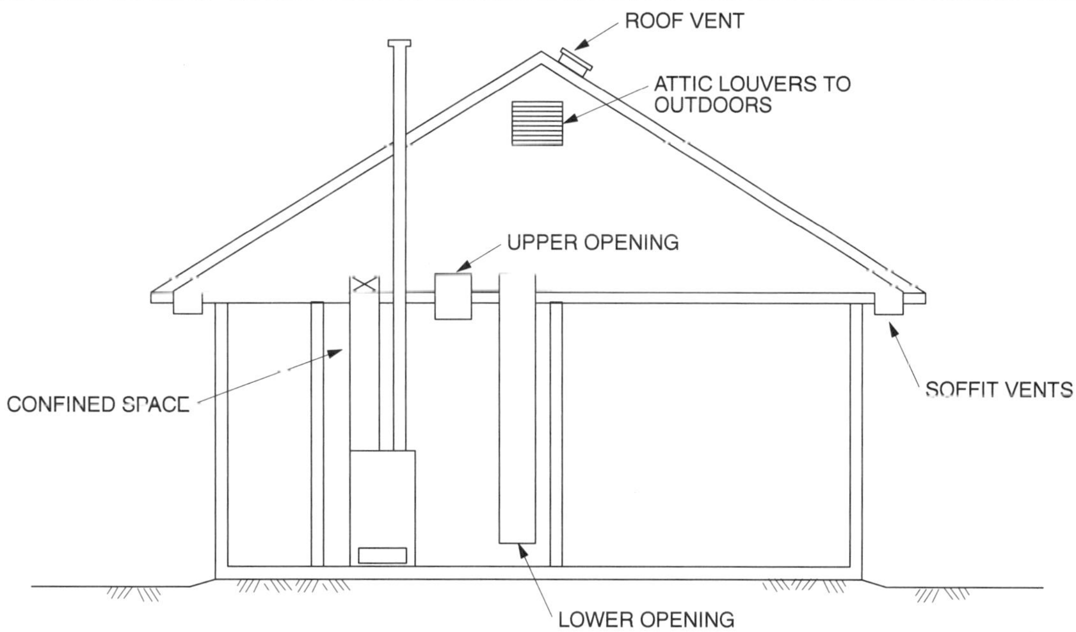

Figure M1703.1(2)
OUTDOOR AIR—AIR FROM VENTILATED ATTIC

- cause freezing of plumbing piping or other water-containing components;
- cause objectionable cold drafts that encourage the occupants to block or cover the openings.

In all cases, combustion air openings should be located to reduce the likelihood they will be accidentally or intentionally blocked or covered. A ceiling transfer opening that connects a furnace room with an attic is an example of a combustion air opening that will probably be intentionally blocked by an occupant because of the drafts that can occur in cold climates.

The building occupants typically do not understand the need for, or the importance of, combustion air openings.

M1703.2 Two openings or ducts. Outside combustion air shall be supplied through openings or ducts, as illustrated in Figures M1703.2(1), M1703.2(2), M1703.2(3) and M1703.2(4). One opening shall be within 12 inches (305 mm) of the top of the enclosure, and one within 12 inches (305 mm) of the bottom of the enclosure. Openings are permitted to connect to spaces directly communicating with the outdoors, such as ventilated crawl spaces or ventilated attic spaces. The same duct or opening shall not serve both combustion air openings. The duct serving the upper opening shall be level or extend upward from the appliance space.

❖ Outside air must be introduced through at least two openings, one of which must be near the top of the room and the other near the bottom. These locations induce flow through the space where the appliances are located.

Where combustion air is supplied through upper and lower combustion air openings, air circulation cannot occur if a single duct serves both combustion air openings. The lower opening functions as an inlet and the upper opening as an air outlet; thus, independent ducts are required.

M1703.2.1 Size of openings. Where communicating with the outdoors by means of vertical ducts, each opening shall have a free area of at least 1 square inch per 4,000 Btu/h (550 mm^2/kW) of total input rating of all appliances in the space. Where horizontal ducts are used, each opening shall have a free area of at least 1 square inch per 2,000 Btu/h (1.00 mm^2/kW) of total input of all appliances in the space. Ducts shall be of the same minimum cross-sectional area as the required free area of the openings to which they connect. The minimum cross-sectional dimension of rectangular air ducts shall be 3 inches (76 mm).

❖ All vertical openings are sized by the same area ratio, regardless of whether they are ducted or open directly to the outdoors, in either attics or crawl spaces.

The unobstructed free area for vertical openings must be sized to provide 1 square inch of opening per 4,000 Btu/h (550 mm^2/kW) total input rating for all appliances located within the room. Vertical ducts allow gravity to assist air movement, which accounts for the decrease in required duct and opening sizes compared to the sizing criteria for horizontal ducts and openings. This opening area-to-Btu/h ratio is deemed adequate in overcoming the static frictional losses for vertical ducts and openings only; it does not apply to vertical ducts that contain horizontal runs or offsets. Where horizontal sections are located within the duct system, the ducts must be sized as horizontal ducts in accordance with this section.

The opening-area-to-Btu/h ratio depends on whether the opening to the outside is through a horizontal duct or directly through a wall. See Commentary Figures M1703.2.1(1) and M1703.2.1(2). When the opening is through a horizontal duct, more area is necessary to offset the static friction losses of the air flowing horizontally through the duct. For this reason, each horizontal duct must have an unobstructed free area of at least 1 square inch for each 2,000 Btu/h (1100 mm^2/kW) total input capacity compared to 1 square inch for each 4,000 Btu/h (550 mm^2/kW) total input capacity for a direct opening through a wall as required by this section. The sizing criteria are based on the gravity flow of combustion air through the ducts, openings, and appliance enclosures.

M1703.3 Attic combustion air. Combustion air obtained from an attic area, as illustrated in Figure M1703.2(3), shall be in accordance with the following:

1. The attic ventilation shall be sufficient to provide the required volume of combustion air.
2. The combustion air opening shall be provided with a metal sleeve extending from the appliance enclosure to at least 6 inches (152 mm) above the top of the ceiling joists and ceiling insulation.
3. An inlet air duct within an outlet air duct shall be an acceptable means of supplying attic combustion air to an appliance room provided that the inlet duct extends at least 12 inches (305 mm) above the top of the outlet duct in the attic space, as illustrated in Figure M1703.3.
4. The end of ducts that terminate in an attic shall not be screened.

❖ It is common practice to provide ducts or direct openings to naturally ventilated attic and crawl spaces for the purpose of obtaining combustion air. For example, an appliance could be located in a room over a crawl space and under an attic space. The upper combustion air opening could open to the attic, and the lower combustion-air opening could open to the crawl space; ducts would be unnecessary. Mechanical ventilation of attics and crawl spaces is not equivalent to an opening to the outdoors because exhaust fans in such spaces could interfere with the movement of combustion air by creating pressure differentials.

Attic ventilation openings must be sized to provide both combustion air and the attic ventilation required by Section R806.2. The attic height must be sufficient to allow unimpeded airflow. The opening sleeve (duct) extension prevents blockage of the opening by loose insulation materials that could fall into or cover the opening. Attics with mechanical ventilation cannot be a source of combustion air.

M1703.4 Under-floor combustion air. Combustion air obtained from under-floor areas, as illustrated in Figure M1703.2(4), shall have free opening areas to the outside equivalent to not less than twice the required combustion air opening.

❖ To obtain the required airflow from the crawl space, openings to the outdoors must remain open (unobstructed) year-round and must be of adequate size to provide both combustion air and the required ventila-

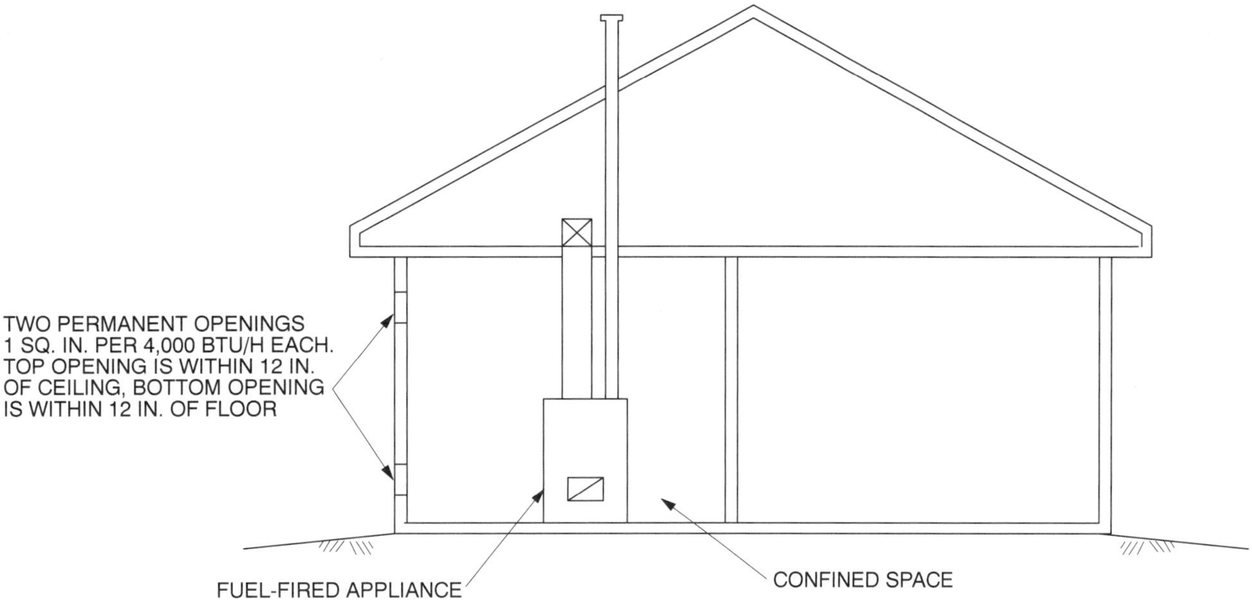

For SI: 1 inch = 25.4 mm, 1 square inch = 645.16 mm^2, 1 btu/h = 0.2931 W.

Figure M1703.2.1(1)
OUTDOOR AIR—DIRECT WALL OPENINGS

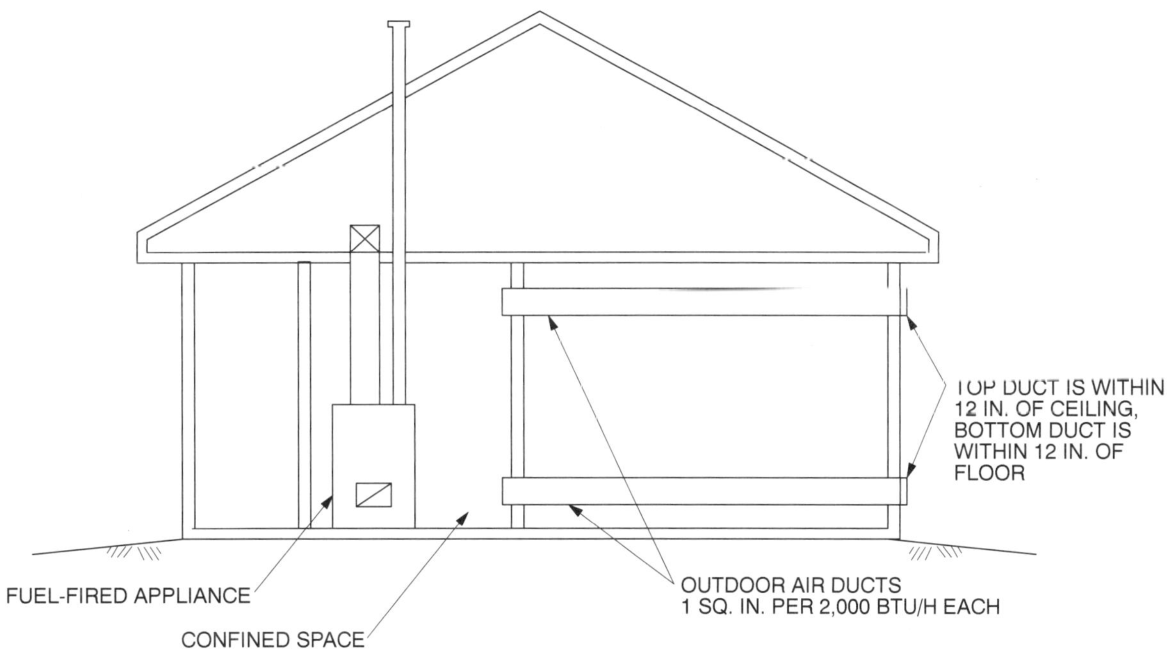

For SI: 1 inch = 25.4 mm, 1 square inch = 645.16 mm^2, 1 btu/h = 0.2931 W.

Figure M1703.2.1(2)
OUTDOOR AIR—HORIZONTAL DUCT OPENINGS

tion. The crawl space openings to the outdoors must have a total area equal to or greater than twice the required area of the combustion air opening to the crawl space.

M1703.5 Opening requirements. Outside combustion air openings shall be covered with corrosion-resistant screen or equivalent protection having not less than 0.25-inch (6.4 mm) openings, and not greater than 0.5-inch (12.7 mm) openings.

❖ The opening sizes for louvers and grilles and the mesh sizes for screens are required to be within the specified range as indicated in the table. The opening size must be large enough to inhibit blockage by debris and to prevent significant resistance to airflow; yet, it must be small enough to keep out what is intended to be kept out. A screen of such mesh size would restrict the passage of rodents and large insects and would be resistant to blockage by lint, debris and plant fibers.

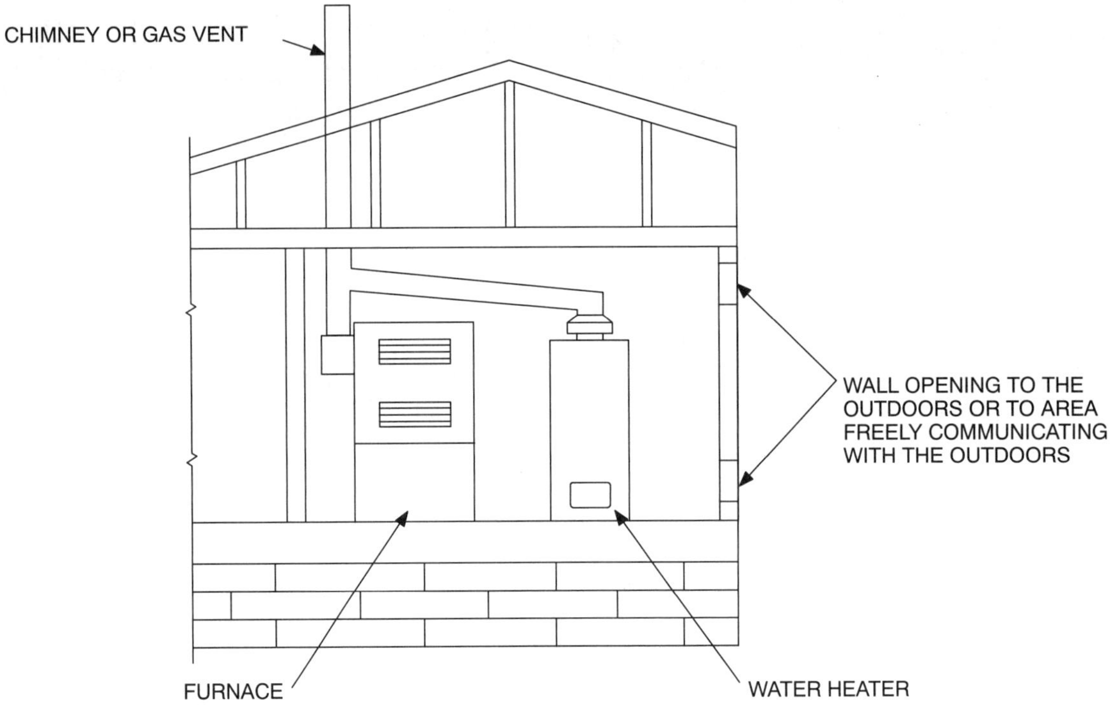

For SI: 1 square inch = 645.16 mm^2, 1 British thermal unit per hour = 0.2931 W.

Note: Each opening shall have a free area of not less than 1 square inch per 4,000 Btu/h of the total input rating of all appliances in the enclosure.

FIGURE M1703.2(1)
APPLIANCES LOCATED IN CONFINED SPACES—ALL AIR TAKEN FROM OUTDOORS THROUGH TWO OPENINGS

❖ See the commentary for Section M1703.2.

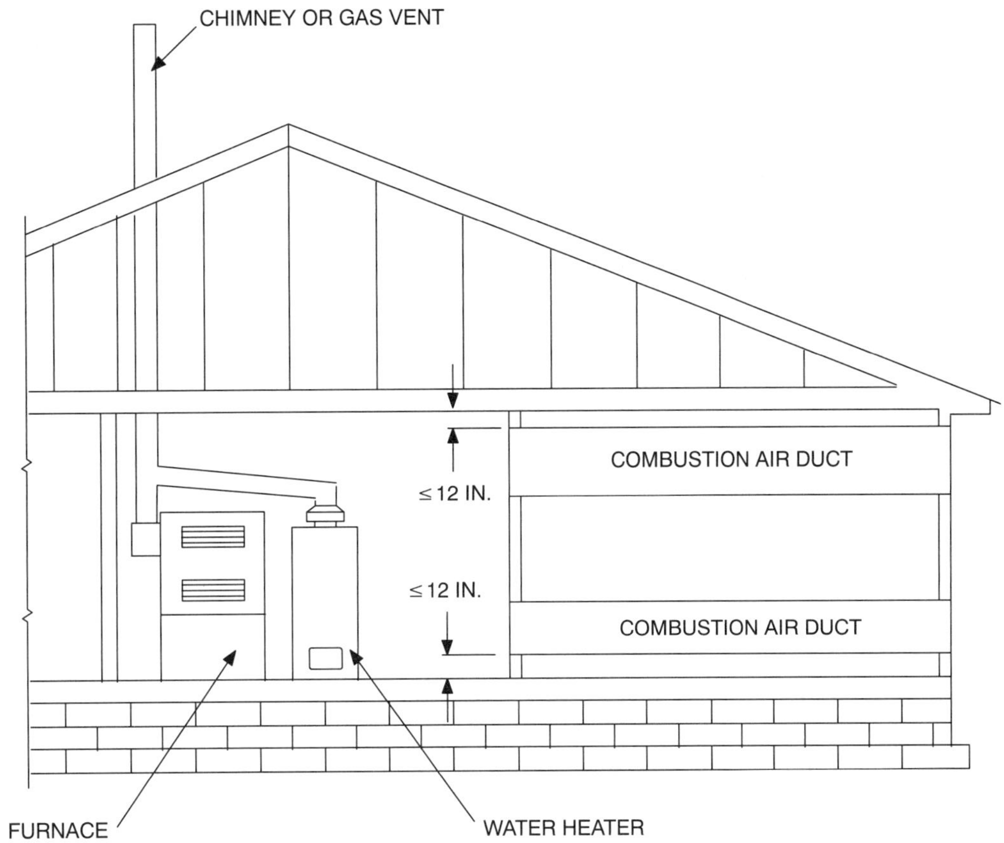

For SI: 1 inch = 25.4 mm, 1 square inch = 645.16 mm^2, 1 British thermal unit per hour = 0.2931 W.

Note: Each opening shall have a free area of at least 1 square inch per 2,000 Btu/h of the total input of all appliances in the space.

FIGURE M1703.2(2)
APPLIANCES LOCATED IN CONFINED SPACES—ALL AIR TAKEN FROM OUTDOORS THROUGH HORIZONTAL DUCTS

❖ See the commentary for Section M1703.2.

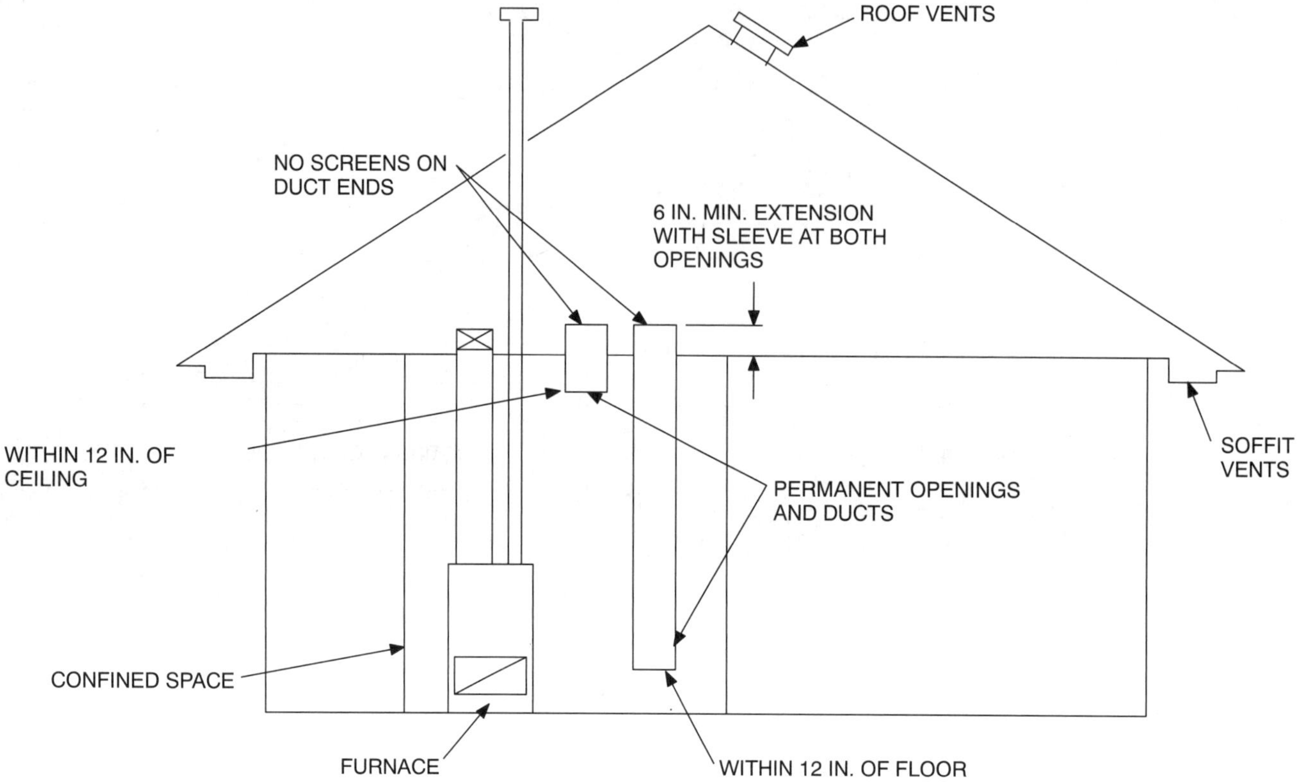

For SI: 1 inch = 25.4 mm, 1 British thermal unit per hour = 0.2931 W, 1 square inch = 645.16 mm^2.

Note: Each opening shall have a free area of at least 1 square inch per 4,000 Btu/h of the total input of all appliances in the space. The attic must be sufficiently ventilated to provide the required combustion air.

FIGURE M1703.2(3)
APPLIANCES LOCATED IN CONFINED SPACES—ALL AIR TAKEN FROM OUTDOORS THROUGH VENTILATED ATTIC

❖ See the commentary for Section M1703.2.

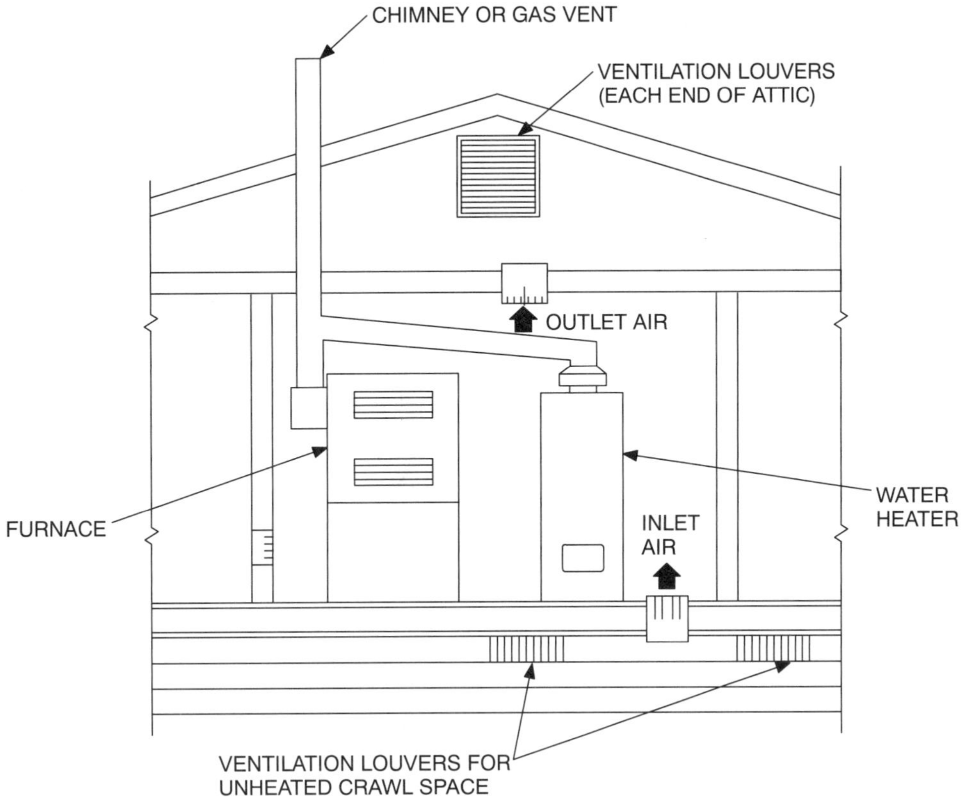

For SI: 1 square inch = 645.16 mm^2, 1 British thermal unit per hour = 0.2931 W.

Note: The inlet and outlet air openings shall have a free area of not less than 1 square inch per 4,000 Btu/h of the total input rating of all appliances in the enclosure.

FIGURE M1703.2(4)
APPLIANCES LOCATED IN CONFINED SPACES—INLET AIR TAKEN FROM VENTILATED CRAWL SPACE AND OUTLET AIR TO VENTILATED ATTIC

❖ See the commentary for Section M1703.2.

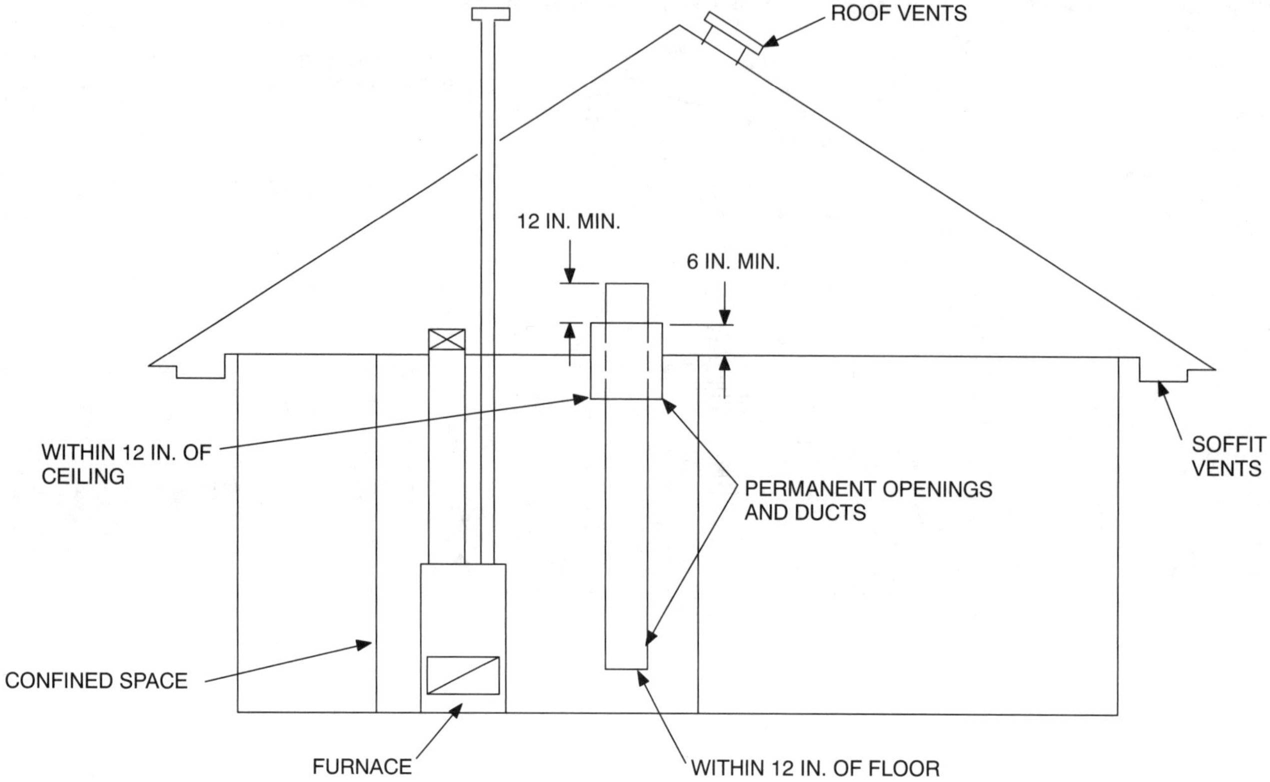

For SI: 1 inch = 25.4 mm, 1 British thermal unit per hour = 0.2931 W, 1 square inch = 645.16 mm^2.

Note: Each opening shall have a free area of at least 1 square inch per 4,000 Btu/h of the total input of all appliances in the space. The attic must be sufficiently ventilated to provide the required combustion air.

FIGURE M1703.3
APPLIANCES LOCATED IN CONFINED SPACES—ALL AIR TAKEN FROM OUTDOORS THROUGH VENTILATED ATTIC, INLET DUCT WITHIN OUTLET DUCT

❖ See the commentary for Section M1703.3.

Bibliography

The following resource material is referenced in this chapter or is relevant to the subject matter addressed in this chapter.

ANSI Z223.1-02, *National Fuel Gas Code*. New York: American National Standards Institute, 2002.

Chapter 18: Chimneys and Vents

General Comments

Chapter 18 regulates the design, construction, installation, maintenance, repair and approval of chimneys, vents and their connections to fuel-burning appliances. A properly designed chimney or vent system is needed to conduct the flue gases produced by a fuel-burning appliance to the outdoors. In the case of natural draft appliances, the chimney or vent system serving the appliance is expected to produce a draft at the appliance connection.

Draft is produced by the temperature difference between combustion gases (flue gases) and the ambient atmosphere. It is measured as pressure in inches of water column (kPa) and is a negative pressure—meaning that it is less than the atmospheric pressure at the point of measurement. Because hotter gases are less dense, they are buoyant and will rise in the chimney, causing negative pressures to develop that are directly proportional to the height of the chimney or vent and directly proportional to the temperature difference between the flue gases and the ambient air. Draft produced by a chimney or vent is the motivating force that conveys combustion products from natural draft appliances to the outside atmosphere.

The type of vent or chimney used in a given installation is generally based on the type of fuel burned by the appliance, the temperature of the flue gases produced by the combustion process, and the appliance category. Because the code classifies appliances into three temperature categories (low, medium, and high heat), the fuel type and characteristics of the flue gas must be analyzed for the design of a chimney or vent system. Higher efficiency appliances continue to produce lower flue gas temperatures, and the diversity of mechanical venting systems continues to expand.

Because of new technology and higher efficiency appliances/equipment, venting means are becoming increasingly diversified and complex. Many of today's appliances are designed to be vented by more than one type of venting system.

Purpose

This chapter contains provisions that would minimize the hazards associated with the products of combustion from fuel-burning appliances.

SECTION M1801 GENERAL

M1801.1 Venting required. Fuel-burning appliances shall be vented to the outdoors in accordance with their listing and label and manufacturer's installation instructions except appliances listed and labeled for unvented use. Venting systems shall consist of approved chimneys or vents, or venting assemblies that are integral parts of labeled appliances. Gas-fired appliances shall be vented in accordance with Chapter 24.

❖ Every fuel-burning appliance must be vented by a chimney or vent, unless the appliance is listed and labeled for unvented use. Section G2445 addresses unvented gas-fired heaters. This chapter addresses other than gas-fired appliances.

M1801.2 Draft requirements. A venting system shall satisfy the draft requirements of the appliance in accordance with the manufacturer's installation instructions, and shall be constructed and installed to develop a positive flow to convey combustion products to the outside atmosphere.

❖ Every fuel-burning appliance must be vented by a chimney or vent unless the appliance is direct vented, integrally vented, or mechanically vented. Although chimneys and vents share a basic purpose, they are different creatures with different code requirements. Some code sections pertain to chimneys only, some to vents only, and some to both. The code is frequently misapplied because people refer to vents as chimneys and vice versa.

The combustion byproducts produced by the burning of fuels consist primarily of carbon dioxide, nitrogen, water vapor, and small amounts of carbon monoxide, nitrous oxide, and other compounds that vary with the type and purity of the fuel. Some of the compounds, especially carbon monoxide (CO), are poisonous, and some are corrosive. The amount of carbon monoxide in combustion gases varies with the type of fuel, the type of burner, and the condition and adjustment of the burner. See the commentary in Chapter 17 for a further discussion of fuel combustion.

Carbon monoxide poisoning as a result of malfunctioning fuel-burning appliances or malfunctioning chimneys or vents, or the lack thereof, has caused many fatalities. The products of combustion must be conducted to the outdoors so that operation of the appliance does not pose a threat of asphyxiation to the building occupants. Fuel-burning appliances have widely varying operating characteristics; the method of venting the combustion by-products must therefore be designed for the particular type of appliance served. Flue gases produced by appliances differ primarily in temperature and chemical composition. Methods of venting differ in the temperature and pressure ranges over which they operate and whether or not they can accommodate condensation.

A mismatch between an appliance and a chimney or vent can result in a hazardous operating condition. For example, an appliance that produces high-temperature flue gases can cause vent material to degrade or melt. It can also cause excessively high temperatures on the outer surface of the vent. On the other hand, an appliance that produces low-temperature flue gases can cause vent material to corrode and deteriorate because of condensation.

M1801.3 Existing chimneys and vents. Where an appliance is permanently disconnected from an existing chimney or vent, or where an appliance is connected to an existing chimney or vent during the process of a new installation, the chimney or vent shall comply with Sections M1801.3.1 through M1801.3.4.

❖ Existing chimneys and vents should be reevaluated for continued suitability whenever the conditions of use change. Size is of primary importance. Other considerations are the presence of liner obstructions, combustible deposits in the liner, the structural condition of the liner, the provision of a cleanout, and clearances to combustibles. Whenever a new appliance is connected or an exiting appliance is disconnected from a chimney or vent, the chimney or vent is subject to the provisions that include requirements for inspection, cleaning, possible repair, the installation of a cleanout if one is not already provided and the establishment of clearances to combustibles required for new chimneys or vent installations.

Chimneys, vents, and the appliances served are all designed to function together as a system. Any change to an existing chimney or vent system will have an impact on the performance of that system. Something as simple as disconnecting an appliance from a chimney can upset the system balance and cause the venting system to fail to produce a draft for the remaining appliances and could cause the venting system to produce harmful condensation.

M1801.3.1 Size. The chimney or vent shall be resized as necessary to control flue gas condensation in the interior of the chimney or vent and to provide the appliance, or appliances served, with the required draft. For the venting of oil-fired appliances to masonry chimneys, the resizing shall be done in accordance with NFPA 31.

❖ The combined input from multiple appliances, especially older lower-efficiency appliances, can maintain chimney or vent temperatures that are high enough to provide the necessary draft and avoid condensation. Changing an existing configuration by disconnecting and eliminating an appliance or by substituting a higher-efficiency appliance can cause a decrease in flue gas temperature, resulting in poor draft or condensation. Also, the elimination of one or more draft-hood-equipped appliances will reduce the amount of dilution air in a venting system, increasing the likelihood of condensation. Often, a chimney or vent must be replaced with a smaller-size chimney or vent or be resized by installation of a liner system. The requirements of NFPA 31 must be followed when resizing a masonry chimney that vents an oil-fired appliance.

A common scenario involves removing chimney-vented furnaces or boilers and leaving a water heater as the only appliance vented to the chimney. In such cases, the chimney would typically be grossly oversized for the water heater, could fail to produce adequate draft, and could be subject to continuous condensation. This scenario has received much attention and has created the phrase "orphaned water heaters."

M1801.3.2 Flue passageways. The flue gas passageway shall be free of obstructions and combustible deposits and shall be cleaned if previously used for venting a solid or liquid fuel-burning appliance or fireplace. The flue liner, chimney inner wall or vent inner wall shall be continuous and free of cracks, gaps, perforations, or other damage or deterioration that would allow the escape of combustion products, including gases, moisture and creosote.

❖ Cleaning the chimney or flue is necessary to allow examination of the flue liner surface, to eliminate fuel for a possible chimney fire, to reduce frictional resistance to flow, and to prevent any residual deposits from falling into lower parts of the venting system. Installation or removal of an appliance without an inspection of the existing flue is prohibited. If the flue serves (or served) a solid- or liquid-fuel-burning appliance or fireplace, cleaning is required before the required inspection to remove any combustible deposits. In that case, the addition or removal of an appliance without both cleaning and inspection of the existing flue would be a code violation. If inspection shows that the flueway of the vent or chimney is incapable of preventing the leakage of combustion gases and liquids, the chimney or vent would have to be repaired or relined. Such repair would have to meet the requirements of NAPA 31 where an oil-fired appliance is vented through the chimney flue.

M1801.3.3 Cleanout. Masonry chimneys shall be provided with a cleanout opening complying with Section R1001.14.

❖ A cleanout opening must be installed, if one is not already provided, for existing masonry chimney flues whenever another appliance is installed or removed.

M1801.3.4 Clearances. Chimneys and vents shall have air-space clearance to combustibles in accordance with this code and the chimney or vent manufacturer's installation instructions.

> **Exception:** Masonry chimneys equipped with a chimney lining system tested and listed for installation in chimneys in contact with combustibles in accordance with UL 1777, and

installed in accordance with the manufacturer's instruction, shall not be required to have a clearance between combustible materials and exterior surfaces of the masonry chimney. Noncombustible firestopping shall be provided in accordance with this code.

❖ This section requires that the existing chimney or vent be inspected for clearances to combustibles for the entire length of the chimney or vent. Clearances to combustibles for masonry chimneys are found in this code. The required clearances for a listed factory-built chimney or vent would be in accordance with the existing chimney or vent manufacturer's installation instructions. The clearances to combustibles for vents and factory-built chimneys are typically indicated on the vent or chimney. If the required clearances are not indicated on the vent or chimney, the manufacturer must be contacted and the original installation instructions consulted.

Existing masonry chimneys commonly do not have the clearances required by this code, especially if the structure is fairly old. Listed chimney lining systems, which provide insulation value to compensate for the lack of clearance, are available for use in such chimneys. The exception gives the criteria for these tested lining systems, citing UL 1777 as the test standard. Fireblocking of any annular spaces where the chimney passes through ceilings or floors must be installed in accordance with this code.

M1801.4 Space around lining. The space surrounding a flue lining system or other vent installed within a masonry chimney shall not be used to vent any other appliance. This shall not prevent the installation of a separate flue lining in accordance with the manufacturer's installation instructions and this code.

❖ When a chimney lining system is installed within a chimney, an annular space will be created between the liner and the chimney. Such a space is usually irregular in shape and is not conducive to the proper flow of flue gases. Also, vent gases could degrade the lining system.

M1801.5 Mechanical draft systems. A mechanical draft system shall be used only with appliances listed and labeled for such use. Provisions shall be made to prevent the flow of fuel to the equipment when the draft system is not operating. Forced draft systems and all portions of induced draft systems under positive pressure during operation shall be designed and installed so as to prevent leakage of flue gases into a building.

❖ Commonly used chimney and vent systems are not designed for positive pressure (above atmospheric). Such chimneys and vents are intended to produce a draft and thus operate with negative internal pressures. Specialized vents and chimneys are available that are designed for positive pressure applications. The misapplication of materials can result in leakage of combustion gases and can cause chimney or vent deterioration from condensation corrosion.

M1801.6 Direct-vent appliances. Direct-vent appliances shall be installed in accordance with the manufacturer's installation instructions.

❖ This section requires that the existing chimney or vent be inspected for clearances to combustibles for the entire length of the chimney or vent. Clearances to combustibles for masonry chimneys are found in this code. The required clearances for a listed factory-built chimney or vent would be in accordance with the existing chimney or vent manufacturer's installation instructions. The clearances to combustibles for vents and factory-built chimneys are typically indicated on the vent or chimney. If the required clearances are not indicated on the vent or chimney, the manufacturer must be contacted and the original installation instructions consulted.

M1801.7 Support. Venting systems shall be adequately supported for the weight of the material used.

❖ Vent manufacturers supply support parts, and manufacturers' instructions provide detailed requirements for support of vent systems. Improper support can cause strain on vent components, appliance connectors, and fittings resulting in vent, appliance, or connector damage and loss of required clearance to combustibles and proper pitch. Frequently, venting installations suffer from lack of or improperly installed supports, brackets, and hangers. See Commentary Figure M1801.7.

M1801.8 Duct penetrations. Chimneys, vents and vent connectors shall not extend into or through supply and return air ducts or plenums.

❖ In residential construction, stud and joist spaces are sometimes used in lieu of ducts for return air and combustion air. Areas under a floor are sometimes used as plenums. A combustion products vent, vent connector or exhaust duct is not permitted to extend into or through an air plenum or space used as an air duct. The combustion vent could leak flue gases into the circulating air, creating a health hazard. The restriction does not apply to a space that is being used to convey combustion air.

M1801.9 Fireblocking. Vent and chimney installations shall be fireblocked in accordance with Section R602.8.

❖ Fireblocking with noncombustible materials must be provided in wood-frame construction at all openings around vents, pipes, ducts, chimneys and fireplaces at ceiling and floor levels. See the commentary for Section R602.8.

M1801.10 Unused openings. Unused openings in any venting system shall be closed or capped.

❖ Unused openings in chimneys or vents can allow vent gases to escape into the building, cause loss of draft or

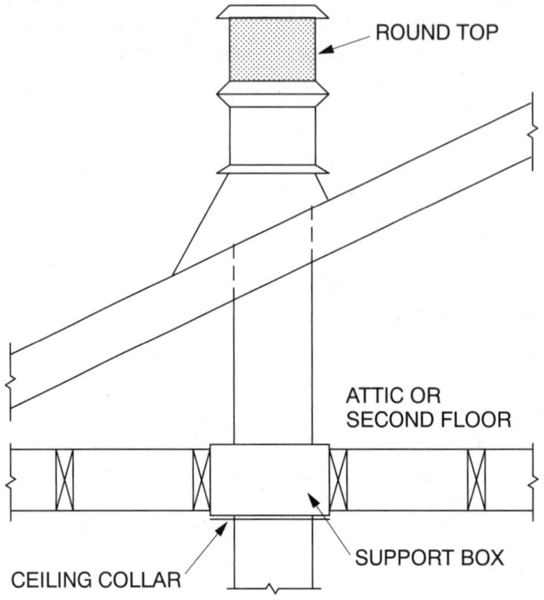

Figure M1801.7
VENT SUPPORT

allow the entry of birds and rodents. The method used to close an unused opening must not create a protrusion that can restrict flow through the vent or chimney passageway.

M1801.11 Multiple-appliance venting systems. Two or more listed and labeled appliances connected to a common natural draft venting system shall comply with the following requirements:

1. Appliances that are connected to common venting systems shall be located on the same floor of the dwelling.

   Exception: Engineered systems as provided for in Section G2427.

2. Inlets to common venting systems shall be offset such that no portion of an inlet is opposite another inlet.
3. Connectors serving appliances operating under a natural draft shall not be connected to any portion of a mechanical draft system operating under positive pressure.

❖ If more than one opening is provided into a single chimney or vent passageway, such openings must be located at different heights or otherwise be arranged so that the flow of combustion products from one connector cannot enter another connector instead of going up the flueway. This provision intends to prevent the flow or spilling of flue gases back into the building through an adjacent appliance flue connection.

Each solid-fuel-burning appliance or fireplace must be provided with a dedicated independent chimney or vent. Solid-fuel appliances cannot share a common chimney with any other appliance or fireplace. Solid-fuel-burning appliances produce creosote deposits on the interior walls of chimney liners. The creosote is highly combustible. In the event of a chimney fire, other connections to the chimney could allow fire to break out of the chimney and into the building. Also, chimney passageways can become partially blocked by creosote formation, thus forcing vent gases from other appliances to discharge back into the building.

M1801.12 Multiple solid fuel prohibited. A solid fuel-burning appliance or fireplace shall not connect to a chimney passageway venting another appliance.

❖ Each solid fuel-burning appliance or fireplace must be provided with a dedicated independent chimney or vent. Solid fuel appliances cannot share a common chimney with any other appliance or fireplace. Solid fuel-burning appliances produce creosote deposits on the interior walls of chimney liners. The creosote is highly combustible. In the event of a chimney fire, other connections to the chimney could allow fire to break out of the chimney and into the building. Also, chimney passageways can become partially blocked by creosote formation, thus forcing vent gases from other appliances to discharge back into the building.

SECTION M1802
VENT COMPONENTS

M1802.1 Draft hoods. Draft hoods shall be located in the same room or space as the combustion air openings for the appliances.

❖ Draft hoods are installed in gas venting systems to prevent downdrafts from entering gas appliances. Draft hoods also allow dilution air to be drawn into the vents, inhibiting the formation of moisture condensed from the flue gases. Air drawn into the venting system must be replaced to prevent a negative pressure that could produce spillage of vent gases from developing in the dwelling. Draft hoods must therefore be located in the same room or space as combustion air openings.

M1802.2 Vent dampers. Vent dampers shall comply with Sections M1802.2.1 and M1802.2.2.

❖ Dampers control the rate of combustion by regulating the amount of draft. When the appliance is not operating, they allow flue gas passageways to be closed to reduce the loss of conditioned air through the chimney. This section allows the use of both manually operated and automatic dampers.

M1802.2.1 Manually operated. Manually operated dampers shall not be installed except in connectors or chimneys serving solid-fuel-burning appliances.

❖ Manual dampers must be installed only in solid-fuel-burning appliances to control the rate of combustion by regulating the amount of draft and to reduce the loss of conditioned air through the vent or chimney when the appliance is not operating by allowing the vent or chimney to be closed. The damper is usually opened manually during the process of starting a fire. If the damper is not opened, smoke will spill back into the room, alerting the user. For gas or liquid-fuel-fired appliances, however, the user may not be aware of a partially or completely closed damper, and a hazardous condition could develop. The code therefore prohibits the use of manual dampers for all but solid-fuel-burning appliances. Note that this section does not prohibit the use of fixed baffles that are components of approved appliance venting systems.

M1802.2.2 Automatically operated. Automatically operated dampers shall conform to UL 17 and be installed in accordance with the terms of their listing and label. The installation shall prevent firing of the burner when the damper is not opened to a safe position.

❖ Automatic dampers can be used with gas- or liquid-fuel-fired natural draft appliances. Because damper failure can result in a hazardous condition, automatic dampers must be listed and labeled, and installed by a qualified installer so that the manufacturer's installation instructions and the terms of the listing are followed. An improperly installed or malfunctioning damper can cause the appliance to malfunction and discharge the combustion products back into the building.

M1802.3 Draft regulators. Draft regulators shall be provided for oil-fired appliances required to be connected to a chimney. Draft regulators provided for solid-fuel-burning appliances to reduce draft intensity shall be installed and set in accordance with the manufacturer's installation instructions.

❖ Oil-fired appliances must be fitted with a draft regulator unless the appliance is listed and labeled for use without the draft regulator. Draft regulators may be installed on solid-fuel-burning appliances to reduce the draft intensity. When required, draft regulators must be installed in the same room or enclosure as the appliance so that no difference in pressure between the air at the regulator and the combustion air supply will exist. Draft regulators are of the barometric type as compared to the draft hood used with gas appliances. See Commentary Figure M1802.3 for illustrations of draft hoods and draft regulators.

M1802.3.1 Location. Where required, draft regulators shall be installed in the same room or enclosure as the appliance so that

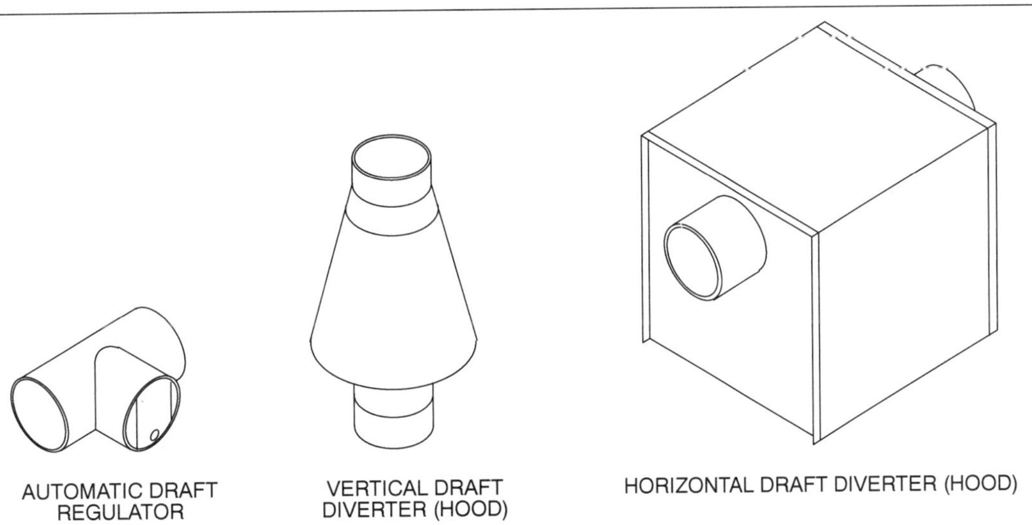

Figure M1802.3
DRAFT HOODS AND DRAFT REGULATORS

no difference in pressure will exist between the air at the regulator and the combustion air supply.

❖ To maintain proper pressure balance where combustion air openings exist, draft regulators must be installed in the same room as the appliances.

SECTION M1803
CHIMNEY AND VENT CONNECTORS

M1803.1 General. Connectors shall be used to connect fuel-burning appliances to a vertical chimney or vent except where the chimney or vent is attached directly to the appliance.

❖ Unless the chimney or vent is connected directly to the appliance, a connector is necessary. A connector is defined as a pipe used to connect an approved fuel-burning appliance to a vent or chimney, including any necessary fittings to make the connection or a change in direction.

M1803.2 Connectors for oil and solid fuel appliances. Connectors for oil and solid-fuel-burning appliances shall be constructed of factory-built chimney material, Type L vent material or single-wall metal pipe having resistance to corrosion and heat and thickness not less than that of galvanized steel as specified in Table M1803.2.

❖ Connectors for oil and solid fuel-burning appliances must be constructed of the materials listed in this section to insure that they are capable of tolerating the high temperatures and corrosive nature of the vent gases being conveyed.

TABLE M1803.2
THICKNESS FOR SINGLE-WALL METAL PIPE CONNECTORS

DIAMETER OF CONNECTOR (inches)	GALVANIZED SHEET METAL GAGE NUMBER	MINIMUM THICKNESS (inch)
Less than 6	26	0.019
6 to 10	24	0.024
Over 10 through 16	22	0.029

For SI: 1 inch = 25.4 mm.

❖ This table provides the minimum thickness of single-wall connectors of varying sizes.

M1803.3 Installation. Vent and chimney connectors shall be installed in accordance with the manufacturer's installation instructions and within the space that the appliance is located. Appliances shall be located as close as practical to the vent or chimney. Connectors shall be as short and straight as possible and installed with a slope of not less than $^1/_4$ inch (6.4 mm) rise per foot of run. Connectors shall be securely supported and joints shall be fastened with sheet metal screws or rivets. Devices that obstruct the flow of flue gases shall not be installed in a connector unless listed and labeled or approved for such installations.

❖ A connector must be supported for the design and weight of the material employed. Proper support is necessary to maintain the clearances required by Section M1803.3.4, to maintain the required pitch, and to prevent physical damage and separation of joints. The joints between connectors and appliance draft hood outlets and flue collars must be secured with screws, rivets, or other approved means. The joints between pipe sections also must be secured with screws, rivets or other approved means. Some connectors are available with a proprietary-type fastening method. The special proprietary joints do not require screws, rivets or other fasteners if they provide equivalent resistance to disengagement or displacement. This section addresses connectors, not chimneys and vents, and most chimney and vent manufacturers do not require or might even prohibit the use of screws or rivets for securing joints for chimney and vent pipe sections and fittings.

The ideal chimney or vent configuration is a vertical system, even though it is not always practical. This section requires all portions of a chimney or vent connector to rise vertically a minimum $^1/_4$ inch per each foot (0.02 mm/m) of its horizontal length. The connector slope is intended to induce the flow of flue gases by employing the natural buoyancy of the hot gases. Connector slope can promote the priming of a cold venting system and can partially compensate for short connector vertical rise.

M1803.3.1 Floor, ceiling and wall penetrations. A chimney connector or vent connector shall not pass through any floor or ceiling. A chimney connector or vent connector shall not pass through a wall or partition unless the connector is listed and labeled for wall pass-through, or is routed through a device listed and labeled for wall pass-through and is installed in accordance with the conditions of its listing and label. Connectors for oil-fired appliances listed and labeled for Type L vents, passing through walls or partitions shall be in accordance with the following:

1. Type L vent material for oil appliances shall be installed with not less than listed and labeled clearances to combustible material.
2. Single-wall metal pipe shall be guarded by a ventilated metal thimble not less than 4 inches (102 mm) larger in diameter than the vent connector. A minimum 6 inches (152 mm) of clearance shall be maintained between the thimble and combustibles.

❖ Floors, ceilings and walls in dwellings are often constructed of combustible materials and must be protected from exposure to high temperature in venting components. Type B and Type L vents must be installed with clearances to combustibles in accordance with their listing and label. Because single-wall metal pipe does not have the insulating qualities of Type B and Type L vents, it must be installed within a ventilated metal thimble when passing through combustible floors, ceilings and walls. The thimble must be at least 4 inches (102 mm) larger in diameter through the connector.

M1803.3.2 Length. The horizontal run of an uninsulated connector to a natural draft chimney shall not exceed 75 percent of the height of the vertical portion of the chimney above the connector. The horizontal run of a listed connector to a natural draft chimney shall not exceed 100 percent of the height of the vertical portion of the chimney above the connector.

❖ The appliance and chimney or vent must be located in close proximity so as to keep the connector length to a minimum. A connector must be limited in length because of flow resistance of the connector and heat loss through the connector wall.

M1803.3.3 Size. A connector shall not be smaller than the flue collar of the appliance.

Exception: Where installed in accordance with the appliance manufacturer's installation instructions.

❖ The size of the connector must be at least equal to the diameter of the appliance flue collar. A lesser size could restrict the flow of flue gases and result in spillage of flue gases into the dwelling.

M1803.3.4 Clearance. Connectors shall be installed with clearance to combustibles as set forth in Table M1803.3.4. Reduced clearances to combustible materials shall be in accordance with Table M1306.2 and Figure M1306.1.

❖ Connectors, vents and chimneys must be provided with minimum clearances to combustibles. These clearances can be reduced if the combustible material is protected by one of the methods of protection provided in Table M1306.2. The clearance requirements in Table M1803.3.4 apply to unlabeled single-wall connectors serving a listed appliance. Connectors that are listed and labeled must maintain a clearance to combustibles as required by the connector manufacturer's installation instructions. If the appliance or connector manufacturer's installation instructions specify larger clearances than those prescribed by this section, the manufacturer's installation instructions govern.

TABLE M1803.3.4
CHIMNEY AND VENT CONNECTOR CLEARANCES TO COMBUSTIBLE MATERIALS[a]

TYPE OF CONNECTOR	MINIMUM CLEARANCE (inches)
Single-wall metal pipe connectors:	
Oil and solid-fuel appliances	18
Oil appliances listed for use with Type L vents	9
Type L vent piping connectors:	
Oil and solid-fuel appliances	9
Oil appliances listed for use with Type L vents	3[b]

For SI: 1 inch = 25.4 mm.

a. These minimum clearances apply to unlisted single-wall chimney and vent connectors. Reduction of required clearances is permitted as in Table M1306.2.

b. When listed Type L vent piping is used, the clearance shall be in accordance with the vent listing.

❖ See the commentary for Section M1803.3.4.

M1803.3.5 Access. The entire length of a connector shall be accessible for inspection, cleaning and replacement.

❖ Because connectors are exposed to high temperatures and possible condensation, the entire length of the connector must be accessible to allow observation of any degradation of the connector and to allow replacement, if needed.

M1803.4 Connection to fireplace flue. Connection of appliances to chimney flues serving fireplaces shall comply with Sections M1803.4.1 through M1803.4.4.

❖ This section regulates installations where fireplace chimneys are used to vent appliances such as room heaters and fireplace-insert heaters.

M1803.4.1 Closure and accessibility. A noncombustible seal shall be provided below the point of connection to prevent entry of room air into the flue. Means shall be provided for access to the flue for inspection and cleaning.

❖ Without an air-tight connection between the chimney flue and the appliance chimney connector, chimney draft will draw room air into the flue, thus weakening the draft drawn through the appliance. Because the chimney flue is no longer open to the fireplace firebox, the chimney connection must be designed to provide access to the chimney for inspection and cleaning.

M1803.4.2 Connection to factory-built fireplace flue. A different appliance shall not be connected to a flue serving a factory-built fireplace unless the appliance is specifically listed for such an installation. The connection shall be made in conformance with the appliance manufacturer's instructions.

❖ Factory-built fireplaces may or may not be tested for uses other than that as a traditional fireplace. Connecting an appliance or fireplace-insert heater could result in abnormally high fireplace or chimney temperatures, leading to component deterioration and hazardous operation. The fireplace manufacturer will provide specific details telling what appliance connections are allowed. Usually, the manufacturer's instructions will prohibit any modifications and any use of the fireplace other than that for which it is designed.

M1803.4.3 Connection to masonry fireplace flue. A connector shall extend from the appliance to the flue serving a masonry fireplace such that the flue gases are conveyed directly into the flue. The connector shall be accessible or removable for inspection and cleaning of both the connector and the flue. Listed direct connection devices shall be installed in accordance with their listing.

❖ The appliance chimney connector must extend up through the fireplace damper and smoke chamber and terminate in the chimney flue. This helps produce an adequate draft and reduces the possibility of combustible creosote deposits forming within the fireplace throat and smoke chamber.

M1803.4.4 Size of flue. The size of the fireplace flue shall be in accordance with Section M1805.3.1.

❖ The size of the connector must be equal to the diameter of the appliance to prevent a restriction in the flow of the flue gases. See the commentary for Section M1805.3.1.

SECTION M1804 VENTS

M1804.1 Type of vent required. Appliances shall be provided with a listed and labeled venting system as set forth in Table M1804.1.

❖ Vents and chimneys are distinct systems. The provisions of this section apply only to vents. The vents addressed in this section are natural draft venting systems that produce draft by employing the same principles that produce draft in chimneys.

Vents regulated by this section are factory fabricated and must be listed and labeled. The labeling requirement applies to all components of the system, including the sections of pipe, fittings, terminal caps, supports and spacers.

A vent system can only be used with oil-burning and pellet-burning appliances that are listed and labeled for use with a vent. Solid-fuel-burning appliances must not be connected to a vent system, because such appliances produce flue gas temperatures that are much higher than the temperatures for which vents are designed. The burning of solid fuels also produces creosote. Creosote formation leaves a combustible deposit on the surface of the flue passageway that, if ignited, can produce a fire with temperatures higher than 2,000°F (1,093°C). Such temperatures far exceed the maximum safe temperature of a vent system.

Type L vents and pellet vents must be tested in accordance with UL 641 and are designed to vent oil-burning and pellet-fuel-burning appliances. Oil-burning appliances produce a higher flue gas temperature than gas-fired appliances. Type L vents are typically double-wall, air-insulated vent piping systems constructed of galvanized and stainless steel. They are designed for natural draft applications only and must not be used to convey combustion gases under a positive pressure. For example, a Type L vent must not be used with Category III or IV gas-fired appliances and must not be used on the discharge side of an exhauster or power-vented appliance. Positive pressures will cause a Type L vent to leak combustion gases into the building, and the leakage can cause condensation to form within the vent system pipe, fittings and joints, resulting in corrosion damage to the vent and eventual vent failure.

TABLE M1804.1
VENT SELECTION CHART

TYPE L OIL	PELLET VENTS
Oil-burning appliances listed and labeled for use with Type L vents.	Pellet fuel burning
	Pellet fuel-burning appliances listed and labeled for use with pellet vents

❖ See the commentary for Section 1804.1.

M1804.2 Termination. Vent termination shall comply with Sections M1804.2.1 through M1804.2.6.

❖ The standards for vents and appliances require that venting instructions be supplied by the manufacturer of the appliance and the manufacturer of the vent system. Such instructions are part of the labeling requirements. Any deviation from them is a violation of the code.

The clearance to combustibles for a vent system is determined by the testing agency and stated on the component labels and in the manufacturer's instructions. Not all vents have the same required clearances to combustibles. The clearances are determined by vent performance during testing in accordance with the applicable standard. Different vent materials and designs impact the vent's ability to control the amount of heat transmitted to surrounding combustibles. Installation of the vent system in accordance with the clearances listed on the vent's label and in the manufacturer's installation instructions is critical.

The termination of a natural draft vent must comply with the requirements of the manufacturer's installation instructions and Sections M1804.2.1 through M1804.2.5. A vent used in conjunction with a mechanical exhauster must meet the termination requirements established in Section M1804.2.6.

The vent system must be physically protected to prevent damage to the vent and to prevent combustibles from coming into contact with or being placed too close to the vents. This protection is usually provided by enclosing the vent in chases, shafts or cavities during the building construction. Physical protection is not required in the room or space where the vent originates (at the appliance connection) and is not required in such locations as attics that are not occupied or used for storage. For example, assume that a vent is installed in an existing building and that it extends from the basement, through a first-floor closet, through the attic, and through the roof. The portion of the vent passing through the closet must be protected from damage. The means of physical damage protection must also be designed to maintain separation between the vent and any combustible storage. To prevent the passage of fire and smoke through the annular space around a vent penetration through a floor or ceiling, the vent must be fireblocked with a noncombustible material in accordance with this code. Vent manufacturers provide factory-built components and installation instructions for fireblocking penetrations.

M1804.2.1 Through the roof. Vents passing through a roof shall extend through flashing and terminate in accordance with the manufacturer's installation requirements.

❖ Manufacturers' installation instructions contain the requirements for the termination of vents through the roof. See Commentary Figure M1804.2.1.

M1804.2.2 Decorative shrouds. Decorative shrouds shall not be installed at the termination of vents except where such shrouds are listed and labeled for use with the specific venting system and are installed in accordance with the manufacturer's installation instructions.

❖ Decorative shrouds must not interfere with the operation of the venting system.

M1804.2.3 Natural draft appliances. Vents for natural draft appliances shall terminate at least 5 feet (1524 mm) above the highest connected appliance outlet, and natural draft gas vents serving wall furnaces shall terminate at an elevation at least 12 feet (3658 mm) above the bottom of the furnace.

❖ To maintain adequate venting action, the termination of gravity vents must be at least 5 feet (1524 mm) above the vent collar of the appliance and at least 12 feet (3658 mm) above the bottom of a wall furnace.

M1804.2.4 Type L vent. Type L venting systems shall conform to UL 641. Such vents shall terminate with a listed and labeled cap in accordance with the vent manufacturer's installation instructions not less than 2 feet (610 mm) above the roof and not less than 2 feet (610 mm) above any portion of the building within 10 feet (3048 mm).

❖ The termination heights are consistent with the requirements of NAPA 31.

Consideration must be given to vent location in the design/installation phase of a building so that the vent termination will not require long extensions of vent pipe exposed to the outdoors. Vent piping exposed to the outdoors encourages condensation in cold weather, could require guy wires or braces depending on

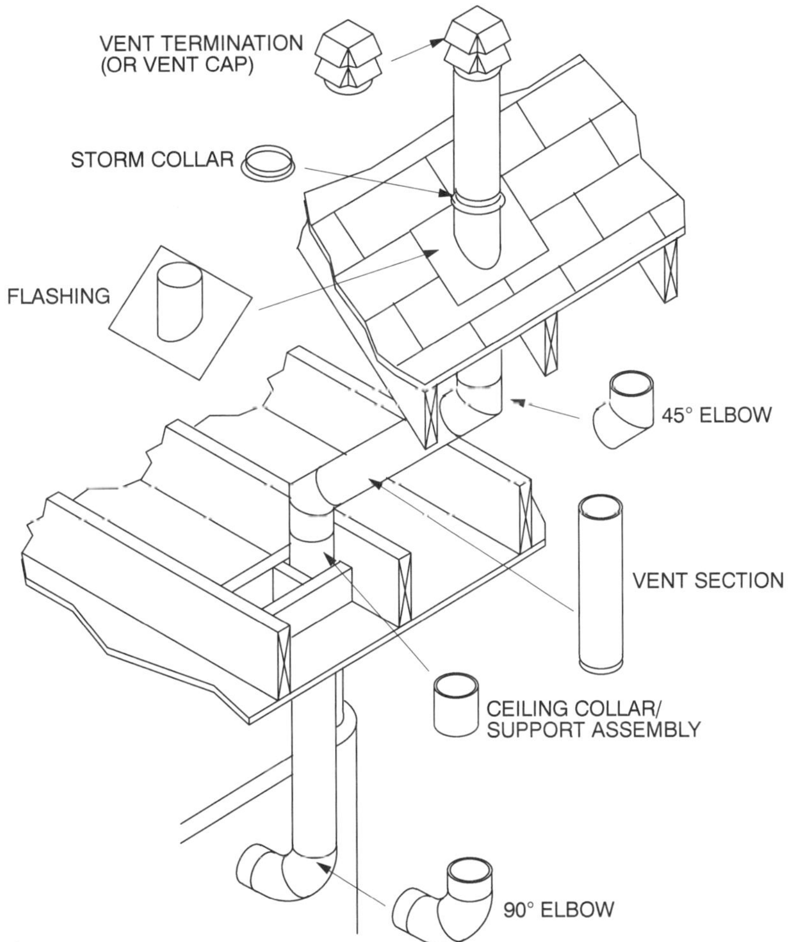

For SI: 1 degree = .01745 rad.

Figure M1804.2.1
VENT TERMINATION THROUGH A ROOF

height, and is considered aesthetically unattractive. Another good reason to plan for vent location during the design phase of a building is to allow straight vertical runs of vent and avoid difficult offsets. Good planning can also prevent vents from passing through the roof on the street side of the building. Quite often, difficult vent offsets in attics are needed for vents to penetrate the roof on the back side of a building.

M1804.2.5 Direct vent terminations. Vent terminals for direct-vent appliances shall be installed in accordance with the manufacturer's installation instructions.

❖ Direct-vent appliances include appliances that have a closed combustion chamber and a conduit for bringing all combustion air directly from the outdoors. Direct-vent appliances are often referred to as "sealed combustion" or "separated combustion" appliances.

The code does not provide prescriptive termination requirements for these appliances and relies on the manufacturers' installation instructions.

M1804.2.6 Mechanical draft systems. Mechanical draft systems shall be installed in accordance with their listing, the manufacturer's installation instructions and, except for direct vent appliances, the following requirements:

1. The vent terminal shall be located not less than 3 feet (914 mm) above a forced air inlet located within 10 feet (3048 mm).
2. The vent terminal shall be located not less than 4 feet (1219 mm) below, 4 feet (1219 mm) horizontally from, or 1 foot (305 mm) above any door, window or gravity air inlet into a dwelling.
3. The vent termination point shall not be located closer than 3 feet (914 mm) to an interior corner formed by two walls perpendicular to each other.
4. The bottom of the vent terminal shall be located at least 12 inches (305 mm) above finished ground level.
5. The vent termination shall not be mounted directly above or within 3 feet (914 mm) horizontally from an oil tank vent or gas meter.
6. Power exhauster terminations shall be located not less than 10 feet (3048 mm) from lot lines and adjacent buildings.
7. The discharge shall be directed away from the building.

❖ The appliance installations addressed in this section use auxiliary or integral fans and blowers to force the flow of combustion products to the outdoors. This section applies to externally installed power exhausters, integrally power-exhausted appliances, and venting systems equipped with draft inducers. This section does not address direct-vent appliances.

The seven requirements in this section provide guidance in locating the vent terminal to prevent drawing the vented products of combustion back into the building or creating a hazardous condition.

M1804.3 Installation. Type L and pellet vents shall be installed in accordance with the terms of their listing and label and the manufacturer's installation instructions.

❖ The standards for vents and appliances require that venting instructions be supplied by the manufacturer of the appliance and the manufacturer of the vent system. Such instructions are part of the labeling requirements. Any deviation from them is a violation of the code.

The clearance to combustibles for a vent system is determined by the testing agency and stated on the component labels and in the manufacturer's instructions. Not all vents have the same required clearances to combustibles. The clearances are determined by vent performance during the testing in accordance with the applicable standard. Different vent materials and designs impact the vent's ability to control the amount of heat transmitted to surrounding combustibles. Installation of the vent system in accordance with the clearances listed on the vent's label and in the manufacturer's installation instructions is critical.

Physical protection of the vent system is required to prevent damage to the vent and to prevent combustibles from coming into contact with or being placed too close to the vents. This protection is typically provided by enclosing the vent in chases, shafts, or cavities in the building construction. Physical protection is not required in room or space where the vent originates (at the appliance connection) and would not be required in locations such as attics that are not occupied or used for storage. For example, assume that a vent is installed in an existing building and that it extends from the basement, through a first-floor closet, through the attic, and through the roof. The portion of the vent passing through the closet must be protected from damage. The means of physical damage protection must also be designed to maintain separation between the vent and any combustible storage. To prevent the passage of fire and smoke through annular space around a vent penetration through a floor or ceiling, the vent must be fireblocked with a noncombustible material in accordance with this code. Vent manufacturers provide installation instructions and factory-built components for fireblocking penetrations.

M1804.3.1 Size of single appliance venting systems. An individual vent for a single appliance shall have a cross-sectional area equal to or greater than the area of the connector to the appliance, but not less than 7 square inches (4515 mm^2) except where the vent is an integral part of a listed and labeled appliance.

❖ This section sets the minimum vent area to be allowed on gravity vent appliances at 7 square inches (4515 mm^2) [which calculates to a minimum 3-inch (76 mm) diameter vent]. Obviously, if the vent is an integral part of the appliance, lesser sizes may be accepted. Section M1803.3.3 requires that the size of the vent or connector not be less than the appliance flue collar. Therefore, when the appliance flue collar is larger than 3 inches (76 mm) in diameter, the connector and vent must be increased in size accordingly.

SECTION M1805 MASONRY AND FACTORY-BUILT CHIMNEYS

M1805.1 General. Masonry and factory-built chimneys shall be built and installed in accordance with Sections R1001 and R1002, respectively. Flue lining for masonry chimneys shall comply with Section R1001.8.

❖ Masonry and factory-built chimneys must be constructed in accordance with Sections R1001, R1002 and R1001.7.

M1805.2 Masonry chimney connection. A chimney connector shall enter a masonry chimney not less than 6 inches (152 mm) above the bottom of the chimney. Where it is not possible to locate the connector entry at least 6 inches (152 mm) above the bottom of the chimney flue, a cleanout shall be provided by installing a capped tee in the connector next to the chimney. A connector entering a masonry chimney shall extend through, but not beyond the wall and shall be flush with the inner face of the liner. Connectors, or thimbles where used, shall be firmly cemented into the masonry.

❖ The appliance chimney connector must extend up through the fireplace damper and smoke chamber and terminate in the chimney flue. This helps produce an adequate draft and reduces the possibility of combustible creosote deposits forming within the fireplace throat and smoke chamber.

M1805.3 Size of chimney flues. The effective area of a natural draft chimney flue for one appliance shall be not less than the area of the connector to the appliance. Chimney flues connected to more than one appliance shall be not less than the area of the largest connector plus 50 percent of the areas of additional chimney connectors.

Exception: Chimney flues serving oil-fired appliances sized in accordance with NFPA 31.

❖ This section requires a chimney or vent serving a single appliance to have a minimum cross-sectional area equal to the cross-sectional area of the appliance outlet connection. The maximum size of a chimney or vent is also an important design consideration. Oversized chimneys and vents can fail to produce sufficient draft and can cause condensation of the flue gases. Existing chimneys and vents might become oversized if an appliance that was previously connected is removed. In excessively large chimneys and vents, the flue gases are cooled by expansion and because of increased heat loss through the larger flue passage surface area. The cooling effect reduces the temperature differences necessary to produce draft and can cause the flue gases to reach the dew point temperature. The resulting condensation is highly corrosive and can cause deterioration of the chimney, vent, connected appliances, and the building.

For connection to more than one appliance, this section requires that the area of the flue be increased by 50 percent of the area of the additional connector. The reason is based on a diversity factor arising from the fact that all appliances are not likely to be firing simultaneously at their maximum rate. Any design conflicting with the sizing criteria of this section must be considered an engineered system.

M1805.3.1 Size of chimney flue for solid fuel appliance. Except where otherwise specified in the manufacturer's installation instructions, the cross-sectional area of a flue connected to a solid-fuel-burning appliance shall be not less than the area of the flue collar or connector, and not larger than three times the area of the flue collar.

❖ So that flue gases properly exhaust, the cross-sectional area of the flue connected to a solid fuel-burning appliance must not be less than the area of the flue collar or connector. For proper draft, and to allow the chimney to be heated sufficiently to produce and maintain this draft, this section puts a maximum size limitation on the chimney, not to exceed three (3) times the area of the flue collar. See the commentary for Section M1805.3.

Bibliography

The following resource materials are referenced in this chapter or are relevant to the subject matter addressed in this chapter.

ANSI Z223.1–02, *National Fuel Gas Code*. New York: American National Standards Institute, 2002.

ASHRAE–00, *HVAC Systems and Equipment Handbook*. Atlanta: American Society for Heating, Refrigerating and Air-Conditioning Engineers, 2000.

ASTM C27–98, *Standard Classification of Fireclay and High-Alumina Refractory Brick*. West Conshohocken, PA: ASTM International, 1998.

ASTM C199–84, *Test Method for Pier Test for Refractory Mortars*. West Conshohocken, PA: ASTM International, 1984.

ASTM C315–00, *Specification for Clay Flue Linings*. West Conshohocken, PA: ASTM International, 2000.

Burning Solid Fuel Safely, 2nd edition. Country Club Hills, IL: Building Officials and Code Administrators International, 1991.

Certified Appliance and Accessories Directory. Washington, DC: American Gas Association, 1996.

Chimney and Gas Vent Sizing Handbook, Publication A99001. Dallas: Selkirk, 1992.

Fire Protection Handbook, 18th edition. Quincy, MA: National Fire Protection Association, 1996.

Grimm, Nils R. and Robert C. Rosaler. *Handbook of HVAC Design*. New York: McGraw-Hill, 1990.

NFPA 54-96, *National Gas Code*. Quincy, MA: National Fire Protection Association, 1996.

NFPA 97-96, *Terms Relating to Chimneys, Vents, and Heat-Producing Appliances*. Quincy, MA: National Fire Protection Association, 1996.

NFPA 211–00, *Chimneys, Fireplaces, Vents, and Solid Fuel–Burning Appliances*. Quincy, MA: National Fire Protection Association, 2000.

UL 103–98, *Chimneys, Factory–Built, Residential Type and Building Heating Appliances*. Northbrook, IL: Underwriters Laboratories, 1998.

UL 127–99, *Factory–Built Fireplaces*. Northbrook, IL: Underwriters Laboratories, 1999.

UL 641–95, *Low–Temperature Venting Systems, Type L*. Northbrook, IL: Underwriters Laboratories, 1995.

UL 737–96, *Fireplace Stoves*. Northbrook, IL: Underwriters Laboratories, 1996.

UL 959–01, *Medium–Heat Appliance Factory–Built Chimneys*. Northbrook, IL: Underwriters Laboratories, 2001.

UL 1777–98, *Chimney Liners*. Northbrook, IL: Underwriters Laboratories, 1998.

Chapter 19: Special Fuel-burning Equipment

General Comments

This chapter regulates the installation of fuel-burning appliances that are not covered in other chapters. Section M1901, "Ranges and Ovens," covers the installation of and clearances to such appliances. Section M1902, Sauna Heaters, addresses the proper installation and controls necessary for safe operation. Section M1903 addresses the testing and installation of stationary fuel cell power plants.

The chapters of the code are dedicated to single subjects, however, because the two subjects in this chapter do not contain the volume of text necessary to warrant individual chapters, they have been combined into a single chapter. This chapter is a collection of requirements for various appliances. The only commonality is that the subjects use energy to perform some task or function.

Purpose

This chapter provides regulations for the proper installation of ranges, ovens, and sauna heaters. The intent is to provide a reasonable level of protection for the occupants of the dwelling.

SECTION M1901 RANGES AND OVENS

M1901.1 Clearances. Freestanding or built-in ranges shall have a vertical clearance above the cooking top of not less than 30 inches (762 mm) to unprotected combustible material. Reduced clearances are permitted in accordance with the listing and labeling of the range hoods or appliances.

❖ Freestanding and built-in ranges must provide required clearances to unprotected combustible materials. Cooking range tops produce considerable heat. The vertical clearance called out will protect the occupants from the ignition of any unprotected combustible materials that are commonly found in residential kitchens.

Vented or unvented range hoods are often installed over ranges. Note, however, that this code does not require such hoods. When installed, range hoods must conform to the manufacturer's installation instructions. Reduced clearances to these hoods are found in most manufacturer's listings of their appliances.

M1901.2 Cooking appliances. Household cooking appliances shall be listed and labeled and shall be installed in accordance with the manufacturer's installation instructions. The installation shall not interfere with combustion air or access for operation and servicing.

❖ This section requires the listing and labeling of all cooking appliances and installation in accordance with the manufacturers' installation instructions. Cooking appliances listed for commercial use must not be installed in residential kitchens. Section G2447.2 prohibits gas-fired commercial cooking appliances from being installed in dwelling units. Commercial cooking appliances do not have the same safety features as residential appliances, such as push-to-turn knobs and insulated doors. There are cooking appliances on the market that have the size and features of commercial appliances, but they are listed for residential installation.

SECTION M1902 SAUNA HEATERS

M1902.1 Locations and protection. Sauna heaters shall be protected from accidental contact by persons with a guard of material having a low thermal conductivity, such as wood. The guard shall have no substantial effect on the transfer of heat from the heater to the room.

❖ Sauna heaters produce steam by evaporating water that passes through or over a heating element. Guards must be installed around the source of steam because occupants of the space of the sauna are usually unclothed. These guards must be made of a material that has low thermal conductivity, such as wood, to protect the occupants from burns.

M1902.2 Installation. Sauna heaters shall be installed in accordance with the manufacturer's installation instructions.

❖ Sauna heaters must be listed, labeled and installed in accordance with the manufacturer's installation instructions. The listing and label must indicate compliance with an approved national standard.

M1902.3 Combustion air. Combustion air and venting for a nondirect vent-type heater shall be provided in accordance with Chapters 17 and 18, respectively.

❖ Combustion air for fuel-burning appliances must not be taken from within the sauna room itself. Sauna rooms are usually very small, confined spaces that do not contain an adequate volume of combustion air. For other than direct vent appliances, combustion air must be provided in accordance with Chapters 17 and 18 of this code.

M1902.4 Controls. Sauna heaters shall be equipped with a thermostat that will limit room temperature to not greater than 194°F (90°C). Where the thermostat is not an integral part of the heater, the heat-sensing element shall be located within 6 inches (152 mm) of the ceiling.

❖ The room must be kept at a temperature not to exceed 194°F (90°C). This temperature has been determined to be the maximum safe level, beyond which occupants can be injured. So that the thermostat will measure the highest temperature of the room and thus maintain an acceptable and safe temperature range throughout the space, the thermostat for the appliance must either be integral with the heater or be installed within the top 6 inches (76 mm) of the room.

SECTION M1903 STATIONARY FUEL CELL POWER PLANTS

M1903.1 General. Stationary fuel cell power plants having a power output not exceeding 1,000 kW, shall be tested in accordance with ANSI Z21.83 and shall be installed in accordance with the manufacturer's installation instructions and NFPA 853.

❖ The private generation of electricity for commercial and residential use is an emerging technology that is expected to gain widespread acceptance as future technological improvements result in lower costs. Fuel cells have been included in this code to provide guidance to the building community for those future installations.

The code requires fuel cell power plants with an output of 1,000 kW or less to be listed in accordance with ANSI Z21.83 and requires the units to be installed in accordance with the manufacturers' instructions and NFPA 853. NFPA 853 provides guidance on location of the power plants, fuel system supplies, ventilation and exhaust of the space containing the unit and fire protection requirements. While fuel cell technology has been available since the 1800s, its first practical use was by NASA to generate electricity and water for the Gemini and Apollo spacecraft and today's space shuttle. Because the installation of fuel cells in privately owned buildings is relatively new, the following discussion of the fuel cell process is included to provide a basic understanding of the technology.

A fuel cell is an electrochemical device that converts the chemical energy of a hydrocarbon fuel (hydrogen, natural gas, coal-bed methane, methanol, gasoline, etc.) and an oxidant (air or oxygen) into useable electricity. The fuel cell consists of a fuel electrode (anode) and an oxidant electrode (cathode) separated by an ion-conducting membrane (electrolyte). When a hydrocarbon fuel is introduced into the fuel cell, the catalyst surface of the membrane converts the hydrogen gas molecules into hydrogen ions and electrons. The hydrogen ions pass through the membrane to react with oxygen and electrons to form water as a reaction by-product. The electrons, which cannot pass through the membrane, must travel around it through an external circuit, thereby creating direct current (dc) electricity. See Commentary Figure M1903.1(1).

When fueled by pure hydrogen, the only by-products of the fuel cell process are heat and water. Low levels of CO_2 are produced when other fuels are used. Unlike traditional fossil-fuel electric generating stations that burn the fuels, the electrochemical fuel cell process produces no particulate matter, nitrogen oxides or sulfur oxides.

When the waste heat is captured for use in heating water or a cogeneration process, fuel cell efficiency can reach 80 percent.

A typical fuel cell consists of three sections: a fuel processor, a fuel stack (power generator), and a power conditioner. See Commentary Figure M1903.1(2).

The fuel processor section has two components: (1) the fuel reformer, which processes a hydrocarbon fuel, such as methane, into a hydrogen-rich formate gas and (2) a carbon monoxide cleanup system to reduce CO concentrations to acceptable levels.

The fuel cell stack consists of many individual fuel cells that operate as previously described to generate electricity and usable heat. The power output of the unit increases based on the number and arrangement of individual cells stacked in series and/or in parallel in the unit. There are several types of fuel cell technologies, including phosphoric acid (PA), molten carbonate (MC), solid oxide (SO) and proton exchange membrane (PEM), which are named based on the type of electrolyte employed in the membrane section of the cell.

The power conditioner (inverter) section converts the direct current (dc) produced in the fuel cell stack into alternating current (ac). Batteries are used to smooth out power surge demands, such as HVAC start-ups, and to provide additional power when demand exceeds the peak output of the cell stack.

Although currently available fuel cell units are not cost competitive with established electrical suppliers, the initial and operating costs are expected to decrease as more manufacturers enter the market. Where there are constraints on transmission expansion, in remote locations or in areas where reduced emission of pollutants is critical, fuel cells could be seriously considered as an energy source. Fuel cells can also be attractive where the quality and dependability of commercial power is a concern.

Bibliography

The following resource materials are referenced in this chapter or are relevant to the subject matter addressed in this chapter.

ANSI Z21.83 - 1998, *Fuel Cell Power Plants*. New York: American National Standards Institute, 1998.

NFPA 31-01, *Installation of Oil-Burning Equipment.* Quincy, MA: National Fire Protection Association, 2001.

NFPA 853 - 00, *Installation of Stationary Fuel Power Plants*. Quincy, MA: National Fire Protection Association, 2000.

UL 2162-94, *Commercial Wood-Fired Baking Ovens*. Northbrook, IL: Underwriters Laboratories, 1994.

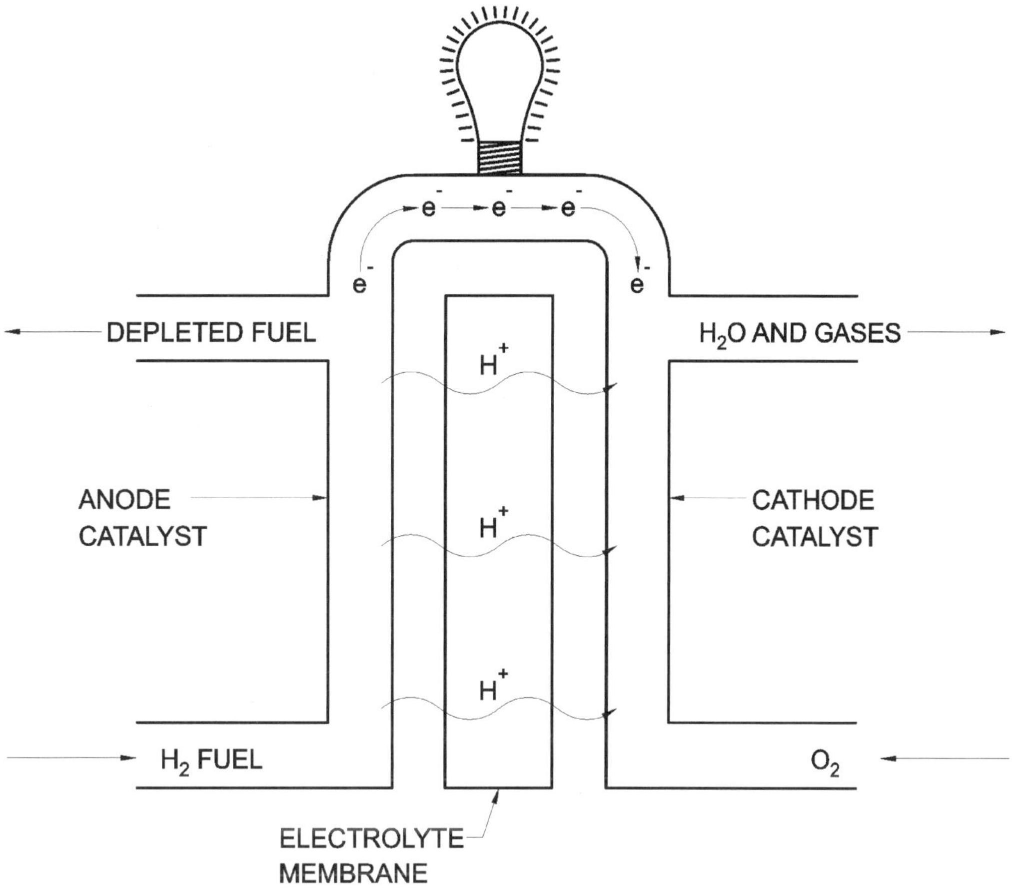

Figure M1903.1(1)
TYPICAL FUEL CELL COMPONENTS

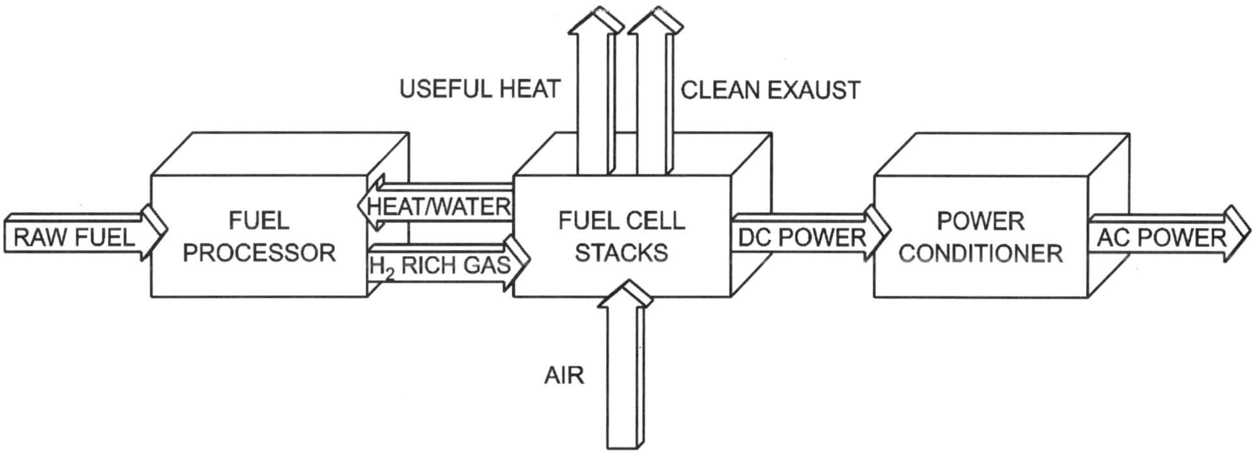

Figure M1903.1(2)
GENERIC FUEL CELL SYSTEM

Chapter 20: Boilers and Water Heaters

General Comments

This chapter applies to all boilers and water heaters. A water heater is any appliance that heats potable water and supplies it to the plumbing hot water distribution system. However, some water heaters are also used for space heating. These are listed for combination potable water heating and space heating applications. A boiler either heats water or generates steam for space heating and is generally a closed system.

Purpose

This chapter regulates the installation of boilers and water heaters. Its purpose is to protect the occupants of the dwelling from the potential hazards associated with such appliances.

SECTION M2001 BOILERS

M2001.1 Installation. In addition to the requirements of this code, the installation of boilers shall conform to the manufacturer's instructions. The manufacturer's rating data, the nameplate and operating instructions of a permanent type shall be attached to the boiler. Boilers shall have all controls set, adjusted and tested by the installer. A complete control diagram together with complete boiler operating instructions shall be furnished by the installer. Solid and liquid-fuel-burning boilers shall be provided with combustion air as required by Chapter 17.

❖ This section governs the installation and commissioning of boilers and their control systems. The mechanical equipment requirements for approval, labeling, installation, maintenance, repair, and alteration are regulated by Chapters 12 and 13.

Complete operating instructions must be permanently affixed to the boiler upon completion of installation by the contractor. Boiler systems can be complex and generally require coordination among several pieces of equipment. The proper operating procedures, set points, etc., must be specified in these instructions. This is usually carried out in the control system design documentation, which includes diagrams, system schematics and control sequence descriptions.

M2001.1.1 Standards. Oil-fired boilers and their control systems shall be listed and labeled in accordance with UL 726. Electric boilers and their control systems shall be listed in accordance with UL 834. Boilers shall be designed and constructed in accordance with the requirements of ASME CSD-1 and as applicable, the ASME *Boiler and Pressure Vessel Code*, Sections I and IV. Gas-fired boilers shall conform to the requirements listed in Chapter 24.

❖ The history of boilers shows that they are potentially dangerous if not properly designed and constructed. Along with the code, several industry standards are referenced for the design and construction of boilers, and equipment manufacturers produce boilers in accordance with these referenced standards. Only the design and construction of boilers are regulated by these references. Design and construction requirements for specific types of boilers are found in the respective sections of these referenced standards.

M2001.2 Clearance. Boilers shall be installed in accordance with their listing and label.

❖ Since boilers, related appliances, and mechanical equipment require inspection, observation, routine maintenance, repairs, and possible replacement, working clearance around such equipment and appliances is required. The manufacturers' installation instructions state the access requirements. See Commentary Figure M2001.2 for an illustration of a typical installation.

M2001.3 Valves. Every boiler or modular boiler shall have a shutoff valve in the supply and return piping. For multiple boiler or multiple modular boiler installations, each boiler or modular boiler shall have individual shutoff valves in the supply and return piping.

> **Exception:** Shutoff valves are not required in a system having a single low-pressure steam boiler.

❖ Shutoff valves are required to facilitate maintenance and repairs by isolating a boiler from the system that it serves. Without isolation valves, the entire system of piping and components would have to be taken out of service, drained and relieved of pressure before a boiler or water-side/steam-side component could be removed for servicing, repair or replacement. Where systems have multiple boilers, isolation valves permit one or more boilers to be removed from service without affecting the remaining units, thus preserving partial system operation. A modular boiler unit is a field-assembled boiler composed of multiple interdependent modules. Modular units are designed to permit close matching of boiler capacity to building loads

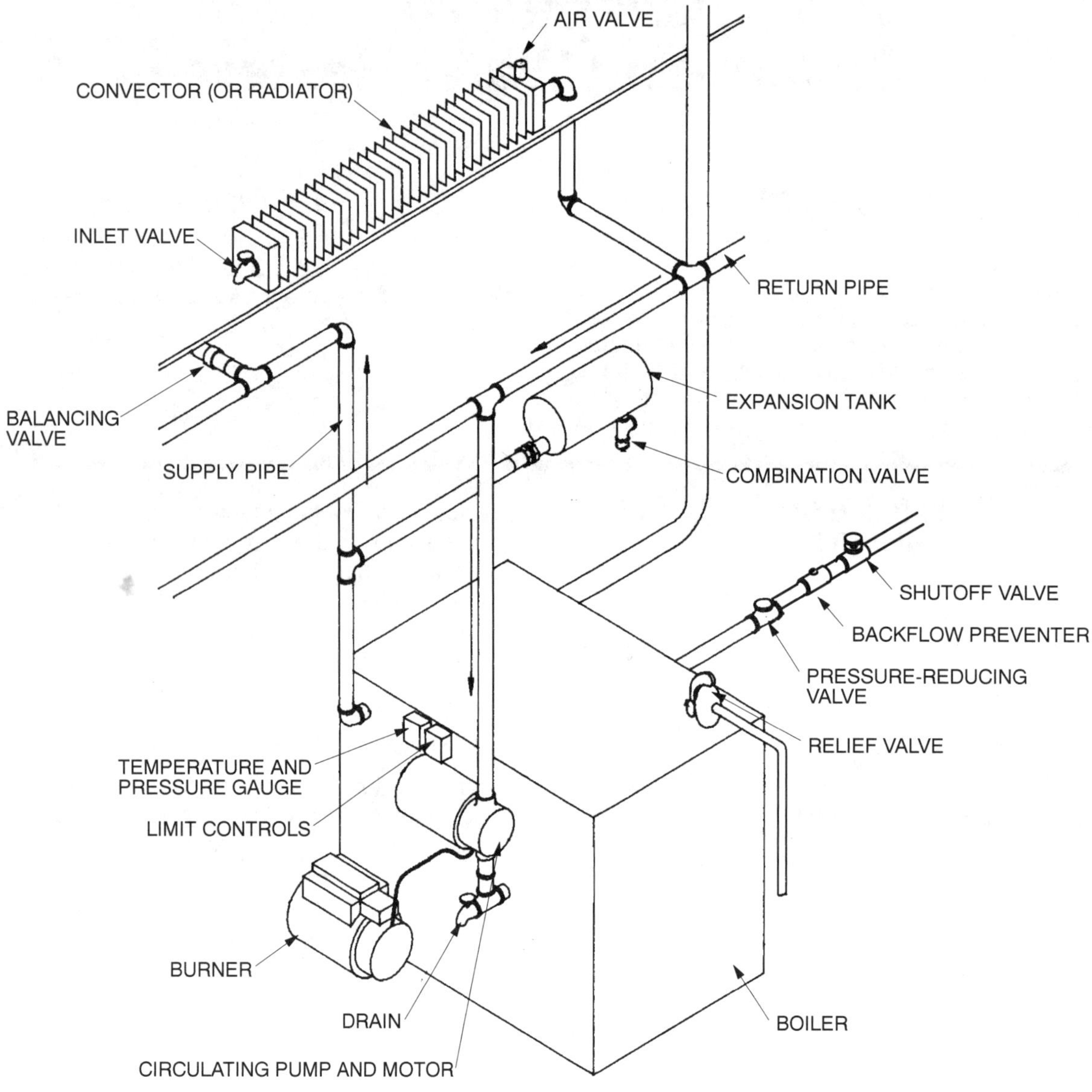

Figure M2001.2
RESIDENTIAL BOILER AND ASSOCIATED EQUIPMENT

and to allow for future expansion of capacity by the addition of modules to a boiler "package." Each independent boiler or modular boiler, whether single or in groups, must be valved individually. See Figure M2001.3.

The exception states that shutoff valves are not required for systems with a single low-pressure steam boiler. Shutoff valves would not be of value for a single low-pressure steam boiler installation because of a negligible amount of heat transfer medium to relieve from the system. Also, maintaining a partial capacity system operation is impossible with only a single unit.

M2001.4 Flood-resistant installation. In areas prone to flooding as established in Table R301.2(1), boilers, water heaters, and their control systems shall be located or installed in accordance with Section R323.1.5.

❖ The local jurisdiction must fill in Table R301.2(1) upon adoption of this code, including the FLOOD HAZARDS information. Boilers, water heaters and associated control systems that are located in flood hazard areas must be installed above the design flood elevation or must be designed and installed to prevent the entrance of water into the components and to resist the forces of the flood waters on the components. See the commentary for Section 323.1.5.

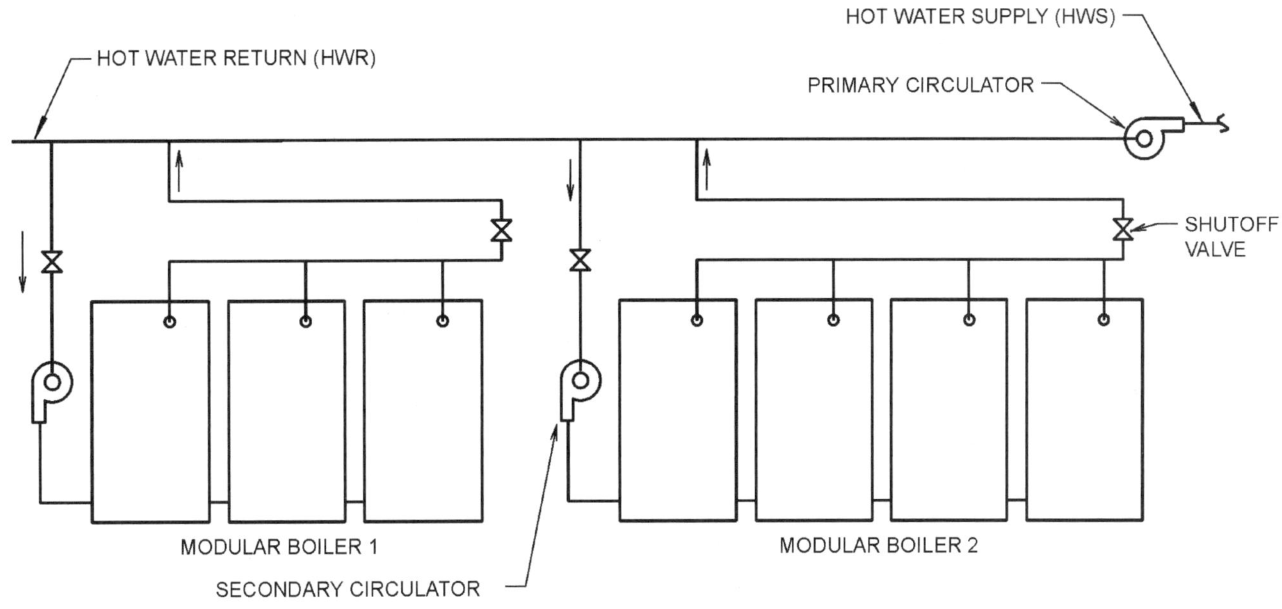

Figure M2001.3
MULTIPLE MODULAR UNITS

SECTION M2002
OPERATING AND SAFETY CONTROLS

M2002.1 Safety controls. Electrical and mechanical operating and safety controls for boilers shall be listed and labeled.

❖ This section requires the listing and labeling of electrical and mechanical operating and safety controls.

M2002.2 Hot water boiler gauges. Every hot water boiler shall have a pressure gauge and a temperature gauge, or combination pressure and temperature gauge. The gauges shall indicate the temperature and pressure within the normal range of the system's operation.

❖ All boilers operate within certain pressure and temperature limitations. Problems and hazards can develop if these limitations are exceeded. Pressure and temperature gauges are necessary to allow boiler owners and service technicians to check and monitor system operating conditions. Gauges indicate hazardous conditions and causes of problems or malfunctions. They also help prevent future hazards or problems from developing by signaling the need for system repairs and service. To allow reasonably accurate readings, gauges must have a range and scale of values that are proportioned to the boiler's normal operating range. For example, a pressure gauge with a range of 0 to 250 psi (0 to 1724 kPa) would barely deflect the needle when installed on a 10-psi (69 kPa) system, and pressure readings would be inaccurate. Hot water boiler gauges are often equipped with combination pressure and temperature gauges. See Commentary Figure M2002.2.

M2002.3 Steam boiler gauges. Every steam boiler shall have a water-gauge glass and a pressure gauge. The pressure gauge shall indicate the pressure within the normal range of the sys-

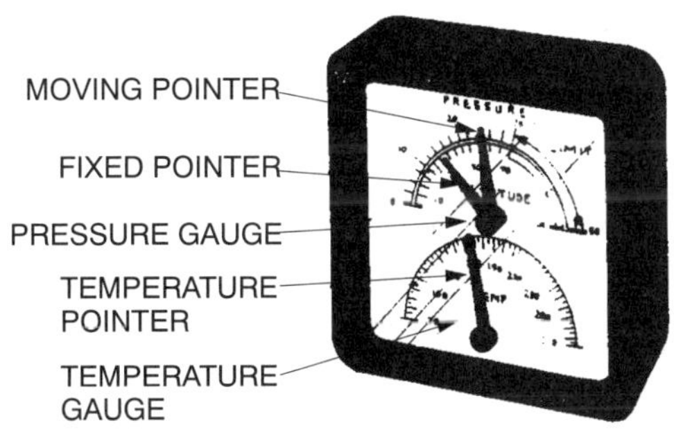

Figure M2002.2
BOILER TEMPERATURE AND PRESSURE GAUGE

tem's operation. The gauge glass shall be installed so that the midpoint is at the normal water level.

❖ All steam boilers operate within certain pressure and water level limits that must be observed. However, unlike with hot water boilers, water temperature is not a concern for steam systems. Whenever steam is present, the water temperature in a steam boiler will be a function of steam pressure and is easily determined when the steam pressure is known. A steam pressure gauge indicates both normal and abnormal operating conditions and the presence of a hazardous condition and is necessary to observe and monitor boiler operation. The water level in a boiler fluctuates within a narrow margin between too low and too high. Because proper boiler operation is dependent on the water level being maintained within the high and low parameters, a water-gauge glass is necessary to visually indicate the actual water level in the boiler.

A water-gauge glass, commonly referred to as a "sight glass," indicates the following: boiler flooding, low water conditions, condensate return operation, unstable water levels resulting from foaming and surging, boiler water condition, and the amount of automatic or manual makeup water feed. See Commentary Figure M2002.3. If either or both of the gauge-glass cocks are closed, the water-level reading will be erroneous, and the boiler could be empty even though the gauge glass reads "normal."

Ideally, a steam boiler pressure gauge should indicate the normal system operating pressure at approximately the first quarter to midpoint of the scale. For example, a 0 to 100 psi (0 to 690 kPa) pressure gauge would be a very poor indicator for a low-pressure [15 psi (103 kPa) or less] steam heating boiler. See the commentary for Section M2002.2. Low-pressure steam gauges are often designed to read both pressure and vacuum because some steam systems will develop a partial vacuum as steam condenses in the distribution system.

A water-gauge glass is always installed on a boiler with the bottom of the glass connected below the normal water level and the top of the glass connected above the normal water level. When connected in this manner, the boiler water level will be accurately represented in the column of water in the gauge glass. It has been an industry tradition to equate the midpoint of a gauge glass with the "normal" water level in a boiler. The coincidence of the gauge-glass midpoint and normal water level also allows the greatest amount of indicator range above and below normal, thus allowing clear indication of abnormal water levels.

M2002.4 Pressure-relief valve. Boilers shall be equipped with pressure-relief valves with minimum rated capacities for the equipment served. Pressure-relief valves shall be set at the maximum rating of the boiler. Discharge shall be piped to drains by gravity to within 18 inches (457 mm) of the floor or to an open receptor.

❖ Boilers must be protected from overpressure by one or more relief valves. A relief valve is designed to open in direct proportion to the boiler pressure force acting on its closure disk. The higher the pressure, the greater the force, and the wider the valve opens. Boilers installed without a relief valve can produce devastating explosions and have been responsible for loss of life and damage to property. Relief valves also play an important role in protecting aging equipment that may have deteriorated or weakened because of prolonged use. Relief valves must be installed directly into the openings provided for such purpose by the boiler manufacturer and must be discharged to a safe location. The discharge is a threat to the occupants and the building because of the high temperature produced when the valves discharge. Because relief valves discharge is a symptom of a boiler malfunction, the discharge must be observable so as to provide a warning to initiate remedial action. Relief valve discharge piping must never be located where it is subject to freezing or installed in a such a way that it functions as a "trap" by allowing discharge water to remain in the pipe.

M2002.5 Boiler low-water cutoff. All steam and hot water boilers shall be protected with a low-water cutoff control. The low-water cutoff shall automatically stop the combustion operation of the appliance when the water level drops below the lowest safe water level as established by the manufacturer.

❖ If a steam or hot water boiler is operated (fired) without water or with water below the minimum level, overheating and severe boiler damage can occur. In addition to damage or destruction of the boiler, a severe fire or explosion hazard can result from the overheating or failure of the combustion chamber enclosure. Burner malfunction can also occur as a result of leakage and destruction of the boiler heat exchange surfaces. The heat exchange surfaces of fuel-fired boilers absorb radiated heat from the flames and heat from the combustion gases produced by the combustion of fuel. This heat is conducted to the water at a rate that prevents the temperature of the heat exchanger material from exceeding the maximum for which it was designed. As the water level drops in a boiler, a greater area of heat exchange surface becomes dry and thus subject to over-heating. All boiler manufacturers specify and mark on the unit the minimum water level to be maintained in the boiler, below which damage would be possible. All boilers are subject to a loss of water. This is especially true for steam boilers because they are dependent on the timely return of condensate water and the proper operation and sequencing of controls for the return and makeup water feed systems.

A low-water cutoff is an essential control device designed to prevent boiler operation when the water level is too low. Because steam boilers (as opposed to hot water boilers, which are completely filled) maintain a water level, a low-water condition is much more likely to occur in a steam system. However, the hazards resulting from low-water levels are the same for both boilers. Also, a flash steam explosion can occur if makeup water is introduced into an over-heated steam or hot water boiler. For these reasons, hot water boilers are also required to be protected by a low-water cutoff control. This section requires low-water cutoffs without regard to heat energy input. Some low-water cutoff controls are intended to be field installed as part of the boiler "trim" (appurtenances), and others are factory installed directly into tapped openings provided on the boiler.

Traditional low-water cutoff controls employ float-actuated mechanisms. Electronic probe-type controls, which are commonly used for hot water boiler protection, detect the water level with an electrode that relies on the slight conductivity of the boiler water to complete a relay circuit. In the case of coil-type (watertube) boilers requiring forced circulation to prevent coil or tube overheating, a flow-sensing device, which detects flow and verifies that the boiler and system are full of water, is required. Low-water cutoff controls do not sense flow and, therefore, cannot protect a forced circulation coil-type/watertube boiler from overheating due to loss of circulation.

Electric boilers are generally protected with factory-installed water-level sensing devices to prevent heating element failure, which could occur if the elements were not submerged.

The sole purpose of a low-water cutoff control is to stop (cut off) the heat input to the boiler whenever the water level is dangerously low. These devices automatically interrupt the power supply to the burner controls or heating elements to cause boiler shutdown.

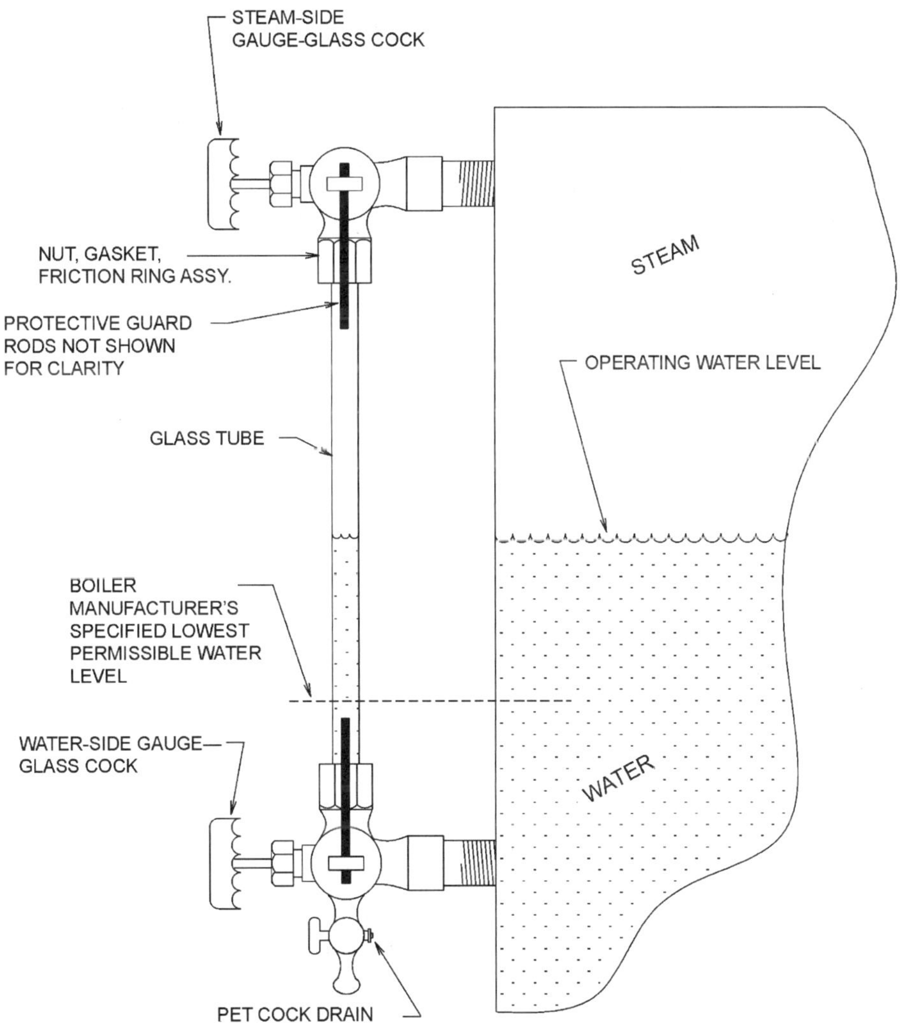

Figure M2002.3
STEAM BOILER SIGHT GLASS

SECTION M2003
EXPANSION TANKS

M2003.1 General. Hot water boilers shall be provided with expansion tanks. Nonpressurized expansion tanks shall be securely fastened to the structure or boiler and supported to carry twice the weight of the tank filled with water. Provisions shall be made for draining nonpressurized tanks without emptying the system.

❖ Hot water heating systems are closed systems that are completely filled with water. When the water is heated, it expands and, because it is in a closed system, quickly develops hydrostatic pressure. Expansion tanks are used to absorb additional system water volume caused by expansion, thus avoiding wide variation in system pressure and possible discharge of pressure relief valves. Expansion tanks are either open tank reservoirs (nonpressurized) or sealed vessels (pressurized).

M2003.1.1 Pressurized expansion tanks. Pressurized expansion tanks shall be consistent with the volume and capacity of the system. Tanks shall be capable of withstanding a hydrostatic test pressure of two and one-half times the allowable working pressure of the system.

❖ Because they are subjected to the same pressures as the boiler system, closed-type expansion tanks must have a pressure rating equal to or exceeding that of the boiler system operating pressure. The code requires pressurized tanks to withstand a hydrostatic test pressure of $2^1/_2$ times the allowable working pressure of the system. Closed-type expansion tanks contain a cushion of air that compresses as water expands into the tank. The pressure in a hot water system will increase several pounds per square inch as the temperature increases. However, the compression of the air in a properly sized expansion tank will prevent the pressure from exceeding the maximum system operating pressure. Open-type expansion tanks are elevated reservoirs holding the expanded hot water at atmospheric pressure.

M2003.2 Minimum capacity. The minimum capacity of expansion tanks shall be determined from Table M2003.2.

❖ The capacity of nonpressurized and pressurized expansion tanks is very important to the proper operation of the boiler system. Refer to the table providing minimum capacities of expansion tanks as they relate to the hot water system volume. Note that the system volume in gallons includes the volume of the water in the boiler, piping, radiators, and convectors. The volume of expansion tanks is not included in the system volume.

TABLE M2003.2
EXPANSION TANK MINIMUM CAPACITY[a] FOR FORCED HOT-WATER SYSTEMS

SYSTEM VOLUME[b] (gallons)	PRESSURIZED DIAPHRAGM TYPE	NONPRESSURIZED TYPE
10	1.0	1.5
20	1.5	3.0
30	2.5	4.5
40	3.0	6.0
50	4.0	7.5
60	5.0	9.0
70	6.0	10.5
80	6.5	12.0
90	7.5	13.5
100	8.0	15.0

For SI: 1 gallon = 3.785 L, 1 pound per square inch gauge = 6.895 kPa, °C = [(°F)-32]/1.8.

a. Based on average water temperature of 195°F, fill pressure of 12 psig and a maximum operating pressure of 30 psig.

b. System volume includes volume of water in boiler, convectors and piping, not including the expansion tank.

❖ This table gives the minimum capacities of two types of expansion tanks. One is the pressurized diaphragm type, which is described in Section M2003.1.1. The other is the nonpressurized type. The capacities of the tanks vary depending on the volume of the system under consideration.

SECTION M2004 WATER HEATERS USED FOR SPACE HEATING

M2004.1 General. Water heaters utilized both to supply potable hot water and to provide hot water for space heating shall be installed in accordance with this chapter, Chapter 24, Chapter 28 and the manufacturer's installation instructions.

❖ Water heaters serving the dual purpose of supplying potable hot water and being a heat source for a hot water space-heating system must be listed and labeled for that dual application. This section does not address water heaters used solely for space-heating applications, but rather addresses water heaters that serve a secondary purpose of space heating. Water heater labels indicate whether or not the appliance is suitable for space heating.

Chapter 28 contains additional requirements for the proper installation of such water heaters.

SECTION M2005 WATER HEATERS

M2005.1 General. Water heaters shall be installed in accordance with the manufacturer's installation instructions and the requirements of this code. Water heaters installed in an attic shall conform to the requirements of Section M1305.1.3. Gas-fired water heaters shall conform to the requirements in Chapter 24. Domestic electric water heaters shall conform to UL 174 or UL 1453. Commercial electric water heaters shall conform to UL 1453. Oiled-fired water heaters shall conform to UL 732.

❖ This section recognizes that water heaters and hot water storage tanks must be considered as both mechanical appliances (equipment) and plumbing appliances; therefore, they must comply with both the mechanical and plumbing sections of this code. Manufacturer's installation instructions are thoroughly evaluated by the listing and labeling agency to establish that a safe installation is prescribed. The listing and labeling agency can require the manufacturer to alter, delete or add information in the installation instructions as necessary to achieve compliance with the applicable standards and code requirements.

When a water heater is tested to obtain a listing and label, the approved agency installs the water heater in accordance with the manufacturer's installation instructions. The water heater is then tested under these conditions; thus, the installation instructions become an integral part of the listing and labeling process.

The manufacturer's installation instructions must be available to the code official because they are an enforceable extension of the code and are necessary for determining that the water heater has been properly installed. Simply put, the listing and labeling process indicates that the water heater and its installation instructions are in compliance with applicable standards. Therefore, an installation in accordance with the manufacturer's installation instructions is required, except where the code requirements are more stringent. It is necessary for an inspector to carefully and completely read and comprehend the manufacturer's instructions in order to properly perform an installation inspection.

Even if an installation appears to be in compliance with the manufacturer's instructions, the installation cannot be complete or approved until all associated components, connections, and systems that serve the water heater are also in compliance with the applicable code provisions.

A water heater installation is complex in that it has a fuel or power supply; a chimney or vent connection, if fuel fired; a combustion air supply, if fuel fired; connections to the plumbing potable water distribution system; and controls and devices to prevent a multitude of potential hazards from conditions such as excessively high temperatures, pressures and ignition failure.

M2005.2 Prohibited locations. Fuel-fired water heaters shall not be installed in a room used as a storage closet. Water heaters located in a bedroom or bathroom shall be installed in a sealed enclosure so that combustion air will not be taken from the living space. Direct-vent water heaters are not required to be installed within an enclosure.

❖ The intent of this section is to prevent fuel-fired appliances from being installed in rooms and spaces where the combustion process could pose a threat to the occupants. Potential threats include depleted oxygen levels; elevated levels of carbon dioxide, nitrous oxides and carbon monoxide, and other combustion gases; ignition of combustibles; and elevated levels of flammable gases.

In small rooms such as bedrooms and bathrooms, the doors are usually closed when the room is occupied, and combustion gases could build up to life-threatening levels. Occupants would not be aware of impending danger.

M2005.2.1 Water heater access. Access to water heaters that are located in an attic or underfloor crawl space is permitted to be through a closet located in a sleeping room or bathroom where ventilation of those spaces is in accordance with this code.

❖ Mechanical equipment and appliances require regular maintenance. Proper access to enable repairs or replacement of equipment is a must. In dwellings, the accessway may be through a closet ceiling or floor. Even though Section M2005.2 prohibits such appliances from being installed in a sleeping room or bathroom, access to them may be through a closet serving such rooms.

All water heaters have a far shorter working life than the building in which they are located; therefore, they must be replaced one or more times during the lifetime of the building. This future need has often been overlooked, and water heaters have been installed in locations that would necessitate the dismantling or destruction of a permanent portion of the structure to install a new water heater.

M2005.3 Electric water heaters. Electric water heaters shall also be installed in accordance with the applicable provisions of Chapters 33 through 42.

❖ This section refers the user back to the electrical chapters in the code for electric water heaters.

M2005.4 Supplemental water-heating devices. Potable water heating devices that utilize refrigerant-to-water heat exchangers shall be approved and installed in accordance with the manufacturer's installation instructions.

❖ Waste heat from refrigeration systems can be used to heat potable water and can provide substantial energy savings over time. This section allows the use of such systems if the equipment is approved and installed in accordance with the manufacturers' installation.

SECTION M2006 POOL HEATERS

M2006.1 General. Pool and spa heaters shall be installed in accordance with the manufacturer's installation instructions. Oil-fired pool heaters shall be tested in accordance with UL 726. Electric pool and spa heaters shall be tested in accordance UL 1261.

❖ This section addresses specialized types of water heaters used with swimming pools, recreational or therapeutic spas, and hot tubs. Pool and spa heaters are similar in design to hot water boilers. Usually, these heaters are of the water-tube type and are designed for either indoor or outdoor installation.

M2006.2 Clearances. In no case shall the clearances be such as to interfere with combustion air, draft hood or flue terminal relief, or accessibility for servicing.

❖ The manufacturer's installation instructions can contain recommendations that are either more or less restrictive than the code for combustion air requirements, draft hood or terminal relief, or accessibility for servicing the appliance. In either case, the more restrictive requirement governs.

M2006.3 Temperature and pressure-limiting devices. Pool heaters shall have temperature and pressure-relief valves.

❖ Pool heaters must be provided with temperature and pressure-relief devices to protect against the development of high temperatures and pressures.

M2006.4 Bypass valves. Where an integral bypass system is not provided as a part of the pool heater, a bypass line and valve shall be installed between the inlet and outlet piping for use in adjusting the flow of water through the heater.

❖ Unless a similar system is integrated within the heater, a bypass line and valve must be provided to allow control of the flow rate of water through the heater, which in turn allows control of the temperature rise of the water circulated through the unit.

Bibliography

The following resource material is referenced in this chapter or is relevant to the subject matter addressed in this chapter.

ASHRAE-00, *HVAC Systems and Equipment Handbook*. Atlanta: American Society of Heating, Refrigerating, and Air-Conditioning Engineers, 2000.

ASME-98, *Boiler and Pressure Vessel Code,* Sections I, II, IV, V and VI. New York: American Society of Mechanical Engineers, 1998.

ASME B31-89, *Pressure Piping Code*. New York: American Society of Mechanical Engineers, 1989.

ASME CSD-1-98, *Controls and Safety Devices for Automatically Fired Boilers*. New York: American Society of Mechanical Engineers, 1998.

AWS D1.1-98, *Structural Welding Code—Steel*. Miami, FL: American Welding Society, 1998.

Grimm, Nils R. and Robert C. Roaler. *Handbook of HVAC Design*. New York: McGraw-Hill, 1990.

NFPA 8501-97, *Single Burner Boiler Operation*. Quincy, MA: National Fire Protection Association, 1997.

UL 726-98, *Oil-Fired Boiler Assemblies*—with Revisions through January 1999. Northbrook, IL: Underwriters Laboratories, 1998.

Chapter 21: Hydronic Piping

General Comments

Hydronic piping includes piping, fittings, and valves used in building space conditioning systems. Applications include hot water, chilled water, steam, steam condensate, brines, and water/antifreeze mixtures. This chapter regulates installation, alteration, and repair of all hydronic piping systems.

Section M2101 contains the installation requirements that apply to all hydronic systems. Section M2102 discusses baseboard convectors. Section M2103 addresses floor heating systems. Section M2104 covers low temperature piping applications. Section M2105 contains the provisions for ground source heat pump system loop piping.

Purpose

The requirements herein are intended to affect the reliability, serviceability, energy efficiency, and safety of hydronic systems.

SECTION M2101 HYDRONIC PIPING SYSTEMS INSTALLATION

M2101.1 General. Hydronic piping shall conform to Table M2101.1. Approved piping, valves, fittings and connections shall be installed in accordance with the manufacturer's installation instructions. Pipe and fittings shall be rated for use at the operating temperature and pressure of the hydronic system. Used pipe, fittings, valves or other materials shall be free of foreign materials.

❖ Table M2101.1 lists the piping materials that are permitted for use in hydronic systems. This section does not prohibit the use of other materials that provide equivalent performance as approved by the building official. Materials that have previously been used for other than hydronic applications can contain residual contaminants that are harmful to the hydronic system. Before any used pipe, fitting, valve, or other material is used, it must be examined to determine that it provides at least equivalent performance to that required for new materials for the intended application.

TABLE M2101.1
HYDRONIC PIPING MATERIALS

MATERIAL	USE CODE[a]	STANDARD[b]	JOINTS	NOTES
Brass pipe	1	ASTM B 43	Brazed, welded, threaded, mechanical and flanged fittings	
Brass tubing	1	ASTM B 135	Brazed, soldered and mechanical fittings	
Chlorinated poly (vinyl chloride) (CPVC) pipe and tubing	1, 2, 3	ASTM D 2846	Solvent cement joints, compression joints and threaded adapters	
Copper pipe	1	ASTM B 42, B 302	Brazed, soldered and mechanical fittings threaded, welded and flanged	
Copper tubing (type K, L or M)	1, 2	ASTM B 75, B 88, B 251, B 306	Brazed, soldered and flared mechanical fittings	Joints embedded in concrete
Cross-linked polyethylene (PEX) tubing	2, 3	ASTM F 876, F 877	Mechanical compression	Install in accordance with manufacturer's instructions.
Cross-linked polyethylene/ aluminum/cross-linked polyethylene (PEX-AL-PEX) pressure pipe	1, 2	ASTM F 1281 or CAN/ CSA B137.10	Mechanical, crimp/insert	Install in accordance with manufacturer's instructions.
Plastic fittings PEX		ASTM F 1807		

(continued)

TABLE M2101.1—continued
HYDRONIC PIPING MATERIALS

MATERIAL	USE CODE[a]	STANDARD[b]	JOINTS	NOTES
Polybutylene (PB) pipe and tubing	1, 2, 3	ASTM D 3309	Heat-fusion, crimp/insert and compression	Joints in concrete shall be heat-fused.
Polyethylene (PE) pipe, tubing and fittings (for ground source heat pump loop systems)	1, 2, 4	ASTM D 2513; ASTM D 3350; ASTM D 2513; ASTM D 3035; ASTM D 2447; ASTM D 2683; ASTM F 1055; ASTM D 2837; ASTM D 3350; ASTM D 1693	Heat-fusion	
Soldering fluxes	1	ASTM B 813	Copper tube joints	
Steel pipe	1, 2	ASTM A 53; A 106	Brazed, welded, threaded, flanged and mechanical fittings	Joints in concrete shall be welded. Galvanized pipe shall not be welded or brazed.
Steel tubing	1	ASTM A 254	Mechanical fittings, welded	

For SI: °C = [(°F)-32]/1.8.
a. Use code:
1. Above ground.
2. Embedded in radiant system.
3. Temperatures below 180°F only.
4. Low temperature (below 130°F) applications only.

b. Standards as listed in Chapter 43.

❖ This table contains approved materials and their associated standards. The notes provided in the last column are intended to assist the user.

M2101.2 System drain down. Hydronic piping systems shall be installed to permit the system to be drained. When the system drains to the plumbing drainage system, the installation shall conform to the requirements of Chapters 25 through 32 of this code.

❖ To facilitate system repairs and maintenance, hydronic piping systems must be sloped and arranged to allow the transfer-medium fluids or condensate to be drained from the system. Each trapped section of the system piping must be provided with drain cocks, unions, or some other means of opening the system for the purpose of draining it. Drainage discharge to the plumbing system must be by indirect connection in accordance with this code.

M2101.3 Protection of potable water. The potable water system shall be protected from backflow in accordance with the provisions listed in Section P2902.

❖ Hydronic systems require means of supplying initial fill water and makeup water to replace any water lost because of evaporation, leakage, or intentional draining. Where direct connections are made to the potable water supply, such connections must be isolated from the potable water source in compliance with the plumbing provisions of this code. This provision protects the potable water system from contamination where a direct connection is made to a hydronic system. Because hydronic systems are usually pressurized and can contain nonpotable water and fluids, antifreeze, and conditioning or cleaning chemicals, protection must be provided so that backflow into the potable water system will not occur.

M2101.4 Pipe penetrations. Openings through concrete or masonry building elements shall be sleeved.

❖ Concrete and masonry penetrations must be sleeved to protect the pipe from abrasion.

M2101.5 Contact with building material. A hydronic piping system shall not be in direct contact with any building material that causes the piping material to degrade or corrode.

❖ All piping must be protected to avoid external degradation resulting from corrosion, abrasion, and chemical reactions. Allowing the pipe to come into contact with certain materials can cause corrosion or degradation. For example, some plastic pipe materials can interact with petroleum-based materials, and concrete can be corrosive to some metals. If hydronic piping is to be installed in an environment or soils that are corrosive, protection for the piping must be provided. Coating, wrappings, and sleeves are forms of protection for piping installed in such an environment.

M2101.6 Drilling and notching. Wood-framed structural members shall be drilled, notched or altered in accordance with the provisions of Sections R502.6, R602.6, R602.6.1 and R802.6. Holes in cold-formed, steel-framed, load-bearing members shall only be permitted in accordance with Sections R506.2, R603.2 and R804.2. In accordance with the provisions of Sections R505.3.5, R603.3.4 and R804.3.5, cutting and notching of flanges and lips of cold-formed, steel-framed, load-bearing members shall not be permitted.

❖ The code recognizes that it may sometimes be necessary to cut or notch framing members. Experience and testing has shown that proper drilling and notching techniques must be followed when mechanical equipment, ducts, piping, and related electrical wiring and cables are installed through framing members. Notches that are too deep or holes that are too large or too close together will significantly weaken these members. The code refers to the building chapters (Chapters 5: Floors, 6: Wall Construction, and 8: Roof/ Ceiling Construction) that elaborate on such requirements and limitations. Reference to these sections is imperative for these installations so that the structure will not be weakened beyond its design parameters. Also, note that this section for notching and cutting is limited in scope to those framing members cited in the code and does not cover engineered framing members such as trusses and premanufactured I-joists.

M2101.7 Prohibited tee applications. Fluid in the supply side of a hydronic system shall not enter a tee fitting through the branch opening.

❖ Commentary Figure M2101.7 depicts a prohibited "bullhead" tee arrangement. Installation of such a connection creates excessive pressure drop, poor system performance, and possible cavitation (air bubbles) that induce erosion of metal pipes. Supply side flow must not enter a tee from the branch opening.

M2101.8 Expansion, contraction and settlement. Piping shall be installed so that piping, connections and equipment shall not be subjected to excessive strains or stresses. Provisions shall be made to compensate for expansion, contraction, shrinkage and structural settlement.

❖ Changes in temperatures cause dimensional changes in all materials to varying degrees. The greatest amount of expansion and contraction in piping occurs along its length. For systems operating at high temperatures, such as steam and hot water, the amount of expansion is substantial, and significant movements can occur even in short runs of piping. Larger movements can be expected in longer lengths of piping. Accordingly, larger forces can develop in restrained piping. Piping systems must be capable of accommodating the forces resulting from thermal expansion and contraction. Inadequate provisions to accommodate these forces can result in failure of pipe and supports, joint damage, and leakage.

M2101.9 Piping support. Hangers and supports shall be of material of sufficient strength to support the piping, and shall be fabricated from materials compatible with the piping material. Piping shall be supported at intervals not exceeding the spacing specified in Table M2101.9.

❖ As with all piping systems, the support of the piping is as important as any other part of the overall design. Proper supports are necessary to maintain piping alignment and slope, to support the weight of the piping and fluid within, to control movement, and to resist imposed loads. Table M2101.9 provides hanger spacing intervals required for different types of piping material.

M2101.10 Tests. Hydronic piping shall be tested hydrostatically at a pressure of not less than 100 pounds per square inch (psi) (689 kPa) for a duration of not less than 15 minutes.

❖ Hydronic piping systems must be tested to determine that the system is leak free and capable of withstanding system operating pressures. The piping under test must be isolated from any component that cannot safely tolerate the test pressure. The prescribed test pressure of 100 psi (690 kPa) will typically exceed the maximum operating pressure of boilers and related components as well as the setting of the system's pressure relief devices. The code official will normally observe all required tests. Maintaining the test pressure for at least 15 minutes provides time for any leaks to be detected.

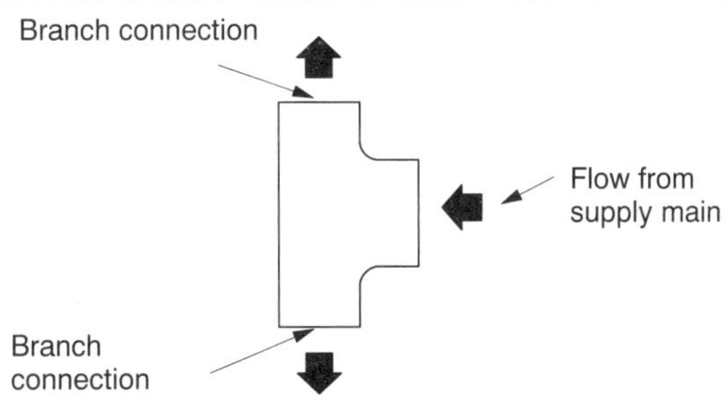

Figure M2101.7
BULLHEAD TEE ARRANGEMENT

TABLE M2101.9
HANGER SPACING INTERVALS

PIPING MATERIAL	MAXIMUM HORIZONTAL SPACING (feet)	MAXIMUM VERTICAL SPACING (feet)
ABS	4	10
CPVC ≤ 1 inch pipe or tubing	3	5
CPVC ≥ $1^1/_4$ inch	4	10
Copper or copper alloy pipe	12	10
Copper or copper alloy tubing	6	10
PB pipe or tubing	2.67	4
PE pipe or tubing	2.67	4
PEX tubing	2.67	4
PVC	4	10
Steel pipe	12	15
Steel tubing	8	10

For SI: 1 inch = 25.4 mm, 1 foot = 304.8 mm.

❖ See the commentary for M2101.9.

SECTION M2102 BASEBOARD CONVECTORS

M2102.1 General. Baseboard convectors shall be installed in accordance with the manufacturer's installation instructions. Convectors shall be supported independently of the hydronic piping.

❖ Convectors must be installed in accordance with the manufacturer's installation instructions. This section requires that convectors be supported independently of the hydronic piping so as not to create a strain on the piping system and to allow for independent expansion and contraction.

SECTION M2103 FLOOR HEATING SYSTEMS

M2103.1 Piping materials. Piping for embedment in concrete or gypsum materials shall be standard-weight steel pipe, copper tubing, cross-linked polyethylene/aluminum/cross-linked polyethylene (PEX-AL-PEX) pressure pipe, chlorinated polyvinyl chloride (CPVC) or polybutylene with a minimum rating of 100 psi at 180°F (690 kPa at 82°C).

❖ Piping embedded in concrete or covered by permanent construction materials or assemblies is not accessible. Therefore, this section is more restrictive than Section M2101.1, as it allows only five materials for embedded hydronic piping.

M2103.2 Piping joints. Piping joints that are embedded shall be installed in accordance with the following requirements:

1. Steel pipe joints shall be welded.
2. Copper tubing shall be joined with brazing material having a melting point exceeding 1,000°F (538°C).
3. Polybutylene pipe and tubing joints shall be installed with socket-type, heat-fused polybutylene fittings.
4. CPVC tubing shall be joined using solvent cement joints.

❖ Joints for these types of installations must be of the type that are least susceptible to failures. For that reason, steel pipe joints must be welded, copper pipe joints brazed, etc.

M2103.3 Testing. Piping or tubing to be embedded shall be tested by applying a hydrostatic pressure of not less than 100 psi (690 kPa). The pressure shall be maintained for 30 minutes, during which all joints shall be visually inspected for leaks.

❖ All hydronic piping systems must be tested to determine that the system is leak free and capable of withstanding the pressures to which it will be subjected. The maximum system design pressure is the maximum pressure at which the system is expected to operate. Such pressure will typically coincide with the setpoint pressure of the system's pressure relief devices. It is customary for the code official to observe all required tests. This testing is especially important where the piping is to be encased and is thereby inaccessible.

SECTION M2104 LOW TEMPERATURE PIPING

M2104.1 Piping materials. Low temperature piping for embedment in concrete or gypsum materials shall be as indicated in Table M2101.1.

❖ Table M2102.1 lists piping materials that may be used in low temperature piping, along with their applicable referenced standards. This section does not prohibit the use of other approved materials that provide equivalent performance.

M2104.2 Piping joints. Piping joints (other than those in Section M2103.2) that are embedded shall comply with the following requirements:

1. Cross-linked Polyethylene (PEX) tubing shall be installed in accordance with manufacturer's instructions.
2. Polyethylene tubing shall be installed with heat fusion joints.

❖ All joining methods for PEX tubing are considered mechanical joints that must be installed in accordance with the manufacturer's installation instructions. Polyethylene pipe and fittings are joined by heat-fusion joining methods, which provide dependability for underground piping.

M2104.2.1 Polyethylene plastic pipe and tubing for ground source heat pump loop systems. Joints between polyethylene plastic pipe and tubing or fittings for ground source heat pump loop systems shall be heat fusion joints conforming to Section M2104.2.1.1, electrofusion joints conforming to Section M2104.2.1.2 or stab-type insertion joints conforming to Section M2104.2.1.3.

❖ See the commentary for Section M2104.2.

M2104.2.1.1 Heat-fusion joints. Joints shall be of the socket-fusion, saddle-fusion or butt-fusion type, fabricated in accordance with the piping manufacturer's instructions. Joint surfaces shall be clean and free of moisture. Joint surfaces shall be heated to melt temperatures and joined. The joint shall be undisturbed until cool. Fittings shall be manufactured in accordance with ASTM D 2683.

❖ Heat fusion is analogous to welding of metals. Specially designed tools, usually electrically heated, are used to heat the pipe ends and/or fitting sockets to the melting temperature and fuse the materials together. See Commentary Figure M2104.2.1.1.

M2104.2.1.2 Electrofusion joints. Joints shall be of the electrofusion type. Joint surfaces shall be clean and free of moisture, and scoured to expose virgin resin. Joint surfaces shall be heated to melt temperatures for the period of time specified by the manufacturer. The joint shall be undisturbed until cool. Fittings shall be manufactured in accordance with ASTM F 1055.

❖ This is another method of joining plastic pipe for ground source heat pump loop systems.

M2104.2.1.3 Stab-type insert fittings. Joint surfaces shall be clean and free of moisture. Pipe ends shall be chamfered and inserted into the fitting to full depth. Fittings shall be manufactured in accordance with ASTM D 2513.

❖ This method, like the method of electrofusion, is intended to be used for ground source heat pump loop systems.

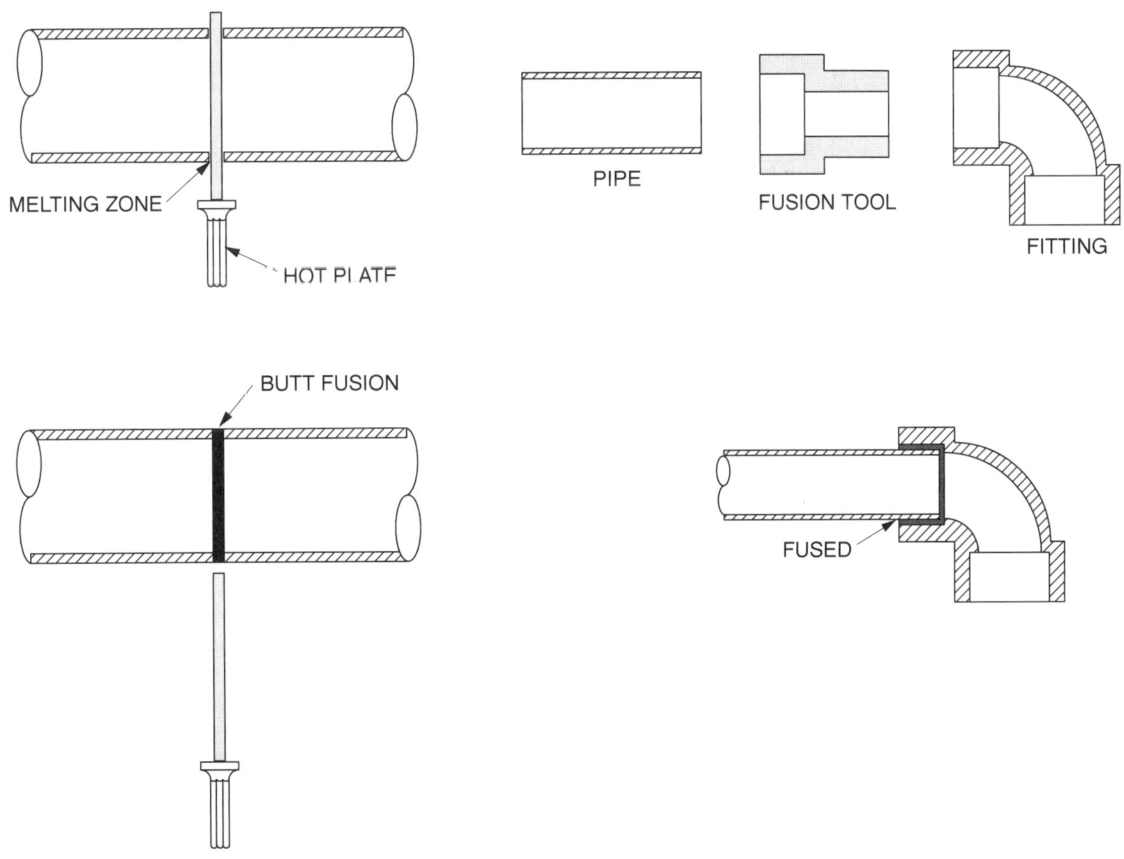

Figure M2104.2.1.1
HEAT-FUSED JOINT

SECTION M2105 GROUND SOURCE HEAT PUMP SYSTEM LOOP PIPING

M2105.1 Testing. The assembled loop system shall be pressure tested with water at 100 psi (690 kPa) for 30 minutes with no observed leaks before connection (header) trenches are backfilled. Flow rates and pressure drops shall be compared to calculated values. If actual flow rate or pressure drop figures differ from calculated values by more than 10 percent, the problem shall be identified and corrected.

❖ Groundwater heat pump systems are either closed loop or open loop. Closed-loop systems circulate water or an antifreeze solution through plastic pipes buried in the ground. Some loop systems, although not as common, use surface water bodies or underground water as the source of heat energy. The heat-transfer fluid absorbs heat from the ground (earth) and transfers it to the building in the heating season, then reverses to absorb heat from the building and transfer it to the ground in the cooling season.

Ground source loop systems use significant lengths of piping installed in multiple circuits (loops) to provide the required area of heat-exchange surface. Also, multiple loops have the advantage over a single loop system because a single piping failure will not cause failure of the entire heat pump system.

The significant difference between testing ground source loop systems and other hydronic systems is that ground source loops may be buried prior to testing. The connection headers (manifolds) that connect the loops (circuits) of piping must remain exposed until completion of testing. Because loops (circuits) are typically installed without joints, all of the potential sources of leaks will be located at the point of connection between the ends of the loops and the header. For this reason, the loops can be covered during testing while the headers remain exposed. Also, it is difficult to dig trenches for large loop fields without backfilling each trench before digging the adjacent trenches.

Depending on the lot, it may be necessary to backfill "as you go" to allow room for the excavator machine. Requiring that each loop be inspected prior to covering could require time-consuming multiple inspections at the same property if the excavator had to backfill each trench before digging the next.

Bibliography

The following resource materials are referenced in this chapter or are relevant to the subject matter addressed in this chapter.

ASHRAE-00, *HVAC Systems and Equipment Handbook*. Atlanta: American Society of Heating, Refrigerating and Air-Conditioning Engineers, 2000.

ASHRAE 90A-80, *Energy Conservation in New Building Design*. Atlanta: American Society of Heating, Refrigerating and Air-Conditioning Engineers, 1980.

ASHRAE/IES 90.1-01, *Energy Code for Commercial and High-Rise Residential Buildings* — Based on ASHRAE/IES. Atlanta: American Society of Heating, Refrigerating and Air-Conditioning Engineers, 2001.

ASME-98, *Boiler and Pressure Vessel Code*, Sections I, II, IV, V & VI. New York: American Society of Mechanical Engineers, 1998.

ASME B1.20.1-83--with revisions through 1999, *Pipe Threads, General Purpose (inch)*. New York: American Society of Mechanical Engineers, 1983.

ASME B31-89, *Pressure Piping Code*. New York: American Society of Mechanical Engineers, 1989.

ASTM A 53-01, *Specification for Pipe, Steel, Black and Hot-Dipped, Zinc-Coated Welded and Seamless*. West Conshohocken, PA: ASTM International, 2001.

ASTM A 106-99e1, *Specification for Seamless Carbon Steel Pipe for High-Temperature Service*. West Conshohocken, PA: ASTM International, 1999.

ASTM A 254-97, *Specification for Copper Brazed Steel Tubing*. West Conshohocken, PA: ASTM International, 1997.

ASTM B 32-00, *Specification for Solder Metal*. West Conshohocken, PA: ASTM International, 2000.

ASTM D 1785-99 *Specification for Poly (Vinyl Chloride)(PVC) Plastic Pipe, Schedules 40, 80 and 120*. West Conshohocken, PA: ASTM International, 1999.

ASTM D 2235-01, *Specification for Solvent Cement for Acrylonitrile-Butadiene-Styrene (ABS) Plastic Pipe and Fittings*. West Conshohocken, PA: ASTM International, 2001.

ASTM D 2241-01, *Specification for Poly (Vinyl Chloride)(PVC) Pressure-Rated Pipe (SDR-Series)*. West Conshohocken, PA: ASTM International, 2001.

ASTM D 2282-99, *Specification for Acrylonitrile-Butadiene-Styrene (ABS) Plastic Pipe (SDR-PR)*. West Conshohocken, PA: ASTM International, 1999.

ASTM D 2846/D 2846M—99, *Specification for Chlorinated Poly (Vinyl Chloride) (CPIC) Plastic Hot and Cold Water Distribution Systems*. West Conshohocken, PA: ASTM International, 1999.

ASTM D 3309-96a, *Specification for Polybutylene (PB) Plastic Hot Water Distribution Systems*. West Conshohocken, PA: ASTM International, 1996.

ASTM E 84-01, *Test Method for Surface Burning Characteristics of Building Materials*. West Conshohocken, PA: ASTM International, 2001.

ASTM F 441/F 441M-99, *Specification for Chlorinated Poly (Vinyl Chloride)(CPIC) Plastic Pipe, Schedules 40 and 80*. West Conshohocken, PA: ASTM International, 1999.

ASTM F 442/F 442M-99, *Specification for Chlorinated Poly (Vinyl Chloride)(CPIC) Plastic Pipe (SDR-PR)*. West Conshohocken, PA: ASTM International, 1999.

ASTM F 493-97, *Specification for Solvent Cements for Chlorinated Poly (Vinyl Chloride)(CPIC) Plastic Pipe and Fittings*. West Conshohocken, PA: ASTM International, 1997.

ASTM F 1476-95a, *Specification for Performance of Gasketed Mechanical Couplings for Use in Piping Applications*. West Conshohocken, PA: ASTM International, 1995.

AWS A5.8-92, *Specifications for Filler Metals for Brazing*. Miami: American Welding Society, 1992.

FS WW-P-325B-76, *Federal Specification for Pipe, Bends, Traps, Caps and Plugs; Lead (for Industrial Pressure, and Soil and Waste Applications)*. Washington, DC: General Services Administration, 1976.

MSS SP-69-96, *Pipe Hangers and Supports*. Vienna, VA: Manufacturers Standardization Society of the Valve and Fittings Industry, 1996.

Chapter 22: Special Piping and Storage Systems

General Comments

This chapter regulates the design and installation of fuel oil storage and piping systems. The regulations include reference to construction standards for above-ground and underground storage tanks, material standards for piping systems (both above-ground and underground) and extensive requirements for the proper assembly of system piping and components.

The importance of adequate regulations for the storage, handling and use of combustible liquid fuel cannot be overstated. A fuel-oil piping system includes piping, valves and fittings. If installed, pumps, reservoirs, regulators, strainers, filters, relief valves, oil preheaters, controls and gauges are also included in the fuel-oil piping system. The piping materials and the method of joining sections of pipe must be approved, and all other components included in the piping system must be labeled and installed in accordance with the manufacturer's instructions. Over the past ten years, public scrutiny of flammable and combustible liquid storage installations has increased with the public's awareness of the consequences of release of these liquids into the environment. Environmental studies have shown that improper installation of these storage systems is a major contributing factor in system failure. Improper use of such systems also accounts for many fire losses. Although the hazards are well known, accidents involving combustible liquids remain one of the most common fire scenarios in the United States.

Fuel oil is considered a Class II combustible liquid, having a flash point of 100°F (38°C) or higher. Though the classification boundaries are somewhat arbitrary, flammable and combustible liquids are distinguished by their flash points. The flash point is that temperature at which the liquid produces sufficient vapor to form an ignitable vapor-air mixture above its surface. Because Class I flammable liquids all have a flash point below 100°F (38°C), it is prudent to assume that such liquids may be capable of igniting when unconfined under normal environmental conditions. On the other hand, combustible liquids, including fuel oil, are materials with flash points above 100°F (38°C) and must usually be heated above their flash points or, in the case of extremely high flash-point liquids, above their boiling points before they will ignite.

Combustible liquids possess significant characteristics other than their flash points. These include ignition temperature, autoignition temperature, flammable (explosive) range, viscosity, vapor density, vapor pressure, boiling point, evaporation rate, specific gravity and water solubility. Once the liquid is ignited, these variables have little influence over the material's heat release rate. Factors such as evaporation rate, viscosity and water solubility may profoundly affect how these fires are extinguished.

Generally, combustible liquids have low specific gravities, high vapor densities and narrow flammable ranges. These characteristics mean that liquids usually float on water, vapors most often hug the ground, and ignitable vapor-air mixtures are generally confined to a range between 6 and 15 percent in air. Thus, smothering is difficult and ignition sources near the ground are more likely to pose a hazard.

Purpose

The purpose of this chapter is to prevent fires, leaks and spills involving fuel oil storage and piping systems, whether inside or outside structures and above or underground.

SECTION M2201 OIL TANKS

M2201.1 Materials. Supply tanks shall be listed and labeled and shall conform to UL 58 for underground tanks and UL 80 for inside tanks.

❖ Shop-fabricated tank(s) must be listed and labeled by an approved agency. The code specifies the applicable national standards governing the manufacture of these tanks.

M2201.2 Above-ground tanks. The maximum amount of fuel oil stored above ground or inside of a building shall be 660 gallons (2498 L). The supply tank shall be supported on rigid noncombustible supports to prevent settling or shifting.

❖ All above-ground oil storage tanks used for residential fuel applications, both inside and outside the building, are limited to 660 gallons (2498 L) capacity. This maximum size is consistent with the *International Fire Code* in regard to additional requirements for larger tanks and the *International Building Code* in regard to additional requirements for larger tanks classified as high-hazard uses in buildings.

In addition, all tanks must be secured and supported in place to prevent displacement from settlement and seismic activity.

M2201.2.1 Tanks within buildings. Supply tanks for use inside of buildings shall be of such size and shape to permit installation and removal from dwellings as whole units. Supply tanks larger than 10 gallons (38 L) shall be placed not less than 5 feet (1524 mm) from any fire or flame either within or external to any fuel-burning appliance.

❖ This section limits the size of a tank, in addition to the capacity limitation of Section M2201.2, by taking into consideration the maintenance aspect of replacing a tank. If a tank is required to be cut apart to be removed from a residence, it could create a hazardous condition.

In addition, this section limits the location of a tank with respect to an open flame, so as to prevent ignition of spilled or leaking fuel.

M2201.2.2 Outside above-ground tanks. Tanks installed outside above ground shall be a minimum of 5 feet (1524 mm) from an adjoining property line. Such tanks shall be suitably protected from the weather and from physical damage.

❖ This section takes into account the location of tanks installed outside a building and above ground. Protection of a tank from physical damage and deterioration caused by local weather conditions is the main concern. The location of the tank must be evaluated with respect to possible impact damage (vehicles, equipment, etc.) and protected in an approved manner. Also, exterior tanks are to be protected from local weather conditions to the satisfaction of the enforcing authority. Such protection usually consists of high quality exterior grade paint. In areas subject to extreme weather conditions, additional protection may be required.

M2201.3 Underground tanks. Excavations for underground tanks shall not undermine the foundations of existing structures. The clearance from the tank to the nearest wall of a basement, pit or property line shall not be less than 1 foot (305 mm). Tanks shall be set on and surrounded with noncorrosive inert materials such as clean earth, sand or gravel well tamped in place. Tanks shall be covered with not less than 1 foot (305 mm) of earth. Corrosion protection shall be provided in accordance with Section M2203.7.

❖ These requirements exist so that a tank will not shift or move because of settlement of the surrounding earth. Shifting stresses a tank and piping and can cause leakage of oil.

M2201.4 Multiple tanks. Cross connection of two supply tanks shall be permitted in accordance with Section M2203.6.

❖ A maximum of two interconnected tanks may be allowed in lieu of one, as long as the total capacity does not exceed 660 gallons (2498 L). See the commentary for Section M2203.6.

M2201.5 Oil gauges. Inside tanks shall be provided with a device to indicate when the oil in the tank has reached a predetermined safe level. Glass gauges or a gauge subject to breakage that could result in the escape of oil from the tank shall not be used.

❖ This section prohibits the use of any gauging device, including sight glasses, that would release tank contents in the event of the gauging device breaking. This helps prevent a hazardous material spill and possible fuel oil vapor ignition.

M2201.6 Flood-resistant installation. In areas prone to flooding as established by Table R301.2(1), tanks shall be installed at or above the design flood elevation established in Section R323 or shall be anchored to prevent flotation, collapse and lateral movement under conditions of the design flood.

❖ The local jurisdiction must fill in Table R301.2(1) upon adoption of this code, including the "Flood Hazards" information. Tanks that are located in flood hazard areas must be installed above the design flood elevation or must be designed and installed to resist the lateral and buoyant forces of the flood waters. See the commentary for Section 323.1.5.

SECTION M2202
OIL PIPING, FITTING AND CONNECTIONS

M2202.1 Materials. Piping shall consist of steel pipe, copper tubing or steel tubing conforming to ASTM A 539. Aluminum tubing shall not be used between the fuel-oil tank and the burner units.

❖ Aluminum is prohibited because it corrodes more readily when it is exposed to moisture or is adjacent to some types of construction material.

M2202.2 Joints and fittings. Piping shall be connected with standard fittings compatible with the piping material. Cast iron fittings shall not be used for oil piping. Unions requiring gaskets or packings, right or left couplings, and sweat fittings employing solder having a melting point less than 1,000°F (538°C) shall not be used for oil piping. Threaded joints and connections shall be made tight with a lubricant or pipe thread compound.

❖ All joining and connecting methods and materials must be compatible with the pipe and the fuel oil used. Lubricants and tape are designed to lubricate the threads for proper mating and to fill in small imperfections within these threads. Lubricants must be compatible with the pipe and fuel oil used, or they will deteriorate and allow fuel oil leaks. Gaskets and packing materials in unions can lose their resiliency and deteriorate over time, which can also lead to fuel oil leaks. Solder with low melting point can quickly fail in a fire. Cast-iron fittings are very brittle and can be damaged by impact.

M2202.3 Flexible connectors. Flexible metal hose used where rigid connections are impractical or to reduce the effect of jarring and vibration shall be listed and labeled in accordance with UL 536 and shall be installed in compliance with its label and the manufacturer's installation instructions. Connectors made from combustible materials shall not be used inside of buildings or above ground outside of buildings.

❖ Flexible metal hoses are used to reduce the transfer of vibrations and where rigid connectors are impractical. These connectors must be listed and labeled in accordance with UL 536 for their intended use so that they are compatible with the environment and the type of fuel used. Combustible connectors are prohibited for installations inside the building or above ground.

SECTION M2203 INSTALLATION

M2203.1 General. Piping shall be installed in a manner to avoid placing stresses on the piping, and to accommodate expansion and contraction of the piping system.

❖ Changes in temperatures can result in substantial longitudinal movement of piping, especially in long runs. Provisions must be made to allow such expansion or contraction without creating stresses on the system. Otherwise, leakage or failure of pipe, pipe supports and pipe joints can occur.

M2203.2 Supply piping. Supply piping used in the installation of oil burners and appliances shall be not smaller than $^3/_8$-inch (9.5 mm) pipe or $^3/_8$-inch (9.5 mm) outside diameter tubing. Copper tubing and fittings shall be a minimum of Type L.

❖ This section requires that the fuel oil supply piping be of a size capable of delivering the maximum fuel demand of the appliance it supplies at the recommended working pressure of the appliance burner. For purposes of flow volume and pressure control, pumps may supply more fuel oil than an appliance consumes, in which case the excess oil must be routed back to the tank through a return line. Because the amount returning to the tank is less than what was supplied, the return line pipe diameter may be smaller than that of the supply pipe.

M2203.3 Fill piping. Fill piping shall terminate outside of buildings at a point at least 2 feet (610 mm) from any building opening at the same or lower level. Fill openings shall be equipped with a tight metal cover.

❖ Requiring fill pipe terminations to be outside buildings and at least 2 feet (610 mm) from building openings that are at the same level or lower than the fill pipe reduces the possibility that liquid or vapor spillage during delivery will enter the building and come into contact with a source of ignition. Likewise, fill pipes must be arranged to reduce the possibility of liquid spilling onto the ground or pavement upon completion of the filling operation, which could result in vapors traveling to an ignition source. One method of managing such spills is to install a spill containment device at the end of the fill pipe. These devices are usually equipped with a watertight cover and are essentially reservoirs designed to catch any spilled fuel oil and retain it for proper disposal. Some models of these devices have a manual drain valve that allows the captured liquid to drain through the fill pipe and into the tank, while other models have a liquid level alarm to warn responsible parties on the premises that the reservoir needs to be emptied.

M2203.4 Vent piping. Vent piping shall be not smaller than $1^1/_4$-inch (32 mm) pipe. Vent piping shall be laid to drain toward the tank without sags or traps in which the liquid can collect. Vent pipes shall not be cross connected with fill pipes, lines from burners or overflow lines from auxiliary tanks. The lower end of a vent pipe shall enter the tank through the top and shall extend into the tank not more than 1 inch (25.4 mm).

❖ Because vent piping can have oil in it under certain circumstances, this section requires that it be laid to drain back to the tank without sags or traps. This is so that if oil reaches the vent, it can drain back to the tank without leaving pockets of oil that could block the vent. The vent must enter the tank through the top and not extend more than 1 inch (25.4 mm) into the tank, again to protect against the vent being blocked by fuel oil.

M2203.5 Vent termination. Vent piping shall terminate outside of buildings at a point not less than 2 feet (610 mm), measured vertically or horizontally, from any building opening. Outer ends of vent piping shall terminate in a weather-proof cap or fitting having an unobstructed area at least equal to the cross-sectional area of the vent pipe, and shall be located sufficiently above the ground to avoid being obstructed with snow and ice.

❖ The 2-foot (610 mm) minimum clearance reduces the possibility of vapors entering the building. Protection of the open end of the vent pipe with an approved vent cap or fitting is necessary to prevent rainwater from entering the tank through the vent.

The vent pipe must terminate above ground at a level that will reduce the possibility of the opening being obstructed by snow accumulation.

M2203.6 Cross connection of tanks. Cross connection of two supply tanks, not exceeding 660 gallons (2498 L) aggregate capacity, with gravity flow from one tank to another, shall be acceptable providing that the two tanks are on the same horizontal plane.

❖ Cross connection of two supply tanks, with gravity flow from one tank to another, is acceptable provided that the two tanks are installed on the same horizontal plane. The capacity of both tanks combined must not exceed 660 gallons (2498 L). Two smaller supply tanks connected together, instead of a single larger tank, allow more flexibility of location, such as with indoor storage tanks.

M2203.7 Corrosion protection. Underground tanks and buried piping shall be protected by corrosion-resistant coatings or special alloys or fiberglass-reinforced plastic.

❖ To minimize the possibility of a hazardous fuel oil spill, underground tanks and buried piping must be protected by corrosion-resistant coating, use of special alloys, or fiberglass reinforced plastic. Materials installed underground and left unprotected are subject to moisture, soil chemistry and stresses that can cause rapid deterioration.

SECTION M2204
OIL PUMPS AND VALVES

M2204.1 Pumps. Oil pumps shall be positive displacement types that automatically shut off the oil supply when stopped. Automatic pumps shall be listed and labeled in accordance with UL 343 and shall be installed in accordance with their listing.

❖ Pumps must be of the positive-displacement type to disallow gravity or siphon delivery of oil when the pump is not operating. Otherwise, oil could drain through the burners and into the combustion chamber, creating a hazard.

M2204.2 Shutoff valves. A readily accessible manual shutoff valve shall be installed between the oil supply tank and the burner. Where the shutoff valve is installed in the discharge line of an oil pump, a pressure-relief valve shall be incorporated to bypass or return surplus oil.

❖ Valves can be used to stop the flow of oil if a leak occurs and can also be used during maintenance or repair of the burners. When the shutoff valve is installed downstream of a pump, the pump becomes capable of overpressurizing the piping system if it pumps against a closed valve. Therefore, if the pump has no built-in bypass, a pressure relief valve must be installed. The fuel oil discharged by the relief valve must be routed back to the tank through a return line to prevent a spill.

M2204.3 Maximum pressure. Pressure at the oil supply inlet to an appliance shall be not greater than 3 pounds per square inch (psi) (20.7 kPa).

❖ Higher pressures could cause malfunctioning of the appliance and/or leaks if the internal parts of the appliance are not designed to withstand higher pressures.

M2204.4 Relief valves. Fuel-oil lines incorporating heaters shall be provided with relief valves that will discharge to a return line when excess pressure exists.

❖ Fuel-oil systems are sometimes designed to incorporate a heater that decreases the viscosity of the oil and improves flow within the system. Because oil will expand when heated, it can reach pressures beyond those of the design operating pressure. Relief valves are necessary to prevent overpressures.

Bibliography

The following resource materials are referenced in this chapter or are relevant to the subject matter addressed in this chapter.

ASME B1.20.1-93 (R1998), *Pipe Threads, General Purpose (inch)*. New York: American Society of Mechanical Engineers, 1993.

ASTM A 53-01, *Specification for Pipe, Steel, Black and Hot-Dipped, Zinc-Coated Welded and Seamless*. West Conshohocken, PA: ASTM International, 2001.

ASTM A 106-99e, *Specification for Seamless Carbon Steel Pipe for High-Temperature Service*. West Conshohocken, PA: ASTM International, 1999.

ASTM A 254-97, *Specification for Copper Brazed Steel Tubing*. West Conshohocken, PA: ASTM International, 1997.

ASTM A 539-99, *Specification for Electric-Resistance-Welded Coiled Steel Tubing for Gas and Fuel Oil Lines*. West Conshohocken, PA: ASTM International, 1999.

ASTM B 42-98, *Specification for Seamless Copper Pipe, Standard Sizes*. West Conshohocken, PA: ASTM International, 1998.

ASTM B 43-98, *Specification for Seamless Red Brass Pipe, Standard Sizes*. West Conshohocken, PA: ASTM International, 1998.

ASTM B 75-99, *Specification for Seamless Copper Tube*. West Conshohocken, PA: ASTM International, 1999.

ASTM B 88-99e1, *Specification for Seamless Copper Water Tube.* West Conshohocken, PA: ASTM International, 1999.

ASTM B 135-00, *Specification for Seamless Brass Tube*. West Conshohocken, PA: ASTM International, 2000.

ASTM B 280-99, *Specification for Seamless Copper Tube for Air Conditioning and Refrigeration Field Service*. West Conshohocken, PA: ASTM International, 1999.

ASTM B 302-00, *Specification for Threadless Copper Pipe*. West Conshohocken, PA: ASTM International, 2000.

ASTM D 56-01, *Test Method for Flash Point by Tag Closed Tester*. West Conshohocken, PA: ASTM International, 2001.

ASTM D 93-00, *Test Method for Flash Point of Pensky-Martens Closed Tester*. West Conshohocken, PA: ASTM International, 2000.

AWS B2.1-84, *Standard for Welding Procedure and Performance Qualification.* Miami: American Welding Society, 1984.

AWS B2.2-91, *Standard for Brazing Procedure and Performance Qualification.* Miami: American Welding Society, 1991.

Benedetti, R.P. *Flammable and Combustible Liquids Handbook,* 4th ed. Quincy, MA: National Fire Protection Association, 1990.

Gas and Oil Equipment Directory. Northbrook, IL: Underwriters Laboratories, 1998.

NACE RP-01-69, *Recommended Practice: Control of External Corrosion on Metallic Buried, Partially Buried, or Submerged Metallic Piping Systems.* Houston: National Association of Corrosion Engineers, 1983.

NFPA 13-99, *Installation of Sprinkler Systems*. Quincy, MA: National Fire Protection Association, 1999.

NFPA 31-01, *Installation of Oil Burning Equipment*. Quincy, MA: National Fire Protection Association, 2001.

UL 443-95, *Standard for Safety for Steel Auxiliary Tanks for Oil-Burner Fuel*. Northbrook, IL: Underwriters Laboratories, 1995.

Chapter 23: Solar Systems

General Comments

Chapter 14 contains requirements for the construction, alteration, and repair of all systems and components of solar energy systems used for space heating or cooling, and domestic hot water heating or processing. The provisions of this chapter are limited to those necessary to achieve safe installations that are relatively hazard free.

A solar energy system can be designed to handle 100 percent of the energy load of a building, although this is rarely accomplished. Because solar energy is a low-intensity energy source and dependent on the weather, it is usually necessary to supplement a solar energy system with traditional energy sources.

Solar heating and cooling systems are classified as either passive or active systems. A passive solar energy system uses solar collectors, a gas or liquid heat-transfer medium, and distribution piping or ductwork. This system does not use circulators or fans but relies on natural (gravity) flow. The flow of energy (heat) is by natural convection, conduction, and radiation. Commentary Figures 23(1) and 23(2) show typical passive solar heating applications.

A complete active solar energy system includes solar collectors, a gas or liquid heat transfer medium, piping, ducts, circulators, fans, and controls to move the medium to the load or to an energy storage system. Commentary Figure 23(3) shows a typical active solar system used for space and domestic water heating, supplemented by an auxiliary furnace and water heater.

Many design variations exist, such as hybrid systems with components of both active and passive systems. Solar systems can act in harmony with the architectural features of the building and can be very complex, or they can be as simple as glazing. See Commentary Figure 23(4). Solar designs can be used to work with, and improve the performance of, conventional space heating systems. For example, a simple passive collection system can greatly improve the performance of an air-to-air or water-to-air heat pump system.

The energy collected by a solar system can be used as it is collected or transferred to some form of thermal storage system. Typical thermal storage systems consist of large quantities of a dense mass, such as stone, masonry, or water. Storage mass can also consist of materials in which a phase change occurs, such as chemical compounds that store the latent heat required to liquefy the compounds that are solid at room temperature.

If the energy collected by a solar system is not used or stored for later use, and unless the collection system is designed to withstand the higher temperatures and pressures resulting from the static (no-flow) mode of operation, the collection system must either stop collecting or dump the energy to the building exterior.

The structural loading effect that the collectors and supports have on the building's roof system also must be considered. Collectors can add considerable weight, affect snow accumulation, and increase wind loads and uplift forces. As with any roof-mounted equipment or appliances, the installation also must comply with the applicable provisions of the building portions of this code.

Purpose

This chapter establishes provisions for the safe installation, operation, and repair of solar energy systems. Although these systems use components similar to those of conventional mechanical equipment, many of these provisions are unique to solar energy systems.

SECTION M2301 SOLAR ENERGY SYSTEMS

M2301.1 General. This section provides for the design, construction, installation, alteration, and repair of equipment and systems utilizing solar energy to provide space heating or cooling, hot water heating, and swimming pool heating.

❖ The provisions within the *International Residential Code* (IRC) cover the design, installation, construction, alteration, and repair of equipment for solar systems that heat water for domestic hot water usage, space heating, and, on rare occasions, space cooling. The IRC also covers the installation of collector systems that provide hot water for swimming pools. There are five major components in active solar heating systems:

- Collector(s) to capture solar energy. Usually, systems used for space heating will have many more collectors than systems designed for domestic water heating only.
- Circulation system to move a fluid between the collectors to a storage tank. The circulation system will either be a direct (open loop) system or an indirect (closed loop) system. A direct system pumps domestic water directly from the storage tank (usually a hot water heater) through the collectors and back to the storage tank. An indirect system circulates an antifreeze fluid, oil, refrigerant or other fluid with a high boiling temperature and low freezing temperature through the collectors and back through a heat exchanger. The

heat exchanger keeps the collection fluid separate from the domestic water.

- Storage tank(s) to collect heated water. This may be a standard water heater or a separate unfired hot water storage tank located next to the water heater.
- Backup heating system. For domestic water heating this will usually be a hot water heater. Backup space heating systems may consist of electric baseboard heating or a gas or electric heater.
- Control system to regulate the overall system operation. The control system will include valves and sensors to provide freeze protection to the collector panels and mixing valves and thermal shutoff devices to protect from excessively high temperatures.

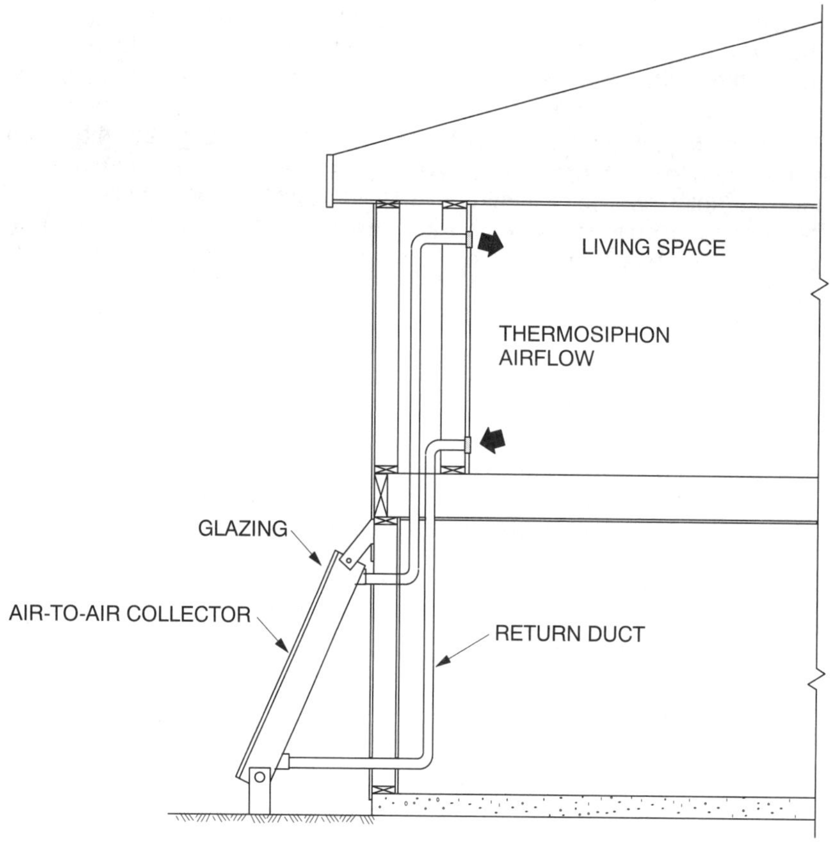

Figure 23(1)
PASSIVE SOLAR SPACE HEATING

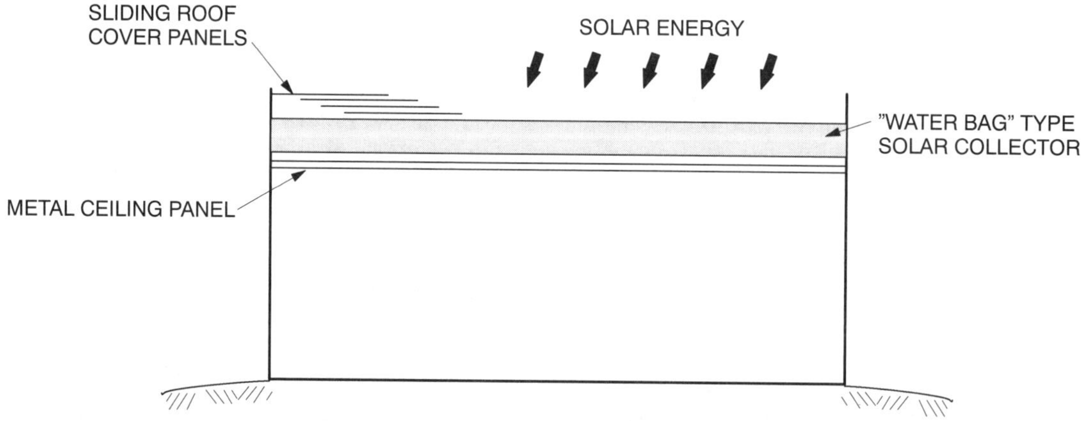

Figure 23(2)
PASSIVE SOLAR HEATING AND COOLING APPLICATIONS

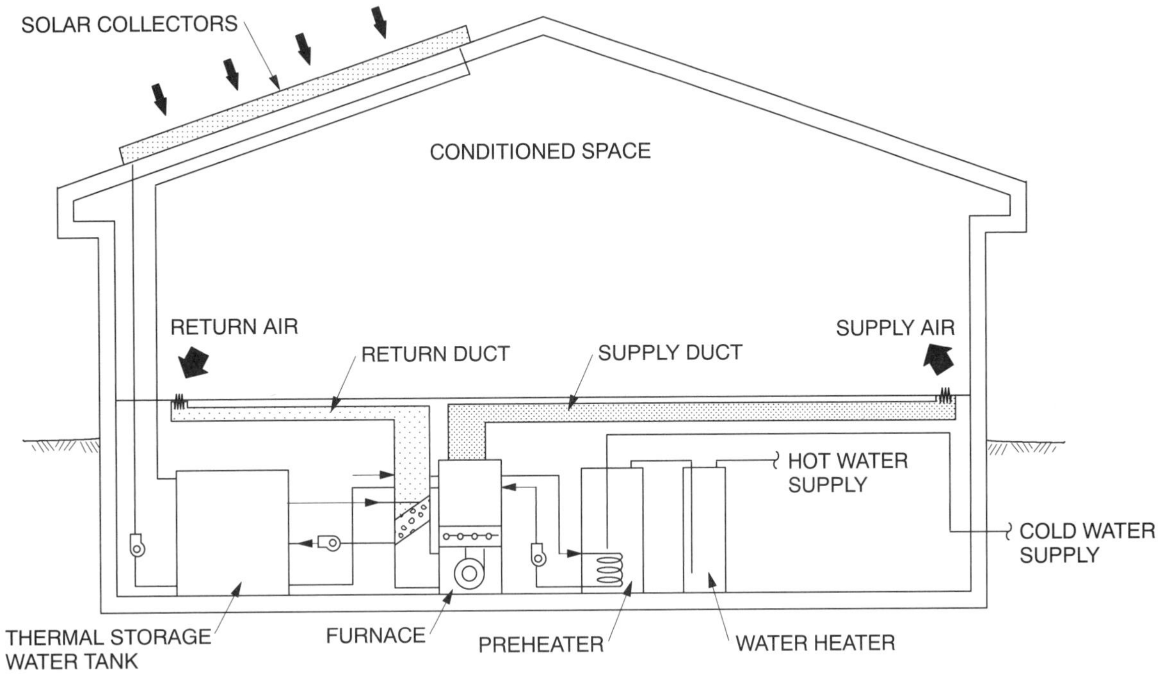

Figure 23(3)
ACTIVE SPACE AND DOMESTIC WATER HEATING SYSTEM

M2301.2 Installation. Installation of solar energy systems shall comply with Sections M2301.2.1 through M2301.2.9.

❖ The requirements of this chapter address the potential hazards to life and property associated with solar installations only. The system must be installed in accordance with the manufacturer's installation instructions. Components of the solar system must also conform to other applicable sections of the code.

M2301.2.1 Access. Solar energy collectors, controls, dampers, fans, blowers and pumps shall be accessible for inspection, maintenance, repair and replacement.

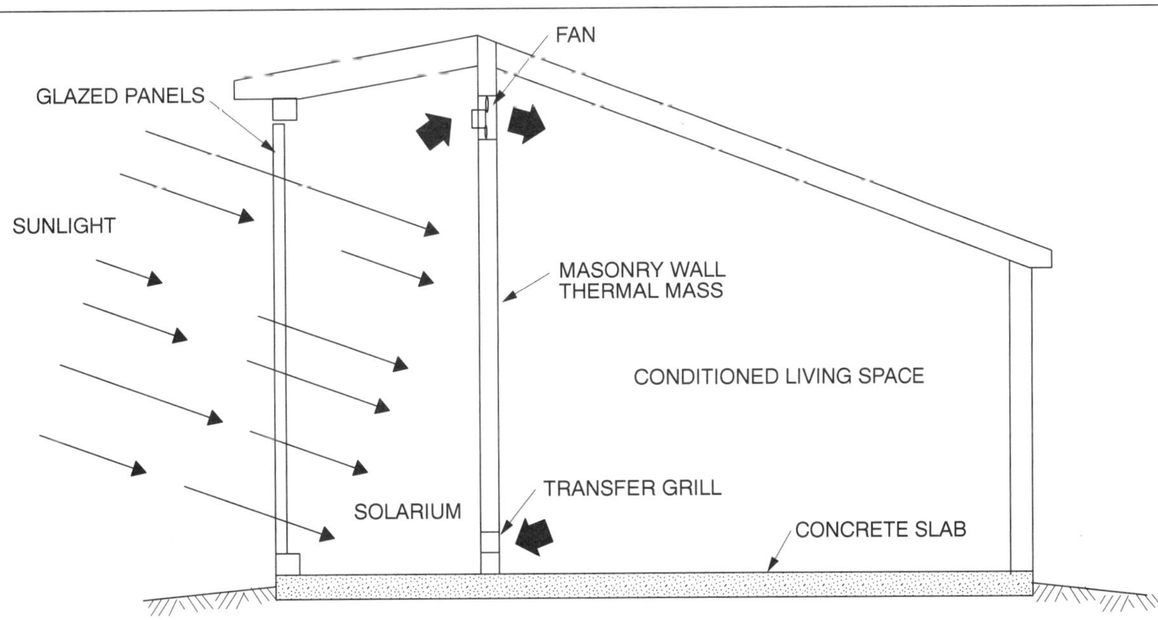

Figure 23(4)
PASSIVE COLLECTION WITH MECHANICAL DISTRIBUTION

❖ This section requires that solar collectors, controls, dampers, fans, blowers, and pumps associated with the solar system be accessible for inspection, maintenance, repair, and replacement. The collection system pumps and controls are often located in the garage or pool equipment building, allowing easy access for the inspector, and are mounted in close proximity to each other. While not a code requirement, it is beneficial that operation and maintenance manuals are available to the owner once the system is installed.

M2301.2.2 Roof-mounted collectors. The roof shall be constructed to support the loads imposed by roof-mounted solar collectors. Roof-mounted solar collectors that serve as a roof covering shall conform to the requirements for roof coverings in Chapter 9 of this code. Where mounted on or above the roof coverings, the collectors and supporting structure shall be constructed of noncombustible materials or fire-retardant-treated wood equivalent to that required for the roof construction.

❖ Roof mounted solar collectors can be designed to be installed integral with the roof, thereby serving as the roof covering. They can also be mounted on or above the roof covering. Systems that are installed as a roof covering must be watertight and properly flashed, in addition to meeting the other requirements applicable to the installation in Chapter 9 of this code. Systems installed on or above the roof covering must be securely fastened to the roof/rafter system. Often, solar collector systems are installed with manufactured metal mounting brackets that can be attached to the roofing system or wood blocking, or with sleepers using lag bolts or other secure mounting systems. The manufacturer's mounting specifications must be followed so that the panels are properly mounted. This is especially important in areas with high wind loads. Also, the weight of the collector system must be checked (when operational and filled with fluid) to confirm that the roof structure is designed to meet the increased load.

Most manufactured panel systems used for domestic hot water are constructed using a metal housing to hold the collection system. This housing must meet the requirements for noncombustible materials. Panel systems constructed on site by the owner/builder that utilize a housing made of wood may or may not meet the requirements for noncombustible materials.

M2301.2.3 Pressure and temperature relief. System components containing fluids shall be protected with pressure- and temperature-relief valves. Relief devices shall be installed in sections of the system so that a section cannot be valved off or isolated from a relief device.

❖ Pressure and temperature relief valves are important to prevent pipes bursting because of high collector temperatures. Flat plate collectors used for domestic water systems are designed with glass coverings and an insulated housing to increase the efficiency of the collection system. The collection surface, or absorber plate, can easily heat up to 200°F (93°C) on a freezing winter day and up to 400°F (204°C) on a summer day when the water is stagnant and not circulated through the collector. These valves are often installed as part of the collector system. These valves must also be installed on the storage tank and back-up water heater.

M2301.2.4 Vacuum relief. System components that might be subjected to pressure drops below atmospheric pressure during operation or shutdown shall be protected by a vacuum-relief valve.

❖ This section requires solar systems that might be subjected to pressure drops below atmospheric pressure during operation or shutdown to be protected by a vacuum relief valve. Vacuum relief valves can be used in conjunction with freeze protection valves that drain the collector panels when the temperature reaches close to 32°F (0°C) or a drain down system that automatically drains the collector fluid when the system is shut off. The vacuum relief valve must be mounted higher than the collector panels.

M2301.2.5 Protection from freezing. System components shall be protected from damage resulting from freezing of heat-transfer liquids at the winter design temperature provided in Table R301.2(1). Freeze protection shall be provided by heating, insulation, thermal mass and heat transfer fluids with freeze points lower than the winter design temperature, heat tape or other approved methods, or combinations thereof.

Exception: Where the winter design temperature is greater than 32°F (0°C).

❖ In climates where the design winter temperature is 32°F (0°C) or less, the system must be protected from freezing. The greatest damage that can be done to the collector system is for water to freeze in the collection system, causing the piping to break. To meet the intent of the code, an indirect (closed loop) system may be installed. See Section M2301.1. Various types of collection fluids that have low freezing temperatures can be used in this system. Usually, a double wall heat exchanger is utilized with this system to transfer the heat from the collector loop to the domestic water loop. This system can be designed to drain the fluid back into the storage tank during periods of nonuse. For a direct (open loop) system, hot water from the storage tank can be circulated through the collector loop to protect the system.

M2301.2.6 Expansion tanks. Expansion tanks in solar energy systems shall be installed in accordance with Section M2003 in closed fluid loops that contain heat transfer fluid.

❖ This section requires pressure tanks to be installed in indirect (closed loop) systems. Manufactured systems that are pre-engineered based on the number of panels in the system, storage tank size and circulation pump size often come with a specified expansion tank presized for the system. The expansion tank is usually installed next to the pumps, storage tank, and controls

M2301.2.7 Roof and wall penetrations. Roof and wall penetrations shall be flashed and sealed to prevent entry of water, rodents and insects in accordance with Chapter 9 of this code.

❖ Roof and wall penetrations must be sealed during and after installation of the solar systems. At a minimum, there will be roof penetrations for the supply and return piping going to and from the collector panels. Systems mounted above the roof covering will require penetrations into the rafter system or blocking to attach the collectors to the roof, and care must be taken to seal the holes with a waterproof sealant. Chapter 9 provides flashing and waterproofing requirements for several roof types. Wall penetrations below roof overhangs and similar locations where it is unlikely that water will enter must still be sealed to prevent the entry of rodents and insects.

M2301.2.8 Solar loop isolation. Valves shall be installed to allow the solar collectors to be isolated from the remainder of the system. Each isolation valve shall be labeled with the open and closed position.

❖ This section requires valves to be installed so that the solar collectors are isolated from the remainder of the system. Gate valves must be installed on both the inlet and outlet side of each bank of collectors (if more than one) so that the collector panel(s) can be removed for repair. Although it is not a requirement of the code, it is advantageous to install valves so that the circulation pump and storage tank can each be isolated and so that the solar loop can be bypassed completely.

M2301.2.9 Maximum temperature limitation. Systems shall be equipped with means to limit the maximum water temperature of the system fluid entering or exchanging heat with any pressurized vessel inside the dwelling to 180°F (82°C). This protection is in addition to the required temperature- and pressure-relief valves required by Section M2301.2.3.

❖ To prevent potential problems with clearances to combustibles and to prevent burn injuries due to human contact with piping or pressure vessels, the system fluid temperature is limited to 180°F (82°C) if it enters any pressurized vessel inside a dwelling. The method of limiting the temperature is to be determined by the designer or system installer. Possible methods include a system to bypass the solar collector or a rejection heat exchanger outside the dwelling.

M2301.3 Labeling. Labeling shall comply with Sections M2301.3.1 and M2301.3.2.

❖ The designer, installer, owner, and building official must know the quality, design, and application limitations of the collector and thermal storage units. This section requires that they be labeled. The label is a form of quality assurance and provides specifications that apply only to the equipment, as well as necessary information for its use.

M2301.3.1 Collectors. Collectors shall be listed and labeled to show the manufacturer's name, model number, serial number, collector weight, collector maximum allowable temperatures and pressures, and the type of heat transfer fluids that are compatible with the collector. The label shall clarify that these specifications apply only to the collector.

❖ Solar collectors must be listed and labeled to show the manufacturer's name, model number, serial number, collector weight, collector maximum allowable temperatures and pressures, and the type of heat transfer fluids that are compatible with the collector. The Solar Rating and Certification Corporation (SRCC) is one entity that provides a third party certification that will meet the requirements of this provision. The *Directory of SRCC Certified Solar Collector System Ratings, Directory of SRCC Certified Water Heating System Ratings* and *Summary of SRCC Certified Solar Collector and Water Heating System Ratings* are published by the SRCC on an annual basis and will list all the information required for this provision except for the serial number, which can be found on the collector.

M2301.3.2 Thermal storage units. Pressurized thermal storage units shall be listed and labeled to show the manufacturer's name, model number, serial number, storage unit maximum and minimum allowable operating temperatures and pressures, and the type of heat transfer fluids that are compatible with the storage unit. The label shall clarify that these specifications apply only to the thermal storage unit.

❖ Pressurized thermal storage units must bear a listing and label that discloses the manufacturer's name, model, serial number, and important operating characteristics such as operating temperatures and pressures, and the type of heat transfer fluid. The information is important for servicing, repairs and inspection.

M2301.4 Prohibited heat transfer fluids. Flammable gases and liquids shall not be used as heat transfer fluids.

❖ Flammable gases or flammable liquids must not be used as heat transfer fluids. Solar energy systems generate large amounts of heat that could ignite flammable materials in the event of a malfunction.

M2301.5 Backflow protection. All connections from the potable water supply to solar systems shall comply with Section P2902.4.5.

❖ The potable water supply connected to a solar system must be equipped with a backflow preventer with an intermediate atmospheric vent. Refer to Section P2902.4.5 for a discussion of this requirement.

Bibliography

The following resource materials are referenced in this chapter or are relevant to the subject matter addressed in this chapter.

Directory of SRCC Certified Solar Collector Ratings. Cocoa, FL: Solar Rating and Certification Corp., 2000.

Directory of SRCC Certified Solar Water Heating System Ratings. Cocoa, FL: Solar Rating and Certification Corp., 2000.

Summary of SRCC Certified Solar Collector and Water Heating System Ratings. Cocoa, FL: Solar Rating and Certification Corp., 2000.

Part VI — Fuel Gas

Chapter 24: Fuel Gas

The code text of this chapter is excerpted from the 2003 edition of the *International Fuel Gas Code* and has been modified where necessary to make such text conform to the scope of application of the *International Residential Code for One- and Two-Family Dwellings.* The section numbers appearing in parentheses after each section number represent the location of the corresponding text in the *International Fuel Gas Code.*

General Comments

This chapter covers all installations of gas piping, gas appliance installation, and gas appliance venting systems. It is extracted from the *International Fuel Gas Code* and is identical in intent. This chapter contains its own gas-specific coverage of combustion air, clearance reduction methods, chimneys and vents, and appliance installation. The dual section numbering system allows this text to be cross-referenced with the IFGC. Chapters 12, 13, 14, and 20 also contain requirements applicable to gas appliance installations.

The *International Fuel Gas Code* itself is segregated by section number into two categories: code and standard. In that document, code sections are identified as IFGC; standards sections are identified as IFGS. The IFGS is a copyrighted work of the American Gas Association.

The commentary text of this chapter is produced and copyrighted by the International Code Council.

Purpose

Chapter 24 intends to protect occupants and their property from fire, explosion, and health hazards that could result from the improper installation of gas piping systems, gas appliances, and appliance venting systems.

SECTION G2401 (101) GENERAL

G2401.1 (101.2) Application. This chapter covers those fuel gas piping systems, fuel gas utilization equipment and related accessories, venting systems and combustion air configurations most commonly encountered in the construction of one- and two-family dwellings and structures regulated by this code.

Exception: As an alternative to the provisions of this code, fuel-gas piping systems, fuel-gas utilization equipment and related accessories in existing buildings that are undergoing repairs, alterations, changes in occupancy or construction of additions shall be permitted to comply with the provisions of the *International Existing Building Code.*

Coverage of piping systems shall extend from the point of delivery to the outlet of the equipment shutoff valves (see "point of delivery"). Piping systems requirements shall include design, materials, components, fabrication, assembly, installation, testing, inspection, operation, and maintenance. Requirements for gas utilization equipment and related accessories shall include installation, combustion, and ventilation air and venting and connections to piping systems.

The omission from this chapter of any material or method of installation provided for in the *International Fuel Gas Code* shall not be construed as prohibiting the use of such material or method of installation. Fuel gas piping systems, fuel gas utilization equipment and related accessories, venting systems and combustion air configurations not specifically covered in these chapters shall comply with the applicable provisions of the *International Fuel Gas Code.*

Gaseous hydrogen systems shall be regulated by Chapter 7 of the *International Fuel Gas Code.*

This chapter shall not apply to the following:

1. Liquified natural gas (LNG) installations.
2. Temporary LP-Gas piping for buildings under construction or renovation that is not to become part of the permanent piping system.
3. Except as provided in Section G2412.1.1, gas piping, meters, gas pressure regulators, and other appurtenances used by the serving gas supplier in the distribution of gas, other than undiluted LP-Gas.
4. Portable LP-gas equipment of all types that is not connected to a fixed fuel piping system.
5. Portable fuel cell appliances that are neither connected to a fixed piping system nor interconnected to a power grid.
6. Installation of hydrogen gas, LP-gas and compressed natural gas (CNG) systems on vehicles.

❖ This section describes the types of fuel gas systems to which the code is intended to apply and specifically lists those systems to which the code does not apply. The applicability of the code spans the initial design of fuel gas systems, through the installation and construction phases, and into the maintenance of operating systems. Chapter 24 of the IRC covers fuel gas systems and is a duplication of the applicable IFGC text. The provisions of IRC Chapter 24 and the IFGC are identical. The exception refers to the IEBC as an option to the provisions of the IRC with regard to existing buildings. See IEBC Section 503.3.1.

SECTION G2402 (201) GENERAL

G2402.1 (201.1) Scope. Unless otherwise expressly stated, the following words and terms shall, for the purposes of this chapter, have the meanings indicated in this chapter.

❖ In the application of the code, the terms used have the meanings given in Section G2403.

G2402.2 (201.2) Interchangeability. Words used in the present tense include the future; words in the masculine gender include the feminine and neuter; the singular number includes the plural and the plural, the singular.

❖ While the definitions contained in Chapter 2 are to be taken literally, gender and tense are considered to be interchangeable; thus, any grammatical inconsistencies within the code text will not hinder the understanding or enforcement of the requirements.

G2402.3 (201.3) Terms defined in other codes. Where terms are not defined in this code and are defined in the ICC *Electrical Code, International Building Code, International Fire Code, International Mechanical Code* or *International Plumbing Code*, such terms shall have meanings ascribed to them as in those codes.

❖ When a word or term appears in the code that is not defined in this chapter, other references may be used to find its definition. These include the *International Building Code,* the ICC *Electrical Code,* the *International Fire Code,* the *International Mechanical Code,* or the *International Plumbing Code.* These codes contain additional definitions (some parallel and duplicative) that may be used in the enforcement of this code or in the enforcement of the other codes by reference.

SECTION G2403 (202) GENERAL DEFINITIONS

❖ This section contains definitions of terms that are associated with the subject matter of this chapter. It is important to emphasize that these terms are not exclusively related to this chapter but are applicable everywhere the term is used in the code. Definitions of terms are necessary for the understanding and application of the code requirements.

AIR CONDITIONING, GAS FIRED. A gas-burning, automatically operated appliance for supplying cooled and/or dehumidified air or chilled liquid.

❖ Gas fired air conditioners are not specifically referred to in the code but would include systems fueled with natural or propane gas including absorption type units and reciprocating units powered by internal combustion engines.

AIR, EXHAUST. Air being removed from any space or piece of equipment and conveyed directly to the atmosphere by means of openings or ducts.

❖ Exhaust air may be from a space or a piece of equipment. Exhaust air systems are terminated outside the building, in some cases after the exhaust air has been treated to remove any harmful emissions. Exhaust air is not recirculated.

AIR-HANDLING UNIT. A blower or fan used for the purpose of distributing supply air to a room, space or area.

❖ In addition to blowers, air-handling units may contain heat exchangers, filters, and means to control air volume.

AIR, MAKEUP. Air that is provided to replace air being exhausted.

❖ Makeup air is not to be confused with combustion air. Makeup air replaces the air being exhausted through such systems as bathroom and toilet exhausts, kitchen exhaust hoods, hazardous exhaust systems, and clothes dryer exhaust systems. Refer to Section G2439.4 for specific requirements for makeup air. Exhaust systems cannot function at design capacity without adequate volumes of makeup air to replace the air being exhausted.

ALTERATION. A change in a system that involves an extension, addition or change to the arrangement, type or purpose of the original installation.

❖ An alteration is any modification or change made to an existing installation. For example, increasing the size of piping for a portion of the system to accommodate different appliances would be an alteration.

ANODELESS RISER. A transition assembly in which plastic piping is installed and terminated aboveground outside of a building.

❖ As the name implies, these riser assemblies accomplish corrosion protection of the steel riser by means other than cathodic protection involving a sacrificial anode. Some anodeless risers allow the termination of plastic piping aboveground by encasing the piping in a steel conduit equipped with a plastic-to-steel transition fitting. See Commentary Figure G2403(1).

APPLIANCE (EQUIPMENT). Any apparatus or equipment that utilizes gas as a fuel or raw material to produce light, heat, power, refrigeration or air conditioning.

❖ An appliance is a manufactured component or assembly of components that converts one source of energy into a different form of energy to serve a specific purpose. The term "appliance" generally refers to residential- and commercial-type utilization equipment that is manufactured in standardized sizes or types. The term "appliance" is generally not associated with industrial-type equipment. For the application of the IRC code

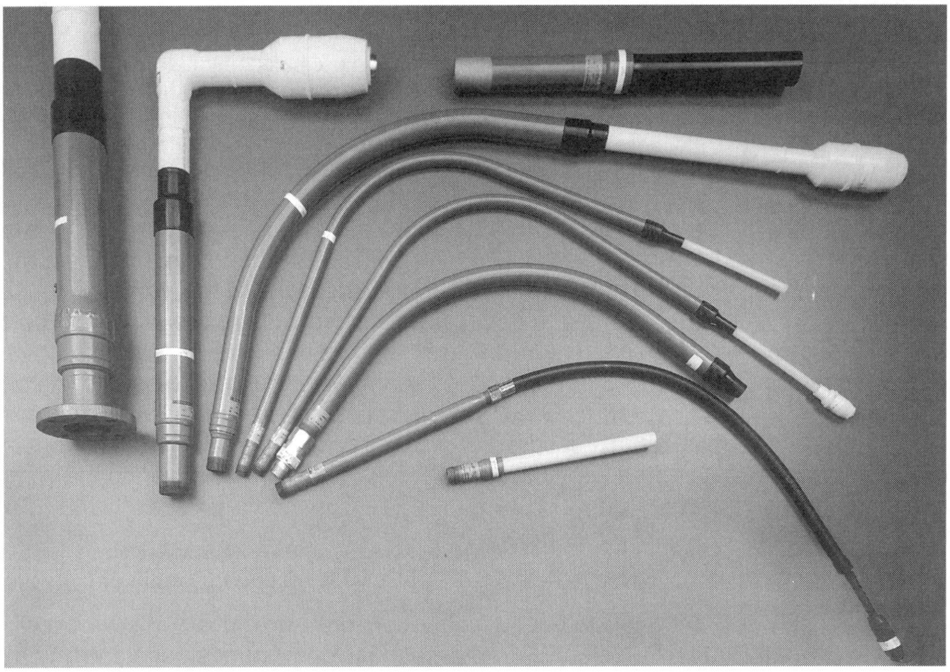

Photo courtesy of Perfection Corporation

Figure G2403(1)
ANODELESS RISERS

provisions, the terms "appliance" and "equipment" are interchangeable.

Examples of appliances regulated by this code include furnaces, boilers, water heaters, room heaters, decorative gas log sets, cooking equipment, clothes dryers, pool, spa and hot tub heaters, unit heaters, ovens, and similar gas-fired equipment.

APPLIANCE, FAN-ASSISTED COMBUSTION. An appliance equipped with an integral mechanical means to either draw or force products of combustion through the combustion chamber or heat exchanger.

❖ Fan-assisted appliances are a specific type of Category I appliance, typically furnaces and boilers. Such appliances are not to be confused with Category II, III, or IV appliances; power-vented appliances; or appliances served by exhausters. As the name implies, such appliances use a fan (blower) to "assist" the combustion process by helping the flue gases to overcome the internal flow resistance of the appliance heat exchanger. Some fan-assisted appliances are field-convertible to Category III appliances.

APPLIANCE, AUTOMATICALLY CONTROLLED. Appliances equipped with an automatic burner ignition and safety shut-off device and other automatic devices, which accomplish complete turn-on and shut-off of the gas to the main burner or burners, and graduate the gas supply to the burner or burners, but do not affect complete shut-off of the gas.

❖ With respect to the IRC, an automatically controlled appliance is an appliance that is cycled through its operation by controls such as thermostats, pressure switches, and timers, without manual intervention. A residential water heater is automatically controlled whereas a four-burner cooktop is not.

APPLIANCE, UNVENTED. An appliance designed or installed in such a manner that the products of combustion are not conveyed by a vent or chimney directly to the outside atmosphere.

❖ Direct-fired makeup air heaters, gas-fired infrared radiant unit heaters, gas-fired cooking equipment, and small gas-fired room heaters are examples of appliances that are sometimes unvented. With the exception of direct-fired equipment, the appliances listed herein can also be of the vented type, depending on the design and type of appliance and the particular application. Unvented appliances must be designed, installed, and operated so as to avoid both the depletion of the oxygen supply and the accumulation of toxic combustion byproducts. See Sections G2425.8, G2445, and G2451.

APPLIANCE, VENTED. An appliance designed and installed in such a manner that all of the products of combustion are conveyed directly from the appliance to the outside atmosphere through an approved chimney or vent system.

❖ The majority of fuel-fired appliances are designed to vent the products of combustion to the outdoors through one or more specific types of vent or chimney. See Section G2425.2.

ATMOSPHERIC PRESSURE. The pressure of the weight of air and water vapor on the surface of the earth, approximately 14.7 pounds per square inch (psia) (101 kPa absolute) at sea level.

❖ This is the "standard" atmospheric pressure at sea level. The actual atmospheric pressure varies due to climate conditions. Barometric pressure is a measure of the actual atmospheric pressure, usually given as inches of mercury. One standard atmosphere (14.7 psi) is equal to 29.9 inches of mercury at 32°F. Absolute pressure includes atmospheric pressure. Gage pressure is the pressure above atmospheric pressure. If not stated, it is assumed that stated pressure is the pressure above atmospheric pressure (gage pressure). A vacuum (partial or complete) is a pressure below atmospheric pressure.

AUTOMATIC IGNITION. Ignition of gas at the burner(s) when the gas controlling device is turned on, including reignition if the flames on the burner(s) have been extinguished by means other than by the closing of the gas controlling device.

❖ As opposed to manual ignition, automatic ignition is accomplished by pilot burners, spark ignitors, hot surface ignitors, and similar means.

BAROMETRIC DRAFT REGULATOR. A balanced damper device attached to a chimney, vent connector, breeching or flue gas manifold to protect combustion equipment by controlling chimney draft. A double-acting barometric draft regulator is one whose balancing damper is free to move in either direction to protect combustion equipment from both excessive draft and backdraft.

❖ These units automatically open or close depending on the difference between the internal vent pressure and the atmospheric pressure. See the commentary for the definition of "Atmospheric Pressure." Excessive negative pressure in the vent will cause excessive draft; therefore, the draft regulator will open by means of the higher atmospheric pressure on the exterior of the damper allowing air to enter the vent, increasing the internal vent pressure. With excessive positive internal vent pressure a double-acting regulator will open to relieve the pressure lowering the vent internal pressure. See Commentary Figure G2403(2).

BOILER, LOW-PRESSURE. A self-contained gas-fired appliance for supplying steam or hot water.

❖ Low-pressure boilers operate at pressures less than or equal to 15 pounds per square inch (psi) (103 kPa) for steam and 160 psi (1103 kPa) for water. High-pressure boilers operate at pressures exceeding those pressures.

Boilers are usually manufactured of steel, cast iron, or copper and are used to transfer heat from the combustion of a fuel or from an electric-resistance element to water to make steam or pressurized hot water for heating or other process or power purposes.

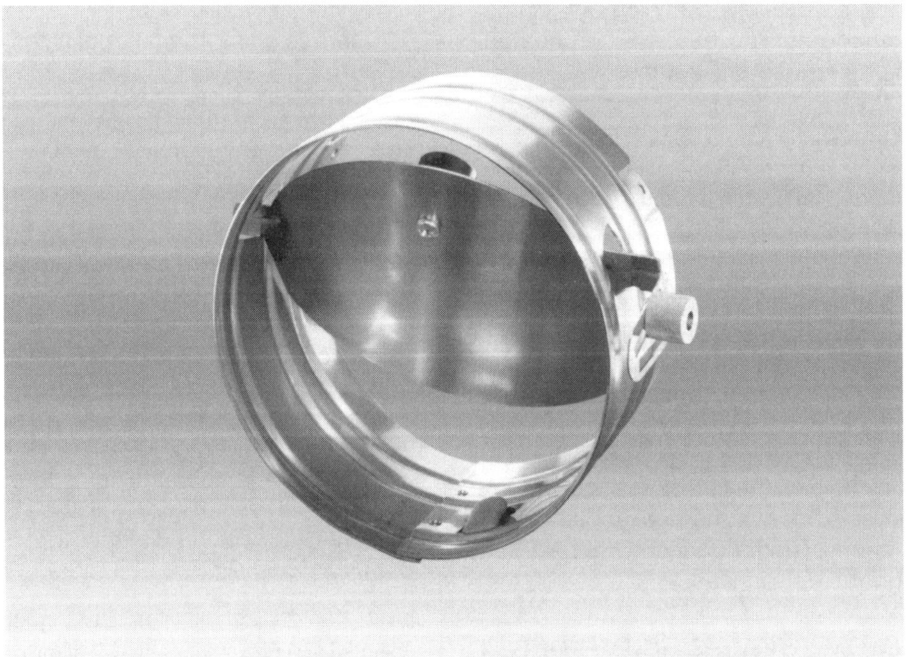

Photo courtesy of Tjernlund Products, Inc.

Figure G2403(2)
BAROMETRIC DRAFT REGULATOR

Boilers are usually installed in closed systems where the heat transfer medium is recirculated and retained within the system.

Boilers must be labeled and installed in accordance with the manufacturer's instructions and the applicable sections of the code. Boilers are rated in accordance with standards published by the American Society of Mechanical Engineers (ASME), the Hydronics Institute, the Steel Boiler Institute (SBI), the Canadian Standards Association (CSA), and the American Boiler Manufacturers Association (ABMA). Boilers can be classified by working temperature, working pressure, type of fuel used (or electric boilers), materials of construction, and whether or not the heat transfer medium changes phase from a liquid to a vapor. See Commentary Figure G2403(3).

Hot water heating boiler. A boiler in which no steam is generated, from which hot water is circulated for heating purposes and then returned to the boiler, and that operates at water pressures not exceeding 160 psig (1100 kPa gauge) and at water temperatures not exceeding 250°F (121°C) at or near the boiler outlet.

❖ Hot water heating boilers are normally part of a closed system where the heated water is circulated through

WEIL-McLAIN

LGB BOILER

Boiler Model No. LGB-8
Series No. 1
Type of Gas .... NATURAL
Input, Btu/hr. 910000
Max. Inlet Gas Press., w.c. 13.0
Min. Inlet Gas Press., w.c. 7.0
Manifold Gas Press., w.c. 3.5
Minimum Input............ 455000
A.G.A. Gross Output, Btu/hr...737100

NET I=B=R Output

Water, Mbh 641
Steam Sq. Ft............. 2304
Steam, Mbh 553

Electrical Input Less Than 12
Amperes, 120 Volts, & 60 Hertz.

CERTIFIED BY
WEIL-McLAIN
A Marley Company
Michigan City, Indiana 46360

MAXIMUM W.P., WATER 50 PSI
STEAM 15 PSI

MIN. RELIEF VALVE CAP. 738 LB/HR OR MBH

Design Certified Under ANS Z21.13 1991
Low Pressure Boiler

550—223—125

Figure courtesy of Weil-McLain

Figure G2403(3)
LABEL FOR LOW PRESSURE BOILER

fan coils or radiators of various types, including convectors, finned tube units, and baseboard units.

Hot water supply boiler. A boiler, completely filled with water, which furnishes hot water to be used externally to itself, and that operates at water pressures not exceeding 160 psig (1100 kPa gauge) and at water temperatures not exceeding 250°F (121°C) at or near the boiler outlet.

❖ Hot water supply boilers are normally part of open systems where the heated water is supplied and used externally to the boiler. Large domestic (potable) water heating systems often use hot water supply boilers.

Steam heating boiler. A boiler in which steam is generated and that operates at a steam pressure not exceeding 15 psig (100 kPa gauge).

❖ Steam heating boilers are normally part of a closed system where low-pressure steam can be circulated through a variety of terminal units including natural convection units, forced-convection units, and radiant panel systems.

BRAZING. A metal joining process wherein coalescence is produced by the use of a nonferrous filler metal having a melting point above 1,000°F (538°C), but lower than that of the base metal being joined. The filler material is distributed between the closely fitted surfaces of the joint by capillary action.

❖ Brazing is the act of producing a brazed joint and is often referred to as silver soldering. Silver soldering is more accurately described as silver brazing and employs high silver-bearing alloys primarily composed of silver, copper, and zinc. Silver soldering (brazing) typically requires temperatures in excess of 1,000°F (538°C), and such solders are classified as "hard" solders.

Confusion has always been present with respect to the distinction between "silver solder" and "silver-bearing" solder. Silver solders are unique and can be further subdivided into soft and hard categories, which are determined by the percentages of silver and the other component elements of the particular alloy. The distinction is that silver-bearing solders (melting point less than 600°F [316°C]) are used in soft-soldered joints, and silver solders (melting point greater than 1,000°F [538°C]) are used in silver-brazed joints.

Brazed joints are considered to have superior strength and stress resistance and, because of the high melting point, are less likely to fail when exposed to fire.

Brazing with a filler metal conforming to AWS A5.8 produces a strong joint that will perform under extreme service conditions. The surfaces to be brazed must be cleaned so as to be free from oxides and impurities. Flux should be applied as soon as possible after the surfaces have been cleaned. Flux helps to remove residual traces of oxides, to promote wetting, and to protect the surfaces from oxidation during heating. Care should be taken to prevent flux from entering the piping system during the brazing operation, since flux that remains may corrode the pipe or contaminate the system.

Air and any other residual products should be removed from the pipe being brazed by purging the piping with a nonflammable gas such as carbon dioxide or nitrogen. Purging the system has several benefits, such as preventing oxidation from occurring on the inside of the pipe and preventing the creation of toxic gases that can result from the chemical breakdown of other products. Additionally, purging will eliminate the possibility of an explosion from any other flammable gas in the pipe, which could ignite when mixed with air.

BTU. Abbreviation for British thermal unit, which is the quantity of heat required to raise the temperature of 1 pound (454 g) of water 1°F (0.56°C) (1 Btu = 1055 J).

❖ Btu (J) is a unit of energy measurement. Fuel-fired appliances and equipment are rated based on their Btu/h (W) input or output.

BURNER. A device for the final conveyance of the gas, or a mixture of gas and air, to the combustion zone.

❖ There are several types of burners that vary in the method of providing combustion air and the pressures of fuel/air mixtures. Burner types include atmospheric ribbon, ported, porous ceramic, slotted, and cone types and atmospheric or powered inshot and upshot gun burners.

Induced-draft. A burner that depends on draft induced by a fan that is an integral part of the appliance and is located downstream from the burner.

❖ See the commentary for the definition of "Draft/Mechanical" or induced draft.

Power. A burner in which gas, air or both are supplied at pressures exceeding, for gas, the line pressure, and for air, atmospheric pressure, with this added pressure being applied at the burner.

❖ Power burners are typically used on larger appliances and employ a blower for introducing combustion air under pressure.

CHIMNEY. A primarily vertical structure containing one or more flues, for the purpose of carrying gaseous products of combustion and air from an appliance to the outside atmosphere.

❖ Chimneys differ from vents in their materials of construction and the type of appliance they are designed to serve. Chimneys are capable of venting much higher temperature flue gases than vents. See Commentary Figure G2403(4).

Factory-built chimney. A listed and labeled chimney composed of factory-made components, assembled in the field in accordance with manufacturer's instructions and the conditions of the listing.

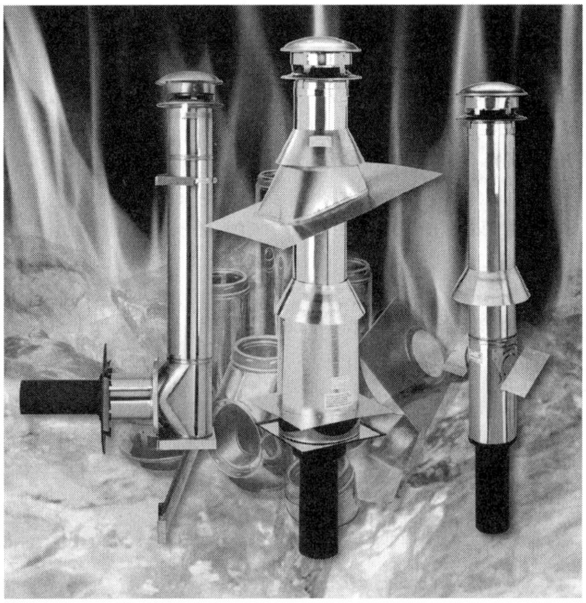

Photo courtesy of Selkirk L.L.C.

Figure G2403(4)
FACTORY-BUILT CHIMNEYS

❖ A factory-built chimney is a manufactured listed and labeled chimney that has been tested by an approved agency to determine its performance characteristics. Factory-built chimneys are manufactured in two basic designs: a double-wall mineral fiber insulated design and a triple-wall air-cooled design. Some chimneys employ both convection cooling and mineral fiber insulation. Both designs use stainless steel inner liners to resist the corrosive effects of combustion products. See Commentary Figure G2403(4).

Masonry chimney. A field-constructed chimney composed of solid masonry units, bricks, stones or concrete.

❖ Masonry chimneys can have one or more flues within them and are field constructed of brick, stone, concrete, or fire-clay materials. Masonry chimneys can stand alone or be part of a masonry fireplace.

CLEARANCE. The minimum distance through air measured between the heat-producing surface of the mechanical appliance, device or equipment and the surface of the combustible material or assembly.

❖ Clearances between sources of heat and combustibles are always airspace clearances.

CLOTHES DRYER. An appliance used to dry wet laundry by means of heated air.

❖ There are two types of clothes dryers.

Type 1. Factory-built package, multiple production. Primarily used in the family living environment. Usually the smallest unit physically and in function output.

❖ Type 1 clothes dryers are used in dwelling units and in residential settings.

CODE. These regulations, subsequent amendments thereto, or any emergency rule or regulation that the administrative authority having jurisdiction has lawfully adopted.

❖ The adopted regulations are generally referred to as "the code" and include not only the IRC but also any adopted modifications to the code and all other related rules and regulations promulgated and enacted by the jurisdiction.

CODE OFFICIAL. The officer or other designated authority charged with the administration and enforcement of this code, or a duly authorized representative.

❖ The statutory power to enforce the code is normally vested in a building department (or the like) of a state, county, or municipality whose designated enforcement officer is termed the "code official."

COMBUSTION. In the context of this code, refers to the rapid oxidation of fuel accompanied by the production of heat or heat and light.

❖ The primary components of combustion are fuel, oxygen, and heat. The code regulates many aspects of combustion technology, including the process of combustion and providing sufficient air, the use of energy produced from combustion, the safe venting of the products of combustion, combustion efficiency, the containment and control of combustion, and the fuel supplies for combustion equipment and appliances.

COMBUSTION AIR. Air necessary for complete combustion of a fuel, including theoretical air and excess air.

❖ The process of combustion requires a specific amount of oxygen to initiate and sustain the combustion reaction. Combustion air includes primary air, secondary air, draft hood dilution air, and excess air. Combustion air includes the amount of atmospheric air required for complete combustion of a fuel and is related to the molecular composition of the fuel being burned, the design of the fuel-burning equipment, and the percentage of oxygen in the combustion air. Too little combustion air will result in incomplete combustion of a fuel and the possible formation of carbon deposits (soot), carbon monoxide, toxic alcohols, ketone, aldehydes, nitrous oxides and other by-products. The required amount of combustion air is usually stated in terms of cubic feet per minute (m^3/s) or pounds per hour (kg/h).

COMBUSTION CHAMBER. The portion of an appliance within which combustion occurs.

❖ Combustion chambers are either open to the atmosphere or are isolated (sealed) from the atmosphere in which the appliance is installed. Combustion chambers are often referred to as "fireboxes."

COMBUSTION PRODUCTS. Constituents resulting from the combustion of a fuel with the oxygen of the air, including the inert gases, but excluding excess air.

❖ Such products include water, carbon dioxide, carbon monoxide, nitrous oxides, and various trace compounds.

CONCEALED LOCATION. A location that cannot be accessed without damaging permanent parts of the building structure or finish surface. Spaces above, below or behind readily removable panels or doors shall not be considered as concealed.

❖ The space above a "drop-in" tile suspended ceiling system, for example, would not be considered as a concealed location.

CONCEALED PIPING. Piping that is located in a concealed location. (See "Concealed location.")

❖ This refers to gas piping in locations meeting the definition of "Concealed location."

CONDENSATE. The liquid that condenses from a gas (including flue gas) caused by a reduction in temperature or increase in pressure.

❖ Condensate forms when the temperature of a vapor is lowered to its dew point temperature. Air conditioning systems produce condensate when an airstream contacts cooling coils. The moisture in the air condenses on the cold surface of the coils and the air is "dehumidified." High efficiency (84 percent and up) fuel burning appliances produce condensate from the combustion gases. Condensate also forms within improperly designed chimneys and vents when the products of combustion (which contain water vapor) contact the colder inner walls of the vent or chimney. If the temperature of the products of combustion is lowered to the dew point temperature of the water vapor, condensate will form on the inside walls of the vent or chimney. Condensed steam in hydronic systems is also referred to as "condensate."

CONNECTOR. The pipe that connects an approved appliance to a chimney, flue or vent.

❖ In most cases, appliances are not located directly in line with the vertically rising chimney or vent; therefore, a vent connector is necessary to connect the appliance flue outlet to the vent. Vent connectors can be single- or double-wall pipes and are usually made from galvanized steel, stainless steel, or aluminum sheet metal. Connectors are also made of corrugated aluminum. In many installations, the vent connectors must be constructed of the same material as the vent, as is typically done with Type B vent systems. There are many requirements in the code regulating the location, length, size, and type of connector. These requirements are in Sections G2426 and G2427.

CONTROL. A manual or automatic device designed to regulate the gas, air, water or electrical supply to, or operation of, a mechanical system.

❖ A control is a device designed to respond to changes in temperature, pressure, liquid or gas flow rates, current, voltage, resistance, humidity, and/or liquid levels.

CONVERSION BURNER. A unit consisting of a burner and its controls for installation in an appliance originally utilizing another fuel.

❖ Typical conversion burners are designed to convert an appliance such as a boiler from solid fuel fired to oil or gas-fired.

CUBIC FOOT. The amount of gas that occupies 1 cubic foot (0.02832 m^3) when at a temperature of 60°F (16°C), saturated with water vapor and under a pressure equivalent to that of 30 inches of mercury (101 kPa).

❖ This definition applies specifically to a cubic foot of gas.

DAMPER. A manually or automatically controlled device to regulate draft or the rate of flow of air or combustion gases.

❖ Dampers act as restrictors or valves in a gaseous flow stream. See Section G2427.13.

DECORATIVE GAS APPLIANCE, VENTED. A vented appliance wherein the primary function lies in the aesthetic effect of the flames.

❖ Such appliances include so-called "gas fireplaces" that are designed to simulate a wood-burning fireplace. The title of the standard for vented decorative gas-fired appliances (Z21.50) has been recently changed to "Vented Gas Fireplaces."

DECORATIVE GAS APPLIANCES FOR INSTALLATION IN VENTED FIREPLACES. A vented appliance designed for installation within the fire chamber of a vented fireplace, wherein the primary function lies in the aesthetic effect of the flames.

❖ Such appliances include gas log sets for installation in solid fuel-burning fireplaces.

DEMAND. The maximum amount of gas input required per unit of time, usually expressed in cubic feet per hour, or Btu/h (1 Btu/h = 0.2931 W).

❖ Demand refers to the load that appliances place on the fuel-gas distribution system.

DESIGN FLOOD ELEVATION. The elevation of the "design flood," including wave height, relative to the datum specified on the community's legally designated flood hazard map.

❖ This term is used in Section G2404.7 and refers to the "worst case" flood that could be anticipated, such as might occur once every 100 years. The design flood elevation is the height to which flood waters will rise during passage or occurrence of the design flood. The datum specified on the flood hazard map is important because it may differ from that used locally for other purposes.

DILUTION AIR. Air that is introduced into a draft hood and is mixed with the flue gases.

❖ Dilution air is associated only with draft hood-equipped Category I appliances. Fan-assisted Category I appliances do not involve dilution air, as all of the air brought in by the fan passes directly through the combustion chamber of the appliance. Dilution air lowers the dew point of vent gases, reducing the possibility that condensation will occur inside the vent or chimney.

DIRECT-VENT APPLIANCES. Appliances that are constructed and installed so that all air for combustion is derived directly from the outside atmosphere and all flue gases are discharged directly to the outside atmosphere.

❖ Such appliances are provided with independent exhaust and intake pipes or are provided with concentric pipes that vent combustion gases through the inner pipe and convey combustion air in the annular space between the inner and outer pipe walls. See Commentary Figures G2403(5) and G2403(6).

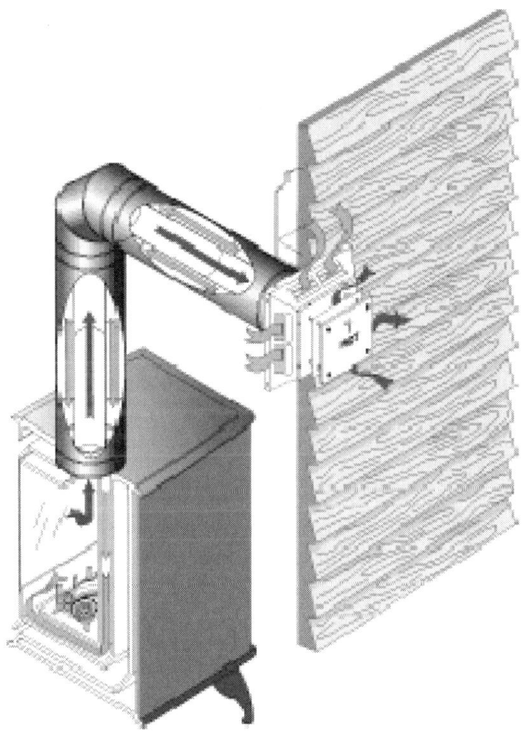

Photo courtesy of Simpson Dura-Vent Company, Incorporated

Figure G2403(5)
DIRECT-VENT APPLIANCE

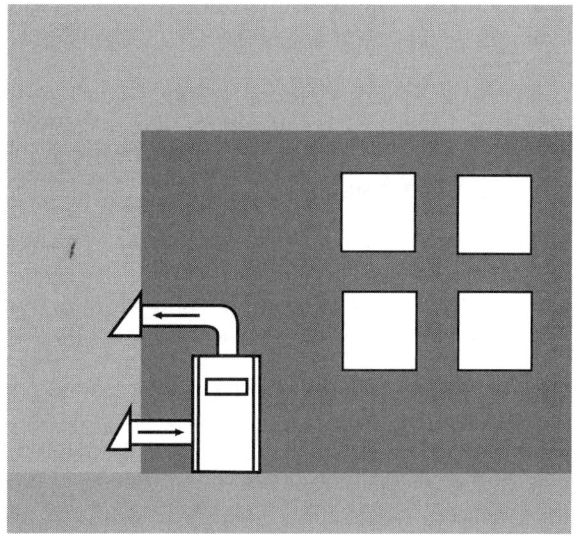

Figure courtesy of Lochinvar Corporation

Figure G2403(6)
DIRECT-VENT APPLIANCE

DRAFT. The pressure difference existing between the equipment or any component part and the atmosphere, that causes a continuous flow of air and products of combustion through the gas passages of the appliance to the atmosphere.

❖ Draft is the negative static pressure measured relative to atmospheric pressure that is developed in chimneys and vents and in the flue-ways of fuel-burning appliances. Draft can be produced by hot flue-gas buoyancy ("stack effect"), mechanically by fans and exhausters, or by a combination of both natural and mechanical means.

Mechanical or induced draft. The pressure difference created by the action of a fan, blower or ejector that is located between the appliance and the chimney or vent termination.

❖ Induced draft systems use a fan or blower to boost or "induce" draft in a venting system that produces insufficient natural draft. Draft induction implies that the combustion gases are "pulled" through a passageway or conduit. Draft inducers produce negative pressures on the inlet (upstream) side of the fan or blower. They are separate field-installed units located between an appliance and its venting system. Draft inducers are used with natural draft venting systems to overcome the resistance of vent or chimney connectors and to compensate for the inability of the chimney or vent to produce sufficient and reliable draft.

Draft induction is also a design principle used in many fan-assisted appliances. Draft inducer fans or blowers that are integral with fuel-fired appliances are necessary to overcome the internal resistance of the heat exchanger flue passageways. To attain higher thermal transfer efficiencies, some heat exchanger designs use flue passageways that retain the combustion gases longer over a greater surface area, thus extracting more heat. Such heat exchanger designs cannot rely upon natural (gravity) venting to overcome the higher resistance to flow and, therefore, must rely upon mechanical means to pull the combustion gases through the heat exchanger.

Induced draft systems are not to be confused with systems using power exhausters or other self-venting equipment.

Mechanical draft systems produce positive pressures on the discharge side of the fan or blower and can be said to "push" combustion gases through a passageway or conduit.

Natural draft. The pressure difference created by a vent or chimney because of its height, and the temperature difference between the flue gases and the atmosphere.

❖ Natural draft systems do not use mechanical devices such as fans or blowers but instead rely on the principle of buoyancy to carry the products of combustion to the atmosphere. Because of the difference in temperature and the resultant difference in density between the hot products of combustion and the ambient atmosphere, the gases within the chimney or flue will rise, creating a buoyant "draft." The phenomenon of natural draft is sometimes referred to as "stack effect" and is measured in inches of water column (kPa). The amount of draft is affected by the height of the chimney or vent and also by the ability of the chimney or vent to maintain the temperature differential between the combustion gases and the ambient air.

DRAFT HOOD. A nonadjustable device built into an appliance, or made as part of the vent connector from an appliance, that is designed to (1) provide for ready escape of the flue gases from the appliance in the event of no draft, backdraft, or stoppage beyond the draft hood, (2) prevent a backdraft from entering the appliance, and (3) neutralize the effect of stack action of the chimney or gas vent upon operation of the appliance.

❖ Draft hoods are integral to or supplied with natural draft atmospheric burner gas-fired appliances other than fan-assisted appliances. When classified, appliances equipped with draft hoods are classified by the manufacturer as Category I appliances. Because of minimum efficiency standards and the popularity of mechanical and other special proprietary venting systems, draft-hood-equipped appliances are becoming increasingly rare in the marketplace. See Commentary Figure G2403(7).

DRAFT REGULATOR. A device that functions to maintain a desired draft in the appliance by automatically reducing the draft to the desired value.

❖ Excessive draft reduces the combustion and thermal efficiency of an appliance. Draft regulators automatically adjust the draft by admitting air into the vent thereby reducing the draft. See Commentary Figure G2403(2).

DRIP. The container placed at a low point in a system of piping to collect condensate and from which the condensate is removable.

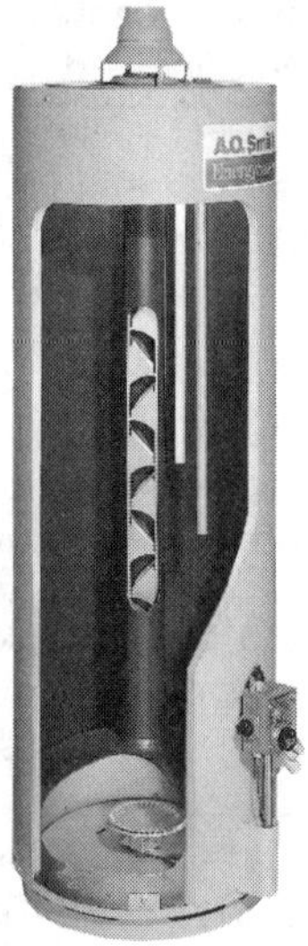

Photo courtesy of A.O. Smith Water Products

Figure G2403(7)
DRAFT HOOD EQUIPPED APPLIANCE

❖ Such reservoirs, also referred to as "drip legs," are made up of pipe and fittings and are intended to collect liquid in piping systems where condensables are possible. A "drip leg" is distinct from a "sediment trap" even though they may be constructed identically.

DUCT FURNACE. A warm-air furnace normally installed in an air-distribution duct to supply warm air for heating. This definition shall apply only to a warm-air heating appliance that depends for air circulation on a blower not furnished as part of the furnace.

❖ Duct furnaces are vented appliances and consist of burners, heat exchangers, and related controls. Duct furnaces depend on an external air handler or blower.

EQUIPMENT. See "Appliance."

❖ See the commentaries for "Appliance (Equipment)"; "Appliance, Fan-Assisted Combustion"; "Appliance, Automatically Controlled"; "Appliance, Unvented"; and "Appliance, Vented."

FIREPLACE. A fire chamber and hearth constructed of noncombustible material for use with solid fuels and provided with a chimney.

❖ Fireplaces burn solid fuels (wood, coal, etc.) and are not referred to as appliances in the code. The *International Residential Code* covers construction of masonry fireplaces.

Masonry fireplace. A hearth and fire chamber of solid masonry units such as bricks, stones, listed masonry units or reinforced concrete, provided with a suitable chimney.

❖ Masonry fireplaces must be constructed in accordance with the requirements found in this code. These specific requirements are based on tradition and field experience and describe the conventional fireplace that has proven to be reliable where properly constructed, used, and maintained.

Factory-built fireplace. A fireplace composed of listed factory-built components assembled in accordance with the terms of listing to form the completed fireplace.

❖ Factory-built fireplaces are solid fuel-burning units having a fire chamber that is intended to be either open to the room or, if equipped with doors, operated with the doors either open or closed. Fireplaces are not referred to as appliances. The term "fireplace" describes a complete assembly, which includes the hearth, the fire chamber, and a chimney. A factory-built fireplace is composed of factory-built components representative of the prototypes tested and is installed in accordance with the manufacturer's installation instructions to form the completed fireplace.

FLAME SAFEGUARD. A device that will automatically shut off the fuel supply to a main burner or group of burners when the means of ignition of such burners become inoperative, and when flame failure occurs on the burner or group of burners.

❖ Such devices are primary safety controls and are provided on all automatically operated fuel-fired appliances.

FLOOD HAZARD AREA. The greater of the following two areas:

1. The area within a floodplain subject to a 1 percent or greater chance of flooding in any given year.
2. This area designated as a flood hazard area on a community's flood hazard map, or otherwise legally designated.

❖ FEMA prepares flood insurance rate maps that delineate the land area that is subject to inundation by the 1-percent annual change flood. Some states and local jurisdictions develop and adopt maps of flood hazard areas that are more extensive than the areas shown on FEMA's map. For the purpose of the code, the flood hazard area within which the requirements are to be applied is the greater of the two delineated areas. See Section G2404.7.

FLOOR FURNACE. A completely self-contained furnace suspended from the floor of the space being heated, taking air for combustion from outside such space and with means for observing flames and lighting the appliance from such space.

❖ Such units supply heat through a floor grille placed directly over the unit's heat exchanger.

FLUE, APPLIANCE. The passage(s) within an appliance through which combustion products pass from the combustion chamber of the appliance to the draft hood inlet opening on an appliance equipped with a draft hood or to the outlet of the appliance on an appliance not equipped with a draft hood.

❖ Appliance flues are the passages through an appliance from the combustion chamber, through the heat exchanger and through the vent outlet.

FLUE COLLAR. That portion of an appliance designed for the attachment of a draft hood, vent connector or venting system.

❖ The flue collar size will be a determining factor in vent connector sizing.

FLUE GASES. Products of combustion plus excess air in appliance flues or heat exchangers.

❖ The exact composition of flue gases will depend on the fuel being burned. The primary components of flue gases are nitrogen, carbon dioxide, water vapor, particulates, and a myriad of compounds and trace elements that vary with the nature of the fuel and purity of the combustion air. Carbon monoxide is also a component of flue gas.

FLUE LINER (LINING). A system or material used to form the inside surface of a flue in a chimney or vent, for the purpose of protecting the surrounding structure from the effects of combustion products and for conveying combustion products without leakage to the atmosphere.

❖ Flue liners must be resistant to heat and the corrosive action of the products of combustion. They provide insulation value to retard the transfer of heat to the chimney structure and limit exposure of the chimney structure to the harmful effects of combustion products. Flue liners are generally made of fire-clay tile, refractory brick, poured-in-place refractory materials, stainless steel alloys, and aluminum. Flue liners used with gas-fired appliances are typically made of stainless steel or aluminum. See Commentary Figure G2403(8).

FUEL GAS. Fuel gases include: a natural gas, manufactured gas, liquified petroleum gas, hydrogen gas and mixtures of these gases.

❖ The nature of fuel gases makes proper design, installation, and selection of materials and devices necessary to minimize the possibility of fire or explosion. Bringing fuel gases into a building is in itself a risk. The provisions of the code are intended to reduce that risk to a level comparable to that associated with other energy sources such as electricity.

The two most commonly used fuel gases are natural gas and liquefied petroleum gas (LP-gas or LPG). These fuel gases have the following characteristics or properties.

Natural gas: The principal constituent of natural gas is methane (CH_4). It can also contain small quantities of nitrogen, carbon dioxide, hydrogen sulfide, water vapor, other hydrocarbons (such as ethane and propane), and various trace elements. Natural gas is

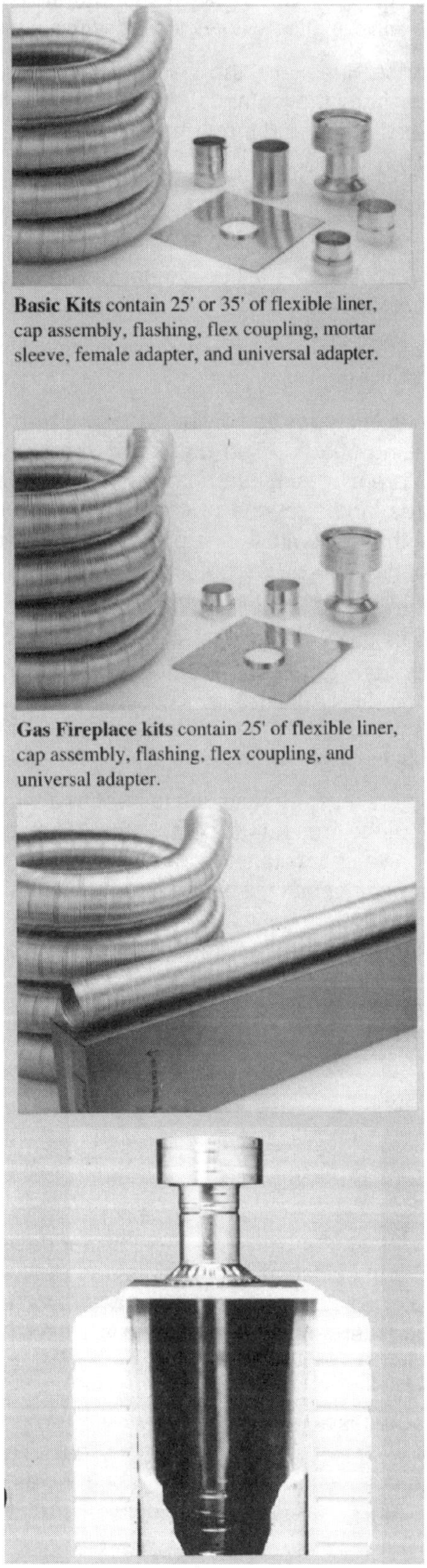

Photo courtesy of Simpson Dura-Vent Company, Incorporated

Figure G2403(8)
CHIMNEY FLUE LINER SYSTEM FOR GAS APPLIANCE

colorless, tasteless, and odorless; however, an odorant is added to the gas so that it can be readily detected. Natural gas is lighter than air (specific gravity of 0.60 typical) and has the tendency to rise when escaping to the atmosphere. Natural gas has a rather narrow flammability range (approximately 3 to 15 percent volume in air) above and below which the gas-to-air mixture ratio will be too rich or too lean to support combustion. The heating value of natural gas is approximately 1,050 Btu per cubic foot (10.8 kW/m^3).

LP-gas: Liquefied petroleum gases include commercial propane and commercial butane. LP-gas vapors are heavier than air (specific gravity of 1.52 typical) and tend to accumulate in low areas and near the floor. The ranges of flammability for LP-gases are narrower than those of natural gas (approximately 2 to 10 percent volume in air). Like natural gas, LP-gases are odorized to make them detectable.

The heating value of propane is approximately 2,500 Btu per cubic foot (25.9 kW/m^3) of gas. The heating value of butane is approximately 3,300 Btu per cubic foot (34.2 kW/m^3) of gas.

FUEL GAS UTILIZATION EQUIPMENT. See "Appliance."

❖ See the commentary for "Appliance (equipment)."

FURNACE. A completely self-contained heating unit that is designed to supply heated air to spaces remote from or adjacent to the appliance location.

❖ The single most distinguishing characteristic of furnaces is that they use air as the heat transfer medium. Furnaces can be fueled by gas, oil, solid fuel, or electricity and use fans, blowers, or gravity (convection) to circulate air to and from the unit. In the context of the code, the primary usage of the term "furnace" refers to heating appliances that combine a combustion chamber with related burner and control components, one or more heat exchangers, a flue gas conveying system, and air-handling fans/blowers.

FURNACE, CENTRAL FURNACE. A self-contained appliance for heating air by transfer of heat of combustion through metal to the air, and designed to supply heated air through ducts to spaces remote from or adjacent to the appliance location.

❖ The term "central furnace" has been adopted to identify a furnace that supplies conditioned air to remote locations from a central location. There are many types of furnaces. Most of these variations are a result of architectural design related to available space; desired location of the unit, and size of the area to be conditioned.

FURNACE PLENUM. An air compartment or chamber to which one or more ducts are connected and which forms part of an air distribution system.

❖ The term "plenum" is no longer defined or used in this chapter. A furnace plenum is a box made of sheet metal or other duct material. Such box is attached directly to a furnace to facilitate connection of supply and return ducts to the furnace air inlet and outlet. This term does not refer to cavities within a building such as stud and joist spaces and spaces above dropped ceilings.

GAS CONVENIENCE OUTLET. A permanently mounted, manually operated device that provides the means for connecting an appliance to, and disconnecting an appliance from, the gas supply piping. The device includes an integral, manually operated valve with a nondisplaceable valve member and is designed so that disconnection of an appliance only occurs when the manually operated valve is in the closed position.

❖ Such devices are listed and labeled and designed to mate with specialized appliance connectors. A gas outlet is the point at which the gas distribution system connects to a gas appliance. Such outlets are intended for use with portable appliances such as space heaters and outdoor cooking appliances.

GAS PIPING. An installation of pipe, valves or fittings installed on a premises or in a building and utilized to convey fuel gas.

❖ Gas piping includes all the components, fittings, and piping needed to deliver the fuel gas from the point of delivery to the appliance or equipment connection. The point of delivery may be a regulator or meter that is typically installed by the gas utility. The point of delivery may be located at the user's property line, immediately outside the structure, or in some instances in the structure.

GAS UTILIZATION EQUIPMENT. An appliance that utilizes gas as a fuel or raw material or both.

❖ See the commentary for "Appliance (Equipment)."

HAZARDOUS LOCATION. Any location considered to be a fire hazard for flammable vapors, dust, combustible fibers or other highly combustible substances. The location is not necessarily categorized in the building code as a high-hazard use group classification.

❖ The environment in which mechanical equipment and appliances operate plays a significant role in the safe performance of the equipment installation. Locations that may contain ignitable or explosive atmospheres are classified as hazardous locations for the installation of mechanical equipment and appliances. For example, repair garages can be classified as hazardous locations because they can contain gasoline vapors from vehicles stored within them as well as other volatile chemicals. Public and private garages are not considered hazardous locations although the presence of motor vehicles and the storage of paint, varnish, thinner, lawn- and home-maintenance products, and other chemicals are a concern as evidenced by Sections G2408.2 and G2408.3.

HOUSE PIPING. See "Piping system."

❖ House piping is the distribution piping downstream of the point of delivery. House piping is an antiquated term.

IGNITION PILOT. A pilot that operates during the lighting cycle and discontinues during main burner operation.

❖ Commonly referred to as "interrupted pilots," such pilots are ignited by a spark or hot surface device. Generally an electronic circuit controls the ignition system. When main burner ignition has been proven, the ignition pilot is shut off, and the main burner flame is supervised for the entire burn cycle.

IGNITION SOURCE. A flame spark or hot surface capable of igniting flammable vapors or fumes. Such sources include appliance burners, burner ignitors and electrical switching devices.

❖ This definition is important in the application of Section G2408.2 regarding the elevation of ignition sources. By means of this definition, Section G2408.2 applies to unintentional ignition sources such as electrical switching devices as well as intentional ignition sources such as pilot lights, spark ignitors, and hot surface ignitors.

In the context of Section G2408.2, an "ignition source" is something capable of igniting flammable vapors that are present in the atmosphere in the locations listed in Section G2408.2.

INFRARED RADIANT HEATER. A heater which directs a substantial amount of its energy output in the form of infrared radiant energy into the area to be heated. Such heaters are of either the vented or unvented type.

❖ Such heaters include low-intensity tubular types and high-intensity ceramic burner element types. They heat objects (such as people) by direct radiation and do not heat the ambient air.

JOINT, FLARED. A metal-to-metal compression joint in which a conical spread is made on the end of a tube that is compressed by a flare nut against a mating flare.

❖ Because the pipe end is expanded in a flared joint, only annealed and bending tempered soft-drawn copper tubing may be flared. Commonly used flaring tools employ a screw yoke and block assembly or an expander tool that is driven into the tube with a hammer. The flared tubing end is compressed between a fitting seat and a threaded nut to form a metal-to-metal seal.

JOINT, MECHANICAL. A general form of gas-tight joints obtained by the joining of metal parts through a positive-holding mechanical construction, such as flanged joint, threaded joint, flared joint or compression joint.

❖ Mechanical joints can take many forms, but most share a common characteristic, which is applying a radial pressure to the pipes or fittings they join. Mechanical joining means can be proprietary, which means that the manufacturer of the fittings and devices must provide adequate instructions for assembling the joint.

JOINT, PLASTIC ADHESIVE. A joint made in thermoset plastic piping by the use of an adhesive substance which forms a continuous bond between the mating surfaces without dissolving either one of them.

❖ Unlike solvent welded joints, adhesive (glue) joints are surface bonded.

LIQUEFIED PETROLEUM GAS or LPG (LP-GAS). Liquefied petroleum gas composed predominately of propane, propylene, butanes or butylenes, or mixtures thereof that is gaseous under normal atmospheric conditions, but is capable of being liquefied under moderate pressure at normal temperatures.

❖ Liquefied petroleum gases are usually obtained as a byproduct of oil refinery operations or by stripping natural gas and are commercially available as butane, propane, or a mixture of the two. Propane has a boiling point of -40°F (-40°C) at atmospheric pressure and a heating value of approximately 2,500 Btu per cubic foot (25.9 m^3). Butane has a boiling point of 32°F (0°C) at atmospheric pressure and a heating value of approximately 3,200 Btu per cubic foot (33.1 kW/m^3). LP-gas is odorized so that gas leakage can be detected.

LIVING SPACE. Space within a dwelling unit utilized for living, sleeping, eating, cooking, bathing, washing and sanitation purposes.

❖ The code uses this term to define locations that are not to be considered as part of a private garage for the purpose of requiring the elevation of ignition sources. See Section G2408.2.

LOG LIGHTER, GAS-FIRED. A manually operated solid-fuel ignition appliance for installation in a vented solid-fuel-burning fireplace.

❖ Log lighters are simple, manually operated burners used to start wood fires. Log lighters are not considered as decorative but rather as functional appliances. The heat produced by the log lighter flame raises the temperature of the wood fuel to its ignition temperature. A log lighter is designed to be turned off manually after the wood fire is capable of sustaining combustion.

MAIN BURNER. A device or group of devices essentially forming an integral unit for the final conveyance of gas or a mixture of gas and air to the combustion zone, and on which combustion takes place to accomplish the function for which the appliance is designed.

❖ With the exception of small amounts of fuel for the pilot or ignition burners, the main burner consumes the entire fuel input into an appliance. Common types of atmospheric burners are the drilled port, slotted port, ribbon, and single port.

MECHANICAL EXHAUST SYSTEM. Equipment installed in and made a part of the vent, which will provide a positive induced draft.

❖ Confusion may occur with respect to the definition of this term. The definition in the *International Mechanical Code* for the same term is different and is as follows: "A system for removing air from a room or space by mechanical means." The only place this term can be found in this code is in Section G2425.8, and the *International Mechanical Code* definition correctly defines its use in that section.

This term, as defined in this code, is currently not used in the body of the code. See Sections G2427.3.3 and G2429.1.

METER. The instrument installed to measure the volume of gas delivered through it.

❖ The meter is the actual point of commerce and is usually associated with the point of delivery. This code does not require or regulate meters. The gas supplying utility generally installs meters for natural gas. Occasionally in locations where there are several tenants (large office buildings, mobile home parks, and apartment complexes for example), the gas utility installs one meter for the property owner and the owner installs individual tenant meters. Liquefied petroleum may be metered to individual tenants from a storage tank in the same manner. See Commentary Figure G2403(9).

Photo courtesy of Washington Gas

Figure G2403(9)
TYPICAL RESIDENTIAL GAS METER SETTING

MODULATING. Modulating or throttling is the action of a control from its maximum to minimum position in either predetermined steps or increments of movement as caused by its actuating medium.

❖ In the context of this code, modulating is associated with burner control valves. One type of automatic control is a modulating control that can infinitely vary the controlled variable over the range being controlled. Another common type of control is a two-position action control, which simply turns the variable being controlled, such as fuel gas, on and off. And another type of control is the floating action control, which regulates the controlled device by adjusting the controlled variable between the open or closed position by a sensing element, which must react faster than the actuating drive time or the system would be essentially the same as a two-position control.

OFFSET (VENT). A combination of approved bends that make two changes in direction bringing one section of the vent out of line, but into a line parallel with the other section.

❖ An offset results in the lateral displacement of a vertical vent. Vent offsets are measured in degrees of angle from vertical and are limited in venting systems, as they restrict the flow of flue gas. An offset angle is measured between a line that is an extension of the original vertical vent and the new line of direction of the vent. See Commentary Figure G2403(10).

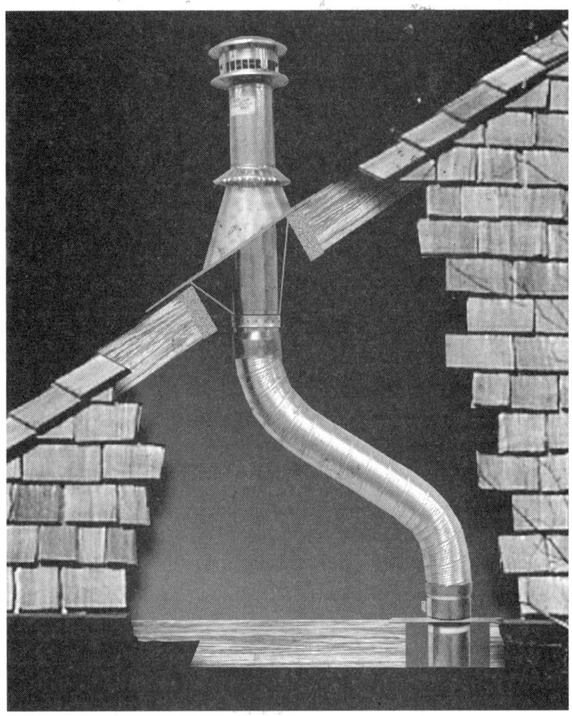

Photo courtesy of Selkirk L.L.C.

Figure G2403(10)
VENT OFFSET

OUTLET. A threaded connection or bolted flange in a pipe system to which a gas-burning appliance is attached.

❖ A gas outlet is analogous to an electrical receptacle outlet. See Section G2415.13 for requirements regarding the location of gas piping outlets.

OXYGEN DEPLETION SAFETY SHUTOFF SYSTEM (ODS). A system designed to act to shut off the gas supply to the main and pilot burners if the oxygen in the surrounding atmosphere is reduced below a predetermined level.

❖ An Oxygen Depletion Safety Shutoff System (ODS) is intended to protect the occupants from carbon monoxide build-up. The ODS measures the oxygen level in the air and shuts off the appliance if the level falls below a present level, typically 18 percent. Monitoring oxygen levels will inversely monitor carbon monoxide levels because of the natural relationship between a decreasing oxygen level and an increasing carbon monoxide level. See Commentary Figure G2403(11).

Photo by D. F. Noyes Studios, courtesy of DESA International

Figure G2403(11)
OXYGEN DEPLETION SAFETY SHUTOFF

PILOT. A small flame that is utilized to ignite the gas at the main burner or burners.

❖ Pilot burners (pilot lights) can be continuously burning or can be intermittent or interrupted depending upon the type of appliance.

PIPING. Where used in this code, "piping" refers to either pipe or tubing, or both.

❖ Piping includes tubing and pipe used to convey fuel gases. See the definition for "Piping System."

Pipe. A rigid conduit of iron, steel, copper, brass or plastic.

Tubing. Semirigid conduit of copper, aluminum, plastic or steel.

PIPING SYSTEM. All fuel piping, valves, and fittings from the outlet of the point of delivery to the outlets of the equipment shutoff valves.

❖ A piping system includes tubing, pipe, fittings, valves, and line pressure regulators used to convey fuel gas from the point of delivery to the appliance. This definition is revised in the 2003 edition of the IFGC to state that the piping system ends at the outlets of the appliance shutoff valves. This revision allows the last six feet or less of gas conduit between the shutoff valve and the appliance inlet connection to be sized as a connector. See Section G2420.5 and the definition for "Point of Delivery."

PLASTIC, THERMOPLASTIC. A plastic that is capable of being repeatedly softened by increase of temperature and hardened by decrease of temperature.

❖ Polyvinyl chloride (PVC) is a type of thermoplastic.

POINT OF DELIVERY. For natural gas systems, the point of delivery is the outlet of the service meter assembly, or the outlet of the service regulator or service shutoff valve where a meter is not provided. Where a valve is provided at the outlet of the service meter assembly, such valve shall be considered to be downstream of the point of delivery. For undiluted liquefied petroleum gas systems, the point of delivery shall be considered the outlet of the first stage pressure regulator that provides utilization pressure, exclusive of line gas regulators, in the system.

❖ It is necessary to understand where the point of delivery begins with respect to fuel gas systems, since that is the location where the enforcement of Chapter 24 begins. For LP applications, the point of delivery is the outlet of the second-stage regulator in low pressure systems or the first regulator that reduces pressure to 2 psi or less in elevated pressure systems. NFPA 58 regulates all piping and components upstream of the second stage regulator. Any valve located on the outlet of the service meter assembly is considered as part of the piping system under the scope of this chapter.

PRESSURE DROP. The loss in pressure due to friction or obstruction in pipes, valves, fittings, regulators and burners.

❖ The pressure in a fuel gas system is reduced as the gas flows in the system. This is caused by a loss of energy due to friction and turbulence. Fuel gas systems are designed so that the pressure drop does not result in the system pressure falling below the minimum pressure required for proper equipment operation.

PRESSURE TEST. An operation performed to verify the gas-tight integrity of gas piping following its installation or modification.

❖ Installed piping systems are tested on the job site to verify that the system is void of any leaks.

READY ACCESS (TO). That which enables a device, appliance or equipment to be directly reached, without requiring the removal or movement of any panel, door or similar obstruction. (See "Access.")

❖ Ready access can be described as the capability of being quickly reached or approached for the purpose of operation, inspection, observation, or emergency action. Ready access does not require the removal or movement of any door, panel, or similar obstruction, nor does it require surmounting physical obstructions or obstacles, including differences in elevation.

REGULATOR. A device for controlling and maintaining a uniform gas supply pressure, either pounds-to-inches water column (MP regulator) or inches-to-inches water column (appliance regulator).

❖ In the context of the code, such devices are fuel-gas pressure regulating devices.

REGULATOR, GAS APPLIANCE. A pressure regulator for controlling pressure to the manifold of gas equipment.

❖ These regulators are integral to the appliance and serve to stabilize the appliance input pressure and provide fuel gas to the appliance at the desired pressure. Most appliances come equipped with appliance regulators, as the appliances are designed to operate at a pressure lower than the delivery pressure, even when low pressure gas (0.5 psig or lower) is supplied.

REGULATOR, LINE GAS PRESSURE. A device placed in a gas line between the service pressure regulator and the equipment for controlling, maintaining or reducing the pressure in that portion of the piping system downstream of the device.

❖ Line gas pressure regulators are used to reduce the gas pressure supplied by the service (point of delivery) regulator. Such devices are typically located at the connection to each gas appliance and are commonly used as "pounds to inches" regulators in systems having elevated (14 inches wc to 5 psi typical) supply pressures.

REGULATOR, PRESSURE. A device placed in a gas line for reducing, controlling and maintaining the pressure in that portion of the piping system downstream of the device.

❖ See the definition and commentary for "Regulator."

REGULATOR, SERVICE PRESSURE. A device installed by the serving gas supplier to reduce and limit the service line gas pressure to delivery pressure.

❖ The utility company (gas supplier) provides this regulator to reduce the gas pressure from the utility distribution mains to that required for delivery to the meter and customer. Such devices are usually located with the meter and are outdoors. See Commentary Figure G2403(12).

Photo courtesy of Washington Gas

Figure G2403(12)
SERVICE REGULATOR

RELIEF OPENING. The opening provided in a draft hood to permit the ready escape to the atmosphere of the flue products from the draft hood in the event of no draft, back draft or stoppage beyond the draft hood, and to permit air into the draft hood in the event of a strong chimney updraft.

❖ Dilution air, a component of combustion air, enters the venting system via the draft hood opening. Combustion gases spilling from (exiting) a draft hood relief opening create an abnormal and potentially hazardous condition except for brief periods of spillage that might occur at the start of an appliance firing cycle. Typical periods of spillage at start-up last only a few seconds.

RELIEF VALVE (DEVICE). A safety valve designed to forestall the development of a dangerous condition by relieving either pressure, temperature or vacuum in the hot water supply system.

❖ A relief valve is a safety device designed to prevent the development of a potentially damaging or dangerous condition in a closed system. This is usually associated with a closed system that has a heat source as part of the system. Requirements for the installation of relief valves in water distribution systems and water heaters are provided in Chapter 28.

RELIEF VALVE, PRESSURE. An automatic valve which opens and closes a relief vent, depending on whether the pressure is above or below a predetermined value.

❖ Pressure relief valves are intended to prevent harm to life and property. These valves are safety devices used to relieve abnormal pressures and to prevent the overpressurization of the vessel or system to which the valves are connected. Pressure relief valves are designed to operate (discharge) only when an abnormal pressure exists in a system. They are not to be used as regulating valves, nor are they intended to control or regulate flow or pressure. Pressure relief valves are set at the factory to begin opening at a predetermined pressure and are rated according to the maximum energy discharge per unit of time, usually in units of British thermal units per hour (Btu/h)(W).

Pressure relief valves must be properly sized and rated for the boiler, pressure vessel, or system served; otherwise, a system malfunction could cause the pressure within the system to continue to rise, thereby creating a hazardous condition. Pressure relief valves do not open fully when they reach the factory-preset pressure; rather, the valves open an amount proportional to the forces produced by the pressure in the system. The valves open fully at a certain percentage above the preset pressure, then close when the pressure drops below the preset pressure. Pressure relief valves are sometimes combined in the same body with temperature relief valves, creating combination temperature and pressure relief valves.

RELIEF VALVE, TEMPERATURE.

Reseating or self-closing type. An automatic valve which opens and closes a relief vent, depending on whether the temperature is above or below a predetermined value.

Manual reset type. A valve which automatically opens a relief vent at a predetermined temperature and which must be manually returned to the closed position.

❖ Temperature relief valves are designed to open in response to excessive temperatures and to discharge heated water to limit the temperature of the water in the vessel, tank, or system. Water heater control failure or improper adjustments could allow excessive temperature rises, which can heat water above the temperature at which it would vaporize at atmospheric pressure. Water heated to above 212°F (100°C) is referred to as superheated water and will flash to steam when the pressure is reduced below its vapor pressure. Very hot water and the presence of steam are hazardous; however, more importantly, higher temperatures produce higher pressures and superheated water increases the potential for and the magnitude of an explosion.

RELIEF VALVE, VACUUM. A valve that automatically opens and closes a vent for relieving a vacuum within the hot water supply system, depending on whether the vacuum is above or below a predetermined value.

❖ The vacuum relief valve is intended to prevent the possible reduction in pressure within the system (water heater tanks being of main concern) to below atmospheric pressure (that is, a partial vacuum). Many tanks are not designed to resist external pressures exceeding internal pressures; therefore, a vacuum relief valve is necessary to prevent atmospheric pressure from possibly collapsing or otherwise damaging the tank. A partial vacuum could also cause water to be siphoned from the tank thereby creating an overheating hazard in a water heater.

The vacuum relief valve operates by automatically opening the system to the atmosphere when a partial vacuum is created thereby permitting air to enter and maintaining the internal pressure at atmospheric pressure.

RISER, GAS. A vertical pipe supplying fuel gas.

❖ A vertical gas pipe that distributes gas from a building main to upper floor levels is commonly referred to as a "riser."

ROOM HEATER, UNVENTED. See "Unvented room heater."

❖ See the commentary for "Unvented room heater" and Commentary Figure G2403(13).

Photo courtesy of Empire Comfort Systems, Inc.

Figure G2403(13)
UNVENTED ROOM HEATER

ROOM HEATER, VENTED. A free-standing gas-fired heating unit used for direct heating of the space in and adjacent to that in which the unit is located. [See also "Vented room heater."]

❖ These heaters are typically small space-heating appliances designed to heat a single room.

SAFETY SHUTOFF DEVICE. See "Flame safeguard."

❖ See the commentary for "Flame safeguard."

SHAFT. An enclosed space extending through one or more stories of a building, connecting vertical openings in successive floors, or floors and the roof.

❖ A shaft is an enclosed, vertical passage that passes through stories and floor levels within a building. Shafts are sometimes referred to as "chases" and are usually constructed of fire-resistance-rated assemblies.

SPECIFIC GRAVITY. As applied to gas, specific gravity is the ratio of the weight of a given volume to that of the same volume of air, both measured under the same condition.

❖ Analogous to density, specific gravity values are used to describe how a gaseous substance will behave in, for example, air or piping systems.

THERMOSTAT

Electric switch type. A device that senses changes in temperature and controls electrically, by means of separate components, the flow of gas to the burner(s) to maintain selected temperatures.

Integral gas valve type. An automatic device, actuated by temperature changes, designed to control the gas supply to the burner(s) in order to maintain temperatures between predetermined limits, and in which the thermal actuating element is an integral part of the device.

1. Graduating thermostat. A thermostat in which the motion of the valve is approximately in direct proportion to the effective motion of the thermal element induced by temperature change.
2. Snap-acting thermostat. A thermostat in which the thermostatic valve travels instantly from the closed to the open position, and vice versa.

❖ Thermostats most often consist of one or more thermally actuated off/on switches, but the term also includes types that transmit varying voltage and signals. A thermostat is also referred to as a controller. The set point or desired unit of temperature is adjusted or set at the thermostat. The thermostat transmits a signal to the controlled device or appliance.

TRANSITION FITTINGS, PLASTIC TO STEEL. An adapter for joining plastic pipe to steel pipe. The purpose of this fitting is to provide a permanent, pressure-tight connection between two materials that cannot be joined directly one to another.

❖ The joining of piping of dissimilar materials such as plastic and steel requires specialized transition fittings. For example, such fittings are used with underground plastic fuel-gas piping and steel meter setting risers. See Commentary Figure G2403(14).

UNIT HEATER

High-static pressure type. A self-contained, automatically controlled, vented appliance having integral means for cir-

Photo courtesy of Perfection Corporation

Figure G2403(14)
TRANSITION FITTINGS

culation of air against 0.2 inch (15 mm H_2O) or greater static pressure. Such appliance is equipped with provisions for attaching an outlet air duct and, where the appliance is for indoor installation remote from the space to be heated, is also equipped with provisions for attaching an inlet air duct.

Low-static pressure type. A self-contained, automatically controlled, vented appliance, intended for installation in the space to be heated without the use of ducts, having integral means for circulation of air. Such units are allowed to be equipped with louvers or face extensions made in accordance with the manufacturer's specifications.

❖ Unit heaters are similar to warm-air furnaces, except that ducts are not usually associated with unit heaters. Such heaters are typically suspended from ceilings or roof structures.

UNVENTED ROOM HEATER. An unvented heating appliance designed for stationary installation and utilized to provide comfort heating. Such appliances provide radiant heat or convection heat by gravity or fan circulation directly from the heater and do not utilize ducts.

❖ As opposed to vented heaters, unvented heaters discharge all products of combustion into the space being heated. See Commentary Figure G2403(13).

VALVE. A device used in piping to control the gas supply to any section of a system of piping or to an appliance.

❖ The code requires valves to perform various functions in a fuel gas system as described in the definitions below. These valves are designed for the specific use intended and many are factory installed on equipment and are part of the listed equipment. However, there are valves that are not listed or manufactured in accordance with the standards listed in the code and if allowed must be evaluated in accordance with Section 105 of the *International Fuel Gas Code.*

Automatic. An automatic or semiautomatic device consisting essentially of a valve and operator that control the gas supply to the burner(s) during operation of an appliance. The operator shall be actuated by application of gas pressure on a flexible diaphragm, by electrical means, by mechanical means or by other approved means.

Automatic gas shutoff. A valve used in conjunction with an automatic gas shutoff device to shut off the gas supply to a water heating system. It shall be constructed integrally with the gas shutoff device or shall be a separate assembly.

Equipment shutoff. A valve located in the piping system, used to isolate individual equipment for purposes such as service or replacement.

❖ This definition has been revised to clarify that the equipment/appliance shutoff valves are not considered to be "emergency" valves.

Individual main burner. A valve that controls the gas supply to an individual main burner.

Main burner control. A valve that controls the gas supply to the main burner manifold.

Manual main gas-control. A manually operated valve in the gas line for the purpose of completely turning on or shutting off the gas supply to the appliance, except to pilot or pilots that are provided with independent shutoff.

Manual reset. An automatic shutoff valve installed in the gas supply piping and set to shut off when unsafe conditions occur. The device remains closed until manually reopened.

Service shutoff. A valve, installed by the serving gas supplier between the service meter or source of supply and the customer piping system, to shut off the entire piping system.

VENT. A pipe or other conduit composed of factory-made components, containing a passageway for conveying combustion products and air to the atmosphere, listed and labeled for use with a specific type or class of appliance.

❖ In code terminology, vents are distinguished from chimneys and usually are constructed of factory-made listed and labeled components intended to function as a system. Type B and BW vents are constructed of galvanized steel and aluminum sheet metal and are double wall and air insulated. Such vents are designed to vent gas-fired appliances and equipment, which are equipped with draft hoods, or which are specifically listed (labeled) for use with Type B or BW vents. Type L vents are typically constructed of sheet steel and stainless steel. They are double-wall and air insulated and are designed to vent gas-and oil-fired appliances and equipment. Some appliances are designed for use with corrosion-resistant special vents, such as those made of plastic pipe and special alloys of stainless steel. See Commentary Figure G2403(15).

Photo courtesy of Selkirk L.L.C.

Figure G2403(15)
TYPE B VENT

Special gas vent. A vent listed and labeled for use with listed Category II, III and IV gas appliances.

❖ Such vents include high-temperature plastic pipe, stainless steel pipe, and low-temperature (PVC and CPVC) plastic pipes.

Type B vent. A vent listed and labeled for use with appliances with draft hoods and other Category I appliances that are listed for use with Type B vents.

❖ See the commentary for "Vent."

Type BW vent. A vent listed and labeled for use with wall furnaces.

❖ See the commentary for "Vent."

Type L vent. A vent listed and labeled for use with appliances that are listed for use with Type L or Type B vents.

❖ See the commentary for "Vent."

VENT CONNECTOR. (See "Connector.")

❖ See the commentary for "Connector."

VENTED GAS APPLIANCE CATEGORIES. Appliances that are categorized for the purpose of vent selection are classified into the following four categories:

❖ Gas appliances are categorized for the purpose of matching the appliance to the required type of venting means.

Category I. An appliance that operates with a nonpositive vent static pressure and with a vent gas temperature that avoids excessive condensate production in the vent.

❖ Such appliances include draft hood-equipped and fan-assisted types.

Category II. An appliance that operates with a nonpositive vent static pressure and with a vent gas temperature that is capable of causing excessive condensate production in the vent.

❖ The condition of non-positive pressure (draft dependent) combined with low flue gas temperatures makes it difficult to design such appliances. At the time of this publication, no such appliances are known to exist in the marketplace.

Category III. An appliance that operates with a positive vent static pressure and with a vent gas temperature that avoids excessive condensate production in the vent.

❖ Such appliances include mid-efficiency (approximately 78 to 83 percent) appliances that are typically side-wall-vented with stainless steel special vents.

Category IV. An appliance that operates with a positive vent static pressure and with a vent gas temperature that is capable of causing excessive condensate production in the vent.

❖ Such appliances include high-efficiency (84 percent and higher) condensing-type furnaces, boilers and water heaters.

VENTED ROOM HEATER. A vented self-contained, free-standing, nonrecessed appliance for furnishing warm air to the space in which it is installed, directly from the heater without duct connections.

❖ These appliances are typically small space heaters that discharge all products of combustion to the outdoors.

VENTED WALL FURNACE. A self-contained vented appliance complete with grilles or equivalent, designed for incorporation in or permanent attachment to the structure of a building, mobile home or travel trailer, and furnishing heated air circulated by gravity or by a fan directly into the space to be heated through openings in the casing. This definition shall exclude floor furnaces, unit heaters and central furnaces as herein defined.

❖ Wall furnaces are designed to occupy very little room area and provide space heating for one or more rooms in small occupancies such as dwelling units, cottages, and the like.

VENTING SYSTEM. A continuous open passageway from the flue collar or draft hood of an appliance to the outside atmosphere for the purpose of removing flue or vent gases. A venting system is usually composed of a vent or a chimney and vent connector, if used, assembled to form the open passageway.

❖ Venting systems fall into one of two categories: natural draft or mechanical draft systems. Natural draft chimneys and vents do not rely on any mechanical means to convey combustion products to the outdoors. Draft is produced by the temperature difference between the vent gases (combustion gases) and the ambient atmosphere. Hot gases are less dense and more buoyant; therefore, they are displaced by cooler (more dense) ambient gases causing the hotter gases to rise and produce a draft.

Mechanical draft systems use fans or other mechanical means to cause the removal of flue or vent gases and also fall into one of two categories, based on positive or non-positive static vent pressure. The positive vent pressure systems are referred to as forced-draft or power venting systems, and the non-positive vent pressure systems are referred to as induced-draft venting systems.

- **Natural-draft venting system.** Natural draft chimneys and vents do not rely on any mechanical means to convey combustion products to the outdoors. Draft is produced by the temperature difference between the combustion gases (flue gases) and the ambient atmosphere. Hot gases are less dense and more buoyant; therefore, they rise as they are displaced by more dense ambient air.
- **Mechanical draft venting system.** Such systems do not depend on draft and employ fans or blowers.
- **Forced-draft venting system.** Power exhausters and some power burner systems are examples of forced-draft systems. Vents and chimneys must be listed for positive pressure applications where used with forced-draft systems.
- **Induced-draft venting system.** Induced-draft venting is commonly accomplished with field-installed inducer fans designed to supplement natural draft chimneys or vents.
- **Power draft venting system.** See the commentary for "Forced-draft venting system."

WALL HEATER, UNVENTED TYPE. A room heater of the type designed for insertion in or attachment to a wall or partition. Such heater does not incorporate concealed venting arrangements in its construction and discharges all products of combustion through the front into the room being heated.

❖ See the definition and commentary for "Unvented Room Heater."

WATER HEATER. Any heating appliance or equipment that heats potable water and supplies such water to the potable hot water distribution system.

❖ A water heater is a closed pressure vessel or heat exchanger that is provided with a heat source and that provides potable (drinkable) water to the building's hot water distribution system. Large commercial water heaters are sometimes referred to as "hot water supply boilers." Water heaters can be of the storage type with an integral storage vessel; circulating type for use with an external storage vessel; tankless instantaneous type without storage capacity; and point-of-use type with or without storage capacity. This chapter and Chapter 28 regulate water heaters because these appliances have elements and installation requirements relative to both mechanical and plumbing systems.

SECTION G2404 (301) GENERAL

G2404.1 (301.1) Scope. This section shall govern the approval and installation of all equipment and appliances that comprise parts of the installations regulated by this code in accordance with Section G2401.

❖ This section states that this chapter governs the approval and installation of all gas-fired equipment and appliances that are regulated by the code. Section G2401.1 establishes the scope of application of the code. See the commentary for Section G2401.1.

G2404.2 (301.1.1) Other fuels. The requirements for combustion and dilution air for gas-fired appliances shall be governed by Section G2407. The requirements for combustion and dilution air for appliances operating with fuels other than fuel gas shall be regulated by Chapter 17.

❖ Chapters 17 and 24 each have combustion air provisions that are specific to the fuels addressed in the respective chapters.

G2404.3 (301.3) Listed and labeled. Appliances regulated by this code shall be listed and labeled unless otherwise approved in accordance with Section R104.11. The approval of unlisted appliances in accordance with Section R104.11 shall be based upon approved engineering evaluation.

❖ Gas-fired appliances must be listed and labeled by an approved agency to show that they comply with the applicable national standards. The code requires listing and labeling for appliances such as boilers, furnaces, space heaters, cooking appliances, clothes dryers, rooftop HVAC units, etc. The code also requires listing for system components as specifically stated in the text addressing such components. The label is the primary, if not the only, assurance to the installer, the inspector, and the end user that a representative sample of an appliance model has been tested and evaluated by an approved agency and has been determined to perform safely and efficiently when installed and operated in accordance with its listing.

The presence of a label is part of the information that the code official is to consider in the approval of appliances. The only exception to the labeling requirement occurs when the code official approves a specific appliance in accordance with the authority granted in Section R104.11.

Caution should be exercised when the official considers the approval of unlisted and unlabeled appliances. Approval of unlabeled appliances must be based on documentation that demonstrates compliance with applicable standards or, where no product standards exist, that the appliance is appropriate for the intended use and will provide the same level of performance as would be provided by listed and labeled appliances. A fundamental principle of the code is the reliance on the listing and labeling process to assure appliance performance; approvals granted in accordance with Section R104.11 must be well justified with supporting documentation. To the code official, the installer, and the end-user, very little is known about the performance of an appliance that is not tested and built to an appliance standard.

G2404.4 (301.8) Vibration isolation. Where means for isolation of vibration of an appliance is installed, an approved means for support and restraint of that appliance shall be provided.

❖ Where vibration isolation connections are used in ducts and piping and where equipment is mounted with vibration dampers, support is required for the ducts, piping, and equipment to maintain positioning and alignment and to prevent stress and strain on the vibration connectors and dampers.

G2404.5 (301.9) Repair. Defective material or parts shall be replaced or repaired in such a manner so as to preserve the original approval or listing.

❖ Repair work must not alter the nature of appliances and equipment in a way that would invalidate the listing or conditions of approval. For example, replacement of safety control devices with different devices could alter the design and operation of an appliance from that intended by the manufacturer and the listing agency.

G2404.6 (301.10) Wind resistance. Appliances and supports that are exposed to wind shall be designed and installed to resist the wind pressures determined in accordance with this code.

❖ Installations of equipment and appliances that are subject to wind forces must be designed to resist those forces. The wind pressures must be based on the wind provisions in the building code for the site. The wind pressure requirements are based on the exposure of the building and wind speeds for that region.

G2404.7 (301.11) Flood hazard. For structures located in flood-hazard areas, the appliance, equipment and system installations regulated by this code shall be located at or above the design flood elevation and shall comply with the flood-resistant construction requirements of Section R323.

Exception: The appliance, equipment and system installations regulated by this code are permitted to be located below the design flood elevation provided that they are designed and installed to prevent water from entering or accumulating within the components and to resist hydrostatic and hydrodynamic loads and stresses, including the effects of buoyancy, during the occurrence of flooding to the design flood elevation and shall comply with the flood-resistant construction requirements of Section R323.

❖ In areas designated as a special flood-hazard in accordance with the requirements of this code, equipment and appliances must be protected from water or must be elevated above the anticipated flood level.

Exposure to water can cause deterioration of system components and serious appliance and equipment malfunctions. For example, appliance manufacturers require the replacement of appliances that have been submerged in floodwaters. Components such as pressure regulators, gas controls, motors, and electronic circuitry can all be ruined by exposure to water. The exception allows appliances and system components to be located below the design flood elevation if they are designed for submersion in flood waters. Appliances and system components designed for submersion may be nonexistent, and it would be extremely difficult to install effective barriers to keep out flood waters.

G2404.8 (301.12) Seismic resistance. Where earthquake loads are applicable in accordance with this code, the supports shall be designed and installed for the seismic forces in accordance with this code.

❖ This code requires building systems to be designed for specified seismic forces. This section references the detailed equipment and component seismic support requirements contained in this code to bring these requirements to the attention of the design professional and the permit applicant.

Equipment and piping must be braced for earthquake loads as stated in this code. The failure of the supports for these components has been shown to be a threat to health and safety in geographical areas where moderate- to high-magnitude earthquakes occur. The code specifies the geographical locations where earthquake design is required for certain piping and equipment and the size of the components that must be braced.

G2404.9 (301.14) Rodent proofing. Buildings or structures and the walls enclosing habitable or occupiable rooms and spaces in which persons live, sleep or work, or in which feed, food or foodstuffs are stored, prepared, processed, served or sold, shall be constructed to protect against the entry of rodents.

❖ This section states the requirements to prevent rodent infestation of a building. Efforts must be made to protect the annular spaces around openings in exterior walls. The annular spaces can be sealed by any effective method, but the primary methods used are to fill the annular space with a sealant material or to place a metal collar around the penetrating pipe, duct, etc. The collar material must be durable for weather exposure and strong enough to prevent rodents from chewing through and entering the building.

Effort must also be made to cover ventilation and combustion air openings with wire cloth to prevent the entry of rodents through the opening.

SECTION G2405 (302) STRUCTURAL SAFETY

G2405.1 (302.1) Structural safety. The building shall not be weakened by the installation of any gas piping. In the process of installing or repairing any gas piping, the finished floors, walls, ceilings, tile work or any other part of the building or premises, which are required to be changed or replaced shall be left in a safe structural condition in accordance with the requirements of this code.

❖ The installation of fuel gas systems must not adversely affect the structural integrity of the building components. This code dictates the structural safety requirements that must be applied to any structural portion of the building that is penetrated, altered, or removed during the installation, replacement, or repair of fuel gas systems.

G2405.2 (302.4) Alterations to trusses. Truss members and components shall not be cut, drilled, notched, spliced or otherwise altered in any way without the written concurrence and approval of a registered design professional. Alterations resulting in the addition of loads to any member (e.g., HVAC equipment, water heaters) shall not be permitted without verification that the truss is capable of supporting such additional loading.

❖ A truss is an engineered system of components that functions as a structural unit. Trusses are susceptible to weakening by almost any type of alteration such as boring, notching, and cutting. Unlike dimensional solid lumber members (e.g., studs, rafters, and joists), trusses are not allowed to be altered in a code-prescribed way. The only exception is attainment of written approval of a specific alteration from a registered design professional.

The placement of a water heater in an attic, for example, can result in abnormal loads on roof trusses that must be accounted for structurally. Subjecting trusses to loads that they were not designed to carry is considered as an alteration and is prohibited without a structural analysis to verify that the truss can bear the additional weight.

SECTION G2406 (303) APPLIANCE LOCATION

G2406.1 (303.1) General. Appliances shall be located as required by this section, specific requirements elsewhere in this code and the conditions of the equipment and appliance listing.

❖ Section G2406 is a consolidation of the code's generally applicable location requirements and limitations. The listing for an appliance or equipment will often contain location requirements parallel with or in addition to these sections.

G2406.2 (303.3) Prohibited locations. Fuel-fired appliances shall not be located in, or obtain combustion air from, any of the following rooms or spaces:

1. Sleeping rooms.
2. Bathrooms.
3. Toilet rooms.
4. Storage closets.

Exceptions:

1. Direct-vent appliances that obtain all combustion air directly from the outdoors.
2. Vented room heaters, wall furnaces, vented decorative appliances and decorative appliances for installation in vented solid fuel-burning fireplaces, provided that the room meets the required volume criteria of Section G2407.5.
3. A single listed wall-mounted unvented room heater equipped with an oxygen depletion safety shutoff system and installed in a bathroom provided that the input rating does not exceed 6,000 Btu/h (1.76 kW) and the bathroom meets the required volume criteria of Section G2407.5.
4. A single listed wall-mounted unvented room heater equipped with an oxygen depletion safety shutoff system and installed in a bedroom provided that the input rating does not exceed 10,000 Btu/h (2.93 kW) and the bedroom meets the required volume criteria of Section G2407.5.
5. Appliances installed in an enclosure in which all combustion air is taken directly from the outdoors, in accordance with Section G2407.6. Access to such an enclosure shall be through a weather-stripped door, equipped with an approved self-closing device.

❖ The intent of this section is to prevent fuel-fired appliances from being installed in rooms and spaces where the combustion process could pose a threat to the occupants. Potential threats include depleted oxygen levels; elevated levels of carbon dioxide, nitrous oxides, carbon monoxide, and other combustion gases; ignition of combustibles; and elevated levels of flammable gases.

In small rooms such as bedrooms and bathrooms, the doors are typically closed when the room is occupied, which could allow combustion gases to build up to life-threatening levels. In bedrooms, sleeping occupants would not be alert or aware of impending danger.

If an appliance obtains combustion air from a room or space, it communicates with the atmosphere in that room or space whether or not it is installed in that room or space. An appliance might be in a room, closet, or alcove and obtain combustion air from an adjacent room, so Section G2406.2 is worded to address both the actual location of an appliance and its source of combustion air. In other words, an appliance in a closet accessed from a bedroom is no different from an appliance located within the bedroom. It is not the intent of this section to prevent combustion air from being taken from a bedroom, bathroom, etc. as evidenced in Exceptions 2, 3, and 4. For example, the volume of a bedroom could be added to the volume of other rooms for the purpose of providing indoor combustion air for an appliance not installed in a location prohibited by this section, provided that openings are installed to conjoin the space volumes in accordance with Section G2407.5.3. The point being made is that if an appliance obtains combustion air from a room, the appliance must be considered to be in such a room. In other words, the appliance combustion chamber would be open to the room if the appliance is able to draw combustion air from such room.

Exception 1 recognizes that direct-vent appliances have sealed combustion chambers and obtain all combustion air directly from the outdoors. The appliance combustion chambers do not communicate with the room atmosphere.

Exception 2 requires the room to be able to supply the necessary combustion air by infiltration as specified in Section G2407.5.

Exceptions 3 and 4 provide for the installation of a single wall-mounted unvented room heater to be installed in bathrooms and bedrooms provided that the heaters are equipped with oxygen depletion safety shutoff systems, are limited in Btu input rating, and the space is capable of providing indoor combustion air in accordance with Section G2407.5. These exceptions specify "wall-mounted" heaters and the nature of this type of heater is such that it is less likely to come in contact with bed linens, towels, clothing, and other combustibles. Exceptions 3 and 4 would not apply to room heaters that stand on the floor or fasten to a fireplace hearth or ventless firebox hearth. See Commentary Figure G2406.2.

Exception 5 would allow installation of fuel-fired appliances within a separate dedicated space that is accessed from the rooms and spaces listed in this section. A separated space containing the appliance must be open to the outdoors in accordance with Section G2407.6, and the access door to the space must be sealed in accordance with the requirements for exterior doors to prevent communication between atmospheres in the separated spaces. The door must also be self-closing; occupants cannot be relied on to keep it closed. The enclosure should not be used for storage or any other purpose. This exception can be used to prevent the hardship of relocating an appliance when an existing appliance installed in a prohibited location needs to be replaced.

Photo by D. F. Noyes Studios courtesy of DESA International

Figure G2406.2
UNVENTED WALL-MOUNTED ROOM HEATER

G2406.3 (303.6) Outdoor locations. Equipment installed in outdoor locations shall be either listed for outdoor installation or provided with protection from outdoor environmental factors that influence the operability, durability and safety of the equipment."

❖ Appliances installed outdoors must be specifically listed for outdoor installation or be protected from outdoor conditions that will affect the "operability, durability, and safety of the equipment."

The concern is for weather and ambient temperatures. The manufacturer's instructions and listing must be consulted to determine whether a particular appliance is designed for or can be made suitable for outdoor installation. For example, furnaces cannot be installed outdoors in cold climates regardless of any weatherproof enclosure, unless the heat exchanger, burner assemblies, and venting system are designed for exposure to temperatures below normal indoor temperatures. Cold ambient temperatures can cause harmful condensation to occur on heat exchanger surfaces of fuel-fired appliances.

Additionally, there may be local ordinances that govern the outdoor installation of appliances. Before installing an appliance outdoors, consult local zoning regulations, ordinances, and subdivision covenants. Many of these regulations strictly limit the location of outdoor mechanical equipment and appliances. Also, for roof installations, the roof structure must be able to support all imposed static and dynamic structural loads. Substantiating data on the structural adequacy of the entire installation must be submitted and approved.

SECTION G2407 (304) COMBUSTION, VENTILATION AND DILUTION AIR

G2407.1 (304.1) General. Air for combustion, ventilation and dilution of flue gases for gas utilization equipment installed in buildings shall be provided by application of one of the methods prescribed in Sections G2407.5 through G2407.9. Where the requirements of Section G2407.5 are not met, outdoor air shall be introduced in accordance with one of the methods prescribed in Sections G2407.6 through G2407.9. Direct-vent appliances, gas appliances of other than natural draft design and vented gas appliances other than Category I shall be provided with combustion, ventilation and dilution air in accordance with the equipment manufacturer's instructions.

Exception: Type 1 clothes dryers that are provided with makeup air in accordance with Section G2439.4.

❖ The provisions of Section G2407 describe requirements for the combustion air necessary for the complete combustion of fuel gas, dilution of flue gases, and ventilation of gas-fired appliances and the space in which they are installed. An inadequate combustion air supply to gas-fired appliances can compromise safety by causing incomplete combustion resulting in appliance malfunction and production of excess carbon monoxide.

Complete combustion of fuel gas is essential for the proper operation of gas-fired appliances, control of harmful emissions, and for achieving maximum fuel efficiency. Combustion air provides, among other things, the oxygen necessary for the complete and efficient burning of fuel. If insufficient quantities of oxygen are supplied, the combustion process will be incomplete, creating dangerous by-products and wasting energy in the form of unburned fuel (hydrocarbons). The by-products of incomplete combustion are poisonous, corrosive, and combustible and can cause serious appliance or equipment malfunctions, which pose fire or explosion hazards.

Although not implied in the term, combustion air also serves other purposes in addition to providing oxygen. Combustion air ventilates and cools appliances and the rooms or spaces that enclose them. Combustion air also plays an important role in producing and controlling draft in vents and chimneys. An insufficient combustion air supply could cause the appliance to overheat and/or discharge combustion by-products into the building. The depletion of available oxygen promotes incomplete fuel combustion, which in turn almost guarantees serious consequences.

Despite the fact that an adequate combustion air supply is extremely important, it is one aspect of gas-fired equipment installations that is often overlooked, ignored, or compromised. Depending on the appliance type, location, and building construction, supplying combustion air can either be easy or can involve complex designs and extraordinary methods. In any case, the importance of a proper combustion air supply cannot be overemphasized.

The methods of supplying combustion air range from simple (inherently more dependable) methods to more complex methods. This section offers five methods for providing combustion air. See Commentary Figure G2407.1(1).

1. All indoor air
2. All outdoor air
3. Combination indoor and outdoor air
4. Mechanical combustion air supply
5. Engineered design

The last sentence clarifies that this section is intended to apply only to natural draft atmospheric burner design appliances, Category I appliances, and non-direct vent appliances. See the commentary for Section G2407.8. For example, the application of Section G2407.6 with large power burner equipped boilers could result in excessively large openings in an outside wall, causing environmental problems within the boiler room. Direct-vent appliances provide their own combustion air by means of an outdoor air intake pipe or duct. See Commentary Figures G2407.1(2) and G2407.1(3).

The exception recognizes that clothes dryers receive the required combustion air via the makeup air that compensates for the exhaust air flow from the appliance.

G2407.2 (304.2) Appliance/equipment location. Equipment shall be located so as not to interfere with proper circulation of combustion, ventilation and dilution air.

❖ The provisions of Sections G2407 rely on the natural (gravity) movement of air; therefore, installations must allow unimpeded flow of combustion air from the source to the appliances being supplied. Other mechanical systems, appliances, or equipment in the same room or building can adversely affect the combustion air supply. For example, when placed too close to the combustion air openings, gas-fired appliances or equipment will restrict the free circulation of air.

G2407.3 (304.3) Draft hood/regulator location. Where used, a draft hood or a barometric draft regulator shall be installed in the same room or enclosure as the equipment served so as to prevent any difference in pressure between the hood or regulator and the combustion air supply.

❖ It is important that draft hoods and barometric draft regulators be located in the same pressure zone as the appliance combustion chamber. Draft hoods and barometric regulators both serve to stabilize draft through an appliance and in some cases act to relieve back-pressure/downdraft in the venting system. It is the draft in a Category I appliance that moves combustion air into the appliance and conveys the combustion gases to the venting system. Pressure differences between the atmospheres in which the draft hood/regulator and the combustion chamber are located could interfere with the function of the draft hood/regulator resulting in poor combustion, excess draft, insufficient draft, combustion gas spillage, and/or hazardous appliance operation. A draft hood/regulator is installed under the assumption that the combustion chamber it serves sees the same atmospheric pressure.

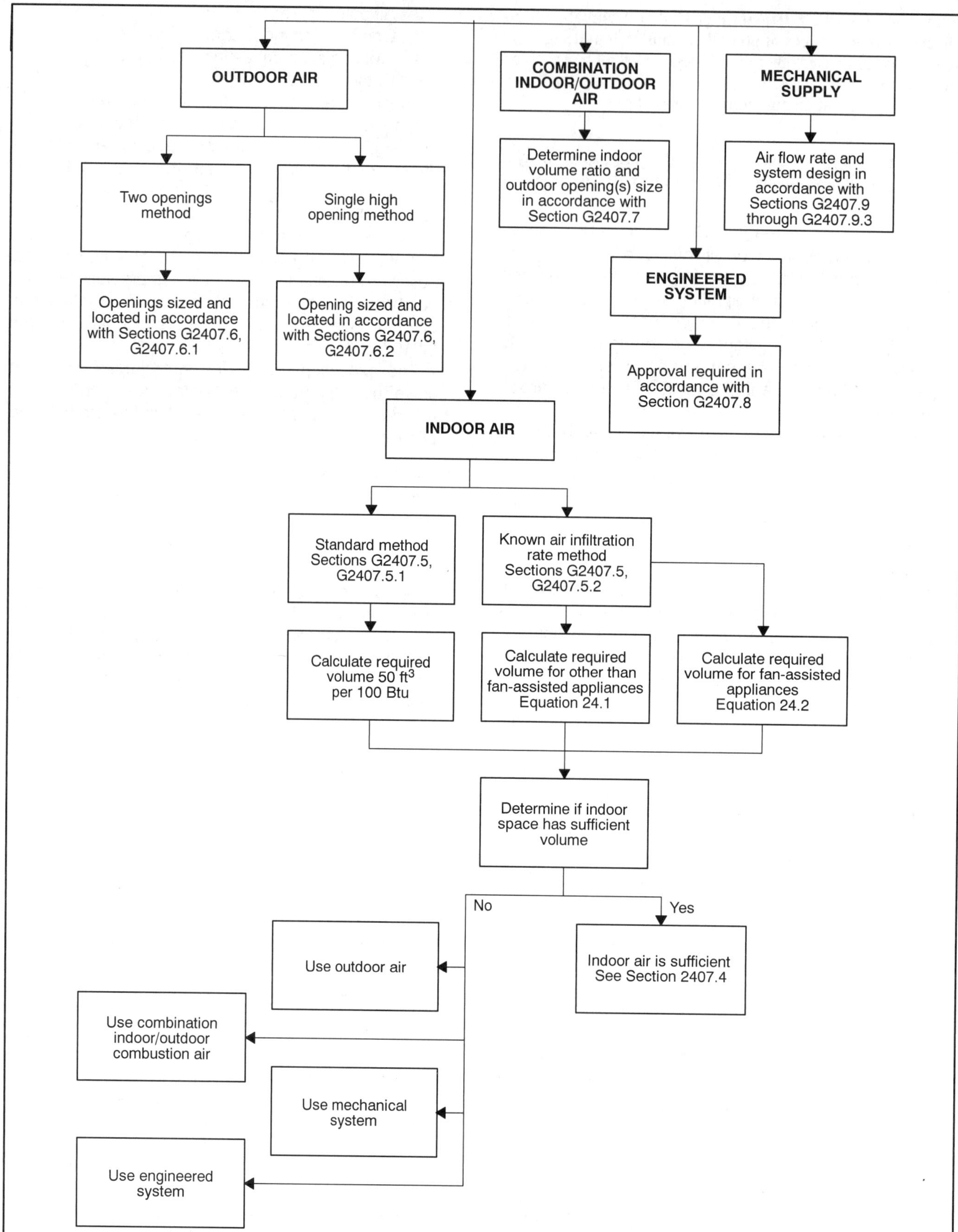

Figure G2407.1(1)
COMBUSTION AIR SUPPLY METHODS

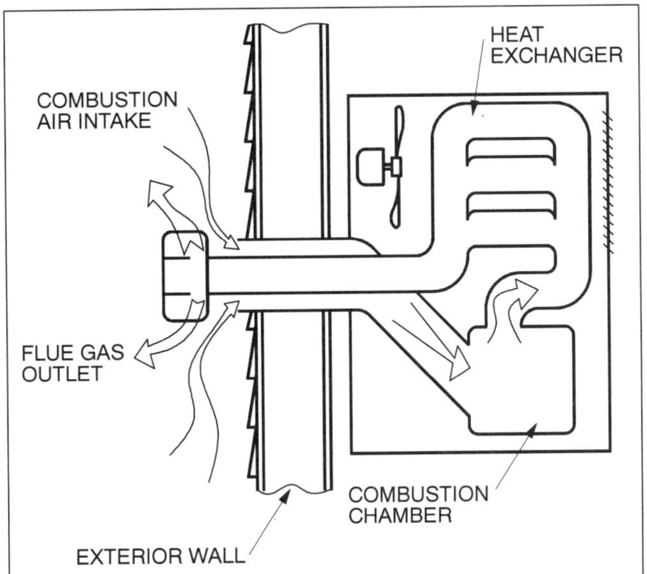

Figure G2407.1(2)
DIRECT-VENT APPLIANCE

G2407.4 (304.4) Makeup air provisions. Air requirements for the operation of exhaust fans, kitchen ventilation systems, kitchen ventilation systems, clothes dryers and fireplaces shall be considered in determining the adequacy of a space to provide combustion air requirements.

❖ The introduction of makeup air is critical to the proper operation of all fuel-burning appliances located in areas subject to the effects of exhaust systems. Too little makeup air will cause excessive negative pressures to develop, not only reducing the exhaust airflow but also interfering with appliance venting. Too little makeup air can cause loss of draft in appliance vents and chimneys or cause combustion by-products to discharge into the building. It is the draft produced in an appliance that causes combustion air to enter the combustion chamber of natural draft atmospheric burner appliances. It is the draft in the venting system that causes the combustion gases and dilution air to move through the venting system toward the vent terminal. A lack of combustion air can result in incomplete fuel combustion, appliance malfunction, and flue gas spillage.

A significant reduction in building pressure could be created by fireplaces, exhaust fans, ventilation systems, clothes dryers, and similar appliances and equipment. Gas-fired appliances are often in competition with other equipment or systems for the available combustion air and makeup air that flows through (infiltrates) the building envelope or is otherwise introduced into a building, room, or space. The competition between powered exhaust equipment and natural draft gas-fired appliance venting systems is an unfair contest at best. Eventually, the powered equipment will starve the natural draft appliances unless provisions are made to compensate for the effect of the powered exhaust equipment/appliances. Providing makeup air will prevent negative pressures within a space, thus negating the effect of exhaust fans and clothes dryers. In many cases, makeup air provisions would be required since this is the only way to prevent exhaust systems from negatively affecting the appliance venting systems that draw combustion air from the same space from which the exhaust systems remove air.

Natural draft appliances also compete among themselves for combustion air. The appliance venting system that produces the strongest draft, such as a solid-fuel appliance or fireplace, can cause combustion air shortages for the appliance venting systems that produce a weaker draft.

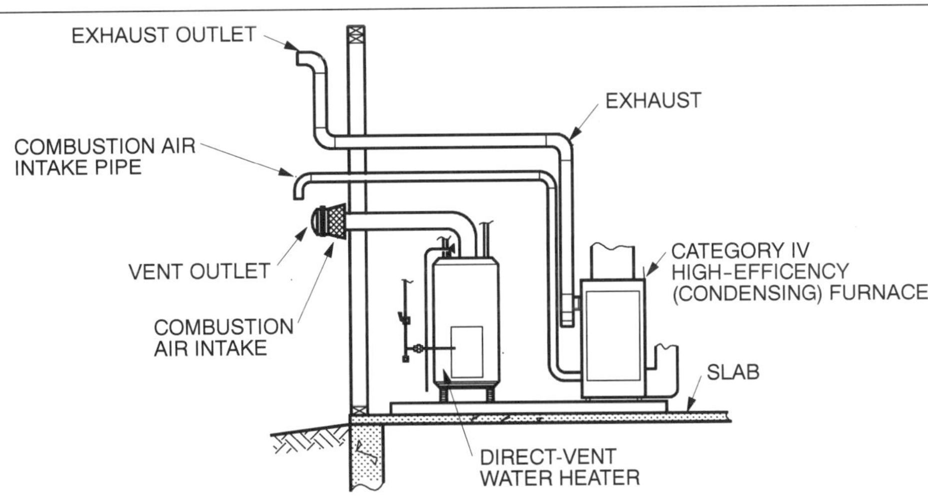

Figure G2407.1(3)
DIRECT COMBUSTION AIR CONNECTION TO THE OUTDOORS

Exhaust fans and similar equipment and appliances can produce significant negative building pressures that can interfere with the operation of vents and chimneys. This interference can cause reverse flow as outdoor air enters the building through the vents and chimneys because of the pressure difference. Any such interference with vents or chimneys could cause products of combustion to be discharged into the building and, therefore, must be avoided. The provisions of Section G2407 could be woefully adequate or nullified where compensating features are not provided to mitigate the effect of appliances or equipment removing air from any room or space. When the system depends on infiltration as the source of combustion air, it will probably be necessary to provide makeup air to offset the deficiency in many buildings. It is obvious that natural draft appliance installations cannot compete with exhaust fans and clothes dryers for the available infiltration air. Without makeup air, the gas-fired appliance venting systems and the mechanical exhaust systems will indeed compete for the same infiltration air, which could be insufficient to satisfy either need. See Commentary Figure G2407.4. Note that exhaust fans, clothes dryers and similar equipment that removes air from a space will have no negative impact on unvented appliances since they have no venting system. Direct-vent appliances are also immune.

G2407.5 (304.5) Indoor combustion air. The required volume of indoor air shall be determined in accordance with Section G2407.5.1 or G2407.5.2, except that where the air infiltration rate is known to be less than 0.40 air changes per hour (ACH), Section G2407.5.2 shall be used. The total required volume shall be the sum of the required volume calculated for all appliances located within the space. Rooms communicating directly with the space in which the appliances are installed through openings not furnished with doors, and through combustion air openings sized and located in accordance with Section G2407.5.3, are considered to be part of the required volume.

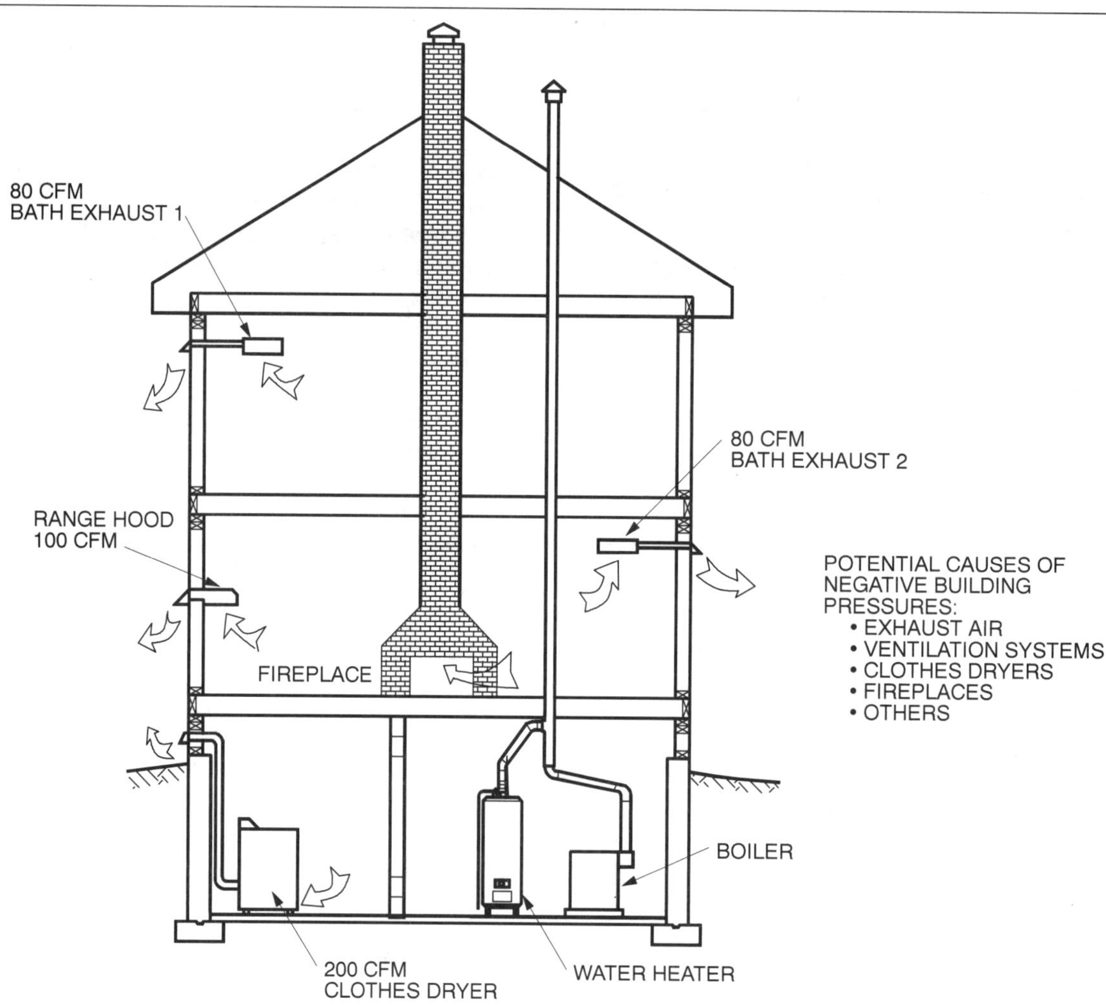

For SI: 1 cubic foot per minute = 0.0004719 m^3/s

Figure G2407.4
DEPLETION OF COMBUSTION AIR

❖ The terms "confined space," "unconfined space," and "unusually tight construction" are no longer used in the code. The required room volume is determined by two methods—one method based on the actual air infiltration rate of the building and the other based on the familiar fixed ratio of 50 ft^3 per 1000 Btu/h. Air taken directly from inside the building is an acceptable source of combustion air, provided that a sufficient volume of air is available for the appliances served and provided that the building construction allows sufficient infiltration. The provisions of this section rely on building envelope infiltration as the only source of combustion air.

The method given in Section G2407.5.1 is based on a conservative assumed air infiltration rate of at least 0.40 air changes per hour (ACH); therefore, if the ACH is known to be less than 0.40, this method cannot be used. The air infiltration rate for most buildings is not known, and it is possible for the ACH rate of a building to be actually less than 0.40. However, research indicates that buildings having less than 0.40 ACH are rare. It is believed that 80 to 95 percent of U.S. homes have infiltration rates of 0.35 ACH or greater. A rate of 0.40 ACH or less is the quantitative expression of "tight construction." Testing also indicates that the 50 ft^3/1000 Btu/h convention is somewhat liberal; thus, a built-in safety factor exists. If the designer suspects that the building in question is extraordinarily tight (less than 0.40 ACH), calculations and/or testing should be done to verify the ACH rate. A conservative approach could be to use the method in Section G2407.5.2 with a conservative ACH rate such as 0.25, 0.30, or 0.35. Commentary Table G2407.5.2 provides calculated required volumes for appliances up to 300,000 Btu/h input rating based on the standard method and known ACH method for multiple ACH rates.

The standard method can be calculated for each appliance and then all the volumes are added, or it can be calculated by adding all of the appliance inputs first and then calculating the total volume. The known ACH rate method must be performed in at least two distinct calculations where both fan-assisted and draft hood-equipped appliances are present.

To increase the available volume, the code allows rooms and spaces to be coupled together by doorways without doors and by openings installed in accordance with Section G2407.5.3.

G2407.5.1 (304.5.1) Standard method. The minimum required volume shall be 50 cubic feet per 1,000 Btu/h (4.8 m^3/kW).

❖ This is the optional default method intended to be used when the ACH rate is unknown. If the actual ACH rate is known to be greater than 0.40, the method of Section G2407.5.2 will yield smaller required volumes.

If the ACH rate is known to be less than 0.40, this method must not be used because it is based on an assumed ACH rate of at least 0.40. The rate of 50 ft^3/1,000 Btu/h is a carryover from the previous editions of the code and was used for buildings not of unusually tight construction.

G2407.5.2 (304.5.2) Known air-infiltration-rate method. Where the air infiltration rate of a structure is known, the minimum required volume shall be determined as follows:

For appliances other than fan assisted, calculate volume using Equation 24-1.

$$\text{Required Volume}_{\text{other}} \geq \frac{21\ \text{ft}^3}{ACH}\left(\frac{I_{other}}{1,000\ \text{Btu/hr}}\right)$$

(Equation (24-1)

For fan-assisted appliances, calculate volume using Equation 24-2.

$$\text{Required Volume}_{\text{fan}} \geq \frac{15\ \text{ft}^3}{ACH}\left(\frac{I_{fan}}{1,000\ \text{Btu/hr}}\right)$$

(Equation (24-2)

where:

I_{other} = All appliances other than fan assisted (input in Btu/h)

I_{fan} = Fan-assisted appliance (input in Btu/h).

ACH = Air change per hour (percent of volume of space exchanged per hour, expressed as a decimal).

For purposes of this calculation, an infiltration rate greater than 0.60 ACH shall not be used in Equations 24-1 and 24-2.

❖ This method considers the actual or calculated ACH rate and requires space volumes commensurate with such rate. This method can also be used when the ACH rate is unknown by simply picking a conservative ACH rate (e.g., 0.40 ACH or less) representing the lowest anticipated ACH rate for the given building. Equation 24-1 is for draft hood-equipped appliances and reflects recent research and test results that show that such appliances need less air flow than previously assumed. The standard ratio of 50 ft^3/1,000 Btu/h that has been in codes for many years was based on an assumed combustion air flow need of 25 ft^3/1,000 Btu/h with an assumed ACH rate of 0.50. The 25 ft^3/1,000 Btu/h air flow consisted of the air required for stoichiometric combustion (10 ft^3/ft^3 of natural gas) plus excess air to assure complete combustion, plus draft hood dilution air, plus a safety factor allowance. The volume of 25 ft^3 divided by an ACH rate of 0.50 yields the familiar 50 ft^3/1,000 Btu/h ratio. Equation 24-1 is based on a total air flow need of 21 ft^3 instead of 25 ft^3 because research supports a less conservative revised volume.

Equation 24-2 is for fan-assisted appliances only and accounts for the fact that fan-assisted appliances have no draft hood and, therefore, do not need dilution air. Fan-assisted appliances need only air for stoichiometric combustion (10 ft^3/ft^3 of natural gas) and excess air (5 ft^3/ft^3 of natural gas). Dividing by the ACH rate adjusts the required volume to correspond to the available infiltration rate.

TABLE 304.5.2
CALCULATED VOLUMES

Standard Method Required Volume All Appliances		Known Air infiltration Rate Method, Minimum Space Volume for Appliance Other Than Fan-Assisted, For Specified Infiltration Rates (ACH*)				Known Air Infiltration Rate Method, Minimum Space Volume for Fan-Assisted Appliance, For specified Infiltration Rates (ACH*)			
Appliance Input Btu/hr	Space Volume (ft³)	Appliance Input Btu/hr	Space Volume (ft³) 0.25 ACH	Space Volume (ft³) 0.30 ACH	Space Volume (ft³) 0.35 ACH	Appliance Input Btu/hr	Space Volume (ft³) 0.25 ACH	Space Volume (ft³) 0.30 ACH	Space Volume (ft³) 0.35 ACH
5,000	250	5,000	420	350	300	5,000	300	250	214
10,000	500	10,000	840	700	600	10,000	600	500	429
15,000	750	15,000	1,260	1,050	900	15,000	900	750	643
20,000	1,000	20,000	1,680	1,400	1,200	20,000	1,200	1,000	857
25,000	1,250	25,000	2,100	1,750	1,500	25,000	1,500	1,250	1.071
30,000	1,500	30,000	2,520	2,100	1,800	30,000	1,800	1,500	1,286
35,000	1,750	35,000	2,940	2,450	2,100	35,000	2,100	1,750	1,500
40,000	2,000	40,000	3,360	2,800	2,400	40,000	2,400	2,000	1,714
45,000	2,250	45,000	3,780	3,150	2,700	45,000	2,700	2,250	1,929
50,000	2,500	50,000	4,200	3,500	3,000	50,000	3,000	2,500	2,143
55,000	2,750	55,000	4,620	3,850	3,300	55,000	3,300	2,700	2,357
60,000	3,000	60,000	5,040	4,200	3,600	60,000	3,600	3,000	2,571
65,000	3,250	65,000	5,460	4,550	3,900	65,000	3,900	3,250	2,786
70,000	3,500	70,000	5,800	4,900	4,200	70,000	4,200	3,500	3,000
75,000	3,750	75,000	6,300	5,250	4,500	75,000	4,500	3,750	3,214
80,000	4,000	80,000	6,720	5,600	4,800	80,000	4,800	4,000	3,429
85,000	4,250	85,000	7,140	5,950	5,100	85,000	5,100	4,250	3,643
90,000	4,500	90,000	7,500	6,300	5,400	90,000	5,400	4,500	3,857
95,000	4,750	95,000	7,900	6,650	5,700	95,000	5,700	4,750	4,071
100,000	5,000	100,000	8,400	7,000	6,000	100,000	6,000	5,000	4,286
105,000	5,250	105,000	8,820	7,350	6,300	105,000	6,300	5,250	4,500
110,000	5,500	110,000	9,240	7,700	6,600	110,000	6,600	5,500	4,714
115,000	5,700	115,000	9,660	8,050	6,900	115,000	6,900	5,750	4,929
120,000	6,000	120,000	10,080	8,400	7,200	120,000	7,200	6,000	5,143
125,000	6,250	125,000	10,500	8,750	7,500	125,000	7,500	6,250	5,357
130,000	6,500	130,000	10,920	9,100	7,800	130,000	7,800	6,500	5,571
135,000	6,750	135,000	11,340	9,450	8,100	135,000	8,100	6,700	5,786
140,000	7,000	140,000	11,760	9,800	8,400	140,000	8,400	7,000	6,000
145,000	7,250	145,000	12,180	10,150	8,700	145,000	8,700	7,250	6,214
150,000	7,500	150,000	12,600	10,500	9,000	150,000	9,000	7,500	6,429
160,000	8,000	160,000	13,440	11,200	9,600	160,000	9,600	8,000	6,857
170,000	8,500	170,000	14,280	11,900	10,200	170,000	10,200	8,500	7,286
180,000	9,000	180,000	15,120	12,600	10,800	180,000	10,800	9,000	7,714
190,000	9,500	190,000	15,960	13,300	11,400	190,000	11,400	9,500	8,143
200,000	10,000	200,000	16,800	14,000	12,000	200,000	12,000	10,000	8,571
210,000	10,500	210,000	17,640	14,700	12,600	210,000	12,600	10,500	9,000
220,000	11,000	220,000	18,480	15,400	13,200	220,000	13,200	11,000	9,429
230,000	11,500	230,000	19,320	16,100	13,800	230,000	13,800	11,500	9,857
240,000	12,000	240,000	20,160	16,800	14,400	240,000	14,400	12,000	10,286
250,000	12,500	250,000	21,000	17,500	15,000	250,000	15,000	12,500	10,714
260,000	13,000	260,000	21,840	18,200	15,600	260,000	15,600	13,000	11,143
270,000	13,500	270,000	22,680	18,900	16,200	270,000	16,200	13,500	11,571
280,000	14,000	280,000	23,520	19,600	16,800	280,000	16,800	14,000	12,000
290,000	14,500	290,000	24,360	20,300	17,400	290,000	17,400	14,500	12,429
300,000	15,000	300,000	25,200	21,000	18,000	300,000	18,000	15,000	12,857

*ACH=Air Change per Hour

*ACH=Air Change per Hour

Table courtesy of American Gas Association

Table G2407.5.2
CALCULATED VOLUMES

For example, if the ACH rate is 0.35, 42.8 ft^3 would be required for each 1,000 Btu/h (ft^3 natural gas), and if the ACH rate is 0.50, 30 ft^3 would be required for each 1,000 Btu/h. The greater the ACH rate, the less room volume that would be needed.

Research and testing results indicate that the minimum volume for draft hood-equipped appliances must be 35 ft^3/1,000 Btu/h, and for fan-assisted appliances it must be 25 ft^3/1,000 Btu/h. This is why this section limits the infiltration rate to a maximum of 0.60 ACH, which provides a lower limit safety factor for the required volume, thereby preventing appliances from being installed in exceedingly small spaces. Commentary Table G2407.5.2 is based on the methodology of Sections G2407.5.1 and G2407.5.2 and is provided for convenience.

G2407.5.3 (304.5.3) Indoor opening size and location. Openings used to connect indoor spaces shall be sized and located in accordance with Sections G2407.5.3.1 and G2407.5.3.2 (see Figure G2407.5.3).

❖ Sections G2407.5.3.1 and G2407.5.3.2 prescribe the size and location of openings used to conjoin (couple) spaces for the purpose of increasing the available volume. The last sentence of Section G2407.5 refers to this. Such openings must be permanently open and not able to be closed except where interlocked motorized dampers are used in accordance with Section G2407.10.

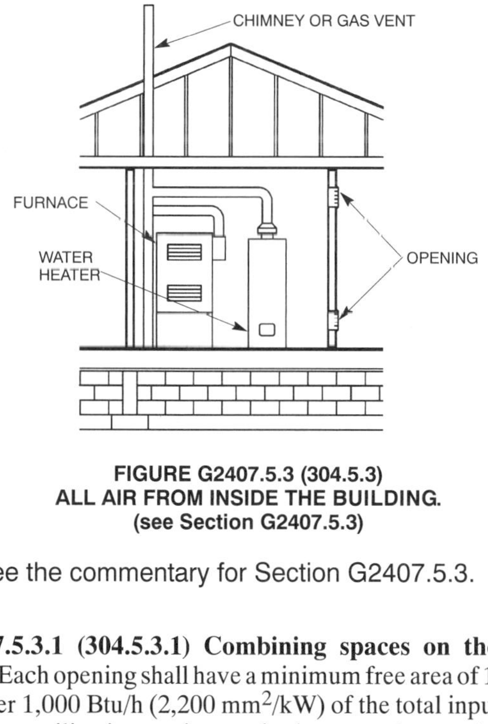

FIGURE G2407.5.3 (304.5.3)
ALL AIR FROM INSIDE THE BUILDING.
(see Section G2407.5.3)

❖ See the commentary for Section G2407.5.3.

G2407.5.3.1 (304.5.3.1) Combining spaces on the same story. Each opening shall have a minimum free area of 1 square inch per 1,000 Btu/h (2,200 mm^2/kW) of the total input rating of all gas utilization equipment in the space, but not less than 100 square inches (0.06 m^2). One opening shall commence within 12 inches (305 mm) of the top and one opening shall commence within 12 inches (305 mm) of the bottom of the enclosure. The minimum dimension of air openings shall be not less than 3 inches (76 mm).

❖ This section is applicable to adjacent spaces on the same floor level and provides for the familiar "high and low" openings. The opening configuration creates a thermosiphon air flow with the bottom opening acting as an inlet and the top opening acting as an outlet. This provision originated as a means to conjoin a small appliance enclosure (such as a furnace or boiler room) with other spaces, thereby effectively increasing the volume of such enclosures.

G2407.5.3.2 (304.5.3.2) Combining spaces in different stories. The volumes of spaces in different stories shall be considered as communicating spaces where such spaces are connected by one or more openings in doors or floors having a total minimum free area of 2 square inches per 1,000 Btu/h (4402 mm^2/kW) of total input rating of all gas utilization equipment.

❖ This section is a new concept and allows spaces on different floor levels to be conjoined, thereby effectively increasing the available volume for providing combustion air. Providing a high and low opening configuration in accordance with the previous section would serve no purpose when vertically connecting adjacent stories; thus a single opening is permitted. The opening or openings are permitted to be in a floor or in a door opening to an unenclosed stairway that connects the two stories. This method of conjoining spaces is particularly useful for dwelling units where the volume of a basement in which the appliances are located can be conjoined with the open spaces on upper stories by means of a louvered door at the basement stairs. The opening size requirement is simply the addition of the areas of the two openings required by Section G2407.5.3.1. A combustion air opening in a floor would be considered as a transfer opening in a horizontal assembly.

G2407.6 (304.6) Outdoor combustion air. Outdoor combustion air shall be provided through opening(s) to the outdoors in accordance with Section G2407.6.1 or G2407.6.2. The minimum dimension of air openings shall not be less than 3 inches (76 mm).

❖ This section provides two methods for providing combustion air from the outdoors: the traditional method by means of two direct openings or ducts to the outdoors and a newer method by means of one opening or duct to the outdoors.

Openings to spaces that are naturally ventilated with outdoor air, such as attic or crawl spaces, are considered as an acceptable alternative to a direct connection to the outside. Attic and crawl spaces can be acceptable sources of combustion air only where such spaces are provided with adequate natural ventilation openings directly to the outdoors. Attic and crawl spaces ventilated by mechanical means are not an acceptable source of combustion air.

Combustion air ducts and openings that penetrate components of wall, floor, ceiling, and roof assemblies must be installed as required by this code.

When designing combustion air installations, consideration should be given to the effect that openings to the outdoors can have on appliances, plumbing systems, and building occupants. For example, depending on the location, openings to the outdoors can:

- Cause drafts that can blow out pilot lights or otherwise interfere with appliance ignition and operation.
- Cause freezing of plumbing piping or other water-containing components.
- Cause objectionable cold drafts that encourage the occupants to block or cover the openings.

In all cases, combustion air openings should be located to reduce the likelihood they will be accidentally or intentionally blocked or covered. A ceiling transfer opening that connects a furnace room with an attic is an example of a combustion air opening that is likely to be intentionally blocked by an occupant because of the drafts that can occur in cold climates. Building occupants typically do not understand the need for or the importance of combustion air openings.

No side dimension of a square or rectangular opening and no diameter of a round opening can be less than 3 inches. The smallest allowed square opening area would be 9 square inches, and the smallest allowed round opening area would be 7 square inches. See the commentary for Section G2407.10.

G2407.6.1 (304.6.1) Two-permanent-openings method. Two permanent openings, one commencing within 12 inches (305 mm) of the top, and one commencing within 12 inches (305 mm) of the bottom, of the enclosure shall be provided. The openings shall communicate directly, or by ducts, with the outdoors or spaces (crawl or attic) that freely communicate with the outdoors.

Where directly communicating with the outdoors, or where communicating to the outdoors through vertical ducts, each opening shall have a minimum free area of 1 square inch per 4,000 Btu per hour (5.5 cm^2 per kW) of total input rating of all equipment in the enclosure. [See Figures G2407.6.1(1) and G2407.6.1(2).]

Where communicating with the outdoors through horizontal ducts, each opening shall have a minimum free area of 1 square inch per 2,000 Btu per hour (11 cm^2 per kW) of total input rating of all equipment in the enclosure. [See Figure G2407.6.1(3).]

❖ Two openings located as prescribed in this section will induce a convective air current in the room or space by admitting cooler, more dense air in the lower opening and allowing the escape of warmer, less dense air through the upper opening. The farther the openings are apart, the greater the temperature differential and the greater the convective force behind the current. A component of combustion air is cooling (ventilation) air for the appliance enclosure. The two- opening method was created to ventilate the appliance enclosure in addition to providing combustion air. This ventilation cools the appliances and would help remove any combustion gases that spilled from the appliances.

G2407.6.2 (304.6.2) One-permanent-opening method. One permanent opening, commencing within 12 inches (305 mm) of the top of the enclosure, shall be provided. The equipment shall have clearances of at least 1 inch (25.4 mm) from the sides and back and 6 inches (152 mm) from the front of the appliance. The opening shall directly communicate with the outdoors or shall communicate through a vertical or horizontal duct to the outdoors or spaces (crawl or attic) that freely communicate with the outdoors [see Figure G2407.6.2], and shall have a minimum free area of 1 square inch per 3,000 Btu per hr (7 cm^2 per kW) of the total input rating of all equipment located in the enclosure, and not less than the sum of the areas of all vent connectors in the space.

❖ Research has shown that for modern appliances, a single opening to the outdoors will perform as well as the traditional two opening method. The one opening method described in this section depends on a reduced pressure being created in the enclosure by the draft created by the venting system. This reduced pressure causes combustion air to enter the enclosure through the single opening. It is important that the opening be properly sized considering both sizing criteria: the square-inch-area-per-Btu/h ratio and the area minimum based on the sum of the areas of all vent connectors in the enclosure. This method allows for fewer openings, fewer ducts, and less objection by the owners/occupants.

G2407.7 (304.7) Combination indoor and outdoor combustion air. The use of a combination of indoor and outdoor combustion air shall be in accordance with Sections G2407.7.1 through G2407.7.3.

❖ This method of providing combustion air is a combined application of Section G2407.5 and Section G2407.6. This method allows credit for the amount of infiltration that exists and makes up for the shortage with supplemental outdoor air. In other words, in addition to obtaining combustion air directly from the outdoors, this method relies on building infiltration for a portion of the total combustion air. In spaces where the volume is insufficient to satisfy the method of Section G2407.5 or where smaller outdoor air openings than required by Section G2407.6 are desired, this method allows infiltration and outdoor openings to supplement each other. If the appliances are enclosed in a small room such as a closet, openings to adjacent spaces as prescribed by Section G2407.5.3 must be provided to couple the appliance room volume with any other space volume counted on to provide combustion air. Frequently, sufficient volume cannot be obtained in the appliance enclosure or by opening the appliance enclosure to adjacent spaces. This method of combining indoor and outdoor air is an alternative solution. Simply stated, this section uses ratios of what is required to what is actually provided so that when combined, the indoor air component and the outdoor air component add up to the whole required.

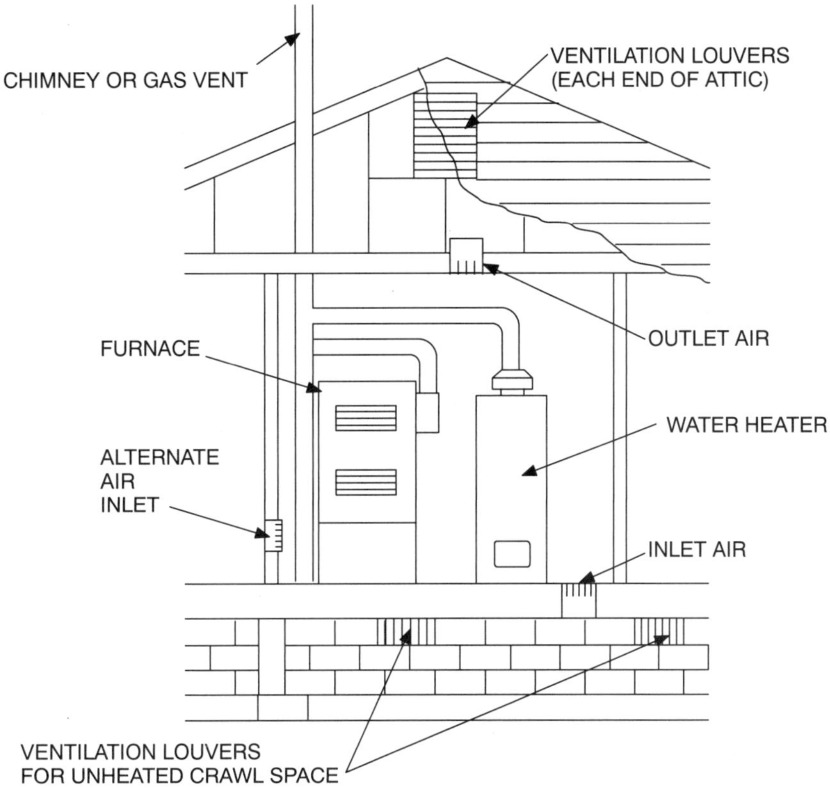

FIGURE G2407.6.1(1) [304.6.1(1)]
ALL AIR FROM OUTDOOR—INLET AIR FROM VENTILATED CRAWL SPACE AND OUTLET AIR TO VENTILATED ATTIC (see Section G2407.6.1)

❖ See the commentary for Section G2407.6.1.

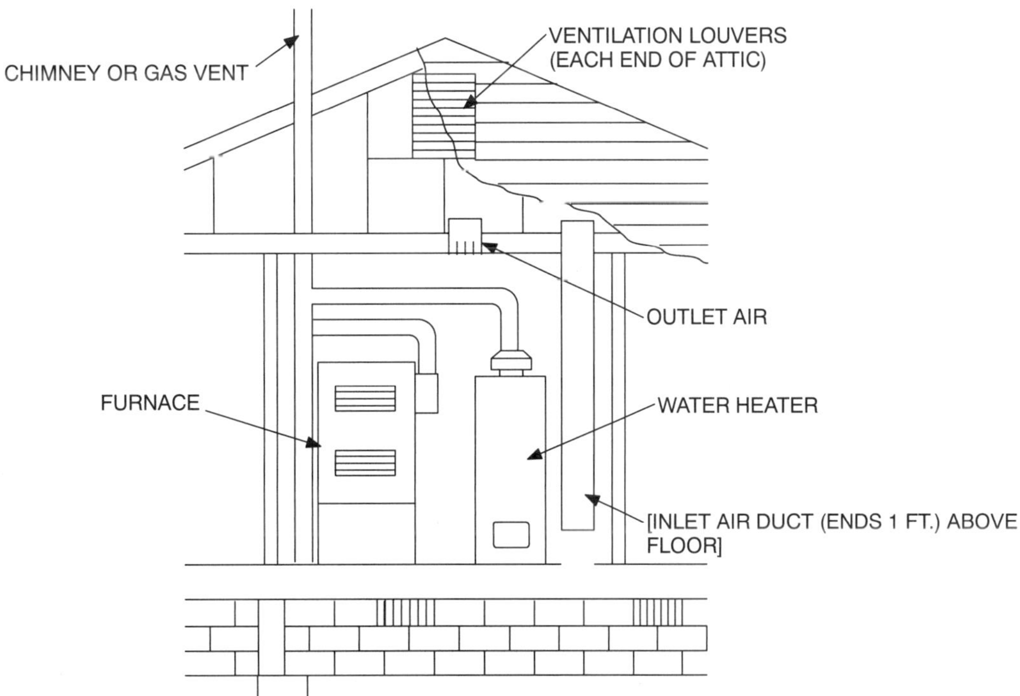

For SI: 1 foot = 304.8 mm.

FIGURE G2407.6.1(2) [304.6.1(2)]
ALL AIR FROM OUTDOORS THROUGH VENTILATED ATTIC (see Section G2407.6.1)

❖ See the commentary for Section G2407.6.1.

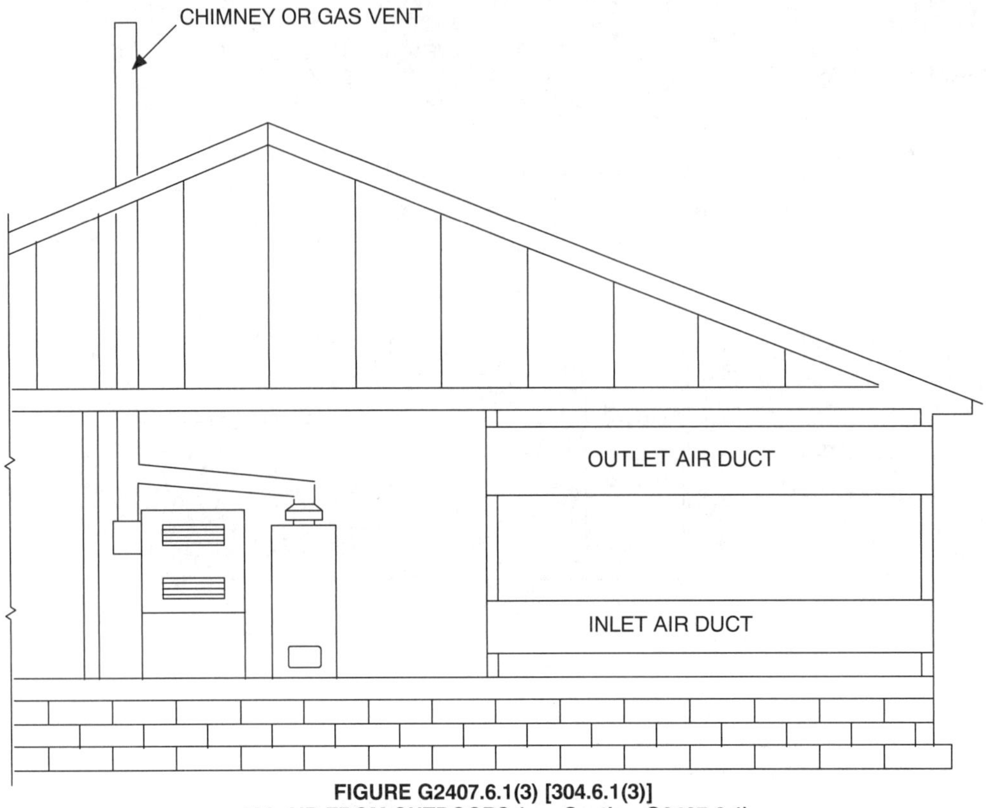

FIGURE G2407.6.1(3) [304.6.1(3)]
ALL AIR FROM OUTDOORS (see Section G2407.6.1)

❖ See the commentary for Section G2407.6.1.

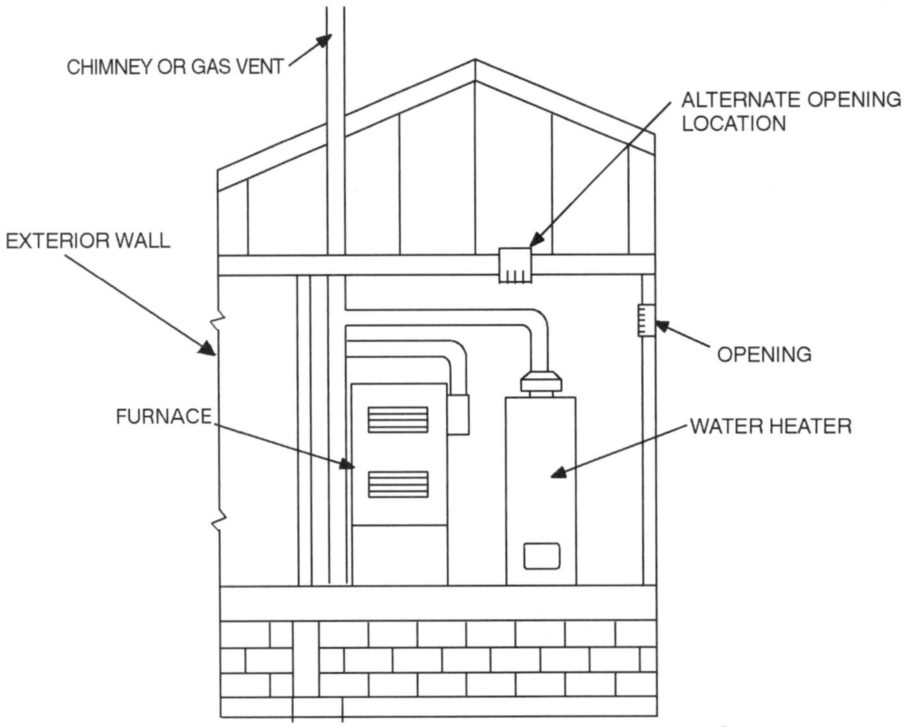

FIGURE G2407.6.2 (304.6.2)
SINGLE COMBUSTION AIR OPENING—ALL AIR FROM OUTDOORS (see Section G2407.6.2)

❖ See the commentary for Section G2407.6.2.

G2407.7.1 (304.7.1) Indoor openings. Where used, opening connecting the interior spaces shall comply with Section G2407.5.3.

❖ See the commentary for Section G2407.5.3.

G2407.7.2 (304.7.2) Outdoor opening location. Outdoor opening(s) shall be located in accordance with Section G2407.6.

❖ See the commentary for Section G2407.6.

G2407.7.3 (304.7.3) Outdoor opening(s) size. The outdoor opening(s) size shall be calculated in accordance with the following:

1. The ratio of interior spaces shall be the available volume of all communicating spaces divided by the required volume.
2. The outdoor size reduction factor shall be one minus the ratio of interior spaces.
3. The minimum size of outdoor opening(s) shall be the full size of outdoor opening(s) calculated in accordance with Section G2407.6, multiplied by the reduction factor. The minimum dimension of air openings shall be not less than 3 inches (76 mm).

❖ While the principle has not changed, this method has been simplified compared to the same provision in the 2000 edition of the code. The intent is still the same; that is, the fraction of the required indoor volume plus the fraction of outdoor openings must be equal to or greater than one. The indoor volume method and the outdoor air method can both be stand-alone methods; therefore, half of one and half of the other will work, as will $^3/_4$ of one and $^1/_4$ of the other, etc.

This section is expressed in the following equation:

$$\left[1 - \left(\frac{\text{available indoor volume}}{\text{volume required by Section G2407.5}}\right)\right] \times \text{full size openings required by Section G2407.6} = \text{reduced size outdoor air openings}$$

Examples:

Given:

7,500 feet^3 of indoor volume is required if all indoor air is to be used.

4,950 feet^3 of indoor volume is available.

One opening to the outdoors is desired and required to be 50 in^2 if all outdoor air is to be used.

$$\left[1 - \left(\frac{4950 \text{ ft}^3}{7500 \text{ ft}^3}\right)\right] \times 50 \text{ in}^2 = \text{reduced outdoor opening size}$$

$$0.34 \times 50 \text{ in}^2 = 17 \text{ in}^2 \text{ outdoor opening size}$$

Either of the provisions of Section G2407.6 when combined with the provision of this section can be used to satisfy the equation.

Example 1:

Room A is within a dwelling unit and contains a fuel-fired furnace and a fuel-fired water heater with input ratings of 150,000 Btu/h and 50,000 Btu/h, respectively.

The size of Room A is 20 feet by 30 feet, and adjacent Room B is also 20 feet by 30 feet. Both rooms have 8-foot-high ceilings. Rooms A and B communicate through two openings located in accordance with Section G2407.5.3, each having dimensions of 10 inches by 20 inches. Two 3-inch-diameter (round) direct openings to the outdoors are provided in Room A and are located in accordance with Section G2407.6.1.

Question: Do the openings to the outdoors when combined with the volumes of Rooms A and B meet the combustion air demand of the appliances?

For **SI:** 1 inch = 25.4 mm, 1 foot = 305 mm, 1 Btu/h = 0.2931 W, 1 cubic foot = 0.0283 m^3, 1 square inch = 645 mm^2.

Given:

The air infiltration rate of the building is unknown.

Volume of Room A = 4,800 ft^3

Volume of Room B = 4,800 ft^3

Area of 3-inch-diameter direct openings = 7.07 in^2

Area of each opening between Rooms A and B = 200 in^2

Total input rating of appliances = 200,000 Btu/h

Step 1: Determine whether Rooms A and B meet the volume requirements of Section G2407.5. Available volume of Rooms A and B = 4,800 ft^3 + 4,800 ft^3 = 9,600 ft^3

The required volume is

$$\frac{200,000 \text{ Btu/h}}{1000 \text{ Btu/h}} \times 50 \text{ ft}^3 = 10,000 \text{ ft}^3$$

Rooms A and B combined do not meet the volume requirements of Section G2407.5.

Step 2: Determine whether the openings between Rooms A and B meet the requirements of Section G2407.5.3.

The minimum required area is

$$\frac{200,000 \text{ Btu/h}}{1000 \text{ Btu/h}} \times 1 \text{ in}^2 = 200 \text{ in}^2$$

Actual area of each opening = 200 in^2

The area of each opening meets the size requirement of Section G2407.5.3.

Step 3: Determine the required area of each direct outdoor opening in accordance with Section G2407.6.1

Required area of each opening =

$$\frac{200,000 \text{ Btu/h}}{4000 \text{ Btu/h}} \times 1 \text{ in}^2 = 50 \text{ in}^2$$

Step 4: Determine whether the 3-inch-diameter direct openings combined with the volumes of Rooms A and B comply with Section G2407.7.3.

$$\left[1 - \left(\frac{9600\ \text{ft}^3}{10{,}000\ \text{ft}^3}\right)\right] \times 50 = (1 - 0.96) \times 50 =$$

$$0.04 \times 50 = 2\ \text{in}^2$$

The actual area of each direct opening (7.07 in^2) exceeds the required area of 2 in^2.

Therefore, the combination of the 3-inch-diameter direct openings and the volumes of Rooms A and B satisfies the combustion air demand of the appliances.

G2407.8 (304.8) Engineered installations. Engineered combustion air installations shall provide an adequate supply of combustion, ventilation and dilution air and shall be approved.

❖ Potential sources of combustion air are the indoor atmosphere in which the appliance is located, outdoor openings and ducts that communicate with the outdoor atmosphere, a combination of these, or any other approved arrangement for the dependable introduction of outdoor air into the building or directly to the fuel-burning appliance.

The provisions of Section G2407 are intended to be applied only to natural draft atmospheric burner appliances and equipment. Gas-fired appliances, especially those that are not of the atmospheric-burner type, are commonly provided with combustion air by engineered systems. For example, special-engineered systems are usually designed for appliances equipped with power burners. Power burners force combustion air into the burner under pressure and are, therefore, not dependent upon gravity airflow. In some applications, engineered combustion air supplies have advantages and can be preferable to large gravity (natural) openings to the outdoors. All engineered designs must be evaluated for code compliance, approved, and installed in accordance with the appliance manufacturer's instructions. See Commentary Figures G2407.9(1) through G2407.9(3).

G2407.9 (304.9) Mechanical combustion air supply. Where all combustion air is provided by a mechanical air supply system, the combustion air shall be supplied from the outdoors at a rate not less than 0.35 cubic feet per minute per 1,000 Btu/h (0.034 m^3/min per kW) of total input rating of all appliances located within the space.

❖ This method provides combustion air by means of a fan/blower that runs when any of the served appliances are in a firing cycle. The fan/blower must be sized to supply the required air flow based on the simultaneous operation of all appliances that are served by such fans/blowers. The total Btu/h input rating of all fuel-burning appliances located in the room or space must be used because the potential exists for all of the appliances to be operating at the same time. This method is used where gravity openings to the outdoors are impractical or undesirable. A small fan-powered intake opening can substitute for comparatively large gravity openings, especially where freezing temperatures are a problem or there is no room for the required size gravity openings. Commentary Figures G2407.9(1), (2), and (3) illustrate factory engineered mechanical combustion air supply units.

The prescribed air flow rate of 0.35 cfm/1000 Btu/h is equivalent to 1 cfm/2867 Btu/h and is based on a combustion air need of 21 ft^3 per ft^3 of natural gas (1000 Btu/h). The rate is derived as follows:

$$\frac{2857\ \text{Btu/h}}{1000\ \text{Btu/ft}^3} = 2.857\ \text{ft}^3\ \text{of gas per hour}$$

2.857 ft^3 of gas per hour X 21 ft^3 of air per ft^3 of gas = 60 ft^3 of air per/hour = 60 CFH

$$\frac{60\ \text{CFH}}{60\ \text{min/hr}} = 1\ \text{CFM}$$

Thus, 1 cfm is required for each 2857 Btu/h or 0.35 cfm/1000 Btu/h.

G2407.9.1 (304.9.1) Makeup air. Where exhaust fans are installed, makeup air shall be provided to replace the exhausted air.

❖ This section expresses the same concern as Section G2407.4. See the commentary for Section G2407.4. If the air that is supplied by the combustion air fan becomes makeup air for an exhaust fan, the combustion air will not be available to the appliance(s) for which the air was intended.

G2407.9.2 (304.9.2) Appliance interlock. Each of the appliances served shall be interlocked with the mechanical air supply system to prevent main burner operation when the mechanical air supply system is not in operation.

❖ This section requires an interlock circuit as opposed to simple parallel operation wiring. In other words, the combustion air fan/blower must start when initiated by the means that controls the appliance operation cycles (e.g., thermostat), and only after the combustion air fan/blower is proven to be operating will the appliance be allowed to start a firing cycle. If the combustion air fan/blower fails during a firing cycle of an appliance, the appliance firing cycle must be terminated. See Commentary Figure G2407.9.2.

The following steps must occur in this order:

1. Call for heat
2. Combustion air fan/blower starts
3. Combustion air fan/blower is proven to be operating
4. Proving controls allow appliance to fire

Photo courtesy of Tjernlund Products, Inc.

Figure G2407.9(1)
MECHANICAL COMBUSTION AIR/MAKEUP AIR SUPPLY SYSTEM

Photo courtesy of Tjernlund Products, Inc.

Figure G2407.9(2)
MECHANICAL COMBUSTION AIR SUPPLY SYSTEM

Photo courtesy of Tjernlund Products, Inc.

Figure G2407.9(3)
MECHANICAL COMBUSTION AIR/MAKEUP AIR SUPPLY SYSTEM

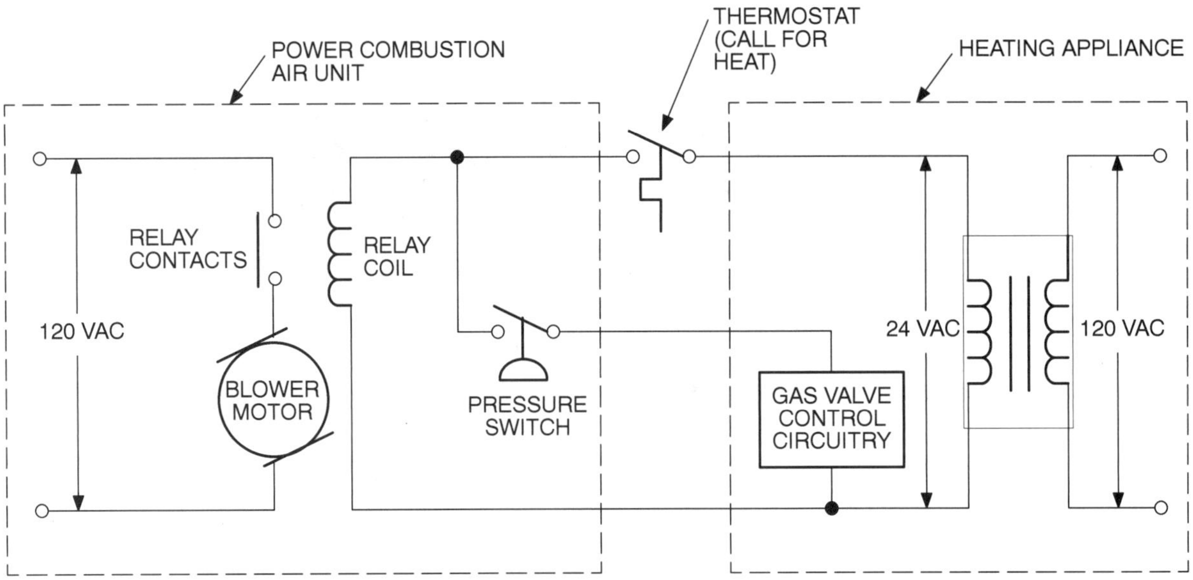

Figure G2407.9.2
APPLIANCE INTERLOCK

G2407.9.3 (304.9.3) Combined combustion air and ventilation air system. Where combustion air is provided by the building's mechanical ventilation system, the system shall provide the specified combustion air rate in addition to the required ventilation air.

❖ If combustion air is mechanically supplied through an HVAC system, such system must deliver the required combustion air rate above and beyond any outdoor air rate provided for ventilation of the occupied spaces. The HVAC system must also be interlocked with the appliance(s) as required by Section G2407.9.2. Neither continuous operation nor parallel operation of the HVAC system satisfies the requirement for a true interlock as required by Section G2407.9.2.

G2407.10 (304.10) Louvers and grilles. The required size of openings for combustion, ventilation and dilution air shall be based on the net free area of each opening. Where the free area through a design of louver, grille or screen is known, it shall be used in calculating the size opening required to provide the free area specified. Where the design and free area of louvers and grilles are not known, it shall be assumed that wood louvers will have 25-percent free area and metal louvers and grilles will have 75-percent free area. Screens shall have a mesh size not smaller than $^1/_4$ inch. Nonmotorized louvers and grilles shall be fixed in the open position. Motorized louvers shall be interlocked with the equipment so that they are proven to be in the full open position prior to main burner ignition and during main burner operation. Means shall be provided to prevent the main burner from igniting if the louvers fail to open during burner startup and to shut down the main burner if the louvers close during operation.

❖ This section recognizes that louvers and grilles are usually installed over air inlets and outlets to prevent rain, snow, and animals from entering the building. When louvers or grilles are used, the solid portion of the louver or grille must be considered when determining the unobstructed (net clear) area of the opening.

Combustion air openings are sized based on there being a free, unobstructed area for the passage of air into the space where the fuel-burning appliances are located. Louvers or grilles placed over these openings reduce the area of the openings because of the area occupied by the solid portions of the grille or louver. The reduction in area must be considered because only the unobstructed area can be credited toward the required opening size.

The reduction in opening area caused by the presence of grilles or louvers will always require openings to be larger than determined from the sizing ratios of this chapter and larger than any duct of the minimum required size that might connect to these openings.

This section does not apply to grilles, louvers, or screens that are an integral component of a labeled appliance that is installed in accordance with the manufacturer's instructions. Where the manufacturer states the free, unobstructed area of a grille or louver, it is acceptable to consider such area as actual without requiring further calculation.

The code also requires the free area of screens to be considered in the opening sizing calculations. If it is unknown, it will have to be calculated since there is no stated default area for screens as there is for louvers and grilles. The code prohibits screens having mesh sizes smaller than $^1/_4$ inch (6.35 mm). Screens with a mesh size of $^1/_4$ inch and larger are not insect screens and would be suitable only for keeping out large debris, rodents, and other small animals. Smaller size mesh is likely to become obstructed by lint, plant fibers, cottonwood seeds, etc. At best, placing screens in combustion air openings is risky, and it is preferable to rely on grilles and louvers.

Metal louvers are constructed so that the solid portion of the louver occupies approximately 25 percent of the opening area, leaving 75 percent of this area unobstructed. Wood louvers occupy approximately 75 percent of the opening area, leaving 25 percent of the area unobstructed. When the required amount of unobstructed area is determined, the effect that a louver has on that area must be taken into account to make certain the minimum amount of unobstructed opening is achieved. The material type, thickness, and spacing all have a significant effect on the actual net free area of any louver or grille.

To prevent freezing of pipes, to save energy, and to maintain comfortable temperatures in appliance rooms, outdoor openings are often provided with motorized dampers that close during the appliance(s)'s off-cycle. The damper motors and/or damper blades must be supervised so that an interlock can be established between the dampers and the appliances, thereby preventing appliance operation when the dampers are closed.

Combustion air openings must not be closable unless they are opened and closed automatically by a mechanism that is interlocked with the appliances. See Commentary Figure G2407.10. Damper supervision is easily accomplished by damper motor end switches or blade position travel switches that monitor the position of the drive motor shaft, the damper/motor linkage hardware, or the damper blades. Such switches provide feedback to the appliance control circuitry to allow appliance firing only when the dampers are proven to be fully open.

G2407.11 (304.11) Combustion air ducts. Combustion air ducts shall comply with all of the following:

1. Ducts shall be of galvanized steel complying with Chapter 16 of this code or of equivalent corrosion-resistant material approved for this application.

   Exception: Unobstructed stud and joist spaces shall not be prohibited from conveying combustion air, provided that not more than one required fireblock is removed.

2. Ducts shall terminate in an unobstructed space allowing free movement of combustion air to the appliances.
3. Ducts shall serve a single enclosure.
4. Ducts shall not serve both upper and lower combustion air openings where both such openings are used. The separation between ducts serving upper and lower combustion air openings shall be maintained to the source of combustion air.
5. Ducts shall not be screened where terminating in an attic space.
6. Horizontal upper combustion air ducts shall not slope downward toward the source of combustion air.
7. The remaining space surrounding a chimney liner, gas vent, special gas vent or plastic piping installed within a masonry, metal or factory-built chimney shall not be used to supply combustion air.

   Exception: Direct-vent gas-fired appliances designed for installation in a solid fuel-burning fireplace where installed in accordance with the listing and the manufacturer's instructions.

8. Combustion air intake openings located on the exterior of a building shall have the lowest side of such openings

located not less than 12 inches (305 mm) vertically from the adjoining grade level.

❖ This section addresses the construction and installation of combustion air ducts.

Item 1 requires combustion air ducts to be constructed of galvanized sheet steel or an approved equivalent material. The intent is to cause ducts to be constructed of a material that is resistant to corrosion and physical damage, which would allow such ducts to remain in place undamaged for the life of the installation. It is doubtful that the authors of item 1 ever anticipated the use of flexible ducts and the higher friction losses associated with such non-smooth-wall ducts. It is reasonable to assume that all of the sizing criteria in Section G2407 is based on smooth wall ducts. The exception expresses a concern for fire safety where stud cavities are used as a duct.

Item 2 is intended to prevent blockages and minimize resistance (friction loss) to airflow.

Item 3 prohibits a duct from serving more than one appliance enclosure because the duct and opening sizing criteria are based on a single duct serving a single opening in an appliance location. Multiple openings supplied by a single duct could result in unpredictable airflow and might not create the convective movements of air intended by the combustion air methods.

Item 4 recognizes that where both upper and lower combustion air openings are required, air circulation cannot occur if a single duct serves both combustion air openings. The lower opening functions as an air inlet and the upper opening as an air outlet, thus independent ducts are required. See Commentary Figure G2407.11.

Item 5 is intended to prevent obstruction or blockage by insulation materials.

Item 6 is intended to prevent trapping or sloping of the upper combustion air opening in a way that would impede the convective movement of air. The horizontal duct for the upper openings must slope upward as it travels away from the appliance(s) and toward the outdoors.

Item 7: When a vent or chimney liner is installed within a chimney, an annular space will exist between the vent or liner and the interior chimney walls. This space is prohibited from being used as a conduit for conveying combustion air because of poor flow characteristics, the tendency for reverse flow to occur as the combustion air is warmed by the liner or vent, and the adverse cooling effect on the vent or liner. The exception recognizes that there could be appliances that are designed and listed for the type of installation that this item intends to prohibit.

Item 8 intends to prevent combustion air openings from being obstructed by snow, leaves and vegetation. No part of a combustion air opening is allowed to be less than 12 inches above grade.

G2407.12 (304.12) Protection from fumes and gases. Where corrosive or flammable process fumes or gases, other than products of combustion, are present, means for the disposal of such fumes or gases shall be provided. Such fumes or gases include carbon monoxide, hydrogen sulfide, ammonia, chlorine and halogenated hydrocarbons.

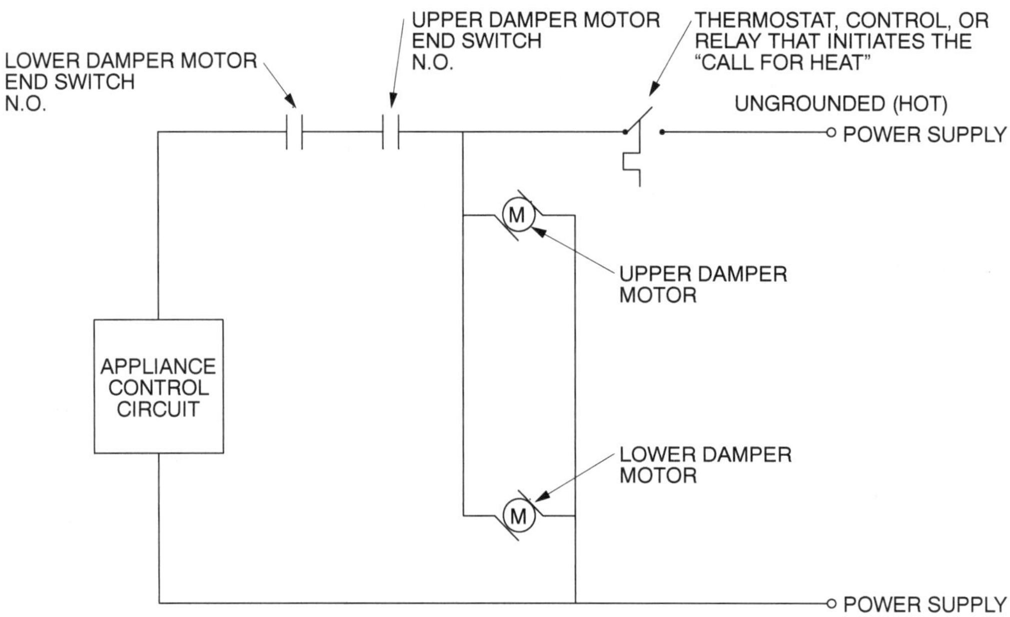

Figure G2407.10
SIMPLIFIED SAMPLE SCHEMATIC OF COMBUSTION AIR DAMPER/APPLIANCE INTERLOCK (SHOWN FOR ILLUSTRATIVE PURPOSE ONLY)

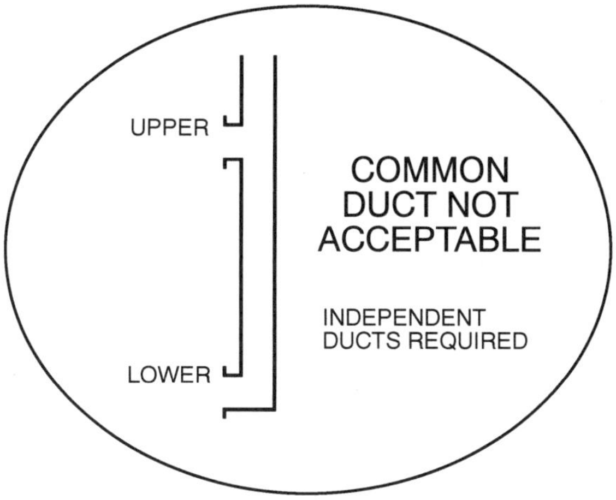

Figure G2407.11
COMMON DUCT PROHIBITED

In barbershops, beauty shops and other facilities where chemicals that generate corrosive or flammable products, such as aerosol sprays, are routinely used, nondirect-vent-type appliances shall be located in an equipment room separated or partitioned off from other areas with provisions for combustion air and dilution air from the outdoors. Direct-vent appliances shall be installed in accordance with the appliance manufacturer's installation instructions.

❖ In many occupancies, the routine use of chemicals contaminates the indoor combustion air. These contaminants can combine with water vapor in the combustion gases, producing acids and other corrosive compounds that can destroy appliance components, vents, chimneys, and connectors. For example, it is common to see vents and vent connectors for draft hood-equipped water heaters in beauty shops that have deteriorated to the point of structural failure. Other examples are swimming pool heaters exposed to chlorine, boilers exposed to laundry and dry cleaning chemicals, and boilers exposed to refrigerant leakage. Some contaminants are more toxic to life after they have passed through an appliance combustion chamber. The highly poisonous chemical phosgene (carbonyl chloride) can be produced when common refrigerants contaminate combustion air.

Obviously, choosing direct-vent appliances will eliminate this problem in most cases and will not require isolation enclosures for the appliances. It is possible that some atmospheres are unsuitable for any appliance, direct-vent or otherwise. The code provides an alternative to direct-vent appliances that requires the appliances to be isolated from the contaminated atmosphere by reasonably air-tight enclosures and requires all outdoor combustion air to be provided for the appliances.

SECTION G2408 (305) INSTALLATION

G2408.1 (305.1) General. Equipment and appliances shall be installed as required by the terms of their approval, in accordance with the conditions of listing, the manufacturer's instructions and this code. Manufacturer's installation instructions shall be available on the job site at the time of inspection. Where a code provision is less restrictive than the conditions of the listing of the equipment or appliance or the manufacturer's installation instructions, the conditions of the listing and the manufacturer's installation instructions shall apply.

Unlisted appliances approved in accordance with Section 2404.3 shall be limited to uses recommended by the manufacturer and shall be installed in accordance with the manufacturer's instructions, the provisions of this code and the requirements determined by the code official.

❖ Manufacturers' installation instructions are evaluated by the listing agency verifying compliance with the appropriate standard. The listing agency can require that the manufacturer alter, delete, or add information to the instructions as necessary to achieve compliance with applicable standards and code requirements. Manufacturers' installation instructions are an enforceable extension of the code and must be in the hands of the code official when an inspection takes place.

When an appliance is tested to obtain a listing and label, the approved agency installs the appliance in accordance with the manufacturer's installation instructions. The appliance is tested under these conditions; thus, the installation instructions become an integral part of the labeling process.

The listing and labeling process assures that the appliance and its installation instructions are in compliance with applicable standards. Therefore, an installation in accordance with the manufacturer's instructions is required, except where the code requirements are more stringent. An inspector must carefully and completely read and comprehend the manufacturer's instructions to properly perform an installation inspection.

In some cases, the code will specifically address an installation requirement that is also addressed in the manufacturer's installation instructions. The code requirements may be the same or may exceed the requirements in the manufacturer's installation instructions, or the manufacturer's installation instructions could contain requirements that exceed those in the code. In all such cases, the more restrictive requirements would apply.

In addition to the installation requirements for the appliance or equipment itself, this section also regulates the connections to appliances and equipment by requiring compliance with all other applicable sections of this code and other ICC codes. These connections include but are not limited to fuel and electrical supplies, control wiring, hydronic piping, chimneys, vents, and ductwork. Some overlap or coincidence of connection requirements may occur in the code and the manufacturer's installation instructions. In the unlikely event that the code was less restrictive (more lenient) than the manufacturer's instructions regarding an installation issue, the manufacturer's instructions must prevail. Where differences or conflicts occur, the requirements that provide the greatest level of safety always apply.

Even if an installation appears to be in compliance with the manufacturer's instructions, the installation cannot be complete or approved until all associated components, connections, and systems that serve the appliance or equipment are also in compliance with the applicable provisions of the code. For example, a gas-fired boiler installation must not be approved if the boiler is connected to a deteriorated, undersized, improper or otherwise unsafe chimney or vent. Likewise, the same installation must not be approved if the existing gas piping is in poor condition or if the electrical supply circuit is inadequate or unsafe.

In the case of replacement installations, the intent of this section is to require all new work associated with the installation to comply with the code without necessarily requiring full compliance for the existing, unchanged portions of the related ductwork, piping, electrical, venting, and similar mechanical systems. In the case of an appliance replacement, such as a furnace, it is the expectation of both the manufacturer's instructions and the code that the new furnace be connected to a code-complying venting system, fuel supply system, and electrical circuit, all of which are part of the furnace installation. If the ductwork system was such that the furnace temperature rise across the heat exchanger exceeded the maximum allowable temperature rise, the ductwork system might have to be altered as part of the furnace installation. Existing mechanical systems are accepted on the basis that they are free from hazard although not necessarily compliant with current codes. The code is not retroactive except where specifically stated that it applies to existing systems.

Manufacturer's installation instructions are often updated and changed for various reasons such as changes in the appliance or equipment design, revisions to the product standards, or as a result of field experience related to existing installations. The code official should stay abreast of any changes by reviewing the manufacturer's instructions for every installation.

Equipment and appliances must be installed in accordance with the manufacturer's installation instructions. The manufacturer's label and installation instructions must be consulted in determining whether or not an appliance or piece of equipment can be installed and operated in a particular hazardous location. Without the manufacturer's installation instructions, an inspector cannot fully perform his or her job; thus, the manufacturer's installation instructions must be available on the job site when the equipment is being inspected.

G2408.2 (305.3) Elevation of ignition source. Equipment and appliances having an ignition source shall be elevated such that the source of ignition is not less than 18 inches (457 mm) above the floor in hazardous locations and private garages. For the purpose of this section, rooms or spaces that are not part of the living space of a dwelling unit and that communicate directly with a private garage through openings shall be considered to be part of the private garage.

Exception: Elevation of the ignition source is not required for appliances that are listed as flammable vapor resistant and for installation without elevation.

❖ To reduce the fire/explosion hazard in hazardous areas, the potential sources of ignition must be elevated above the surface supporting the equipment or appliance. It is not the intent of this section for the installer to measure the elevation from the surface of a dedicated appliance stand or other structure built to support the appliance where that stand or structure does not afford room for the storage of flammable liquids. Some flammable and combustible liquids typically associated with hazardous locations give off vapors that are denser than air and tend to collect near the floor. The 18-inch (457 mm) height requirement is intended to reduce the possibility of fire or explosion by keeping ignition sources elevated above the anticipated level of accumulated vapors. The 18-inch (457 mm) value is a minimum requirement and must be increased when required by the manufacturer's installation instructions. This section will effectively prohibit the installation of most furnaces, boilers, space heaters, clothes dryers, and water heaters directly on the floor of residential garages.

The accumulation of flammable vapors more than 18 inches (457 mm) deep is unlikely in most ventilated locations; therefore, maintaining all possible sources of ignition at least 18 inches (457 mm) above the floor will substantially reduce the risk of explosion and fire. See Commentary Figure G2408.2(1). The installation of appliances and equipment having an ignition source within Group H occupancies or control areas where combustible, flammable, or explosive materials are used, handled, or dispensed is prohibited regardless of elevation.

In the context of this section, a source of ignition could be a pilot flame, burner, burner igniter, or electrical component capable of producing a spark. The term "ignition source" is defined and can be interpreted as meaning an intentional source of ignition such as for a burner or an unintentional source of ignition for any flammable vapors that may be present. See the definition of "Ignition source."

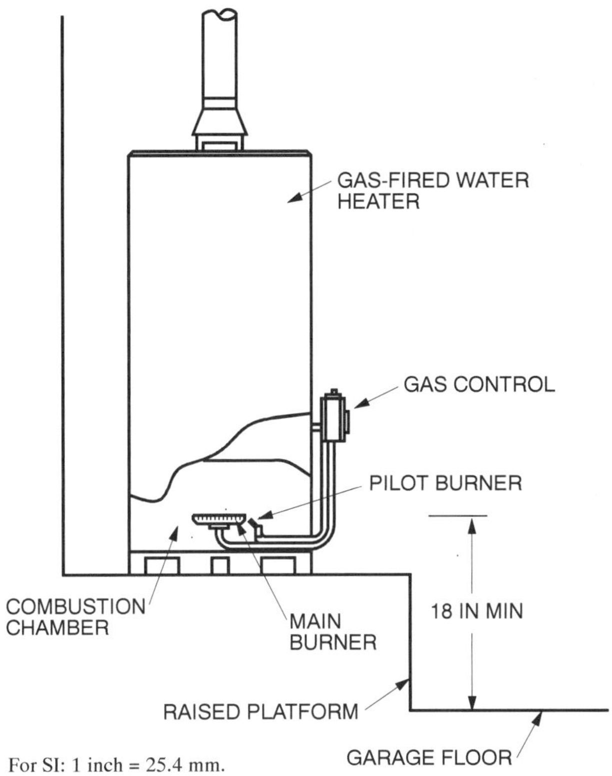

For SI: 1 inch = 25.4 mm.

Figure G2408.2(1)
GAS-FIRED WATER HEATER INSTALLATION IN A GARAGE

An appliance installed in a closet or room that is accessible only from the garage must be considered as part of the garage for application of this section. Even though the room may be separated from the garage by walls and a door, there is no practical means of making the door vapor-tight, nor is there any assurance that the door will remain closed during normal use. An appliance room that is accessed only from the outdoors or only from the living space would not be considered as part of the garage. Rooms such as utility rooms or laundry rooms that communicate both with the garage and the living space and that are considered as part of the living space are not part of the garage. See the definition of "Living space." If a room opens to the garage and such room is not living space, it is part of the garage for the purpose of this section.

The exception recognizes that new technology exists that allows appliances (water heaters at this point in time) to be tested and listed as being flammable-vapor ignition resistant and suitable for installation without the 18-inch elevation requirement. This new technology was developed for conventional draft hood-equipped water heaters, and now the standard for these appliances, Z21.10.1, mandates that by specified effective dates, all water heaters must comply with the flammable-vapor ignition criteria in the standard. The technology involves variously configured flame arrestor screens having precisely engineered openings (slots). All combustion air for the appliance is passed through the flame arrestor screen. In passing through the small slots in the screen, the combustion air velocity is increased (Venturi principle) such that the velocity exceeds the velocity at which flames travel through a flammable vapor. If the combustion air contains flammable vapors, the vapors will enter the appliance and be ignited within the combustion chamber. The flames will be contained (trapped) within the combustion chamber because the arrestor screen will not allow flames to exit the combustion chamber. The result is that the vapor ignition is limited to the appliance interior and not allowed to spread to the atmosphere exterior to the appliance. During such an event, a thermal sensing device shuts down and locks out the appliance. The appliance manufacturers typically require the appliance to be replaced after a flammable-vapor ignition event. The appliance standard includes testing requirements to determine if the arrestor screen is subject to blockage by debris including lint, plant fibers, dust particles, etc. If the arrestor screen openings become restricted by debris, the appliance will be deprived of combustion air and incomplete combustion and serious appliance malfunction can result. If an appliance shuts down as a result of the arrestor screen being blocked by debris, thus restricting the combustion air supply, a thermal device can be reset or replaced to restore operation of the appliance.

If a water heater equipped with flammable-vapor resistant technology has installation instructions or a label that states that the appliance must be elevated so that the source of ignition is at least 18 inches above the floor, it must be installed as directed. See Commentary Figures G2408.2(1) through G2408.2(5).

G2408.3 (305.4) Private garages. Appliances located in private garages shall be installed with a minimum clearance of 6 feet (1829 mm) above the floor.

> **Exception:** The requirements of this section shall not apply where the appliances are protected from motor vehicle impact and installed in accordance with Section G2408.2.

❖ Appliances located in a private garage or carport must be protected from vehicle impact. This section is applicable to appliances located in an area where motor vehicles can be operated and includes appliances under which a vehicle can pass and those located anywhere in a vehicle's path where impact is possible. The 6-foot (1829 mm) minimum height requirement is intended to provide adequate clearance above the typical automobile. With the popularity of conversion vans and recreational vehicles, which can be much higher than other automobiles, the 6-foot (1829 mm) minimum installation height above the floor might not provide adequate clearance. The garage door height can be used as a guide in determining the maximum vehicle height.

Figure G2408.2(2)
SOURCES OF IGNITION ELEVATED IN PRIVATE GARAGE

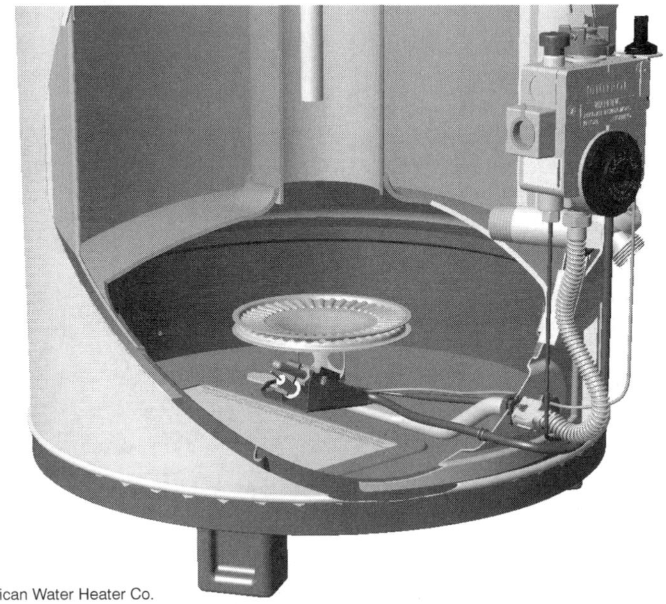

Photo courtesy of American Water Heater Co.

Figure G2408.2(3)
EXAMPLE OF FLAMMABLE VAPOR IGNITION-RESISTANT WATER HEATER TECHNOLOGY

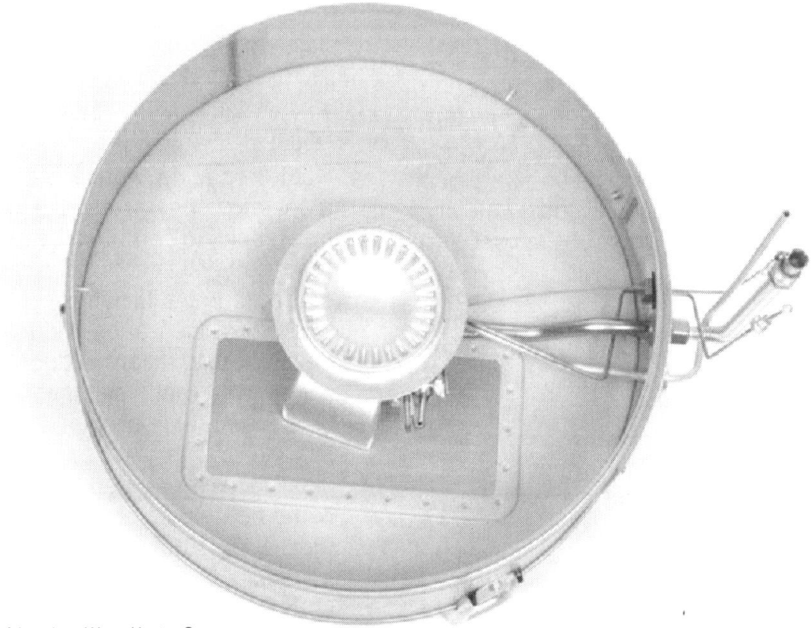

Photo courtesy of American Water Heater Co.

Figure G2408.2(4)
EXAMPLE OF FLAMMABLE VAPOR IGNITION-RESISTANT WATER HEATER TECHNOLOGY

Photo courtesy of Bradford White Corp. (Bradford White Defender Safety System[TM])

Figure G2408.2(5)
EXAMPLE OF FLAMMABLE VAPOR IGNITION-RESISTANT WATER HEATER TECHNOLOGY

G2408.4 (305.7) Clearances from grade. Equipment and appliances installed at grade level shall be supported on a level concrete slab or other approved material extending above adjoining grade or shall be suspended a minimum of 6 inches (152 mm) above adjoining grade.

❖ This section addresses outdoor and crawl space installations where an appliance is resting on the earth and not supported by a building. The intent is to maintain level support and protect the appliance from corrosion and deterioration. The slab must be concrete or a material having comparable strength and longevity. Outdoor appliance installations must be designed to tolerate movement where frost heave will cause a slab to rise and fall.

G2408.5 (305.8) Clearances to combustible construction. Heat-producing equipment and appliances shall be installed to maintain the required clearances to combustible construction as specified in the listing and manufacturer's instructions. Such clearances shall be reduced only in accordance with Section G2409. Clearances to combustibles shall include such considerations as door swing, drawer pull, overhead projections or shelving and window swing. Devices, such as door stops or limits and closers, shall not be used to provide the required clearances.

❖ Section G2409 allows clearances to be reduced where protective assemblies are installed to reduce the transfer of heat from the source to the combustible material/assembly. Clearances must be measured from shelving; overhead projections; and the possible positions of doors, windows, and drawers. Doorstops and closers are easily defeated and subject to failure.

SECTION G2409 (308) CLEARANCE REDUCTION

G2409.1 (308.1) Scope. This section shall govern the reduction in required clearances to combustible materials and combustible assemblies for chimneys, vents, appliances, devices and equipment.

❖ Heat-producing appliances and mechanical equipment must be installed with the required minimum clearances to combustible materials indicated by their listing label. It is not uncommon to encounter practical or structural difficulties in maintaining clearances. Therefore, clearance reduction methods have been developed to allow reduction of the minimum prescribed clearance distance while achieving equivalent protection. An important understanding is that all prescribed clearances to combustibles are airspace clearances measured from the heat source to the face of the nearest combustible surface. See Commentary Figure G2409.1. When using the assemblies described in Table G2409.2, the clearance is measured as described in Note "b." In the case of listed equipment, the required clearances are intended to be clear airspace, and therefore the space is not to be filled with insulation or any other material. This is especially important where clearances are required from appliances and equipment that rely on the airspace for convection cooling to maintain their proper operation.

The provisions contained in this section are based on the principles of heat transfer. Mechanical equipment or appliances producing heat will by themselves become hot, and many appliances have hot exterior surfaces by design. The heat energy is then radiated to objects surrounding the appliances or equipment. When mechanical equipment and appliances are tested, the minimum clearances are established so that radiant and, to a lesser extent, convective heat transfer do not represent an ignition hazard to adjacent surfaces and objects. This distance is called the "required clearance" to combustible materials. Appliance and equipment labels are required to specify minimum clearances in all directions.

This section permits the use of materials and systems to act as radiation shields, allowing a reduction in the amount of heat energy transferred to surrounding objects and reducing the required clearances between gas-fired appliances and equipment and combustible material to combustibles and does not change the fact that the adjacent surface is combustible.

Plaster and gypsum by themselves are classified as noncombustible materials. Under continued exposure to heat, however, these materials will gradually decompose as water molecules are driven out of the material. Plaster on wood lath, plasterboard, sheetrock, and drywall are considered combustible materials when the code provisions are applied.

Additionally, gypsum wallboard has a paper face that has a flame spread index that is measurable in the ASTM E 84 test. This alone identifies the need to classify gypsum wallboard as a combustible material for the purpose of requiring a separation from heat-producing equipment and appliances.

The code specifically prohibits the reduction of clearances in certain applications. For examples, see Sections G2409.2, G2409.3, and G2409.4. Also see the definition of "noncombustible materials."

Applying gypsum wallboard to a wood frame wall does not reduce the required clearance, nor does the type of gypsum board make any difference (that is, class X, fire code, or standard).

G2409.2 (308.2) Reduction table. The allowable clearance reduction shall be based on one of the methods specified in Table G2409.2 or shall utilize an assembly listed for such application. Where required clearances are not listed in Table G2409.2, the reduced clearances shall be determined by linear interpolation between the distances listed in the table. Reduced clearances shall not be derived by extrapolation below the range of the table. The reduction of the required clearances to combustibles for listed and labeled appliances and equipment shall be in accordance with the requirements of this section except that such clearances shall not be reduced where reduction is specifically prohibited by the terms of the appliance or equipment listing [see Figures G2409.2(1) and G2409.2(2)].

❖ Another option for reducing the required clearances to combustible materials is to use one of the on-site field-constructed methods specified in Table G2409.2. Note that the table refers to 24 gage (0.024-inch thick) sheet metal.

G2409.3 (308.3) Clearances for Indoor Air-Conditioning Equipment. Clearance requirements for indoor air-conditioning equipment shall comply with Sections G2409.3.1 through G2409.3.5.

❖ The code emphasizes requirements for clearances to combustibles because of the potential fire hazard posed when such clearances are not observed. Maintaining an appropriate distance from the outer surfaces of an appliance or piece of equipment to combustible materials reduces the possibility of ignition of combustible materials.

The minimum clearances to combustibles are specified in the manufacturer's installation instructions for a labeled appliance. Since an approved agency tests appliances in accordance with these instructions, the clearances required are necessary for correct installation and operation of the appliance.

Reduction of the required clearances to combustibles is allowed only when the combustibles are protected by one of the methods outlined in Section G2409 and clearance reduction is not prohibited by Section G2409 or the appliance and equipment listing. See the commentary for Section G2409. The manufacturer's specified minimum clearances and the clearances specified in Section G2409 are all airspace clearances, and such spaces cannot be filled with insulation or any other material, even if the material is noncombustible. In some cases, the manufacturer's installation instructions will specify absolute minimum clearances that must not be reduced by any clearance reduction method.

The most common wall covering material—gypsum wallboard—is a combustible finish material for the purpose of this chapter. As a result, gypsum wallboard as well as all other combustible wall finishes must be separated from an appliance or equipment in accordance with the prescribed clearance to combustibles. Clearances to combustibles also apply to furnishings, window treatments, and moveable items that can be placed within the required clearance range of appliances and equipment.

G2409.3.1 (308.3.1) Equipment installed in rooms that are large in comparison with the size of the equipment. Air-conditioning equipment installed in rooms that are large in comparison with the size of the equipment shall be installed with clearances per the terms of their listing and the manufacturer's instructions.

❖ See the definition of "Room Large in Comparison with Size of Equipment."

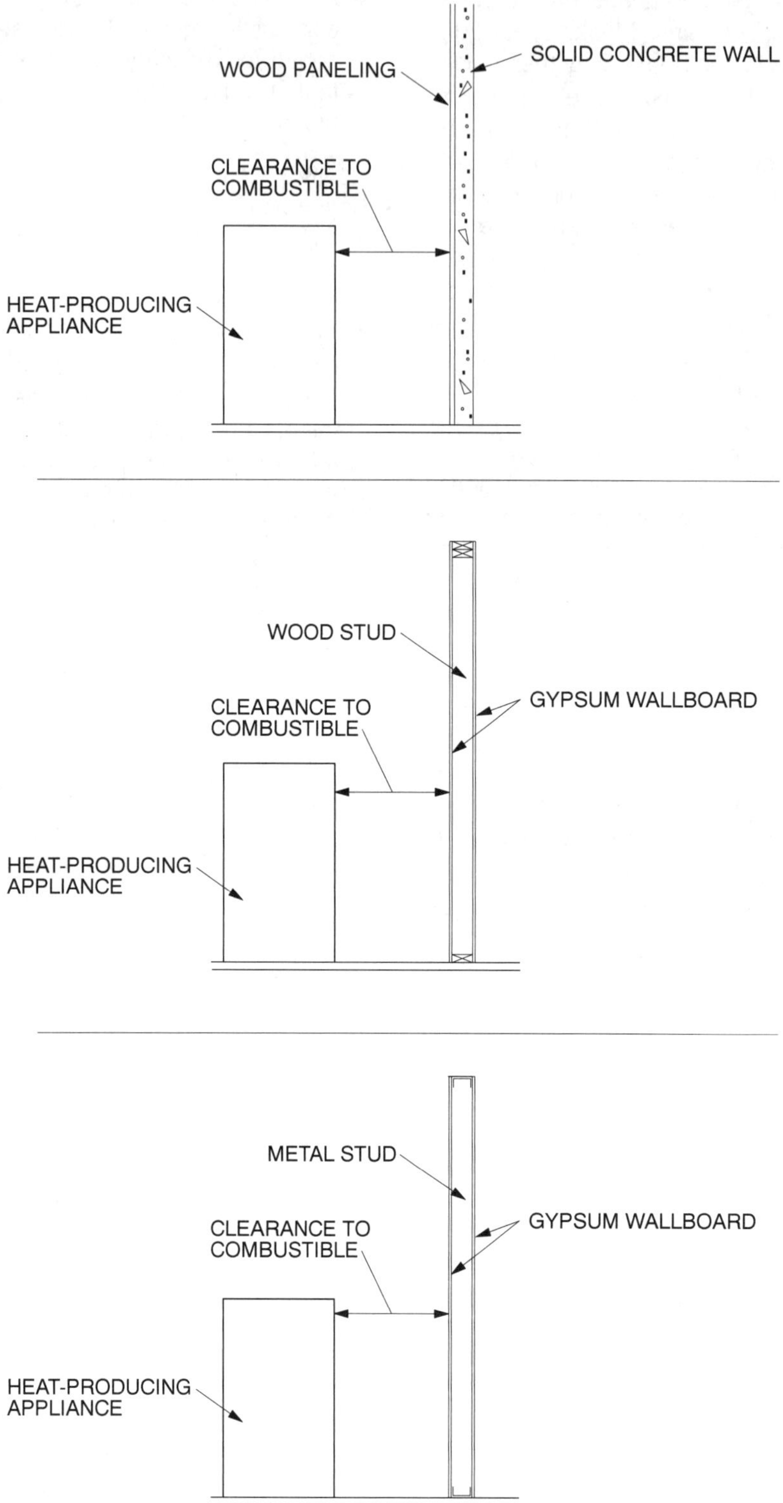

Figure G2409.1
CLEARANCE TO COMBUSTIBLES—EXAMPLES OF HOW TO MEASURE

TABLE G2409.2 (308.2)
REDUCTION OF CLEARANCES WITH SPECIFIED FORMS OF PROTECTION[a, b, c, d, e, f, g, h, i, j, k]

Type of protection applied to and covering all surfaces of combustible material within the distance specified as the required clearance with no protection [*see* Figures G2409.2(1) and G2409(2)]	WHERE THE REQUIRED CLEARANCE WITH NO PROTECTION FROM APPLIANCE, VENT CONNECTOR OR SINGLE WALL METAL PIPE IS:									
	36 inches		18 inches		12 inches		9 inches		6 inches	
	Allowable clearances with specified protection (inches)									
	Use column 1 for clearances above appliance or horizontal connector. Use column 2 for clearances from appliance, vertical connector and single-wall metal pipe.									
	Above column 1	Sides and rear column 2	Above column 1	Sides and rear column 2	Above column 1	Sides and rear column 2	Above column 1	Sides and rear column 2	Above column 1	Sides and rear column 2
3.5-in. thick masonry wall without ventilated air space	—	24	—	12	—	9	—	6	—	5
0.5-in. insulation board over 1-in. glass fiber or mineral wool batts	24	18	12	9	9	6	6	5	4	3
0.024 inch (24 gage) sheet metal over 1-in. glass fiber or mineral wool batts reinforced with wire on rear face with ventilated air space	18	12	9	6	6	4	5	3	3	3
3.5-in. thick masonry wall with ventilated air space	—	12	—	6	—	6	—	6	—	6
0.024 inch (24 gage) sheet metal with ventilated air space	18	12	9	6	6	4	5	3	3	2
0.5-in. thick insulation board with ventilated air space	18	12	9	6	6	4	5	3	3	3
0.024 inch (24 gage) sheet metal with ventilated air space over 0.024 inch (24 gage) sheet metal with ventilated air space	18	12	9	6	6	4	5	3	3	3
1-in. glass fiber or mineral wool batts sandwiched between two sheets 0.024 inch (24 gage) sheet metal with ventilated air space	18	12	9	6	6	4	5	3	3	3

For SI: 1 inch = 25.4 mm.

NOTES:

a. Reduction of clearances from combustible materials shall not interfere with combustion air, draft hood clearance and relief, and accessibility of servicing.

b. All clearances shall be measured from the outer surface of the combustible material to the nearest point on the surface of the appliance, disregarding any intervening protection applied to the combustible material.

c. Spacers and ties shall be of noncombustible material. No spacer or tie shall be used directly opposite appliance or connector.

d. Where clearance reduction systems use ventilated air space, adequate provision for air circulation shall be provided as described. [See Figure G2409.2(2).]

e. There shall be at least 1 inch between clearance reduction systems and combustible walls and ceilings for reduction systems using ventilated air space.

f. Where a wall protector is mounted on a single flat wall away from corners, it shall have a minimum 1-inch air gap. To provide air circulation, the bottom and top edges or only the side and top edges, or all edges shall be left open.

g. Mineral wool batts (blanket or board) shall have a minimum density of 8 pounds per cubic foot (128 kg per m) and a minimum melting point of 1,500°F (816°C).

h. Insulation material used as part of clearance reduction system shall have a thermal conductivity of 1.0 Btu inch per square foot per hour-°F (0.144 Watts per meter-Kelvin) or less.

i. There shall be at least 1 inch between the appliance and the protector. In no case shall the clearance between the appliance and the combustible surface be reduced below that allowed in Table G2409.2.

j. All clearances and thicknesses are minimum; larger clearances and thicknesses are acceptable.

k. Listed single-wall connectors shall be installed in accordance with the terms of their listing and the manufacturer's instructions.

❖ The column headings of Table G2409.2 list required clearances without protection. The numbers to the right of each method indicate the permissible reduced clearance measured from the heat-producing appliances/equipment to the face of the combustible material.

The rationale behind the methods of protection listed in Table G2409.2 is the ability of the protection to reduce radiant heat transmission from the appliance and equipment to the combustible material so that the temperature rise of the combustible material will remain below the maximum allowed.

Although the materials referred to in Table G2409.2 are common construction materials, confusion often arises over what satisfies the requirement for "insulation board" (item 6 in the table), sometimes referred to as an inorganic insulating board, noncombustible mineral board, or noncombustible insulating board. These products are not made of carbon-based compounds. Carbon-based compounds are those found in cellulose (wood), plastics, and other materials manufactured from raw materials that once existed as living organisms. Cement board materials must have a specified maximum "C" (conductance) value in addition to being noncombustible.

Footnote "h" specifies a maximum thermal conductivity of 1.0 Btu /in x h x ft^2 x degree F. Conductivity is the amount of heat in Btu that will flow each hour through a one foot square slab of material, one inch thick, with a one degree Fahrenheit temperature difference between both sides and is usually identified by the symbol *K*. Tables of *K* values usually do not include in the dimensions for conductivity the area term, and it must be understood that the value must be multiplied by the area to obtain the total Btu value.

Thermal conductance (overall) is the time rate of heat flow through a body not taking thickness into account and is usually identified by the symbol *C* (Btu / h x square feet x degree F).

Thermal resistance (overall) is the reciprocal of overall thermal conductance and is usually identified by the symbol *R* (h x square feet x degree F / Btu).

This translates into a minimum required insulation *R*-value of 1.0 (ft^2 x hr x degree F)/Btu per inch of insulating material. When implemented, the methods in Table G2409.2 control heat transmission by reflecting heat radiation, retarding thermal conductance, and providing convective cooling. Where sheet metal materials or metal plates are specified, the effectiveness of the protection can be enhanced by the reflective surface of the metal. Painting or otherwise covering the surface would reduce the metal's ability to reflect radiant heat and depending on the color could increase heat absorption; however, the code does not require that the metal be reflective or unpainted. The airspace between the protected surface and the clearance-reduction assembly allows convection air currents to cool the protection assembly by carrying away heat that has been conducted through the assembly. Where a clearance-reduction assembly is required to be spaced 1 inch off the wall, the top, bottom, and sides of the assembly must remain open as required by Notes "d" and "f" to permit unrestricted airflow (convection currents). If such openings were not provided, the air-cooling effect would not take place, and the protection assembly would not be as effective in limiting the temperature rise on the protected surfaces. Ideally, the protection assembly should be open on all sides to provide maximum ventilation. Figures G2409.2(1) and G2409.2(2) show assemblies incorporating airspace.

Spacers must be noncombustible. Spacers should not be placed directly behind the heat source because the location would increase the amount of heat conduction through the spacer, thus creating a "hot spot." Figure G2409.2(2) specifically shows a noncombustible spacer arrangement.

The performance of a protective assembly when applied to a horizontal surface, such as a ceiling, will differ substantially from the same assembly placed in a vertical plane. Obviously, temperatures at a ceiling surface will be higher because of natural convection and because the air circulation between the method of protection and the protected ceiling surface will be substantially reduced or nonexistent. It is for these reasons that Table G2409.2 is divided into two application groups.

The manufacturer's instructions or label for many appliances will state an absolute minimum clearance, regardless of any clearance reduction method used. Those clearance requirements take precedence over Table G2409.2.

The methods in Table G2409.2 are intended for permanent installations properly supported to prevent displacement or deformation. Movement could adversely affect the performance of the protection method, thus posing a potential fire hazard.

The assemblies in Table G2409.2 are the product of experience and testing. To achieve predictable and dependable performance, the components of the various assemblies cannot be mixed, matched, combined, or otherwise rearranged to comprise new assemblies, and materials cannot be substituted for those prescribed in the table. Any alterations or substitutions could have an effect on the assembly, and its performance must be tested and approved.

Note "b" states the method of measuring the reduced clearance. The intent is to measure the reduced clearance from the heat source to the combustible material, disregarding any intervening protection assembly.

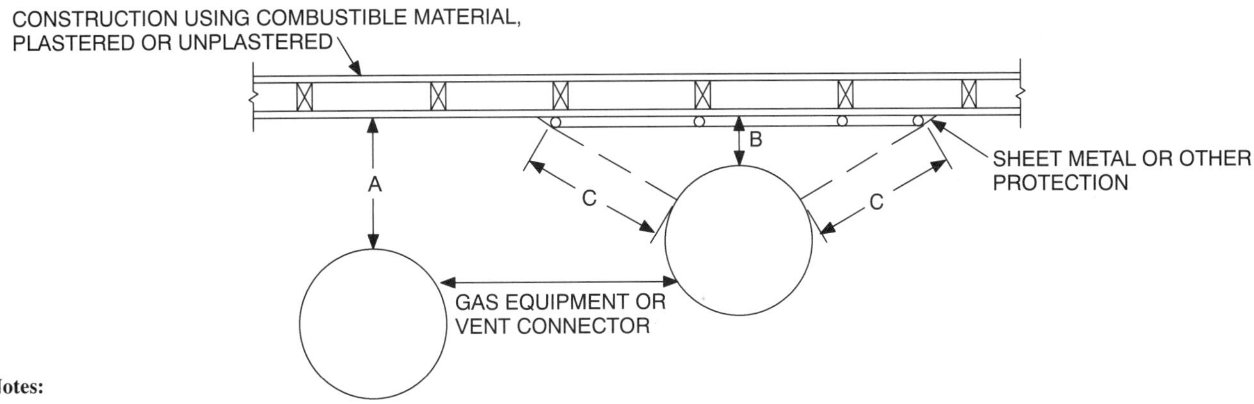

Notes:

"A" equals the clearance with no protection.

"B" equals the reduced clearance permitted in accordance with Table G2409.2. The protection applied to the construction using combustible material shall extend far enough in each direction to make "C" equal to "A."

FIGURE G2409.2(1) [308.2(1)]
EXTENT OF PROTECTION NECESSARY TO REDUCE CLEARANCES FROM GAS EQUIPMENT OR VENT CONNECTORS

❖ See the commentary for Section G2409.2.

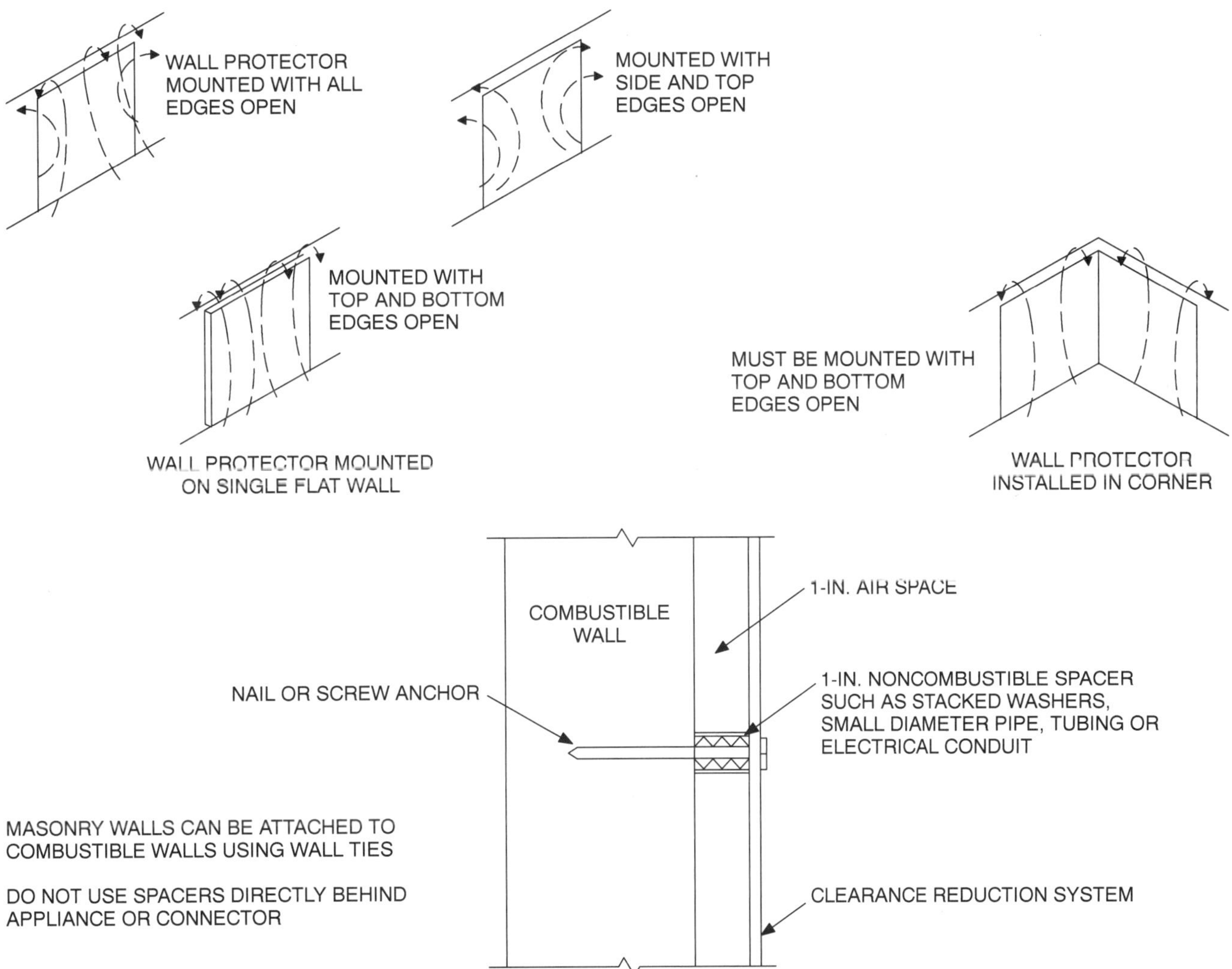

For SI: 1 inch = 25.4 mm.

FIGURE G2409.2(2) [308.2(2)]
WALL PROTECTOR CLEARANCE REDUCTION SYSTEM

❖ See the commentary for Section 2409.2.

G2409.3.2 (308.3.2) Equipment installed in rooms that are not large in comparison with the size of the equipment. Air-conditioning equipment installed in rooms that are not large (such as alcoves and closets) in comparison with the size of the equipment shall be listed for such installations and installed in accordance with the manufacturer's instructions. Listed clearances shall not be reduced by the protection methods described in Table G2409.2, regardless of whether the enclosure is of combustible or noncombustible material.

❖ See the definition of "Room Large in Comparison with Size of Equipment."

Indirectly, the definition along with this section defines an alcove and a closet with respect to size.

An appliance must be listed for closet/alcove installation if it is installed in a space that does not meet the requirements of a Room Large in Comparison with Size of Equipment as defined in Section G2403. Simply stated, if the room or space volume is less than the stated minimum volume in the definition, that room or space is considered to be a closet or alcove.

Fuel-fired appliances installed in small rooms or spaces can pose a potential fire hazard because of the heat emanating from the appliance and the inadequate ventilation afforded by the boundaries of small enclosures. For appliances installed in closets or alcoves, the clearances stated in the listing (label) are necessary for ventilation cooling of the appliance and must not be reduced as provided for in this section.

G2409.3.3 (308.3.3) Clearance reduction. Air-conditioning equipment installed in rooms that are large in comparison with the size of the equipment shall be permitted to be installed with reduced clearances to combustible material provided the combustible material or equipment is protected as described in Table G2409.2.

❖ See the definition of "Room Large in Comparison with Size of Equipment."

G2409.3.4 (308.3.4) Plenum clearances. Where the furnace plenum is adjacent to plaster on metal lath or noncombustible material attached to combustible material, the clearance shall be measured to the surface of the plaster or other noncombustible finish where the clearance specified is 2 inches (51 mm) or less.

❖ Where the air-conditioning equipment plenum is adjacent to surfaces covered by noncombustible material as described in this section, and the required clearance from combustibles is 2 inches or less, the required clearance must be measured to the surface of the adjacent material. Measuring to the combustible material would reduce the clearance between the equipment plenum and adjacent surface. Reducing this clearance would affect the cooling of the equipment and could cause overheating.

G2409.3.5 (308.3.5) Clearance from supply ducts. Air-conditioning equipment shall have the clearance from supply ducts within 3 feet (914 mm) of the furnace plenum be not less than that specified from the furnace plenum. No clearance is necessary beyond this distance.

❖ Ducts within 3 feet of a supply plenum to which they connect will reach nearly the same temperature as the plenum.

G2409.4 (308.4) Central heating boilers and furnaces. Clearance requirements for central heating boilers and furnaces shall comply with Sections G2409.4.1 through G2409.4.6. The clearance to this equipment shall not interfere with combustion air, draft hood clearance and relief, and accessibility for servicing.

❖ See the commentary for Section G2409.3.

G2409.4.1 (308.4.1) Equipment installed in rooms that are large in comparison with the size of the equipment. Central heating furnaces and low-pressure boilers installed in rooms large in comparison with the size of the equipment shall be installed with clearances per terms of their listing and the manufacturer's instructions.

❖ See the definition of "Room Large in Comparison with Size of Equipment."

G2409.4.2 (308.4.2) Equipment installed in rooms that are not large in comparison with the size of the equipment. Central heating furnaces and low-pressure boilers installed in rooms that are NOT large (such as alcoves and closets) in comparison with the size of the equipment shall be listed for such installations. Listed clearances shall not be reduced by the protection methods described in Table G2409.2 and illustrated in Figures G2409.2(1) and G2409.2(2), regardless of whether the enclosure is of combustible or noncombustible material.

❖ See the definition of "Room Large in Comparison with Size of Equipment."

Indirectly, the definition along with this section defines an alcove and a closet with respect to size.

An appliance must be listed for closet/alcove installation if it is installed in a space that does not meet the requirements of a Room Large in Comparison with Size of Equipment as defined in Section G2403. Simply stated, if the room or space volume is less than the stated minimum volume in the definition, that room or space is considered to be a closet or alcove.

Fuel-fired appliances installed in small rooms or spaces can pose a potential fire hazard because of the heat emanating from the appliance and the inadequate ventilation afforded by the boundaries of small enclosures. For appliances installed in closets or alcoves, the clearances stated in the listing (label) are necessary for ventilation cooling of the appliance and must not be reduced as provided for in this section.

G2409.4.3 (308.4.3) Clearance reduction. Central heating furnaces and low-pressure boilers installed in rooms that are large in comparison with the size of the equipment shall be permitted to be installed with reduced clearances to combustible material provided the combustible material or equipment is protected as described in Table G2409.2.

❖ See the definition of "Room Large in Comparison with size of Equipment."

G2409.4.4 (308.4.5) Plenum clearances. Where the furnace plenum is adjacent to plaster on metal lath or noncombustible material attached to combustible material, the clearance shall be measured to the surface of the plaster or other noncombustible finish where the clearance specified is 2 inches (51 mm) or less.

❖ See the commentary for Section G2409.3.4.

G2409.4.5 (308.4.6) Clearance from supply ducts. Central heating furnaces shall have the clearance from supply ducts within 3 feet (914 mm) of the furnace plenum be not less than that specified from the furnace plenum. No clearance is necessary beyond this distance.

❖ See the commentary for Section G2409.3.5.

G2409.4.6 (308.4.4) Clearance for servicing equipment. Front clearance shall be sufficient for servicing the burner and the furnace or boiler.

❖ Because mechanical equipment and appliances require routine maintenance, repairs, and possible replacement, access is required. Additionally, access recommendations or requirements are usually stated in the manufacturer's installation instructions.

The intent of this section is to assure access to components that require observation, inspection, adjustment, servicing, repair, or replacement such as controls, gauges, burners, filters, blowers, and motors. Access is also necessary to conduct operating procedures such as startup or shutdown.

The code defines access as a component that is being able to be reached, but first may require the removal of a panel, door, or similar obstruction. An appliance or piece of equipment is not accessible if any portion of the structure's permanent finish materials, such as drywall, plaster, paneling, built-in furniture, cabinets, or any other similar permanently affixed building component must be removed.

SECTION G2410 (309) ELECTRICAL

G2410.1 (309.1) Grounding. Gas piping shall not be used as a grounding electrode.

❖ This section prohibits metal underground gas piping from being used as a grounding electrode. Metal gas piping must be bonded to the electrical service grounding electrode system, but it cannot serve as a grounding electrode. See Section 250.104(B) of the 2002 NFPA 70 for clarification of gas piping bonding means.

G2410.2 (309.2) Connections. Electrical connections between gas utilization equipment and the building wiring, including the grounding of the equipment, shall conform to Chapters 33 through 42.

❖ Field-installed power wiring and control wiring for appliances and equipment must be installed in accordance with the electrical chapters of this code, which are based on the 2002 edition of NFPA 70.

The power wiring includes all the wiring, disconnects, overcurrent protection devices, starters, and related hardware used to supply electrical power to the appliance or equipment. The control wiring includes all the wiring, devices, and related hardware that connect the main unit to all external controls and accessories, such as temperature and pressure sensors, thermostats, exhausters, equipment contactors, interlock controls, and remote damper motors. The internal factory wiring of appliances and equipment is not covered by this section unless it is specifically addressed in the electrical chapters; however, such wiring is covered by the testing and review performed by an approved agency as part of the labeling process.

The mechanical or electrical code official responsible for the inspection of appliances and equipment must be familiar with the applicable sections of the electrical chapters.

SECTION G2411 (310) ELECTRICAL BONDING

G2411.1 (310.1) Gas pipe bonding. Each above-ground portion of a gas piping system that is likely to become energized shall be electrically continuous and bonded to an effective ground-fault current path. Gas piping shall be considered to be bonded where it is connected to gas utilization equipment that is connected to the equipment grounding conductor of the circuit supplying that equipment.

❖ This section is consistent with Section 250.104(B) of the 2002 NFPA 70. In the majority of cases, the gas piping system will be connected to at least one appliance that uses electrical power. Such an appliance will be connected to an equipment grounding conductor that will serve to bond the gas piping to the grounding electrode system for the building. Where the appliance branch circuits include grounding conductors properly connected to the appliance and one or more such appliances are connected to the gas piping system, the gas piping system is automatically bonded as required by this section.

SECTION G2412 (401) GENERAL

G2412.1 (401.1) Scope. This section shall govern the design, installation, modification and maintenance of piping systems. The applicability of this code to piping systems extends from the point of delivery to the connections with the equipment and includes the design, materials, components, fabrication, assembly, installation, testing, inspection, operation and maintenance of such piping systems.

❖ This section regulates aspects of fuel gas distribution systems, including the design, installation, testing, repair, and maintenance. The applicability of this section is limited to the consumer side of the public utility company's gas distribution system.

The code governs all piping and system components from the end point of the gas purveyor's (utility company) service line to the supplied appliances and equipment. Typically, the gas utility company service line terminates at the service pressure regulator and meter setting. In other words, the code does not apply to piping and components that are owned by the gas utility company. The construction of utility-owned gas piping is governed by the federal Department of Transportation regulations wherever such piping is located. However, in cases where the utility-owned piping runs through, in, or on a building, it must do so in a way that does not jeopardize the structural integrity or fire safety of the building. Therefore, the relationship of the utility-owned piping to the building (prohibited locations, penetrations, etc.) is governed by the building portion of this code. The *International Fuel Gas Code* contains a reference (see Section 401.1.1) to the *International Building Code* to make sure this is recognized by the user. No other provision within this chapter applies to utility-owned gas piping.

Although LP-gas storage systems generally remain under the ownership of the gas supplier, such systems are governed by the code due to the reference to NFPA 58. See the Commentary for Section G2412.2. An LP-gas piping system begins at the outlet side of the first stage pressure regulator. The piping between the first and second stage (final) regulator is governed by NFPA 58, and the piping downstream of the second stage regulator is governed by this code. The first stage pressure regulator reduces the storage container pressure to 10 pounds per square inch gauge (psig) (138 kPa) or less.

The termination point of gas utility services varies depending on the particular company's policies, but typically the service terminates at the property line or outdoors, immediately adjacent to the building or structure served by the service. See Commentary Figures G2412.1(1) and G2412.1(2).

G2412.1.1 (401.1.1) Utility piping systems located within buildings. Utility service piping located within buildings shall be installed in accordance with the structural safety and fire protection provisions of this code.

❖ Piping and components upstream of the point of delivery are designed, installed, tested, owned, and maintained by the gas utility company and are regulated by the federal DOT regulations, not this code. This code does, however, regulate two aspects of the installation related to protection of the structure from a structural and fire safety standpoint. The first is to regulate the piping penetrations through structural components and through fire-resistance-rated assemblies. The second is to regulate piping support relative to structural loads and seismic forces. This section recognizes that utility piping within buildings is under the coincidental jurisdiction of the DOT and the local government.

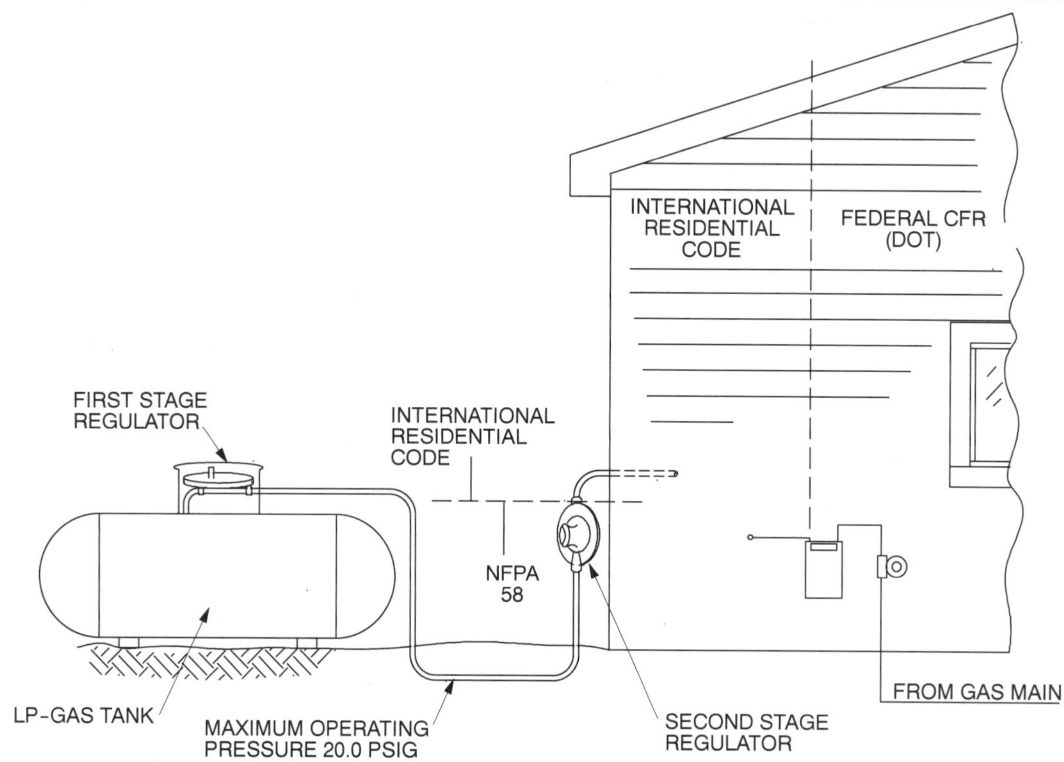

Figure G2412.1(1)
SCOPE OF THE CODES

Illustration courtesy of NORMAC

Figure G2412.1(2)
GAS SERVICE POINT OF DELIVERY

G2412.2 (401.2) Liquefied petroleum gas storage. The storage system for liquefied petroleum gas shall be designed and installed in accordance with the *International Fire Code* and NFPA 58.

❖ LP-gas tanks, containers, cylinders, storage vessels and related components must comply with NFPA 58. The storage system includes the storage vessel of any type and the vessel appurtenances such as shutoff valves, pressure gauges, liquid level indicators, pressure regulators, pressure relief valves, and filling connections. The first stage and second stage pressure regulators and interconnecting piping are also addressed by the standard and could therefore considered to be part of the LP-gas storage system. If the piping between the first stage regulator and the point of delivery (second stage or 2-psi regulator) is not placed under the control of NFPA 58, it would be unregulated since Section G2412.1 states that the code coverage begins at the point of delivery. See the definition of "point of delivery." 2-psi regulators reduce the outlet pressure of the first stage regulator to 2 psi, and a line pressure regulator downstream of the 2-psi regulator reduces the pressure from 2 psi to a maximum of 14 inches WC. Second stage regulators reduce the outlet pressure of the first stage regulator to a maximum of 14 inches WC. NFPA 58 addresses all aspects of LP-gas storage systems such as container design and construction, container appurtenances, safety devices, container location, and installation. A referenced standard is only enforceable in the context in which it is referenced. Any portions of NFPA 58 that extend beyond the scope of the storage system are not applicable, because of the context of the reference in this section.

G2412.3 (401.3) Modifications to existing systems. In modifying or adding to existing piping systems, sizes shall be maintained in accordance with this chapter.

❖ The gas volume demand on any portion of a distribution system cannot be increased beyond the capacity of the piping serving that portion. Modifications and additions to existing piping systems can have a detrimental effect on the existing piping system's ability to serve the connected loads. Commonly, additional gas utilization equipment or appliances will be added to existing distribution piping without any regard for the effect of the extra load. To prevent excessive pressure losses, existing piping must be able to supply adequately additional loads, or piping must be increased in size.

G2412.4 (401.4) Additional appliances. Where an additional appliance is to be served, the existing piping shall be checked to determine if it has adequate capacity for all appliances served. If inadequate, the existing system shall be enlarged as required or separate piping of adequate capacity shall be provided.

❖ This section parallels Section G2412.8 and makes it clear that when any appliances are added to an existing gas piping system, the system must be carefully evaluated to verify that adding an appliance will not create a dangerous low gas pressure condition within the gas piping system. The system evaluation must be conducted on the completed system as it will be configured after the appliance is added, and if the results indicate that the existing piping system cannot adequately serve the gas demand of the new appliance the system must be revised accordingly. Undersized (overloaded) piping will cause pressure losses that can result in serious appliance malfunction.

G2412.5 (401.5) Identification. For other than black steel pipe, exposed piping shall be identified by a yellow label marked "Gas" in black letters. The marking shall be spaced at intervals not exceeding 5 feet (1524 mm). The marking shall not be required on pipe located in the same room as the equipment served.

❖ The intent of this section is to prevent gas piping from being mistaken for other piping. Such a case of mistaken identity could lead to a dangerous condition if gas piping were to be cut or opened unintentionally. It is not uncommon for fuel-gas piping systems and various other gas or liquid piping systems in a building to be constructed of the same materials and thus have the same appearance. The method of identifying fuel gas piping must be permanent, legible, and conspicuous. Black steel pipe is exempt from the identification requirements because it is the traditional gas pipe material and is generally recognized as such. CSST is manufactured in compliance with the marking requirements of this section, and some copper tubing is manufactured with a similar yellow plastic jacket. See Commentary Figures G2414.5 and G2414.5.3(1). Note that black steel pipe that has been painted another color is still black steel pipe.

G2412.6 (401.6) Interconnections. Where two or more meters are installed on the same premises, but supply separate consumers, the piping systems shall not be interconnected on the outlet side of the meters.

❖ If gas distribution systems with independent meters were connected together downstream of the meters, it would be impossible to determine the actual consumption of each of the interconnected systems. As the result of building renovations, remodeling, additions, or intentional cross-connections between separately metered systems, it is possible for a consumer to be paying for fuel gas consumed by another consumer. Accurate consumption metering can only occur if each metered distribution system is independent of all other metered distribution systems on the premises. Also, a system thought to be shut off could be backfed from an interconnected system, thereby maintaining pressure in a system that was intended to be shut off.

G2412.7 (401.7) Piping meter identification. Piping from multiple meter installations shall be marked with an approved permanent identification by the installer so that the piping system supplied by each meter is readily identifiable.

❖ This provision allows service and emergency personnel to readily locate the shutoff valve that serves each piping system. The identification also helps to prevent interconnections between systems and between spaces when additions and/or alterations are made in the piping systems.

G2412.8 (401.8) Minimum sizes. All pipe utilized for the installation, extension and alteration of any piping system shall be sized to supply the full number of outlets for the intended purpose and shall be sized in accordance with Section G2413.

❖ Undersized gas piping systems are not capable of delivering the required volume of fuel at the required pressure. Inadequate gas pressure can cause hazardous operation of appliances. Appliances depend on proper gas pressure to maintain the design Btu (J) input rate, to maintain the required minimum manifold pressure, and to produce the required gas velocity in the burners. If the gas pressure to an appliance is too low, the result can be incomplete combustion, burner malfunction and flashback, soot production, and a potential fire, explosion, or asphyxiation hazard. Gas piping must be sized to maintain the required minimum gas pressures at the appliance inlet.

When one is designing a gas piping distribution system, it must be assumed that all connected appliances and equipment will operate simultaneously, except as provided for in Section G2413.2. In other words, the system is designed for maximum demand. The code contains provisions for sizing gas piping systems using an exact equation or design tables. The tabular method is most often used because it is simpler and less likely to produce errors.

Factors that affect the sizing of gas piping systems include the specific gravity of the gas, the length of the pipe and the number of fittings installed, the maximum gas demand, the allowable pressure loss through the system, and any diversity factor that is applicable.

The design of gas piping installations should also take into account the possibility of future increases in gas demand (load). The further the gas meter or point of delivery is from the building, the more extensive will be the disruption of the property in the event that a larger yard line between the meter and the building becomes necessary in the future due to increased gas demand within the building, a change in fuel gas type, or a change in service pressure.

SECTION G2413 (402) PIPE SIZING

G2413.1 (402.1) General considerations. Piping systems shall be of such size and so installed as to provide a supply of gas sufficient to meet the maximum demand without undue loss of pressure between the point of delivery and the gas utilization equipment.

❖ This section is a summation of the provisions of Sections G2412.8 and G2413.5 and is intended to give guidance to the designer. See the Commentary for Sections G2412.8 and G2413.5. "Undue loss of pressure" is not defined but must be considered as pressure loss (drop) in excess of the allowable loss described in Section G2413.5.

G2413.2 (402.2) Maximum gas demand. The volume of gas to be provided in cubic feet per hour shall be determined directly from the manufacturer's input ratings of the gas utilization equipment served. Where input rating is not indicated, the gas supplier, equipment manufacturer or a qualified agency shall be contacted, or the rating from Table G2413.2 shall be used for estimating the volume of gas to be supplied.

The total connected hourly load shall be used as the basis for piping sizing assuming that all equipment could be operating at full capacity simultaneously. Where a diversity of load can be established, pipe sizing shall be permitted to be based on such loads.

❖ To determine the demand volume required by an appliance in cubic feet (m^3) of gas per hour, the input rating of the appliance in British thermal units per hour (Btu/h) (W) as specified by the appliance manufacturer must be known. If the average heating value of the gas is known [about 1,000 Btu/ft^3 (10.4 kW/m^3) for natural gas], the volume of gas required per hour can be calculated. The heating value can be obtained for the gas supplier. The input rating of all appliances is required to be shown on the appliance label. See the Commentary for Section 301.5. When the appliances' input rating is unknown at the time of the piping system design, the gas demand will have to be estimated based on input from appliance manufacturers, gas utilities, or other source. See Table G2413.2. However, when the actual appliance demands are determined upon appliance selection, the approximate calculations done in accordance with this section must be checked to verify that the piping system is, in fact, adequate to supply the known demand for all appliances. The designer must be aware of the fact that a system designed in accordance with an estimated load could be undersized relative to the actual connected load.

TABLE G2413.2 (402.2)
APPROXIMATE GAS INPUT FOR TYPICAL APPLIANCES

APPLIANCE	INPUT BTU/H (Approx.)
Space Heating Units	
Hydronic boiler	
Single family	100,000
Multifamily, per unit	60,000
Warm-air furnace	
Single family	100,000
Multifamily, per unit	60,000
Space and Water Heating Units	
Hydronic boiler	
Single family	120,000
Multifamily, per unit	75,000
Water Heating Appliances	
Water heater, automatic instantaneous	
Capacity at 2 gal./minute	142,800
Capacity at 4 gal./minute	285,000
Capacity at 6 gal./minute	428,400
Water heater, automatic storage, 30- to 40-gal. tank	35,000
Water heater, automatic storage, 50-gal. tank	50,000
Water heater, domestic, circulating or side-arm	35,000

(Continued)

TABLE G2413.2 (402.2)—continued
APPROXIMATE GAS INPUT FOR TYPICAL APPLIANCES

APPLIANCE	INPUT BTU/H (Approx.)
Cooking Appliances	
Built-in oven or broiler unit, domestic	25,000
Built-in top unit, domestic	40,000
Range, free-standing, domestic	65,000
Other Appliances	
Barbecue	40,000
Clothes dryer, Type 1 (domestic)	35,000
Gas fireplace, direct vent	40,000
Gas light	2,500
Gas log	80,000
Refrigerator	3,000

For SI: 1 gallon = 3.785 L, 1 gallon per minute = 1 British thermal unit per hour = 0.293 W, .

In all cases, the piping must have the capacity to supply the actual connected load of appliances. As also stated in Section G2411.8, the load on the piping system must be based on the simultaneous operation of all appliances at full output. The only exception to this would be the demonstration of actual load diversity. For example, the designer could have access to data that demonstrates that the total load for cooking ranges or water heaters in a multiple family dwelling (apartment building) is less than the sum of all of the appliance demands at full output. The code official would have to be convinced that any such diversity factor has been established. Any known diversity factors would likely come from the serving gas supplier.

The following is an example of how gas demand is calculated for a representative appliance. An appliance has a nameplate (label) input rating of 120,000 Btu/h (35.2 kW). Using natural gas with an average heating value of 1,000 Btu/ft^3 (10.4 kW/m^3), the volume of gas required for the appliance is: 120,000 ÷ 1,000 = 120 cubic feet per hour (3.52 m^3/h).

G2413.3 (402.3) Sizing. Gas piping shall be sized in accordance with one of the following:

1. Pipe sizing tables or sizing equations in accordance with Section G2413.4.
2. The sizing tables included in a listed piping system's manufacturer's installation instructions.
3. Other approved engineering methods.

❖ Using the sizing tables is generally simpler and less time consuming than using the equations found in Section G2413.4. The gas system pressure, allowable pressure drop, maximum gas demand, and the specific gravity of the fuel gas must be known in order to use tables. The allowable pressure drop is the designer's choice because the intent of the code is to provide the appliance or equipment with the required rate of gas flow at the required minimum pressure. The tables are based on a given pressure drop and cannot be used if the designer wishes to use any other pressure drop. The sizing tables are based on using smooth bore pipe or tube and can be used with the piping materials listed in Section G2414. Pipe materials with higher resistance to flow cannot be sized by these tables. CSST is sized by CSST-specific tables. The pipe sizes in the pipe sizing tables are based on the traditional steel pipe size designation as used in the material standards. More information may be obtained from a variety of pipe design publications, including the Piping Handbook by Sabin Crocker and the ASHRAE Handbook of HVAC Systems and Equipment. Each of the tables presents different piping material and design parameters and represents the most common system design applications. The tables assume that the system being sized is constructed of a single type pipe or tubing material throughout; thus, a single table cannot be used for systems constructed of multiple materials. For example, it is common to find hybrid systems composed of both steel pipe and CSST. Such systems would have to be sized from two different tables based on the longest run of the entire system.

The code does not limit sizing methodology to the tables or equations. Manufacturers of gas piping/tubing may provide sizing charts, tables, or slide calculators that, in effect, become part of the installation instructions. Such is the case for CSST, which is a listed piping system.

Section G2413.4 provides a sizing method which uses an exact thermodynamic flow equation to determine the required pipe sizes. Although this is an exact method, the degree of accuracy it affords may not always prove valuable, because pipe sizes are standardized and the pipe diameter calculated may not correspond to an available pipe size. For example, a calculated pipe diameter of 0.80 inches (20.3 mm) will require the selection of the next larger standard pipe size [1-inch (25.4 mm) pipe].

There are two equations to be used, depending on the gas pressure within the piping system.

As stated in Appendix A, A.2.2, and A.2.3, an equivalent length of pipe should be added to the length of any piping run having four or more fittings. See Appendix A, Table A.2.2. Many designers simply add 50 percent of the actual length of piping as an all-inclusive fitting allowance (i.e., actual pipe length x 1.5). This is a conservative estimate and is used because the necessary number and type of fittings is usually not known when the system is being designed.

G2413.4 (402.4) Sizing tables and equations. Where Tables G2413.4(1) through G2413.4(6) are used to size piping or tubing, the pipe length shall be determined in accordance with Section G2413.4.1, G2413.4.2 or G2413.4.3.

Where Equations 24-3 and 24-4 are used to size piping or tubing, the pipe or tubing shall have smooth inside walls and the pipe length shall be determined in accordance with Section G2413.4.1, G2413.4.2 or G2413.4.3.

1. Low-pressure gas equation [Less than 15 pounds per square inch (psi) (10.3 kPa)]:

$$D = \frac{Q^{0.381}}{19.17\left(\frac{\Delta H}{C_r \times L}\right)^{0.206}} \quad \textbf{(Equation 24-3)}$$

2. High-pressure gas equation [1.5 psi (10.3 kPa) and above]:

$$D = \frac{Q^{0.381}}{18.93\left[\frac{(P_1^2 - P_2^2) \times Y}{C_r \times L}\right]^{0.206}} \quad \textbf{(Equation 24-4)}$$

where:

D = Inside diameter of pipe, inches (mm).

Q = Input rate appliance(s), cubic feet per hour at 60°F (16°C and 30-inch mercury column

P_1 = Upstream pressure, psia (P_1 + 14.7)

P_2 = Downstream pressure, psia (P_2 + 14.7)

L = Equivalent length of pipe, feet

ΔH = Pressure drop, inch water column (27.7 inch water column = 1 psi)

TABLE G2413.4 (402.4)
C_r AND Y VALUES FOR NATURAL GAS AND UNDILUTED PROPANE AT STANDARD CONDITIONS

GAS	EQUATION FACTORS	
	C_r	Y
Natural gas	0.6094	0.9992
Undiluted propane	1.2462	0.9910

For SI: 1 cubic foot = 0.028 m^3, 1 foot = 305 mm, 1 inch water column = 0.1249 kPa, 1 pound per square inch = 6.895 kPa, 1 British thermal unit per hour = 0.293 W.

❖ The length of pipe or tube to be used with the tables and equations is determined by the sizing method being used. Section G2413.4.1 involves only one length whereas Sections G2413.4.2 and G2413.4.3 involve multiple lengths.

Example 1:

Determine the required pipe sizes for the gas distribution system in Commentary Figure G2413.4 using the equation method outlined in this section. The system is designed to operate at less than 0.5 pounds per square inch (psig), using steel pipe. Assume the following:

Fuel gas: Natural

Specific gravity: 0.60

Heating value: 1,000 Btu per cubic foot

Maximum allowable pressure drop: $^1/_2$ inch of water column

Negligible number of fittings

For SI: 1 inch = 25.4 mm, 1 foot = 304.8 mm, 1 Btu/h = 0.2931 W, 1 Btu/ft^3 = 10.35 W/m^3, 1 inch water column = 0.2488 kPa, 1 cubic foot = 0.20832 m^3/h, 1 pound per square inch gauge = 6.895 kPa.

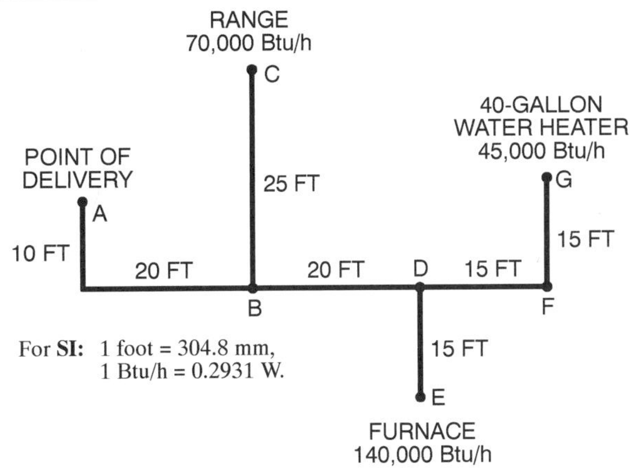

Figure G2413.4
GAS PIPING LAYOUT

To use equations in this section, the gas volume of each appliance must be known.

Gas volume of an appliance = Input rating of an appliance (Btu/h)/Heating value of the fuel gas (Btu/ft^3)

APPLIANCE	INPUT RATING (Btu/h)	GAS VOLUME (ft^3/h)
Range	70,000	70
Furnace	140,000	140
Water heater	45,000	45

Using the equation for maximum demand quantity *(Q)*, we can solve for the minimum pipe diameter *(D)* by rearranging it to read as follows:

$D = Q^{0.381}/19.17\ [(\Delta H) / (Cr \times L)]^{0.206}$

Starting at the most remote appliance from the point of delivery (Point G - water heater), determine the minimum pipe size for pipe segment D.

$D = (45)^{0.381}/19.17\ [(^1/_2) / (0.609 \times 80)]^{0.206} = 0.574$ inches

Repeat the calculation for each pipe segment. The gas piping system is sized as follows:

Pipe segment	Flow rate *(Q)*	Calculated pipe Diameter *(D)* (in.)	Nominal pipe size (in.)
A-B	255	1.11	1$^1/_4$
B-C	70	0.679	$^3/_4$
B-D	185	0.983	1
D-E	140	0.884	1
D-G	45	0.574	$^1/_2$

Most gas pipe sizing applications will be completed using the sizing tables. In the event that circumstances warrant it, however, the pipe sizing may be done by calculation using the provisions of Appendix A or other approved methods. If in a particular design the piping length, desired pressure loss, system supply pressure, specific gravity, or pipe size does not fall within the parameters of a table, an approved alternate sizing procedure must be employed.

In lieu of the methods specified in this section, piping sizes may be determined by the use of accurate gas flow, computer, or pressure drop charts. Such sizing tools must be acceptable to the code official. Calculators, charts, and computer programs can be designed to size gas piping for any pressure drop since the intent of the code is to make sure that the supply pressure at the appliance is greater than the minimum pressure required for proper operation of the appliance. See Section G2413.5.

TABLE G2413.4(1) [402.4(2)]
SCHEDULE 40 METALLIC PIPE

Gas	Natural
Inlet Pressure	0.5 psi or less
Pressure Drop	0.5 inch WC
Specific Gravity	0.60

	PIPE SIZE (in.)					
Nominal	$^1/_4$	$^3/_8$	$^1/_2$	$^3/_4$	1	$1^1/_4$
Actual ID	0.364	0.493	0.622	0.824	1.049	1.380
Length (ft)	**Maximum Capacity in Cubic Feet of Gas per Hour**					
10	43	95	175	360	680	1,400
20	29	65	120	250	465	950
30	24	52	97	200	375	770
40	20	45	82	170	320	660
50	18	40	73	151	285	580
60	16	36	66	138	260	530
70	15	33	61	125	240	490
80	14	31	57	118	220	460
90	13	29	53	110	205	430
100	12	27	50	103	195	400
125	11	24	44	93	175	360
150	10	22	40	84	160	325

For SI: 1 inch = 25.4 mm, 1 foot = 304.8 mm, 1 cubic foot = 0.0283 m^3/h, 1 pound per square inch = 6.895 kPa, 1-inch water column = 0.2488 kPa.

❖ See the commentary for Section G2413.4.

TABLE G2413.4(2) [402.4(3)]
SCHEDULE 40 METALLIC PIPE

Gas	Natural
Inlet Pressure	2.0 psi
Pressure Drop	1.0 psi
Specific Gravity	0.60

	PIPE SIZE (in.)			
Nominal	$^1/_2$	$^3/_4$	1	$1^1/_4$
Actual ID	0.622	0.824	1.049	1.380
Length (ft)	**Maximum Capacity in Cubic Feet of Gas per Hour**			
10	1,506	3,041	5,561	11,415
20	1,065	2,150	3,932	8,072
30	869	1,756	3,211	6,591
40	753	1,521	2,781	5,708
50	673	1,360	2,487	5,105
60	615	1,241	2,270	4,660
70	569	1,150	2,102	4,315
80	532	1,075	1,966	4,036
90	502	1,014	1,854	3,805
100	462	934	1,708	3,508
125	414	836	1,528	3,138
150	372	751	1,373	2,817

For SI: 1 inch = 25.4 mm, 1 foot = 304.8 mm, 1 cubic foot = 0.0283 m^3/h, 1 pound per square inch = 6.895 kPa.

❖ See the commentary for Section G2413.4.

Gas	Natural
Inlet Pressure	0.5 psi or less
Pressure Drop	0.5 inch WC
Specific Gravity	0.60

TABLE G2413.4(3) [402.4(8)]
SEMI-RIGID COPPER TUBING

TUBE SIZE (in.)							
Nominal	K & L	1/4	3/8	1/2	5/8	3/4	1
	ACR	3/8	1/2	5/8	3/4	7/8	1 1/8
Outside		0.375	0.500	0.625	0.750	0.875	1.125
Inside		0.305	0.402	0.527	0.652	0.745	0.995
Length (ft)		Maximum Capacity in Cubic Feet of Gas per Hour					
10		27	55	111	195	276	590
20		18	38	77	134	190	406
30		15	30	61	107	152	326
40		13	26	53	92	131	279
50		11	23	47	82	116	247
60		10	21	42	74	105	224
70		9.3	19	39	68	96	206
80		8.6	18	36	63	90	192
90		8.1	17	34	59	84	180
100		7.6	16	32	56	79	170
125		6.8	14	28	50	70	151
150		6.1	13	26	45	64	136

Note: Table capacities are based on Type K copper tubing inside diameter (shown), which has the smallest inside diameter of the copper tubing products.
For SI: 1 inch = 25.4 mm, 1 foot = 304.8 mm, 1 cubic foot per hour = 0.0283 m^3/h, 1 pound per square inch = 6.895 kPa, 1-inch water column = 0.2488 kPa.

❖ See the commentary for Section G2413.4.

Gas	Natural
Inlet Pressure	2.0 psi or less
Pressure Drop	1.0 inch WC
Specific Gravity	0.60

TABLE G2413.4(4) [402.4(11)]
SEMI-RIGID COPPER TUBING

TUBE SIZE (in.)							
Nominal	K & L	1/4	3/8	1/2	5/8	3/4	1
	ACR	3/8	1/2	5/8	3/4	7/8	1 1/8
Outside		0.375	0.500	0.625	0.750	0.875	1.125
Inside		0.305	0.402	0.527	0.652	0.745	0.995
Length (ft)		Maximum Capacity in Cubic Feet of Gas per Hour					
10		245	506	1,030	1,800	2,554	5,455
20		169	348	708	1,237	1,755	3,749
30		135	279	568	993	1,409	3,011
40		116	239	486	850	1,206	2,577
50		103	212	431	754	1,069	2,284
60		93	192	391	683	969	2,069
70		86	177	359	628	891	1,904
80		80	164	334	584	829	1,771
90		75	154	314	548	778	1,662
100		71	146	296	518	735	1,570
125		63	129	263	459	651	1,391
150		57	117	238	416	590	1,260

Note: Table capacities are based on Type K copper tubing inside diameter (shown), which has the smallest inside diameter of the copper tubing products.
For SI: 1 inch = 25.4 mm, 1 foot = 304.8 mm, 1 cubic foot per hour = 0.0283 m^3/h, 1 pound per square inch = 6.895 kPa.

❖ See the commentary for Section G2413.4.

Gas	Natural
Inlet Pressure	0.5 psi or less
Pressure Drop	0.5 inch WC
Specific Gravity	0.60

TABLE G2413.4(5) [402.4(14)] CORRUGATED STAINLESS STEEL TUBING (CSST)

	TUBE SIZE (EHD*)										
Flow Designation	13	15	18	19	23	25	30	31	37	46	62
Length (ft)	Maximum Capacity in Cubic Feet of Gas per Hour										
5	46	63	115	134	225	270	471	546	895	1,790	4,142
10	32	44	82	95	161	192	330	383	639	1,261	2,934
15	25	35	66	77	132	157	267	310	524	1,027	2,398
20	22	31	58	67	116	137	231	269	456	888	2,078
25	19	27	52	60	104	122	206	240	409	793	1,860
30	15	25	47	55	96	112	188	218	374	723	1,698
40	13	21	41	47	83	97	162	188	325	625	1,472
50	12	19	37	42	75	87	144	168	292	559	1,317
60	11	17	34	38	68	80	131	153	267	509	1,203
70	10	16	31	36	63	74	121	141	248	471	1,114
80	11	15	29	33	60	69	113	132	232	440	1,042
90	10	14	28	32	57	65	107	125	219	415	983
100	9	13	26	30	54	62	101	118	208	393	933
150	7	10	20	23	42	48	78	91	171	320	762
200	6	9	18	21	38	44	71	82	148	277	661
250	5	8	16	19	34	39	63	74	133	247	591
300	5	7	15	17	32	36	57	67	95	226	540

Note: Table includes losses for four 90-degree bends and two end fittings. Tubing runs with larger numbers of bends and/or fittings shall be increased by an equivalent length of tubing to the following equation: $L = 1.3n$ where L is additional length (ft) of tubing and n is the number of additional fittings and/or bends.

*EHD—Equivalent Hydraulic Diameter, which is a measure of the relative hydraulic efficiency between different tubing sizes. The greater the value of EHD, the greater the gas capacity of the tubing.

For SI: 1 inch = 25.4 mm, 1 foot = 304.8 mm, 1 cubic foot per hour = 0.0283 m^3/h, 1 pound per square inch = 6.895 kPa, 1-inch water column = 0.2488 kPa, 1 degree = 0.01745 rad.

❖ See the commentary for Section G2413.4.

Gas	Natural
Inlet Pressure	2.0 psi or less
Pressure Drop	1.0 inch WC
Specific Gravity	0.60

TABLE G2413.4(6) [402.4(17)] CORRUGATED STAINLESS STEEL TUBING (CSST)

	TUBE SIZE (EHD*)										
Flow Designation	13	15	18	19	23	25	30	31	37	46	62
Length (ft)	Maximum Capacity in Cubic Feet of Gas per Hour										
10	270	353	587	700	1,098	1,372	2,592	2,986	4,509	9,599	21,637
25	166	220	374	444	709	876	1,620	1,869	2,887	6,041	13,715
30	151	200	342	405	650	801	1,475	1,703	2,642	5,509	12,526
40	129	172	297	351	567	696	1,273	1,470	2,297	4,763	10,855
50	115	154	266	314	510	624	1,135	1,311	2,061	4,255	9,715
75	93	124	218	257	420	512	922	1,066	1,692	3,467	7,940
80	89	120	211	249	407	496	892	1,031	1,639	3,355	7,689
100	79	107	189	222	366	445	795	920	1,4e71	2,997	6,881
150	64	87	155	182	302	364	646	748	1,207	2,442	5,624
200	55	75	135	157	263	317	557	645	1,049	2,111	4,874
250	49	67	121	141	236	284	497	576	941	1,886	4,362
300	44	61	110	129	217	260	453	525	862	1,720	3,983
400	38	52	96	111	189	225	390	453	749	1,487	3,452
500	34	46	86	100	170	202	348	404	552	1,329	3,089

Notes to Table G2413.4(6):

1. Table does not include effect of pressure drop across the line regulator. Where regulator loss exceeds $^3/_4$ psi, DO NOT USE THIS TABLE. Consult with regulator manufacturer for pressure drops and capacity factors. Pressure drops across a regulator may vary with flow rate.
2. CAUTION: Capacities shown in table might exceed maximum capacity for a selected regulator. Consult with regulator or tubing manufacturer for guidance.
3. Table includes losses for four 90-degree bends and two end fittings. Tubing runs with larger numbers of bends and/or fittings shall be increased by an equivalent length of tubing to the following equation: $L = 1.3n$ where L is additional length (ft) of tubing and n is the number of additional fittings and/or bends.

*EHD—Equivalent Hydraulic Diameter, which is a measure of the relative hydraulic efficiency between different tubing sizes. The greater the value of EHD, the greater the gas capacity of the tubing.

For SI: 1 inch = 25.4 mm, 1 foot = 304.8 mm, 1 cubic foot = 0.0283 m^3/h, 1 pound per square inch = 6.895 kPa, 1 degree = 0.01745 rad.

❖ See the commentary for Section G2413.4.

G2413.4.1 (402.4.1) Longest length method. The pipe size of each section of gas piping shall be determined using the longest length of piping from the point of delivery to the most remote outlet and the load of the section.

❖ This section provides a step-by-step approach to proper application of the tables using the traditional longest length method. When using the tables, the maximum pipe length from the point of delivery to the farthest outlet must be determined, including any allowance for the equivalent length of fittings. The designer will use the column (or row) in the table that equals the determined length or the next higher column (or row) if the determined length is between table values. The system will be designed using only values taken from this column (or row). This is known as the longest run method. Basing the sizing on the most demanding circuit (longest run) compensates for pressure losses throughout the entire system. See Commentary Figure G2413.4.2 and the examples that are part of the commentary for Section G2413.4.2.

G2413.4.2 (402.4.2) Branch length method. Pipe shall be sized as follows:

1. Pipe size of each section of the longest pipe run from the point of delivery to the most remote outlet shall be determined using the longest run of piping and the load of the section.
2. The pipe size of each section of branch piping not previously sized shall be determined using the length of piping from the point of delivery to the most remote outlet in each branch and the load of the section.

❖ This sizing method is a variation of the longest length method, and the results can be less conservative. Since less headroom is built into this method, it is especially important to account for the equivalent length of fittings installed in the system. This method involves multiple piping lengths within a system for application of the tables or equations, whereas the longest length method involves only one piping length per system. See Commentary Figure G2413.4.2 and the example below.

Example (refer to Commentary Figure G2413.4.2):

1. In accordance with Section G2413.4.2, item 1, determine the size of sections A, B, C, D and E (constructed of Schedule 40 steel pipe) based on the load of each section and the longest run length of 85 feet. Because the system pressure is less than 0.5 psi (14 in. w.c.), Table G2413.4(1) is chosen. Because the longest run length is between rows in the table, the 90-foot row must be chosen.

PIPE SECTION	LOAD MBH	SIZE (Inches)
A	230	$1^1/_4$
B	170	1
C	135	1
D	95	$^3/_4$
E	20	$^3/_8$

2. In accordance with Section G2413.4.2, item 2, determine the size of branch sections F, G, H, I and J (constructed of CSST) based on the load of each section and the length of piping and tubing from the point of delivery to the outlet on that section. Table G2413.4(5) is chosen because the branch sections are constructed of CSST and the pressure is less than 0.5 psi (14 in. w.c.). Where a length falls between entries in the table, use the next longer length row.

CSST BRANCH	LOAD MBH	LENGTH OF PIPING AND TUBING (Feet)	SIZE EHD
F	20	85	18
G	60	25	19
H	35	40	18
I	40	60	23
J	75	75	30

G2413.4.3 (402.4.3) Hybrid pressure. The pipe size for each section of higher pressure gas piping shall be determined using the longest length of piping from the point of delivery to the most remote line pressure regulator. The pipe size from the line pressure regulator to each outlet shall be determined using the length of piping from the regulator to the most remote outlet served by the regulator.

❖ Hybrid pressure piping systems convey gas at different pressures in different parts of the system. One or more pressure regulators are used to reduce the pressure in portions of the system. In some cases, a single gas piping system must deliver widely varying pressure to loads throughout a building. The most common use of hybrid pressure systems allows small, economically sized piping to carry large loads over long distances by conveying gas at high pressures. Near the load being supplied, pressure regulators reduce the higher pressure to suit the load. For example, rooftop units on buildings are typically designed for a natural gas pressure of 14 inches WC or less. Running hundreds of feet of pipe across expansive roofs would require large, expensive pipe at such low pressure. At 5 psi pressure, small, less expensive pipe can be run over long distances, and regulators at each rooftop unit reduce the 5 psi pressure to less than $^1/_2$ psi, typically 8 inches WC.

G2413.5 (402.5) Allowable pressure drop. The design pressure loss in any piping system under maximum probable flow conditions, from the point of delivery to the inlet connection of the equipment, shall be such that the supply pressure at the equipment is greater than the minimum pressure required for proper equipment operation.

❖ Between the point of delivery and the load, gas pressure losses will occur because of pipe friction. The design of the system must control these losses such that the pressure at the load (appliance) connection point will be in excess of the minimum required pressure as dictated by the appliance manufacturer. As stated in the commentary to Section G2412.8, inadequate gas pressures can result in dangerous appliance operation. This section speaks of "maximum probable flow conditions," which must be taken to mean the total connected load of all appliances operating at full capacity except where a diversity of load factor is applied.

The designer can choose any pressure loss for a piping system, provided that the goal of this section is met. For example, an appliance has a minimum required inlet pressure of 5 inches of water column, and the gas pressure at the point of delivery is 8 inches of water column. The maximum allowable pressure drop through the piping system would be just under 3 inches WC because the pressure at the appliance connection must be greater than 5 inches WC. 5.1 inches WC is greater than 5.0 inches WC and would satisfy the intent of this section. Sizing tables such as Table G2413.4(1) and G2413.4 (3) are based on a very small pressure drop (0.5 inches WC) and thus are quite conservative sizing methods that account for the friction losses through a limited number of fittings used in a piping system. The tables are optional sizing methods placed in the code for convenience. Piping systems are permitted to be designed for the exact pressure losses allowed by this section.

G2413.6 (402.6) Maximum design operating pressure. The maximum design operating pressure for piping systems located inside buildings shall not exceed 5 psig (34 kPa gauge) except where one or more of the following conditions are met:

1. The piping system is welded.
2. The piping is located in a ventilated chase or otherwise enclosed for protection against accidental gas accumulation.
3. The piping is a temporary installation for buildings under construction.

❖ This section is specific to indoor piping systems. The pressure is limited to 5 psig to minimize the consequences of a leak. Also, the higher the pressure, the

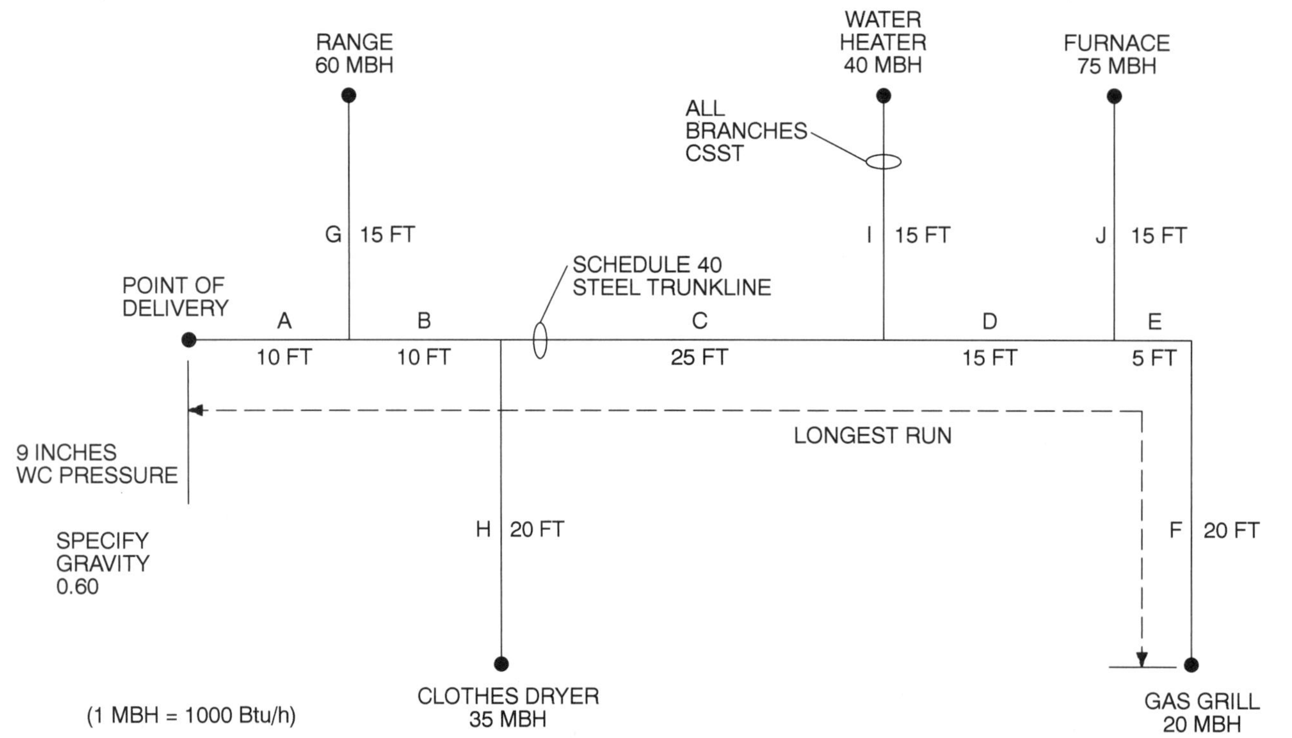

Figure G2413.4.2
BRANCH PIPING SYSTEM WITH SCHEDULE 40 STEEL PIPE TRUNKLINE AND CSST BRANCHES

more difficult it is to contain any gas, and thus leakage is more likely. Item 1 recognizes the superior strength and integrity of steel piping with welded joints. Item 2 allows the option of higher pressures where the piping is enclosed so as to prevent leakage from entering the building. Item 3 allows higher pressures in specific occupancies where the risks are acceptable because of low occupant density, the presence of trained maintenance and emergency response personnel, and/or the well ventilated nature of such buildings. Gas utilization equipment in some of the occupancies listed might require gas to be supplied at higher than 5 psig pressure. Item 4 recognizes the unique nature of temporary piping on construction sites where small piping is desirable, thus making higher pressures necessary.

G2413.6.1 (402.6.1) Liquefied petroleum gas systems. The operating pressure for undiluted LP- Gas systems shall not exceed 20 psig (140 kPa gauge). Buildings having systems designed to operate below -5°F (-21°C) or with butane or a propane-butane mix shall be designed to either accommodate liquid LP-Gas or prevent LP-Gas vapor from condensing back into a liquid.

❖ The maximum operating pressure for LP-gas distribution systems applies to indoor and outdoor installations with the same concerns as expressed in the Commentary to Section G2413.5. In low-temperature environments, LP gases might condense into the liquid state within the piping system. If the liquid fuel enters system components or appliances, dangerous malfunctions could result. At atmospheric pressure, 14.7 psia (sea level). propane will condense at 44°F and butane will condense at 15°F. Within the piping system the pressure could be higher than one atmosphere and would therefore condense at a higher temperature. For example, at 20 psig, propane will condense at 5°F, hence the 5 degree threshold requirement of this section. The exception allows higher pressures in the specified occupancies if the buildings that house such occupancies are constructed in accordance with NFPA 58. Chapter 7 of NFPA 58 regulates buildings that house LP-gas distribution facilities and contains provisions for construction type, height, explosion venting, ventilation, heating systems and fire rated separations, among others.

For SI: 1 cubic foot = 0.028 m^3, 1 foot = 0.305 m,
1-inch water column = 0.249 kPa, 1 psig = 6.894 kPa,
1,000 Btu/h = 0.2931 W, 90 degrees = 1.6 rad

SECTION G2414 (403) PIPING MATERIALS

G2414.1 (403.1) General. Materials used for piping systems shall comply with the requirements of this chapter or shall be approved.

❖ This section dictates what materials and components are permitted to be used to construct gas distribution systems and also specifies the allowable applications of such materials.

G2414.2 (403.2) Used materials. Pipe, fittings, valves or other materials shall not be used again unless they are free of foreign materials and have been ascertained to be adequate for the service intended.

❖ Ideally, gas pipe installations should be constructed of new materials. However, this section recognizes that there are occasions when used piping materials may be perfectly acceptable for an installation provided that they meet a number of very strict criteria that intend to reduce the potential for poor performance. To judge equivalency of reused piping materials, compliance with approval criteria for new materials should be used.

G2414.3 (403.3) Other materials. Material not covered by the standards specifications listed herein shall be investigated and tested to determine that it is safe and suitable for the proposed service, and, in addition, shall be recommended for that service by the manufacturer and shall be approved by the code official.

❖ This section echoes the intent of Section R104.11.

G2414.4 (403.4) Metallic pipe. Metallic pipe shall comply with Sections G2414.4.1 and G2414.4.2.

❖ This section addresses the traditional metal pipes used in piping installations.

G2414.4.1 (403.4.1) Cast iron. Cast-iron pipe shall not be used.

❖ Because cast iron pipe is brittle compared to malleable metals, it is more likely to fail under stress and therefore is not suitable for conveying fuel gases.

G2414.4.2 (403.4.2) Steel. Steel and wrought-iron pipe shall be at least of standard weight (Schedule 40) and shall comply with one of the following:

1. ASME B36.10, 10M; or
2. ASTM A 53; or
3. ASTM A 106.

❖ Steel pipe must be Schedule 40 or heavier, must comply with one of the listed standards, and must be black iron or galvanized. Contrary to popular belief, natural gas does not adversely react with the zinc coating on galvanized pipe.

G2414.5 (403.5) Metallic tubing. Seamless copper, aluminum alloy or steel tubing shall be permitted to be used with gases not corrosive to such material.

❖ Aluminum and steel tubing are used in the manufacturing of appliances but are rarely used for gas distribution. Copper tubing is widely used in gas distribution systems and as an alternative to CSST in parallel distribution manifold systems. As with copper pipe, copper tubing can not be used with gases that will cause it to corrode. See Section G2414.5.2. Commentary Figure G2414.5 illustrates a type of sheathed copper tubing that is designed for gas distribution.

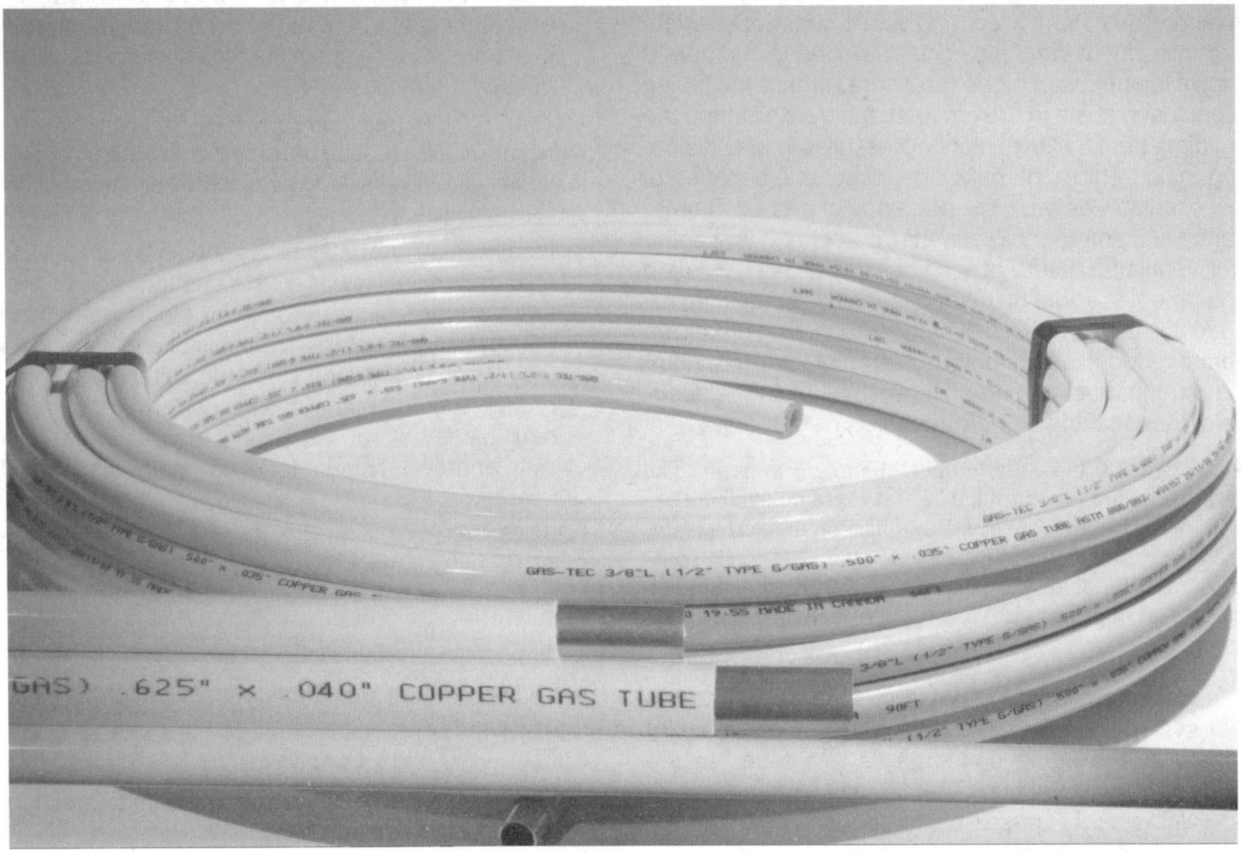

Photo courtesy of Kamco Products

Figure G2414.5
SHEATHED COPPER TUBING FOR GAS DISTRIBUTION

G2414.5.1 (403.5.1) Steel tubing shall comply with ASTM A 539 or ASTM A 254.

❖ Steel tubing is used only in industrial applications and in the construction of appliances.

G2414.5.2 (403.5.2) Copper tubing. Copper tubing shall comply with standard Type K or L of ASTM B 88 or ASTM B 280.

Copper and brass tubing shall not be used if the gas contains more than an average of 0.3 grains of hydrogen sulfide per 100 standard cubic feet of gas (0.7 milligrams per 100 liters).

❖ ASTM B88 applies to copper water tubing, and Type K and L refers to wall thickness. Type M tubing is the thinnest wall and is not allowed for gas service. The serving gas supplier must be contacted to determine the hydrogen sulfide content of the gas.

G2414.5.3 (403.5.4) Corrugated stainless steel tubing. Corrugated stainless steel tubing shall be tested and listed in compliance with the construction, installation and performance requirements of ANSI LC 1.

❖ ANSI LC-1 provides material and performance criteria and installation requirements for corrugated stainless steel tubing (CSST) gas distribution systems. CSST systems are used as an alternative to more traditional gas piping systems for the distribution of natural gas within buildings.

Corrugated stainless steel tubing is a semirigid stainless steel tubing with a plastic jacket and is available in coil lengths varying from 100 to 250 feet (30 480 to 76 200 mm) and in the following standard sizes: 3/8, $^1/_2$, $^3/_4$, and 1 inch (9.5, 12.7, 19.1, and 25.4 mm) (larger sizes may be available). See Commentary Figure G2414.5.3(1). Special proprietary fittings are required for connection of the tubing. See Commentary Figure G2414.5.3(2) through G2414.5.3(4). CSST can be used in low-pressure systems [up to 0.5 psig (14 inches of water column) (3.45 kPa)] and in higher pressure systems operating at pressures up to 5 psig (3.45 kPa). Commentary Figure G2414.5.3(5) shows a typical manifold installation using CSST and shows a multiport manifold with a pressure regulator and shutoff valve installed. To prevent damage or puncture by screws and nails, specialized steel shield plates and protective flexible conduit are used to protect CSST systems that pass through or near structural members. Such protection must employ the parts specified by the CSST manufacturer. See Commentary Figures G2414.5.3(1), (6), (9) and (10).

CSST systems are usually designed as parallel distribution (manifold) systems, whereas traditional gas distribution systems are typically designed as branch

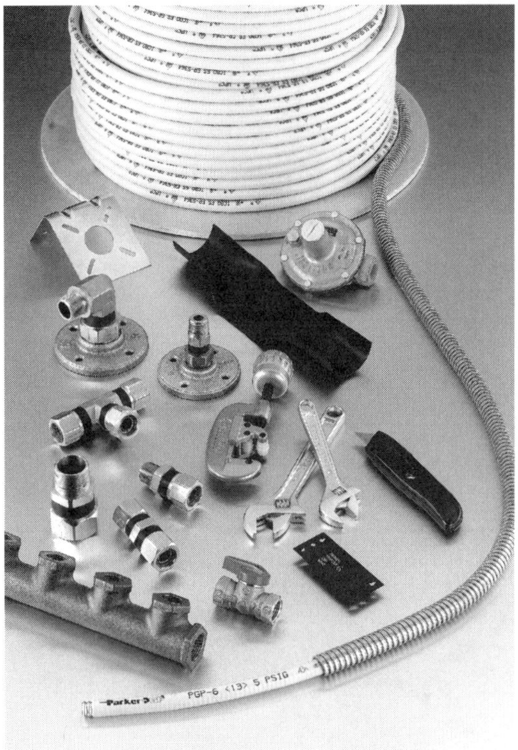

Photo courtesy of Parker Hannifin

Figure G2414.5.3(1)
CSST AND SYSTEM COMPONENTS

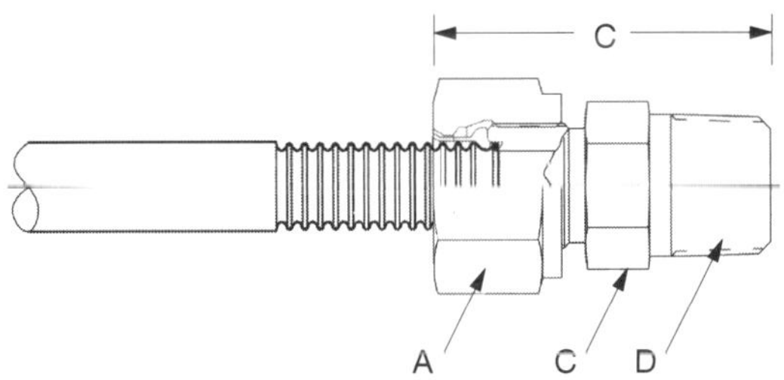

Figure courtesy of Parker Hannifin

Figure G2414.5.3(2)
CSST FITTING

(series) systems. See Commentary Figure G2414.5.3(7).

Medium-pressure CSST systems, typically 2 psig (13.8 kPa) maximum, are designed to allow the use of small diameter tubing for long runs. Significant distribution pressure losses can be compensated for by providing sufficiently high supply pressures. Medium-pressure distribution systems require the installation of pressure regulators, commonly called pounds-to-inches regulators, to reduce the supply pressure to the pressure required for delivery to the utilization equipment. See the Commentary for Section G2421 and Commentary Figure G2414.5.3 (8). Systems with pressures over 2 psi that supply appliances designed for a maximum inlet pressure of 14 inches WC must be provided with over-pressure protection devices (OPD) in addition to the "pounds-to-inches" line pressure regulator.

Section G2415.3 prohibits tubing fittings in concealed locations but does not apply to CSST fittings. In accordance with the manufacturer's installation instructions and the material listing, CSST fittings can be concealed in the building construction. The tubing, fittings, shields, and outlet terminals are all part of a system and must be installed together as a system. The manufacturer's installation instructions are to be considered as a complete system installation manual, which also includes the proprietary sizing criteria. Note that unless specific to CSST, the sizing tables in Section G2414 do not apply to and cannot be used for CSST systems.

CSST is not allowed to be installed underground (directly buried) unless it is of a type specially designed and listed for underground applications. Some manufacturers' installation instructions allow CSST to be "indirectly" buried by having the tubing installed within a non-metallic, watertight conduit such as PVC, PE, or ABS plastic pipe. See Sections G2415.6 and 2415.11 regarding piping buried in slabs and underground beneath buildings. CSST is allowed by the manufacturer's installation instructions to connect directly to fixed-in-place, nonportable, nonmoveable appliances such as furnaces, boilers, and water heaters. CSST is not intended for direct connection to moveable appliances such as clothes dryers or ranges nor is it intended to connect to appliances suspended on chains such as radiant heaters. It is also not allowed to connect directly to log lighters installed in solid fuel burning fireplaces. Direct connection is allowed for gas fireplace appliances in accordance with the CSST and fireplace manufacturers' instructions. At range and clothes dryer locations, the CSST must terminate at a CSST termination fitting assembly, and a listed gas appliance connector must be used to connect the appliance to the termination fitting assembly. CSST must not be exposed to impact, vibration, or repeated movement. "Drops" to fixed-in-place, nonmoveable appliances must be well supported and routed to avoid being disturbed. In other words "flying runs" of CSST are prohibited. CSST, like copper tubing, is not flexible, but rather is semirigid, allowing it to be formed into shape as it is installed. CSST, like any metallic tubing, must not be repeatedly bent in the same location as this can result in kinking, work hardening, and structural failure.

Generally, the sale of CSST materials is restricted; only individuals who have been specially trained and certified are allowed to purchase and install CSST systems.

Traditionally, the utility company (gas supplier) service regulator delivers natural gas to the building at a

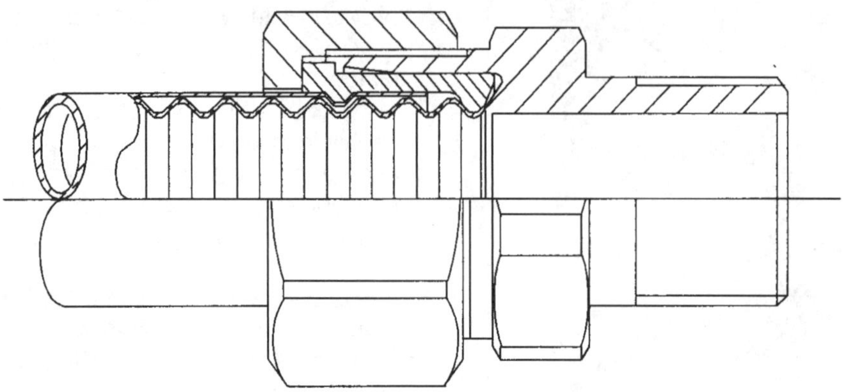

Figure courtesy of Gastite Division, Titeflex Corp.

Figure G2414.5.3(3)
CSST FITTING

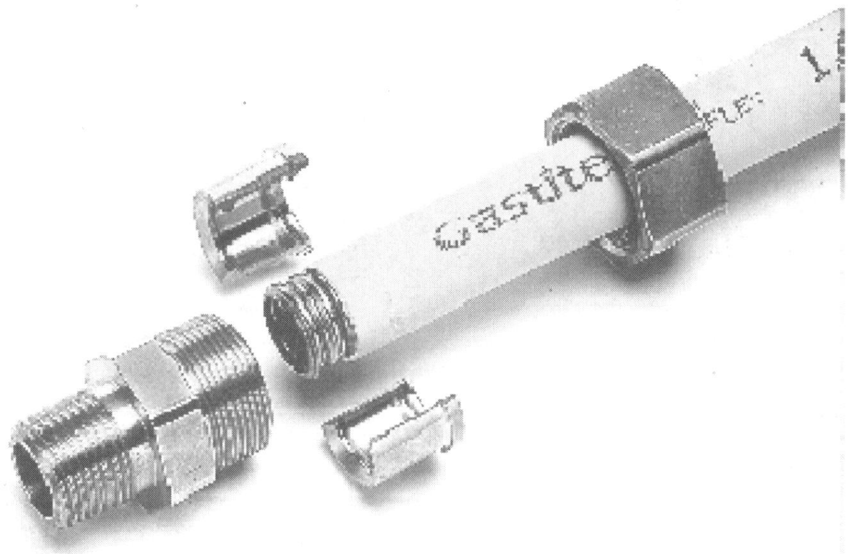

Photo courtesy of Gastite Division, Titeflex Corp.

Figure G2414.5.3(4)
CSST FITTING

pressure of approximately 6 to 10 inches of water column (1.49 to 2.49 kPa). Where medium-pressure (2 psig [13.8 kPa]) systems are used, the service regulator is set to deliver gas at a pressure of 2 psig (13.8 kPa). A medium-pressure gas distribution system must be properly designed to deliver gas to the connected appliances at pressures not exceeding the maximum inlet (supply) pressure for which the appliances are designed. A hazardous condition could develop if gas is supplied to appliances or equipment at pressures greater than the design inlet pressure of such appliances and equipment.

ANSI LC-1/CSA 6.26 addresses CSST systems and contains installation requirements. Installers and inspectors must have copies of the standard and the CSST manufacturer's installation instructions in order to properly install and inspect CSST systems. Manufacturers' instructions contain detailed requirements for protection hardware designed to prevent damage to installed CSST systems. As with most tubing systems, the most common installation errors involve abuse of the CSST tubing, lack of support, and improperly installed puncture protection devices or the omission of protection devices. CSST manufacturers publish design and installation guides that are very helpful and should be on the bookshelves of inspectors, installers, and designers. See Commentary Figures G2414.5.3(9) and G2414.5.3(10). Some CSST manufacturers offer preassembled manifolds in a rough-in cabinet to simplify installation. See Commen-

Photo courtesy of Titeflex Corporation

Figure G2414.5.3(5)
MULTIPORT MANIFOLD INSTALLATION USING CSST

Photo courtesy of Gastite Division, Titeflex Corp.

Figure G2414.5.3(6)
TOP PLATE TUBING PROTECTION SHIELD

tary Figure G2414.5.3(11). CSST transitions to rigid pipe appliance connections and listed appliance connectors must be made with fittings that prevent stress and rotational forces from being applied to the CSST. Commentary Figures G2414.5.3(12) through G2414.5.3(14) illustrate typical fitting assemblies used to make the transition from CSST to appliance connections or connectors.

G2414.6 (403.6) Plastic pipe, tubing and fittings. Plastic pipe, tubing and fittings shall be used outside underground only and shall conform with ASTM D 2513. Pipe to be used shall be marked "gas" and "ASTM D 2513."

❖ The installation of plastic pipe and tubing is limited to areas that are both outside the building and underground because of the potential hazard associated with the use of a material that has lower resistance to physical damage and heat compared to metallic pipe. Plastic pipe and tubing are widely used for underground gas distribution systems because of their ease of installation and inherent resistance to corrosion.

G2414.6.1 (403.6.1) Anodeless risers. Anodeless risers shall comply with the following:

1. Factory-assembled anodeless risers shall be recommended by the manufacturer for the gas used and shall be leak-tested by the manufacturer in accordance with written procedures.
2. Service head adapters and field-assembled anodeless risers incorporating service head adapters shall be recommended by the manufacturer for the gas used by the manufacturer and shall be designed certified to meet the requirements of Category I of ASTM D 2513, and U.S. Department of Transportation, Code of Federal Regulations, Title 49, Part 192.281(e). The manufacturer shall provide the user qualified installation instructions as prescribed by the U.S. Department of Transportation, Code of Federal Regulations, Title 49, Part 192.283(b).

❖ This section is an exception to the absolute prohibition of plastic gas piping aboveground. Plastic pipe is allowed to rise from underground within a steel riser conduit constructed as an anodeless riser assembly. Such risers are coated to resist corrosion and do not employ a sacrificial anode for the purpose of providing cathodic corrosion protection, hence the name "anodeless riser." The steel riser protects the PE plastic pipe from physical damage and terminates the plastic pipe with a threaded steel pipe transition fitting. A sweeping 90-degree bend is placed on the underground end of rigid risers to prevent the risers from rotating in the ground when threaded connections are made at the terminal end. The plastic pipe could be damaged by rotational and bending stresses transmitted by the riser; therefore, a horizontal portion of steel piping is provided at the bottom of the riser to limit such movement. These risers can eliminate the need for an underground plastic-to-metal transition fitting and simplify installation. They are used extensively by utility companies for service laterals and can be used for other installations such as supply runs to out-buildings, pool heaters, gas grills, and yard lights.

Service head adapters are also a type of terminal transition fitting that allow plastic pipe to be pulled through a retired steel gas service lateral and terminated at the point where the concentric piping passes through the building foundation wall. The adapter makes the transition from plastic pipe to steel pipe and seals the end of the retired steel service lateral. See

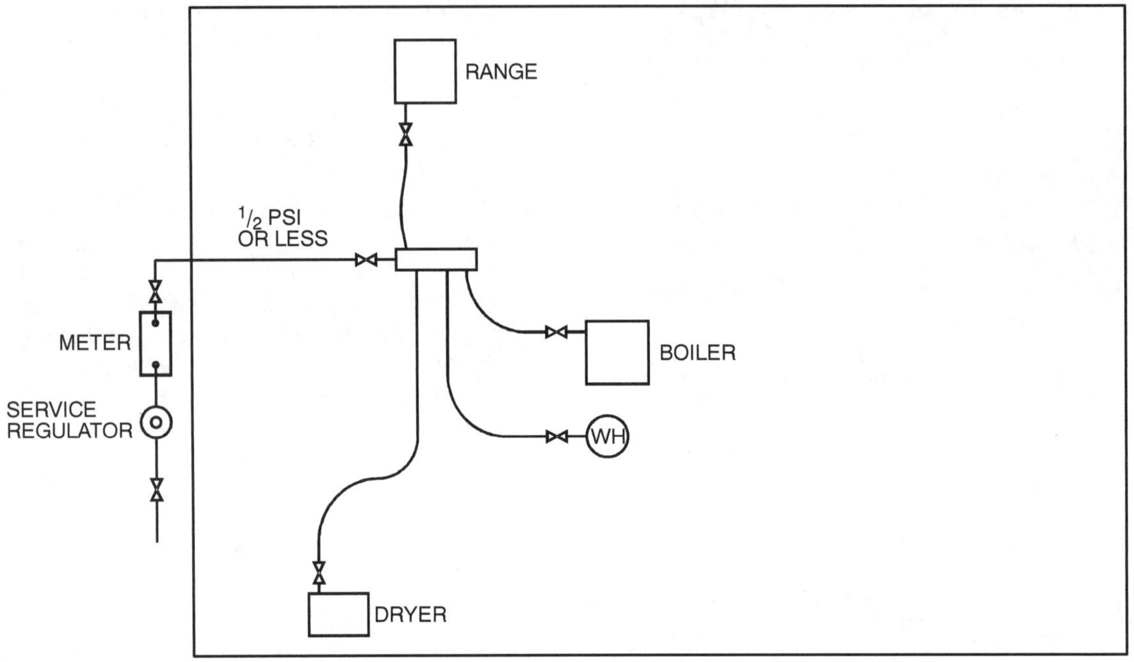

For SI: 1 pound per square inch = 6.895 kPa.

Figure G2414.5.3(7)
LOW-PRESSURE GAS PIPING SYSTEM USING CSST

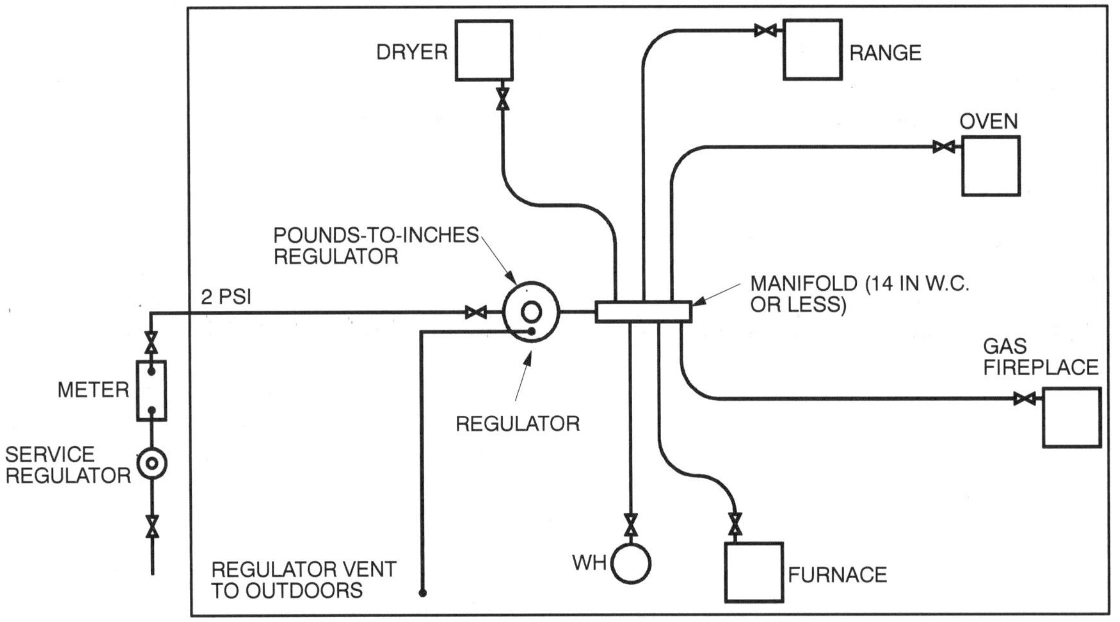

For SI: 1 pound per square inch = 6.896 kPa.

Figure G2414.5.3(8)
MEDIUM-PRESSURE GAS PIPING SYSTEM USING CSST

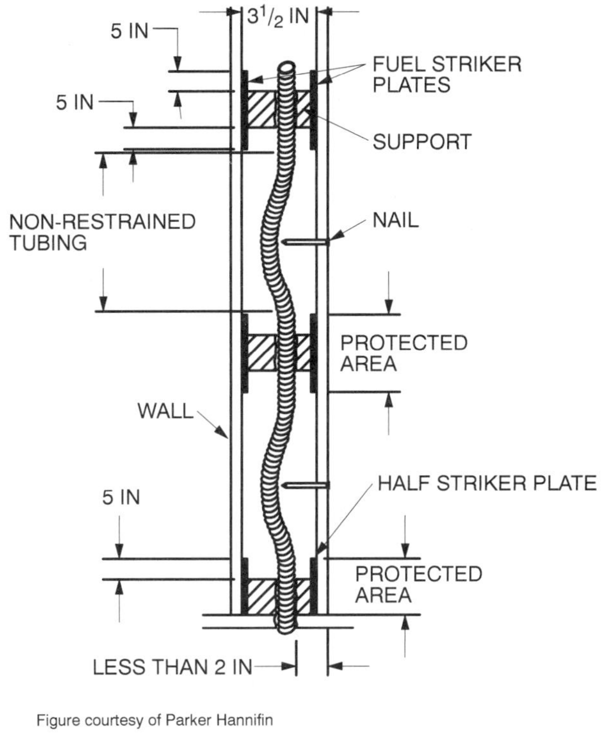

Figure G2414.5.3(9)
EXCERPT FROM CSST MANUFACTURER'S INSTALLATION INSTRUCTIONS

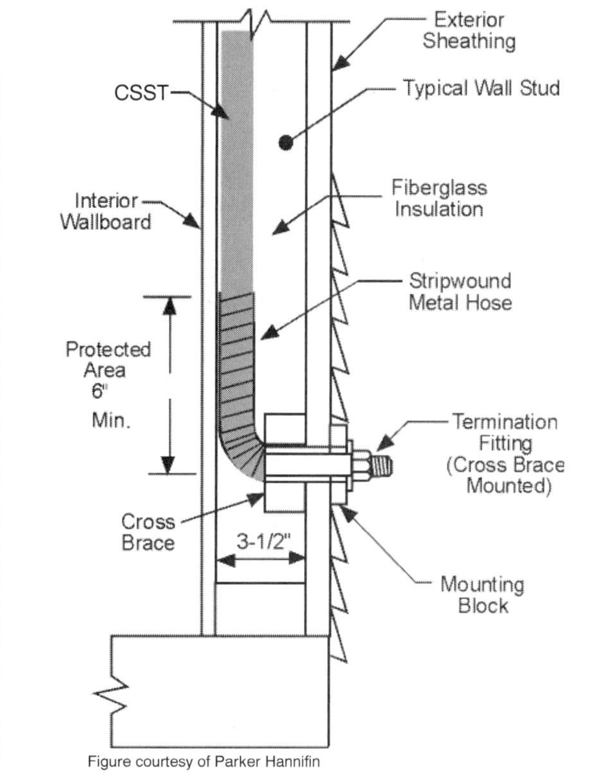

Figure G2414.5.3(10)
EXCERPT FROM CSST MANUFACTURER'S INSTALLATION INSTRUCTIONS

Commentary Figures G2403(1), G2414.6.1(1), G2414.6.1(2) and G2414.6.1(3).

G2414.6.2 (403.6.2) LP-Gas systems. The use of plastic pipe, tubing and fittings in undiluted liquefied petroleum gas piping systems shall be in accordance with NFPA 58.

❖ NFPA 58 limits plastic piping for LP-gas service to only polyethylene pipe and tubing. The PE pipe and tubing must be marked as being in compliance with ASTM D2513 and is allowed for LP-gas service only if so recommended by the material manufacturer.

G2414.7 (403.7) Workmanship and defects. Pipe or tubing and fittings shall be clear and free from cutting burrs and defects in structure or threading, and shall be thoroughly brushed, and chip and scale blown.

Defects in pipe or tubing or fittings shall not be repaired. Defective pipe, tubing or fittings shall be replaced. (See Section G2417.1.2.)

❖ This section stresses that proper installation procedures are necessary for a safe installation.

In the context of this section, good workmanship is an enforceable requirement. Burrs result from the cutting of pipe or tube and cause flow restriction. Pipe scale, dirt, construction site debris, metal chips from cutting and threading operations, and excess thread compound are potentially harmful contaminates that might cause damage to valves, regulators, and appliances. Damaged and/or improperly cut threads can cause leakage or joint failure. Defective materials must be replaced because attempts to repair them are always makeshift at best, and any cost savings would not be worth the risk.

G2414.8 (403.8) Protective coating. Where in contact with material or atmosphere exerting a corrosive action, metallic piping and fittings coated with a corrosion-resistant material shall be used. External or internal coatings or linings used on piping or components shall not be considered as adding strength.

❖ Protective coatings in the form of wrappings, tapes, enamels, epoxies, sleeves, casings, and factory-applied coverings are common methods of isolating metallic piping materials from the atmosphere or from the earth in which they are buried. Application of these materials under strictly controlled factory conditions improves the quality control process and reduces the likelihood of coating defects. Corrosion can be caused by weather exposure, burial in the soil, or construction materials that are in contact with the pipe. Cinders (also known as dross) are the waste residue from steel production and carry the waste products skimmed from the molten metal during the smelting process. Cinders are also the waste residue of coal-fired appliances and equipment. Such materials are extremely corrosive to metallic piping and may not be used in

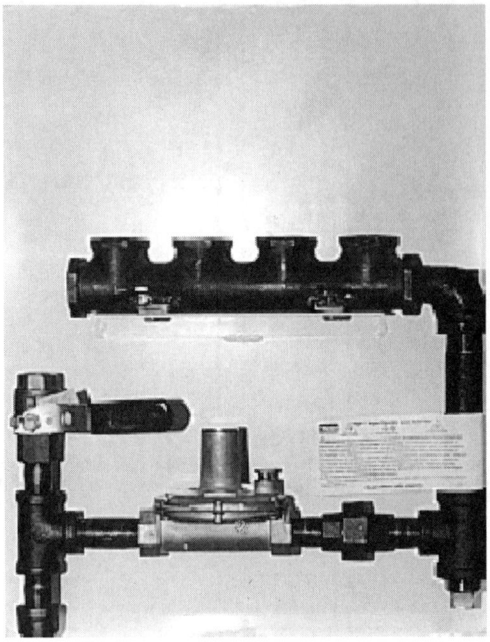

Photo courtesy of Parker Hannifin

**Figure G2414.5.3(11)
MANIFOLD FOR 2 PSI PARALLEL DISTRIBUTION SYSTEM**

Photo courtesy of Gastite Division, Titeflex Corp.

**Figure G2414.5.3(12)
STUB-OUT TERMINATION TRANSITION FITTINGS**

any form (backfill or cinder blocks, for example) where in direct contact with the pipe. Protection is usually provided by a factory-applied coating or by field wrapping the pipe with a protective covering, such as a coal tar-based or plastic wrapping. Galvanized piping in contact with soil would require additional coating or protection. The zinc coating does not usually provide long-term protection in underground applications. Where possible, a piping material is chosen that is not subject to the type of corrosion of the application.

Corrosion can also be caused by galvanic action that takes place where dissimilar metals are joined together in a current-carrying medium, such as soil or water. For example, if steel and copper pipe are joined together in a medium that conducts electrical current, the steel pipe will corrode at an accelerated rate because of the electrochemical process between the dissimilar metals. In this case, the soil acts as an electrolyte; the steel will be a sacrificial anode, and the copper will be a cathode. Because a cell is created, current will

Photo courtesy of Gastite Division, Titeflex Corp.

**Figure G2414.5.3(13)
VALVE BOX TERMINATION TRANSITION FITTING**

Photo courtesy of Gastite Division, Titeflex Corp.

**Figure G2414.5.3(14)
TERMINATION TRANSITION FITTING**

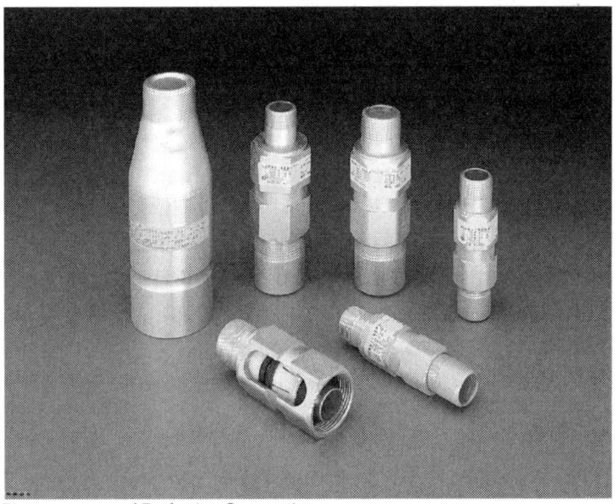

Photo courtesy of Perfection Corporation

Figure G2414.6.1(1)
SERVICE HEAD ADAPTERS

flow through the metal junction and the soil, resulting in the gradual deterioration of the steel. To protect against galvanic corrosion, dielectric fittings and couplings are used to join the piping together, thus breaking the circuit of the cell.

This section is commonly interpreted as requiring black steel pipe that is exposed to the weather to be primed and painted as the minimum form of protection.

G2414.9 (403.9) Metallic pipe threads. Metallic pipe and fitting threads shall be taper pipe threads and shall comply with ASME B1.20.1.

❖ Threads must be taper cut in accordance with the stated standard. The standard regulates tapered and straight threads, but this section specifies tapered threads. Tapered pipe threads, when made up, form a metal-to-metal (interference fit) seal. Pipe-joint compound or PTFE tape is to be applied to male threads only to decrease the possibility of tape fragments or compound entering the piping system. Such debris in piping systems can block orifices, restrict flow, or interfere with the operation of controls. The primary purpose of pipe-thread compounds is to act as a lubricant to allow proper tightening and to achieve a metal-to-metal seal. They also fill in small imperfections on the threaded surfaces. This section prohibits the use of straight tapped fittings such as couplings because the threaded joint will not provide the same seal as a joint made with taper-tapped fittings. Steel taper threaded couplings are easily identified, as they look exactly like the run of a tee fitting. See Commentary Figure G2414.9. The common designation "NPT" is misinterpreted as "nominal pipe threads" and actually stands for "national standard, pipe, tapered."

G2414.9.1 (403.9.1) Damaged threads. Pipe with threads that are stripped, chipped, corroded or otherwise damaged shall not be used. If a weld opens during the operation of cutting or threading, that portion of the pipe shall not be used.

❖ Damaged or improperly cut threads and defective welded-seam pipe must be eliminated from any gas piping system because of the potential for leaks.

G2414.9.2 (403.9.2) Number of threads. Field threading of metallic pipe shall be in accordance with Table G2414.9.2.

❖ The table sets forth the required number of threads per unit length of pipe based on the pipe size. These specifications are intended to be both maximums and mini-

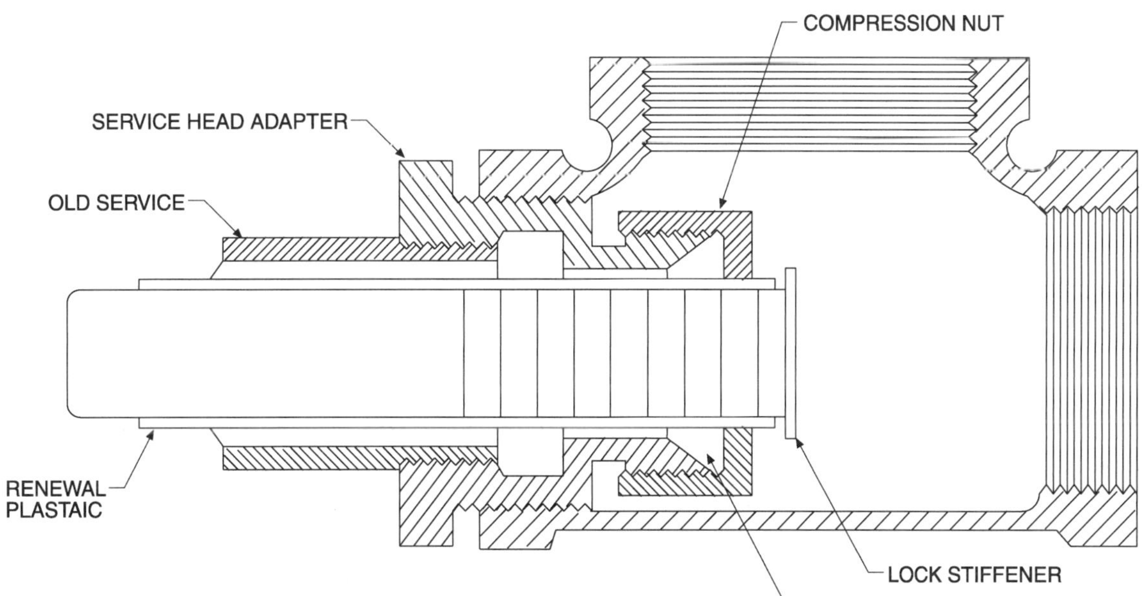

Figure courtesy of NORMAC

Figure G2414.6.1(2)
SERVICE HEAD ADAPTOR

Photo courtesy of NORMAC

Figure G2414.6.1(3)
FLEXIBLE ANODELESS RISER

mums since both undercutting and overcutting can result in faulty joints. Proper threading of pipe requires skill, practice, and the proper well-maintained tools.

TABLE G2414.9.2 (403.9.2)
SPECIFICATIONS FOR THREADING METALLIC PIPE

IRON PIPE SIZE (inches)	APPROXIMATE LENGTH OF THREADED PORTION (inch)	APPROXIMATE NO. OF THREADS TO BE CUT
$^1/_2$	$^3/_4$	10
$^3/_4$	$^3/_4$	10
1	$^7/_8$	10
$1^1/_4$	1	11
$1^1/_2$	1	11

For SI: 1 inch = 25.4 mm.

❖ See the commentary for Section G2413.9.2.

G2414.9.3 (403.9.3) Thread compounds. Thread (joint) compounds (pipe dope) shall be resistant to the action of liquefied petroleum gas or to any other chemical constituents of the gases to be conducted through the piping.

❖ Joint compounds in both paste and tape forms are commonly misapplied and used for the wrong application. Some compounds chemically react with the gas being conveyed, which could result in leakage. The label on the compound container will specify the applications for which the compound is suitable. Thread compounds act as a lubricant for the threads during assembly and also act as a sealant for the life of the joint. Care must be taken to keep all compounds out of the piping system interior as such contamination can cause damage to components and appliances. Pipe-joint compound or tape is limited to application on the male threads only to decrease the possibility of tape fragments or compound entering the piping system. Such debris in piping systems can block orifices, restrict flow, or interfere with the operation of safety controls. It is important to leave the first one or two threads on the end of the male threads bare to help prevent the compound from entering the piping system.

Once made from lead compounds and linseed oil, joint compound formulations now commonly contain PTFE (Teflon). The primary purpose of pipe-thread compounds is to act as a lubricant to allow proper tightening and to achieve a metal-to-metal seal. They also fill in small imperfections on the threaded surfaces. Pipe-thread compounds and tapes must be compatible with both the piping material and the contents of the piping.

LP-gases can be a solvent for some pipe joint compound formulations. Therefore, a type must be chosen that will not react with the gas.

G2414.10 (403.10) Metallic piping joints and fittings. The type of piping joint used shall be suitable for the pressure-temperature conditions and shall be selected giving consideration to joint tightness and mechanical strength under the service conditions. The joint shall be able to sustain the maximum end force due to the internal pressure and any additional forces due to temperature expansion or contraction, vibration, fatigue, or to the weight of the pipe and its contents.

❖ This section is in performance language and requires the designer and/or installer to choose the type of joining means that is appropriate for the application. Besides being leak-tight, the joints must be able to withstand the conditions of service. There are many joining methods addressed in the code, but not all of them are suitable for all applications. For example, PE pipe for LP-gas use is limited to a maximum pressure of 30 psig, and most tubing materials would be unsuitable for uses involving equipment vibration that is transferred to the tubing. Also, some fittings might not be suitable for exposure to extreme changes in temperature.

G2414.10.1 (403.10.1) Pipe joints. Pipe joints shall be threaded, flanged, brazed or welded. Where nonferrous pipe is brazed, the brazing materials shall have a melting point in excess of 1,000°F (538°C). Brazing alloys shall not contain more than 0.05-percent phosphorus.

❖ Joints in piping materials are allowed to be made by four methods: threaded (screwed), brazed, welded, and flanged. Flanged joints must also involve threading or welding as the means of attaching the flanges to the pipe. The limit on phosphorus content is related to

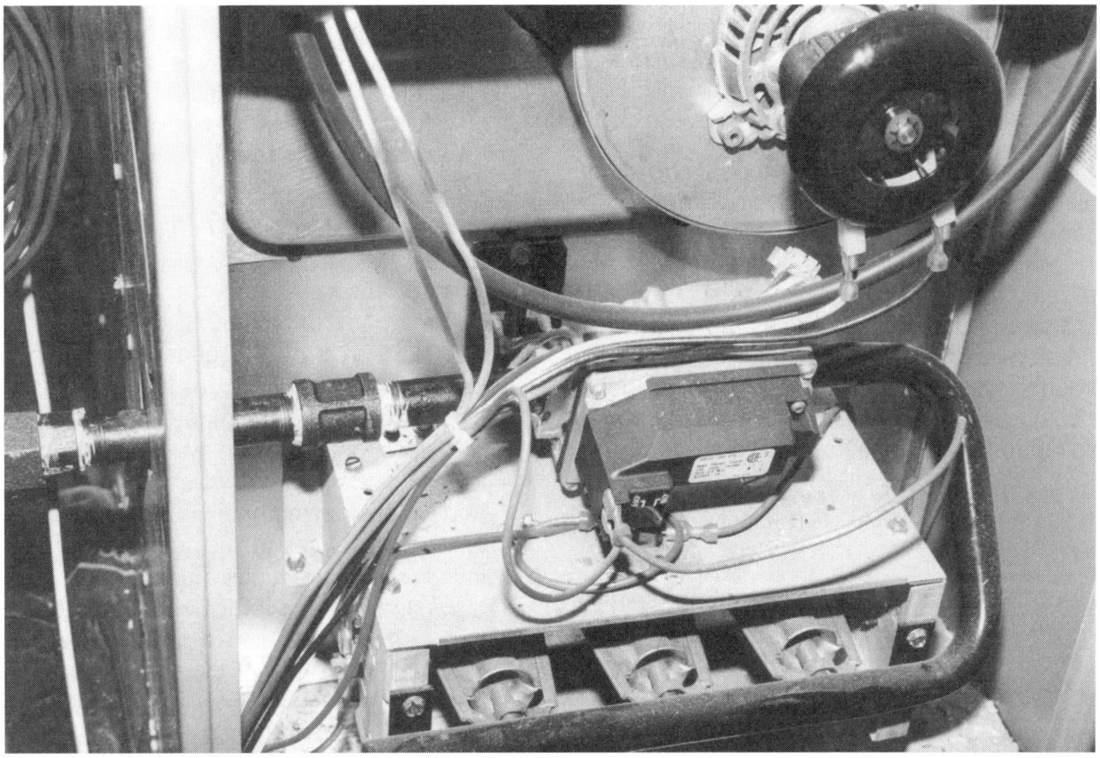

Figure G2414.9
TAPERED THREAD COUPLING IN CONNECTION TO FURNACE

corrosion that can occur from the reaction of phosphorus with a trace contaminant (sulphur) in the gas. See the commentary for Section G2413.10.2.

G2414.10.2 (403.10.2) Tubing joints. Tubing joints shall either be made with approved gas tubing fittings or be brazed with a material having a melting point in excess of 1,000°F (538°C). Brazing alloys shall not contain more than 0.05-percent phosphorus.

❖ Tubing joints are allowed to be made by two methods: brazing and tubing fittings. Tubing fittings include flare, compression, and CSST proprietary fittings.

Brazing is the act of producing a brazed joint and is often referred to as silver soldering. Silver soldering is more accurately described as silver brazing and employs high silver-bearing alloys primarily composed of silver, copper, and zinc. Silver soldering (brazing) typically requires temperatures in excess of 1,000°F (538°C), and such solders are classified as "hard" solders. See the definition of "Brazed joint."

Confusion has always been present with respect to the distinction between "silver solder" and "silver-bearing" solder. Silver solders are unique and can be further subdivided into soft and hard categories, determined by the percentages of silver and the other component elements of the particular alloy. The distinction is that silver-bearing solders (melting point less than 600°F [316°C]) are used in soft-soldered joints, and silver solders (melting point greater than 1,000°F [538°C]) are used in silver-brazed joints.

Brazed joints are considered to have superior strength and stress resistance and, because of the high melting point, are less likely to fail when exposed to fire.

Brazing with a filler metal conforming to AWS A5.8 produces a strong joint that will perform under extreme service conditions. The surfaces to be brazed must be cleaned so as to be free from oxides and impurities. Flux should be applied as soon as possible after the surfaces have been cleaned. Flux helps to remove residual traces of oxides, to promote wetting, and to protect the surfaces from oxidation during heating. Care should be taken to prevent flux from entering the piping system during the brazing operation, since flux that remains may corrode the pipe or contaminate the system.

Air should be removed from the pipe being brazed by purging the piping with a nonflammable gas such as carbon dioxide or nitrogen. Purging the system has several benefits such as preventing oxidation from occurring on the inside of the pipe. Mechanical joints are usually proprietary joints that are developed and marketed by individual manufacturers. Many types of mechanical joints employ a sleeve or ferrule that is compressed around the circumference of the pipe or tube. Mechanical joints must be specifically designed for and compatible with the type of pipe or tube to be joined.

G2414.10.3 (403.10.3) Flared joints. Flared joints shall be used only in systems constructed from nonferrous pipe and tubing where experience or tests have demonstrated that the joint is suitable for the conditions and where provisions are made in the design to prevent separation of the joints.

❖ Flared joints are typically used with copper and aluminum tubing and are prohibited for steel tubing. They require the use of specialized tools. Because the pipe end is expanded in a flared joint, only annealed and bending tempered (drawn) copper tubing may be flared. Commonly used flaring tools employ a screw yoke and block assembly or an expander tool that is driven into the tube with a hammer. The flared tubing end is compressed between a fitting seat and a threaded nut to form a metal-to-metal seal.

G2414.10.4 (403.10.4) Metallic fittings. Metallic fittings, including valves, strainers and filters shall comply with the following:

1. Fittings used with steel or wrought-iron pipe shall be steel, brass, bronze, malleable iron, ductile iron or cast iron.
2. Fittings used with copper or brass pipe shall be copper, brass or bronze.
3. Cast-iron bushings shall be prohibited.
4. Special fittings. Fittings such as couplings, proprietary-type joints, saddle tees, gland-type compression fittings, flared, flareless or compression-type tubing fittings shall be: used within the fitting manufacturer's pressure-temperature recommendations; used within the service conditions anticipated with respect to vibration, fatigue, thermal expansion or contraction; installed or braced to prevent separation of the joint by gas pressure or external physical damage; and shall be approved.

❖ Item 1 allows cast iron fittings to be used with steel, but cast iron pipe is prohibited. Malleable iron (steel) fittings are used almost exclusively. Item 4 refers to special fittings of the mechanical type. None of the joining methods addressed in the code use a nonmetallic or elastomeric sealing element except those used outdoors only.

G2414.11 (403.11) Plastic piping, joints and fittings. Plastic pipe, tubing and fittings shall be joined in accordance with the manufacturers' instructions. Such joints shall comply with the following:

1. The joints shall be designed and installed so that the longitudinal pull-out resistance of the joints will be at least equal to the tensile strength of the plastic piping material.
2. Heat-fusion joints shall be made in accordance with qualified procedures that have been established and proven by test to produce gas-tight joints at least as strong as the pipe or tubing being joined. Joints shall be made with the joining method recommended by the pipe manufacturer. Heat fusion fittings shall be marked "ASTM D2513."
3. Where compression type mechanical joints are used, the gasket material in the fitting shall be compatible with the plastic piping and with the gas distributed by the system. An internal tubular rigid stiffener shall be used in conjunction with the fitting. The stiffener shall be flush with the end of the pipe or tubing and shall extend at least to the outside end of the compression fitting when installed. The stiffener shall be free of rough or sharp edges and shall not be a force fit in the plastic. Split tubular stiffeners shall not be used.
4. Plastic piping joints and fittings for use in liquefied petroleum gas piping systems shall be in accordance with NFPA 58.

❖ This section provides general requirements for the proper joining of plastic gas pipe materials. While plastic pipe materials offer many advantages such as ease of installation, corrosion resistance, lighter weight, and lower cost, such materials require different handling and installation methods than steel or other metal pipe or tubing. In all cases, the pipe material manufacturer's installation instructions must be strictly adhered to in order to reduce the likelihood of creating improper joints, which could lead to gas leakage or piping or joint failure. Dissimilar types of plastic materials must not be joined by heat fusion methods. Due to the resulting reduction in wall thickness, plastic pipe and tubing must not be field-threaded as a joining method. All approved joints must be designed to account for the forces acting upon the piping system such as expansion, contraction, and external loads imposed by burial in underground installations. The longitudinal pull-out resistance of the joint should be approximately the same as the tensile strength of the materials being joined.

Heat-fusion joints for plastic pipe are analogous to welded joints of steel pipe. Heat fusion must be performed in accordance with the pipe manufacturer's instructions. The process involves heating the pipe and fittings with a special iron. When the parts to be joined reach their melting points, they are assembled and allowed to fuse. Electrical fusion fittings are available and employ an integral heating element that is connected to a power supply unit that allows the fitting to be heated at the required rate and duration.

Mechanical compression joints employ an elastomeric seal that must be compatible not only with the piping material but also with the gas that is conveyed in the system. The manufacturer's instructions normally require an internal stiffener insert to be used in conjunction with the fitting to provide additional support. The stiffener insert prevents the pipe from deforming under the compression force exerted by the seal and compression nut assembly.

Category 1 fittings in accordance with ASTM D 2513 are designed to provide a gas-tight seal and resist axial (pull-out) forces equal to the yield strength of the piping material. See Commentary Figures G2414.11(1) through G2414.11(4).

Photo courtesy of Perfection Corporation

Figure G2414.11(1)
MECHANICAL JOINTS

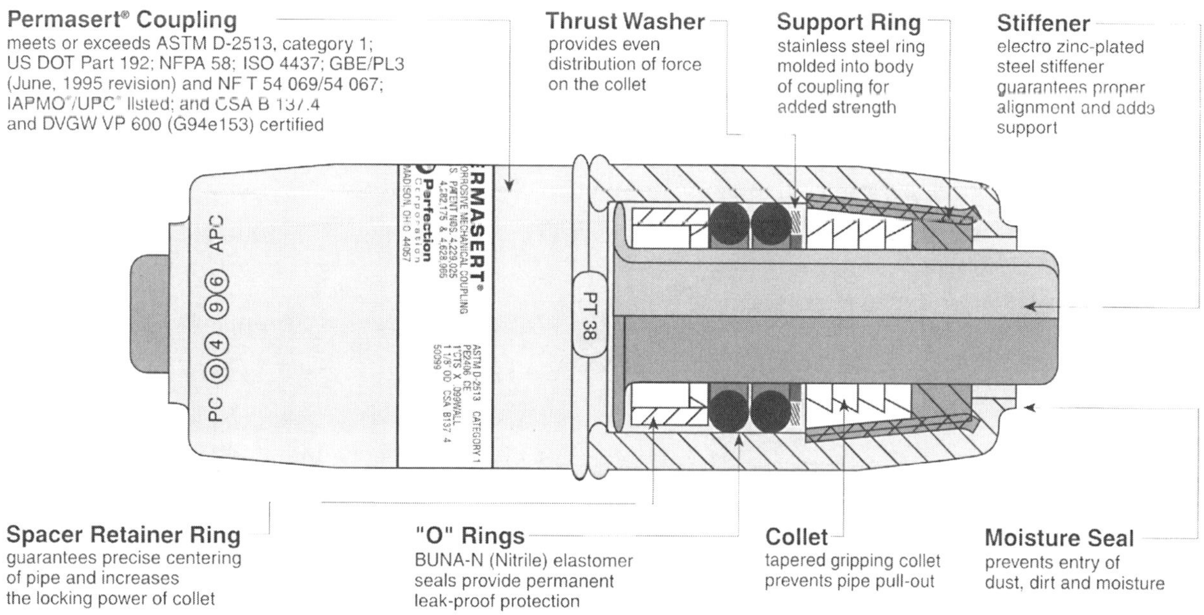

Figure courtesy of Perfection Corporation

Figure G2414.11(2)
MECHANICAL JOINT

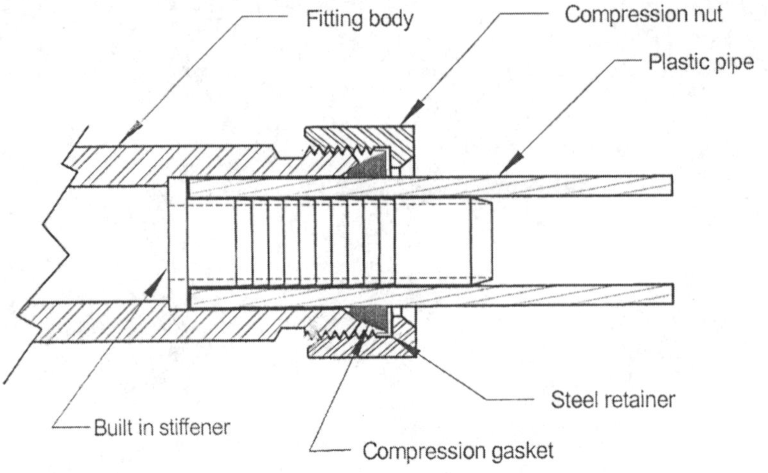

Figure courtesy of NORMAC

Figure G2414.11(3)
PLASIC PIPE TRANSITION FITTING

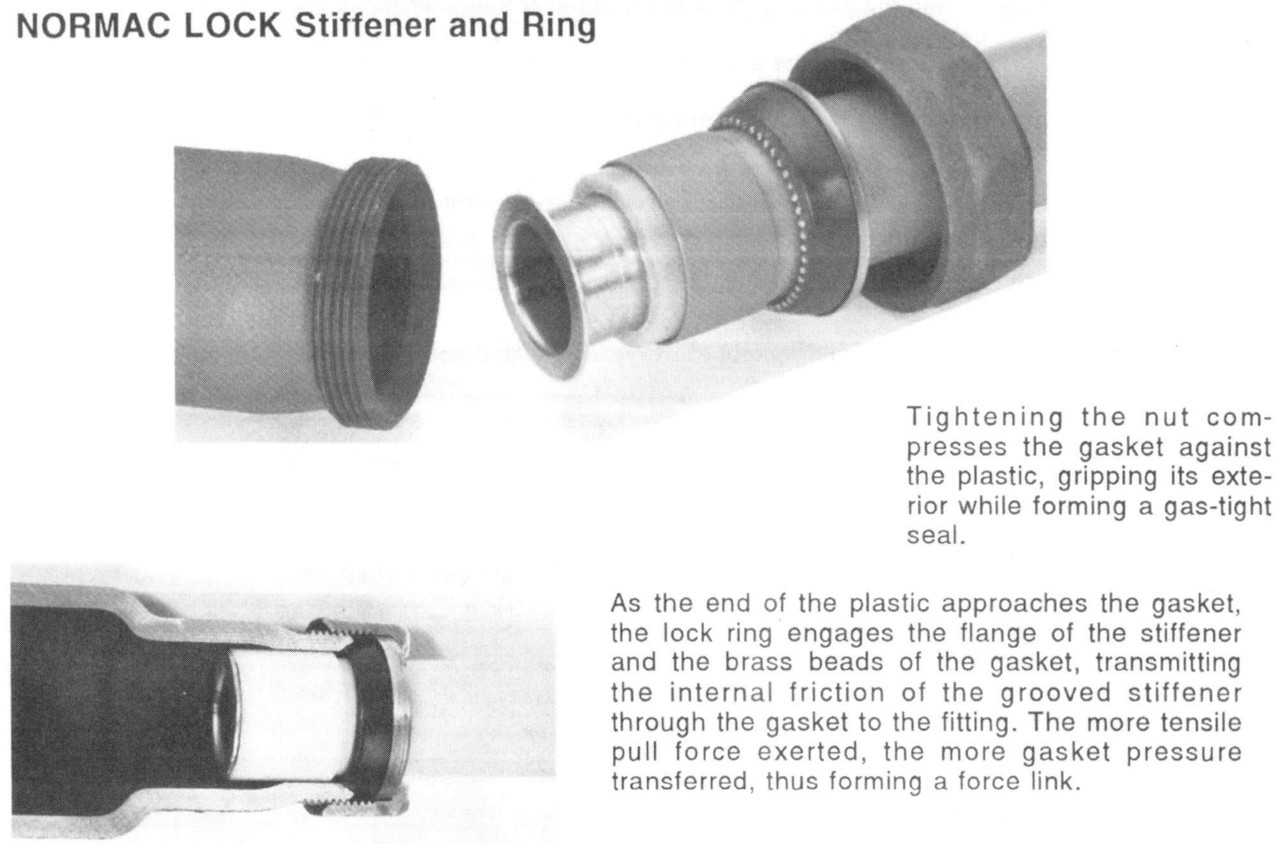

Figure courtesy of NORMAC

Figure G2414.11(4)
PLASTIC PIPE TRANSITION FITTING

SECTION G2415 (404) PIPING SYSTEM INSTALLATION

G2415.1 (404.1) Prohibited locations. Piping shall not be installed in or through a circulating air duct, clothes chute, chimney or gas vent, ventilating duct, dumbwaiter or elevator shaft.

❖ An air supply duct or areas that create a shaft that provides a path for the gas to travel are considered to be locations where the problem of a gas leak could be compounded by the gas or any resultant fire spreading throughout the building. Also, in some locations, the gas piping could be subject to deterioration because of temperature or corrosive conditions such as the ones present in air ducts or chimneys.

Locating fuel gas piping in certain areas or atmospheres may cause corrosion of the pipe, which in turn may cause a leak in the piping system. Within a supply air duct, the conditioned air may cause moisture to condense on or within the pipe, thereby causing corrosion of the pipe. Fuel gas piping located within a chimney or a vent will be subject to high temperatures and the corrosion effects of flue gases.

Additionally, piping located in a dumbwaiter or elevator shaft may be subject to an additional hazard resulting in mechanical damage to the pipe. Damage from a dumbwaiter or elevator impacting the fuel gas piping may not only cause a leak that would allow the gas to escape and travel up the shaft but may also put any occupants within an elevator in immediate danger. The elevator codes also intend to prohibit alien piping within elevator shafts.

Gas piping is not prohibited in concealed spaces used to convey environmental air (e.g. above ceiling and below floor plenums, stud cavities, and joist spaces). Gas piping is often installed in return air plenums above a suspended ceiling with no history of problems. It is assumed that the piping system will not leak in a ceiling air plenum or any other location, for that matter. In fact, when viewed logically, it would be more hazardous for a leak to occur in a location where the air is static and an explosive gas/air mixture could be created. A circulating air duct is not considered to be a plenum. Such ducts are supply and return air conduits constructed as components of a ductwork system. On the other hand, a furnace plenum is part of a ductwork system and would be treated as a circulating air duct. See the definition of "Furnace Plenum."

G2415.2 (404.2) Piping in solid partitions and walls. Concealed piping shall not be located in solid partitions and solid walls, unless installed in a chase or casing.

❖ Similar to the alternative installation requirements in Section G2416.6, this section allows gas piping to be installed within solid walls or partitions only if the piping is installed within a chase or casing to protect the pipe from stress and from corrosive effects of the wall material, such as concrete.

G2415.3 (404.3) Piping in concealed locations. Portions of a piping system installed in concealed locations shall not have unions, tubing fittings, right and left couplings, bushings, compression couplings and swing joints made by combinations of fittings.

Exceptions:

1. Tubing joined by brazing.
2. Fittings listed for use in concealed locations.

❖ A concealed location is a location that requires the removal of permanent construction in order to gain access to the piping within. See the definition of "Concealed location." The space above a dropped ceiling having readily removable lay-in panels or other locations that have a removable access panel are not considered to be concealed locations for the purposes of this section. Concealed locations include wall, floor, and ceiling cavities bounded by permanent finish materials, such as gypsum board, masonry, or paneling. Unions and mechanical joint tubing fittings are not permitted in concealed locations because they are more likely to loosen and leak than other joining means and fittings. Tubing fittings include flare, compression, and similar proprietary-type fittings, all of which are mechanical joints. Joints for CSST systems are mechanical tubing joints; however, such joints use specialized proprietary fittings that are listed for concealment as part of the CSST system. Such fittings undergo stringent testing to determine their ability to remain leak-tight.

Right and left couplings are somewhat archaic and were used as a union is used. The coupling is made with right-hand threads on one end and left-hand threads on the other and with the corresponding male pipe threads can be used to connect pipe and fittings within a section of existing piping. If the coupling were concealed, people would not be aware of such a fitting in a piping system and would not realize that they were loosening a joint as they applied the traditional tightening torque to an exposed section of piping. Bushings have been known to split after assembly and are not trusted in concealed locations.

Swing joints were used to control the forces that develop in piping systems due to expansion, contraction, and movement in structural components. If the joint "swings" (rotates), a leak could develop. Brazing is a commonly used method of joining copper tubing and is as reliable as welded joints for steel pipe.

G2415.4 (404.4) Piping through foundation wall. Underground piping, where installed below grade through the outer foundation or basement wall of a building, shall be encased in a protective pipe sleeve. The annular space between the gas piping and the sleeve shall be sealed.

❖ Because of the likelihood of the fuel gas piping settling after installation and because of the abrasive nature of foundation wall materials, such as concrete, an approved sleeve is required where piping passes through foundation walls. If the elevation of the underground piping is above the frost-depth line, additional protection may be necessary because of the vertical movement of the pipe caused by frost-heave action. The preferred method for bringing outdoor gas piping into the building is to penetrate the exterior wall at a point above grade. See Commentary Figure G2415.4.

The pipe and protective sleeve must be adequately sealed to prevent the possibility of water or insect entry into the building. A gas pipe that enters or exits the building above grade is preferable from the standpoint that any gas leakage in the underground piping will not be channeled into the building via the surrounding soil and the wall penetration. Gas leakage can travel in the annular space that exists between the outside wall of the pipe and the soil that surrounds it. With an above-ground wall penetration, any outdoor piping leakage will probably vent to the outdoors instead of finding its way into the building. Underground penetrations of building foundations must be sealed to eliminate the possibility of gas entering the building.

G2415.5 (404.5) Protection against physical damage. In concealed locations, where piping other than black or galvanized steel is installed through holes or notches in wood studs, joists, rafters or similar members less than 1 inch (25.4 mm) from the nearest edge of the member, the pipe shall be protected by shield plates. Shield plates shall be a minimum of 0.0625-inch-thick (1.6 mm) steel, shall cover the area of the pipe where the member is notched or bored, and shall extend a minimum of 4 inches (102 mm) above sole plates, below top plates and to each side of a stud, joist or rafter.

❖ This section is intended to minimize the possibility that nails or screws will be driven into the gas pipe or tube. Because nails and screws sometimes miss the stud, rafter, joist, or sole or top plates, the shield plates must extend parallel to the pipe or tube not less than 4 inches (105 mm) beyond the member on each side and not less than 4 inches (105 mm) above or below sole or top wall plates, respectively. Commentary Figure G2415.5 shows typical shield plates. Black and galvanized steel pipe (Schedule 40) each have wall thicknesses greater than the required thickness for the shield plates, which makes these piping materials inherently resistant to nail and screw penetrations. This section does not apply to CSST tubing because CSST systems have their own protection requirements dictated by the manufacturers' instructions and ANSI LC-1. The protection requirements for CSST systems are, overall, more stringent than those of this section. Before studs, rafters, joists, or other structural members are drilled or notched, the building portion of this code should be consulted.

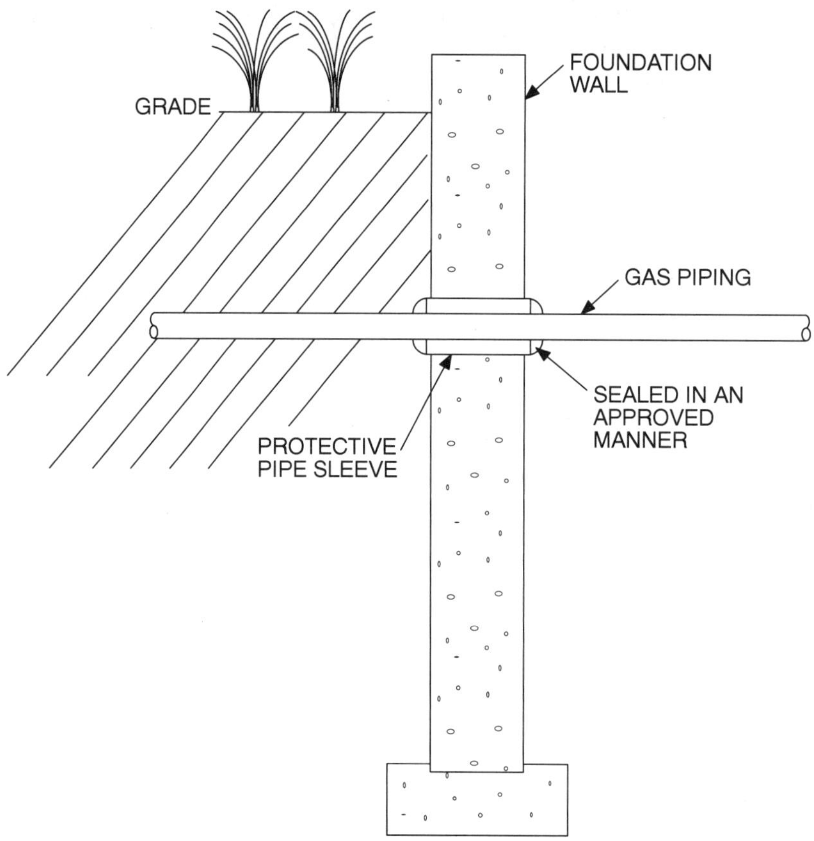

Figure G2415.4
PIPING THROUGH FOUNDATION WALLS

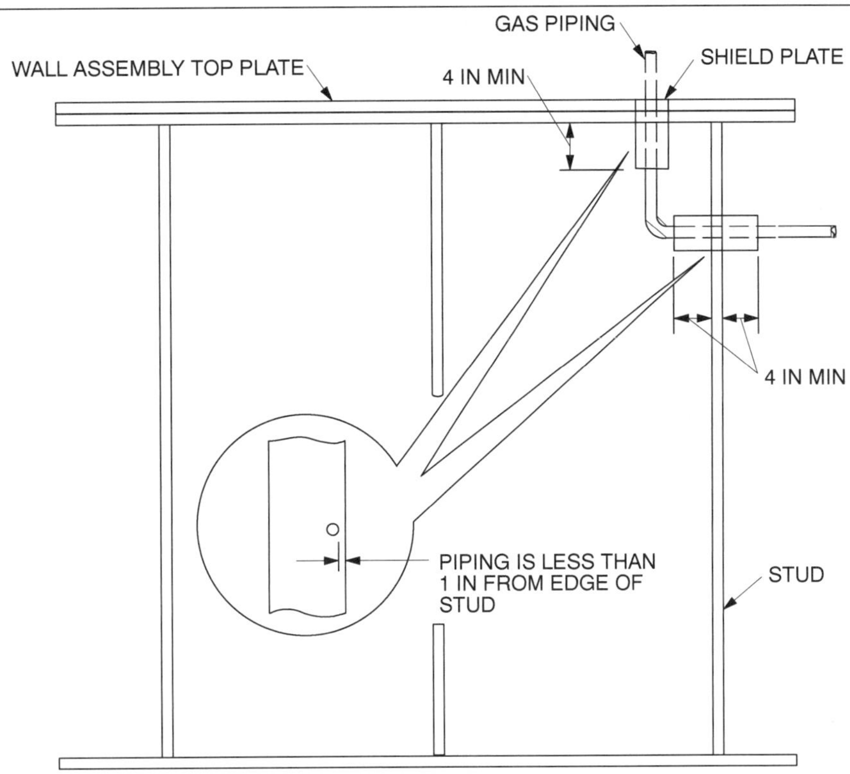

For SI: 1 inch = 25.4 mm.

Figure G2415.5
PIPE PROTECTION

G2415.6 (404.6) Piping in solid floors. Piping in solid floors shall be laid in channels in the floor and covered in a manner that will allow access to the piping with a minimum amount of damage to the building. Where such piping is subject to exposure to excessive moisture or corrosive substances, the piping shall be protected in an approved manner. As an alternative to installation in channels, the piping shall be installed in a casing of schedule 40 steel, wrought iron, PVC or ABS pipe with tightly sealed ends and joints. Both ends of such casing shall extend not less than 2 inches (51 mm) beyond the point where the pipe emerges from the floor.

❖ Piping must not be installed in any solid concrete or masonry floor construction. The potential for pipe damage due to slab settlement, cracking, or the corrosive action of the floor material makes it imperative that one of the installation methods in this section be used. This section does not intend to allow any direct encasement of gas piping in solid concrete or masonry floor systems.

Gas piping installed within a solid floor system must be safeguarded by installation in a sealed casing or in a floor channel with a removable cover for pipe access. See Commentary Figures G2415.6 (1) and (2). Either of these methods should provide reasonable protection of the pipe from the effects of settling, cracking, and corrosive materials being in contact with the pipe.

Because casings constructed of steel pipe could corrode where installed within a concrete slab on grade, consideration should be given to corrosion protection for the steel casing, or an alternate material should be chosen.

G2415.7 (404.7) Above-ground piping outdoors. All piping installed outdoors shall be elevated not less than $3^1/_2$ inches (152 mm) above ground and where installed across roof surfaces, shall be elevated not less than $3^1/_2$ inches (152 mm) above the roof surface. Piping installed above ground, outdoors, and installed across the surface of roofs shall be securely supported and located where it will be protected from physical damage. Where passing through an outside wall, the piping shall also be protected against corrosion by coating or wrapping with an inert material. Where piping is encased in a protective pipe sleeve, the annular space between the gas piping and the sleeve shall be sealed.

❖ Gas piping in any location must be properly supported and protected from physical damage. Protection from damage is especially important where piping is run outdoors near grade or across roof surfaces. See Sections G2418 and G2424. Piping passing through an outside wall must be protected where the material of the wall could corrode or abrade the piping (concrete, masonry, or stucco, for example). To help protect piping from corrosion resulting from exposure to moisture, the piping must always be located at least $3^1/_2$

inches above the earth and above roof surfaces. $3^1/_2$ inches corresponds to the width of pressure treated 4-inch x 4-inch nominal lumber, which is commonly used to support piping run across roof surfaces.

Where piping is supported on 4-inch x 4-inch nominal wood blocks or similar supports, the piping must be attached to the blocking with suitable clamps or straps. It is common to see wood blocking that has been displaced or piping that has been displaced because the piping was not properly fastened to the blocking. Section G2424.1 will dictate the minimum required spacing for such blocking.

G2415.8 (404.8) Protection against corrosion. Metallic pipe or tubing exposed to corrosive action, such as soil condition or moisture, shall be protected in an approved manner. Zinc coatings (galvanizing) shall not be deemed adequate protection for gas piping underground. Ferrous metal exposed in exterior locations shall be protected from corrosion in a manner satisfactory to the code official. Where dissimilar metals are joined underground, an insulating coupling or fitting shall be used. Piping shall not be laid in contact with cinders.

❖ See the commentary for Section G2414.8.

G2415.8.1 (404.8.1) Prohibited use. Uncoated threaded or socket welded joints shall not be used in piping in contact with soil or where internal or external crevice corrosion is known to occur.

❖ Pits, crevices, and other surface imperfections can create concentrated corrosion cells where the corro-

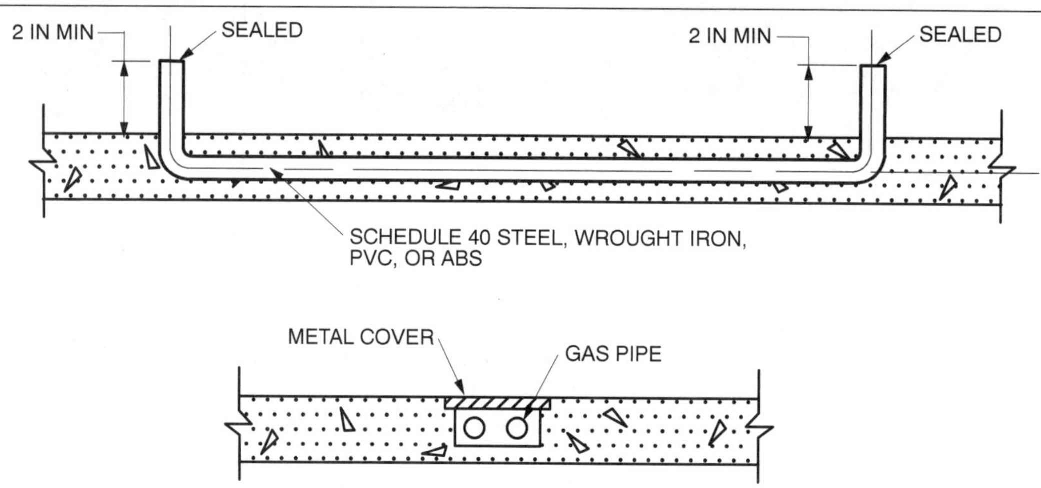

For SI: 1 inch = 25.4 mm.

Figure G2415.6(1)
PIPING IN SOLID FLOORS

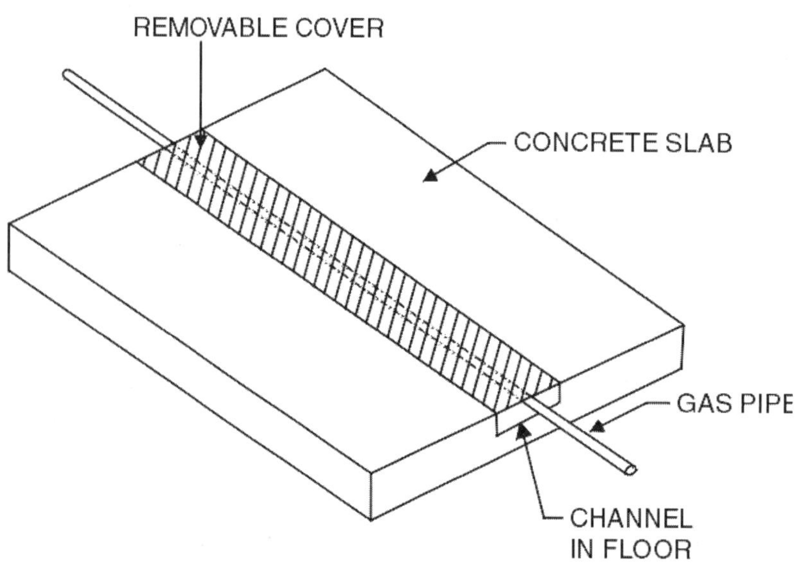

Figure G2415.6(2)
PIPING IN CONCRETE SLABS

sion effect is amplified at such sites. Pipe or fitting failure could occur more quickly than if the corrosion was uniformly distributed over the entire surface of the pipe or fitting. See the commentary for Section G2415.8.2.

G2415.8.2 (404.8.2) Protective coatings and wrapping. Pipe protective coatings and wrappings shall be approved for the application and shall be factory applied.

Exception: Where installed in accordance with the manufacturer's installation instructions, field application of coatings and wrappings shall be permitted for pipe nipples, fittings and locations where the factory coating or wrapping has been damaged or necessarily removed at joints.

❖ Protective coatings in the form of wrappings, tapes, enamels, epoxies, sleeves, casings, and factory-applied coverings are a common method of isolating metallic piping materials from the atmosphere or from the earth in which they are buried. Corrosion can be caused by the soil or construction materials that are in contact with the pipe. Protection is usually provided by a factory-applied coating or by field-wrapping the pipe with a protective covering, such as a coal-tar-based or plastic wrapping. Where possible, a piping material is chosen that is not subject to the type of corrosion of the application. Application of these materials under strictly controlled factory conditions improves the quality control process and reduces the likelihood of coating defects. In the event that coatings or wrappings must be applied in the field, the exception states that they must be installed in accordance with the manufacturers' installation instructions. Coatings and wrappings should be applied only by persons trained and experienced in such work in order to provide improved reliability of the coating/wrapping process. Field application is only permitted in cases where the factory-applied material may have been damaged during transit or installation of the pipe, where the applied coating was removed for pipe welding or threading of the pipe, or for short sections (nipples) of pipe used in the installation. Corrosion can focus on the pipe or fitting where there are flaws in the coating or wrapping; therefore, improperly applied coatings and wrappings can cause pipe or fitting failures to occur faster than failures would occur in unprotected metals. It is evident why plastic piping is used extensively in underground installations.

G2415.9 (404.9) Minimum burial depth. Underground piping systems shall be installed a minimum depth of 12 inches (305 mm) below grade, except as provided for in Section G2415.9.1.

❖ The depth of 12 inches (457 mm) is considered sufficient to avoid possible harm to the pipe from the use of hand tools such as spades and shovels.

G2415.9.1 (404.9.1) Individual outside appliances. Individual lines to outside lights, grills or other appliances shall be installed a minimum of 8 inches (203 mm) below finished grade, provided that such installation is approved and is installed in locations not susceptible to physical damage.

❖ Gas piping is permitted to be installed within 8 inches (203 mm) of the ground surface where it serves individual outdoor appliances and is not likely to be subjected to damage such as might occur from vehicular traffic, gardening, future excavation, etc. Each individual installation must be reviewed and approved by the code official. The intent is to allow shallow installations for small distribution lines that serve outdoor gas lights, cooking appliances, pool heaters, and similar loads only where the piping is unlikely to be disturbed.

G2415.10 (404.10) Trenches. The trench shall be graded so that the pipe has a firm, substantially continuous bearing on the bottom of the trench.

❖ Where trenches have non-uniform depth or peaks and valleys in the bottom, the piping could lack continuous support and could be subjected to stresses from the backfill and surface loads.

G2415.11 (404.11) Piping underground beneath buildings. Piping installed underground beneath buildings is prohibited except where the piping is encased in a conduit of wrought iron, plastic pipe, or steel pipe designed to withstand the superimposed loads. Such conduit shall extend into an occupiable portion of the building and, at the point where the conduit terminates in the building, the space between the conduit and the gas piping shall be sealed to prevent the possible entrance of any gas leakage. If the end sealing is capable of withstanding the full pressure of the gas pipe, the conduit shall be designed for the same pressure as the pipe. Such conduit shall extend not less than 4 inches (102 mm) outside the building, shall be vented above grade to the outdoors, and shall be installed so as prevent the entrance of water and insects. The conduit shall be protected from corrosion in accordance with Section G2415.8.

❖ This section prohibits the installation of gas piping beneath buildings in order to reduce the potential for an inaccessible pipe failure due to the settling of the structure. The prohibition also reduces the potential for corrosion-caused failure of piping embedded in soil or fill. See Section G2415.6 for gas pipe installations in floor slabs. Where underground installation under a building is unavoidable, the piping must be encased in another pipe so as to channel any leakage to the outdoors. See Commentary Figure G2415.11.

G2415.12 (404.12) Outlet closures. Gas outlets that do not connect to appliances shall be capped gas tight.

Exception: Listed and labeled flush-mounted-type quick-disconnect devices and listed and labeled gas convenience outlets shall be installed in accordance with the manufacturer's installation instructions.

❖ Unused fuel gas outlets must be capped or plugged gas-tight, regardless of whether or not a shutoff valve is provided at the outlet. A closed valve alone is not dependable and poses an unnecessary risk of leakage. The exception recognizes that listed gas outlet devices are built with an inherent safety feature that will automatically shut off the gas flow if the mating connector is disengaged or that will not allow the appliance connector to be disengaged until the integral shutoff valve is manually closed.

G2415.13 (404.13) Location of outlets. The unthreaded portion of piping outlets shall extend not less than 1 inch (25.4 mm) through finished ceilings and walls and where extending through floors, outdoor patios and slabs, shall not be less than 2 inches (51 mm) above them. The outlet fitting or piping shall be securely supported. Outlets shall not be placed behind doors. Outlets shall be located in the room or space where the appliance is installed.

Exception: Listed and labeled flush-mounted-type quick-disconnect devices and listed and labeled gas convenience outlets shall be installed in accordance with the manufacturer's installation instructions.

❖ This section regulates the location, installation, and termination of gas piping system outlets in order to reduce the likelihood of physical damage and to provide clearances for the use of tools. When making connections to piping outlets, "back-up" wrenches are used to prevent piping from rotating, loosening, or being damaged. Sufficient pipe length is necessary to allow the application of tools. Gas outlet devices that have been tested and labeled for installation methods other than those addressed in this section must be installed in accordance with the terms of their testing and listing as contained in the manufacturers' installation instructions. Gas outlets for CSST systems must be installed with the termination fitting designed specifically for that purpose and provided by the CSST manufacturer.

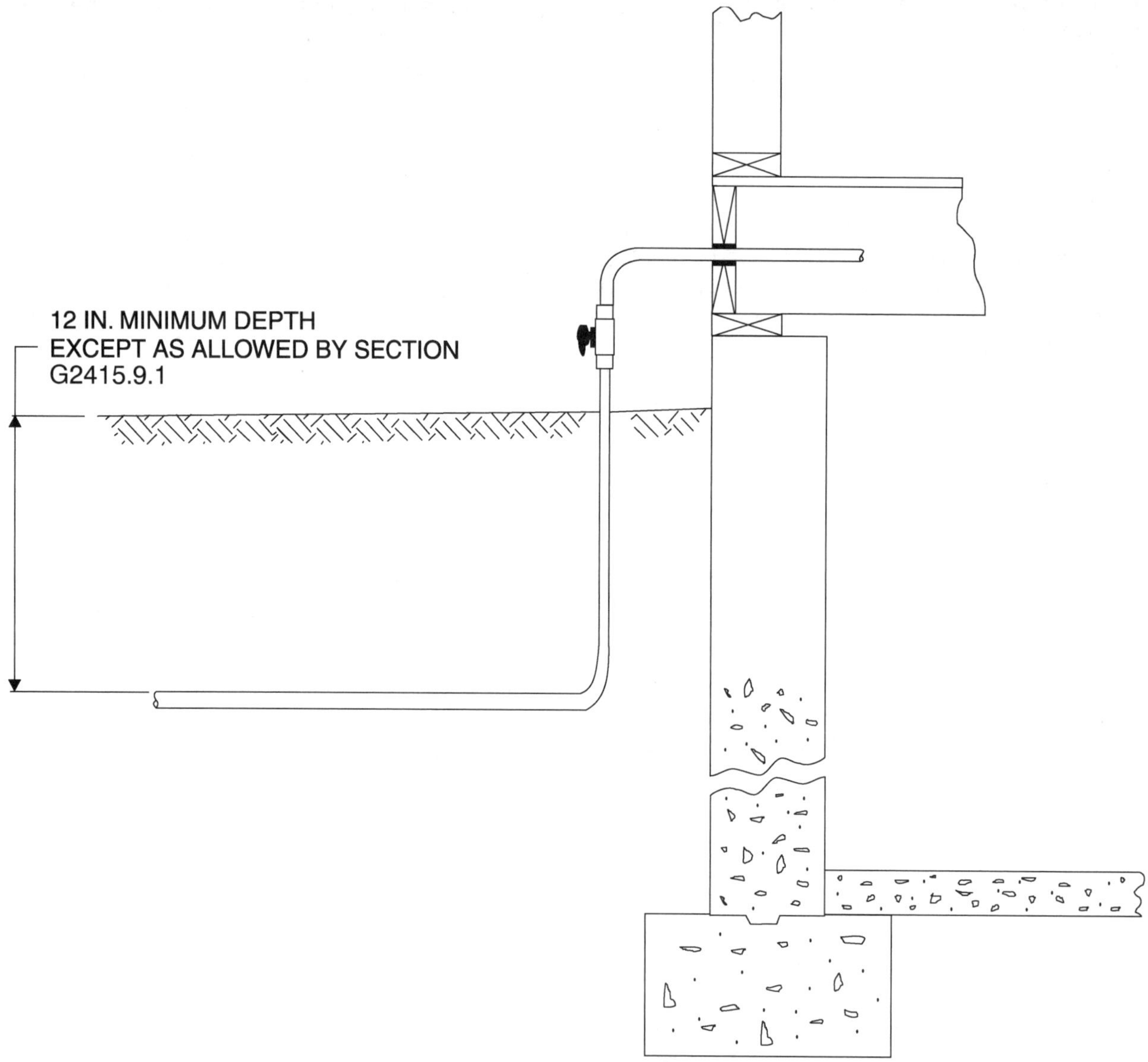

Figure G2415.11
MINIMUM BURIAL DEPTH

G2415.14 (404.14) Plastic pipe. The installation of plastic pipe shall comply with Sections G2415.14.1 through G2415.14.3.

❖ This section places restrictions on the location and operating pressures for plastic pipe in addition to providing installation requirements specific to plastic pipe.

G2415.14.1 (404.14.1) Limitations. Plastic pipe shall be installed outside underground only. Plastic pipe shall not be used within or under any building or slab or be operated at pressures greater than 100 psig (689 kPa) for natural gas or 30 psig (207 kPa) for LP gas.

Exceptions:

1. Plastic pipe shall be permitted to terminate aboveground outside of buildings where installed in premanufactured anodeless risers or service head adapter risers that are installed in accordance with that manufacturer's installation instructions.
2. Plastic pipe shall be permitted to terminate with a wall head adapter within buildings where the plastic pipe is inserted in a piping material for fuel gas use in buildings.

❖ Because of the potential hazard associated with the use of a material that has lower resistance to physical damage and heat as compared to metallic pipe, the installation of plastic pipe and tubing is limited to areas that are both outside of the building and underground. Plastic pipe and tubing are widely used for underground gas distribution systems because of their ease of installation and inherent resistance to corrosion. Polyethylene (PE) pipe is the only allowable plastic pipe for use with LP-gas, and the code user is directed to the referenced standard, NFPA 58, which requires that PE piping materials comply with ASTM D 2513. The exception makes it clear that if an equivalent level of physical protection is provided, plastic pipe may terminate above ground outside of the building. It is common practice for gas utility companies to install premanufactured riser assemblies at their meter settings. Such risers typically consist of a steel pipe with a corrosion resistant coating in which a length of PE pipe is pre-installed for coupling to the service lateral. A 90-degree sweeping bend is provided on the underground end of the riser assembly. Plastic piping is not allowed in or under a concrete slab. The code official might determine that it is not the intent of this section to prevent an outdoor slab from being placed over an existing customer owned (private) underground gas line, such as in the case of sidewalks and open patios. The prohibition on piping within a slab is applicable to all slabs. Exceptions 1 and 2 relate to Section G2414.6.1. See the commentary for that section and Commentary Figure G2415.14.1.

G2415.14.2 (404.14.2) Connections. Connections made outside and underground between metallic and plastic piping shall be made only with transition fittings categorized as Category I in accordance with ASTM D 2513.

❖ Because plastic piping is permitted to be installed only underground and outside, the same requirement holds for joining plastic and metallic piping. Mechanical compression joints employ an elastomeric seal that must be compatible not only with the piping material but also with the gas that is conveyed in the system. The manufacturers' instructions normally require an internal stiffener insert to be used in conjunction with the fitting to provide additional support. The stiffener insert prevents the pipe from deforming under the compression force exerted by the seal and compression nut assembly.

Category 1 fittings are designed to provide a gas-tight seal and resist axial (pull-out) forces equal to the yield strength of the piping material. See Commentary Figure G2414.11(3).

G2415.14.3 (404.14.3) Tracer. A yellow insulated copper tracer wire or other approved conductor shall be installed adjacent to underground nonmetallic piping. Access shall be provided to the tracer wire or the tracer wire shall terminate above ground at each end of the nonmetallic piping. The tracer wire size shall not be less than 18 AWG and the insulation type shall be suitable for direct burial.

❖ To avoid piping damage and a hazardous condition, the location of underground piping must be known prior to excavation work being performed in the vicinity of such piping. In the past, gas utility companies, installers, and contractors relied on metal detectors to locate gas piping underground. With the advent of nonmetallic (plastic) piping, this is no longer possible. A tracer wire provides a means for locating nonmetallic gas piping without having to excavate, thereby avoiding the possibility of damaging the pipe. An electrical current can be passed through the wire to allow a metal detector to locate the piping. Insulated wire is required because copper wire or other approved conductors may be susceptible to corrosion in some soils.

Yellow is the standardized color used to identify gas piping. Tapes are also available that contain a conducting strip and also serve as a visual warning.

G2415.15 (404.15) Prohibited devices. A device shall not be placed inside the piping or fittings that will reduce the cross-sectional area or otherwise obstruct the free flow of gas.

Exception: Approved gas filters.

❖ Devices such as filters and flow measuring instruments could create considerable pressure drop and are prohibited except where the piping system is designed to tolerate such losses. An unaccounted for restriction in the piping could make an otherwise properly designed system fail to supply the required pressure.

G2415.16 (404.16) Testing of piping. Before any system of piping is put in service or concealed, it shall be tested to ensure that it is gas tight. Testing, inspection and purging of piping systems shall comply with Section G2417.

❖ A pressure test is required after every installation, alteration, addition, or repair to the fuel gas piping system. The location of a leak may be difficult to determine, especially if it is concealed in the building construction. If a leak is found, the leaking component must be repaired or replaced before the system is concealed or put into operation. Section G2417 specifies testing pressures based on the type of system, the design working pressure, or other parameters. The testing duration is based on the total cubic feet of pipe volume, and the piping system must sustain the test pressure for the duration without exhibiting any sign of leakage.

SECTION G2416 (405) PIPING BENDS AND CHANGES IN DIRECTION

G2416.1 (405.1) General. Changes in direction of pipe shall be permitted to be made by the use of fittings, factory bends or field bends.

❖ This section permits changes in direction of gas piping to be made through the use of pipe fittings compatible with the piping material or by bending the pipe, if such a procedure is recommended by the pipe manufacturer. If bending is the chosen method, it must be accomplished in accordance with Sections G2416.2 and G2416.3. In practice, steel gas piping is rarely bent to accomplish changes in direction because pipe failure at a seam is a possibility for steel pipe that is bent.

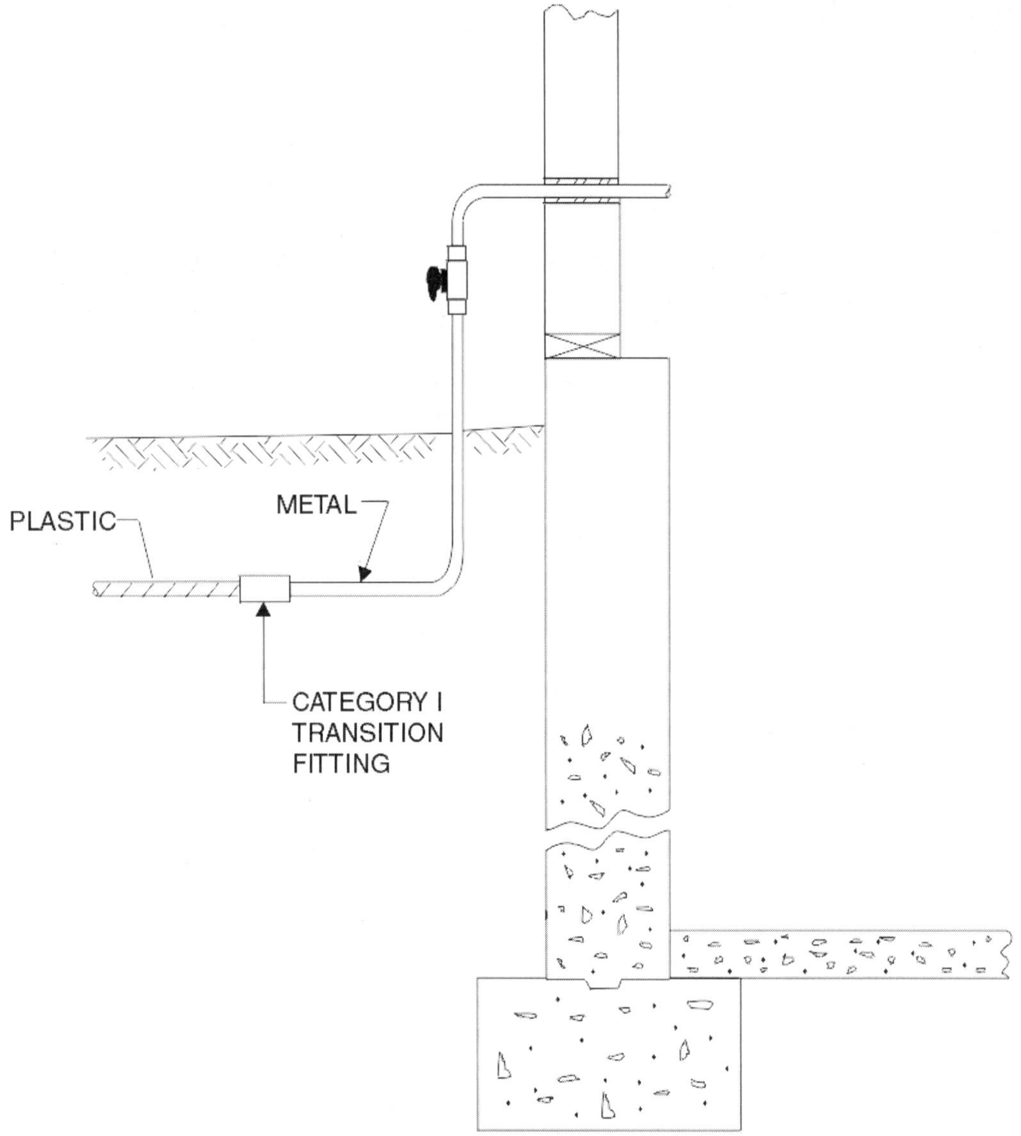

Figure G2415.14.1
INSTALLATION OF PLASTIC TO METALLIC PIPE

G2416.2 (405.2) Metallic pipe. Metallic pipe bends shall comply with the following:

1. Bends shall be made only with bending equipment and procedures especially intended for that purpose.
2. All bends shall be smooth and free from buckling, cracks or other evidence of mechanical damage.
3. The longitudinal weld of the pipe shall be near the neutral axis of the bend.
4. Pipe shall not be bent through an arc of more than 90 degrees (1.6 rad).
5. The inside radius of a bend shall be not less than six times the outside diameter of the pipe.

❖ Pipe is bent to accomplish changes in direction without the use of fittings. Pipe bending must done with the appropriate bending tools and materials. Some materials are not tempered or intended to be bent; therefore, consideration must be given to choosing the proper materials. Bending tools are designed to make bends without damaging the pipe. All pipe intended to be bent has a minimum bend radius which must be observed to avoid damage to the pipe. For example, bending welded seam pipe with the seam located outside of the neutral axis of the bend may result in a split seam because of the stresses induced from the bend. In other words, the welded seam must not be oriented in a bend so as to be placed in tension or compression. If pipe is to be bent, it should be stated in the pipe specifications that the pipe is suitable for bending. The bending of rigid gas piping is not commonly done, because of the perceived risk of pipe stress failures at the bend and because bending can be more labor intensive than using fittings.

G2416.3 (405.3) Plastic pipe. Plastic pipe bends shall comply with the following:

1. The pipe shall not be damaged and the internal diameter of the pipe shall not be effectively reduced.
2. Joints shall not be located in pipe bends.
3. The radius of the inner curve of such bends shall not be less than 25 times the inside diameter of the pipe.
4. Where the piping manufacturer specifies the use of special bending equipment or procedures, such equipment or procedures shall be used.

❖ Due to the material characteristics of plastic pipe, the code defers to the manufacturers' installation and bending instructions to achieve the performance level contemplated by this section. For example, coiled plastic pipe and tubing have limitations regarding bending beyond or against the natural curvature of the coil. A joint in a bend would be subject to bending stresses and pipe and fitting misalignments that could cause joint structural failure and/or leakage. Thus, a one-inch I.D. plastic pipe would have a minimum bending radius of 25 inches measured to the inner curve of the bend.

SECTION G2417 (406) INSPECTION, TESTING AND PURGING

G2417.1 (406.1) General. Prior to acceptance and initial operation, all piping installations shall be inspected and pressure tested to determine that the materials, design, fabrication, and installation practices comply with the requirements of this code.

❖ Before any gas piping system is put to use, it must be inspected, tested, and approved. See Sections G2417.1.1 through G2417.1.4. Pressure testing is distinct from leakage testing as covered in Section G2417.6.

G2417.1.1 (406.1.1) Inspections. Inspection shall consist of visual examination, during or after manufacture, fabrication, assembly or pressure tests as appropriate.

❖ The inspection of piping installations is intended to be a visual observation of the system and the testing procedure. The designer may require that, for example, welded joints in very high-pressure applications be examined by a method that is capable of discovering internal defects that are not detectable by visual observation.

G2417.1.2 (406.1.2) Repairs and additions. In the event repairs or additions are made following the pressure test, the affected piping shall be tested.

Minor repairs or additions are not required to be pressure tested provided that the work is inspected and connections are tested with a noncorrosive leak-detecting fluid or other approved leak-detecting methods.

❖ Some time after the initial test of the piping, a repair could be made or a branch could be added or extended. The repaired, added, or extended portion of the piping system needs to be tested without requiring the unchanged portions to be tested again. Minor work is allowed to be done to an existing system without pressure testing the minor work, provided that such work is visually inspected and tested by a leak-detecting method. Leak-detecting methods include bubble test fluids and electronic sensors. Some bubble test fluids could be corrosive to some piping; therefore, a non-corrosive fluid designed for the purpose should always be used. Methods other than bubble test fluids must be acceptable to the code official.

G2417.1.3 (406.1.3) New branches. Where new branches are installed from the point of delivery to new appliances, only the newly installed branches shall be required to be pressure tested. Connections between the new piping and the existing piping shall be tested with a noncorrosive leak-detecting fluid or other approved leak-detecting methods.

❖ Often, because it is convenient or because the existing piping is fully loaded, a new piping branch will be run from the point of delivery to serve a new appliance installation. The new branch is typically taken from a tee fitting installed immediately downstream of the meter. The entire new branch piping must be tested, including the tee fitting in the existing piping from which the new branch is supplied. The point of connection to the existing piping is allowed to be tested by means other than pressure testing; all other piping in the new branch must be isolated from the existing piping and pressure tested as required for new work. Section G2417.1.4 requires the new branch piping to be disconnected from the existing piping during pressure testing except where an assembly consisting of two closed valves in series with an intermediate open valved port is installed to isolate the new piping from the existing piping. See Commentary Figure G2417.1.4.

G2417.1.4 (406.1.4) Section testing. A piping system shall be permitted to be tested as a complete unit or in sections. Under no circumstances shall a valve in a line be used as a bulkhead between gas in one section of the piping system and test medium in an adjacent section, unless two valves are installed in series with a valved "tell-tale" located between these valves. A valve shall not be subjected to the test pressure unless it can be determined that the valve, including the valve closing mechanism, is designed to safely withstand the test pressure.

❖ Depending on the progression of a job, it may be desirable to test portions of a system as they are completed. It is also possible that portions of a system will be put in service before the entire system is completed. To prevent a test medium from leaking into piping containing fuel gas or vice versa, portions of piping under test must be isolated from portions that are in service. A single valve cannot be depended on to establish this isolation because all valves have some "leak-through" allowance, and also the valve could be accidentally opened. To eliminate such risk, two valves in series must be used, and a valved open port (tell-tale) must be installed between the valves. The open port will release any gases that have passed through either isolation valve and also will serve to indicate any such leakage or accidental opening of an isolation valve. See Commentary Figure G2417.1.4.

G2417.1.5 (406.1.5) Regulators and valve assemblies. Regulator and valve assemblies fabricated independently of the piping system in which they are to be installed shall be permitted to be tested with inert gas or air at the time of fabrication.

❖ Assemblies of valves, regulators, and/or over-pressure devices might be factory or shop assembled and are allowed to be tested independently of the piping system in which they will be installed.

G2417.2 (406.2) Test medium. The test medium shall be air, nitrogen, carbon dioxide or an inert gas. Oxygen shall not be used.

❖ The gas that is forced into the piping system for pressure testing must be chemically unreactive in order to prevent undesirable reactions with the piping system components and the fuel to be conveyed in the piping. Oxygen is not inert, and any residual amount could form an explosive mixture with the fuel gas. The typical test gas is air, which is also not inert but is readily available and free. Nitrogen and carbon dioxide are considered to be inert for the purpose of testing piping systems and are inexpensive compared to most truly inert gases.

G2417.3 (406.3) Test preparation. Pipe joints, including welds, shall be left exposed for examination during the test.

Exception: Covered or concealed pipe end joints that have been previously tested in accordance with this code.

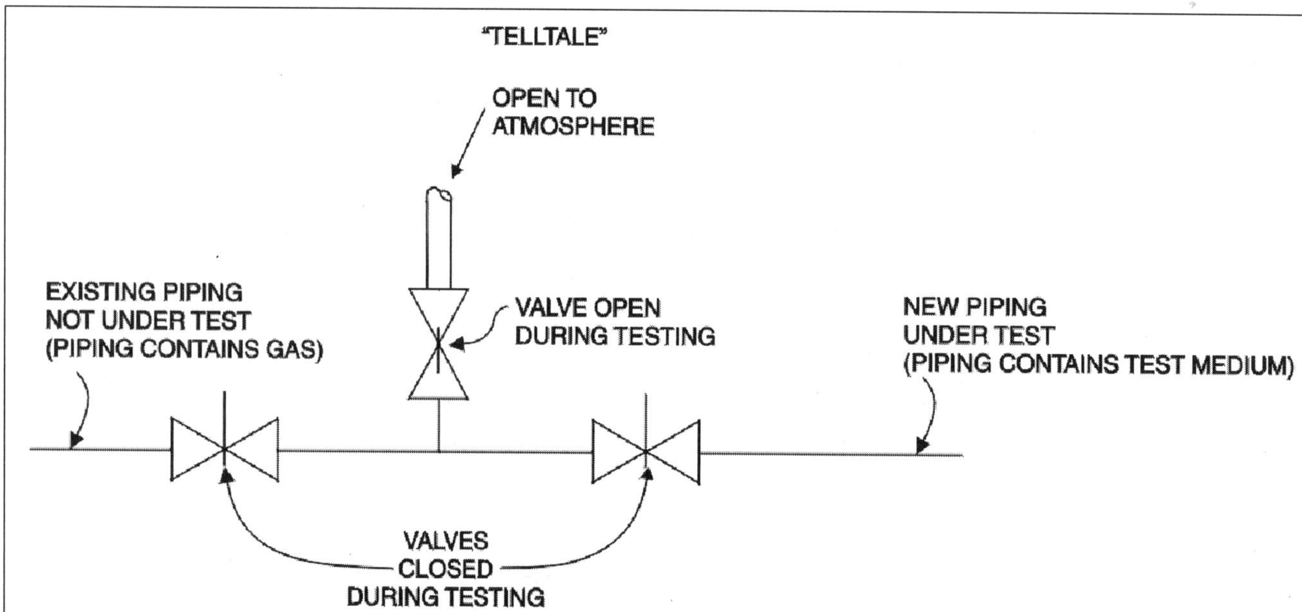

Figure G2417.1.4
PIPING ISOLATION ARRANGEMENT FOR TESTING

❖ This section is consistent with Section R109.1.2 and requires all piping joints to remain exposed during testing to allow defects to be located. This section refers only to joints, but Section R109.1.2 will require the piping and all joints to be left exposed until inspected and approved.

G2417.3.1 (406.3.1) Expansion joints. Expansion joints shall be provided with temporary restraints, if required, for the additional thrust load under test.

❖ If the piping system employs expansion joints to accommodate piping movement, such joints might need to be restrained to resist the additional thrust loading caused by the movement of the pressurized test gas. Generally, testing does not produce noticeable dynamic thrust forces, because of the slow rate at which the test gas is put in and taken out of the piping system. The thrust force produced by the static pressure of the test gas is of concern. Expansion joints are rarely used because expansion loops and offsets perform the same function.

G2417.3.2 (406.3.2) Equipment isolation. Equipment that is not to be included in the test shall be either disconnected from the piping or isolated by blanks, blind flanges or caps.

❖ Some equipment can be damaged by test pressure and is therefore disconnected from the piping system or is otherwise isolated by blank-offs placed between the flanges.

Caps and blind flanges are used to cap piping that has been disconnected from the equipment.

G2417.3.3 (406.3.3) Equipment disconnection. Where the piping system is connected to equipment or components designed for operating pressures of less than the test pressure, such equipment or equipment components shall be isolated from the piping system by disconnecting them and capping the outlet(s).

❖ This requirement is often overlooked in the field, which could lead to appliance damage and serious consequences. Appliances are typically designed for a maximum inlet gas pressure of ½ psig or less, above which damage to controls and regulators could occur. The test pressure will be at least 3 psig and often times much higher in accordance with general practice or local traditions or requirements. Installers routinely pressurize piping systems to pressures between 50 and 100 psig, often at the insistence of the local authority. It is obvious that there is a good potential to expose appliances to extreme pressures during a test, which is why the code requires the precaution of disconnecting the appliances from the supply piping. Disconnection is normally accomplished by opening a union or disconnecting the appliance connector. The appliance shutoff valve must not be used as a means of appliance isolation since the valve can leak through and/or can be accidentally opened or left open. See Section G2417.3.4 and Commentary Figure G2417.3.3.

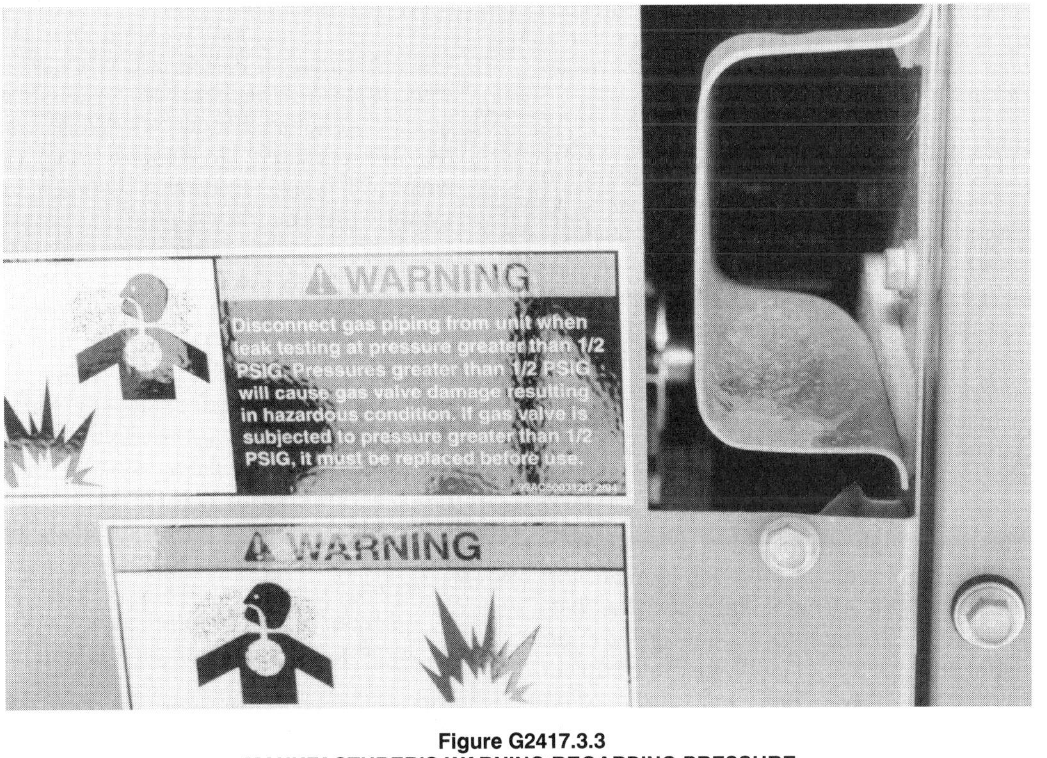

Figure G2417.3.3
MANUFACTURER'S WARNING REGARDING PRESSURE TESTING OF GAS PIPING

G2417.3.4 (406.3.4) Valve isolation. Where the piping system is connected to equipment or components designed for operating pressures equal to or greater than the test pressure, such equipment shall be isolated from the piping system by closing the individual equipment shutoff valve(s).

❖ If the appliances and other system components are not subject to damage by the test pressure, the equipment shutoff valve can serve to isolate the equipment during the test. Unlike Section G2417.3.3, this section does not require disconnection of the appliance/equipment. See the commentary for Section G2417.3.3.

G2417.3.5 (406.3.5) Testing precautions. All testing of piping systems shall be done with due regard for the safety of employees and the public during the test. Prior to testing, the interior of the pipe shall be cleared of all foreign material.

❖ Compressed gases store energy in direct proportion to the pressure. This stored energy can be destructive and dangerous to personnel if released quickly due to joint, component, or pipe failure or an intentional action. Debris will be found in all piping systems after installation. This debris can be harmful to appliances and system components and can act as projectiles during testing and purging operations. Piping is typically purged before testing to flush out foreign materials.

G2417.4 (406.4) Test pressure measurement. Test pressure shall be measured with a manometer or with a pressure measuring device designed and calibrated to read, record or indicate a pressure loss caused by leakage during the pressure test period. The source of pressure shall be isolated before the pressure tests are made. Mechanical gauges used to measure test pressures shall have a range such that the highest end of the scale is not greater than five times the test pressure.

❖ Test pressures must be measured with a manometer or other pressure-measuring device that is designed and calibrated to indicate a pressure loss due to leakage during the pressure test period. Pressure-measuring devices must be graduated such that small variations in test pressure will be readily detectable. Small pressure variations might go undetected in instruments having broader ranges. Gauges should have scale increments that are small enough to allow detection of small changes in pressure. Mechanical gauges generally achieve their best accuracy near mid-range of their scale. All test instruments should have recently been calibrated and their accuracy verified before starting the test. Although no piping system can be absolutely leak free, the intent of the code is make the system free of leaks that are measurable with the testing procedures and instruments prescribed in the code and attainable in the field. Spring-type mechanical pressure gauges must have a high-end scale reading of not greater than 5 times the test pressure. If the test pressure was 3 psig, for example, the gauge would have to have a maximum range of zero to 15 psig. Such gauges are a wise choice because the test reading will be higher in the range of the gauge where the gauge accuracy is better for such instruments. Also, the gauge can have much smaller increments than the typical 100 psig gauge, thus allowing greater discernability in detecting a leak. See Commentary Figure G2417.4(1). All spring-type mechanical gauges are susceptible to inaccuracies due to conditions of use and mishandling and should be calibrated regularly. Manometers are very accurate and do not need to be calibrated. In all cases, the intent of this section is to require an instrument that is capable of indicating the very small pressure changes that occur as a result of a detectable leak. Test instrument sensitivity is especially important considering the short duration of the test.

The source of pressure in the piping system under test must be disconnected from the system to prevent such source from invalidating the test. For example, the compressed air tank or inert gas cylinder could be adding gas to the piping system at the same rate as a leak is letting gas escape; the test would then be meaningless.

Commentary Figures G2417.4(2), (3), and (4) show examples of test instruments that can be considerably more accurate than mechanical spring gauges. Electronic instruments can provide high resolution readings, making very small leaks detectable and making leaks detectable within short time periods. Commentary Figures G2417.4 (3) and (4) show instruments that are designed to give accurate pressure readings and give positive indication of a leak. The liquid manometer-type instrument, Commentary Figure G2417.4 (3), functions as a manometer indicating the test pressure and readily can indicate leakage. A separate feature of the instrument allows small leaks to be quickly recognized as follows. While the system is under test pressure, an integral valve can be closed, which will isolate the liquid reservoir from the piping system, thereby trapping the test pressure in the manometer's liquid reservoir. A small diameter vertical dip tube, having its open end submerged in the liquid, rises above the liquid level in the reservoir. The top end of the dip tube is connected to the piping system under test. If there is no leakage, the pressures at the upper and lower (submerged) ends of the dip tube are equal, and liquid will not rise in the dip tube. If leakage occurs, the pressure in the piping system will fall below the pressure trapped in the reservoir at the start of the test procedure, and liquid will be forced up into the small diameter dip tube indicating the leak. The use of a small diameter dip tube allows a small change in liquid level in the reservoir to be seen as a comparatively large movement of liquid in the dip tube.

G2417.4.1 (406.4.1) Test pressure. The test pressure to be used shall be not less than one and one-half times the proposed maximum working pressure, but not less than 3 psig (20 kPa gauge), irrespective of design pressure. Where the test pressure exceeds 125 psig (862 kPa gauge), the test pressure shall not ex-

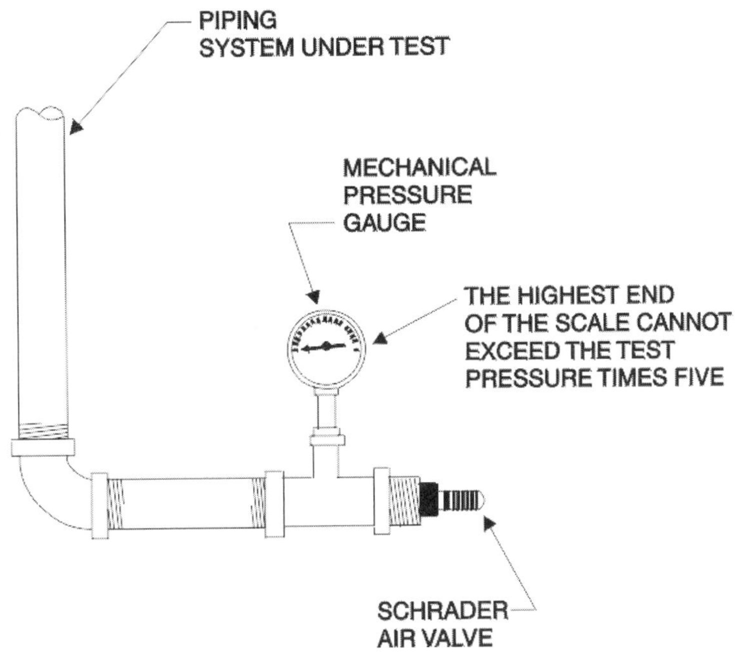

Figure G2417.4(1)
TYPICAL SETUP FOR AIR PRESSURE TESTING

ceed a value that produces a hoop stress in the piping greater than 50 percent of the specified minimum yield strength of the pipe.

❖ The minimum test pressure will never be less than 3 psig. The majority of residential and small commercial occupancies are served by piping systems with a pressure of less than ½ psig (14 inches water column), and for such systems the test pressure must be 3 psig or higher. The test pressure must also be not less than 1.5 times the pressure at which the piping system is designed to operate. If the piping system is designed to operate at 5 psig, the test pressure must be not less than 1.5 x 5 = 7.5 psig. In very high-pressure applications, the test pressure might approach the structural limits of the piping material, and therefore the test pressure must be limited. Hoop stress is the result of internal pressure that tends to expand the pipe walls outward along the circumference of the pipe.

Photo courtesy of Dwyer Instruments, Inc.

Figure G2417.4(2)
ELECTRONIC PRESSURE GAUGE

G2417.4.2 (406.4.2) Test duration. The test duration shall be not less than 10 minutes.

❖ As the piping system becomes larger, the test duration must be longer so any leak can be detected. For example, a leakage rate of one cubic foot per hour would produce a more easily detected pressure drop in a small 10 cubic foot volume system than it would in a large 500 cubic foot volume system. It would require a longer test period to detect the leak in the large volume system since the leakage of one cubic foot represents a much smaller fraction of the total gas volume in the system. If a system has volume of 501 cubic feet, the minimum test duration would be 1 hour. To put this in perspective, 500 cubic feet of piping system internal volume would equate to 83,333 feet of 1-inch schedule 40 steel pipe or 9,747 feet of 3-inch schedule 40 steel pipe.

Because 10 cubic feet equates to 1,667 feet of 1-inch pipe, it is obvious that single-family dwellings fall in the 10 cubic foot or less volume category. A ten-minute test period is not much time to detect a pressure drop, which makes it even more important to use

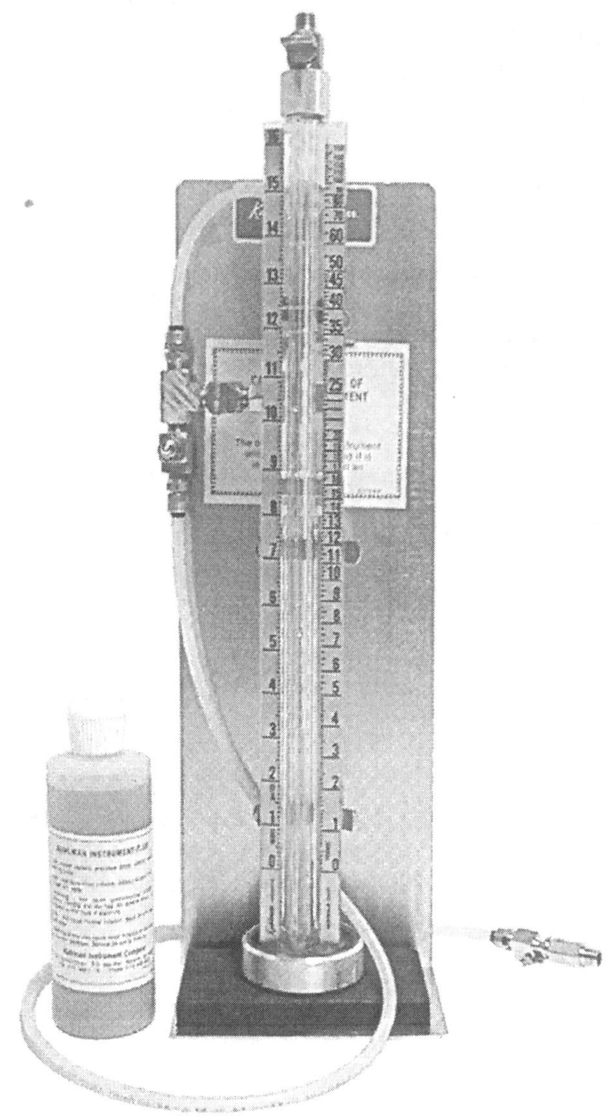

Photo courtesy of Kuhlman Company

Figure G2417.4(3)
LEAK DETECTOR INSTRUMENT

a pressure instrument that is capable of indicating very small drops in pressure. See the commentary for Section G2417.4.

G2417.5 (406.5) Detection of leaks and defects. The piping system shall withstand the test pressure specified without showing any evidence of leakage or other defects. Any reduction of test pressures as indicated by pressure gauges shall be deemed to indicate the presence of a leak unless such reduction can be readily attributed to some other cause.

❖ Any pressure drop, no matter how small, is considered to be a failure of the test, except where it can be demonstrated that the drop was caused by a change in temperature or other such cause. This would be practically impossible to demonstrate in the field, especially for short test durations. A true leak should show up as a continuous drop in test pressure. If the rate of pressure drop slows as the test progresses it may be due to the cooling of the test medium, which causes it to contract. For example, warm air coming from a compressor will contract when introduced into a cold piping system, thereby causing a test instrument to indicate a drop in pressure. To help eliminate any guess work, the temperature of the test medium should be allowed to stabilize within the piping system before the testing period is started.

G2417.5.1 (406.5.1) Detection methods. The leakage shall be located by means of an approved combustible gas detector, a noncorrosive leak detection fluid or an equivalent nonflammable solution. Matches, candles, open flames or other methods that could provide a source of ignition shall not be used.

❖ Once the pressure measuring instrument indicates leakage, the leak or leaks must be located so that repairs or replacements can be performed. Electronic sensors and bubble fluids are used to find leaks that are not readily found by human senses. Many people who have worked with and/or inspected gas piping installations can recall instances of personal injury, fires, and close calls that were caused by searching for leaks with an open flame.

G2417.5.2 (406.5.2) Corrections. Where leakage or other defects are located, the affected portion of the piping system shall be repaired or replaced and retested.

❖ Where a leak is detected by a method named in Section G2417.5.1, such defect must be corrected. Corrections include tightening of fittings or threaded joints and replacing defective pipe, tubing, or fittings.

G2417.6 (406.6) System and equipment leakage test. Leakage testing of systems and equipment shall be in accordance with Sections G2417.6.1 through G2417.6.4.

❖ Leakage testing is different from pressure testing. The previous subsections in Section G2417 addressed the pressure-testing of newly installed or altered piping systems for the purpose of locating defects in the installation or materials. This section addresses leakage tests that are intended to discover open outlets, defective appliance connections, and defects that have developed since the initial installation. See Section G2417.6.2.

G2417.6.1 (406.6.1) Test gases. Fuel gas shall be permitted to be used for leak checks in piping systems that have been tested in accordance with Section G2417.

❖ Pressure testing of piping systems is done with air or an inert gas, and leakage testing is done with the fuel gas at whatever pressure the system is designed to operate. See Section G2417.6.2.

G2417.6.2 (406.6.2) Before turning gas on. Before gas is introduced into a system of new gas piping, the entire system shall be inspected to determine that there are no open fittings or ends and that all valves at unused outlets are closed and plugged or capped.

❖ This section requires a precautionary measure be taken to prevent fuel gas from escaping from the piping system when the gas is turned on in the system for the first time. Since the piping system has never been used, it is not to be fully trusted as being gas tight. Because anything could have been done to the system since it was pressure tested, an additional visual check is justified before the gas is turned on. The condition of appliances may also be unknown; thus, all appliance shutoff valves must be closed until such appliances are ready to be put in service. Often, parties other than those who installed and tested the piping system are involved with connecting to the outlets in the system. This makes it likely that what was gas tight at the initial pressure testing is no longer gas tight as a result of incomplete or improper connections at the system outlets.

G2417.6.3 (406.6.3) Test for leakage. Immediately after the gas is turned on into a new system or into a system that has been initially restored after an interruption of service, the piping system shall be tested for leakage. If leakage is indicated, the gas supply shall be shut off until the necessary repairs have been made.

❖ The section is a logical progression of the previous section. After the inspector makes a visual check for the status of all outlets and equipment shutoff valves, the gas is introduced into the system, and then the system is tested for leakage. Such testing is typically accomplished by examining all of the outlets, observing a gas meter for indication of flow, monitoring the system pressure for any drop, or any combination of these. Since all outlet valves are closed and all unused outlets are capped, any gas flow or pressure drop indicates leakage, and the gas must be turned off. This section also allows the leakage test to be conducted with the appliance shutoff valves open and the appliances shut down to prevent their operation. Testing in this manner would verify that the appliance gas controls were not faulty. A leakage test is also required after gas service is restored.

G2417.6.4 (406.6.4) Placing equipment in operation. Gas utilization equipment shall be permitted to be placed in operation after the piping system has been tested and determined to be free of leakage and purged in accordance with Section G2417.7.2.

❖ Gas appliances cannot be put into service until the gas piping system that supplies them has been pressure tested, tested for leakage, and purged of air. See Section G2417.7.2.

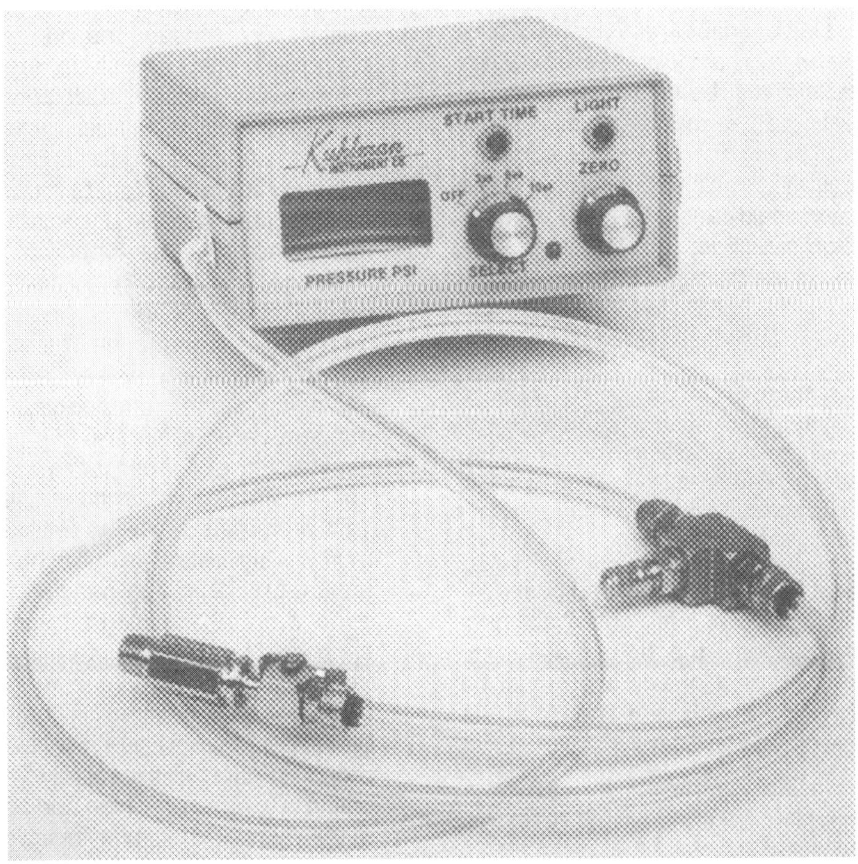

Photo courtesy of Kuhlman Company

Figure G2417.4(4)
ELECTRONIC LEAK DETECTOR INSTRUMENT

G2417.7 (406.7) Purging. Purging of piping shall comply with Sections G2417.7.1 through G2417.7.4.

❖ Purging is intended to prevent a flammable gas/air mixture from being created in the piping. A flammable mixture could be a fire/explosion hazard in the piping, room, space, or in an appliance combustion chamber.

G2417.7.1 (406.7.1) Removal from service. When gas piping is to be opened for servicing, addition or modification, the section to be worked on shall be turned off from the gas supply at the nearest convenient point, and the line pressure vented to the outdoors, or to ventilated areas of sufficient size to prevent accumulation of flammable mixtures.

❖ If gas piping is opened for some reason, the residual gas in the affected piping must be disposed of in such a manner that will not create a fire/explosion hazard.

G2417.7.2 (406.7.2) Placing in operation. When piping full of air is placed in operation, the air in the piping shall be displaced with fuel gas. The air can be safely displaced with fuel gas provided that a moderately rapid and continuous flow of fuel gas is introduced at one end of the line and air is vented out at the other end. The fuel gas flow should be continued without interruption until the vented gas is free of air. The point of discharge shall not be left unattended during purging. After purging, the vent shall then be closed.

❖ Piping that has been left open for some period of time because it is out of service or has never been in service will contain air. This air must be eliminated from the piping to avoid creating a flammable gas/air mixture that could ignite in the piping or an appliance combustion chamber. Gas/air mixtures can also be out of the flammability range and could cause delayed ignition in appliance combustion chambers resulting in personnel injury and/or appliance damage. In small volume systems, the air can be purged by introduction of fuel gas at a rate sufficient to flush the air from the system.

G2417.7.3 (406.7.3) Discharge of purged gases. The open end of piping systems being purged shall not discharge into confined spaces or areas where there are sources of ignition unless precautions are taken to perform this operation in a safe manner by ventilation of the space, control or purging rate, and elimination of all hazardous conditions.

❖ Obviously, mixtures of fuel gas, air, and inert gases resulting from the purging of piping are potentially hazardous and must be discharged to the outdoors or to adequately ventilated areas. Consideration must be given to sources of ignition as well as ventilation where purged gases are discharged indoors.

G2417.7.4 (406.7.4) Placing equipment in operation. After the piping has been placed in operation, all equipment shall be purged and then placed in operation, as necessary.

❖ For the same reason that piping systems must be purged of foreign gases, the appliances served must also be purged before being placed in operation.

SECTION G2418 (407) PIPING SUPPORT

G2418.1 (407.1) General. Piping shall be provided with support in accordance with Section G2418.2.

❖ Often, piping support is inadequate, poorly installed, or overlooked entirely, despite the fact that support is critical in preventing stresses and strains in piping, fittings, joints, connectors, and appliance gas trains. See Section G2418.2.

G2418.2 (407.2) Design and installation. Piping shall be supported with pipe hooks, metal pipe straps, bands, brackets or hangers suitable for the size of piping, of adequate strength and quality, and located at intervals so as to prevent or damp out excessive vibration. Piping shall be anchored to prevent undue strains on connected equipment and shall not be supported by other piping. Pipe hangers and supports shall conform to the requirements of MSS SP-58 and shall be spaced in accordance with Section G2424. Supports, hangers, and anchors shall be installed so as not to interfere with the free expansion and contraction of the piping between anchors. All parts of the supporting equipment shall be designed and installed so they will not be disengaged by movement of the supported piping.

❖ As with all piping systems, the support of the fuel gas system is as important as any other part of the overall design. Proper supports are necessary to maintain piping alignment and slope, to support the weight of the pipe and its contents, to control movement, and to resist dynamic loads, such as thrust. Inadequate support can cause piping to fail under its own weight, resulting in fire, explosion, or property damage. Building design must take into consideration the structural loads created by the support of piping systems.

Hangers or supports must not react with or be detrimental to the pipe they support. Hangers or supports for metallic pipe must be of a material that is compatible with the pipe to prevent any corrosive action. For example, copper, copper-clad, or specially coated hangers are required if the piping system is constructed of copper tubing. Hangers and supports are constructed of noncombustible materials to prevent premature failure in a fire. This section is commonly interpreted to prohibit combustible hooks, bands, brackets, and hangers. Gas piping must not depend on other piping for support; thus, it must be supported from the structure independently. Gas controls for appliances can be stressed and damaged by unsupported piping drops that bear on or produce bending moments in such controls. A simple check for proper support at an appliance connected with rigid piping is to open the union and observe any movement of the piping. If the piping drops or moves in any direction, it is evident that stress is being applied to the appliance controls. Eliminating such stress is one of the advantages of using flexible gas connectors. See Section G2424.

SECTION G2419 (408) DRIPS AND SLOPED PIPING

G2419.1 (408.1) Slopes. Piping for other than dry gas conditions shall be sloped not less than 0.25 inch in 15 feet (6.4 mm in 4572 mm) to prevent traps.

❖ The required minimum slope (0.14 percent) is intended for gas known to contain enough water vapor to cause condensate to form inside the piping. Such conditions are rarely, if ever, encountered today.

G2419.2 (408.2) Drips. Where wet gas exists, a drip shall be provided at any point in the line of pipe where condensate could collect. A drip shall also be provided at the outlet of the meter and shall be installed so as to constitute a trap wherein an accumulation of condensate will shut off the flow of gas before the condensate will run back into the meter.

❖ Drips, often referred to as "drip legs," are distinct from sediment traps. Modern gas supplies and distribution systems are typically dry; thus, drips would only be required as advised by the gas supplier.

G2419.3 (408.3) Location of drips. Drips shall be provided with ready access to permit cleaning or emptying. A drip shall not be located where the condensate is subject to freezing.

❖ A drip is not the same as a sediment trap as they are required for different reasons. See Section G2419.4.

G2419.4 (408.4) Sediment trap. Where a sediment trap is not incorporated as a part of the gas utilization equipment, a sediment trap shall be installed downstream of the equipment shutoff valve as close to the inlet of the equipment as practical. The sediment trap shall be either a tee fitting with a capped nipple in the bottom opening of the run of the tee or other device approved as an effective sediment trap. Illuminating appliances, ranges, clothes dryers, and outdoor grills need not be so equipped.

❖ In addition to the code requirement, most appliance manufacturers require the installation of a sediment trap (dirt leg) to protect the appliance from debris in the gas. Sediment traps are necessary to protect appliance gas controls from the dirt, soil, pipe chips, pipe joint tapes and compounds, and construction site debris that enters the piping during installation and repairs. Hazardous appliance operation could result from debris entering gas controls and burners. Despite the fact that utilities supply clean gas, debris can enter the piping prior to and during installation on the utility side of the system and on the customer side.

Sediment traps are designed to cause the gas flow to change direction 90 degrees at the sediment collection point, thus causing the solid or liquid contaminants to drop out of the gas flow. See Commentary Figure G2419.4(1). The nipple and cap must not be placed in the branch opening of a tee fitting, because this would not create a change in direction of flow and would allow debris to pass over the collection point. Commentary Figure G2419.4(2) illustrates an improper sediment trap that is prohibited by this section. The code does not specify a minimum length for the capped nipple. Three inches minimum is customary. The sediment trap must be downstream of the appliance shutoff valve and as close to the appliance inlet as practical. Requiring the sediment trap to be downstream of the appliance shutoff valve makes sure that the sediment trap is within 6 feet of the appliance inlet. If there is 6 feet or less of piping between a sediment trap and the appliance inlet served, the intent of the code has been met, regardless of the shutoff valve location. Manufactured sediment traps are available that have the configuration of a straight section of pipe and are equipped with cleanout openings. Although it would be wise to install sediment traps at all appliance connections, they are not mandated by code for gas lights, ranges, clothes dryers, and outdoor grills. These appliances are susceptible to harm from debris in gas, especially ranges and clothes dryers, and the appliance manufacturer may require sediment traps where the code does not. The code's logic is that these appliances are attended while in use therefore, the user would be aware of a problem.

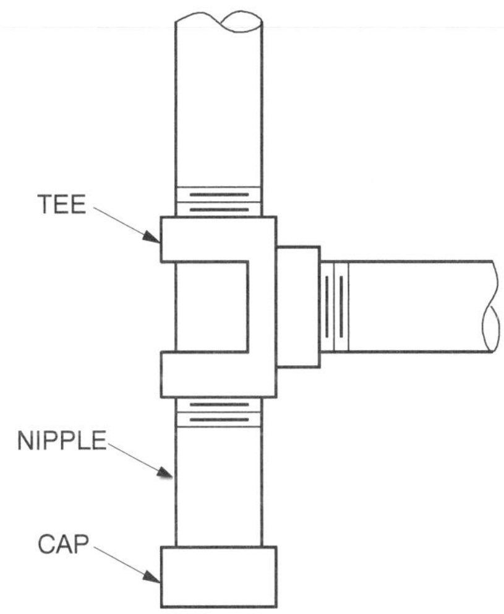

Figure G2419.4(1) SEDIMENT TRAP

SECTION G2420 (409) GAS SHUTOFF VALVES

G2420.1 (409.1) General. Piping systems shall be provided with shutoff valves in accordance with this section.

❖ This section addresses shutoff valves for meters, buildings, tenant spaces, appliances, and fireplaces.

G2420.1.1 (409.1.1) Valve approval. Shutoff valves shall be of an approved type. Shutoff valves shall be constructed of materials compatible with the gas piping. Shutoff valves installed in a portion of a piping system operating above 0.5 psig shall comply with ASME B16.33. Shutoff valves installed in a portion of a piping system operating at 0.5 psig or less shall comply with ANSI Z21.15 or ASME Bl6.33.

❖ The code official must approve the type of shutoff valve used. For example, he or she might not approve a valve that requires a wrench to operate. All shutoff valves must be tested to the specified standard and so identified. Z21.15 is applicable to appliance shutoff valves and appliance connector shutoff valves. Ball valves are used almost exclusively today because of their low pressure losses, ease of operation, and reliability. See Commentary Figures G2420.1.1(1) and G2420.1.1(2).

G2420.1.2 (409.1.2) Prohibited locations. Shutoff valves shall be prohibited in concealed locations and furnace plenums.

❖ This section does not prohibit shutoff valves in spaces above ceilings and below floors used for conveying environmental air. See the definition of "Concealed location." A valve in a concealed location would be of no value and a potential leak source. The concern for valves in furnace plenums is temperature extremes and lack of access.

G2420.1.3 (409.1.3) Access to shutoff valves. Shutoff valves shall be located in places so as to provide access for operation and shall be installed so as to be protected from damage.

❖ As expressed in the previous section, a shutoff valve has no utility if it is installed where it cannot be accessed. Location is especially important for valves installed for emergency use such as a ball valve installed indoors upstream of all appliances to allow the building owner to shut off all gas in the event of an emergency. Outdoor valves are particularly subject to damage.

G2420.2 (409.2) Meter valve. Every meter shall be equipped with a shutoff valve located on the supply side of the meter.

❖ This valve is typically provided and owned by the gas supplier because it is upstream of the point of delivery for utility-owned meters. Such valves require a wrench to operate and are capable of being locked off.

G2420.3 (409.3.2) Individual buildings. In a common system serving more than one building, shutoff valves shall be installed outdoors at each building.

❖ This section requires the shutoff valve for each building to be located outside of the building, thus allowing quick access in an emergency.

G2420.4 (409.4) MP regulator valves. A listed shutoff valve shall be installed immediately ahead of each MP regulator.

❖ See the definition of "Regulator, Medium-pressure." This regulator (also known as "pounds-to-inches" regulator) is placed between the service regulator and the appliance regulator and is typically associated with manifold parallel distribution systems using CSST or copper tubing. A shutoff valve ahead of the MP regulator will allow the regulator to be conveniently serviced, adjusted, or replaced. The valve also serves to isolate the higher pressure system from the lower pressure system. For purposes of sizing, the piping downstream of an MP regulator is considered as a separate system, with the MP regulator being the source of supply for such system.

G2420.5 (409.5) Equipment shutoff valve. Each appliance shall be provided with a shutoff valve separate from the appliance. The shutoff valve shall be located in the same room as the appliance, not further than 6 feet (1829 mm) from the

Figure G2419.4(2)
PROHIBITED SEDIMENT TRAP CONFIGURATION

appliance, and shall be installed upstream from the union, connector or quick disconnect device it serves. Such shutoff valves shall be provided with access.

> **Exception:** Shutoff valves for vented decorative appliances and decorative appliances for installation in vented fireplaces shall not be prohibited from being installed in an area remote from the appliance where such valves are provided with ready access. Such valves shall be permanently identified and shall serve no other equipment.

❖ Each appliance, regardless of type, size, or location, must have its own dedicated shutoff valve to isolate the appliance from its gas supply. It is not the intent of this section to allow a single shutoff valve to control more than one appliance. Equipment/appliance shutoff valves are not considered to be emergency valves, but rather are considered to be service valves that allow the appliance to be serviced, repaired, replaced, or shut down. See the definition of "Valve, Equipment shutoff." The location requirements of this section are intended to make the valve conspicuous and easy to operate. The code does not require the equipment shutoff valve to be provided with "ready access." Because such a valve is not an emergency valve, there is no reason to require ready access, and simply providing access is sufficient. Common practice has been to install the appliance shutoff valve behind ranges and clothes dryers and in the control compartment of decorative gas fireplace appliances. This section allows such practice since the valves are accessible. To avoid the use of an exposed valve or concealed "key-handle" valve, it is accepted practice to locate the shutoff valve for gas fireplace appliances in the control compartment in the bottom of the unit, behind the control access panel. See Commentary Figure G2420.5.

This section calls for a separate shutoff valve, which means a valve other than the valve that is integral with the appliance gas control. The intent is that the separate shutoff be located such that there is no more than 6 feet of piping between the outlet of the shutoff valve and the inlet of the appliance. This section works with the definition of "piping system" and Section G2413.3 to allow the piping downstream of the shutoff valve to be considered as and sized as a connector.

The exception allows shutoff valves for the stated appliances to be located remote from the appliance, provided that such valves are permanently tagged as to their function, serve only the single appliance, and are provided with ready access. In the exception, "ready access" is intended because the shutoff valve will be more difficult to locate being remote from the appliance served. The exception provides an alternative to using concealed "key-handle" type valves located in the wall or floor adjacent to the appliance. Unvented appliances are not addressed by the exception. The shutoff valve for a log lighter is also the appliance operating valve; therefore, such valve cannot be located remotely. Because the remote shutoff valve allowed by the exception could be more than 6 feet from the appliance, the piping from the shutoff valve outlet to within 3 feet of the appliance inlet must be treated as and sized as gas piping, not as a connector.

Photo courtesy of Dormont Manufacturing Company

Figure G2420.1.1(1)
TYPICAL BALL-TYPE GAS SHUTOFF VALVE

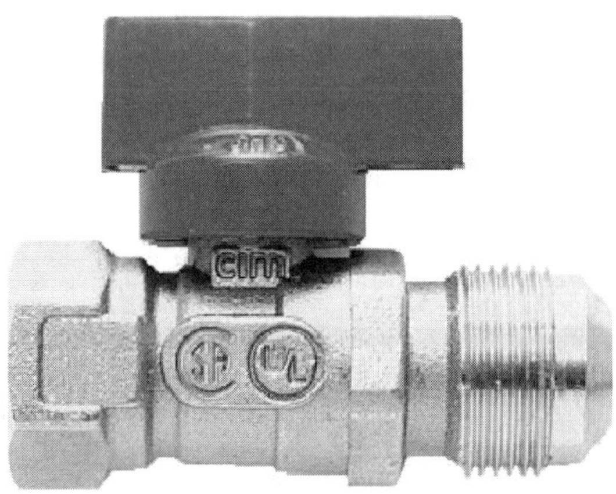

Photo courtesy of Dormont Manufacturing Company

Figure G2420.1.1(2)
BALL-TYPE GAS SHUTOFF VALVE DESIGNED TO MATE WITH A CONNECTOR

G2420.5.1 (409.5.1) Shutoff valve in fireplace. Equipment shutoff valves located in the firebox of a fireplace shall be installed in accordance with the appliance manufacturer's instructions.

❖ This section addresses appliances such as decorative gas log sets that are installed in solid fuel-burning fireplaces. The manual shutoff is permitted to be in the firebox of the fireplace provided that it is allowed by the appliance manufacturer's instructions and the valve is located as directed by those instructions. There is typically an area adjacent to the appliance controls or inlet connection that is shielded from radiant heat, and this "cool zone" is where the manual shutoff valve must be located. A shutoff valve must never be located in the firebox if solid fuel is burned.

SECTION G2421 (410) FLOW CONTROLS

G2421.1 (410.1) Pressure regulators. A line pressure regulator shall be installed where the appliance is designed to operate at a lower pressure than the supply system. Access shall be provided to pressure regulators. Pressure regulators shall be protected from physical damage. Regulators installed on the exterior of the building shall be approved for outdoor installation.

❖ All gas supply systems employ regulators at some point in the distribution system, and the type and location of these devices depends on the design pressures of the system and utilization equipment. A line pressure regulator reduces the pressure from the service delivery pressure to the pressure of the distribution system. Additional pressure regulators could be required at other points in the system or at the appliance to reduce the pressure even further. See Commentary Figure G2421.1(1). With the increased use of medium-pressure gas distribution systems, systems are designed with one or more regulators in addition to the service regulators. Appliances and gas utilization equipment are also equipped with factory-installed regulators that control burner input (manifold) pressure.

Pressure regulators are extremely important devices that are relied on to prevent overpressurization of appliances and equipment. Typical gas appliance controls are not designed to tolerate a pressure of more than 14 inches of water column (3.5 kPa). The standards that regulate the design and function of appliance gas controls require that the controls be tested for resistance to permanent damage resulting from exposure to excessive gas supply pressure. Each function of the gas controls must be tested for exposure to 2½ psi for a period of one minute, after which the control must function normally. Excess pressures could lead to severe appliance malfunction. Depending on the application, gas pressure regulators may contain relief mechanisms that are designed to vent gas from the downstream side in the event that the downstream (load) side pressure exceeds a predetermined maximum. This feature is a safeguard intended to prevent the overpressurization of the piping and utilization equipment supplied by the regulator. Regulators with relief devices must be installed outdoors, or the relief vent must be piped to the outdoors to prevent gas from being discharged into the building.

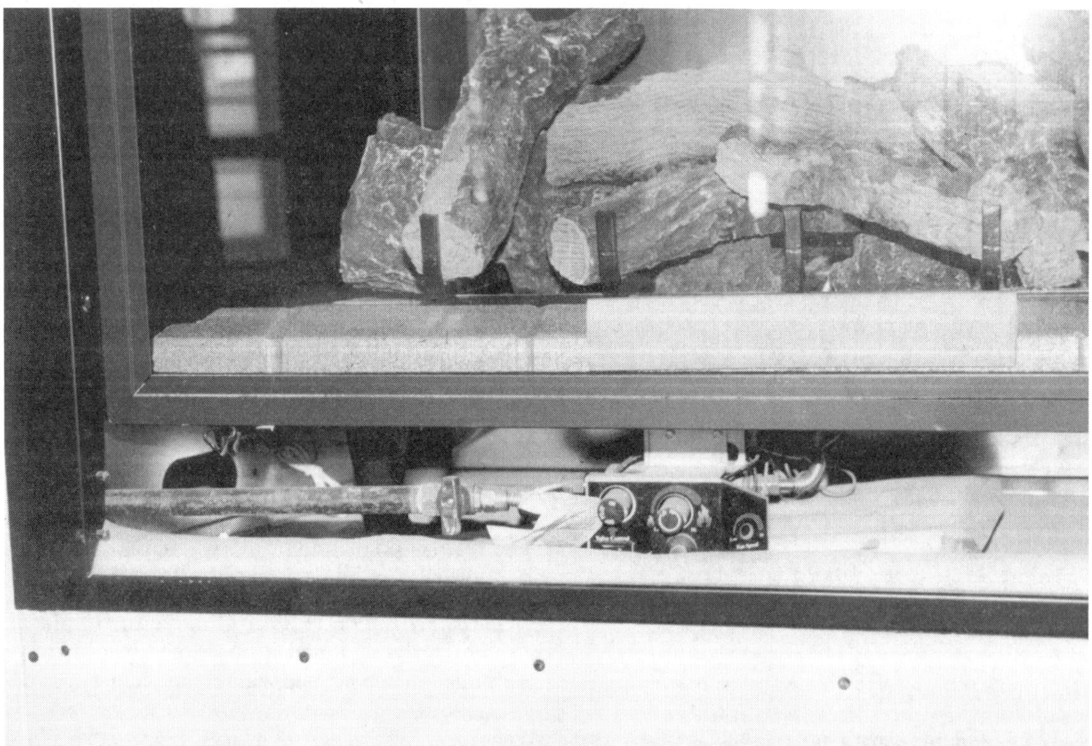

Figure G2420.5
SHUTOFF VALVE IN CONTROL COMPARTMENT OF GAS FIREPLACE

Access to pressure regulators is required for the maintenance, adjustment, and servicing of the devices. The definition of "Access" appears in Section G2403. Protecting regulators from physical damage essentially means that regulators are to be installed in locations that are not prone to impact from equipment or vehicles, or they are to be shielded from such impact. Because of the range of environmental conditions encountered outside the building, regulators must be evaluated and approved for installation outdoors. Regulators exposed to the weather must be protected from moisture and precipitation. Figure G2421.1(2) shows a service regulator with discharge piping connected to the integral relief valve port. This regulator establishes the pressure at the point of delivery. Figure G2421.1(3) illustrates the basic operating principle of gas pressure regulators. Outlet pressure acts on the flexible diaphragm and adjustment spring causing the valve to close on a rise in outlet pressure and open on a fall in outlet pressure. The vent allows the diaphragm to move freely by admitting or emitting air above the diaphragm. Atmospheric pressure and the adjustment spring force act together to establish the outlet pressure setting of the regulator.

G2421.2 (410.2) MP regulators. MP pressure regulators shall comply with the following:

1. The MP regulator shall be approved and shall be suitable for the inlet and outlet gas pressures for the application.
2. The MP regulator shall maintain a reduced outlet pressure under lockup (no-flow) conditions.
3. The capacity of the MP regulator, determined by published ratings of its manufacturer, shall be adequate to supply the appliances served.
4. The MP pressure regulator shall be provided with access. Where located indoors, the regulator shall be vented to the outdoors or shall be equipped with a leak-limiting device, in either case complying with Section G2421.3.
5. A tee fitting with one opening capped or plugged shall be installed between the MP regulator and its upstream shutoff valve. Such tee fitting shall be positioned to allow connection of a pressure measuring instrument and to serve as a sediment trap.
6. A tee fitting with one opening capped or plugged shall be installed not less than 10 pipe diameters downstream of the MP regulator outlet. Such tee fitting shall be positioned to allow connection of a pressure measuring instrument.

❖ This section sets forth the minimum requirements for medium-pressure (MP) gas regulators installed in 2-psi (13.8 kPa) gas piping systems or portions thereof. These requirements intend to keep the regulators accessible for service and repair, protect them from physical damage, require that they are suitable for the use intended, are properly sized and installed, and, if installed indoors, are properly vented to the outdoors. See the Commentary for Section G2421.3. Item 2 requires a "lockup"-type regulator capable of holding the approximate outlet pressure setpoint indefinitely under no flow conditions. Section G2420.4 requires a gas shutoff valve immediately upstream of the MP regulator to allow isolation of the regulator during testing, re-

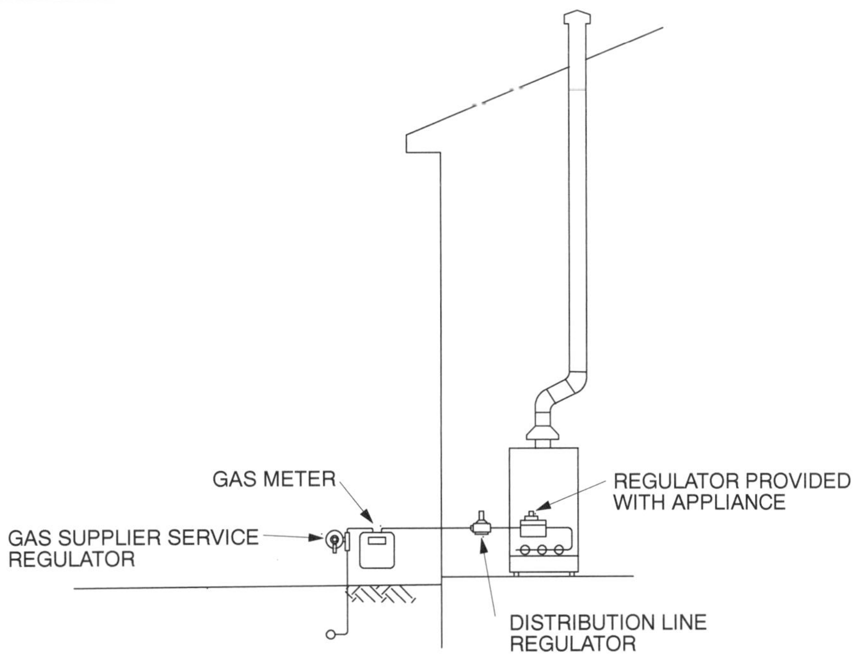

Figure G2421.1(1)
GAS PRESSURE REGULATORS

Photo courtesy of Washington Gas

Figure G2421.1(2)
SERVICE REGULATOR

pair, or replacement of the regulator. Items 5 and 6 require a tee fitting upstream and downstream of the MP regulator to facilitate the connection of pressure-measuring instruments for calibrating the regulator. Such fittings should provide a $^1/_8$-inch threaded tap-

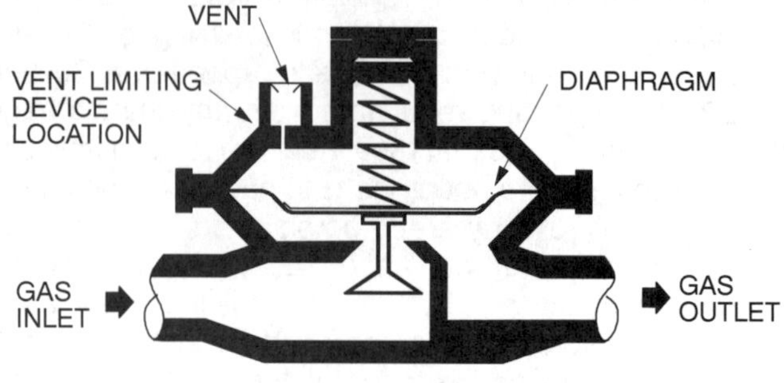

Figure G2421.1(3)
GAS PRESSURE REGULATOR

ping for connection of common gauges and manometers. Item 5 also requires a sediment trap on the inlet side of the regulator. Because a lot of faith is placed in the integrity of this regulator, it deserves the protection from debris afforded by a simple tee, nipple, and cap arrangement. See Section G2419.4.

Where a line pressure regulator serves appliances rated for $^1/_2$ psi or less inlet pressure and the regulator's inlet pressure is over 2 psi, a downstream overpressure protection device must be provided in accordance with ANSI Z21.80, the standard for line pressure regulators. See Commentary Figure G2421.2. The OPD is designed to limit the downstream (outlet) pressure to 2 psi.

G2421.3 (410.3) Venting of regulators. Pressure regulators that require a vent shall have an independent vent to the outside of the building. The vent shall be designed to prevent the entry of water or foreign objects.

Photo courtesy of Maxitrol Company

Figure G2421.2
PRESSURE REGULATOR/OVERPRESSURE PROTECTION DEVICE ASSEMBLIES

Exception: A vent to the outside of the building is not required for regulators equipped with and labeled for utilization with approved vent-limiting devices installed in accordance with the manufacturer's instructions.

❖ All regulators employ synthetic or natural rubber diaphragms, one side of which is vented to the atmosphere. The devices must vent one side of the diaphragm to the atmosphere in order to allow unimpeded movement of the diaphragm within its enclosure. Thus, these devices, in effect, see atmospheric pressure as a reference point for controlling the fuel-gas pressure. Regulators cannot operate without the vent. If the diaphragm were to rupture or perforate, gas would be vented to the exterior of the device. For this reason, regulators installed indoors must be vented to the outdoors or must comply with the exception to this section. The vent termination outdoors must be protected from blockage and the entry of foreign objects, insects, and water, which could cause regulator failure. See Commentary Figures G2421.3(1) and G2421.3(2).

This section also requires regulators to be vented individually to the outdoors in order to prevent a potentially hazardous situation that may arise if more than one regulator were served by a single vent. The hazard could occur if the diaphragm of one regulator failed and allowed an increase of pressure within the vent, which would then cause all of the other regulators using the common vent to operate at a higher discharge pressure. ANSI Z223.1 permits engineered vent manifold systems to be used to vent more than one regulator, provided that the manifold is designed to allow unimpeded air movement and to minimize back pressure in the event of diaphragm failure.

The exception permits regulators to be installed indoors without a vent to the outdoors where the regulator is equipped with and labeled for use with a factory-installed vent-limiting device. Vent-limiting devices, as the name implies, are designed to allow the vent to "breathe" under normal operating conditions, and in the event of diaphragm failure, the vent-limiting devices will limit gas escapement to a rate near one cubic foot per hour for natural gas at 2 psi pressure. A vent-limiting device is a ball check valve with a fixed orifice and is designed to allow air inhalation and escapement while restricting gas leakage to a small rate. Even small leakage rates can be hazardous if inadequate ventilation exists to prevent buildup of explosive concentrations. Although a standard is not referenced in this section, ANSI Z21.18 is appropriate for evaluating such vent-limiting devices. The standard limits the volume of gas that a vent-limiting device is allowed to discharge into a room or space. Vent-limiting devices should be installed in locations that are ventilated sufficiently to dissipate any gas discharged from the device.

Photo courtesy of Washington Gas

Figure G2421.3(1)
REGULATOR WITH PROTECTION

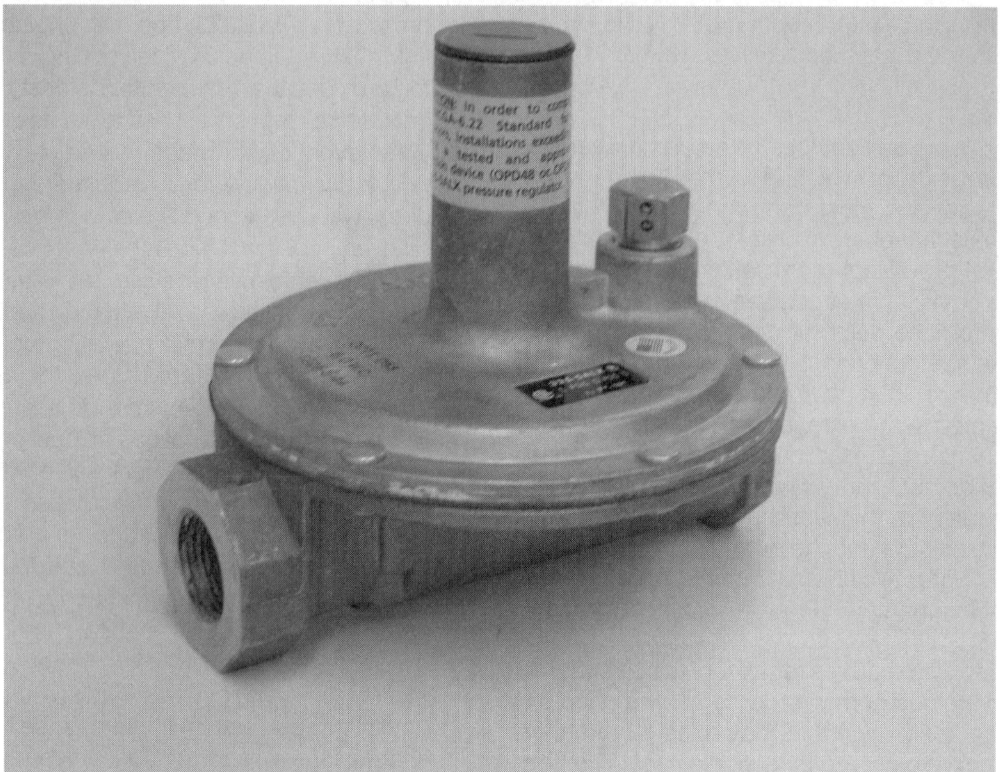

Photo courtesy of Mixitrol Company

Figure G2421.3(2)
PRESSURE REGULATOR WITH VENT-LIMITING DEVICE

SECTION G2422 (411) APPLIANCE CONNECTIONS

G2422.1 (411.1) Connecting appliances. Appliances shall be connected to the piping system by one of the following:

1. Rigid metallic pipe and fittings.
2. Listed and labeled appliance connectors installed in accordance with the manufacturer's installation instructions and located entirely in the same room as the appliance.
3. Listed and labeled quick-disconnect devices used in conjunction with listed and labeled appliance connectors.
4. Listed and labeled gas convenience outlets used in conjunction with listed and labeled appliance connectors.

❖ Connectors are used to connect the appliance or equipment to the gas distribution system outlet. The pipe sizing methods in this chapter size distribution piping up to but not beyond the appliance shutoff valve outlet, and this section sizes the connection between the shutoff valve outlet and the appliance. See Commentary Figure G2422.1(1). The choice of a connection type must take into consideration such factors as appliance movement, vibration, ambient conditions, and susceptibility to physical damage. Between the shutoff valve and appliance, a means of disconnecting the piping must also be provided by a union fitting or similar arrangement. Flared or ground-joint connections that are part of an approved, labeled, semirigid ("flexible") connector are permitted to serve as a union.

Item 2 refers to copper or aluminum tubing, which is rarely used today as an appliance connector. Such connectors are field fabricated. Item 3 refers to the commonly used, so called "flexible" connectors typically made of corrugated stainless steel or brass. Note that although CSST is allowed to directly connect to specific appliances (fixed/non-moveable), CSST is not to be used as a substitute for a connector where a connector is required. CSST manufacturer's instructions specify where CSST can directly connect to appliances and where it cannot. Such connectors are listed and labeled with the label typically being a stamped metal ring placed around the connector tubing, or the label information is stamped into the fittings on the end of the connector. Most connectors manufactured today have attached plastic labels that state the installation and sizing instructions. These connectors are not designed for repeated movement that causes bending of the metal. Such connectors should not be reused after the initial installation. Reuse is typically prohibited by the manufacturer's instructions. Repeated bending and/or vibration can cause metal fatigue, stress cracking, and gas leakage. See Commentary Figures G2422.1(2) and G2422.1(3). Item 6 refers to specialized connectors for moveable appliances. Such connectors are commonly used for restaurant cooking appliances that are routinely moved for cleaning or equipment line order changes.

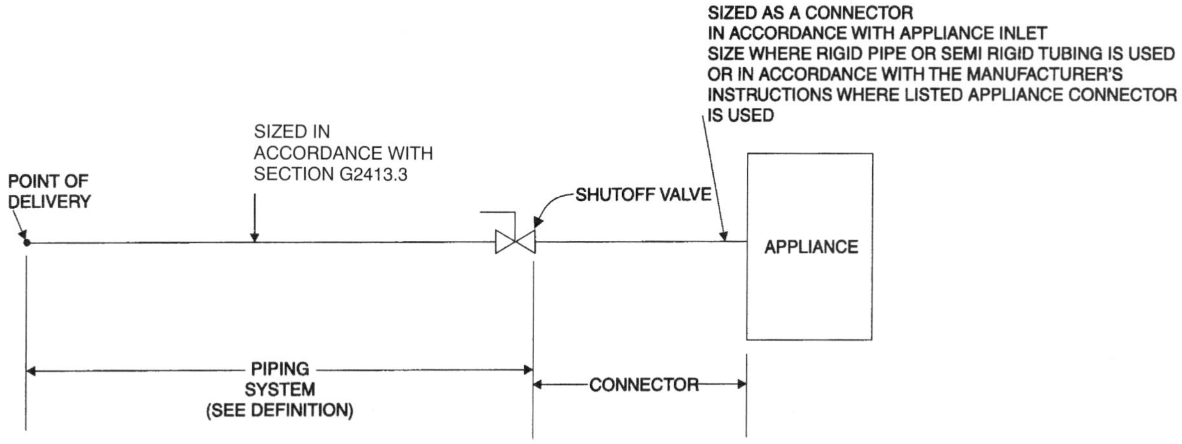

Figure G2422.1(1)
PIPING SYSTEM AND APPLIANCE CONNECTION SIZING

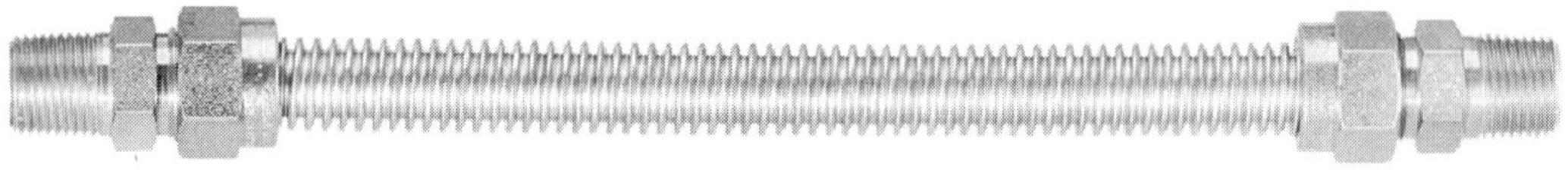

Photo courtesy of Dormont Manufacturing Company

Figure 2422.1(2)
SEMI-RIGID (FLEXIBLE) APPLIANCE CONNECTOR

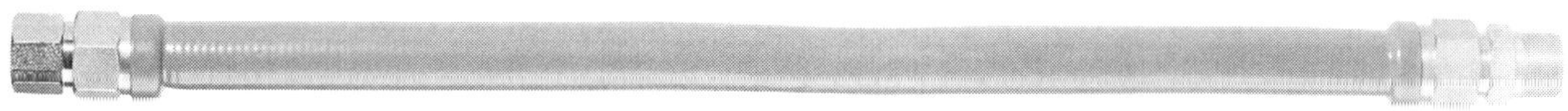

Photo courtesy of Dormont Manufacturing Company

Figure 2422.1(3)
OUTDOOR MOBILE HOME CONNECTOR

The appliance must be provided with restraint cables or chains designed to limit movement of the appliance to protect the connector from damage. The restraints limit the travel distance of the appliance to that which can be safely tolerated by the connector. The typical flexible gas appliance connector is not designed to withstand vibration and repeated movement such as might occur for connections to appliances suspended on chains or long hangers. For example, infrared heaters are commonly hung by chains and can move frequently because of air movement in the space. In this application, the heater should be connected with a connector configuration or specialized connector that can tolerate such movement, or the appliance should be stabilized. Infrared tube-type heaters expand and contract during normal operation and can cause connector flexing, which can damage the connector if it is not properly installed to accommodate such movement.

The piping system is sized up to the outlet of the appliance shutoff valve. The connector occurs between the shutoff valve outlet and the appliance and is sized by the connector manufacturer or is based on the appliance input rating, gas inlet size, and length of the connector. For example, a boiler is supplied by branch piping required to be 1 inch based on the sizing methods in Section G2413. The branch piping is 1 inch up to the boiler shutoff valve, but the boiler inlet opening is only $^1/_2$ inch. The piping (connector) downstream of the shutoff valve is sized as a connector because it is 6 feet or less in length. See the commentary for Section G2420.5.

G2422.1.1 (411.1.1) Protection from damage. Connectors and tubing shall be installed so as to be protected against physical damage.

❖ Appliance connectors, while sound, are not constructed or tested to withstand the same rigors of service as the gas piping system with respect to resisting physical damage. Therefore, they must be installed so as to minimize the potential for damage. For example, "flexible" and semirigid tubing connectors should not be used where subject to excessive vibration or impact by occupants, vehicles, animals, doors, stored materials, etc. Impact, repeated movement, and vibration can cause connector failure.

It is not the intent of this section to require that appliance connectors be hidden from view behind or under an appliance in order to consider them as protected from damage. For typical range and clothes dryer installations, the connectors are physically protected by the appliances themselves. However, for water heaters, furnaces, boilers, unit heaters, infrared heaters, etc. the connector will be located in the open and thus will need protection from damage by proper location, placement, elevation, or guards. Protection from damage is commonly afforded by locating the connector out of harm's way such as between a wall and an appliance or 8 feet or more above the floor.

G2422.1.2 (411.1.2) Appliance fuel connectors. Connectors shall have an overall length not to exceed 3 feet (914 mm), except for range and domestic clothes dryer connectors, which shall not exceed 6 feet (1829 mm) in length. Connectors shall not be concealed within, or extended through, walls, floors, partitions, ceilings or appliance housings. A shutoff valve not less than the nominal size of the connector shall be installed ahead of the connector in accordance with Section G2420.5. Connectors shall be sized to provide the total demand of the connected appliance.

Exception: Fireplace inserts factory equipped with grommets, sleeves or other means of protection in accordance with the listing of the appliance.

❖ Flexible appliance fuel connectors are primarily intended for use with cooking ranges, clothes dryers and similar appliances where the gas connection is located behind the appliance and some degree of flexibility is necessary to facilitate the hook-up. However, flexible connectors are allowed to be used with any appliance where they are not subject to damage. See the commentary for Section G2422.1.1. Rigid connections are not practical for such movable appliances. Flexible connectors are typically constructed of stainless steel and are labeled with tags or metal rings placed over the tubing. The manufacturer's installation instructions dictate the installation requirements that are intended to protect the connector from damage and prevent leakage. Most flexible connectors are intended to be installed only once and are not designed for repeated bending and forming. For other than commercial flexible connections that are designed for repeated movement, a new connector should be used each time an appliance is reinstalled. Repeated flexing or bending can cause metal fatigue and connector failure. Flexible connectors must be protected from physical damage and are not allowed to pass through any walls, floors, ceilings or appliance housings. See the Exception.

Flexible connectors must bear the label of an approved agency and are limited to a maximum of 3 feet (914 mm) in length, except when used in a domestic range or dryer installation where a 6-foot (1829 mm) maximum length is permitted. Longer lengths for range and dryer installations allow the connector to be coiled or otherwise arranged to allow limited movement of the appliance without stressing the connector. Flexible connectors are evaluated in accordance with ANSI Z21.24 or ANSI Z21.69. Multiple connectors are not permitted to be connected together to circumvent the maximum length requirement.

Connectors must also be chosen on the basis that they will provide the total gas demand of the appliance served at the required minimum pressure, as indicated in the manufacturer's instructions. Item 2 of Section G2422.1 specifically allows semirigid metallic tubing to be a maximum of 6 feet (1829 mm) long and to enter an appliance housing where the entry opening is designed to protect the tubing.

The exception allows connectors to enter the housing of a fireplace insert appliance where factory provided means protect the connector from abrasion, corrosion, and sharp metal edges. None of the connectors addressed in Section G2422.1 are allowed to pass through walls, floors, ceilings or partitions. Connectors are installed downstream of the shutoff valve, and such valves must be located in the same room as the appliances. All of the connectors addressed in Section G2422.1 must be sized for the connected load.

G2422.1.3 (411.1.3) Movable appliances. Where appliances are equipped with casters or are otherwise subject to periodic movement or relocation for purposes such as routine cleaning and maintenance, such appliances shall be connected to the supply system piping by means of an approved flexible connector designed and labeled for the application. Such flexible connectors shall be installed and protected against physical damage in accordance with the manufacturer's installation instructions.

❖ Commercial appliances such as cooking equipment commonly employ heavy-duty, commercial-type flexible gas connectors. Such connectors are designed to tolerate repeated movement to allow appliances on wheels or casters to be moved for cleaning operations or relocation. Such connectors are commonly equipped with quick-disconnect couplings for ease of disconnection and reconnection and restraint cables to limit the amount of appliance movement, thereby preventing stress and damage to the connector.

Flexible connectors designed for specific applications are also used for the connection of equipment and appliances that produce significant vibrations. For example, gas-burning combustion engines in generator sets and HVAC equipment are connected with specially designed vibration dampening flexible connectors. Some pulse-combustion heating appliances employ flexible connectors to control the transmission of noise to gas distribution piping.

SECTION G2423 (413) CNG GAS-DISPENSING SYSTEMS

G2423.1 (413.1) General. Service stations for CNG fuel shall be in accordance with this section and the *International Fire Code.* The operation of CNG service stations shall be regulated by the *International Fire Code.*

❖ Section 2208 of the IFC regulates CNG service stations.

SECTION G2424 (415) PIPING SUPPORT INTERVALS

G2424.1 (415.1) Interval of support. Piping shall be supported at intervals not exceeding the spacing specified in Table G2424.1. Spacing of supports for CSST shall be in accordance with the CSST manufacturer's instructions.

❖ See the commentary for Table G2424.1.

TABLE G2424.1 (415.1) SUPPORT OF PIPING

STEEL PIPE, NOMINAL SIZE OF PIPE (inches)	SPACING OF SUPPORTS (feet)	NOMINAL SIZE OF TUBING SMOOTH-WALL (inch O.D.)	SPACING OF SUPPORTS (feet)
$^1/_2$	6	$^1/_2$	4
$^3/_4$ or 1	8	$^5/_8$ or $^3/_4$	6
$1^1/_4$ or larger (horizontal)	10	$^7/_8$ or 1 horizontal	8
$1^1/_4$ or larger (vertical)	Every floor level	1 or larger (vertical)	Every floor level

For SI: 1 inch = 25.4 mm, 1 foot = 304.8 mm.

❖ This table is two tables side by side. The first two columns are for steel pipe, and the second two columns are for tubing materials. See Section G2418. Support for CSST is dictated by the manufacturer's installation instructions.

SECTION G2425 (501) GENERAL

G2425.1 (501.1) Scope. This section shall govern the installation, maintenance, repair and approval of factory-built and masonry chimneys, chimney liners, vents and connectors serving gas-fired appliances.

❖ Section G2425 contains provisions for the installation, maintenance, repair, and approval of all residential, commercial, and industrial chimney and venting systems that convey the products of combustion from a gas-fired appliance to the outside atmosphere. Venting systems for fuel-fired appliances other than gas-fired appliances, such as oil and solid-fuel appliances, are covered in Chapter 18. The construction of masonry chimneys is regulated by the *International Building Code.* See Section G2425.3.

G2425.2 (501.2) General. Every appliance shall discharge the products of combustion to the outdoors, except for appliances exempted by Section G2425.8.

❖ All appliances except those listed in Section G2425.8 must be vented to convey the potentially harmful combustion by-products to the outdoor atmosphere. See Section G2425.8.

G2425.3 (501.3) Masonry chimneys. Masonry chimneys shall be constructed in accordance with Section G2427.5 and Chapter 10.

❖ A masonry chimney is a field-constructed assembly that can consist of solid masonry units, reinforced concrete, rubble stone, fire-clay liners, and mortars. A masonry chimney is permitted to serve low-heat, medium-heat, and high-heat appliances. Chapter 10 outlines the general code requirements regarding the construction details for all masonry chimneys, including those serving masonry fireplaces.

G2425.4 (501.4) Minimum size of chimney or vent. Chimneys and vents shall be sized in accordance with Section G2427.

❖ Section G2427 contains the requirements for sizing Category I appliance venting systems. Sizing of venting systems for all other categories is dictated by the appliance manufacturer.

G2425.5 (501.5) Abandoned inlet openings. Abandoned inlet openings in chimneys and vents shall be closed by an approved method.

❖ Abandoned inlet openings result from appliances being disconnected or reconnected at different elevations. Unused openings in chimneys and vents can allow combustion gases to enter the building, can cause loss of draft at other appliances connected to the chimney or vent, can allow conditioned air to escape resulting in energy losses, and can allow the entry of birds and rodents. For example, an appliance that was connected to a chimney has been replaced with an appliance that vents through an exterior wall. The opening in the chimney left by the disconnection of the appliance must be properly closed to prevent a hazardous condition. Combustion gases from appliances connected to chimneys and vents can escape through openings in the chimney and vent. Such openings can affect the draft of the remaining appliances, causing them to spill combustion gases into the building.

The method used to close an unused opening must not create a protrusion in the flue passageway, which could restrict flue gas flow or provide a surface for the accumulation of creosote or soot. It is not uncommon to find multiple openings in chimneys that were used at one time but have since been abandoned and covered, closed in some makeshift fashion, or left open. Inlet openings that are not closed could allow the escape of vent gases and could affect the draft produced by the chimney or vent. Excess dilution air will be admitted through unused openings, which could weaken draft by cooling the vent gases and bypassing the flow through appliance connectors. It is rare to find unused openings in vents because vent openings are easily closed or adapted to different appliance connections; however, the hazard of abandoned openings is the same as for chimneys. Closure of vent and factory-built chimney openings is easily accomplished by eliminating tees and wyes or by using metal caps. Closure of masonry chimney openings may require the replacement of masonry, especially where the chimney has been structurally weakened by the openings.

G2425.6 (501.6) Positive pressure. Where an appliance equipped with a mechanical forced draft system creates a positive pressure in the venting system, the venting system shall be designed for positive pressure applications.

❖ Commonly used chimney and vent systems are not designed for positive (above atmospheric) pressure. Such chimneys and vents are intended to produce a draft and thus operate with negative internal pressures. Specialized vents and chimneys are available that are designed for positive pressure applications. See Commentary Figure G2425.6. The misapplication of materials can result in leakage of combustion gases and can cause chimney or vent deterioration from condensate corrosion. Masonry chimneys, Type B vents, Type L vents, and most factory-built chimneys are designed for negative pressure applications, meaning that the venting system is producing draft. Such venting systems will leak if subjected to positive internal pressures. Leakage can not only allow combustion gases to enter the building; it can cause rapid deterioration of the chimney or vent due to condensation.

G2425.7 (501.7) Connection to fireplace. Connection of appliances to chimney flues serving fireplaces shall be in accordance with Sections G2425.7.1 through G2425.7.3.

❖ This section regulates installations where fireplace chimneys are used as the venting means for appliances such as room heaters and fireplace insert-type heaters.

G2425.7.1 (501.7.1) Closure and access. A noncombustible seal shall be provided below the point of connection to prevent entry of room air into the flue. Means shall be provided for access to the flue for inspection and cleaning.

❖ Without an air-tight connection between the chimney flue and the appliance chimney connector, chimney draft will draw room air into the flue, thus weakening the draft drawn through the appliance and promoting creosote accumulation. The chimney connection must be designed for access for inspection and cleaning of the chimney because the chimney flue is no longer open to the fireplace firebox.

Photo courtesy of Selkirk L.L.C.

Figure G2425.6
VENT DESIGNED FOR CATEGORY III AND IV (POSITIVE PRESSURE) APPLICATIONS

G2425.7.2 (501.7.2) Connection to factory-built fireplace flue. An appliance shall not be connected to a flue serving a factory-built fireplace unless the appliance is specifically listed for such installation. The connection shall be made in accordance with the appliance manufacturer's installation instructions.

❖ Factory-built fireplaces may or may not be tested for any use other than as a traditional fireplace. Installing an appliance or fireplace insert-type heater could result in abnormally high fireplace or chimney temperatures, component deterioration, and hazardous operation. The fireplace manufacturer will provide specific instructions as to what is or is not allowed regarding appliance installations. Typically the manufacturer's instructions will prohibit any modifications and any use of the fireplace other than that for which it was designed.

G2425.7.3 (501.7.3) Connection to masonry fireplace flue. A connector shall extend from the appliance to the flue serving a masonry fireplace such that the flue gases are exhausted directly into the flue. The connector shall be accessible or removable for inspection and cleaning of both the connector and the flue. Listed direct connection devices shall be installed in accordance with their listing.

❖ The appliance chimney connector must extend up through the fireplace damper and smoke chamber and terminate in the chimney flue. This helps produce an adequate draft and reduces the possibility of forming combustible creosote deposits within the fireplace throat and smoke chamber.

G2425.8 (501.8) Equipment not required to be vented. The following appliances shall not be required to be vented:

1. Ranges.
2. Built-in domestic cooking units listed and marked for optional venting.
3. Hot plates and laundry stoves.
4. Type 1 Clothes dryers (Type 1 clothes dryers shall be exhausted in accordance with the requirements of Section G2439).
5. Refrigerators.
6. Counter appliances.
7. Room heaters listed for unvented use.

Where the appliances and equipment listed in Items 5 through 7 above are installed so that the aggregate input rating exceeds 20 Btu per hour per cubic foot (207 watts per m^3) of volume of the room or space in which such appliances and equipment are installed, one or more shall be provided with venting systems or other approved means for conveying the vent gases to the outdoor atmosphere so that the aggregate input rating of the remaining unvented appliances and equipment does not exceed the 20 Btu per hour per cubic foot (207 watts per m^3) figure. Where the room or space in which the equipment is installed is directly connected to another room or space by a doorway, archway or other opening of comparable size that cannot be closed, the volume of such adjacent room or space shall be permitted to be included in the calculations.

❖ The specified appliances do not have to be vented unless required by the last paragraph of this section. The last paragraph of this section is intended to address items 5 through 7 and not items 1 through 4. Section G2404.3 requires all appliances to be listed.

Item 1 recognizes that residential-type cooking appliances are listed for unvented operation. Item 6 refers to gas-fired cooking and food service appliances designed for use on countertop surfaces.

Item 4 clarifies that clothes dryers are not vented, but instead are exhausted. The products of combustion for gas-fired clothes dryers are discharged to the outdoors along with the moisture exhaust; in effect, the dryer is "vented" but not in the sense that other appliances are vented. Clothes dryer exhaust ducts are not referred to as vents because they are exhaust ducts, not a type of vent. Clothes dryers, both gas and electric, are required to exhaust to the outdoors.

20 Btu per hour per cubic foot of room volume is the same ratio as 50 cubic feet per 1000 Btu per hour on which Section G2407.5.1 is based. In other words, this section requires the appliances that are not vented to be located in a space that meets the volume requirement for indoor combustion air specified in Section G2407.5.1. If the total appliance input rating exceeds the ratio of 20 Btu/h per cu ft of space volume, one or more appliances would have to be vented or removed to keep the total at or below the limit. Maintaining a minimum space volume will help control the level of combustion products in the space by means of dilution with infiltration air.

G2425.9 (501.9) Chimney entrance. Connectors shall connect to a masonry chimney flue at a point not less than 12 inches (305 mm) above the lowest portion of the interior of the chimney flue.

❖ This provision is intended to prevent blockage of the connector opening by debris that has collected at the bottom of the flue passage. The 12-inch (305 mm) deep space below the connector functions as a trap (dirt leg) for debris such as dead animals, leaves, flaking tile, mortar, and soot. This section is not intended to apply to factory-built chimneys, because such chimneys are protected from the entry of debris by the required cap.

G2425.10 (501.10) Connections to exhauster. Appliance connections to a chimney or vent equipped with a power exhauster shall be made on the inlet side of the exhauster. Joints on the positive pressure side of the exhauster shall be sealed to prevent flue-gas leakage as specified by the manufacturer's installation instructions for the exhauster.

❖ When a mechanical draft device or power exhauster is installed, the resulting "forced draft" creates a positive pressure inside the chimney or vent on the discharge or outlet side of the fan. A hazardous condition will result if an appliance is connected on the outlet or discharge side of a power exhauster. The appliance connector provides an alternate path for the flue gases to travel back into the building. Also, the appliance connected on the discharge side (outlet) of the exhauster will not be provided with draft, and its combustion products will spill into the building interior.

A hazardous condition can also occur when a vent connector serving an appliance vented by natural draft is connected to the vent of an appliance equipped with integral power-venting means. This section does not prohibit Category I fan-assisted appliances from being common vented with draft hood (natural draft) appliances as addressed in Section G2428.

Any vent, chimney, or connector piping into which an exhauster discharges must be rated and approved for positive pressure and properly installed to prevent leakage of combustion gases. For example, Type B vents cannot be used on the outlet side of an exhauster or a power vented appliance regardless of any appliance manufacturer's recommendations. Type B vent material cannot be sealed to prevent leakage when exposed to positive pressure. Positive pressure can force combustion products into the annular space between the pipe walls where condensation could occur, causing deterioration of the piping.

G2425.11 (501.11) Masonry chimneys. Masonry chimneys utilized to vent appliances shall be located, constructed and sized as specified in the manufacturer's installation instructions for the appliances being vented and Section G2427.

❖ Manufacturers' instructions for today's gas-fired appliances specify very limited conditions under which the appliance is allowed to vent to a masonry chimney. The conditions include chimney size, state of structural condition, location, and construction. Because of being oversized, unlined, in poor structural condition, or partially exposed to the outdoors, masonry chimneys are usually not allowed to serve gas-fired appliances meeting current minimum energy-efficiency requirements. Once common for 75-percent or less efficiency appliances, masonry chimneys serving to vent modern gas-fired appliances are becoming rare. In most cases a masonry chimney must be relined with a retrofit metal liner system, or a vent such as Type B vent will have to be installed within the flueway of the chimney. Masonry chimneys have a large thermal mass that has to be heated to above the dew point of the combustion gases to avoid condensation. The flue gas temperatures discharged from many modern appliances makes this impossible.

G2425.12 (501.12) Residential and low-heat appliances flue lining systems. Flue lining systems for use with residential-type and low-heat appliances shall be limited to the following:

1. Clay flue lining complying with the requirements of ASTM C 315 or equivalent. Clay flue lining shall be installed in accordance with Chapter 10.
2. Listed chimney lining systems complying with UL 1777.
3. Other approved materials that will resist, without cracking, softening or corrosion, flue gases and condensate at temperatures up to 1,800°F (982°C).

❖ Flue lining systems for masonry chimneys include clay tile, poured-in-place refractory materials, and stainless steel pipe.

G2425.13 (501.13) Category I appliance flue lining systems. Flue lining systems for use with Category I appliances shall be limited to the following:

1. Flue lining systems complying with Section G2425.12.
2. Chimney lining systems listed and labeled for use with appliances with draft hoods and other Category I gas appliances listed and labeled for use with Type B vents.

❖ Chimney lining systems for chimneys serving gas-fired appliances include stainless steel and aluminum piping (uninsulated or insulated) designed for both new construction and for relining existing chimneys. Type B vent is also used as a chimney liner, although it is not considered to be a lining system, and in such cases the chimney is actually a masonry chase containing a vent.

Relining of existing chimneys is quite common now because of the incompatibility of mid-efficiency appliances and masonry chimneys. Relining systems allows existing chimneys to be reused, thereby saving expense and retaining desirable architectural features of buildings. See Commentary Figures G2425.13(1) through G2425.13(4).

G2425.14 (501.14) Category II, III and IV appliance venting systems. The design, sizing and installation of vents for Category II, III and IV appliances shall be in accordance with the appliance manufacturer's installation instructions.

❖ Venting systems for such appliances are special vent systems specified by the appliance manufacturer and are not addressed further in the code. See Commentary Figure G2425.14.

G2425.15 (501.15) Existing chimneys and vents. Where an appliance is permanently disconnected from an existing chimney or vent, or where an appliance is connected to an existing chimney or vent during the process of a new installation, the chimney or vent shall comply with Sections G2425.15.1 through G2425.15.4.

❖ Existing chimneys and vents need to be reevaluated for continued suitability whenever the conditions of use change. Size, which is covered in Section G2425.15.1, is of primary importance. Other considerations are the presence of liner obstructions, combustible deposits in the liner, the structural condition of the liner, the provision of a cleanout, and clearances to combustibles, which are addressed in Sections G2425.15.2 through G2425.15.4. It is the intent of this section that whenever a new appliance is connected or an existing appliance is disconnected from a chimney or vent, the chimney or vent is subject to the requirements in Sections G2425.15.2 through G2425.15.4, which include requirements for inspection, cleaning,

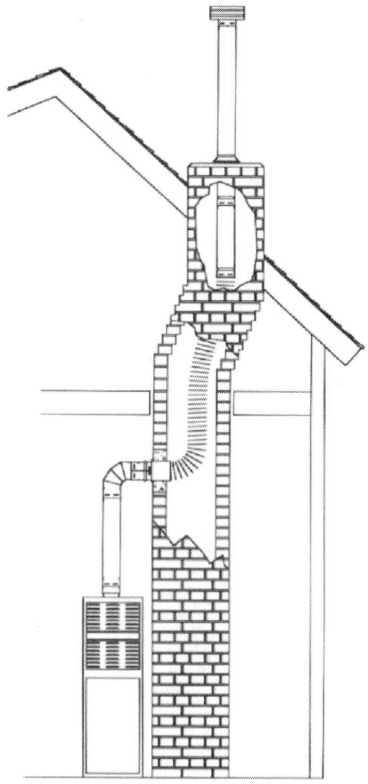

Figure courtesy of Simpson Dura-Vent Company, Incorporated

Figure G2425.13(1)
MASONRY CHIMNEY RELINING SYSTEM

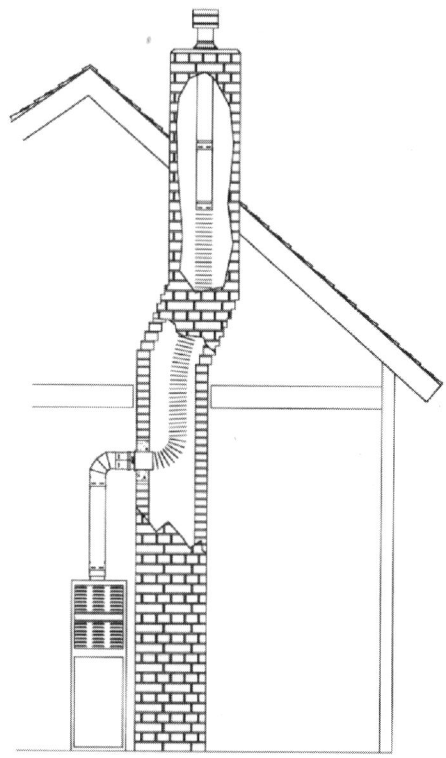

Figure courtesy of Simpson Dura-Vent Company, Incorporated

Figure G2425.13(2)
MASONRY CHIMNEY RELINING SYSTEM

possible repair, the installation of a cleanout if one is not already provided, and the establishment of clearances to combustibles in accordance with the requirements for new chimneys or vent installations.

Chimneys, vents, and the appliances served are all designed to function together as a system, and any change to an existing chimney or vent system will have an impact on the performance of such a system. Something as simple as disconnecting an appliance from a chimney can upset the system balance and cause the venting system to fail to produce a draft for the remaining appliances and could cause the venting system to produce harmful condensation. See Commentary Figure G2425.15.

G2425.15.1 (501.15.1) Size. The chimney or vent shall be resized as necessary to control flue gas condensation in the interior of the chimney or vent and to provide the appliance or appliances served with the required draft. For Category I appliances, the resizing shall be in accordance with Section G2426.

❖ The combined input from multiple appliances, especially older lower-efficiency appliances, can maintain sufficiently high chimney or vent temperatures to provide the necessary draft and to avoid condensation. Changing an existing configuration by disconnecting and eliminating an appliance or by substituting a higher-efficiency appliance can cause a decrease in flue-gas temperature resulting in condensation or poor draft. Also, the elimination of one or more draft-hood-equipped appliances will reduce the amount of dilution air in a venting system, thus increasing the likelihood of condensation. Often, it is necessary to resize a chimney or vent by replacing it with a smaller-size chimney or vent or by installing a liner system. See Commentary Figure G2425.15.

A common scenario involves removing chimney-vented furnaces or boilers and leaving a water heater as the only appliance vented to the chimney. In such cases, the chimney would typically be grossly oversized for the water heater, could fail to produce adequate draft, and could be subject to continuous condensation. This scenario has received much attention and has created the phrase "orphaned water heaters."

G2425.15.2 (501.15.2) Flue passageways. The flue gas passageway shall be free of obstructions and combustible deposits and shall be cleaned if previously used for venting a solid or liquid fuel-burning appliance or fireplace. The flue liner, chimney inner wall or vent inner wall shall be continuous and shall be free of cracks, gaps, perforations, or other damage or deterioration that would allow the escape of combustion products, including gases, moisture and creosote.

❖ Cleaning of the chimney or flue is necessary to examine the flue liner surface, to eliminate fuel for a possible chimney fire, to reduce frictional resistance to

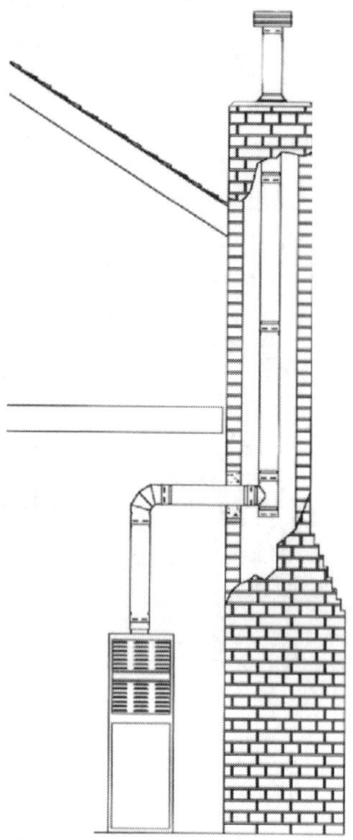

Figure courtesy of Simpson Dura-Vent Company, Incorporated

Figure G2425.13(3)
MASONRY CHIMNEY RELINED BY INSTALLATION OF TYPE B VENT SYSTEM

Photo courtesy of Simpson Dura-Vent Company, Incorporated

Figure G2425.13(4)
MASONRY CHIMNEY RELINING SYSTEM

flow, and to prevent any residual deposits from falling into lower parts of the venting system. Installation or removal of an appliance without an inspection of the existing flue is prohibited. If the flue served a solid or liquid fuel-burning appliance or fireplace, cleaning is required in order to make the required inspection and to remove any combustible deposits. In that case, the addition or removal of an appliance without both cleaning and inspection of the existing flue would be a code violation. If inspection shows that the flueway of the vent or chimney is incapable of preventing the leakage of combustion gases, the chimney or vent would have to be repaired or relined.

G2425.15.3 (501.15.3) Cleanout. Masonry chimney flues shall be provided with a cleanout opening having a minimum height of 6 inches (152 mm). The upper edge of the opening shall be located not less than 6 inches (152 mm) below the lowest chimney inlet opening. The cleanout shall be provided with a tight-fitting, noncombustible cover.

❖ This section requires a cleanout to be installed in a chimney to facilitate cleaning and inspection. A fireplace inherently provides access to its chimney through the firebox, throat, and smoke chamber. The cleanout cover and opening frame must be of an approved material such as cast iron, precast cement, or other noncombustible material and must be arranged to remain tightly closed. A loose fitting or unsecured

Photo courtesy of Selkirk L.L.C.

Figure G2425.14
SPECIAL VENT SYSTEM FOR CATEGORY II, III, AND IV APPLIANCES

cleanout can allow air to flow into the chimney, affecting draft at the appliance or fireplace, and can result in an energy loss in the building. Under certain conditions, combustion products could escape into the building through an ill-fitting cleanout door or cover. The requirement for placing the cleanout at least 6 inches (152 mm) below the lowest connection to the chimney is intended to minimize the possibility of combustion products exiting the chimney through the cleanout. It is the intent of this section that a cleanout opening be installed, if one is not already provided, for existing masonry chimney flues whenever another appliance is installed or removed.

G2425.15.4 (501.15.4) Clearances. Chimneys and vents shall have airspace clearance to combustibles in accordance with Chapter 10 and the chimney or vent manufacturer's installation instructions. Noncombustible firestopping or fireblocking shall be provided in accordance with Chapter 10.

Exception: Masonry chimneys equipped with a chimney lining system tested and listed for installation in chimneys in contact with combustibles in accordance with UL 1777, and installed in accordance with the manufacturer's instructions, shall not be required to have clearance between combustible materials and exterior surfaces of the masonry chimney.

❖ The effect of this requirement is to require that the existing chimney or vent be inspected for clearances to combustibles for the entire length of the chimney or vent. Clearances to combustibles for masonry chimneys are found in Chapter 10. The required clearances for a listed factory-built chimney or vent would be in accordance with the existing chimney or vent manufacturer's installation instructions. The clearances to combustibles for vents and factory-built chimneys are typically indicated on the vent or chimney. If the required clearances are not indicated on the vent or chimney, the manufacturer must be contacted and the original installation instructions consulted.

Existing masonry chimneys commonly do not have the clearances required by code, especially if the structure is fairly old. Listed chimney lining systems, which provide insulation value to compensate for the lack of clearance, are available for use in such chimneys. The exception gives the criteria for these tested lining systems, citing UL 1777 as the test standard. Fireblocking of any annular spaces where the chimney passes through ceilings or floors is required to be installed in accordance with Chapter 10.

SECTION G2426 (502) VENTS

G2426.1 (502.1) General. All vents, except as provided in Section G2427.7, shall be listed and labeled. Type B and BW vents shall be tested in accordance with UL 441. Type L vents shall be tested in accordance with UL 641. Vents for Category II and III appliances shall be tested in accordance with UL 1738. Plastic vents for Category IV appliances shall not be required to be listed and labeled where such vents are as specified by the appliance manufacturer and are installed in accordance with the appliance manufacturer's installation instructions.

❖ Vents regulated by this section are factory fabricated and must be listed and labeled. The labeling requirement applies to all components of the system such as

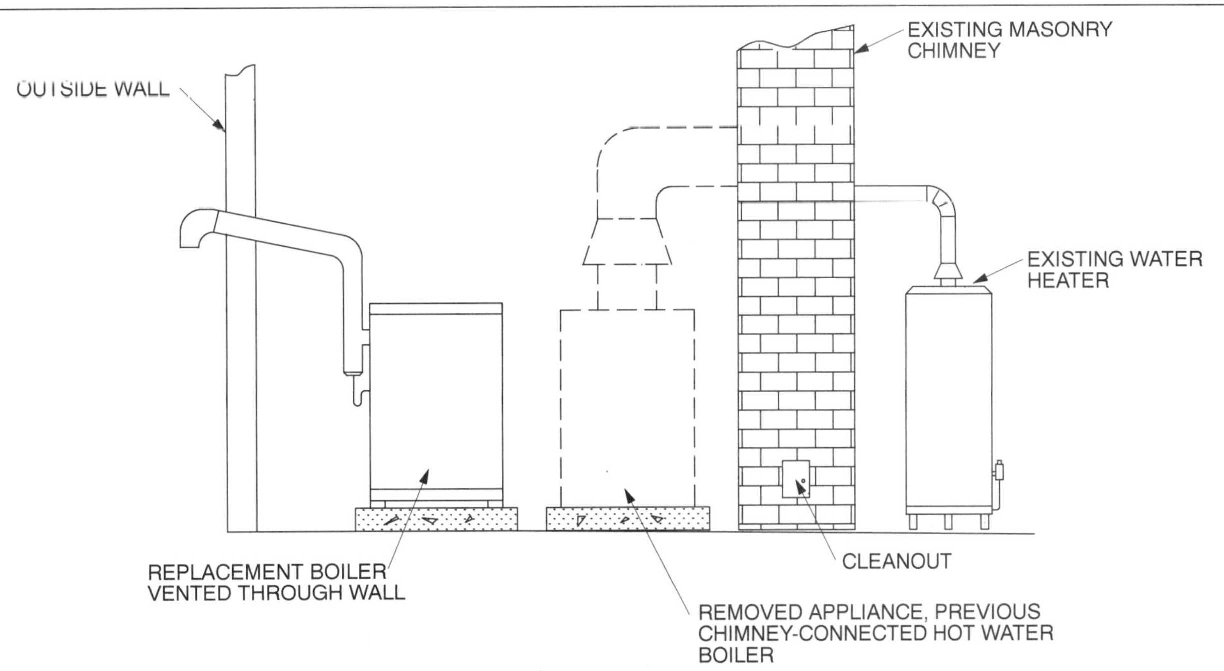

Figure G2425.15
APPLIANCE REPLACEMENTS AFFECTING EXISTING APPLIANCES

the sections of pipe, fittings, terminal caps, supports, and spacers.

The provision for unlisted plastic vents is necessary to allow for the venting systems commonly specified in the installation instructions for Category IV condensing appliances. Unlisted plastic pipe (commonly PVC or CPVC plumbing pipe) vents are distinct from listed high-temperature plastic special gas vents that were designed primarily for Category III appliances. Unlisted plastic pipe can be used only when specified by the appliance manufacturer and must be installed as directed in the appliance installation instructions. Unlisted plastic pipe is used only for high efficiency Category IV gas-fired appliances, and such pipe functions as an exhaust pipe for combustion gases and as a combustion air intake pipe.

Gas appliances (furnaces and boilers) are categorized so that the proper venting system can be selected. Four categories have been developed based on the pressure produced in the vent passageway and whether or not the flue gas temperatures approach the dew point making condensation likely. See the definition for "Vented Gas Appliance Categories."

Category I appliances operate under a negative or neutral vent pressure, are noncondensing appliances, and may generate vent gases with temperatures up to 550°F (288°C) for draft hood appliances and up to 470°F for fan-assisted appliances. The Category I appliance can use a draft hood or an integral vent blower or draft inducer (fan assisted). See Commentary Figure G2426.1(1).

Category II appliances also operate under negative or neutral vent pressure, but the vent gas temperature and percent of CO_2 yield a flue loss of less than 17 percent, meaning that condensation of flue gases can occur. Vent systems serving Category II appliances fall under the special vent provisions of Sections G2425.14, G2427.4, and G2427.6.8.2.

Category III appliances operate with a positive vent pressure and can produce condensation, although they are not considered "condensing appliances." The positive pressure in the vent system necessitates air-tight flue passageways to avoid leakage. Vent systems serving Category III appliances fall under the special vent provisions of Sections G2425.14, G2427.4, and G2427.6.8.2 and are typically constructed of high-temperature plastic or stainless steel pipe. The type of stainless steel specified today is typically AL29-4C, a special alloy known for its corrosion resistance.

Category IV appliances operate with a positive vent pressure and condense flue gases. Condensing (high-efficiency) appliances require air-tight vent passageways and a method of collecting and draining condensation. The vent system serving this type of appliance is usually constructed of plastic, such as PVC or CPVC. Vent systems serving Category IV appliances also fall under the special vent provisions of Sections G2425.14, G2427.4, and G2427.6.8.2. See Commentary Figure G2426.1(2).

A vent system can only be used in conjunction with gas-burning appliances that are listed and labeled for use with a vent. Appliances equipped with draft hoods and Category I (including fan-assisted) appliances both depend on the vent to produce a draft in order to convey the combustion products to the outdoors.

Solid fuel-burning appliances must not be connected to a vent system, because such appliances produce flue gas temperatures that are much higher than the temperatures for which vents are designed. The burning of solid fuels also produces creosote. Creosote formation leaves a combustible deposit on the surface of the flue passageway and, if ignited, can produce a fire with temperatures exceeding 2,000°F (1093°C), far exceeding the safe temperature limitations of a vent system.

Type B vents are designed for venting noncondensing gas appliances equipped with a draft hood and fan-assisted appliances that operate with a nonpositive vent pressure and that are listed and labeled for use with a Type B vent. Type B vents are capable of withstanding continuous firing at flue temperatures not exceeding 400°F (204°C) above room temperature. This section requires a Type B vent to be tested in accordance with UL 441. Type B vents must not be used with incinerators, appliances readily converted to the use of solid or liquid fuels, or combination gas/oil burning appliances. Type B vents are permitted to serve multiple gas-fired appliance installations. See Commentary Figure G2426.1(3).

Type BW vents are also required to be tested in accordance with UL 441 and are only permitted to serve labeled gas-fired wall furnaces. A Type BW vent is an oval-shaped vent designed and tested to be installed in a stud wall cavity and is capable of withstanding continuous firing at flue temperatures not exceeding 400°F (204°C) above room temperature. Type BW vents must not be used to vent incinerators, appliances readily converted to the use of solid or liquid fuels, and combination gas/oil burning appliances.

Type L vents are required to be tested in accordance with UL 641 and are designed to vent oil-burning appliances. Oil-burning appliances produce a higher flue gas temperature than gas-fired appliances. Type L vents can also be used with appliances listed and labeled for Type B vent systems, meaning that Type L vents can substitute for Type B vents, but Type B vents cannot serve appliances that require Type L vents.

Type B and BW vents are double-walled, air-insulated vent piping systems constructed of galvanized steel outer walls and aluminum inner walls. See Commentary Figure G2426.1(4). Type L vents are typically double-wall, air-insulated vent piping systems constructed of galvanized and stainless steel.

Type B, BW, and L vents are designed for natural draft applications only and must not be used to convey combustion gases under positive pressure. For example, a Type B vent must not be used with Category III or IV gas-fired appliances and must not be used on the discharge side of an exhauster or power-vented appliance. Positive pressures will cause a Type B vent to leak combustion gases into the building, and the leakage can cause condensation to form within the vent system pipe, fittings, and joints, resulting in corrosion damage to the vent and eventual vent failure. Fan-assisted Category I appliances are not considered power vented and do not produce positive vent pressures when vented to a properly sized draft-producing vent such as a Type B vent system. See the commentary for Section G2428. UL 1738 covers factory-built vents for Category II, III, and IV appliances. Such vents can be constructed of metallic or plastic materials including aluminum alloys, steel (coated, galvanized, stainless, or aluminized), PVC, and CPVC.

G2426.2 (502.2) Connectors required. Connectors shall be used to connect appliances to the vertical chimney or vent, except where the chimney or vent is attached directly to the appliance. Vent connector size, material, construction and installation shall be in accordance with Section G2427.

❖ Connectors designed, constructed, and installed in accordance with Section G2427 are required except where a chimney or vent serves a single appliance and connects directly to the appliance. In such cases, a connector does not exist. The use of single-wall connectors for fan-assisted appliances is limited by the sizing tables because of the high heat input necessary (minimum capacity column) to offset the heat loss of single-wall pipe.

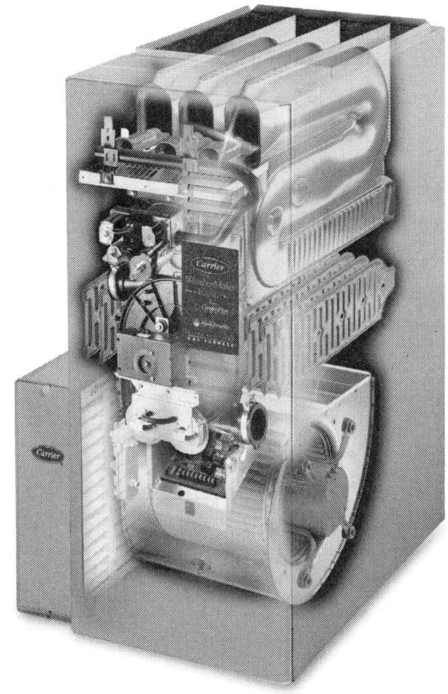

Photo courtesy of Carrier Corporation

Figure G2426.1(2)
CATEGORY IV APPLIANCE (FURNACE)

Figure G2426.1(1)
CATEGORY I FAN-ASSISTED FURNACE

G2426.3 (502.3) Vent application. The application of vents shall be in accordance with Table G2427.4.

❖ See Section G2427.4.

G2426.4 (502.4) Insulation shield. Where vents pass through insulated assemblies, an insulation shield constructed of not less than 26 gage sheet (0.016 inch) (0.4 mm) metal shall be installed to provide clearance between the vent and the insulation material. The clearance shall not be less than the clearance to combustibles specified by the vent manufacturer's installation instructions. Where vents pass through attic space, the shield shall terminate not less than 2 inches (51 mm) above the insulation materials and shall be secured in place to prevent displacement. Insulation shields provided as part of a listed vent system shall be installed in accordance with the manufacturer's installation instructions.

❖ Loose insulation in attic floor assemblies or roof assemblies can fall against vents and create a fire hazard. They can also cause abnormally high vent temperatures. Even though clearances to combustible construction such as wood framing are maintained, the continued heating of combustible insulation materials that have fallen against a vent could be a source of ignition. This section applies regardless of the combustibility of the insulation. Shields constructed in the field must be sufficient to serve their purpose and must be securely attached to building construction, because shields that are merely resting in place could be inadvertently dislodged or removed during maintenance activities.

G2426.5 (502.5) Installation. Vent systems shall be sized, installed and terminated in accordance with the vent and appliance manufacturer's installation instructions and Section G2427.

❖ The standards to which vents and appliances comply require that venting instructions be supplied by the manufacturer of the appliance and the manufacturer of the vent system. Such instructions are part of the labeling requirements, and any deviation from them is in violation of the code.

The instructions for Type B vents and for gas appliances approved for use with Type B vents may be specific to a certain appliance or manufacturer's B-vent material. The clearance to combustibles for a vent system is determined by the testing agency and is stated on the component labels and in the manufacturer's instructions. Not all vents have the same required clearances to combustibles. The clearances are determined by the vent's performance during the testing in accordance with the applicable standard. Different vent materials and designs impact the vent's ability to control the amount of heat transmitted to surrounding combustibles. It is critical that the vent system be installed in accordance with the clearances listed on the vent's label and in the manufacturer's installation instructions.

Photo courtesy of Simpson Dura-Vent Company, Incorporated

Figure G2426.1(3)
TYPE B GAS VENT SERVING TWO CATEGORY I APPLIANCES

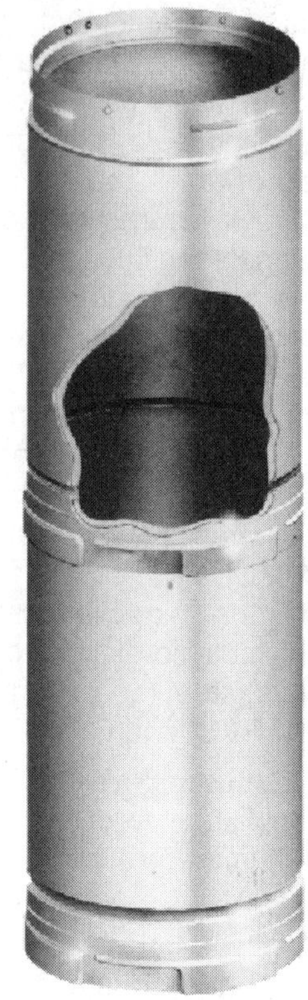

Photo courtesy of Simpson Dura-Vent Company, Incorporated

Figure G2426.1(4)
TYPE B VENT SECTIONS WITH AIR SPACE BETWEEN PIPE WALLS

The termination of a natural draft vent must comply with the requirements of the manufacturer's installation instructions and Sections G2427.6.3, G2427.6.5 and G2427.6.6. A vent used in conjunction with a mechanical exhauster must meet the termination requirements established in Sections G2427.3.3 and G2427.8.

Physical protection of the vent system is required to prevent damage to the vent and to prevent combustibles from coming into contact with or being placed too close to the vents. Such protection is typically provided by enclosing the vents in chases, shafts, or cavities in the building construction. Physical protection is not required in the room or space where the vent originates (at the appliance connection) and would not be required in such locations as attics that are not occupied or used for storage. For example, assume that a vent is installed in an existing building and that it extends from the basement, through a first-floor closet, through the attic, and through the roof. The portion of the vent passing through the closet must be protected from damage. The means of physical damage protection should also be designed to maintain separation between the vent and any combustible storage. To prevent the passage of fire and smoke through the annular space around a vent penetration through a floor or ceiling, the vent must be fireblocked with a noncombustible material in accordance with this code. Vent manufacturers provide installation instructions and factory-built components for fireblocking penetrations.

G2426.6 (502.6) Support of vents. All portions of vents shall be adequately supported for the design and weight of the materials employed.

❖ Vent manufacturers supply support parts, and manufacturers' instructions provide detailed requirements for support of vent systems. Improper support can cause strain on vent components, appliance connectors, and fittings resulting in vent, appliance, or connector damage, as well as loss of required clearance to combustibles and proper slope. Frequently, venting installations suffer from lack of or improperly installed supports, brackets, and hangers.

SECTION G2427 (503) VENTING OF EQUIPMENT

G2427.1 (503.1) General. This section recognizes that the choice of venting materials and the methods of installation of venting systems are dependent on the operating characteristics of the equipment being vented. The operating characteristics of vented equipment can be categorized with respect to (1) positive or negative pressure within the venting system; and (2) whether or not the equipment generates flue or vent gases that may condense in the venting system. See Section G2403 for the definition of these vented appliance categories.

❖ This section is in itself commentary and does not state any requirements. Section G2427 contains general sizing, design, and installation requirements for venting systems. Section G2428 covers vent system sizing for Category I appliances.

G2427.2 (503.2) Venting systems required. Except as permitted in Section G2427.2.1, Section G2427.2.2 and Section G2425.8, all equipment shall be connected to venting systems.

❖ All appliances must be vented by a venting system except those allowed to be unvented in accordance with Section G2425.8. The appliances addressed in Sections G2427.2.1 and G2427.2.2 are provided with a venting system.

G2427.2.1 (503.2.3) Direct-vent equipment. Listed direct-vent equipment shall be considered properly vented where installed in accordance with the terms of its listing, the manufacturer's instructions and Section G2427.8 Item 3.

❖ Direct-vent appliances are vented by a natural or mechanical draft system that is designed as part of the appliance and that usually requires some assembling on the job site. Except for the vent terminal provisions of Section G2427.8, item 3, the code depends on the appliance manufacturers' installation instructions for all aspects of venting direct-vent appliances. Direct-vent appliances include some Category IV condensing furnaces and boilers and power and gravity vented direct-vent water heaters, unit heaters, space heaters, and packaged terminal units.

Direct-vent appliances have the distinct advantage of providing their own combustion air supply in addition to their own dedicated venting means. See the definition of "Direct-vent appliance" as well as Commentary Figures G2403(5) and G2427.2.1(1) through G2427.2.1(3).

G2427.2.2 (503.2.4) Equipment with integral vents. Equipment incorporating integral venting means shall be considered properly vented when installed in accordance with its listing, the manufacturer's instructions and Section G2427.8, Items 1 and 2.

❖ As it does with direct-vent appliances, the code trusts the manufacturers' installation instructions for the installation of appliances having integral venting means. Integral vent appliances such as rooftop HVAC units, outdoor makeup air heaters, and outdoor heaters for swimming pools have a built-in natural or mechanical venting means. The natural draft integral venting means is simply a short vertical vent with a factory supplied vent cap assembly. The mechanical integral vent appliances use supervised blowers to discharge the combustion gases directly to the atmosphere. The typical means of blower supervision is one or more pressure switches or a centrifugal switch mounted to the blower motor shaft. Integral vent

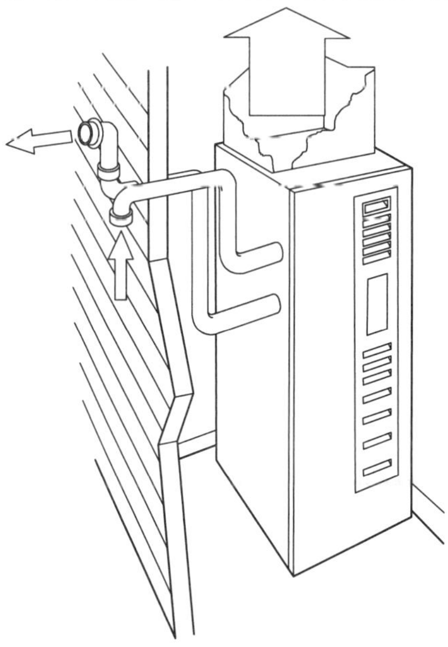

Figure courtesy of Carrier Corporation

Figure G2427.2.1(1) DIRECT-VENT APPLIANCE

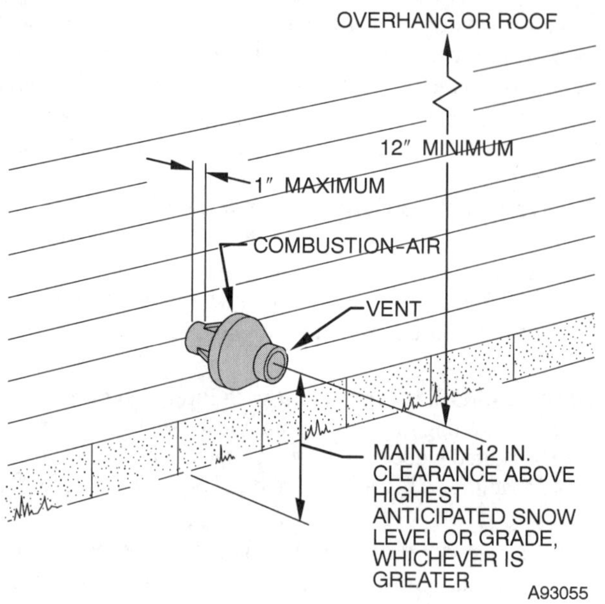

Figure courtesy of Carrier Corporation

Figure G2427.2.1(2)
EXAMPLE OF MANUFACTURER'S DETAIL FOR DIRECT-VENT APPLIANCE CONCENTRIC VENT-AIR INTAKE TERMINAL

appliances must not be confused with unvented appliances and appliances not required to be vented in accordance with Section G2425.8. Integral vent appliances are required to be vented to the outside atmosphere and thus are always designed for outdoor use only. See Section G2427.8, items 1 and 2, and Commentary Figures G2427.2.2(1) through G2427.2.2(3).

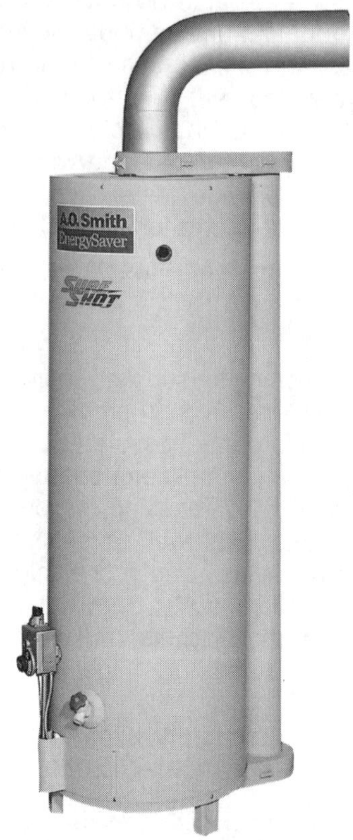

Photo courtesy of A.O. Smith Water Products

Figure G2427.2.1(3)
GRAVITY DIRECT-VENT WATER HEATER

G2427.3 (503.3) Design and construction. A venting system shall be designed and constructed so as to develop a positive flow adequate to remove flue or vent gases to the outdoor atmosphere.

❖ This section is stated in performance language and requires venting systems to convey the products of combustion to the outdoors. The reference to a "positive flow" does not imply that the venting system is under a positive pressure, but rather implies that the vent produces a dependable flow at whatever pressure it is designed to operate under.

G2427.3.1 (503.3.1) Equipment draft requirements. A venting system shall satisfy the draft requirements of the equipment in accordance with the manufacturer's instructions.

❖ This section is stated in performance language and requires venting systems to produce the draft necessary for the operation of the appliance being vented. An appliance manufacturer will specify the required type of venting means and, to assure the required amount of draft, may specify the size and height of a vent or chimney or may specify a mechanical draft system.

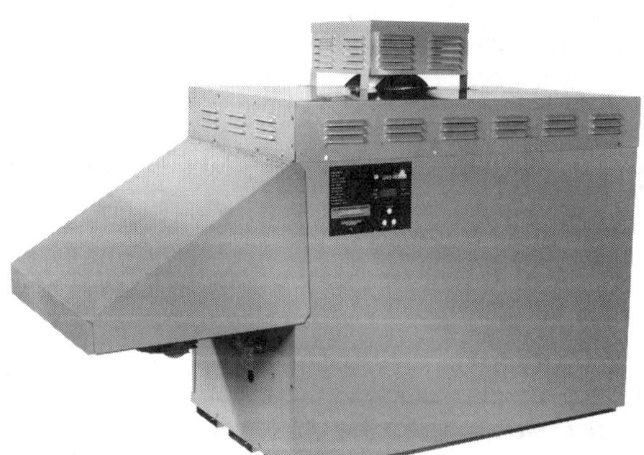

Photo courtesy of A.O. Smith Water Products

Figure G2427.2.2(1)
HOT WATER BOILER WITH INTEGRAL VENT FOR OUTDOOR INSTALLATION

G2427.3.2 (503.3.2) Design and construction. Gas utilization equipment required to be vented shall be connected to a venting system designed and installed in accordance with the provisions of Sections G2427.4 through G2427.15.

❖ As defined in Section G2403, a venting system includes natural, induced, and forced- draft systems, which are addressed in Sections G2427.3.3 through G2427.15.

G2427.3.3 (503.3.3) Mechanical draft systems. Mechanical draft systems shall comply with the following:

1. Mechanical draft systems shall be listed and shall be installed in accordance with the terms of their listing and both the appliance and the mechanical draft system manufacturer's instructions.
2. Equipment, except incinerators, requiring venting shall be permitted to be vented by means of mechanical draft systems of either forced or induced draft design.
3. Forced draft systems and all portions of induced draft systems under positive pressure during operation shall be designed and installed so as to prevent leakage of flue or vent gases into a building.

Photo courtesy of Reznor/Thomas & Betts Corporation

Figure G2427.2.2(2)
OUTDOOR DUCT FURNACE WITH INTEGRAL POWER VENTING

Photo courtesy of Carrier Corporation

Figure G2427.2.2(3)
ROOFTOP HVAC UNIT WITH INTEGRAL POWER VENT

4. Vent connectors serving equipment vented by natural draft shall not be connected into any portion of mechanical draft systems operating under positive pressure.
5. When a mechanical draft system is employed, provision shall be made to prevent the flow of gas to the main burners when the draft system is not performing so as to satisfy the operating requirements of the equipment for safe performance.
6. The exit terminals of mechanical draft systems shall be not less than 7 feet (2134 mm) above grade where located adjacent to public walkways and shall be located as specified in Section G2427.8, Items 1 and 2.

❖ The appliances and equipment installations addressed in this section employ auxiliary or integral fans and blowers to force the flow of combustion products to the outdoors.

This section applies to externally installed power exhausters, integrally power-exhausted appliances, and venting systems equipped with draft inducers. This section does not address fan-assisted Category I appliances (except in item 3) and does not address direct-vent appliances. Category III and IV appliances are subject to the requirements of this section if such appliances are not listed as direct-vent appliances. Some Category I appliances are field convertible to Category III. Item 3 applies to Category I fan-assisted appliances since they are vented by natural draft.

Power exhausters are field-installed pieces of equipment that are independent of but used in conjunction with other appliances. See Commentary Figures G2427.3.3(1) through G2427.3.3(11).

Power exhausters are typically used where other means of venting are impractical, impossible, or uneconomical. Power exhausters are typically designed for use with gas-fired and oil-fired natural draft appliances. Some are also designed for use with fan-assisted appliances and appliances equipped with integral forced draft (power vent) fans. Power exhausters produce negative pressures at their inlet connection and produce positive pressures at their outlet (dis-

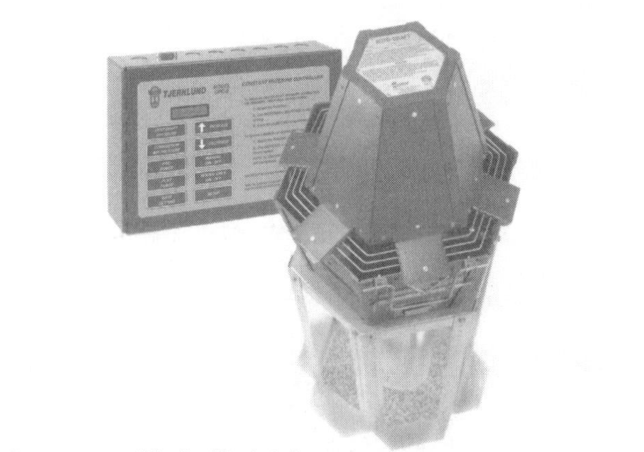

Photo courtesy of Tjernlund Products, Inc.

Figure G2427.3.3(1)
VERTICAL DRAFT INDUCER

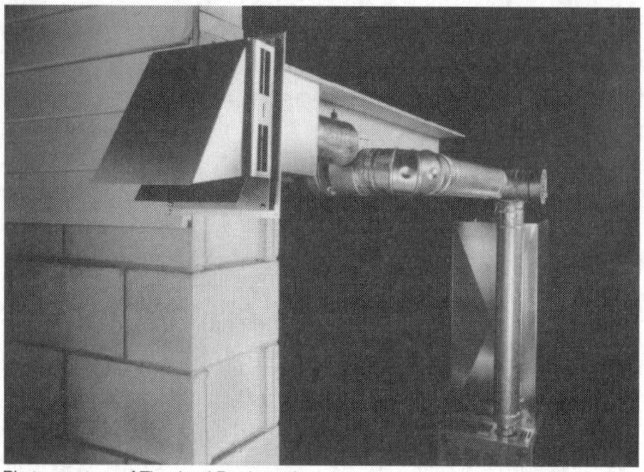

Photo courtesy of Tjernlund Products, Inc.

Figure G2427.3.3(2)
HORIZONTAL POWER EXHAUSTER

charge connection). The most common installation locates the exhauster at the point of termination of the vent or chimney. In such installations, the vent or chimney between the appliance and the exhauster operates under negative pressure. See Commentary Figures G2427.3.3(2) through G2427.3.3(7).

Induced draft systems employ separate field-installed units designed to boost draft in a natural (gravity) draft chimney or vent. See Commentary Figures G2427.3.3(1), (8) and (9).

Power-vented (self-venting) appliances are equipped with factory-installed integral blowers that force the combustion products through the special venting systems that are addressed in Sections G2425.14 and G2427.4. Power-vented appliances include mid-efficiency to high-efficiency furnaces, boilers, water heaters, and heating units that are designed for through-the-wall or through-the-roof venting. See Commentary Figure G2427.3.3(10). Appliances for outdoor installation are typically power vented, discharge directly to the atmosphere, and are addressed in Section G2427.2.2.

Although not specifically referred to as an appliance, a power exhauster (mechanical draft system) is considered a mechanical appliance appurtenance and therefore must bear the label of an approved agency in accordance with item 1. See the definition of "Appliance." A power exhauster is an essential component of the appliance installation it serves and must be installed in accordance with the manufacturer's installation instructions for both the power exhauster and the appliance served. Mechanical draft devices that are an integral part of an appliance are covered by the appliance listing.

Vent or chimney systems installed downstream (discharge side) of a power exhauster must be designed and approved for positive-pressure applications. For example, a Type B vent cannot be used on the discharge side of an exhauster, because such pipe is not

Photo courtesy of Exhausto, Inc.

Figure G2427.3.3(3)
HORIZONTAL POWER EXHAUSTERS

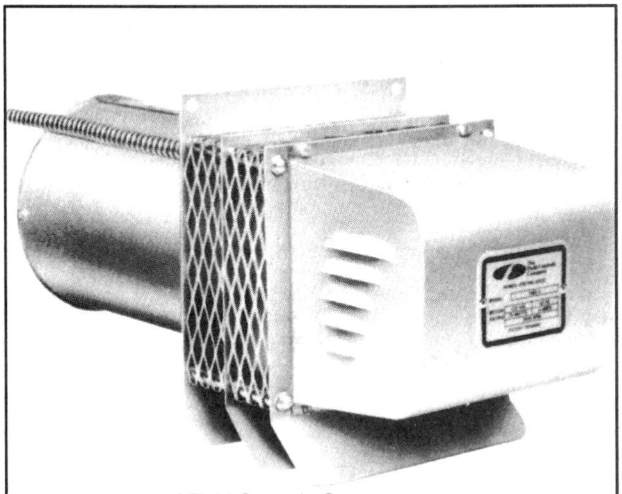

Photo courtesy of Field Controls Company

Figure G2427.3.3(4)
POWER EXHAUSTER OUTDOOR TERMINAL-TYPE UNIT

designed for positive pressure applications. The application and installation of exhausters (power vents) must be in accordance with the manufacturers' installation instructions for the exhausters and the appliance(s) served by the exhausters. See the commentary for Section G2425.6.

There are three distinct variations in use of blowers at the combustion chamber inlet working in conjunction with the fuel burner.

1. Blowers that supply turbulent combustion air to aid fuel-air mixing in a combustion chamber that is under negative pressure. Residential pressure-atomizing oil burners are examples. To obtain proper negative over-fire draft (which optimizes combustion) also requires steady negative (below atmospheric) pressure at the flue outlet. This will be produced by a natural draft chimney and may be controlled by a barometric draft regulator.
2. Blowers that supply sufficient combustion air and pressure to produce flow through the combustion chamber, but the combustion process does not need additional vent or chimney draft. This permits use of gravity or neutral draft venting products such as Type B vent for gas conversion burners and chimneys for equipment burning oil, coal, or wood. A draft regulator may be used for such equipment to prevent excess draft from affecting combustion efficiency.

Photo courtesy of Tjernlund Products, Inc.

Figure G2427.3.3(5)
EXHAUSTER AND CONTROL PACKAGE

3. Blowers with enough power to overcome internal flue passage pressure losses (i.e., in fire-tube boilers) that also produce positive pressure at the outlet. This outlet pressure must be added to gravity draft as the motive force for flow in a chimney. If positive outlet pressure exists, the use of a sealed pressure-tight chimney is required.

The above types of equipment usually have integral blower/burner systems, and all could be considered forced combustion systems. Only those described in the third paragraph truly produce forced chimney draft.

When power-venting equipment is installed, it becomes an essential part of the appliance it serves. The appliance relies on the power-venting equipment (exhauster) to provide sufficient draft for proper appliance operation and for venting of the combustion products. Power exhausters are electrically interlocked with the appliance or appliances they serve to assure that the appliances will not operate if there is insufficient draft to vent the products of combustion. If an exhauster fails, the appliances served by the exhauster would discharge the products of combustion into the building unless shut off by controls that monitor draft and draft hood spillage. An improper or missing interlock can cause the appliances to malfunction and spill harmful flue gases into the building space. The electrical interlock (interconnection) is typically accomplished with controls provided by the exhauster manufacturer. See Commentary Figure G2427.3.3(11).

The usual sequence of operation is as follows: The call for heat from the appliance operating control starts the exhauster, and pressure controls start the appliance only after adequate draft has been proven to exist. In some cases, temperature sensors are used in addition to pressure controls. The temperature sensors are used to sense draft hood spillage or unusual vent system temperatures and shut off the appliance.

Item 6 addresses the concern for pedestrian safety relative to exposure to high-temperature combustion gases. See Section G2427.8.

G2427.3.4 (503.3.5) Circulating air ducts and furnace plenums. No portion of a venting system shall extend into or pass through any circulating air duct or furnace plenum.

❖ If a vent, chimney, or a connector for such passes through or extends into a duct or furnace plenum, it is possible to subject the venting system to negative or positive pressures and/or temperature extremes that

Photo courtesy of Tjernlund Products, Inc.

Figure G2427.3.3(6)
TYPICAL POWER EXHAUSTER

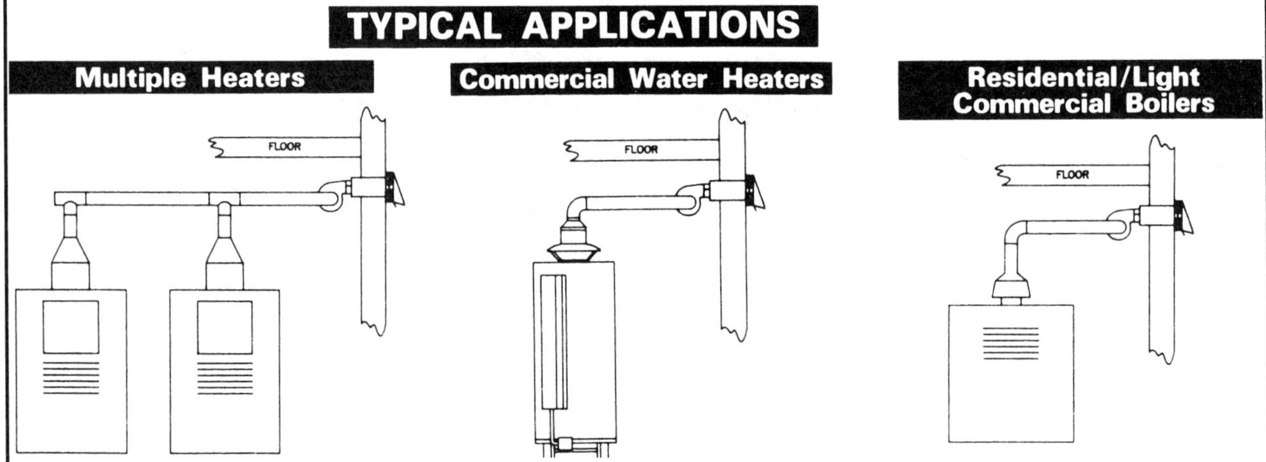

Illustration courtesy of Tjernlund Products, Inc.

Figure G2427.3.3(7)
TYPICAL EXHAUSTER APPLICATIONS

Photo courtesy of A.O. Smith Water Products

Figure G2427.3.3(10)
POWER VENTED WATER HEATER

Photo courtesy of Tuernlund Products, Incorporated

Figure G2427.3.3(8)
DRAFT INDUCER

Photo courtesy of Field Controls Company

Figure G2427.3.3(9)
DRAFT INDUCER

could cause the venting system to deteriorate, fail to produce the required draft, produce condensation, and/or leak combustion gases. A furnace plenum is a component of the ductwork system constructed as a junction box for multiple duct connections. Consider that pressure differentials can also occur between a space such as an air ceiling plenum and occupied spaces, thus creating a condition similar to what would occur if a venting system component was exposed to the interior of a ductwork plenum.

G2427.4 (503.4) Type of venting system to be used. The type of venting system to be used shall be in accordance with Table G2427.4.

❖ The table lists various types of vents and the corresponding types of appliances that can be served by the vents. The vent system must be tested and specifically approved for use with the approved appliance. If the vent system is not a tested and labeled component of the appliance, the material must be approved for use with the appliance and installed in accordance with the manufacturer's installation instructions.

Gas-burning appliances (listed as Category II, III, or IV) require special vent systems that are specific to the type of appliance. Special vent systems are typically associated with mid-efficiency to high-efficiency appliances and include vent materials such as polyvinyl chloride (PVC), chlorinated polyvinyl chloride (CPVC), and special alloys of stainless steel. See Commentary Figure G2425.14.

A primary consideration for the design of a special vent system that serves a high-efficiency appliance is the selection of a material that is capable of withstanding the corrosive effects of condensate. Flue gas condensate is acidic and corrosive and can deteriorate many vent materials. Sulfur in the fuel and halogens

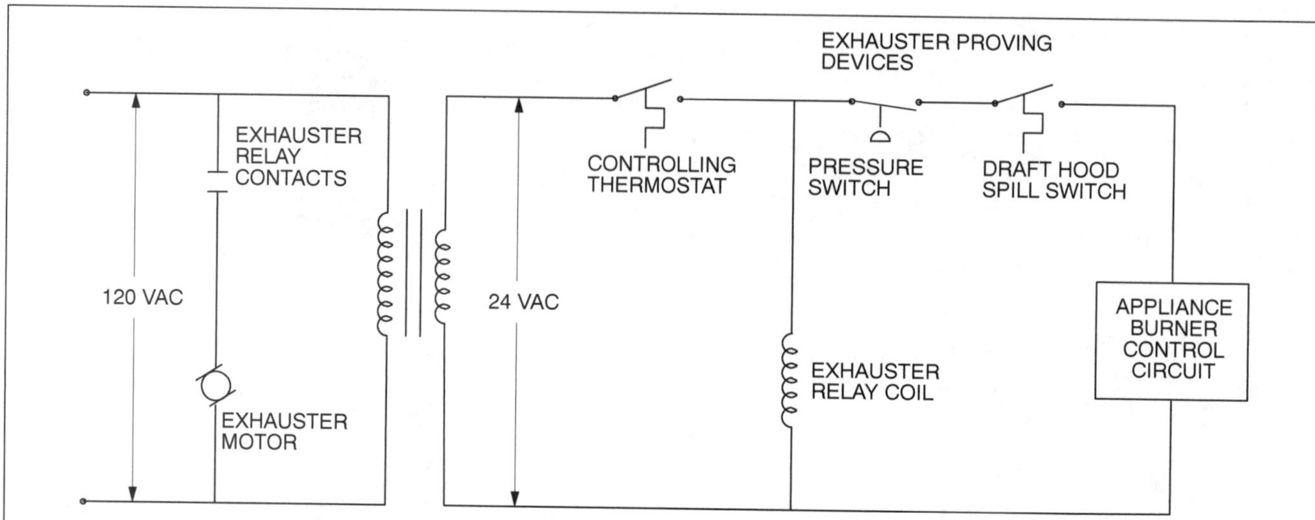

Figure G2427.3.3(11)
SIMPLIFIED SAMPLE INTERLOCK CIRCUIT DIAGRAM FOR APPLIANCE SERVED BY POWER EXHAUSTER (SHOWN FOR ILLUSTRATIVE PURPOSE ONLY)

carried in the combustion air can combine with the combustion products to create acids. To prevent the corrosion of a vent and possible escape of flue gas, metallic flue passageways can be protected with appropriate coatings; however, any imperfections, pinholes, or discontinuities in the coating can create areas for condensate to collect and accelerate the deterioration of the chimney or vent. The solution to this problem has been to use a material such as PVC, CPVC, or high-temperature plastic that is resistant to acidic condensation and compatible with the temperature of the appliance flue gases. Plastic pipe is a suitable material because it is impervious to corrosive condensation and can easily be designed to drain off accumulated condensate. The flue gas temperatures of most condensing-type appliances are not high enough to affect such materials adversely.

Special vent systems must be designed and installed in compliance with the manufacturers' instructions. The manufacturers' instructions for appliances requiring special vent systems will specify installation requirements that are specific to that type of vent and appliance. For example, a maximum developed length, a maximum number of directional fittings, fitting turn radius, and specific support requirements will be specified for special vents employing plastic pipe. Special vent systems also have special vent termination requirements and typically incorporate condensate collection and drainage fittings and connections.

The type and size of the vent must be as dictated by the manufacturer's installation instructions for the appliance. The design and installation instructions provided by the vent manufacturer must also be consulted when designing a vent system for any particular application. Commentary Figure G2427.4 depicts some of the variables that affect vent system sizing. In addition to Section G2428, the vent manufacturer's design and application handbook and installation instructions contain information and tables for the sizing and application of the vent systems.

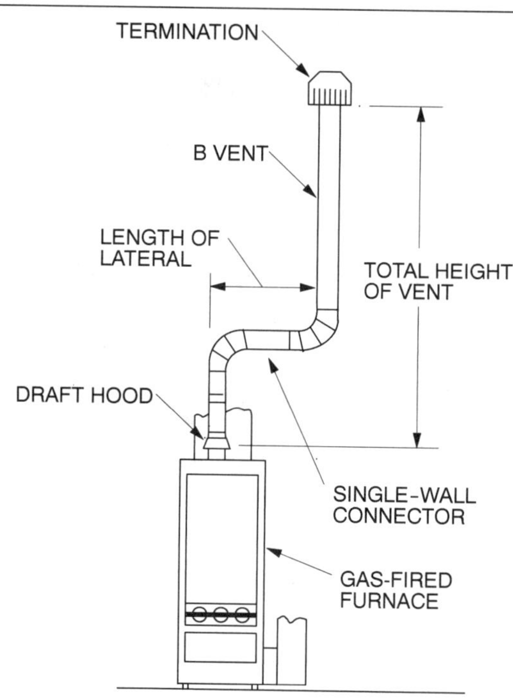

Figure G2427.4
DETERMINING VENT HEIGHT

G2427.4.1 (503.4.1) Plastic piping. Plastic piping used for venting equipment listed for use with such venting materials shall be approved.

❖ Plastic pipe has been used as a venting material for Category III and IV appliances for many years and has recently been used to vent specialized power vented appliances that are equipped with draft hoods and blowers that introduce enough dilution air to keep the vent gases within the service temperature range of the plastic pipe. See Sections G2426.1 and G2427.4.

G2427.4.2 (503.4.2) Special gas vent. Special gas vents shall be listed and installed in accordance with the terms of the special gas vent listing and the manufacturers' instructions.

❖ Special gas vents include high temperature plastic pipe, stainless steel alloy pipe, and positive pressure pipe that are specified by the appliance manufacturer for use with a specific appliance. This section does not require plastic pipe to be listed as a venting system component if such pipe is specified by the appliance manufacturer for venting appliances such as Category IV condensing furnaces and boilers. The installation instructions for Category IV appliances often specify PVC, CPVC, and ABS plastic plumbing pipe for venting the appliances, and such pipe is not listed as a component of a venting system. See Section G2427.4.1.

G2427.5 (503.5) Masonry, metal, and factory-built chimneys. Masonry, metal and factory-built chimneys shall comply with Sections G2427.5.1 through G2427.5.9.

❖ See the commentary for Sections G2427.5.1 through G2427.5.9.

G2427.5.1 (503.5.1) Factory-built chimneys. Factory-built chimneys shall be installed in accordance with their listing and the manufacturers' instructions. Factory-built chimneys used to vent appliances that operate at positive vent pressure shall be listed for such application.

❖ All prefabricated chimney systems must bear the label of an approved agency. A label is required on all components of the chimney system such as the pipe sections, shields, fireblocks, fittings, termination caps, and supports. The label states information such as the type of appliance the chimney was tested for use with, a reference to the manufacturer's installation instructions, and the minimum required clearances to combustibles. See Commentary Figure G2403(4).

Most factory-built chimneys are either of the double-walled fiber-insulated design or the triple-walled air-cooled design. Factory-built chimneys are constructed of stainless steel inner liners with stainless steel outer walls. Factory-built chimneys, like vent systems, are composed of components that must be installed as a complete system. Components from different manufacturers are not designed to be mixed and installed together.

The manufacturer's instructions contain sizing criteria and requirements for every aspect of a factory-built chimney installation. The requirements include component assembly, clearances to combustibles, support, terminations, connections, protection from damage, and fireblocking.

Any vent or chimney that conveys vent gases under positive pressure must be factory designed for such application. There is no allowable method of taping or sealing a chimney or vent that will convert a nonpositive pressure chimney or vent into a positive pressure chimney.

TABLE G2427.4 (503.4)
TYPE OF VENTING SYSTEM TO BE USED

GAS UTILIZATION EQUIPMENT	TYPE OF VENTING SYSTEM
Listed Category I equipment Listed equipment equipped with draft hood Equipment listed for use with Type B gas vent	Type B gas vent (G2427.6) Chimney (G2427.5) Single-wall metal pipe (G2427.7) Listed chimney lining system for gas venting (G2427.5.3). Special gas vent listed for this equipment (G2427.4.2)
Listed vented wall furnaces	Type B-W gas vent (G2427.6, G2436)
Category II equipment	As specified or furnished by manufacturers of listed equipment (G2427.4.1, G2427.4.2)
Category III equipment	As specified or furnished by manufacturers of listed equipment (G2427.4.1, G2427.4.2)
Category IV equipment	As specified or furnished by manufacturers of listed equipment (G2427.4.1, G2427.4.2)
Unlisted equipment	Chimney (G2427.5)
Decorative appliance in vented fireplace	Chimney
Direct-vent equipment	See G2427.2.1
Equipment with integral vent	See G2427.2.2

❖ Table G2427.4 lists the currently available types of appliances and the types of venting systems to be used with such appliances. In the first row, the three types of appliances can be matched up with any of the five types of venting systems, if allowed by the appliance manufacturer's installation instructions. The 4th row up from the bottom addresses unlisted appliances that are not allowed by the code except under the alternative approval provisions of Section R104.11.

G2427.5.2 (503.5.3) Masonry chimneys. Masonry chimneys shall be built and installed in accordance with this code and shall be lined with approved clay flue lining, a listed chimney lining system, or other approved material that will resist corrosion, erosion, softening or cracking from vent gases at temperatures up to 1,800°F (982°C).

Exception: Masonry chimney flues serving listed gas appliances with draft hoods, Category I appliances, and other gas appliances listed for use with Type B vent shall be permitted to be lined with a chimney lining system specifically listed for use only with such appliances. The liner shall be installed in accordance with the liner manufacturers' instructions and the terms of the listing. A permanent identifying label shall be attached at the point where the connection is to be made to the liner. The label shall read: "This chimney liner is for appliances that burn gas only. Do not connect to solid or liquid fuel-burning appliances or incinerators."

❖ A chimney liner might be listed for use with solid, liquid, and gaseous fuel-fired appliances. The label required by this section is intended to warn the unknowing installer who sees a masonry chimney and thinks that it can serve a solid or liquid fuel-fired appliance. The label would be applicable to chimney liner systems that are designed for gas-fired appliances only. See Section G2425.3.

G2427.5.3 (503.5.4) Chimney termination. Chimneys shall extend at least 3 feet (914 mm) above the highest point where it passes through a roof of a building and at least 2 feet (610 mm) higher than any portion of a building within a horizontal distance of 10 feet (3048 mm). (See Figure G2427.5.3.) Chimneys shall extend at least 5 feet (1524 mm) above the highest connected equipment draft hood outlet or flue collar. Decorative shrouds shall not be installed at the termination of factory-built chimneys except where such shrouds are listed and labeled for use with the specific factory-built chimney system and are installed in accordance with the manufacturers' installation instructions.

❖ Low-heat and residential-type chimneys must extend at least 3 feet (914 mm) above the roof, measured from the highest point of the roof penetration. They must also be at least 2 feet (610 mm) higher than any portion of the roof within a 10-foot (3048 mm) horizontal distance. See Commentary Figure G2427.5.3.

The 2-foot (610 mm) termination requirement is intended to prevent wind and pressure zones from reducing the amount of draft produced by the chimney. Wind and wind-induced eddy currents can react with building structural surfaces, thereby creating air pressure zones that can diminish chimney draft and cause reverse flow (backdraft) in the chimney. Loss of draft will cause the appliance or fireplace served by the chimney to discharge combustion products into the building interior. Locating the chimney outlet well into the undisturbed wind stream and away from the cavity and wake (eddy) zones around the building can counteract the adverse effects and also prevent the reentry of flue gases into the building through openings and fresh air intakes. Terminating a chimney in the eddy current area recirculates the combustion products and allows them to enter the building via infiltration, wall openings, and air intakes. A chimney terminal properly located above the eddy current area allows the wind to carry the combustion products away from the building. Factory-built chimneys are provided with termination caps that are part of the labeled chimney system and that must be installed. Like vent systems, the chimney cap is designed to keep out precipitation and animals and minimize the adverse effects of wind.

Decorative shrouds are designed to be aesthetically pleasing and to hide chimney terminations from view. These shrouds have, in some cases, caused fires resulting from overheated combustible construction and the accumulation of debris and animal nesting. Shrouds can also interfere with the functioning of a vent or chimney. The listing of the shroud must state the specific make and model of chimney with which the shroud is intended to be used. At this time, the author is unaware of any shrouds that are listed for use with a specific make and model of chimney. Most often, shrouds are field-constructed of sheet metal, masonry, or stucco on frame.

G2427.5.4 (503.5.5) Size of chimneys. The effective area of a chimney venting system serving listed appliances with draft hoods, Category I appliances, and other appliances listed for use with Type B vents shall be determined in accordance with one of the following methods.

1. The provisions of Section G2428.
2. For sizing an individual chimney venting system for a single appliance with a draft hood, the effective areas of the vent connector and chimney flue shall be not less than the area of the appliance flue collar or draft hood outlet, nor greater than seven times the draft hood outlet area.
3. For sizing a chimney venting system connected to two appliances with draft hoods, the effective area of the chimney flue shall be not less than the area of the larger draft hood outlet plus 50 percent of the area of the smaller draft hood outlet, nor greater than seven times the smallest draft hood outlet area.
4. Chimney venting systems using mechanical draft shall be sized in accordance with approved engineering methods.
5. Other approved engineering methods.

❖ Chimneys are sized in Section G2428, or they could be sized by items 2, 3 or 4 or by an engineering method acceptable to the code official. Some appliances are not categorized; however, appliances with draft hoods and appliances listed for use with Type B vents would fit in Category I if they were categorized. The manufacturers' instructions for some Category I appliances, such as fan-assisted furnaces and boilers, may not allow the appliance to be connected to a masonry chimney except under specific limited conditions.

Item 2 applies only to a single appliance that is factory equipped with a draft hood. The "seven times rule" prevents the chimney from being too large, which

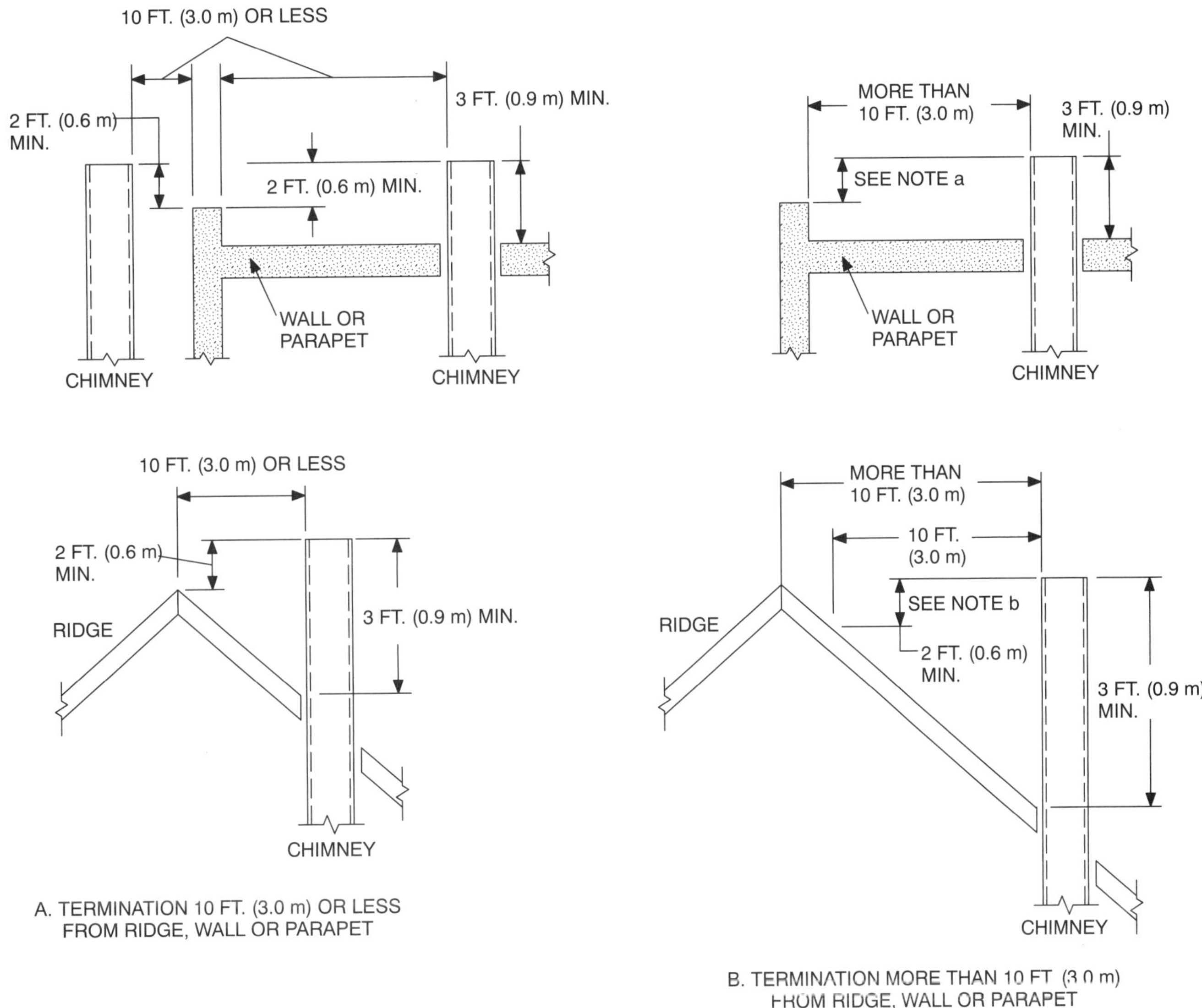

For SI: 1 inch = 25.4 mm, 1 foot = 304.8 mm.

Notes:

a. No height above parapet required when distance from walls or parapet is more than 10 feet.

b. Height above any roof surface within 10 feet horizontally.

FIGURE G2427.5.3 (503.5.4)
TYPICAL TERMINATION LOCATIONS FOR CHIMNEYS AND SINGLE-WALL METAL PIPES

❖ See the commentary for Sections G2426.5.1 through G2426.5.9.

could result in poor draft and condensation problems. Like vents, chimneys that are too large in area can be as problematic as chimneys that are too small in area.

Item 3 is limited to only one venting arrangement. It can only be used for a chimney that serves two appliances, both of which must be factory-equipped with draft hoods. The method is based on the areas of the chimneys and draft hood outlets, not on the diameters.

Item 4 recognizes that some manufacturers of mechanical draft equipment (exhausters) provide their own sizing methodology specific to their product. Recall that item 1 of Section G2427.3.3 requires such equipment to be listed.

G2427.5.5 (503.5.6) Inspection of chimneys. Before replacing an existing appliance or connecting a vent connector to a chimney, the chimney passageway shall be examined to ascertain that it is clear and free of obstructions and it shall be cleaned if previously used for venting solid or liquid fuel-burning appliances or fireplaces.

❖ If a chimney is going to serve a new appliance installation or a replacement appliance installation, the chimney must be inspected to determine if it is still service-

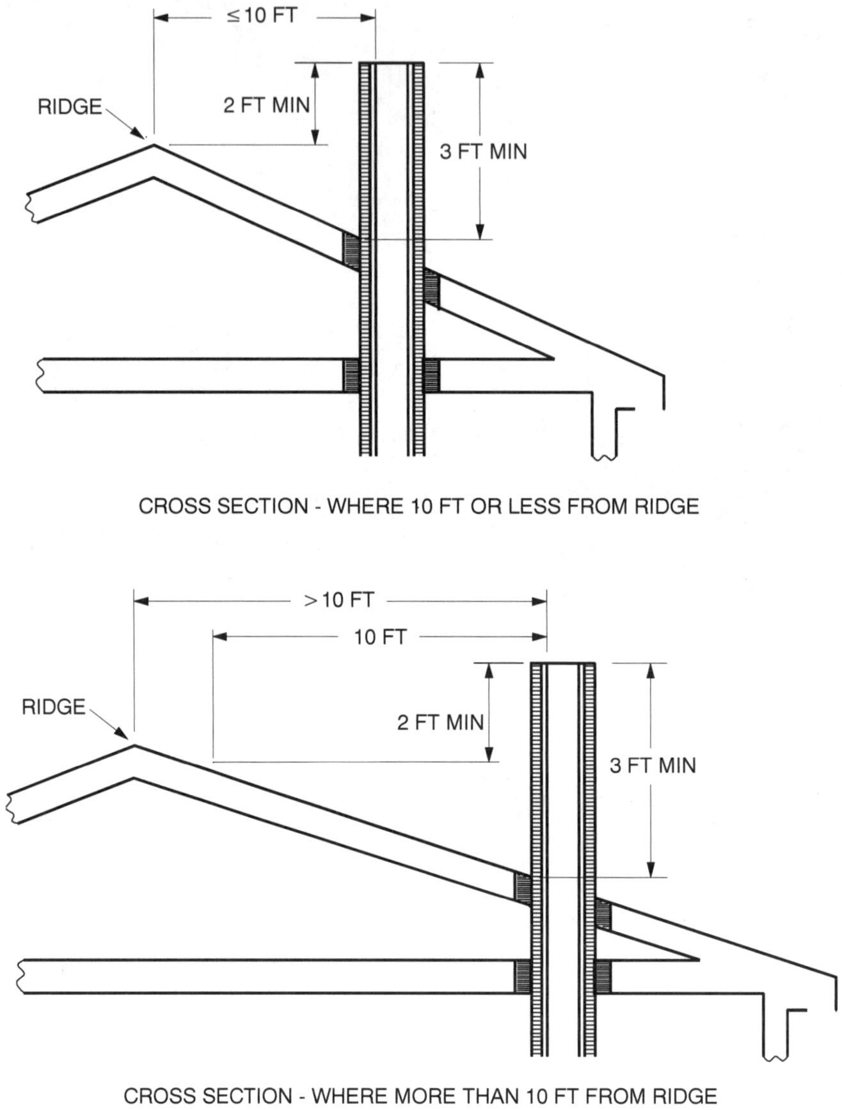

For SI: 1 foot = 304.8 mm.

Figure G2427.5.3
LOW-HEAT CHIMNEY TERMINATIONS

able and free of deposits. Chimneys can become obstructed by debris such as leaves, animal carcasses, loose mortar, and pieces of masonry and liner. Combustible deposits can accumulate on chimney walls used to vent liquid and solid fuel-fired appliances, and this section mandates that the chimney be cleaned if it served such appliances in the past. Masonry chimneys are especially susceptible to internal deterioration and should be inspected regularly, in addition to the occasion of connecting a new or replacing an old appliance.

G2427.5.5.1 (503.5.6.1) Chimney lining. Chimneys shall be lined in accordance with Chapter 10.

Exception: Existing chimneys shall be permitted to have their use continued when an appliance is replaced by an appliance of similar type, input rating and efficiency.

❖ The liner forms the flue passageway and is the actual conductor of all products of combustion. The chimney liner must be able to withstand exposure to high temperatures and corrosive chemicals. The chimney lining protects the masonry construction of the chimney walls and allows the chimney to be constructed gas tight. This section regulates liners and relining systems. Liners are often used to reline an existing chimney to salvage a masonry chimney or allow connection of higher-efficiency appliances.

Flue lining systems for masonry chimneys include clay tile, poured-in-place refractory materials, and stainless steel pipe. Chimney lining systems for chimneys serving gas-fired appliances include stainless steel and aluminum piping (uninsulated or insulated) designed for both new construction and for relining existing chimneys. See Commentary Figures G2427.5.5.1(1) and G2427.5.5.1(2). Type B vent is also used as a chimney liner, although it is not considered to be a lining system, and in such cases the chimney is actually a masonry chase containing a vent.

Relining of existing chimneys is quite common now because of the incompatibility of mid-efficiency appliances and masonry chimneys. Relining systems allow existing chimneys to be reused, thereby saving expense and retaining desirable architectural features of buildings.

The intended purpose of the exception is to allow an unlined chimney to serve a new appliance that is installed to replace a previously served appliance, provided that the new appliance does not create any different operating conditions in the chimney. That is, the volume, temperature, water vapor content, and dilution air content of the new appliance flue gases must be the same as the old appliance.

The exception assumes that the existing chimney and the existing appliances operated as intended by this code. If there was evidence that the previous chimney and appliance system was not working as required by this code, the exception is not intended to be applicable. Because this exception would allow an installation that is in conflict with the appliance manufacturer's installation instructions, it is in effect negated because the manufacturer's instructions prevail. See Sections G2408.1 and G2427.5.5.3.

G2427.5.5.2 (503.5.6.2) Cleanouts. Cleanouts shall be examined to determine they will remain tightly closed when not in use.

❖ A cleanout door or panel must be capable of securely sealing the opening to prevent air from entering the chimney, which would reduce draft for the appliances served. Under certain conditions, a faulty cleanout could also cause vent gases to leak from the chimney.

G2427.5.5.3 (503.5.6.3) Unsafe chimneys. Where inspection reveals that an existing chimney is not safe for the intended application, it shall be repaired, rebuilt, lined, relined, or replaced with a vent or chimney to conform to M1805 and Chapter 10, and it shall be suitable for the equipment to be vented.

❖ If a chimney is inspected and found to be unsafe for any reason, it must be made safe or replaced. See Section G2425.3. Some chimneys and appliances are not compatible as dictated by the appliance and/or factory-built chimney manufacturer. Masonry chimneys usually become unsafe as a result of deteriorated mortar joints and liners. Other causes include structural instability, cracks, faulty cleanouts, deposit formation, and moisture damage.

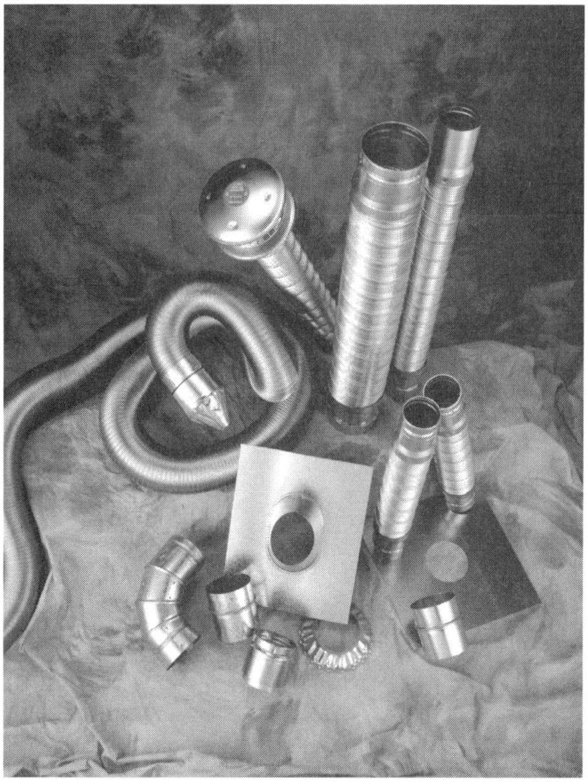

Photo courtesy of Selkirk L.L.C.

Figure G2427.5.5.1(1)
CHIMNEY LINING SYSTEM FOR GAS APPLIANCES

G2427.5.6 (503.5.7) Chimney serving equipment burning other fuels. Chimneys serving equipment burning other fuels shall comply with Sections G2427.5.6.1 through G2427.5.6.4.

❖ See Sections G2427.5.6.1 through G2427.5.6.4.

G2427.5.6.1 (503.5.7.1) Solid fuel-burning appliances. Gas utilization equipment shall not be connected to a chimney flue serving a separate appliance designed to burn solid fuel.

❖ Each solid fuel-burning appliance or fireplace must be provided with a dedicated independent chimney or a dedicated independent flue in multiple-flue chimney constructions. Solid fuel-burning appliances and fireplaces cannot share a common chimney or flue-way with any other appliance or fireplace. For example, a solid-fuel appliance cannot be connected to an existing masonry chimney that serves other appliances or a fireplace.

Solid fuel-burning appliances produce creosote deposits on the interior walls of chimney liners. The creosote formation is highly combustible and creates a potential fire hazard. Because of the potential for a chimney fire, other connections to the chimney can allow fire to break out of the chimney into the building. Also, chimney passageways can become blocked by the creosote formations. If other appliances were vented to such chimneys, combustion products would be discharged into the building.

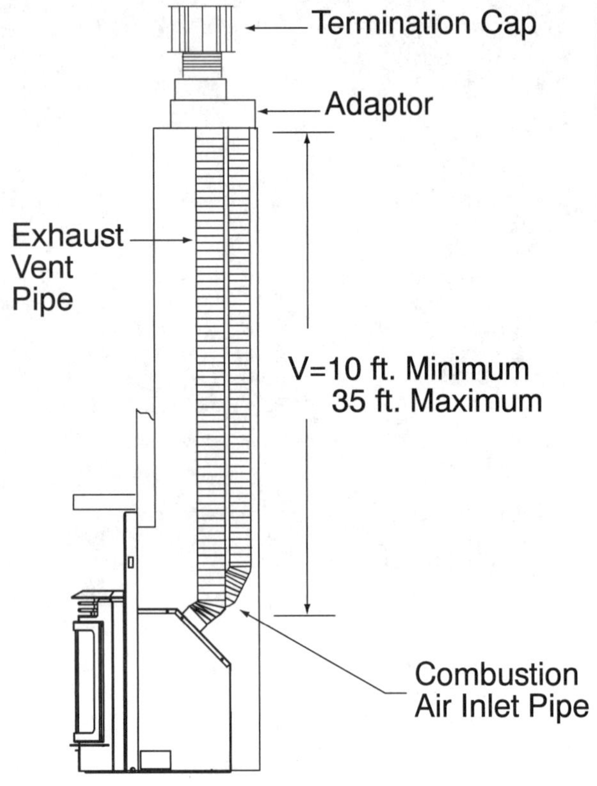

Figure G2427.5.5.1(2)
DIRECT-VENT GAS APPLIANCE FOR INSTALLATION IN SOLID FUEL FIREPLACE

This section is also intended to prevent the possibility of creosote leaking out of the chimney through appliance connectors.

Combination (dual fuel) gas or oil and solid fuel-burning appliances are designed to be connected to a single chimney passageway. Such dual-fuel appliances must be listed, labeled, and installed in accordance with the manufacturer's instructions.

G2427.5.6.2 (503.5.7.2) Liquid fuel-burning appliances. Where one chimney flue serves gas utilization equipment and equipment burning liquid fuel, the equipment shall be connected to separate openings or shall be connected through a single opening where joined by a suitable fitting located as close as practical to the chimney. Where two or more openings are provided into one chimney flue, they shall be at different levels. Where the gas utilization equipment is automatically controlled, it shall be equipped with a safety shutoff device.

❖ Gas-fired and oil-fired appliances are allowed to share a chimney or flue. The code does not provide a sizing method for such arrangements; therefore, the system would have to be engineered or otherwise approved by the code official. Placing the openings at different levels in the chimney flue will minimize interference between the flows of the chimney connectors. The last sentence is consistent with the design of all automatically controlled appliances listed to today's appliance standards as required by this code.

G2427.5.6.3 (503.5.7.3) Combination gas- and solid fuel-burning appliances. A combination gas- and solid fuel-burning appliance equipped with a manual reset device to shut off gas to the main burner in the event of sustained backdraft or flue gas spillage shall be permitted to be connected to a single chimney flue. The chimney flue shall be sized to properly vent the appliance.

❖ A dual-fuel appliance, gas and solid fuel, can be served by a single chimney or flue if the appliance is equipped with a safety control that monitors chimney spillage. A solid fuel fire cannot be turned on and off like other fuel fires; thus, it is possible that the gas burner could be operated while the solid fuel fire is still burning, and the chimney could be overloaded causing spillage of combustion gases. The chimney size would have to be engineered or would have to rely on the appliance manufacturer's instructions.

G2427.5.6.4 (503.5.7.4) Combination gas- and oil fuel-burning appliances. A listed combination gas- and oil-burning appliance shall be permitted to be connected to a single chimney flue. The chimney flue shall be sized to properly vent the appliance.

❖ See Section G2427.5.6.3.

G2427.5.7 (503.5.8) Support of chimneys. All portions of chimneys shall be supported for the design and weight of the materials employed. Factory-built chimneys shall be supported and spaced in accordance with their listings and the manufacturers' instructions.

❖ See Section G2426.6. Chimneys, including factory-built, are very heavy compared to vents and need substantial support to carry their weight and prevent displacement. Factory-built chimneys require special support fittings at offsets to prevent the elbows from being damaged by bearing the weight of the chimney sections above the offset. Chimneys, like piping systems, often suffer from lack of adequate support.

G2427.5.8 (503.5.9) Cleanouts. Where a chimney that formerly carried flue products from liquid or solid fuel-burning appliances is used with an appliance using fuel gas, an accessible cleanout shall be provided. The cleanout shall have a tight-fitting cover and be installed so its upper edge is at least 6 inches (152 mm) below the lower edge of the lowest chimney inlet opening.

❖ If an existing chimney has no cleanout and was previously used for oil or solid-fuel appliances, a cleanout must be added. The cleanout allows access to inspect and clean the chimney and monitor deposits on the interior walls. The cleanout placement must be such that vent gases will not be in contact with the cleanout door/cover.

G2427.5.9 (503.5.10) Space surrounding lining or vent. The remaining space surrounding a chimney liner, gas vent, special gas vent or plastic piping installed within a masonry chimney flue shall not be used to vent another appliance. The insertion of another liner or vent within the chimney as provided in this code and the liner or vent manufacturer's instructions shall not be prohibited.

The remaining space surrounding a chimney liner, gas vent, special gas vent, or plastic piping installed within a masonry, metal or factory-built chimney, shall not be used to supply combustion air. Such space shall not be prohibited from supplying combustion air to direct-vent appliances designed for installation in a solid fuel-burning fireplace and installed in accordance with the listing and the manufacturer's instructions.

❖ If a vent or a chimney liner is installed within a chimney flue, there will be space between the vent or liner and the chimney walls. This space must not be used to convey vent gases because of its irregular size and geometry and because the vent gases could damage the vent or liner installed in the chimney. A metallic vent or liner could be corroded by surrounding vent gases and plastic pipes could be overheated. Liner systems and vent systems are not listed for use within an atmosphere of vent gases. The exception is not actually an exception to the prohibition of the main paragraph; rather, it explains the obvious fact that installing multiple liners or vents within a chimney is not the same as using the annular space between a vent or liner and the inside chimney walls.

The space between a liner, vent, or pipe and the interior chimney walls cannot be used as a duct for conveying combustion air. Such space has poor air flow characteristics, and cold combustion air moving over a liner or vent can cause condensation to occur in the venting system. An exception is made for direct-vent appliances that are listed for such application in solid fuel-burning fireplaces.

G2427.6 (503.6) Gas vents. Gas vents shall comply with Sections G2427.6.1 through G2427.6.10. (See Section G2403, Definitions.)

❖ See Sections G2427.6.1 through G2427.6.10.

G2427.6.1 (503.6.1) Installation, general. Gas vents shall be installed in accordance with the terms of their listings and the manufacturer's instructions.

❖ See Sections G2426.1 and G2426.5. Gas vents, such as Type B vents, are provided with installation instructions available at the place of purchase or enclosed within the boxed items. The manufacturers also will provide, upon request, very detailed and informative design manuals that discuss sizing and venting fundamentals. If you install, inspect, or design Type B vent systems, these design manuals are invaluable.

G2427.6.2 (503.6.2) Type B-W vent capacity. A Type B-W gas vent shall have a listed capacity not less than that of the listed vented wall furnace to which it is connected.

❖ A Type BW vent is an oval shaped version of a Type B vent designed to fit within a nominal 2 x 4-inch framed wall and serve a wall furnace. See Section G2436.

G2427.6.3 (503.6.3) Roof penetration. A gas vent passing through a roof shall extend through the roof flashing, roof jack or roof thimble and shall be terminated by a listed termination cap.

❖ This section duplicates the typical vent manufacturers' instructions and emphasizes that a vent is a system of components that are all necessary for proper functioning. The cap must be as specified by the vent manufacturer and is a listed component. A vent cap not only keeps out moisture, debris, and animals, it also serves to prevent wind interference that could negatively affect the vent's ability to produce the required draft. Vent caps are tested for flow resistance and performance in wind.

The emphasis here is that the vent system must extend through the roof all the way to the cap, using components that are listed as part of the listed venting system.

G2427.6.4 (503.6.4) Offsets. Type B or Type L vents shall extend in a generally vertical direction with offsets not exceeding 45 degrees, except that a vent system having not more than one 60-degree offset shall be permitted. Any angle greater than 45 degrees from the vertical is considered horizontal. The total horizontal distance of a vent plus the horizontal vent connector serving drafthood-equipped appliances shall not be greater than 75 percent of the vertical height of the vent.

Exception: Systems designed and sized as provided in Section G2428 or in accordance with other approved engineering methods.

Vents serving Category I fan-assisted appliances shall be installed in accordance with the appliance manufacturer's instructions and Section G2428 or other approved engineering methods.

❖ This section addresses vent offsets differently than Section G2428 and is in no way related to Section G2428. This section would be applied only if Section G2428 is not applied; therefore, it is rarely applied, because Section G2428 is necessary for sizing the majority of today's Category I venting systems. The only time that a Type B vent system would not be designed in accordance with Section G2428 is if it were an engineered system or if it were designed in accordance with the exceptions of Section G2427.6.8.1, applicable only to draft hood-equipped appliances.

This section regulates vent offsets, but without the tables and design requirements of Section G2428 there are only very limited design requirements in Section 503 to accompany the offset provisions. In other words, this section would be only one of many parts necessary to design a vent system. The exception to this section will almost always circumvent the main paragraph because in all but the rarest of cases the vent system will either be engineered or designed in accordance with Section G2428. This section is not intended to be applied in conjunction with the provisions of Section G2428.

The last paragraph states that this section is not applicable to fan-assisted appliances.

G2427.6.5 (503.6.6) Gas vent termination. A gas vent shall terminate in accordance with one of the following:

1. Above the roof surface with a listed cap or listed roof assembly. Gas vents 12 inches (305 mm) in size or smaller with listed caps shall be permitted to be terminated in accordance with Figure G2427.6.5, provided that such vents are at least 8 feet (2438 mm) from a vertical wall or similar obstruction. All other gas vents shall terminate not less than 2 feet (610 mm) above the highest point where they pass through the roof and at least 2 feet (610 mm) higher than any portion of a building within 10 feet (3048 mm).
2. As provided for direct-vent systems in Section G2427.2.1.
3. As provided for equipment with integral vents in Section G2427.2.2.
4. As provided for mechanical draft systems in Section G2427.3.3.

❖ This section duplicates typical vent manufacturers' instructions and emphasizes that a vent is a system of components that are all necessary for proper functioning. The cap must be as specified by the vent manufacturer and is a listed component. A vent cap not only keeps out moisture, debris, and animals, it also serves to prevent wind interference that could negatively affect the vent's ability to produce the required draft. Vent caps are tested for flow resistance and performance in wind.

This section is consistent with typical vent manufacturers' instructions and only applies to vents equipped with listed caps as required by Section G2427.6.3. The relationship between vent terminal height and roof pitch is based on mitigating the effects of wind and maintaining a minimum separation between the vent terminal and the roof surface.

See Section G2428.2.9 for a discussion of vent terminations that extend above the roof higher than required by the code.

Figure G2427.6.5 requires greater vent height above the roof as the roof approaches being a vertical surface. The greater the roof pitch, the greater the effect of wind striking the roof surface and the closer the cap becomes (horizontally) to the roof surface. See Figure G2427.6.5.

It is a common misapplication for code users to apply chimney termination height requirements to vents, thereby causing vents to extend above roofs much higher than required in many cases. For example, a 6-inch (152 mm) diameter Type B vent is required to extend above a simple gabled roof only 1 foot (305 mm) for roofs having up to 6/12 pitch, provided that the terminal is not within 8 feet (2438 mm) of a vertical surface such as an upper-story exterior wall. Consideration should be given to vent location in a building in the design/installation phase so that the vent termination will not be difficult or require long extensions of vent pipe exposed to the outdoors. Vent piping exposed to the outdoors encourages condensation in cold weather, could require guy wires or braces depending on height, and is considered aesthetically unattractive. Another good reason to plan for vent location during the design phase of a building is to allow straight vertical runs of vent to avoid offsets.

Good planning can also prevent vents from passing through the roof on the street side of the building. Quite often, vent offsets in attics are installed for the sole purpose of penetrating the roof on the back side of a building.

Vent termination heights are affected by the pitch of the roof and the proximity of walls and other vertical or near-vertical structural components. Where vent terminal locations within 8 feet (2438 mm) of a gable end, upper story, or other vertical surface cannot be avoided, vent terminations must be regulated in a manner similar to chimneys (i.e., 2-feet [610 mm] minimum height and not less than 2 feet [610 mm] above any part of building within 10 feet [3048 mm]). Figure G2427.6.5 does not apply to vent sizes larger than 12 inches (305 mm). This is consistent with manufacturers' instructions and relates to the testing of gas vents to the referenced standard.

It is the intent of Section G2428.2.9 that vents exposed to the outdoors below the roof penetration be sized by a method other than the tables in Section G2428. This means that if a vent is within 8 feet of a vertical wall or is larger in size than 12 inches, it must extend 2 feet above any part of the building within 10 feet horizontally, and this could expose a substantial run of vent to the outdoors. In such cases, the tables of Section G2428 would not apply, and the vent system would have to be engineered.

G2427.6.5.1 (503.6.6.1) Decorative shrouds. Decorative shrouds shall not be installed at the termination of gas vents except where such shrouds are listed for use with the specific gas venting system and are installed in accordance with the manufacturer's installation instructions.

❖ Decorative shrouds have become a popular architectural feature. They are designed to be aesthetically pleasing and to hide chimney and vent terminations. These shrouds have, in some cases, caused fires resulting from overheated combustible construction and the accumulation of debris and animal nesting. Shrouds can also interfere with the functioning of a chimney or vent system. Shrouds are allowed only where they are listed and labeled for use with the specific factory-built chimney system.

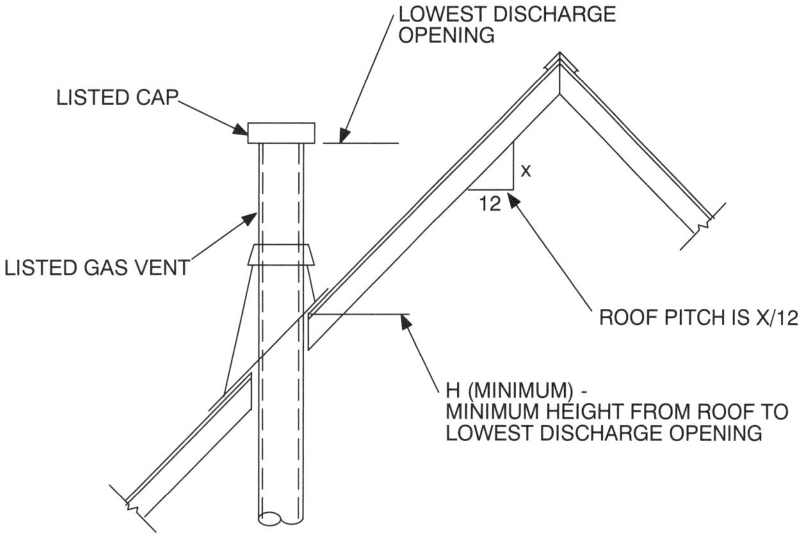

ROOF PITCH	H (minimum) ft	m
Flat to 6/12	1.0	0.30
Over 6/12 to 7/12	1.25	0.38
Over 7/12 to 8/12	1.5	0.46
Over 8/12 to 9/12	2.0	0.61
Over 9/12 to 10/12	2.5	0.76
Over 10/12 to 11/12	3.25	0.99
Over 11/12 to 12/12	4.0	1.22
Over 12/12 to 14/12	5.0	1.52
Over 14/12 to 16/12	6.0	1.83
Over 16/12 to 18/12	7.0	2.13
Over 18/12 to 20/12	7.5	2.27
Over 20/12 to 21/12	8.0	2.44

For SI: 1 foot = 304.8 mm.

FIGURE G2427.6.5 (503.6.6)
GAS VENT TERMINATION LOCATIONS FOR LISTED CAPS 12 INCHES OR LESS IN SIZE AT LEAST 8 FEET FROM A VERTICAL WALL

❖ See the commentary for Section G2427.6.5.

G2427.6.6 (503.6.7) Minimum height. A Type B or a Type L gas vent shall terminate at least 5 feet (1524 mm) in vertical height above the highest connected equipment draft hood or flue collar. A Type B-W gas vent shall terminate at least 12 feet (3658 mm) in vertical height above the bottom of the wall furnace.

❖ The amount of draft produced by a vent is directly related to the height of the vent. A minimum height must be established to produce the minimum draft necessary for the appliance served. This is made evident by looking at the tables in Section G2428 that indicate increasing vent capacity as the vent height increases. This is due to the increase in draft and vent flow velocity. The tables in Section G2428 and some vent manufacturers' instructions indicate a minimum height of 6 feet (1829 mm), as this is the lowest entry for height in the sizing tables.

This section is often violated for appliances installed near the roof such as suspended unit heaters and appliances installed under shed roofs.

G2427.6.7 (503.6.8) Exterior wall penetrations. A gas vent extending through an exterior wall shall not terminate adjacent to the wall or below eaves or parapets, except as provided in Sections G2427.2.1 and G2427.3.3.

❖ The termination locations prohibited by this section would result in poor draft or backdraft because of the effects of wind. Also, termination in such locations could result in a fire hazard, damage to the structure, and entry of vent gases into the building. If a gas vent (typically Type B) penetrates an exterior wall, the vent would have to extend vertically to above the roof in accordance with Section G2427.6.5; however, this exterior extension of vent presents a problem in itself. Gas vents, such as Type B vents, are not designed to be exposed to the outdoors, except for the short section of vent that passes through the roof. Sections G2428.2.9 and G2428.3.16 reinforce this by stating that the vent sizing tables in Section G2428 do not apply to vents exposed to the outdoors below the roof line. Vents exposed to the outdoors will suffer from poor draft and condensation production in most climates because of the heat loss through the exposed vent walls. Vents sized in accordance with Section G2428 cannot extend up the exterior side of a building as implied by this section and Section G2427.6.5. This type of installation could only occur if the vent system was engineered to account for the outdoor exposure. This is also consistent with typical vent manufacturers' design manuals and installation instructions. Even though it is common to see gas vents run up the exterior wall of a building, such installations are not allowed unless they are engineered and allowed by the vent manufacturer, neither of which is likely.

G2427.6.8 (503.6.9) Size of gas vents. Venting systems shall be sized and constructed in accordance with Section G2428 or other approved engineering methods and the gas vent and gas equipment manufacturers' instructions.

❖ Gas vents are to be sized by either Section G2428 or an engineering method acceptable to the code official and, in all cases, in accordance with the vent manufacturer's instructions. See Sections G2427.6.8.1, G2427.6.8.2 and G2427.6.8.3.

G2427.6.8.1 (503.6.9.1) Category I appliances. The sizing of natural draft venting systems serving one or more listed appliances equipped with a draft hood or appliances listed for use with Type B gas vent, installed in a single story of a building, shall be in accordance with one of the following methods.

Exceptions:

1, The provisions of Section G2428.
2. For sizing an individual gas vent for a single, draft hood-equipped appliance, the effective area of the vent connector and the gas vent shall be not less than the area of the appliance draft hood outlet, nor greater than seven times the draft hood outlet area.
3. For sizing a gas vent connected to two appliances with draft hoods, the effective area of the vent shall be not less than the area of the larger draft hood plus 50 percent of the smaller draft hood outlets area, nor greater than seven times the smaller draft hood outlet area.
4. Approved engineering practices.

❖ This section reiterates the intent of the preceding section, is specific to Category I appliances, and applies only to appliances installed within the same story. Multiple-story applications are addressed in the *International Fuel Gas Code.* The last two sentences of this section explain that Category I appliances are vented by natural draft and, depending upon the design and efficiency, are either equipped with a draft hood or a fan-assist3ed combustion system. Because of the higher thermal efficiencies and lack of dilution air, fan-assisted appliances must have their vents designed differently than draft hood-equipped appliances.

Item 3 allows a very limited application of the old "fifth-percent rule." Item 2 applies only to a single draft hood-equipped appliance. Item 3 applies only to a common vent system serving two draft-hood-equipped appliances. Neither item 2 nor 3 can be applied to the venting of fan-assisted appliances, because the old alternate method (fifty-percent rule) was not created with fan-assisted appliances in mind and will not work with such appliances. Besides being limited to draft hood appliances, items 2 and 3 should be further limited to simple venting system configurations having short laterals; few, if any changes in direction (fittings); short connector lengths with maximum vertical rise; and tall vent heights. Vent manufacturers emphatically warn against the use of the alternative sizing rule in item 3 because this method does not account for many factors now know to be important in vent sizing.

G2427.6.8.2 (503.6.9.2) Category II, Category III, and Category IV appliances. The sizing of gas vents for Category II, Category III and Category IV equipment shall be in accordance with the equipment manufacturer's instructions.

❖ Venting systems for such appliances are special vent systems specified by the appliance manufacturer and are not addressed further in the code. See Section G2427.4.

G2427.6.8.3 (503.6.9.3) Mechanical draft. Chimney venting systems using mechanical draft shall be sized in accordance with approved engineering methods.

❖ The manufacturers of exhausters (mechanical draft systems) provide sizing criteria for the vents and chimneys served by such equipment. The code's sizing criteria are based on natural draft chimneys and vents; therefore, engineered sizing methods must be used for other than natural draft. See Section G2427.3.3.

G2427.6.9 (503.6.11) Support of gas vents. Gas vents shall be supported and spaced in accordance with their listings and the manufacturer's instructions.

❖ Type B vents must be supported to prevent the weight of the system from bearing on appliance3s and vent fittings. Fittings such as adjustable elbows are not designed to carry the weight of upper or lower sections of vent pipe, nor are appliance draft hoods and flue collars. Support bases, brackets, and spacers are available from the vent manufacturer and are used to support the weight of the vent and maintain alignment and clearance to combustibles. On most job sites, the vents will be supported by various field-constructed brackets and straps made of sheet metal scraps, duct slips and drives, and plumbers' perforated straps. The installations should be carefully examined to determine that such field-constructed supports are adequate. Sheet metal screws must not penetrate the inner liner of any Type B vent component because they are designed to bed installed without the use of any additional fasteners that penetrate the vent walls.

G2427.6.10 (503.6.12) Marking. In those localities where solid and liquid fuels are used extensively, gas vents shall be permanently identified by a label attached to the wall or ceiling at a point where the vent connector enters the gas vent. The determination of where such localities exist shall be made by the code official. The label shall read:

"This gas vent is for appliances that burn gas. Do not connect to solid or liquid fuel-burning appliances or incinerators."

❖ Type B vents installed in localities where oil and solid fuel-fired appliances are common must be labeled to warn the installers and occupants that Type B vents cannot be used to vent any appliances other than gas-fired. The label is not placed on the vent; rather it is to be placed on a wall or ceiling where it would be conspicuous to someone attempting to connect to the vent. The code official determines if their locality is the intended target for this provision. It is assumed that ignorance of gas vents will be more likely in regions where oil and solid fuel have been commonly used for fuel.

G2427.7 (503.7) Single-wall metal pipe. Single-wall metal pipe vents shall comply with Sections G2427.7.1 through G2427.7.12.

❖ Although discouraged or prohibited by designers, code officials, and appliance and vent manufacturers alike, the code recognizes the use of unlisted single-wall metal pipe as a vent. All other types of vents are listed systems as required by Section G2426.1. As evidenced by Sections G2427.7.1 through G2427.7.12, single-wall metal pipe is restricted to very limited applications, and extraordinary installation precautions are necessary. Single-wall metal pipe was once common but is rarely used today. Appliance manufacturers' instructions will typically prohibit the use of a single-wall metal vent. See Section G2427.7.8.

G2427.7.1 (503.7.1) Construction. Single-wall metal pipe shall be constructed of galvanized sheet steel not less than 0.0304 inch (0.7 mm) thick, or other approved, noncombustible, corrosion-resistant material.

❖ The galvanized sheet steel thickness specified is equivalent to 22 gage, which is heavier than the pipe commonly used for connectors for most residential solid-fuel-fired appliances.

G2427.7.2 (503.7.2) Cold climate. Uninsulated single-wall metal pipe shall not be used outdoors in cold climates for venting gas utilization equipment.

❖ This section limits the application of single-wall vents to regions such as the southwestern and southeastern parts of the United States, unless such pipe is insulated to compensate for the high heat loss of single-wall pipe. The code provides no guidance as to what type of insulation is required or what R-value is required. At the minimum, insulation applied directly to the pipe would have to be a noncombustible material, the R-value would have to exceed that of Type B vent, and the insulation would have to be permanently attached and protected from the weather. Insulating single-wall pipe could be considerably more difficult and expensive than using a venting system designed for outdoor use. Air insulated double-wall vent pipe (Type B) is not designed to be used outdoors in cold climates. When applying Section G2428, single-wall vents are also prohibited from being exposed to the outdoors below the point of roof penetration.

G2427.7.3 (503.7.3) Termination. Single-wall metal pipe shall terminate at least 5 feet (1524 mm) in vertical height above the highest connected equipment draft hood outlet or flue collar. Single-wall metal pipe shall extend at least 2 feet (610 mm) above the highest point where it passes through a roof of a building and at least 2 feet (610 mm) higher than any portion of a building within a horizontal distance of 10 feet (3048 mm). (See Figure G2426.5.) An approved cap or roof assembly shall be attached to the terminus of a single-wall metal pipe. [Also see Section 2427.7.8, Item 3.]

❖ Single-wall metal vents are not allowed to terminate as allowed by Section G2427.6.5 and must comply with the termination requirements for chimneys, except that the minimum height above the roof penetration is 2 feet, whereas it is 3 feet for chimneys. See Figure G2427.5.3. The minimum height and vent cap requirements are consistent with Sections G2427.6.5 and G2427.6.6.

G2427.7.4 (503.7.4) Limitations of use. Single-wall metal pipe shall be used only for runs directly from the space in which the equipment is located through the roof or exterior wall to the outdoor atmosphere.

❖ Single-wall metal pipe is further restricted by this section to only two applications. It can run from the appliance directly to the outdoors through an exterior wall, or it can run from the appliance directly to the outdoors through a roof. Single-wall metal pipe cannot pass through any floors, or interior walls, or partitions and cannot pass through attics or concealed spaces. This requires the appliance served to be in a space with an exterior wall or in the same story as the roof.

G2427.7.5 (503.7.5) Roof penetrations. A pipe passing through a roof shall extend without interruption through the roof flashing, roof jack or roof thimble. Where a single-wall metal pipe passes through a roof constructed of combustible material, a noncombustible, nonventilating thimble shall be used at the point of passage. The thimble shall extend at least 18 inches (457 mm) above and 6 inches (152 mm) below the roof with the annular space open at the bottom and closed only at the top. The thimble shall be sized in accordance with Section G2427.10.15.

❖ A thimble is a sheet metal assembly designed to provide clearance between a vent and combustible materials, analogous to a thimble that protects one's fingers from a sewing needle. It is essentially a spacer that provides an annular space around the vent. Such annular space is usually ventilated by punched holes in the faces of the thimble assembly. Some thimbles are constructed with insulation in addition to an annular space.

A roof jack is an assembly that passes through the roof and provides a thimble and weather protection all in one. Roof jacks are flashing/thimble combinations and typically mount to a roof curb.

G2427.7.6 (503.7.6) Installation. Single-wall metal pipe shall not originate in any unoccupied attic or concealed space and shall not pass through any attic, inside wall, concealed space or floor. The installation of a single-wall metal pipe through an exterior combustible wall shall comply with Section G2427.10.15.

❖ A single-wall metal vent must be exposed for its entire length, except for the section of pipe that is within a thimble or roof jack assembly. Single-wall vents can only penetrate exterior walls and roofs. A roof/ceiling assembly cannot be penetrated if there is an attic or a concealed space between the ceiling and the roof. See Section G2427.7.4.

G2427.7.7 (503.7.7) Clearances. Minimum clearances from single-wall metal pipe to combustible material shall be in accordance with Table G2427.7.7. The clearance from single-wall metal pipe to combustible material shall be permitted to be reduced where the combustible material is protected as specified for vent connectors in Table G2409.2.

❖ Consideration should be given to the fact that single-wall vents will have high-surface temperatures compared to insulated vents such as Type B vents. Such surface temperatures could be an ignition or burn injury hazard. See the commentary for Table 2426.7.7.

G2427.7.8 (503.7.8) Size of single-wall metal pipe. A venting system constructed of single-wall metal pipe shall be sized in accordance with one of the following methods and the equipment manufacturer's instructions:

1. For a draft hood-equipped appliance, in accordance with Section G2428.
2. For a venting system for a single appliance with a draft hood, the areas of the connector and the pipe each shall not be less than the area of the appliance flue collar or draft hood outlet, whichever is smaller. The vent area shall not be greater than seven times the draft hood outlet area.
3. Other approved engineering methods.

❖ As can be seen from items 1, 2, and 3 and Tables 504.2(5) and 504.3(5) of the *International Fuel Gas Code*, single-wall metal pipe vents are limited to use with draft hood-equipped appliances, except where the system is engineered.

G2427.7.9 (503.7.9) Pipe geometry. Any shaped single-wall metal pipe shall be permitted to be used, provided its equivalent effective area is equal to the effective area of the round pipe for which it is substituted and provided the minimum internal dimension of the pipe is not less than 2 inches (51 mm).

❖ Single-wall metal pipe does not have to be round; it could be oval, square, rectangular, etc. Round pipe has the best flow characteristics as evidenced by the fact that a round chimney liner will have a greater capacity than a rectangular liner of equal area.

G2427.7.10 (503.7.10) Termination capacity. The vent cap or a roof assembly shall have a venting capacity not less than that of the pipe to which it is attached.

❖ Section G2427.7.3 requires an approved vent cap or roof termination assembly, and this section requires such terminus to be sized and designed so as not to reduce the capacity of the vent. An unlisted untested cap or roof assembly will have an unknown capacity unless it is engineered.

G2427.7.11 (503.7.11) Support of single-wall metal pipe. All portions of single-wall metal pipe shall be supported for the design and weight of the material employed.

❖ Single-wall vents, like all vents and chimneys, must be supported to prevent structural failure and joint separation and to maintain the required clearances.

G2427.7.12 (503.7.12) Marking. Single-wall metal pipe shall comply with the marking provisions of Section G2427.6.10.

❖ See the commentary for Section G2427.6.10.

TABLE G2427.7.7 (503.7.7)
CLEARANCES FOR CONNECTORS

MINIMUM DISTANCE FROM COMBUSTIBLE MATERIAL				
Equipment	**Listed Type B gas vent material**	**Listed Type L vent material**	**Single-wall metal pipe**	**Factory-built chimney sections**
Listed equipment with draft hoods and equipment listed for use with Type B Gas Vents	as listed	as listed	6 inches	as listed
Residential boilers and furnaces with listed gas conversion burner and with draft hood	6 inches	6 inches	9 inches	as listed
Residential appliances listed for use with Type L vents	not permitted	as listed	9 inches	as listed
Unlisted residential appliances with draft hood	not permitted	6 inches	9 inches	as listed
Residential and low-heat equipment other than those above	not permitted	9 inches	18 inches	as listed

For SI: 1 inch = 25.4 mm.
NOTE: These clearances shall apply unless the listing of an appliance or connector specifies different clearances, in which case the listed clearances shall apply.

❖ This is a vent connector clearance table; the third column applies to single-wall vents as well as single-wall vent connectors.

G2427.8 (503.8) Venting system termination location. The location of venting system terminations shall comply with following (see Appendix C):

1. A mechanical draft venting system shall terminate at least 3 feet (914 mm) above any forced air inlet located within 10 feet (3048 mm).

   Exceptions:

   1. This provision shall not apply to the combustion air intake of a direct vent appliance.
   2. This provision shall not apply to the separation of the integral outdoor air inlet and flue gas discharge of listed outdoor appliances.

2. A mechanical draft venting system, excluding direct-vent appliances, shall terminate at least 4 feet (1219 mm) below, 4 feet (1219 mm) horizontally from, or 1 foot (305 mm) above any door, operable window, or gravity air inlet into any building. The bottom of the vent terminal shall be located at least 12 inches (305 mm) above grade.
3. The vent terminal of a direct-vent appliance with an input of 10,000 Btu per hour (3 kW) or less shall be located at least 6 inches (152 mm) from any air opening into a building, and such an appliance with an input over 10,000 Btu per hour (3 kW) but not over 50,000 Btu per hour (14.7 kW) shall be installed with a 9 inch (230 mm) vent termination clearance, and an appliance with an input over 50,000 Btu/hr shall have at least a 12-inch (305 mm) vent termination clearance. The bottom of the vent terminal and the air intake shall be located at least 12 inches (305 mm) above grade.
4. Through-the-wall vents for Category II and Category IV appliances and noncategorized condensing appliances shall not terminate over public walkways or over an area where condensate or vapor could create a nuisance or hazard or could be detrimental to the operation of regulators, relief valves, or other equipment. Where local experience indicates that condensate is a problem with Category I and Category III appliances, this provision shall also apply.

❖ This section addresses the terminations of mechanical drafts systems and direct-vent appliances. To prevent vent gases from entering the building, item 1 requires forced air (mechanical) intakes to be at least 10 feet away from a mechanical draft termination, or such termination must be at least 3 feet above the air intake. The natural buoyancy of vent gases is the justification for allowing a closer distance where the terminal is above the air intake. A combustion air intake for a direct-vent appliance is not what item 1 intends to address. Direct-vent appliance exhaust and intake opening locations are dictated by the appliance listing and manufacturer's instructions. Commentary Figures G2427.8(1) and G2427.8(2) are extracted from Z21.47/CSA 2.3–2001, the standard for gas-fired central furnaces, and dictate the content of furnace installation instructions. Note the distinction between direct-vent and other than direct-vent in the figures. Item 1 also does not apply to appliances such as rooftop units that have built-in outdoor air intakes and combustion gas exhaust outlets. The code assumes that the design and listing of the outdoor appliance accounts for the location of such openings.

Item 2 addresses gravity air inlets in the exterior envelope of a building, including doors, operable (openable) windows, and intake louvers and grilles. The appliances addressed by this item employ auxiliary or integral fans and blowers to force the flow of combustion products to the outdoors. This item applies to externally installed power exhausters, integrally power-exhausted appliances, and venting systems equipped with draft inducers. This item does not address fan-assisted Category I appliances and does not address direct-vent appliances.

Power exhausters are field-installed pieces of equipment that are independent of, but used in conjunction with, other appliances. See Commentary Figure G2427.8(3) relative to power exhausters.

This item is consistent with appliance installation instructions that will provide additional details for the location of the venting system terminals, including but not limited to clearance to grade, inside corners, outside corners, decks, porches, balconies, soffits, roof eaves, utility meters, combustion air intakes, other appliance vents, service regulators, sidewalks, and driveways. See Commentary Figures G2427.8(1) and G2427.8(2).

There are significant differences between the requirements of items 2 and 3 as they relate to non-direct-vent and direct-vent appliances, respectively. A Category IV, 90-percent efficient furnace or boiler would be subject to item 2 if installed without a combustion air intake pipe to the outdoors but would be subject to item 3, not item 2, if installed with a combustion air intake pipe to the outdoors. This is because such a furnace can be listed as both a direct-vent and nondirect-vent appliance, and as such can be installed either way. Less restrictive clearances, improved operation efficiency, longer appliance life, immunity from the effects of building exhaust systems, and ease of providing combustion air are some of the advantages of direct-vent appliances. For example, according to item 2, a sidewall vent terminal must be 4 feet or more to either side of an openable window, or must be 4 feet or more below the bottom of the window or must be 1 foot or more above the top of the window.

Item 3 is consistent with appliance manufacturers' instructions. Research has shown that the distances associated with the input ranges are necessary to allow the combustion gases to dissipate in the atmosphere, thus avoiding reentry into the building through openings. This item is applicable only to direct-vent appliances. For example, a Category IV, 90-percent efficient furnace installed without an outdoor air intake pipe for combustion air would be subject to item 2, not 3. See Commentary Figures G2427.8(4), G2427.8(5) and G2427.8(6).

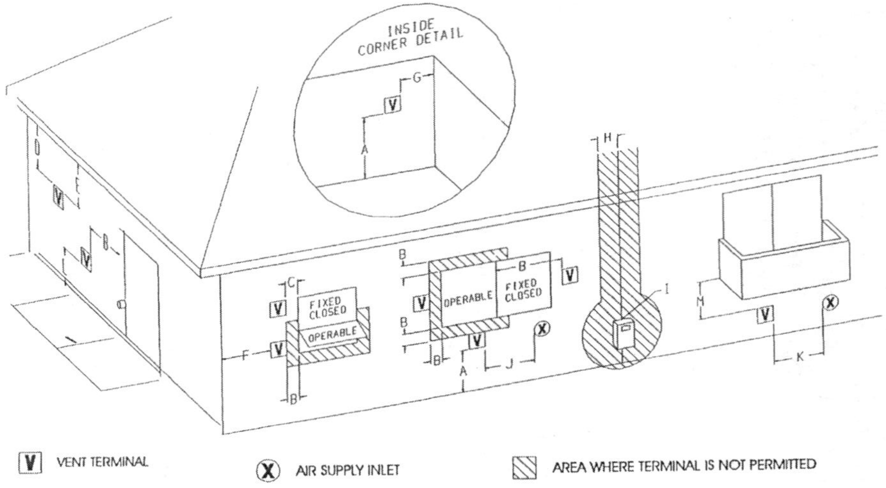

		Canadian Installations[1]	US Installations[2]
A=	Clearance above grade, veranda, porch, deck, or balcony	12 inches (30 cm)	12 inches (30 cm)
B=	Clearance to window or door that may be opened	6 inches (15 cm) for appliances $\le$ 10,000 Btuh (3 kW), 12 inches (30 cm) for appliances > 10,000 Btuh (3 kW) and $\le$ 100,000 Btuh (30 kW), 36 inches (91 cm) for appliances > 100,000 Btuh (30 kW)	6 inches (15 cm) for appliances $\le$ 10,000 Btuh (3 kW), 9 inches (23 cm) for appliances > 10,000 Btuh (3 kW) and $\le$ 50,000 Btuh (15 kW), 12 inches (30 cm) for appliances > 50,000 Btuh (15 kW)
C=	Clearance to permanently closed window	*	*
D=	Vertical clearance to ventilated soffit located above the terminal within a horizontal distance of 2 feet (61 cm) from the center line of the terminal	*	*
E=	Clearance to unventilated soffit	*	*
F=	Clearance to outside corner	*	*
G=	Clearance to inside corner	*	*
H=	Clearance to each side of center line extended above meter/regulator assembly	3 feet (91 cm) within a height 15 feet (4.6 m) above the meter/regulator assembly	*
I=	Clearance to service regulator vent outlet	3 feet (91 cm)	*
J=	Clearance to nonmechanical air supply inlet to building or the combustion air inlet to any other appliance	6 inches (15 cm) for appliances $\le$ 10,000 Btuh (3 kW), 12 inches (30 cm) for appliances > 10,000 Btuh (3 kW) and $\le$ 100,000 Btuh (30 kW), 36 inches (91 cm) for appliances > 100,000 Btuh (30 kW)	6 inches (15 cm) for appliances $\le$ 10,000 Btuh (3 kW), 9 inches (23 cm) for appliances > 10,000 Btuh (3 kW) and $\le$ 50,000 Btuh (15 kW), 12 inches (30 cm) for appliances > 50,000 Btuh (15 kW)
K=	Clearance to a mechanical air supply inlet	6 feet (1.83 m)	3 feet (91 cm) above if within 10 feet (3 m) horizontally
L=	Clearance above paved sidewalk or paved driveway located on public property	7 feet (2.13 m) †	*
M=	Clearance under veranda, porch deck, or balcony	12 inches (30 cm) ‡	*

[1] In accordance with the current CSA B149.1 *Natural Gas and Propane Installation Code*

[2] In accordance with the current ANSI Z223.1 / NFPA 54 *National Fuel Gas Code*

† A vent shall not terminate directly above a sidewalk or paved driveway that is located between two single family dwellings and serves both dwellings.

‡ Permitted only if veranda, porch, deck, or balcony is fully open on a minimum of two sides beneath the floor.

* For clearances not specified in ANSI Z223.1/NFPA 4 or CSA B149.1, the following statement shall be included:

"Clearance in accordance with local installation codes, the requirements of the gas supplier and the manufacturer's installation instructions."

Figure G2427.8(1)
DIRECT-VENT TERMINAL CLEARANCES

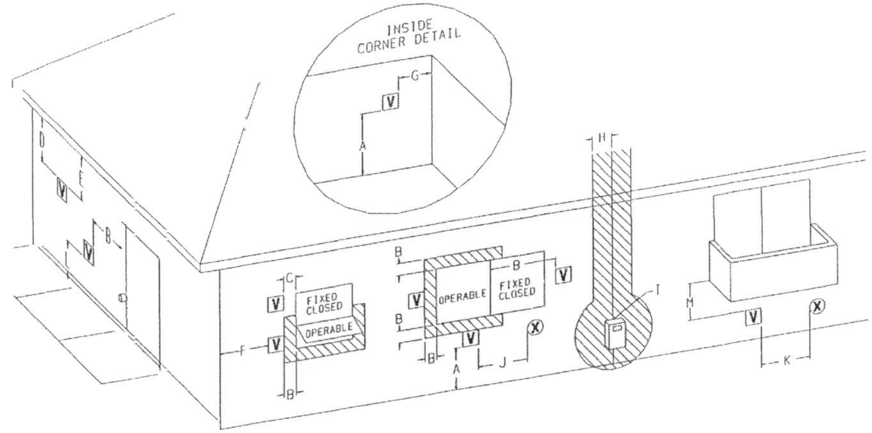

V VENT TERMINAL

X AIR SUPPLY INLET

AREA WHERE TERMINAL IS NOT PERMITTED

		Canadian Installations[1]	US Installations[2]
A=	Clearance above grade, veranda, porch, deck, or balcony	12 inches (30 cm)	12 inches (30 cm)
B=	Clearance to window or door that may be opened	6 inches (15 cm) for appliances ≤ 10,000 Btuh (3 kW), 12 inches (30 cm) for appliances > 10,000 Btuh (3 kW) and ≤ 100,000 Btuh (30 kW), 36 inches (91 cm) for appliances > 100,000 Btuh (30 kW)	4 feet (1.2 m) below or to side of opening; 1 foot (300 mm) above opening
C=	Clearance to permanently closed window	*	*
D=	Vertical clearance to ventilated soffit located above the terminal within a horizontal distance of 2 feet (61 cm) from the center line of the terminal	*	*
E=	Clearance to unventilated soffit	*	*
F=	Clearance to outside corner	*	*
G=	Clearance to inside corner	*	*
H=	Clearance to each side of center line extended above meter/regulator assembly	3 feet (91 cm) within a height 15 feet (4.6 m) above the meter/regulator assembly	*
I=	Clearance to service regulator vent outlet	3 feet (91 cm)	*

		Canadian Installations[1]	US Installations[2]
J=	Clearance to nonmechanical air supply inlet to building or the combustion air inlet to any other appliance	6 inches (15 cm) for appliances ≤ 10,000 Btuh (3 kW), 12 inches (30 cm) for appliances > 10,000 Btuh (3 kW) and ≤ 100,000 Btuh (30 kW), 36 inches (91 cm) for appliances > 100,000 Btuh (30 kW)	4 feet (1.2 m) below or to side of opening; 1 foot (300 mm) above opening
K=	Clearance to a mechanical air supply inlet	6 feet (1.83 m)	3 feet (91 cm) above if within 10 feet (3 m) horizontally
L=	Clearance above paved sidewalk or paved driveway located on public property	7 feet (2.13 m) †	7 feet (2.13 m)
M=	Clearance under veranda, porch deck, or balcony	12 inches (30 cm) ‡	*

[1] In accordance with the current CSA B149.1 *Natural Gas and Propane Installation Code*

[2] In accordance with the current ANSI Z223.1 / NFPA 54 *National Fuel Gas Code*

† A vent shall not terminate directly above a sidewalk or paved driveway that is located between two single family dwellings and serves both dwellings.

‡ Permitted only if veranda, porch, deck, or balcony is fully open on a minimum of two sides beneath the floor.

* For clearances not specified in ANSI Z223.1/NFPA 4 or CSA B149.1, the following statement shall be included:

"Clearance in accordance with local installation codes, the requirements of the gas supplier and the manufacturer's installation instructions."

Figure G2427.8(2)
OTHER THAN DIRECT-VENT TERMINAL CLEARANCES

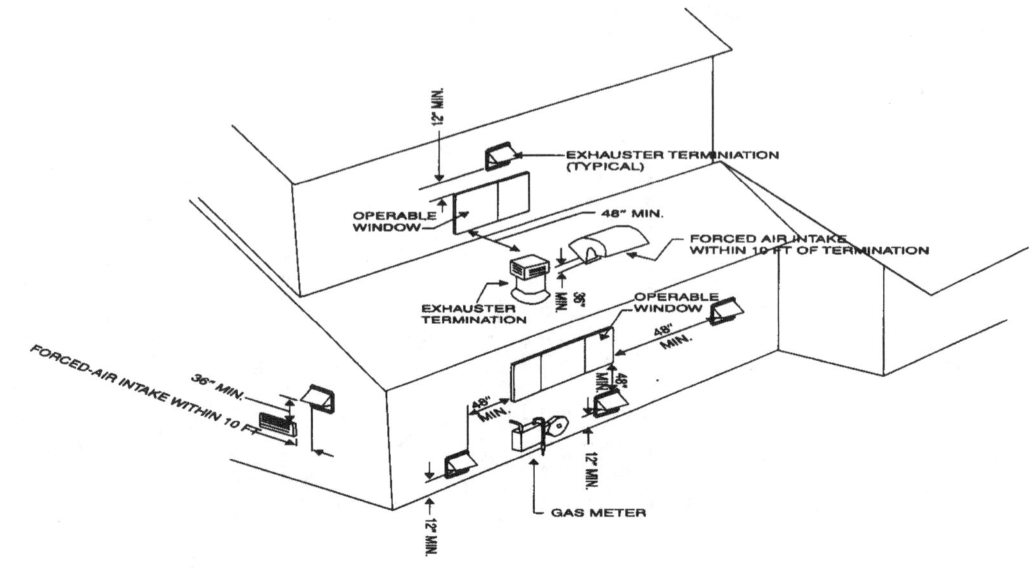

Figure G2427.8(3)
MECHANICAL DRAFT VENTING SYSTEM EXHAUSTER TERMINATIONS

Item 4 addresses the potential problem of condensation relative to its possible effect on pedestrians, service equipment, etc. Condensate is corrosive and could create a hazard when it freezes. Condensate ice could be a slip hazard and could obstruct regulator and relief valve vents. Category I and III appliances do not produce as much condensate as the other two categories, but they can also cause condensate problems in cold climates. It is common to see large masses of ice build up on or near the vent terminals of Category I and III vents. Such ice masses can fall or slide from roofs and injure someone or cause damage to a building, vehicle, or equipment.

G2427.9 (503.9) Condensation drain. Provision shall be made to collect and dispose of condensate from venting systems serving Category II and Category IV equipment and noncategorized condensing appliances in accordance with G2427.8, Item 4. Where local experience indicates that condensation is a problem, provision shall be made to drain off and dispose of condensate from venting systems serving Category I and Category III equipment in accordance with G2427.8, Item 4.

❖ See Section G2427.8, Item 4.

G2427.10 (503.10) Vent connectors for Category I gas utilization equipment. Vent connectors for Category I equipment shall comply with Sections G2427.10.1 through G2427.10.15.

❖ See Sections G2427.10.1 through G2427.10.15.

G2427.10.1 (503.10.1) Where required. A vent connector shall be used to connect equipment to a gas vent, chimney or single-wall metal pipe, except where the gas vent, chimney or single-wall metal pipe is directly connected to the equipment.

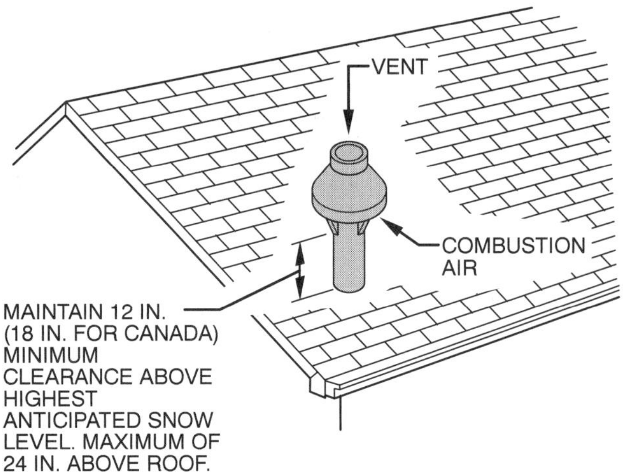

For SI: 1 inch = 25.4 mm.

Figure courtesy of Carrier Corporation

Figure G2427.8(4)
EXAMPLE OF MANUFACTURER'S DETAIL FOR DIRECT-VENT APPLIANCE TERMINAL

❖ Unless the chimney or vent is connected directly to the appliance, a "Connector," defined in Section 2403 as a pipe used to connect an approved fuel-burning appliance to a chimney or vent, is necessary. This includes the necessary fittings to make a connection or change in direction. This is usually accomplished with a single-wall metal pipe, but it is also common practice to use listed and labeled chimney and vent pipe or listed factory-built single-wall or double-wall bendable connectors.

There are many factors that affect the design and configuration of a connector. The most important of

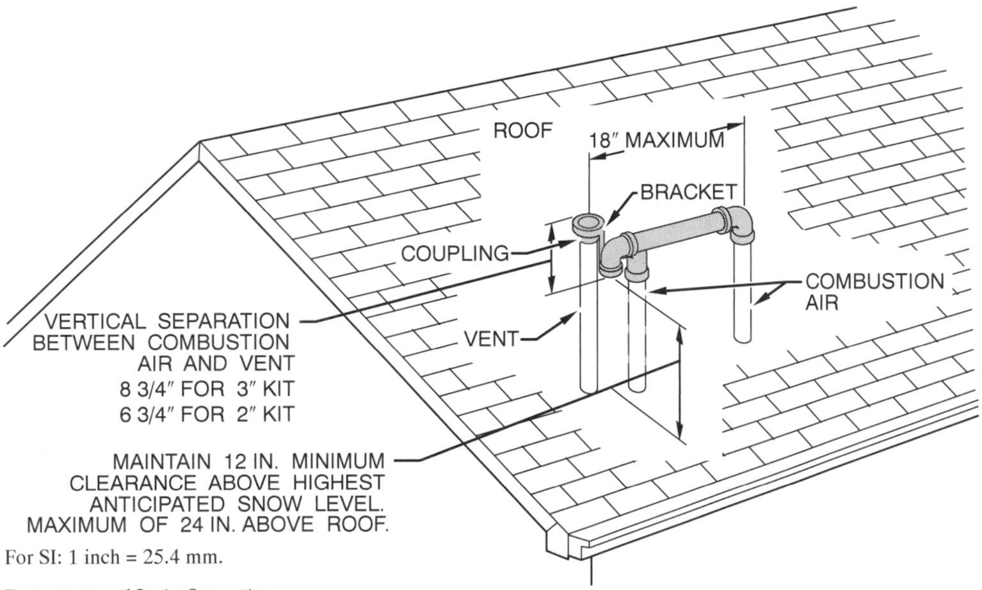

For SI: 1 inch = 25.4 mm.

Figure courtesy of Carrier Corporation

Figure G2427.8(5)
EXAMPLE OF MANUFACTURER'S DETAIL FOR DIRECT-VENT APPLIANCE TERMINALS

these is the appliance location with respect to the chimney or vent system. This impacts the connector size, length, and rise. Another important factor is the number of appliances being vented.

The appliance manufacturer's installation instructions may prohibit the use of single-wall connectors. For example, a single-wall vent connector may not be appropriate for fan-assisted appliances, because of the possibility of condensation, corrosion, and leakage. Single-wall vent connectors are also not permitted in attic or crawl space installations that are subject to cold temperatures. The cold temperatures increase the possibility of condensation, which leads to accelerated material failure. If the vent or chimney extends all the way to a single appliance vent outlet, no connector is needed.

G2427.10.2 (503.10.2) Materials. Vent connectors shall be constructed in accordance with Sections G2427.10.2.1 through G2427.10.2.4.

❖ See Sections G2427.10.2.1 through G2427.10.2.4.

G2427.10.2.1 (503.10.2.1) General. A vent connector shall be made of noncombustible corrosion-resistant material capable of withstanding the vent gas temperature produced by the equipment and of sufficient thickness to withstand physical damage.

❖ See Sections G2427.10.2.2 through G2427.10.2.4.

G2427.10.2.2 (503.10.2.2) Vent connectors located in unconditioned areas. Where the vent connector used for equipment having a draft hood or a Category I appliance is located in or passes through attics, crawl spaces or other unconditioned spaces, that portion of the vent connector shall be listed Type B or Type L or listed vent material or listed material having equivalent insulation properties.

Exception: Single-wall metal pipe located within the exterior walls of the building in areas having a local 99-percent winter design temperature of 5°F (-15°C) or higher shall be permitted to be used in unconditioned spaces other than attics and crawl spaces.

❖ Compared to double-wall pipe, single-wall pipe has a much higher heat loss and when installed in an unconditioned space, condensation can occur, deteriorating the pipe. A typical scenario is an attic, garage, or crawl space appliance installation where a connector would be exposed to cold temperatures, practically ensuring the formation of condensation in the connector. Except as allowed in the exception, this section effectively prohibits the use of single-wall connectors in garages, attics, and crawl spaces where such spaces are unheated. The term "unconditioned" is not defined and the debate over the intent has created new text in the exception, which is climate related. Attics and crawl spaces are unconditioned spaces without doubt; however, the uncertainty has been with residential basements and garages, which are common locations for gas-fired appliances. In many climates, the garage winter temperatures will be cold enough to cause the combustion gases in a connector to reach their dew point, meaning that condensate will form and linger in the connector, causing its rapid destruction. Most residential garages and some basements are considered to be unconditioned insofar as they are not directly or indirectly heated like the living space.

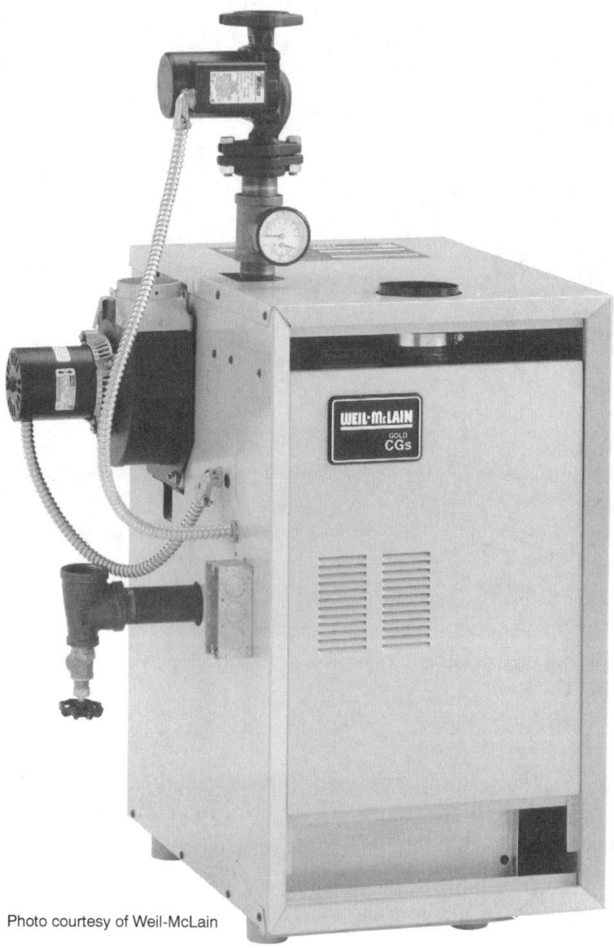

Photo courtesy of Weil-McLain

Figure G2427.8(6)
DIRECT-VENT WATER BOILER

The typical residential garage is not heated (i.e., conditioned) and in most climates, will certainly have low ambient temperatures. The exception allows single-wall metal pipe in unconditioned spaces (other than attics and crawl spaces) if the climate is as described. The winter design temperatures can be found in the ASHRAE *Handbook of Fundamentals*. To put the climate criterion in perspective, Philadelphia, PA., Louisville, KY., Fall River, MA., and Providence, RI. all have 99 percent winter design temperatures of 5°F or higher. The exception is applicable only to vent connectors that are inside a building (i.e. located within the exterior walls). See Commentary Figure G2427.10.2.2.

G2427.10.2.3 (503.10.2.3) Residential-type appliance connectors. Where vent connectors for residential-type appliances are not installed in attics or other unconditioned spaces, connectors for listed appliances having draft hoods and for appliances having draft hoods and equipped with listed conversion burners shall be one of the following:

1. Type B or Type L vent material; or

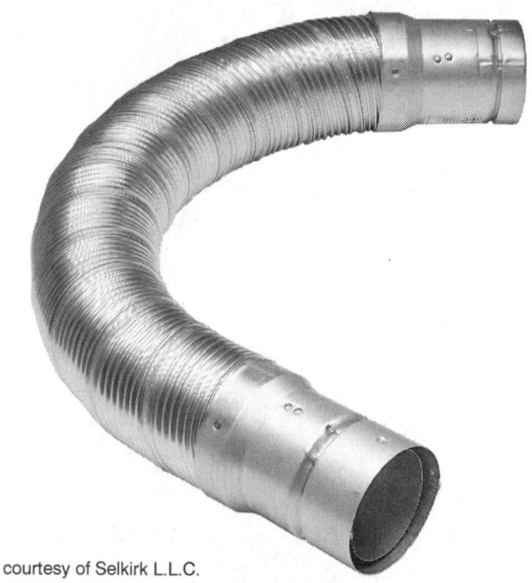

Photo courtesy of Selkirk L.L.C.

Figure G2427.10.2.2
LISTED BENDABLE DOUBLE-WALL CONNECTOR

2. Galvanized sheet steel not less than 0.018 inch (0.46 mm) thick; or
3. Aluminum (1100 or 3003 alloy or equivalent) sheet not less than 0.027 inch (0.69 mm) thick; or
4. Stainless steel sheet not less than 0.012 inch (0.31 mm) thick; or
5. Smooth interior wall metal pipe having resistance to heat and corrosion equal to or greater than that of Items 2, 3 or 4 above; or
6. A listed vent connector.

❖ This section applies to draft-hood-equipped appliances only and lists the allowable materials, including sheet metals, listed connectors, and listed vent system pipe. The minimum thickness for single-wall galvanized steel pipe is equivalent to 26 gage. Installers have field-installed insulation on vent connectors in an attempt to control heat loss and the resultant condensation; however, this is prohibited because of the combustibility of most insulation materials and the possibility of excessive connector temperatures. Listed double-wall (air insulated) Type B connectors are available as an option to using Type B vent pipe.

G2427.10.2.4 (503.10.2.4) Low-heat equipment. A vent connector for low-heat equipment shall be a factory-built chimney section or steel pipe having resistance to heat and corrosion equivalent to that for the appropriate galvanized pipe as specified in Table G2427.10.2.4. Factory-built chimney sections shall be joined together in accordance with the chimney manufacturers' instructions.

❖ Residential appliances are also low-heat by definition, and the provisions of Section G2427.10.2.3 apply as well.

**TABLE G2427.10.2.4 (503.10.2.4)
MINIMUM THICKNESS FOR GALVANIZED STEEL VENT CONNECTORS FOR LOW-HEAT APPLIANCES**

DIAMETER OF CONNECTOR (inches)	MINIMUM THICKNESS (inch)
less than 6	0.019
6 to less than 10	0.023
10 to 12 inclusive	0.029
14 to 16 inclusive	0.034
over 16	0.056

For SI: 1 inch = 25.4 mm, 1 square inch = 645.16 mm^2.

❖ See the commentary for Sections G2427.10.3.1 through G2427.10.3.5.

G2427.10.3 (503.10.3) Size of vent connector. Vent connectors shall be sized in accordance with Sections G2427.10.3.1 through G2427.10.3.5.

❖ See Sections G2427.10.3.1 through G2427.10.3.5.

G2427.10.3.1 (503.10.3.1) Single draft hood and fan-assisted. A vent connector for equipment with a single draft hood or for a Category I fan-assisted combustion system appliance shall be sized and installed in accordance with Section G2428 and other approved engineering methods.

❖ The sizing tables in Section G2428 will determine connector size with the fundamental requirement that a connector not be smaller than the appliance vent connection, except as allowed by Sections G2428.2.2 and G2428.3.21. The tables will often require a connector to be larger than the appliance vent outlet. See Section G2428.

G2427.10.3.2 (503.10.3.2) Multiple draft hood. For a single appliance having more than one draft hood outlet or flue collar, the manifold shall be constructed according to the instructions of the appliance manufacturer. If there are no instructions, the manifold shall be designed and constructed in accordance with approved engineering practices. As an alternate method, the effective area of the manifold shall equal the combined area of the flue collars or draft hood outlets and the vent connectors shall have a minimum 1 foot (305 mm) rise.

❖ Some appliances are designed with more than one venting (flue) outlet. For example, some large furnaces were built with dual heat exchangers and burner sections and thus have two draft hoods or flue collar outlets. The manufacturer's instructions must be followed. However, in the event that the manufacturer's instructions are not provided, this section specifies required sizing criteria. Section G2408.1 requires the manufacturer's instructions to be on the job site.

G2427.10.3.3 (503.10.3.3) Multiple appliances. Where two or more appliances are connected to a common vent or chimney, each vent connector shall be sized in accordance with Section G2428 or other approved engineering methods.

As an alternative method applicable only where all of the appliances are draft hood-equipped, each vent connector shall have an effective area not less than the area of the draft hood outlet of the appliance to which it is connected.

❖ The sizing tables in Section G2428 will determine connector size with the fundamental requirement that a connector not be smaller than the appliance vent connection, except as allowed by Sections G2428.2.2 and G2428.3.17. The tables will often require a connector to be larger than the appliance vent outlet. See Section G2428. The alternative method paragraph of this section is incompatible with Section G2428 and is intended only for use with the alternative sizing method of Section G2427.6.8.1.

G2427.10.3.4 (503.10.3.4) Common connector/manifold. Where two or more gas appliances are vented through a common vent connector or vent manifold, the common vent connector or vent manifold shall be located at the highest level consistent with available headroom and the required clearance to combustible materials and shall be sized in accordance with Section G2427 or other approved engineering methods.

As an alternate method applicable only where there are two draft-hood equipped appliances, the effective area of the common vent connector or vent manifold and all junction fittings shall be not less than the area of the larger vent connector plus 50 percent of the area of the smaller flue collar outlet.

❖ Vent manufacturers' instructions provide detailed design criteria for manifolds. Connectors serving gas-fired appliances are allowed to join together to form a common connector manifold rather than connecting independently to the vertical common vent. See Commentary Figures G2427.10.3.4(1) and G2427.10.3.4(2).

Various manufacturers' instructions require a 10-percent to 20-percent reduction in the capacity of a common vent where a manifold is used. See the commentary for Section G2428.3.4. A manifold is defined as a vent or connector that is a lateral (horizontal) extension of the lower end of a common vent. Because a manifold serves multiple appliances and thus conveys vent gases from two or more appliances, manifolds must be sized by the common vent tables.

This section requires that manifolds be installed as high as possible, respective of ceiling height and clearance to combustibles requirements. This will provide the most appliance connector rise, which is always beneficial for venting performance. Connector rise takes advantage of the energy of hot gases discharging directly from the appliance and develops flow velocity and draft in addition to that produced by the common vent. See Commentary Figures G2427.10.3.4(1) and G2427.10.3.4(2). The alternative method in the second paragraph is intended for use only with the alternative sizing method of Section G2427.6.8.1 and is not compatible with Section G2428.

G2427.10.3.5 (503.10.3.5) Size increase. Where the size of a vent connector is increased to overcome installation limitations

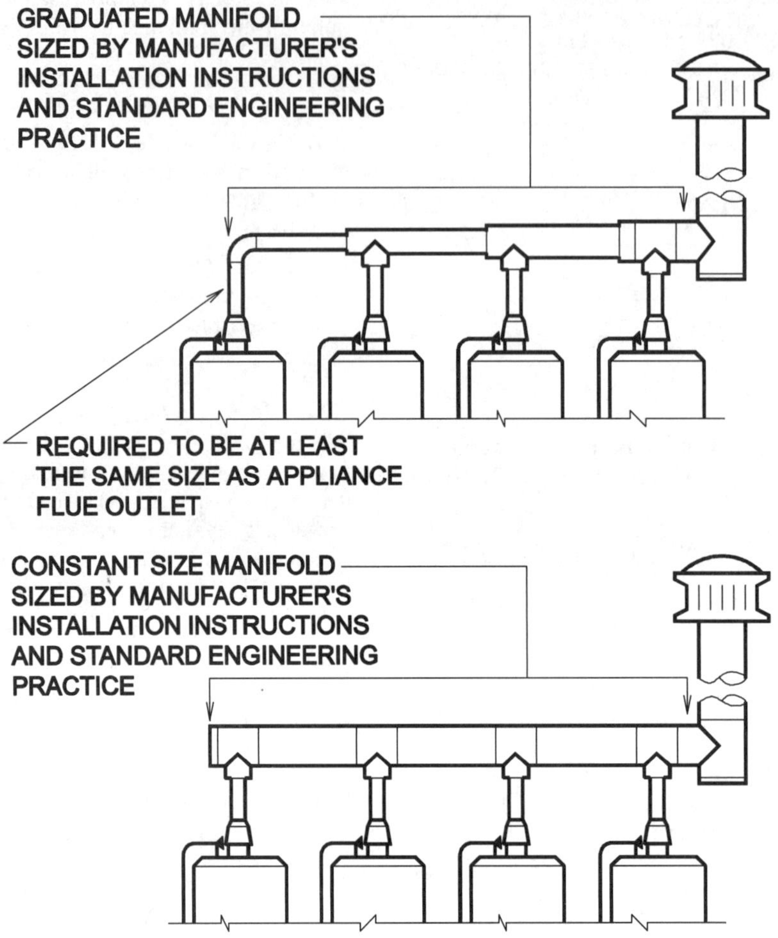

Figure courtesy of Selkirk, LLC.

Figure G2427.10.3.4(1)
MANIFOLD SIZING

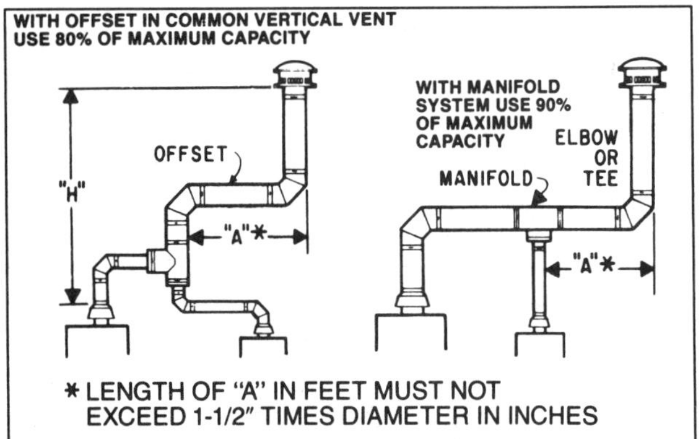

Figure courtesy of Selkirk, LLC.

Figure G2427.10.3.4(2)
COMMON VENT OFFSET AND MANIFOLD

and obtain connector capacity equal to the equipment input, the size increase shall be made at the equipment draft hood outlet.

❖ If a connector is required to be larger than the appliance vent outlet, the size increase must occur at the appliance connection by use of an appropriate increaser fitting. For example, it is common for a water heater with a 3-inch draft hood outlet to be connected to a 4-inch connector by a 3 x 4-inch increaser fitting, as dictated by the vent tables in Section G2428.

G2427.10.4 (503.10.4) Two or more appliances connected to a single vent. Where two or more vent connectors enter a common gas vent, chimney flue or single-wall metal pipe, the smaller connector shall enter at the highest level consistent with the available headroom or clearance to combustible material. Vent connectors serving Category I appliances shall not be connected to any portion of a mechanical draft system operating under positive static pressure, such as those serving Category III or Category IV appliances.

❖ A common vent is somewhat oversized when only a single appliance is operating; therefore, every effort is made to achieve proper vent/chimney operation during all possible operating circumstances. Placing the smallest appliance connection at the highest elevation in the common vent/chimney will allow the greatest connector rise for the smaller appliance connector and will take advantage of any draft priming effect caused by the lower connector. For example, it has been common practice to connect a domestic water heater connector above the furnace or boiler connector in a chimney or vent system.

Category I appliances require a vent that produces a draft and therefore cannot be connected to any venting system that produces a positive pressure. For example, a Category I appliance and a Category III appliance cannot share the same vent unless they both connect to the negative pressure (inlet) side of a mechanical draft system in accordance with the appliance and draft system installation instructions. See Section G2425.10.

G2427.10.5 (503.10.5) Clearance. Minimum clearances from vent connectors to combustible material shall be in accordance with Table G2427.7.7.

Exception: The clearance between a vent connector and combustible material shall be permitted to be reduced where the combustible material is protected as specified for vent connectors in Table G2409.2.

❖ All flue gas passageways must be provided with minimum clearances to ignitable materials. The clearance requirements in Table G2427.7.7 apply to unlisted single wall connectors. The clearances for connectors that are listed and labeled for such use must maintain a clearance to combustibles as required by the connector manufacturers installation instructions. If the appliance or connector manufacturer's installation instructions specify larger clearances than those prescribed by this section, the manufacturer's installation instructions govern. Table G2427.7.7 dictates the required air-space clearances between vent connectors and combustible materials and assemblies. The provisions of Section G2409 would allow connector clearances to be reduced where the prescribed clearance cannot be provided or is impractical. Lack of vent connector clearance is a very common code violation. Connectors usually lack the required clearances to wood joists, plastic plumbing piping, building and piping insulation materials, and gypsum board. A major disadvantage of single-wall connectors is the large clearance required to combustibles. Listed connectors usually require far less clearance, in some cases only one inch (25.4 mm).

G2427.10.6 (503.10.6) Flow resistance. A vent connector shall be installed so as to avoid turns or other construction features that create excessive resistance to flow of vent gases.

❖ A vent connector must have no more bends or fittings than absolutely necessary to accomplish the installation because changes in direction add to the flow resistance. Excessive resistance to flow could be defined as any resistance in excess of what would be caused by the two changes in direction anticipated by the vent sizing tables in Section G2428. This means that the installer and designer will have to plan the installations of vents and appliances so that the connectors can be as short and direct as possible. It is quite common to see installations with three, four, or more changes of direction in a connector because the appliance and vent location were not planned well. Section G2428 will penalize connector installations that contain more than two changes in direction (fittings). Listed bendable connectors offer the advantage of long sweeping changes in direction, which can have better flow characteristics than short sweep fittings. See Commentary Figure G2427.10.2.2. If a connector could have been installed with less fittings than the installer used, such installation could be rejected under this section. See Section G2427.10.9 and the commentary for Section G2428.3.11.

G2427.10.7 (503.10.7) Joints. Joints between sections of connector piping and connections to flue collars or hood outlets shall be fastened by one of the following methods:

1. Sheet-metal screws
2. Vent connectors of listed vent material, assembled and connected to flue collars and draft hood outlets in accordance with the manufacturers' instructions.
3. Other approved means.

❖ Single-wall connectors (unlisted) have been traditionally fastened with sheet metal screws and rivets. All connector joints must be fastened, including the joint at the appliance draft hood or flue collar. A displaced connector could result in a life threatening condition; therefore, connectors must be fastened and supported well. Appliance connectors are often located where they can be impacted or otherwise disturbed by building occupants, making proper fastening even more important. Item 2 speaks of listed connectors such as Type B vent material and factory-built corrugated (bendable) connectors, which employ adapter fittings or integral collars that are mechanically fastened to the appliance with screws or rivets. See Commentary Figure G2427.10.2.2.

G2427.10.8 (503.10.8) Slope. A vent connector shall be installed without any dips or sags and shall slope upward toward the vent or chimney at least $^1/_4$ inch per foot (21 mm/m).

Exception: Vent connectors attached to a mechanical draft system installed in accordance with the manufacturer's instructions.

❖ The ideal chimney or vent configuration is a totally vertical system, even though it is not always practical. This section requires all portions of a chimney or vent connector to rise vertically a minimum of a $^1/_4$ inch per each foot (0.02 mm/m) of its horizontal length. The connector slope is intended to induce the flow of flue gases by employing the natural buoyancy of the hot gases. Connector slope can promote the priming of a cold venting system and can partially compensate for short connector vertical rise. Low points, dips, and sags could also trap condensate and accelerate corrosion of the connector. The exception allows connectors without slope where connected to mechanical draft systems because the connector is on the negative pressure (inlet) side of the exhauster, and slope would provide no benefit.

G2427.10.9 (503.10.9) Length of vent connector. A vent connector shall be as short as practical and the equipment located as close as practical to the chimney or vent. Except as provided for in Section G2427.10.3, the maximum horizontal length of a single-wall connector shall be 75 percent of the height of the chimney or vent. Except as provided for in Section G2427.10.3, the maximum horizontal length of a Type B double-wall connector shall be 100 percent of the height of the chimney or vent. For a chimney or vent system serving multiple appliances, the maximum length of an individual connector, from the appliance outlet to the junction with the common vent or another connector, shall be 100 percent of the height of the chimney or vent.

❖ This section is applied in conjunction with Section G2428.3.2. The appliance and chimney or vent must be located so as to keep the connector length as short as practicable. This section establishes the maximum allowable length for uninsulated chimney and vent connectors, insulated chimney and vent connectors, and individual connectors for a chimney or vent system serving multiple appliances. These requirements are based on the heat loss of the connector and the ability of the vent or chimney system to produce a draft. An insulated connector (double wall) reduces the amount of heat transfer through the connector pipe; thus, the flue gas is maintained at a higher temperature. See the commentary for Section G2428.3.2. The amount of draft is directly related to the chimney or vent height and the difference in temperature between the flue gases and the ambient air. An uninsulated connector is a run of single-wall pipe.

Commentary Figure G2427.10.9 shows that the total chimney or vent height is measured from the top of the highest appliance flue outlet connection to the termination point of the chimney or vent. The length limitations of this section are based on the total vertical rise of the chimney or vent and on the developed length of horizontal connectors within the chimney or vent system. The venting tables will often prohibit the use of single-wall connectors with fan-assisted appliances, and most appliance and vent manufacturers recommend against the use of single-wall (uninsulated) connectors.

Connectors are limited in length because of the flow resistance of the connector pipe and because heat loss through the connector is directly related to its length. For a venting system to work, the draft produced by the vertical vent or chimney must be able to overcome the resistance to flow created by the connector. The longer the connector, the longer it takes to prime a cold venting system and develop draft. As stated in Section G2428.3.2, connectors should always be as short as the installation conditions will permit.

The sizing methodology and tables in Section G2428 for connectors in single-appliance installations might allow connector lengths greater than the general limits given in this section or might require a shorter connector length than would be allowed by these limits, depending on the combination of appliance type and input, connector size, and type and vent height. Because the sizing tables in Section G2428 are based on specific computer modeling of the installation configuration, the limitations or allowances in the single-appliance tables supersede what would be required in this section.

For multiple-appliance systems, the sizing methods and tables do not contain the same limitations on connector length. This section sets a reasonable length limitation for connectors in multiple-appliance systems (100 percent of the height of the chimney or vent), which still provides liberal installation flexibility without permitting excessively long connectors. This section functions as an absolute cap for the connector length provisions of Sections G2428.3.2 and G2428.3.3. As discussed in the commentary to Section G2427.10.6, the appliance and vent location must be well planned to allow the shortest horizontal run of connector piping. The code official must exercise good judgment in determining what is "short as practical" and "close as practical," because this section could require an appliance, vent, or chimney to be relocated to comply with the intent of the first sentence. See Sections G2428.3.2 and G2428.3.3.

This section refers to Section G2427.10.3 as if the section contains an exception to the length limits. However, Section G2427.10.3 is devoted to connector sizing only and does not address length. The intent of the reference is to allow connector length to be determined by design as part of an engineered venting system.

G2427.10.10 (503.10.10) Support. A vent connector shall be supported for the design and weight of the material employed to maintain clearances and prevent physical damage and separation of joints.

❖ A connector is required to be supported for the design and weight of the material employed. Proper support is necessary to maintain the clearances required by Section G2427.10.5, to maintain the required slope, and to prevent physical damage and separation of joints. The joints between connectors and appliance draft-hood outlets and flue collars are required to be secured with screws, rivets, or other approved means. The joints between pipe sections are also required to be secured with screws, rivets, or other approved means. Some connectors are available with a proprietary-type fastening method, which does not require screws, rivets, or other fasteners if it provides equivalent resistance to disengagement or displacement. This section ad-

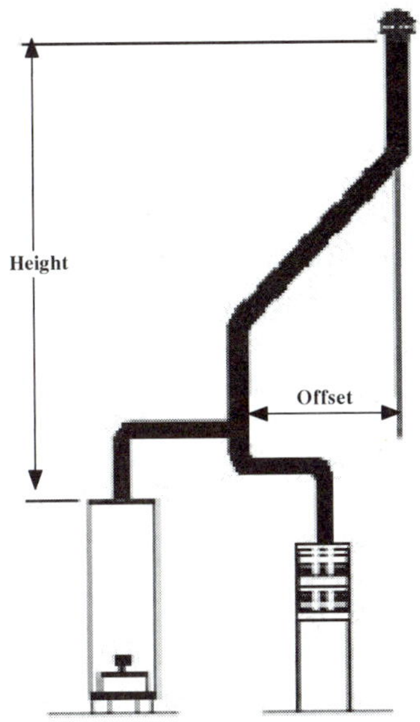

Figure courtesy of Simpson Dura-Vent Company, Incorporated

Figure G2427.10.9
VENT HEIGHT

dresses connectors, not chimneys and vents, and most chimney and vent manufacturers do not require or might even prohibit the use of screws or rivets for securing joints for chimney and vent pipe sections and fittings.

G2427.10.11 (503.10.11) Chimney connection. Where entering a flue in a masonry or metal chimney, the vent connector shall be installed above the extreme bottom to avoid stoppage. A thimble or slip joint shall be permitted to be used to facilitate removal of the connector. The connector shall be firmly attached to or inserted into the thimble or slip joint to prevent the connector from falling out. Means shall be employed to prevent the connector from entering so far as to restrict the space between its end and the opposite wall of the chimney flue. (See Section G2425.9.)

❖ All of the precautions taken to provide adequate chimney design may be ineffective if the connection between the appliance and the chimney is not properly accomplished. Improper connections can lead to appliance or connector failure and the leakage of harmful flue gases into the building. This section provides requirements for the connection between a chimney and the appliance connector. The connection to a vent is properly accomplished by vent fittings. The same holds true for factory-built chimneys.

Chimney connectors are required to pass through a masonry chimney wall to the inner face of the liner, but not beyond. A connector that extends into a chimney passageway can restrict the flow of flue gases and provide a ledge on which creosote, soot, and debris can accumulate. The joint between the connector and the chimney must be fastened in a manner that will prevent separation. If the connector enters a masonry chimney, it must be cemented in place with an approved material such as refractory mortar or other heat-resistant cement.

This section also permits the use of thimbles at a masonry chimney opening to provide for easy removal of the connector to facilitate cleaning. When a thimble is installed, it is to be permanently cemented in place with an approved high-temperature cement. A connector must be attached to a thimble in an approved manner to prevent displacement. A collar on the connector or a similar arrangement must be provided to limit the penetration of the connector into the chimney. Otherwise, it is quite possible that a connector will be shoved blindly into a chimney, reaching to the back wall of the chimney, and thus obstruct both the connector and the chimney.

G2427.10.12 (503.10.12) Inspection. The entire length of a vent connector shall be provided with ready access for inspection, cleaning and replacement.

❖ A connector cannot be concealed by any form of building construction, whether or not such construction is removable. Connectors need to be periodically inspected for corrosion damage, structural integrity, joint separation, and clearances to combustibles. See Section G2427.10.15.

G2427.10.13 (503.10.13) Fireplaces. A vent connector shall not be connected to a chimney flue serving a fireplace unless the fireplace flue opening is permanently sealed.

❖ A fireplace chimney cannot serve any appliance unless the fireplace is retired by permanently sealing off the chimney from the fireplace. If this is not done, the fireplace can affect draft for the connected appliance, and/or vent gases could escape into the building through the fireplace opening.

G2427.10.14 (503.10.14) Passage through ceilings, floors or walls. A vent connector shall not pass through any ceiling, floor or fire-resistance-rated wall. A single-wall metal pipe connector shall not pass through any interior wall.

Exception: Vent connectors made of listed Type B or Type L vent material and serving listed equipment with draft hoods and other equipment listed for use with Type B gas vents shall be permitted to pass through walls or partitions constructed of combustible material if the connectors are installed with not less than the listed clearance to combustible material.

❖ A single-wall connector has more than double the heat loss of a listed metal vent or chimney system. This produces high temperatures on the outside surface and cooling of the flue gases on the inside of the connector. The cooling of flue gases produces condensation, which can cause connector deterioration, while the excessive surface temperatures pose a potential fire hazard. For these reasons, a single-wall connector can only be located within the room or space in which the appliance is located. All connectors are prohibited from passing through any ceiling, floor, or fire-resistance-rated wall. Connectors that pass through walls, floors, or ceilings might not be readily observable, and a potential fire hazard or connector deterioration could go undetected. Also, such pass-throughs increase the likelihood that the connector will be excessively long, exposed to low ambient temperatures, or subject to contact with combustibles. Connectors that pass through a wall, floor, or ceiling would be out of sight and therefore out of mind, meaning that they would not get the attention needed to avoid hazardous conditions. The exception allows connectors made of Type B or L vent to pass through walls because such vents are allowed to be concealed within construction and pass through assemblies. Clearances to vents must be observed, regardless of their use as vents or as connectors. Although the exception appears to allow connectors to penetrate only combustible walls, it is the intent to allow penetration of noncombustible walls also. See Section G2427.10.15 for exterior wall penetrations.

G2427.10.15 (503.10.15) Single-wall connector penetrations of combustible walls. A vent connector made of a single-wall metal pipe shall not pass through a combustible exterior wall unless guarded at the point of passage by a ventilated metal thimble not smaller than the following:

1. For listed appliances equipped with draft hoods and appliances listed for use with Type B gas vents, the thimble shall be not less than 4 inches (102 mm) larger in diameter than the vent connector. Where there is a run of not less than 6 feet (1829 mm) of vent connector in the open between the draft hood outlet and the thimble, the thimble shall be permitted to be 2 inches (51 mm) larger in diameter than the vent connector.
2. For unlisted appliances having draft hoods, the thimble shall be not less than 6 inches (152 mm) larger in diameter than the vent connector.
3. For residential appliances and low-heat appliances, the thimble shall be not less than 12 inches (305 mm) larger in diameter than the vent connector.

Exception: In lieu of thimble protection, all combustible material in the wall shall be removed from the vent connector a sufficient distance to provide the specified clearance from such vent connector to combustible material. Any material used to close up such opening shall be noncombustible.

❖ This section applies to exterior (outside) walls. Section G2427.10.2.2 does not allow a single-wall connector to extend outdoors; thus, this section has no apparent application for vent connectors. The only application for the provisions of this section would be for single-wall vents under Section G2427.7.6. See also Sections G2427.10.14, G2427.7.2, and G2427.7.6. Item 1 dictates the construction of thimbles used to protect the combustible wall. Ventilated thimbles have holes in the faces of the thimble to allow air to cool the thimble by convection. Item 2 would rarely apply, because unlisted appliances are not allowed by this code unless specifically approved by the code official under Section R104.11. Item 3 would rarely apply, because residential incinerators are believed to be extinct, and gas-fired appliances without draft hoods and appliances not listed for use with Type B vent are either factory/industrial types or Category II, III, or IV.

The exception requires the combustible wall to be made noncombustible at the point of penetration. The noncombustible portion of the wall must extend from the connector in all directions a distance equal to the required airspace clearance for the connector.

G2427.11 (503.11) Vent connectors for Category II, Category III and Category IV gas utilization equipment. Vent connectors for category II, III and IV appliances shall be as specified for the venting system in accordance with Section G2427.4.

❖ See Sections G2427.4, G2427.4.1, and G2427.4.2.

G2427.12 (503.12) Draft hoods and draft controls. The installation of draft hoods and draft controls shall comply with Sections G2427.12.1 through G2427.12.7.

❖ See Sections G2427.12.1 through G2427.12.7.

G2427.12.1 (503.12.1) Equipment requiring draft hoods. Vented equipment shall be installed with draft hoods.

Exception: Dual oven type combination ranges; incinerators; direct-vent equipment; fan-assisted combustion system appliances; equipment requiring chimney draft for operation; single firebox boilers equipped with conversion burners with inputs greater than 400,000 Btu per hour (117 kw); equipment equipped with blast, power or pressure burners that are not listed for use with draft hoods; and equipment designed for forced venting.

❖ Draft hoods are integral to or are supplied with natural draft atmospheric burner gas-fired appliances. See Commentary Figure G2427.12.1. With the development of higher appliance efficiencies, fan-assisted combustion, and Category III and IV appliances, fewer draft hood appliances will be used. Draft-hood-equipped appliances are the least efficient appliances and are typically found on boilers and conventional tank-type and tankless water heaters. A draft-hood-equipped appliance will allow a continuous flow of air into the venting system, which represents an energy loss except where an automatic vent damper is installed to close off the vent in the off cycle. Without an automatic vent damper, conditioned air is taken from the space in which the appliance is located and residual heat is robbed from the appliance itself, both of which contribute to energy loss. Not all appliances are categorized, but if they were, draft hood-equipped appliances would fall under Category I. Draft hoods have multiple functions. They serve as a relief opening in the event of vent or chimney downdraft, they stabilize/regulate the amount of draft that occurs in the appliance combustion chamber, and they allow the introduction of dilution air into the chimney or vent.

G2427.12.2 (503.12.2) Installation. A draft hood supplied with or forming a part of listed vented equipment shall be

Photo courtesy of A. O. Smith Water Products

Figure G2427.12.1
DRAFT HOOD EQUIPPED APPLIANCE

installed without alteration, exactly as furnished and specified by the equipment manufacturer.

G2427.12.2.1 (503.12.2.1) Draft hood required. If a draft hood is not supplied by the equipment manufacturer where one is required, a draft hood shall be installed, shall be of a listed or approved type and, in the absence of other instructions, shall be of the same size as the equipment flue collar. Where a draft hood is required with a conversion burner, it shall be of a listed or approved type.

❖ In the unlikely event that an appliance manufacturer does not supply the required draft hood, a listed or approved draft hood must be obtained and installed.

G2427.12.2.2 (503.12.2.2) Special design draft hood. Where it is determined that a draft hood of special design is needed or preferable for a particular installation, the installation shall be in accordance with the recommendations of the equipment manufacturer and shall be approved.

❖ There could be unusual circumstances, such as where a horizontal type draft hood is needed because of low ceilings and/or insufficient connector rise.

G2427.12.3 (503.12.3) Draft control devices. Where a draft control device is part of the equipment or is supplied by the equipment manufacturer, it shall be installed in accordance with the manufacturer's instructions. In the absence of manufacturers' instructions, the device shall be attached to the fluecollar of the equipment or as near to the equipment as practical.

❖ Draft hoods also serve to control draft; however, this section would be redundant with Section G2427.12.2 if it were not addressing draft controls such as barometric dampers. There are two types of draft controls that are installed in the appliance vent outlet or vent connector: the passive draft hood and the active barometric damper. Barometric dampers consist of a damper plate mounted on an axle that rotates in low friction bearing points such as needle or knife edge bearings. The device is mounted in a tee fitting arrangement and connects with the appliance vent connector. It admits air in varying amounts into the vent connector to control the amount of draft through the appliance combustion chamber. Such devices respond to varying chimney or vent draft caused by changes in indoor and outdoor temperatures, changes in atmospheric (barometric) pressure, and changes in wind speed and direction. Barometric dampers are tuned by adjustable balancing weights attached to the damper plate. The damper manufacturer will provide instructions on how and where to install the device. See Commentary Figure G2403(2).

G2427.12.4 (503.12.4) Additional devices. Equipment requiring controlled chimney draft shall be permitted to be equipped with a listed double acting barometric draft regulator installed and adjusted in accordance with the manufacturer's instructions.

❖ Like Section G2427.12.3, this section addresses barometric dampers, specifically, the double acting type. These dampers open inward to admit air into the vent system like all other barometric dampers but can also open outward in the event of backdraft to allow vent gases to be spilled into the room or space in which the appliance is located, just as a draft hood would do. The manufacturer's installation instructions may require or recommend that a sensor (spill switch) be installed that will sense prolonged vent gas spillage from the damper and shut down the appliance. In such cases, a manufacturer's recommendation should be interpreted as an enforceable requirement. Some Category I fan-assisted furnaces can be fitted with a factory supplied conversion part that is, in effect, a retrofit draft hood. This will lower the annual fuel utilization efficiency, (AFUE) of the appliance installation by converting it to a draft-hood-equipped appliance installation that will now introduce dilution air (conditioned room air) into the vent. This is done solely to allow fan-assisted furnaces to connect to existing masonry chimneys that would otherwise not be allowed to serve such appliances. Such conversion parts employ a sensor (spill switch) in the relief/air inlet opening. The manufacturer of the appliance and the draft hood conversion kit will provide specific installation instructions that must be followed for the conversion. Such draft hood conversion kits are listed for use only with the appliance with which they were tested. In some cases, the manufacturer's instructions will contain specific chimney sizing tables that must be used instead of tables in the code.

G2427.12.5 (503.12.6) Location. Draft hoods and draft regulators shall be installed in the same room or enclosure as the equipment in such a manner as to prevent any difference in pressure between the hood or regulator and the combustion air supply.

❖ Draft regulation devices are designed to maintain a constant draft through the appliance combustion chamber and flue passages for the purpose of maintaining proper combustion conditions and achieving the highest possible efficiency. If the draft control "sees" a different ambient pressure than the appliance, it could allow too much or too little draft through the appliance. The positive, negative, or neutral pressures within a building must act equally on both the appliance and the draft control to allow the draft control to be adjusted and perform its intended function.

G2427.12.6 (503.12.7) Positioning. Draft hoods and draft regulators shall be installed in the position for which they were designed with reference to the horizontal and vertical planes and shall be located so that the relief opening is not obstructed by any part of the equipment or adjacent construction. The equipment and its draft hood shall be located so that the relief opening is accessible for checking vent operation.

❖ Draft hoods and barometric dampers are both affected by their orientation. There are two basic types of draft hoods: vertical draft hoods and horizontal draft hoods, which are not interchangeable. Barometric dampers can be adjusted to connect to either vertical or horizontal vent connectors; however, they are always positioned so that the closed damper plate is in a vertical plane. The manufacturers' instructions must always be followed.

The relief opening on all draft controls is also the dilution air inlet, and it must be unobstructed to allow air to enter under normal conditions and allow vent gases to spill in the event of backdraft and to allow testing of the vent system for proper draft. For example, for many years, appliance venting/draft has been checked by observing the movement of smoke introduced near the relief/inlet opening.

G2427.12.7 (503.12.8) Clearance. A draft hood shall be located so its relief opening is not less than 6 inches (152 mm) from any surface except that of the equipment it serves and the venting system to which the draft hood is connected. Where a greater or lesser clearance is indicated on the equipment label, the clearance shall be not less than that specified on the label. Such clearances shall not be reduced.

❖ The specified clearance is necessary to assure the unimpeded flow of air into the draft hood and flow of vent gas spillage from the hood. The clearance also protects materials from the hot gases that can spill from the draft hood and allows access for testing and observation.

G2427.13 (503.13) Manually operated dampers. A manually operated damper shall not be placed in the vent connector for any equipment. Fixed baffles shall not be classified as manually operated dampers.

❖ Manual dampers are associated with solid-fuel appliances but not with gas-fired appliances. A manual damper requires manual operation by a human, but this operation cannot be relied upon since humans forget. If a damper is left closed or partially closed during appliance operation, a severe hazard could result from vent gas spillage and/or appliance malfunction. See Section G2427.14.

G2427.14 (503.14) Automatically operated vent dampers. An automatically operated vent damper shall be of a listed type.

❖ Automatic vent dampers are intended for use with gas-fired natural draft appliances. The installation of an automatic vent damper must be in strict accordance with the manufacturer's installation instructions. Because automatic vent damper failure can result in a hazardous condition, automatic dampers must be listed and labeled. An automatic vent damper is installed on the draft hood outlet of an individual gas-fired appliance. See Commentary Figure G2427.14. Such dampers must not serve more than one appliance. The manufacturer's installation instructions require the damper to be installed by a qualified installer in accordance with the terms of the listing and the manufacturer's instructions. Because the purpose of a vent damper is to close or restrict the flue passageway of an appliance, it is imperative that the device be properly installed to minimize the possibility of failure. A malfunctioning or improperly installed vent damper device can cause the appliance to malfunction and could cause the discharge of products of combustion directly into the building interior.

Automatic vent dampers are energy-saving devices designed to close off or restrict an appliance flue passageway when the appliance is not operating and is in its "off" cycle. Such devices save energy by trapping residual heat in a heat exchanger after the burners shut off and by preventing the escape of conditioned room air up the chimney or vent. Thus, the appliance efficiency can be boosted, and building air infiltration can be reduced.

Such devices can be field-installed additions to existing equipment and can be a factory-supplied component of an appliance. See Commentary Figure G2428.2.1. A common application for such devices today is for mid-efficiency (80 to 83 percent) hot water boilers, which are draft-hood-equipped. There are three types of automatic vent dampers: electrical-, mechanical-, and thermal-actuating. Electrical- and mechanical-actuating automatic dampers are required to open automatically prior to main burner ignition and must remain open until the completion of the burner cycle. They are also required to be interlocked electrically with the appliance control circuitry to prevent the operation of the appliance in the event of damper failure. Thermal-actuating dampers open in response to flue gas temperature after burner ignition and close after burner shutdown. Thermal actuated dampers employ bimetal components that convert heat energy into mechanical energy. Some thermal-actuated dampers require the installation of a draft hood spillage sensor that would shut off the appliance if the damper failed to open. The most common type of vent damper employs an electric motor to rotate a damper blade. The device employs switches that prove the damper position and allow the appliance to start only after the damper is opened and verified. The sequence of operation is the same as that described in the commentary to Section G2427.3.3.

G2427.15 (503.15) Obstructions. Devices that retard the flow of vent gases shall not be installed in a vent connector, chimney or vent. The following shall not be considered as obstructions:

1. Draft regulators and safety controls specifically listed for installation in venting systems and installed in accordance with the terms of their listing.
2. Approved draft regulators and safety controls that are designed and installed in accordance with approved engineering methods.
3. Listed heat reclaimers and automatically operated vent dampers installed in accordance with the terms of their listing.
4. Approved economizers, heat reclaimers, and recuperators installed in venting systems of equipment not required to be equipped with draft hoods, provided the gas utilization equipment manufacturer's instructions cover the installation of such a device in the venting system and performance in accordance with Section G2427.3 and Section G2427.3.1 is obtained.
5. Vent dampers serving listed appliances installed in accordance with Sections G2428.2.1 and G2428.3.1 or other approved engineering methods.

❖ Manuals dampers, flow restricting orifices, and similar devices are not accounted for in the vent design methods of this code. The tables in Section G2428 assume that no such devices are installed in the vent. Item 3 addresses the automatic vent dampers covered in Section G2427.14. Such dampers open fully before the appliance is allowed to fire and offer negligible flow resistance because the damper plate is parallel to the vent gas flow. Item 4 addresses devices that extract heat from the vent gases to save energy that would

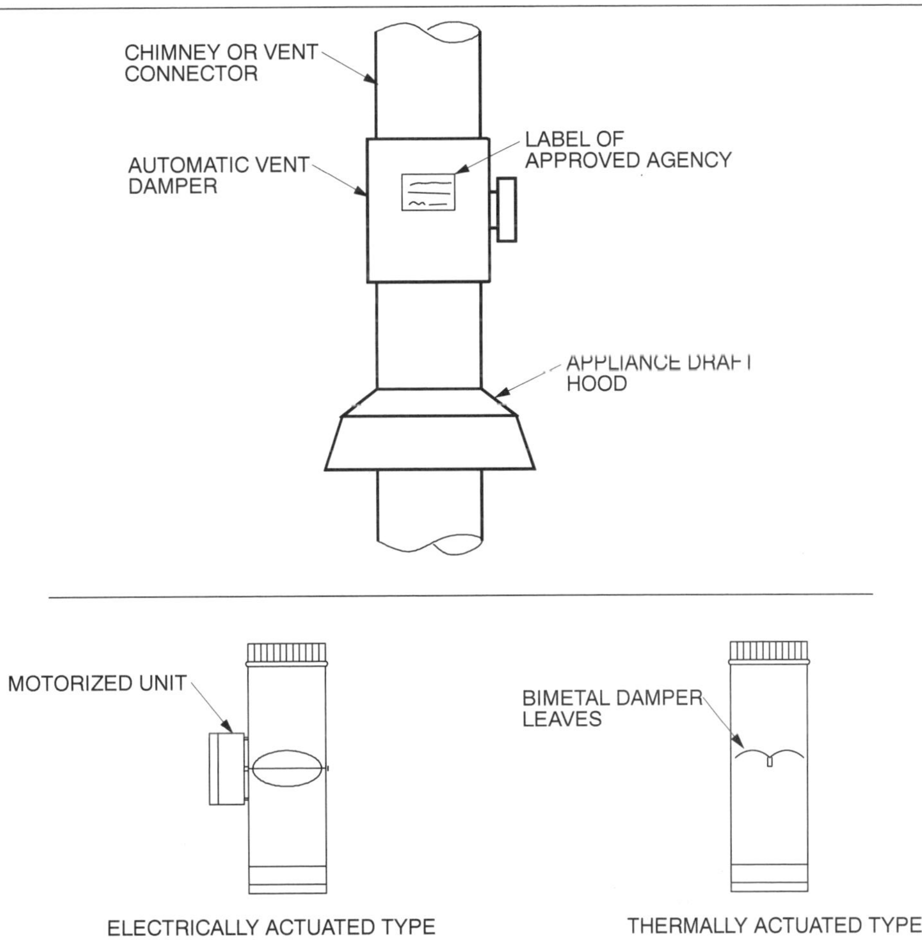

Figure G2427.14
AUTOMATIC VENT DAMPERS

otherwise be lost to the outside atmosphere. Appliances without draft hoods, excluding Category I fan-assisted, Category II, III, and IV appliances, usually have higher vent gas temperatures than those with draft hoods; therefore, waste heat can be reclaimed with less chance of diminishing draft. See Section G2428.2.1.

SECTION G2428 (504) SIZING OF CATEGORY I APPLIANCE VENTING SYSTEMS

G2428.1 (504.1) Definitions. The following definitions apply to tables in this section.

❖ The definitions in this section are necessary for application of the vent sizing tables and are exclusively related to Section G2428.

As stated in the main section title, this section applies to Category I appliances. See the definition of "Vented Appliance Categories." However, it also applies to draft hood-equipped appliances, which may or may not be categorized at all, even though they could be Category I because they meet the definition of Category I appliances. For example, a draft hood water heater is not required to be categorized by its governing standard, but if it were categorized it would be a Category I appliance.

A gas-fired appliance with a draft hood depends on gravity flow for supply of combustion air, for flow of combustion products through the heat exchanger, and for proper gas venting. The draft hood is designed to maintain proper combustion in the event of draft fluctuations. Its relief opening serves as a flue product exit in the event of a blocked vent or downdraft. The relief opening, however, also allows dilution air to enter the vent during normal appliance/vent operation. When this dilution air is obtained from within the heated space, there is a loss of seasonal efficiency, which can be reduced by installation of a vent damper designed to reduce the flow of air through the vent when the burner is not firing.

To improve annual fuel utilization efficiency (AFUE), many appliance manufacturers have designed fan-assisted combustion using mechanical means (blowers or fans) to obtain either induced or forced flow of combustion air and combustion products. Greater efficiency results from three major effects. First, heat exchange improves because of higher internal flow velocity, which enhances heat transfer by creating turbulent flow and the "scrubbing" of heat exchanger surfaces. Second, induced draft through the appliance combustion chambers and heat exchangers allows the heat exchangers to be constructed with longer passes and thus greater surface area and thus longer fuel gas retention time. This allows more heat to be extracted from the combustion gases. Third, there is no longer a draft hood to cause heated air loss up the vent, both when the appliance is operating and when it is not. See Commentary Figures G2428.1(1) through G2428.1(4).

A fan-assisted combustion appliance uses pressure-sensing and heat-sensing controls to monitor proper venting. The controls must prove that the appliance flue outlet is at neutral or negative pressure and that adequate flow of flue gases exists. Without a draft hood and dilution air to dilute the combustion products, gases entering the vent will have a higher water vapor content than those from a draft hood appliance. See Commentary Figure G2428.1(1). There will likely be a longer period of condensation in the vent (wet time), particularly where it connects to the appliance. These differences between fan-assisted and draft hood appliances are reflected in the tables by the lack of minimum capacities for draft hood appliances. The minimums for fan-assisted appliances are based on the heat input needed to control the duration of wet time or condensation during appliance start-up and operation.

The tables do not apply to:

1. Wall furnaces (recessed heaters), which require Type BW vents.
2. Decorative gas appliances, which generally require a specific vent size and are best individually vented.
3. Category II, III, or IV gas appliances. If permitted by the appliance manufacturer, some Category III appliances can also be vented in accordance with Category I conditions. Specifically, if the vent for a Category III appliance is sized and configured (connector rise, total height, etc.) for the input, the vent will operate under nonpositive pressure. This allows the use of Type B gas vent. For a Category III appliance to operate as a Category I, appliance manufacturers' instructions must be followed, and the vent size may need to be larger than the appliance flue collar.
4. Gas-fired appliances listed for use only with chimneys or dual-fuel appliances such as oil/gas, wood/gas, or coal/gas.

Dual-fuel appliances require chimneys sized in accordance with the appliance manufacturers' instructions and in accordance with the chimney manufacturers' instructions. Capacities in the tables assume cold starts such as occur with an intermittent ignition system (no standing pilot). The type of ignition system does not affect maximum capacities, but it does affect minimums. A standing pilot on an appliance will keep vent gas temperatures slightly above outdoor ambient temperature, whereas a tank of hot water plus water heater pilot operation will maintain flue gas temperature prior to the draft hood at approximately the same temperature as the stored water. This flow of heat into a vent aids in priming as well as reducing wet time. In some cases, the fan-assisted appliance manufacturer and/or the sizing tables in the code will require a draft hood-type appliance to be connected to the venting system. This is because the relatively inefficient draft hood appliance contributes heat to the venting system necessary to make it function.

The tables also assume that there are no adverse or building depressurization effects. A strong wind or mechanical ventilation/exhaust might cause a downdraft with draft hood appliances that will prevent the vent from priming properly.

The input capacity values in the tables were computed for typical natural gas. They can be used for LP gases, such as propane and butane, and mixtures of these with air. With LP gases, the maximum capacity remains the same, but because these have less hydrogen and produce less water vapor, the possibility of condensation is somewhat less. To simplify matters, assume that minimum capacities are the same regardless of fuel type.

Oversized or excessively long vents can cause condensation and should be avoided. The tables define limits that minimize wet time in the indoor portions, as well as in the upper exposed end of the vent.

The maximum vent lateral lengths for FAN appliances were calculated on the basis of several assumed conditions, including flue gas temperature and composition, and an outdoor ambient temperature of 42°F (6°C), chosen as a representative value. Colder assumed temperatures or greater outdoor exposure of the vent would lead to shorter maximum allowable lengths or to greater possibilities of condensation. In order to minimize condensation, it is essential to operate closer to maximum than minimum capacity and also to use the smallest allowable vent size.

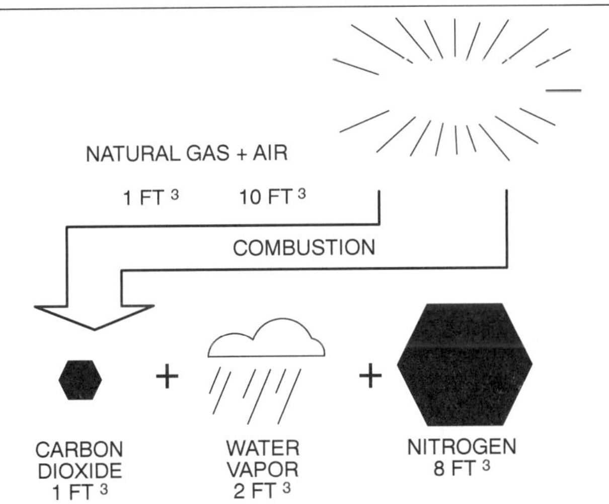

Figure courtesy of Simpson Dura-Vent Company, Inc.

Figure G2428.1(1)
COMBUSTION PROCESS FOR NATURAL GAS

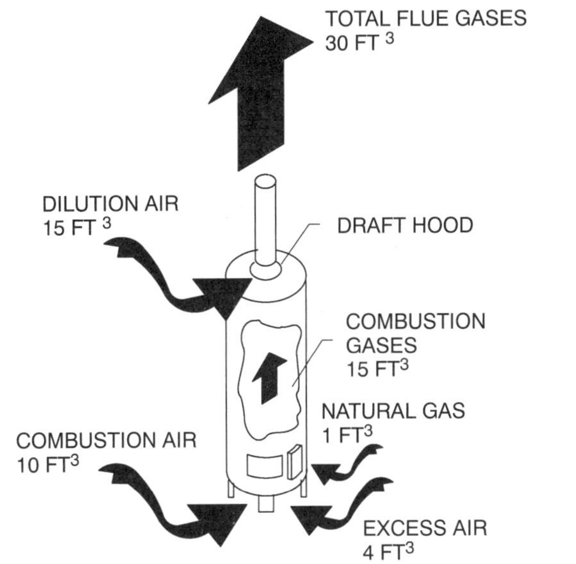

Figure courtesy of Simpson Dura-Vent Company, Incorporated

Figure G2428.1(2)
FLUE GAS COMPOSITION

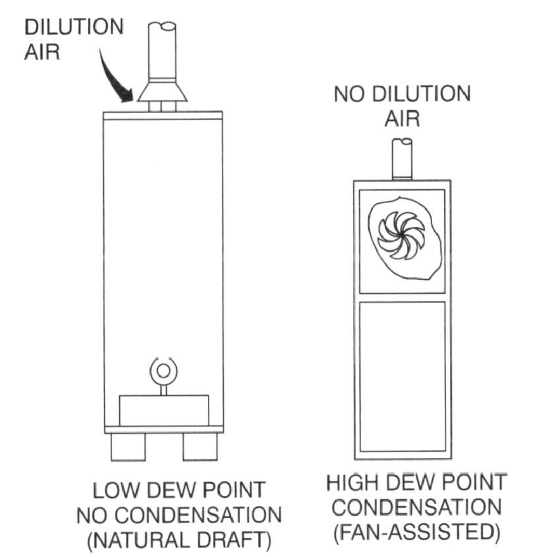

Figure courtesy of Simpson Dura-Vent Company, Incorporated

Figure G2428.1(3)
DRAFT HOOD APPLIANCE COMPARED TO FAN-ASSISTED APPLIANCE

APPLIANCE CATEGORIZED VENT DIAMETER/AREA. The minimum vent area/diameter permissible for Category I appliances to maintain a nonpositive vent static pressure when tested in accordance with nationally recognized standards.

❖ The minimum Category I appliance vent area/diameter necessary to maintain a non-positive vent static pressure when tested in accordance with the applicable standard.

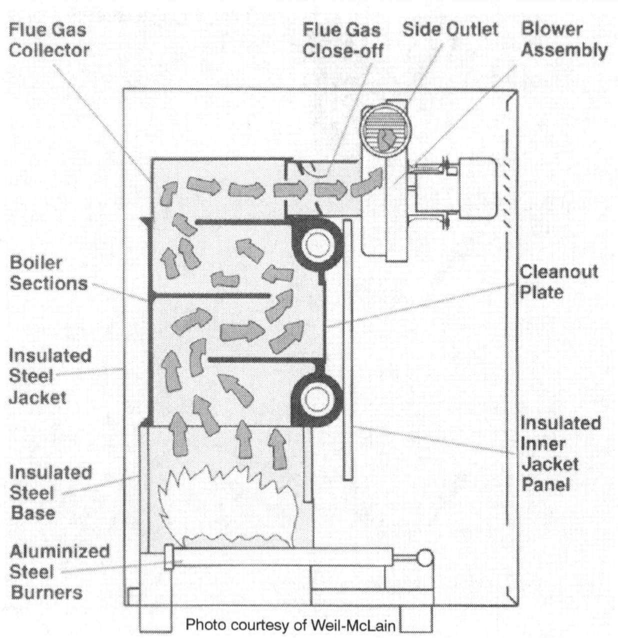

Photo courtesy of Weil-McLain

Figure G2428.1(4)
BASIC PRINCIPLE OF FAN-ASSISTED COMBUSTION

The appliance categorized vent diameter/area is determined by the appliance manufacturer and is typically the vent outlet collar provided on the appliance.

FAN-ASSISTED COMBUSTION SYSTEM. An appliance equipped with an integral mechanical means to either draw or force products of combustion through the combustion chamber or heat exchanger.

❖ An appliance equipped with an integral mechanical means to either draw or force products of combustion through the combustion chamber or heat exchanger. See Commentary Figure G2428.1(5).

FAN MIN. The minimum input rating of a Category I fan-assisted appliance attached to a vent or connector.

❖ The minimum appliance input rating of a Category I appliance with a fan-assisted combustion system that is capable of being attached to the vent.

FAN MAX. The maximum input rating of a Category I fan-assisted appliance attached to a vent or connector.

❖ The maximum appliance input rating of a Category I appliance with a fan-assisted combustion system that is capable of being attached to the vent.

NAT MAX. The maximum input rating of a Category I draft-hood equipped appliance attached to a vent or connector.

❖ The maximum input rating of a Category I appliance equipped with a draft hood that is capable of being attached to the vent. There are no minimum appliance input ratings for draft hood-equipped appliances.

Photo courtesy of Weil-McLain

Figure G2428.1(5)
FAN-ASSISTED WATER BOILER

FAN + FAN. The maximum combined appliance input rating of two or more Category I fan-assisted appliances attached to the common vent.

❖ The maximum combined input rating of two or more fan-assisted appliances attached to the common vent.

FAN + NAT. The maximum combined appliance input rating of one or more Category I fan-assisted appliances and one or more Category I draft hood-equipped appliances attached to the common vent.

❖ The maximum combined input rating of one or more fan-assisted appliances and one or more draft hood-equipped appliances attached to the common vent.

NA. Vent configuration is not permitted due to potential for condensate formation or pressurization of the venting system, or not applicable due to physical or geometric restraints.

❖ Vent configuration is prohibited.

NAT + NAT. The maximum combined appliance input rating of two or more Category I draft hood- equipped appliances attached to the common vent.

❖ The maximum combined input rating of two or more draft hood-equipped appliances attached to the common vent.

G2428.2 (504.2) Application of single appliance vent Tables G2428.2(1) through G2428.2(2). The application of Tables G2428.2(1) through G2428.2(5) shall be subject to the requirements of Sections G2428.2.1 through G2428.2.15.

❖ The tables do not apply to Type BW vents; vents for decorative gas appliances; vents for Category II, III or IV appliances; vents for dual-fuel appliances; vents for appliances not listed for use with Type B vent; and appliances listed only for connection to chimneys. Category I appliances can be a fan-assisted design or can be a draft hood-equipped design and each such design has different vent system design considerations. A noteworthy difference between fan-assisted and draft hood-equipped designs is that the vents for fan-assisted appliances have both a maximum and a minimum vent capacity. In the past, lower-efficiency draft hood-type appliances were dominant, and vent sizing was based only on the maximum venting capacity of the vent. In other words, the only concern was to make sure that the vent was big enough. This is no longer the case with the advent of mid-efficiency fan-assisted-type appliances. Today, it is possible for a vent to be either too small or too large. An undersized vent will have insufficient capacity and can allow flue gas spillage or cause positive pressure to occur within the vent. An oversized vent can fail to produce sufficient draft and can be subject to the continuous formation of water vapor condensation on the interior vent surfaces.

A typical individual vent is shown in Commentary Figure G2428.2(1) for a FAN furnace with an input rating of 150,000 Btu/h (44 kW) and a 5-inch (127 mm) outlet connection. Procedure for a Type B vent in Table G2428.2(1): Go down the height column to 20 feet (6096 mm) and across on the 10-foot (3048 mm) lateral line.

For this furnace, the MAX capacity under the 5-inch (127 mm) size heading is 229,000 Btu/h (67.1 kW) and the MIN capacity is 50,000 Btu/h (14.7 kW). A 5-inch (127 mm) size Type B vent will be correct. See Commentary Figure G2428.2(1). In applying the tables, extreme care must be taken to observe the titles of the table and to apply the conditions and instructions set forth in Sections G2428.2.1 through G2428.2.15. To determine the proper size for an individual vent, apply the table as follows:

1. Determine the total vent height and the length of lateral, based on the appliance and vent location and the height to top of vent, as indicated in Commentary Figure G2428.2(1). If gas appliances, such as a furnace, boiler, or water heater, have not been chosen or installed, estimate the height beginning at 6 feet (1829 mm) above the floor. For attic or horizontal furnaces, floor furnaces, room heaters, and small boilers, the height location of the draft hood outlet or vent collar should be known.
2. Read down the height column to a height equal to or less than the estimated total height.
3. Select the horizontal row for the appropriate lateral (L) length.
4. Read across to the first column under the type of appliance (FAN or NAT) that shows an appliance input rating equal to or greater than the nameplate sea level input rating of the appliance to be vented.

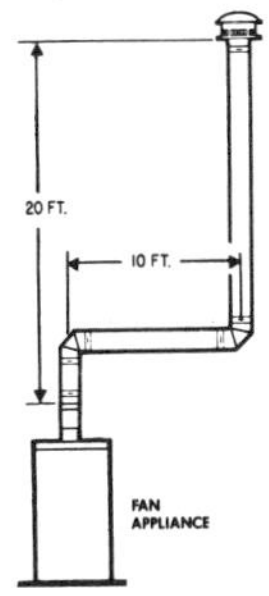

For SI: 1 foot = 304.8 mm

Figure courtesy of Selkirk, LLC.

Figure G2428.2(1)
EXAMPLE OF INDIVIDUAL VENT DIMENSIONS

Commentary Figure G2428.2(2) illustrates the measurement of height and laterals for draft hood appliances. Some vent manufacturers include special requirements regarding certain types of appliances in their installation/design instructions. The following example is taken from the 1992 *Chimney and Gas Vent Sizing Handbook* by Selkirk Metalbestos:

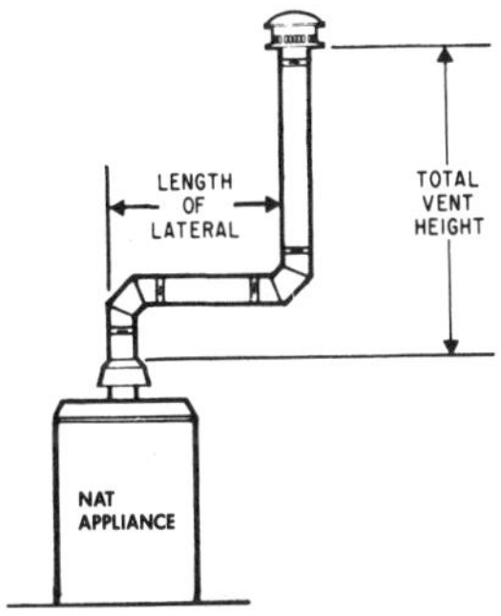

Figure courtesy of Selkirk, LLC.

Figure G2428.2(2)
INDIVIDUAL VENT—LATERAL AND HEIGHT

1. Room heaters, floor furnaces, wall furnaces: If the appliance has a draft hood, assume an adjusted input 40 percent greater than name-plate value and design vent for this increased input. Example: The vent for a 50,000 Btu/h room heater will be 10 feet high with a 2-foot lateral. Adjusted input is 1.4 x 50,000 = 70,000. The NAT column in the tables shows that a 4-inch vent with a 81,000 Btu/h capacity will be adequate. If the appliance has a fan-assisted burner, use the FAN column without adjustment.
2. Central heating boilers, gravity and forced-air furnaces, duct furnaces and unit heaters, conversion burners with draft hoods: Use full tabulated capacities, depending on the type of appliance (NAT or FAN).
3. Cooking equipment: domestic, restaurant, and commercial: Use outlet size or five times input to determine individual vent size.
4. Decorative appliances: Use outlet size or five times input to determine individual vent size.

Appendix B contains vent sizing examples.

TABLE G2428.2(1) [504.2(1)]
TYPE B DOUBLE-WALL GAS VENT

Number of Appliances	Single
Appliance Type	Category I
Appliance Vent Connection	Connected directly to vent

HEIGHT H (feet)	LATERAL L (feet)	VENT DIAMETER — D (inches): 3			4			5			6			7			8			9		
		Appliance input rating in thousands of Btu/h																				
		FAN		NAT	FAN		NAT	FAN		NAT	FAN		NAT	FAN		NAT	FAN		NAT	FAN		NAT
		Min	Max	Max	Min	Max	Max	Min	Max	Max	Min	Max	Max	Min	Max	Max	Min	Max	Max	Min	Max	Max
6	0	0	78	46	0	152	86	0	251	141	0	375	205	0	524	285	0	698	370	0	897	470
	2	13	51	36	18	97	67	27	157	105	32	232	157	44	321	217	53	425	285	63	543	370
	4	21	49	34	30	94	64	39	153	103	50	227	153	66	316	211	79	419	279	93	536	362
	6	25	46	32	36	91	61	47	149	100	59	223	149	78	310	205	93	413	273	110	530	354
8	0	0	84	50	0	165	94	0	276	155	0	415	235	0	583	320	0	780	415	0	1,006	537
	2	12	57	40	16	109	75	25	178	120	28	263	180	42	365	247	50	483	322	60	619	418
	5	23	53	38	32	103	71	42	171	115	53	255	173	70	356	237	83	473	313	99	607	407
	8	28	49	35	39	98	66	51	164	109	64	247	165	84	347	227	99	463	303	117	596	396
10	0	0	88	53	0	175	100	0	295	166	0	447	255	0	631	345	0	847	450	0	1,096	585
	2	12	61	42	17	118	81	23	194	129	26	289	195	40	402	273	48	533	355	57	684	457
	5	23	57	40	32	113	77	41	187	124	52	280	188	68	392	263	81	522	346	95	671	446
	10	30	51	36	41	104	70	54	176	115	67	267	175	88	376	245	104	504	330	122	651	427
15	0	0	94	58	0	191	112	0	327	187	0	502	285	0	716	390	0	970	525	0	1,263	682
	2	11	69	48	15	136	93	20	226	150	22	339	225	38	475	316	45	633	414	53	815	544
	5	22	65	45	30	130	87	39	219	142	49	330	217	64	463	300	76	620	403	90	800	529
	10	29	59	41	40	121	82	51	206	135	64	315	208	84	445	288	99	600	386	116	777	507
	15	35	53	37	48	112	76	61	195	128	76	301	198	98	429	275	115	580	373	134	755	491
20	0	0	97	61	0	202	119	0	349	202	0	540	307	0	776	430	0	1,057	575	0	1,384	752
	2	10	75	51	14	149	100	18	250	166	20	377	249	33	531	346	41	711	470	50	917	612
	5	21	71	48	29	143	96	38	242	160	47	367	241	62	519	337	73	697	460	86	902	599
	10	28	64	44	38	133	89	50	229	150	62	351	228	81	499	321	95	675	443	112	877	576
	15	34	58	40	46	124	84	59	217	142	73	337	217	94	481	308	111	654	427	129	853	557
	20	48	52	35	55	116	78	69	206	134	84	322	206	107	464	295	125	634	410	145	830	537
30	0	0	100	64	0	213	128	0	374	220	0	587	336	0	853	475	0	1,173	650	0	1,548	855
	2	9	81	56	13	166	112	14	283	185	18	432	280	27	613	394	33	826	535	42	1,072	700
	5	21	77	54	28	160	108	36	275	176	45	421	273	58	600	385	69	811	524	82	1,055	688
	10	27	70	50	37	150	102	48	262	171	59	405	261	77	580	371	91	788	507	107	1,028	668
	15	33	64	NA	44	141	96	57	249	163	70	389	249	90	560	357	105	765	490	124	1,002	648
	20	56	58	NA	53	132	90	66	237	154	80	374	237	102	542	343	119	743	473	139	977	628
	30	NA	NA	NA	73	113	NA	88	214	NA	104	346	219	131	507	321	149	702	444	171	929	594
50	0	0	101	67	0	216	134	0	397	232	0	633	363	0	932	518	0	1,297	708	0	1,730	952
	2	8	86	61	11	183	122	14	320	206	15	497	314	22	715	445	26	975	615	33	1,276	813
	5	20	82	NA	27	177	119	35	312	200	43	487	308	55	702	438	65	960	605	77	1,259	798
	10	26	76	NA	35	168	114	45	299	190	56	471	298	73	681	426	86	935	589	101	1,230	773
	15	59	70	NA	42	158	NA	54	287	180	66	455	288	85	662	413	100	911	572	117	1,203	747
	20	NA	NA	NA	50	149	NA	63	275	169	76	440	278	97	642	401	113	888	556	131	1,176	722
	30	NA	NA	NA	69	131	NA	84	250	NA	99	410	259	123	605	376	141	844	522	161	1,125	670

For SI: 1 inch = 25.4 mm, 1 foot = 304.8 mm, 1,000 Btu = 0.293 kW, 1 square inch = 645.16 mm^2.

❖ See the commentary for Section G2428.2.

Number of Appliances	Single
Appliance Type	Category I
Appliance Vent Connection	Single-wall metal connector

TABLE G2428.2(2) [504.2(2)]
TYPE B DOUBLE-WALL GAS VENT

HEIGHT H (feet)	LATERAL L (feet)	VENT DIAMETER — D (inches)																										
		3			4			5			6			7			8			9			10			12		
		Appliance input rating in thousands of Btu/h																										
		FAN		NAT	FAN		NAT	FAN		NAT	FAN		NAT	FAN		NAT	FAN		NAT	FAN		NAT	FAN		NAT	FAN		NAT
		Min	Max	Max	Min	Max	Max	Min	Max	Max	Min	Max	Max	Min	Max	Max	Min	Max	Max	Min	Max	Max	Min	Max	Max	Min	Max	Max
6	0	38	77	45	59	151	85	85	249	140	126	373	204	165	522	284	211	695	369	267	894	469	371	1,118	569	537	1,639	849
	2	39	51	36	60	96	66	85	156	104	123	231	156	159	320	213	201	423	284	251	541	368	347	673	453	498	979	648
	4	NA	NA	33	74	92	63	102	152	102	146	225	152	187	313	208	237	416	277	295	533	360	409	664	443	584	971	638
	6	NA	NA	31	83	89	60	114	147	99	163	220	148	207	307	203	263	409	271	327	526	352	449	656	433	638	962	627
8	0	37	83	50	58	164	93	83	273	154	123	412	234	161	580	319	206	777	414	258	1,002	536	360	1,257	658	521	1,852	967
	2	39	56	39	59	108	75	83	176	119	121	261	179	155	363	246	197	482	321	246	617	417	339	768	513	486	1,120	743
	5	NA	NA	37	77	102	69	107	168	114	151	252	171	193	352	235	245	470	311	305	604	404	418	754	500	598	1,104	730
	8	NA	NA	33	90	95	64	122	161	107	175	243	163	223	342	225	280	458	300	344	591	392	470	740	486	665	1,089	715
10	0	37	87	53	57	174	99	82	293	165	120	444	254	158	628	344	202	844	449	253	1,093	584	351	1,373	718	507	2,031	1,057
	2	39	61	41	59	117	80	82	193	128	119	287	194	153	400	272	193	531	354	242	681	456	332	849	559	475	1,242	848
	5	52	56	39	76	111	76	105	185	122	148	277	186	190	388	261	241	518	344	299	667	443	409	834	544	584	1,224	825
	10	NA	NA	34	97	100	68	132	171	112	188	261	171	237	369	241	296	497	325	363	643	423	492	808	520	688	1,194	788
15	0	36	93	57	56	190	111	80	325	186	116	499	283	153	713	388	195	966	523	244	1,259	681	336	1,591	838	488	2,374	1,237
	2	38	69	47	57	136	93	80	225	149	115	337	224	148	473	314	187	631	413	232	812	543	319	1,015	673	457	1,491	983
	5	51	63	44	75	128	86	102	216	140	144	326	217	182	459	298	231	616	400	287	795	526	392	997	657	562	1,469	963
	10	NA	NA	39	95	116	79	128	201	131	182	308	203	228	438	284	284	592	381	349	768	501	470	966	628	664	1,433	928
	15	NA	NA	NA	NA	NA	72	158	186	124	220	290	192	272	418	269	334	568	367	404	742	484	540	937	601	750	1,399	894
20	0	35	96	60	54	200	118	78	346	201	114	537	306	149	772	428	190	1,053	573	238	1,379	750	326	1,751	927	473	2,631	1,346
	2	37	74	50	56	148	99	78	248	165	113	375	248	144	528	344	182	708	468	227	914	611	309	1,146	754	443	1,689	1,098
	5	50	68	47	73	140	94	100	239	158	141	363	239	178	514	334	224	692	457	279	896	596	381	1,126	734	547	1,665	1,074
	10	NA	NA	41	93	129	86	125	223	146	177	344	224	222	491	316	277	666	437	339	866	570	457	1,092	702	646	1,626	1,037
	15	NA	NA	NA	NA	NA	80	155	208	136	216	325	210	264	469	301	325	640	419	393	838	549	526	1,060	677	730	1,587	1,005
	20	NA	NA	NA	NA	NA	NA	186	192	126	254	306	196	309	448	285	374	616	400	448	810	526	592	1,028	651	808	1,550	973
30	0	34	99	63	53	211	127	76	372	219	110	584	334	144	849	472	184	1,168	647	229	1,542	852	312	1,971	1056	454	2,996	1,545
	2	37	80	56	55	164	111	76	281	183	109	429	279	139	610	392	175	823	533	219	1,069	698	296	1,346	863	424	1,999	1,308
	5	49	74	52	72	157	106	98	271	173	136	417	271	171	595	382	215	806	521	269	1,049	684	366	1,324	846	524	1,971	1,283
	10	NA	NA	NA	91	144	98	122	255	168	171	397	257	213	570	367	265	777	501	327	1,017	662	440	1,287	821	620	1,927	1,243
	15	NA	NA	NA	115	131	NA	151	239	157	208	377	242	255	547	349	312	750	481	379	985	638	507	1,251	794	702	1,884	1,205
	20	NA	NA	NA	NA	NA	NA	181	223	NA	246	357	228	298	524	333	360	723	161	133	955	615	570	1,216	768	780	1,841	1,166
	30	NA	NA	NA	NA	NA	NA	NA	NA	NA	NA	NA	NA	389	477	305	461	670	426	541	895	574	704	1,147	720	937	1,759	1,101
50	0	33	99	66	51	213	133	73	394	230	105	629	361	138	928	515	176	1,292	704	220	1,724	948	295	2,223	1,189	428	3,432	1,818
	2	36	84	61	53	181	121	73	318	205	104	495	312	133	712	443	168	971	613	209	1,273	811	280	1,615	1,007	401	2,426	1,509
	5	48	80	NA	70	174	117	94	308	198	131	482	305	164	696	435	204	953	602	257	1,252	795	347	1,591	991	496	2,396	1,490
	10	NA	NA	NA	89	160	NA	118	292	186	162	461	292	203	671	420	253	923	583	313	1,217	765	418	1,551	963	589	2,347	1,455
	15	NA	NA	NA	112	148	NA	145	275	174	199	441	280	244	646	405	299	894	562	363	1,183	736	481	1,512	934	668	2,299	1,421
	20	NA	NA	NA	NA	NA	NA	176	257	NA	236	420	267	285	622	389	345	866	543	415	1,150	708	544	1,473	906	741	2,251	1,387
	30	NA	NA	NA	NA	NA	NA	NA	NA	NA	315	376	NA	373	573	NA	442	809	502	521	1,086	649	674	1,399	848	892	2,159	1,318

For SI: 1 inch = 25.4 mm, 1 foot = 304.8 mm, 1,000 Btu = 0.293 kW.

❖ See the commentary for Section G2428.2.

G2428.2.1 (504.2.1) Vent obstructions. These venting tables shall not be used where obstructions, as described in the exceptions to Section G2427.15, are installed in the venting system. The installation of vents serving listed appliances with vent dampers shall be in accordance with the appliance manufacturer's instructions or in accordance with the following:

1. The maximum capacity of the vent system shall be determined using the "NAT Max" column.
2. The minimum capacity shall be determined as if the appliance were a fan-assisted appliance, using the "FAN Min" column to determine the minimum capacity of the vent system. Where the corresponding "FAN Min" is "NA," the vent configuration shall not be permitted and an alternative venting configuration shall be utilized.

❖ An engineered vent sizing method or appliance and vent/chimney manufacturer's instructions must be used to size vents where devices (obstructions) are installed in a venting system. The tables do not account for the flow resistance caused by such devices, nor do they account for the reduction of flue gas temperatures resulting from heat reclaim exchangers. Section G2427.15 lists several items that are not to be viewed as obstructions, including automatic vent dampers. This section recognizes the use of the vent sizing tables with appliances equipped with automatic vent damper devices. By referring to the appliance manufacturer's installation instructions, this section implies that the appliances are factory equipped with automatic vent dampers. Item 1 recognizes that during the firing cycle, an appliance with an automatic vent damper affects a venting system no differently than a draft hood appliance without such device. Item 2 addresses the effect that appliances equipped with automatic vent dampers have on venting systems during the appliance off cycle. The appliance acts as if it were an idle fan-assisted appliance because a closed vent damper will allow little or no air to enter the vent system, just as an idle fan-assisted appliance allows no air to enter the vent system. The lack of dilution air entering the venting system is a primary reason that the minimum input columns were created in the vent sizing tables. Boilers and waters heaters with draft hoods commonly utilize automatic vent dampers to boost the efficiency of the appliance by trapping residual heat of the firing cycle, reducing standby losses, and reducing the amount of conditioned air that escapes up the vent via the draft hood. See Commentary Figure G2428.2.1.

G2428.2.2 (504.2.2) Minimum size. Where the vent size determined from the tables is smaller than the appliance draft hood outlet or flue collar, the smaller size shall be permitted to be used provided all of the following are met:

1. The total vent height (H) is at least 10 feet (3048 mm).
2. Vents for appliance draft hood outlets or flue collars 12 inches (305 mm) in diameter or smaller are not reduced more than one table size.
3. Vents for appliance draft hood outlets or flue collars larger than 12 inches (305 mm) in diameter are not reduced more than two table sizes.
4. The maximum capacity listed in the tables for a fan-assisted appliance is reduced by 10 percent (0.90 by maximum table capacity).
5. The draft hood outlet is greater than 4 inches (102 mm) in diameter. Do not connect a 3-inch-diameter (76 mm) vent to a 4-inch-diameter (102 mm) draft hood outlet. This provision shall not apply to fan-assisted appliances.

❖ If a vent complies with all of items 1 through 5 and is sized in accordance with the appropriate table, it is allowed to be smaller than the appliance connection. This is primarily due to the increase in capacity of the vent because of its height and because vent capacity is based on the Btu/h input rating of the appliance and not the size of the appliance flue collar. This code section suggests that flue collars are sized for worst case vent designs and are thus somewhat oversized for vents with higher venting capacities. Item 4 is necessary to prevent positive pressure from developing in the vent serving fan-assisted appliances. A 6-inch draft hood connected to a 5-inch Type B vent, for example, might look like an installation error, but this section could allow it in some cases, and operating a vent nearer its maximum capacity is always desirable. See Section G2428.2.10.

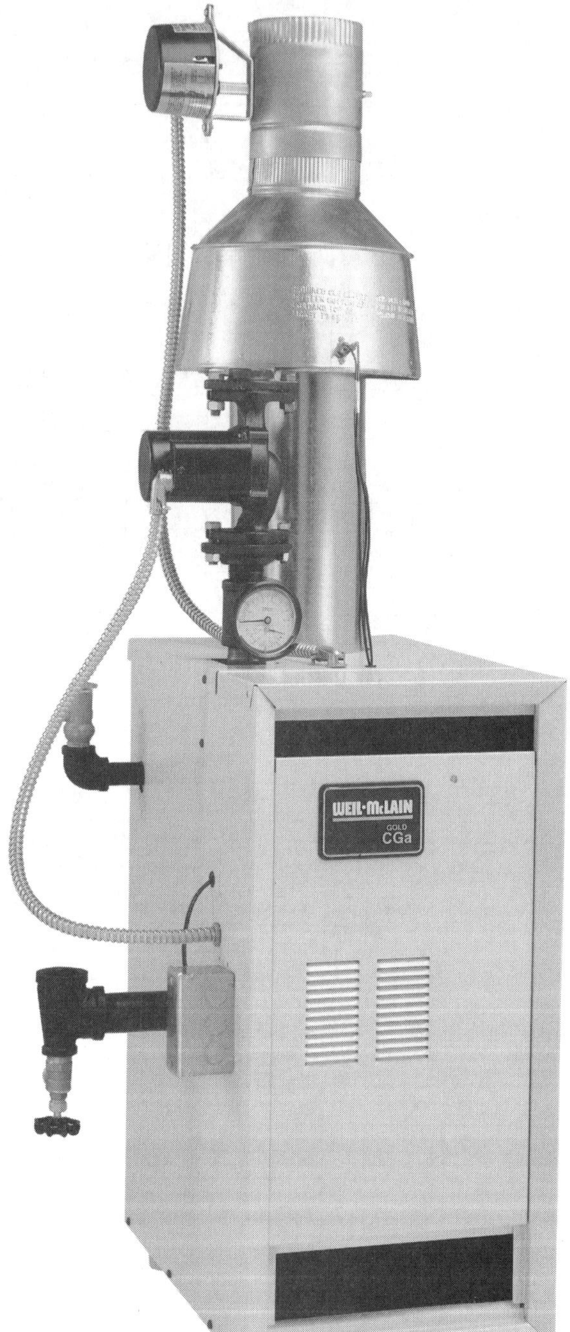

Photo courtesy of Weil-McLain

Figure G2428.2.1
CATEGORY I NATURAL-DRAFT WATER BOILER WITH AUTOMATIC VENT DAMPER

G2428.2.3 (504.2.3) Vent offsets. Single-appliance venting configurations with zero lateral lengths in Tables G2428.2(1) and G2428.2(2) shall have no elbows in the venting system. For vent configurations with lateral lengths, the venting tables include allowance for two 90-degree (1.57 rad) turns. For each elbow up to and including 45-degrees (0.79 rad), the maximum capacity listed in the venting tables shall be reduced by 5 percent. For each elbow greater than 45 degrees (0.79 rad) up to and including 90 degrees (1.57 rad), the maximum capacity listed in the venting tables shall be reduced by 10 percent.

❖ A vent system with a zero lateral length is an ideal system as it is entirely vertical and thus has the least flow resistance and greatest draft. To create a lateral run of vent or vent connector, one or two elbows (adjustable fittings) are necessary. See Commentary Figure G2428.2(2). The tables account for two 90-degree (1.6 rad) changes in direction, which will offer considerable resistance to the flow of flue gases. Depending on the appliance flue collar or draft hood outlet orientation, a single elbow can create a lateral, but the tables assume that a vertical outlet orientation and two elbows are used. Two 45-degree (0.79 rad) fittings joined together, for example, are the equivalent of a single 90-degree (1.6 rad) fitting but would offer slightly less resistance to flow because of the larger turning radius. The capacity reduction penalty for each change in direction in excess of the two "free" 90-degree fittings accounts for the increased flow resistance caused by the additional fittings (elbows). For example, a vent employing two 90-degree fittings and two 45-degree fittings would have a 10 percent capacity reduction penalty. A vent employing two 90-degree fittings and two 30-degree fittings would also have a 10-percent capacity penalty. All changes of direction, regardless of the angle, that are in addition to the two "free" 90-degree fittings are accounted for by a capacity reduction penalty. A fitting adjusted to 45 degrees or less will cause a 5-percent penalty to be applied, and a fitting adjusted to greater than 45 degrees up to 90 degrees will cause a 10-percent penalty to be applied.

Whenever possible, an individual vent should be located directly over the appliance outlet. If the appliance flue outlet is horizontal, one 90-degree (1.6 rad) elbow should be used with the vent directly over it. A straight vertical vent is more easily supported and has less flow resistance.

There is no need to offset the vertical vent to include a tee and bracket. The use of a tee for cleanout or inspection purposes is quite unnecessary for three reasons. First, using a B-vent cap keeps debris and birds out. Second, clean-burning gas does not produce any deposits needing removal. Third, Type B vent joints are easily opened for someone to inspect the inside of the piping.

Should an offset be needed, the use of two elbows will provide somewhat greater capacity than an elbow and a tee. A tee has greater flow resistance than an elbow. The two 90-degree turns that are accounted for by the tables for laterals can occur anywhere within the vent system. This means that if the turns are used where the appliance connects to the system, the rest of the vent must be entirely vertical, or else the penalties of this section must be applied. The two "free" 90-degree turns can be composed of any number and angle of fittings that total not more than 180 degrees. Simply put, there are free turns up to a maximum of 180 degrees, and after that each additional turn will cause a capacity penalty to apply.

Vent offsets should always be avoided where possible, especially offsets in the upper portion of the vent and in cold spaces such as attics and truss spaces. Offsets can be avoided with simple planning of the vent system and appliance locations. The attic offset is prevalent because the vent and appliance location is often not planned, and the building owners and designers do not want the vent to terminate on the street side of a roof for aesthetic reasons. This type of offset causes the most problems for several reasons. The typical attic will have winter temperatures close to the outdoor temperature; therefore, offsets in attics will cause more vent pipe to be exposed to the cold attic. See the commentary for Section G2428.2.9. This can increase the production of condensate in the vent, and the fittings used will provide points at which the condensate can leak from the inner vent lining. Most condensate leakage from vents can be traced to a fitting used in an offset. The code (Sections G2428.2.9 and G2428.3.16) assumes that a vent will not be exposed to outdoor temperatures except for the portion that extends above the roof. The typical attic today is nearly equivalent to the outdoors with regard to temperature. Studies have shown that offsets in the upper portion of vent systems can produce slight positive pressures because of the flow resistance, and this encourages water vapor to escape from the vent and condense between the vent walls. Offsets in attics usually involve long runs of pipe that increase the surface area of vent pipe exposed to cold temperature. It is quite common to find that attic offsets exceed the horizontal length allowed by Section G2428.3.5.

G2428.2.4 (504.2.4) Zero lateral. Zero (0) lateral (L) shall apply only to a straight vertical vent attached to a top outlet draft hood or flue collar.

❖ The zero lateral rows in the tables refer only to an entirely vertical vent run without offsets, lateral runs, and directional fittings. This would be the ideal vent. Unfortunately, it is seldom seen because of lack of planning and coordination of vent and appliance locations. Architectural features of buildings usually dictate that vents take less than desirable paths to get from the appliances to the outdoors.

G2428.2.5 (504.2.5) High altitude installations. Sea level input ratings shall be used when determining maximum capacity for high altitude installation. Actual input, derated for altitude, shall be used for determining minimum capacity for high altitude installation.

❖ Appliances installed at elevations above sea level must have their Btu/h (W) input reduced (derated) because the atmosphere is less dense and oxygen levels decrease as altitude increases. The actual volume of flue gases entering a vent (i.e., primary and secondary combustion air, excess air, and combustion by-products) does not significantly change for different elevations; therefore, the maximum vent capacity is based on the sea-level input rating. The minimum capacity of a vent is dependent upon heat input to the vent to control condensation and maintain the required draft. Therefore, the actual heat input (as derated) is used to determine the minimum capacity of a vent.

G2428.2.6 (504.2.6) Multiple input rate appliances. For appliances with more than one input rate, the minimum vent capacity (FAN Min) determined from the tables shall be less than the lowest appliance input rating, and the maximum vent capacity (FAN Max/NAT Max) determined from the tables shall be greater than the highest appliance rating input.

❖ To improve efficiency, many appliances are equipped for modulating input or high-low two-stage input. The venting system must be capable of operating properly throughout the entire firing range of the appliance. The venting system must be capable of handling the maximum input of the appliance and also must be compatible with the appliance when the appliance is operating at its lowest input. Such appliances will usually operate at the "low-fire" input most of their total operating time. To satisfy the minimum input requirements for fan-assisted appliance vents, the vent design must be based on the lowest Btu/h rating of the appliance because this is what the vent "sees" as the connected load.

The lowest appliance input rate is used for minimum vent capacity, and the highest input rate must be used for maximum vent capacity. This requirement can be easily overlooked if the appliance label is not carefully studied, because the appliance will usually be described by its maximum input, and any minimum input will have to be obtained from the label.

G2428.2.7 (504.2.7) Liner system sizing. Listed corrugated metallic chimney liner systems in masonry chimneys shall be sized by using Table G2428.2(1) or G2428.2(2) for Type B vents with the maximum capacity reduced by 20 percent (0.80 × maximum capacity) and the minimum capacity as shown in Table G2428.2(1) or G2428.2(2). Corrugated metallic liner systems installed with bends or offsets shall have their maximum capacity further reduced in accordance with Section G2428.2.3. The 20-percent reduction for corrugated metallic chimney liner systems includes an allowance for one long-radius 90-degree (1.57 rad) turn at the bottom of the liner.

❖ A liner system installed within a masonry chimney is sized and designed as if the chimney/liner system combination were a Type B vent. Corrugated metal liners are constructed of semi-rigid piping having a higher flow resistance than smooth-wall rigid pipe. A capacity reduction penalty compensates for the poorer flow characteristics of corrugated pipe. Bends in semi-rigid liners count as elbows and are regulated by Section G2428.2.3. Note that the sweeping 90-degree bend in the liner at the point where it exits the bottom of the chimney and connects to the appliance(s) is already accounted for in the 20-percent capacity penalty, and no further capacity adjustment is required.

G2428.2.8 (504.2.8) Vent area and diameter Where the vertical vent has a larger diameter than the vent connector, the vertical vent diameter shall be used to determine the minimum vent capacity, and the connector diameter shall be used to determine the maximum vent capacity. The flow area of the vertical vent shall not exceed seven times the flow area of the listed appliance categorized vent area, flue collar area, or draft hood outlet area unless designed in accordance with approved engineering methods.

❖ The maximum capacity of a venting system must be based on the element of the system having the least capacity (i.e., the weakest link in the chain). Likewise, the minimum Btu/h (W) input required for proper vent system operation must be based on the largest element of the system because the largest element will naturally have the highest minimum input requirement. A connector that is smaller than the vent will set the maximum capacity limit for the system because it has the greater flow resistance and smaller cross-sectional area. The minimum Btu/h input is based on the larger vent size because the vent will require more heat to produce adequate draft and avoid condensation. The ratio of appliance vent-connection area to vertical vent area is intended to prevent oversizing of the vent, which can result in poor draft and condensation problems. Because the tables will allow vents to be larger than needed, this section serves as a limit to prevent gross oversizing. For example, in Table G2428.2(1) it can be seen that a fan-assisted appliance having an input rating near the maximum end of the minimum/maximum range for any 4-inch venting system configuration can be served by not only a 4-inch vent but also a 5-inch and sometimes a 6-inch or 7-inch vent. Consistent with vent manufacturers' instructions, a venting system should always be designed to utilize the smallest size vent allowed by the tables. Vents that are larger than they need to be usually will not perform as well as accurately sized vents and are more likely to suffer from poor draft and condensate problems.

G2428.2.9 (504.2.9) Chimney and vent locations. Tables G2428.2(1) and G2428.2(2) shall be used for chimneys and vents not exposed to the outdoors below the roof line. A Type B vent or listed chimney lining system passing through an unused masonry chimney flue shall not be considered to be exposed to the outdoors.

Exception: The installation of vents serving listed appliances shall be permitted to be in accordance with the appliance manufacturer's instructions and the terms of the listing.

❖ Chimneys that have one or more walls exposed to the outdoors and vents exposed to the outdoors can develop condensation and weak draft problems in cold weather. Commentary Figure G2428.2.9 illustrates why masonry chimneys, including interior chimneys, are more prone to draft and condensation problems. The tables assume that the only portion of the chimney or vent that is exposed to the outdoors is the portion that extends above the roof penetration. Even though an attic is technically not outdoors, the temperature in an attic can be very close to the outdoor temperature. In some climates, the actual attic temperature can be far lower than the assumed temperature used in the computer modeling on which the vent sizing tables are based. Therefore, the portion of a venting system that passes through an attic should be kept to an absolute minimum. If at all possible, the installation should be planned to avoid the need for vent offsets in an attic. This section states that an exterior chimney lined with a listed liner or Type B vent is to be sized as if it were a Type B vent. The walls of an exterior chimney will provide some insulating effect for a properly installed Type B vent or lining system installed within the chimney. Thus, such installations are allowed to be sized as interior chimneys or vents. A chimney lined with a Type B vent is not considered to be a chimney but is treated as a Type B vent installed within a masonry chase. See Section G2428.2.3.

This section intends that the listed tables be applicable only to vents and chimneys that are not exposed to the outdoors below the roof penetration. If the vent or chimney is exposed to the outdoors below the roof penetration, the listed sizing tables do not apply. The intent is that vents and chimneys be enclosed within the building envelope to lessen exposure to low ambient temperatures and wind. It is these conditions that were considered in the development of the vent sizing tables. The sizing tables are based on an assumed outdoor temperature of 42°F and the assumption that all portions of a vent or chimney below the roof penetration are exposed to an ambient temperature of 60°F. This makes it obvious that long runs of vent piping through attics and long extensions above roofs should be avoided in cold climates. See Section G2428.3.5.

This section prohibits the use of the vent sizing tables for the installation of a Type B vent run up the outside of an exterior wall of a building. This is consistent with vent manufacturers' instructions. Outdoor vents obviously suffer from poorer draft and increased condensation formation because they are exposed to the elements, low temperatures being the main factor. "Stack effect" in the building interior can cause back drafts (reverse flow) to occur in any vent or chimney, particularly vents exposed to the outdoors. For the same reasons, vents should extend above the roof no more than required by the code because the more vent pipe that extends above the roof, the more vent pipe will be exposed to the cold and wind. In some cases, designers and installers will intentionally raise a vent above a roof more than Section G2427.6.5 requires for the purpose of increasing the vent capacity. In doing so, additional vent pipe is exposed to the outdoors, thus inviting condensation and possibly diminishing draft. Section G2427.6.5 could also require a vent to extend well above a lower roof penetration, thus causing the sizing tables of this section to be inapplicable to such a vent. The tables for exterior masonry chimneys obviously apply to chimneys that are exposed to the outdoors below the roof line, and the list of conditions intends to compensate for the cold masonry. Conditions such as requiring double-wall connectors and draft hood equipped appliances are intended to put as much heat into the chimney as possible because it will be needed to produce draft and minimize condensation in a less than ideal chimney. Exterior chimneys require greater heat input to make them work. An alternative to the continued use of an exterior chimney would be relining or abandoning the chimney and using high-efficiency appliances.

The exception addresses the possibility of an appliance that was tested and listed for venting applications other than allowed by this section. An example might be an appliance that is field-converted to a draft hood appliance for the purpose of utilizing a masonry chimney. Some Category I fan-assisted furnaces can be fitted with a factory supplied conversion part that is, in effect, a retrofit draft hood. This will lower the annual fuel utilization efficiency, (AFUE) of the appliance installation by converting it to a draft-hood-equipped appliance installation that will now introduce dilution air (conditioned room air) into the vent. This is done solely to allow fan-assisted furnaces to connect to existing masonry chimneys that would otherwise not be allowed to serve such appliances. Such conversion parts employ a sensor (spill switch) in the relief/air inlet opening. The manufacturer of the appliance and the draft hood conversion kit will provide specific installation instructions that must be followed for the conversion. Such draft hood conversion kits are listed for use only with the appliance with which they were tested. In some cases, the manufacturer's instructions will contain specific chimney sizing tables that must be used instead of tables in the code.

G2428.2.10 (504.2.10) Corrugated vent connector size. Corrugated vent connectors shall be not smaller than the listed appliance categorized vent diameter, flue collar diameter or draft hood outlet diameter.

❖ Corrugated vent connectors are becoming more common since they can save installation time, and their use avoids the use of fittings. This section is consistent with Section G2428.3.17 and applies to both draft

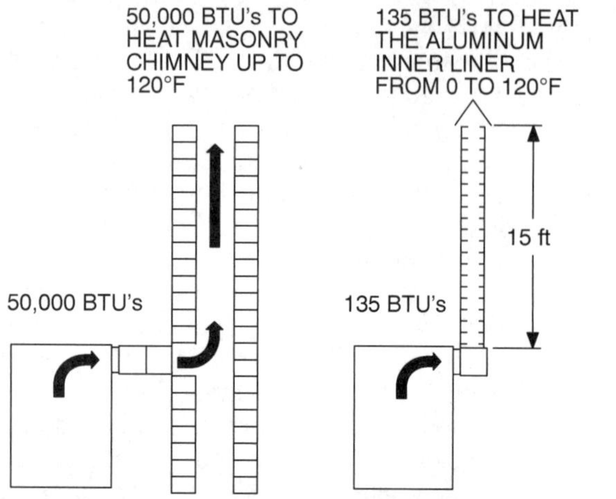

Figure courtesy of Simpson Dura-Vent Company, Incorporated

Figure G2428.2.9
MASONRY CHIMNEY COMPARED TO TYPE B VENT

hood-equipped and fan-assisted appliances. Corrugated materials have higher resistance to flow than smooth wall materials; therefore, to assure the required capacity, full size connectors must be used.

G2428.2.11 (504.2.11) Vent connector size limitation. Vent connectors shall not be increased in size more than two sizes greater than the listed appliance categorized vent diameter, flue collar diameter or draft hood outlet diameter.

❖ The sizing tables of this section will often require a connector to be larger than the appliance vent connection. The appliance manufacturer's instructions may also require a connector to be increased in size. The increase in size compensates for flow resistance in the connector and prevents positive pressurization. A grossly oversized connector will allow the vent gases to expand and cool too much, thus negatively affecting draft and contributing to the formation of condensation. Connectors are sometimes increased for the sole purpose of gaining more allowable length (Section G2428.3.2), and this section will limit such practice.

G2428.2.12 (502.2.12) Component commingling. In a single run of vent or vent connector, different diameters and types of vent and connector components shall be permitted to be used, provided that all such sizes and types are permitted by the tables.

❖ In the unlikely event that an installer wants to mix, for example, Type B vent with single-wall pipe in a single connector (e.g., Type B vent from draft hood switching to single wall and connecting to Type B vent again), such mixed system must comply with the code based on each type of material as if constructed entirely of that material. There may be some unique situations that would justify commingling of different types and/or sizes of components, but many things must be considered such as clearances, heat loss, ambient temperatures, joining methods, and flow characteristics.

G2428.2.13 (504.2.13) Table interpolation. Interpolation shall be permitted in calculating capacities for vent dimensions that fall between the table entries. (See Example 3, Appendix B.)

❖ Because Tables G2428.2(1) and G2428.2(2) do not contain values for every possible height of a vent or length of a lateral, interpolation could be necessary. The maximum and minimum capacity of vents having heights not listed in the tables can be determined using the following equations.

Interpolated Maximum Vent Capacity = $(A/B \times C) + D$

Interpolated Minimum Vent Capacity = $E - (A/B \times F)$

where:

A = The difference between the design height entry and the next lower height entry in the applicable table.

B = The difference between the closest consecutive height entries in the applicable table.

C = The difference between the maximum capacity column consecutive entries in the applicable table.

D = The lower maximum capacity entry in the applicable table.

E = The higher minimum capacity entry in the applicable table.

F = The difference between the minimum capacity column consecutive entries in the applicable table.

For SI: 1 inch = 25.4 mm, 1 foot = 304.8 mm, 1 Btu/h = 0.2931 W

Example 1:

Determine the minimum and maximum vent capacity for a single fan-assisted appliance given the following:

Vertical design height of vent	17 feet
Diameter of vent	6 inches
Vent type	Type B
Length of lateral	10 feet

Using Table G2428.2(1):

Interpolated Maximum Vent Capacity = [(17 ft - 15 ft)/(20 ft - 15 ft) x (351,000 Btu/h - 315,000 Btu/h)] + 315,000 Btu/h

= (2/5 x 36,000 Btu/h) + 315,000 Btu/h)]

= 329,000 Btu/h

Interpolated Minimum Vent Capacity = 64,000 Btu/h - [(17 ft - 15 ft)/(20 ft - 15 ft) x (64,000 Btu/h - 62,000 Btu/h)]

= 64,000 Btu/h – (2.5 x 2,000 Btu/h)

= 63,200 Btu/h

Example 2:

Determine the maximum vent capacity for a single fan assisted furnace with:

Input rating of furnace	210,000 Btu/h
Vertical design height of vent	12½ feet
Diameter of vent	6 inches

Vent type Type B
Length of lateral 2 feet
Using Table G2428.2(1):

Step 1: Find the maximum vent capacity (MAX) at the first height entry greater than $12^1/_2$ feet, i.e., 15 feet.

MAX_{15} = 226,000 Btu/h

Step 2: Find the maximum vent capacity (MAX) at the next height entry lower than $12^1/_2$ feet, i.e., 10 feet.

MAX_{10} = 194,000 Btu/h

Step 3: Determine the difference between the two maximum vent capacities.

MAX_{15} - MAX_{10} = 226,000 Btu/h - 194,000 Btu/h = 32,000 Btu/h

Step 4: Determine the maximum vent capacity for a 12½-foot high vent using:

$MAX_{12½} = [(A/B) \times (MAX_{15} - MAX_{10})] + MAX_{10}$

where:

A = The difference between the next higher table vent height and the design vent height.

B = The difference between the consecutive closest table vent heights.

$MAX_{12½}$ = [(15 ft - $12^1/_2$ ft)/(15 ft - 10 ft) x 32,000 Btu/h] + 194,000 Btu/h
= [($2^1/_2$ ft)/(5 ft) x 32,000 Btu/h] + 194,000 Btu/h
= 16,000 Btu/h + 194,000/h
= 210,000 Btu/h

When a vent height falls between height entries in a table, the code user can choose to interpolate or use the closest table height entry that is lower than the actual height of the vent. These are the only allowable options. This example also illustrates how the tables may be used to reduce vent size if there is adequate height and capacity to do so. See the commentary for Section G2428.2.2. It is important to consider interpolation before assuming that an "in between" height vent installation is not in compliance with the code. For example, a vent with a 25-foot height could fail to comply with a table when it is viewed as a 20-foot vent, but that same vent could quite possibly comply with the table when viewed as what it actually is, a 25-foot height vent. This interpolation process can also be used to estimate in-between capacities for intermediate lengths of laterals, as well as in-between minimum capacities. There is no way to estimate in-between capacity between a zero lateral (straight vertical vent) and a 2-foot lateral (which has two elbows). If the vent has just one 90-degree turn, the 2-foot lateral capacity applies.

G2428.2.14 (504.2.14) Extrapolation prohibited. Extrapolation beyond the table entries shall not be permitted.

❖ Projecting capacities below the lowest or above the highest boundary entries of the tables is not allowed, because the tables are based on the known operational limits of venting systems. The tables cannot predict the operating characteristics of venting systems having heights shorter or taller than the table limits; therefore, any such system must be an engineered system.

G2428.2.15 (504.2.15) Engineering calculations. For vent heights less than 6 feet (1829 mm) and greater than shown in the tables, engineering methods shall be used to calculate vent capacities.

❖ This section parallels the previous section by stating that vents having heights not within the tables must be designed by an engineered method. In other words, extrapolation above or below the table height entries is not allowed. See Section G2428.2.13.

G2428.3 (504.3) Application of multiple appliance vent Tables G2428.3(1) through G2428.3(4). The application of Tables G2428.3(1) through G2428.3(4) shall be subject to the requirements of Sections G2428.3.1 through G2428.3.22.

❖ The application of Tables G2428.3(1) through G2428.3(4) shall be subject to the requirements of Sections G2428.3.1 through G2428.3.22. This part of Section G2428 regulates vents and chimneys that serve more than one appliance. Such venting systems can be described as combined vents and include multiple connectors. A "combined vent" is a vent for two or more appliances at one level served by a "common vent." "Least total height" is the vertical distance from the highest appliance outlet (draft hood or flue collar) to the lowest discharge opening of the vent top. This height dimension is illustrated in Commentary Figure G2428.3(1) for a typical (FAN + NAT) system. The same height measurement applies to FAN+FAN and NAT+NAT systems. Least total height is used for vent sizing of all connected appliances on one level. "Connector rise" for any appliance is the vertical distance from its outlet connection to the level at which it joins the common vent, as shown in Commentary Figure G2428.3(1).

A "connector" for purposes of designing a combined vent is that part of the vent piping between the appliance outlet and its junction with or interconnection to the rest of the system. A chimney is referred to as a venting means, although a chimney and a vent are distinct. An improperly sized common vent or chimney could fail to vent the combustion products of the appliances served. There are many factors that influence the sizing of a vent or chimney, including total height, offsets, lateral lengths, the type of appliance, the appliance energy input, and the appliance connector configurations. The "common vent" is that portion of the system serving two or more connected appliances. If connectors are joined before reaching the vertical vent, the run between the last entering connector and the vertical portion is also treated as part of the common vent. In Commentary Figure G2428.3(1), the vertical common vent is cross-hatched beginning at the interconnection tee.

For each connector, the correct size must be found from the applicable tables based on its appliance input, rise, and least total vent height. For draft hood (NAT) appliances, the outlet size may be too small if there is not enough rise; therefore, connector design involves choosing the correct size and verifying if use of a connector the same size as the outlet is within input rating limits.

For the common vent, the capacity table shows maximum combined ratings only. The size of the common vent is thus based on least total height and the combination of attached appliances, FAN+FAN, FAN+NAT, or NAT+NAT.

To find each connector size, use Table G2428.3(1) for Type B vent connectors or Table G2428.3(2) for single-wall metal connectors and proceed as follows:

1. Determine "least total height" for the system.
2. Determine connector rise for each appliance.
3. Enter the applicable vent connector table at the least total height. Continue across on the line for appliance connector rise to a MAX input rating equal to or greater than that of the appliance. For a FAN appliance, this input rating should also be greater than shown in the MIN column. Read the connector size at the top of the column. If Table G2428.3(2) for single wall shows "NA," use a double-wall connector in accordance with Table G2427.3(1). In some cases, the table will dictate a connector size that is larger than the appliance vent outlet (flue collar).

To find the common vent size, proceed as follows:

1. Add all appliances Btu input ratings to get the total Btu input.
2. If one or both connectors are single-wall metal, use Table G2428.3(2) for the common vent.
3. Enter the common vent table at the same least total height used for connectors.
4. Continue across and stop at the first applicable column of combined appliance input rating equal to or greater than the total. If Table G2428.3(2) for the common vent shows "NA," the entire system must be double-wall Type B vent.
5. Read the size of the common vent at the top of the applicable column (NAT+NAT, FAN+NAT, or FAN+FAN).
   (a) Regardless of table results for size, the common vent must be at least as large as the largest connector.
   (b) For NAT+NAT appliance combinations, if both connectors are the same size, the common vent must be at least one size larger.

Example 1

Commentary Figure G2428.3(1) shows a two-appliance "FAN+NAT" system, combining a draft hood water heater (NAT appliance) with a fan-assisted combustion Category I (FAN) furnace. The system is designed in the following steps using Table G2428.3(1) for double-wall connectors.

1. For the water heater, enter the vent connector table at a least total height of 15 feet (4572 mm) and a connector rise of 1 foot (305 mm). Read across to the MAX Btu/h rating for a NAT appliance vent higher than 35,000 Btu/h (10.3 kW). The table shows that a 4-inch (102 mm) size connector is needed. This size must be used beginning at the draft hood, regardless of draft hood size, which might be 3 inches (76 mm). See Commentary Figure G2428.3(1).
2. For the 105,000 Btu/h (30.8 kW) furnace, enter the vent connector table at the same least total height [15 feet (4572 mm)] at a connector rise of 3 feet (914 mm). Read across to 163,000 under

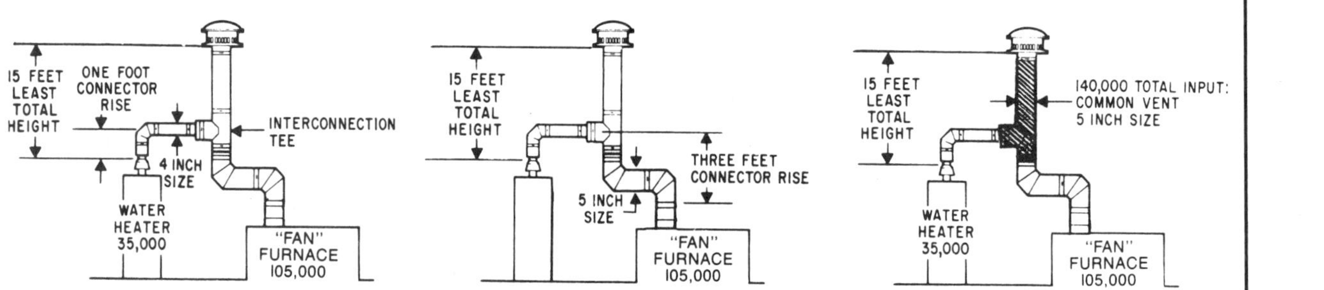

Figure G2428.3(1)
COMBINED VENT DIMENSIONS

the FAN MAX column. The MIN is 51,000. Therefore, 5 inches (127 mm) is the correct connector size. See Commentary Figure G2428.3(1).

3. For the common vent, the sum of the two ratings is 140,000 Btu/h (41.0 kW). Enter the common vent table at 15 foot (4572 mm) least total height. For a FAN+NAT combination, the maximum input of 5-inch (127 mm) vent is 164,000 (48.1 kW), so 5 inches (127 mm) is the proper size, as shown in Commentary Figure G2428.3(1).

Both draft hood and fan-assisted appliances may be common vented in any combination, as indicated by the headings in the common vent table. Appliance types include the following:

- Central heating furnaces;
- Central heating boilers (hot water and steam);
- Water heaters;
- Unit heaters;
- Duct furnaces;
- Room heaters*; and
- Floor furnaces*.

*If these have draft hoods, an appropriate design input increase of 40 percent is recommended by some vent manufacturers in order to use the tables. See the commentary for Section G2428.2.

The common vent tables do not apply to the following:

1. Gas cooking appliances, which should be vented into an appropriate exhaust hood or vented individually in accordance with the appliance manufacturer's instructions.
2. Forced draft, and commercial or industrial hot water or steam boilers without draft hoods. For this equipment, see the boiler and vent/chimney manufacturer's instructions.
3. Categories II, III, and IV gas-burning equipment, for which the equipment manufacturers' venting instructions must be used.
4. Gas-fired incinerators.

The manufacturers of chimney and vent systems provide installation and design information as part of their installation instructions. The common chimney or vent must be designed to properly vent the products of combustion when any one, any combination or all of the connected appliances operate. See Commentary Figure G2428.3(2). Solid fuel-burning appliances must not connect to common chimneys (see Section G2427.5.6.1).

Combined vent systems for two or more gas appliances of either type (FAN or NAT) must be designed to prevent draft hood spillage for natural draft (NAT) appliances and to avoid positive pressure for fan-assisted (FAN) appliances. The connector and common vent tables have been computed by examining the most critical situations for any operating combination. A common vent must function properly when all, one, or any combination of the connected appliances are firing. It is particularly demanding of a vent system to operate properly when only a single appliance having the lowest input rating is firing, thus creating the greatest potential for poor draft and condensation. Oversizing of venting systems should be avoided.

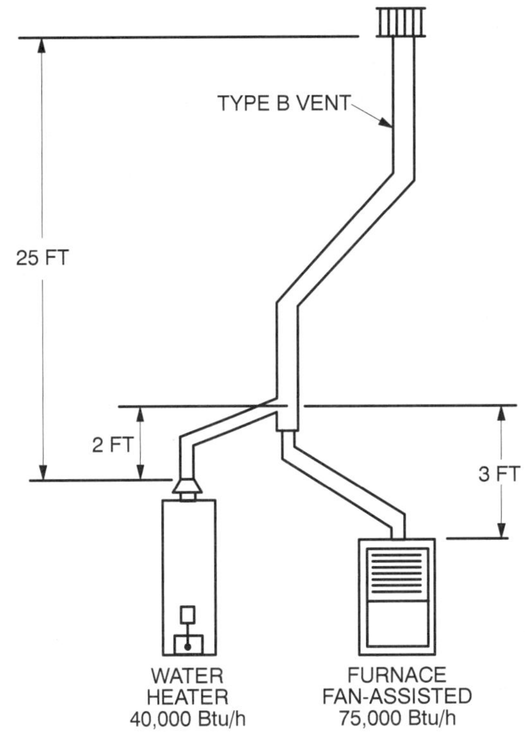

For SI: 1 foot = 304.8 mm.

Figure G2428.3(2)
FAN AND NAT VENT SIZING EXAMPLE

The connector tables are based on the most critical condition for that particular appliance when operating by itself, whereas the common vent tables show sizes that assure adequate capacity and draft, whether one or all appliances are operating simultaneously.

All parts of a combined vent must be checked for capacity. For connectors, the size must be determined from the tables, particularly for low-height vents or where headroom restricts available connector rise.

The connector in a combined vent system is defined as the piping from a draft hood or flue collar to the junction of the common vent or to a junction in a vent manifold. Proper connector design is vital to obtaining adequate capacity. The connector must produce its share of the total draft for its NAT or FAN appliance and must deliver enough heat to the common vent so that it can contribute the balance of draft needed.

For application of the codes' installation provisions, the Type B gas vent connector is a "gas vent." It is essential, however, for system design purposes to use the word "connector" so that its rise and configuration may be explained and tabulated.

The connector tables for combined vents show MIN and MAX capacities only for FAN appliances because it has been found that no minimums are needed for NAT appliances with Type B gas vent connectors.

Important factors in connector design include the following:

- Connector material;
- Connector length;
- Connector rise;
- Number of fittings;
- Appliance location as it affects the piping arrangement;
- Number of attached appliances or different connector sizes; and
- Connection to an offset or manifold rather than directly to the vertical common vent.

Example 2

This example is a common scenario. See Figure G2428.3(2).

Given:

- Single-family house with Type B vent serving two appliances
- Category I fan-assisted furnace, 80-percent efficient, 75,000 Btu/h input (4-inch vent collar)
- Furnace connector single-wall pipe with 3-foot rise
- Draft hood equipped water heater, 40,000 Btu/h (3-inch draft hood outlet)
- Water heater connector single-wall with 2-foot rise
- Vent height 25 feet
- A 45-degree offset occurs in the common vent in the attic

Step 1. Choose the applicable table. Table G2428.3(2) addresses two or more Category I appliances with single-wall connectors.

Step 2. Using the top portion of the table, size the appliance connectors. Because the table has no row for 25 feet height, the 20 feet row must be used, or interpolation must be used to calculate the capacities at 25 feet. To simplify the procedure, first proceed using the 20 feet row until such time that interpolation is necessary.

A. It can be seen in the “NAT” columns that a 4-inch connector is required for the water heater because the water heater input of 40,000 Btu/h exceeds the maximum capacity (36,000) of a 3-inch connector with a 2-foot rise.

B. It can be seen in the “FAN” columns that the furnace input of 75,000 Btu/h is less than the minimum capacity of 4-inch and larger connectors with a 3-foot rise. This means that single-wall pipe cannot be used for the furnace connector. The furnace connector must be double-wall (Type B or equivalent) and must be sized by Table G2428.3(1). In Table G2428.3(1), it can be seen that a 4-inch double-wall connector is suitable for the furnace because the 75,000 Btu/h inputs fits within the range of 35,000 to 110,000 Btu/h for a 3-foot rise.

The furnace connector has been determined to be 4-inch double-wall pipe, and the water heater connector has been determined to be 4-inch single-wall pipe.

Step 3. Using the bottom portion of Table G2428.3(2), the common vent is sized. (Because both single and double-wall connectors are used within the same system, Section G2428.3.18 requires the single-wall table to be used.) The water heater and furnace inputs total 115,000 Btu/h. Under the 4-inch vent heading, FAN + NAT column, it can be seen that 4-inch Type B vent has a capacity of 118,000 Btu/h, which exceeds the appliance total of 115,000. However, the system contains an offset, which will reduce this capacity by 20 percent in accordance with Section G2428.3.5.

0.80 x 118,000 = 94,400 Btu/h

The actual capacity of the 4-inch common vent is 94,400 Btu/h, which is less than the 115,000 appliance total; therefore a 5-inch common vent is required.

Verify 5-inch capacity: 0.80 x 177 = 141,600 Btu/h

By looking at the 30-feet height rows, it is apparent that interpolation for the actual height of 25 feet would not have changed the outcome of this system design.

Appendix B contains additional sizing examples.

G2428.3.1 (504.3.1) Vent obstructions. These venting tables shall not be used where obstructions, as described in Section G2427.15, are installed in the venting system. The installation of vents serving listed appliances with vent dampers shall be in accordance with the appliance manufacturer’s instructions or in accordance with the following:

1. The maximum capacity of the vent connector shall be determined using the NAT Max column.
2. The maximum capacity of the vertical vent or chimney shall be determined using the FAN+NAT column when the second appliance is a fan-assisted appliance, or the NAT+NAT column when the second appliance is equipped with a draft hood.
3. The minimum capacity shall be determined as if the appliance were a fan-assisted appliance.
   - 3.1. The minimum capacity of the vent connector shall be determined using the FAN Min column.
   - 3.2. The FAN+FAN column shall be used when the second appliance is a fan-assisted appliance, and the FAN+NAT column shall be used when the second appliance is equipped with a draft hood, to determine whether the vertical vent or chimney configuration is not permitted (NA). Where the vent configuration is NA, the vent configuration shall not be permitted and an alternative venting configuration shall be utilized.

❖ See the commentary for Section G2428 .2.1. This section is a bit more complex than Section G2428.2.1 because the tables for multiple appliances are two-part tables. The difference lies in steps 2 and 3.2 dealing with the common vent portion of the sizing tables. Step

2 determines the maximum capacity of the common vent by treating the vent damper equipped appliance as a draft hood appliance coupled with either another draft hood appliance or a fan-assisted appliance. Step 3.2 is conducted as a test wherein the vent damper equipped appliance is treated as a fan-assisted appliance. If the table says NA, the vent system must be redesigned, regardless of the fact that it may have been allowed by the table under step 2.

TABLE G2428.3(1) [504.3(1)] TYPE B DOUBLE-WALL VENT

Number of Appliances	Two or more
Appliance Type	Category I
Appliance Vent Connection	Type B double-wall connector

Vent Connector Capacity

VENT HEIGHT H (feet)	CONNECTOR RISE R (feet)	TYPE B DOUBLE-WALL VENT AND CONNECTOR DIAMETER — D (inches)																															
		3			4			5			6			7			8			9			10										
		Appliance input rating limits in thousands of Btu/h																															
		FAN		NAT	FAN		NAT	FAN		NAT	FAN		NAT	FAN		NAT	FAN		NAT	FAN		NAT	FAN		NAT								
		Min	Max	Max	Min	Max	Max	Min	Max	Max	Min	Max	Max	Min	Max	Max	Min	Max	Max	Min	Max	Max	Min	Max	Max								
6	1	22	37	26	35	66	46	46	106	72	58	164	104	77	225	142	92	296	185	109	376	237	128	466	289								
	2	23	41	31	37	75	55	48	121	86	60	183	124	79	253	168	95	333	220	112	424	282	131	526	345								
	3	24	44	35	38	81	62	49	132	96	62	199	139	82	275	189	97	363	248	114	463	317	134	575	386								
8	1	22	40	27	35	72	48	49	114	76	64	176	109	84	243	148	100	320	194	118	408	248	138	507	303								
	2	23	44	32	36	80	57	51	128	90	66	195	129	86	269	175	103	356	230	121	454	294	141	564	358								
	3	24	47	36	37	87	64	53	139	101	67	210	145	88	290	198	105	384	258	123	492	330	143	612	402								
10	1	22	43	28	34	78	50	49	123	78	65	189	113	89	257	154	106	341	200	125	436	257	146	542	314								
	2	23	47	33	36	86	59	51	136	93	67	206	134	91	282	182	109	374	238	128	479	305	149	596	372								
	3	24	50	37	37	92	67	52	146	104	69	220	150	94	303	205	111	402	268	131	515	342	152	642	417								
15	1	21	50	30	33	89	53	47	142	83	64	220	120	88	298	163	110	389	214	134	493	273	162	609	333								
	2	22	53	35	35	96	63	49	153	99	66	235	142	91	320	193	112	419	253	137	532	323	165	658	394								
	3	24	55	40	36	102	71	51	163	111	68	248	160	93	339	218	115	445	286	140	565	365	167	700	444								
20	1	21	54	31	33	99	56	46	157	87	62	246	125	86	334	171	107	436	224	131	552	285	158	681	347								
	2	22	57	37	34	105	66	48	167	104	64	259	149	89	354	202	110	463	265	134	587	339	161	725	414								
	3	23	60	42	35	110	74	50	176	116	66	271	168	91	371	228	113	486	300	137	618	383	164	764	466								
30	1	20	62	33	31	113	59	45	181	93	60	288	134	83	391	182	103	512	238	125	649	305	151	802	372								
	2	21	64	39	33	118	70	47	190	110	62	299	158	85	408	215	105	535	282	129	679	360	155	840	439								
	3	22	66	44	34	123	79	48	198	124	64	309	178	88	423	242	108	555	317	132	706	405	158	874	494								
50	1	19	71	36	30	133	64	43	216	101	57	349	145	78	477	197	97	627	257	120	797	330	144	984	403								
	2	21	73	43	32	137	76	45	223	119	59	358	172	81	490	234	100	645	306	123	820	392	148	1,014	478								
	3	22	75	48	33	141	86	46	229	134	61	366	194	83	502	263	103	661	343	126	842	441	151	1,043	538								

Common Vent Capacity

VENT HEIGHT H (feet)	TYPE B DOUBLE-WALL COMMON VENT DIAMETER — D (inches)																				
	4			5			6			7			8			9			10		
	Combined appliance input rating in thousands of Btu/h																				
	FAN +FAN	FAN +NAT	NAT +NAT	FAN +FAN	FAN +NAT	NAT +NAT	FAN +FAN	FAN +NAT	NAT +NAT	FAN +FAN	FAN +NAT	NAT +NAT	FAN +FAN	FAN +NAT	NAT +NAT	FAN +FAN	FAN +NAT	NAT +NAT	FAN +FAN	FAN +NAT	NAT +NAT
6	92	81	65	140	116	103	204	161	147	309	248	200	404	314	260	547	434	335	672	520	410
8	101	90	73	155	129	114	224	178	163	339	275	223	444	348	290	602	480	378	740	577	465
10	110	97	79	169	141	124	243	194	178	367	299	242	477	377	315	649	522	405	800	627	495
15	125	112	91	195	164	144	283	228	206	427	352	280	556	444	365	753	612	465	924	733	565
20	136	123	102	215	183	160	314	255	229	475	394	310	621	499	405	842	688	523	1,035	826	640
30	152	138	118	244	210	185	361	297	266	547	459	360	720	585	470	979	808	605	1,209	975	740
50	167	153	134	279	244	214	421	353	310	641	547	423	854	706	550	1,164	977	705	1,451	1,188	860

For SI: 1 inch = 25.4 mm, 1 foot = 304.8 mm, 1,000 Btu = 0.293 kW.

❖ See the commentary for Section G2428.3.

TABLE G2428.3(2) [504.3(2)]
TYPE B DOUBLE-WALL VENT

Number of Appliances	Two or more
Appliance Type	Category I
Appliance Vent Connection	Single-wall metal connector

Vent Connector Capacity

VENT HEIGHT H (feet)	CONNECTOR RISE R (feet)	SINGLE-WALL METAL VENT CONNECTOR DIAMETER — D (inches)																							
		3			4			5			6			7			8			9			10		
		Appliance input rating in thousands of Btu/h																							
		FAN		NAT	FAN		NAT	FAN		NAT	FAN		NAT	FAN		NAT	FAN		NAT	FAN		NAT	FAN		NAT
		Min	Max	Max	Min	Max	Max	Min	Max	Max	Min	Max	Max	Min	Max	Max	Min	Max	Max	Min	Max	Max	Min	Max	Max
6	1	NA	NA	26	NA	NA	46	NA	NA	71	NA	NA	102	207	223	140	262	293	183	325	373	234	447	463	286
	2	NA	NA	31	NA	NA	55	NA	NA	85	168	182	123	215	251	167	271	331	219	334	422	281	458	524	344
	3	NA	NA	34	NA	NA	62	121	131	95	175	198	138	222	273	188	279	361	247	344	462	316	468	574	385
8	1	NA	NA	27	NA	NA	48	NA	NA	75	NA	NA	106	226	240	145	285	316	191	352	403	244	481	502	299
	2	NA	NA	32	NA	NA	57	125	126	89	184	193	127	234	266	173	293	353	228	360	450	292	492	560	355
	3	NA	NA	35	NA	NA	64	130	138	100	191	208	144	241	287	197	302	381	256	370	489	328	501	609	400
10	1	NA	NA	28	NA	NA	50	119	121	77	182	186	110	240	253	150	302	335	196	372	429	252	506	534	308
	2	NA	NA	33	84	85	59	124	134	91	189	203	132	248	278	183	311	369	235	381	473	302	517	589	368
	3	NA	NA	36	89	91	67	129	144	102	197	217	148	257	299	203	320	398	265	391	511	339	528	637	413
15	1	NA	NA	29	79	87	52	116	138	81	177	214	116	238	291	158	312	380	208	397	482	266	556	596	324
	2	NA	NA	34	83	94	62	121	150	97	185	230	138	246	314	189	321	411	248	407	522	317	568	646	387
	3	NA	NA	39	87	100	70	127	160	109	193	243	157	255	333	215	331	438	281	418	557	360	579	690	437
20	1	49	56	30	78	97	54	115	152	84	175	238	120	233	325	165	306	425	217	390	538	276	546	664	336
	2	52	59	36	82	103	64	120	163	101	182	252	144	243	346	197	317	453	259	400	574	331	558	709	403
	3	55	62	40	87	107	72	125	172	113	190	264	164	252	363	223	326	476	294	412	607	375	570	750	457
30	1	47	60	31	77	110	57	112	175	89	169	278	129	226	380	175	296	497	230	378	630	294	528	779	358
	2	51	62	37	81	115	67	117	185	106	177	290	152	236	397	208	307	521	274	389	662	349	541	819	425
	3	54	64	42	85	119	76	122	193	120	185	300	172	244	412	235	316	542	309	400	690	394	555	855	482
50	1	46	69	34	75	128	60	109	207	96	162	336	137	217	460	188	284	604	245	364	768	314	507	951	384
	2	49	71	40	79	132	72	114	215	113	170	345	164	226	473	223	294	623	293	376	793	375	520	983	458
	3	52	72	45	83	136	82	119	221	123	178	353	186	235	486	252	304	640	331	387	816	423	535	1,013	518

Common Vent Capacity

VENT HEIGHT H (feet)	TYPE B DOUBLE-WALL VENT AND DIAMETER — D (inches)																				
	4			5			6			7			8			9			10		
	Combined appliance input rating in thousands of Btu/h																				
	FAN +FAN	FAN +NAT	NAT +NAT	FAN +FAN	FAN +NAT	NAT +NAT	FAN +FAN	FAN +NAT	NAT +NAT	FAN +FAN	FAN +NAT	NAT +NAT	FAN +FAN	FAN +NAT	NAT +NAT	FAN +FAN	FAN +NAT	NAT +NAT	FAN +FAN	FAN +NAT	NAT +NAT
6	NA	78	64	NA	113	99	200	158	144	304	244	196	398	310	257	541	429	332	665	515	407
8	NA	87	71	NA	126	111	218	173	159	331	269	218	436	342	285	592	473	373	730	569	460
10	NA	94	76	163	137	120	237	189	174	357	292	236	467	369	309	638	512	398	787	617	487
15	121	108	88	189	159	140	275	221	200	416	343	274	544	434	357	738	599	456	905	718	553
20	131	118	98	208	177	156	305	247	223	463	383	302	606	487	395	824	673	512	1,013	808	626
30	145	132	113	236	202	180	350	286	257	533	446	349	703	570	459	958	790	593	1,183	952	723
50	159	145	128	268	233	208	406	337	296	622	529	410	833	686	535	1,139	954	689	1,418	1,157	838

For SI: 1 inch = 25.4 mm, 1 foot = 304.8 mm, 1,000 Btu = 0.293 kW.

❖ See the commentary for Section G2428.3.

TABLE G2428.3(3) [504.3(3)] MASONRY CHIMNEY

Number of Appliances	Two or more
Appliance Type	Category I
Appliance Vent Connection	Type B double-wall connector

Vent Connector Capacity

VENT HEIGHT H (feet)	CONNECTOR RISE R (feet)	TYPE B DOUBLE-WALL VENT CONNECTOR DIAMETER — D (inches)																							
		3			4			5			6			7			8			9			10		
		Appliance input rating limits in thousands of Btu/h																							
		FAN		NAT	FAN		NAT	FAN		NAT	FAN		NAT	FAN		NAT	FAN		NAT	FAN		NAT	FAN		NAT
		Min	Max	Max	Min	Max	Max	Min	Max	Max	Min	Max	Max	Min	Max	Max	Min	Max	Max	Min	Max	Max	Min	Max	Max
6	1	24	33	21	39	62	40	52	106	67	65	194	101	87	274	141	104	370	201	124	479	253	145	599	319
	2	26	43	28	41	79	52	53	133	85	67	230	124	89	324	173	107	436	232	127	562	300	148	694	378
	3	27	49	34	42	92	61	55	155	97	69	262	143	91	369	203	109	491	270	129	633	349	151	795	439
8	1	24	39	22	39	72	41	55	117	69	71	213	105	94	304	148	113	414	210	134	539	267	156	682	335
	2	26	47	29	40	87	53	57	140	86	73	246	127	97	350	179	116	473	240	137	615	311	160	776	394
	3	27	52	34	42	97	62	59	159	98	75	269	145	99	383	206	119	517	276	139	672	358	163	848	452
10	1	24	42	22	38	80	42	55	130	71	74	232	108	101	324	153	120	444	216	142	582	277	165	739	348
	2	26	50	29	40	93	54	57	153	87	76	261	129	103	366	184	123	498	247	145	652	321	168	825	407
	3	27	55	35	41	105	63	58	170	100	78	284	148	106	397	209	126	540	281	147	705	366	171	893	463
15	1	24	48	23	38	93	44	54	154	74	72	277	114	100	384	164	125	511	229	153	658	297	184	824	375
	2	25	55	31	39	105	55	56	174	89	74	299	134	103	419	192	128	558	260	156	718	339	187	900	432
	3	26	59	35	41	115	64	57	189	102	76	319	153	105	448	215	131	597	292	159	760	382	190	960	486
20	1	24	52	24	37	102	46	53	172	77	71	313	119	98	437	173	123	584	239	150	752	312	180	943	397
	2	25	58	31	39	114	56	55	190	91	73	335	138	101	467	199	126	625	270	153	805	354	184	1011	452
	3	26	63	35	40	123	65	57	204	104	75	353	157	104	493	222	129	661	301	156	851	396	187	1067	505
30	1	24	54	25	37	111	48	52	192	82	69	357	127	96	504	187	119	680	255	145	883	337	175	1,115	432
	2	25	60	32	38	122	58	54	208	95	72	376	145	99	531	209	122	715	287	149	928	378	179	1,171	484
	3	26	64	36	40	131	66	56	221	107	74	392	163	101	554	233	125	746	317	152	968	418	182	1,220	535
50	1	23	51	25	36	116	51	51	209	89	67	405	143	92	582	213	115	798	294	140	1,049	392	168	1,334	506
	2	24	59	32	37	127	61	53	225	102	70	421	161	95	604	235	118	827	326	143	1,085	433	172	1,379	558
	3	26	64	36	39	135	69	55	237	115	72	435	180	98	624	260	121	854	357	147	1,118	474	176	1,421	611

Common Vent Capacity

VENT HEIGHT H (feet)	MINIMUM INTERNAL AREA OF MASONRY CHIMNEY FLUE (square inches)																							
	12			19			28			38			50			63			78			113		
	Combined appliance input rating in thousands of Btu/h																							
	FAN +FAN	FAN +FAN	NAT +NAT	FAN +FAN	FAN +FAN	NAT +NAT	FAN +FAN	FAN +FAN	NAT +NAT	FAN +FAN	FAN +FAN	NAT +NAT	FAN +FAN	FAN +FAN	NAT +NAT	FAN +FAN	FAN +FAN	NAT +NAT	FAN +FAN	FAN +FAN	NAT +NAT	FAN +FAN	FAN +FAN	NAT +NAT
6	NA	74	25	NA	119	46	NA	178	71	NA	257	103	NA	351	143	NA	458	188	NA	582	246	1,041	853	NA
8	NA	80	28	NA	130	53	NA	193	82	NA	279	119	NA	384	163	NA	501	218	724	636	278	1,144	937	408
10	NA	84	31	NA	138	56	NA	207	90	NA	299	131	NA	409	177	606	538	236	776	686	302	1,226	1,010	454
15	NA	NA	36	NA	152	67	NA	233	106	NA	334	152	523	467	212	682	611	283	874	781	365	1,374	1,156	546
20	NA	NA	41	NA	NA	75	NA	250	122	NA	368	172	565	508	243	742	668	325	955	858	419	1,513	1,286	648
30	NA	NA	NA	NA	NA	NA	NA	270	137	NA	404	198	615	564	278	816	747	381	1,062	969	496	1,702	1,473	749
50	NA	NA	NA	NA	NA	NA	NA	NA	NA	NA	NA	NA	NA	620	328	879	831	461	1,165	1,089	606	1,905	1,692	922

For SI: 1 inch = 25.4 mm, 1 foot = 304.8 mm, 1,000 Btu = 0.293 kW, 1 square inch = 645.16 mm^2.

❖ See the commentary for Section G2427.3.

TABLE G2428.3(4) [504.3(4)] MASONRY CHIMNEY

Number of Appliances	Two or more
Appliance Type	Category I
Appliance Vent Connection	Single-wall metal connector

Vent Connector Capacity

VENT HEIGHT H (feet)	CONNECTOR RISE R (feet)	SINGLE-WALL METAL VENT CONNECTOR DIAMETER — D (inches)																							
		3			4			5			6			7			8			9			10		
		Appliance input rating limits in thousands of Btu/h																							
		FAN		NAT	FAN		NAT	FAN		NAT	FAN		NAT	FAN		NAT	FAN		NAT	FAN		NAT	FAN		NAT
		Min	Max	Max	Min	Max	Max	Min	Max	Max	Min	Max	Max	Min	Max	Max	Min	Max	Max	Min	Max	Max	Min	Max	Max
6	1	NA	NA	21	NA	NA	39	NA	NA	66	179	191	100	231	271	140	292	366	200	362	474	252	499	594	316
	2	NA	NA	28	NA	NA	52	NA	NA	84	186	227	123	239	321	172	301	432	231	373	557	299	509	696	376
	3	NA	NA	34	NA	NA	61	134	153	97	193	258	142	247	365	202	309	491	269	381	634	348	519	793	437
8	1	NA	NA	21	NA	NA	40	NA	NA	68	195	208	103	250	298	146	313	407	207	387	530	263	529	672	331
	2	NA	NA	28	NA	NA	52	137	139	85	202	240	125	258	343	177	323	465	238	397	607	309	540	766	391
	3	NA	NA	34	NA	NA	62	143	156	98	210	264	145	266	376	205	332	509	274	407	663	356	551	838	450
10	1	NA	NA	22	NA	NA	41	130	151	70	202	225	106	267	316	151	333	434	213	410	571	273	558	727	343
	2	NA	NA	29	NA	NA	53	136	150	86	210	255	128	276	358	181	343	489	244	420	640	317	569	813	403
	3	NA	NA	34	97	102	62	143	166	99	217	277	147	284	389	207	352	530	279	430	694	363	580	880	459
15	1	NA	NA	23	NA	NA	43	129	151	73	199	271	112	268	376	161	349	502	225	445	646	291	623	808	366
	2	NA	NA	30	92	103	54	135	170	88	207	295	132	277	411	189	359	548	256	456	706	334	634	884	424
	3	NA	NA	34	96	112	63	141	185	101	215	315	151	286	439	213	368	586	289	466	755	378	646	945	479
20	1	NA	NA	23	87	99	45	128	167	76	197	303	117	265	425	169	345	569	235	439	734	306	614	921	387
	2	NA	NA	30	91	111	55	134	185	90	205	325	136	274	455	195	355	610	266	450	787	348	627	986	443
	3	NA	NA	35	96	119	64	140	199	103	213	343	154	282	481	219	365	644	298	461	831	391	639	1,042	496
30	1	NA	NA	24	86	108	47	126	187	80	193	347	124	259	492	183	338	665	250	430	864	330	600	1,089	421
	2	NA	NA	31	91	119	57	132	203	93	201	366	142	269	518	205	348	699	282	442	908	372	613	1,145	473
	3	NA	NA	35	95	127	65	138	216	105	209	381	160	277	540	229	358	729	312	452	946	412	626	1,193	524
50	1	NA	NA	24	85	113	50	124	204	87	188	392	139	252	567	208	328	778	287	417	1,022	383	582	1,302	492
	2	NA	NA	31	89	123	60	130	218	100	196	408	158	262	588	230	339	806	320	429	1,058	425	596	1,346	545
	3	NA	NA	35	94	131	68	136	231	112	205	422	176	271	607	255	349	831	351	440	1,090	466	610	1,386	597

Common Vent Capacity

VENT HEIGHT H (feet)	MINIMUM INTERNAL AREA OF MASONRY CHIMNEY FLUE (square inches)																							
	12			19			28			38			50			63			78			113		
	Combined appliance input rating in thousands of Btu/h																							
	FAN +FAN	FAN +NAT	NAT +NAT	FAN +FAN	FAN +NAT	NAT +NAT	FAN +FAN	FAN +NAT	NAT +NAT	FAN +FAN	FAN +NAT	NAT +NAT	FAN +FAN	FAN +NAT	NAT +NAT	FAN +FAN	FAN +NAT	NAT +NAT	FAN +FAN	FAN +NAT	NAT +NAT	FAN +FAN	FAN +NAT	NAT +NAT
6	NA	NA	25	NA	118	45	NA	176	71	NA	255	102	NA	348	142	NA	455	187	NA	579	245	NA	846	NA
8	NA	NA	28	NA	128	52	NA	190	81	NA	276	118	NA	380	162	NA	497	217	NA	633	277	1,136	928	405
10	NA	NA	31	NA	136	56	NA	205	89	NA	295	129	NA	405	175	NA	532	234	771	680	300	1,216	1,000	450
15	NA	NA	36	NA	NA	66	NA	230	105	NA	335	150	NA	400	210	677	602	280	866	772	360	1,359	1,139	540
20	NA	NA	NA	NA	NA	74	NA	247	120	NA	362	170	NA	503	240	765	661	321	947	849	415	1,495	1,264	640
30	NA	NA	NA	NA	NA	NA	NA	NA	135	NA	398	195	NA	558	275	808	739	377	1,052	957	490	1,682	1,447	740
50	NA	NA	NA	NA	NA	NA	NA	NA	NA	NA	NA	NA	NA	612	325	NA	821	456	1,152	1,076	600	1,879	1,672	910

For SI: 1 inch = 25.4 mm, 1 foot = 304.8 mm, 1,000 Btu = 0.293 kW, 1 square inch = 645.16 mm^2.

❖ See the commentary for Section G2428.3.

G2428.3.2 (504.3.2) Connector length limit. The vent connector shall be routed to the vent utilizing the shortest possible route. Except as provided in Section G2428.3.3, the maximum vent connector horizontal length shall be 1.5 feet (457 mm) for each inch (18 mm per mm) of connector diameter as shown in Table G2428.3.2.

❖ Connectors are necessary for chimney and vent systems that serve multiple appliances. To reduce friction loss and flue gas heat loss and to conserve available draft, connectors must be kept to a minimum length. Thoughtful placement of appliances and chimneys or vents can reduce connector lengths and achieve optimum venting performance. Unlike Tables G2428.2(1) and G2428.2(2), Tables G2428.3(1) through G2428.3(4) do not provide a range of permissible lateral lengths. Therefore, Section G2427.10.9 places an overall maximum limit on connector length. Section G2427.10.9 provides a necessary limiting function because Tables G2428.3(1) through G2428.3(4) would appear to allow (unintentionally) a connector to be considerably longer than the height of the vent.

Increasing the length of the connector increases the required minimum capacity of a vent and decreases its maximum capacity. Some vent manufacturers appear to measure only the horizontal (lateral) run of a connector in their design manuals. The more conservative approach would be to consider the friction loss and heat loss for the entire run of a connector throughout its developed length. See the commentary for Sections G2427.10.6 and G2427.10.9. The vent sizing tables were developed with the assumption that vent connectors are no longer than 18 inches per inch of connector diameter. This section establishes the basic rule, which is modified by Sections G2428.3.3 and G2427.10.9. Simply put, connector length is limited by this section to 18 inches per inch of diameter without capacity penalties or is limited in length by Section G2427.10.9 if the capacity penalties of Section G2428.3.3 are applied and the minimum capacity is met for fan-assisted appliances. Section G2428.3.3 allows longer connector lengths, but this section and the others mentioned still require connectors to take the shortest route to the vent.

TABLE G2428.3.2 (504.3.2)
MAXIMUM VENT CONNECTOR LENGTH

CONNECTOR DIAMETER	CONNECTOR HORIZONTAL
Maximum (inches)	Length (feet)
3	4.5
4	6
5	7.5
6	9
7	10.5
8	12
9	13.5

For SI: 1 inch = 25.4 mm, 1 foot = 304.8 mm.

❖ See the commentary for Section G2428.3.2.

G2428.3.3 (504.3.3) Connectors with longer lengths. Connectors with longer horizontal lengths than those listed in Section G2428.3.2 are permitted under the following conditions:

1. The maximum capacity (FAN Max or NAT Max) of the vent connector shall be reduced 10 percent for each additional multiple of the length listed above. For example, the maximum length listed above for a 4-inch (102 mm) connector is 6 feet (1829 mm). With a connector length greater than 6 feet (1829 mm), but not exceeding 12 feet (3658 mm), the maximum capacity must be reduced by 10 percent (0.90 maximum vent connector capacity). With a connector length greater than 12 feet (3658 mm), but not exceeding 18 feet (5486 mm), the maximum capacity must be reduced by 20 percent (0.80 maximum vent capacity).
2. For a connector serving a fan-assisted appliance, the minimum capacity (FAN Min) of the connector shall be determined by referring to the corresponding single appliance table. For Type B double-wall connectors, Table G2428.2(1) shall be used. For single-wall connectors, Table G2428.2(2) shall be used. The height (H) and lateral (L) shall be measured according to the procedures for a single appliance vent, as if the other appliances were not present.

❖ Section G2427.10.9 still applies and serves to limit the application of this section and Section G2428.3.2. For fan-assisted appliances, both items 1 and 2 apply. Item 1 allows the maximum length permitted by Section G2428.3.2 to be increased, i.e., doubled, tripled, etc., provided that the vent connector capacity is reduced by 10 percent if the maximum connector length is doubled, by 20 percent if tripled, by 30 percent if quadrupled, etc. Regardless of the length permitted, connectors should always be as short as practicable.

Item 2 applies to fan-assisted appliances and allows connector lengths to be as long as the lateral entries in Table G2428.2(1) or G2428.2(2), provided that the limits of Section G2427.10.9 are not exceeded and the appliance complies with the minimum input requirement of the respective table.

When horizontal lengths in excess of those stated in Section G2428.3.2 are necessary, the minimum capacity of the system is determined by referring to the corresponding single appliance table. In this case, for each appliance the entire vent connector and common vent from appliance to vent termination is treated as a single appliance vent (of the same size as the common vent), as if the others were not present. Any appliance failing to meet the minimum input may be prone to creating excessive condensation and/or insufficient draft within the vent system if such appliance is operated by itself. In such case, options may include relocation of appliances, selective sequential or simultaneous operation of appliances, or separate vent installations. This section is intended to be used only where a longer connector cannot be avoided as required by Sections G2427.10.6 and G2427.10.9 and the connector takes the shortest possible route to the vent. The longer the connector, the greater the flow resistance and the greater the heat loss, both of which negatively affect vent system operation. The capacity penalties account for the increased flow resistance and item 2 accounts for the increased heat loss. Item 2 addresses the worst case condition when a single appliance is operating and treats a connector serving a fan-assisted appliance as if the appliance were being independently vented, hence, the reference to the single appliance tables. Recall that the single appliance tables limit lateral length, which is, in effect, a connector length limitation.

In many cases, the sizing tables will prevent someone from taking advantage of the allowances of this section for connectors. For example, a single-wall connector serving a fan-assisted appliance may not be able to meet the minimum input requirement (FAN Min) if the capacity penalty of this section forces the connector to be increased in size. Note that Section G2427.10.9 will not allow a connector length to exceed the vent or chimney height in any case.

G2428.3.4 (504.3.4) Vent connector manifold. Where the vent connectors are combined prior to entering the vertical portion of the common vent to form a common vent manifold, the size of the common vent manifold and the common vent shall be determined by applying a 10-percent reduction (0.90 × maximum common vent capacity) to the common vent capacity part of the common vent tables. The length of the common vent connector manifold (L_M) shall not exceed $1^1/_2$ feet for each inch (457 mm per mm) of common vent connector manifold diameter (D) (see Appendix B Figure B-11).

❖ This section inflicts a 10-percent capacity penalty on the common vent as a result of the lateral (offset) effect of a manifold. A connector manifold is a horizontal extension of the lower end of a common vent. See Commentary Figure G2427.10.3.4(2). The capacity reduction accounts for the additional turn in the common vent. The manifold portion of a vent is sized based on the common vent portion of the tables. The length of a connector manifold is measured from the vertical common vent to the most upstream interconnection fitting.

G2428.3.5 (504.3.5) Common vertical vent offset. Where the common vertical vent is offset, the maximum capacity of the common vent shall be reduced in accordance with Section G2428.3.6. The horizontal length of the common vent offset (L_o) shall not exceed $1^1/_2$ feet for each inch (457 mm per mm) of common vent diameter.

❖ Offsets, as opposed to a single change in direction, require two fittings (elbows). See Commentary Figures G2403(10) and G2427.10.3.4(2). An offset offers resistance to flow as does any change of direction in a vent chimney or connector, and the capacity reduction penalty compensates for such flow resistance. Common vent locations should be well planned in a building to avoid the need for an offset. Offsets also introduce the need for additional supports to prevent stress on fittings (elbows). The term Offset (Vent) is defined in Section G2403. See the commentary for Sections G2428.2.3 and G2428.2.9. Offsets should be avoided because of the flow resistance, capacity reduction, additional support requirements, and possibility of condensation leakage. The length limitation of this section is absolute with no allowance for greater lengths. This provision is commonly violated in attic space offsets with some violations resulting in offsets that are as long or longer than the vent height. The length of an offset is measured in a horizontal plane, not along the developed length of the vent pipe. For example, a 45-degree offset with 7 feet of pipe between the fittings will have an offset length of 5 feet for application of this section. (For a 45-degree offset, actual pipe developed length is approximately 1.4 times the horizontal distance between the parallel vertical sections of pipe.) See Commentary Figure G2428.3.5.

G2428.3.6 (504.3.6) Elbows in vents. For each elbow up to and including 45 degrees (0.79 rad) in the common vent, the maximum common vent capacity listed in the venting tables shall be reduced by 5 percent. For each elbow greater than 45 degrees (0.79 rad) up to and including 90 degrees (1.57 rad), the maximum common vent capacity listed in the venting tables shall be reduced by 10 percent.

❖ This section applies to the common vent portion of a vent system. See the commentary for Section G2428.2.3. For example, a common vent may have two 30-degree (1.6 rad) changes of direction at its base and might also have an offset constructed with 45-degrees fittings in an attic space, thus having a total capacity reduction of 20 percent. If that same vent system had two 45-degree changes of direction at the base and an offset constructed with 60-degree fittings in the attic, the total capacity penalty would be 30 percent. In Figure G2428.3.6(1), an offset is created in the common vent by the use of two fittings (elbows) adjusted to 45-degree angles. The common vent capacity would be reduced by 10 percent total (5 percent for each fitting/elbow). If the fittings were adjusted to angles greater than 45 degrees, the total capacity reduction would be 10 percent per fitting for a total of 20 percent. See Commentary Figures G2428.3.6(2) and G2428.3.6(3).

G2428.3.7 (504.3.7) Elbows in connectors. The vent connector capacities listed in the common vent sizing tables include allowance for two 90-degree (1.57 rad) elbows. For each additional elbow up to and including 45 degrees (0.79 rad), the maximum vent connector capacity listed in the venting tables shall be reduced by 5 percent. For each elbow greater than 45 degrees (0.79 rad) up to and including 90 degrees (1.57 rad), the maximum vent connector capacity listed in the venting tables shall be reduced by 10 percent.

❖ In Figure G2428.3.6(1), the connector for the fan-assisted appliance is offset with two 60-degree fittings (elbows). Because the tables include an allowance for two "free" 90-degree fittings, there would be no capacity reduction required for this connector. The draft-hood-equipped appliance has a connector with a total of four 90-degree fittings, including the tee in the common vent. This would require a total capacity reduction of 20 percent. Sections G2427.10.6 and G2427.10.9 both intend to prohibit connectors with excessive length and fittings/elbows. Excess fittings and length are often the result of poor planning, improper appliance location, and obstructions such as ducts, structural members, and other appliances. Capacity reduction penalties can force an increase in connector size, and in the case of fan-assisted appliances, increasing the connector size might raise the minimum required input rating above the appliance rating, thus forcing the system to be redesigned.

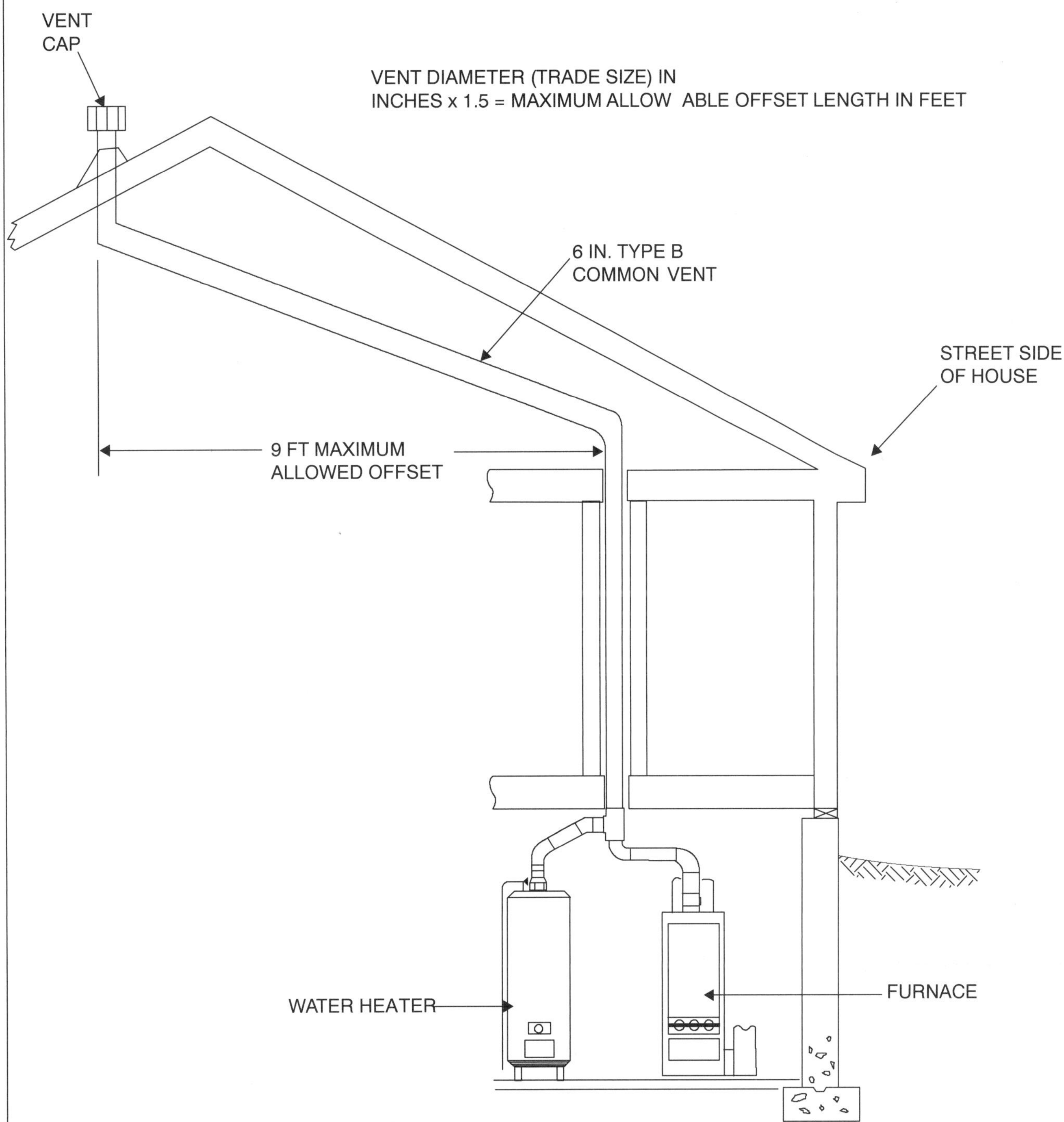

Figure G2428.3.5
COMMON VENT OFFSET IN ATTIC TO AVOID A STREET-SIDE ROOF PENETRATION

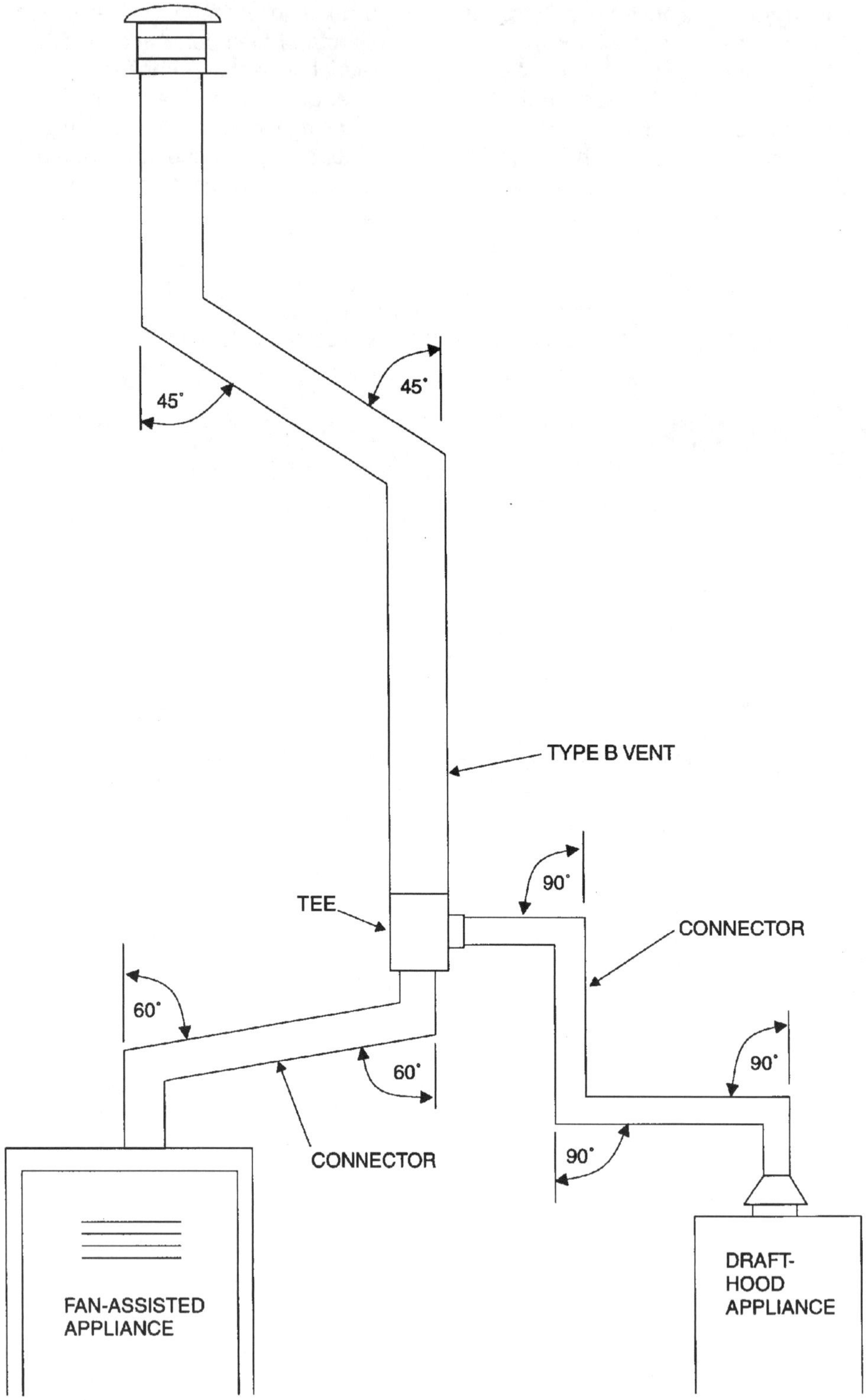

Figure G2428.3.6(1)
OFFSET IN COMMON VENT AND CONNECTORS

Illustration courtesy of Simpson Dura-Vent Co., Inc.

Figure G2428.3.6(2)
TYPE B VENT ADJUSTABLE 45-DEGREE ELBOW FITTING

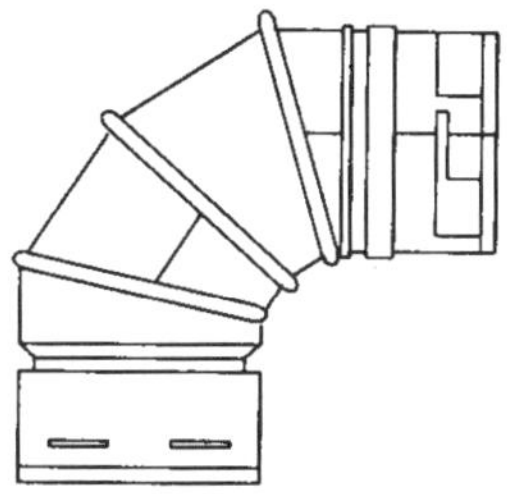

Illustration courtesy of Simpson Dura-Vent Co., Inc.

Figure G2428.3.6(3)
TYPE B VENT ADJUSTABLE 90-DEGREE ELBOW FITTING

G2428.3.8 (504.3.8) Common vent minimum size. The cross-sectional area of the common vent shall be equal to or greater than the cross-sectional area of the largest connector.

❖ This requirement is consistent with vent manufacturers. instructions and is necessary to prevent overloading of the common vent. It is apparent that a vent's maximum capacity could be exceeded where multiple connectors are involved, especially if one of the connectors is larger than the common vent itself. Vent manufacturers may also require that the common vent be increased one size where more than one appliance connector and the common vent are the same size.

G2428.3.9 (504.3.9) Common vent fittings. At the point where tee or wye fittings connect to a common vent, the opening size of the fitting shall be equal to the size of the common vent. Such fittings shall not be prohibited from having reduced-size openings at the point of connection of appliance vent connectors.

❖ This section specifically allows tee and wye fittings with reduced-size openings to be used to join appliance vent connectors to the common vent. The fittings must be the same size as the common vent at the point where they join but are allowed to have smaller branch openings and smaller openings on the opposite end of the run of the fitting. A common question regarding interconnection fittings is whether or not the code intends to allow single-wall interconnection fittings such as tees and wyes. Such fittings have commonly been used with single-wall connectors to join two appliances to a common Type B vent (e.g., 5 x 4 x 3-inch wyes, 6 x 6 x 4- tees, etc.). Vent manufacturers' opinion on this issue is that because the interconnection fitting is part of the common vent, it should be constructed as required for the common vent, meaning Type B vent tees and wyes. Certainly, double-wall vent connectors are not allowed to discharge into a single-wall fitting. However, the code is silent with regard to single-wall connectors discharging into single-wall fittings. Double-wall tees and wyes are preferable because they have far less heat loss and less required clearance to combustibles. See Commentary Figures G2428.3.9(1) and (2).

Photo courtesy of Simpson Dura-Vent Company, Incorporated

Figure G2428.3.9(1)
TYPE B VENT REDUCING TEE

Photo courtesy of Simpson Dura-Vent Company, Incorporated

Figure G2428.3.9(2)
TYPE B VENT REDUCING WYE

G2428.3.10 (504.3.10) High altitude installations. Sea level input ratings shall be used when determining maximum capacity for high altitude installation. Actual input, derated for altitude, shall be used for determining minimum capacity for high altitude installation.

❖ See the commentary for Section G2428.2.5.

G2428.3.11 (504.3.11) Connector rise measurement. Connector rise (R) for each appliance connector shall be measured from the draft hood outlet or flue collar to the centerline where the vent gas streams come together.

❖ See Commentary Figures G2428.3.11 and G2428.3(2). It is desirable to achieve the greatest connector rise as conditions will permit.

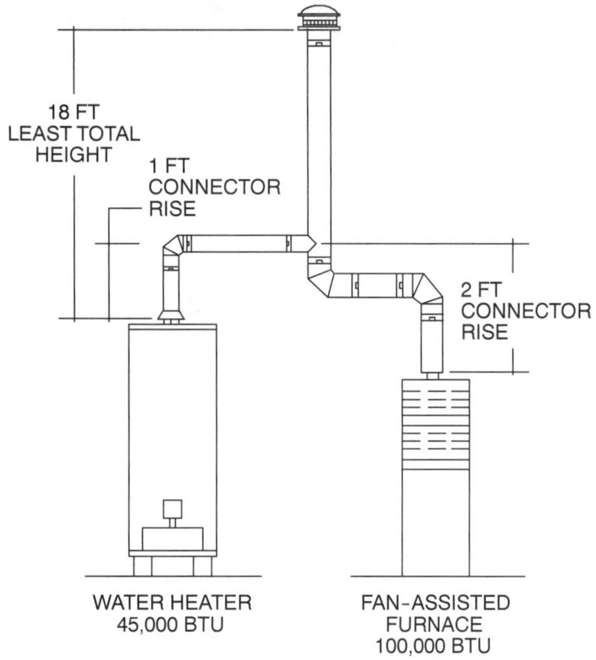

Figure courtesy of Simpson Dura-Vent Company, Incorporated

Figure G2428.3.11
CONNECTOR RISE

G2428.3.12 (504.3.12) Vent height measurement. For multiple units of equipment all located on one floor, available total height (H) shall be measured from the highest draft hood outlet or flue collar up to the level of the outlet of the common vent.

❖ See Commentary Figure G2428.3(2). The vent design is based on the "worst case" (most conservative) height so that a vent is not given credit for being taller than it actually is.

G2428.3.13 (504.3.17) Vertical vent maximum size. Where two or more appliances are connected to a vertical vent or chimney, the flow area of the largest section of vertical vent or chimney shall not exceed seven times the smallest listed appliance categorized vent areas, flue collar area, or draft hood outlet area unless designed in accordance with approved engineering methods.

❖ Vent manufacturers' instructions address ratios of areas of appliance connectors connecting to a common vent. This section relates the area ratio limitation to the smallest appliance connection and the area of the vertical common vent. The concern is that the smaller or smallest appliance may not be able to provide enough heat to the common vent to develop sufficient draft and prevent a condensation problem. It can be seen that the common vent tables would allow gross oversizing if it were not for the limit of this section. For example, if the table permits a 5-inch common vent, it would also permit a 6, 7, 8, 9, and 10-inch common vent. This section serves to limit the size of the common vent, whereas the table does not. The larger the common vent, the more difficult it becomes to maintain draft and avoid condensation, especially when the smallest appliance is firing by itself.

G2428.3.14 (504.3.18) Multiple input rate appliances. For appliances with more than one input rate, the minimum vent connector capacity (FAN Min) determined from the tables shall be less than the lowest appliance input rating, and the maximum vent connector capacity (FAN Max or NAT Max) determined from the tables shall be greater than the highest appliance input rating.

❖ See Section G2428.2.6.

G2428.3.15 (504.3.19) Liner system sizing. Listed, corrugated metallic chimney liner systems in masonry chimneys shall be sized by using Table G2427.3(1) or G2427.3(2) for Type B vents, with the maximum capacity reduced by 20 percent (0.80 maximum capacity) and the minimum capacity as shown in Table G2427.3(1) or G2427.3(2). Corrugated metallic liner systems installed with bends or offsets shall have their maximum capacity further reduced in accordance with Sections G2428.3.5 and G2428.3.6. The 20-percent reduction for corrugated metallic chimney liner systems includes an allowance for one long-radius 90-degree (1.57 rad) turn at the bottom of the liner.

❖ See the commentary for Section G2428.2.7.

G2428.3.16 (504.3.20) Chimney and vent location. Tables G2428.3(1), G2428.3(2), G2428.3(3) and G2428.3(4) shall be used for chimneys and vents not exposed to the outdoors below the roof line. A Type B vent or listed chimney lining system passing through an unused masonry chimney flue shall not be considered to be exposed to the outdoors.

> **Exception:** Vents serving listed appliances installed in accordance with the appliance manufacturer's instructions and the terms of the listing.

❖ See the commentary for Section G2428.2.9.

G2428.3.17 (504.3.21) Connector maximum and minimum size. Vent connectors shall not be increased in size more than two sizes greater than the listed appliance categorized vent diameter, flue collar diameter, or draft hood outlet diameter. Vent connectors for draft hood-equipped appliances shall not be smaller than the draft hood outlet diameter. Where a vent connector size(s) determined from the tables for a fan-assisted appliance(s) is smaller than the flue collar diameter, the use of the smaller size(s) shall be permitted provided that the installation complies with all of the following conditions:

1. Vent connectors for fan-assisted appliance flue collars 12 inches (305 mm) in diameter or smaller are not reduced by more than one table size [e.g., 12 inches to 10 inches (305 mm to 254 mm) is a one-size reduction] and those larger than 12 inches (305 mm) in diameter are not reduced more than two table sizes [e.g., 24 inches to 20 inches (610 mm to 508 mm) is a two-size reduction].
2. The fan-assisted appliance(s) is common vented with a draft hood-equipped appliance(s).
3. The vent connector has a smooth interior wall.

❖ Using a connector that is larger than the appliance outlet connection may sometimes be necessary to comply with the requirements of the sizing tables for multiple-appliance systems. However, having a connector significantly larger than the appliance outlet connection could cause reduced flow velocity, excessive loss of heat, and could result in the allowance of excessively long connector lengths.

In accordance with Section G2428.2.2, a venting system serving a single appliance can, under certain conditions, be smaller than the appliance outlet connection. However, this is prohibited by this section for common venting systems serving more than one appliance. See the commentary for Section G2428.2.11. The reason for this prohibition is that a common vent or chimney, being large enough for multiple appliances, may not be able to produce adequate draft or avoid condensation where a single appliance is operating with a reduced-size vent connector.

Recent research involving computerized modeling has shown that connectors for fan-assisted appliances in multiple-appliance systems can be reduced under the conditions similar to those for single-appliance installations. Therefore, this section makes an exception for systems with fan-assisted appliances commonly vented with appliances equipped with a draft hood. See the commentary for Section G2428.2.2. The allowance for a connector to be smaller than an appliance flue collar is applicable only to fan-assisted appliances. A reduction in size for corrugated metal connectors is not allowed.

G2428.3.18 (504.3.22) Component commingling. All combination of pipe sizes, single-wall, and double-wall metal pipe shall be allowed within any connector run(s) or within the common vent, provided all of the appropriate tables permit all of the desired sizes and types of pipe, as if they were used for the entire length of the subject connector or vent. Where single-wall and Type B double-wall metal pipes are used for vent connectors within the same venting system, the common vent must be sized using Table G2428.3(2) or G2428.3(4) as appropriate.

❖ Since this section parallels Section G2428.2.12, see the commentary for that section. This section addresses the installation that employs both single-wall connectors and double-wall connectors in the same venting system. The penalty for such mixing is that the common vent must be sized from the more restrictive single-wall connector tables. Double-wall pipe is not designed to be interchangeable with single-wall pipe, and all such transitions would have to made with fittings designed for the transition.

G2428.3.19 (504.3.23) Multiple sizes permitted. Where a table permits more than one diameter of pipe to be used for a connector or vent, all the permitted sizes shall be permitted to be used.

❖ This section broadly states that any size pipe permitted by the sizing tables is allowed to be used. However, do not overlook Sections G2428.3.8, G2428.3.13 and G2428.3.17, all of which contain provisions that can override this section. Vent manufacturers stress that the smallest size pipe allowed by the code should be used rather than any larger sizes that may be permitted. Operating a vent or chimney at nearer to its maximum capacity will produce stronger draft and provide better protection against the formation of condensation.

G2428.3.20 (504.3.24) Table interpolation. Interpolation shall be permitted in calculating capacities for vent dimensions that fall between table entries. (See Example 3, Appendix B.)

❖ See Section G2428.2.13.

G2428.3.21 (504.3.25) Extrapolation prohibited. Extrapolation beyond the table entries shall not be permitted.

❖ See Section G2428.2.14.

G2428.3.22 (504.3.26) Engineering calculations. For vent heights lower than 6 feet (1829 mm) and greater than shown in the tables, engineering methods shall be used to calculate vent capacities.

❖ See Section G2428.2.15.

SECTION G2429 (505) DIRECT-VENT, INTEGRAL VENT, MECHANICAL VENT AND VENTILATION/EXHAUST HOOD VENTING

G2429.1 (505.1) General. The installation of direct-vent and integral vent appliances shall be in accordance with Section G2427. Mechanical venting systems shall be designed and installed in accordance with Section G2427.

❖ See Sections G2427.2.1, G2427.2.2, G2427.3.3, and G2427.8.

SECTION G2430 (506) FACTORY-BUILT CHIMNEYS

G2430.1 (506.1) Listing. Factory-built chimneys for building heating appliances producing flue gases having a temperature not greater than 1,000°F (538°C), measured at the entrance to the chimney, shall be listed and labeled in accordance with UL 103 and shall be installed and terminated in accordance with the manufacturer's installation instructions.

❖ Factory-built chimneys can be used with gas-fired appliances if allowed by the chimney manufacturer. The sizing tables in Section G2428 do not apply to factory-built chimneys; therefore, sizing must be engineered or as specified by the chimney and appliance manufacturers' instructions. Factory-built chimneys will be too large for many residential appliances. See Figure G2430.1(1) and G2430.1(2).

Photo courtesy of Selkirk L.L.C.

Figure G2430.1(1) FACTORY-BUILT CHIMNEY

G2430.2 (506.2) Support. Where factory-built chimneys are supported by structural members, such as joists and rafters, such members shall be designed to support the additional load.

❖ Factory-built chimneys are supported by the building structure and can impose considerable weight on structural components. Joists, rafters, and other structural members must be designed to support the additional loading. Structural evaluation may be necessary, especially where framing for a chimney requires cutting and heading of joists and rafters.

SECTION G2431 (601) GENERAL

G2431.1 (601.1) Scope. Sections G2432 through G2453 shall govern the approval, design, installation, construction, maintenance, alteration and repair of the appliances and equipment specifically identified herein.

❖ This chapter regulates all aspects of appliances and equipment to the extent found in each of Sections G2432 through G2453.

SECTION G2432 (602) DECORATIVE APPLIANCES FOR INSTALLATION IN FIREPLACES

G2432.1 (602.1) General. Decorative appliances for installation in approved solid fuel burning fireplaces shall be tested in accordance with ANSI Z21.60 and shall be installed in accordance with the manufacturer's installation instructions. Manually lighted natural gas decorative appliances shall be tested in accordance with ANSI Z21.84.

❖ These appliances include gas log sets that are designed to simulate wood fires. See Commentary Figure G2432.1. Sections G2404 through G2411 addresses the requirements for testing, labeling, and installing mechanical equipment and appliances. The gas-burning appliance must be tested to the standard or standards appropriate for the equipment. The testing agency is responsible for determining the appropriate standard to be used to test the equipment. In the case of decorative gas log sets, ANSI Z21.60 or Z21.84 is the appropriate test standard.

Labeling is the code official's assurance that the subject product is a representative duplication of the product that the testing agency tested in the laboratory. The label indicates that an independent agency has conducted inspections at the plant to verify that all units conform to the specifications that the quality control manual sets forth for fabricating the gas appliances. Information that must be contained on the label is described in Section M1303.1 and includes the manufacturer's identification, the third-party inspection agency's identification, the model number, the serial number, the input ratings, and the type of fuel intended for use with the appliance.

The code requires that the appliance be installed in accordance with the manufacturer's installation instructions. This requirement is also linked to the laboratory testing of the appliance because the laboratory used the same installation instructions to install the prototype appliance being tested. When the appliance has been tested and evaluated for code compliance and judged to meet the performance and construction requirements of the applicable standard, the installation instructions become an integral part of the labeling requirements and must be strictly adhered to.

Photo courtesy of Simpson Dura-Vent Co., Inc.

Figure G2430.1(2)
FACTORY-BUILT AIR-COOLED CHIMNEY SYSTEM

Photography by D. F. Noyes Studio, courtesy of DESA International

Figure G2432.1
VENTED GAS LOG SET

The intent of this section is to regulate gas-burning appliances that are accessory to, and designed for installation in, vented solid fuel-burning fireplaces. Gas-fired decorative log sets and log lighters are examples of accessory appliances that are designed for installation in solid fuel-burning fireplaces. Gas log sets provide some radiant heat; however, their primary function is to create an aesthetically pleasing simulation of a wood log fire. This section addresses vented appliances and does not address "unvented gas log sets" (room heaters).

Decorative gas-burning appliances, such as some gas log sets, are designed to simulate wood fires by intentionally causing incomplete combustion as necessary to yield yellow or yellow-tipped flames. This incomplete combustion produces an increase in the amount of carbon monoxide produced. The highly toxic carbon monoxide is odorless and colorless, and the accompanying products of combustion produced by a gas-burning appliance do not have a strong odor or an odor that is readily recognized by an untrained person.

If a decorative gas appliance were operated with the fireplace damper closed, the carbon monoxide levels could be dangerously high before the building occupants became aware of the hazard. To prevent harm and the possible asphyxiation of the occupants, it is imperative that the fireplace damper be open whenever the appliance is burning. The manufacturer's installation instructions will specify the minimum free area of damper opening required to vent the appliance combustion products. The damper area is proportional to the appliance's input rating. The fireplace damper plate must be removed or permanently fixed in a position that provides the opening area required by the appliance manufacturer's installation instructions.

G2432.2 (602.2) Flame safeguard device. Decorative appliances for installation in approved solid fuel-burning fireplaces, with the exception of those tested in accordance with ANSI Z21.84, shall utilize a direct ignition device, an ignitor or a pilot flame to ignite the fuel at the main burner, and shall be equipped with a flame safeguard device. The flame safeguard device shall automatically shut off the fuel supply to a main burner or group of burners when the means of ignition of such burners becomes inoperative.

❖ In order to eliminate the hazards associated with manually igniting a decorative gas appliance, the code requires that the main burner be ignited by a supervised means of direct ignition, such as a standing pilot. The exception to this requirement is for manually lighted appliances listed to ANSI Z21.84. The use of automatic ignition controls prevents the operator of the appliance from being exposed to the potential hazard of trying to ignite the main burner with a match or other manual lighting device. The flame safeguard device consists of a control valve assembly with an integral means of pilot or igniter supervision and is designed to prevent or shut off the flow of gas to the appliance main burner in the event that the source of ignition is extinguished or otherwise fails. The typical flame safeguard device used with gas log set appliances is a combination manual control valve, pressure regulator, pilot feed, and magnetic pilot safety mechanism with a thermocouple generator. If the pilot flame is extinguished, the drop in thermocouple output voltage will cause the control valve to "lock out" in the closed position, thereby preventing the flow of gas to the main burner and the pilot burner. It is not the intent of this section to require flame safeguard devices for manually operated log lighter appliances used to kindle wood fires.

A flame safeguard, also known as a safety shutoff device, functions to automatically shut off the fuel supply to the main burner or burners of an appliance when the source of ignition becomes inoperative. Gas-burning equipment may operate using standing pilot ignition, which is a small flame kept constantly burning for the sole purpose of igniting the main burner when the thermostat calls for heat or a manual gas valve is opened by the operator. If the pilot is not lit, for whatever reason, the flame safeguard will automatically activate to prevent the flow of fuel to the main burner or burners of the appliance. Appliances that use automatic sparking devices or other means for ignition are subject to the same requirements for providing flame safeguards, and the method of operation of the flame safeguard will differ depending on the type of ignition system in use. All automatically operated gas-burning equipment and appliances are equipped with some means of monitoring the means of ignition or the main burner flame because if fuel is introduced into a combustion chamber and the fuel is not ignited and burned, a dangerous condition could result.

The most commonly used type of flame safeguard device employs a thermocouple and an electro-mechanical device that function together to supervise a standing pilot flame. In the event of pilot failure, the thermocouple and flame safeguard device will function to prevent the flow of fuel to the main burners, and if the safeguard control is of the 100-percent shutoff type, it will also shut off the flow of fuel to the pilot burner.

Modern appliances seldom employ standing pilots and instead employ electric-spark-ignited intermittent pilots or direct ignition devices such as hot surface igniters and glow coils. The methods of detecting the presence of a pilot flame include thermocouples, millivolt (power pile) generators, liquid-filled capillary tubes, bimetal mechanisms, and flame rectification circuits. Where only the main burner is monitored to verify ignition and continued combustion, the methods of detection include flame rectification, infrared detectors and ultraviolet detectors. The majority of today's appliances rely on electronic flame rectification or radiant energy detectors (infrared or ultraviolet) as the means of proving ignition of a pilot burner or the main burner(s). The flame rectification method of sensing the presence of flames is very common today and involves two electrodes: a flame rod and the burner itself. An AC voltage is applied between the flame rod and the burner. If a flame is present and impinges on both of these electrodes, a few microamps of current will flow through the gases in the flame, but almost entirely in one direction because of the relative size of the surface areas of the two electrodes. This unidirectional flow rectifies the AC current to DC current, which is then recognized by electronic circuitry, thereby proving the presence of a flame. Manually lighted natural gas decorative appliances tested and listed to Z21.84 do not have to be equipped with a flame safeguard device.

Flame safeguard devices are commonly integral with a combination gas control. Such controls typically incorporate multiple solenoid valves, a manual valve, an appliance gas pressure regulator, a pressure operated valve and flame safeguard devices all in one device body. It is also common for flame safeguard devices to be incorporated into an electronic integrated circuit control module that is separate from the combination gas control. For example, spark ignition and hot surface ignition systems incorporate the means of ignition and ignition supervision in an electronic control module that may also contain components and circuitry for control of combustion fans, furnace blowers, combustion chamber purging and monitoring of multiple system pressure and temperature sensors located throughout an appliance. Commentary Figure G2432.2(1) illustrates a type of combination gas control designed for use with an outboard intermittent ignition control module. Commentary Figure G2432.2(2) illustrates the internal workings that are typical of such combination gas controls. Such controls have redundant valve mechanisms as an additional safety feature.

G2432.3 (602.3) Prohibited installations. Decorative appliances for installation in fireplaces shall not be installed where prohibited by Section G2406.2.

❖ Section G2406.2 lists various locations where gas appliances cannot be installed, such as bedrooms and bathrooms. Those prohibitions also apply to decorative appliances. See Section G2406.2 and its exceptions.

SECTION G2433 (603)
LOG LIGHTERS

G2433.1 (603.1) General. Log lighters shall be tested in accordance with CSA 8 and shall be installed in accordance with the manufacturer's installation instructions.

❖ Log lighters are simple, manually operated burners used to start wood fires. Log lighters are considered functional rather than decorative appliances. The heat produced by the log lighter flames raises the temperature of the wood fuel to its ignition temperature. The log lighters are designed to be turned off manually after the wood fire is capable of sustaining combustion.

The code, rather than providing prescriptive installation requirements, requires the log lighters to be installed in accordance with the manufacturer's installation instructions.

Photo courtesy of Emerson Climate Technologies, White-Rodgers

Figure G2432.2(1)
COMBINATION GAS CONTROL

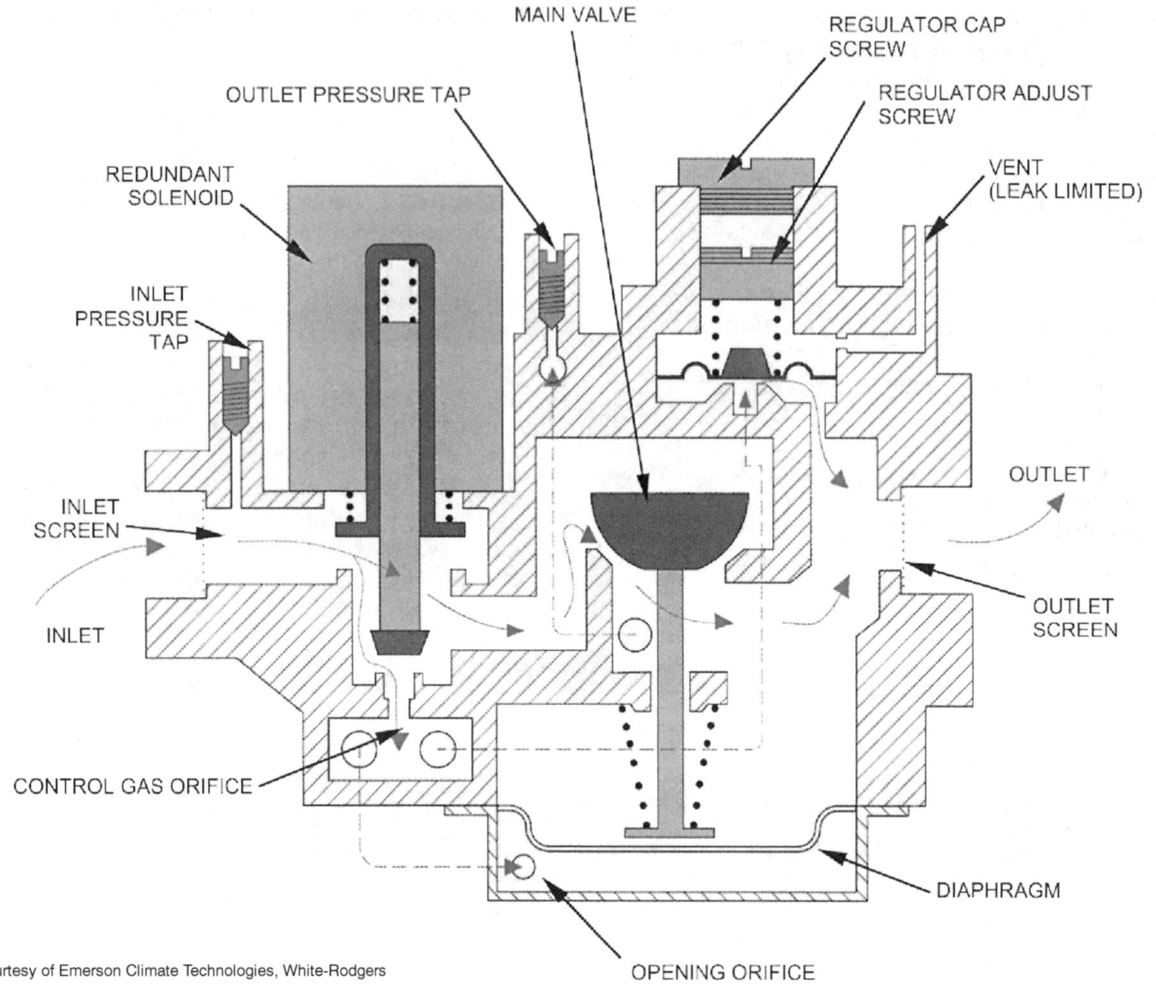

Figure courtesy of Emerson Climate Technologies, White-Rodgers

Figure G2432.2(2)
COMBINATION GAS CONTROL FOR APPLIANCES
(shown for illustrative purposes only)

SECTION G2434 (604) VENTED GAS FIREPLACES (DECORATIVE APPLIANCES)

G2434.1 (604.1) General. Vented gas fireplaces shall be tested in accordance with ANSI Z21.50, shall be installed in accordance with the manufacturer's installation instructions and shall be designed and equipped as specified in Section G2432.2.

❖ Unlike the appliances addressed in Section G2432, these appliances are self-contained and do not rely on a fireplace to contain or vent them. Such appliances are referred to as gas fireplaces because they are designed to simulate a solid fuel-burning fireplace. The standard, Z21.50, that regulates such appliances has recently been retitled as "Vented Gas Fireplaces." Therefore, these appliances will be referred to as gas fireplaces even though they do not fall under the definition of "fireplace." See Commentary Figures G2434.1(1), G2434.1(2), and G2434.1(3).

These appliances are designed for various methods of venting, including direct-venting through the wall or roof and conventional venting with Type B vent or factory-supplied vent material. Section G2406.2 controls the type of appliance that is allowed in bathrooms, toilet rooms, and bedrooms. Direct-vent appliances have the advantage of a closed combustion chamber that does not communicate with the room in which they are installed, as well as the advantage of an outdoor combustion air supply.

G2434.2 (604.2) Access. Panels, grilles, and access doors that are required to be removed for normal servicing operations shall not be attached to the building.

❖ Access to the appliance for maintenance and repair may be through panels, grilles, or doors, but such panels must be removable as intended by the manufacturer of the appliance and not a permanent part of the structure. See the definition of "Access" in Section G2403.

Photography by D. F. Noyes Studio, courtesy of DESA International

Figure G2434.1(1)
VENTED DECORATIVE APPLIANCE

Photography by D. F. Noyes Studio, courtesy of DESA International

Figure G2434.1(2)
DIRECT-VENT DECORATIVE APPLIANCE

SECTION G2435 (605) VENTED GAS FIREPLACE HEATERS

G2435.1 (605.1) General. Vented gas fireplace heaters shall be installed in accordance with the manufacturer's installation instructions, shall be tested in accordance with ANSI Z21.88 and shall be designed and equipped as specified in Section G2432.2.

❖ These appliances are very similar to those addressed in Section G2434, the main difference being that vented gas fireplace heaters are designed with more emphasis on space heating while maintaining the decorative features. Such heaters are required to comply with minimum thermal efficiency requirements.

Figure G2434.1(3)
DIRECT-VENT GAS FIREPLACE

SECTION G2436 (608) VENTED WALL FURNACES

G2436.1 (608.1) General. Vented wall furnaces shall be tested in accordance with ANSI Z21.49 or Z21.86/CSA 2.32 and shall be installed in accordance with the manufacturer's installation instructions.

❖ Wall furnaces are a type of room heater usually designed to be installed within a 2- by 4-inch stud cavity in frame construction. They are typically used in cottages, room additions, and in homes in mild climates. Some units are designed to serve a single room, and others are designed as through-the-wall units to serve adjacent rooms. Wall furnaces are ductless; however, some units are listed for use with a surface-mounted supply outlet extension. Wall furnaces can be either gravity or forced air type.

G2436.2 (608.2) Venting. Vented wall furnaces shall be vented in accordance with Section G2427.

❖ Because wall furnaces are designed to fit within a wall cavity, a special type of oval-shaped vent is required to fit within the same wall. Type BW vent is specially designed for wall furnace applications. See the commentary for Section G2427.

G2436.3 (608.3) Location. Vented wall furnaces shall be located so as not to cause a fire hazard to walls, floors, combustible furnishings or doors. Vented wall furnaces installed between bathrooms and adjoining rooms shall not circulate air from bathrooms to other parts of the building.

❖ Wall furnaces, like all room heaters, can present a fire hazard if improperly located. The heat discharged or directly radiated from such units can ignite nearby wall or floor surfaces, furniture, trim items, window treatments, and doors. See the commentary for Section G2436.4. A through-the-wall unit serving both a bathroom and an adjacent room must not be capable of recirculating air between such spaces.

G2436.4 (608.4) Door swing. Vented wall furnaces shall be located so that a door cannot swing within 12 inches (305 mm) of an air inlet or air outlet of such furnace measured at right angles to the opening. Doorstops or door closers shall not be installed to obtain this clearance.

❖ A combustible door that swings close to a wall furnace could be a fire hazard if at some position the door would be within 12 inches (305 mm) of an air inlet or outlet. Because door closers and door stops are easily defeated, they must not be depended upon to secure the required clearance. If the door swing cannot comply with this section, the door would have to be removed or rehung to change its swing direction.

G2436.5 (608.5) Ducts prohibited. Ducts shall not be attached to wall furnaces. Casing extension boots shall not be installed unless listed as part of the appliance.

❖ Most wall furnaces are not designed to force air through ducts, especially those gravity types that do not employ fans. Attachment of ducts would add resistance to the flow of air through the furnace, thereby causing abnormally high temperature rise across the heat exchanger and abnormally high temperatures on furnace surfaces. Some wall furnaces are listed and designed for use with supply duct extensions intended for wall mounting and intended to improve heat distribution. Such duct extensions are factory-built and supplied only by the furnace manufacturer.

G2436.6 (608.6) Access. Vented wall furnaces shall be provided with access for cleaning of heating surfaces, removal of burners, replacement of sections, motors, controls, filters and other working parts, and for adjustments and lubrication of parts requiring such attention. Panels, grilles and access doors that are required to be removed for normal servicing operations shall not be attached to the building construction.

❖ All access panels and doors must be removable as intended by the manufacturer of the furnace.

SECTION G2437 (609) FLOOR FURNACES

G2437.1 (609.1) General. Floor furnaces shall be tested in accordance with ANSI Z21.48 or Z21.86/CSA 2.32 and shall be installed in accordance with the manufacturer's installation instructions.

❖ Floor furnaces are vented appliances and are installed in an opening in the floor. Such units supply heat to the room by gravity convection and direct radiation and usually serve as the sole source of space heating. Such furnaces are common in cottages, small homes, seasonally occupied structures, and rural homes. Because the floor grille can become hot, extreme care must be exercised to prevent occupants, especially children, from contacting the grille by walking or falling on it. Also, care must be taken to avoid a fire hazard caused by placement of materials or furnishings on or near the furnace floor grille. See Commentary Figure G2437.1.

Photo courtesy of Empire Comfort Systems Inc.

**Figure G2437.1
FLOOR FURNACE**

G2437.2 (609.2) Placement. The following provisions apply to floor furnaces:

1. Floors. Floor furnaces shall not be installed in the floor of any doorway, stairway landing, aisle or passageway of any enclosure, public or private or in an exitway from any such room or space.
2. Walls and corners. The register of a floor furnace with a horizontal warm air outlet shall not be placed closer than 6 inches (152 mm) to the nearest wall. A distance of at least 18 inches (457 mm) from two adjoining sides of the floor furnace register to walls shall be provided to eliminate the necessity of occupants walking over the warm air discharge. The remaining sides shall be permitted to be placed not closer than 6 inches (152 mm) to a wall. Wall-register models shall not be placed closer than 6 inches (152 mm) to a corner.

3. Draperies. The furnace shall be placed so that a door, drapery, or similar object cannot be nearer than 12 inches (305 mm) to any portion of the register of the furnace.
4. Floor construction. Floor furnaces shall not be installed in concrete floor construction built on grade.
5. Thermostat. The controlling thermostat for a floor furnace shall be located within the same room or space as the floor furnace or shall be located in an adjacent room or space that is permanently open to the room or space containing the floor furnace.

❖ 1. Floor furnaces must not be located so as to interfere with or impede egress. In an emergency egress situation, a floor furnace could be a tripping hazard and could collapse under the live load of many occupants.
2. The floor furnace register must be at least 6 inches from walls to avoid creating a fire hazard by raising the wall temperature to a combustible level. The same reasoning applies to the 6-inch distance from wall registers to a corner.
The 18-inch clearance from adjoining sides of a register to walls provides space for the occupants to walk around the grill, which can reach temperatures high enough to cause burns.
3. As expressed in Section G2436.3, the furnace location must not create a fire hazard by being too close to wall surfaces, trim items, furnishings, and window treatments.
4. A floor furnace installed in a slab on grade would have to be in a pit, would be subject to flooding and corrosion, and would be inaccessible for service and inspection.
5. If the controlling thermostat does not sense the air temperature in the room in which the furnace is installed, dangerous overheating could result. A thermostat isolated from the source of heat it controls would not respond to the condition in the space served by the furnace.

G2437.3 (609.3) Bracing. The floor around the furnace shall be braced and headed with a support framework designed in accordance with Chapter 5.

❖ The framing around the floor opening must be capable of supporting the floor system, the anticipated floor loads, and the weight of the furnace. The structural requirements of this code must be complied with.

G2437.4 (609.4) Clearance. The lowest portion of the floor furnace shall have not less than a 6-inch (152 mm) clearance from the grade level; except where the lower 6-inch (152 mm) portion of the floor furnace is sealed by the manufacturer to prevent entrance of water, the minimum clearance shall be reduced to not less than 2 inches (51 mm). Where these clearances cannot be provided, the ground below and to the sides shall be excavated to form a pit under the furnace so that the required clearance is provided beneath the lowest portion of the furnace. A 12-inch (305 mm) minimum clearance shall be provided on all sides except the control side, which shall have an 18-inch (457 mm) minimum clearance.

❖ This section specifies clearances between the ground and the furnace, which apply to crawl space installations. The clearances allow access for service and inspection and help prevent corrosion of the furnace assembly.

G2437.5 (609.5) First floor installation. Where the basement story level below the floor in which a floor furnace is installed is utilized as habitable space, such floor furnaces shall be enclosed as specified in Section G2437.6 and shall project into a nonhabitable space.

❖ Where a floor furnace is installed above a habitable basement space, the furnace must project into a nonhabitable space and be separated from that space by noncombustible construction. See the commentary for Section G2436.6.

G2437.6 (609.6) Upper floor installations. Floor furnaces installed in upper stories of buildings shall project below into nonhabitable space and shall be separated from the nonhabitable space by an enclosure constructed of noncombustible materials. The floor furnace shall be provided with access, clearance to all sides and bottom of not less than 6 inches (152 mm) and combustion air in accordance with Section G2407.

❖ Where the floor furnace is installed in an upper floor of a building, the furnace must project into a nonhabitable space, similar to the installation in Section G2437.5, and must be separated from that space by noncombustible construction. Access and clearance must be provided for maintenance and servicing. An adequate source of combustion air must also be provided to the furnace.

SECTION G2438 (613) CLOTHES DRYERS

G2438.1 (613.1) General. Clothes dryers shall be tested in accordance with ANSI Z21.5.1 and shall be installed in accordance with the manufacturer's installation instructions.

❖ This section addresses clothes dryer appliances, Types 1 and 2. See the definition of "Clothes Dryer."

Dryers are tested to the appropriate safety standard for the appliance, and the manufacturer's installation instructions convey the information needed to duplicate the installation configuration that was tested and found to meet the requirements of the safety standard. The manufacturer's installation instructions are evaluated by the agency responsible for testing, listing, and labeling the appliance, and therefore such instructions will prescribe an installation that is consistent with the appliance installation that was tested. Clothes dryers must be exhausted in accordance with Section G2439 and the manufacturer's installation instructions. See Commentary Figure G2438.1.

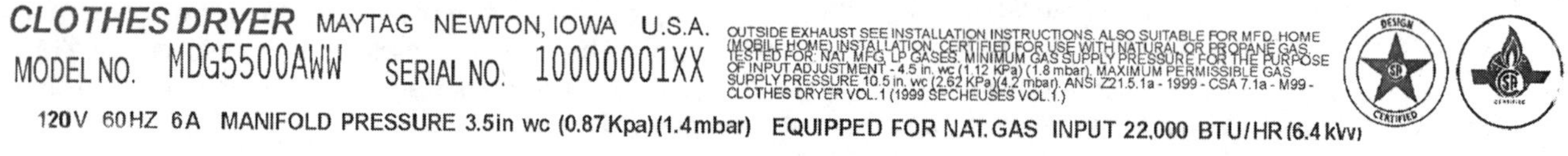

Figure courtesy of Maytag Corporation

Figure G2438.1
TYPE 1 CLOTHES DRYER LABEL

SECTION G2439 (614) CLOTHES DRYER EXHAUST

G2439.1 (614.1) Installation. Clothes dryers shall be exhausted in accordance with the manufacturer's instructions. Dryer exhaust systems shall be independent of all other systems and shall convey the moisture and any products of combustion to the outside of the building.

❖ All clothes dryer exhaust systems must convey the moisture and any products of combustion directly to the exterior of the building (outdoors). The code does not use the term "dryer vent." Gas-fired clothes dryers are not "vented" as are other fuel-fired appliances, but rather, are exhausted.

Clothes dryer exhaust ducts must be installed in accordance with the dryer manufacturer's installation instructions and the provisions of this section. Dryers are designed and built to meet industry safety standards.

The clothes dryer manufacturer's installation instructions control the type of exhaust duct material allowed and the method of installation. For example, typical dryer installation instructions will require metallic duct materials and will impose more stringent length limitations for flexible metallic ducts than for rigid ducts because of the poorer flow characteristics of flexible duct materials.

Because clothes dryer exhaust contains high concentrations of combustible lint, debris, and water vapor, dryer exhaust systems must be independent of all other systems. This requirement prevents the fire hazards associated with such an exhaust system from extending into or affecting other systems or other areas in the building. Additionally, this section intends to prevent toxic products of combustion from entering the building through other systems.

Dryer exhaust ducts must be independent of other dryer exhaust ducts unless connected to an engineered exhaust system specifically designed to serve multiple dryers. Type I domestic dryers are not designed for connection to a common exhaust duct serving multiple dryers. For example, connecting multiple Type I dryers to a common duct riser would pressurize the riser and cause exhaust to back up into any dryer that was not operating. Also, the low-flow velocities and duct temperature losses could result in water vapor condensation and the buildup of lint and debris at low points.

Clothes dryer exhaust systems cannot terminate in or discharge to any enclosed space such as an attic or crawl space, regardless of whether or not the space is ventilated through openings to the outdoors. The high levels of moisture in the exhaust air can cause condensation to form on exposed surfaces or in insulation materials. Water vapor condensation can cause structural damage and deterioration of building materials, and contribute to the growth of mold and fungus. Clothes dryer exhausts that discharge to enclosed spaces will also cause an accumulation of combustible lint and debris, creating a significant fire hazard. An improperly installed clothes dryer exhaust system not only reduces dryer efficiency and increases running time but can also cause a significant increase in exhaust temperature, in turn causing the dryer to cycle on its high limit control, which is an unsafe operating condition.

G2439.2 (614.2) Duct penetrations. Ducts that exhaust clothes dryers shall not penetrate or be located within any fireblocking, draftstopping or any wall, floor/ceiling or other assembly required by this code to be fire-resistance-rated, unless such duct is constructed of galvanized steel or aluminum of the thickness specified in the mechanical provisions of this code and the fire-resistance rating is maintained in accordance with this code. Fire dampers shall not be installed in clothes dryer exhaust duct systems.

❖ Rigid clothes dryer exhaust ducts are permitted to penetrate assemblies that are not fire-resistance-rated and building elements not used as fireblocking or draftstopping. In all other cases, ducts must be constructed of galvanized steel or aluminum of the thickness specified in this code, and the penetration must be protected to maintain the fire-resistance rating and integrity of the assembly or element being penetrated. Because of the strength and rigidity differences between steel and aluminum, aluminum ducts generally must be of a heavier (thicker) gage than steel ducts for a given application.

The metal thickness requirements of this code practically necessitate rigid pipe and all but rule out the use of flexible duct where the duct must penetrate fireblocking, draftstopping, or a fire-resistance-rated assembly. Where penetrating fireblocking or draftstopping, the exhaust duct must be constructed of galvanized steel or aluminum, and the annular space around the duct must be fireblocked in accordance with this code.

Where penetrating a fire-resistance-rated assembly, the penetration must be protected. The requirements of this section combined with Section G2439.5.1 make a compelling case for always placing clothes dryers against outside walls to avoid long duct runs and penetrations of other than exterior walls.

G2439.3 (614.4) Exhaust installation. Dryer exhaust ducts for clothes dryers shall terminate on the outside of the building and shall be equipped with a backdraft damper. Screens shall not be installed at the duct termination. Ducts shall not be connected or installed with sheet metal screws or other fasteners that will obstruct the flow. Clothes dryer exhaust ducts shall not be connected to a vent connector, vent or chimney. Clothes dryer exhaust ducts shall not extend into or through ducts or plenums.

❖ Exhaust ducts must connect directly to terminals that pass through the building envelope to the outdoor atmosphere. Attics and crawl spaces are not considered to be outdoors, and exhaust ducts cannot terminate in those spaces. See the commentary for Section G2439.1. Backdraft dampers must be provided in dryer exhaust ducts to avoid outdoor air infiltration during periods when the dryer is not operating and to prevent the entry of animals. These dampers should be designed and installed to provide an adequate seal when in a closed position so as to minimize air leakage (infiltration). Backdraft dampers are usually of the gravity types that are opened by the energy of the exhaust discharge. Some dryer manufacturers prohibit the use of magnetic backdraft dampers because of the extra resistance that the exhaust flow must overcome.

Exhaust terminal opening size is also governed by the dryer manufacturer's instructions. Full-opening terminals present less resistance to flow and might be mandated by the dryer manufacturer. A "full opening" is considered to be an opening having no dimension less than the diameter of the exhaust duct. Dryer exhaust flow must not be restricted by screens or fastening devices such as sheet metal screws. Any type of screen would become completely blocked with fibers in a very short time. Consider that the filter screen integral with the appliance becomes restricted with lint in each cycle of operation. These restrictions and projections will promote the accumulation of combustible lint and debris in the exhaust duct, thereby creating a potential fire hazard and causing flow resistance. Pop-rivets and ¼-inch-long screws are commonly used to join dryer exhaust ducts because they have no ability to catch lint and create an obstruction. Duct tape should not be relied upon as the sole means of joining dryer exhaust ducts, because adhesives can deteriorate with age and when exposed to high temperatures, causing joints to separate. However, sealing joints is still desirable to limit leakage of lint, fibers, moisture vapor, and combustion products into the occupied space.

G2439.4 (614.5) Makeup air. Installations exhausting more than 200 cfm (0.09 m^3/s) shall be provided with makeup air. Where a closet is designed for the installation of a clothes dryer, an opening having an area of not less than 100 square inches (0.0645 m^2) for makeup air shall be provided in the closet enclosure, or makeup air shall be provided by other approved means.

❖ Makeup air must be provided to compensate for the air exhausted by the dryer exhaust system. A typical domestic clothes dryer will exhaust approximately 200 cfm (0.09 m^3/s). For closet installations, an opening with a minimum area of 100 square inches (0.0645 m^2) must be provided in the closet door or in the closet enclosure. Where louvers or grilles are used, the solid portion of the louver or grille should be evaluated in accordance with Section G2407.10. Makeup air is necessary to prevent the room or space housing the dryer(s) from developing a negative pressure with respect to adjacent spaces or to the outdoors, which could result in the improper and dangerous operation of the dryer and other fuel-burning appliances. The required amount of makeup air should be approximately equal to the amount exhausted and is normally provided by infiltration of air from outdoors or through openings to the outdoors.

The make-up air not only provides the air that is to be exhausted from the dryer, but also provides combustion air for gas-fired appliances. Commercial dryer manufacturers. installation instructions will prescribe outdoor air opening requirements for the purpose of providing the necessary make-up and combustion air. See the Commentary for Section G2407.4.

To simplify the installation, commercial and multiple dryer installations are commonly provided with transfer or ducted openings to the outdoors that introduce outdoor air into the immediate vicinity of the appliances. Commercial units usually obtain make-up air from an enclosed accessible space (plenum) behind the units. This enclosure provides maintenance access; houses the unit's power, piping, and duct connections; and isolates the make-up air from the conditioned occupiable spaces. Make-up air is rarely provided by the building's HVAC system because if it were, the HVAC system would have to be interlocked with the dryers to avoid starving any one dryer of make-up air if the HVAC system were not operating. Also, providing make-up air through the HVAC system would waste energy because large volumes of conditioned air are exhausted through the dryers. Additionally, cold outdoor air will normally have low moisture content and thus serves well as make-up air for drying operations. (The warmed outdoor air will have low relative humidity.)

G2439.5 (614.6) Clothes dryer ducts. Exhaust ducts for domestic clothes dryers shall be constructed of metal and shall have a smooth interior finish. The exhaust duct shall be a minimum nominal size of 4 inches (102 mm) in diameter. The entire exhaust system shall be supported and secured in place. The male end of the duct at overlapped duct joints shall extend in the direction of airflow. Clothes dryer transition ducts used to connect the appliance to the exhaust duct system shall be metal and limited to a single length not to exceed 8 feet (2438 mm) in length and shall be listed and labeled for the application. Transition ducts shall not be concealed within construction.

❖ Section G2408.1 states that gas-fired equipment and appliances must be installed in accordance with the manufacturer's installation instructions for the listed and labeled equipment. An installation in accordance with the manufacturer's installation instructions is required, except where the code requirements are more stringent. In the case of this section and Section G2439.5.1, the code specifically addresses dryer exhaust ducts, an installation requirement that is also addressed in the dryer manufacturer's installation instructions. The code requirements for the items addressed in these sections, such as maximum total developed length of dryer exhaust ducts, equivalent lengths of directional fittings, and minimum nominal size to name a few, may parallel or exceed the applicable requirements in the manufacturer's installation instructions. The manufacturer's installation instructions could also contain requirements that exceed those in the code.

In all cases, the more restrictive requirements would apply. Therefore, the clothes dryer exhaust duct size must be determined by the clothes dryer manufacturer but must not be less than 4 inches (102 mm) in diameter.

Dryer exhaust ducts cannot be larger than 4 inches in diameter unless specifically allowed by the manufacturer. Note that enlarging a duct will cause a reduction in flow velocity, and dryer exhaust ducts must maintain sufficient flow velocity to transport lint and fibers through the duct to the discharge terminal.

The duct length, the number and degree of directional fittings, the smoothness of the duct interior wall, and the type of exhaust outlet terminal all contribute to the overall friction loss of a clothes dryer exhaust system. When the friction loss is high enough to restrict the required exhaust flow, the duct system must be redesigned to allow the required flow rate. An exhaust duct may have to be shortened or rerouted to compensate for the flow resistance.

Excessive friction losses will also result in reduced flow velocities, which means that the exhaust ducts would be much more likely to collect debris. An improperly designed exhaust system will result in poor dryer performance and poor energy efficiency and can cause the appliance to cycle on its limit control, which can be hazardous. Joints in the dryer exhaust system must be reasonably airtight, must have a smooth interior finish, and must run in the direction of airflow. For example, the male end of each section of duct must point away from the dryer. This permits the dryer exhaust system to function as intended so that joints and connections do not serve as collection points for lint and debris. Proper support is required and necessary for maintaining alignment of the dryer exhaust duct system and to prevent excessive stress on ducts and duct joints. A sagging duct will increase the internal resistance to airflow, reduce the efficiency of the system, and cause the accumulation of lint and debris.

Section G2439.5 specifically addresses transition duct connectors. Within the context of this section, a transition duct is a flexible connector used as a transition between the dryer outlet and the connection point to the exhaust duct system. Transition duct connectors must be listed and labeled as transition ducts for clothes dryer application. Transition ducts are currently listed to comply with UL 2158A. Such transition ducts are not considered flexible air ducts or flexible air connectors subject to the material requirements of the *International Mechanical Code*.

Transition ducts are flexible ducts constructed of a metalized (foil) fabric supported on a spiral wire frame. They are more fire resistant than the typical plastic spiral duct. Transition duct connectors are limited to 8 feet (1829 mm) in length and must be installed in accordance with their listing and the manufacturer's instructions. These duct connectors must not be concealed by any portion of the structure's permanent finish materials such as drywall, plaster, paneling, built-in furniture or cabinets, or any other similar permanently affixed building component; they must remain entirely within the room in which the appliance is installed. Transition duct connectors cannot be joined together to extend beyond the 8-foot maximum length limit. Transition ducts are to be cut to length as needed to avoid excess duct and unnecessary bends. Transition duct connectors are necessary for domestic dryers because of appliance movement, vibration, and outlet location. In many cases, connecting a domestic clothes dryer directly to rigid duct would be difficult.

G2439.5.1 (614.6.1) Maximum length. The maximum length of a clothes dryer exhaust duct shall not exceed 25 feet (7620 mm) from the dryer location to the outlet terminal. The maximum length of the duct shall be reduced $2^1/_2$ feet

(762 mm) for each 45 degree (0.79 rad) bend and 5 feet (1524 mm) for each 90 degree (1.6 rad) bend.

> **Exception**: Where the make and model of the clothes dryer to be installed is known and the manufacturer's installation instructions for such dryer are provided to the code official, the maximum length of the exhaust duct, including any transition duct, shall be permitted to be in accordance with the dryer manufacturer's installation instructions.

❖ The maximum length of 25 feet and the associated reductions for changes in direction can be very restrictive and make it necessary in most cases to locate the clothes dryer close to an exterior wall. Locating the dryer in the middle of a house or apartment creates a design problem for the architect and the contractor. Exceeding the length requirement of this section will contribute to excessive pressure loss and potential fire hazards. See the commentary for Section G2439.5.

The length reduction for changes in direction apply to all changes in direction in the exhaust duct, including the first 90-degree turn required to run the duct inside a wall cavity. Any change in direction made with the 8-foot transition duct connector does not require a reduction in length.

The exception allows the 25-foot duct length to be exceeded if the manufacturer's installation instructions specify a longer duct allowance. The manufacturer's instructions must be made available to the code official to allow verification of the allowable length and to allow the rough-in installation to be inspected for compliance with such instructions. The code official will also have to verify that the specific clothes dryer is installed. The dryer chosen for application of the exception to this section must be installed before final approval can be given.

The exception could create a problem when the occupants move and take the dryer with them. The next occupant must either buy a dryer with an equivalent duct length allowance or install a duct that is compatible with the dryer to be used. If the new occupant is not aware of the duct requirements, he or she might install a standard dryer designed for only 25 feet of duct length. This could result in a buildup of lint in the duct and a potential fire hazard. The code official is usually not aware of the occupancy change and has no method of assuring that the proper dryer and duct are installed.

The code does not currently recognize the installation of booster fans that are marketed for the purpose of extending exhaust duct lengths. The concerns have been that dryers are not currently listed for use with external booster fans and there is no interlock to prevent dryer operation in the event of booster fan failure. Clothes dryer exhaust duct length problems could be eliminated if designers would give thought to the location of laundry facilities.

G2439.5.2 (614.6.2) Rough-in-required. Where a compartment or space for a clothes dryer is provided, an exhaust duct system shall be installed.

❖ If the design of a residence includes a laundry room or clothes dryer compartment, this section requires the exhaust system to be installed prior to the final inspection. It should also be shown on the plans submitted for review. This will allow the plans examiner to advise the designer or builder of an exhaust duct problem prior to construction.

SECTION G2440 (615) SAUNA HEATERS

G2440.1 (615.1) General. Sauna heaters shall be installed in accordance with the manufacturer's installation instructions.

❖ Sauna heaters are used in steam baths and similar rooms to generate heat and steam. They must be installed in accordance with the manufacturer's instructions to ensure they are installed as designed and tested. The code relies on these installation instructions rather than providing prescriptive requirements that might contradict the manufacturer.

G2440.2 (615.2) Location and protection. Sauna heaters shall be located so as to minimize the possibility of accidental contact by a person in the room.

❖ Sauna heaters produce steam by passing water over a heated surface. The choice of locations must consider the fact that a heat-producing appliance will be exposed within a small visually obscured room with unclothed occupants.

G2440.2.1 (615.2.1) Guards. Sauna heaters shall be protected from accidental contact by an approved guard or barrier of material having a low coefficient of thermal conductivity. The guard shall not substantially affect the transfer of heat from the heater to the room.

❖ Guards are required to protect the occupants from being burned. The guards must be constructed of a material that is a poor conductor of heat (e.g., wood) so that the guard itself will not present a burn hazard. The design of the guard must protect the occupants but not have a negative impact on the flow of heat into the room from the heater.

G2440.3 (615.3) Access. Panels, grilles and access doors that are required to be removed for normal servicing operations, shall not be attached to the building.

❖ Access panels, covers, and doors must not be made unusable by trim, woodwork, or room enclosures that impede or interfere with access.

G2440.4 (615.4) Combustion and dilution air intakes. Sauna heaters of other than the direct-vent type shall be installed with the draft hood and combustion air intake located outside the sauna room. Where the combustion air inlet and the draft hood are in a dressing room adjacent to the sauna room, there shall be provisions to prevent physically blocking the combustion air inlet and the draft hood inlet, and to prevent physical contact with the draft hood and vent assembly, or warning notices shall be posted to avoid such contact. Any warning notice shall be easily readable, shall contrast with its background, and the wording shall be in letters not less than 0.25 inch (6.4 mm) high.

❖ Combustion air and dilution air must not be taken from the sauna room, because of the excessive water vapor in the air and the fact that the sauna room would probably be incapable of providing the required volume of combustion air. If a draft hood or combustion air inlet is obstructed, the appliance could malfunction and threaten the occupants. For example, a combustion air-starved heater would produce high levels of carbon monoxide that could enter the sauna room in the event of venting failure. In general, the code expresses concern for the use of fuel-fired appliances in small closed rooms, especially where the occupants are sleeping or would have impaired senses or a diminished ability to recognize danger.

G2440.5 (615.5) Combustion and ventilation air. Combustion air shall not be taken from inside the sauna room. Combustion and ventilation air for a sauna heater not of the direct-vent type shall be provided to the area in which the combustion air inlet and draft hood are located in accordance with Section G2407.

❖ This section would require the heater to be either a direct vent type or a separated combustion type in which the combustion chamber does not communicate with the sauna room. See the commentary for Section G2440.4.

G2440.6 (615.6) Heat and time controls. Sauna heaters shall be equipped with a thermostat which will limit room temperature to 194°F (90°C). If the thermostat is not an integral part of the sauna heater, the heat-sensing element shall be located within 6 inches (152 mm) of the ceiling. If the heat-sensing element is a capillary tube and bulb, the assembly shall be attached to the wall or other support, and shall be protected against physical damage.

❖ A thermostat is required to limit the temperature in the sauna for fire safety purposes. The control must sense the warmest air near the ceiling and must be protected from physical damage.

G2440.6.1 (615.6.1) Timers. A timer, if provided to control main burner operation, shall have a maximum operating time of 1 hour. The control for the timer shall be located outside the sauna room.

❖ Timers are not required by the code but provide an extra level of protection when installed. To protect both the occupants and the building, timers must limit the heater operating time to 1 hour. Resetting the timer would require the occupant to exit the sauna, thus lessening the chances of overexposure.

G2440.7 (615.7) Sauna room. A ventilation opening into the sauna room shall be provided. The opening shall be not less than 4 inches by 8 inches (102 mm by 203 mm) located near the top of the door into the sauna room.

❖ A ventilation opening is required to allow the escape of steam and heat and to provide ventilation for the occupants. The opening is required near the ceiling to take advantage of the natural tendency of heat and steam to rise, facilitating the exhaust.

SECTION G2441 (617) POOL AND SPA HEATERS

G2441.1 (617.1) General. Pool and spa heaters shall be tested in accordance with ANSI Z21.56 and shall be installed in accordance with the manufacturer's installation instructions.

❖ Pool and spa heaters are specialized water heaters very similar in design to hot water supply boilers and are utilized with swimming pools, recreational or therapeutic spas, and hot tubs. These heaters are usually of the water tube type and are designed for either indoor or outdoor installation.

SECTION G2442 (618) FORCED-AIR WARM-AIR FURNACES

G2442.1 (618.1) General. Forced-air warm-air furnaces shall be tested in accordance with ANSI Z21.47 or UL 795 and shall be installed in accordance with the manufacturer's installation instructions.

❖ Forced-air warm-air furnaces are considered to be central heating units and consist of burners or heating elements, heat exchangers, blowers, and associated controls. Forced-air furnaces are made in many different configurations, including upflow, counterflow (down flow), horizontal flow, multi-position flow, and indoor and outdoor units. Commentary Figure G2426.1(1) is an example of a multi-position (i.e., up, down and horizontal flow) forced-air warm-air furnace.

G2442.2 (618.2) Forced-air furnaces. The minimum unobstructed total area of the outside and return air ducts or openings to a forced air warm-air furnace shall be not less than 2 square inches for each 1,000 Btu/h (4402 mm^2/W) output rating

capacity of the furnace and not less than that specified in the furnace manufacturer's installation instructions. The minimum unobstructed total area of supply ducts from a forced-air warm-air furnace shall be not less than 2 square inches for each 1,000 Btu/h (4402 mm^2/W) output rating capacity of the furnace and not less than that specified in the furnace manufacturer's installation instructions.

> **Exception:** The total area of the supply air ducts and outside and return air ducts shall not be required to be larger than the minimum size required by the furnace manufacturer's installation instructions.

❖ The aggregate area of all ducts or openings that convey supply air from the furnace or return air back to the furnace must be adequate to allow the required airflow through the furnace. A furnace that is "starved" for return air or is restricted by an inadequate supply air duct size will produce an abnormal temperature rise across the heat exchanger, which is both a fire hazard and detrimental to the furnace. The furnace output rating is not usually indicated on the label and would be determined as approximately the input rating in Btu/h (W) times the efficiency rating of the furnace. As an example, 100,000 Btu/h (29 310 W) input times 0.80 (80 percent efficiency) is 80,000 Btu/h (23 448 W) output. Return or supply air openings required by this section must not be less than that specified by the furnace manufacturer's installation instructions.

The exception intends to clarify that if the furnace installation instructions specify a lesser return or supply area than this section, then that lesser area is permitted. The code specified minimum area applies where the furnace manufacturer does not specify a minimum area.

G2442.3 (618.3) Dampers. Volume dampers shall not be placed in the air inlet to a furnace in a manner that will reduce the required air to the furnace.

❖ Dampers are usually avoided in return air ducts and openings because of the risk of starving the furnace for return air. If such dampers are installed, the total unrestricted return air duct or opening area must be as required by the code with all of the dampers in the fully closed position.

G2442.4 (618.4) Circulating air ducts for forced-air warm-air furnaces. Circulating air for forced-air-type, warm-air furnaces shall be conducted into the blower housing from outside the furnace enclosure by continuous air-tight ducts.

❖ This section parallels the manufacturer's installation instructions that require return air ducts to extend from the furnace cabinet to the exterior of any closet, alcove, or furnace room that encloses a furnace. If return air were drawn from within a furnace enclosure, negative pressures could be produced in such a room, and negative pressure could cause combustion byproducts to be drawn from the combustion chamber or venting system and introduced into the circulating airflow.

Negative and positive pressure differentials between the interior and exterior of a furnace enclosure could also cause hazardous burner and venting system malfunctions.

G2442.5 (618.5) Prohibited sources. Outside or return air for a forced-air heating system shall not be taken from the following locations:

1. Closer than 10 feet (3048 mm) from an appliance vent outlet, a vent opening from a plumbing drainage system or the discharge outlet of an exhaust fan, unless the outlet is 3 feet (914 mm) above the outside air inlet.
2. Where there is the presence of objectionable odors, fumes or flammable vapors; or where located less than 10 feet (3048 mm) above the surface of any abutting public way or driveway; or where located at grade level by a sidewalk, street, alley or driveway.
3. A room or space, the volume of which is less than 25 percent of the entire volume served by such system. Where connected by a permanent opening having an area sized in accordance with Section G2442.2, adjoining rooms or spaces shall be considered as a single room or space for the purpose of determining the volume of such rooms or spaces.

Exception: The minimum volume requirement shall not apply where the amount of return air taken from a room or space is less than or equal to the amount of supply air delivered to such room or space.

4. A room or space containing an appliance where such a room or space serves as the sole source of return air.

Exception: This shall not apply where:

1. The appliance is a direct-vent appliance or an appliance not requiring a vent in accordance with Section G2425.8.
2. The room or space complies with the following requirements:
   - 2.1. The return air shall be taken from a room or space having a volume exceeding 1 cubic foot (28 316.85 mm^3) for each 10 Btu/h (9.6 L/W) of combined input rating of all fuel-burning appliances therein.
   - 2.2. The volume of supply air discharged back into the same space shall be approximately equal to the volume of return air taken from the space.
   - 2.3. Return-air inlets shall not be located within 10 feet (3048 mm) of any appliance firebox or draft hood in the same room or space.
3. Rooms or spaces containing solid-fuel burning appliances, provided that return-air inlets are located not less than 10 feet (3048 mm) from the firebox of such appliances.

5. A closet, bathroom, toilet room, kitchen, garage, mechanical room, boiler room or furnace room.

❖ This section prohibits outdoor air and return air from being taken from locations that are potential sources of contamination, odor, flammable vapors, or toxic substances and also from locations that would negatively affect the operation of the furnace itself or other fuel-burning appliances.

The intent of item 4 is to prevent the system from being starved for return air by the placement of the main or only return air intake in an area not meeting the volume requirements of this item. Items 4 and 5 are airflow balance related, and items 1, 2, 3, 5, and 6 are contaminant related. It is not the intent of these items to prohibit the common practice of installing return air intakes in bedrooms and similarly sized rooms that typically have a volume that is far less than 25 percent of the total volume of the space served by the furnace.

The return air system must be able to convey the required air flow to the furnace regardless of the position of any doors to any rooms in the building served by the furnace.

It is the intent of this section is to avoid arrangements that cause an air pressure imbalance. Air pressure imbalances can cause fuel-fired appliances to spill combustion products into the occupied space. Pressure imbalances can be avoided by making sure that the amount of supply air discharge to a room or space is approximately equal to the amount of return air taken from the room or space.

G2442.6 (618.6) Screen. Required outdoor air inlets shall be covered with a screen having $^1/_4$ inch (6.4 mm) openings. Required outdoor air inlets serving a nonresidential portion of a building shall be covered with screen having openings larger than $^1/_4$ inch (6.4 mm) and not larger than 1 inch (25.4 mm).

❖ The inlet openings for outdoor air must be covered with screen or mesh material with a mesh opening size as specified. The screen openings must be small enough to keep out insects, rodents, birds, etc., while being large enough to prevent blockage of air flow by lint, debris, and plant fibers.

G2442.7 (618.7) Return-air limitation. Return air from one dwelling unit shall not be discharged into another dwelling unit.

❖ This section prohibits a forced-air heating/cooling system from serving more than one dwelling unit. Any arrangement in which dwelling units share all or part of an air distribution system would allow a communication of atmospheres between the units. This type of communication would spread odors, smoke, allergens, contaminants, and disease-causing organisms from one dwelling unit to another and therefore must be avoided.

SECTION G2443 (619)
CONVERSION BURNERS

G2443.1 (619.1) Conversion burners. The installation of conversion burners shall conform to ANSI Z21.8.

❖ The referenced standard is an installation standard that, in addition to the manufacturer's instructions, would govern the installation of conversion burners. Conversion burners are an assembly of components including burners, gas controls, blowers, safety devices, and supporting means. These units are designed to convert an existing appliance from another fuel to gas, commonly from fuel-oil or coal. Conversion of an existing appliance to a different fuel can involve much more than installation of a conversion burner. It can also include the addition of safety controls and limits, combustion air and secondary air provisions, fuel gas piping installation, chimney or vent alterations, and other modifications to the existing appliance and its control system.

SECTION G2444 (620)
UNIT HEATERS

G2444.1 (620.1) General. Unit heaters shall be tested in accordance with ANSI Z83.8 and shall be installed in accordance with the manufacturer's installation instructions.

❖ Unit heaters are ductless warm-air space heaters that are self-contained and usually suspended from a ceiling or roof structure. Garages, workshops, warehouses, factories, gymnasiums, mercantile spaces, and similar large open buildings are the most common locations for unit heaters. See Commentary Figures G2444.1(1) through G2444.1(5).

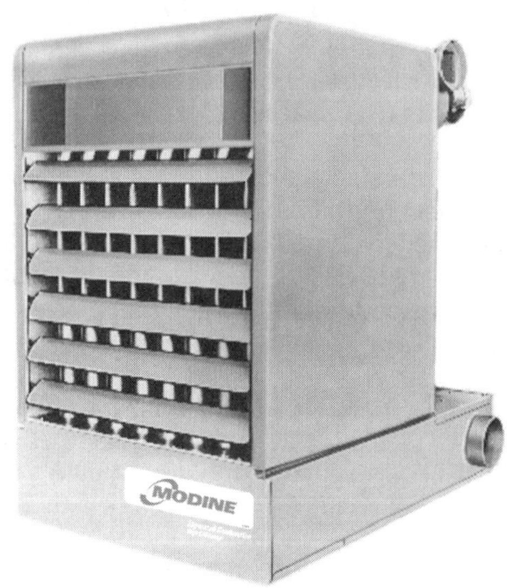

Photo courtesy of Modine Manufacturing Company

Figure G2444.1(1)
DIRECT-VENT UNIT HEATER

G2444.2 (620.2) Support. Suspended-type unit heaters shall be supported by elements that are designed and constructed to accommodate the weight and dynamic loads. Hangers and brackets shall be of noncombustible material.

❖ As with all suspended fuel-fired appliances, a support failure can result in a fire, explosion, or injury to building occupants. The supports themselves must be properly designed. Equally important are the structural members to which the supports are attached, such as rafters, beams, joists, and purlins. All brackets, pipes, rods, angle iron, structural members, and fasteners must be designed for the dead and dynamic loads of the suspended appliance.

G2444.3 (620.3) Ductwork. Ducts shall not be connected to a unit heater unless the heater is listed for such installation.

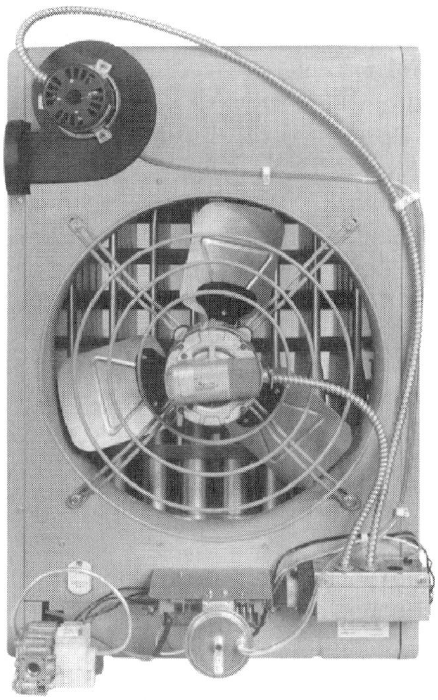

Photo courtesy of Modine Manufacturing Company

Figure G2444.1(2)
POWER-VENTED UNIT HEATER

❖ Unit heaters are usually not designed to move air through ductwork. Unless specifically listed for the application, the fans or blowers on unit heaters are designed only for moving air across the heat exchanger without the added friction of ductwork. The addition of ductwork could create a hazard by restricting airflow through the heater.

G2444.4 (620.4) Clearance. Suspended-type unit heaters shall be installed with clearances to combustible materials of not less than 18 inches (457 mm) at the sides, 12 inches (305 mm) at the bottom and 6 inches (152 mm) above the top where the unit heater has an internal draft hood or 1 inch (25.4 mm) above the top of the sloping side of the vertical draft hood.

Floor-mounted type unit heaters shall be installed with clearances to combustible materials at the back and one side only of not less than 6 inches (152 mm). Where the flue gases are vented horizontally, the 6-inch (152 mm) clearance shall be measured from the draft hood or vent instead of the rear wall of the unit heater. Floor-mounted-type unit heaters shall not be installed on combustible floors unless listed for such installation.

Clearance for servicing all unit heaters shall be in accordance with the manufacturer's installation instructions.

Exception: Unit heaters listed for reduced clearance shall be permitted to be installed with such clearances in accordance with their listing and the manufacturer's instructions.

❖ The required clearances to combustible materials for suspended and floor mounted unit heaters are specified in this section and must be adhered to unless the unit is listed for less clearance. The clearance reduction methods detailed in Section G2409 may be applied where applicable. See the commentary for Section G2409 for a complete discussion concerning clearance requirements and clearance reduction methods.

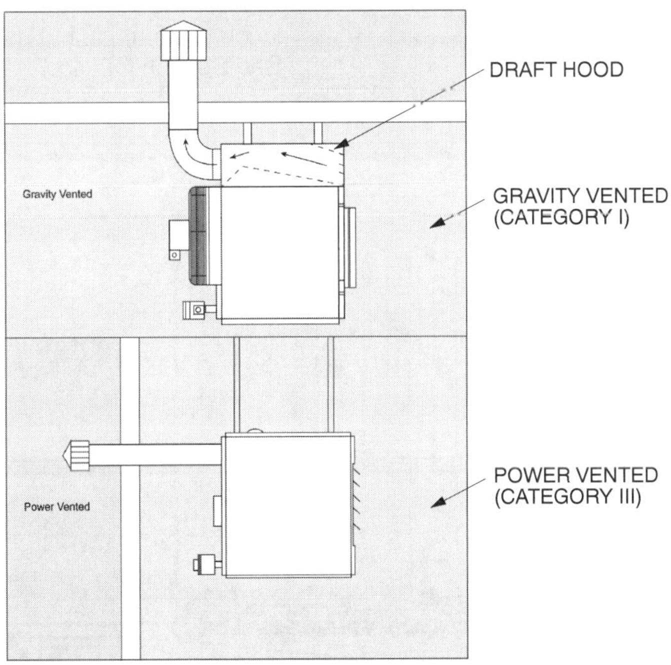

Figure courtesy of Reznor/Thomas & Betts Corporation

Figure G2444.1(3)
UNIT HEATER VENTING (TYPICAL)

Photo courtesy of Reznor/Thomas & Betts Corporation

Figure G2444.1(4)
DRAFT HOOD-EQUIPPED UNIT HEATER

SECTION G2445 (621) UNVENTED ROOM HEATERS

G2445.1 (621.1) General. Unvented room heaters shall be tested in accordance with ANSI Z 21.11.2 and shall be installed in accordance with the conditions of the listing and the manufacturer's installation instructions.

❖ Unvented room heaters are limited-size gas-fired space heaters that discharge the combustion by-products into the space being heated. Like all appliances regulated by this code, unvented gas-fired room heaters must be listed and labeled, and their installation must comply with the manufacturer's installation instructions. See Commentary Figures G2445.1(1) and G2445.1(2).

G2445.2 (621.2) Prohibited use. One or more unvented room heaters shall not be used as the sole source of comfort heating in a dwelling unit.

❖ Unvented room heaters are designed to supplement a central heating system to allow zone heating of particular rooms and spaces. Unvented room heaters are not intended to be used continuously as would occur if they were the only source of heat in a building. One or more unvented room heaters used as the sole source of heat would not provide adequate heat distribution in most building arrangements and, depending on the heating load, could require continuous operation of the appliance. Unvented gas log heaters are listed as room heaters and are referred to in UL 127 regarding their installation in factory-built fireplaces. See Commentary Figure G2445.2. Fireplaces built to the current edition of UL 127 will have a label and installation instructions that will either allow or disallow the installation of an unvented gas log in the fireplace firebox. See Section G2445.7.

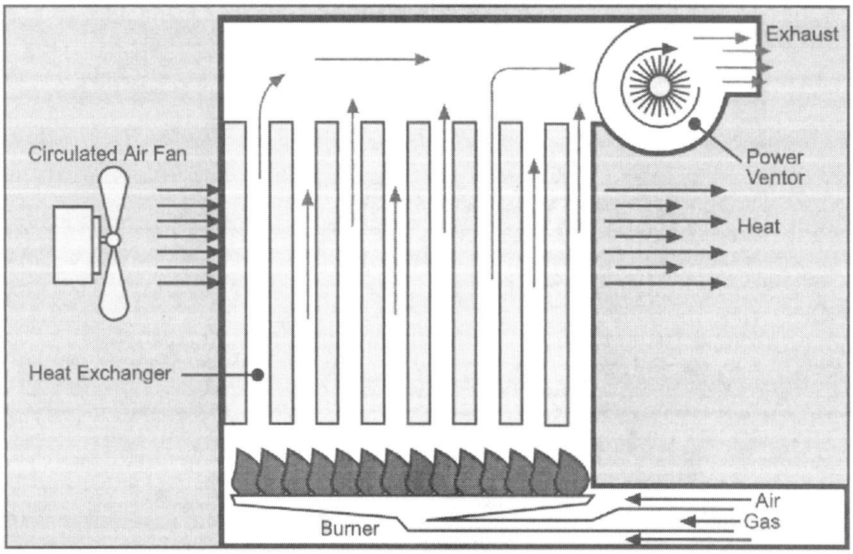

Figure courtesy of Reynor/Thomas & Betts Corp.

Figure G2444.1(5)
FUNDAMENTAL COMPONENTS OF A UNIT HEATER

Photo courtesy of Empire Comfort Systems Inc.

Figure G2445.1(1)
UNVENTED ROOM HEATER

Photography by D. F. Noyes Studio, courtesy of DESA International

Figure G2445.2
UNVENTED GAS LOG HEATER

Photography by D. F. Noyes Studio, courtesy of DESA International

Figure G2445.1(2)
FIREBOX FOR UNVENTED ROOM HEATER

G2445.3 (621.3) Input rating. Unvented room heaters shall not have an input rating in excess of 40,000 Btu/h (11.7 kW).

❖ The input rating limitation is consistent with the industry standard for such appliances and allows unvented room heaters to be categorized in the supplemental room heater classification.

G2445.4 (621.4) Prohibited locations. The location of unvented room heaters shall comply with Section G2406.2.

❖ In accordance with this section and Section G2406.2, unvented gas-fired heaters are prohibited in assembly, educational, and institutional use groups and in sleeping rooms, bathrooms, and toilet rooms in all occupancies. See the commentary for Section G2406.2 for exceptions. These heaters are considered to be an unacceptable risk in such occupancies because of occupant density, occupant age, occupant physical condition and awareness, and small room volumes.

G2445.5 (621.5) Room or space volume. The aggregate input rating of all unvented appliances installed in a room or space shall not exceed 20 Btu/h per cubic foot (0.21 kW/m^3) of volume of such room or space. Where the room or space in which the equipment is installed is directly connected to another room or space by a doorway, archway or other opening of comparable size that cannot be closed, the volume of such adjacent room or space shall be permitted to be included in the calculations.

❖ The purpose of the Btu/h (W) input to room volume ratio is to limit the accumulation of combustion by-products in the building interior. Combustion by-products include CO_2, CO, NO_2, and H_2O. The required room volume would allow dilution of the combustion by-products by infiltration.

The aggregate input rating would include the input of all unvented appliances in the room or space such as room heaters, refrigerators, and cooking appliances. See the commentary for Section G2425.8.

G2445.6 (621.6) Oxygen-depletion safety system. Unvented room heaters shall be equipped with an oxygen-depletion-sensitive safety shutoff system. The system shall shut off the gas supply to the main and pilot burners when the oxygen in the surrounding atmosphere is depleted to the percent concentration specified by the manufacturer, but not lower than

18 percent. The system shall not incorporate field adjustment means capable of changing the set point at which the system acts to shut off the gas supply to the room heater.

❖ Since unvented heaters are not vented to the outdoors, they are required by the appliance standard to incorporate this extra safety feature. This safety system consists of a special pilot burner device that is incorporated with the appliance's flame safeguard device. See the Commentary for Section G2432.2. The oxygen-depletion sensor is basically a pilot burner that is extremely sensitive to the oxygen content in the combustion air. If the oxygen content (approximately 20 percent normal) drops to a predetermined level, the pilot flame will become "lazy," shifting from a stable horizontal flame to a less stable, more vertical flame, which is incapable of sufficiently heating the thermocouple or thermopile generator, resulting in main gas control valve shutdown and lockout. The predetermined oxygen level that activates burner shutdown is above the level at which incomplete combustion would start to occur. The monitoring of oxygen levels in this manner is an indicator of approaching insufficient combustion air and is designed to prevent the formation of excess carbon monoxide. If the oxygen level reaches a low enough percentage of air, the level of oxygen will have an inverse relationship with the level of carbon monoxide. This means that as burners are increasingly starved for oxygen, the amount of carbon monoxide produced by the flames is increased. This relationship is valid for appliances in good working order. An improperly maintained, maladjusted, or defective burner could produce abnormal (elevated) amounts of carbon monoxide regardless of the oxygen level in the combustion air, and, in such cases, the oxygen-depletion sensor device will react only to the oxygen level because it cannot directly detect burner malfunction or abnormal carbon monoxide production.

G2445.7 (621.7) Unvented log heaters. An unvented log heater shall not be installed in a factory-built fireplace unless the fireplace system has been specifically tested, listed and labeled for such use in accordance with UL 127.

❖ Because the fireplace chimney damper can be closed while operating an unvented log heater (room heater), the firebox might reach surface temperatures higher than would occur with the damper open. The lack of chimney draft with the resultant dilution air introduction into the firebox would allow higher temperatures to occur in the fireplace components. To prevent surface temperatures from exceeding those allowed by UL 127, the standard now requires factory-built fireplace manufacturers to either test their units for use with unvented log heaters and meet the standard criteria or provide instructions and labels that prohibit such use.

If a factory-built fireplace has not been tested for use with an unvented appliance, the installation instructions and labels must state that such use is prohibited. Fireplaces manufactured before the UL standard added coverage for unvented log heaters could not have been tested for such use, and the manufacturers did not have the opportunity to provide instructions and labels relative to unvented log heaters for use with their units. The installation instructions for fireplace units that were built prior to unvented heater coverage being added to the UL 127 standard stated that the fireplace was permitted for use only with solid fuel and decorative gas logs listed to Z21.60. Therefore, regardless of when the fireplace unit was manufactured, unvented log heaters can only be installed in factory-built fireplaces that specifically state that they are tested for use with unvented log heaters. The fireplace manufacturers will currently indicate whether or not their units are intended for use with unvented log heaters. If listed for the application, an unvented log heater can be used as a vented decorative appliance by opening the fireplace chimney damper.

SECTION G2446 (622) VENTED ROOM HEATERS

G2446.1 (622.1) General. Vented room heaters shall be tested in accordance with ANSI Z21.11.1 or ANSI Z21.86/CSA 2.32, shall be designed and equipped as specified in Section G2432.2 and shall be installed in accordance with the manufacturer's installation instructions.

❖ This section addresses vented gas-fired space/room heaters, including direct-vent and vent- or chimney-connected appliances, which are limited-size room heaters designed to connect to a vent or chimney. As required by the appliance standard, such appliances are equipped with safety controls that will prevent gas flow to the burners in the event of ignition system failure. See Commentary Figures G2446.1(1) and G2446.1(2).

SECTION G2447 (623) COOKING APPLIANCES

G2447.1 (623.1) Cooking appliances. Cooking appliances that are designed for permanent installation, including ranges, ovens, stoves, broilers, grills, fryers, griddles, hot plates and barbecues, shall be tested in accordance with ANSI Z21.1 or ANSI Z21.58 and shall be installed in accordance with the manufacturer's installation instructions.

❖ This section addresses cooking appliances in all occupancies that are designed for permanent installation, including ranges, ovens, stoves, broilers, grills, fryers, griddles, and barbecues. These appliances must be installed in accordance with the listing and the manufacturer's installation instructions.

The code intends to regulate the design, construction, and installation of cooking appliances that are designed for permanent installation. The code does not regulate portable appliances such as electrically

Photo courtesy of Quadra-Fire

Figure G2446.1(1)
VENTED GAS FIREPLACE HEATER

Photo courtesy of Empire Comfort Systems Incorporated

Figure G2446.1(2)
VENTED ROOM HEATER

heated counter-top appliances. Appliances that are not readily moveable to another location because of a gas fuel supply connection would be considered permanently installed, even if they were on casters. Line equipment under a Type I hood, for example, is usually on casters and connected with quick-disconnect-type fuel supply lines to allow movement for routine cleaning. Such equipment would be considered permanently installed.

G2447.2 (623.2) Prohibited location. Cooking appliances designed, tested, listed and labeled for use in commercial occupancies shall not be installed within dwelling units or within any area where domestic cooking operations occur.

❖ Commercial cooking appliances are tested and labeled to different standards than those listed for domestic use. Commercial cooking appliances generally are not insulated to the same level as domestic cooking appliances, have higher surface operating temperatures than domestic appliances, and require a much greater clearance to combustible material. The safety measures inherent to household cooking appliances, such as child-safe push-to-turn knobs and insulated oven doors, are not usually provided in commercial cooking appliances.

Commercial cooking appliances also have a greater ventilation air requirement for safe operation than household-type cooking appliances. As such, installation of commercial-type cooking appliances in dwellings is prohibited.

G2447.3 (623.3) Domestic appliances. Cooking appliances installed within dwelling units and within areas where domestic cooking operations occur shall be listed and labeled as household-type appliances for domestic use.

❖ Cooking appliances used in dwelling units or in areas where domestic cooking operations occur require a greater degree of user protection and as such must be listed and labeled as household-type appliances for domestic use. See the commentary for Section G2447.2. To satisfy residential consumer demand for commercial appliances in the home, some manufacturers are producing listed household-type appliances that have the appearance of commercial application cooking appliances. See Commentary Figures G2447.3(1) and (2).

G2447.4 (623.4) Range installation. Ranges installed on combustible floors shall be set on their own bases or legs and shall be installed with clearances of not less than that shown on the label.

❖ This section requires the installation of a domestic range using the legs or base provided by the manufacturer where the range is resting on a combustible surface. These supports were part of the test set-up when the range was tested and as such must be included in the installation to ensure compliance with the listing. Clearances behind and to the sides of the range must also comply with the listing.

SECTION G2448 (624) WATER HEATERS

G2448.1 (624.1) General. Water heaters shall be tested in accordance with ANSI Z 21.10.1 and ANSI Z 21.10.3 and shall be installed in accordance with the manufacturer's installation instructions.

❖ Water heaters are recognized by the codes as both plumbing and mechanical appliances. The purpose of this section is to identify those aspects of water heaters specifically related to fuel gas.

The standards listed are specific to gas-fired water heaters. ANSI Z21.10.1 is for water heaters with an input rating less than or equal to 75,000 Btu per hour. ANSI Z21.10.3 is for water heaters with an input rating greater than 75,000 Btu per hour and for circulating and instantaneous water heaters.

G2448.1.1 (624.1.1) Installation requirements. The requirements for water heaters relative to sizing, relief valves, drain pans and scald protection shall be in accordance with this code.

❖ This section refers the user to the plumbing provisions of this code for generic requirements related to water heaters and not related to the fuel source. These include connections to the potable water system, safety devices such as relief valves, drain pans for protection of the structure, and sizing of the water heaters.

G2448.2 (624.2) Water heaters utilized for space heating. Water heaters utilized both to supply potable hot water and provide hot water for space-heating applications shall be listed and labeled for such applications by the manufacturer and shall be installed in accordance with the manufacturer's installation instructions and this code.

❖ Water heaters serving the dual purpose of supplying potable hot water and serving as a heat source for a space-heating system must be listed and labeled for that dual application. This section does not address water heaters used solely for space-heating applications, but rather addresses water heaters that serve a secondary purpose of space heating. The label will indicate whether the water heater is suitable for space heating. See Commentary Figures G2448.2(1) and (2).

SECTION G2449 (627) AIR CONDITIONING EQUIPMENT

G2449.1 (627.1) General. Air conditioning equipment shall be tested in accordance with ANSI Z21.40.1 or ANSI Z21.40.2 and shall be installed in accordance with the manufacturer's installation instructions.

❖ Gas-fired air conditioning systems include absorption types and internal combustion engine driven machines. The referenced ANSI standards provide requirements for gas-fired air conditioning equipment. ANSI Z21.40.1 regulates gas-fired absorption systems and ANSI Z21.40.2 regulates work activated gas-fired air conditioning and heat pump systems. The code relies on the manufacturer's installation instructions and the referenced standards for installation requirements.

G2449.2 (627.2) Independent piping. Gas piping serving heating equipment shall be permitted to also serve cooling equipment where such heating and cooling equipment cannot be operated simultaneously. (See Section G2413.)

❖ Where gas supply piping serves both the heating and cooling equipment, this section permits the piping to be sized based on the larger gas demand of the two systems rather than sizing based on the total demand of the two systems. This provision is not applicable unless the heating and cooling systems are incapable of operating simultaneously.

Photo courtesy of Maytag Corp.

Figure G2447.3(1) DOMESTIC (HOUSEHOLD TYPE) RANGE

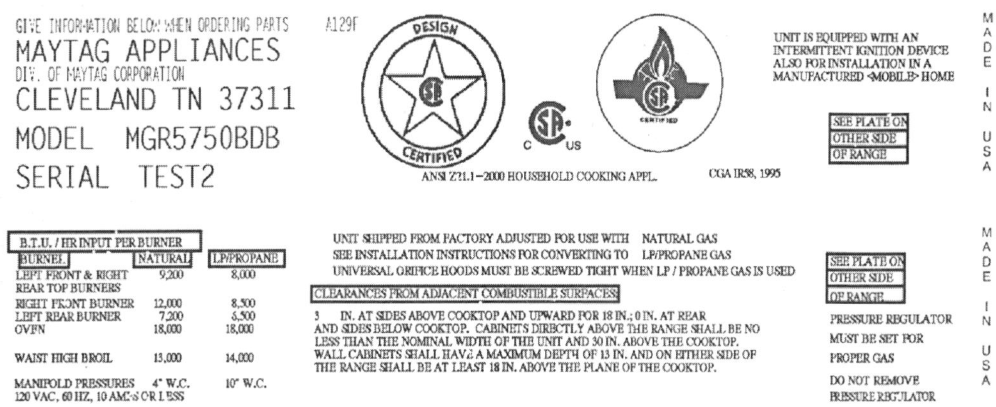

Illustration courtesy of Maytag Corp.

Figure G2447.3(2)
HOUSEHOLD-TYPE COOKING APPLIANCE LABEL

G2449.3 (627.3) Connection of gas engine-powered air conditioners. To protect against the effects of normal vibration in service, gas engines shall not be rigidly connected to the gas supply piping.

❖ Because of the inherent vibration of a gas engine, it is necessary to install a connector between the engine and the gas supply pipe in such a manner that the vibration does not place undue stress on the connector or have an adverse effect on the shutoff valve or the gas piping. Vibration transmitted to the connector and piping could eventually cause joint failure and/or connector failure. The connector must be designed and approved for such application.

G2449.4 (627.6) Installation. Air conditioning equipment shall be installed in accordance with the manufacturer's instructions. Unless the equipment is listed for installation on a combustible surface such as a floor or roof, or unless the surface is protected in an approved manner, equipment shall be installed on a surface of noncombustible construction with noncombustible material and surface finish and with no combustible material against the underside thereof.

❖ The intent of this section is to prohibit the installation of gas-fired air conditioning equipment on combustible surfaces unless specifically listed for such an installation or the surface is protected in a manner approved by the code official.

SECTION G2450 (628) ILLUMINATING APPLIANCES

G2450.1 (628.1) General. Illuminating appliances shall be tested in accordance with ANSI Z21.42 and shall be installed in accordance with the manufacturer's installation instructions.

❖ Illuminating appliances must be listed and labeled as required for other appliances regulated by this code. These appliances include gas lamps designed for outdoor use.

G2450.2 (628.2) Mounting on buildings. Illuminating appliances designed for wall or ceiling mounting shall be securely attached to substantial structures in such a manner that they are not dependent on the gas piping for support.

❖ Gas piping is not designed to act as a support for any appliance. All appliances must be independently supported so as not to strain or stress the fuel supply connection.

G2450.3 (628.3) Mounting on posts. Illuminating appliances designed for post mounting shall be securely and rigidly attached to a post. Posts shall be rigidly mounted. The strength and rigidity of posts greater than 3 feet (914 mm) in height shall be at least equivalent to that of a 2.5-inch (64 mm) diameter post constructed of 0.064-inch (1.6 mm) thick steel or a 1-inch (25.4 mm) Schedule 40 steel pipe. Posts 3 feet (914 mm) or less in height shall not be smaller than a $^3/_4$-inch (19 mm) Schedule 40 steel pipe. Drain openings shall be provided near the base of posts where there is a possibility of water collecting inside them.

❖ Regardless of the intended method of mounting and support of an appliance, the installation must be secure and must not transmit any loading to the fuel supply connection. A requirement is added to provide a drain at the base of the post to prevent collection of water that can deteriorate the post material.

G2450.4 (628.4) Appliance pressure regulators. Where an appliance pressure regulator is not supplied with an illuminating appliance and the service line is not equipped with a service pressure regulator, an appliance pressure regulator shall be installed in the line to the illuminating appliance. For multiple installations, one regulator of adequate capacity shall be permitted to serve more than one illuminating appliance.

❖ Outdoor lighting appliances generally require low pressure gas flow. Some of these lighting appliances are provided from the manufacturer with a built-in pressure regulator set for the proper operating pressure. If, however, the regulator is not provided, this section requires the installation of one in the gas line serving the appliance. This regulator must be appropriate for the pressure required by the manufacturer. One pressure regulator with the required capacity may be provided to serve more than one lighting appliance if the piping pressure losses permit such application.

SECTION G2451 (630) INFRARED RADIANT HEATERS

G2451.1 (630.1) General. Infrared radiant heaters shall be tested in accordance with ANSI Z 83.6 and shall be installed in accordance with the manufacturer's installation instructions.

❖ This section addresses radiant heaters including ceramic element and steel tube-type designs. Infrared heaters are produced in both vented and unvented types and function by creating a very hot surface area from which heat energy is directly radiated. Infrared radiant heaters are usually suspended from ceilings or roofs. Such heaters are typically used for "spot" heating in spaces that are otherwise unconditioned or that are not conditioned to human comfort levels. Radiant heaters have the advantage of being able to heat objects and personnel without having to heat the surrounding air. See Commentary Figures G2451.1(1) through G2451.1(3).

The code relies on the manufacturer's installation instructions for installation requirements. Maintaining the clearance to combustibles for radiant heaters is of paramount importance. Direct radiation is a very effective method of transferring heat energy, and improperly located combustible materials can be readily ignited. See Commentary Figure G2451.1(4).

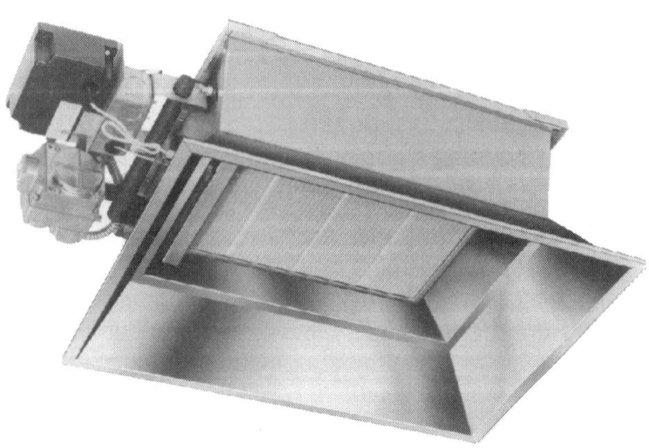

Photo courtesy of Reznor/Thomas & Betts Corporation

**Figure G2451.1(1)
CERAMIC BURNER INFRARED RADIANT HEATER**

Photo courtesy of Reznor/Thomas & Betts Corporation

**Figure G2451.1(2)
TUBE-TYPE INFRARED RADIANT HEATER**

Photo courtesy of Reznor/Thomas & Betts Corporation

**Figure G2451.1(3)
TUBE-TYPE INFRARED RADIANT HEATER**

G2451.2 (630.2) Support. Infrared radiant heaters shall be safely and adequately fixed in an approved position independent of gas and electric supply lines. Hanger and brackets shall be of noncombustible material.

❖ Supports must prevent radiant heaters from falling. A fall would mean losing the required clearance to combustibles; putting tension or pressure on electrical, fuel, and vent connections; and dislocating so as to redirect the radiant output to where a fire hazard would result. Often, such heaters are improperly hung and restrained such that heater movement (swinging) puts stress on flexible (semi-rigid) gas connectors. Gas connectors must be designed and approved for applications in which the appliance moves because of expansion and contraction or lack of restraints. To accommodate expansion/contraction movement, some appliance manufacturers require flexible connectors to be coiled or shaped in a manner that allows limited

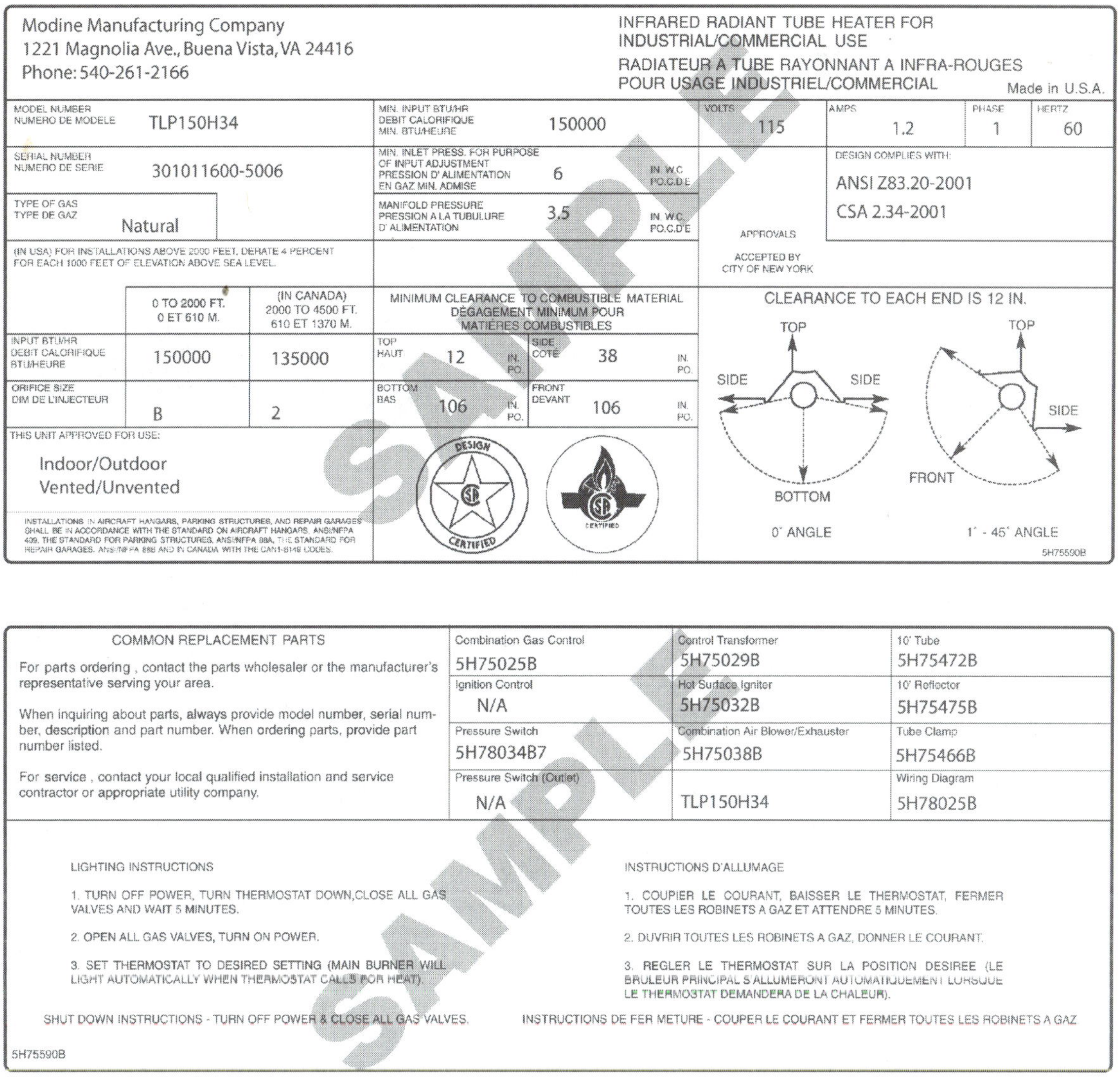

Figure courtesy of Modine Manufacturing Co.

Figure G2451.1(4)
SAMPLE LABEL FOR INFRARED RADIANT TUBE HEATER

movement without stressing the metal of the connector, similar to expansion loops and offsets in rigid piping systems.

SECTION G2452 (631) BOILERS

G2452.1 (631.1) Standards. Boilers shall be listed in accordance with the requirements of ANSI Z21.13 or UL 795. If applicable, the boiler shall be designed and constructed in accordance with the requirements of ASME CSD-1 and as applicable, the ASME *Boiler and Pressure Vessel Code,* Sections I, II, IV, V and IX and NFPA 85.

❖ The scope of this section includes boilers in all occupancies including power plants; factories; industrial plants; schools; institutional occupancies such as hospitals, commercial laundries, and hotels; and residential structures. Boilers are defined in Section 202 of this code. See Commentary Figure G2452.1.

Boilers are potentially dangerous if not properly designed, constructed, and operated, more so than many other appliances because of the potential explosion hazard associated with pressure vessels. In addition to this code, several industry standards are referenced. Manufactured gas-fired boilers are required to be listed to either ANSI Z21.13 or UL 795. Figure G2403(3) shows a typical label with the required information for a listed boiler. Only the design and construction requirements of boilers are regulated by the referenced standards. The requirements for specific types of boilers can be found in the respective sections of the referenced standards.

Photo courtesy of Weil-McLain

Figure G2452.1
STEAM HEATING BOILER WITH GAUGE GLASS, LOW-WATER CUTOFF CONTROL, PRESSURE GAUGE, PRESSURE LIMIT CONTROL AND AUTOMATIC VENT DAMPER

G2452.2 (631.2) Installation. In addition to the requirements of this code, the installation of boilers shall be in accordance with the manufacturer's instructions and this code. Operating instructions of a permanent type shall be attached to the boiler. Boilers shall have all controls set, adjusted and tested by the installer. A complete control diagram together with complete boiler operating instructions shall be furnished by the installer. The manufacturer's rating data and the nameplate shall be attached to the boiler.

❖ This section governs the installation and commissioning of boilers and their control systems. The mechanical equipment requirements for approval, labeling, installation, maintenance, repair, and alteration are regulated by the mechanical portion of this code and the manufacturer's instructions. The mechanical portion of this code contains installation requirements for boilers including provisions for shutoff valves, pressure relief valves, safety valves, expansion tanks and low water cutoff controls. See Commentary Figures G2452.1, G2452.2(1), G2452.2(2), and G2452.2(3).

Complete operating instructions must be permanently affixed to the boiler upon completion of the

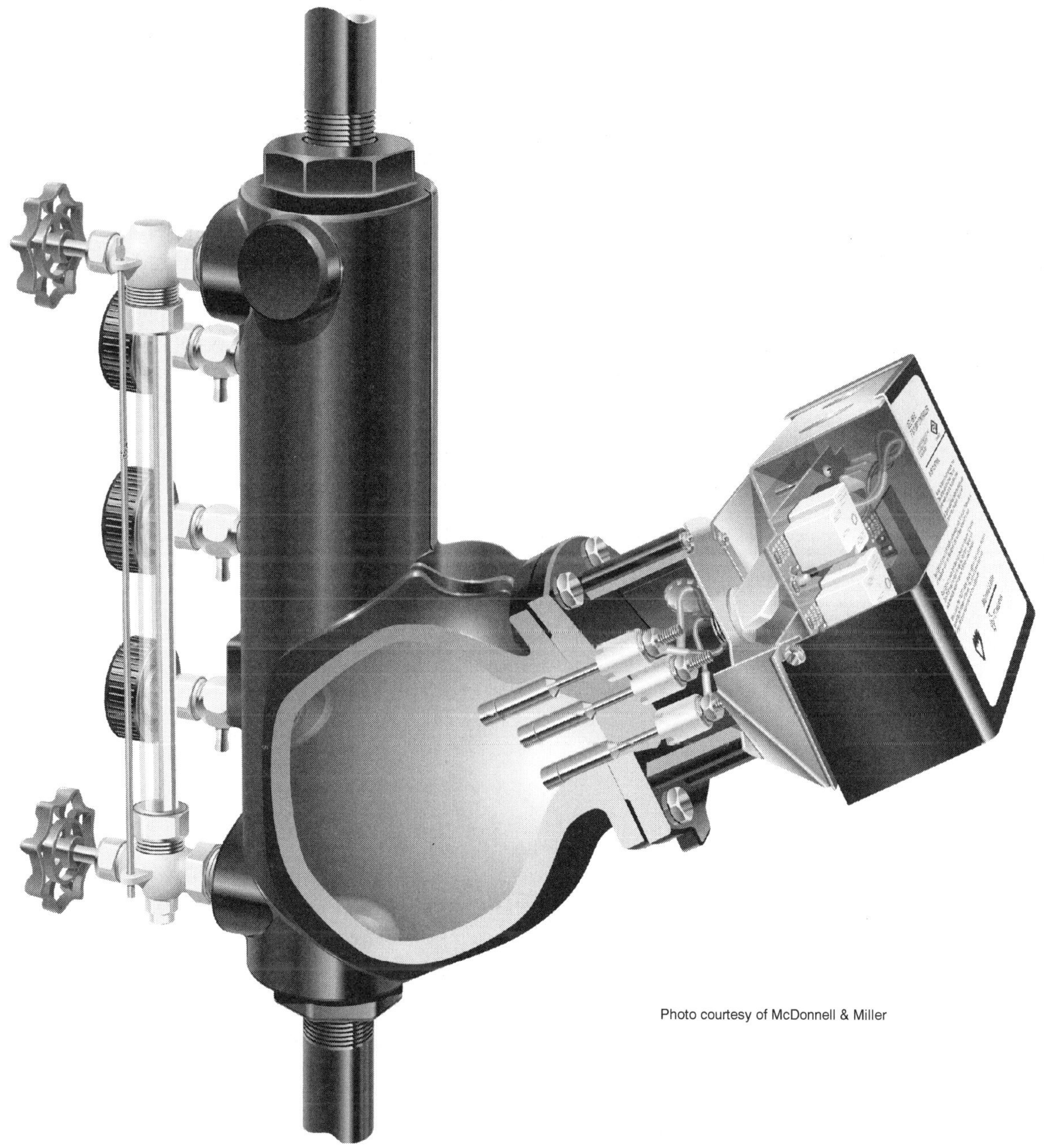

Photo courtesy of McDonnell & Miller

Figure G2452.2(1)
ELECTRONIC PROBE-TYPE LOW WATER CUTOFF CONTROL WITH GAUGE GLASS AND TRI-COCKS FOR STEAM BOILER APPLICATION

installation. Boiler systems can be complex and generally require coordination of several pieces of equipment. The proper operating procedures, set points, etc., must be specified and included in these instructions. This is usually done in the control system design documentation, including such things as diagrams, system schematics, and control sequence descriptions. Typically, an operating and maintenance (O&M) manual is provided to the building owner/operator upon completion of the project. Along with operating and control procedures, the O&M manual will clearly identify routine maintenance and calibration information. It is also typical for operation sequences and calibration information to be displayed under glass in a conspicuous place in the boiler room for use by operating and service personnel. The intent is to provide the owner/operator with all operating and control information necessary to properly operate and maintain the boiler and its associated controls and equipment.

Photo courtesy of McDonnell & Miller

Figure G2452.2(2)
ELECTRONIC PROBE-TYPE LOW WATER CUTOFF CONTROL FOR WATER BOILER APPLICATION

G2452.3 (631.3) Clearance to combustible material. Clearances to combustible materials shall be in accordance with Section G2409.4.

❖ Boilers, like all gas-fired equipment and appliances, operate at high temperatures that require adequate separation from combustibles. This section refers the user to Section G2409 of this code for requirements for clearances to combustibles and allowable clearance reduction methods.

SECTION G2453 (634)
CHIMNEY DAMPER OPENING AREA

G2453.1 (634.1) Free opening area of chimney dampers. Where an unlisted decorative appliance for installation in a vented fireplace is installed, the fireplace damper shall have a permanent free opening equal to or greater than specified in Table G2453.1.

❖ This section intends to ensure that an adequate fireplace damper opening is provided for all unlisted decorative appliances. The minimum damper opening size is part of the listing for listed appliances, but the installer and the code official might not have access to this information for an unlisted appliance. Section G2404.3 requires all appliances to be listed. The free opening size is a function of the height of the chimney and the appliance input rating. As noted in footnote .a,. the opening areas in the table correspond to the cross-sectional area of round chimney flues having nominal sizes from 3 to 8 inches, and the 64 square inches represents the cross-sectional area of a standard 8-inch by 8-inch chimney tile.

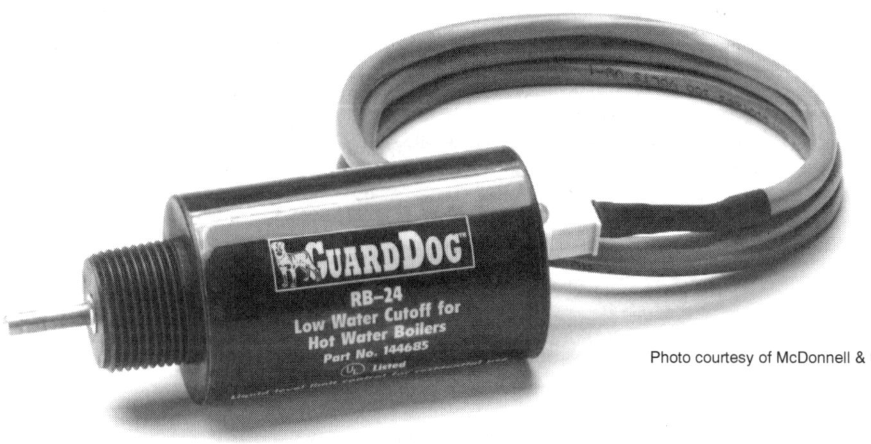

Photo courtesy of McDonnell & Miller

Figure G2452.2(3)
ELECTRONIC PROBE-TYPE LOW WATER CUTOFF CONTROL FOR WATER BOILER APPLICATION

TABLE G2453.1 (634.1)
FREE OPENING AREA OF CHIMNEY DAMPER FOR VENTING FLUE GASES FROM UNLISTED DECORATIVE APPLIANCES FOR INSTALLATION IN VENTED FIREPLACES

CHIMNEY HEIGHT (feet)	MINIMUM PERMANENT FREE OPENING (square inches)[a]						
	8	13	20	29	39	51	64
	Appliance input rating (Btu per hour)						
6	7,800	14,000	23,200	34,000	46,400	62,400	80,000
8	8,400	15,200	25,200	37,000	50,400	68,000	86,000
10	9,000	16,800	27,600	40,400	55,800	74,400	96,400
15	9,800	18,200	30,200	44,600	62,400	84,000	108,800
20	10,600	20,200	32,600	50,400	68,400	94,000	122,200
30	11,200	21,600	36,600	55,200	76,800	105,800	138,600

For SI: 1 foot = 304.8 mm, 1 square inch = 645.16 mm^2, 1,000 Btu per hour = 0.293 kW.

a. The first six minimum permanent free openings (8 square inches to 51 square inches) correspond approximately to the cross-sectional areas of chimneys having diameters of 3 inches through 8 inches, respectively. The 64-square inch opening corresponds to the cross-sectional area of standard 8-inch by 8-inch chimney tile.

❖ See the commentary for Section G2453.1.

Bibliography

The following resource material is referenced in this chapter or is relevant to the subject matter addressed in this chapter.

ANSI LC1-97, *Gas Piping Systems Using Corrugated Stainless Steel Tubing*. New York: American National Standards Institute, 1997.

ANSI Z21.15-97, *Manually Operated Gas Valves for Appliances, Appliance Connector Valves, and Hose End Valves*. New York: American National Standards Institute, 1997.

ANSI Z21.18-2000, *Gas Appliance Pressure Regulators*. New York: American National Standards Institute, 2000.

ANSI Z21.47-01, *Gas-Fired Central Furnaces*. New York: American National Standards Institute, 2001.

ANSI Z83.8-96, *Gas Unit Heaters*. New York: American National Standards Institute, 1996.

ANSI Z223.1-02, *National Fuel Gas Code.* New York: American National Standards Institute, 2002.

ASCE 7-02, *Minimum Design Loads for Buildings and Other Structures*. Reston, VA: American Society of Civil Engineers, 2002.

ASHRAE Handbook of HVAC Systems and Equipment. Atlanta, GA: American Society of Heating, Refrigerating and Air-Conditioning Engineers, 1999.

ASTM D 2513-01a, *Specification for Thermoplastic Gas Pressure Pipe, Tubing, and Fittings*. West Conshohocken, PA: ASTM International, 2001.

ASTM E 136-99, *Standard Test Method for Behavior of Materials in a Vertical Tube Furnace at 750°C*. West Conshohocken, PA: ASTM International, 1999.

AWS A5.8-92, *Filler Metals for Brazing and Braze Welding*. Miami, FL: American Welding Society, 1992.

Chimney and Gas Vent Sizing Handbook. Dallas, TX: Selkirk Metalbestos, 1992.

Crocker, Sabin and R. C. King. *Piping Handbook*. Columbus, OH: McGraw-Hill, 1973

GPSA Engineering Data Book. Tulsa, OK: Gas Processors Suppliers Association, 1998.

NFPA 58-01, *Liquefied Petroleum Gases Code*. Quincy, MA: National Fire Protection Association, 2001.

NFPA 70, *National Electrical Code*. Quincy, MA: National Fire Protection Association, 2002.

UL 641-95, *Type L Low Temperature Venting Systems*. Northbrook, IL: Underwriters Laboratories, 1995.

UL 795-99, *Commercial-Industrial Gas Heating Equipment*. Northbrook, IL: Underwriters Laboratories, 1999.

UL 1738-00, *Venting Systems for Gas Burning Appliances: Categories II, III and IV*. Northbrook, IL: Underwriters Laboratories, 2000.

UL 1777-98, *Chimney Liners*. Northbrook, IL: Underwriters Laboratories, 1998.

Part VII — Plumbing

Chapter 25: Plumbing Administration

General Comments

Chapter 25 contains administrative requirements that apply to the installation of plumbing systems and to the ensuing plumbing related chapters of the code. Chapter 25 refers the user to the applicable provisions of Chapter 1 and addresses the code requirements' applicability to additions, alterations, and repairs. This chapter also outlines the building official's authority and discretion regarding existing construction that is used in conjunction with new construction.

Chapter 25 lists the types and phases of various inspections as well as it details the permit holder's responsibilities as to when and how to perform tests.

Purpose

The requirements of Chapter 25 do not precede the administrative provisions of Chapter 1. Rather, the administrative guidelines of Chapter 25 pertain to plumbing installations that are best referenced and located within the plumbing chapters. This chapter's purpose is to address how to apply the plumbing provisions of this code to specific types or phases of construction. This chapter also serves to outline the responsibilities of the applicant, installer, and inspector with regard to testing plumbing installations.

SECTION P2501 GENERAL

P2501.1 Scope. The provisions of this chapter shall establish the general administrative requirements applicable to plumbing systems and inspection requirements of this code.

❖ This section states that the provisions of Chapter 25 set forth the general administration requirements for plumbing systems.

P2501.2 Application. In addition to the general administration requirements of Chapter 1, the administrative provisions of this chapter shall also apply to the plumbing requirements of Chapters 25 through 32.

❖ Section P2501.2 refers to the applicability of the administrative requirements of Chapter 1 and of plumbing Chapters 25 through 32.

[EB] SECTION P2502 EXISTING PLUMBING SYSTEMS

P2502.1 Existing building sewers and drains. Existing building sewers and drains shall be used in connection with new systems when found by examination and/or test to conform to the requirements prescribed by this document.

❖ It is necessary to examine the existing drainage and sewer piping when adding to or altering an existing system. The building official may require the system to be tested to determine whether the system leaks. Where additional load requires a pipe size increase, the change must be in accordance with Sections P3005.4.1 and P3005.4.2 based on drainage fixture unit values.

P2502.2 Additions, alterations or repairs. Additions, alterations, renovations or repairs to any plumbing system shall conform to that required for a new plumbing system without requiring the existing plumbing system to comply with all the requirements of this code. Additions, alterations or repairs shall not cause an existing system to become unsafe, insanitary or overloaded.

Minor additions, alterations, renovations and repairs to existing plumbing systems shall be permitted in the same manner and arrangement as in the existing system, provided that such repairs or replacement are not hazardous and are approved.

❖ Additions or alterations to the existing system can place additional loads on the system. Such loads might necessitate changing all or part of the existing system. Repairs to an existing plumbing system are permitted without replacing the entire system provided that such additions or alterations do not cause a health hazard or an insanitary condition. Minor additions, alterations, or repairs are permitted when installed in the same manner as the existing system provided that the building official approves such repairs.

SECTION P2503 INSPECTION AND TESTS

P2503.1 Inspection required. New plumbing work and parts of existing systems affected by new work or alterations shall be inspected by the building official to ensure compliance with the requirements of this code.

❖ This section establishes the building official's authority to conduct inspections of plumbing work to determine code compliance. The code requires inspection and approval of new work and existing systems that could be affected by new work. This provision allows the building official to inspect existing systems as part of the overall approval process.

P2503.2 Concealment. A plumbing or drainage system, or part thereof, shall not be covered, concealed or put into use until it has been tested, inspected and approved by the building official.

❖ This section informs the applicant that the installation must remain exposed for inspection, testing, and approval of the plumbing system. The majority of a plumbing installation is hidden within finished construction. Thus, the code requires inspections to be conducted prior to covering or concealing the installation.

P2503.3 Responsibility of permittee. Test equipment, materials and labor shall be furnished by the permittee.

❖ It is the responsibility of the permittee, not the building official, to provide the necessary equipment, material, and labor to perform the required tests to the plumbing system.

P2503.4 Building sewer testing. The building sewer shall be tested by insertion of a test plug at the point of connection with the public sewer and filling the building sewer with water, testing with not less than a 10-foot (3048 mm) head of water and be able to maintain such pressure for 15 minutes.

❖ This section requires that the building sewer be plugged at the point of connection to the public sewer and tested with 10 feet (3048 mm) of water for 15 minutes without leaking. While such a test provides a relatively small amount of pressure (approximately 4.33 psi [29.89 kPa]), it is sufficient in nonpressure piping applications.

P2503.5 DWV systems testing. Rough and finished plumbing installations shall be tested in accordance with Sections P2503.5.1 and P2503.5.2.

❖ Section P2503.5 indicates that drainage and venting systems are to be tested once during the rough-in phase and again after the plumbing fixtures have been set and traps installed.

P2503.5.1 Rough plumbing. DWV systems shall be tested on completion of the rough piping installation by water or air with no evidence of leakage. Either test shall be applied to the drainage system in its entirety or in sections after rough piping has been installed, as follows:

1. Water test. Each section shall be filled with water to a point not less than 10 feet (3048 mm) above the highest fitting connection in that section, or to the highest point in the completed system. Water shall be held in the section under test for a period of 15 minutes. The system shall prove leak free by visual inspection.
2. Air test. The portion under test shall be maintained at a gauge pressure of 5 pounds per square inch (psi) (34 kPa) or 10 inches of mercury column (2488 Pa). This pressure shall be held without introduction of additional air for a period of 15 minutes.

❖ This section describes the procedures for performing a water or air test of the DWV system. The water test is performed during the rough-in phase to determine whether the plumbing system complies with the code and operates as intended. This test allows any defects to be identified and repaired before the system is concealed within the building. Except for the top 10 feet (3048 mm) of the system, each section must be tested with a minimum 10-foot (3048 mm) head of water for a period of 15 minutes. The system must be tight at all points.

As an alternative, this section allows an air pressure test of the DWV system to be conducted, introducing air until the system can maintain 5 pounds per square inch (34 kPa) pressure throughout for a period of at least 15 minutes without introducing additional air. Testing DWV systems with air is inherently more dangerous to personnel than testing with water because of the energy release that can occur if the system fails or ruptures. Increasing the test pressure beyond the material's safety limits poses a hazard. Some manufacturers of plastic DWV materials do not recommend testing with air. It is best to consult the manufacturer's installation instructions for proper testing procedures.

P2503.5.2 Finished plumbing. After the plumbing fixtures have been set and their traps filled with water, their connections shall be tested and proved gas tight and/or water tight as follows:

1. Water tightness. Each fixture shall be filled and then drained. Traps and fixture connections shall be proven water tight by visual inspection.
2. Gas tightness. When required by the local administrative authority, a final test for gas tightness of the DWV system shall be made by the smoke or peppermint test as follows:
    2.1. Smoke test. Introduce a pungent, thick smoke into the system. When the smoke appears at vent terminals, such terminals shall be sealed and a pressure equivalent to a 1-inch water column (249 Pa) shall be applied and maintained for a test period of not less than 15 minutes.
    2.2. Peppermint test. Introduce 2 ounces (59 mL) of oil of peppermint into the system. Add 10 quarts (9464 mL) of hot water and seal all vent terminals. The odor of peppermint shall not be detected at any trap or other point in the system.

❖ Section P2503.5.2 requires a visual inspection of the finished plumbing system. When there is any doubt that the finished system is water tight, this section allows for the entire system to be subjected to a smoke or peppermint test. The smoke test involves filling all traps with water and then introducing a thick pungent smoke produced by a smoke machine. When smoke appears at stack openings, they must be capped, and a pressure equivalent to 1 inch of water column (249 Pa) must be maintained for a period of 15 minutes. The pressure developed for the smoke test must not exceed 1 inch (249 Pa) of water column, because any higher pressure could affect the water seal in the traps, thereby rendering the test invalid. Detection of any smoke or peppermint odor indicates a leak that must be repaired.

P2503.6 Water-supply system testing. Upon completion of the water-supply system or a section thereof, the system, or portion completed, shall be tested and proved tight under a water pressure not less than the working pressure of the system; or, for piping systems other than plastic, by an air test of not less than 50 psi (344 kPa). The water used for tests shall be obtained from a potable water source.

❖ The water pressure required to test water-supply systems must be equal to or greater than the working pressure of the system. Air testing is permitted for all systems except for those made of plastic because of the hazard described in the commentary for Section P2503.5.1.

P2503.7 Inspection and testing of backflow prevention devices. Inspection and testing of backflow prevention devices shall comply with Section P2503.7.1 and Section P2503.7.2.

❖ A device must be tested in accordance with Section P2503.7.2 to determine whether it is operating properly. A visual inspection will determine if the unit is installed according to its listing and the manufacturer's installation instructions. See the commentary for Section P2902.

P2503.7.1 Inspections. Inspections shall be made of all backflow prevention assemblies to determine whether they are operable.

❖ This section requires inspection of all backflow assemblies to ascertain whether they are installed correctly and are operating in accordance with the manufacturer's installation instructions. The commentary for Section P2902 provides further discussion of installation.

P2503.7.2 Testing. Reduced pressure principle backflow preventers, double check valve assemblies, double-detector check valve assemblies and pressure vacuum breaker assemblies shall be tested at the time of installation, immediately after repairs or relocation and at least annually.

❖ Four types of assemblies require testing: reduced pressure principal backflow assemblies, double check valve assemblies, double-detector check valve assemblies, and pressure vacuum breaker assemblies. See the commentary for Section P2902. Other backflow prevention devices such as atmospheric vacuum breakers and air gaps are used within a water distribution system to protect against cross-contamination. These devices and installations are required to be tested, inspected, and approved. For the purpose of testing their adequacy under certain working conditions, the four assemblies referenced in this section are provided with test-cocks (or ports) that are used in conjunction with specific testing equipment and procedures. These tests are required at the time of installation, repair, and relocation, and at least on an annual basis thereafter.

P2503.8 Test gauges. Gauges used for testing shall be as follows:

1. Tests requiring a pressure of 10 psi or less shall utilize a testing gauge having increments of 0.10 psi or less.
2. Tests requiring a pressure of greater than 10 psi but less than or equal to 100 psi shall utilize a testing gauge having increments of 1 psi or less.
3. Tests requiring a pressure of greater than 100 psi shall utilize a testing gauge having increments of 2 psi or less.

❖ It is common to see test gauges that register in one or two psi increments being used in a DWV inspection that requires only a five psi test. These gauges are slow to respond, difficult to read and interpret, and not appropriate at these lower pressures.

In each case, the user needs to ensure that the proper pressure gauge is selected with respect to indicating range and design. Most standard dial-type pressure gauges use a bourdon tube sensing element generally made of a copper alloy (brass) or stainless steel. The construction is simple, and operation does not require any additional power source. The C-shaped or spirally wound bourdon tube flexes when pressure is applied, producing a rotational movement that in turn causes the pointer to indicate the measured pressure. It is recommended to install shut-off devices between the measuring point and the pressure gauge. This will allow an exchange of the pressure gauge and checks on the gauge's zero setting while the system remains operative.

Bibliography

The following resource materials are referenced in this chapter or are relevant to the subject matter addressed in this chapter.

ASSE 5013-98, *Performance Requirements for Testing Reduced Pressure Principle Backflow Preventers (RP) and Reduced Pressure Fire Protection Principle Backflow Preventers (RFP)*. Westlake, OH: American Society of Sanitary Engineers, 1998.

ASSE 5015-98, *Performance Requirements for Testing Double Check Backflow Prevention Assemblies (DC) and Double Check Fire Protection Backflow Prevention Assemblies (DCF)*. Westlake, OH: American Society of Sanitary Engineers, 1998.

ASSE 5020-98, *Performance Requirements for Testing Pressure Vacuum Breaker Assembly (PVBA)*. Westlake, OH: American Society of Sanitary Engineers, 1998.

ASSE 5047-98, *Performance Requirements for Testing Reduced Pressure Detector Fire Protection Backflow Prevention Assemblies (RPDF)*. Westlake, OH: American Society of Sanitary Engineers, 1998.

ASSE 5048-98, *Performance Requirements for Testing Double Check Detector Fire Protection Backflow Prevention Assemblies (DCDF)*. Westlake, OH: American Society of Sanitary Engineers, 1998.

ASSE 5052-98, *Performance Requirements for Testing Hose Connection Backflow Preventers*. Westlake, OH: American Society of Sanitary Engineers, 1998.

ASSE 5056-98, *Performance Requirements for Testing Spill Resistant Vacuum Breaker*. Westlake, OH: American Society of Sanitary Engineers, 1998.

Legal Aspects of Code Administration. Country Club Hills, IL: International Code Council, Inc., 2001.

Chapter 26:
General Plumbing Requirements

General Comments

This chapter provides for the proper installation of piping systems used in conjunction with plumbing fixtures, drains, and appliances. It mandates the removal of liquid wastes and sewage within a building and specifies approved sources of potable water and proper collection of wastes.

Chapter 26 outlines approved methods for maintaining structural integrity when installing the plumbing system as well as for protecting the piping system itself. This chapter also provides guidelines for installing and inspecting underground piping and for preventing the undermining of foundation footings.

Chapter 26 includes a table as well as specifications for providing proper horizontal and vertical support to piping systems using diverse materials. This chapter also addresses the water tightness of a structure's exterior membrane penetrations. Chapter 26 includes provisions for adhering to standards that regulate materials used in piping systems for durability and health effects.

Purpose

Chapter 26 contains requirements for plumbing systems that do not correlate to the provisions set forth in the remaining plumbing related chapters. In other words, if specific provisions do not demand that a requirement be located in another chapter, then the requirement is located in this chapter.

SECTION P2601
GENERAL

P2601.1 Scope. The provisions of this chapter shall govern the installation of plumbing not specifically covered in other chapters applicable to plumbing systems.

❖ This section includes requirements that are not specifically addressed elsewhere but that relate to the overall plumbing system.

P2601.2 Connection. Plumbing fixtures, drains and appliances used to receive or discharge liquid wastes or sewage shall be connected to the sanitary drainage system of the building or premises in accordance with the requirements of this code. This section shall not be construed to prevent indirect waste systems.

❖ Plumbing fixtures, drains, appurtenances, and appliances that receive or discharge liquid waste or sewage must be directly connected to the sanitary drainage system unless specifically required to be indirectly connected. An example of an indirect waste connection would be a standpipe receiving the discharge of a clothes washer. See the commentary for Section P2718.

P2601.3 Flood hazard area. In areas prone to flooding as established by Table R301.2(1), plumbing fixtures, drains, and appliances shall be located or installed in accordance with Section R323.1.5.

❖ See the commentary for Section R323.1.5.

SECTION P2602
INDIVIDUAL WATER SUPPLY AND SEWAGE DISPOSAL

P2602.1 General. The water-distribution and drainage system of any building or premises where plumbing fixtures are installed shall be connected to a public water-supply or sewer system, respectively, if available. When either a public water-supply or sewer system, or both, are not available, or connection thereto is not feasible, an individual water-supply or individual (private) sewage-disposal system, or both, shall be provided.

❖ To comply with Section P2601, this section requires that both the water distribution system and the drainage system be connected to a public water supply and a public sewer system, respectively. The code allows for connection to an individual water supply and a private sewage disposal when a public connection is not feasible. Usually, the local jurisdictional authorities (i.e. building department, water purveyor, sewer utility district, etc.) determine feasibility. The most common individual water supply is a private well.

P2602.2 Flood-resistant installation. In areas prone to flooding as established by Table R301.2(1):

1. Water supply systems shall be designed and constructed to prevent infiltration of floodwaters.
2. Pipes for sewage disposal systems shall be designed and constructed to prevent infiltration of floodwaters into the systems and discharges from the systems into floodwaters.

❖ This section requires that plumbing systems be designed in such a manner as to prevent or minimize infiltration of floodwaters into them. The purpose of these requirements is to prevent damage to the system, guard against contamination of the water supply through floodwater entering the system, and to prevent sewage from escaping and further contaminating floodwaters.

Potable water service pipes, both in the lower levels of the building and those buried below ground, are to be water tight to prevent floodwater from entering them. While a water pipe under normal conditions does not leak, it could develop a leak when it is subjected to the additional stress from floodwater. This means that the entire piping system subject to floodwater contact is to be designed accordingly by, for example, having hangers that are capable of resisting not only the normal downward load of the pipe and its contents, but also lateral or upward forces on the pipe caused by floodwaters.

Sanitary drainage and vent piping must be designed so that the additional stress of the floodwater does not cause it to fail and thus contaminate the floodwater.

SECTION P2603
STRUCTURAL AND PIPING PROTECTION

[EB] P2603.1 General. In the process of installing or repairing any part of a plumbing and drainage installation, the finished floors, walls, ceilings, tile work or any other part of the building or premises that must be changed or replaced shall be left in a safe structural condition in accordance with the requirements of the building portion of this code.

❖ Where new plumbing components are installed within existing construction, structural components that are modified must comply with the minimum structural requirements of the code's building chapters.

P2603.2 Drilling and notching. Wood-framed structural members shall not be drilled, notched or altered in any manner except as provided in Sections R502.6, R602.5, R602.6, R802.7 and R802.7.1. Holes in cold-formed steel-framed load-bearing members shall only be permitted in accordance with Sections R506.2, R603.2 and R804.2. In accordance with the provisions of Sections R603.3.4 and R804.3.5 cutting and notching of flanges and lips of cold-formed steel-framed load-bearing members shall not be permitted.

❖ The code recognizes that it may sometimes be necessary to cut or notch framing members. Experience and testing has shown that proper drilling and notching techniques must be followed when piping is installed through framing members. Notches that are too deep or holes that are too large or too close together will significantly weaken these members. The code refers to the building chapters (Chapters 5, Floors; 6, Wall Construction; and 8, Roof/Ceiling Construction) that elaborate on such requirements and limitations. Reference to these sections is imperative for these installations so that the structure will not be weakened beyond its design parameters. Also, note that this section for notching and cutting is limited in scope to those framing members cited in the code and does not cover engineered framing members such as trusses and premanufactured I-joists.

P2603.2.1 Protection against physical damage. In concealed locations, where piping, other than cast-iron or galvanized steel, is installed through holes or notches in studs, joists, rafters or similar members less than 1.5 inches (38 mm) from the nearest edge of the member, the pipe shall be protected by shield plates. Protective shield plates shall be a minimum of 0.062-inch-thick (1.6 mm) steel, shall cover the area of the pipe where the member is notched or bored and shall extend a minimum of 2 inches (51 mm) above sole plates and below top plates.

❖ Section P2603.2.1 intends to minimize the possibility of drainage and water pipe damage caused by nails, screws, or other fasteners. Because nails sometimes miss the wood member, shields must protect the pipe for the full width of the member on each side and must extend a minimum 2 inches (51mm) above sole plates and below top wall plates. Cast iron and galvanized steel pipe have a wall thickness greater than that required for the shield plates. See Commentary Figure P2603.2.1.

P2603.3 Breakage and corrosion. Pipes passing through or under walls shall be protected from breakage. Pipes passing through concrete or cinder walls and floors, cold-formed steel framing or other corrosive material shall be protected against external corrosion by a protective sheathing or wrapping or other means that will withstand any reaction from lime and acid of concrete, cinder or other corrosive material. Sheathing or wrapping shall allow for expansion and contraction of piping to prevent any rubbing action. Minimum wall thickness of material shall be 0.025 inch (0.64 mm).

❖ Metallic pipes made of brass, copper, cast iron, and steel are subject to corrosion when exposed to lime and acid contained in concrete, cinder, or other corrosive material such as soil. Protective sheathing having a minimum thickness of 0.025 inch (0.64 mm) must be placed over the piping before installing it in these corrosive environments. The protective coating will act as a barrier when the piping expands and contracts, thereby preventing rubbing. Section P2603.3 requires pipes passing through or under walls to be protected from breakage. The following two sections provide the means for such protection.

P2603.4 Sleeves. Annular spaces between sleeves and pipes shall be filled or tightly caulked as approved by the building official. Annular spaces between sleeves and pipes in fire-rated assemblies shall be filled or tightly caulked in accordance with the building portion of this code.

❖ A sleeve or a relieving arch is required where piping passes through or under building construction such as walls. The annular space between the sleeve and the pipe must be filled with a non-hardening material that protects the pipe from being loaded by any forces from the building material, while at the same time creating a permanent barrier to prevent the intrusion of insects and rodents. The material used, which typically consists of caulk with coal tar or with an asphaltic compound, must create a moisture-proof seal. See Commentary Figure P2603.4(1). Verify that the filler material is compatible with the pipe and sleeve material. See Commentary Figure P2603.4(2). When the penetration of fire-resistance-rated construction creates annular space, the filler material must be rated the same as the assembly being penetrated.

P2603.5 Pipes through footings or foundation walls. Any pipe that passes under a footing or through a foundation wall shall be provided with a relieving arch; or there shall be built into the masonry wall a pipe sleeve two pipe sizes greater than the pipe passing through.

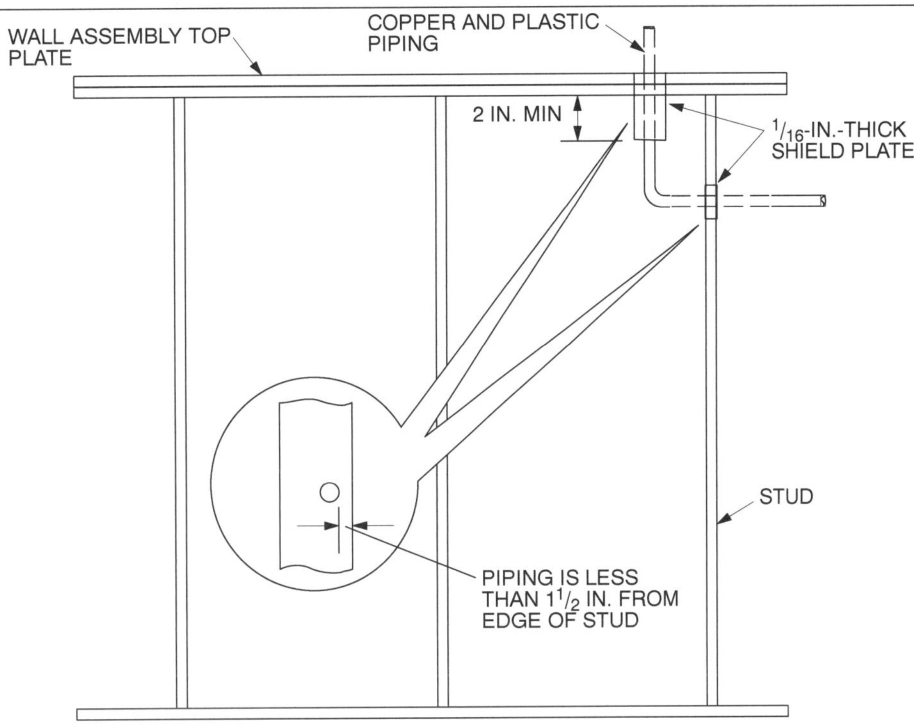

For SI: 1 inch = 25.4 mm.

Figure P2603.2.1
PIPE PROTECTION

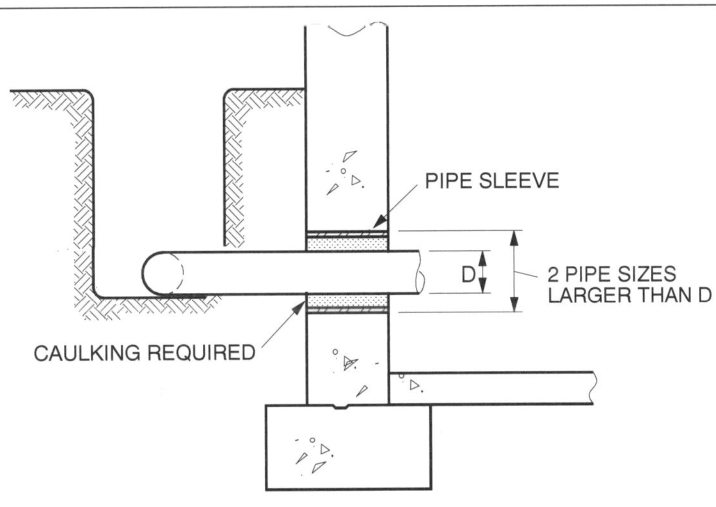

Figure P2603.4(1)
PIPE PROTECTION

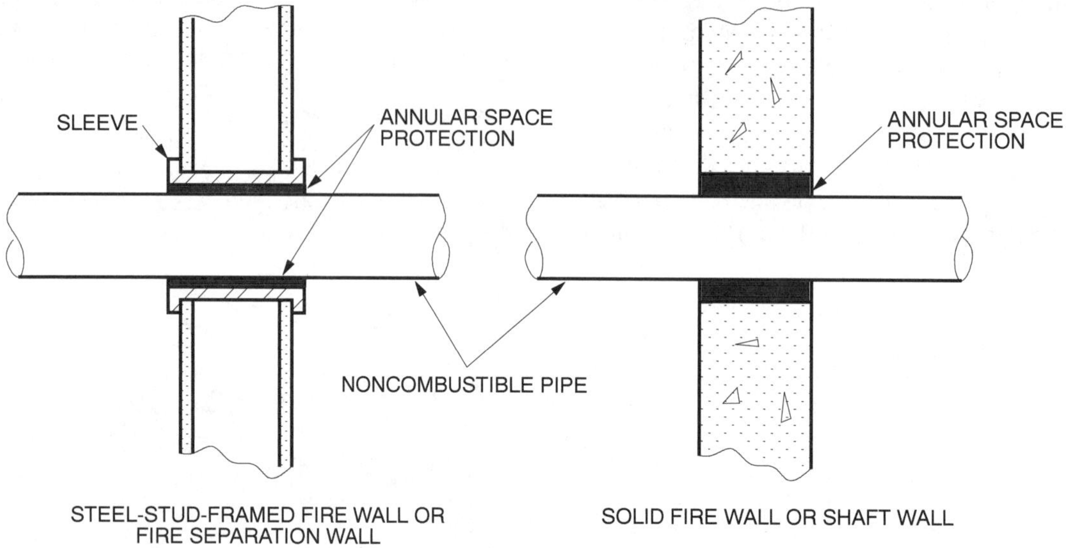

Figure P2603.4(2)
ANNULAR SPACE PROTECTION

❖ Any pipe installed within or under a footing or foundation wall must be protected structurally from any transferred loading from the foundation wall or footing. A relieving arch or a pipe sleeve can be used to provide this protection. Where a sleeve is used, it must be two pipe sizes larger than the enclosed pipe. This space will allow for differential movement of the pipe. See Commentary Figure P2603.5.

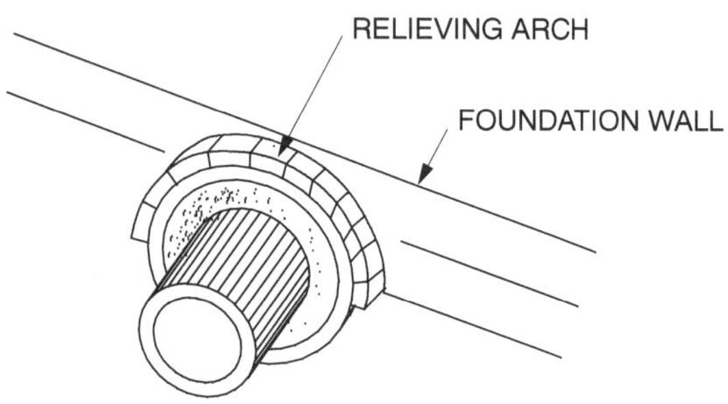

Figure P2603.5
PIPE PROTECTION

P2603.6 Freezing. In localities having a winter design temperature of 32°F (0°C) or lower as shown in Table R301.2(1) of this code, a water, soil or waste pipe shall not be installed outside of a building, in exterior walls, in attics or crawl spaces, or in any other place subjected to freezing temperature unless adequate provision is made to protect it from freezing by insulation or heat or both. Water service pipe shall be installed not less than 12 inches (305 mm) deep or less than 6 inches (152 mm) below the frost line.

❖ Water, soil, or waste pipes must be protected from freezing, whether installed inside or outside the building. Where pipe installation occurs in an exterior wall or unheated space, such as a crawl space or attic, adequate protection must be provided in the form of insulation or heat or both. Where the temperature of the air surrounding the insulation remains low for a significant period of time, insulation alone will not provide adequate protection from freezing without the addition of heat. In such conditions, the water in the pipe will freeze regardless of the amount of insulation used. Conditions differ significantly between occupied and unoccupied buildings because of heat added in occupied buildings for the comfort of the occupants. When a building remains vacant for an extended period of time, heat from another source must be supplied to offset heat loss. A common method used to supply make-up heat is to place electric resistance heat tape or cable inside the insulation next to the water pipe.

Hose bibbs and wall hydrants located on the exterior wall must be protected from freezing when installed in areas subject to freezing temperatures. This protection can be accomplished by installing devices such as frost-proof hose bibbs that locate the valve seat within the heated space and allow residual water within the hydrant to drain after the valve is closed. Frostproof hose bibbs cannot be installed where the valve seat is located in an unheated garage or storage room. The valve seat must extend through to the heated side of the exterior wall. The valve casing on these devices is available in different lengths to accommodate various wall thicknesses. See the commentary for Section P2903.10.

Table R301.2(1) supplies a method for determining areas that have a winter design temperature of 32°F (0°C) or less. Water service pipe installed outside or underground must be at least 12 inches (305 mm) below grade or 6 inches (152 mm) below the frost depth, whichever is greater, to protect against freezing. The 12-inch (305 mm) minimum cover above the pipe protects it from external damage. See Commentary Figure P2603.6.

P2603.6.1 Sewer depth. Building sewers that connect to private sewage disposal systems shall be a minimum of [NUMBER] inches (mm) below finished grade at the point of septic tank connection. Building sewers shall be a minimum of [NUMBER] inches (mm) below grade.

❖ A building sewer is not subject to the same constraints as a water pipe because a sewer carries heat to the surrounding soil as the drainage flows through it. A building sewer also has an intermittent flow. As a result, the frost depth is higher in the immediate area of the building sewer; therefore, the burial depth of a building sewer does not have to be below the recorded frost depth. A typical burial depth for a building sewer ranges from 6 to 18 inches (152 to 457 mm).

The installation depth of the sewer pipe connecting to a septic tank could be different in order to facilitate the proper operation of the private sewage disposal system. The connection at the septic tank is typically 18 inches (457 mm) below grade; however, the code official is to determine the depths necessary for the locality.

SECTION P2604 TRENCHING AND BACKFILLING

P2604.1 Trenching and bedding. Piping shall be installed in trenches so that the piping rests on solid and continuous bearing. When over excavated, the trench shall be backfilled to the proper grade with compacted earth, sand, fine gravel or similar granular material. Piping shall not be supported on rocks or blocks at any point. Rocky or unstable soil shall be over excavated by two or more pipe diameters and brought to the proper grade with suitable compacted granular material.

❖ A trench must be sufficiently wide to allow proper pipe alignment. Piping should be supported on solid, continuous bedding for its entire length. Improper backfill material will subject the pipe to uneven earth loading that can lead to damage to the pipe or joints. Where a trench contains unstable material, the code requires overexcavation by at least two pipe diameters below the pipe and placement of suitable, compacted granular material below the pipe to ensure proper and even support. Manufacturer's installation instructions often include information on burial techniques.

P2604.2 Common trench. See Section P2904.4.1.

❖ See the commentary for Section P2904.4.1.

P2604.3 Backfilling. Backfill shall be free from discarded construction material and debris. Backfill shall be free from rocks, broken concrete and frozen chunks until the pipe is covered by at least 12 inches (305 mm) of tamped earth. Backfill shall be placed evenly on both sides of the pipe and tamped to retain proper alignment. Loose earth shall be carefully placed in the trench in 6-inch (152 mm) layers and tamped in place.

❖ This section requires that backfill material be devoid of construction material or debris, rocks, broken concrete, frozen chunks, and other rubble. Twelve inches (305 mm) of clean tamped earth, sand, or fine gravel must be used to cover the pipe. Loose earth must be

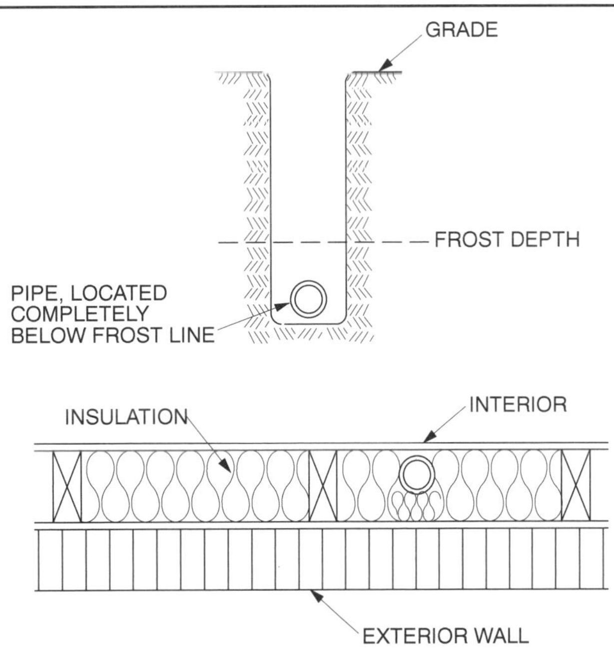

Figure P2603.6
PROTECTION AGAINST FREEZING

placed carefully in the trench in 6-inch (152 mm) layers and tamped in place to reach the minimum 12-inch (305 mm) cover above the pipe. See Commentary Figure P2604.3.

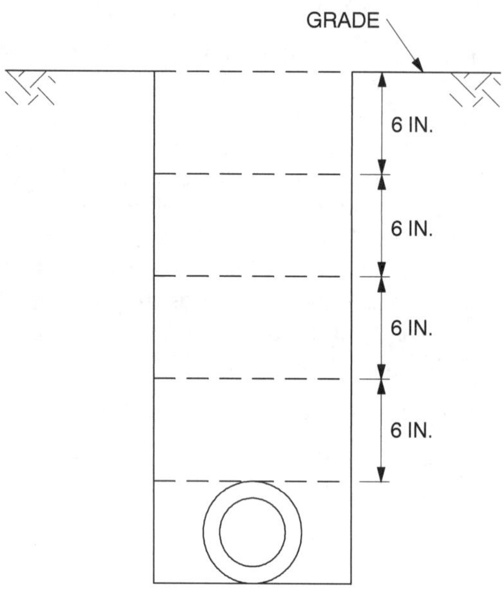

For SI: 1 inch = 25.4 mm.

Figure P2604.3
BACKFILLING 6-IN. TAMPED LAYER

P2604.4 Protection of footings. Trenching installed parallel to footings shall not extend below the 45-degree (0.79 rad) bearing plane of the bottom edge of a wall or footing (see Figure P2604.4).

❖ Section P2603.7 provides for the parallel installation of pipes alongside a footing. The code includes Section P2604.4 among its trenching and backfilling provisions to address the need to avoid trenching below the 45-degree (0.79 rad) load-bearing plane supporting a footing or wall. See Figure P2604.4.

SECTION P2605 SUPPORT

P2605.1 General. Support for piping shall be provided in accordance with the following:

1. Piping shall be supported so as to ensure alignment and prevent sagging, and allow movement associated with the expansion and contraction of the piping system.
2. Piping in the ground shall be laid on a firm bed for its entire length, except where support is otherwise provided.
3. Hangers and anchors shall be of sufficient strength to maintain their proportional share of the weight of pipe and contents and of sufficient width to prevent distortion to the pipe. Hangers and strapping shall be of approved material that will not promote galvanic action. Rigid support sway bracing shall be provided at changes in direction greater than 45 degrees (0.39 rad) for pipe sizes 4 inches (102 mm) and larger.
4. Piping shall be supported at distances not to exceed those indicated in Table P2605.1.

❖ The intent of supporting the pipe is to maintain proper slope and alignment and to prevent sagging, while allowing movement of the pipe due to expansion and contraction. This section refers to support of piping installed in a trench, as discussed in the commentary for Section P2604.1. Hangers and supports for piping must be capable of supporting the load imposed by the piping system and must not be detrimental to the pipes

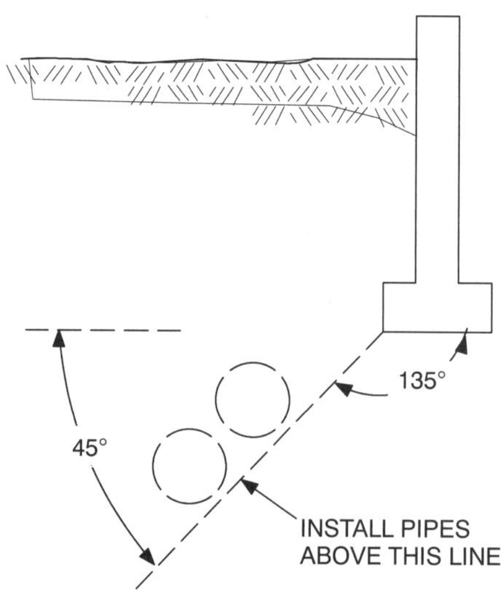

For SI: 1 degree = 0.018 rad.

FIGURE P2604.4
PIPE LOCATION WITH RESPECT TO FOOTINGS

❖ See the commentary for Section P2604.4.

they support. Using hanger or strapping material that is not compatible with the piping material can result in corrosion caused by galvanic action. Galvanic action occurs when dissimilar metals are in contact with one another and moisture is sufficient to carry an electrical current. Section P2605.1 also contains lateral support requirements. Pipes are subject to the velocity and momentum resulting from the flow of waste and from changes in direction. Hangers alone may not be sufficient to resist such forces, and, thus, rigid support sway bracing is required for pipe sizes 4 inches (102 mm) or larger at changes of direction greater than 45 degrees (0.79 rad). See Commentary Figure P2605.1(1).

TABLE P2605.1
PIPING SUPPORT

PIPING MATERIAL	MAXIMUM HORIZONTAL SPACING (feet)	MAXIMUM VERTICAL SPACING
ABS pipe	4	10[b]
Aluminum tubing	10	15
Brass pipe	10	10
Cast-iron pipe	5[a]	15
Copper or copper alloy pipe	12	10
Copper or copper alloy tubing ($1^1/_4$ inch diameter and smaller)	6	10
Copper or copper alloy tubing ($1^1/_2$ inch diameter and larger)	10	10
Cross-linked polyethylene (PEX) pipe	2.67 (32 inches)	10[b]
Cross-linked polyethylene/aluminum/cross-linked polyethylene (PEX-AL-PEX) pipe	2.67 (32 inches)	4[b]
CPVC pipe or tubing (1 inch in diameter and smaller)	3	10[b]
CPVC pipe or tubing ($1^1/_4$ inch in diameter and larger)	4	10[b]
Lead pipe	Continuous	4
PB pipe or tubing	2.67 (32 inches)	4
Polyethylene/aluminum/polyethylene (PE-AL-PE) pipe	2.67 (32 inches)	4[b]
PVC pipe	4	10[b]
Stainless steel drainage systems	10	10[b]
Steel pipe	12	15

For SI: 1 inch = 25.4 mm, 1 foot = 304.8 mm.

a. The maximum horizontal spacing of cast-iron pipe hangers shall be increased to 10 feet where 10-foot lengths of pipe are installed.

b. Midstory guide for sizes 2 inches and smaller.

❖ The code refers to Table P2605.1 for piping support requirements. The maximum amount of sag occurs at a point equidistant from the supports and when the pipe is full of water. The recommended spacing of supports in Table P2605.1 is designed to limit sag and pipe stress and to maintain slope and alignment. As the pipe's diameter increases, so does the pipe's strength and resistance to sag. The spacing in the table is calculated for a small-diameter pipe. See Commentary Figures P2605.1(2) and P2605.1(3). Consult the pipe manufacturer's installation instructions for more specific spacing instructions.

The spacing of supports indicated in Table P2605.1 determines the minimum number of supports required. Extra supports should be used as necessary because of fittings, connections, change in direction, or connection to a vertical pipe. See Commentary Figure P2605.1(4).

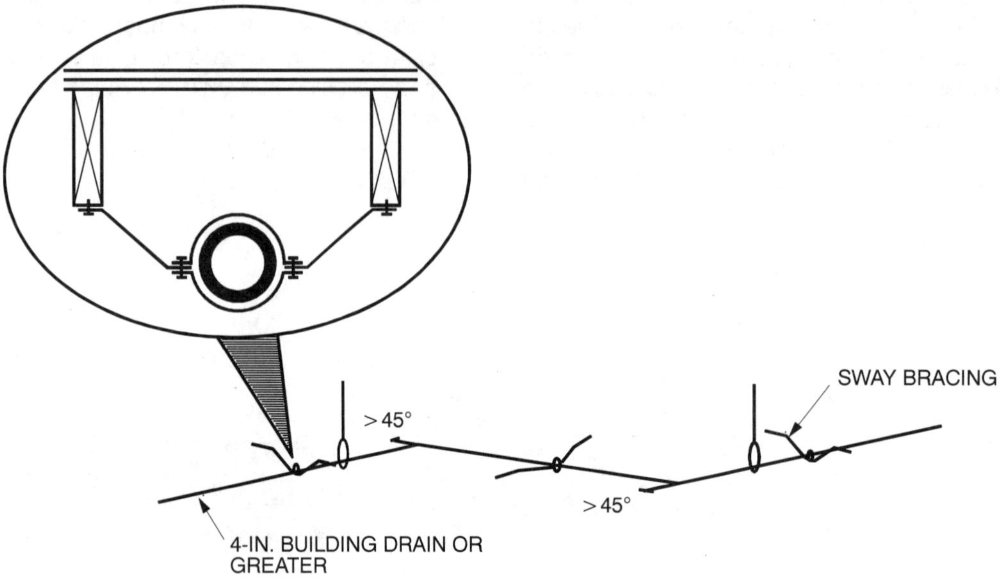

For SI: 1 inch = 25.4 mm, 45 degrees = 0.79 rad.

Figure P2605.1(1)
SWAY BRACING

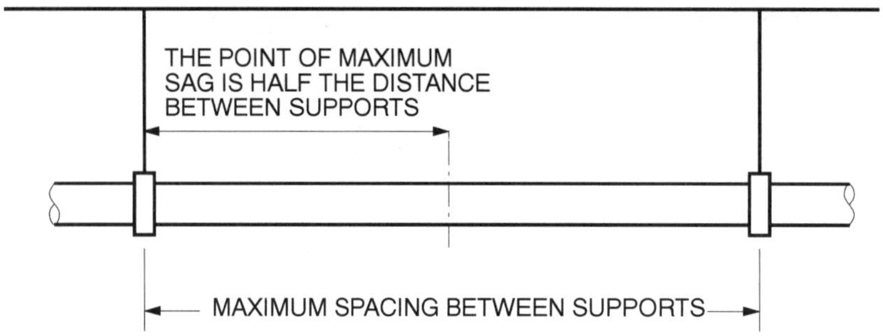

Figure P2605.1(2)
POINT OF SAG BETWEEN HANGERS

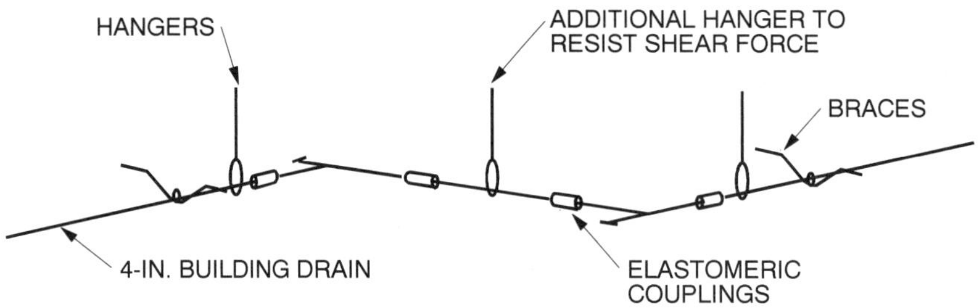

For SI: 1 inch = 25.4 mm.

Figure P2605.1(3)
LOCATION OF HANGERS

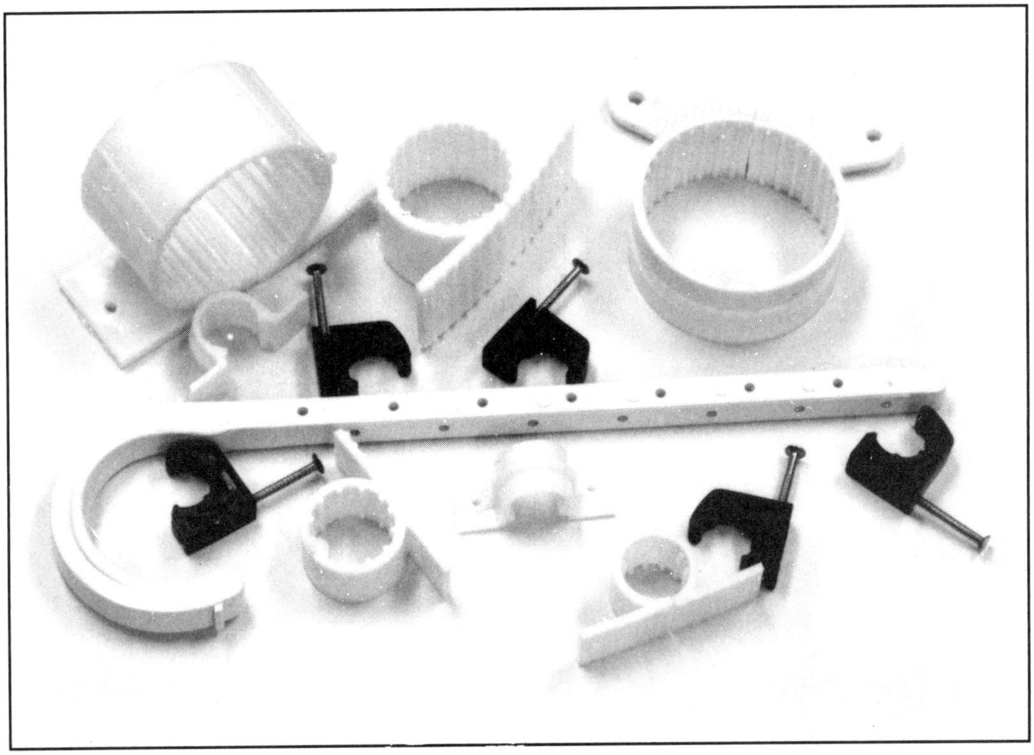

Photo courtesy of A. O. Smith Water Products

Figure P2605.1(4)
TYPICAL PIPE SUPPORTS

SECTION P2606
WATERPROOFING OF OPENINGS

P2606.1 General. Roof and exterior wall penetrations shall be made water tight. Joints at the roof, around vent pipes, shall be water tight by the use of lead, copper or galvanized iron flashings or an approved elastomeric material. Counterflashing shall not restrict the required internal cross-sectional area of any vent.

❖ Roof and exterior wall penetrations must be watertight when penetrated by piping. Use of lead, copper, or galvanized flashing is a common and approved method of waterproofing roof penetrations. Use of approved elastomeric material is another acceptable means of waterproofing.

When using flashing materials such as lead that wrap over the top of the pipe, take care to avoid reducing the required cross-sectional area of the vent terminal. See Commentary Figure P2606.1.

Because exterior walls are not usually subjected to the same severe exposure as a roof, the code does not require the space around pipes that penetrate exterior walls to be sealed with flashing; rather, it requires that the space be made watertight. Openings in exterior walls can be sealed with an approved method that will prevent moisture and foreign matter from penetrating into the structure.

SECTION P2607
WORKMANSHIP

P2607.1 General. Valves, pipes and fittings shall be installed in correct relationship to the direction of the flow. Burred ends shall be reamed to the full bore of the pipe.

❖ Drainage piping, fittings, and valves must be installed in the correct direction to assist the flow of waste. Table P3005.1 regulates the installation of fittings according to the change of direction. Pipe ends that are burred because of cutting procedures must be reamed to the full bore of the pipe to eliminate turbulence in water piping and rough edges that can cause blockage in drainage piping.

Burred ends on the pipe's outer edges must be removed to create the desired joint. A chamfering tool can be used to remove the burred ends. See Commentary Figure P2607.1.

SECTION P2608
MATERIALS EVALUATION AND LISTING

P2608.1 Identification. Each length of pipe and each pipe fitting, trap, fixture, material and device utilized in a plumbing system shall bear the identification of the manufacturer.

❖ The manufacturer is given the option of determining the type of marking for the material. If there is no appli-

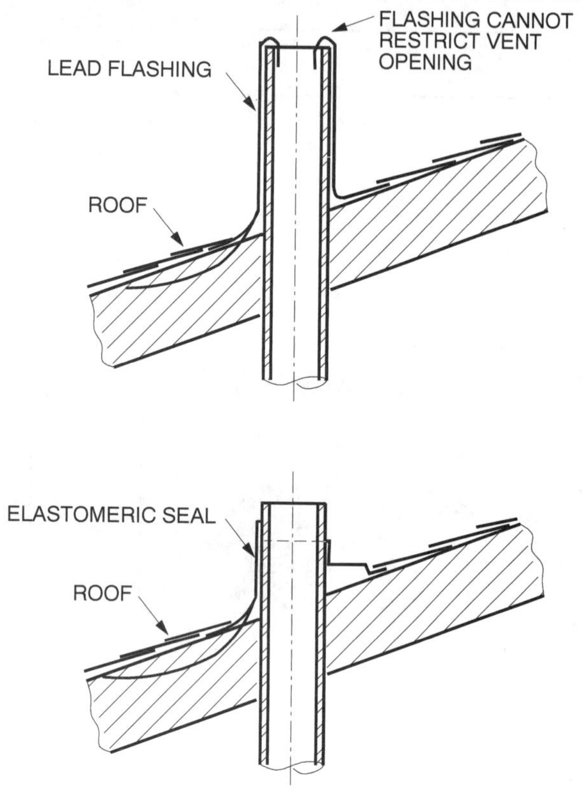

Figure P2606.1
ROOF PENETRATIONS

cable standard or the applicable standard does not require a material to be identified, identification of the manufacturer is still required by the code. Where the code indicates compliance with an approved standard, the manufacturer must comply with the requirements for marking in accordance with the applicable standard.

P2608.2 Installation of materials. All materials used shall be installed in strict accordance with the standards under which the materials are accepted and approved. In the absence of such installation procedures, the manufacturer's installation instructions shall be followed. Where the requirements of referenced standards or manufacturer's installation instructions do not conform to the minimum provisions of this code, the provisions of this code shall apply.

❖ Plumbing components and materials are to be installed in accordance with the installation requirements of the applicable standard listed in the code.

Where a standard is not provided, the manufacturer's instructions must be followed. For example, since there are very few standards available that regulate the installation of valves, the manufacturer's installation instructions must be used to install these components.

P2608.3 Plastic pipe, fittings and components. All plastic pipe, fittings and components shall be third-party certified as conforming to NSF 14.

❖ All plastic piping, fittings, and plastic pipe-related components, including solvent cements, primers, tapes, lubricants, and seals used in plumbing systems, must be tested and certified as conforming to NSF 14. This includes all water service, water distribution, and drainage piping and fittings as well as plastic piping system components, including but not limited to pipes, fittings, valves, joining materials, gaskets, and appurtenances. This section does not apply to components that include plastic parts such as brass valves with a plastic stem or to fixture fittings such as fixture stop valves. Plastic piping systems, fittings, and related components intended for use in the potable water supply system must comply with NSF 61 in addition to NSF 14.

P2608.4 Third-party testing and certification. All plumbing products and materials shall comply with the referenced standards, specifications and performance criteria of this code and shall be identified in accordance with Section P2608.1. When required by Table P2608.4, plumbing products and materials shall either be tested by an approved third-party testing agency or certified by an approved third-party certification agency.

❖ This section has been revised to require that all plumbing products and materials comply with the referenced standards. However, the provisions contained in Section R104.11 regarding the evaluation and approval of alternative materials, methods, and equipment are still applicable; see the commentary Section R104.11. Additionally, the code has been revised to include specific requirements for plumbing products in regard to third-party certification and testing. Table P2608.4 lists these requirements. "Third-party certified" indicates that the minimum level of quality required by the appropriate standard is maintained and the product is often referred to as "listed." "Third-party tested" is a product that has been tested by an approved testing laboratory and found to be in compliance with the appropriate standard. While the code does not specifically state the identification or marking requirements except for the manufacturer's identification, the applicable referenced standard states the minimum information required. The identification or marking requirements typically include the name of the manufacturer, product name or serial number, installation specifications, applicable tests and standards, testing agency, and labeling agency. Commentary Figure P2608.4 shows an example of a typical label for PVC plastic pipe.

P2608.5 Water supply systems. All water service pipes, water distribution pipes and the necessary connecting pipes, fittings, control valves, faucets and all appurtenances used to dispense water intended for human ingestion shall be evaluated and listed as conforming to the requirements of NSF 61.

❖ Piping material that comes in contact with the potable water source must comply with the requirements of NSF 61. The purpose is to control potential adverse health effects produced by indirect additives, products, and materials that come in contact with potable water. Plastic piping systems, fittings, and related components intended for use with the potable water supply system must comply with NSF 14 in addition to NSF 61.

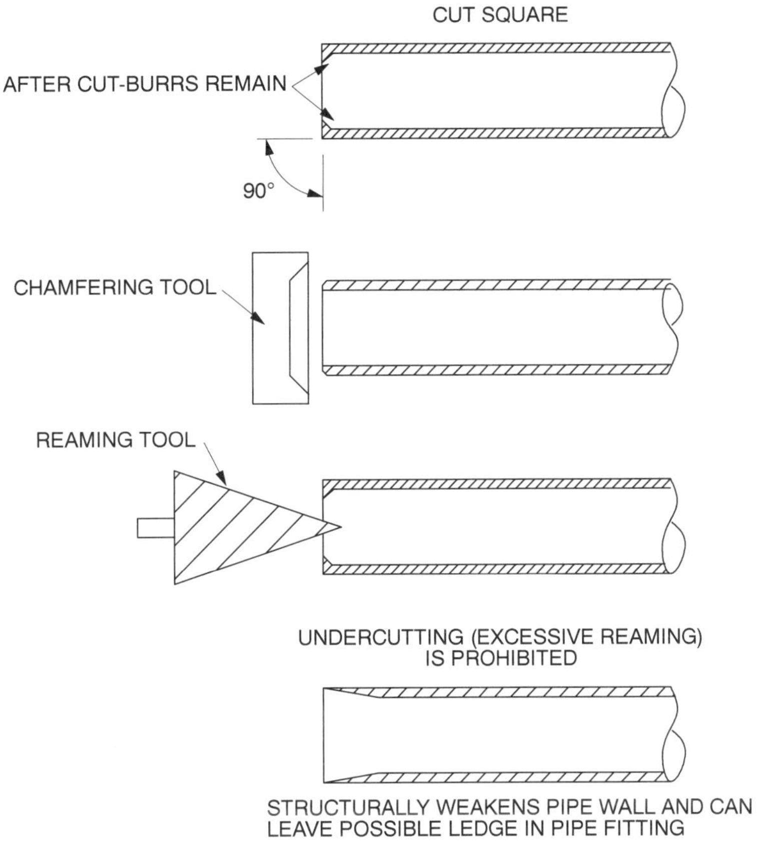

For SI: 1 degree = 0.079 rad.

Figure P2607.1
PREPARATION OF PIPE END

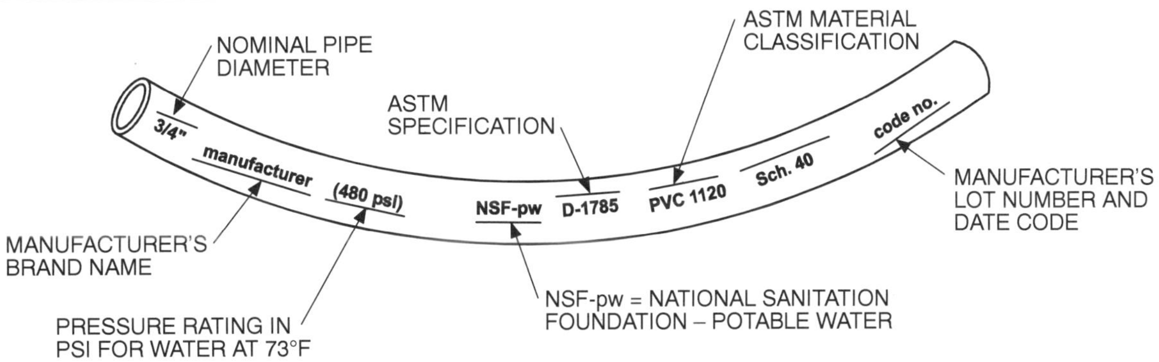

For SI: C = [(F) - 32]/1.8.

Figure P2608.4
SAMPLE MARKING OF PVC PIPE

TABLE P2608.4
PRODUCTS AND MATERIALS REQUIRING THIRD-PARTY TESTING AND THIRD-PARTY CERTIFICATION

PRODUCT OR MATERIAL	THIRD-PARTY CERTIFIED	THIRD-PARTY TESTED
Backflow prevention devices	Required	—
Plumbing appliance	Required	—
Plumbing fixtures	—	Required
Potable water supply system components and potable water fixture fittings	Required	—
Sanitary drainage and vent system components	Plastic pipe, fittings, and pipe related components	All others
Special waste system components	—	Required
Storm drainage system components	Plastic pipe, fittings, and pipe related components	All others
Subsoil drainage system components	—	Required
Waste fixture fittings	Plastic pipe, fittings, and pipe related components	All others
Water distribution system safety devices	Required	—

❖ See the commentary for Section P2608.4.

Bibliography

The following resource materials are referenced in this chapter or are relevant to the subject matter addressed in this chapter.

FEMA 348, *Protecting Building Utilities From Flood Damage: Principles and Practices for the Design and Construction of Flood Resistant Building Utility Systems.* Washington, DC: Federal Emergency Management Agency, 1999.

NSF 14-99, *Plastic Piping System Components and Related Materials.* Ann Arbor, MI: NSF International, 1999.

NSF 61-01, *Drinking Water System Components—Health Effects*. Ann Arbor, MI: NSF International, 2001.

Chapter 27: Plumbing Fixtures

General Comments

Chapter 27 addresses each type of fixture individually, including the accessory parts or installation criteria specific to a given fixture.

Chapter 27 outlines the fixtures, faucets, and fixture fittings and the standards for their manufacture. In particular, it covers the materials used in the manufacture of tailpieces and traps, their fittings, and the strainer requirements for the fixture outlet. This chapter also regulates fixture tailpieces and specifies the minimum tailpiece size required for each fixture. Chapter 27 lists access requirements for inspection and repair of slip-joint connections.

This chapter also details installation criteria for material, methods, and location of fixtures and their parts.

The majority of the chapter is devoted to detailing specific provisions relating to waste receptors; kitchen fixtures and appliances; bathroom and toilet room fixtures; laundry facilities; valves and fixture fittings; and macerating toilet systems.

Purpose

Fixtures must be of the proper type, approved for the purpose of installation, and installed properly to promote usability and a safe, sanitary condition. Chapter 27 provides the standards to which fixtures must be manufactured and installed.

SECTION P2701
FIXTURES, FAUCETS AND FIXTURE FITTINGS

P2701.1 Quality of fixtures. Plumbing fixtures, faucets and fixture fittings shall be constructed of approved materials, shall have smooth impervious surfaces, shall be free from defects and concealed fouling surfaces, and shall conform to the standards cited in this code. Plumbing fixtures shall be provided with an adequate supply of potable water to flush and keep the fixtures in a clean and sanitary condition without danger of backflow or cross connection.

❖ Plumbing fixtures, faucets, and fixture fittings must be manufactured using only approved materials. They are to be constructed as required by a nationally recognized standard specific to the individual item. A list of those standards appears in Table P2701.1. These items and devices are required to have smooth, impervious surfaces and to be free of concealed fouling surfaces. Fixtures that do not thoroughly scour the used surface or area will harbor bacteria and germs such that the fixtures become a source of contamination. Most fixtures are required by their respective referenced standards to be self-cleaning. Plumbing fixtures must be supplied with adequate flow of potable water to flush and keep them clean and sanitary. The water supply must be installed as to avoid any possibility of backflow or cross connection that may contaminate the water supply piping or the public or private water supply system.

TABLE P2701.1
PLUMBING FIXTURES, FAUCETS AND FIXTURE FITTINGS

MATERIAL	STANDARD
Air gap fittings for use with plumbing fixtures, appliances and appurtenances	ASME A112.1.3
Bathtub/Whirlpool Pressure Sealed Doors	ASME A112.19.15
Diverters for faucets with hose spray anti-syphon type, residential application	ASSE 1025
Enameled cast-iron plumbing fixtures	ASME A112.19.1M
Floor drains	ASME A112.6.3
Framing-affixed supports for off-the-floor water closets with concealed tanks	ASME A112.6.2
Handheld showers	ASSE 1014
Home laundry equipment	ASSE 1007
Hose connection vacuum breaker	ASSE 1052
Hot water dispensers, household storage type, electrical	ASSE 1023
Household dishwashing machines	ASSE 1006
Household disposers	ASSE 1008

(continued)

TABLE P2701.1
PLUMBING FIXTURES, FAUCETS AND FIXTURE FITTINGS—continued

MATERIAL	STANDARD
Hydraulic performance for water closets and urinals	ASME A112.19.6
Individual pressure balancing valves for individual fixture fittings	ASSE 1066
Individual shower control valves anti-scald	ASSE 1016
Macerating toilet systems and related components	ASME A112.3.4
Nonvitreous ceramic plumbing fixtures	ASME A112.19.9M
Plastic bathtub units	ANSI Z124.1
Plastic lavatories	ANSI Z124.3
Plastic shower receptors and shower stall	ANSI Z124.2
Plastic sinks	ANSI Z124.6
Plastic water closet bowls and tanks	ANSI Z124.4
Plumbing fixture fittings	ASME A112.18.1M
Porcelain enameled formed steel plumbing fixtures	ASME A112.19.4M
Pressurized flushing devices for plumbing fixtures	ASSE 1037
Specification for copper sheet and strip for building construction	ASTM B 370
Suction fittings for use in swimming pools, wading pools, spas, hot tubs and whirlpool bathtub appliances	ASME A112.19.8M
Stainless steel plumbing fixtures (residential)	ASME A112.19.3M
Temperature-actuated, flow reduction valves to individual fixture fittings	ASSE 1062
Thermoplastic accessible and replaceable plastic tube and tubular fittings	ASTM F 409
Trench drains	ASME A112.6.3
Trim for water closet bowls, tanks and urinals	ASME A112.19.5
Vacuum breaker wall hydrant-frost resistant automatic draining type	ASSE 1019
Vitreous china plumbing fixtures	ASME A112.19.2M
Wall-mounted and pedestal-mounted, adjustable and pivoting lavatory and sink carrier systems	ASME A112.19.12
Water closet flush tank fill valves	ASSE 1002
Whirlpool bathtub appliances	ASME A112.19.7M

SECTION P2702 FIXTURE ACCESSORIES

P2702.1 Plumbing fixtures. Plumbing fixtures, other than water closets, shall be provided with approved strainers.

❖ This section requires plumbing fixtures, except water closets, to be equipped with approved strainers. This requirement prevents items such as toothbrushes, rings, cosmetic devices, and large food waste particles from entering the drainage system and causing detrimental and costly problems. This requirement does not include clothes washer standpipes or other indirect waste receptors, as they are exempt from these provisions and are addressed in other sections in this chapter.

There are several other fixtures common in the plumbing industry that do not require strainers, such as urinals. However, the code contains provisions that apply primarily to one- and two-family dwellings. Thus, the code does not address fixtures uncommon to this type of construction.

P2702.2 Material for tail pieces and traps. Continuous wastes, waste and overflow fittings, traps and tail pieces constructed of seamless drawn brass shall be of No. 20 gage (0.8 mm) minimum thickness.

❖ Section P2702.2 requires tail pieces, traps, waste and overflow, and continuous waste (piping connecting two or more fixtures to the same trap) to be minimum 20 gage (0.8 mm) thickness when constructed of seamless drawn brass.

This provision does not preclude the use of approved plastic material, which is addressed in Section P2702.3.

P2702.3 Plastic tubular fittings. Plastic tubular fittings shall conform to ASTM F 409 listed in Table P2701.1.

❖ The code allows the same devices manufactured of seamless drawn brass as specified in Section P2702.2 to be manufactured of plastic. Plastic tubular fittings must conform to ASTM F 409. Table P2701.1 lists this standard and the products to which it pertains.

P2702.4 Carriers for wall-hung water closets. Carriers for wall-hung water closets shall conform to ASME A112.6.1 or ASME A112.6.2.

❖ A wall-hung water closet differs greatly in structural design from a floor-mounted water closet because the structural load is transferred to the wall; the floor provides no support. The hanger or carrier for wall-hung water closets must distribute the structural load to the building structural elements without transferring any load to the plumbing piping system. These off-the-floor water closet support assemblies consist of hardware to mount and connect the water closet, the fitting and conduit component means for conveying waste, the structural anchor assembly carrier, and the necessary gasketing.

The referenced standards ASME A112.6.1 or ASME A112.6.2 contain design and test requirements for the loading of the hanger. See commentary Figure P2702.4.

SECTION P2703 TAIL PIECES

P2703.1 Minimum size. Fixture tail pieces shall be not less than 1.5 inches (38 mm) in diameter for sinks, dishwashers, laundry tubs, bathtubs and similar fixtures, and not less than 1.25 inches (32 mm) in diameter for bidets, lavatories and similar fixtures.

❖ This section covers the sizing requirements for fixture waste outlets. Lavatories and bidets must have minimum 1.25-inch (32 mm) tailpieces. Sinks, dishwashers, laundry tubs, and bathtubs must have minimum 1.5-inch (38 mm) tailpieces. These standards apply to similar types of fixtures.

SECTION P2704 ACCESS TO CONNECTIONS

P2704.1 General. Slip joints shall be made with an approved elastomeric gasket and shall only be installed on the trap outlet, trap inlet and within the trap seal. Fixtures with concealed slip-joint connections shall be provided with an access panel or utility space at least 12 inches (305 mm) in its smallest dimension or other approved arrangement so as to provide access to the slip connections for inspection and repair.

❖ Slip joints may be installed on the trap inlet, on the trap outlet, and within the trap seal. Fixtures such as bathtubs and bidets commonly employ slip joints in concealed spaces, which require access in accordance with this section.

Slip joints usually connect lavatories and sinks. Having a sink cabinet door that opens to a minimum 12-inch by 12-inch (305 mm) space will satisfy the access requirement. The code does not require access through finished construction where joints are accessible in an underfloor space or similar area.

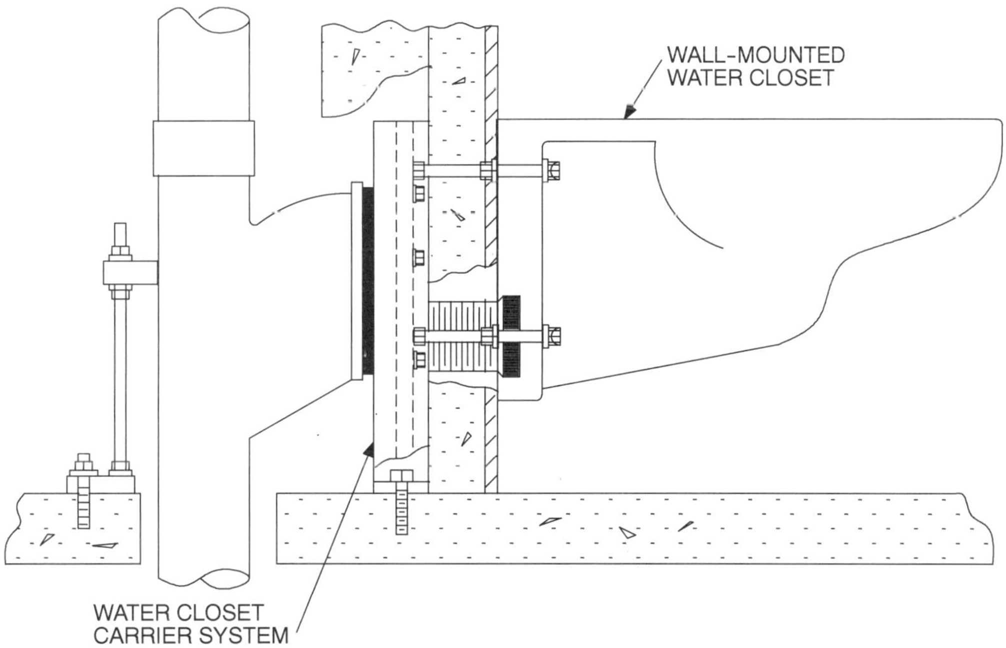

Figure P2702.4
WALL-MOUNTED WATER CLOSET

SECTION P2705
INSTALLATION

P2705.1 General. The installation of fixtures shall conform to the following:

1. Floor-outlet or floor-mounted fixtures shall be secured to the drainage connection and to the floor, when so designed, by screws, bolts, washers, nuts and similar fasteners of copper, brass or other corrosion-resistant material.
2. Wall-hung fixtures shall be rigidly supported so that strain is not transmitted to the plumbing system.
3. Where fixtures come in contact with walls and floors, the contact area shall be watertight.
4. Plumbing fixtures shall be functionally accessible.
5. The centerline of water closets or bidets shall not be less than 15 inches (381 mm) from adjacent walls or partitions or not less than 15 inches (381 mm) from centerline of a bidet to the outermost rim of an adjacent water closet. There shall be at least 21 inches (533 mm) clearance in front of the water closet, bidet or lavatory to any wall, fixture or door.
6. The location of piping, fixtures or equipment shall not interfere with the operation of windows or doors.
7. In areas prone to flooding as established by Table R301.2(1), plumbing fixtures shall be located or installed in accordance with Section R323.1.5.
8. Integral fixture fitting mounting surfaces on manufactured plumbing fixtures or plumbing fixtures constructed on site, shall meet the design requirements of ASME A112.19.2 or ASME A112.19.3.

❖ Floor-outlet fixtures such as water closets and bidets are to be held in place by bolts or screws that attach to the floor and to the floor flange. These bolts or screws must be corrosion-resistant, because they might be exposed to moisture. Screws, bolts, washers, nuts, and similar fasteners are usually made of copper, brass, or plastic.

Wall hung fixtures must be rigidly supported using approved carriers securely fastened to the wall structure so as not to strain the piping system. Connections to the wall must be watertight. This requirement not only applies to wall hung fixtures, but it also applies to fixtures installed in contact with walls.

Plumbing fixtures require space around them for use and cleaning purposes. This section requires that water closets or bidets be at least 15 inches (381 mm) from adjacent side walls or partitions and not less than 15 inches (381 mm) from the centerline of a bidet to the outermost rim of an adjacent water closet. The code requires a clearance of at least 21 inches (533 mm) in front of a water closet, bidet, or lavatory to any wall, fixture, or door. Plumbing fixture locations must not interfere with the operation of any window or door, especially where a fixture such as a water closet is adjacent to a shower or tub-shower enclosure. See Commentary Figure P2705.1.

Plumbing systems must be located at or above the design flood elevation, and this section directs the reader to Section 323.1.5.

Integral fixture fitting mounting surfaces on manufactured plumbing fixtures or plumbing fixtures constructed on-site must meet the design requirements of ASME A112.19.2 or ASME A112.19.3. Not all plumbing fixture standards limit the allowable depression for fixture fitting mounting surfaces. These standards ensure that fixture fittings or on-site construction of plumbing fixtures meet the air gap requirements of fixture fitting standards and provide backflow protection when mounted on a plumbing fixture.

SECTION P2706
WASTE RECEPTORS

P2706.1 General. Every waste receptor shall be of an approved type. Plumbing fixtures or other receptors receiving the discharge of indirect waste pipes shall be shaped and have a capacity to prevent splashing or flooding and shall be readily accessible for inspection and cleaning. Waste receptors and standpipes shall be trapped and vented and shall connect to the building drainage system. A removable strainer or basket shall cover the waste outlet of waste receptors. Waste receptors shall be installed in ventilated spaces. Waste receptors shall not be installed in bathrooms or in any inaccessible or unventilated space such as a closet. Ready access shall be provided to waste receptors.

Exception: Open hub waste receptors shall be permitted in the form of a hub or pipe extending not less than 1 inch (25.4 mm) above a water-impervious floor, and are not required to have a strainer.

❖ Waste receptors are typically used to receive discharge from fixtures or appliances required to discharge indirectly, such as condensate drains, clothes washers, and dishwashers. These appliances discharge waste either to dedicated waste receptors or to a tailpiece serving other plumbing fixtures. Appliances must be designed and installed to prevent splashing or flooding. They must be located in areas that are readily accessible for monitoring and cleaning; in ventilated spaces; and not in frequently accessed areas, such as closets. The receptor is to be equipped with a strainer unless an open hub receptor is installed at least 1 inch (25.4 mm) above an impervious floor. The receptor, whether independent or part of a fixture connection with the drainage system (e.g. lavatory tailpiece), must be trapped, vented, and connected to the drainage system. See Commentary Figures P2706.1(1), P2706.1(2), and P2706.1(3).

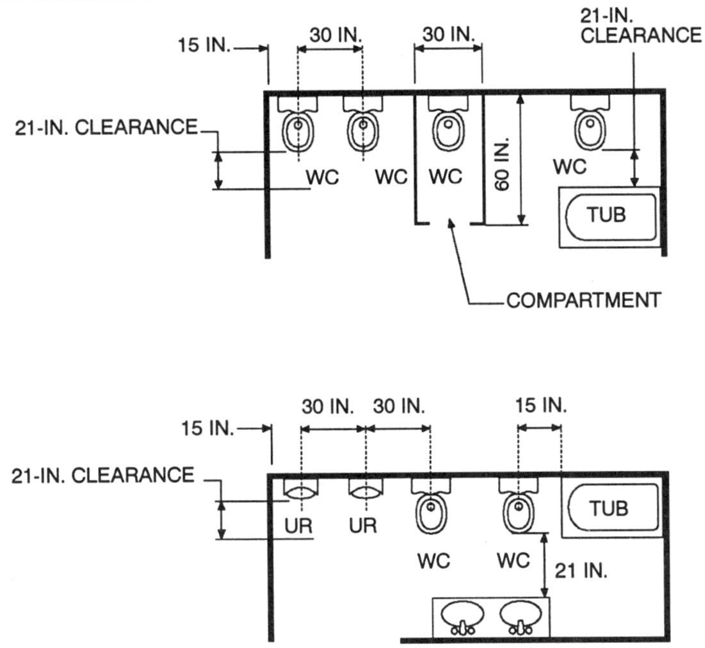

For SI: 1 inch = 25.4 mm.

Figure P2705.1
FIXTURE CLEARANCES

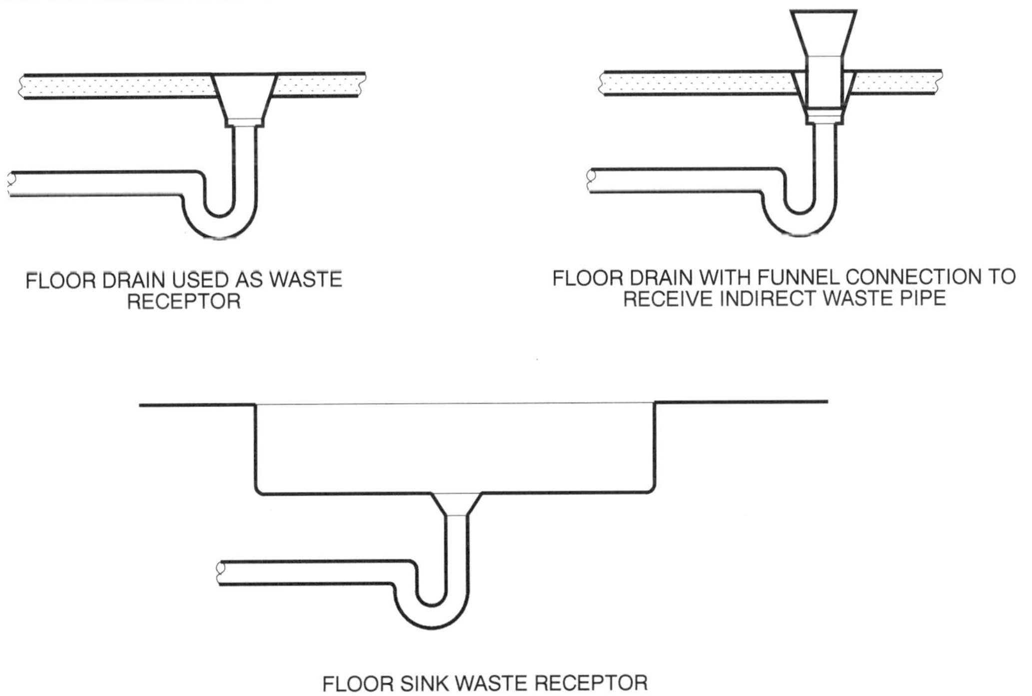

Figure P2706.1(1)
WASTE RECEPTORS

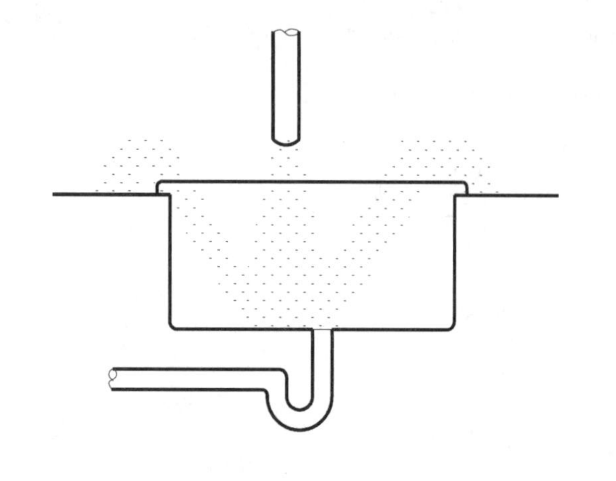

Figure P2706.1(2)
SHALLOW FLAT BOTTOM WASTE RECEPTOR

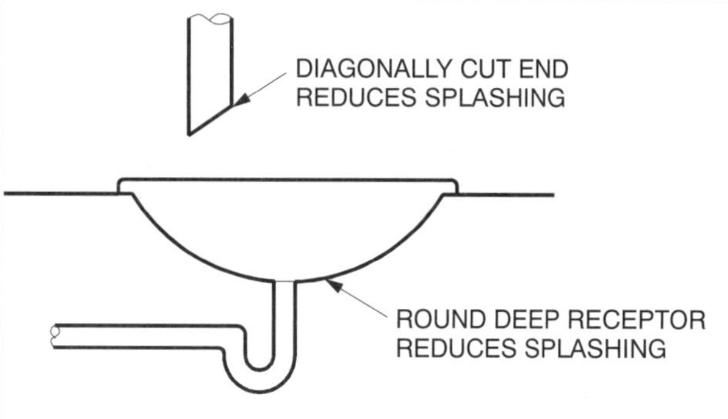

Figure P2706.1(3)
SIZE AND SHAPE OF WASTE RECEPTOR

P2706.2 Standpipes. Standpipes shall extend a minimum of 18 inches (457 mm) and a maximum of 42 inches (1067 mm) above the trap weir. Access shall be provided to all standpipe traps and drains for rodding.

❖ A standpipe may serve as an indirect waste receptor. The established minimum receptor height aims to provide a minimal retention capacity and head pressure to prevent overflow. This height limitation is especially necessary when an indirect waste pipe receives a high rate of discharge, such as the pumped discharge from a clothes washer. See Commentary Figure P2706.2.

The maximum height requirement is a measurement above the trap weir and intends to control the waste flow velocity at the trap inlet. Excessive velocity promotes trap siphonage. While the standpipe itself can be concealed within construction, the point of indirect connection at the standpipe inlet must be accessible to drain cleaning equipment.

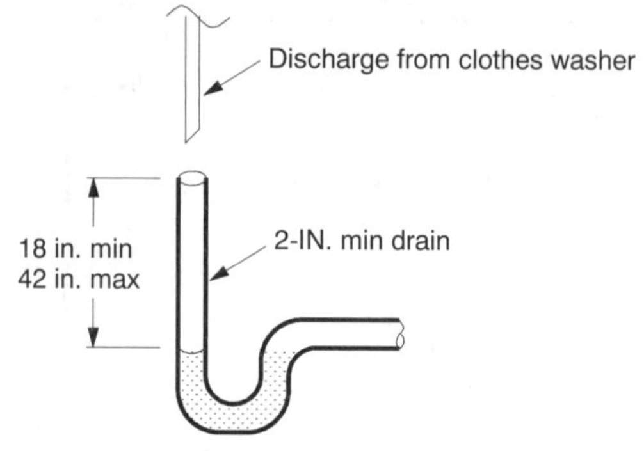

For SI: 1 inch = 25.4 mm.

Figure P2706.2
STANDPIPES

P2706.2.1 Laundry tray connection. A laundry tray waste line is permitted to connect into a standpipe for the automatic clothes washer drain. The standpipes shall not be less than 30 inches (762 mm) as measured from the crown weir. The outlet of the laundry tray shall be a maximum horizontal distance of 30 inches (762 mm) from the standpipe trap.

❖ The discharge from a laundry tray may connect to a standpipe provided that the standpipe is at least 30 inches (762 mm) high and that the laundry tray is not more than 30 inches (762 mm) horizontal from the standpipe trap. These requirements aim to keep the flood level rim of the laundry tray above the height of the standpipe outlet and to make it possible for the laundry tray and the clothes washer to be used simultaneously. The 30 inches (762 mm) horizontal distance limitation is intended to maintain a sanitary condition by minimizing bacterial buildup.

P2706.3 Prohibited waste receptors. Plumbing fixtures that are used for domestic or culinary purposes shall not be used to receive the discharge of an indirect waste.

Exceptions:

1. A kitchen sink trap is acceptable for use as a receptor for a dishwasher.
2. A laundry tray is acceptable for use as a receptor for a clothes washing machine.

❖ For sanitary reasons, this section prohibits discharge of an indirect waste receptor into a sink used for domestic or culinary purposes. However, this provision does not apply to the indirect connection of a dishwasher to a kitchen sink, since a dishwasher essentially performs the same function as a culinary sink. By the same logic, this section also permits the discharge from a clothes washing machine into a laundry tray.

SECTION P2707
DIRECTIONAL FITTINGS

P2707.1 Directional fitting required. Approved directional-type branch fittings shall be installed in fixture tailpieces receiving the discharge from food waste disposal units or dishwashers.

❖ Waste disposal units or dishwashers must connect to a tailpiece of a kitchen sink using a directional-type branch fitting to prevent blockage or backup of discharge under pressure. One option is to use a tee equipped with a baffle, a wye branch, or a similar type of directional fitting that would guide the discharge in the direction of flow.

SECTION P2708 SHOWERS

P2708.1 General. Shower compartments shall have at least 900 square inches (0.581 m^2) of interior cross-sectional area. Shower compartments shall not be less than 30 inches (762 mm) in minimum dimension measured from the finished interior dimension of the shower compartment, exclusive of fixture valves, shower heads, soap dishes, and safety grab bars or rails. The minimum required area and dimension shall be measured from the finished interior dimension at a height equal to the top of the threshold and at a point tangent to its centerline and shall be continued to a height not less than 70 inches (1778 mm) above the shower drain outlet.

Exception: Fold-down seats shall be permitted in the shower, provided the required 900-square-inch (0.581 m^2) dimension is maintained when the seat is in the folded-up position.

❖ A minimum 900 square inches (0.58 m^2) of cross-sectional area is required within a shower to allow bending over to clean one's body extremities. The 30-inch (762 mm) minimum dimension is meant to facilitate safety and comfort based on typical body movement while showering. The shower walls must extend to a height of 6 feet (1829 mm) above the floor level of the room (see Section R307.2) or 70 inches (1778 mm) above the shower floor and be constructed of smooth, noncorrosive, and nonabsorbent materials to protect the building materials from water damage. The height of the shower head above the floor is not regulated by the code. The standard design practice is to locate the shower head between 70 and 80 inches (1778 and 2032 mm) above the floor. Hinged doors must open outward to allow easy maneuvering in and out of a shower and proper egress. The code permits fold-down seats to be installed and to encroach into the required minimum dimension as long as that dimension is maintained when the seats are folded.

This section requires that walls form a watertight joint with each other and with the shower floor in accordance with the provisions of the code's building chapters.

Refer to Section R702.4 for provisions on use of water-resistant gypsum backing board and installation of non-absorbent finish material, such as ceramic tile. See Commentary Figure P2708.1.

P2708.2 Water-supply riser. The water supply riser from the shower valve to the shower head outlet shall be secured to the permanent structure.

❖ The shower pipe riser must be attached to a building structural element to protect the pipe from leaks due to fractures caused by showerhead movement (see Commentary Figure P2708.2). A standard practice for installing the riser pipe is to place a drop-ear elbow at the top of the riser. A drop-ear elbow has two wing connections permitting the elbow to be screwed into a backing structural member. When drop-ear elbows are not used, a pipe strap is installed at the top of the riser. When the shower riser is exposed, the manufacturer typically provides a clamp or other means of support matching the finish of the exposed piping. The clamp or other support is firmly secured to a structural member and attached to the riser.

P2708.3 Shower control valves. Showers and tub/shower combinations shall be equipped with control valves of the pressure balance, the thermostatic mixing or the combination pressure balance/thermostatic mixing valve types with high limit stops in accordance with ASSE 1016. The high limit stops shall be set to limit water temperature to a maximum 120°F (49°C).

❖ Every shower must have a control valve capable of protecting an individual from being scalded when taking a shower. Many incidents have been documented involving severe burns that result from sudden surges of high-temperature water flowing from a showerhead. These devices are also required to protect against thermal shock. Thermal shock is a change in discharge temperature great enough to produce a potentially hazardous situation. ASSE 1016 requires control valves to protect against rapid temperature fluctuations by automatically maintaining the discharge temperature to ±3°F (16°C) of the selected discharge temperature.

The three types of valves available are:

1. A balanced pressure valve that senses the pressure at the inlet to both the hot and cold water supply. If the pressure changes on one side, it automatically adjusts the pressure on the other side.
2. A thermostatic valve that senses the temperature of the outlet water, and adjusts the mixture ratio of the inlet water to maintain a constant set point temperature.
3. A combination balanced-pressure/thermostatic valve that adjusts to pressure and temperature fluctuations.

To help prevent accidental scalding, ASSE 1016 requires the valves to have a maximum set point feature. The valve can be set and adjusted to prevent the temperature from rising above 120°F (49°C). The risk of being scalded by shower water dramatically increases when temperatures rise above 120°F (49°C). See Commentary Figure P2708.3.

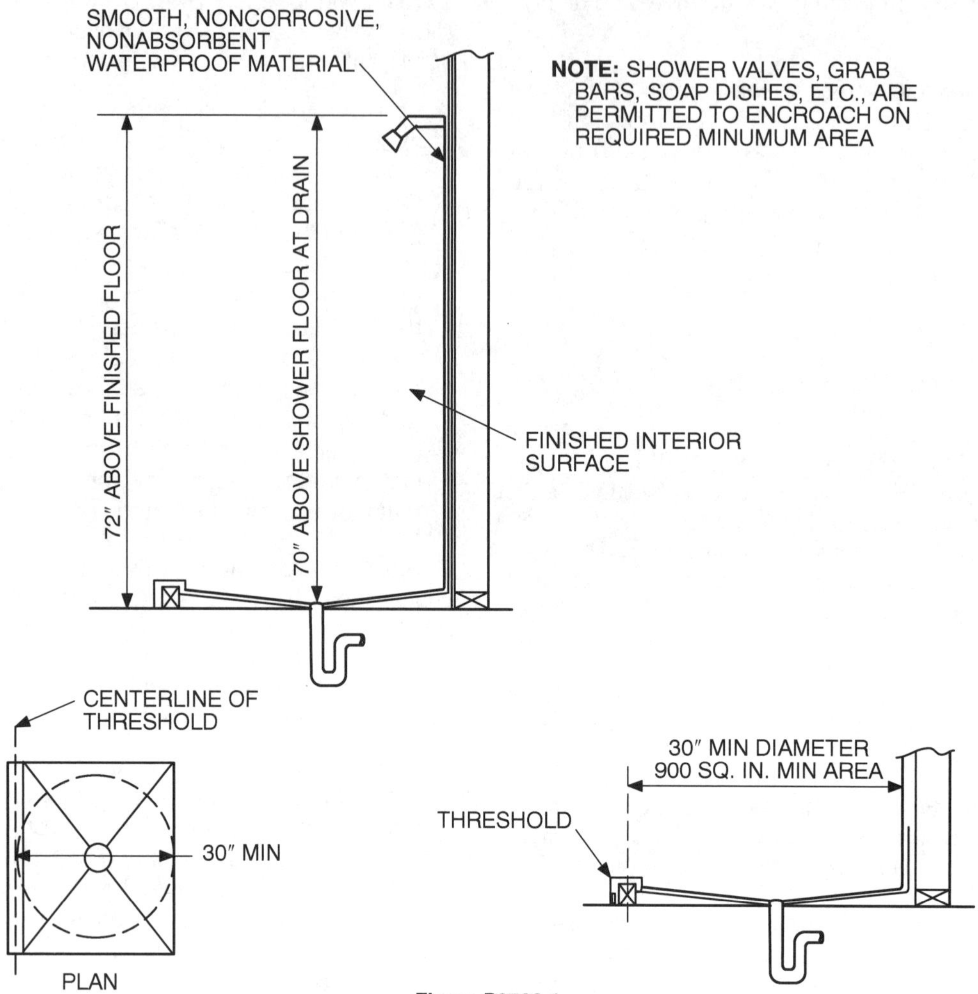

Figure P2708.1
SHOWER COMPARTMENT

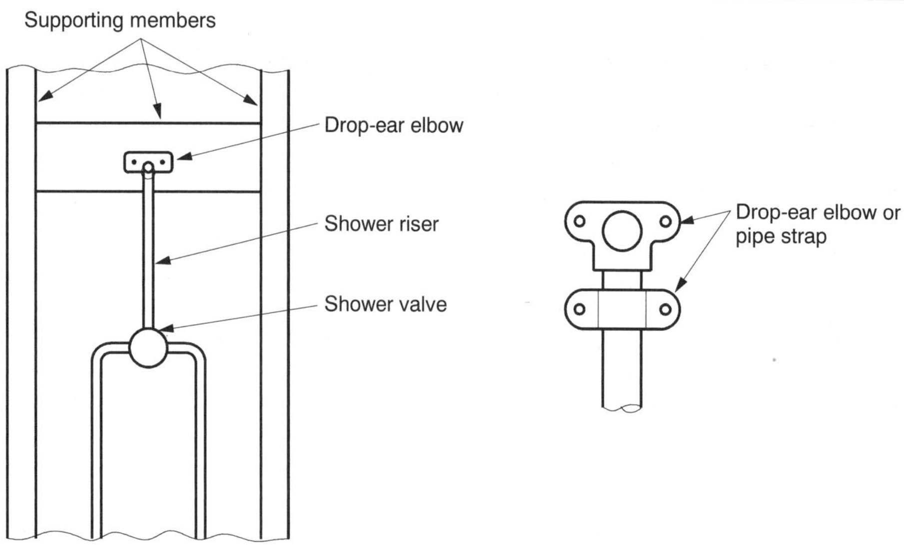

Figure P2708.2
SECURING SHOWER RISER

The high-limit stop consists of an adjustable set screw or cam that is set manually to limit the travel of the faucet control handle. The high-limit stop must be set and adjusted using a thermometer. Any change to the water heater thermostat will affect the faucet discharge temperature, thereby requiring readjustment of the high-limit stop. Seasonal fluctuations in the cold water supply temperature will also affect the discharge temperature. The high-limit stop setting needs to be set to accommodate cold water temperature variations; otherwise, seasonal readjustments might be necessary.

SECTION P2709 SHOWER RECEPTORS

P2709.1 Construction. Shower receptors shall have a finished curb threshold not less than 1 inch (25.4 mm) below the sides and back of the receptor. The curb shall be not less than 2 inches (51 mm) and not more than 9 inches (229 mm) in depth when measured from the top of the curb to the top of the drain. The finished floor shall slope uniformly toward the drain not less than one-fourth unit vertical in 12 units horizontal (2-percent slope) nor more than 0.5 inch (12.7 mm), and floor drains shall be flanged to provide a water-tight joint in the floor.

❖ This section contains minimum provisions for proper scouring and draining of shower receptors. While shower receptors and enclosures complying with ANSI Z 124.2 meet these requirements, Section P2709 includes minimum dimensions for receptors built onsite. See Commentary Figure P2709.1.

P2709.2 Lining required. The adjoining walls and floor framing, enclosing on-site built-up shower receptors shall be lined with sheet lead, copper or a plastic liner material that complies with ASTM D 4068. The lining material shall extend not less than 3 inches (76 mm) beyond or around the rough jambs and not less than 3 inches (76 mm) above finished thresholds. Hot mopping shall be permitted in accordance with Section P2709.2.1.

❖ Lining may be lead, copper, or plastic liner material that complies with ASTM D 4068. Extending the material at least 3 inches (76 mm) beyond rough jambs and 3 inches (76 mm) above the finished threshold creates a watertight pan beneath the finished floor of the shower receptor.

P2709.2.1 Hot-mopping. Shower receptors lined by hot mopping shall be built-up with not less than three layers of standard grade Type 15 asphalt-impregnated roofing felt. The bottom layer shall be fitted to the formed subbase and each succeeding

TIME AND TEMPERATURE RELATIONSHIP TO SERIOUS BURNS			
		Adults (skin thickness of 2.5 mm)	Children (skin thickness of .56 mm)
Water temperature		Time required for a third-degree burn to occur	
155°F	68°C	1 second	0.5 second
148°F	64°C	2 seconds	1 second
140°F	60°C	5 seconds	1 second
133°F	56°C	15 seconds	4 seconds
127°F	52°C	1 minute	10 seconds
124°F	51°C	3 minutes	1.5 minute
120°F	48°C	5 minutes	2.5 minutes
100°F	37°C	Safe temperature for bathing	Safe temperature for bathing

For SI: °C = [(°F) - 32]/1.8

Figure P2708.3
TEMPERATURE BURN CHART

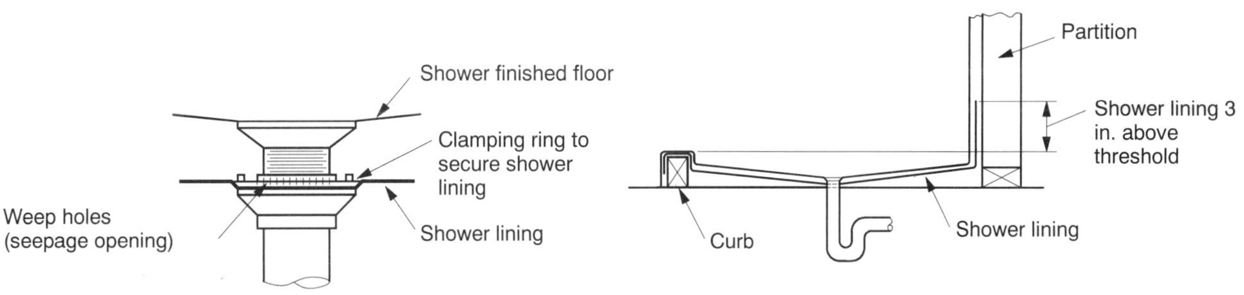

For SI: 1 inch = 25.4 mm.

Figure P2709.1
SHOWER DRAIN AND PAN CONNECTION

layer thoroughly hot-mopped to that below. All corners shall be carefully fitted and shall be made strong and water tight by folding or lapping, and each corner shall be reinforced with suitable webbing hot-mopped in place. All folds, laps and reinforcing webbing shall extend at least 4 inches (102 mm) in all directions from the corner and all webbing shall be of approved type and mesh, producing a tensile strength of not less than 50 pounds per inch (893 kg/m) in either direction.

❖ Hot mopping is permitted if three separate layers of Type 15 asphalt-impregnated roofing felt are thoroughly hot mopped in succession. Corners must be carefully fitted, with folds extending at least 4 inches (102 mm) and with webbing that produces a minimum tensile strength of 50 pounds per inch (893 kg/m) in either direction.

P2709.3 Installation. Lining materials shall be pitched one-fourth unit vertical in 12 units horizontal (2-percent slope) to weep holes in the subdrain by means of a smooth, solidly formed subbase, shall be properly recessed and fastened to approved backing so as not to occupy the space required for the wall covering, and shall not be nailed or perforated at any point less than 1 inch (25.4 mm) above the finished threshold.

❖ This section requires that the lining material (just below the finished floor) maintain a gradient to the drain of at least one-fourth unit vertical in 12 units horizontal (2-percent slope) to the weep holes in the subdrain. See Commentary Figure P2709.1. No punctures or perforations in the lining material are permitted at any point less than 1 inch (25.4 mm) above finished threshold.

P2709.3.1 Materials. Lead and copper linings shall be insulated from conducting substances other than the connecting drain by 15-pound (6.80 kg) asphalt felt or its equivalent. Sheet lead liners shall weigh not less than 4 pounds per square foot (19.5 kg/m^2). Sheet copper liners shall weigh not less than 12 ounces per square foot (3.82 kg/m^2). Joints in lead and copper pans or liners shall be burned or silver brazed, respectively. Joints in plastic liner materials shall be jointed per the manufacturer's recommendations.

❖ Sheet lead liners must weigh at least 4 pounds per square foot (19.5 kg/m^2). Copper liners must weigh at least 12 ounces per square foot (3.82 kg/m^2). Copper joints must be brazed, and plastic joints must be made in accordance with the manufacturer's recommendations.

P2709.4 Receptor drains. An approved flanged drain shall be installed with shower subpans or linings. The flange shall be placed flush with the subbase and be equipped with a clamping ring or other device to make a water-tight connection between the lining and the drain. The flange shall have weep holes into the drain.

❖ As illustrated in Commentary Figure P2709.1, a flanged drain must be installed flush with the sub-base and equipped with a clamping ring to create a watertight connection between the lining and the drain.

SECTION P2710
SHOWER WALLS

P2710.1 Bathtub and shower spaces. Shower walls shall be finished in accordance with Section R307.2.

❖ Section P2710.1 directs the reader to Section R307.2, which contains provisions for using water-resistant gypsum wallboard and installing non-absorbent finish materials, such as ceramic tile.

SECTION P2711
LAVATORIES

P2711.1 Approval. Lavatories shall conform to ANSI Z124.3, ASME A112.19.1, ASME A112.19.2, ASME A112.19.3, ASME A112.19.4, ASME A112.19.9, CSA B45.1, CSA B45.2, CSA B45.3 or CSA B45.4.

❖ Lavatories are available in enameled cast iron, vitreous china, stainless steel, porcelain-enameled formed steel, plastic, and nonvitreous ceramic. Lavatories also come in a variety of shapes and sizes such as wall-mounted, hanger-mounted, pedestal, rimmed, above-counter basins, and undermounted types. Referenced standards previously required lavatories to have an overflow; however, that is not currently the case. The provision of an overflow and its location is an option of the manufacturer. The reason for eliminating the overflow requirement was the lack of use, which resulted in the growth of bacteria and microorganisms. Where a lavatory has an overflow, the standards require the cross-sectional area of the overflow to be a minimum of $1^1/_8$ square inches (726 mm^2). This is designed to prevent too small an overflow, which would promote rapid bacterial growth.

P2711.2 Cultured marble lavatories. Cultured marble vanity tops with an integral lavatory shall conform to ANSI Z124.3 or CSA B45.5.

❖ Vanity tops with an integral lavatory that is all plastic must conform to ANSI Z124.3 or CSA B45.5. Lavatory units can be constructed of a variety of synthetic materials including ABS, PVC, gel-coated fiberglass-reinforced plastic, acrylic, cultured marble cast-filled fiberglass, polyester, and cultured marble acrylic. Cultured marble vanity tops are impregnated with fire-retardant chemicals, reducing the fuel contribution of the unit during a fire. They are also designed to resist the effect of a burning cigarette left unattended on the vanity top.

P2711.3 Lavatory waste outlets. Lavatories shall have waste outlets not less than 1.25 inch (32 mm) in diameter. A strainer, pop-up stopper, crossbar or other device shall be provided to restrict the clear opening of the waste outlet.

❖ The waste outlet must be the same size as the fixture trap described in Section P3201.7. Section P2711.1 establishes the minimum size requirement for lavatory waste outlets as well as the restriction for the waste opening. Each lavatory opening must have a restriction device that prevents items such as jewelry, toiletry items, and cosmetics from inadvertently entering and potentially damaging the drainage system.

P2711.4 Movable lavatory systems. Moveable lavatory systems shall comply with ASME A112.19.12.

❖ Moveable and pivoting lavatories allow the user to adjust the location of a lavatory upward and downward, from side-to-side, and front-to-back in order to maximize the ease of use of the fixture. Some products also pivot.

SECTION P2712 WATER CLOSETS

P2712.1 Approval. Water closets shall conform to the water consumption requirements of Section P2903.2 and shall conform to ANSI Z124.4, ASME A112.19.2, CSA B45.1, CSA B45.4 or CSA B45.5. Water closets shall conform to the hydraulic performance requirements of ASME A112.19.6. Water closets tanks shall conform to ANSI Z124.4, ASME A112.19.2, ASME A112.19.9, CSA B45.1, CSA B45.4 or CSA B45.5. Water closets that have an invisible seal and unventilated space or walls that are not thoroughly washed at each discharge shall be prohibited. Water closets that permit backflow of the contents of the bowl into the flush tank shall be prohibited.

❖ A water closet is referred to by many other names, including toilet and commode. The term "water closet" is the technical expression that derives from the first attempt to bring plumbing indoors. Early water closets consisted of a room that was the size of a closet, containing a cistern and a chair. Today, the term "water closet" applies to the name of the fixture itself.

The most common word used in the United States to identify a water closet is "toilet." However, the word "toilet" more accurately describes the room in which water closets are installed.

The water closet is the fixture primarily responsible for the modernization of plumbing. The fixture effectively disposes of human waste in a sanitary manner. A water closet that does not thoroughly scour the used surface will harbor bacteria and germs that pose health risks to occupants due to contamination. This would also be the case if the contents of the bowl were to enter the flush tank, where the potable water source is located. The code requires fixtures to be self-cleaning in accordance with their respective referenced standards.

Water closets must be of the water-conservation type. The respective fixture standard permits a maximum average of 1.6 gallons (6.1 L) of water per flushing cycle. No cycle during the testing can exceed 2.0 gallons (7.6 L) per flush. To determine conformance, a representative model of the water closet must be tested for its flushing efficiency. The flushing test utilizes polypropylene balls and polyethylene granule discs. There is also a test for the ability to remove an ink stain on the inside surface of the water closet bowl, and a dye test measuring the exchange rate of water in the bowl. Another test measures the capacity of the drain line attached to the water closet to carry away the solids from the bowl after being flushed.

There are three styles of water closets: close coupled, one piece, and flushometer valve. The close-coupled water closet has a bowl and separate gravity-type tank or flushometer tank that is supported by the bowl. A one-piece water closet is constructed with the gravity-type tank or flushometer tank and the bowl as one integral unit. A flushometer valve water closet is a bowl with a flushometer valve. See Commentary Figures P2712.1(1), P2712.1(2), P2712.1(3), P2712.1(4) and P2712.1(5).

P2712.2 Flushing devices required. Water closets shall be provided with a flush tank, flushometer tank or flushometer valve designed and installed to supply water in sufficient quantity and flow to flush the contents of the fixture, to cleanse the fixture and refill the fixture trap in accordance with ASME A112.19.2 and ASME A112.19.6.

❖ There are three device classifications for flushing water closets. The most common in residential buildings is the flush- tank water closet, which relies on the volume of water in the tank to cleanse and flush the contents of the bowl. Flushometer tanks were introduced more recently when water-conservation requirements were established for water closets. The concern was that the tank capacity requirements [typically 1.6 gal-

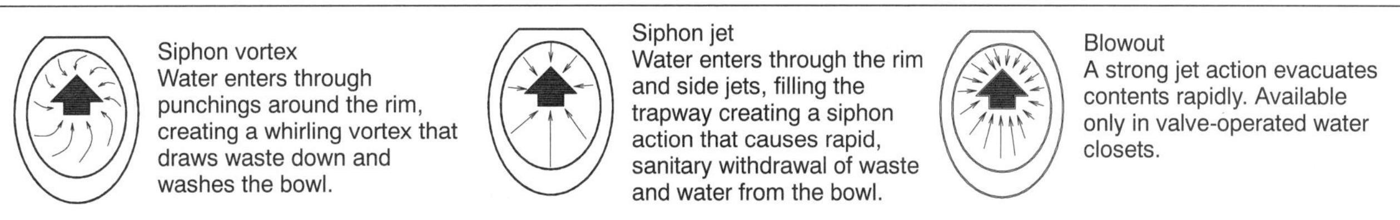

Figure P2712.1(1)
FLUSHING ACTION FOR CERTAIN WATER CLOSETS

lon (6.056 L) per flush] might not be sufficient in volume to flush the contents of the bowl. The flushometer tank uses air pressure that is stored behind a membrane to push the water from a bladder-type container into the bowl under higher pressure and velocity. This type requires higher residual water pressure to fill the bladder after each flush, since the fill water flows against pressure higher than that of the atmosphere. See Commentary Figure P2712.2. Flushometer valve water closets are common in non-residential buildings and are rarely found in single-family residences. They are designed for more frequent use than is common in non-residential occupancies.

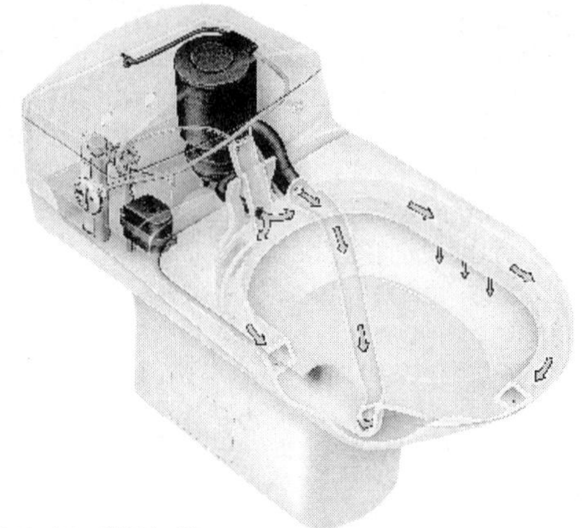

Illustration courtesy of Kohler Co.

Figure P2712.1(2)
WATER CLOSET
(PRESSURE-ASSISTED ONE PIECE)

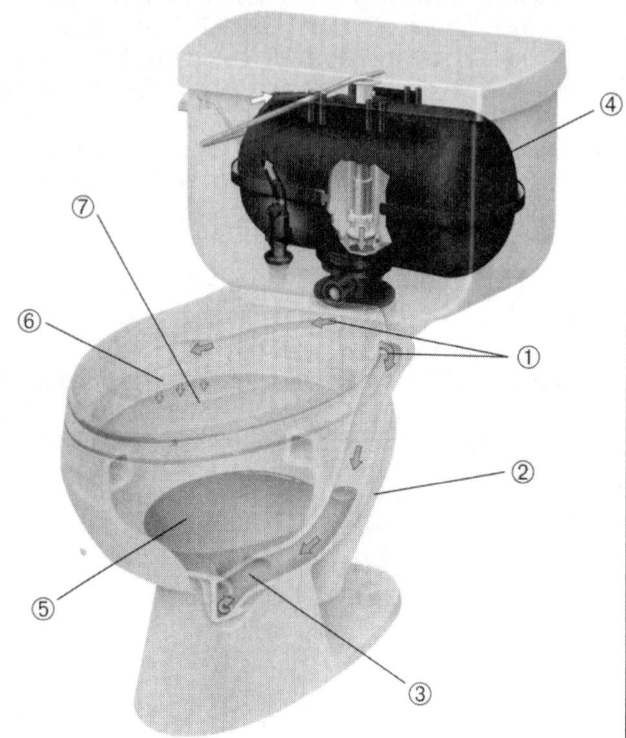

Illustration courtesy of Kohler Co.

Figure P2712.1(3)
WATER CLOSET
(PRESSURE-ASSISTED TWO PIECE)

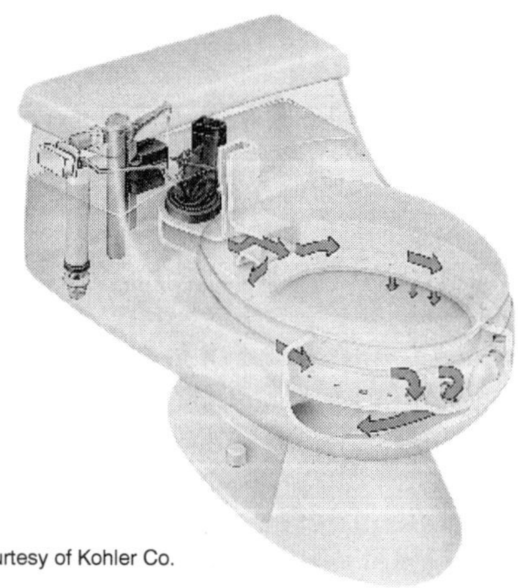

Illustration courtesy of Kohler Co.

Figure P2712.1(4)
WATER CLOSET
(GRAVITY-FED LOW-PROFILE TANK)

P2712.3 Water supply for flushing devices. An adequate quantity of water shall be provided to flush and clean the fixture served. The water supply to flushing devices equipped for manual flushing shall be controlled by a float valve or other automatic device designed to refill the tank after each discharge and to completely shut off the water flow to the tank when the tank is filled to operational capacity. Provision shall be made to automatically supply water to the fixture so as to refill the trap after each flushing.

❖ This section requires that water be supplied in quantities sufficient to cleanse and refill the fixture trap. A float valve that opens when the tank water drops below the fill level and closes when the tank is filled to operational capacity must be used to control manual flushing. In addition, there must be a connection to the bowl to refill the trap after each flushing. This requirement is usually fulfilled by installing a small rubber hose connecting to the overflow pipe that will discharge water to refill the trap, even after the flap at the bottom of the tank closes once the tank has discharged its contents into the bowl.

P2712.4 Flush valves in flush tanks. Flush valve seats in tanks for flushing water closets shall be at least 1 inch (25.4 mm) above the flood-level rim of the bowl connected thereto, except an approved water closet and flush tank combination designed so that when the tank is flushed and the fixture is clogged or partially clogged, the flush valve will close tightly so that water will not spill continuously over the rim of the bowl or backflow from the bowl to the tank.

❖ The flush valve seat (flap) must be installed at least 1 inch (25.4 mm) above the flood-level rim of the bowl to protect the potable water source from contamination caused by the submerged inlet in the flush tank containing nonpotable water. Another means of protection is to use a design that will close the flush valve tightly when the bowl is clogged or partially clogged, thus preventing the bowl contents from flowing back to the tank. This type of design applies to special one-piece water closets.

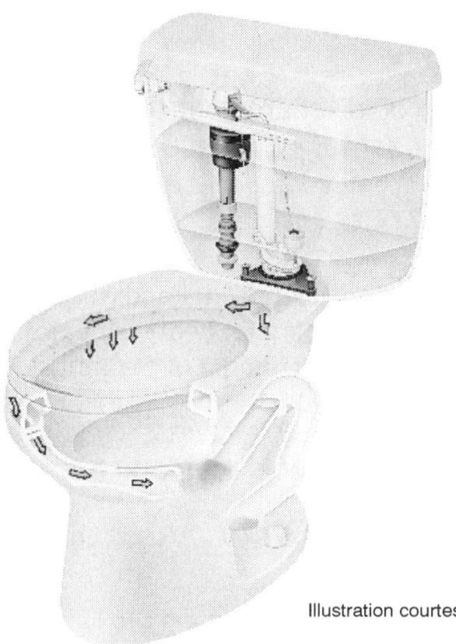

Illustration courtesy of Kohler Co.

Figure P2712.1(5)
WATER CLOSET
(GRAVITY-FED SIPHON JET)

Photo courtesy of Sloan Flushmate, Div. of Sloan Valve Co.

Figure P2712.2
FLUSHOMETER TANK

P2712.5 Overflows in flush tanks. Flush tanks shall be provided with overflows discharging to the water closet connected thereto and such overflow shall be of sufficient size to prevent flooding the tank at the maximum rate at which the tanks are supplied with water according to the manufacturer's design conditions.

❖ The overflow in the flush tank prevents flooding of the tank and submersion of the fill valve. The fill valve antisiphon device must be located above water at all times to maintain protection against backflow. The overflow discharges into the water closet, ultimately discharging to the drainage system.

When the flush tank is a separate unit (a two-piece water closet), the overflow is always above the flood level rim of the water closet or urinal. If the water closet drain is obstructed, the water discharges onto the floor before submerging the fill valve antisiphon device. In certain styles of one-piece water closets, the overflow opening is below the water closet flood level rim; therefore, openings are provided in the tank to serve as a secondary overflow to prevent the flooding of the fill valve.

P2712.6 Access. All parts in a flush tank shall be accessible for repair and replacement.

❖ Parts within a flush tank must be accessible for repair and replacement.

P2712.7 Water closet seats. Water closets shall be equipped with seats of smooth, nonabsorbent material and shall be properly sized for the water closet bowl type.

❖ A water closet seat must be designed to prevent the accumulation of dirt or bacterial growth and to facilitate cleaning. The seat must also be designed to suit the bowl. For example, an elongated bowl requires an elongated seat.

P2712.8 Flush tank lining. Sheet copper used for flush tank linings shall have a minimum weight of 10 ounces per square foot (3.18 kg/m^2).

❖ Whereas sheet copper was the material commonly used to line plumbing fixtures in the past, it is rarely used today.

If used, sheet copper must weigh at least 10 ounces per square foot (3.1 kg/m2). ASTM B 152 regulates cold-rolled tempered sheet copper, hot-rolled tempered sheet copper, and annealed copper sheet.

P2712.9 Electro-hydraulic water closets. Electro-hydraulic water closets shall conform to ASME A112.19.13.

❖ An electro-hydraulic water closet is a siphonic or wash-down water closet with a nonmechanical trap seal. The water closet incorporates electrical motor(s), pump(s), and controllers to facilitate the flushing action.

SECTION P2713
BATHTUBS

P2713.1 Bathtub waste outlets and overflows. Bathtubs shall have outlets and overflows at least 1.5 inches (38 mm) in diameter, and the waste outlet shall be equipped with an approved stopper.

❖ The minimum diameter required of a bathtub waste outlet is 1.5 inches (38 mm). An overflow waste outlet is required to remove displaced water when a bather enters the bathtub or to protect against accidental flooding due to an unattended filling operation. Because a bathtub has a much higher threshold (flood level rim) than a shower receptor or enclosure, bathtubs equipped with a shower riser do not require an increase in trap and drain size to the 2-inch (51 mm) minimum. See the commentary for Table P3201.7.

P2713.2 Bathtub enclosures. Doors within a bathtub enclosure shall conform to ASME A112.19.15.

❖ Bathtub/whirlpool bathtubs with pressure sealed doors allow the bather to enter the fixture at approximately the same level as the floor. The door can be opened allowing the bather easier access to the bathtub. Once inside the fixture, the door is closed and the water is turned on. Sensor switches, which note the presence of water entering the vessel, activate an air pump that pressurizes the door seal, thereby keeping the water within the bathing vessel. Upon completion of bathing, the bather can open the door only when the bath water is drained from the vessel.

SECTION P2714 SINKS

P2714.1 Sink waste outlets. Sinks shall be provided with waste outlets not less than 1.5 inches (38 mm) in diameter. A strainer, crossbar or other device shall be provided to restrict the clear opening of the waste outlet.

❖ Section P2714.1 requires that kitchen-type sinks be provided with waste outlets at least 1.5 inches (38 mm) in diameter. The minimum opening must be increased to at least 3.5 inches (89 mm) when the sink is equipped with a food-waste grinder. Strainers or crossbars are required to restrict the flow. This requirement prevents items from entering the drainage system.

P2714.2 Movable sink systems. Movable sink systems shall comply with ASME A112.19.12.

❖ See the commentary for Section P2711.4.

SECTION P2715 LAUNDRY TUBS

P2715.1 Laundry tub waste outlet. Each compartment of a laundry tub shall be provided with a waste outlet not less than 1.5 inches (38 mm) in diameter and a strainer or crossbar to restrict the clear opening of the waste outlet.

❖ Laundry tubs must be provided with a minimum 1.5-inch (38 mm) waste outlet equipped with a strainer or cross bar to catch debris or other materials that could be discharged into the sink, causing the drain to become clogged.

SECTION P2716 FOOD WASTE GRINDER

P2716.1 Food waste grinder waste outlets. Food waste grinders shall be connected to a drain of not less than 1.5 inches (38 mm) in diameter.

❖ The outlet, or discharge, of the food-waste grinder must be at least 1.5 inches (38 mm) in diameter, which is consistent with kitchen-type sink requirement in Section P2714.1.

P2716.2 Water supply required. Food waste grinders shall be provided with an adequate supply of water at a sufficient flow rate to ensure proper functioning of the unit.

❖ A supply of cold water is necessary to promote proper food-waste grinder operation and to carry food waste into the drain.

SECTION P2717 DISHWASHING MACHINES

P2717.1 Protection of water supply. The water supply for dishwashers shall be protected by an air gap or integral backflow preventer.

❖ Residential or domestic dishwashers must conform to ASSE 1006. This standard includes protection of the potable water supply against backflow. Units conforming to this standard possess backflow protection consisting of an internal, integral backflow prevention device. Where a unit does not meet this standard, protection must be provided by means of an air gap.

P2717.2 Sink and dishwasher. A sink and dishwasher are permitted to discharge through a single 1.5-inch (38 mm) trap. The discharge pipe from the dishwasher shall be increased to a minimum of 0.75 inch (19.1 mm) in diameter and shall be connected with a wye fitting to the sink tailpiece. The dishwasher waste line shall rise and be securely fastened to the underside of the counter before connecting to the sink tailpiece.

❖ Dishwashing machines may connect directly to the drainage system and do not require an indirect connection by drainage air gap or air break. The direct connection is acceptable, because the potable water supply is adequately protected against backflow. Many domestic dishwasher manufacturers require the waste line to be looped as high as possible and to be securely fastened to the underside of the sink rim or countertop to minimize the potential of waste backflow into the dishwasher unit. See Commentary Figure P2717.2.

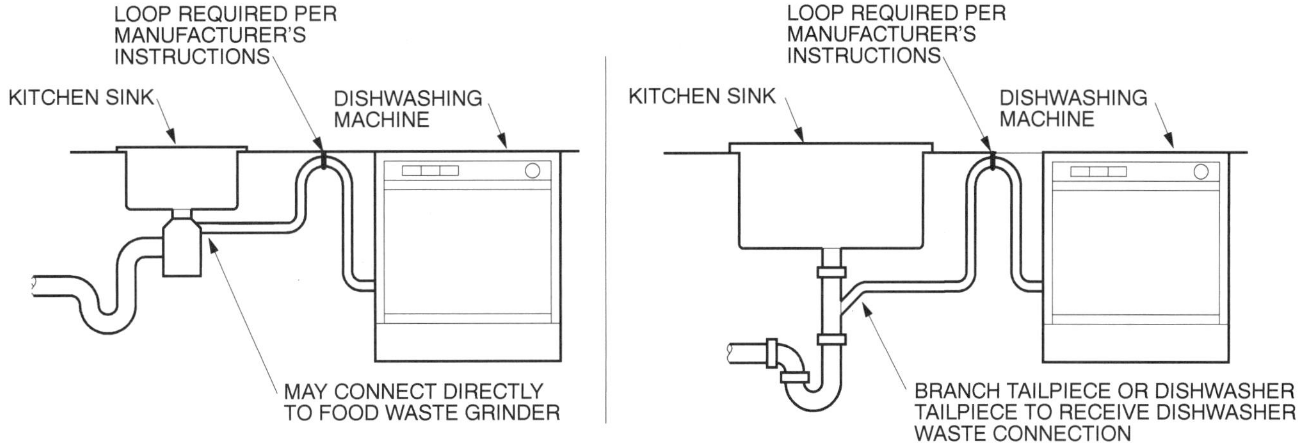

Figure P2717.2
DISHWASHER WASTE CONNECTIONS

P2717.3 Sink, dishwasher and food grinder. The combined discharge from a sink, dishwasher, and waste grinder is permitted to discharge through a single 1.5 inch (38 mm) trap. The discharge pipe from the dishwasher shall be increased to a minimum of 0.75 inch (19.1 mm) in diameter and shall connect with a wye fitting between the discharge of the food-waste grinder and the trap inlet or to the head of the food grinder. The dishwasher waste line shall rise and be securely fastened to the underside of the counter before connecting to sink tail piece or food grinder.

❖ This section permits the discharge from a sink, dishwasher, and food waste grinder to be intercepted by a single, 1.5-inch (38 mm) trap. The discharge piping from the dishwasher must be increased to a minimum $^3/_4$ inch (19.1 mm) and connect to a branch tail piece of a kitchen sink or directly to the food waste grinder. For the same reason discussed in the previous section, the dishwasher waste must rise and be securely fastened to the underside of the counter before connecting to the branch tail piece or food waste grinder. See Commentary Figure P2717.2.

SECTION P2718 CLOTHES WASHING MACHINE

P2718.1 Waste connection. The discharge from a clothes washing machine shall be through an air break.

❖ ASSE 1007 requires that an automatic clothes washer have an internal air gap or backflow preventor to protect the potable water and that the waste discharge through an air break. See the commentary for Section R202 for the definition of "Air break." Such waste usually discharges to a standpipe meeting the provisions of Section P2706.2 or to a laundry tray in accordance with Section P2706.3, Exception 2.

SECTION P2719 FLOOR DRAINS

P2719.1 Minimum size. Floor drains shall have waste outlets not less than 2 inches (51 mm) in diameter and shall be provided with a removable strainer with an open area of at least two-thirds of the cross-sectional area of the drain line to which it connects.

❖ Whereas the code does not specifically require floor drains, designers often include floor drains in single family dwellings to provide for emergency drainage in case a room or space floods or to meet specific drainage needs (i.e., relief drains for water heaters, backflow preventer discharge, etc.).

Floor drains must conform to ASME A112.6.3 as listed in Table P2701.1. This standard requires the floor drains to be constructed of cast iron, brass, or bronze. However, it is not uncommon to see floor drains made of plastic material. Plastic floor drains must be approved for use in accordance with Section R104.11. Because floor drains are viewed as emergency drains, Table P3004.1 does not assign them a fixture unit value. Therefore, Section P2719.1 clearly states that the outlet (hence, the trap and drain also) must be a minimum of 2 inches (51 mm) in diameter. In addition, the strainer, although it provides some restriction to prevent materials from accidentally entering or damaging the drainage system. The strainer must be removable for drain cleaning purposes. See Commentary Figures P2719.1(1) and P2719.1(2).

SECTION P2720 WHIRLPOOL BATHTUBS

P2720.1 Access panel. A door or panel of sufficient size shall be installed to provide access to the pump for repair and/or replacement.

❖ The code does not dictate the size of the access door or panel. The factor for determining the access door or

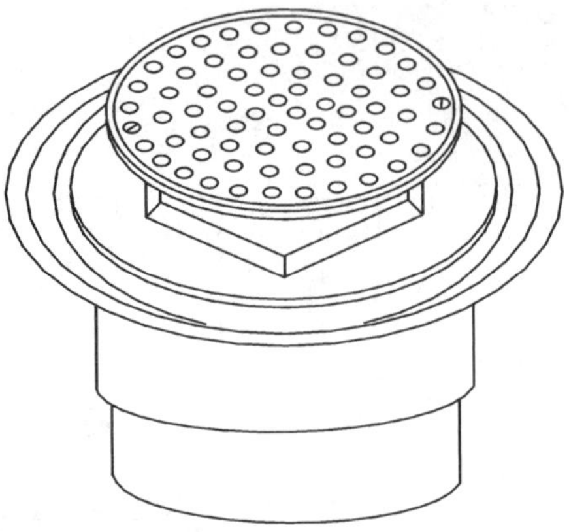

Figure P2719.1(1)
TYPICAL FLOOR DRAIN WITHOUT INTEGRAL TRAP

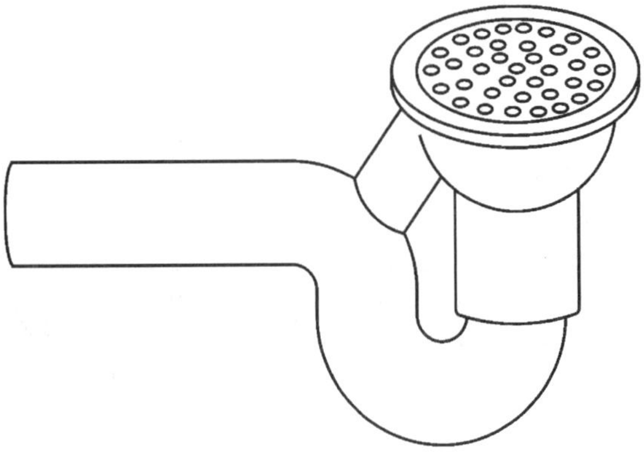

Figure P2719.1(2)
TYPICAL FLOOR DRAIN WITH INTEGRAL TRAP

panel size is the size of the pump plus an area sufficient to repair or to remove and replace the pump. The access door or panel must be located such that the pump can be reached for service. This access must not be confused with the access requirement for fixture connections in Section P2704, where the whirlpool bathtub is provided with a waste and overflow using slip joints.

P2720.2 Piping drainage. The circulation pump shall be accessibly located above the crown weir of the trap. The pump drain line shall be properly graded to ensure minimum water retention in the volute after fixture use. The circulation piping shall be installed to be self-draining.

❖ The pump and suction lines for the whirlpool bathtub must be designed to drain completely after each use. This improves the whirlpool bathtub's sanitation. Wastewater left standing in the pump body or in the suction and discharge piping will promote bacterial growth. Therefore, the pump is required to be located above the crown weir of the trap, and the piping must be properly graded to ensure minimum water retention.

P2720.3 Leak testing. Leak testing and pump operation shall be performed in accordance with the manufacturer's installation instructions.

❖ The code consistently requires testing for plumbing installations, and whirlpool bathtub installations are by no means exempt. Testing the installed unit verifies water carrying capabilities and pump operation to determine whether the unit has been damaged during shipping or installation in such a way that would compromise the integrity of the unit.

P2720.4 Manufacturer's instructions. The product shall be installed in accordance with the manufacturer's installation instructions.

❖ Unlike most fixtures, a whirlpool bathtub comprises numerous parts that can vary widely from one manufacturer to another. For this reason, the installer must follow carefully the manufacturer's installation instructions with regard to shipping, installation, testing, and use.

SECTION P2721
BIDET INSTALLATIONS

P2721.1 Water supply. The bidet shall be equipped with either an air-gap-type or vacuum-breaker-type fixture supply fitting.

❖ Most bidets present a submerged outlet hazard, because the design typically employs a water supply nozzle located below the flood level rim of the fixture. Hence, they must be equipped with a vacuum breaker on the fixture supply to protect against back siphonage. Bidets that are equipped with a faucet type fixture supply are required to have an air gap protection above the flood level rim of the fixture.

SECTION P2722
FIXTURE FITTING

P2722.1 General. Fixture supply valves and faucets shall comply with ASME A112.18.1 as listed in Table P2701.1. Faucets and fixture fittings that supply drinking water for human ingestion shall conform to the requirements of NSF 61, Section 9. Flexible water connectors shall conform to the requirements of Section P2904.7.

❖ Plumbing fixture fittings must comply with ASME A112.18.1, as listed in Table P2701.1. When the water supply is for human consumption, this section also requires fixture fittings to comply with NSF 61, Section 9. This standard contains toxicity requirements to prevent minor toxins from exceeding appropriate limits. Since NSF 61, Section 9 addresses water for human consumption, it does not apply to faucets or fixtures such as shower valves or hose bibbs. See Commentary Figures P2722.1(1), P2722.1(2), P2722.1(3) and P2722.1(4).

P2722.2 Hot water. Fixture fittings, faucets and diverters shall be installed and adjusted so that the flow of hot water from the fittings corresponds to the left-hand side of the fitting.

Exception: Shower or tub/shower mixing valves conforming to ASSE 1016, where the flow of hot water corresponds to the markings on the device.

❖ For safety reasons, the code mandates that hot water correspond to the left side of the fixture. It is an accepted practice and is human nature to equate the left side of the faucet with hot water. The code intends to protect an individual from scalding when turning on what the individual incorrectly assumes to be cold water.

Some manufacturers design a fixture fitting that can be reversed inside the faucet if a plumber inadvertently (or intentionally in the case of back-to-back faucets) pipes the water supplies in reverse. This type of installation is acceptable.

Several single-handle shower valves complying with ASSE 1016 have identifiable "hot" designations on the control valve that do not correspond with the left side of the valve. This section does not prohibit using these devices. These devices operate by rotating a single handle or knob away from the "off" position to initiate water flow. Moving the handle further in the same direction increases the temperature of flowing water. Because the rotating operation is limited to one direction, the user cannot mistakenly activate water flow at the maximum hot-water setting without first rotating the handle or knob through the cold water setting. In addition, ASSE 1016 requires control valves to have settings that indicate clearly the direction or means of adjusting the temperature.

P2722.3 Individual pressure-balancing in-line valves for individual fixture fittings. Where individual pressure-balanc-

Photo courtesy of Delta Faucet Company

Figure P2722.1(1)
FAUCETS AND FIXTURE FITTINGS

ing in-line valves for individual fixture fittings are installed, the valves shall comply with ASSE 1066. Such valves shall be installed in an accessible location and shall not be utilized alone as a substitute for the balanced pressure, thermostatic or combination shower valves required in Section P2708.3.

❖ These valves must comply with ASSE 1066, which applies to automatic pressure balancing in-line valves that are used to equalize incoming hot and cold water line pressures for the purpose of minimizing mixed water temperature variations due to pressure fluctuations when used in conjunction with a mixing valve or two-handle valve set. They are not designed to limit the maximum outlet temperature at the point of use. These devices are intended for use in individual plumbing fix-

Photo courtesy of Delta Faucet Company

Figure P2722.1(2)
FAUCETS AND FIXTURE FITTINGS

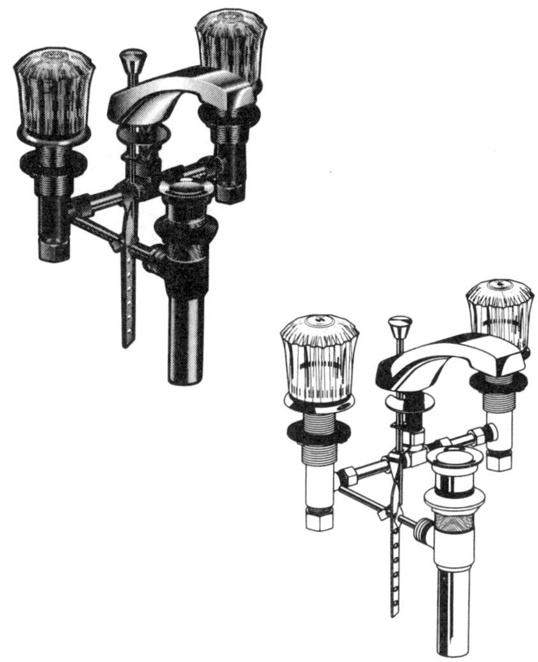

Illustration courtesy of Gerber Plumbing Fixture Corp.

Figure P2722.1(3)
FAUCETS AND FIXTURE FITTINGS

ture fittings such as bath faucets, sinks, and lavatory faucets. These devices may not be used as a substitute for shower or tub/shower valves required in Section P2708.3.

SECTION P2723 MACERATING TOILET SYSTEMS

P2723.1 General. Macerating toilet systems shall be installed in accordance with manufacturer's installation instructions.

❖ A macerating toilet system collects the waste from a single water closet, lavatory and bathtub located in the same room. The system then grinds the waste and discharges it through a small discharge line to the drainage system. The concept of this system is similar to that of a building drain installation using a sewage ejector pump.

P2723.2 Drain. The minimum size of the drain from the macerating toilet system shall be 0.75 inch (19.1 mm) in diameter.

❖ Since a macerating toilet alters the contents significantly, a $^{3}/_{4}$-inch (19.1 mm) discharge line is adequate for handling the pumped drainage.

SECTION P2724 SPECIALTY TEMPERATURE CONTROL DEVICES AND VALVES

P2724.1 Temperature-actuated, flow-reduction devices for individual fixtures. Temperature-actuated, flow-reduction devices, where installed for individual fixture fittings, shall conform to ASSE 1062. Such valves shall not be used alone as a substitute for the balanced pressure, thermostatic or combination shower valves required for showers in Section P2708.3.

❖ Temperature-actuated flow reduction valves for individual fixture fittings must comply with the provisions set forth in ASSE 1062. These valves reduce flow to .25 gpm (.946 L/m) or less automatically in response to outlet temperatures greater than a preset temperature not to exceed 120°F (49°C). This limits a person's exposure to high temperature water discharge from an individual fixture fitting such as a bath and utility faucet, lavatory, or sink faucet. These devices may not be used as a substitute for shower or tub/shower valves required in Section P2708.3.

Bibliography

The following resource materials are referenced in this chapter or are relevant to the subject matter addressed in this chapter.

ANSI Z124.1-95, *Plastic Bathtub Units*. New York: American National Standards Institute, 1995.

ANSI Z124.2-95, *Plastic Shower Receptors and Shower Stalls*. New York: American National Standards Institute, 1995.

ANSI Z124.3-95, *Plastic Lavatories*. New York: American National Standards Institute, 1995.

ANSI Z124.4-96, *Plastic Water Closet Bowls and Tanks*. New York: American National Standards Institute, 1996.

ANSI Z124.6-97, *Plastic Sinks*. New York: American National Standards Institute, 1997.

ASME A112.6.1M-1997, *Floor Affixed Supports for Off-the-floor Plumbing Fixtures for Public Use*. New York: American Society of Mechanical Engineers, 1997.

ASME A112.6.2-2000, *Framing-Affixed Supports for Off-the-floor Water Closets with Concealed Tanks*. New York: American Society of Mechanical Engineers, 2000.

ASME A112.6.3-2001, *Floor and Trench Drains*. New York: American Society of Mechanical Engineers, 2001.

ASME A112.18.1-2000, *Plumbing Fixture Fittings*. New York: American Society of Mechanical Engineers, 2000.

ASME A112.19.1M-1994 (R1999), *Enameled Cast Iron Plumbing Fixtures*. New York: American Society of Mechanical Engineers, 1999.

Photo courtesy of Delta Faucet Company

Figure P2722.1(4)
FAUCETS AND FIXTURE FITTINGS

ASME A112.19.2M-1998, *Vitreous China Plumbing Fixtures*. New York: American Society of Mechanical Engineers, 1998.

ASME A112.19.3M-1987 (R1996), *Stainless Steel Plumbing Fixtures (Designed for Residential Use)*. New York: American Society of Mechanical Engineers, 1996.

ASME A112.19.4M-1994 (R1999), *Porcelain Enameled Formed Steel Plumbing Fixtures*. New York: American Society of Mechanical Engineers, 1999.

ASME A112.19.5-1999, *Trim for Water-Closet Bowls, Tanks, and Urinals*. New York: American Society of Mechanical Engineers, 1999.

ASME A112.19.6-1995, *Hydraulic Performance Requirements for Water Closets and Urinals*. New York: American Society of Mechanical Engineers, 1995.

ASME A112.19.7M-1995, *Whirlpool Bathtub Appliances*. New York: American Society of Mechanical Engineers, 1995.

ASME A112.19.8M-1987 (R1996), *Suction Fittings for Use in Swimming Pools, Wading Pools, Spas, Hot Tubs, and Whirlpool Bathtub Appliances*. New York: American Society of Mechanical Engineers, 1996.

ASME A112.19.9M-1998, *Non-vitreous Ceramic Plumbing Fixtures*. New York: American Society of Mechanical Engineers, 1998.

ASME A112.19.12-2000, *Wall Mounted and Pedestal Mounted, Adjustable and Pivoting Lavatory and Sink Carrier Systems*. New York: American Society of Mechanical Engineers, 2000.

ASME A112.19.13-2001, *Electrohydraulic Water Closets*. New York: American Society of Mechanical Engineers, 2001.

ASME A112.19.15-2001, *Bathtub/Whirlpool Bathtubs with Pressure Sealed Doors*. New York: American Society of Mechanical Engineers, 2001.

ASSE 1002-99, *Performance Requirements for Anti-Siphon Fill Valves (Ballcocks) for Gravity Water Closet Flush Tanks*. Westlake, OH: American Society of Sanitary Engineers, 1999.

ASSE 1006-89, *Performance Requirements for Residential Use Dishwashers*. Westlake, OH: American Society of Sanitary Engineers, 1989.

ASSE 1007-92, *Performance Requirements for Home Laundry Equipment*. Westlake, OH: American Society of Sanitary Engineers, 1992.

ASSE 1008-89, *Performance Requirements for Household Food Waste Disposer Units*. Westlake, OH: American Society of Sanitary Engineers, 1989.

ASSE 1014-90, *Performance Requirements for Handheld Showers*. Westlake, OH: American Society of Sanitary Engineers, 1990.

ASSE 1016-96, *Performance Requirements for Individual Thermostatic, Pressure Balancing and Combination Control Valves for Individual Fixture Fittings*. Westlake, OH: American Society of Sanitary Engineers, 1996.

ASSE 1019-97, *Performance Requirements for Vacuum Breaker Wall Hydrants, Freeze Resistant, Automatic Draining Type*. Westlake, OH: American Society of Sanitary Engineers, 1997.

ASSE 1025-78, *Performance Requirements for Diverters for Plumbing Faucets with Hose Spray, Anti-Siphon Type, Residential Applications*. Westlake, OH: American Society of Sanitary Engineers, 1978.

ASSE 1037-90, *Performance Requirements for Pressurized Flushing Devices for Plumbing Fixtures*. Westlake, OH: American Society of Sanitary Engineers, 1990.

ASSE 1052-94, *Performance Requirement for Hose Connection Backflow Preventers*. Westlake, OH: American Society of Sanitary Engineers, 1994.

ASSE 1062-97, *Performance Requirements for Temperature Actuated, Flow Reduction Valves to Individual Fixture Fittings*. Westlake, OH: American Society of Sanitary Engineers, 1997.

ASSE 1066-97, *Performance Requirements for Individual Pressure Balancing Valves for Individual Fixture Fittings*. Westlake, OH: American Society of Sanitary Engineers, 1997.

ASTM B 152/152M-00, *Specification for Copper Sheet, Strip Plate and Rolled Bar*. West Conshohocken, PA: ASTM International, 2000.

ASTM D 4068-01, *Specification for Chlorinated Polyethylene (CPE) Sheeting for Concealed Water-Containment Membrane*. West Conshohocken, PA: ASTM International, 2001.

ASTM F 409-99a, *Specification for Thermoplastic Accessible and Replaceable Plastic Tube and Tubular Fittings*. West Conshohocken, PA: ASTM International, 1999.

CSA B45.1-99, *Ceramic Plumbing Fixtures*. Toronto, Ontario: Canadian Standards Association, 1999.

CSA B45.2-99, *Enameled Cast Iron Plumbing Fixtures*. Toronto, Ontario: Canadian Standards Association, 1999.

CSA B45.3-99, *Porcelain Enameled Steel Plumbing Fixtures*. Toronto, Ontario: Canadian Standards Association, 1999.

CSA B45.4-99, *Stainless Steel Plumbing Fixtures*. Toronto, Ontario: Canadian Standards Association, 1999.

CSA B45.5-99, *Plastic Plumbing Fixtures*. Toronto, Ontario: Canadian Standards Association, 1999.

NSF 61-2001, *Drinking Water System Components-Health Effects*. Ann Arbor, MI: NSF International, 2001.

Chapter 28: Water Heaters

General Comments

Chapter 28 regulates the design and installation of water heaters and additional safety devices. It provides the criteria necessary to meet installation and location provisions for various types of water heaters based on their energy or fuel sources. Chapter 28 also contains specifications designed to protect the water heater from leaking and causing structural damage or from igniting flammable vapors in garages.

This chapter includes information pertinent to combination water heating and space heating systems as well as requirements designed to maintain the potability of the domestic water distribution system. Chapter 28 contains additional provisions specific to the types and installations of relief valves and their drains.

Purpose

The code requires that hot water be provided to fixtures and equipment used for bathing, washing, and culinary purposes. A water heater is any appliance that heats potable water and supplies it to the hot water distribution system and, hence, to these fixtures.

Improper water heater installations can prove extremely dangerous. There are some documented cases in which water heaters have overheated and exploded, causing them to launch through the roof or upper floors of a dwelling and land hundreds of feet away. As past incidents have demonstrated, explosions can cause severe property damages and even injuries and deaths. For this reason, among others, the code establishes installation provisions for both the installer and the inspector. The manufacturer's installation instructions must be used in conjunction with the code for safe and efficient water heater installation. The safety concerns addressed in the code and the manufacturer's instructions are also designed to minimize the wear and tear on the water heater and the hot water piping system.

SECTION P2801 GENERAL

P2801.1 Required. Each dwelling shall have an approved automatic water heater or other type of domestic water-heating system sufficient to supply hot water to plumbing fixtures and appliances intended for bathing, washing or culinary purposes. Storage tanks shall be constructed of noncorrosive metal or shall be lined with noncorrosive material.

❖ This section requires each dwelling to have an automatic domestic water heater to supply hot water enough for bathing, washing, and culinary purposes. This requirement corresponds to Section R306.4, which establishes a general requirement for dwellings to provide both hot and cold water to such fixtures. This section also requires hot water storage tanks either to be constructed of non-corrosive metal or to be lined with non-corrosive material to prevent deterioration of the water heater's interior.

P2801.2 Installation. Water heaters shall be installed in accordance with this chapter and Chapters 20 and 24.

❖ This section refers to Chapter 20, Boilers/Water Heaters, and Chapter 24, Fuel Gas, for provisions applicable to water heaters.

P2801.3 Location. Water heaters and storage tanks shall be located and connected so as to provide access for observation, maintenance, servicing and replacement.

❖ This section provides for access to all components of a water heater that require observation, inspection, adjustment, servicing, repair, and replacement. Access is also required to conduct startup and shutdown operations.

Consult the manufacturer's installation instructions for clearance specifications for access and sidewall distances. As provided by Section R102.4, the manufacturer's instructions are by default part of this code, since they compliment the standard under which the water heater is listed.

P2801.4 Prohibited locations. Water heaters shall be located in accordance with Chapter 20.

Exceptions:

1. Direct-vent water heaters.
2. Appliances installed in a dedicated enclosure in which all combustible air is taken directly from the outdoors, in accordance with Section M1703. Access to such enclosure shall be through a solid door, weather-stripped in accordance with the exterior door air leakage requirements of the *International Energy Conservation Code* and equipped with an approved self-closing device.

❖ This section refers to Chapter 20 for water heater location requirements. Section M2005.2 prohibits fuel-fired appliances in rooms or spaces where the combustion process could be hampered and could pose a threat to the occupants. Two exceptions provide at least some water heater access from these rooms.

The first exception exempts direct-vent water heaters, which are installed in a sealed, weather-stripped compartment and draw clean air for combustion from outdoors. A direct-vent water heater should not be confused with a sealed combustion chamber-type appliance. Although a direct-vent appliance draws air from outdoors rather than from the room, the heat source is not sealed off completely. Direct-vent water heaters are not exempt from the requirements of Section P2801.6.

The second exception provides for appliances installed in an enclosure in which all combustion air is taken from the outdoors, and the enclosure is equipped with a solid weatherstripped door and a self-closing device. See the commentary for Section M1703.

P2801.5 Required pan. Where water heaters or hot water storage tanks are installed in locations where leakage of the tanks or connections will cause damage, the tank or water heater shall be installed in a galvanized steel pan having a minimum thickness of 24 gage (0.016 inch) (0.4 mm) or other pans for such use. Listed pans shall comply with CSA LC3.

❖ Where leaks from a water heater or storage tank or the associated connections will cause damage to the structure or surrounding property, the tank must be installed in a pan. The pan must be galvanized steel or other material listed for such use. Where listed pans are used they must conform to the requirements of CSA LC3. Such a pan is intended to catch water from a leaking tank, a leaking connection to the tank, or condensate from the tank. The relief valve piping is prohibited from terminating into a safe pan. See Commentary Figure P2801.5. The $^3/_4$-inch-diameter (19.1 mm) pan drain is not capable of collecting and removing the pressurized discharge of the relief valve at full flow conditions.

P2801.5.1 Pan size and drain. The pan shall be not less than 1.5 inches (38 mm) deep and shall be of sufficient size and shape to receive all dripping and condensate from the tank or water heater. The pan shall be drained by an indirect waste pipe having a minimum diameter of $^3/_4$ inch (19 mm) or the outlet diameter of the relief valve, whichever is larger.

❖ The pan is intended to catch condensate from the tank or water leaking from the tank or its connection. The pan must be at least 1.5 inches (38 mm) deep and of a size and shape sufficient to meet the intent of this section.

The pan drain must not connect directly to the drainage system. An indirect waste pipe is necessary to prevent backflow from the drainage system into the pan. The indirect waste pipe must have a minimum $^3/_4$-inch (19.1 mm) diameter, but the diameter must not be less than that of the required relief valve outlet. The relief valve piping is prohibited from terminating into a safe pan and must be piped to the drainage system or safe location.

P2801.5.2 Pan drain termination. The pan drain shall extend full-size and terminate over a suitably located indirect waste receptor or shall extend to the exterior of the building and termi-

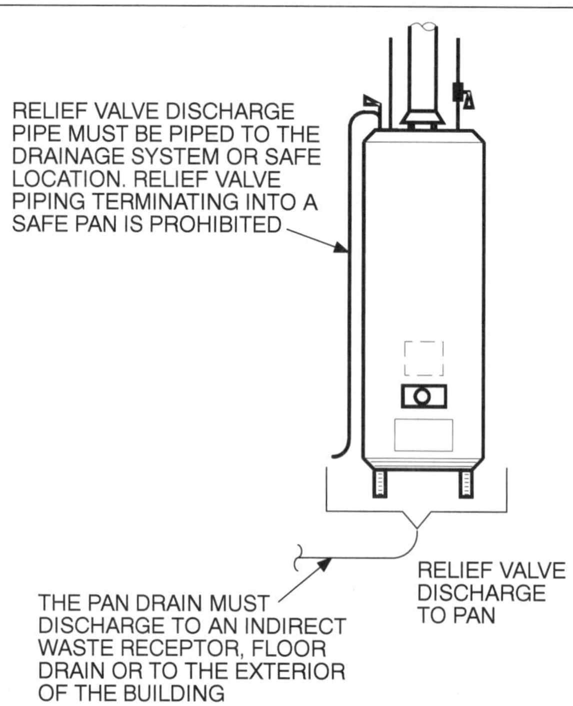

Figure P2801.5
WATER HEATER IN WASTE RECEPTOR SERVING AS PAN

nate not less than 6 inches (152 mm) and not more than 24 inches (610 mm) above the adjacent ground surface.

❖ The pan drain's size must not be reduced, because a reduction would impede the flow. The pan drain must terminate to an indirect waste receptor or a floor drain, or must extend to the building exterior. An air gap or air break must be provided to prevent backflow when the pan drain terminates into an indirect waste receptor or a floor drain.

When the pan drain terminates to the building exterior, the termination must be at least 6 inches (152 mm) but not more than 24 inches (610 mm) above the adjacent ground surface. Locating the termination within this range above the ground surface serves to prevent the pan drain from becoming blocked by vegetation, snow, and ice, while ensuring that the pan drain is low enough not to create a nuisance or safety concern.

P2801.6 Water heaters installed in garages. Water heaters having an ignition source shall be elevated such that the source of ignition is not less than 18 inches (457 mm) above the garage floor.

❖ Unless properly elevated, a fuel-burning water heater installed in a garage poses a serious fire hazard. A garage is commonly used as a storage area for gasoline and common household items such as paint, paint thinner, and solvents. Fumes from these flammable liquids tend to collect near the floor and produce vapors denser than air.

The code language now refers to water heater ignition sources instead of ignition devices. This distinction clarifies that this section addresses all types of water heaters, not just gas- and oil-fired water heaters. This section applies equally to gas-fired water heaters of the direct-vent type. If the water heater were a true sealed combustion chamber type, the elevation requirement would not apply. However, this type of water heater is seldom seen in domestic or most commercial uses. The 18-inch (457 mm) minimum height requirement is intended to prevent an energy or heat source from becoming a source of ignition. The ignition source could be a pilot flame, burner, burner igniter, or electrical component capable of producing a spark or an arc. The 18-inch (457 mm) elevation must be increased where required by the manufacturer's installation instructions. This section effectively prohibits the installation of water heaters directly on the floor of residential garages. The accumulation of flammable vapors more than 18 inches (457 mm) deep is unlikely in most residential garages; therefore, maintaining the water heater ignition source at least 18 inches (457mm) above the floor substantially reduces the risk of explosion and fire.

SECTION P2802 WATER HEATERS USED FOR SPACE HEATING

P2802.1 Protection of potable water. Piping and components connected to a water heater for space heating applications shall be suitable for use with potable water in accordance with Chapter 29. Water heaters that will be used to supply potable water shall not be connected to a heating system or components previously used with nonpotable water heating appliances. Chemicals for boiler treatment shall not be introduced into the water heater.

❖ There are approved combined systems that serve the dual purpose of providing both space heating and domestic hot water. In this case, precautions must be taken to protect the potable water from contamination. Thus, any piping (i.e., under floor hydronic piping system) that is used as part of the domestic potable water system must meet the material, installation, and backflow requirements of Chapter 29. Chemicals for boiler treatment will cause contamination and must not be used in a dual, flow-through potable system.

P2802.2 Temperature control. Where a combination water heater-space heating system requires water for space heating at temperatures exceeding 140°F(60°C), a master thermostatic mixing valve complying with ASSE 1017 shall be installed to temper the water to a temperature of 140°F(60°C) or less for domestic uses.

❖ When a water heater has a dual purpose of supplying hot water and serving as a heat source for a hot water space heating system, the maximum outlet water temperature for the potable hot water distribution system is limited to 140°F (60°C). A master thermostatic mixing valve conforming to ASSE 1017 must be provided to limit the water temperature. These valves are used extensively in applications for domestic service to mix hot and cold water to reduce high service water temperature to the building distribution system. These devices are not intended for final temperature control at fixtures and appliances. See the commentary for Section P2708.3.

A water heater used as part of a space heating system must be protected from any conditions that can cause contamination of the potable water supply system. A typical installation might be an under-floor radiant heating system. Since the water heater is part of the potable water system, materials used in the heating system must be approved for use in a potable water system, and all connections must be protected against contamination. In the summer months when the heating system is inactive, a method to prevent stagnation of the water should be employed. A small orifice in the isolation valves to permit a small amount of water to circulate through the system may be provided for this purpose. Chemicals of any type must not be added to the heating system, as this would directly contaminate the potable water supply. Protection of the potable water supply must be in accordance with Section P2902.

SECTION P2803 RELIEF VALVES

P2803.1 Relief valves required. Appliances and equipment used for heating water or storing hot water shall be protected by:

1. A separate pressure-relief valve and a separate temperature-relief valve; or
2. A combination pressure- and temperature-relief valve.

❖ A combination temperature- and pressure-relief valve or separate temperature-relief and pressure-relief valves are required to protect the water heater against damage and potential explosion. Pressure-relief valves are designed to relieve excessive pressures that can develop in a closed system (the water heater and piping). Temperature-relief valves are designed to open in response to excessive temperatures and to discharge heated water, thereby limiting the water temperature in the water heater. In the event of water heater control malfunction, it is possible for the water pressure to exceed the maximum pressure for which the water heater and piping system were designed. Also, water heater control failures could allow excessive temperature rises that could heat the water above the temperature at which it would vaporize. See Commentary Figure P2803.1.

P2803.2 Rating. Relief valves shall have a minimum rated capacity for the equipment served and shall conform to ANSI Z 21.22.

❖ Relief valves must be manufactured and tested in accordance with ANSI Z21.22. Conformance to the standard indicates to the installer, inspector, and occupant that a similar relief valve has been tested and evaluated and will perform as required when properly installed.

P2803.3 Pressure relief valves. Pressure-relief valves shall have a relief rating adequate to meet the pressure conditions for the appliances or equipment protected. In tanks, they shall be installed directly into a tank tapping or in a water line close to the tank. They shall be set to open at least 25 psi (172 kPa) above the system pressure but not over 150 psi (1034 kPa). The relief-valve setting shall not exceed the tanks rated working pressure.

❖ The code requires an independent pressure-relief valve for a water heater or hot water storage tank to be set to open at least 25 psi (172 kPa) above the system pressure, but not over 150 psi (1034 kPa). Regardless, the valve setting must not exceed the tank's rated working pressure. Because the valve operates on pressure, not temperature, it can be placed anywhere in the water distribution system, preferably close to the water heater or the hot water storage tank. Although the code does not require the pressure-relief valve to be installed on the hot water system downstream of the water heater's water supply shutoff valve, it is best to install it there in case the valve is turned off while the energy or fuel source is still operating.

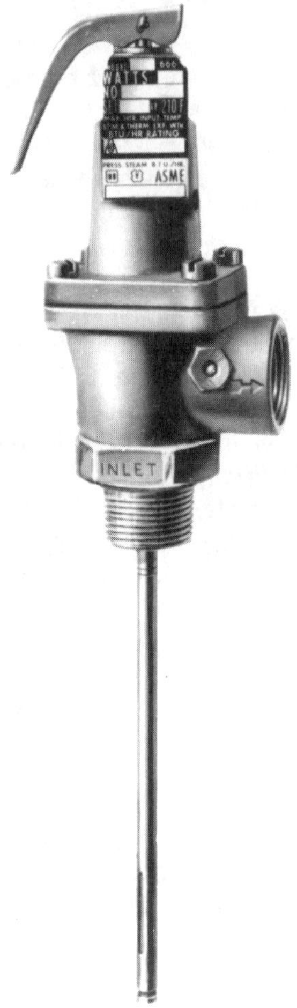

Figure P2803.1
COMBINATION TEMPERATURE AND PRESSURE RELIEF VALVE
Photo courtesy of Watts Regulator Company

P2803.4 Temperature relief valves. Temperature-relief valves shall have a relief rating compatible with the temperature conditions of the appliances or equipment protected. The valves shall be installed such that the temperature-sensing element monitors the water within the top 6 inches (152 mm) of the tank. The valve shall be set to open at a maximum temperature of 210°F (99°C).

❖ The code requires temperature-relief valves to be set to open at a maximum temperature of 210°F (99°C). The highest temperature water settles in the top of the tank in tank-type water heaters. The thermal sensing element of the temperature-relief valve must be installed within 6 inches (152 mm) of the top of the tank. This measure allows the valve to sense the water temperature accurately and to provide an early response to afford maximum protection. The water heater manufacturer's installation instructions usually specify the location, in which case the instructions must be followed. See Commentary Figure P2803.4.

P2803.5 Combination pressure/temperature relief valves. Combination pressure/temperature-relief valves shall comply with all the requirements of separate pressure- and temperature-relief valves.

❖ All of the provisions of Sections P2803.3 and P2803.4 that apply to separate relief valves apply to combination relief valves. Since combination valves operate by sensing changes in temperature, the location requirements of Section P2803.4 apply.

P2803.6 Installation of relief valves. A check or shutoff valve shall not be installed in the following locations:

1. Between a relief valve and the termination point of the relief valve discharge pipe;
2. Between a relief valve and a tank; or
3. Between a relief valve and heating appliances or equipment.

❖ This provision intends to promote proper functioning of the relief valve. A check valve or shutoff valve can restrict or completely cut off the physical connection between the water heater and the relief valve, thus rendering the relief valve ineffective for its designed purpose and causing the installation to be a safety hazard.

P2803.6.1 Requirements of discharge pipe. The outlet of a pressure relief valve, temperature relief valve or combination thereof, shall not be directly connected to the drainage system. The discharge from the relief valve shall be piped full size separately to the floor, to the outside of the building or to an indirect waste receptor located inside the building. In areas subject to freezing, the relief valve shall discharge through an air gap into an indirect waste receptor located within a heated space, or by other approved means. The discharge shall be installed in a manner that does not cause personal injury or property damage and that is readily observable by the building occupants. The discharge from a relief valve shall not be trapped. The diameter of the discharge piping shall not be less than the diameter of the relief valve outlet. The discharge pipe shall be installed so as to drain by gravity flow and shall terminate atmospherically not more than 6 inches (152 mm) above the floor. The outlet end of the discharge pipe shall not be threaded and such discharge pipe shall not have a valve installed.

❖ Relief valves are emergency devices that are not intended to operate continuously. Any discharge must not go unnoticed, because discharge from a relief valve indicates that something is wrong with the system. The termination of a relief valve discharge must be visible for observation so that corrective measures can be taken as necessary.

If a relief valve discharges to a drainage system, the discharge must be an indirect connection through an air gap to prevent backflow from potentially contaminating the potable water system.

The diameter of the discharge pipe must not be reduced or be less than the diameter of the relief valve outlet. Relief valve discharge piping must not be exposed to freezing temperatures, since freezing water could block the pipe and disable the relief valve. The relief valve must first discharge through an air gap into an indirect waste receptor located in a heated space, which, in turn, terminates outside. The discharge pipe must terminate close to the floor level to prevent harm to building occupants. See Commentary Figure P2803.6.1. Water must not be allowed to discharge where it can cause structural decay. Discharge piping must drain by gravity and must not be trapped within the relief piping system. Standing water in the discharge line may become contaminated. To discourage any obstruction in the discharge pipe, the code prohibits the installation of a valve or threads on the outlet end of such pipe.

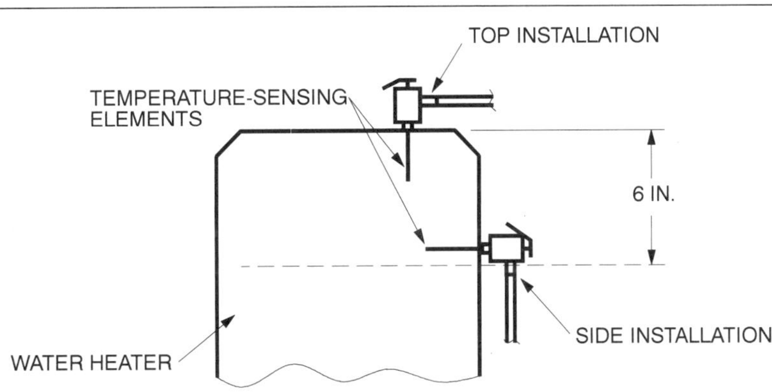

For SI: 1 inch = 25.4 mm.

Figure P2803.4
TEMPERATURE AND PRESSURE RELIEF VALVE INSTALLATION

P2803.6.2 Relief valve drains. Relief valve drains shall comply with Section P2904.5 or ASME A112.4.1.

❖ The discharge pipe for the relief valve must either be of material suitable for potable water systems, as specified in Section P2904.5, or must comply with the provisions of ASME A112.4.1. Although this section allows certain plastic tubing products, the material must be of the rigid type (not coiled) and be suitable for use with hot water distribution piping.

Bibliography

The following resource materials are referenced in this chapter or are relevant to the subject matter addressed in this chapter.

ANSI Z21.10.1-00, *Gas Water Heaters — Volume I – Storage, Water Heaters with Input Ratings of 75,000 Btu per Hour or Less*. New York: American National Standards Institute, 2000.

ANSI Z21.10.3-98, *Gas Water Heaters — Volume III – Storage, Water Heaters with Input Ratings above 75,000 Btu per hour, Circulating and Instantaneous Water Heaters with Z21.10.3a-99 Addendum*. New York: American National Standards Institute, 1998.

ANSI Z21.22-99, *Relief Valves for Hot Water Supply Systems*. New York: American National Standards Institute, 1999.

ASME A112.4.1-1993(R1998), *Water Heater Relief Valve Drain Tubes*. New York: American Society of Mechanical Engineers, 1998.

ASSE 1016-96, *Performance Requirements for Individual Thermostatic, Pressure Balancing and Combination Control Valves for Individual Fixture Fittings*. Westlake, OH: American Society of Sanitary Engineering, 1996.

UL 174-98, *Household Electric Storage Tank Water Heaters—with Revisions through October 1999*. Northbrook, IL: Underwriters Laboratories, Inc., 1998.

UL 732-95, *Oil Fired Storage Tank Water Heaters—with Revisions through January 1999*. Northbrook, IL: Underwriters Laboratories, Inc., 1995.

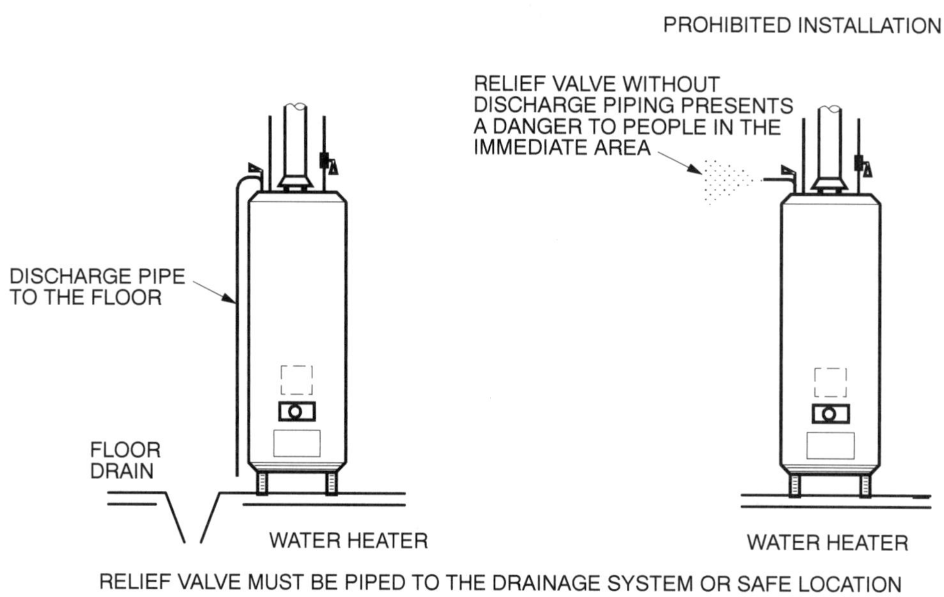

Figure P2803.6.1
RELIEF VALVE DISCHARGE

Chapter 29:
Water Supply and Distribution

General Comments

Chapter 29 regulates the design and installation of water supply systems. The code requires potable water to be provided to dwelling units and contains specific guidelines for maintaining the potability of the water supply. These provisions for protecting the water system from contamination are arguably the most important part of this chapter.

The chapter further outlines the methods that can be used in sizing water supply systems, along with specific precautions that must be adhered to in order to maintain the integrity of the material and its contents. Information specific to the types of materials that are approved for use within a potable water supply system are included, along with provisions relating to the types of connecting components or methods specific to each material.

The type and installation of various drinking water treatment units such as water softeners and water filters are also included.

Purpose

This chapter contains the minimum requirements necessary to design and install the water distribution system. The provisions include materials, installation guidelines, and methods for maintaining the potability of the water system.

A properly functioning water distribution system must be sized such that fluctuations in available pressure are compensated and that the effects of the type of material chosen are taken into consideration. This chapter provides, among other provisions, a prescriptive method for sizing the system to accommodate each fixture according to its intended use, taking into account the possibility of it being used simultaneously with additional fixtures and outlets.

SECTION P2901
GENERAL

P2901.1 Potable water required. Dwelling units shall be provided with a supply of potable water in the amounts and pressures specified in this chapter. In a building where both a potable and nonpotable water-distribution system are installed, each system shall be identified by color marking, metal tag or other appropriate method. Any nonpotable outlet that could inadvertently be used for drinking or domestic purposes shall be posted.

❖ Each dwelling unit must be supplied with sufficient potable water to flush and clean fixtures and be utilized by the occupants of the building. An increased number of installations now utilize nonpotable water (usually reclaimed water) for irrigation and for flushing certain types of fixtures. This section requires that nonpotable water systems be marked by an identifying color. In addition, the potable water system must be marked by a different identifying color so that each system is recognized. In lieu of color, metal tags or other approved methods can be used to identify each system. Nonpotable water outlets must be clearly posted so that they will not inadvertently be used for drinking.

SECTION P2902
PROTECTION OF POTABLE WATER SUPPLY

P2902.1 General. A potable water supply system shall be designed and installed in such a manner as to prevent contamination from nonpotable liquids, solids or gases being introduced into the potable water supply. Connections shall not be made to a potable water supply in a manner that could contaminate the water supply or provide a cross-connection between the supply and source of contamination unless an approved backflow-prevention device is provided. Cross-connections between an individual water supply and a potable public water supply shall be prohibited.

❖ One of the most important missions of a plumbing code is to protect the potable water supply from contamination. Previous incidents indicate that unprotected water supply systems can become contaminated and lead to the spread of contagious diseases. Therefore, the cross-connection control provisions are designed to maintain the water supply system in safe conditions at all times and to protect against actual or potential cross connections. The system must be designed and installed to prevent any contamination from liquid, solids, or gases introduced because of cross connection. This section also prohibits interconnection between a public water supply system and any private system that might be supplying the premises. Private water supply systems are not under the control of a public entity and are thus unregulated. Without monitoring, they are subject to contamination from numerous sources of pollution.

In order to protect the potable water supply, identify potential threats, and determine the appropriate backflow prevention method, an understanding of how backflow prevention can occur is necessary. Backflow depends on atmospheric pressure, water pressures, and variations in the water supply at any moment. The following discusses pressure and backflow.

ATMOSPHERIC PRESSURE: The pressure exerted by the weight of the atmosphere. Atmospheric pressure at sea level is 14.7 psi (101 kPa) and varies with altitude.

GAUGE PRESSURE (PSIG): The pressure read on a gauge. The reference point for the gauge is atmospheric pressure equal to zero. In other words, gauge pressure is the amount of pressure above or below atmospheric pressure.

ABSOLUTE PRESSURE (PSIG): The sum of atmospheric pressure and gauge pressure.

WATER PRESSURE: A function of the weight and height of a column of water. A cubic foot of water weights 62.4 lb (28.3 kg); the pressure exerted at the base of the cube is 62.4 lb/ft^2 (0.433 lb/m^2). Therefore, the height of a column of water determines the pressure that is exerted at its base.

The diameter of the pipe or vessel has no bearing on this pressure. For example, a 100-foot (30 480 mm) high column of water that is 20 feet (6096 mm) in diameter has the same pressure at its base as a 100-foot high (30 480 mm) column of water that is 1 inch (25.4 mm) in diameter.

Columns of water with a free surface (exposed to atmosphere) will have atmospheric pressure in addition to pressure that is due to the weight of water exerted at its base.

BACKFLOW: Flow of liquids in potable water distribution piping that is in reverse of its intended path. There are two types of pressure conditions that cause backflow: backsiphonage and backpressure.

BACKSIPHONAGE: Backflow of water that is due to system pressure falling below atmospheric pressure. Atmosphere supplies the force that reverses flow. Any gauge reading of pressure below atmospheric pressure will be negative.

To illustrate the effect atmospheric pressure has on water, assume that an open tube is inserted in a container of water. Atmospheric pressure acts equally on the surface of the water within the tube and outside the tube.

If the tube is sealed and a vacuum pump is used to evacuate all the air from the tube, a vacuum with a pressure of 0 psig (0 kPa) is created within the tube. Because liquids will flow toward the point of lowest pressure, water in the tube will rise to 33.9 feet (10 333 mm) above the surface of the water in the container. This height is equal to the height that can be supported by the atmospheric pressure. Standard atmospheric pressure is 14.7 psi (101 kPa) and will support a column of water 33.9 feet (103 32 mm) high. This column of water is equal in weight to 14.7 psi (101 kPa) times the area of the tube.

Backsiphonage occurs in the water distribution system when supply pressure falls below atmospheric pressure. The following are practical examples of conditions that can cause backsiphonage.

A simple siphon occurs when pressure is reduced as a result of a difference in water levels at two separate points within a continuous fluid system. See Commentary Figure P2902.1(1).

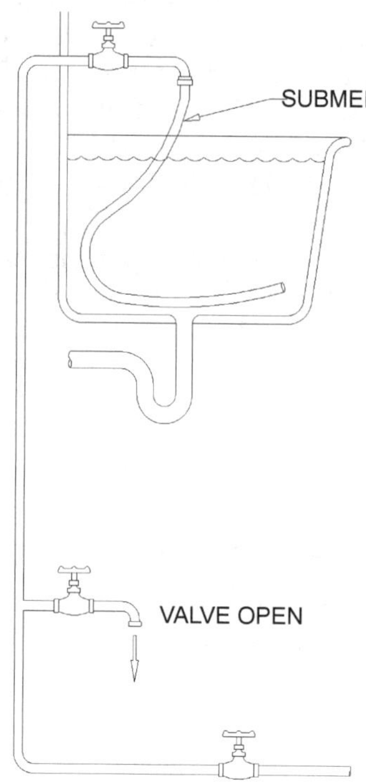

Figure P2902.1(1)
EXAMPLE OF SIMPLE SIPHON

Aspiration or venturi principle is achieved when reduced pressure is created within a fluid system as a result of fluid motion. Intentional or unintentional constriction of flow through an opening or pipe increases velocity and decreases pressure. If this point of reduction is linked to a source of pollution, backsiphonage of the pollutant can occur. See Commentary Figure P2902.1(2).

Supply pressure can be reduced at the pump intake or suction side. Insufficient water pressure or undersized piping supplying the pump can cause a negative pressure and allow contaminated water to be drawn into the system. See Commentary Figure P2902.1(3).

BACKPRESSURE: pressure created in a nonpotable system in excess of the water supply mains, causing backflow. Backpressure may be created by mechanical means such as a pump, by static head

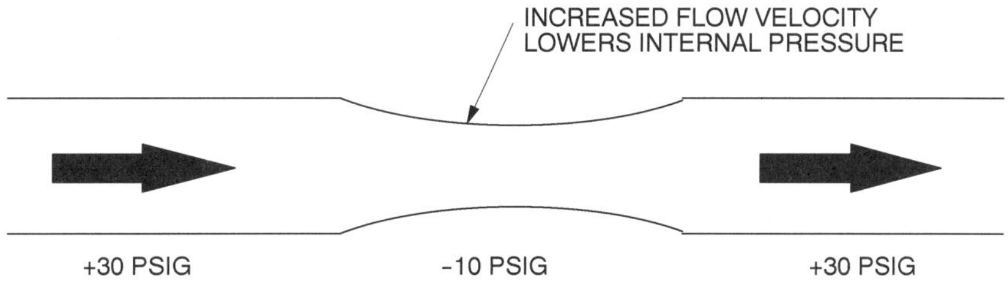

For SI: 1 pound per square inch = 6.895 kPa.

Figure P2902.1(2)
EXAMPLE OF ASPIRATION OR VENTURI PRINCIPLE

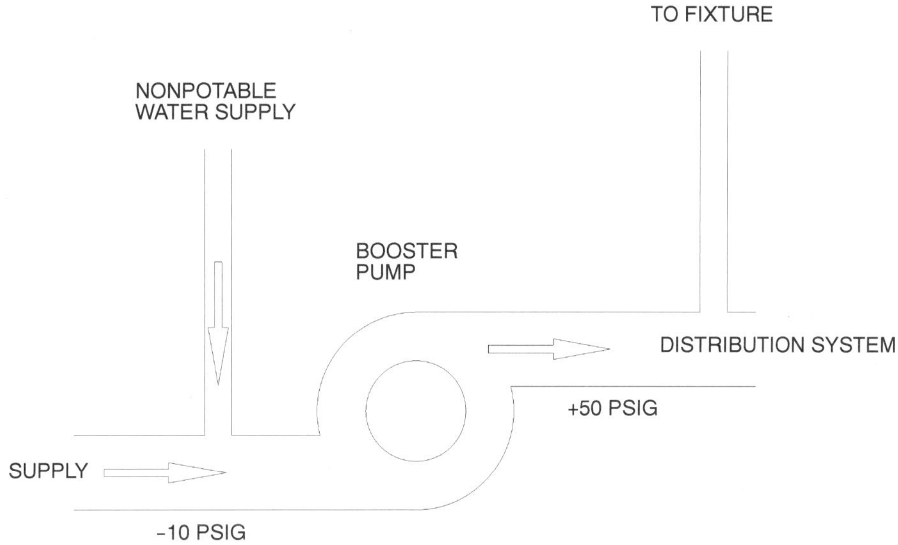

For SI: 1 pound per square inch = 6.895 kPa

Figure P2902.1(3)
EXAMPLE OF PUMP CAUSING NEGATIVE PRESSURE

pressure including an elevated tank, or by thermal expansion from a heat source such as a water heater.

P2902.2 Backflow protection. A means of protection against backflow shall be provided in accordance with Sections P2902.2.1 through P2902.2.6. Backflow prevention applications shall conform to Table P2902.2, except as specifically stated in Sections P2902.2 through P2902.4.

❖ This section refers the reader to six subsections, each detailing a type of approved backflow preventer that can be used, dependent on the type of installation and commensurate with the degree of hazard.

P2902.2.1 Air gaps. Air gaps shall comply with ASME A112.1.2 and air gap fittings shall comply with ASME A112.1.3. The minimum air gap shall be measured vertically from the lowest end of a water supply outlet to the flood level rim of the fixture or receptor into which such potable water outlets discharge. The minimum required air gap shall be twice the diameter of the effective opening of the outlet, but in no case less than the values specified in Table P2902.2.1. An air gap is required at the discharge point of a relief valve or piping. Air gap devices shall be incorporated in dishwashing and clothes washing appliances.

❖ The table referred to in this section has been a mainstay in the plumbing industry since 1940. The values specified have not changed in more than 50 years.

The minimum air gap distance must be twice the diameter of the effective opening. See Commentary Figure P2902.2.1(1).

This section refers the reader to Table P2902.2.1 for distance requirements from side walls. See the commentary for Table P2902.2.1.

Further provisions specify minimum air gap requirements. An air gap is required at the discharge point of a relief valve and within dishwashers and clothes washers. See the commentaries for Sections P2803.6.1, P2717.1 and P2718.1.

TABLE P2902.2
APPLICATION FOR BACKFLOW PREVENTERS

DEVICE	DEGREE OF HAZARD[a]	APPLICATION[b]	APPLICABLE STANDARDS
Air gap	High or low hazard	Backsiphonage or backpressure	ASME A112.1.2
Air gap fittings for use with plumbing fixtures, appliances and appurtenances	High or low hazard	Backsiphonage or backpressure	ASME A112.1.3
Antisiphon-type fill valves for gravity water closet flush tanks	High hazard	Backsiphonage only	ASSE 1002 CSA CAN/CSA B125
Backflow preventer with intermediate atmospheric vents	Low hazard	Backpressure or backsiphonage Sizes $^1/_4''$ - $^3/_4''$	ASSE 1012 CSA CAN/CSA-B64.3
Double check backflow prevention assembly and double check fire protection backflow prevention assembly	Low hazard	Backpressure or backsiphonage Sizes $^3/_8''$ - 16″	ASSE 1015 AWWA C510
Double check detector fire protection backflow prevention assemblies	Low hazard	Backpressure or backsiphonage (Fire sprinkler systems) Sizes 2″ - 16″	ASSE 1048
Dual-check-valve-type backflow preventer	Low hazard	Backpressure or backsiphonage Sizes $^1/_4''$ - 1″	ASSE 1024
Hose connection backflow preventer	High or low hazard	Low head backpressure, rated working pressure backpressure or backsiphonage Sizes $^1/_2''$-1″	ASSE 1052
Hose-connection vacuum breaker	High or low hazard	Low head backpressure or backsiphonage Sizes $^1/_2''$, $^3/_4''$, 1″	ASSE 1011 CSA CAN/CSA-B64.2
Laboratory faucet backflow preventer	High or low hazard	Low head backpressure and backsiphonage	ASSE 1035 CSA B64.7
Pipe-applied atmospheric-type vacuum breaker	High or low hazard	Backsiphonage only Sizes $^1/_4''$ - 4″	ASSE 1001 CSA CAN/CSA-B64.1.1
Pressure vacuum breaker assembly	High or low hazard	Backsiphonage only Sizes $^1/_2''$ - 2″	ASSE 1020
Reduced pressure detector fire protection backflow prevention assemblies	High or low hazard	Backsiphonage or backpressure (Fire sprinkler systems)	ASSE 1047
Reduced pressure principle backflow preventer and reduced pressure principle fire protection backflow preventer	High or low hazard	Backpressure or backsiphonage Sizes $^3/_8''$ - 16″	ASSE 1013 AWWA C511 CSA CAN/CSA B64.4
Spillproof vacuum breaker	High or low hazard	Backsiphonage only Sizes $^1/_4''$-2″	ASSE 1056
Vacuum breaker wall hydrants, frost-resistant, automatic draining type	High or low hazard	Low head backpressure or backsiphonage Sizes $^3/_4''$, 1″	ASSE 1019 CSA CAN/CSA-B64.2.2

For SI: 1 inch = 25.4 mm.

a. Low hazard—See Pollution (Section 202). High hazard—See Contamination (Section 202).

b. See Backpressure (Section 202). See Backpressure, Low Head (Section 202). See Backsiphonage (Section 202).

❖ This table is to be used to assist in the determination of the type of backflow prevention devices to be installed in a water distribution system. The "DEGREE OF HAZARD" is the level of risk to which the water distribution system is subjected, and the "APPLICATION" lists the type of backflow for which the device is rated. This table contains general information about various types of backflow preventers. There are instances throughout Section P2902 where specific provisions override the general information in Table P2902.2. For example, Section P2902.4.3 states that potable water supply to lawn irrigation systems may be protected by an atmospheric-type or pressure-type vacuum breaker that protects against backsiphonage. In addition, where valved sprinklers or zone controls are utilized, a pressure-type vacuum breaker is required because a valve installed downstream from an atmospheric vacuum breaker could render this device inoperative. However, Table P2902.2 lists several devices rated for backsiphonage. Therefore, in order to avoid the apparent conflict, the rule of thumb is that the specific code provision overrides the general classifications given in the table.

TABLE P2902.2.1
MINIMUM AIR GAPS

FIXTURE	MINIMUM AIR GAP Away from a wall[a] (inches)	MINIMUM AIR GAP Close to a wall (inches)
Effective openings greater than 1 inch	Two times the diameter of the effective opening	Three times the diameter of the effective opening
Lavatories and other fixtures with effective opening not greater than $^1/_2$ inch in diameter	1	1.5
Over-rim bath fillers and other fixtures with effective openings not greater than 1 inch in diameter	2	3
Sink, laundry trays, gooseneck back faucets and other fixtures with effective openings not greater than $^3/_4$ inch in diameter	1.5	2.5

For SI: 1 inch = 25.4 mm.

a. Applicable where walls or obstructions are spaced from the nearest inside edge of the spout opening a distance greater than three times the diameter of the effective opening for a single wall, or a distance greater than four times the diameter of the effective opening for two intersecting walls.

❖ The middle column of the table provides minimum air gap dimensions for outlets located away from walls or obstructions. The third column provides more stringent air gap dimensions for outlets located near walls or obstructions. See Commentary Figures P2902.2.1(2) and P2902.2.1(3).

ASME A112.1.2 is the standard for air gaps that require specific dimensions; these are listed in Table P2902.2.1. An air gap is not a device and has no moving parts. It is, therefore, the most effective and dependable method of preventing backflow and should be used where feasible. However, when this method of separation is utilized, there must be an assurance or at least a level of awareness by the user that the air gap cannot be bypassed or defeated by the addition of a hose or extension from the end of the potable water outlet to the source of contamination. The potable water opening or outlet must terminate at an elevation above the level of the source of contamination. The potable water could only be contaminated if the entire area or room becomes flooded to a depth that would submerge the potable water opening or outlet. An air gap installation must always be constructed so that the air gap remains permanently fixed.

ASME A112.1.3 is the standard for air gap fittings for use with plumbing fixtures, appliances and appurtenances. This standard was developed for applications where an air gap fitting is manufactured for appliance applications.

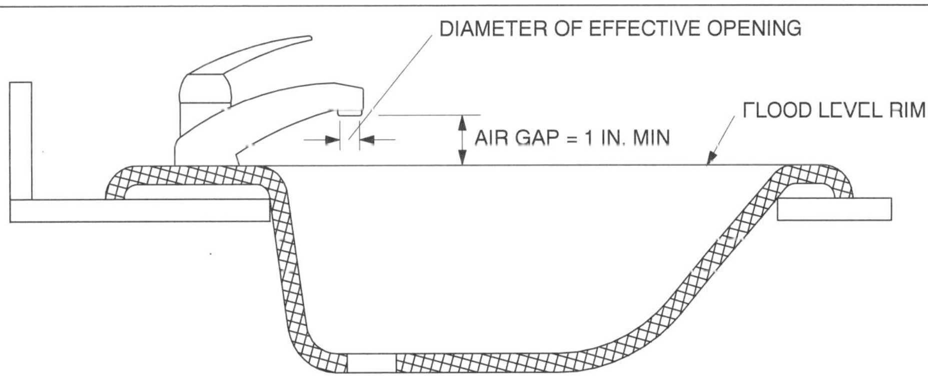

For SI: 1 inch = 25.4 mm.

Figure P2902.2.1(1)
AIR GAP FOR A LAVATORY FAUCET

P2902.2.2 Atmospheric-type vacuum breakers. Pipe-applied atmospheric-type vacuum breakers shall conform to ASSE 1001 or CSA CAN/CSA B64.1.1. Hose connection vacuum breakers shall conform to ASSE 1011, ASSE 1019, ASSE 1035, ASSE 1052, CSA CAN/CSA B64.2, CSA CAN/CSA B64.2.2, CSA B64.7. These devices shall operate under normal atmospheric pressure when the critical level is installed at the required height.

❖ An atmospheric vacuum breaker is designed to prevent siphonic action from occurring downstream of the device. When siphon action or a vacuum is applied to the water supply, the vacuum breaker opens to the atmosphere, allowing air to enter the piping system, thus breaking the vacuum. This prevents contamination of the potable water system by stopping the backsiphonage backflow. When a siphon occurs on the supply inlet of the vacuum breaker (see Commentary Figure

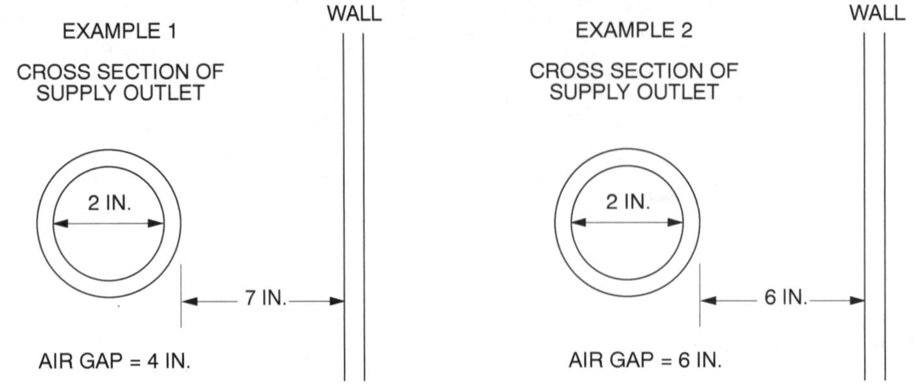

For SI: 1 inch = 25.4 mm.

Figure P2902.2.1(2)
AIR GAPS-LOCATED NEAR SINGLE WALL

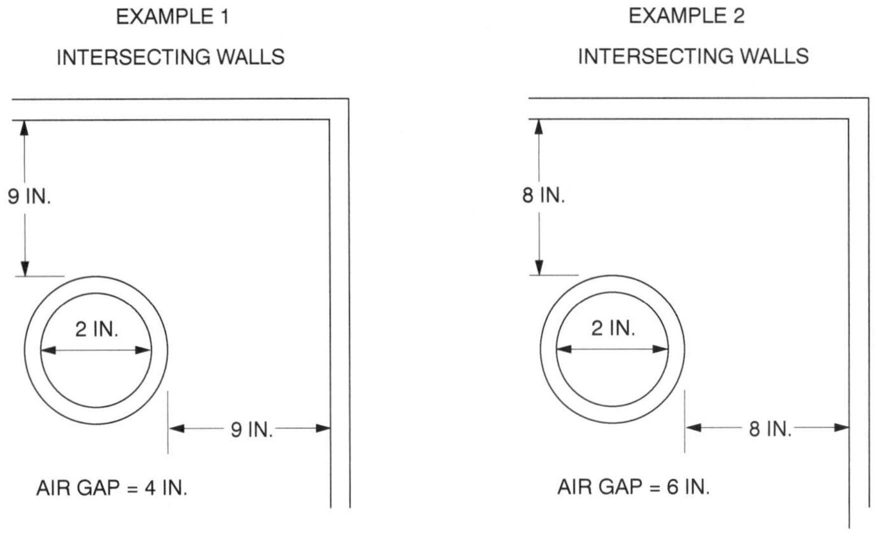

For SI: 1 inch = 25.4 mm.

Figure P2902.2.1(3)
AIR GAPS-LOCATED NEAR INTERSECTING WALLS

P2902.2.2), the disc float drops over the opening. Air enters the atmospheric port opening allowing the remaining water in the piping downstream from the vacuum breaker to drain.

These devices are not effective in preventing backflow resulting from backpressure. The outlet of atmospheric vacuum breakers must remain open to the atmosphere by terminating with a pipe, spout, or similar unobstructed opening. Valves must not be installed downstream of this device, as this would subject the device to supply pressure, thereby rendering it inoperative.

An atmospheric-type vacuum breaker:

- Provides protection against low-hazard or high-hazard contamination.
- Provides protection against backsiphonage only.
- Cannot be installed where it is under continuous pressure from the water supply (12-hour or less intervals).
- Has a critical level of installation of 6 inches (152 mm) above highest point of use downstream.
- Must be accessible for inspection and replacement.

Hose-connection backflow preventers are designed to be installed on the discharge side of a hose-threaded outlet on a potable water system. This two-check device protects against backflow due to backsiphonage and low-head backpressure, under the high-hazard conditions present at a hose-threaded outlet. This device must be used only on systems where sources of backpressure are not introduced and where low-head backpressure does not exceed that generated by an elevated hose equal to or less than 10 feet (3048 mm) in height.

P2902.2.3 Backflow preventer with intermediate atmospheric vent. Backflow preventers with intermediate atmospheric vents shall conform to ASSE 1012 or CSA CAN/CSA B64.3. These devices shall be permitted to be installed where subject to continuous pressure conditions. The relief opening shall discharge by air gap and shall be prevented from being submerged.

❖ A backflow preventer with intermediate atmospheric vent is a nontestable device consisting of two spring-loaded check valves in the closed position and a vent to atmosphere between the two check valves. The design and principle of operation of these backflow preventers are similar to those of a reduced pressure principle backflow preventer. Supply pressure keeps the atmospheric vent closed, but zero supply pressure or backsiphonage will open the inner chamber to the atmosphere.

With a flow through the valve (see Commentary Figure P2902.2.3), the primary check (1) opens away from the diaphragm seal (2). The atmospheric vent remains closed by deflection of the diaphragm seal (2). The secondary check (4) opens away from downstream seal (5), permitting flow of water through the valve.

A backflow preventer with intermediate atmospheric vent:

- Provides protection against low-hazard contamination.
- Provides protection against backsiphonage and backpressure backflow.
- Must have its relief vent opening discharge through an air gap.
- Can be used where it is under continuous pressure from the water supply.
- Cannot be installed below grade where it may be subject to submersion.
- Must be accessible for inspection and replacement.

Provisions must be made at or near the location of the installation to prevent drainage from the relief vent opening causing damage to the structure.

P2902.2.4 Pressure-type vacuum breakers. Pressure-type vacuum breakers shall conform to ASSE 1020 and spill-proof vacuum breakers shall comply with ASSE 1056. These devices are designed for installation under continuous pressure conditions when the critical level is installed at the required height. Pressure-type vacuum breakers shall not be installed in locations where spillage could cause damage to the structure.

❖ A pressure-type vacuum breaker contains one or two independently operating spring-loaded check valves and an independently operating spring-loaded air inlet valve located on the discharge side of the check valve. This device is testable, as it has a shutoff valve and test cock at each end. See Commentary Figure P2902.2.4. While a pressure-type vacuum breaker operates by the same principle as an atmospheric vacuum breaker, a pressure-type vacuum breaker is spring activated. The water pressure acting against the spring closes off the air opening. Because of the spring activation, a valve can be located downstream of a pressure vacuum breaker.

A pressure-type vacuum breaker:

- Provides protection against low-hazard and high-hazard contamination.
- Provides protection against backsiphonage only.

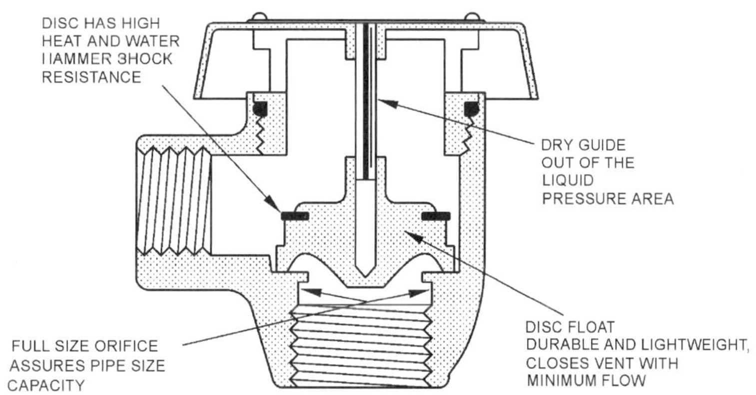

Figure P2902.2.2

ATMOSPHERIC VACUUM BREAKER

Figure courtesy of Watts Regulator Company

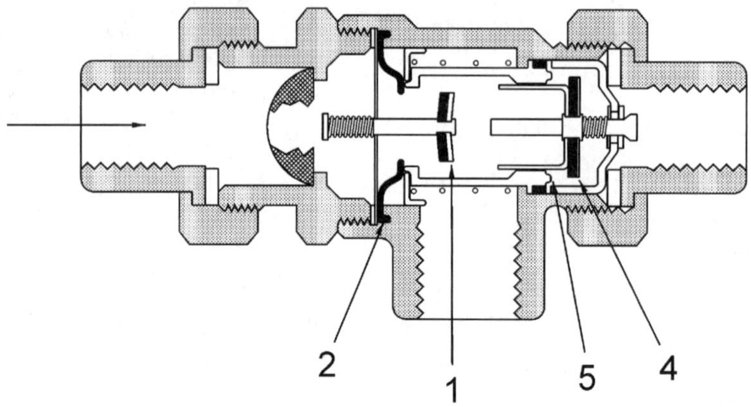

Figure P2902.2.3
DOUBLE CHECK VALVE WITH INTERMEDIATE ATMOSPHERE VENT
Figure courtesy of Watts Regulator Company

- Can be used where it is under continuous pressure from the water supply (valving permitted downstream).
- Has a critical level of installation of 6 inches (152 mm) above flood level rim.
- Must be accessible for testing and inspection.

Provisions must be made at or near the location of the installation to prevent leakage from the air vent opening causing damage to the structure when the device is installed in a building or structure.

ASSE 1056 was developed to specifically address indoor applications offering the same vacuum breaker capabilities of ASSE 1020. However, water discharges each time the valve is pressurized. As with ASSE 1020 valves, backflow protection is achieved against backsiphonage by a check valve backed up with an air inlet vent that opens in response to a loss of supply pressure. Such valves are not for use in any system where backpressure is applied to the device. Where the system is pressurized, the vent closes to prevent flow through the upstream check valve and to eliminate vent spillage.

P2902.2.5 Reduced pressure principle backflow preventer. Reduced pressure principle backflow preventers shall conform to ASSE 1013, AWWA C511 or CSA CAN/CSA B64.4. Reduced pressure detector assembly backflow preventers shall conform to ASSE 1047. These devices are permitted to be installed where subject to continuous pressure conditions. The relief opening shall discharge by air gap and shall be prevented from being submerged.

❖ A reduced pressure principle backflow preventer is considered to be the most reliable mechanical method to prevent backflow. These devices consist of dual, independently acting, spring-loaded check valves that are separated by a chamber or "zone" equipped with a relief valve. See Commentary Figure P2902.2.5. The pressure downstream of the device and in the central chamber between the check valves is maintained at a minimum of 2 psi (13.8 kPa) less than the potable water supply pressure at the device inlet, hence the name "reduced pressure principle." The relief valve located in the central chamber is held closed by the pressure differential between the inlet supply pressure and the central chamber pressure. If the inlet supply pressure decreases, the device will discharge water from the central chamber thus maintaining a lower pressure in the chamber. In the event of backpressure or negative supply pressure, the relief valve will open to the atmosphere and drain any backflow that has leaked through the check valve. The relief vent will also allow air to enter to prevent any siphonage.

During a static or no-flow condition, both check valves and the relief vent remain closed. When there is normal flow, the two check valves open and the relief vent port remains closed. The supply pressure drops across the valve because of the force required to open the spring-loaded check valves. If backsiphonage or backpressure occurs, both check valves close and the relief vent opens to the atmosphere, discharging the water in the intermediate zone. If both check valves are fouled and held in the open position, and there is either a backsiphonage or backpressure or both, the relief vent port opens, discharging the possibly contaminated water to the atmosphere and allowing air to enter to break the siphon. This gives a visual indication of a malfunction of one or both of the check valves or relief valve.

ASSE 1013 is the standard for two types of backflow prevention assemblies, identified as "reduced pressure principle backflow preventers" (RP) and "reduced pressure principle fire protection backflow preventers" (RPF). An RP device is designed for a working pressure of 150 psi (1024 kPa) and an RPF device is designed for a working pressure of 175 psi (1207 kPa). The RP and RPF are identical in their backflow protection. The RPF has specific performance requirements relating to their use on fire protection systems.

ASSE 1047 is the standard for devices known as "reduced pressure detector fire protection backflow prevention assemblies" (RPDF). This device is primarily used for the protection of the public water supply from contaminants found in fire sprinkler systems, and it is used to detect low rates of flow up to 0.126 L/s (2 gpm) within the sprinkler system caused by leakage or unauthorized use. This standard also allows provisions for alarm signaling devices to be included in the assembly.

A reduced pressure principle backflow preventer:

- Provides protection against low- or high-hazard contamination.
- Provides protection against backsiphonage and backpressure backflow.
- Can be used where it is under continuous pressure from the water supply.
- Must have its relief vent opening discharge through an air gap.
- Must be accessible for testing and inspection.
- Cannot be installed below grade where it may be subject to submersion.

Provisions must be made at or near the location of the installation to prevent drainage from the relief vent opening causing damage to the structure when the device is installed in a building or structure.

P2902.2.6 Double check valve assemblies. Double check valve assemblies shall conform to ASSE 1015 or AWWA C510. Double-detector check valve assemblies shall conform to ASSE 1048. These devices shall be capable of operation under continuous pressure.

❖ These devices are designed for low-hazard applications subject to backpressure and backsiphonage applications. The devices consist of two independent spring-loaded check valves in series. Test cocks are provided to permit testing of the devices. See Commentary Figure P2902.2.6. Note that these devices must not be confused with dual check-valve devices or two single check valves placed in series.

Double-detector check-valve assemblies are always permitted as an option where double check-valve assemblies are required because they offer identical protection. Double-detector devices provide the user the ability to monitor water flow through the device to determine if there is a system leak or if water is being intentionally withdrawn from the system downstream of the device.

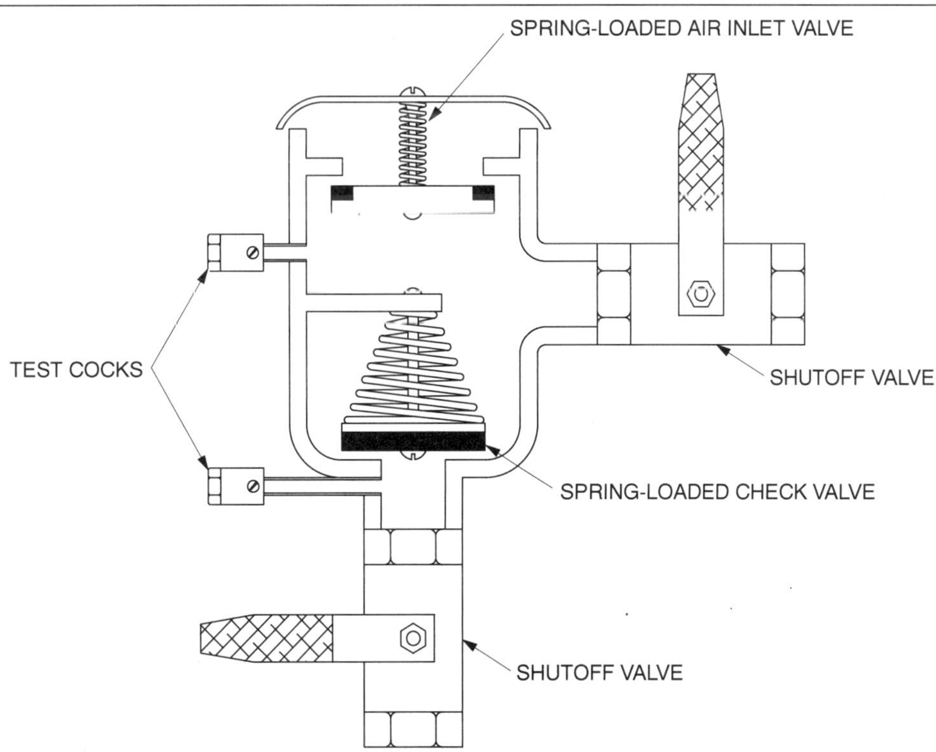

Figure P2902.2.4
PRESSURE VACUUM BREAKER

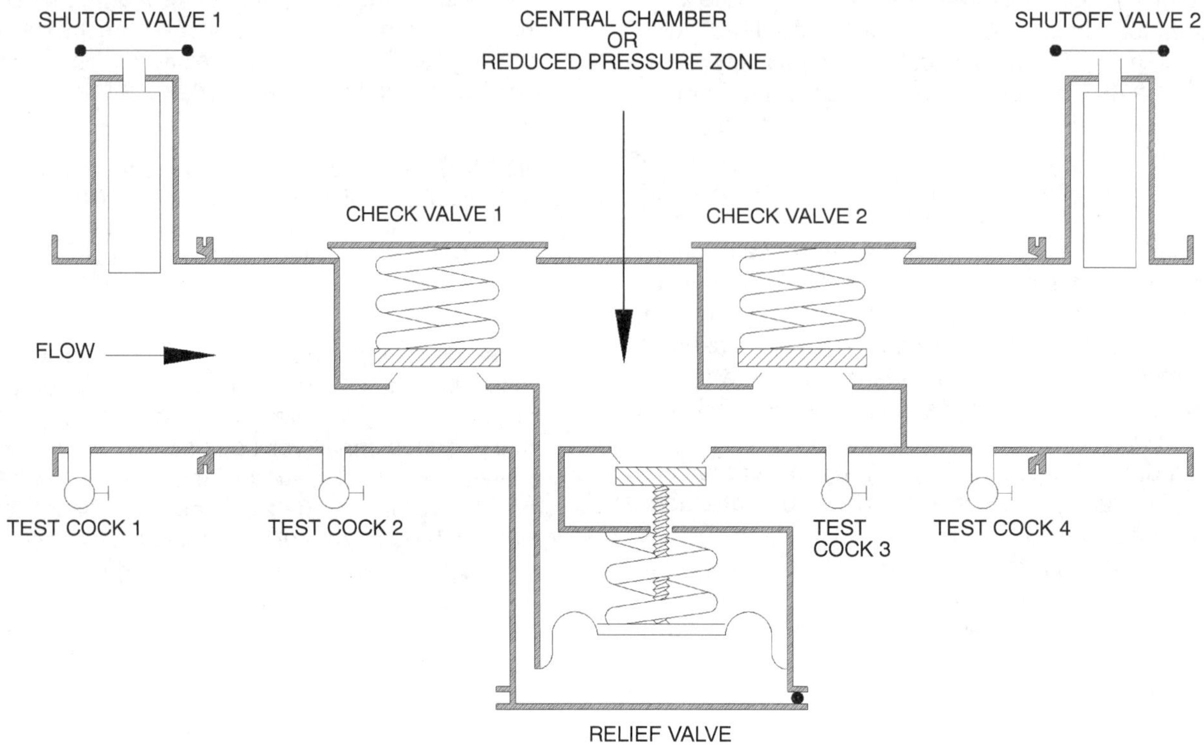

Figure P2902.2.5
REDUCED PRESSURE PRINCIPLE BACKFLOW PREVENTER

Figure courtesy of Watts Regulator Company

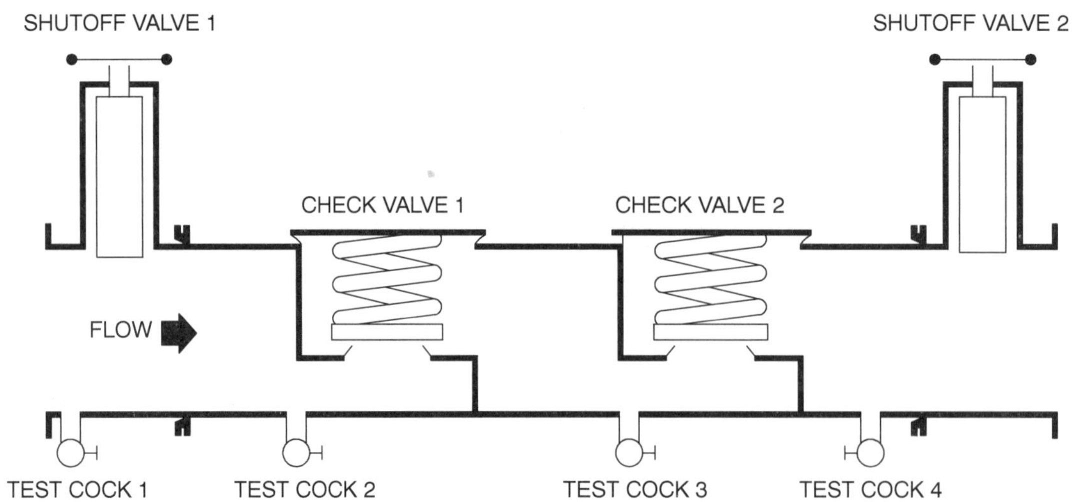

Figure P2902.2.6
DOUBLE CHECK-VALVE ASSEMBLY

P2902.3 Protection of potable water outlets. All potable water openings and outlets shall be protected by an air gap, reduced pressure principle backflow preventer with atmospheric vent, atmospheric-type vacuum breaker, pressure-type vacuum breaker or hose connection backflow preventer.

❖ The potable water system comes in contact with the atmosphere at openings or outlets. These include faucets, hose bibbs, and plumbing appliances. The openings can be protected by an air gap, reduced pressure principle backflow preventer with atmospheric vent, atmospheric-type vacuum breaker, pressure-type vacuum breaker, or hose connection backflow preventer.

P2902.3.1 Fill valves. Flush tanks shall be equipped with an antisiphon fill valve conforming to ASSE 1002 or CSA B125. The fill valve backflow preventer shall be located at least 1 inch (25.4 mm) above the full opening of the overflow pipe.

❖ The fill valve is the refilling component that controls the flow of water into the flush tank. Since the fill valve connects the potable water to the flush tank, it must have adequate protection against backflow. The referenced standards require an antisiphon device in the fill valve. An antisiphon device is a type of vacuum breaker that must be located a minimum of 1 inch (25 mm) above the overflow pipe to prevent its submersion. The 1-inch (25 mm) minimum is equivalent to the vacuum breaker critical level.

P2902.3.2 Deck-mounted and integral vacuum breakers. Approved deck-mounted vacuum breakers and faucets with integral atmospheric or spill-proof vacuum breakers shall be installed in accordance with the manufacturer's installation instructions and the requirements for labeling with the critical level not less than 1 inch (25.4 mm) above the flood level rim.

❖ These devices are installed on the decks of plumbing fixtures such as custom bathtubs, whirlpools, bidets, or sinks. They need to be installed only 1 inch (25.4 mm), instead of 6 inches (152 mm), above the flood rim unless required otherwise by the manufacturer. However, these devices must be tested for that reduced elevation so that they function properly and prevent backsiphonage. Faucets with integral atmospheric or spill-proof vacuum breakers are also required to be installed 1 inch (25.4 mm) above the flood rim of the fixture served. The manufacturer's installation instructions must be followed so that the devices will function as intended.

P2902.3.3 Hose connection. Sillcocks, hose bibbs, wall hydrants and other openings with a hose connection shall be protected by an atmospheric-type or pressure-type vacuum breaker or a permanently attached hose connection vacuum breaker.

Exceptions:

1. This section shall not apply to water heater and boiler drain valves that are provided with hose connection threads and that are intended only for tank or vessel draining.
2. This section shall not apply to water supply valves intended for connection of clothes washing machines where backflow prevention is otherwise provided or is integral with the machine.

❖ Sillcocks, hose bibbs, wall hydrants, and other openings with hose connections must be protected by either an atmospheric or pressure-type vacuum breaker or a permanently attached hose connection vacuum breaker. See Commentary Figure P2902.3.3. There is danger of backsiphonage with any outlet that has a hose connected to it if the hose can be submerged. Therefore, a device that will stop the siphonage action is required. The type of device will depend on whether a shutoff valve is installed upstream or downstream of the device. Where a shutoff valve is installed upstream of the device, and the device is not subjected to prolonged flow conditions, an atmospheric-type vacuum breaker or a hose connection vacuum breaker can be installed. Where a shutoff valve is installed downstream, or the device is subjected to prolonged flow, a pressure-type vacuum breaker is required.

Backflow prevention devices are not required for fixtures that have integral backflow prevention devices, such as dishwashers, clothes washers, and ice machines.

P2902.4 Protection of potable water connections. All connections to the potable water shall conform to Sections P2902.4.1 through P2902.4.5.

❖ This section regulates direct connections between potable and nonpotable systems that result in a cross-connection. There are instances where specific provisions override the previous general provisions, as indicated in Sections P2902.4.1 through P2902.4.5.

P2902.4.1 Connections to boilers. The potable supply to the boiler shall be equipped with a backflow preventer with an intermediate atmospheric vent complying with ASSE 1012 or CSA CAN/CSA B64.3. Where conditioning chemicals are introduced into the system, the potable water connection shall be protected by an air gap or a reduced pressure principle backflow preventer complying with ASSE 1013, CSA CAN/CSA B64.3 or AWWA C511.

❖ Because boilers are pressurized vessels, the potential for backflow caused by backpressure is high. If the water supply pressure drops below the boiler pressure, backflow can occur. When the boiler contains untreated water (without additives), a backflow preventer with an intermediate atmospheric vent may be installed. However, when chemicals are added to the potable water supply, the connection must be protected with a reduced pressure principle backflow prevention device or by an air gap.

If potable water is used for domestic use in conjunction with space heating, such systems must be protected from conditions that can cause contamination.

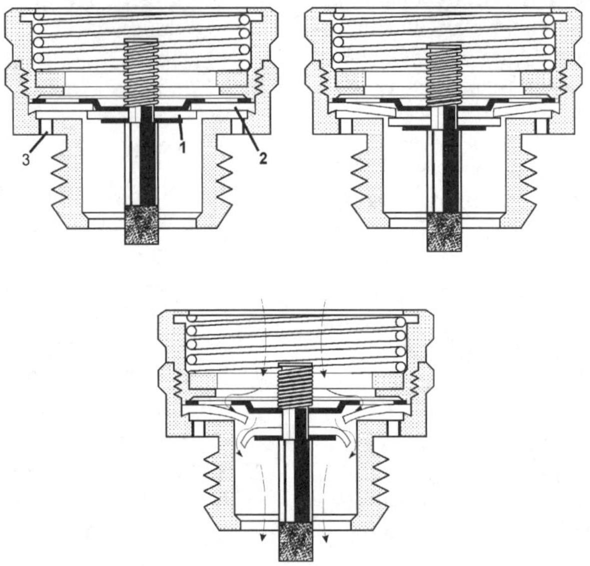

Figure P2902.3.3
HOSE CONNECTION VACUUM BREAKER

Because the water heater or boiler is part of the potable water system, all materials used in the heating system must be approved for use in a potable water system. However, temperature protection is required in accordance with Section P2802.2.

P2902.4.2 Heat exchangers. Heat exchangers utilizing an essentially toxic transfer fluid shall be separated from the potable water by double-wall construction. An air gap open to the atmosphere shall be provided between the two walls. Heat exchangers utilizing an essentially nontoxic transfer fluid shall be permitted to be of single-wall construction.

❖ Heat exchangers are utilized to transfer energy to or from potable water without mixing with the medium carrying the energy to or from the heat exchanger. Because a wall separates the medium from the potable water, care must be exercised so that mixing between the two will not occur, especially when the medium is toxic. If the medium is toxic, double wall heat exchangers with an air gap between the walls vented to atmosphere are required so that any leak from either side will leak to the outside, alerting the occupant that a problem exists. Single wall heat exchangers are also permitted when the heat transfer fluid is not toxic.

P2902.4.3 Lawn irrigation systems. The potable water supply to lawn irrigation systems shall be protected against backflow by an atmospheric-type vacuum breaker, a pressure-type vacuum breaker or a reduced pressure principle backflow preventer. A valve shall not be installed downstream from an atmospheric vacuum breaker. Where chemicals are introduced into the system, the potable water supply shall be protected against backflow by a reduced pressure principle backflow preventer.

❖ Lawn irrigation system outlets can be installed at or below grade, and such systems are not regulated by this code. However, the potable water supply point of connection to the system must be protected against backflow by an atmospheric-type vacuum breaker, a pressure-type vacuum breaker, or a reduced pressure principle backflow preventer. The most common method of protection utilizes pressure-type vacuum breakers. These devices are permitted for all irrigation systems that employ valved sprinklers or zone controls that could render atmospheric-type vacuum breakers inoperative. Where chemicals can be introduced into the irrigation system, the potable water system must be protected by a reduced pressure principle backflow preventer.

P2902.4.4 Connections to automatic fire sprinkler systems. The potable water supply to automatic fire sprinkler systems shall be protected against backflow by a double check-valve assembly or a reduced pressure principle backflow preventer.

Exception: Where systems are installed as a portion of the water distribution system in accordance with the requirements of this code and are not provided with a fire department connection, isolation of the water supply system shall not be required.

❖ Automatic fire sprinkler systems installed in dwellings are either installed as a portion of the water distribution system in accordance with the requirements of this code, in which case there are usually no stagnant portions, or they are completely isolated from the potable water system, in which case there could be areas where water can stagnate and possibly develop pollutants and residue from the piping system. If the entire piping system, including sprinkler piping, is constructed of materials that are approved for potable water, a backflow preventer is not required. However, if the water distribution system and fire sprinkler system share only the main supply (water service) in common, then a double check valve assembly is required to isolate the fire sprinkler system.

P2902.4.4.1 Additives or nonpotable source. Where systems contain chemical additives or antifreeze, or where systems are connected to a nonpotable secondary water supply, the potable water supply shall be protected against backflow by a reduced pressure principle backflow preventer. Where chemical additives or antifreeze is added to only a portion of an automatic fire sprinkler or standpipe system, the reduced pressure principle backflow preventer shall be permitted to be located so as to isolate that portion of the system.

❖ Some fire sprinkler systems incorporate the use of chemicals or additives that can be used to protect the piping system from corrosion or protect it from freezing. In such cases, a reduced pressure principle backflow preventer is required. In cases where sections of the system are chemically treated, only those treated sections need to be isolated.

P2902.4.5 Solar systems. The potable water supply to a solar system shall be equipped with a backflow preventer with intermediate atmospheric vent complying with ASSE 1012 or a reduced pressure principle backflow preventer complying with ASSE 1013. Where chemicals are utilized, the potable water supply shall be protected by a reduced pressure principle backflow preventer.

Exception: Where all solar system piping is a part of the potable water distribution system, in accordance with the requirements of the *International Plumbing Code*, and all components of the piping system are listed for potable water use, cross-connection protection measure shall not be required.

❖ Similar to a fire sprinkler system, solar systems must be equipped with a backflow preventer with an intermediate atmospheric vent conforming to ASSE 1012. If chemicals or additives are used in the solar system, the more reliable reduced pressure principle backflow preventer conforming to ASSE 1013 must be used. Current text requires solar systems to be provided with a form of backflow prevention device. An exception provides for an installation where the solar system piping is a part of the potable water distribution system, and all components of the piping system are listed for potable water use; then cross-connection protection measures are not required.

P2902.5 Access. All backflow prevention devices shall be accessible for inspection and servicing.

❖ Backflow assemblies must be accessible for inspection and servicing. Every backflow preventer must be monitored to confirm that the assembly is providing the necessary backflow protection. The referenced standards usually include specifications for clearance and height limitations with regard to installation and location.

SECTION P2903 WATER-SUPPLY SYSTEM

P2903.1 Water supply system design criteria. The water service and water distribution systems shall be designed and pipe sizes shall be selected such that under conditions of peak demand, the capacities at the point of outlet discharge shall not be less than shown in Table P2903.1.

❖ Each plumbing fixture requires a given flow rate and minimum water supply pressure to properly function. The minimum values are supplied in Table P2903.1. Fixtures not listed in the table must be provided with their minimum requirements as specified by the manufacturer. Minimum flow rate and pressure requirements must be satisfied during peak demand. Peak demand is the maximum amount (flow rate) of water at any given time during the day. Usually, peak demand for residential occupancies is in the morning before 8 A.M. and at night between 9 P.M. and 10 P.M. It is unrealistic to calculate the maximum flow rate if all fixtures are assumed to operate at the same time. In determining the maximum probable flow of peak conditions, the loading effect of each fixture has to be considered. However, this method fails to take into consideration fixtures that are not normally used, effects of water-conserving fixtures, changes in water use habits in the last 60 years, multiple bathroom groups and differences in flow characteristics between various fixtures. When estimating peak demand, this chapter utilizes a sizing method that incorporates the use of water supply fixture units (w.s.f.u.). A w.s.f.u. is a numerical factor on an arbitrary scale that measures the load-producing effects of a single plumbing fixture. Table 2903.1 requires 8 psi (55.152 kPa) residual pressure for most fixtures. Only temperature control showers, flushometer tank water closets, and one-piece tank-type water closets require higher residual pressure to operate properly. This issue becomes important when utilizing the sizing method in this chapter, as the designer will be required to deduct the additional pressure higher than 8 psi (55.152 kPa) from the minimum available street pressure to determine the appropriate sizing table for the proposed system. Note that the lowest pressure in the public main occurs in the summer because of lawn sprinkling and the need to supply make-up water for air-conditioning cooling towers. The lowest pressure must be selected to guarantee a continuous, adequate supply of water.

TABLE P2903.1
REQUIRED CAPACITIES AT POINT OF OUTLET DISCHARGE

FIXTURE AT POINT OF OUTLET	FLOW RATE (gpm)	FLOW PRESSURE (psi)
Bathtub	4	8
Bidet	2	4
Dishwasher	2.75	8
Laundry tub	4	8
Lavatory	2	8
Shower	3	8
Shower, temperature controlled	3	20
Sillcock, hose bibb	5	8
Sink	2.5	8
Water closet, flushometer tank	1.6	15
Water closet, tank, close coupled	3	8
Water closet, tank, one-piece	6	20

For SI: 1 gallon per minute = 3.785 L/m, 1 pound per square inch = 6.895 kPa.

❖ This table specifies the design parameters required for a water distribution system. These minimum values must be satisfied in the design of the system for proper functioning of each plumbing fixture. The design professional should use the table when calculating the pipe sizes as required in Section P2903.7. See the commentary for Section P2903.7.

P2903.2 Maximum flow and water consumption. The maximum water consumption flow rates and quantities for all plumbing fixtures and fixture fittings shall be in accordance with Table P2903.2.

❖ This section conserves water resources by limiting the consumption of water by plumbing fixtures. Water conservation is important not only to conserve our limited water supplies but also to conserve the energy required to pump, transport, or treat water. The relationship between water usage and energy usage is evidenced by the fact that the federal government water conservation requirements are part of the National Energy Policy Act. Table P2903.2 below provides the maximum allowable flow rates for various types of fixtures.

TABLE P2903.2
MAXIMUM FLOW RATES AND CONSUMPTION FOR PLUMBING FIXTURES AND FIXTURE FITTINGS[b]

PLUMBING FIXTURE OR FIXTURE FITTING	PLUMBING FIXTURE OR FIXTURE FITTING
Lavatory faucet	2.2 gpm at 60 psi
Shower head[a]	2.5 gpm at 80 psi
Sink faucet	2.2 gpm at 60 psi
Water closet	1.6 gallons per flushing cycle

For SI: 1 gallon per minute = 3.785 L/m, 1 pound per square inch = 6.895 kPa.

a. A handheld shower spray is also a shower head.

b. Consumption tolerances shall be determined from referenced standards.

❖ The fixtures and faucets listed in this table are those that function in a cyclical manner or that are allowed to consume water continuously for the duration of fixture use. Water is conserved by limiting the flow rate for manually controlled fixtures and by limiting the volume-per-cycle usage for cyclically operating fixtures.

The maximum flow rates for nonmetering showers, lavatories, and sink faucets are dependent on the supply pressure; therefore, a reference pressure is stated for each item in the table. Where the actual supply pressure is less than the pressure in the table, the actual flow rate of the faucet or shower head will be less than the maximum flow rate in the table.

P2903.3 Minimum pressure. Minimum static pressure (as determined by the local water authority) at the building entrance for either public or private water service shall be 40 psi (276 kPa).

❖ To allow for a properly designed system, the code requires a minimum street supply pressure (from the local water purveyor) of 40 psi (275 66 kPa). Private water systems must also be capable of supplying this minimum pressure to the building. Lower pressures would not provide the quantity of flow that is required at the plumbing fixtures to flush, operate and cleanse fixtures and fixture fittings. Sufficient flow is also necessary to fill plumbing fixtures in a reasonable amount of time for convenience purposes.

P2903.3.1 Maximum pressure. Maximum static pressure shall be 80 psi (551 kPa). When main pressure exceeds 80 psi (551 kPa), an approved pressure-reducing valve conforming to ASSE 1003 shall be installed on the domestic water branch main or riser at the connection to the water-service pipe.

❖ The piping system, components and fixtures in a water distribution system are designed to withstand a given pressure. The code established 80 psi (551.52 kPa) as the maximum working pressure of the system. Where the pressure from the public water system or the individual private water supply system exceeds 80 psi (551.52 kPa), a pressure reducing valve conforming to ASSE 1003 must be installed at the point where the water service enters the building.

P2903.4 Thermal expansion. In addition to the required pressure relief valve, an approved device for thermal expansion control shall be installed on any water supply system utilizing storage water heating equipment whenever the building supply pressure exceeds the pressure-reducing valve setting or when any device, such as a pressure-reducing valve, backflow preventer or check valve, is installed that prevents pressure relief through the building supply. The thermal expansion control device shall be sized in accordance with the manufacturer's installation instructions.

❖ As the water in a water heater is heated to higher temperatures, it expands in volume. Usually, the water expands into the water service and the public water main. This prevents an increase in water pressure in the building distribution system. If the water cannot expand, dangerous high pressure will develop, which can damage the piping or the water heater. The code therefore mandates that in addition to the required pressure relief valve, an approved device for thermal expansion control must be installed, especially if the building supply pressure is greater than the relief valve pressure setting. This is because, even if the regulator is equipped with a bypass to let the water expand into the public water main, the high pressure in the water main will not allow the expansion until the pressure on the building side of the regulator reaches this much higher level. A thermal expansion (compression) tank is used under these conditions. The tank must be adequately sized to prevent major fluctuations in the system pressure. When the street pressure is less than that of the relief valve setting but is higher than 80 psi (551.52 kPa), necessitating a pressure reducing valve, the same problem exists because the pressure reducing valve (regulator) will act as a barrier, preventing the hot water from expanding back through the device check valve. In this instance, either the regulator must be equipped with an integral bypass to allow the reverse flow of expanding water to the public main, or an expansion tank must be installed downstream of the pressure reducing valve. Where a backflow prevention device or check valve is installed on a water supply system utilizing storage water heating equipment such that thermal expansion causes an increase in pressure, a device for controlling pressure must be installed. Thermal expansion of water is minimal where tankless water heaters are used. These water heaters are flow activated and shut off when the water usage stops. Usually, an expansion tank or an auxiliary pressure relief device is installed somewhere in the system to prevent major fluctuations in the system pressure, whereas a pressure relief device will allow greater pressure variations. A relief device also wastes water each time it discharges. An antisiphon fill valve with built-in relief valve assembly limits increased system pressure to 80 psi (551.52 kPa) by relieving into the water closet. In all cases where a double check valve backflow preventer or a reduced pressure principle backflow preventer is installed on the water supply, thermal expansion must be addressed. Some water meters are also equipped with check valves that necessitate expansion control. See Commentary Figure P2903.4.

P2903.5 Water hammer. The flow velocity of the water distribution system shall be controlled to reduce the possibility of water hammer. A water-hammer arrestor shall be installed where quick-closing valves are utilized. Water-hammer arrestors shall be installed in accordance with manufacturers' specifications. Water-hammer arrestors shall conform to ASSE 1010.

❖ Water hammer is an effect of hydraulic shock and is a symptom of a dangerous condition. Hydraulic shock happens when fluid flowing through a pipe is subjected to sudden and rapid changes in velocity. Water hammer is a phenomenon that results from the dissipation of the kinetic energy of flowing water. The dissipation of energy occurs in the form of a pressure surge that produces a hydraulic shockwave traveling at extremely high velocity. The pressure surge is sometimes accompanied by a banging noise within the piping system, hence the expression "water hammer." Any time flowing water is abruptly stopped, water hammer occurs, whether heard or not. To prevent any adverse effect from water hammer, the code requires the intensity of the water hammer to be controlled by installing a

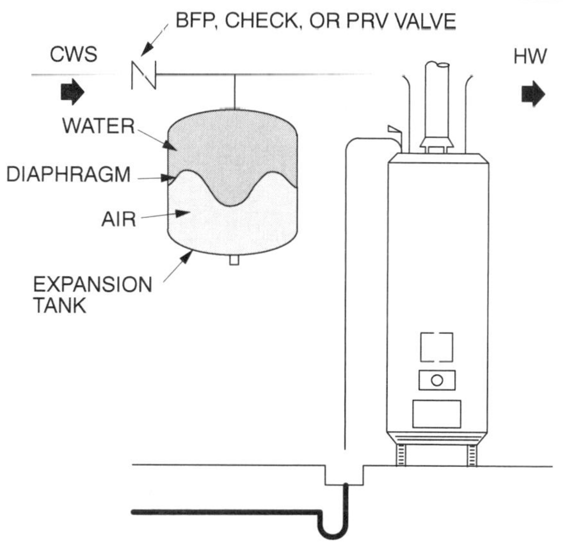

Figure P2903.4
EXAMPLE METHOD OF THERMAL EXPANSION CONTROL

water hammer arrestor close to quick-closing valves. Mechanical water hammer arrestors in conformance with ASSE 1010 alleviate water hammer intensity by absorbing the energy produced by the water. See Commentary Figure P2903.5 and the commentary for the definition of "Quick-closing valves" in Chapter 2.

P2903.6 Determining water-supply fixture units. Supply loads in the building water-distribution system shall be determined by total load on the pipe being sized, in terms of water-supply fixture units (w.s.f.u.), as shown in Table P2903.6, and gallon per minute (gpm) flow rates [see Table P2903.6(1)]. For fixtures not listed, choose a w.s.f.u. value of a fixture with similar flow characteristics.

❖ A prescriptive method for estimating the water demand for a building has evolved through the years and has been accepted as a standard method of sizing of water systems. The method is based upon weighing fixtures in accordance with their water supply load-producing effects on the water distribution system. This system couples the flow characteristics of the various fixtures with probability curves of simultaneous flushing of the fixtures. Table P2903.6 provides water fixture units assigned for various fixtures, and Table P2903.6(1) provides for conversions from water fixture units to gallon per minute (gpm) flow rates. For fixtures not listed, the fixture unit value of a similar fixture may be chosen and allocated.

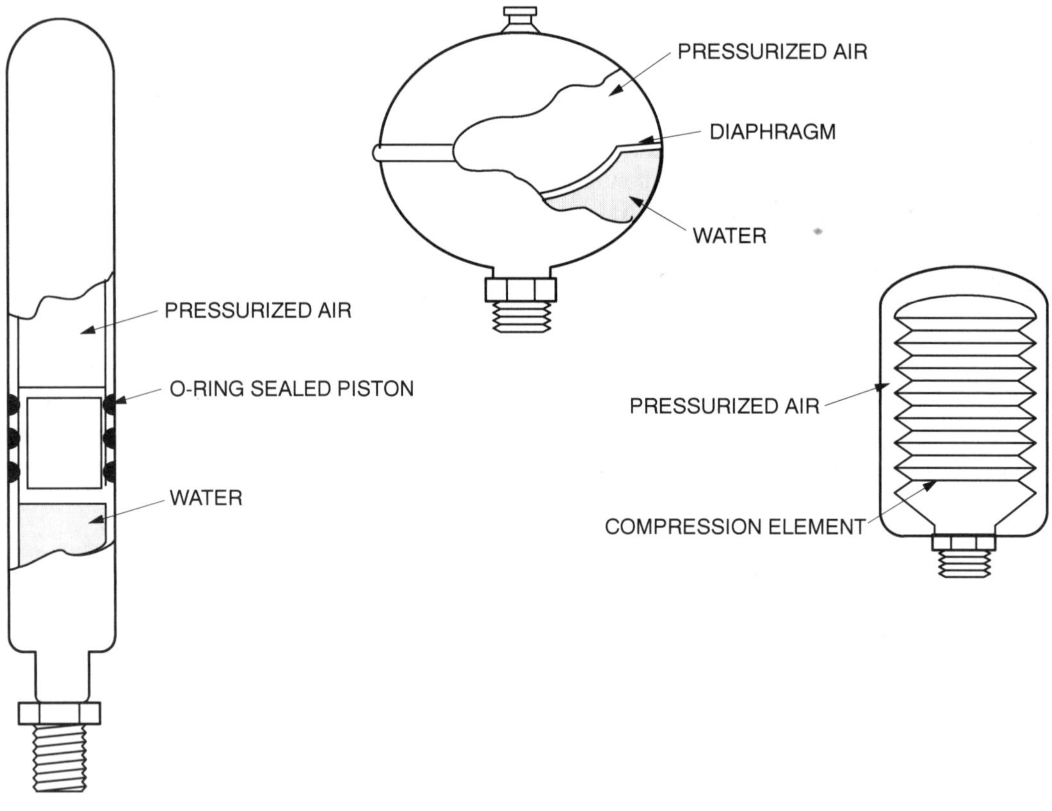

Figure P2903.5
EXAMPLE OF MECHANICAL WATER-HAMMER ARRESTORS

TABLE P2903.6
WATER-SUPPLY FIXTURE-UNIT VALUES FOR VARIOUS PLUMBING FIXTURES AND FIXTURE GROUPS

TYPE OF FIXTURES OR GROUP OF FIXTURES	WATER-SUPPLY FIXTURE-UNIT VALUE (w.s.f.u.)		
	Hot	Cold	Combined
Bathtub (with/without overhead shower head)	1.0	1.0	1.4
Clothes washer	1.0	1.0	1.4
Dishwasher	1.4	—	1.4
Full-bath group with bathtub (with/without shower head) or shower stall	1.5	2.7	3.6
Half-bath group (water closet and lavatory)	0.5	2.5	2.6
Hose bibb (sillcock)[a]	—	2.5	2.5
Kitchen group (dishwasher and sink with/without garbage grinder)	1.9	1.0	2.5
Kitchen sink	1.0	1.0	1.4
Laundry group (clothes washer standpipe and laundry tub)	1.8	1.8	2.5
Laundry tub	1.0	1.0	1.4
Lavatory	0.5	0.5	0.7
Shower stall	1.0	1.0	1.4
Water closet (tank type)	—	2.2	2.2

For SI: 1 gallon per minute = 3.785 L/m.

a. The fixture unit value 2.5 assumes a flow demand of 2.5 gpm, such as for an individual lawn sprinkler device. If a hose bibb/sill cock will be required to furnish a greater flow rate, the equivalent fixture-unit value may be obtained from Table P2903.6 or Table P2903.7.

❖ This table provides w.s.f.u. values for various fixtures and groups of fixtures, which are designed to work in conjunction with the prescriptive sizing methods and tables outlined in Section P2903.7.

Values are given for hot water and cold water usage, along with a combined hot and cold usage, the combined use value being somewhat less than the total of the hot and cold w.s.f.u. values. To estimate the supply demand of the building main and branches of the system, total the corresponding demand in the right column of the table. When sizing branches, add the water supply fixture unit demand passing through the branch with the related hot or cold column of Table P2903.6. Section P2903.7 does not contain provisions for the minimum size of a fixture supply pipe defined in Chapter 2. See the definition of "Fixture Supply." To size the fixture supply pipe when connecting faucets to sinks, lavatories, tubs, dishwashers and hose bibbs, the referenced standard under fixture fittings in Section P2722.1 is used. The following sizes are the recommendation of ASME A112.18.1.

TABLE P2903.6(1)
CONVERSIONS FROM WATER SUPPLY FIXTURE UNIT TO GALLON PER MINUTE FLOW RATES

SUPPLY SYSTEMS PREDOMINANTLY FOR FLUSH TANKS			SUPPLY SYSTEM PREDOMINANTLY FOR FLUSH VALVES		
Load	Demand		Load	Demand	
(Water supply fixture units)	(Gallons per minute)	(Cubic feet per minute)	(Water supply fixture units)	(Gallons per minute)	(Cubic feet per minute)
1	3.0	0.04104	—	—	—
2	5.0	0.0684	—	—	—
3	6.5	0.86892	—	—	—
4	8.0	1.06944	—	—	—
5	9.4	1.256592	5	15.0	2.0052
6	10.7	1.430376	6	17.4	2.326032
7	11.8	1.577424	7	19.8	2.646364
8	12.8	1.711104	8	22.2	2.967696
9	13.7	1.831416	9	24.6	3.288528
10	14.6	1.951728	10	27.0	3.60936
11	15.4	2.058672	11	27.8	3.716304
12	16.0	2.13888	12	28.6	3.823248
13	16.5	2.20572	13	29.4	3.930192
14	17.0	2.27256	14	30.2	4.037136
15	17.5	2.3394	15	31.0	4.14408
16	18.0	2.90624	16	31.8	4.241024

(Continued)

TABLE P2903.6(1)—continued
CONVERSIONS FROM WATER SUPPLY FIXTURE UNIT TO GALLON PER MINUTE FLOW RATES

SUPPLY SYSTEMS PREDOMINANTLY FOR FLUSH TANKS			SUPPLY SYSTEM PREDOMINANTLY FOR FLUSH VALVES		
Load	Demand		Load	Demand	
(Water supply fixture units)	(Gallons per minute)	(Cubic feet per minute)	(Water supply fixture units)	(Gallons per minute)	(Cubic feet per minute)
17	18.4	2.459712	17	32.6	4.357968
18	18.8	2.513184	18	33.4	4.464912
19	19.2	2.566656	19	34.2	4.571856
20	19.6	2.620128	20	35.0	4.6788
25	21.5	2.87412	25	38.0	5.07984
30	23.3	3.114744	30	42.0	5.61356
35	24.9	3.328632	35	44.0	5.88192
40	26.3	3.515784	40	46.0	6.14928
45	27.7	3.702936	45	48.0	6.41664
50	29.1	3.890088	50	50.0	6.684

For SI: 1 gallon per minute = 3.785 L/m, 1 cubic foot per minute = 0.4719 L/s.

❖ This table provides conversions from water supply fixture units to gallon per minute flow rate as required by Section P2903.8.1 for the sizing of manifolds (estimating demand).

P2903.7 Size of water-service mains, branch mains and risers. The minimum size water service pipe shall be $^3/_4$ inch (19.1 mm). The size of water service mains, branch mains and risers shall be determined according to water supply demand [gpm (L/m)], available water pressure [psi (kPa)] and friction loss due to the water meter and developed length of pipe [feet (m)], including equivalent length of fittings. The size of each water distribution system shall be determined according to the procedure outlined in this section or by other design methods conforming to acceptable engineering practice and approved by the administrative authority:

1. Obtain the minimum daily static service pressure [psi (kPa)] available (as determined by the local water authority) at the water meter or other source of supply at the installation location. Adjust this minimum daily static pressure [psi (kPa)] for the following conditions:
   - 1.1. Determine the difference in elevation between the source of supply and the highest water supply outlet. Where the highest water supply outlet is located above the source of supply, deduct 0.5 psi (3.4 kPa) for each foot (0.3 m) of difference in elevation. Where the highest water supply outlet is located below the source of supply, add 0.5 psi (3.4 kPa) for each foot (0.3 m) of difference in elevation.
   - 1.2. Where a water pressure reducing valve is installed in the water distribution system, the minimum daily static water pressure available is 80 percent of the minimum daily static water pressure at the source of supply or the set pressure downstream of the pressure reducing valve, whichever is smaller.
   - 1.3. Deduct all pressure losses due to special equipment such as a backflow preventer, water filter and water softener. Pressure loss data for each piece of equipment shall be obtained through the manufacturer of such devices.
   - 1.4. Deduct the pressure in excess of 8 psi (55 kPa) due to installation of the special plumbing fixture, such as temperature controlled shower and flushometer tank water closet.

   Using the resulting minimum available pressure, find the corresponding pressure range in Table P2903.7.
2. The maximum developed length for water piping is the actual length of pipe between the source of supply and the most remote fixture, including either hot (through the water heater) or cold water branches multiplied by a factor of 1.2 to compensate for pressure loss through fittings.

   Select the appropriate column in Table P2903.7 equal to or greater than the calculated maximum developed length.
3. To determine the size of water service pipe, meter and main distribution pipe to the building using the appropriate table, follow down the selected "maximum developed length" column to a fixture unit equal to, or greater than the total installation demand calculated by using the "combined" water supply fixture unit column of Table P2903.6. Read the water service pipe and meter sizes in the first left-hand column and the main distribution pipe to the building in the second left-hand column on the same row.
4. To determine the size of each water distribution pipe, start at the most remote outlet on each branch (either hot or cold branch) and, working back toward the main distribution pipe to the building, add up the water supply fixture unit demand passing through each segment of the distribution system using the related hot or cold column of Table P2903.6. Knowing demand, the size of each segment shall be read from the second left-hand column of the same table and a maximum developed length column selected in Steps 1 and 2, under the same or next smaller size meter row. In no case does the size of any branch or main need to be larger that the size of the main distribution pipe to the building established in Step 3.

❖ This section provides a prescriptive method of sizing the water distribution system. The following is a step-by-step method to be followed in using this sizing method.

1. Obtain the minimum static service pressure at the house meter from the water purveyor and adjust it as follows:
   - 1.1 Determine the maximum difference in elevation between the meter and the highest water supply outlet and deduct .5 psi (3.477 kPa) for each foot of difference in elevation. Where the highest water supply outlet is located below the source of supply add .5 psi (3.477 kPa) for each foot of difference in elevation. This step is to account for the static pressure loss or gain between the water supply source and the highest supply outlet. The .5 psi (3.477 kPa) is an approximation of the actual water pressure with respect to each foot of head pressure (0.433 psi = 1 foot of head pressure).
   - 1.2 Where the street pressure is over 80 psi (511.52 kPa), the code requires the installation of a pressure reducing valve (regulator) to reduce the pressure to 80 psi (511.52 kPa). See section P2903.3.1. The available street pressure has a maximum and minimum range. The regulator is set to reduce the maximum pressure below 80 psi (511.52 kPa) during peak demand periods. To account for the reduced pressure, a factor of 80 percent of the minimum street pressure is used. This adjustment must be made before any other deductions that are due to elevation, devices, or residual pressure adjustment are made. The corresponding pressure range can be found in Table P2903.7.
   - 1.3 Deduct any pressure losses that are due to devices on the system, such as backflow preventers, water softeners, etc. The sizing tables do not include losses other than the meter. Therefore, any additional losses must be accounted for to allow accurate results.
   - 1.4 The sizing tables have accounted for a residual pressure of 8 psi (55.152 kPa) that must be maintained at fixtures to produce the required flow. Where fixtures require higher than 8 psi (55.152 kPa) to function (such as temperature controlled showers, flushometer tank water closets, and one-piece tank-type water closets), the difference between the required operating pressure and the 8 psi (55.152 kPa) considered in the sizing tables must be deducted from the available pressure.
2. The maximum developed length for the water piping between the source of supply (meter) and the most remote outlet, including either hot (through the water heater) or cold water branches must be calculated. To account for fittings and valve losses, the length must be increased by 20 percent (multiply length by 1.2). This percentage factor is used only as an estimate for friction imposed for fittings. Select the appropriate column equal to or greater than the calculated maximum developed length.
3. To determine the size of the water service pipe, meter, and main distribution pipe using the appropriate table, obtain the total fixture unit count. Locate the row that has a fixture unit count under the length column that equals or exceeds the total fixtures units. Determine the required meter size in the first left hand column and the main distribution pipe size in the second left hand column on the same row.
4. To determine the size of each water distribution pipe, start at the most remote outlet on each branch (hot or cold) and, working back toward the main distribution pipe, add up the water supply fixture unit demand using the related hot or cold column of Table P2903.6. The size of each branch is read from the same table and maximum developed length column selected in steps 1 and 2 under the same or next smaller meter row. In no case does the size of any branch or main need to be larger than the size of the main distribution pipe to the building established by calculating the total "combined" demand.

TABLE P2903.7
MINIMUM SIZE OF WATER METERS, MAINS AND DISTRIBUTION PIPING BASED ON WATER SUPPLY FIXTURE UNIT VALUES

Pressure Range—30 to 39 psi

METER AND SERVICE PIPE (inches)	DISTRIBUTION PIPE (inches)	MAXIMUM DEVELOPMENT LENGTH (feet)									
		40	60	80	100	150	200	250	300	400	500
3/4	1/2[a]	2.5	2	1.5	1.5	1	1	.5	.5	0	0
3/4	3/4	9.5	7.5	6	5.5	4	3.5	3	2.5	2	1.5
3/4	1	32	25	20	16.5	11	9	7.5	6.5	5.5	4.5
1	1	32	32	27	21	13.5	10	8	7	5.5	5
3/4	1 1/4	32	32	32	32	30	24	20	17	13	10.5
1	1 1/4	80	80	70	61	45	34	27	22	16	12
1 1/2	1 1/4	80	80	80	75	54	40	31	25	17.5	13
1	1 1/2	87	87	87	87	84	73	74	56	45	36
1 1/2	1 1/2	151	151	151	151	117	92	79	69	54	43

Pressure Range—40 to 49 psi

METER AND SERVICE PIPE (inches)	DISTRIBUTION PIPE (inches)	MAXIMUM DEVELOPMENT LENGTH (feet)									
		40	60	80	100	150	200	250	300	400	500
3/4	1/2[a]	3	2.5	2	1.5	1.5	1	1	.5	.5	.5
3/4	3/4	9.5	9.5	8.5	7	5.5	4.5	3.5	3	2.5	2
3/4	1	32	32	32	26	18	13.5	10.5	9	7.5	6
1	1	32	32	32	32	21	15	11.5	9.5	7.5	6.5
3/4	1 1/4	32	32	32	32	32	32	32	27	21	16.5
1	1 1/4	80	80	80	80	65	52	42	35	26	20
1 1/2	1 1/4	80	80	80	80	75	59	48	39	28	21
1	1 1/2	87	87	87	87	87	87	87	78	65	55
1 1/2	1 1/2	151	151	151	151	151	130	109	93	75	63

Pressure Range—50 to 60 psi

METER AND SERVICE PIPE (inches)	DISTRIBUTION PIPE (inches)	MAXIMUM DEVELOPMENT LENGTH (feet)									
		40	60	80	100	150	200	250	300	400	500
3/4	1/2[a]	3	3	2.5	2	1.5	1	1	1	.5	.5
3/4	3/4	9.5	9.5	9.5	8.5	6.5	5	4.5	4	3	2.5
3/4	1	32	32	32	32	25	18.5	14.5	12	9.5	8
1	1	32	32	32	32	30	22	16.5	13	10	8
3/4	1 1/4	32	32	32	32	32	32	32	32	29	24
1	1 1/4	80	80	80	80	80	68	57	48	35	28
1 1/2	1 1/4	80	80	80	80	80	75	63	53	39	29
1	1 1/2	87	87	87	87	87	87	87	87	82	70
1 1/2	1 1/2	151	151	151	151	151	151	139	120	94	79

(continued)

TABLE P2903.7—continued
MINIMUM SIZE OF WATER METERS, MAINS AND DISTRIBUTION PIPING BASED ON WATER SUPPLY FIXTURE UNIT VALUES

Pressure Range—greater than 60 psi

METER AND SERVICE PIPE (inches)	DISTRIBUTION PIPE (inches)	MAXIMUM DEVELOPMENT LENGTH (feet)									
		40	60	80	100	150	200	250	300	400	500
3/4	1/2[a]	3	3	3	2.5	2	1.5	1.5	1	1	.5
3/4	3/4	9.5	9.5	9.5	9.5	7.5	6	5	4.5	3.5	3
3/4	1	32	32	32	32	32	24	19.5	15.5	11.5	9.5
1	1	32	32	32	32	32	28	22	17	12	9.5
3/4	1 1/4	32	32	32	32	32	32	32	32	32	30
1	1 1/4	80	80	80	80	80	80	69	60	46	36
1 1/2	1 1/4	80	80	80	80	80	80	76	65	50	38
1	1 1/2	87	87	87	87	87	87	87	87	87	84
1 1/2	1 1/2	151	151	151	151	151	151	151	144	114	94

For SI: 1 inch = 25.4 mm, 1 foot = 304.8 mm, 1 pound per square inch = 6.895 kPa.

a. Minimum size for building supply is 3/4-inch pipe.

❖ These tables are used in conjunction with Section P2903.7. See the commentary for Section P2903.7.

P2903.8 Parallel water-distribution system manifolds. Hot and cold parallel water-distribution system manifolds with individual distribution lines to each fixture or fixture fitting shall be sized and installed in accordance with Sections P2903.8.1 through P2903.8.7

❖ A manifold with parallel water distribution lines is a system that provides a centralized distribution point for all fixture supply piping. Each fixture is supplied with an individual supply line that extends from a common manifold, thus eliminating concealed joints and simplifying the method of installation. This type of system can be used in lieu of a conventional system, provided that the system is installed in accordance with Sections P2903.8.1 through P2903.8.7. Commentary Figures P2903.8(1) and P2903.8(2) show typical details of a manifold system.

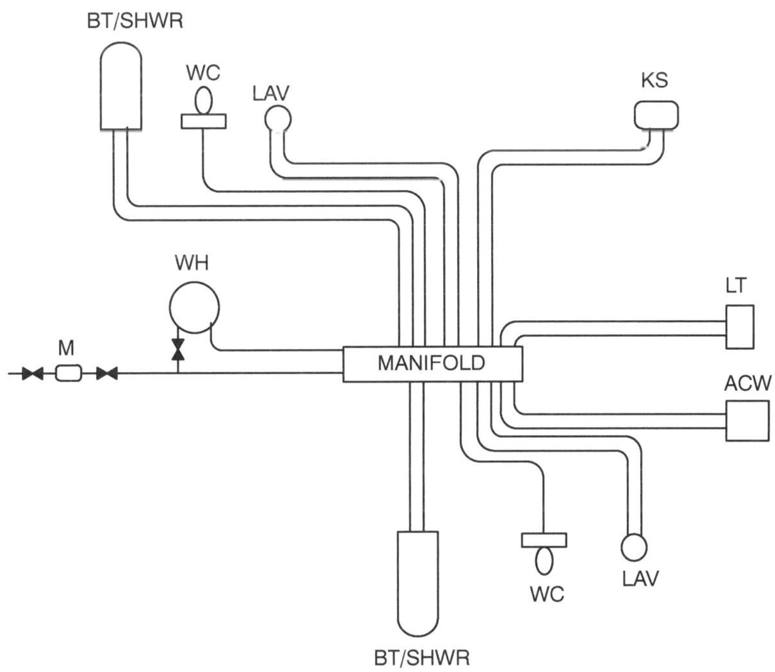

Figure P2903.8(1)
SIMPLIFIED SCHEMATIC OF PARALLEL WATER DISTRIBUTION

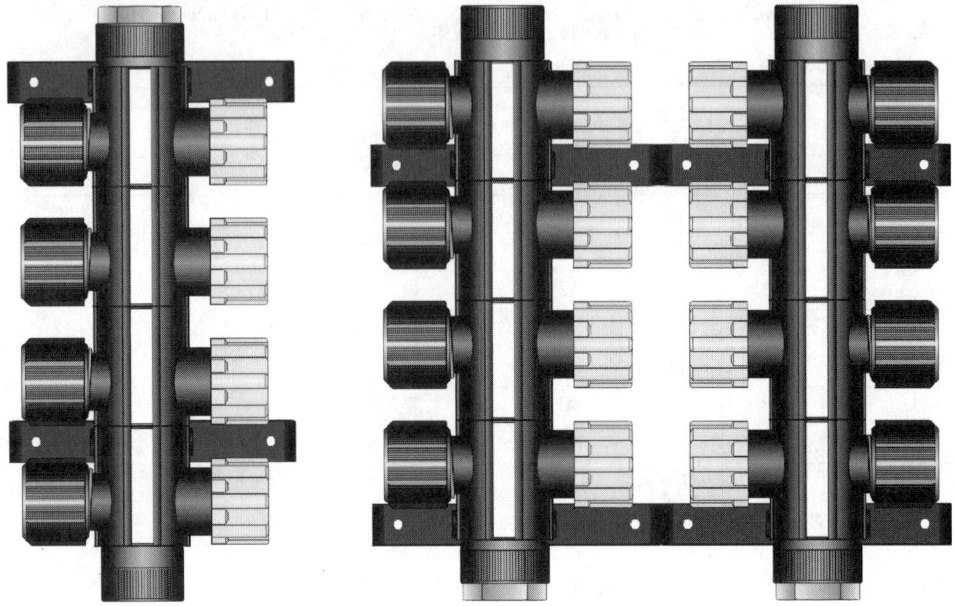

Figure P2903.8(2)
PARALLEL DISTRIBUTION MANIFOLD
(Figure courtesy of Vanguard Plastics)

P2903.8.1 Sizing of manifolds. Manifolds shall be sized in accordance with Table P2903.8.1. Total gallons per minute is the demand for all outlets.

❖ The size of parallel water-distribution system manifolds are regulated by Table P2903.8.1. The total water supply fixture units must be converted into gallon per minute (gpm) demand. See Table P2903.6(1).

TABLE P2903.8.1
MANIFOLD SIZING

PLASTIC		METALLIC	
Nominal Size ID (inches)	Maximum[a] gpm	Nominal Size ID (inches)	Maximum[a] gpm
$^3/_4$	17	$^3/_4$	11
1	29	1	20
$1^1/_4$	46	$1^1/_4$	31
$1^1/_2$	66	$1^1/_2$	44

For SI: 1 inch = 25.4 mm, 1 gallon per minute = 3.785 L/m, 1 foot per second = 0.3048 m/s.

NOTE: See Table P2903.6 for w.s.f.u and Table 2903.6(1) for gallon-per-minute (gpm) flow rates.

a. Based on velocity limitation: plastic—12 fps; metal—8 fps.

❖ Determine the total gallon per minute demand based on the use of either flush tank water closets or flush valve water closets. For example, a typical two-story single-family residence with three baths (extra tub), kitchen sink, dishwasher, laundry tray, clothes washer and two hose bibbs would have a total water supply fixture unit (w.s.f.u.) count of 25.0 and a 21.5 gallon (81 L) per minute demand based on flush tank water closets. In accordance with Table P2903.8.1, a 1 inch (25 mm) water service and manifold size could serve this single-family residence.

P2903.8.2 Minimum size. The minimum size of individual distribution lines shall be $^3/_8$ inch (9.5 mm). Certain fixtures such as one-piece water closets and whirlpool bathtubs shall require a larger size where specified by the manufacturer. If a water heater is fed from the end of a cold water manifold, the manifold shall be one size larger than the water heater feed.

❖ The minimum size of individual distribution lines are determined by the required gallons per minute (gpm), flow rate of each fixture, and the developed length. This section requires a minimum individual distribution line of $^3/_8$ inch (9.525 mm). However, one-piece water closets and whirlpool bathtubs might require a larger size based on the manufacturer's instructions. In addition, where an individual distribution line serves the cold water supply to a water heater, the manifold must be one size larger than the cold water inlet. This section also requires that the manifold be at least one size larger than the water heater inlet line size, if the water heater is supplied by one of the individual branches of the manifold.

P2903.8.3 Maximum length. The maximum length of individual distribution lines shall be 60 feet (18 288 mm) nominal.

❖ Usually, the maximum length of individual distribution lines does not exceed 50 feet (15 240 mm), because such systems are used where the fixtures are in close proximity.

P2903.8.4 Orientation. Manifolds shall be permitted to be installed in a horizontal or vertical position.

❖ This section allows the manifold to be installed either in the horizontal or vertical position, as long as appropriate support is provided and such system is installed in accordance with the manufacturer's instructions.

P2903.8.5 Support and protection. Plastic piping bundles shall be secured in accordance with the manufacturer's installation instructions and supported in accordance with Section P2605. Bundles that have a change in direction equal to or greater than 45 degrees (0.79 rad) shall be protected from chaffing at the point of contact with framing members by sleeving or wrapping.

❖ This section requires that plastic pipe bundles be secured as per the manufacturer's installation instructions and in accordance with the requirements of Section P2605. See the commentaries for Section P2605.1 and Table P2605.1. In addition, changes in direction can create movement of the branches because of expansion and contraction. This section requires that branches with a change in direction of 45 degrees (0.79 rad) or more are protected against chaffing at the point of contact with framing members by the use of sleeves or wrapping. This applies to both single and bundled branches.

P2903.8.6 Valving. Fixture valves, when installed, shall be located either at the fixture or at the manifold. If valves are installed at the manifold, they shall be labeled indicating the fixture served.

❖ Fixture valves can either be installed at the fixture or at the manifold. However, where valves are installed at the manifold, they must be labeled to identify the fixture served.

P2903.8.7 Hose bibb bleed. A readily accessible air bleed shall be installed in hose bibb supplies at the manifold or at the hose bibb exit point.

❖ This section requires a readily accessible air bleed to be installed in hose bibb supplies, either at the manifold or at the hose bibb exit point, to allow the disposal of any trapped air that might be in the system. This added protective requirement is due to the inherent dangers associated with pressurized air in plastic piping systems.

P2903.9 Valves. Valves shall be installed in accordance with Sections P2903.9.1 through P2903.9.3.

❖ This section addresses requirements for fixture shutoff valves. Where such valves are located at the manifold, they must be labeled to identify the fixture served.

P2903.9.1 Service valve. Each dwelling unit shall be provided with an accessible main shutoff valve near the entrance of the water service. The valve shall be of a full-open type having nominal restriction to flow, with provision for drainage such as a bleed orifice or installation of a separate drain valve. Additionally, the water service shall be valved at the curb or property line in accordance with local requirements.

❖ Each dwelling unit must have a full-open valve at the curb or property line as required by the local water purveyor. In addition, this section requires an accessible full-open valve for the dwelling unit at or near the entrance of the water service. In multiple family dwellings, the full-open valve must be located at the point where the water distribution pipe enters each dwelling unit so that in the event of a leak within the unit, the unit can be isolated. See Commentary Figures P2903.9.1(1), P2903.9.1(2) and P2903.9.1(3).

The two valves allow isolation of the meter for the water purveyor and shutoff of the entire building for service.

Full-open valves, such as gate valves and ball valves, reduce the amount of friction loss, as compared to other valves, such as globe valves. Therefore, this section requires full-open type valves so as to minimize friction loss and to coincide with the prescriptive sizing methods of water distribution systems as outlined in this chapter. The bleed orifice or termination of the drain from a drain valve must be above grade to prevent cross-contamination of the potable water system.

P2903.9.2 Water heater valve. A readily accessible full-open valve shall be installed in the cold-water supply pipe to each water heater at or near the water heater.

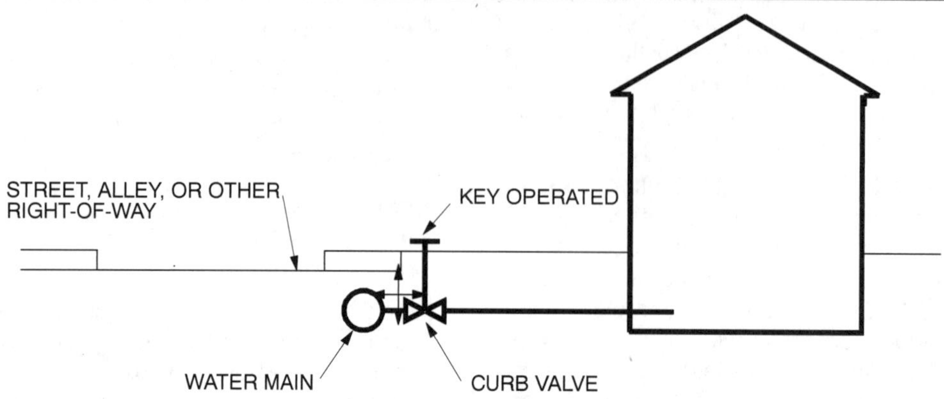

Figure P2903.9.1(1)
CURB VALVE ON WATER SUPPLY

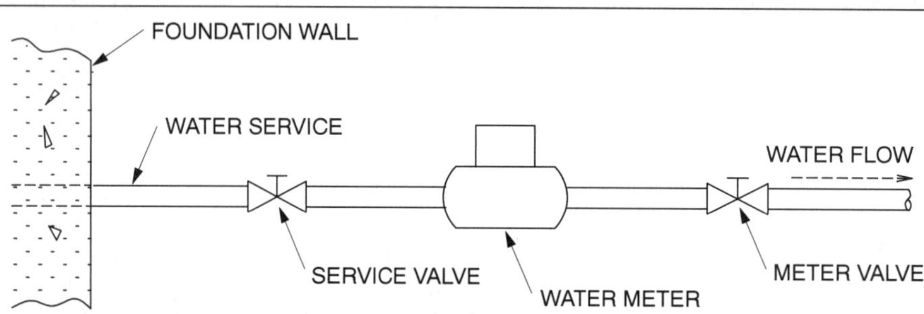

Figure P2903.9.1(2)
VALVE AT ENTRANCE TO THE BUILDING AND ON THE DISCHARGE SIDE OF METER

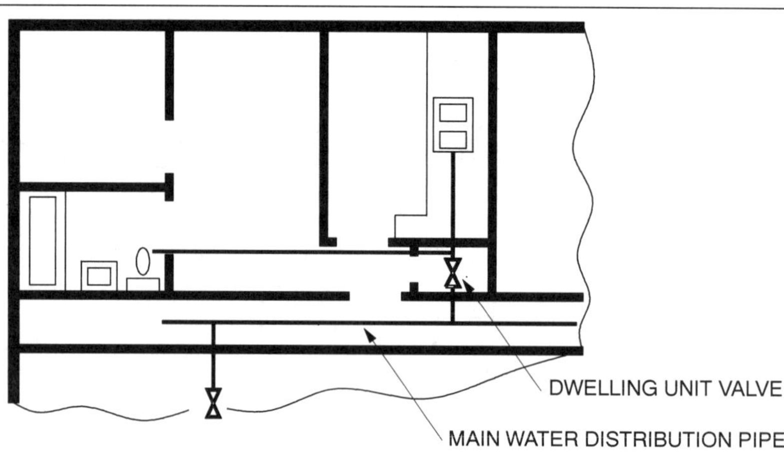

Figure P2903.9.1(3)
DWELLING UNIT VALVE

❖ To allow for maintenance, replacement, or emergency shutoff of a water heater, a readily accessible full open valve is required on the cold water supply to the water heater. See Commentary Figure P2903.9.2. The valve must be located at or near the water heater so that it is readily discernable as serving the water heater, without the use of a label.

P2903.9.3 Valve requirements. Valves serving individual fixtures, appliances, risers and branches shall be provided with access. An individual shutoff valve shall be required on the fix-

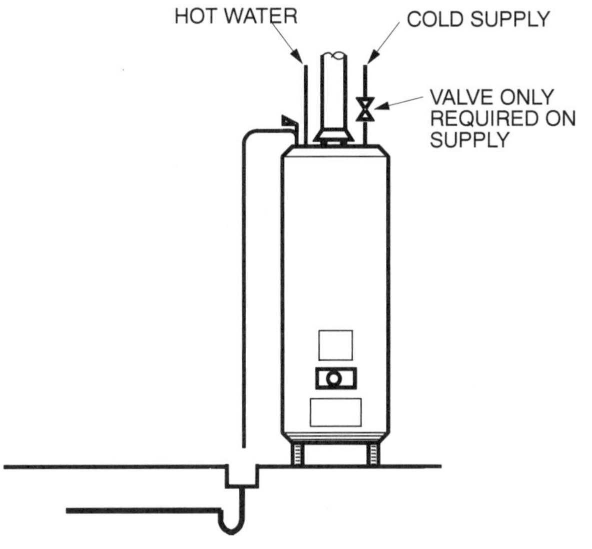

Figure P2903.9.2
WATER HEATER VALVE

ture supply pipe to each plumbing fixture other than bathtubs and showers.

❖ Within the dwelling, a shutoff valve is required on the fixture supply pipe to each plumbing fixture except for bathtubs and showers. Bathtubs and showers are typically not supplied with shutoff valves because such valves would not be accessible for maintenance.

P2903.10 Hose bibb. Hose bibbs subject to freezing, including the "frost-proof" type, shall be equipped with an accessible stop-and-waste-type valve inside the building so that they may be controlled and/or drained during cold periods.

Exception: Frost-proof hose bibbs installed such that the stem extends through the building insulation into an open heated or semi-conditioned space need not be separately valved. (See Figure P2903.10.)

❖ This section requires that hose bibbs that are subject to freezing be equipped with an accessible stop-and-waste type valve that is capable of shutting off water to the piping. This feature allows drainage of water from the piping and hose bibb. An exception to this requirement exists for frost-proof hose bibbs that have stems extending beyond the building insulation to an area that will not subject the water to freezing conditions. To prevent cross-contamination of the potable water system, the bleed orifice or termination of the drain must be above grade. The exception applies only to frost-proof hose bibbs where the point of shutoff is not subject to freezing. See the commentary for Figure P2903.10. If the bibb is subject to freezing, the exception does not apply.

SECTION P2904
MATERIALS, JOINTS AND CONNECTIONS

P2904.1 Soil and groundwater. The installation of water service pipe, water distribution pipe, fittings, valves, appurtenances and gaskets shall be prohibited in soil and groundwater that is contaminated with solvents, fuels, organic compounds or other detrimental materials that cause permeation, corrosion, degradation or structural failure of the water service or water distribution piping material.

❖ This section requires that soil be investigated for contaminants, based on the material selected. When a pipe is buried, the surrounding soil conditions may cause a component of the system to corrode at an accelerated rate. The soil must be evaluated, or historical knowledge of the soil applied, to determine if additional requirements are necessary to protect the piping material. Most soils are corrosive to galvanized steel pipe; therefore, additional protection would be required. The pipe is either coated or wrapped with coal tar, or an elastomeric or epoxy coating is applied to the exterior of the pipe. Some soils affect copper tubing, and soil with cinders will attack unprotected copper pipe or tubing. The pipe or tube must therefore be coated with a protective layer. Thermoplastic piping can degrade quickly in soil that contains heavy concentrations of hydrocarbons.

P2904.1.1 Investigation required. Where detrimental conditions are suspected by or brought to the attention of the building official, a chemical analysis of the soil and groundwater conditions shall be required to ascertain the acceptability of the water service material for the specific installation.

❖ If historical knowledge of the soil or a visual inspection reveals the possibility of contaminants or abrasive compounds, a chemical analysis is required. Based on the results of such analysis, approved materials must be installed.

P2904.1.2 Detrimental condition. When a detrimental condition exists, approved alternate materials or alternate routing shall be required.

❖ If a detrimental soil condition exists, either an alternate material or routing must be used that will be compatible with both the potable water and the soil condition.

P2904.2 Lead content. Pipe and fittings utilized in the water-supply system shall have a maximum of 8 percent lead.

❖ The EPA limits the lead content in water system components to a maximum of 8 percent. This limitation also applies to the various brass alloys used in producing faucets, fittings, and valves. All commercially available faucets, fittings, and valves have less than 8 percent lead.

P2904.3 Polyethylene plastic piping installation. Polyethylene pipe shall be cut square, using a cutter designed for plastic pipe. Except when joined by heat fusion, pipe ends shall be

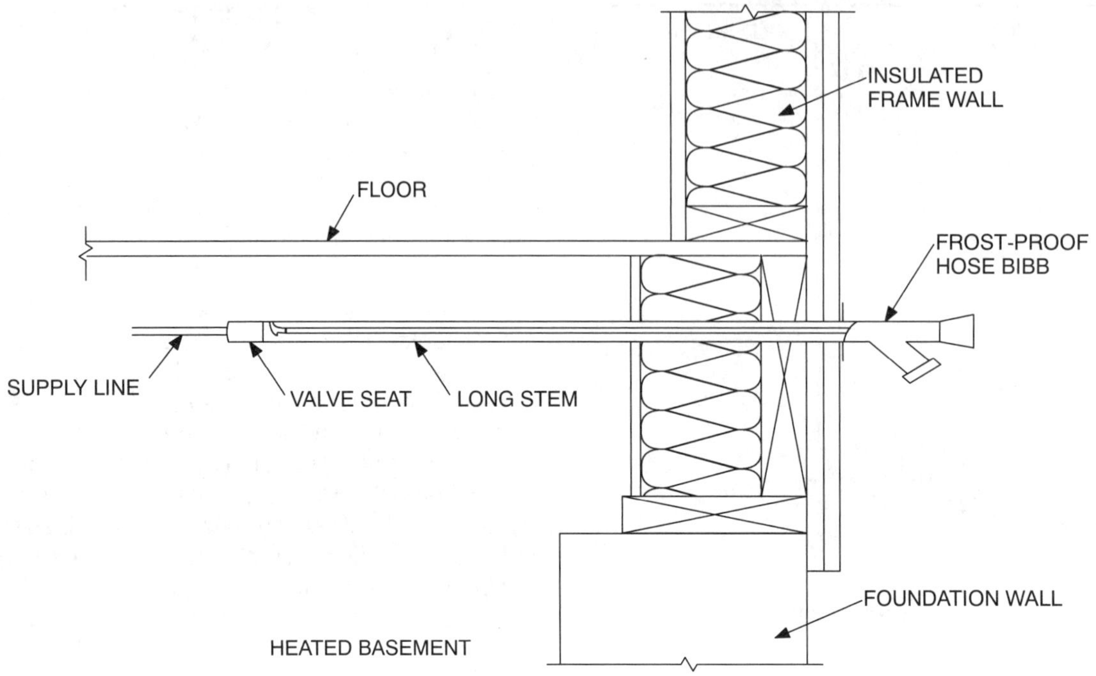

FIGURE P2903.10
TYPICAL FROST-PROOF HOSE BIBB INSTALLATION NOT REQUIRING SEPARATE VALUE

❖ This figure shows a detail of a proper installation of a frost-proof hose bibb through an insulated exterior wall separating a conditioned space from the exterior of the building. The point of shutoff (valve seat) is within the conditioned (heated) space. The hose bibb is installed such that the stem portion drains automatically when the valve seat is in the closed position. This type of installation would not require a separate shutoff valve as specified in Section P2903.10.

chamfered to remove sharp edges. Pipe that has been kinked shall not be installed. For bends, the installed radius of pipe curvature shall be greater than 30 pipe diameters or the coil radius when bending with the coil. Coiled pipe shall not be bent beyond straight. Bends shall not be permitted within ten pipe diameters of any fitting or valve. Stiffener inserts used with compression-type fittings shall not extend beyond the clamp or nut of the fitting. Flared joints shall be permitted where recommended by the manufacturer and made by the use of a tool designed for that operation.

❖ Any properly made pipe joint requires that the pipe ends be cut square for proper alignment, proper insertion depth, and adequate surface area for joining. Because polyethylene plastic pipe and tubing can be kinked and stress weakened, a minimum pipe bending radius is specified so that the structural capacity of the material is not exceeded. The minimum bending radius must not be less than the radius of the pipe coil as it came from the manufacturer or less than 30 pipe diameters, whichever is larger. See Commentary Figure P2904.3.

The pipe can be straightened as it is taken from the coil, but it cannot be bent in a direction opposite the bending direction of the coil. See Commentary Figure P2904.3. To avoid stress on connected fittings or valves, the location of pipe bends is restricted to specific distances from such components.

P2904.4 Water service pipe. Water service pipe shall conform to NSF 61 and shall conform to one of the standards listed in Table P2904.4.1. Water service pipe or tubing, installed underground and outside of the structure, shall have a minimum working pressure rating of 160 psi at 73°F (1100 kPa at 23°C).

❖ Water service pipe must conform to the requirements of NSF 61 and to the requirements for one of the materials listed in Table P2904.4.1. See the definition of "Water service pipe" in Chapter 2. Materials that will be underground and outside of the building must have a working pressure rating of at least 160 psi (1103.04 kPa).

P2904.4.1 Water service installation. Trenching, pipe installation and backfilling shall be in accordance with Section P2604. Water-service pipe is permitted to be located in the same trench with a building sewer provided such sewer is constructed of materials listed for underground use within a building in Section P3002.1. If the building sewer is not constructed of materials listed in Section P3002.1, the water-service pipe shall be separated from the building sewer by a minimum of 5 feet (1524 mm), measured horizontally, of undisturbed or compacted earth or placed on a solid ledge at least 12 inches (305 mm) above and to one side of the highest point in the sewer line.

MINIMUM BENDING RADIUS IS THE LARGER DIMENSION DETERMINED FROM THE FOLLOWING:

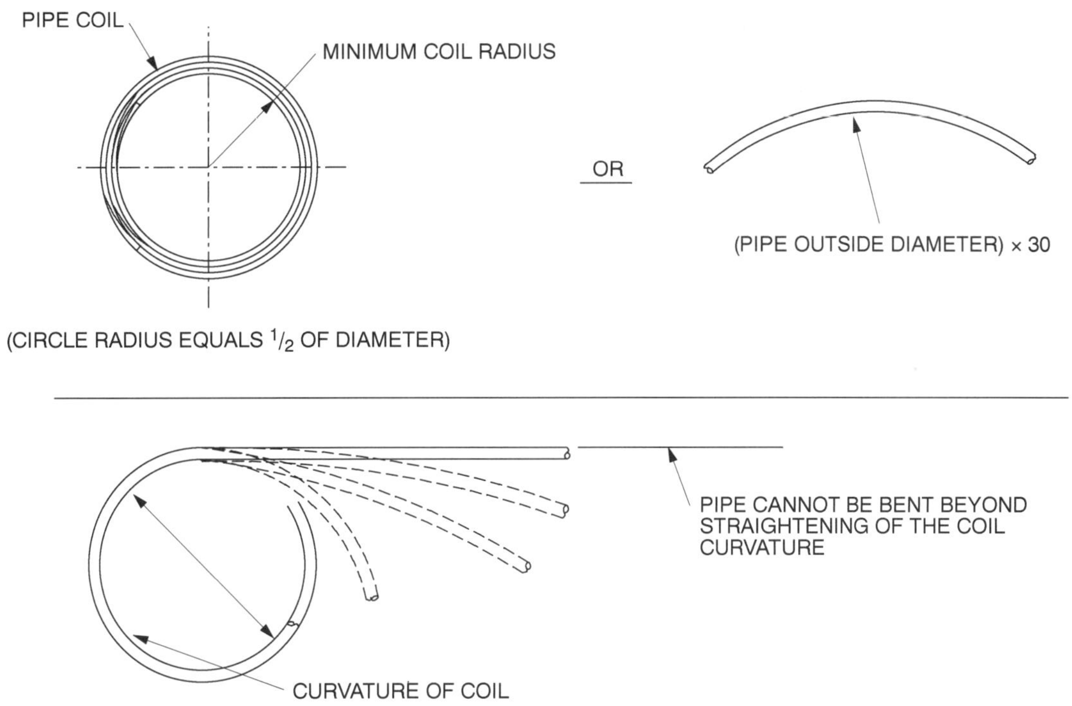

Figure P2904.3
BENDING

Exception: The required separation distance shall not apply where a water service pipe crosses a sewer pipe, provided the water service pipe is sleeved to at least 5 feet (1524 mm), horizontally from the sewer pipe centerline, on both sides of such crossing with pipe materials listed in Tables P2904.4.1, P3002.1 or P3002.2.

❖ The requirement for separation of water service from the building sewer is intended to reduce the possibility of contaminating the potable water supply. When there is a leak in the building sewer, the soil becomes contaminated around the water pipe. If the water service pipe has a subsequent failure, contamination of the potable water supply can occur. The simplest means of reducing the possibility of soils contamination is installing the building sewer and water service pipe in two separate trenches separated horizontally by at least 5 feet (1524 mm) of undisturbed or compacted earth. This section permits the water service to be located in the same trench, provided that the building sewer is constructed of drain, waste, and vent piping in accordance with Table P3002.1. Where the building sewer is not constructed of materials listed in Table P3002.1, the water service must be separated from the building sewer by 5 feet (1524 mm) measured horizontally or placed 12 inches (305 mm) measured vertically above the highest point in the building sewer. See Commentary Figures P2904.4.1(1) and P2904.4.1(2).

The exception provides for the installation where the water service pipe crosses a sewer pipe. Sometimes this installation is unavoidable, especially with existing installations. The water service must be sleeved to a point at least 5 feet horizontally (1524 mm) from the sewer pipe centerline, on both sides of the crossing. Simply stated, the water service must be run in a continuous conduit with piping listed from Table P2904.4.1, Table P3002.1, or Table P3002.2 for a minimum distance of 10 feet (3048 mm) to reduce the possibility of soil contamination of the potable water supply.

P2904.5 Water-distribution pipe. Water-distribution piping within dwelling units shall conform to NSF 61 and shall conform to one of the standards listed in Table P2904.5. All hot-water-distribution pipe and tubing shall have a minimum pressure rating of 100 psi at 180°F (689 kPa at 82°C).

❖ This section refers to Table P2904.5 for approved piping material inside the building for both hot and cold water. The referenced table specifies materials that are suitable for water distribution piping. Hot water distribution piping is required to be rated at 100 psi (689 4 kPa) at 180°F (266°C) (as opposed to 160 psi [1103.04 kPa] at 73°F [23°C] for water service piping or cold water piping). Note that all potable water piping and fittings must conform to NSF 61 so that all toxicity limitations imposed by that standard are met.

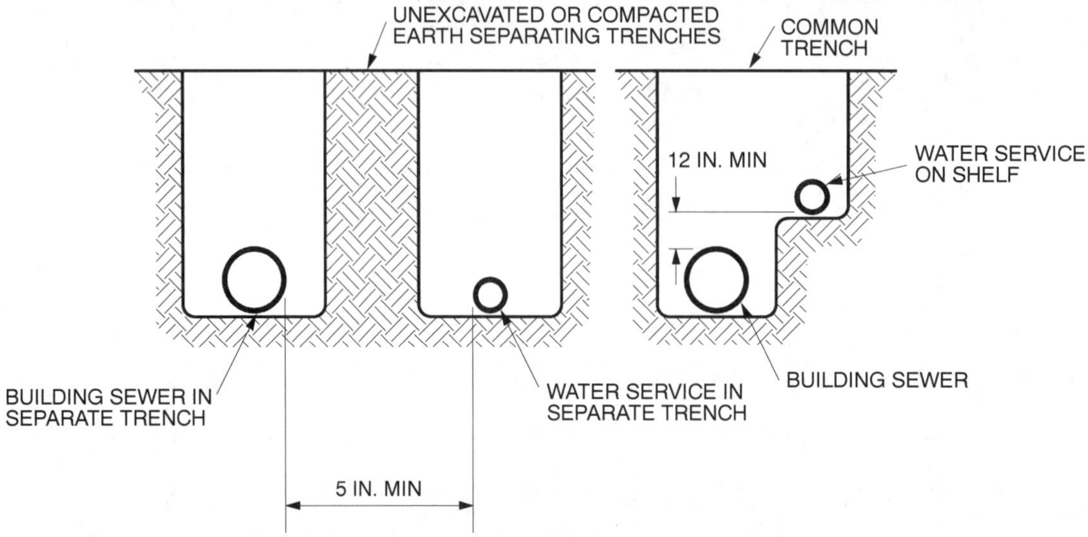

For SI: 1 inch = 25.4 mm, 1 foot = 304.8 mm.

Figure P2904.4.1(1)
SEPARATION OF WATER SERVICE AND BUILDING SEWER

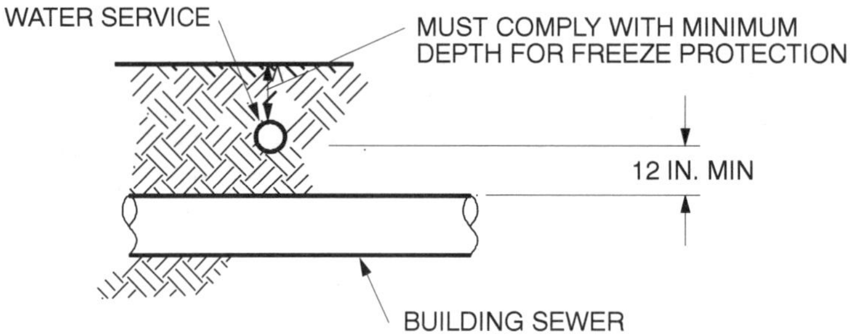

For SI: 1 inch = 25.4 mm.

Figure P2904.4.1(2)
WATER SERVICE CROSSING OVER BUILDING SEWER

TABLE P2904.4.1
WATER SERVICE PIPE

MATERIAL	STANDARD
Acrylonitrile butadiene styrene (ABS) plastic pipe	ASTM D 1527; ASTM D 2282
Asbestos-cement pipe	ASTM C 296
Brass pipe	ASTM B 43
Chlorinated polyvinyl chloride (CPVC) plastic pipe	ASTM D 2846; ASTM F 441; ASTM F 442; CSA B137.6
Copper or copper-alloy pipe	ASTM B 42; ASTM B 302
Copper or copper-alloy tubing (Type K, WK, L, WL, M or WM)	ASTM B 75; ASTM B 88; ASTM B 251; ASTM B 447
Cross-linked polyethylene/aluminum/cross-linked polyethylene (PEX-AL-PEX) pipe	ASTM F 1281; CSA CAN/CSA B137.10
Cross-linked polyethylene (PEX) plastic tubing	ASTM F 876; ASTM F 877; CSA B137.5

(Continued)

TABLE P2904.4.1—continued
WATER SERVICE PIPE

MATERIAL	STANDARD
Ductile iron water pipe	AWWA C151; AWWA C115
Galvanized steel pipe	ASTM A 53
Polybutylene (PB) plastic pipe and tubing	ASTM D 2662; ASTM D 2666; ASTM D 3309; CSA B137.8M
Polyethylene/aluminum/ polyethylene (PE-AL-PE) pipe	ASTM F 1282; CSA CAN/CSA-B137.9M
Polyethylene (PE) plastic pipe	ASTM D 2239; CSA-B137.1
Polyethylene (PE) plastic tubing	ASTM D 2737; CSA B137.1
Polyvinyl chloride (PVC) plastic pipe	ASTM D 1785; ASTM D 2241; ASTM D 2672; CSA B137.3
Stainless steel (Type 304/304L) pipe	ASTM A312; ASTM A778
Stainless steel (Type 316/316L) pipe	ASTM A312; ASTM A778

❖ The referenced table specifies allowable piping materials for water service. It also identifies the standards with which these materials must comply.

ACRYLONITRILE BUTADIENE STYRENE (ABS) PLASTIC PIPE: ABS pipe is rarely manufactured for water service. However, it is approved for use outside the structure. The pipe is black and must be rated for distribution of potable water. ASTM D 1527 is the referenced standard for Schedules 40 and 80 ABS pipe. The schedule number indicates the wall thickness: the higher the schedule number, the thicker the pipe wall. ASTM D 2282 is the referenced standard for SDR-PR pipe. SDR is an abbreviation for standard pipe dimension ratio, which is the ratio of pipe diameter to wall thickness. The PR designation indicates that the pipe is pressure rated.

ASBESTOS-CEMENT PIPE: This material is predominantly used for street water mains. It is used for large-size water service since it is only manufactured in 4-inch through 36-inch (102 mm through 914 mm) diameters.

Asbestos fibers are used to bond portland cement when manufactured into pipe. Because the asbestos is bonded in the pipe, it has not been considered a health hazard. However, very little asbestos-cement pipe is being produced because there is a hazard of breathing asbestos during pipe production and cutting. ASTM C 296 is the referenced standard for asbestos-cement pipe.

BRASS PIPE: Brass pipe is approved for both water service and water distribution. The piping material is a red brass composed of approximately 85 percent copper and 15 percent zinc. The pipe is available in both standard and extra strong weight. Extra strong weight pipe has the same outside diameter as standard weight pipe, but with a larger pipe-wall thickness.

Standard weight pipe is often referred to as "Schedule 40," and extra strong pipe is often referred to as "Schedule 80." Brass pipe dimensions are very similar to steel pipe dimensions. Section P2904.5 requires brass pipe to comply with NSF 61, which restricts lead content to very small concentrations.

COPPER OR COPPER-ALLOY TUBING (TYPE K, WK, L, WL, M, OR WM): This material is approved both for water service and water distribution. The tubing is manufactured in two different tempers, which are identified as drawn (called hard copper with the designation "H") and annealed (called soft copper with the designation "O"). Hard (drawn) copper tubing comes in straight lengths. Soft (annealed) copper tubing comes in both straight lengths and coils.

Annealed copper and tempered drawn copper may be formed by bending. Bending tempered copper is identified on the pipe with the designation "BT." Only bending tempered copper tubing may be joined by flared connections.

Seamless copper water tubing is available in Type K, L, or M. The tubing type indicates the wall thickness. Type K has the greatest wall thickness, followed by Types L and M. The outside diameter of the tubing is the same for all three types; only the inside diameter varies with wall thickness. Welded copper tubing has the designation "WK," "WL," or "WM." The "W" indicates that the tubing is welded-seam. The pipe dimensions correspond to Type K, L, or M seamless tubing.

The types of copper tubing are continuously identified with a colored marking for ease of identification. These colored markings are as follows:

Type K (WK) - Green
Type L (WL) - Blue
Type M (WM) - Red

CHLORINATED POLYVINYL CHLORIDE (CPVC) PLASTIC PIPE: CPVC is approved for both water service and water distribution. CPVC raw material, used to produce pipe, is a polyvinyl chloride that has been chlorinated to improve the material characteristics. The resulting pipe is more resistant to temperature extremes than PVC pipe.

CPVC plastic pipe may be either a white or a milky white (cream) color. The pipe listed is continuously marked with the manufacturer's name, the ASTM standard, and CPVC 4120 or CPVC 41, followed by two additional numbers. The designation CPVC 4120 identifies the quality of the material used to produce the pipe. The pipe is made with Grade 23447 plastic material. The grade number is a code for the quality of the pipe. The first digit indicates that the base resin is of CPVC. The remaining four digits indicate the impact strength, tensile strength, modulus of elasticity, and deflection temperature.

Grade 23445 was previously designated Type IV, Grade 1, which was shortened to CPVC 41. The last two digits in CPVC 4120 indicate the hydrostatic design stress in hundreds of psi. CPVC 4120 has a hydrostatic design stress of 2,000 psi (13 790 kPa).

ASTM D 2846 is the referenced standard for SDR 11 CPVC plastic pipe and CPVC socket fittings. SDR means "standard dimension ratio," which is a ratio of the average outside diameter to the minimum wall thickness. ASTM F 441 is the referenced standard for Schedules 40 and 80 plastic pipe. The dimensions of the pipe are the same as Schedules 40 and 80 steel pipe. The referenced standard for SDR 13.5, 17, 21, 26, and 32.5 CPVC pipe is ASTM F 442. The pipe is designated "PR," indicating that it is pressure rated.

DUCTILE-IRON WATER PIPE: Piping material is identified as either cast iron or ductile iron. Both materials are within the classification of cast iron. Ductile iron is a cast iron that is higher in ductility, hence the name. It is not as brittle as cast-iron pipe. This is necessary to provide a protective barrier between the potable water supply and the ductile-iron pipe to prevent impurities and contaminants from leaching into the water supply.

GALVANIZED STEEL PIPE: There is a wide variety of pipe classified as galvanized steel pipe. The grade of steel (thickness) can differ greatly. These thicknesses are referred to as Schedules 40, 80, and 160. As the schedule number increases, the wall thickness of the pipe increases. The outside diameter, however, remains the same for each pipe size.

Galvanized steel pipe is sometimes called "galvanized pipe" or mistakenly called "wrought-iron pipe." Wrought-iron pipe has not been manufactured in the United States in more than 30 years. All pipe referred to as "wrought-iron pipe" is steel pipe.

POLYBUTYLENE (PB) PLASTIC PIPE AND TUBING: Polybutylene (PB) pipe has been the subject of much media attention in the past few years. This is due primarily to the settlement of a multimillion-dollar lawsuit and the announcement by the primary producer that they have stopped manufacturing polybutylene resin for water supply piping. As a result, the pipe material will no longer be available once existing supplies run out. However, the material remains in the code to address modifications and repairs of existing systems.

Polybutylene is classified as an inert polyolefin material, meaning it is chemically resistant. This is the reason that polybutylene pipe may not be solvent cemented like other plastic pipes. It is joined by either flared, heat fusion, or mechanical joints. Polybutylene is one of the most flexible piping materials acceptable for the transportation of potable water. The minimum bending radius is ten times the diameter of the pipe. Polybutylene raw material is a by-product of the manufacture of ethylene.

Polybutylene pipe is usually blue or gray in color. The blue coloring indicates that the pipe is for water service use only and is not intended for water distribution. The gray color indicates that the pipe is for both water service and water distribution. The pipe is supplied in both straight lengths and rolled coils. The pipe is continuously marked with the manufacturer's name and the ASTM or CSA standard. Piping marked in accordance with the ASTM standard has the following additional markings: PB 2110 or PB 21, followed by two additional digits.

The designation "PB 2110" indicates the quality of the material. The material must be of Type 2, Category 1, Polybutylene, which is identified as PB 21. Type 2 indicates density, and Category 1 shows the extrusion flow rate. The last two digits in PB 2110 indicate the hydrostatic design stress in hundreds of psi. PB 2110 has a hydrostatic design stress of 1,000 psi (6895 kPa). The hydrostatic design stress is not an indication of the pressure rating of the pipe.

ASTM D 2666 is the referenced standard for SDR-PR polybutylene plastic tubing with an SDR of 9 and 13.5. The SDR for tubing is a ratio of the average outside diameter of the tubing and the minimum wall thickness.

POLYETHYLENE (PE) PLASTIC PIPE AND TUBING: PE pipe and tubing is approved for water service. Like polybutylene, polyethylene is an inert polyolefin material. It is resistant to chemical action, making it impossible to be joined by solvent cement. Polyethylene pipe is blue or black when supplied for water service. Orange-colored polyethylene is typically rated for only gas pipe installations. The pipe is continuously marked with the manufacturer's name, the ASTM or CSA standard, and the grade of polyethylene.

Various grades of polyethylene resin may be used to produce pipe or tubing. In accordance with the ASTM standard, the material is identified by the term "PE," followed by four digits. The first digit indicates the type of material based on the density. The second digit is the category identifying the extrusion flow rate. The last two digits show the hydrostatic design stress in hundreds of psi. The hydrostatic design stress is not an indication of the pressure rating of the pipe.

ASTM D 2737 is for polyethylene tubing with an SDR of 7.3, 9, and 11. The SDR for tubing is based on the average outside diameter. The grades of polyethylene permitted are PE 2305, PE 2306, PE 3306, PE 3406, and PE 3408.

CROSS-LINKED POLYETHYLENE (PEX) PLASTIC TUBING: PEX tubing is approved for both water service and water distribution, except that PEX tubing that conforms to ASTM F876 may not be used for water distribution. Cross-linked polyethylene tubing has been used extensively in Europe for many years for hot and cold potable water distribution systems. The material is available in sizes NPS $^1/_4$ inch through 2 inch (6 mm through 51 mm) and a standard dimension ratio (SDR 9). See Commentary Figure P2904.3(1).

Unlike most other pipe materials, PEX can be produced by several different manufacturing processes, such as the Engel, Silane, Monosil, Point a'Mousson, Daoplas, and electron beam methods. Although each manufacturing process can yield tubing that complies with the referenced standards, each process produces a tube that possesses unique characteristics. The cross-linked molecular structuring gives the pipe additional resistance to rupture over a wider range of temperatures and pressures than that of other polyolefin plastics (PE, PB, and polypropylene [PP]). Because of PEX pipe's unique molecular structure and resistance to heat, heat fusion is not permitted as a joining method. Because PEX is a member of the polyolefin plastic family, it is resistant to solvents. Therefore, the pipe cannot be joined by solvent cementing.

PEX pipe is flexible and can be bent using two methods: hot and cold bending. For hot bending, a hot-air gun with a diffuser nozzle is used. The pipe cannot be exposed to an open flame. The hot air meeting the pipe surface must not exceed 338°F (170°C), and the heat-up time must not exceed 5 minutes. The pipe is bent using conventional methods and is allowed to cool to room temperature before removal from the bending tool. The minimum hot-bending radius is two and one-half times the outside diameter. PEX can be bent at room temperature (cold bending) to a minimum radius of six times the outside diameter.

Mechanical connectors and fittings for PEX pipe are proprietary in nature and must only be used with the pipe for which they have been designed and tested. Fitting systems currently available include compression fittings with inserts and ferrules or O-rings, insert fittings utilizing a metallic lock (crimp) ring, and insert fittings with compression collars utilizing a cold flaring tool. It is important to consult the manufacturer's installation instructions to identify the fittings authorized for use with PEX piping.

CROSS-LINKED POLYETHYLENE/ALUMINUM/CROSS-LINKED POLYETHYLENE (PEX-AL-PEX) PIPE: PEX-AL-PEX is approved for both water distribution and water service. PEX-AL-PEX is a composite pipe made of an aluminum tube laminated to interior and exterior layers of cross-linked polyethylene. The layers are bonded together with an adhesive. See Commentary Figure P2904.3(2).

The cross-linked molecular structuring gives the pipe additional resistance to rupture to that of polyethylene. See above commentary for "cross-linked polyethylene tubing." Therefore, the pipe is suitable for hot and cold water distribution and is pressure rated for 125 psi at 180°F (862 kPa at 82°C).

Although it is partially plastic, the PEX-AL-PEX pipe resembles metal tubing in that it can be bent by hand or with a suitable bending device and maintain its shape without fittings or supports. The minimum bending radius specified by manufacturers is five times the outside diameter.

Mechanical joints are the only methods currently available to join PEX-AL-PEX pipe. A number of proprietary mechanical-compression-type connectors have been developed for use with the composite pipe to permit transition to

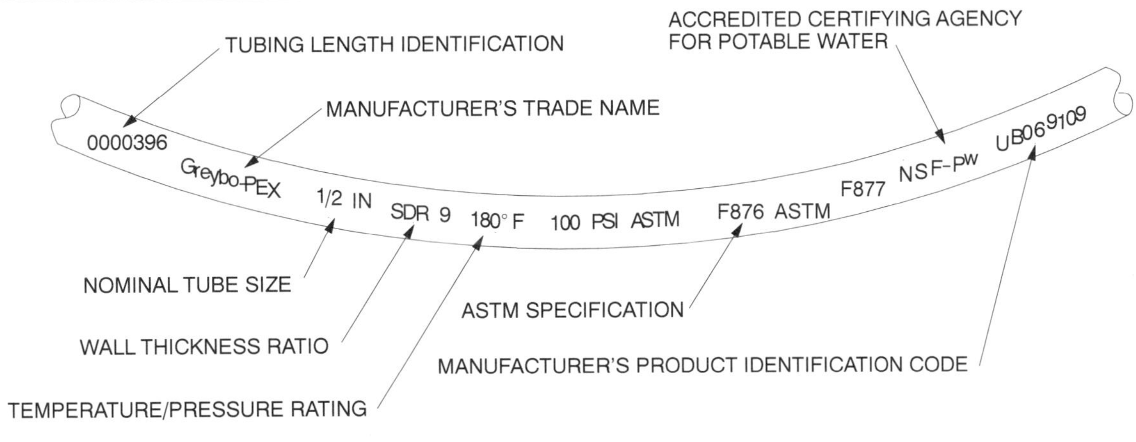

Figure P2904.3(1)
SAMPLE MARKING OF PEX TUBING

other pipes and fittings. The installation of such fittings must be done in accordance with the manufacturer's instructions.

POLYETHYLENE/ALUMINUM/POLYETHYLENE (PE-AL-PE) PIPE: PE-AL-PE is identical to PEX-AL-PEX composite pipe except for the physical properties of the polyethylene. Polyethylene displays the same resistance to temperature and pressure as cross-linked polyethylene does; therefore, it is suitable for hot and cold water distribution and is pressure rated for 125 psi at 180°F (862 kPa at 82°C). See the commentary for PEX-AL-PEX in this section.

POLYVINYL CHLORIDE (PVC) PLASTIC PIPE: PVC: PVC water service pipe is a different material than PVC drainage pipe, although both materials are white in color. The pipe must be continuously marked with the manufacturer's name, the ASTM standard, and the grade of the PVC material.

There are a number of grades of PVC material used to produce pipe. In accordance with the ASTM standards, the compounds are identified as PVC 12454-B, 12454-C, and 14333-D. The first digit indicates the base resin, with the following four digits identifying the impact strength, modulus of elasticity, and deflection temperature. The letter suffix indicates the material's chemical resistance. The compounds were previously identified by type and grade: 12454-B was Type 1; 12454-C was Type 1, Grade 2; and 14333-D was Type 2, Grade 1. The marking on the pipe lists the material grade by the term "PVC," followed by four digits. The first two digits use the previous type and grade numbers to identify the compound. The last two digits indicate the hydrostatic design stress in hundreds of psi. The hydrostatic design stress is not an indication of the pressure rating of the pipe.

ASTM D 1785 is the referenced standard for Schedules 40, 80, and 160 PVC plastic pipe. These dimensions are the same as Schedules 40, 80, and 160 steel pipe. ASTM D 2241 is the referenced standard for SDR-PR PVC plastic pipe. SDR means "standard dimension ratio," which is the ratio between the average outside diameter of the pipe and the minimum wall thickness. "PR" indicates that the pipe is pressure rated. SDRs of the pipe are 13.5, 17, 21, 26, 32.5, 41, and 64.

STAINLESS STEEL: Stainless steel is a family of iron-based alloys that must contain at least 10 percent or more chromium. The presence of chromium creates an invisible surface film that resists oxidation and makes the material corrosion resistant. The fact that stainless steel has a great resistance to corrosion means that stainless will have a very long life compared to mild steel. The benefits of stainless includes corrosion resistance, fire and heat resistance, hygiene (easy cleaning ability), aesthetic appearance, strength-to-weight advantage, ease of fabrication, impact resistance, and long-term value.

Wrought austenitic stainless steels combine a useful combination of outstanding corrosion and heat resistance with good mechanical properties over a wide temperature range.

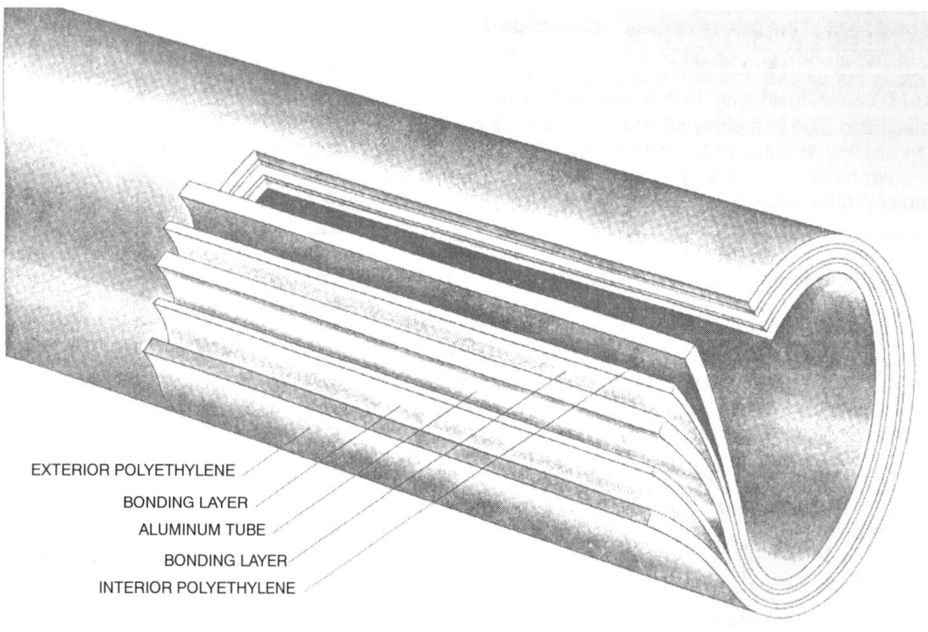

Figure P2904.3(2)
CROSS-LINKED POLYETHYLENE/ALUMINUM/CROSS-LINKED POLYETHYLENE PIPE
(Courtesy of Kitec® used license by IPEX Co., Ltd.)

TABLE P2904.5
WATER DISTRIBUTION PIPE

MATERIAL	STANDARD
Brass pipe	ASTM B 43
Chlorinated polyvinyl chloride (CPVC) plastic pipe and tubing	ASTM D 2846; ASTM F 441; ASTM F 442; CSA B137.6
Copper or copper-alloy pipe	ASTM B 42; ASTM B 302
Copper or copper-alloy tubing (Type K, WK, L, WL, M or WM)	ASTM B 75; ASTM B 88; ASTM B 251; ASTM B 447
Cross-linked polyethylene (PEX) plastic tubing	ASTM F 877; CSA B137.5
Cross-linked polyethylene/aluminum/cross-linked polyethylene (PEX-AL-PEX) pipe	ASTM F 1281; CSACAN/CSA-B137.10
Galvanized steel pipe	ASTM A 53
Polybutylene (PB) plastic pipe and tubing	ASTM D 3309; CSA CAN3-B137.8
Polyethylene/aluminum/polyethylene (PE-AL-PE) composite pipe	ASTM F 1282
Stainless steel (Type 304/304L) pipe	ASTM A312; ASTM A778
Stainless steel (Type 316/316L) pipe	ASTM A312; ASTM A778

❖ See the commentary for Section P2904.5.

P2904.5.1 Under concrete slabs. Inaccessible water distribution piping under slabs shall be copper water tube minimum Type M, brass, ductile iron pressure pipe, cross-linked polyethylene/aluminum/cross-linked polyethylene (PEX-AL-PEX) pressure pipe, chlorinated polyvinyl chloride (CPVC) or polybutylene (PB) or cross-linked polyethylene (PEX) plastic pipe or tubing—all to be installed with approved fittings or bends. The minimum pressure rating for plastic pipe or tubing installed under slabs shall be 100 psi at 180°F (689 kPa at 82°C).

❖ Where piping is installed under a concrete slab and is not accessible, it must be either Type M copper water tube, brass, ductile iron pressure pipe, PEX-AL-PEX, CPVC, PB, or PEX. Where plastic pipe or tube is installed under concrete slabs, it must be pressure rated at 100 psi (689 4 kPa) at 180°F (266°C), and fittings must be approved for such use by the manufacturer.

Because this section is a subsection of Section 2904.5, it applies to the installation beneath concrete slabs within a building.

P2904.6 Fittings. Pipe fittings shall be approved for installation with the piping material installed, and shall conform to the respective pipe standards listed in Table P2904.6. Pipe fittings utilized in the water supply system shall also conform to NSF 61.

❖ Table 2904.6 can be used to determine the applicable approved standards for pipe fittings. In some cases, especially with certain plastic pipe or tubing systems, the manufacturer will specify the type of fittings to be used. Where such proprietary conditions exist, they must be adhered to so as not to violate the conditions of the material listing.

TABLE P2904.6
PIPE FITTINGS

Acrylonitrile butadiene styrene (ABS) plastic	ASTM D 2468
Brass	ASTM F1974
Cast-iron	ASME B16.4; ASME B16.12
Chlorinated polyvinyl chloride (CPVC) plastic	ASTM F 437; ASTM F 438; ASTM F 439
Copper or copper alloy	ASME B16.15; ASME B16.18; ASME B16.22; ASME B16.23 ASME B16.26; ASME B16.29
Fittings for cross-linked polyethylene (PEX) plastic tubing	ASTM F1807; ASTM F1960; ASTM F2080
Gray iron and ductile iron	AWWA C110; AWWA C153
Malleable iron	ASME B16.3
Polyethylene (PE) plastic	ASTM D 2609
Polyvinyl chloride (PVC) plastic	ASTM D 2464; ASTM D 2466; ASTM D 2467; CSA B137.2
Stainless steel (Type 304/304L) pipe	ASTM A312; ASTM A778
Stainless steel (Type 316/316L) pipe	ASTM A312; ASTM A778
Steel	ASME B16.9; ASME B16.11; ASME B16.28

❖ This table specifies acceptable materials for pipe fittings along with the appropriate standards to which the fittings must comply. Each fitting is designed to be installed in a particular system with a given material or combination of material. For example, drainage pattern fittings must be installed in drainage systems, and vent fittings are limited to the venting system. Many fittings are intended to be used only for water distribution systems. Fittings that may be installed in any type of plumbing system are also available.

Fittings must be of the same material as the pipe and must be compatible with the pipe. This avoids any chemical or corrosive action between dissimilar materials. Although a number of standards strictly regulate fittings, many pipe standards include them as well. All pipe fittings must comply with the limitations set forth in NSF 61.

P2904.7 Flexible water connectors. Flexible water connectors, exposed to continuous pressure, shall conform to ASME A112.18.6. Access shall be provided to all flexible water connectors.

❖ ASTM A112.18.6 is the referenced standard for flexible water connectors used in potable water systems under continuous pressure. Plastic materials coming into contact with potable water must comply with the applicable sections of NSF 14. Flexible connectors for delivery of drinking water must comply with NSF 61. Solder and fluxes containing lead in excess of 0.2 percent must not be used. Metal alloys must not exceed 8-percent lead content. Copper alloy components used with the assembled connector must contain not less than 58 percent copper.

Flexible water connectors are designed to function at water pressure up to 125 psi (862 kPa) and supply temperature from 40°F to 180°F (4°C to 82°C).

P2904.8 Joint and connection tightness. Joints and connections in the plumbing system shall be gas tight and water tight for the intended use or required test pressure.

❖ This section simply states that joints in a plumbing system must be air- and water-tight. The degree of tightness will depend on the use of the material and the required test pressure. See the commentary for Section P2503.6.

P2904.9 Plastic pipe joints. Joints in plastic piping shall be made with approved fittings by solvent cementing, heat fusion, corrosion-resistant metal clamps with insert fittings or compression connections. Flared joints for polyethylene pipe are permitted in accordance with Section P2904.3.

❖ Plastic pipe may be joined in numerous ways depending on the material and the methods allowed by the manufacturer. This section provides the different acceptable methods of joining plastic pipe and specifies the various standards applicable in each case. Solvent cementing, heat fusion, clamps with insert fittings, compression connections, and flared joints are all permitted methods of joining plastic pipe. Reference is made to Section P2904.3 for provisions pertaining to the used of flared joints in polyethylene piping systems.

P2904.9.1 Solvent cementing. Solvent-cemented joints shall comply with Sections P2904.9.1.1 through P2904.9.1.3.

❖ Depending on the type of pipe material, standards and methods to be applied are outlined in Sections P2904.9.1(1) through P2904.9.1(3).

P2904.9.1.1 ABS plastic pipe. Solvent cement for ABS plastic pipe conforming to ASTM D 2235 shall be applied to all joint surfaces.

❖ Solvent cementing is the most common method of joining ABS plastic pipe and fittings. It is also one of the most misunderstood methods of joining pipe. The ASTM referenced standard indicates how to handle solvent cement and recommends a practice for making joints.

The pipe and fittings must be approximately the same temperature prior to solvent cementing. The pipe end and socket fitting must be clean, dry, and free of grease, oil, and other foreign materials. The solvent cement must be uniformly applied to the socket of the fitting and to the end of the pipe to the depth of the socket. Excessive solvent cement must be avoided to prevent a bead from forming on the inside of the pipe joint. The joint is made by twisting the pipe into the fitting socket immediately after applying the cement. The pipe must be held in place for a few seconds until the solvent cement begins to set. The pipe will tend to back out unless it is restrained. A small bead of solvent cement forms between the pipe exterior and fitting. The bead must be removed with a clean cloth to avoid the possibility of weakening the pipe wall. A solvent-cement joint must not be exposed to the working pressure for a period of 24 hours. The pipe must not be handled roughly for at least 1 hour after the joint is made.

Because each type of solvent cement is specifically designed for a given piping material, all-purpose solvent cement or universal solvent cement cannot be used to join ABS pipe or fittings unless it conforms to ASTM D 2235. Solvent cement for each plastic pipe material requires a unique mixture or combination of solvents and dissolved plastic resins in order to conform to the standard. Although all-purpose cement may dissolve the surfaces of ABS, PVC, and CPVC, it may be too aggressive or not aggressive enough to keep a joint strong enough to meet the minimum joint strength requirements of the referenced standard. Solvent cement must be identified on the can as complying with the ASTM D 2235 standard.

P2904.9.1.2 CPVC plastic pipe. Solvent cement for CPVC plastic pipe conforming to ASTM F 493 shall be applied to all joint surfaces. The joint shall be made while the cement is wet, and in accordance with ASTM D 2846 or ASTM F 493. Solvent-cement joints shall be permitted above or below ground.

❖ ABS and CPVC solvent-cement joints are different in the type of solvent cement used and the fact that a primer must be applied to all CPVC joint surfaces prior to applying the solvent cement. The primer breaks down the glossy exterior wall of the pipe to permit fusion of the pipe material with the solvent cement.

Failure to apply a primer to a CPVC solvent-cement joint will result in an improper joint and may affect the pipe manufacturer's product warranty. If a clear primer has been used, close examination is required to determine that the primer has been applied. Primer removes the shiny finish on the pipe, leaving a dull-colored surface.

The solvent cement for CPVC must conform to ASTM F 493. The standard recommends that CPVC solvent cement be orange in color to facilitate identification and minimize unintentional use of other cements.

P2904.9.1.3 PVC plastic pipe. A primer complying with ASTM F 656 shall be applied to all PVC solvent cemented joints. Solvent cement for PVC plastic pipe conforming to ASTM D 2564 shall be applied to all joint surfaces.

❖ Like CPVC, PVC plastic pipe requires a primer prior to solvent cementing. ASTM F 656 recommends that PVC primer not be orange in color, insofar as this is the recommended color for CPVC solvent cement. The coloring allows the installer and the inspector to verify that the required primer has been applied.

To distinguish the primer from the solvent cement, the solvent cement must be of a different color than the primer. Solvent cement for PVC is typically clear.

The PVC solvent cement is applied to the pipe and socket fitting. The joint is made with a slight twisting until the pipe reaches the full depth of the socket fitting. It is held in place until the solvent cement begins to set. The installer in these figures is wearing impervious gloves to prevent solvent cement from coming into contact with the skin. The solvent-cementing procedure must be done in accordance with the referenced standard. To minimize the health and fire hazards associated with solvent cements and primers, the manufacturer's instructions for the use and handling of these materials must be followed.

P2904.9.1.4 Cross-linked polyethylene plastic (PEX). Joints between cross-linked polyethylene plastic tubing or fittings shall comply with Section P2904.9.1.4.1 or Section P2904.9.1.4.2.

❖ This section speaks specifically to cross-linked polyethylene plastic (PEX) systems. Two methods of joining such material are outlined in the following subsections. Often, this type of material and the fittings and joining methods are proprietary. The manufacturer's installation instructions and the specified fittings and tools must therefore be adhered to.

P2904.9.1.4.1 Flared joints. Flared pipe ends shall be made by a tool designed for that operation.

❖ Flared joints must be made with tools specific to the design of the connector system. The assembly method must be in accordance with the manufacturer's instructions specific to the type of connectors being used. Flare configurations will vary depending upon the tool employed. Utilizing a cold flaring tool, wherein the tubing outside diameter is expanded mechanically, the flared end is then secured between the fitting. The flared surface serves as the sealing area between the tubing and fitting.

P2904.9.1.4.2 Mechanical joints. Mechanical joints shall be installed in accordance with the manufacturer's instructions. Fittings for cross-linked polyethylene (PEX) plastic tubing as described in ASTM F 1807, ASTM F1960 and ASTM F2080 shall be installed in accordance with the manufacturer's instructions.

❖ ASTM F 1807, ASTM F1960, and ASTM F2080 are the standards for metal insert fittings and copper crimp rings or cold-expansion fittings with PEX reinforcing rings or metal compression sleeves for use with cross-linked polyethylene (PEX) plastic tubing or pipe. These fittings are intended for use in 100 psi (689.5 kPa) cold and hot water distribution systems operating at temperatures up to and including 180°F (82°C). These fittings must be installed in accordance with the manufacturer's instructions. Insert fittings must be joined to PEX tubing by the compression of a copper crimp ring around the outer circumference of the tubing forcing the tubing material into annular spaces formed by ribs on the fittings.

The crimping procedure involves sliding the crimp ring onto the tubing, then inserting the ribbed end of the fitting into the end of the tubing until the tubing contacts the shoulders of the fitting or tube stop. The crimp ring must then be positioned on the tubing so that the edge of the crimp ring is $^1/_8$ to $^1/_4$ inch (3.2 to 6.4 mm) from the end of the tube. The jaws of the crimping tool must be centered over the crimp ring, and the tool must be held so that the crimping jaws are approximately perpendicular to the axis of the barb. The jaws of the crimping tool must be closed around the crimp ring, compressing the crimp ring onto the tubing. The crimp ring must not be crimped more than once.

P2904.10 Stainless steel. Joints between stainless steel pipe and fittings shall comply with Sections P2904.10.1 and P2904.10.2.

❖ Acceptable joints for stainless steel pipe or fittings are identified in the referenced sections.

P2904.10.1 Mechanical joints. Mechanical joints shall be installed in accordance with the manufacturer's instructions.

❖ Mechanical joints such as compression joints, grooved couplings, hydraulic pressed fittings, or flange designs can be used to join stainless steel pipe and fittings and are specified by the manufacturer. Such joints must be assembled and installed in compliance with the manufacturer's instructions.

P2904.10.2 Welded joints. All joint surfaces shall be cleaned. The joint shall be welded autogenously or with an approved filler metal in accordance with ASTM A312.

❖ The two basic methods for welding stainless steels are fusion welding and resistance welding. In fusion welding, heat is provided by an electric arc struck between a electrode and the metal to be welded. In resistance welding, bonding is the result of heat and pressure. Heat is produced by the resistance to the flow of electric current through the parts to be welded, and pressure is applied by the electrodes.

Welds and the surrounding area should be thoroughly cleaned to avoid impairment of corrosion resistance. Weld spatter, flux, or scale may become focal points for corrosive attack if not properly removed, especially in aggressive environments.

P2904.11 Threaded pipe joints. Threaded joints shall conform to American National Taper Pipe Thread specifications. Pipe ends shall be deburred and chips removed. Pipe joint compound shall be used only on male threads.

❖ Threaded pipe joints must conform to the American National Taper Pipe Thread (ASME B1.20.1, referred to under the standards chapter). After cutting the threads, chips must be removed, and the pipe must be reamed so that any remaining material that might obstruct the flow within the pipe or hinder the proper tightness of the joint is removed. It is imperative that the threads not exceed the maximum required for the size of pipe. If there are too many threads, the proper amount of taper will not be in contact with the fitting to ensure tightness. If there are not enough threads, the integrity of the joint is diminished.

Pipe joint compound must be applied to the male threads only. This prevents the compound from entering the piping system when the joints are made.

P2904.12 Soldered joints. Soldered joints in tubing shall be made with fittings approved for water piping and shall conform to ASTM B 828. Surfaces to be soldered shall be cleaned bright. The joints shall be properly fluxed and made with approved solder. Solders and fluxes used in potable water-supply systems shall have a maximum of 0.2 percent lead. Fluxes shall conform to ASTM B 813.

❖ A soldered joint is the most common method of joining copper pipe and tubing. Pipe must be cut square to provide proper alignment, adequate surface area for joining, and an interior free from obstructions.

When the pipe is cut, an edge (burr) is left protruding into the pipe. The pipe must be reamed to properly remove the burr. Chamfering is required to bevel the outer edge of the cut end of the pipe. Undercutting (excessive reaming) can reduce pipe wall thickness. See Commentary Figure P3003.3.4.

ASTM B 828 governs the procedures for making joints by soldering of copper and copper tubing and fittings. Flux must be applied to all joint surfaces. Flux is a chemically active material that removes and excluded oxides from the joint area during heating, allowing the melted solder to spread out onto the surfaces to be joined. The flux material must conform to ASTM B 813, which establishes minimum performance criteria. It also limits the maximum lead content to 0.2 percent. Any solder used in connection with potable water systems must also be limited to a maximum lead content of 0.2 percent.

The solder will flow by capillary action toward the heat. As such, the temperatures of the joint surfaces play a major role in the making of a properly soldered joint. The fitting and pipe must be approximately the same temperature when the solder is applied. Heating the copper liquefies the flux and solder and expands the surface voids of the material, allowing the solder to penetrate. A strong bond is created when the solder solidifies. If the temperatures are incorrect or uneven, the solder will run down the inside of the pipe and fitting or onto the outside of the pipe. The result is a solder-lined pipe.

Lead-bearing solder can sometimes be visually identified because the solder tends to darken in color with age. Test kits are available that can be used to identify lead-bearing solders with absolute certainty.

P2904.13 Flared joints. Flared joints in water tubing shall be made with approved fittings. The tubing shall be reamed and then expanded with a flaring tool.

❖ Outside of polyethylene materials, copper tube is the most common material that can be joined by the use of a flared joint. See the commentaries for Sections P2904.3 and P2904.9.1.4.1 for flared joint requirements pertaining to polyethylene and PEX materials.

Because the pipe end is expanded in a flared joint, only annealed and bending-tempered drawn copper tubing may be flared. There are two types of flaring tools: hammered and a screw yoke and block. The hammer-type flaring tool is inserted into the pipe and hammered. With the screw yoke and block, the end of the pipe is firmly held by the block assembly. The screw yoke is attached, and a flaring surface is pressed into the pipe end, creating the flare. The flared surface of the pipe is compressed against the mating surface of the fitting by the flare nut to form a watertight seal. Commonly, pipe-joint compound is placed on flare-fitting threads to lubricate the assembly. See Commentary Figure P2904.13.

Because hard-drawn pipe is subject to splitting when flared, a flared joint is restricted to soft (annealed) copper tubing. A flared joint is readily dismantled and is, in effect, a type of union connection.

P2904.14 Underground joints. Joints in polybutylene (PB) plastic pipe or tubing underground or under a concrete floor slab shall be installed using heat fusion, in accordance with the manufacturer's installation instructions. Joints in copper pipe or tube installed in a concrete floor slab or under a concrete floor slab on grade shall be installed using wrought-copper fittings and brazed joints.

❖ Because leaks in pipe joints under slabs can present a major repair problem, polybutylene joints must be made of the heat fusion type, and joints in copper tubing using wrought copper fittings must be brazed. Wrought copper fittings must be used with copper pipe or tube when installed under a concrete slab on grade.

Although brazed joints are similar to soldered joints, brazed joints are joined at a higher temperature (in excess of 1,000°F [538°C]). Because of the inherent resistance to elevated temperatures of a brazed joint, it is much stronger than a soldered joint. Brazing is often referred to as "silver soldering." Silver soldering is more accurately described as silver brazing and employs high silver-bearing alloys composed primarily of silver, copper, and zinc. Silver soldering (brazing) usually requires temperatures in excess of 1,000°F (538°C), and such solders are classified as "hard" solders.

Before brazing flux is applied to the joint surfaces, the joint surfaces must be cleaned. This is usually accomplished by polishing the pipe with an emery cloth. The socket fitting is cleaned with a specially designed brush.

The joints are usually heated with an oxyacetylene torch, though an air acetylene torch may be used on a smaller-diameter pipe. The pipe and fitting must be at approximately the same temperature when making the joint. The filler metal flows to the heat with the aid of the flux. The pipe must first be heated above the fitting. When the flux becomes clear, the heat must be directed to the base of the fitting socket. When the flux becomes clear and quiet, the filler metal is applied to the joint. The heat draws the filler metal into the socket. Any remaining flux residue must be removed from the pipe and joint surfaces. See Commentary Figure P2904.14.

P2904.15 Above-ground joints. Joints within the building between copper pipe, polybutylene tubing or CPVC tubing, in any combination with compatible outside diameters, are permitted to be made with the use of approved push-in mechanical fittings of a pressure-lock design.

❖ Push-in mechanical fittings of a pressure-lock design are allowed to be used to join copper pipe, polybutylene tubing, or CPVC tubing when piping is above ground within the building.

In addition, joints between different materials must be installed in accordance with Sections P2904.16.1 through P2404.16.3. See the commentaries for Sections P2904.16.1 through P2904.16.3.

P2904.16 Joints between different materials. Joints between different piping materials shall be made in accordance with Sections P2904.16.1, P2904.16.2 and P2904.16.3 or with a mechanical joint of the compression or mechanical sealing type having an elastomeric seal conforming to ASTM D 1869 or ASTM F 477. Joints shall be installed in accordance with the manufacturer's instructions.

❖ Mechanical joints can be used to join dissimilar piping materials. Because many piping materials have the same outside diameter, a number of fittings are designed to join different piping materials. This section requires mechanical joints to be of the compression or mechanical sealing type having an elastomeric seal conforming to ASTM D 1869 or ASTM F 477. This type

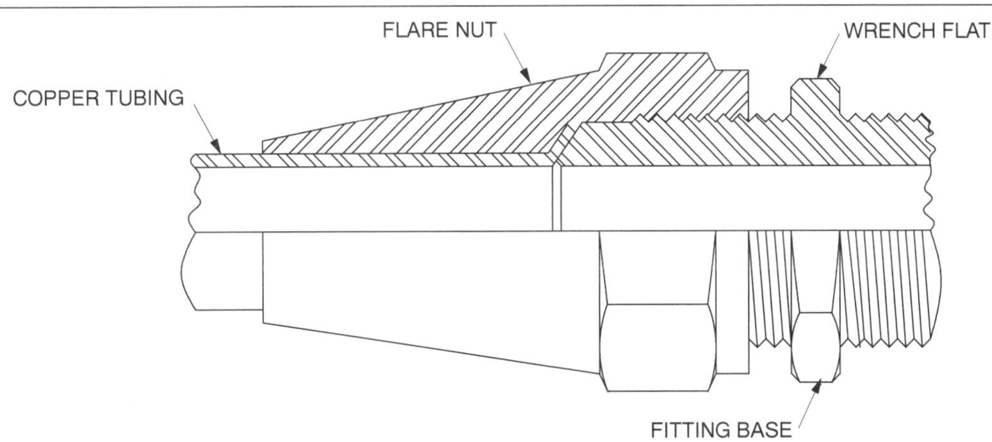

Figure P2904.13
TYPICAL FLARED JOINT

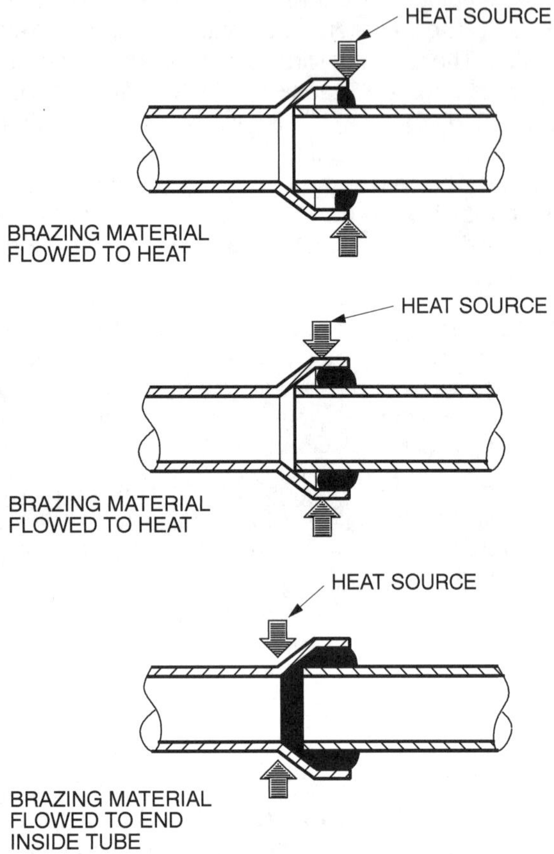

Figure P2904.14
HEAT APPLICATION OF BRAZED JOINT

of joint must be capable of handling the required test pressure and must also be compatible with the material and its intended use. See the commentary for Section P2503.6.

P2904.16.1 Copper or copper-alloy tubing to galvanized steel pipe. Joints between copper or copper-alloy tubing and galvanized steel pipe shall be made with a brass fitting or dielectric fitting. The copper tubing shall be joined to the fitting in an approved manner, and the fitting shall be screwed to the threaded pipe.

❖ This section requires a method to protect against galvanic action when joining copper or copper-alloy tubing with galvanized steel pipe. Galvanic action occurs when two dissimilar metals come into contact with each other, resulting in corrosion. When copper and galvanized steel are coupled, the steel will typically corrode and detach particles to the copper material. To stop galvanic action, this section requires either a dielectric coupling that includes a barrier between the two metals or a brass fitting that acts as a buffer and retards the galvanic action. See Commentary Figure P2904.16.1 for an illustration of the former.

P2904.16.2 Plastic pipe or tubing to other piping material. Joints between different grades of plastic pipe or between plastic pipe and other piping material shall be made with an approved adapter fitting. Joints between plastic pipe and cast-iron hub pipe shall be made by a caulked joint or a mechanical compression joint.

❖ The joining of different types of plastic pipe, and of plastic pipe to other piping materials, is permitted only through the use of adapter fittings specifically designed for such connections. The adaptor fitting must be compatible with the two dissimilar materials being joined. Joints between cast iron hub pipe and plastic pipe must be made by a mechanical compression joint or caulked joint.

The adaptor fittings must be evaluated individually and approved by the building official. Some key items to review when evaluating adaptor fittings are joint tightness, compatibility of the adaptor to the dissimilar

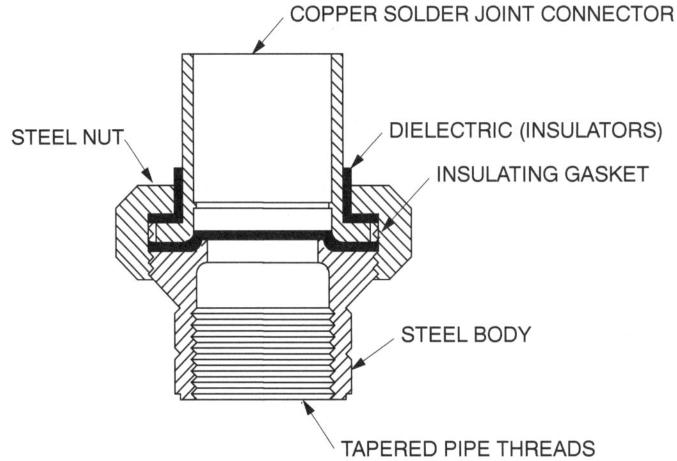

Figure P2904.16.1
DIELECTRIC UNION

materials being joined, and the presence of any ledges, shoulders, or reductions that could retard or obstruct the flow of drainage in the piping.

P2904.16.3 Stainless steel. Joints between stainless steel and different piping materials shall be made with a mechanical joint of the compression or mechanical-sealing type or a dielectric fitting.

❖ Where two dissimilar metals are in contact in the presence of an electrolyte, a battery effect is created, current flows, and one of the metals corrodes. Any metal in this situation will tend to have corrosion accelerated when it is coupled in the presence of an electrolyte. A very important factor to consider in evaluating the potential for galvanic corrosion is the relative surface area of the two different metals in contact.

In theory all metals immersed in an electrolyte, such as water, have a voltage potential. The relative activity of the metal is determined by voltage potential with those metals having negative voltages being the most active and more likely to corrode. For example, magnesium and iron have a negative voltage and corrode very rapidly when exposed to moist atmosphere. Copper and stainless steel have a positive voltage and are less likely to corrode.

Dielectric sleeves may be used with flange bolting or as gaskets for connections between different types of materials to prevent such corrosion. The plastic or rubber gasket engages the thread of the fitting attached, thereby insulating the metal from the water. This prevents corrosion on either part.

P2904.17 Press joints. Press-type mechanical joints in copper tubing shall be made in accordance with the manufacturer's instructions using approved tools which affix the copper fitting with integral O-ring to the tubing.

❖ Press joints are mechanical joints where the copper pipe is inserted into the fitting and a tool presses the fitting to the pipe. The fittings utilize an O-ring and are rated for a working pressure of 200 psi (1379 kPa) with a temperature limitation of 250°F (121°C). The speed of installation far exceeds a solder joining method.

SECTION P2905 CHANGES IN DIRECTION

P2905.1 Bends. Changes in direction in copper tubing are permitted to be made with bends having a radius of not less than four diameters of the tube, providing such bends are made by use of forming equipment that does not deform or create loss in cross-sectional area of the tube.

❖ Annealed copper and tempered drawn copper may be formed by bending. Whereas in most cases copper fittings will be used to change direction, "soft" copper can be bent to make the desired change in direction. Care must be exercised so that the minimum bending radius of four times the diameter of the copper tube is not exceeded. Although such a sharp bend is possible, the use of proper bending tools will be required in order to avoid a kink in the copper that would reduce both the cross-sectional area and the integrity of the material.

SECTION P2906 SUPPORT

P2906.1 General. Pipe and tubing support shall conform to Section P2605.

❖ This section points the reader to Section P2605, which is entitled "Support" and contains provisions pertaining to the horizontal and vertical support of piping materials used within water distribution systems. See the commentaries for Section P2605.1 and Table P2605.1.

SECTION P2907 DRINKING WATER TREATMENT UNITS

P2907.1 Design. Drinking water treatment units shall meet the requirements of either NSF 42, NSF 44 or NSF 53.

❖ Water delivered to a building is expected to be within established guidelines for health, odor, and chemical content. These guidelines are established by the U.S. Department of Health and are applicable to all potable water sources, whether these sources are public municipal water treatment facilities or individual private wells. Although the guidelines make certain that the water is safe to consume, there are occasionally times when the water supply has an objectionable odor or dissolved materials within it that are noticeable to the occupants of the building, particularly in the drinking water.

NSF 42 contains the requirements for units that are designed to improve the perceived quality of the drinking water supplied to the occupants of the building.

These units are designed to reduce the amount of specific contaminants present in the water supply. These contaminants affect the aesthetic quality of the water. Most of the water treatment units are designed to remove certain odors or dissolved materials in the drinking water supply.

The user or installer of the treatment unit is responsible for the selection of the proper unit based upon the offensive characteristics of the water and the unit's reported performance capabilities. Treatment units require periodic maintenance so that they will continue to function as designed. Because of this requirement, the treatment unit must be installed in a location where it is accessible for maintenance.

NSF 53, similar to the NSF 42 standard discussed above, contains the standards for units designed to improve the quality of drinking water by removing dissolved materials or substances that could have a negative effect on the health of the occupants of the building. These materials or substances may be microbiological, chemical, or particulate in nature and are often considered to be either an established or potential health hazard.

NSF 44 contains requirements pertaining to caution exchange water softeners, which utilize charged ions in breaking down and removing particulates.

P2907.2 Reverse osmosis drinking water treatment units. Point-of-use reverse osmosis drinking water treatment units, designed for residential use, shall meet requirements of NSF 58. Waste or discharge from reverse osmosis drinking water treatment units shall enter the drainage system through an air gap or an air gap device that meets the requirements of NSF 58.

❖ Reverse osmosis systems manufactured in accordance with NSF 58 are intended to remove specific contaminants from public or private drinking water supplies considered to be microbiologically safe. The treatment units designed in conformance to NSF 58 are intended to reduce the total dissolved solids (TDS) and other contaminants present in the drinking water. Many of these dissolved materials are considered to be either established or potential health hazards.

The reduction of the objectionable dissolved materials in a reverse osmosis system is accomplished by the use of pressure to force the water through a filter membrane or media in such a manner that the concentration level of the discharged water is less than the concentration of the incoming water. Reverse osmosis treatment systems are intended to be installed as a point of use system. This means that the treatment unit serves the drinking water fixture located immediately adjacent to the treatment system.

The discharge from these units must be by means of an air gap or air gap device meeting the requirements of NSF 58 and Section P2902.2.1. See the commentaries for Section P2902.2.1 and Table P2902.2.1.

P2907.3 Connection tubing. The tubing to and from drinking water treatment units shall be of a size and material as recommended by the manufacturer. The tubing shall comply with NSF 14, NSF 42, NSF 44, NSF 53, NSF 58 or NSF 61.

❖ The NSF standards listed address the connections for these units. The tubing utilized for drinking water treatment units must be of a size and material recommended by the manufacturer. The tubing materials are certified during the evaluation of the equipment for compliance with materials or products that come into contact with drinking water.

Bibliography

The following resource materials are referenced in this chapter or are relevant to the subject matter addressed in this chapter.

ASME A112.1.2-1991 (R1998), *Air Gaps in Plumbing Systems*. New York: American Society of Mechanical Engineers, 1998.

ASME A112.1.3-2000, *Air Gap Fittings for Use with Plumbing Fixtures, Appliances and Appurtenances*. New York: American Society of Mechanical Engineers, 2000.

ASME A112.18.1-2000, *Plumbing Fixture Fittings*. New York: American Society of Mechanical Engineers, 2000.

ASME A112.18.6-1999, *Flexible Water Connectors*. New York: American Society of Mechanical Engineers, 1999.

ASME B1.20.1-1983 (R1999), *Pipe Threads General Purpose (Inch)*, New York: American Society of Mechanical Engineers, 1999.

ASSE 1001-90, *Performance Requirements for Pipe Applied Atmospheric Type Vacuum Breakers*. Westlake, OH: American Society of Sanitary Engineering, 1990.

ASSE 1003-95, *Performance Requirements for Water Pressure Reducing Valves*. Westlake, OH: American Society of Sanitary Engineering, 1995.

ASSE 1010-98, *Performance Requirements for Water Hammer Arresters*. Westlake, OH: American Society of Sanitary Engineering, 1998.

ASSE 1011-95, *Performance Requirements for Hose Connection Vacuum Breakers*. Westlake, OH: American Society of Sanitary Engineering, 1995.

ASSE 1012-95, *Performance Requirements for Backflow Preventers with Intermediate Atmospheric Vent*. Westlake, OH: American Society of Sanitary Engineering, 1995.

ASSE 1013-99, *Performance Requirements for Reduced Pressure Principle Backflow Preventers and Reduced Pressure Fire Protection Principle Backflow Preventers*. Westlake, OH: American Society of Sanitary Engineering, 1999.

ASSE 1019-97, *Performance Requirements for Wall Hydrants, Freezeless, Automatic Draining, Anti-Backflow Types*. Westlake, OH: American Society of Sanitary Engineering, 1997.

ASSE 1020-98, *Performance Requirements for Pressure Vacuum Breaker Assemblies*. Westlake, OH: American Society of Sanitary Engineering, 1998.

ASSE 1047-99, *Performance Requirements for Reduced Pressure Detector Fire Protection Backflow Prevention Assemblies*. Westlake, OH: American Society of Sanitary Engineering, 1999.

ASSE 1052-94, Performance *Requirements for Hose Connection Backflow Preventers*. Westlake, OH: American Society of Sanitary Engineering, 1994.

ASSE 1056-95, *Performance Requirements for Back Siphonage Vacuum Breakers*. Westlake, OH: American Society of Sanitary Engineering, 1995.

ASTM B 813-00e01, *Specification for Liquid and Paste Fluxes for Soldering Applications of Copper and Copper Alloy Tube*. West Conshohocken, PA: ASTM International, 2001.

ASTM B 828-00, *Practice for Making Capillary Joints by Soldering of Copper and Copper Alloy Tube and Fittings*. West Conshohocken, PA: ASTM International, 2000.

ASTM D 1527-99, *Specification for Acrylonite-Butadiene-Styrene (ABS) Plastic Pipe, Schedules 40 and 80*. West Conshohocken, PA: ASTM International, 1999.

ASTM D 1785-99, *Specification for Poly (Vinyl Chloride) (PVC) Plastic Pipe, Schedules 40, 80 and 120*. West Conshohocken, PA: ASTM International, 1999.

ASTM D 1869-95 (2000), *Specification for Rubber Rings for Asbestos-Cement Pipe*. West Conshohocken, PA: ASTM International, 2000.

ASTM D 2282-99, *Specification for Acrylonitrile-Butadiene-Styrene (ABS) Plastic Pipe (SDR-PR)*. West Conshohocken, PA: ASTM International, 1999.

ASTM D 2235-01, *Specification for Solvent Cement for Acrylonitrile-Butadiene-Styrene (ABS) Plastic Pipe and Fittings*. West Conshohocken, PA: ASTM International, 2001.

ASTM D 2666-96a, *Specification for Polybutylene (PB) Plastic Tubing*. West Conshohocken, PA: ASTM International, 1996.

ASTM D 2737-01, *Specification for Polyethylene (PE) Plastic Tubing*. West Conshohocken, PA: ASTM International, 2001.

ASTM D 2846/D 2846M-99, *Specification for Chlorinated Poly (Vinyl Chloride) (CPVC) Plastic Hot- and Cold-Water Distribution Systems*. West Conshohocken, PA: ASTM International, 1999.

ASTM F 438-01, *Specification for Socket-Type Chlorinated Poly (Vinyl Chloride) (CPVC) Plastic Pipe Fittings, Schedule 40*. West Conshohocken, PA: ASTM International, 2001.

ASTM F 441/F441M-99, *Specification for Chlorinated Poly (Vinyl Chloride) (CPVC) Plastic Pipe, Schedules 40 and 80*. West Conshohocken, PA: ASTM International, 1999.

ASTM F 477-99, *Specification for Elastomeric Seals (Gaskets) for joining Plastic Pipe*. West Conshohocken, PA: ASTM International, 1999.

ASTM F 493-97, *Specification for Solvent Cements for Chlorinated Poly (Vinyl Chloride) (CPVC) Plastic Pipe and Fittings*. West Conshohocken, PA: ASTM International, 1997.

ASTM F 656-96a, *Specification for Primers for Use in Solvent Cement Joints of Poly (Vinyl Chloride) (PVC) Plastic Pipe and Fittings*. West Conshohocken, PA: ASTM International, 1996.

ASTM F 876-00, *Specification for Cross-linked Polyethylene (PEX) Tubing*. West Conshohocken, PA: ASTM International, 2000.

ASTM F 1807-99, *Specification for Metal Insert Fittings for Utilizing a Copper Crimp Ring for SDR9 Cross-linked Polyethylene (PEX) Tubing*. West Conshohocken, PA: ASTM International, 1999.

ASTM F 1960-99, *Specification for Cold Expansion Fittings with PEX Reinforcing Rings for Use with Cross-linked Polyethylene (PEX) Tubing*. West Conshohocken, PA: ASTM International, 1999.

ASTM F 2080-01, *Specification for Cold-expansion Fittings with Metal Compression-Sleeves for Cross-linked Polyethylene (PEX) Pipe*. West Conshohocken, PA: ASTM International, 2001.

NSF 14-1999, *Plastic Piping System Components and Related Materials.* Ann Arbor, MI: NSF International, 1999.

NSF 42-2000, *Drinking Water Treatment Units — Aesthetic Effects*. Ann Arbor, MI: NSF International, 2000.

NSF 44-2001, *Residential Cation Exchange Water Softeners*. Ann Arbor, MI: NSF International, 2001.

NSF 53-2001, *Drinking Water Treatment Units — Health Effects*. Ann Arbor, MI: NSF International, 2001.

NSF 58-2001, *Reverse Osmosis Drinking Water Treatment Systems*. Ann Arbor, MI: NSF International, 2001.

NSF 61-2001, *Drinking Water System Components — Health Effects*. Ann Arbor, MI: NSF International, 2001.

Chapter 30: Sanitary Drainage

General Comments

Chapter 30 provides the requirements pertaining to the design, sizing, and installation of sanitary drainage systems.

Material requirements for sanitary drainage systems and applicable standards for the various materials are provided in a table format for ease of reference. This chapter discusses joining methods of similar or different types of material and includes a table whereby the designer can ascertain the drainage fixture unit values as they apply to various fixtures and appliances.

It also addresses the appropriate use of various fittings, provisions for future fixtures or connections, adequate placement of cleanouts, sizing requirements pertaining to horizontal and vertical drain lines, and limitations pertaining to vertical and horizontal offsets in vertical drainage stacks. It includes requirements for drainage fixtures and drains that cannot be drained by gravity to the sewer.

This chapter regulates the use of pumps. It provides the requirement for, and the manufacture of, backwater valves.

Purpose

This chapter provides the minimum requirements necessary for achieving a properly sized sanitary drainage system. It addresses the proper use of fittings for the purpose of directing the flow and its contents effectively. It also outlines appropriated materials and provisions necessary for servicing the drainage system.

A properly working drainage system will depend on adequate air to balance the pressure fluctuations that are common in such systems; hence, there is a need for properly sized and located vents to the atmosphere. Venting requirements are located in Chapter 31.

In the nineteenth century, typhoid fever, cholera, and dysentery were common. The elimination of these diseases in industrialized nations has been due in large part to modern plumbing systems with proper drainage piping. Medical professionals give much of the credit for improvements in health and longevity to the plumbing profession. Medicine alone, without improvements in sanitation practices, would have had only a marginal effect.

SECTION P3001 GENERAL

P3001.1 Scope. The provisions of this chapter shall govern the materials, design, construction and installation of sanitary drainage systems. Plumbing materials shall conform to the requirements of this chapter. The drainage waste and vent (DWV) system shall consist of all piping for conveying wastes from plumbing fixtures, appliances and appurtenances, including fixture traps; above-grade drainage piping; below-grade drains within the building (building drain); below- and above-grade venting systems; and piping to the public sewer or private septic system.

❖ This chapter covers both sanitary drainage and sanitary sewers. Drainage usually consists of all sanitary waste piping and appurtenances within the building, whereas the sanitary sewer consists of piping outside the building discharging to the public sewer or private septic system. This chapter covers the materials, design, construction, and installation of the above two systems.

All portions of the drain, waste, and vent (DWV) system are listed here. Although this chapter primarily addresses drainage and waste lines, venting is also an important part of the entire drainage system in that it provides the air needed for the drainage system to function properly. Vent materials, although including specialized installation criteria, comprise much of the same piping and fittings as does the drainage system within the structure; hence, the materials are compatible.

P3001.2 Protection from freezing. No portion of the above grade DWV system other than vent terminals shall be located outside of a building, in attics or crawl spaces, concealed in outside walls, or in any other place subjected to freezing temperatures unless adequate provision is made to protect them from freezing by insulation or heat or both, except in localities having a winter design temperature above 32°F (0°C) (ASHRAE 97.5 percent column, winter, see Chapter 3).

❖ In areas where freezing can occur, care must be exercised not to install drainage systems where their contents can freeze. To determine whether the locality is subject to freezing, the ASHRAE 97.5 percent column of the winter design condition table is referenced. The reader is referred to Chapter 3. A review of Table R301.2(1) for the winter design temperature will show that this determination can be made by the use of Appendix D of the *International Plumbing Code®* (IPC®). If the temperature is 32°F (0°C) or lower, this section will apply. In this case, all drainage and vent piping must not be allowed outside the building (unless buried and below the frost level), in attics or crawl spaces, concealed in outside walls, or any other place subject to freezing. If the designer has no alternative but to place the piping in these places, provision must be made to either insulate or heat those sections so that the contents will not freeze. The concern is not only that the pipe can be blocked or stopped, but also that the pipe can burst because freezing liquids expand in volume. No amount of insulation by itself (without a heat source) can prevent freezing. Insulation can only delay freezing by slowing the rate of heat loss. Pipe heating devices, such as a self-limiting heat tape can be used. Care is especially necessary where portions of the system that contain standing water, such as a trap, are installed; for example, where the trap is installed within a cantilever and may extend below the floor insulation. See Commentary Figure P2603.6.

P3001.3 Flood-resistant installation. In areas prone to flooding as established by Table R301.2(1), drainage, waste and vent systems shall be located and installed to prevent infiltration of floodwaters into the systems and discharges from the systems into floodwaters.

❖ See the commentary for Section P2602.2.

SECTION P3002 MATERIALS

P3002.1 Piping within buildings. Drain, waste and vent (DWV) piping in buildings shall be as shown in Table P3002.1 except that galvanized wrought-iron or galvanized steel pipe shall not be used underground and shall be maintained not less than 6 inches (152 mm) above ground. Allowance shall be made for the thermal expansion and contraction on plastic piping.

❖ Drain, waste, and vent piping within buildings are allowed to be any of the materials specified in Table P3002.1. So that corrosion will be kept at a minimum, galvanized wrought iron or galvanized steel pipe is restricted from being installed underground or within 6 inches (152 mm) above ground. Where plastic pipe is used, because of the much higher rate of expansion in comparison to metallic pipe, provision for thermal expansion and contraction must be incorporated in the installation.

P3002.2 Building sewer. Building sewer piping shall be as shown in Table P3002.2. Forced main sewer piping shall conform to one of the standards for ABS plastic pipe, cast-iron pipe, copper or copper-alloy tubing, PVC plastic pipe, or pressure-rated pipe listed in Table P3002.2.

❖ Any of the material listed in Table P3002.2 is allowed for sanitary building sewer piping. Forced main sewer piping must conform to one of the standards for ABS plastic pipe, PVC plastic pipe, copper alloy tubing, or pressure rated pipe listed in Table P3002.2. There is a common misunderstanding that high pressure pipe and fittings must be used for a forced main sewer. There is no continuous pressurization of the pipe, only head pressure from the pump, and most building sewer pipe and fittings are rated for such pressure. However, the manufacturers' installation instructions for the pump and discharge piping should be taken into consideration when choosing the type of material that best suits the installation.

P3002.3 Fittings. Fittings shall be approved and compatible with the type of piping being used and shall be of a sanitary or DWV design for drainage and venting. Waterpipe fittings shall be permitted for use in engineer designed systems where the design indicates compliance with Section P3101.2.1.

❖ Each fitting is designed to be installed in a particular system with a given material. Drainage pattern fittings are typically "directional" to guide the flow derived from gravitational force. The required radius of curvature of the fittings will vary with the location where the fittings are to be used. Flows changing direction from horizontal to vertical, vertical to horizontal, and horizontal to horizontal require progressively longer radiuses to allow the flow to change direction without slowing down to a point where solids can separate from the liquid. When the velocity of the flow slows below a certain threshold [typically 2 feet (608.9 mm) per second], solids may separate and adhere to the pipe, thus creating blockages. Therefore, the type of fittings used in drainage systems not only have to be directional but must also be correctly selected for the specific location where they are to be installed. Vent fittings can have less radius than draining fittings because they are only carrying air. Therefore, vent fittings must never be used in drainage systems.

P3002.3.1 Drainage. Drainage fittings shall have a smooth interior waterway of the same diameter as the piping served. All fittings shall conform to the type of pipe used. Drainage fittings shall have no ledges, shoulders or reductions which can retard or obstruct drainage flow in the piping. Threaded drainage pipe fittings shall be of the recessed drainage type, black or galvanized. Drainage fittings shall be designed to maintain one-fourth unit vertical in 12 units horizontal (2-percent slope) grade.

❖ So as not retard the flow in drainage piping, drainage fittings must have a smooth interior water way. Fittings must be free of ledges, shoulders, or reductions that can retard or obstruct the flow within. Recessed threads are required in drainage fittings so that the interior bore remains constant in the transition through the pipe and fitting, thus maintaining a smooth waterway. Drainage fittings must also be "tapped" so that a horizontal pipe connection to the fitting will naturally be sloped at one-quarter unit vertical in twelve units horizontal (2 percent).

TABLE P3002.1
DRAIN, WASTE AND VENT PIPING AND FITTING MATERIALS

MATERIAL	STANDARD
Welded and seamless steel pipe (black or galvanized)	ASTM A 53
Cast-iron soil pipe and fittings (hub and spigot)	ASTM A 74
Cast-iron fittings (threaded)	ASTM A 126
Malleable iron fittings (threaded)	ASTM A 197
Seamless copper pipe, standard sizes	ASTM B 42
Seamless red brass pipe, standard sizes	ASTM B 43
Seamless copper tube	ASTM B 75
Seamless copper water tube Type K, L and M	ASTM B 88
ABS-DWV pipe and fittings	ASTM D 2661
PVC-DWV pipe and fittings	ASTM D 2665
3.25-inch O.D. PVC-DWV pipe and fittings	ASTM D 2949
ABS-DWV sch.40 pipe with cellular core	ASTM F 628
Co-extruded PVC plastic pipe with cellular core	ASTM F 891
Co-extended composite ABS-DWV pipe and fittings	ASTM F 1488
Cast-iron soil pipe and fittings (hubless)	CISPI 301
Copper drainage tube (DWV)	ASTM B 306
Mechanical couplings for drain, waste and vent pipe and sewer pipe	ASTM C1460; ASTM C 1461; CSA CAN/CSA B602
Solvent cement for ABS-DWV pipe and fittings	ASTM D 2235
Solvent cement for PVC-DWV pipe and fittings	ASTM D 2564
Socket bell for PVC-DWV pipe and fittings	ASTM D 2672
Primers for solvent cemented PVC-DWV pipe and fittings	ASTM F 656
Hubless cast-iron soil pipe and fittings	ASTM A 888
Couplings for use in connection with hubless cast-iron soil pipe and fittings	ASTM C1460; ASTM C1461; CISPI 310
Shielded couplings joining hubless cast-iron soil pipe and fittings	ASTM C 1277; ASTM C1460; ASTM C1461
Co-extruded composite ABS DWV schedule 40 IPS pipe (solid or cellular core) solvent cement fittings	ASTM F 1488; ASTM D 2235; ASTM D 2661; ASTM F 628
Co extruded composite PVC DWV schedule 40 IPS pipe (solid or cellular core) solvent cement fittings	ASTM F 1488, ASTM D 2564; ASTM D 2665; ASTM F 891
Co-extruded composite PVC DWV IPS pipe-DR-PS140 or PS200 solvent cement fittings	ASTM F 1488; ASTM D 2564; ASTM D 2665; ASTM F 891

For SI: 1 inch = 25.4 mm.

❖ This table specifies allowable piping and fitting materials for DWV systems. It also identifies the standards to which these materials must comply.

ABS PLASTIC PIPE: There are two types of ABS plastic pipe used for drainage waste and vent pipe. One is a solid wall pipe, and the other has a foam core. The solid wall pipe is produced by either a single extrusion or a simultaneous multiple coextrusion. Foam core pipe is produced by the coextrusion process.

ABS pipe is either black or dark gray and is continuously marked with the manufacturer's name, the ASTM standard, and the term "DWV" (indicating that the pipe is for drainage, waste, or vent systems). When the pipe is produced by the coextrusion process, it is marked "CoeX." The pipe dimensions are the same as standard weight (Schedule 40) steel pipe. As such, it is often called "Schedule 40 ABS drain pipe."

ASTM D 2661 is the referenced standard for both pipe and drainage fittings. ASTM F 628 is the referenced standard for foam core ABS pipe. The foam core is a closed-cell framed ABS plastic.

BRASS PIPE: See the commentary for Table P2904.4.1, Brass Pipe.

CAST-IRON PIPE: There are two pipe styles available in cast iron. The oldest type utilizes a hub for making joints and connections, whereas the other is a hubless pipe joined by mechanical couplings. ASTM A 74 is the referenced standard for hubbed cast iron, and CISPI 301 and ASTM A 888 are the referenced standards for hubless cast iron. Hubbed cast iron is produced in both service and extra-heavy-weight. As the name implies, extra-heavy is the heavier pipe by virtue of a thicker pipe wall. Hubless cast iron has a wall thickness that is approximately equal to service-weight hub pipe. Hubless pipe can be produced with or without a spigot bead.

Cast-iron pipe is also called "soil" pipe or "cast-iron soil pipe." Soil pipe is an expression developed when cast iron was essentially the only material used for soil drainage pipes. The expression is still in use today to distinguish between cast-iron pipe used for drainage and cast-iron pipe used for water.

ASTM A 74 cast iron is manufactured in 5-foot (1524 mm) and 10-foot (3048 mm) lengths, and CISPI 301 and ASTM A 888 are manufactured in 5-, 10-, and 18-foot (1542, 3048, and 5486 mm) lengths. The pipe is marked with the manufacturer's name and is designated hubless, XH for extra heavy, or SV for service.

COPPER OR COPPER-ALLOY PIPE: There are two types of copper pipe: threaded and threadless. Both pipes are seamless and have a chemical composition of 99.9 percent copper. Threaded copper pipe is made in both standard and extra-strong weights. The pipe dimensions are similar to brass pipe and steel pipe. Threadless copper pipe has the same outside diameter as threaded copper pipe, steel pipe, and brass pipe. The wall thickness of the pipe is less than threaded copper pipe; hence the inside diameter is greater. Threadless copper pipe is intended to be joined by brazing and is continuously marked in a gray color. The designation "TP" for threadless pipe is included in the marking.

COPPER OR COPPER-ALLOY TUBING: ASTM B 306 is the referenced standard for Type DWV copper tubing. Type DWV has a thinner pipe wall than Type M copper. The tubing is only produced in a drawn (hard) temper. Type DWV copper tubing is identified by a continuous yellow color marking. See the commentary for Table P3002.1, Copper or Copper-Alloy Tubing, Type K, L, or M.

GALVANIZED STEEL PIPE: A wide variety of pipe is classified as galvanized steel pipe. The grade of steel and the method of manufacturing can differ greatly. The common pipe used in plumbing application is either continuous welded or electric-resistance welded (ERW). Continuous-welded pipe is heated and formed, whereas ERW pipe is cold rolled and welded. Galvanized steel pipe is available in different wall thicknesses, the most common being standard weight, extra strong, and double extra strong. These thicknesses are also referred to as Schedules 40, 80, and 160, respectively. As the schedule number increases, the wall thickness of the pipe increases. The outside diameter, however, remains the same for each pipe size. Steel pipe is hot dipped and zinc coated to produce a galvanized protective coating on both the inside and outside of the pipe. New galvanized pipe is a shiny silver and, as the pipe ages, the coating oxidizes to protect the zinc. As a result, the color turns a dull gray. Galvanized steel pipe is sometimes called "galvanized pipe" or, mistakenly, "wrought-iron pipe." Wrought-iron pipe has not been manufactured in the United States in more than 30 years. All pipe referred to as "wrought-iron pipe" is steel pipe. The steel pipe industry claims to be responsible for nominal pipe dimensions, meaning the pipe is approximately the size by which it is identified. The dimensions date back to 1862 when Robert Briggs of Pascal Iron Works wrote the first pipe standards in the United States. It is believed that the dimensions were based on the pipe Briggs was manufacturing. That is why a $^{3}/_{4}$-inch (19 mm) standard-weight galvanized steel pipe has an outside diameter of 1.05 inches (27 mm) and an average inside diameter of 0.824 inches (21 mm).

POLYVINYL CHLORIDE (PVC) PLASTIC PIPE (TYPE (DWV): PVC plastic pipe for drainage and vent systems is available in solid wall or coextruded with cellular core. The pipe is either white or a milky white (cream) and is continuously marked with the manufacturer's name, the ASTM or CSA standard, and Type DWV. Cellular core pipe is also marked COEX CELLULAR CORE PVC PIPE. The pipe will be double marked if it conforms to the requirements for water service pipe.

ASTM D 2665 is the referenced standard for PVC plastic pipe and fittings. The pipe has the same dimensions as standard-weight (Schedule 40) steel pipe. As such, it is often called "Schedule 40 PVC drain pipe." ASTM F 891 is the referenced standard for coextruded cellular core PVC pipe. The standard specifies various grades of pipe. Only Schedule 40 pipe, also designated as Type DWV, is permitted for above-ground use.

ASTM D 2949 is the referenced standard for 3-inch (76 mm) PVC plastic pipe and fittings only. The pipe has an outside diameter of $3^{1}/_{4}$ inches (83 mm), as compared to an outside diameter of $3^{1}/_{2}$ inches (89 mm) for other 3-inch (76 mm) PVC plastic pipe. The inside diameter is the same as a 3-inch (76 mm) pipe conforming to ASTM D 2665. The resulting pipe wall is $^{1}/_{8}$-inch (3.2 mm) thinner. The $3^{1}/_{4}$-inch (83 mm) outside diameter pipe was developed to permit installation of PVC plastic pipe in a nominal 2 by 4 stud wall cavity. The pipe is often incorrectly identified as Schedule 30 plastic pipe or thin wall pipe. See the commentary for Table 2904.4.1, Polyvinyl Chloride Plastic Pipe.

TABLE P3002.2
BUILDING SEWER PIPING

MATERIAL	STANDARD
Cast-iron pipe and fittings	ASTM A 74
Cast-iron soil pipe and fittings for hubless sanitary system	CISPI 301
Seamless copper tube	ASTM B 75
Copper water tube	ASTM B 88
Concrete sewer, storm drain and culvert pipe	ASTM C 14
Compression joints for vitrified clay pipe and fittings	ASTM C 425
Vitrified clay pipe and fittings	ASTM C 700
Bitumenized fiber drain and sewer pipe	ASTM D 1861
ABS-DWV pipe and fittings	ASTM D 2661
Polyethylene (PE) plastic pipe (SDR-PR)	ASTM F 714
PVC/DWV pipe and fittings	ASTM D 2665
ABS Sewer pipe and fittings	ASTM D 2751
3.25-inch O.D. PVC/DWV pipe and fittings	ASTM D 2949
Type PSM/PVC sewer pipe and fittings	ASTM D 3034
Joints for drain and sewer plastic pipe using flexible elastomeric seals	ASTM D 3212
ABS Schedule 40 DWV pipe with cellular core	ASTM F 628
Co-extruded PVC schedule 40, PS 50 or PS 100 plastic pipe with cellular core	ASTM F 891
Copper drainage tube (DWV)	ASTM B 306
Mechanical couplings for drain waste and vent pipe and sewer pipe	ASTM C 1460; ASTM C 1461; CSA CAN/CSA B602
Solvent cement for ABS-DWV pipe and fittings	ASTM D 2235
Solvent cement for PVC-DWV pipe and fittings	ASTM D 2564
Socket bell for PVC-DWV pipe and fittings	ASTM D 2672
Primers for solvent cemented PVC-DWV pipe and fittings	ASTM F 656
Couplings for hubless cast-iron soil pipe and fittings	ASTM C 1460; ASTM C 1461; CISPI 310
Shielded couplings joining cast-iron soil pipe and fittings	ASTM C 1277; ASTM C 1460; ASTM C 1461
Cast-iron soil pipe and fittings for hubless sanitary system	ASTM A 888
Co-extruded composite ABS DWV schedule 40 IPS pipe (solid or cellular core) solvent cement fittings	ASTM F 1488; ASTM D 2235; ASTM D 2661; ASTM F 628
Co-extruded composite PVC DWV schedule 40 IPS pipe (solid or cellular core) solvent cement fittings	ASTM F 1488; ASTM D 2564; ASTM D 2665; ASTM F 891
Co-extruded composite PVC DWV IPS-DR-PS in PS35, PS50, PS100, PS140, PS 200 solvent cement fittings	ASTM F 1488; ASTM D 2564; ASTM D 2665; ASTM F 891
Co-extruded composite ABS sewer and drain pipe DR-PS in PS35, PS50, PS100, PS140, PS200 solvent cement fittings	ASTM F 1488; ASTM D 2235; ASTM D 2751
Co-extruded composite PVC sewer and drain pipe DR-PS in PS35, PS50, PS100, PS140, PS200 solvent cement fittings	ASTM F 1488; ASTM D 2564; ASTM D 3034; ASTM F 789

For SI: 1 inch = 25.4 mm.

❖ The referenced table specifies allowable piping material for sewers. It also identifies the standards to which these materials must comply.

ABS PLASTIC PIPE: ASTM D 2751 is the referenced standard for ABS sewer pipe and fittings.

The pipe is produced in SDRs of 23.5, 35, and 42. SDR means "standard dimension ratio," which is the ratio between the average outside diameter of the pipe and the minimum wall thickness. See the commentary for Table 3002.1.

CAST-IRON PIPE: See the commentary for Table P3002.1, Cast-Iron Pipe.

CONCRETE PIPE: This pipe is made from portland cement, blast-furnace slag cement of portland-pozzolan cement, and aggregate mixture. Both reinforced and nonreinforced pipe is produced.

ASTM C 14 and CSA A257.1 are the referenced standards for nonreinforced concrete pipe. ASTM C 14 concrete pipe is available in three classes: 1, 2, and 3. The pipe classification relates to the concrete strength based on diameter. Class 1 has the lowest strength requirements, and Class 3 has the highest. The pipe is manufactured in diameters of 4 through 36 inches (102 through 914 mm).

COPPER OR COPPER-ALLOY TUBING TYPE K OR L: See the commentary for Table P3002.1, Copper or Copper-Alloy tubing; type K, L, M, or DWV.

POLYETHYLENE (PE) PLASTIC PIPE (SDR-PR): ASTM F 714 is the referenced standard for polyethylene plastic pipe based on outside diameters of 3.5 inches (90 mm) and larger. This piping material is intended for new construction and insertion renewal of old piping systems used for the transport of municipal sewage, domestic sewage, industrial process liquids, and effluents in both pressure and non-pressure systems. All pipes produced under ASTM F 714 are pressure rated. Polyethylene plastics are defined as plastics or resins prepared by the polymerization of not less than 85 percent ethylene and not less than 95 percent of total olefins with additional compounding ingredients.

POLYVINYL CHLORIDE (PVC) PLASTIC PIPE: ASTM F 891 specifies sewer grade pipe by a PS designation, in addition to Schedule 40 (Type DWV). PS indicates pipe stiffness. The acceptable grades for sewers are PS 50 and 100. The numerical designation indicates the pipe stiffness in pounds per square inch (psi). The higher value indicates the thicker pipe wall.

ASTM D 2655 and D 3034 are sewer-grade PVC pipe and fittings. The materials can be used interchangeably. Both pipes are produced in SDRs of 35 and 41, with a hub end. The PVC material used to produce the pipe may be classified as 12454-B, 12454-C, or 13364-B. The pipe is produced in diameters of 4 through 15 inches (102 through 381 mm).

VITRIFIED CLAY PIPE: The pipe material is surface clay, fire clay, shale, or a combination of these materials. The pipe is classified as standard strength, extra strength, extra quality, and heavy duty. The classification determines the strength of the pipe.

Vitrified clay pipe is produced in diameters of 3 inches (76 mm) and larger. Each length of pipe is marked with the manufacturer's name and the classification of the pipe.

P3002.4 Other materials. Sheet lead, lead bends, lead traps and sheet copper shall comply with Sections P3002.4.1 through P3002.4.3.

❖ This section regulates the material requirements for sheet lead, lead bends, lead traps, and sheet copper. Each of these are addressed in the subsections herein.

P3002.4.1 Sheet lead. Sheet lead for the following uses shall weigh not less than indicated below:

1. Flashing of vent terminals, 3 psf (0.144 kN/m^2).
2. Prefabricated flashing for vent pipes, $2^1/_2$ psf (0.120 kN/m^2).

❖ Sheet lead is ususally used for flashing material, which protects the vent piping penetration through the roof membrane at the point of termination. If such flashes are prefabricated by a manufacturer, they must have a weight not less than 2.5 psf (0.120 kN/m^2).

Where sheet lead is used as the flashing, the minimum weight must not be less than 3 psf (0.144 kN/m2). See the commentary for Section P2606.

P3002.4.2 Lead bends and traps. Lead bends and lead traps shall not be less than 0.125-inch (3.2 mm) wall thickness.

❖ Although lead is not one of the materials referenced in Table P3002.1, on those rare occasions where it must be used for the purpose of repairs to existing installations of the same material the minimum wall thickness of lead bends and traps must not be less than 0.125 inches (3.2 mm).

P3002.4.3 Sheet copper. Sheet copper for the following uses shall weigh not less than indicated below:

1. General use, 12 ounces per square feet (3.82 L/m^2).
2. Flashing for vent pipes, 8 ounces per square feet (2.55 L/m^2).

❖ Sheet copper is used in similar ways as sheet lead, as both a lining material and for flashing material. When used as a flashing material the minimum thickness must not be less than 8 ounces (226.8 g) per square foot (3.82 L/m2). For other general uses (such as lining material), it must not be less than 12 ounces per square foot (2.55 L/m2). See Section P2709.3.1 for requirements pertaining to the use of sheet copper for shower lining material.

SECTION P3003 JOINTS AND CONNECTIONS

P3003.1 Tightness. Joints and connections in the DWV system shall be gas tight and water tight for the intended use or pressure required by test.

❖ Drainage, waste, and vent systems carry waste and sewer gases, and are therefore required to have gas-tight and water-tight joints sufficient to contain the pressure imposed by the test and not allow leakage of gases or liquids during normal operation. See the commentary for Section 2503.5.

P3003.2 Prohibited joints. Running threads and bands shall not be used in the drainage system. Drainage and vent piping shall not be drilled, tapped, burned or welded.

The following types of joints and connections shall be prohibited:

1. Cement or concrete.
2. Mastic or hot-pour bituminous joints.
3. Joints made with fittings not approved for the specific installation.
4. Joints between different diameter pipes made with elastomeric rolling O-rings.
5. Solvent-cement joints between different types of plastic pipe.
6. Saddle-type fittings.

❖ Running threads in drainage and vent piping joints restrict the interior diameter of the piping and violate the requirement for a smooth interior; they are therefore prohibited.

Drilling, tapping, burning, or welding to create fittings for connections is prohibited because these means create sharp corners in lieu of the smooth transitions afforded by directional fittings. Cement mortar joints, which could allow liquids and gas to "sweat" through, are prohibited, as are joints made of mastic and hot-pour bituminous. Couplings used for joining different diameters of pipe made with elastomeric rolling O-rings are prohibited because they create sudden enlargement or reduction in the inside water way. Fittings that either increase or decrease the pipe size must be used in these cases. Saddle type fittings create sharp corners that can contribute to accumulation of solids, thus creating blockage. Finally, because solvent-cements are usually designed for specific types of plastic pipe, they must not be used for joining dissimilar types of plastic pipe and fittings.

P3003.3 Joint requirements, similar piping materials. Joints between similar piping materials shall be made in accordance with Sections P3003.3.1 through P3003.3.7.

❖ The remainder of Section P3003 is divided into two categories: (1) joining similar types of materials and (2) joining different types of materials. This section now addresses the first of these categories: the joining of similar piping and fitting materials.

P3003.3.1 Cast-iron pipe, caulked joints. Lead-caulked joints for cast-iron hub and spigot soil pipe shall be firmly packed with oakum and filled with molten lead not less than 1 inch (25.4 mm) deep and shall not extend more than 0.125 inch (3.2 mm) below the rim of the hub. Lead shall be poured in one operation and shall be caulked tight.

❖ A caulked joint is the oldest form of joining cast-iron pipe. The joint is sealed by the use of tightly packed oakum or hemp. The oakum or hemp expands the first time it becomes wet, thus sealing the joint. Lead is poured in place behind the oakum to prevent the oakum from expanding out of the joint. Quite often, a caulked joint will leak when it is first tested because the oakum has not yet expanded in place. Once the oakum has expanded, the joint will not leak; see Commentary Figure P3003.3.1.

The joint is made by packing the oakum or hemp into the hub fitting around the pipe to be joined. The oakum and joint surfaces must be dry to prevent the oakum from expanding prematurely. Once in place, the oakum is packed tight with a packing iron and hammer. Molten lead is poured in one continuous operation, securing the oakum. If the joint surfaces or oakum are wet when pouring the lead, rapidly expanding steam can cause the lead to pop or explode. After the lead solidifies, it must be packed tight. A properly made caulked joint will bear marks on the lead caused by a finishing iron used to pack the lead. Because caulked joints are labor intensive and involve the risk of worker injury, and because lead is a health hazard, such joints are becoming increasingly rare.

The code does not recognize a substitute for lead in caulked joints. In the case of remodeling and repair work where one or more caulked joints must be made to allow transitions to different types of piping, building officials have on occasion permitted the use of approved lead substitutes such as cement-based compounds. Elastomeric seals are also available for the purpose of making transition joints with hubbed cast-iron pipe. See the commentary for Section P3003.3.2.2.

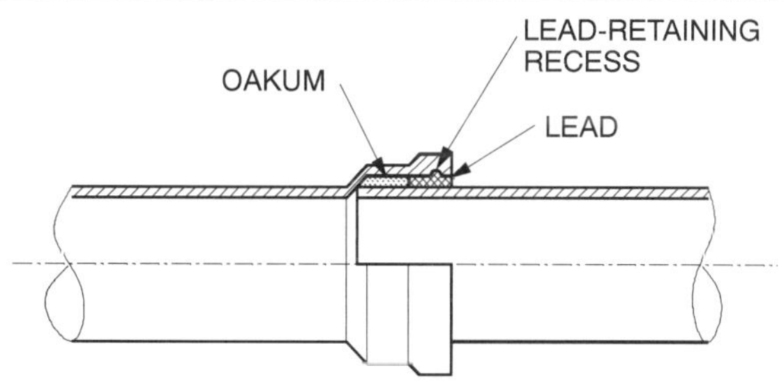

Figure P3003.3.1
CAULKED JOINT FOR HUB AND SPIGOT (CAST IRON)

P3003.3.2 Cast-iron pipe, mechanical joints. Mechanical joints used with cast-iron pipe shall comply with Section P3003.3.2.1 or P3003.3.2.2.

❖ There are two primary methods for mechanically joining cast-iron pipe. These are outlined in the subsections below.

P3003.3.2.1 Hubless pipe. Joints for hubless cast-iron soil pipe shall be made with an approved elastomeric sealing sleeve and stainless steel-retaining sleeve.

❖ Mechanical joint couplings refer to devices that are used with hubless cast-iron drainage, waste, and vent pipe and fittings. Mechanical joint couplings are manufactured in a variety of types, all of which include a metallic shield or sleeve, elastomeric seal, and a clamp assembly that compresses the seal around the outside circumference of the pipe or fitting.

Mechanical joint couplings are usually designed to resist the shear forces that can cause misalignment of pipe and fittings. CSA B602 applies only to the elastomeric seal component of mechanical joint couplings. CISPI 310 and ASTM C 1277 address the entire coupling assembly. The purpose of the center stop is to prevent the pipe or fitting from penetrating too far into the coupling. The center stop is designed to allow proper penetration depth for both pipes and fittings being joined.

The code regulates the design and quality of the elastomeric seals and the overall design of mechanical joint coupling assemblies used to join cast-iron DWV pipe and fittings. The standards CISPI 310 and ASTM C 1277 address couplings that utilize a metal shield. The shield protects the elastomeric seal and contributes shear force resistance and rigidity to the assembled joint.

Mechanical joints and mechanical joint couplings must be installed in accordance with the manufacturer's installation instructions for the device. See Commentary Figure P3003.3.2.1.

P3003.3.2.2 Hub and spigot joints. An approved positive-seal one-piece elastomeric compression-type gasket that is placed in the hub before the spigot is inserted shall be permitted to be used for joining hub and spigot cast-iron soil piping and fittings as an alternate for lead and oakum joints.

❖ The elastomeric gasket is often called a "rubber gasket" or "rubber joint." The gasket is inserted into the hub of the pipe. A lubricant is used to coat the end of the joining pipe and the gasket. The pipe is inserted into the hub by the use of a special tool that pulls the pipe and hub together, compressing the gasket in the annular space between the pipe and the hub.

The pipes must be completely pulled together to properly make the joint. A typical compression joint has two main compression points in the seal. See Commentary Figure P3003.3.2.2.

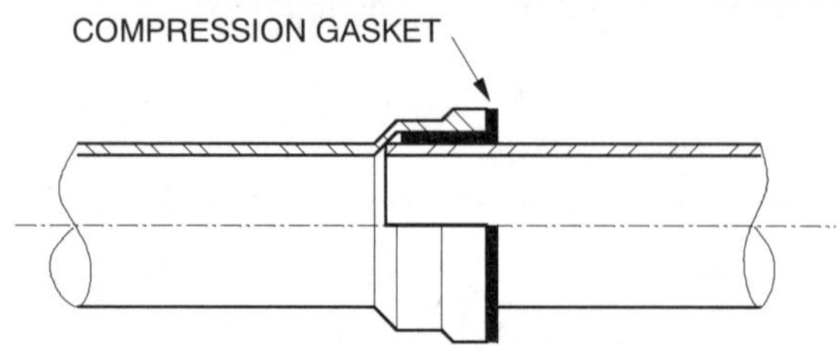

Figure P3003.3.2.2
CAST-IRON COMPRESSION GASKET

P3003.3.3 Threaded pipe joints. Threaded joints shall conform to American National Taper Pipe Thread. Pipe ends shall be reamed or filed to size and all chips removed. Pipe-joint compound or tape shall be applied on the male threads only.

❖ Threads made in the field must conform to the American National Taper Pipe Thread. After cutting the threads, burrs must be removed, and the pipe must be reamed so that any remaining material that might obstruct the flow within the pipe or hinder the proper tightness of the joint is removed. It takes much practice and hands-on training to master the use of pipe threading equipment. When threading steel pipe in the field, one must be sure to align the piping accurately for proper thread alignment. It is imperative that the threads not exceed the maximum required for the size of pipe. If there are too many threads, the proper amount of taper will not be in contact with the fitting to ensure tight-

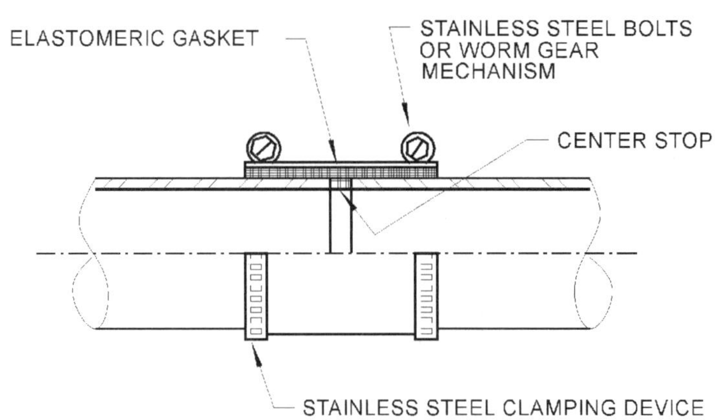

Figure P3003.3.2.1
HUBLESS CAST-IRON DRAINAGE PIPE JOINT

ness. If there are not enough threads, the integrity of the joint is diminished.

Pipe joint compound must be applied to the male threads only. This will prevent the compound from entering the piping system when the joints are made.

P3003.3.4 Soldered joints. Soldered joints in tubing shall be made with fittings approved for DWV piping. Surfaces to be soldered shall be cleaned bright. The joints shall be properly fluxed and made with approved solder. Fluxes shall conform to ASTM B 813. The joint shall be soldered with a solder conforming to ASTM B 32.

❖ A soldered joint is the most common method of joining copper pipe and tubing. Pipe must be cut square to provide proper alignment, adequate surface area for joining, and an interior free from obstructions. When the pipe is cut, an edge (burr) is left protruding into the pipe. The pipe must be reamed to remove the burr properly. Chamfering is required to bevel the outer edge of the cut end of the pipe. Undercutting (excessive reaming) can reduce pipe wall thickness. See Commentary Figure P3003.3.4.

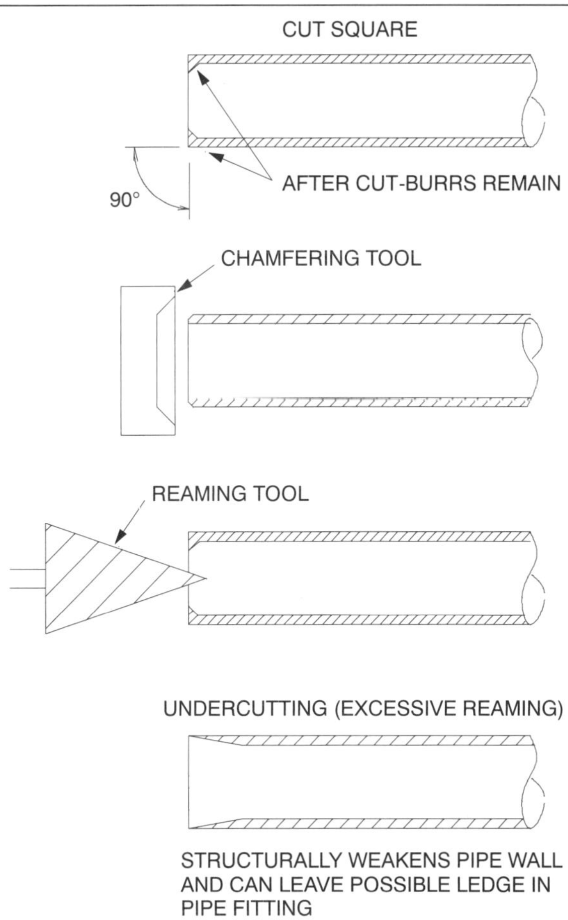

For SI: 1 degree = 0.0175 rad.

Figure P3003.3.4
PREPARATION OF PIPE END

ASTM B 32 covers many grades of solder. The standard also includes tin-silver solders. Previously, the grades used in the plumbing industry were ususally 40-60 (40 percent tin and 60 percent lead), 50-50 (50 percent tin and 50 percent lead) and 95-5 (95 percent tin and 5 percent antimony). While lead-bearing solders are no longer permitted for joining copper tubing used for the distribution of potable water, lead-bearing solders are still acceptable for joining drainage and vent piping. See the commentary for Section P2904.10.

Flux must be applied to all joint surfaces. Flux is a chemically active material that removes and excludes oxides from the joint area during heating, allowing the melted solder to spread out onto the surfaces to be joined. The flux material must conform to ASTM B 813, which establishes minimum performance criteria. It also limits the maximum lead content to 0.2 percent.

The solder will flow by capillary action toward the heat. As such, the temperatures of the joint surfaces play a major role in making a properly soldered joint. The fitting and the pipe need to be approximately the same temperature when the solder is applied. Heating the copper liquefies the flux and solder and expands the surface voids of the material, allowing the solder to penetrate. A strong bond is created when the solder solidifies. If the temperatures are incorrect or uneven, the solder will run down the inside of the pipe and fitting or on the outside of the pipe. The result is a solder-lined pipe. Care must also be exercised when using wrought fittings because there is less of a ledge provided to prevent the flow of solder into the fitting socket. The same care must also be exercised when making a swage joint.

In a swage joint, the end of one pipe is expanded to form a socket fitting. A swaging tool, similar to a hammer-type flaring tool, is used to expand the pipe. Because it is similar to a flared joint, only annealed and bending tempered-drawn copper can be swaged.

P3003.3.5 Clay or cement soil pipe joints. Joints between concrete pipe or fittings shall be made with an elastomeric seal conforming to ASTM C 443, ASTM C 1173, CSA CAN/CSA A257.3 or CSA CAN/CSA-B602. Joints between vitrified clay pipe or fittings shall be made with an elastomeric seal conforming to ASTM C 425, ASTM C 1173 or CSA CAN/CSA-B602.

❖ An acceptable method of joining concrete pipe is by an elastomeric gasket that fits in the hub of the pipe. The gasket is compressed between the hub and the pipe to form a seal. The ASTM and CSA standards regulate the design, material, and performance of the elastomeric gasket. See Commentary Figure P3003.3.5.

P3003.3.6 Plastic pipe joints. Joints in plastic piping shall be made with approved fittings by solvent cementing, elastomeric gaskets or other approved manufactured system.

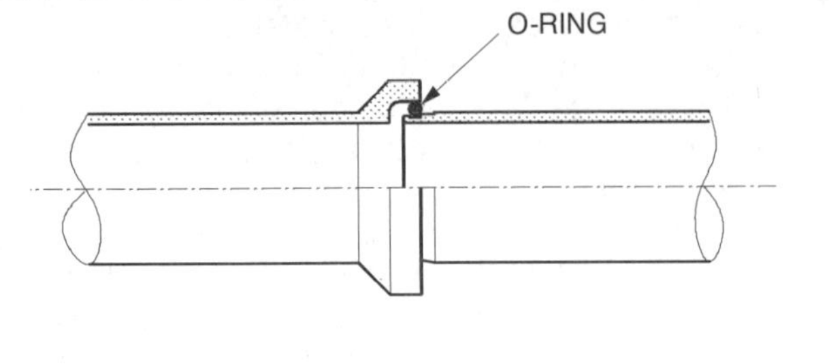

Figure P3003.3.5
CONCRETE MECHANICAL JOINT

❖ Two typical types of plastic joints in DWV systems and sewer pipes are ABS and PVC. Both of these types of materials may involve the use of solvent cement. However, PVC also utilizes a primer. Although solvent cementing is the most common type of joint for plastic DWV, mechanical joints are often used where other joining methods cannot be used. For example, underground PVC pipe can be joined by means of an approved bell and spigot type of mechanical joint; because of water in the trench, solvent cementing is not possible. See Commentary Figure P3003.3.6.

Solvent cementing is the most common method of joining ABS plastic pipe and fittings. ABS solvent cement is a flammable liquid and a source of flammable vapors. It must be kept at a safe distance from any source of ignition and used in a well-ventilated area. Protective goggles and impervious gloves must be worn.

The pipe and fittings need to be approximately the same temperature prior to solvent cementing. The pipe end and socket fitting must be clean, dry, and free from grease, oil, and other foreign materials. The solvent cement must be uniformly applied to the socket of the fitting and to the end of the pipe to the depth of the socket. Excessive solvent cement must be avoided to prevent a puddle of cement from forming on the inside of the pipe joint. Puddling of cement could deform or structurally weaken the fitting or pipe. The joint is made by twisting the pipe into the fitting socket immediately after applying the cement. The pipe must be held in place for a few seconds until the solvent cement begins to set. The pipe will tend to back out unless it is restrained. A small bead of solvent cement forms between the pipe exterior and fittings. The bead must be removed with a clean cloth to avoid the possibility of weakening the pipe wall. A solvent-cement joint must not be exposed to the working pressure for a period of 24 hours. The pipe must not be handled roughly for at least 1 hour after the joint is made.

Because each type of solvent cement is specifically designed for a given piping material, all-purpose solvent or universal solvent cement cannot be used to join ABS pipe or fittings unless it conforms to ASTM D 2235. Solvent cement for each plastic pipe material requires a unique mixture or combination of solvents and dissolved plastic resins in order to conform to the appropriate standard. Although all-purpose cement may dissolve the surfaces of ABS, PVC, and CPVC, it may be too aggressive or not aggressive enough to meet the minimum strength requirements of the appropriate referenced standards. Solvent cement must be identified on its container as complying with the appropriate ASTM standard.

Other methods, such as elastomeric gaskets, can also be used to join plastic pipe.

Slip joints cannot be concealed and are limited to those connections that occur between a fixture waste outlet and its connection to the fixture drain, including the trap adapter on the trap outlet arm. In locations where a slip joint connection would be concealed, an access panel must be provided so that the connection can be serviced. This would include tub waste and overflow connections. See the commentary for Section P2704.

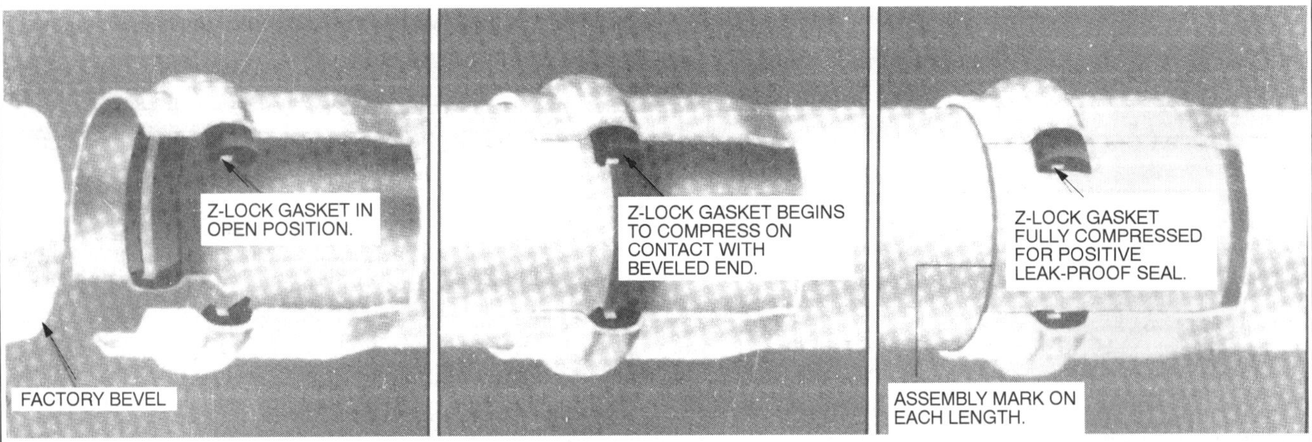

Figure P3003.3.6
PVC MECHANICAL JOINT
(Photo courtesy of Lamson Vylon Pipe)

P3003.3.7 Slip joints. Slip joints shall be made using approved gaskets or compression washers. Ground joint connections that allow adjustment of tubing but provide a rigid joint when made up shall not be considered slip joints. When a ground joint connection is used and the assembled joint does not permit free movement, the joint need not be accessible.

❖ An elastomeric gasket must be used to make the seal for drainage slip joints. See Commentary Figures P3003.3.7(1) and P3003.3.7(2) and the definition of "Slip Joint" in Chapter 2.

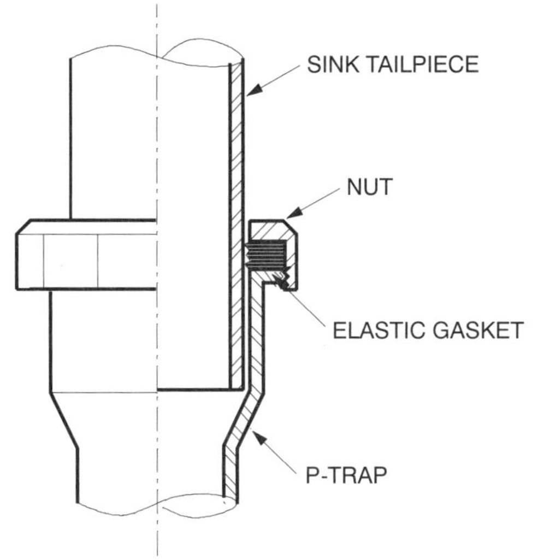

Figure P3003.3.7(1)
SLIP JOINT

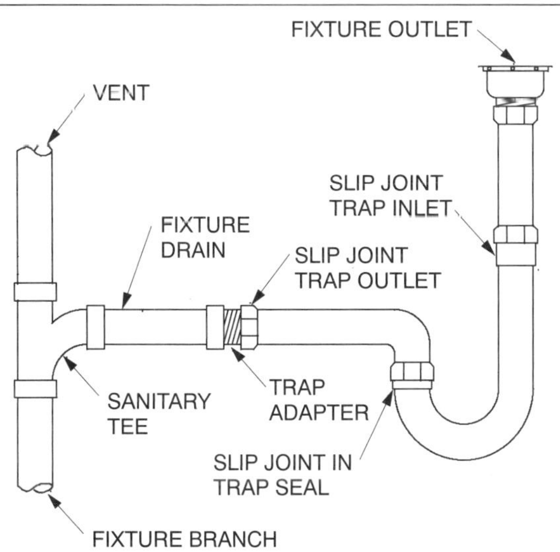

Figure P3003.3.7(2)
SLIP JOINTS

Slip joints cannot be concealed and are limited to those connections that occur between a fixture waste outlet and its connection to the fixture drain, including the trap adapter on the trap outlet arm. In locations where a slip joint connection would be concealed, an access panel must be provided so that the connection can be serviced. This would include tub waste and overflow connections. See the commentary for Section P2704.

P3003.4 Joints between different piping materials. Joints between different piping material shall comply with Sections P3003.4.1 through P3003.4.5.

❖ Mechanical joints are typically used to join dissimilar piping materials; see Commentary Figure P3003.4. Because many piping materials have the same outside diameters, many fittings are designed by manufacturers for use with a number of different piping materials; see Sections P3003.4.1 through P3003.4.5. Some mechanical joints are manufactured specifically for joining pipes having different outside diameters. Transition couplings utilize bushings (inserts) to match the outside diameters of the pipes or are simply designed to fit over different outside diameter pipe ends. ASTM C 1460 or ASTM C 1461 are the standards for the complete coupling assembly, which includes an elastomeric gasket, clamp assembly, or both; see Table P3002.1 and Table P3002.2.

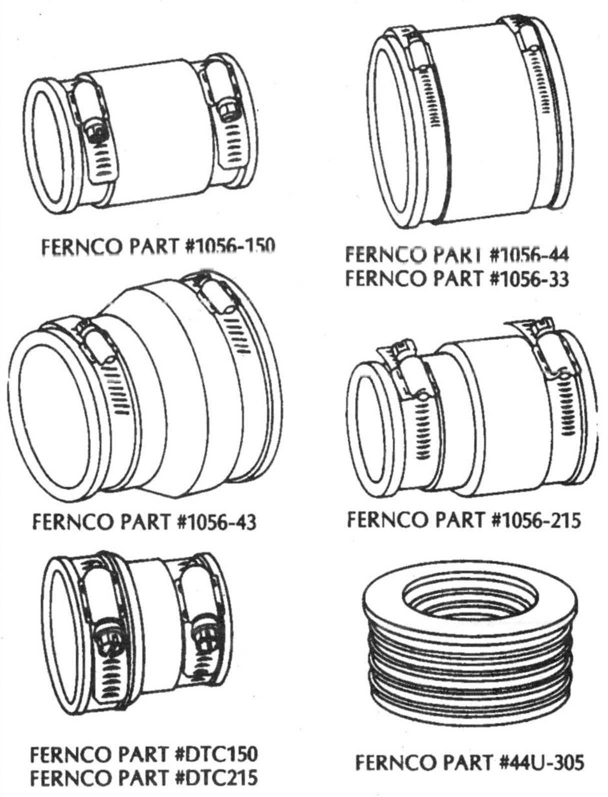

Figure P3003.4
ELASTOMERIC MECHANICAL JOINT
(Figure courtesy of Fernco, Inc.)

P3003.4.1 Hub-type cast-iron or vitrified clay to other piping materials. Joints between hub-type drainage piping and other materials, including steel, plastic and copper shall be made with a lead and oakum joint and an approved caulking ferrule of the other material or by a lead and oakum joint directly with the other material omitting the adapter ferrule. Where the outside diameter of the other material matches that of the hub pipe, an elastomeric gasket designed for use with the hub pipe shall be permitted to be used.

❖ Joints between hub-type drainage piping and other material such as steel, copper, or plastic can be made with lead and oakum joints. For this type of joint, a caulking ferrule of the other type of material can be used. However, with plastic pipe, care must be exercised so that the high temperature used with lead will not negatively affect the plastic and melt it. Because most materials have the same outside diameter, mechanical joints utilizing elastomeric gaskets designed for use with the hub pipe can be used. This method of joining is simple and easy and, unlike the lead and oakum method, does not affect the piping material.

P3003.4.2 Hubless pipe to other piping materials. Joints between hubless drainage piping and any other materials shall be joined with an approved elastomeric sleeve and stainless steel clamp.

❖ An elastomeric sleeve and stainless steel clamp, with or without a shield, can be used to join cast-iron pipe to other hubless pipe material. However, the adapter must be manufactured to fit the outside diameter of the connecting material.

P3003.4.3 Threaded pipe to cast-iron soil pipe. Joints between threaded pipe and cast-iron soil pipe shall be made with approved adapter fittings.

❖ Because cast-iron pipe must not be threaded, an approved adapter fitting must be used to join the cast-iron pipe with other threaded pipe such as steel.

P3003.4.4 Threaded pipe to copper or plastic pipe. Joints from threaded pipe to copper or plastic piping shall utilize approved adapter fittings.

❖ Adapter fittings must be used to connect threaded pipe to copper or plastic pipe. Threading copper in the field is not recommended, because the pipe wall thickness is usually not sufficient to make an appropriate thread. Plastic pipe, unless made of thick wall piping (Schedule 40) has the same problem. Therefore, prefabricated adapter fittings made under factory controlled conditions are most commonly used.

P3003.4.5 Joints between drainage piping and water closets. Joints between drainage piping and water closets or similar fixtures shall be made by means of a closet flange compatible with the drainage system material, securely fastened to a structurally firm base. The inside diameter of the drainage pipe shall not be used as a socket fitting for a four by three closet flange. The joint shall be bolted, with an approved gasket, flange to fixture connection complying with ASME A112.4.3 or setting compound between the fixture and the closet flange.

❖ Joints between the drainage piping and water closets must be made tight by means of closet flanges that are compatible with the piping material. The closet flange must be securely fastened to the floor, which must be firm and structurally sound. It is only fitting that noncorrosive fasteners be used for the purpose of securing the flange to the floor. The joint must be bolted, utilizing a gasket or a setting compound between the water closet and the flange or a flange-to-fixture connection complying with ASME A112.4.3 for a tight seal. The flange must never be used to structurally support the water closet weight; rather, it must be sandwiched between the water closet and the supporting ground. The weight of the water closet must be transferred directly to the floor support system to prevent cracking of the flange, which can lead to leakage of the joint.

SECTION P3004
DETERMINING DRAINAGE FIXTURE UNITS

P3004.1 DWV system load. The load on DWV-system piping shall be computed in terms of drainage fixture unit (d.f.u.) values in accordance with Table P3004.1.

❖ This section regulates the maximum loads that can be imposed on a drainage system. Table P3004.1 is used for the determination of drainage fixture units (d.f.u.) The conventional method of sizing a sanitary drainage system is by drainage fixture unit load values. The fixture unit approach takes into consideration the probability of load on a drainage system. The probability method of sizing drainage systems was developed largely by Dr. Roy B. Hunter. Through his research, Dr. Hunter helped standardize and simplify design principles, which reduced the cost of plumbing systems.

Fixture unit values were determined based on the average rate of discharge by a fixture, the time of a single operation, and the frequency of use or interval between each operation. The theoretical approach considers a large group of fixtures being connected to the plumbing system with only a small fraction of the total number of fixtures in use simultaneously. The probability method has also been effective in the design of smaller plumbing systems because of the overdesign factor added by Dr. Hunter. Dr. Hunter sought an adequate design methodology to provide satisfactory service without interruption or inconvenience to the user.

Because the fixture unit values have a built-in probability factor, they cannot be directly translated into flow rates of discharge. For example, 1 drainage fixture unit is equivalent to a discharge rate of 1 cubic foot per minute (0.0004719 m^3/S) or approximately 7.5 gallons (28.78 L) per minute. Two independent fixtures, however, each with a value of 1 drainage fixture unit, cannot be considered to have a combined discharge rate of 15 gallons (56.78 L) per minute, because the drainage fixture unit value incorporates the element of probability.

TABLE P3004.1
DRAINAGE FIXTURE UNIT (d.f.u.) VALUES FOR VARIOUS PLUMBING FIXTURES

TYPE OF FIXTURE OR GROUP OF FIXTURES	DRAINAGE FIXTURE UNIT VALUE (d.f.u.)[a]
Bar sink	1
Bathtub (with or without shower head and/or whirlpool attachments)	2
Bidet	1
Clothes washer standpipe	2
Dishwasher	2
Floor drain[b]	0
Kitchen sink	2
Lavatory	1
Laundry tub	2
Shower stall	2
Water closet (1.6 gallons per flush)	3
Water closet (greater than 1.6 gallons per flush)	4
Full-bath group with bathtub (with 1.6 gallon per flush water closet, and with or without shower head and/or whirlpool attachment on the bathtub or shower stall)	5
Full-bath group with bathtub (water closet greater than 1.6 gallon per flush, and with or without shower head and/or whirlpool attachment on the bathtub or shower stall)	6
Half-bath group (1.6 gallon per flush water closet plus lavatory)	4
Half-bath group (water closet greater than 1.6 gallon per flush plus lavatory)	5
Kitchen group (dishwasher and sink with or without garbage grinder)	2
Laundry group (clothes washer standpipe and laundry tub)	3
Multiple-bath groups[c]: 1.5 baths 2 baths 2.5 baths 3 baths 3.5 baths	 7 8 9 10 11

For SI: 1 gallon = 3.785 L.

a. For a continuous or semicontinuous flow into a drainage system, such as from a pump or similar device, 1.5 fixture units shall be allowed per gpm of flow. For a fixture not listed, use the highest d.f.u. value for a similar listed fixture.

b. A floor drain itself adds no hydraulic load. However, where used as a receptor, the fixture unit value of the fixture discharging into the receptor shall be applicable.

c. Add 2 d.f.u. for each additional full bath.

❖ This table identifies common plumbing fixtures with their corresponding drainage fixture unit values. The drainage fixture unit values (d.f.u.) are used to determine the load on a drain, which in turn is used to determine the minimum drainage pipe size requirement for the load served. In determining the fixture unit values, each fixture was evaluated for its impact on the sanitary drainage system. The original development of fixture unit values was based on an arbitrary scale established by Dr. Hunter.

The table takes into consideration the unlikely probability of multiple fixtures being used simultaneously. Therefore, sizing can be based on d.f.u. values for individual fixtures or for fixtures combined with adjacent fixtures within the same bathroom or grouping. For this purpose, the table assigns combined d.f.u. values for groups of bathroom fixtures, kitchen fixtures, appliances, and laundry fixtures. These groups are adequately defined within the table itself.

Additional information is provided in the footnotes to the table for the purpose of sizing drains that receive the discharge of pumps or similar devices, floor drains, multiple bathroom groups, and other types of fixtures or drains not specifically listed within the table.

SECTION P3005 DRAINAGE SYSTEM

P3005.1 Drainage fittings and connections. Changes in direction in drainage piping shall be made by the appropriate use of sanitary tees, wyes, sweeps, bends or by a combination of these drainage fittings in accordance with Table P3005.1. Change in direction by combination fittings, heel or side inlets or increasers shall be installed in accordance with Table P3005.1 and Sections P3005.1.1 through P3005.1.4. based on the pattern of flow created by the fitting.

❖ This section refers the user to Table P3005.1 and the appropriate subsections for the determination of the type of drainage fitting that can be used depending on the installation. Consideration is given to the direction of flow, the type of fixtures installed, and the point of connection as it relates to stacks, vents, drains, or other connections. Drainage fittings and connections must provide a smooth transition of flow without creating obstructions or causing interference. The use of proper fittings helps to maintain the required flow velocities and reduces the possibility of stoppage in the drainage system.

Combination fittings are commonly used in drainage systems and must be evaluated for their pattern of flow. Combination fittings include fittings made up of two or more fittings and one-piece fittings such as combination wye and eighth bends or tee-wyes.

TABLE P3005.1
FITTINGS FOR CHANGE IN DIRECTION

TYPE OF FITTING PATTERN	CHANGE IN DIRECTION		
	Horizontal to vertical[c]	Vertical to horizontal	Horizontal to horizontal
Sixteenth bend	X	X	X
Eighth bend	X	X	X
Sixth bend	X	X	X
Quarter bend	X	X[a]	X[a]
Short sweep	X	X[a,b]	X[a]
Long sweep	X	X	X
Sanitary tee	X[c]		
Wye	X	X	X
Combination wye and eighth bend	X	X	X

For SI: 1 inch = 25.4 mm.

a. The fittings shall only be permitted for a 2-inch or smaller fixture drain.

b. Three inches and larger.

c. For a limitation on multiple connection fittings, see Section P3005.1.1.

❖ Where flow is from the horizontal to the horizontal, long-pattern fittings are necessary to prevent excessive reduction in flow velocity. Where flow is from the horizontal to the vertical or from the vertical to the horizontal, a shorter pattern fitting can be used because the acceleration of flow due to gravity will maintain higher flow velocities. An "X" in the column indicates that a particular fitting is acceptable for the described change in direction. For example, the only "X" shown for a sanitary tee is in the column for horizontal to vertical change in direction.

The table has been developed to identify where a particular drainage pattern fitting is allowed to be used. The terminology used to identify drainage pattern fittings was originally developed by the cast-iron soil pipe industry.

The pattern of the fittings is regulated by the fitting standards referenced in Table P3002.1. A fitting of one material will not necessarily look the same as a fitting of another material with the same name.

Note "b" permits the use of short-sweep fittings of 3 inches (76 mm) in diameter or larger where the change in direction is from the vertical to the horizontal. Note "a" permits the use of 2 inch (51 mm) or smaller short-sweep and quarter-bend fittings for drains serving a single fixture where the change in direction is from the vertical to the horizontal or from the horizontal to the horizontal. The intent of Note "c" is to allow short-pattern fittings for fixture drains, which are generally difficult to install using long-pattern fittings in frame construction. Only cast-iron pipe fittings have a separate quarter-bend fitting and short-sweep fitting. For other drainage piping materials, a quarter-bend and short-sweep are the same fitting. A quarter-bend with a very short radius, less than that of a short sweep, is identified as a vent elbow or vent quarter bend. These particular fittings are designed for vent system applications only.

Because there is no waste flow in vent piping and because the airflow in vent piping can be in either direction, the types of fittings used in vent piping are not regulated by this section. For example, a sanitary tee placed on its back with the branch inlet connected to a vertical dry vent riser is a common installation practice and is not prohibited. In this application, the only waste flow is through the straight run of the fitting, and the branch inlet conducts only air into or out of the horizontal drain pipe. Note that the tee must be oriented to allow compliance with Section P3104.3. The restrictions placed on the use of a sanitary tee are based on its inability to guide waste flow entering through the branch inlet. Such restrictions do not apply where the branch inlet is used for a dry vent connection in compliance with Section P3104.

In all cases, the code requires that fittings be installed in a manner that will encourage high-velocity drainage flow and minimize the chances of blockage.

P3005.1.1 Horizontal to vertical (multiple connection fittings). Double fittings such as double sanitary tees and tee-wyes or approved multiple connection fittings and back-to-back fixture arrangements that connect two or more branches at the same level shall be permitted as long as directly opposing connections are the same size and the discharge into directly opposing connections is from similar fixture types or fixture groups. Double sanitary tee patterns shall not receive the discharge of back-to-back water closets and fixtures or appliances with pumping action discharge.

> **Exception:** Back-to-back water closet connections to double sanitary tee patterns shall be permitted where the horizontal developed length between the outlet of the water closet and the connection to the double sanitary tee is 18 inches (457 mm) or greater.

❖ This section addresses the approved use of double (or multiple) inlet fittings that connect more than one horizontal drain to a vertical drain or stack. This is most common with, but is not limited to, back-to-back fixtures. See Commentary Figures P3005.1.1(1) through P3005.1.1(8).

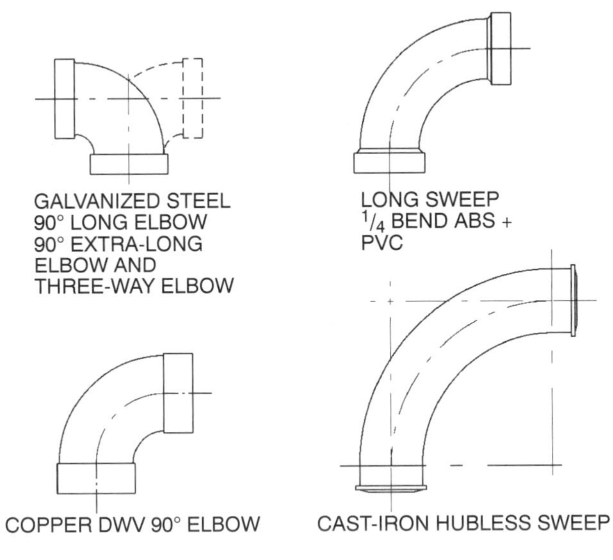

For SI: 1 degree = 0.0175 rad.

Figure P3005.1.1(1)
COMPARISON OF DIFFERENT $1^{1}/_{2}$-INCH LONG SWEEPS FOR VARIOUS PIPING MATERIALS

It is mistakenly thought that the use of a double combination fitting as shown in Commentary Figure P3005.1.1(5) is not permitted for the water closet arrangement, insofar as it will not permit the vent to connect above the trap weir. However, this arrangement is permitted because the vent connection limitation is designed to prevent self-siphonage, and water closets must self-siphon to operate properly.

Double sanitary tees must not be installed where water closets or appliances with pump action are back-to-back. Washing machines and dishwashers are the most typical appliances with pump action in one- and two-family dwellings. The exception is if the distance of the water closet fixture drain (the piping between the level mounting surface of the closet flange where it makes contact with the water closet and the connection to the vent) is 18 inches (457.2 mm) in length or greater. Only in this case may a double sanitary tee be used for back-to-back water closet installations.

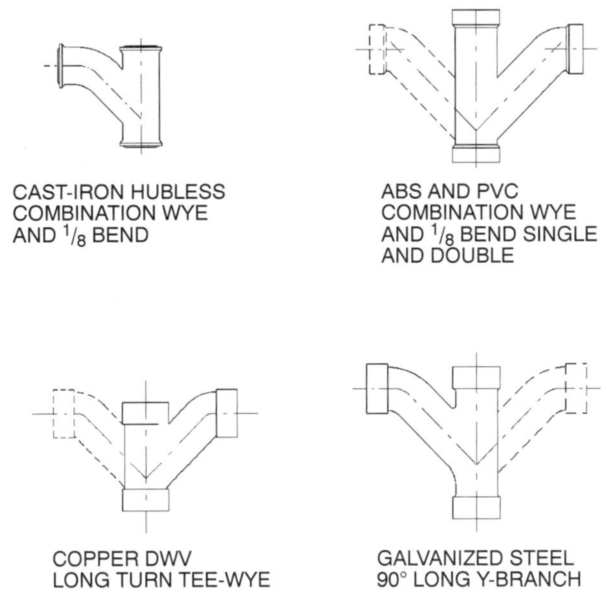

For SI: 1 degree = 0.0175 rad.

Figure P3005.1.1(2)
COMPARISON OF 3-INCH COMBINATION WYE AND EIGHTH BEND FOR VARIOUS PIPING MATERIALS

For all other fixture connections using double sanitary tees and tee-wyes or other multiple connection fittings, the back-to-back fixtures must be similar to one another (two lavatories, two bathtubs, etc.), and each opposing branch connecting to the fitting must be the same size.

P3005.1.2 Heel- or side-inlet quarter bends, drainage. Heel-inlet quarter bends shall be an acceptable means of connection, except where the quarter bends serves a water closet. A low-heel inlet shall not be used as a wet-vented connection. Side-inlet quarter bends shall be an acceptable means of connection for both drainage, wet venting and stack venting arrangements.

❖ This section addresses the use of heel-inlet quarter bends and side-inlet quarter bends for the purpose of drainage. There are two types of heel-inlet quarter bends: a high-inlet and a low-inlet. See Commentary Figure P3005.1.1(6). A heel-inlet quarter bend cannot be used to connect a water closet, and a low-heel inlet quarter bend in particular cannot be used as a wet-vented connection. Side-inlet quarter bends can be used for drainage, wet-venting and stack-venting applications.

P3005.1.3 Heel- or side-inlet quarter bends, venting. Heel-inlet or side-inlet quarter bends, or any arrangement of pipe and fittings producing a similar effect, shall be acceptable as a dry vent where the inlet is placed in a vertical position. The

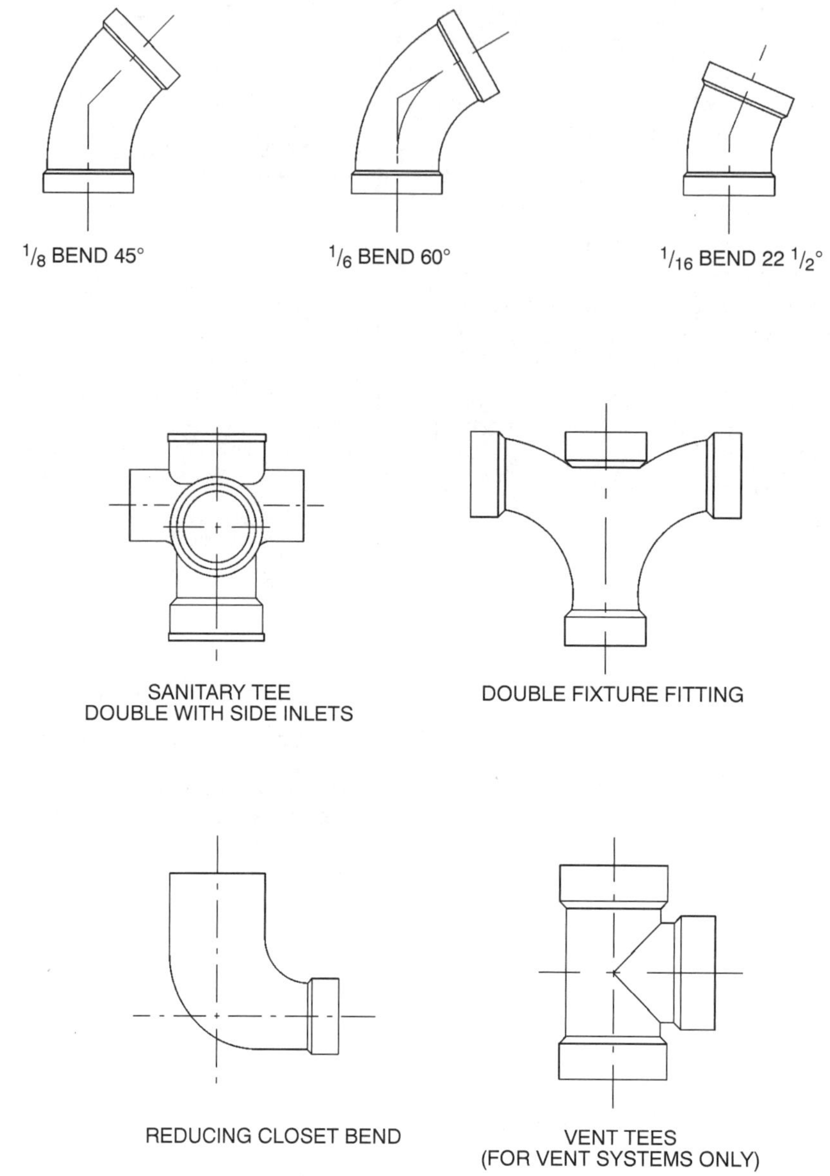

For SI: 1 degree = 0.0175 rad.

Figure P3005.1.1(3)
DRAINAGE PATTERN FITTINGS

inlet is permitted to be placed in a horizontal position only where the entire fitting is part of a dry vent arrangement.

❖ The only application where heel- or side-inlet quarter bends can be used for the purpose of venting a drain or fixture is where the inlet is in the vertical position. As specified in Section P3005.1.2, this inlet cannot serve as a wet vent, so in this application the inlet can only serve as a dry vent.

P3005.1.4 Water closet connection between flange and pipe. One-quarter bends 3 inches (76 mm) in diameter shall be acceptable for water closet or similar connections, provided a 4-inch-by-3-inch (102 mm by 76 mm) flange is installed to receive the closet fixture horn. Alternately, a 4-inch-by-3-inch (102 mm by 76 mm) elbow shall be acceptable with a 4-inch (102 mm) flange.

❖ This section emphasizes that a 3-inch (76 mm) drain is acceptable for a water closet. However, the flange must be 4 inches (102 mm) at the point of connection

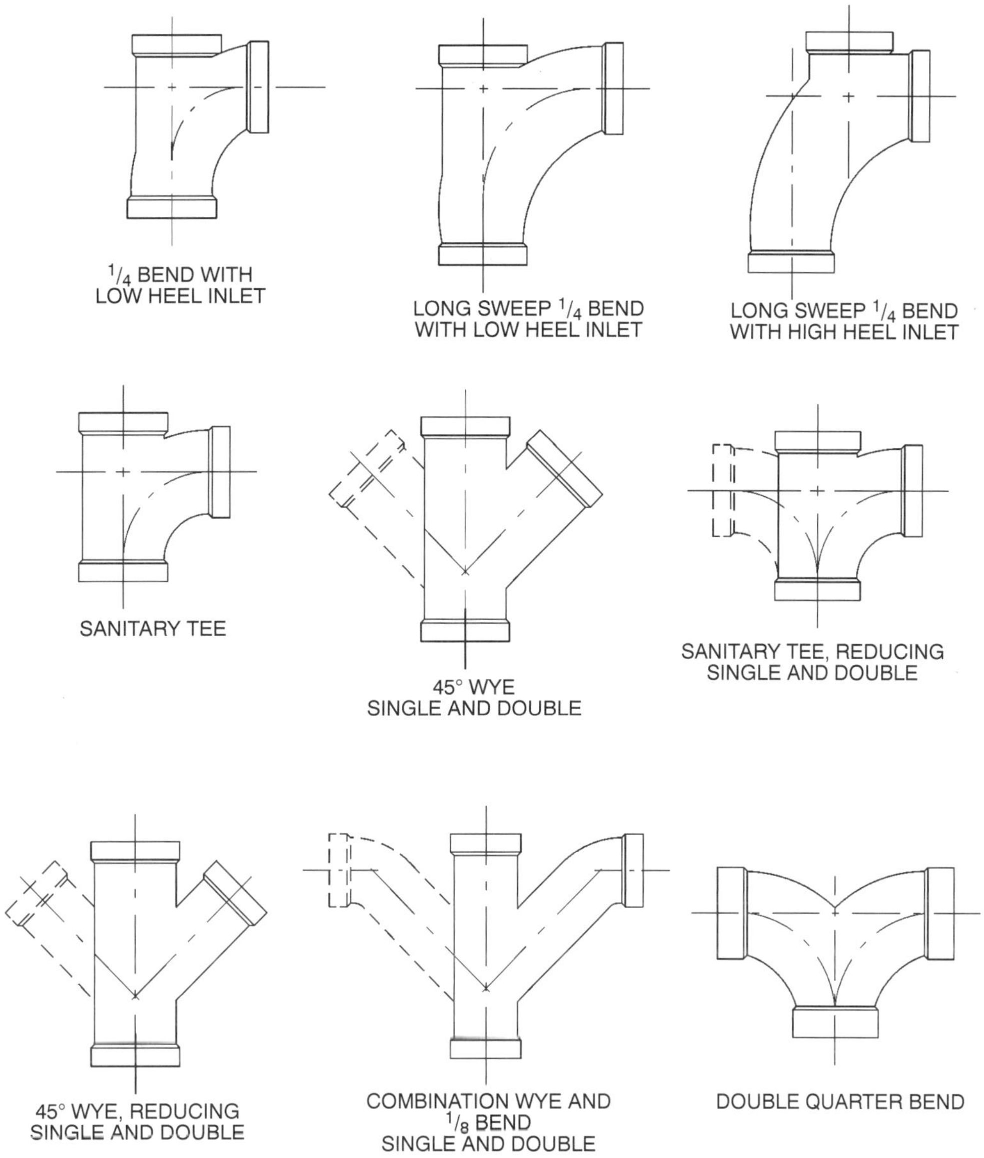

For SI: 1 degree = 0.0175 rad.

Figure P3005.1.1(3)(Cont'd)
DRAINAGE PATTERN FITTINGS

of the water closet in order to encompass the spigot or seat. The reduction from 4 inches (102 mm) to 3 inches (76 mm) by means of a 4-inch by 3-inch (102 mm by 76 mm) closet flange or a 4-inch by 3-inch (102 mm by 76 mm) closet elbow is acceptable and is not to be considered an obstruction, as would be the case with reducing drain sizes downstream of the connection.

P3005.1.5 Dead ends. Dead ends shall be prohibited except where necessary to extend a cleanout or as an approved part of a rough-in more than 2 feet (610 mm) in length.

❖ A dead end is defined as a closed extension of pipe that is 2 feet (610 mm) or more in length. Where such piping is installed in the drainage system and does not receive the discharge of any fixture, solid waste tends to accumulate in the dead end. Flow in a horizontal drain can back up into the dead end because of the potential for solids to drop out of suspension.

The only exceptions are where the piping serves as a cleanout or is an extension to serve an installation. This section must be considered where the installation involves existing construction. Quite often, an existing fixture or set of fixtures can be omitted. Care must be exercised so that the existing drain serving such fixtures is not capped and left as a dead end; hence, this provision will apply.

For SI: 1 degree = 0.0175 rad.

Figure P3005.1.1(4)
DIFFERENCE BETWEEN PVC PLASTIC $^1/_4$-BEND SHORT SWEEP AND $^1/_4$-BEND BENT

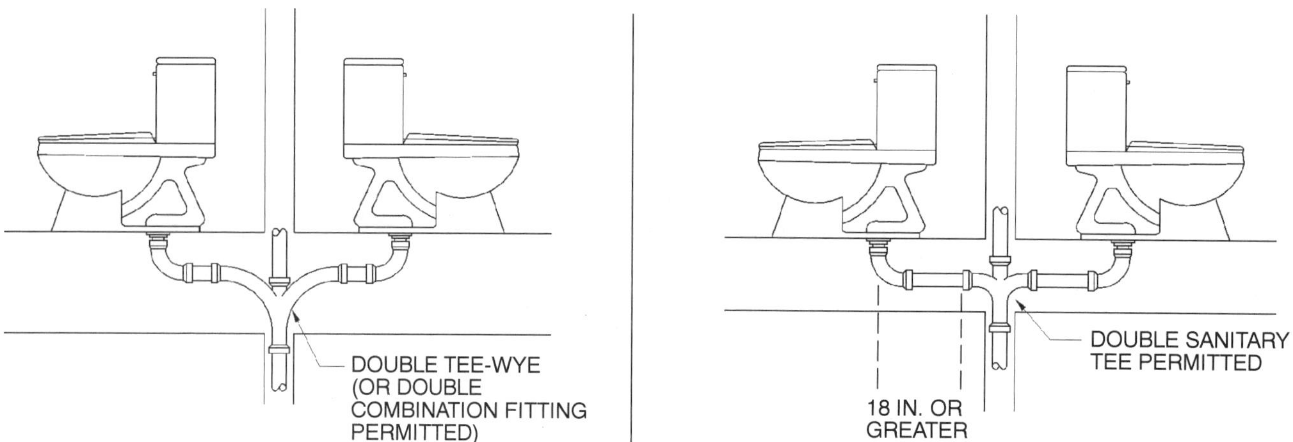

For SI: 1 inch = 25.4 mm.

Figure P3005.1.1(5)
BACK-TO-BACK FIXTURES

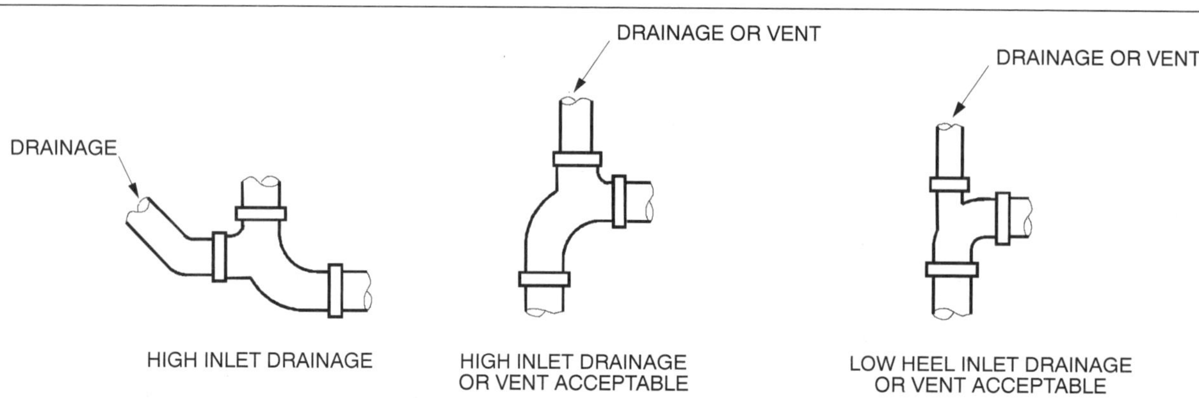

Figure P3005.1.1(6)
HEEL INLET FITTINGS

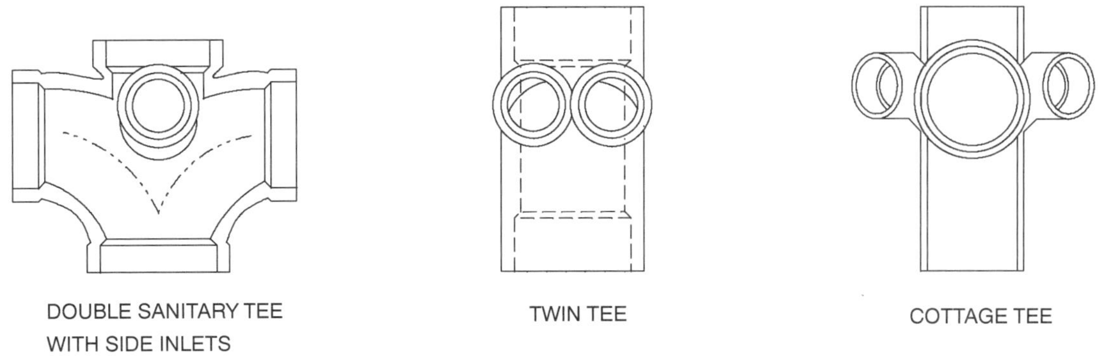

Figure P3005.1.1(7)
COMBINATION TEES

P3005.1.6 Provisions for future fixtures. Where drainage has been roughed-in for future fixtures, the drainage unit values of the future fixtures shall be considered in determining the required drain sizes. Such future installations shall be terminated with an accessible permanent plug or cap fitting.

❖ Where a provision is made for a drain connection to serve future fixtures, the d.f.u. values must be included in determining the size of the drainage system. The piping must be permanently plugged or capped to prevent the escape of sewer gases or the intrusion of foreign substances. The point of future connection at the plug or cap must be accessible. See the definition of "Accessible" in Chapter 2. Note that Section P3104.6 also requires a vent to be roughed-in for drainage piping that will serve fixtures intended to be installed.

P3005.1.7 Change in size. The size of the drainage piping shall not be reduced in size in the direction of the flow. A 4-inch by 3-inch (102 mm by 76 mm) water closet connection shall not be considered as a reduction in size.

❖ One of the fundamental requirements of a drainage system is that the piping cannot be reduced in size in the direction of drainage flow. A size reduction would create an obstruction to flow, possibly resulting in a backup of flow, an interruption of service in the drainage system, or stoppage in the pipe. A 4-inch by 3-inch (102 mm by 76 mm) water closet connection is not considered to be a reduction in pipe size.

P3005.2 Drainage pipe cleanouts. Drainage pipe cleanouts shall comply with Sections P3005.2.1 through P3005.2.11.

Exception: These provisions shall not apply to pressurized building drains and building sewers that convey the discharge of automatic pumping equipment to a gravity drainage system.

❖ This section governs the installation of cleanouts. Even though a code-compliant installation will result in an effective and properly functioning drainage system, the code cannot prevent what might be introduced into the system once the premises are occupied. Hence, there may be times when the drain must be cleaned in order to clear stoppages. This is done with the use of drain cleaning equipment. The following subsections provide appropriate cleanouts strategically located and designed to accommodate such equipment in areas of the installation where it would most likely be needed.

However, an exception provides for pressurized building drains and building sewers that convey the discharge of automatic pumping equipment to a gravity drainage system. The discharge is under pressure; therefore, it is unlikely that a stoppage would occur.

P3005.2.1 Materials. Cleanouts shall be liquid and gas tight. Cleanout plugs shall be brass or plastic.

❖ Because cleanouts are part of the DWV system, they must be tested to prove that they are watertight. Brass or plastic plugs are to be used because they can be threaded and are resistant to corrosion.

P3005.2.2 Spacing. Cleanouts shall be installed not more than 100 feet (30 480 mm) apart in horizontal drainage lines.

❖ Cleanouts must be spaced at a reasonable distance to facilitate cleaning of any portion of the drainage system. This spacing distance is based on the use of modern-day cleaning equipment, and it attempts to minimize the inconvenience and health hazard of creating a mess inside the building as a result of removing a blockage. Inadequate cleanout spacing can require the use of excessively long rodding cables, which complicates the task and increases the likelihood that the rodding cable will jam or break inside the pipe.

Every horizontal branch or building drain must have at least one cleanout, regardless of the length of the horizontal drain. See Commentary Figure P3005.2.2.

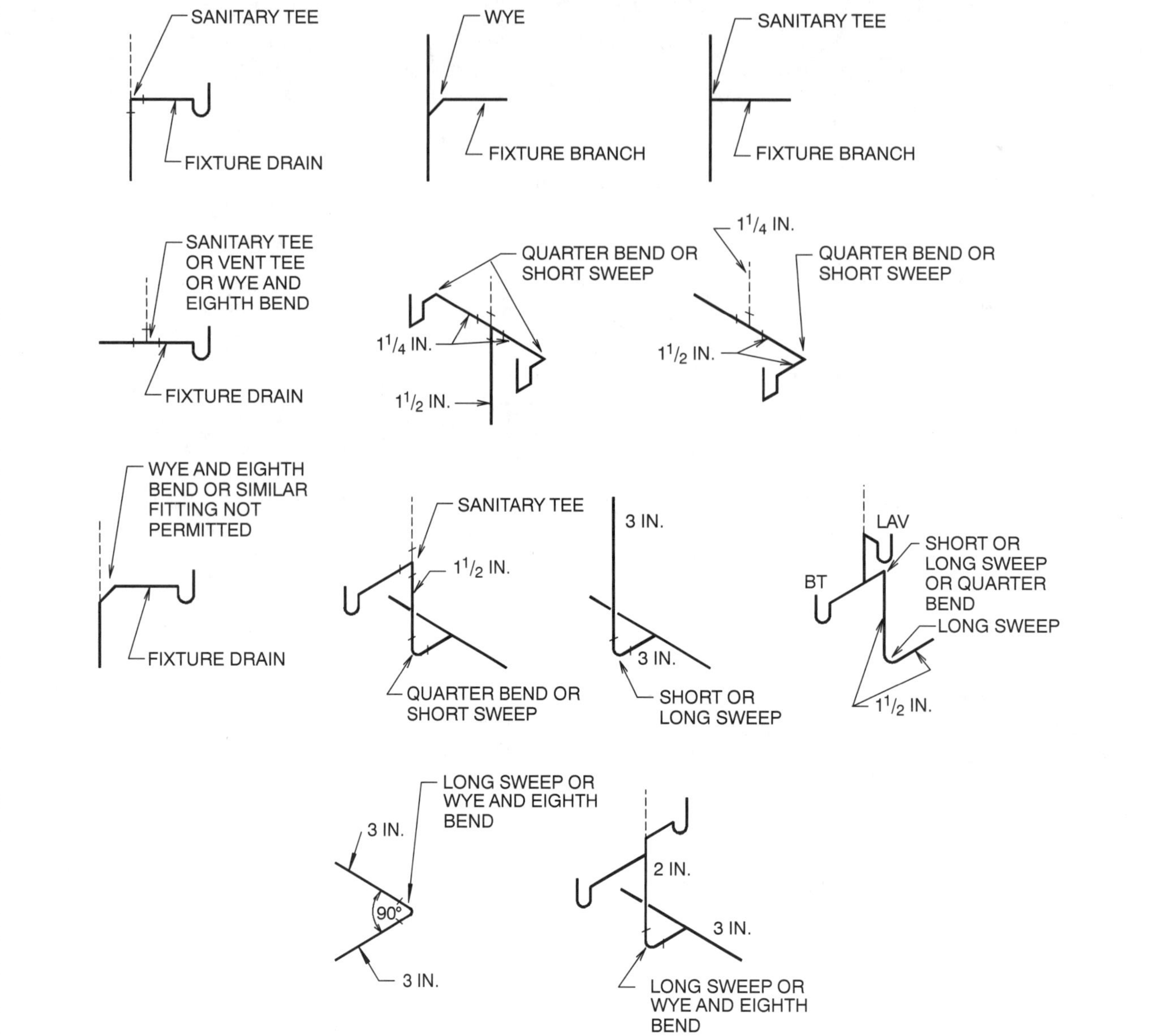

For SI: 1 inch = 5.4 mm, 1 degree = 0.0175 rad.

Figure P3005.1.1(8)
FITTING USES

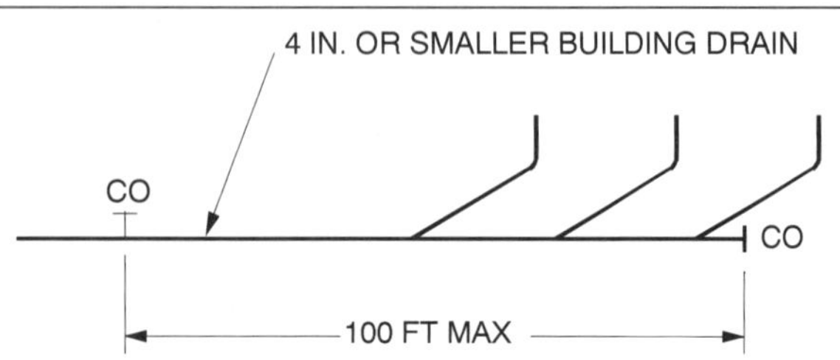

For SI: 1 degree = 0.0175 rad, 1 foot = 304.8 mm.

Figure P3005.2.2
CLEANOUT LOCATIONS

P3005.2.3 Underground drainage cleanouts. When installed in underground drains, cleanouts shall be extended vertically to or above finished grade either inside or outside the building.

❖ Because a cleanout is designed to provide access into the drainage system, the cleanout itself must be accessible, regardless of whether the piping it serves is concealed or in a location not readily accessible. The cleanout must extend up to and be flush with the finished floor level or outside grade or extend to a location that is flush with a finished wall. Note that cleanouts located on walking surfaces must be countersunk to minimize the tripping hazard and to protect the cleanout from damage. See Commentary Figure P3005.2.3(1).

A cleanout must be accessible for use wherever it is located in the drainage system and must not be covered by any permanent construction such as gypsum wall board. If a cleanout is concealed in a floor or wall assembly, an approved access cover must be provided. The access opening must be of suitable size to allow for the removal of the plug and to allow for the drain-cleaning procedure. Cleanouts must terminate flush with the wall or floor surface where an access pit or chase opening is not provided. See Commentary Figure P3005.2.3(2).

Where a cleanout is located in an area subject to vehicular traffic, or even excessive pedestrian traffic, the designer needs to evaluate the cover plate's structural capacity.

Usually, an underfloor space of less than 24 inches (610 mm) will prove difficult for the maneuvering required to operate drain cleaning equipment. While not specifically stated within the code, the installer or designer must give consideration to the location of cleanouts with regard to underfloor spaces and underfloor access.

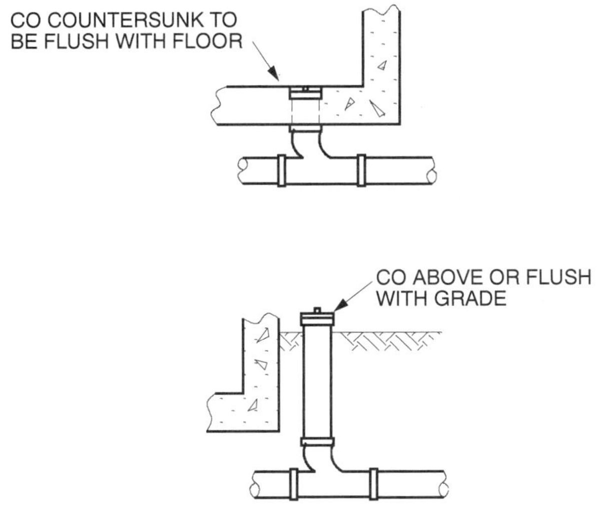

Figure P3005.2.3(1)
CLEANOUT INSTALLATIONS

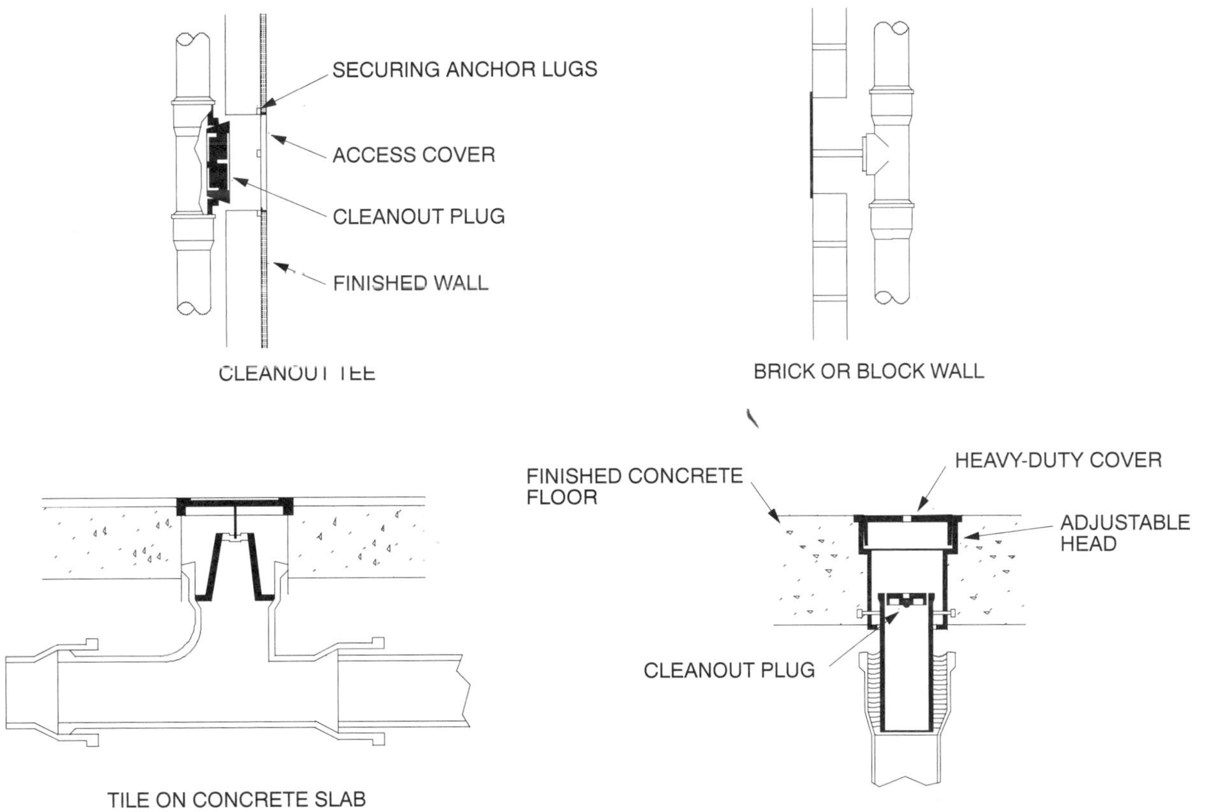

Figure P3005.2.3(2)
CLEANOUTS ON CONCEALED PIPING

P3005.2.4 Change of direction. Cleanouts shall be installed at each change of direction of the drainage system greater than 45 degrees, except not more than one cleanout shall be required in each 40 feet (12 192 mm) of run regardless of change in direction.

❖ The requirement for a cleanout at a change in direction greater than 45 degrees (0.79 rad) is for an individual fitting, not a combination of fittings. If a 90-degree (1.58 rad) change in direction is accomplished with a single fitting, such as a sweep or combination tee-wye, a cleanout is required. If the same change in direction is accomplished with two quarter bends, a cleanout is not required.

Although this requirement might initially appear to be inconsistent, it is logically based on the limitation of the cleaning equipment. Rodding equipment will easily pass through fittings having a change in direction of 45 degrees (0.79 rad) or less, even if multiple fittings are connected together. A fitting with a greater change in direction, however, could inhibit the rodding equipment from passing through the fitting.

A cleanout provides access to the drainage system for cleaning; therefore, cleanouts must be located near where blockages are likely to occur in the drainage system. Cleanout locations at or adjacent to abrupt changes in direction allow such access. See Commentary Figure P3005.2.4.

A single cleanout is permitted to serve up to 40 feet (12 192 mm) of developed length of drainage pipe regardless of the number of changes in direction that occur within that developed length. This provision takes into consideration the improved ability of modern drain-cleaning equipment to pass through fittings over this distance and the technological advances in the manufacturing of pipe and fittings that have improved flow characteristics and decreased the likelihood of stoppages.

P3005.2.5 Accessibility. Cleanouts shall be accessible. Minimum clearance in front of cleanouts shall be 18 inches on 3 inches (457 mm on 76 mm) and larger pipes, and 12 inches (305 mm) on smaller pipes. Concealed cleanouts shall be provided with access of sufficient size to permit removal of the cleanout plug and rodding of the system. Cleanout plugs shall not be concealed by permanent finishing material.

❖ This section regulates access to cleanouts, meaning the measurement from the plug to any obstruction that could hinder access of drain cleaning equipment. See Commentary Figure P3005.2.5. Cleanouts must not be concealed behind finish construction. Where they are concealed, provision must be made for access such that drain cleaning equipment can be used. See Commentary Figure P3005.2.3(2).

P3005.2.6 Base of stacks. Accessible cleanouts shall be provided near the base of each vertical waste or soil stack. Alternatively, such cleanouts may be installed outside the building within 3 feet (914 mm) of the building wall.

❖ The characteristics of the drainage flow in horizontal pipe increases the probability of a stoppage. Also, it is possible for solids to collect at the change of direction from vertical to horizontal. Therefore, a cleanout is required at the base of every drainage stack to provide access to the horizontal piping that serves the stack. See Commentary Figure P3005.2.6.

As an alternative, a cleanout can be placed outside of the structure within 3 feet (914 mm) of the building wall. This will place the cleanout no more than 3 feet (914 mm) horizontally from the stack within the outside wall.

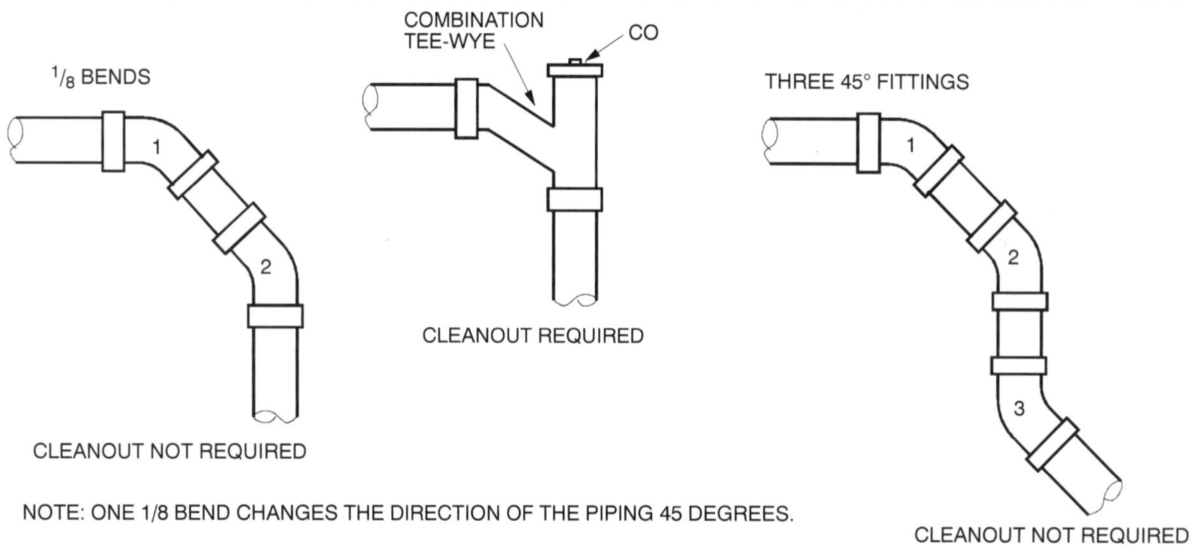

NOTE: ONE 1/8 BEND CHANGES THE DIRECTION OF THE PIPING 45 DEGREES.

For SI: 1 degree = 0.0175 rad.

Figure P3005.2.4
CLEANOUTS AT CHANGE OF DIRECTION

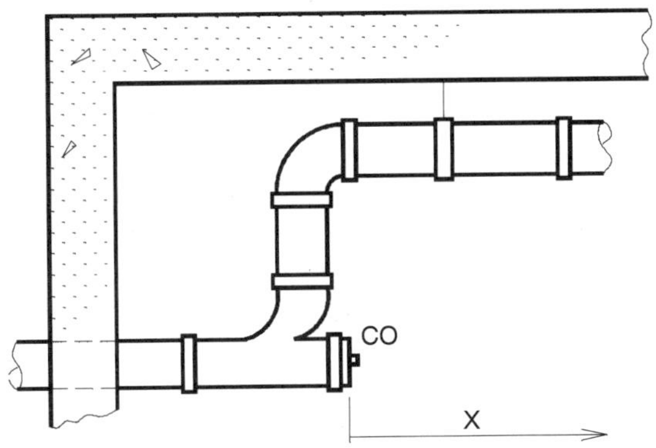

MINIMUM CLEARANCE:
X = MINIMUM 12 IN. FOR $2^1/_2$ IN. AND SMALLER PIPES
X = MINIMUM 18 IN. FOR 3 IN. AND LARGER PIPES

For SI: 1 inch = 25.4 mm.

Figure P3005.2.5
CLEARANCE FOR CLEANOUTS

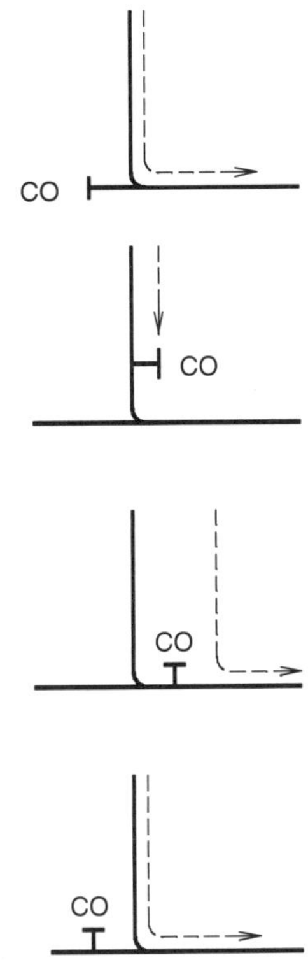

Figure P3005.2.6
CLEANOUTS AT BASE OF STACK

P3005.2.7 Building drain and building sewer junction. There shall be a cleanout near the junction of the building drain and building sewer. This cleanout shall be either inside or outside the building wall, provided it is brought up to finish grade or to the lowest floor level. An accessible interior building drain cleanout or test tee within close proximity to the building drain exit point shall fulfill this requirement.

❖ For stoppages that are known to be outside of a building within the building sewer, it is more convenient to have a cleanout at the junction of the building drain and sewer. The code permits the cleanout to be located either inside or outside the building. The majority of drain stoppages occur in the building sewer because of low-flow velocities and tree roots that penetrate the sewer. See Commentary Figure P3005.2.7.

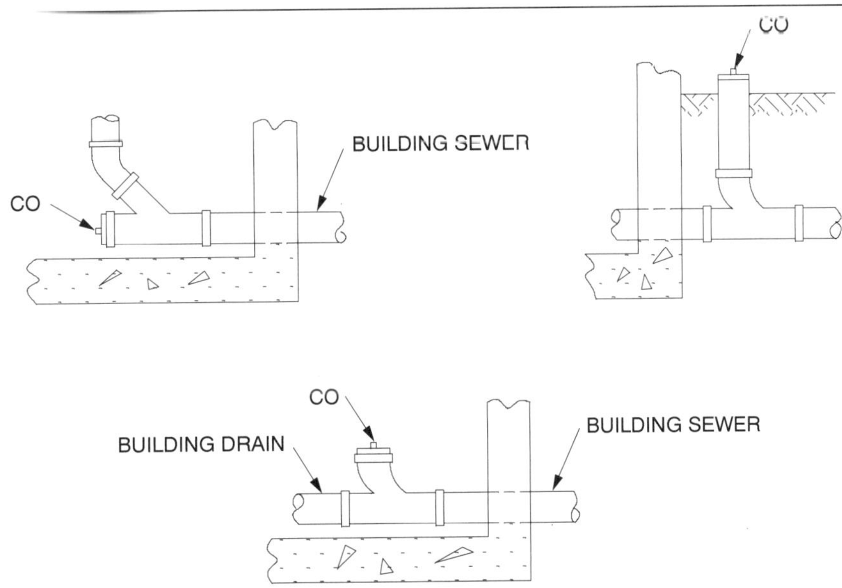

For SI: 1 foot = 304.8 mm.

Figure P3005.2.7
CLEANOUTS AT SEWER JUNCTION

P3005.2.8 Direction of flow. Cleanouts shall be installed so that the cleanout opens to allow cleaning in the direction of the flow of the drainage line.

❖ The normal practice for rodding or cleaning a drainage pipe is in the direction of flow. The cleaning rod (cable) or instrument will follow the normal direction of flow to remove the obstruction at the point of stoppage.

If a cleaning instrument is inserted opposite the direction of flow, the instrument can go in either direction at branch connections; hence, the travel of the instrument cannot be controlled. Fixtures can be damaged if the instrument goes in the wrong direction and into a fixture branch. For example, a floor-mounted water closet can easily be damaged by a cable connected to an electric drain-cleaning machine. Additionally, cleaning a drain to remove an upstream stoppage can result in spillage when the stoppage is cleared and the backed-up waste suddenly flows toward the open cleanout.

P3005.2.9 Cleanout size. Cleanouts shall be sized in accordance with Table P3005.2.9.

Exception: Cast-iron cleanout sizing shall be in accordance with referenced standards in Table P3002.1, ASTM A 74 for hub and spigot fittings or ASTM A 888 or CISPI 301 for hubless fittings.

❖ This section refers the reader to Table 3005.2.9 for the proper size of cleanouts in relation to the size of drain lines being served. An exception provides for the sizing for cast-iron cleanouts and refers the reader to Table P3002.1 and ASTM A 74 for hub and spigot fittings or ASTM A 888 or CISPI 301 for hubless fittings because the standards require a different tap size than the above requirements.

**TABLE P3005.2.9
CLEANOUTS**

PIPE SIZE (inches)	CLEANOUT SIZE (inches)
$1^1/_2$	$1^1/_2$
2	$1^1/_2$
3	$2^1/_2$
4 and larger	$3^1/_2$

For SI: 1 inch = 25.4 mm.

❖ This table provides the size of openings required for cleanouts. These are based on the size of the drain line being served and the type of drain cleaning equipment typically used for rodding of drains in one- and two-family dwellings. See the commentary for the definition of "Cleanout."

P3005.2.10 Cleanout equivalent. A fixture trap or a fixture with integral trap, readily removable without disturbing concealed piping shall be acceptable as a cleanout equivalent.

❖ This section allows for the removal of a trap or a fixture with an integral trap (such as a water closet) to allow for drain cleaning access. It recognizes that the trap is often the first place to look for blockage when a fixture fails to drain properly. When this is the case, either the trap or, with integral type traps, the fixture needs to be removed. When the blockage is located downstream of the trap or fixture, the natural inclination is to rod the drain by accessing the drain at the opening created by the removal of the trap or fixture.

This provision being an acceptable equivalent to a cleanout means that a cleanout fitting with opening is not required to be installed in addition to the access at the trap. This does not negate cleanouts that might be required elsewhere in accordance with Section 3005.2.

P3005.2.11 Connections to cleanouts prohibited. Cleanout openings shall not be used for the installation of new fixtures except where approved and an acceptable alternate cleanout is provided.

❖ So that cleanouts continue to function as a means of clearing the drainage and sewer system, this section prohibits their being used for installing new fixtures or floor drains unless an alternate cleanout is installed.

P3005.3 Horizontal drainage piping slope. Horizontal drain-age piping shall be installed in uniform alignment at uniform slopes not less than one-fourth unit vertical in 12 units horizontal (2-percent slope) for $2^1/_2$-inch (64 mm) diameter and less, and not less than one-eighth unit vertical in 12 units horizontal (1-percent slope) for diameters of 3 inches (76 mm) or more.

❖ The minimum desired velocity in a horizontal drain pipe is approximately 2 feet (610 mm) per second. This velocity is often referred to as the "scouring velocity." This minimum velocity is intended to keep solids in suspension. If velocity is too low because the pipe slope is too shallow, the solids will tend to drop out of suspension, sitting or even sticking to the bottom of the pipe. This will ultimately lead to a drain blockage. The slope requirements of this section produce the minimum required drainage flow velocity. Two percent slope is required for $2^1/_2$-inch (64 mm) diameter and smaller pipe, and one-percent slope is required for 3-inch (76 mm) diameter and larger pipe. Drainage pipe can be installed with greater slopes.

P3005.4 Drain pipe sizing. Drain pipes shall be sized according to drainage fixture unit (d.f.u.) loads. The size of the drainage piping shall not be reduced in size in the direction of flow. The following general procedure is permitted to be used:

1. Draw an isometric layout or riser diagram denoting fixtures on the layout.
2. Assign d.f.u. values to each fixture group plus individual fixtures using Table P3004.1.
3. Starting with the top floor or most remote fixtures, work downstream toward the building drain accumulating d.f.u. values for fixture groups plus individual fixtures for each branch. Where multiple bath groups are being

added, use the reduced d.f.u. values in Table P3004.1, which take into account probability factors of simultaneous use.

4. Size branches and stacks by equating the assigned d.f.u. values to pipe sizes shown in Table P3005.4.1.
5. Determine the pipe diameter and slope of the building drain and building sewer based on the accumulated d.f.u. values, using Table P3005.4.2.

❖ This section outlines in detail the method whereby the designer can effectively size the drainage portion of the DWV system.

The first step is to draw an isometric layout of the drain lines. This drawing must show each fixture (or group of fixtures) and must indicate all points of connection. See Commentary Figure P3005.4.

Step two is to assign d.f.u. values to each fixture or group of fixtures. This information is found in Table P3004.1. See the commentaries for Section P3004.1 and Table P3004.1.

The third step is to add up the d.f.u.'s for each section of piping. Start with the most remote fixture (or group) and follow downstream, adding the total for each section of pipe. See Commentary Figure P3005.4.

Step four is to size the branches and stacks. This requires the use of Table P3005.4, and the size of drainage piping must not be reduced in size in the direction of flow. See the commentaries for Section P3005.4.1, Table P3005.4.1, and definitions of "branch" and "stack."

The fifth and last step is to size the building drain and building sewer based on the d.f.u.'s as determined in the previous steps. This will require the use of Table P3005.4.2. See the commentaries for Section P3005.4.2, Table P3005.4.2, and the definitions of "Building Drain" and "Building Sewer."

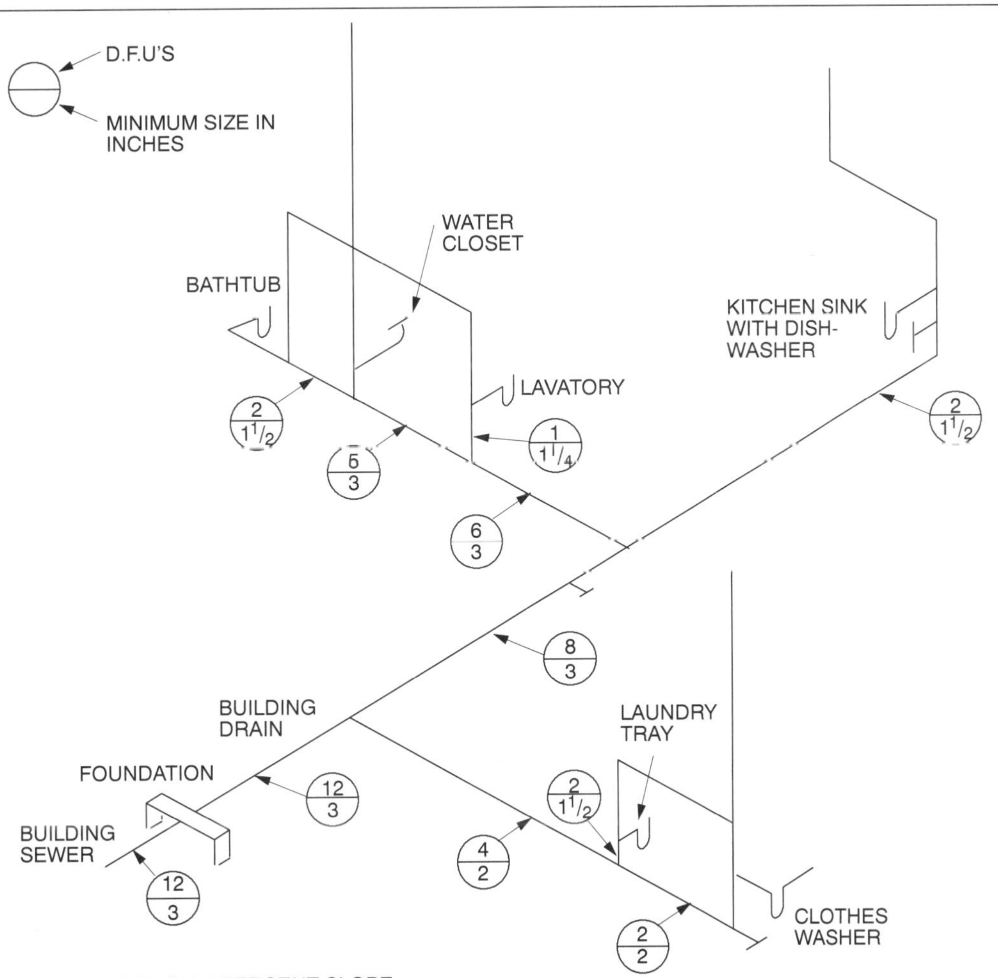

NOTE: BASED ON 2-PERCENT SLOPE

For SI: 1 inch = 25.4 mm.

Figure P3005.4
DRAIN LINE LAYOUT

P3005.4.1 Fixture branch and stack sizing.

1. Branches and stacks shall be sized according to Table P3005.4.1. Below grade drain pipes shall not be less than $1^1/_2$ inches (38 mm) in diameter.
2. Minimum stack size. Drain stacks shall not be smaller than the largest horizontal branch connected, with the following exceptions:
   2.1. A 4-inch by 3-inch (102 mm by 76 mm) closet bend or flange or a 4-inch (102 mm) closet bend into a 3-inch (76 mm) stack tee shall be acceptable (see Section P3005.1.4).

❖ This section outlines the method for sizing branches and stacks. See Chapter 2 for definitions of "branch" and "stack." Refer to Table P3005.4.1 for determining the size of pipe based on the d.f.u. value. It is important to understand the definitions of branch and a stack, especially as they relate to the definition of a building drain. See the commentaries for Chapter 2 definitions of "Branch," "Building Drain," and "Stack." A closer look at these definitions as they relate to Commentary Figure P3005.4 shows that none of the drain lines in the diagram can be defined as a stack or branch. Rather, they are part of the building drain and the building drain branches, which are sized in accordance with Table P3005.4.2. In addition, Table P3201.7 dictates the trap size required for each fixture.

Item 2 of this section requires that the drain stack size not be reduced to less than the largest horizontal branch connected. The only exception is where a 4-inch (102 mm) by 3-inch (76 mm) closet bend is used as provided in Section P3005.1.4. See the commentary for Section P3005.1.4.

A further requirement limits the size of any drain pipe below grade to a minimum diameter of $1^1/_2$ inches (38 mm).

TABLE P3005.4.1
MAXIMUM FIXTURE UNITS ALLOWED TO BE CONNECTED TO BRANCHES AND STACKS

NOMINAL PIPE SIZE (inches)	ANY HORIZONTAL FIXTURE BRANCH	ANY ONE VERTICAL STACK OR DRAIN
$1^1/_4$[a]	—	—
$1^1/_2$[b]	3	4
2[b]	6	10
$2^1/_2$[b]	12	20
3	20	48
4	160	240

For SI: 1 inch = 25.4 mm.

a. $1^1/_4$-inch pipe size limited to a single-fixture drain or trap arm. See Table P3201.7.

b. No water closets.

❖ This table lists the maximum number of fixture units that are allowed to be discharged into a pipe of a given size. As the number of drainage fixture units increases in the direction of flow, the pipe might also have to be increased in size.

Note that a d.f.u. load is not given for the use of $1^1/_4$-inch (32 mm) pipe. Note "a" shows that a $1^1/_4$-inch (32 mm) drain can serve only one fixture drain or trap. Of course, the fixture must be such that it can be served by a $1^1/_4$-inch (32 mm) drain or trap in accordance with Table P3201.7. Note "b" specifies that piping less than 3-inches (76 mm) in diameter cannot serve a water closet.

P3005.4.2 Building drain and sewer size and slope. Pipe sizes and slope shall be determined from Table P3005.4.2 on the basis of drainage load in fixture units (d.f.u.) computed from Table P3004.1.

❖ This section refers the reader to Table P3005.4.1 to determine the size of the building drain, its branches, and the building sewer.

TABLE P3005.4.2
MAXIMUM NUMBER OF FIXTURE UNITS ALLOWED TO BE CONNECTED TO THE BUILDING DRAIN, BUILDING DRAIN BRANCHES OR THE BUILDING SEWER

DIAMETER OF PIPE (inches)	SLOPE PER FOOT		
	$^1/_8$ inch	$^1/_4$ inch	$^1/_2$ inch
$1^1/_2$[a,b]	—	Note a	Note a
2[b]	—	21	27
$2^1/_2$[b]	—	24	31
3	36	42	50
4	180	216	250

For SI: 1 inch = 25.4 mm, 1 foot = 304.8 mm.

a. $1^1/_2$-inch pipe size limited to a building drain branch serving not more than two waste fixtures, or not more than one waste fixture if serving a pumped discharge fixture or garbage grinder discharge.

b. No water closets.

❖ This table provides the appropriate size of pipe to be used, based upon the accumulated d.f.u.'s to be loaded. As indicated in the table, the d.f.u. increases when a greater slope is used, providing the designer with greater flexibility in sizing and layout. Where a d.f.u. load is not given, this is due to a minimum slope required for certain sizes of drains or the limitation on fixtures as outlined in Note "a." As with Table P3005.4.1, Note "b" clarifies that piping with diameters of less than 3 inches (76 mm) cannot serve a water closet.

SECTION P3006
SIZING OF DRAIN PIPE OFFSETS

P3006.1 Vertical offsets. An offset in a vertical drain, with a change of direction of 45 degrees (0.79 rad) or less from the vertical, shall be sized as a straight vertical drain.

❖ An offset diverts a pipe by the combination of two fittings that create a change of direction. It brings one section of the pipe out of line and into a line parallel with the other section. Though the drain line changes direction, it is sized as though it were still vertical. However, the offset (or the section of piping out of line with the run of pipe) portion must not exceed 45 degrees (0.79

rad) with the connecting lines on a vertical installation. Where the offset exceeds 45 degrees (0.79 rad) it is considered horizontal and must be sized in accordance with Section P3006.2 or P3006.3, whichever applies.

P3006.2 Horizontal offsets above the lowest branch. A stack with an offset of more than 45 degrees (0.79 rad) from the vertical shall be sized as follows:

1. The portion of the stack above the offset shall be sized as for a regular stack based on the total number of fixture units above the offset.
2. The offset shall be sized as for a building drain in accordance with Table P3005.4.2.
3. The portion of the stack below the offset shall be sized as for the offset or based on the total number of fixture units on the entire stack, whichever is larger.

❖ This section applies to offsets that are installed on a stack, are located above the lowest branch, and are not defined as vertical offsets. See the commentary for Section P3006.1 and the definition of "Offset." By default they are defined as horizontal offsets.

The portion of the stack above the horizontal offset must be sized for a regular stack in accordance with Table P3005.4.1. The horizontal offset portion must be sized in accordance with Table P3005.4.2. This will then determine the minimum size of the remainder of the stack below the offset, except that additional d.f.u.'s for fixtures and branches connecting to it must be included in the sizing of the stack; this could necessitate a further increase in size beyond that required for the horizontal offset.

P3006.3 Horizontal offsets below the lowest branch. In soil or waste stacks below the lowest horizontal branch, there shall be no change in diameter required if the offset is made at an angle not greater than 45 degrees (0.79 rad) from the vertical. If an offset greater than 45 degrees (0.79 rad) from the vertical is made, the offset and stack below it shall be sized as a building drain (see Table P3005.4.2).

❖ Vertical offsets that are below the lowest horizontal branch connection to the stack are not required to increase in size because of the offset.

The pressure differences created by such offsets mainly affect the horizontal branches downstream of the offset. Horizontal offsets that are below the lowest horizontal branch connection are required to be sized as a building drain in accordance with Table P3005.4.2.

SECTION P3007 SUMPS AND EJECTORS

P3007.1 Sewage ejectors or sewage pumps. A sewage ejector, sewage pump, or grinder pump receiving discharge from a water closet shall have minimum discharge velocity of 1.9 feet per second (0.579 m/s) throughout the discharge piping to the point of connection with a gravity building drain, gravity sewer or pressure sewer system. A nongrinding pump or ejector shall be capable of passing a 1.5-inch-diameter (38 mm) solid ball, and the discharge piping shall be not less than 2 inches (51 mm) in diameter. The discharge piping of grinding pumps shall be not less than 1.25 inches (32 mm) in diameter. A check valve and a gate valve located on the discharge side of the check valve shall be installed in the pump or ejector discharge piping between the pump or ejector and the drainage system. Access shall be provided to such valves. Such valves shall be located above the sump cover or, when the discharge pipe from the ejector is below grade, the valves shall be accessibly located outside the sump below grade in an access pit with a removeable access cover.

Exception: Macerating toilet systems shall be permitted to have the discharge pipe sized in accordance with manufacturer's instructions, but not less than 0.75 inch (19.1 mm) in diameter.

❖ The sewage pump or pneumatic ejector must be sized properly to accommodate the peak flow into the receiver or pit, to prevent excessive cycling of the pumping equipment, and to handle the type of waste that discharges to the sump.

Sewage pumps and ejectors must be able to handle, without creating blockages, the types of solid waste discharged into the sump pit or receiver. Pumps and ejectors serving water closets must be able to handle the solids associated with soil drainage. See Commentary Figures P3007.1(1) and P3007.1(2).

Grinder pumps and grinder ejectors that serve water closets are exempt from the 1.5-inch (38 mm) spherical solids handling requirements of this section. These pumps and ejectors pulverize solids to a near-liquid state and pump the slurry to the drainage system.

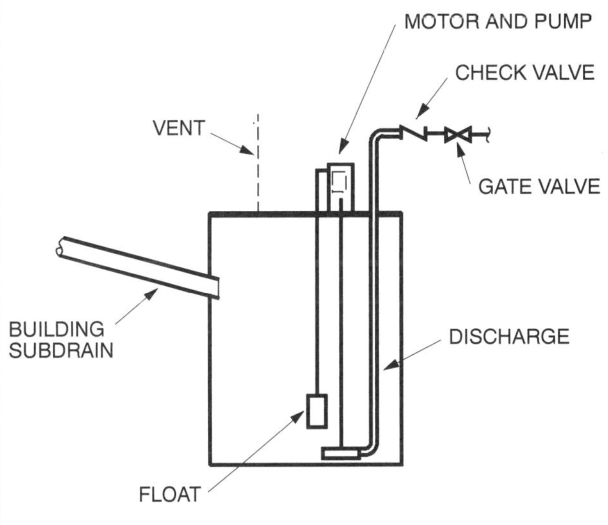

Figure P3007.1(1)
MOTOR AND PUMP

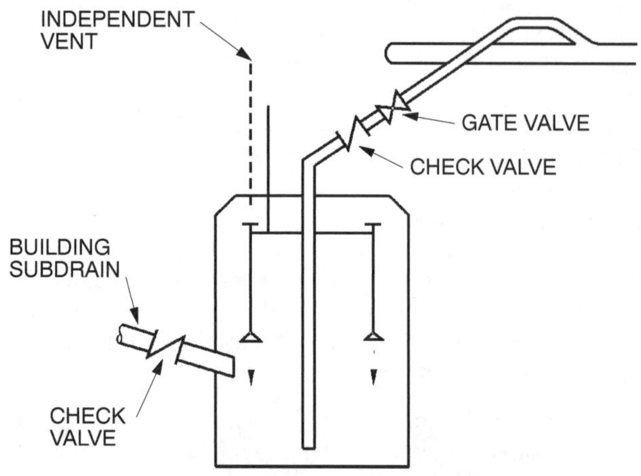

Figure P3007.1(2)
PNEUMATIC EJECTOR

The pulverized solids pass readily through the pump and into the drainage system, thus reducing the possibility of a stoppage in either the pump or the receiving drainage line.

The sewage pump or ejector must have a minimum capacity that is based on the size of the discharge pipe. The required capacity provides a full-flow velocity of at least 2 feet (610 mm) per second in the discharge pipe.

The check valve and the gate valve must be located where they can be accessed for servicing the pump. The gate valve must be downstream of the check valve in the direction of flow or discharge.

Macerating toilet systems are exempted from the requirements of this section and must be installed in accordance with Section P2723. See the commentary for Section P2723.

P3007.2 Building drains below sewer (building subdrains). Building drains which cannot be discharged to the sewer by gravity flow shall be discharged into a tightly covered and vented sump from which the contents shall be lifted and discharged into the building gravity drainage system by automatic pumping equipment.

❖ Only those fixtures or portions of the drainage system that cannot drain by gravity to the building sewer can drain to the pit or sump to be discharged by means of a pump. Such a pump must be automatic and not require the occupant to manually turn the pump to activate it. Usually, a float automatically activates the pump.

The pit or sump must be tightly covered and vented in accordance with Chapter 31. See Commentary Figure P3007.2.

Because the lid is removable, the vent connection must also be designed for disassembly.

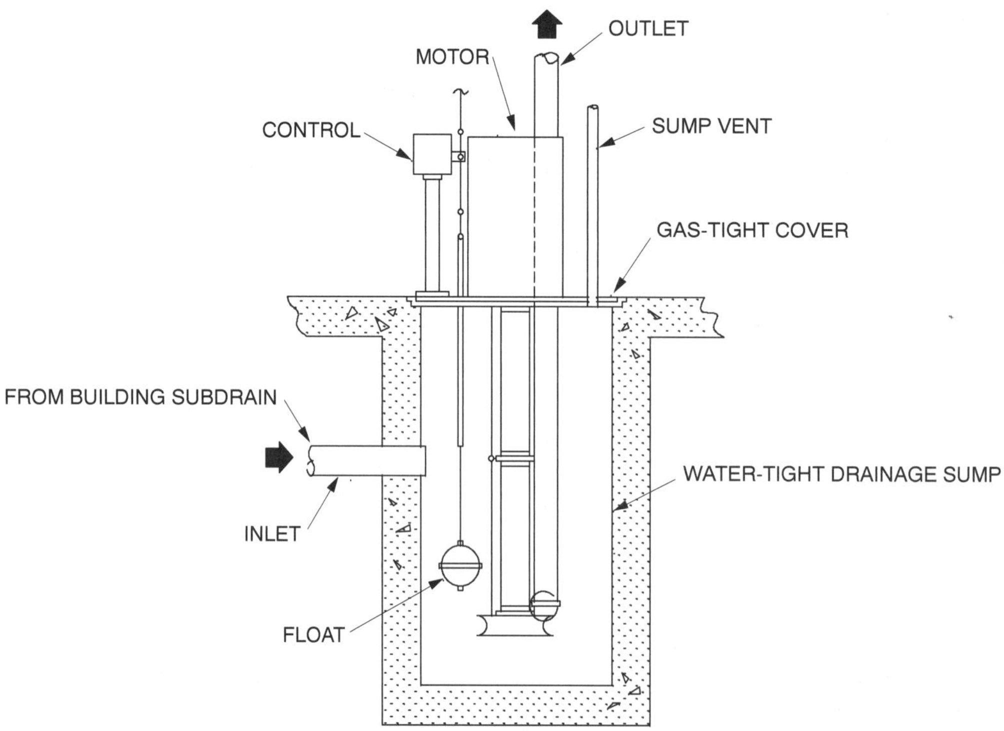

Figure P3007.2
MECHANICAL SEWAGE PUMP

P3007.2.1 Drainage piping. The system of drainage piping below the sewer level shall be installed and vented in a manner similar to that of the gravity system. Only such drains that must be lifted for discharge shall be discharged into sumps. All other drains shall be discharged by gravity.

> **Exception:** Macerating toilet systems shall be permitted as an alternate to the sewage pump or ejector system. The macerating toilet shall comply with ASME A112.3.4 or CSA B45.9 and shall be installed in accordance with the manufacturers' instructions.

❖ As stated in the commentary for Section P3007.2, only those fixtures or portions of the drainage system that cannot drain by gravity to the building sewer can drain to the pit or sump to be discharged by means of a pump. The DWV system that is to be connected to the sump or pit must conform to the requirements of a gravity system. Macerating toilet systems are exempt from the provisions of this section and must be installed per manufacturer's specifications in Section P2723. Macerating toilet systems are unique in that there are many acceptable types available. CSA B45.9 and ASME A112.3.4 are the standards for macerating toilet systems.

SECTION P3008 BACKWATER VALVES

P3008.1 General. Fixtures that have flood level rims located below the elevation of the next upstream manhole cover of the public sewer serving such fixtures shall be protected from backflow of sewage by installing an approved backwater valve. Fixtures having flood level rims above the elevation of the next upstream manhole shall not discharge through the backwater valve. Backwater valves shall be provided with access.

❖ A backwater valve is required in areas where the sewer could back up into the building through the sanitary drainage system. See Commentary Figure P3008.1(1).

Where the plumbing fixtures are located above the next upstream manhole cover from the building sewer connection to the public sewer, the sewer will back up through the street manhole before entering the building.

Public sewers can become blocked or overloaded, which can result in sewage backing up into the manholes and any laterals (taps) connected to the sewer system. The point of overflow for the public sewer will be the top of the manholes.

Fixtures or drains that are located at an elevation below that of the top of the manhole for the relative portion of the sewer system are subject to backflow and must be protected by backwater valves. See Commentary Figure P3008.1(2).

Plumbing fixtures that are not subject to backflow are not permitted to discharge through a backwater valve. In theory, limiting the fixtures that discharge through a backwater valve prevents the waste from upstream fixtures from backing up through downstream fixtures because the waste cannot pass through the backwater valve as the public sewer is blocked or overloaded, and protects the valve from excess wear and potential failure resulting from accumulation of debris.

The last sentence of this section, if taken literally, would apply to fixtures having a flood lever rim a fraction of an inch above the upstream manhole cover. Such fixtures could still be subject to the backup of sewage, which would be insanitary and present the risk of sewage entering the structure through fixture connections and traps.

P3008.2 Construction. Backwater valves shall have noncorrosive bearings, seats and self-aligning discs, and shall be constructed to ensure a positive mechanical seal. Valve access covers shall be water tight.

❖ This section requires that all bearing parts of the backwater valve be made of corrosion-resistant materials so that the valve remains operable for its expected life.

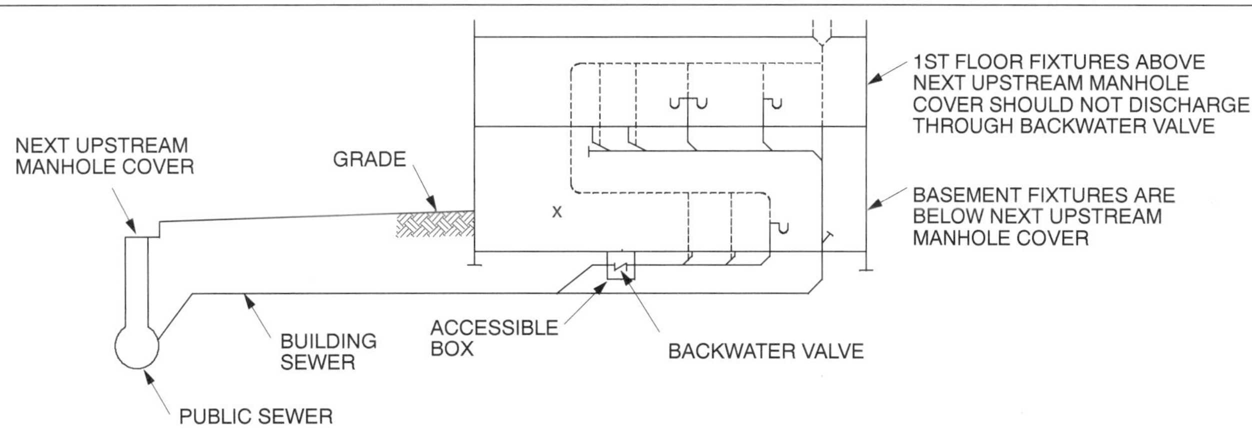

Figure P3008.1(1)
BACKWATER VALVE

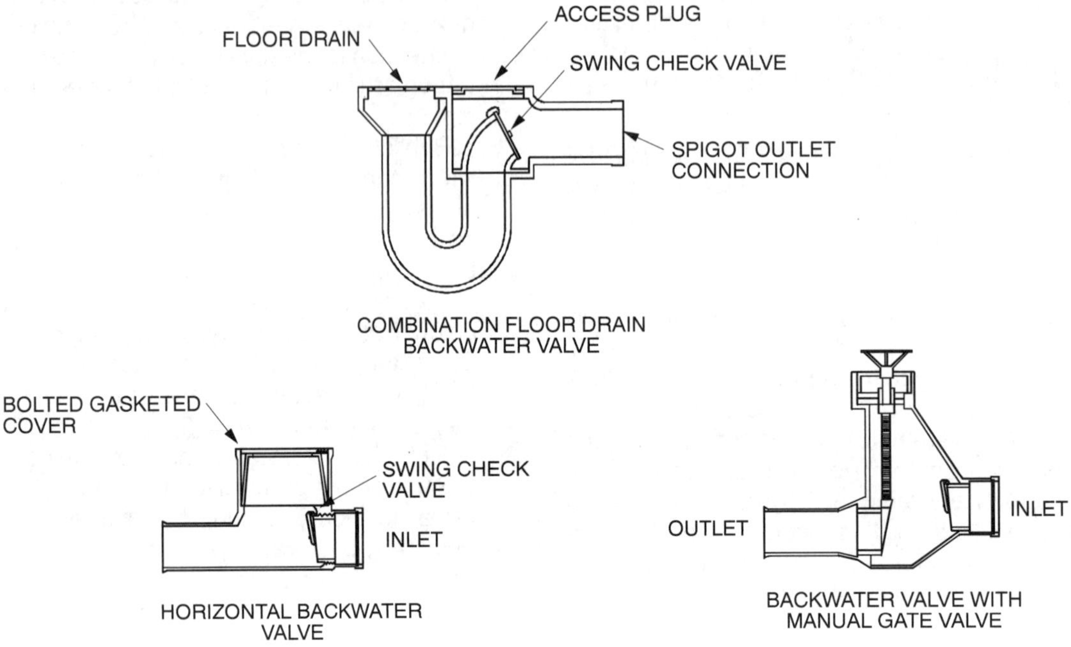

Figure P3008.1(2)
BACKWATER VALVES

The backwater valve incorporates a swing check valve on the inlet side of the valve body that will close when there is a reserve flow. This section requires that the swing check valve have a positive seal against the backflow of drainage. This section also requires that the valve access cover be watertight to prevent any sewage backup from leaking outside the valve.

Bibliography

The following resource materials are referenced in this chapter or are relevant to the subject matter addressed in this chapter.

ASHRAE 2001, *ASHRAE Handbook: Fundamentals*. Atlanta, GA: American Society of Heating, Refrigerating and Air-Conditioning Engineers, Inc., 2001.

ASTM A 53/A53M-01, *Specification for Pipe, Steel, Black and Hot-Dipped, Zinc-Coated Welded and Seamless*. West Conshohocken, PA: ASTM International, 2001.

ASTM A 74-98, *Specification for Cast Iron Soil Pipe and Fittings*. West Conshohocken, PA: ASTM International, 1998.

ASTM A 126-01, *Specification for Gray Iron Castings for Valves, Flanges and Pipe Fittings*. West Conshohocken, PA: ASTM International, 2001.

ASTM A 197/A197M-00, *Specification for Cupola Malleable Iron*. West Conshohocken, PA: ASTM International, 2000.

ASTM A 888-98e1, *Specification for Hubless Cast-Iron Soil Pipe and Fittings for Sanitary and Storm Drain, Waste and Vent Piping Application*. West Conshohocken, PA: ASTM International, 1998.

ASTM B 32-00, *Specification for Solder Metal*. West Conshohocken, PA: ASTM International, 2000.

ASTM B 42-98, *Specification for Seamless Copper Pipe, Standard Sizes*. West Conshohocken, PA: ASTM International, 1998.

ASTM B 43-98, *Specification for Seamless Red Brass Pipe, Standard Sizes*. West Conshohocken, PA: ASTM International, 1998.

ASTM B 75-99, *Specification for Seamless Copper Tube [Metric]*. West Conshohocken, PA: ASTM International, 1999.

ASTM B 88-99e[1], *Specification for Seamless Copper Water Tube*. West Conshohocken, PA: ASTM International, 1999.

ASTM B 306-99, *Specification for Copper Drainage Tube (DWV)*. West Conshohocken, PA: ASTM International, 1999.

ASTM B 813-00e01, *Specification for Liquid and Paste Fluxes for Soldering Applications of Copper and Copper Alloy Tubing*. West Conshohocken, PA: ASTM International, 2001.

ASTM B 828-00, *Practice for Making Capillary Joints by Soldering of Copper and Copper Alloy Tube and Fittings*. West Conshohocken, PA: ASTM International, 2000.

ASTM C 14-99, *Specification for Concrete Sewer, Storm Drain, and Culvert Pipe*. West Conshohocken, PA: ASTM International, 1999.

ASTM C 425-01, *Specification for Compression Joints for Vitrified Clay Pipe and Fittings*. West Conshohocken, PA: ASTM International, 2001.

ASTM C 443-01, *Specification for Joints for Concrete Pipe and Manholes, Using Rubber Gaskets*. West Conshohocken, PA: ASTM International, 2001.

ASTM C 700-00, *Specification for Vitrified Clay Pipe, Extra Strength, Standard Strength and Perforated*. West Conshohocken, PA: ASTM International, 2000.

ASTM C 1173-97, *Specification for Flexible Transition Couplings for Underground Piping System*. West Conshohocken, PA: ASTM International, 1997.

ASTM C 1277-97, *Specification for Shielded Couplings Joining Hubless Cast-Iron Soil Pipe and Fittings*. West Conshohocken, PA: ASTM International, 1997.

ASTM C 1460-00, *Specification for Shielded Transition Couplings for Use with Dissimilar DWV Pipe and Fittings Above Ground*. West Conshohocken, PA: ASTM International, 2000.

ASTM C 1461-00, *Specification for Mechanical Couplings Using Thermoplastic Elastomeric (TPE) Gaskets for Joining Drain, Waste and Vent (DWV) Sewer, Sanitary and Storm Plumbing Systems for Above and Below Ground Use*. West Conshohocken, PA: ASTM International, 2000.

ASTM D 2235-01, *Specification for Solvent Cement for Acrylonitrile-Butadiene-Styrene (ABS) Plastic Pipe and Fittings*. West Conshohocken, PA: ASTM International, 2001.

ASTM D 2564-96a, *Specification for Solvent Cements for Poly (Vinyl Chloride)(PVC) Plastic Piping Systems*. West Conshohocken, PA: ASTM International, 1996.

ASTM D 2661-01, *Specification for Acrylonitrile-Butadiene-Styrene (ABS) Schedule 40 Plastic Drain, Waste, and Vent Pipe and Fittings*. West Conshohocken, PA: ASTM International, 2001.

ASTM D 2665-01, *Specification for Poly (Vinyl Chloride) (PVC) Plastic Drain, Waste, and Vent Pipe and Fittings*. West Conshohocken, PA: ASTM International, 2001.

ASTM D 2751-96a, *Specification for Acrylonitrile-Butadiene-Styrene (ABS) Sewer Pipe and Fittings*. West Conshohocken, PA: ASTM International, 1996.

ASTM D 2949-00a, *Specification for 3.25-in. Outside Diameter Poly (Vinyl Chloride) (PVC) Plastic Drain, Waste, and Vent Pipe and Fittings*. West Conshohocken, PA: ASTM International, 2000.

ASTM D 3034-00, *Specification for Type PSM Poly (Vinyl Chloride)(PVC) Sewer Pipe and Fittings*. West Conshohocken, PA: ASTM International, 2000.

ASTM D 3212-96a, *Specification for Joints for Drain and Sewer Plastic Pipes Using Flexible Elastomeric Seals*. West Conshohocken, PA: ASTM International, 1996.

ASTM F 438-01, *Specification for Socket-Type Chlorinated Poly (Vinyl Chloride) (CPVC) Plastic Pipe Fittings Schedule 40*. West Conshohocken, PA: ASTM International, 2001.

ASTM F 628-01, *Specification for Acrylonitrile-Butadiene-Styrene (ABS) Schedule 40 Plastic Drain, Waste and Vent Pipe with a Cellular Core*. West Conshohocken, PA: ASTM International, 2001.

ASTM F 656-96a, *Specification for Primers for Use in Solvent Cement Joints of Poly (Vinyl Chloride) (PVC) Plastic Pipe and Fittings*. West Conshohocken, PA: ASTM International, 1996.

ASTM F 789-95a, *Specification for Type PS-46 and Type PS-115 Poly (Vinyl Chloride) (PVC) Plastic Gravity Flow Sewer Pipe and Fittings*. West Conshohocken, PA: ASTM International, 1995.

ASTM F 891-00, *Specification for Coextruded Poly (Vinyl Chloride) (PVC) Plastic Pipe with a Cellular Core*. West Conshohocken, PA: ASTM International, 2000.

ASTM F 1488-00, *Specification for Coextruded Composite Pipe*. West Conshohocken, PA: ASTM International, 2000.

CISPI 301-97, *Specification for Hubless Cast-Iron Soil Pipe and Fittings for Sanitary and Storm Drain, Waste and Vent Piping Applications*. Chattanooga, TN: Cast-Iron Soil Pipe Institute, 1997.

CISPI 310-97, *Specification for Coupling for Use in Connection with Hubless Cast-Iron Soil Pipe and Fittings for Sanitary and Storm Drain, Waste and Vent Piping Applications*. Chattanooga, TN: Cast-Iron Soil Pipe Institute, 1997.

CSA A 257.3M-92, *Joints for Circular Concrete Sewer and Culvert Pipe, Manhole Sections, and Fittings Using Rubber Gaskets*. Toronto, Ontario: Canadian Standards Association, 1992.

CSA B602-99, *Mechanical Couplings for Drain, Waste, and Vent Pipe and Sewer Pipe*. Toronto, Ontario: Canadian Standards Association, 1999.

Chapter 31: Vents

General Comments

Venting protects the trap seals within the plumbing system. When installed in accordance with this chapter, the system will function properly. Failure to adhere to these provisions can result in not only the loss of trap seal but also in the introduction of sewer gases into the building.

This chapter includes basic venting requirements and the purpose for such. It also provides installation guidelines that address sizing, penetrations, termination, protection, and provisions for future vents. It outlines maximum distances and sizing limitations for trap and fixture drain sizes.

This chapter details various venting concepts, including individual venting for each fixture, common venting, waste stack venting, circuit venting, combination waste and vent systems, and the use of air admittance valves. Some of these vents are sized such that a drain can serve as a vent, providing design options that may result in minimal material use. This chapter also includes a technique specific to the installation of island sinks.

Purpose

Venting protects the trap seal of each trap. The vents are designed to maintain maximum differential pressures at each trap to less than 1-inch of water column (249 Pa). Waste flow in the drainage system creates pressure fluctuations. Venting is not intended to provide for the circulation of air within the drainage system. If there were no traps in a drainage system, venting would not be required. The system would function adequately because it would be open to the atmosphere at the fixture connections, thereby allowing airflow.

Chapter 31 covers the requirements for vents and venting. Knowing why venting is required makes it easier to understanding this chapter. Venting protects every trap against the loss of a trap seal. All of the provisions set forth in this chapter are geared toward limiting the pressure differentials in the drainage system to a maximum of 1 inch of water column (249 Pa) above or below atmospheric pressure (i.e., positive or negative pressures).

SECTION P3101 VENT SYSTEMS

P3101.1 General. This chapter shall govern the selection and installation of piping, tubing and fittings for vent systems. This chapter shall control the minimum diameter of vent pipes, circuit vents, branch vents and individual vents, and the size and length of vents and various aspects of vent stacks and stack vents. Additionally, this chapter regulates vent grades and connections, height above fixtures and relief vents for stacks and fixture traps, and the venting of sumps and sewers.

❖ This section established the scope of this chapter, which governs the selection and installation of vent systems. The chapter also provides requirements for sizing vents, circuit vents, branch vents, and individual vents. It regulates the size and length of vents as well as providing vent stack and stack vent requirements. In addition, it specifies and regulates vent grades, height above fixtures, relief vents for stacks and fixture traps, and venting of sumps and ejectors.

P3101.2 Trap seal protection. The plumbing system shall be provided with a system of vent piping that will permit the admission or emission of air so that the seal of any fixture trap shall not be subjected to a pneumatic pressure differential of more than 1 inch of water column (249 Pa).

❖ Protection of the trap seal means that during the normal operation of the plumbing system, the water seal remains in the trap. Each trap has a minimum of a 2-inch (51 mm) trap seal depth translating to a hydrostatic pressure equal to a 1-inch water column (498 Pa). If exposed to a 1-inch water column (249 Pa) pressure differential, a 1-inch (25 mm) water seal remains in the trap. The vent methods identified in this chapter limit the pressure differential at trap seals to 1-inch of water column (249 Pa) or less.

A higher pressure on the drainage pipe side of the trap is detrimental when there is excessive pressure or a surge in pressure. This could force the trap seal back into the fixture, possibly resulting in water spilling out of the fixture onto the floor from fixtures such as floor drains and standpipes. A higher pressure on the drainage pipe side of the trap will also result in sewer gases being pushed through the trap seal. For example, bubbles rising through the trap seal of a water closet are an indication of excessive pressure on the drainage side of the trap.

P3101.2.1 Venting required. Every trap and trapped fixture shall be vented in accordance with one of the venting methods specified in this chapter.

❖ This section establishes that all traps and trapped fixtures must be vented. It also indicates that the method of venting can be any of the methods described in the chapter. Proper application of the venting options will provide various options that allow the designer latitude that can be applied to numerous situations encountered in building construction and which may prove cost effective.

P3101.3 Use limitations. The plumbing vent system shall not be utilized for purposes other than the venting of the plumbing system.

❖ The venting system is designed for a specific function and cannot be used for any other purposes. Any other use could interfere with the proper operation of the plumbing system or could create a hazard. For example, vent piping must not serve as a drain for condensate from an air conditioning unit. Such use would introduce waste into piping not intended to convey waste and might allow sewer gases to enter the building through the air-conditioning unit.

P3101.4 Extension outside a structure. In climates where the 97.5-percent value for outside design temperature is 0°F (-18°C) or less (ASHRAE 97.5-percent column, winter, see Chapter 3), vent pipes installed on the exterior of the structure shall be protected against freezing by insulation, heat or both. Vent terminals shall be protected from frost closure in accordance with Section P3103.2.

❖ In areas where the outdoor design temperature (as per the ASHRAE data) is less than 0°F (-18°C), vent pipes installed on the exterior of the building must be protected from freezing by either insulation or heat tape. In addition, where subject to snow or frost, the vent terminals must be protected against closure as outlined in Section P3108.2. See the commentary for Table R301.2(1).

P3101.5 Flood resistance. In areas prone to floodings as established by Table R301.2(1), vents shall be located at or above the design flood elevation established in Section R323.1.

❖ See the commentary for Section P2602.2.

SECTION P3102 VENT STACKS AND STACK VENTS

P3102.1 Stack required. Every building shall have a vent stack or a stack vent. Such vent shall run undiminished in size and as directly as possible from the building drain through to the open air above the roof.

❖ This section requires the stack vent or vent stack (formerly called "main vent") to extend from the building drain to the outdoors and that it must be undiminished in size for its entire run. See Commentary Figure P3102.1. The required stack is either a vent stack or a stack vent and is sized in accordance with Section P3113.1. In no case can the stack be less than one-half of the required size of the building drain. For example, a 3-inch soil stack could have a $1^1/_2$-inch (38 mm) stack vent. This stack vent could be the required stack for the purpose of complying with this section. Historically, the required main vent was a large stack vent that served as a convenient point to connect branch vents.

Present venting methodology depends on vents installed for fixtures and traps to provide the necessary airflow; stack vents serve only a supplemental function.

P3102.2 Vent connection to drainage system. Every vent stack shall connect to the base of the drainage stack. The vent stack shall connect at or below the lowest horizontal branch. Where the vent stack connects to the building drain, the connection shall be located within 10 pipe diameters downstream of the drainage stack. A stack vent shall be a vertical extension of the drainage stack.

❖ Vent stacks are normally used in multiple-floor buildings to relieve pressure fluctuations commonly realized in drainage stacks. The *International Plumbing*

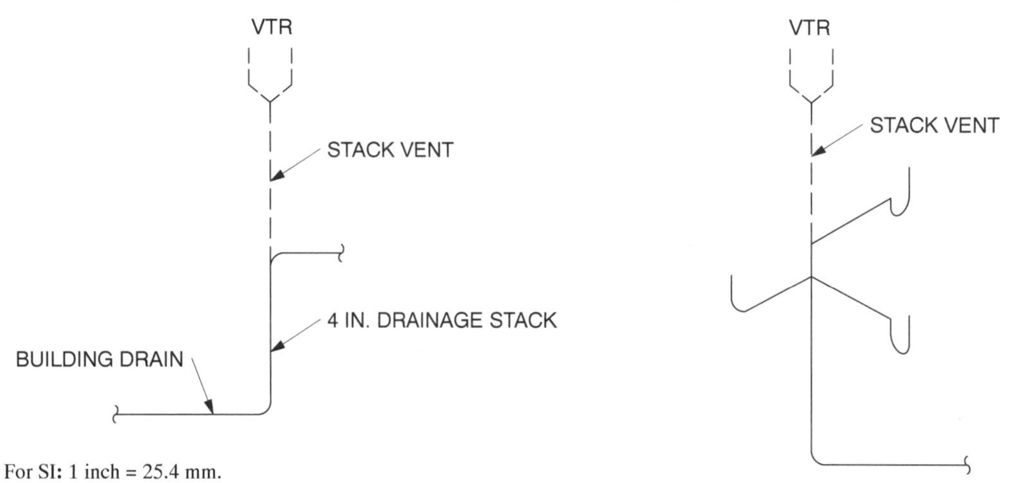

For SI: 1 inch = 25.4 mm.

Figure P3102.1 STACK VENT

Code requires a vent stack where a drainage stack is five or more branch intervals (horizontal branches serving plumbing on each floor level). The code does not specifically require a vent stack. Where the designer chooses to install one, it must connect at or below the lowest horizontal branch. Where the vent stack connects to the building drain, the connection must be made within 10 pipe diameters of the building drain downstream of the stack. A stack vent must continue vertically above the stack and is considered a vertical extension of the stack.

P3102.3 Vent termination. Every vent stack or stack vent shall terminate outdoors to the open air or terminate to a stack-type air admittance valve.

❖ A vent stack or stack vent cannot terminate inside the building, in an attic, or in any other space within the building envelope regardless of how well such space is ventilated. The vent must terminate to the outdoors to permit gases, vapors, and odors from the drainage system to escape into the atmosphere away from the building. However, this section does not prohibit a vent stack or stack vent from terminating to a stack-type air admittance valve, provided that the system complies with Section P3114. In particular, this type of air admittance valve must meet the requirements of ASSE 1050.

SECTION P3103 VENT TERMINALS

P3103.1 Roof extension. All open vent pipes which extend through a roof shall be terminated at least [NUMBER] inches above the roof or [NUMBER] inches above the anticipated snow accumulation, except that where a roof is to be used for any purpose other than weather protection, the vent extensions shall be run at least 7 feet (2134 mm) above the roof.

❖ Where the roof is occupied for any purpose, such as occurs with a sunning or promenade deck, plumbing vents must extend at least 7 feet (2134 mm) above the roof to prevent harmful sewer gasses from contaminating the area. The sewer gases will tend to disperse into the air rather than accumulate near the roof surface. The minimum termination height must be determined by the local jurisdiction based on the local snowfall averages, but it should be no less than 6 inches (152 mm) above the roof to protect the terminal from being blocked by snow and to allow sufficient length of pipe for a proper roof flashing. See Commentary Figure P3103.1.

The vent terminal through the roof must tolerate movement resulting from expansion and contraction of the piping material. This is extremely important when the vent is a piping material with high rates of expansion and contraction, such as acrylonitrile butadiene styrene (ABS) and polyvinyl chloride (PVC).

P3103.2 Frost closure. Where the 97.5-percent value for outside design temperature is 0°F (-18°C) or less, every vent extension through a roof or wall shall be a minimum of 3 inches (76 mm) in diameter. Any increase in the size of the vent shall be made inside the structure a minimum of 1 foot (305 mm) below the roof or inside the wall.

❖ The possibility of frost closure only occurs in areas of the country having cold climate with outdoor design temperature of less than 0°F (-18°C). A vent 3 inches (76 mm) or greater is not likely to completely close up

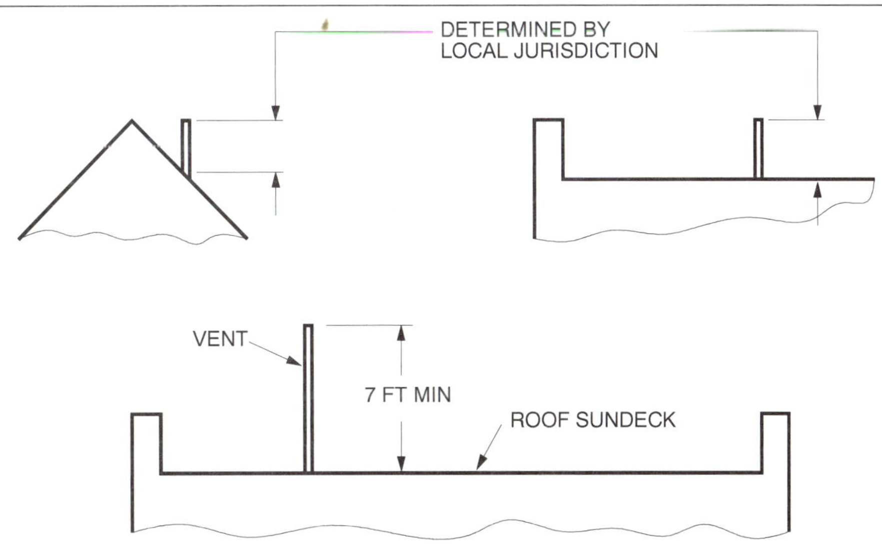

For SI: 1 foot = 304.8 mm.

Figure P3103.1
VENT TERMINAL

from the frost. This section requires vents smaller than 3 inches (76 mm) to be increased in size where frost enclosure is possible. The increase must occur within the structure to protect the smaller pipe from direct exposure to the outside atmosphere. See the commentary for Table R301.2(1).

P3103.3 Flashings and sealing. The juncture of each vent pipe with the roof line shall be made water tight by an approved flashing. Vent extensions in walls and soffits shall be made weather tight by caulking.

❖ Water penetration of the roof or wall at the point where the vent exits can cause decay of the wood and degradation of other building materials. Flashing must be provided at roof penetrations, and caulking must be provided at wall penetrations to protect the interior of the building from moisture or rain and to prevent the intrusion of insects or vermin. See Sections P2606 and P3002.4.1 for additional provisions.

P3103.4 Prohibited use. Vent terminals shall not be used as a flag pole or to support flag poles, TV aerials, or similar items, except when the piping has been anchored in an approved manner.

❖ The use of the vent terminal to support imposed loads is prohibited unless the vent terminal is properly anchored in a manner that is approved by the building official. In any case, manufacturer's instructions must be followed.

P3103.5 Location of vent terminal. An open vent terminal from a drainage system shall not be located less than 4 feet (1219 mm) directly beneath any door, openable window, or other air intake opening of the building or of an adjacent building, nor shall any such vent terminal be within 10 feet (3048 mm) horizontally of such an opening unless it is at least 2 feet (610 mm) above the top of such opening.

❖ To prevent sewer gasses from entering the building through gravity openings or forced air intakes, this section requires that vents be terminated not less than 4 feet (1219 mm) directly beneath any opening into the building or adjacent buildings. When the vent is to terminate above an opening, it must be at least 10 feet (3048 mm) away or 2 feet (610 mm) above that opening. See Commentary Figure P3103.5.

P3103.6 Extension through the wall. Vent terminals extending through the wall shall terminate a minimum of 10 feet (3048 mm) from the lot line and 10 feet (3048 mm) above the highest adjacent grade within 10 feet (3048 mm) horizontally of the vent terminal. Vent terminals shall not terminate under the overhang of a structure with soffit vents. Side wall vent terminals shall be protected to prevent birds or rodents from entering or blocking the vent opening.

❖ Side wall vent terminations provide an alternative to roof penetrations. This alternative might provide sig-

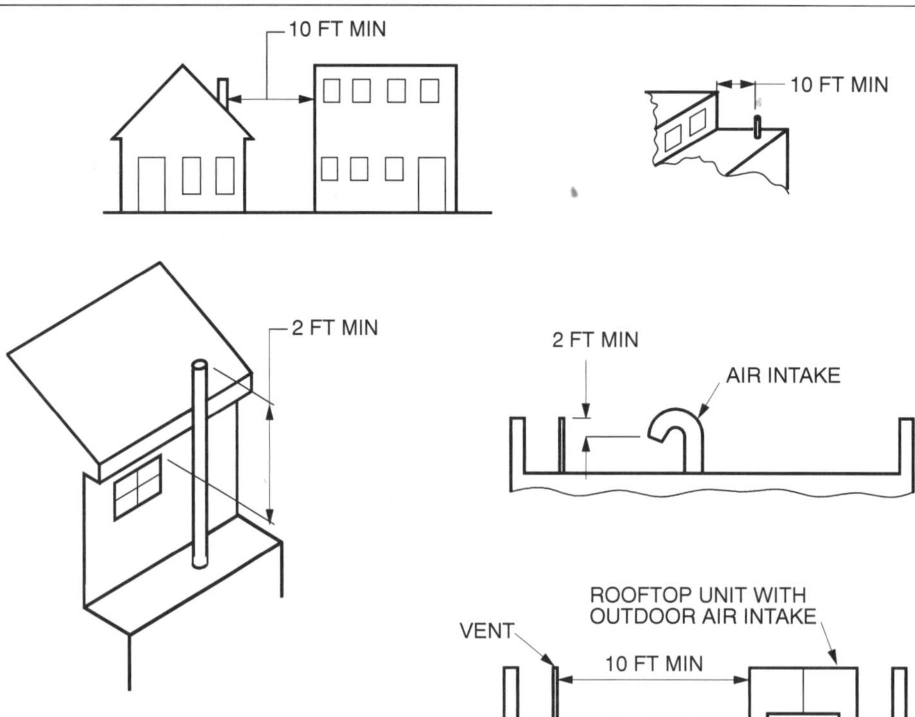

For SI: 1 foot = 304.8 mm.

Figure P3103.5
SEPARATION OF VENT TERMINAL

nificant cost savings and a more aesthetically pleasing installation. For example, a side wall vent may be preferred to penetrating membrane, built-up, slate, or tile roofs. Such roof penetrations are difficult to make leakproof, can be expensive, and are often unsightly.

In multiple-story buildings where remodeling work is done, a side wall vent termination can serve as an alternative to running vent piping through finished or occupied stories above. When a side wall vent is installed, the vent opening must be protected with a screen or louver to prevent birds from building a nest in the pipe. The screen or louver also prevents rodents from entering the vent system. See Commentary Figure P3103.6. Such vents must not terminate where the emissions can cause structural damage or enter the building envelope through soffit vents. Also, such vents must comply with the location requirements of Section P3103.5.

SECTION P3104 VENT CONNECTIONS AND GRADES

3104.1 Connection. All individual branch and circuit vents shall connect to a vent stack, stack vent or extend to the open air.

Exception: Individual, branch and circuit vents shall be permitted to terminate at an air admittance valve in accordance with Section P3114.

❖ This section further emphasizes the requirement that all vents must extend to the outdoors. This can be accomplished by connecting to a vent stack or a stack vent that extends outdoors and terminates to the open air or by extending these vents independently to the outside. Either method prevents sewer gasses from escaping into the building. The exception recognizes the use of air admittance valves in venting systems complying with the provisions of Section P3114.

P3104.2 Grade. All vent and branch vent pipes shall be so graded, connected and supported as to allow moisture and condensate to drain back to the soil or waste pipe by gravity.

❖ Vents commonly convey moisture vapor from the drainage system, and such vapor can condense in the vent piping. Also, rainwater can enter the vent system at the vent terminal. Therefore, the vent must be graded to drain any moisture back into the drainage piping, preventing accumulation of condensate or rainwater. Also, a vent pipe that is not sloped or that has a reverse slope can either be partially or completely filled with water, which lessens the efficiency of the vent or makes it inoperative. See Commentary Figure P3104.2.

P3104.3 Vent connection to drainage system. Every dry vent connecting to a horizontal drain shall connect above the centerline of the horizontal drain pipe.

❖ When drainage enters a pipe, it is assumed that the liquid proceeds down the pipe in the direction of flow. Being a liquid, however, the drainage seeks its own level and can move against the direction of flow before reversing and draining down the pipe. When this occurs, the solids may drop out of suspension to the bottom of the pipe. In a drain, the solids will again move down the pipe with the next discharge of liquid to the drain.

If the horizontal pipe serves only as a dry vent, the solids have no possibility of being brought back into suspension and, over a period of time, the horizontal vent pipe could be completely obstructed by waste. To avoid these blockages, the vent must connect to horizontal drains above the centerline. See Commentary Figure P3104.3. This prevents waste from entering a dry vent by connecting the vent above the flow line of the drain. Such connections will result in vent piping that forms a 45-degree (0.79 rad) or greater angle with the horizontal.

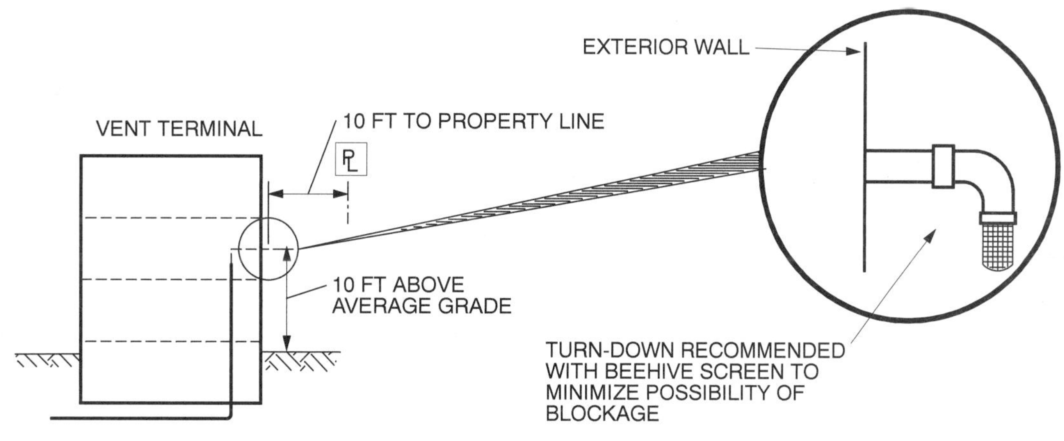

For SI: 1 foot = 304.8 mm.

Figure P3103.6
SIDE WALL VENT

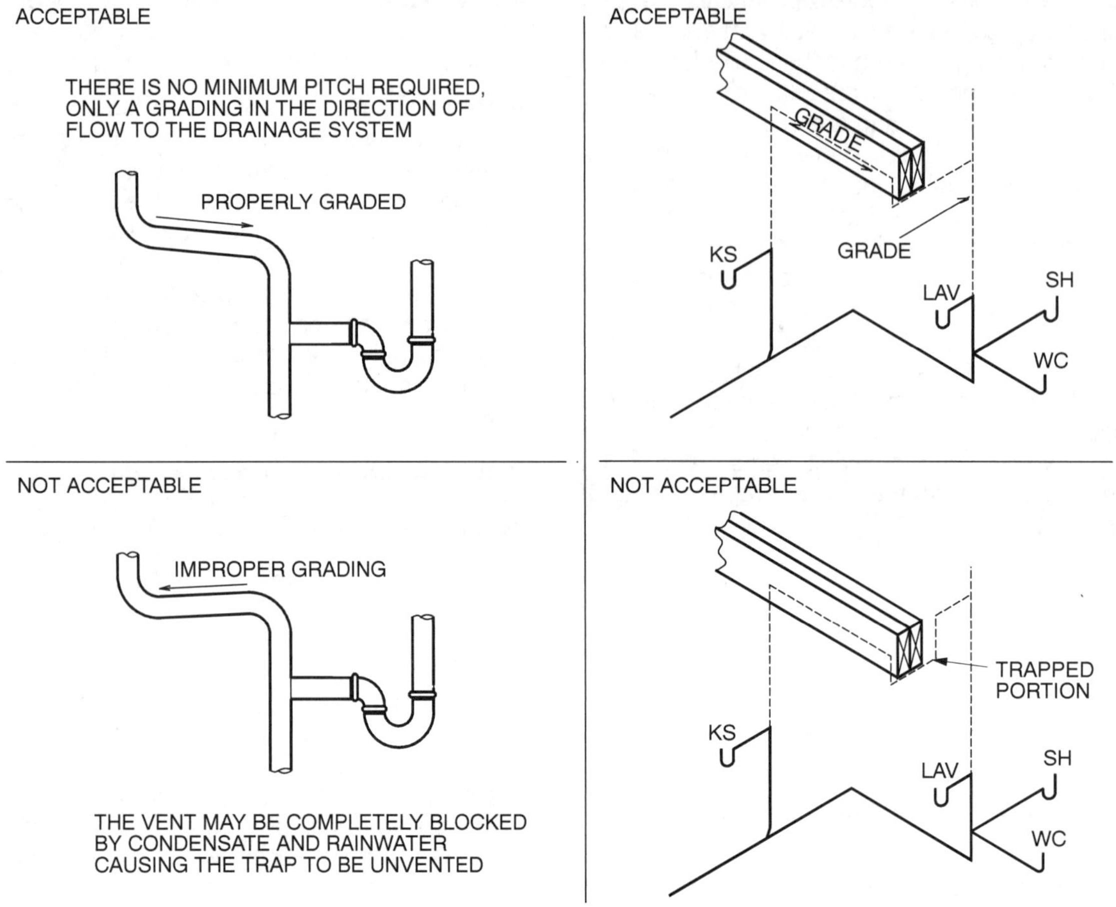

Figure P3104.2
GRADE OF VENTS

P3104.4 Vertical rise of vent. Every dry vent shall rise vertically to a minimum of 6 inches (152 mm) above the flood level rim of the highest trap or trapped fixture being vented.

❖ The provisions of this section prohibit horizontal vent piping from occurring where it will be subject to waste flow. A vertical rise of the vent piping will reduce the possibility of having a blockage in the vent pipe. Drainage system stoppage will flow over the fixture flood rim before reaching the horizontal vent line. Drain stoppage is immediately apparent, but vent stoppage is not; however, the latter can cause trap-seal siphonage and allow sewer gases to enter the building. A vertical rise would meet the definition of a vertical pipe, which is an angle of 45 degrees (0.79 rad) or more with the horizontal.

P3104.5 Height above fixtures. A connection between a vent pipe and a vent stack or stack vent shall be made at least 6 inches (152 mm) above the flood level rim of the highest fixture served by the vent. Horizontal vent pipes forming branch vents shall be at least 6 inches (152 mm) above the flood level rim of the highest fixture served.

❖ This section expresses the same concern as the previous section. In the event of drain stoppage, waste could rise into the vent piping. Over a period of time, the vent piping could become blocked with solids that have settled out of the waste liquids. Additionally, individual, common, or branch vents that improperly connect to vent stacks or stack vents could become drains where the fixture drain or drains served are blocked or restricted. It is entirely possible for a vent to serve unintentionally as a drain, and this condition could go unnoticed for a long period of time. See Commentary Figures P3104.5(1) and P3104.5(2).

P3104.6 Vent for future fixtures. Where the drainage piping has been roughed-in for future fixtures, a rough-in connection for a vent shall be installed a minimum of one-half the diameter of the drain. The vent rough-in shall connect to the vent system or shall be vented by other means as provided in this chapter. The connection shall be identified to indicate that the connection is a vent.

❖ When future fixture rough-in drainage piping is installed, vents must also be installed. The vent rough-in must be tied into the vent system or must extend to a vent terminal. The rough-in connection must be identi-

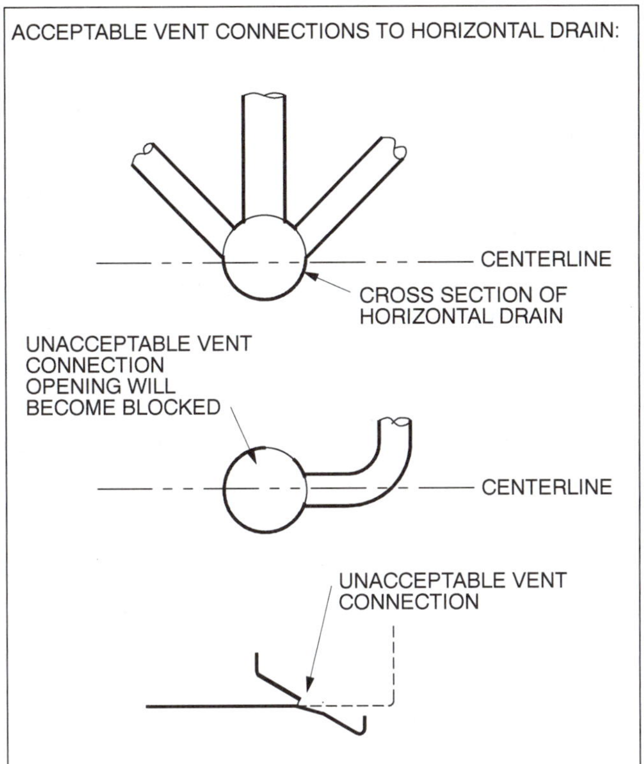

Figure P3104.3
VENT CONNECTIONS

fied as a vent so that the purpose of the original installation is evident when the future plumbing is completed.

SECTION P3105 FIXTURE VENTS

P3105.1 Distance of trap from vent. Each fixture trap shall have a protecting vent located so that the slope and the developed length in the fixture drain from the trap weir to the vent fitting are within the requirements set forth in Table P3105.1.

Exception: The developed length of the fixture drain from the trap weir to the vent fitting for self-siphoning fixtures, such as water closets, shall not be limited.

❖ The distance from a trap to its vent is limited to reduce the possibility of the trap self-siphoning, which is the siphoning caused by the discharge from the fixture the trap serves. This section locates the vent within the flow line of the fixture drain to prevent the opening to the vent from being clogged by the waste flow. See Commentary Figure P3105.1.

An exception is made for the trap-to-vent distance for fixtures such as water closets because these fixtures rely on self-siphonage to operate properly, and the trap is resealed after each use. The water closet must be vented in all cases, and limiting the distance between the vent connection and the water closet serves no purpose.

TABLE P3105.1
MAXIMUM DISTANCE OF FIXTURE TRAP FROM VENT

SIZE OF TRAP (inches)	SLOPE (inch per foot)	DISTANCE FROM TRAP (feet)
$1^1/_4$	$^1/_4$	5
$1^1/_2$	$^1/_4$	6
2	$^1/_4$	8
3	$^1/_8$	12
4	$^1/_8$	16

For SI: 1 inch = 25.4 mm, 1 foot = 304.8 mm, 1 inch per foot = 83.3 mm/m.

❖ The distances listed in Table P3105.1 are based on laboratory testing and prevent the weir of the trap from being located above the highest inlet to the vent. As the total change in elevation of a fixture drain (resulting from the fixture drain's slope) approaches the inside diameter of the drain, the weir of the trap approaches the highest level of the vent connection, and self-siphoning is more likely. See the commentary for Section P3105.2.

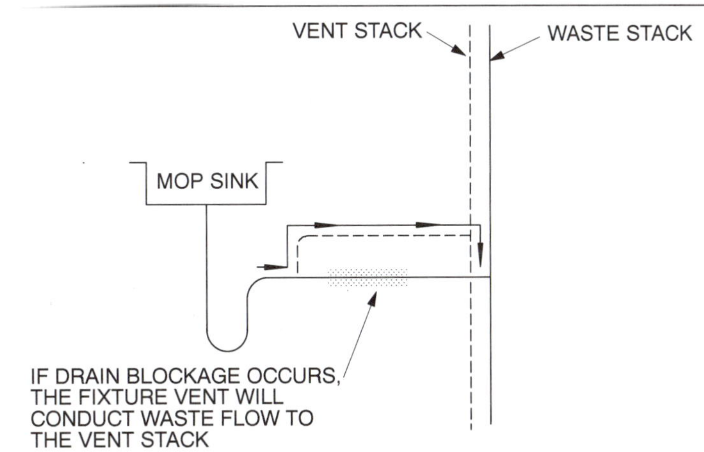

Figure P3104.5(1)
IMPROPERLY CONNECTED VENT SERVING A DRAIN

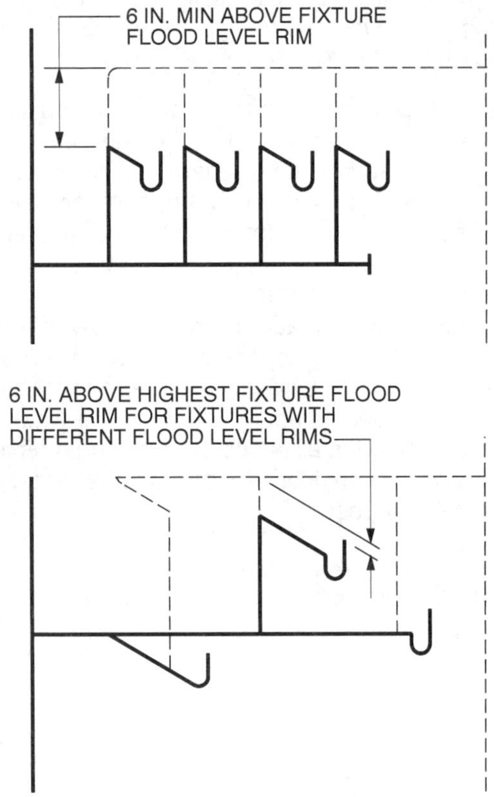

For SI: 1 inch = 25.4 mm.

Figure P3104.5(2)
HORIZONTAL OFFSET OF BRANCH VENT

P3105.2 Fixture drains. The total fall in a fixture drain due to pipe slope shall not exceed one pipe diameter, nor shall the vent pipe connection to a fixture drain, except for water closets, be below the weir of the trap, except as provided in Section P3105.3.

❖ This section's requirements reinforce those of the previous section. They substantiate the basic rule applied in Table P3105.1 and require that the total fall in a fixture drain due to slope does not exceed one pipe diameter. They also require that the vent take-off be above the weir of the trap. Since a water closet depends on trap siphonage in its operation to remove the contents, water closets are exempted from these requirements.

This section does not provide for increasing the fixture drain size alone to accommodate a longer distance. Distance will be determined primarily by trap size, and the fixture drain will be sized accordingly. Chapter 32 outlines trap size limitations.

P3105.3 Vertical leg for waste fixture drains. A vertical leg (see Figure P3105.3) is permitted within a fixture drain of a waste fixture in accordance with the following criteria:

1. Minimum trap diameter shall be in accordance with Table P3201.7.
2. The diameter of Section A shall be equal to the diameter of the trap.
3. The length of Section A shall not be less than 8 inches (203 mm) and in accordance with Table P3105.1.
4. The diameter of Section B shall be one pipe size larger than the diameter of Section A.
5. The length of Section B shall not be more than 36 inches (914 mm).
6. The diameter of Section C shall be one pipe size larger than the diameter of Section B.
7. The total length of Section A and Section C shall not exceed the distance allowed in Table P3105.1.
8. Bends shall be the diameter of the largest connected section.

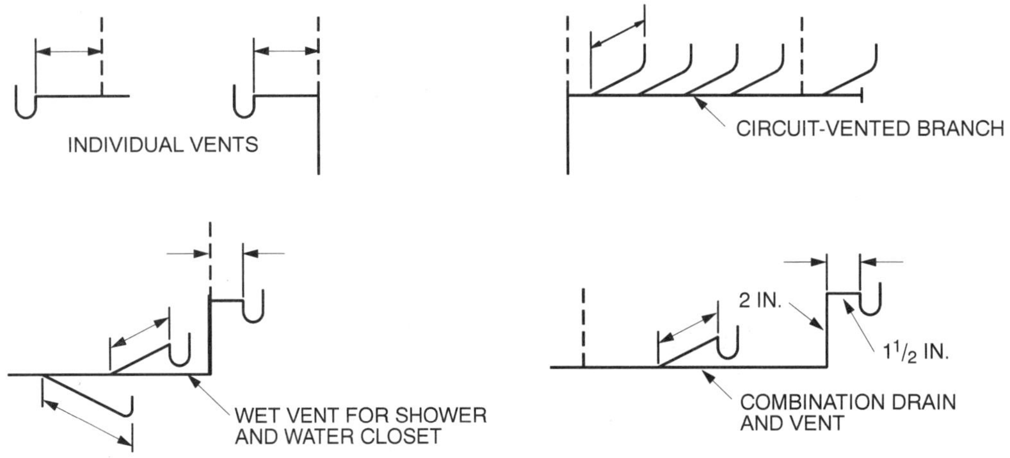

For SI: 1 inch = 25.4 mm.

Figure P3105.1
DISTANCE FROM TRAP TO VENT

❖ This section permits a certain configuration that allows the vent takeoff to be below the weir of the trap. Laboratory tests have shown that when installed following the specific requirements, this trap seal will not siphon. However, strict adherence to these requirements must be observed to ensure proper functioning of the system.

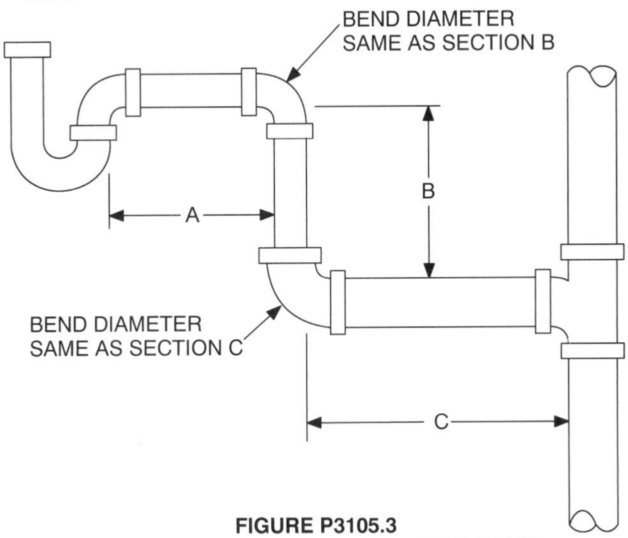

FIGURE P3105.3
VERTICAL LEG FIXTURE DRAIN SCHEMATIC

❖ This figure illustrates the concept of adding a vertical leg within a waste fixture drain installation. Section P3105.3 provides the eight requirements to be adhered to. Refer to Figure P3105.3 as these requirements are explained.

1. The trap size limitations of Table P3201.7 must be applied. This table provides the minimum size of trap for individual or groups of fixtures. See the commentary for Table P3201.7.
2. Section A must be the same diameter as the trap.
3. Section A must adhere to the requirements of Table P3105.1 but must not be less than 8-inches (203 mm). See the commentary for Table P3105.1.
4. Section B, which is the vertical leg, must be a minimum of one pipe size diameter larger than the diameter of Section A. Since Section B is increased, Step 8 will require the elbow at the connection of Sections A and B to be sized equal to Section B, the larger of the two sections.
5. Regardless of the size, the vertical leg portion (Section B) must not exceed 36 inches (914 mm) in length.
6. Section C must now be a minimum of one pipe size larger than that of Section B. Step 8 will require the elbow at the connection of Sections B and C to be sized equal to Section C, the larger of the two sections.
7. This step refers the reader back to the provisions of Table P3105.1 with regard to distance limitations. Sections A and C combined (the horizontal sections) must not exceed the length or distance limitations as specified in the table for the size of trap.
8. This step requires the size of the connecting elbow to be the same as the largest connecting pipe. See the commentary for Steps 4 and 6.

For example, suppose that this figure illustrates a fixture drain installation of a lavatory. According to Table P3207.1, a lavatory will require a minimum $1^1/_4$-inch (32 mm) trap. Therefore, Section A and the trap will be $1^1/_4$ inches (32 mm) in diameter. Further, Section A cannot be less than 8 inches (203 mm) long. So far, we have met the requirements of 1, 2, and 3. Section B must be increased one pipe size; it will now be a minimum of $1^1/_2$ inches (38 mm) in diameter. Due to the increase in size, the elbow connecting Sections A and B must be the same diameter—$1^1/_2$ inches (38 mm). Further, Section B cannot exceed 36 inches (914 mm) in length. This completes the requirements of 4, 5, and 8. Finally, Section C must be sized at least one pipe diameter larger than Section B, requiring it to be a minimum of 2 inches (51 mm) in diameter. Again, according to Step 8, the elbow connecting Sections B and C must be sized equal to the diameter of Section C; therefore, it will be at least 2 inches (51 mm). The length of Section C will be limited to the distance allowed in Table P3105.1 less the distance of Section A. When that is verified, we will have met the requirements of 6, 7, and 8.

P3105.4 Crown vent. A vent shall not be installed within two pipe diameters of the trap weir.

❖ Crown venting is an arrangement where a vent connects at the top of the weir (crown) of a trap. The problem with this type of connection is that the vent opening can become blocked, thus closing the vent and allowing the trap to siphon. The blockage is a result of the action of the drainage flowing through the trap. Flow direction and velocity will force waste up into the vent connection. It has been found that when the vent is connected at least two pipe diameters downstream from the trap weir, this problem can be avoided. See Commentary Figures P3105.4(1) and P3105.4(2).

SECTION P3106
INDIVIDUAL VENT

P3106.1 Individual vent permitted. Each trap and trapped fixture is permitted to be provided with an individual vent. The individual vent shall connect to the fixture drain of the trap or trapped fixture being vented.

❖ The simplest form of venting a trap or trapped fixture is an individual vent for each trap. A single vent pipe is connected between the trap of fixture and the branch connection to the drainage system. With a properly installed individual vent, only the drainage of the fixture served is flowing past the vent. The sizes of individual vents are to be in accordance with Section P3113. See

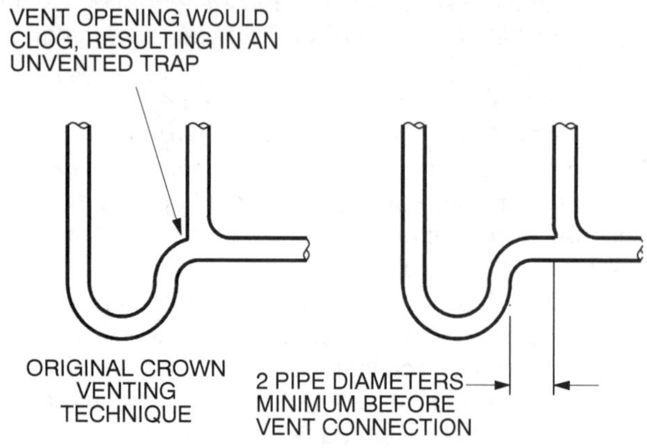

TRAP SIZE	MINIMUM DISTANCE TRAP TO VENT
1 1/4"	2 1/2"
1 1/2"	3"
2"	4"
3"	6"

For SI: 1 inch = 25.4 mm.

Figure P3105.4(1)
CROWN VENTING

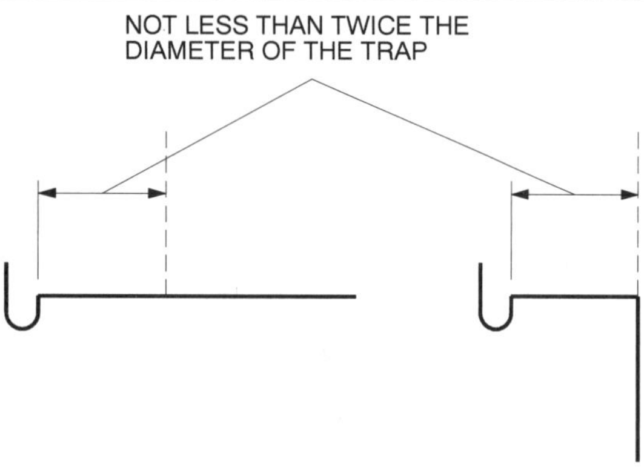

Figure P3105.4(2)
CROWN VENTING

the commentary to Sections P3113.1 and P3113.2. Also, see Commentary Figures P3106.1(1) and P3106.1(2).

SECTION P3107
COMMON VENT

P3107.1 Individual vent as common vent. An individual vent is permitted to vent two traps or trapped fixtures as a common vent. The traps or trapped fixtures being common vented shall be located on the same floor level.

❖ Common venting allows an individual vent installed in accordance with Section P3107.1 to serve two fix-

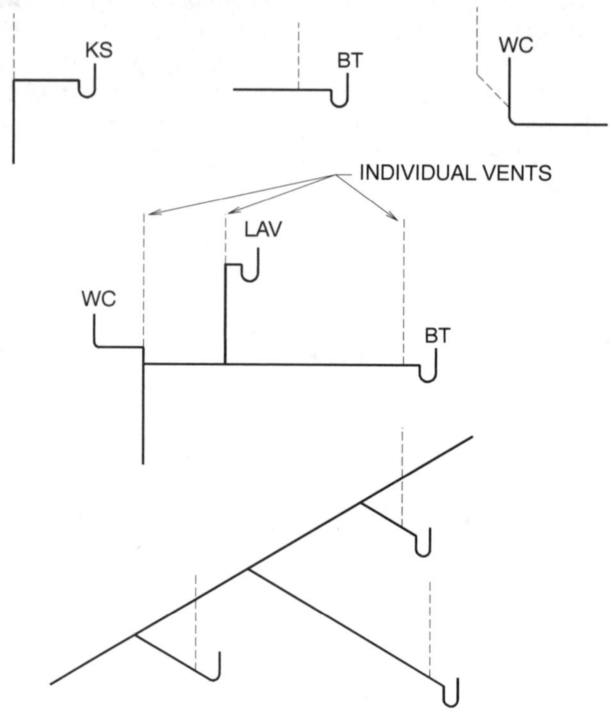

Figure P3106.1(1)
INDIVIDUAL VENTS

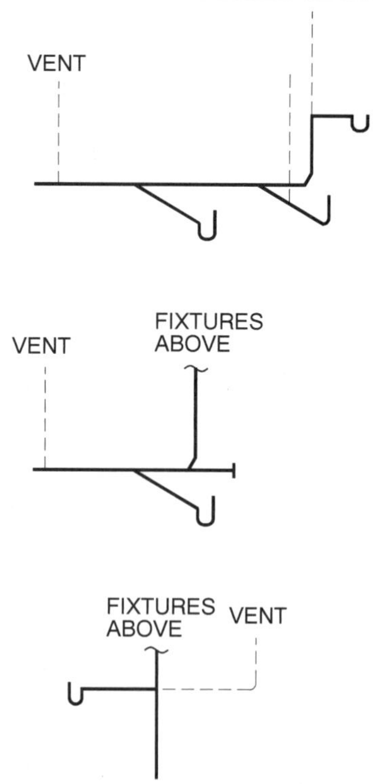

Figure P3106.1(2)
IMPROPERLY INSTALLED INDIVIDUAL VENTS

tures. Sections P3107.2 and P3107.3 contain the provisions that pertain to connection of fixture drains at the same level or at different levels, respectively. In either case, the fixtures being connected by means of the same common vent must be located on the same floor level.

P3107.2 Connection at the same level. Where the fixture drains being common vented connect at the same level, the vent connection shall be at the interconnection of the fixture drains or downstream of the interconnection.

❖ One form of common venting connects two fixtures at the same level. A typical installation consists of two horizontal drains connecting to a vertical drain through a double-pattern fitting. The extension of the vertical drain serves as the vent. Two drains connecting horizontally to a horizontal drain through a double-pattern fitting is also a form of common venting. The vent is allowed to connect at the interconnection of the fixture drains or downstream along the horizontal drain. The distance from the trap to the vent for each fixture drain must be in accordance with Table P3105.1. See Commentary Figure P3107.2.

The fittings used in such installations must be as outlined in Section P3005, in particular those fittings used for connecting back-to-back water closets or fixtures with pumping action discharge. See the commentary for Section P3005.1.1.

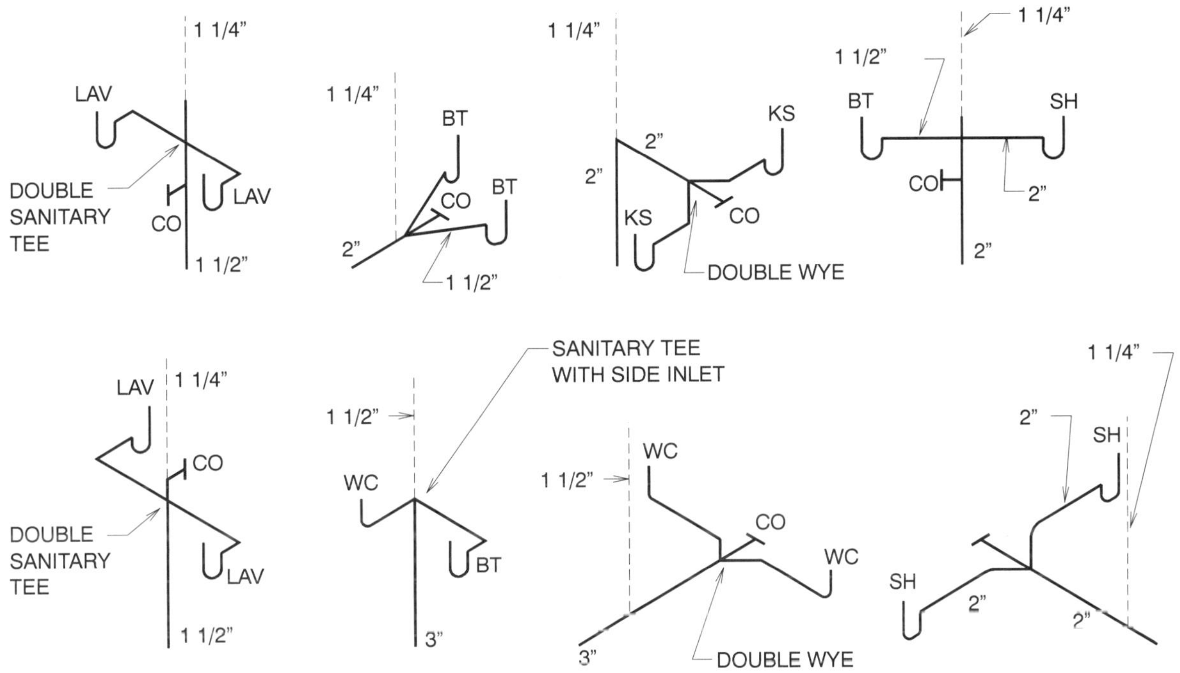

For SI: 1 inch = 25.4 mm.

Figure P3107.2
COMMON VENTS WITH FIXTURE DRAINS CONNECTING AT SAME LEVEL

P3107.3 Connection at different levels. Where the fixture drains connect at different levels, the vent shall connect as a vertical extension of the vertical drain. The vertical drain pipe connecting the two fixture drains shall be considered the vent for the lower fixture drain, and shall be sized in accordance with Table P3107.3. The upper fixture shall not be a water closet.

❖ Two fixtures within a single story can connect at different levels to a vertical drain and be considered common vented, but the vent remains an individual vent serving as a common vent. The drainpipe between the upper and lower fixtures, however, must meet the sizing requirements of Table 3107.3. See the commentary for Table P3107.3. The upper fixture cannot be a water closet because the pressures created by the rapid discharge of a water closet could adversely affect the venting of the lower fixture.

TABLE P3107.3
COMMON VENT SIZES

PIPE SIZE (inches)	MAXIMUM DISCHARGE FROM UPPER FIXTURE DRAIN (d.f.u.)
$1^1/_2$	1
2	4
$2^1/_2$ to 3	6

For SI: 1 inch = 25.4 mm.

❖ This table provides the maximum allowable drainage fixture units (d.f.u.'s) that can be served by the common vent, which is that portion of piping between the upper fixture drain and the lower fixture drain connection. The common vent is sized based on the d.f.u.'s of the upper fixture drain. The drainpipe downstream of the lower fixture drain must be sized in accordance with Section 3005.4 based on the combined d.f.u.'s of both the upper and lower fixture drains connecting to the common vent. See the commentary for Section P3005.4 and Commentary Figure P3107.3.

SECTION P3108
WET VENTING

P3108.1 Wet vent permitted. Any combination of fixtures within two bathroom groups located on the same floor level are permitted to be vented by a wet vent. The wet vent shall be considered the vent for the fixtures and shall extend from the connection of the dry vent along the direction of the flow in the drain pipe to the most downstream fixture drain connection to the horizontal branch drain. Only the fixtures within the bathroom groups shall connect to the wet vented horizontal branch drain. Any additional fixtures shall discharge downstream of the wet vent. [See Figures P3108.1(1), P3108.1(2), P3108.1(3) and P3108.1(4) for typical wet vent configurations.]

❖ A wet vent is a vent pipe that conveys drainage. Wet venting is a method of venting any combination of fixtures within a single or double bathroom group. See the commentary for the definition of "bathroom group." Wet venting employs oversized piping to allow for the flow of air above the waste flow. Reliance is placed on two factors for establishing that there will always be adequate volume within the wet vent pipe to permit required air flow: the low probability of simultaneous fixture discharge and the low-flow velocity that results from requiring the fixtures to be on the same floor level. A single properly sized vent will provide airflow when fixtures are being discharged to relieve pressures that develop in the drain piping.

The wet vent is that portion of the drainage system receiving discharge from a fixture and serving as a vent for other fixtures. Only the fixture types and quantities addressed in this section are allowed to connect to or be served by a wet vent. Because of distances or rough-in heights, more than one fixture might be vented by an individual dry vent. However, this does not prohibit such fixtures from discharging into the wet vent. Fixtures that are part of the two bathroom groups are allowed to discharge to the wet vent, even if such fixtures are not being wet vented. Additional fixtures must connect downstream of the wet vented section. See Figures P3108.1(1) through (5) and their commentaries.

P3108.2 Vent connections. The dry vent connection to the wet vent shall be an individual vent or common vent to the lavatory, bidet, shower or bathtub.

❖ The dry vent is not part of the wet vent system. It is only an extension of the wet vent system to a vent stack, stack vent, or other approved point of termination. The dry extension of the wet vent is sized in accordance with Section P3113. The dry-vent connection must be connected to a lavatory, bidet, shower, or bathtub, thus eliminating a water-closet vent for this purpose.

P3108.3 Size. Horizontal and vertical wet vents shall be of a minimum size as specified in Table P3108.3, based on the fixture unit discharge to the wet vent.

❖ This section refers the reader to Table P3108.3 for sizing the wet-vented section of piping. The table applies to both vertical and horizontal wet venting.

TABLE P3108.3
WET VENT SIZE

WET VENT PIPE SIZE (inches)	FIXTURE UNIT LOAD (d.f.u.)
$1^1/_2$	1
2	4
$2^1/_2$	6
3	12
4	32

For SI: 1 inch = 25.4 mm.

❖ This table provides minimum sizes of wet-vent piping, ranging from 1½ inches (38 mm) through 4 inches (102 mm) in diameter.

P3108.4 Vertical wet vent. A combination of fixtures located on the same floor level are permitted to be vented by a vertical wet vent. The vertical wet vent shall extend from the connection to the dry vent down to the lowest fixture drain connection. Each fixture shall connect independently to the vertical wet vent. All water closet drains shall connect at the same elevation. Other fixture drains shall connect above or at the same elevation as the water closet fixture drains. The dry vent connection to the vertical wet vent shall be an individual or common vent serving one or two fixtures. [See Figures P3108.1(4) and P3108.1(5) for typical vertical wet vent configurations.]

❖ This section contains the requirements for vertical wet venting, which is the same basic idea as that of horizontal wet venting as provided for in Section P3108. In a vertical wet-vent configuration, the connection of a water closet must be the lowest connection to the vertical wet-vented section. In addition, where two water closets are part of the wet vented system and connect to the vertical drain independently, they must connect at the same level. All other connections to the vertical drain must be above or at the same height as the water closet connection. Table P3108.3 shows vertical wet-vent sizing.

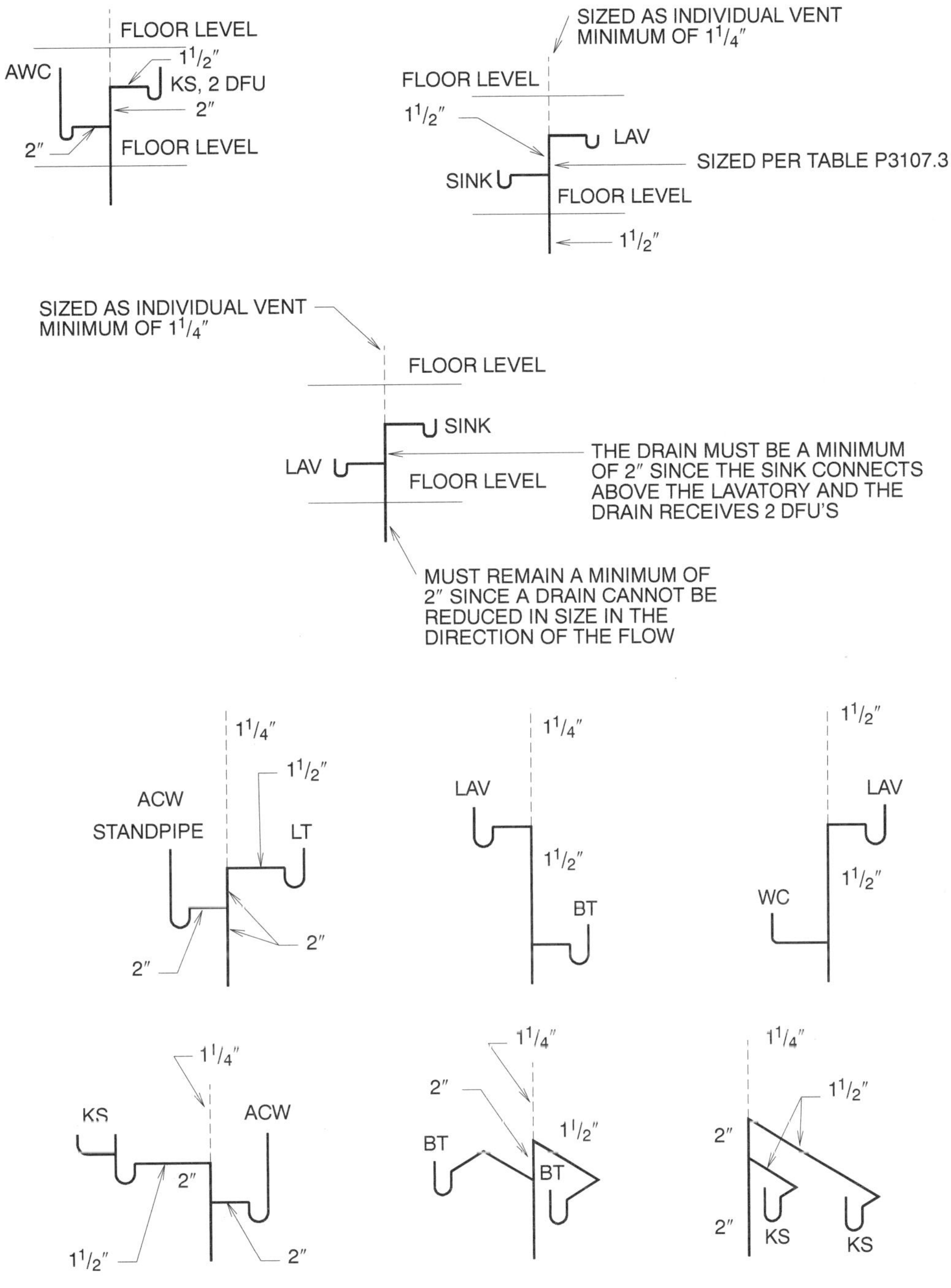

For SI: 1 inch = 25.4 mm.

Figure P3107.3
COMMON VENTS WITH FIXTURE DRAINS CONNECTING AT DIFFERENT LEVELS

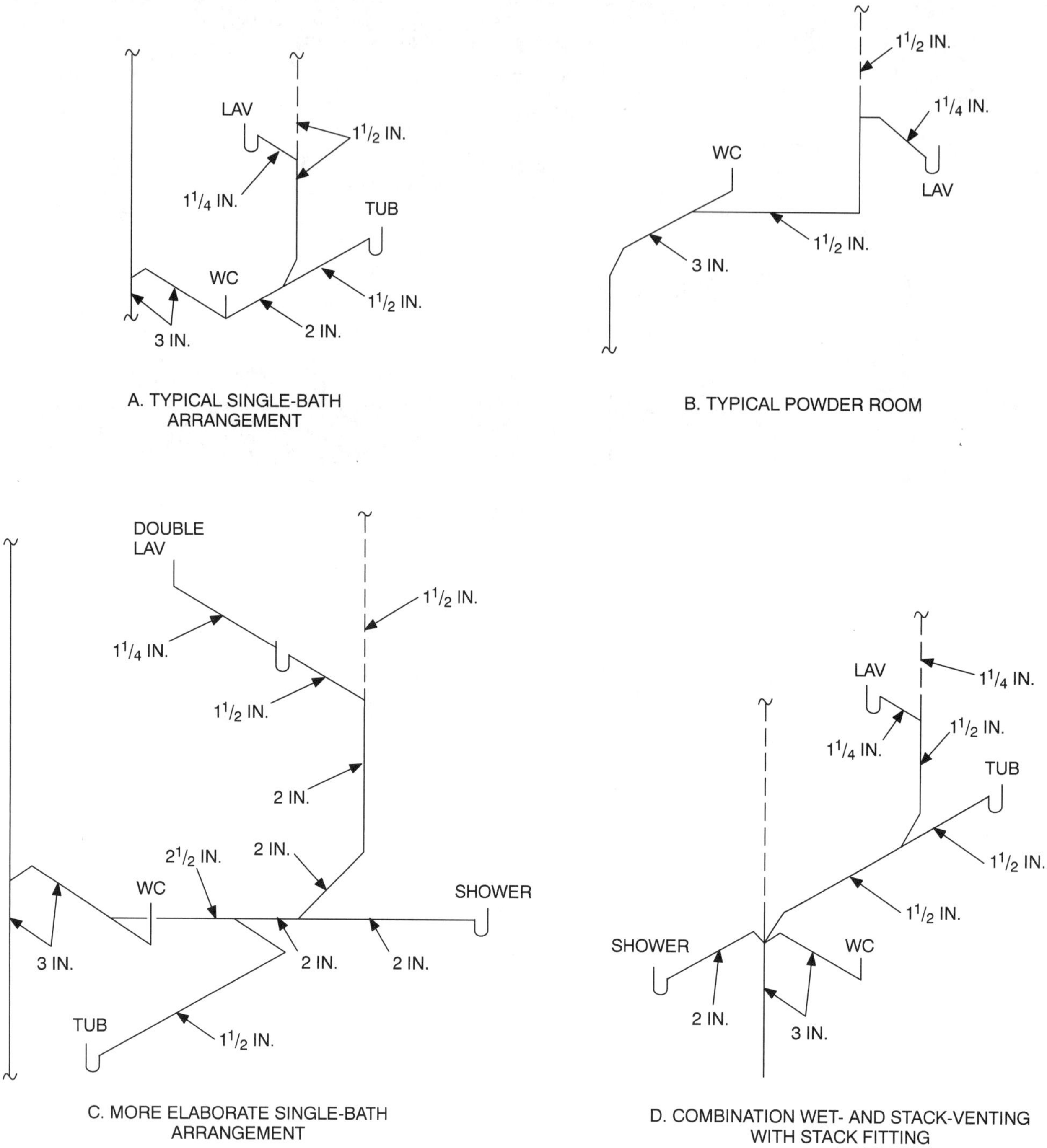

For SI: 1 inch = 25.4 mm.

Note: This figure is for illustration purposes only and the written text shall apply.

FIGURE P3108.1(1)
TYPICAL SINGLE-BATH WET-VENT ARRANGEMENTS

❖ This figure depicts four typical wet-vent installations common to one- and two-family dwellings in a single-bathroom configuration. The dry-vent connection is in accordance with Section P3108.2. As stated in Section P3108.1, the wet vent extends from the connection of the dry vent along the direction of flow in the drain pipe to the most downstream fixture drain connected to the horizontal branch drain. Thus, the dry vent serving the wet vent is properly located in each of these illustrations. The size of the dry vent must conform to Sections P3113. The individual fixture drains must comply with Section P3105. The size of the wet vent section is in accordance with Table P3108.3.

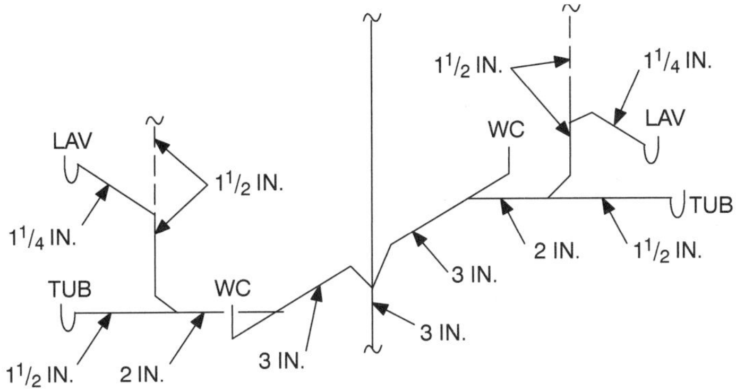

A. TYPICAL BACK-TO-BACK BATHS

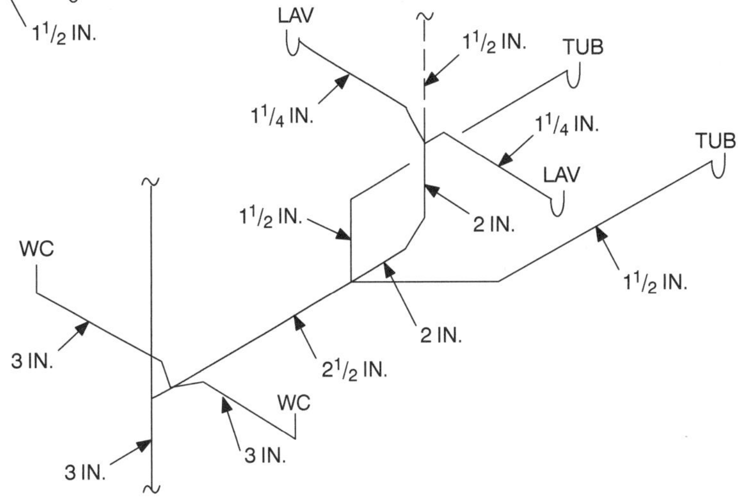

B. DOUBLE BATHS WITH FIXTURES ON COMMON HORIZONTAL BRANCH, COMMON WET VENT

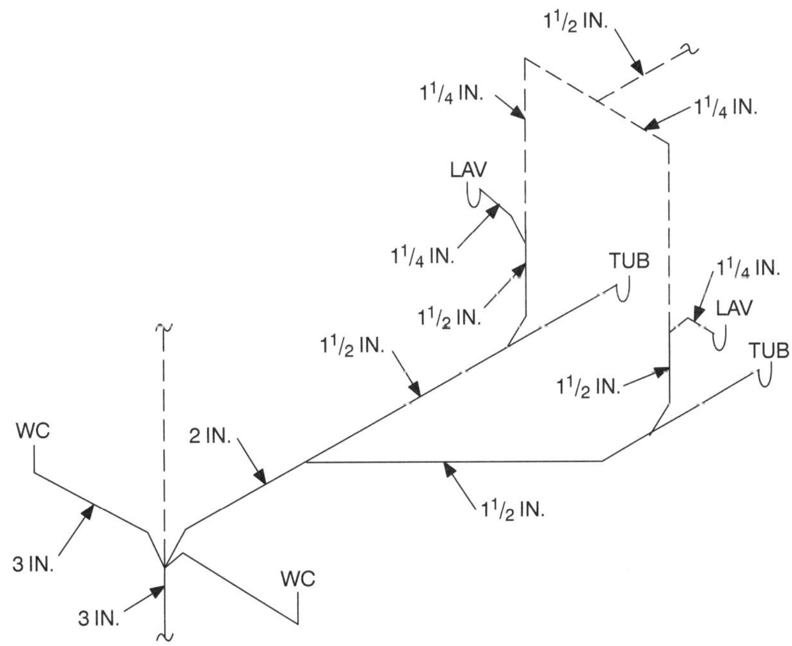

C. DOUBLE BATHS WITH WASTE FIXTURES ON COMMON HORIZONTAL BRANCH, INDIVIDUAL WET VENTS

For SI: 1 inch = 25.4 mm.

Note: This figure is for illustration purposes only and the written text shall apply.

FIGURE P3108.1(2)
TYPICAL DOUBLE-BATH WET-VENT ARRANGEMENTS

❖ This figure provides three illustrations depicting wet venting arrangements involving two bathroom groups. Again, the same limitations discussed in the commentary to Figure P3108.1(1) apply.

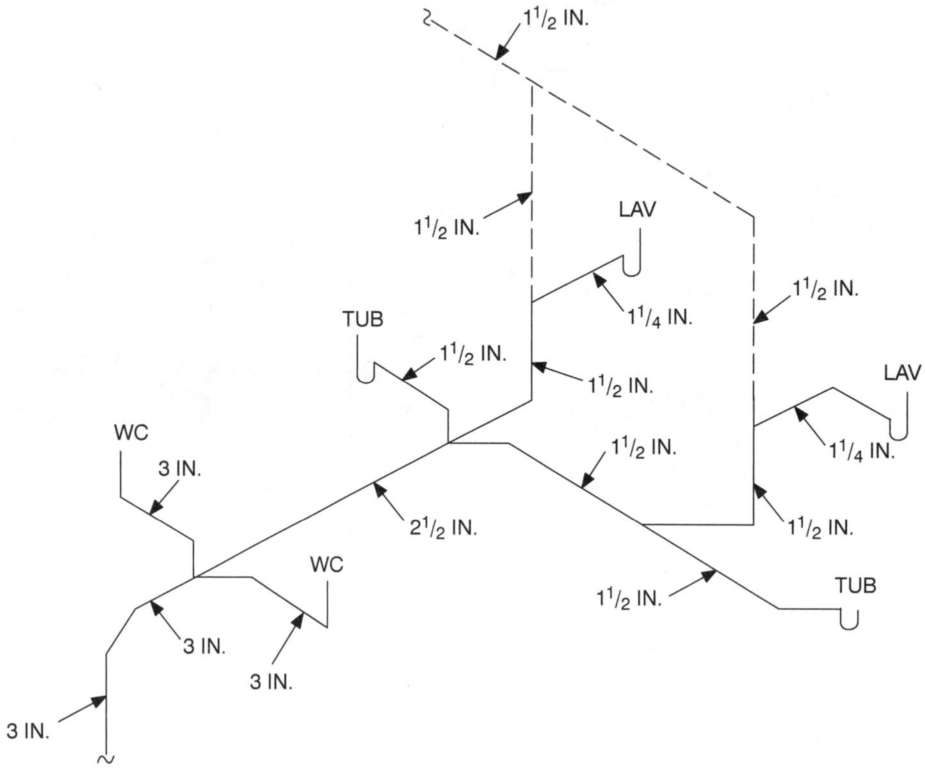

For SI: 1 inch = 25.4 mm.

Note: This figure is for illustration purposes only and the written text shall apply.

FIGURE P3108.1(3)
TYPICAL HORIZONTAL WET VENTING

❖ Figure P3108.1(3) illustrates two bathroom groups served by a wet vent system. A wet-vented system can serve two water closets, two lavatories, two bidets, and two bathtubs or showers. Combinations of fixtures must consist only of fixtures taken from this list. For example, a wet vent cannot serve three lavatories because the fixtures that are within two bathroom groups include only two lavatories. See the definition of "Bathroom Group" in Chapter 2.

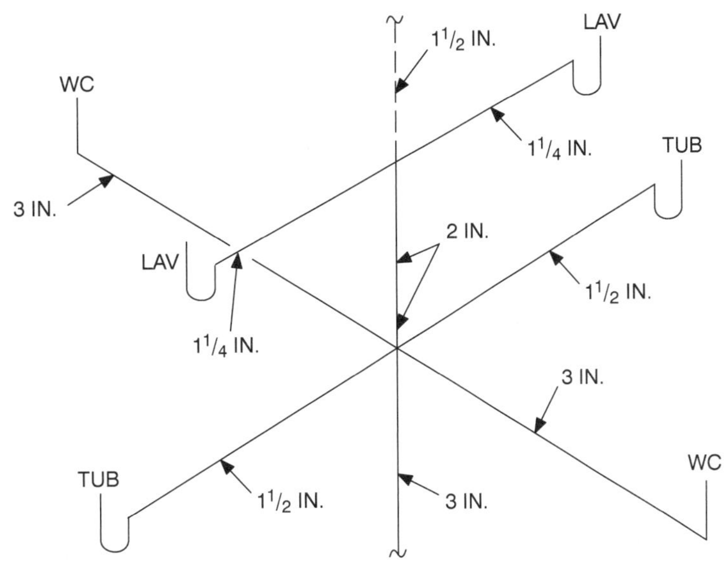

A. VERTICAL WET VENTING

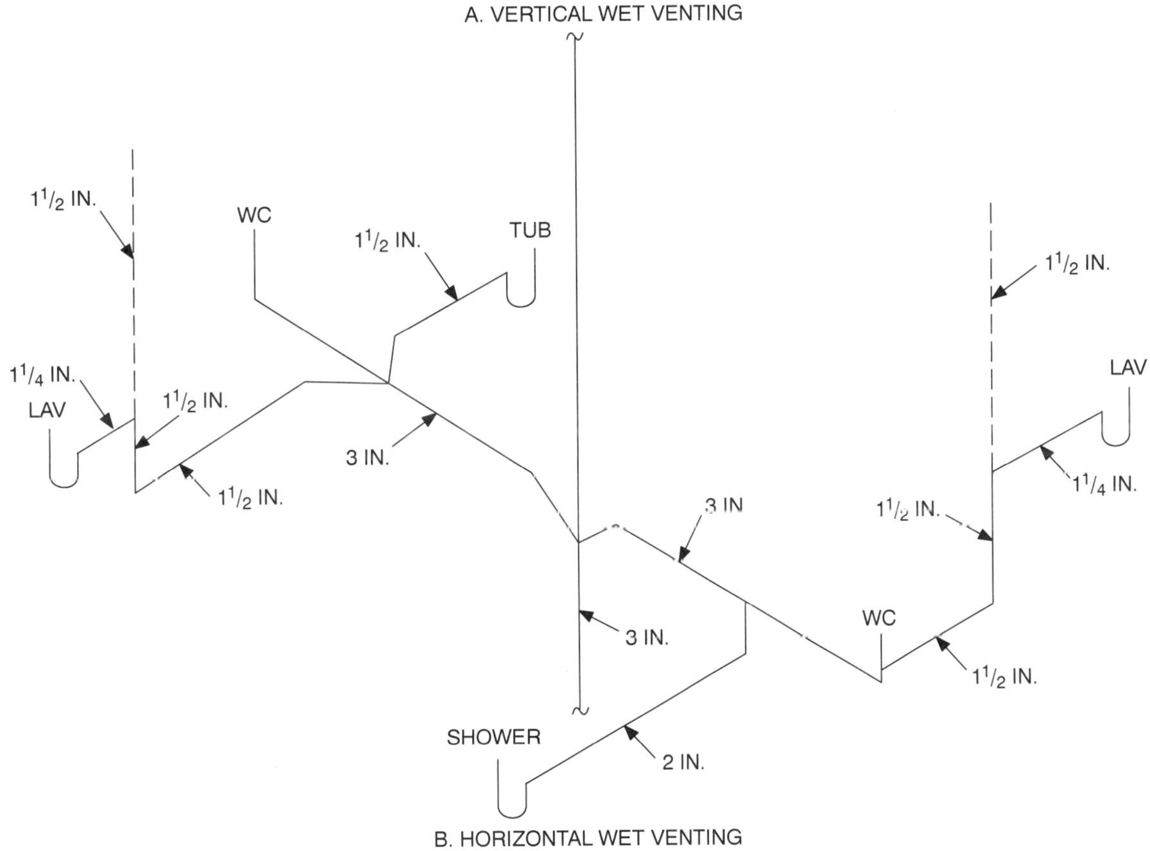

B. HORIZONTAL WET VENTING

For SI: 1 inch = 25.4 mm.

Note: This figure is for illustration purposes only and the written text shall apply.

FIGURE P3108.1(4)
TYPICAL METHODS OF WET VENTING

❖ This figure illustrates two wet-vented configurations: Diagram A is a vertical wet vent, and Diagram B shows both vertical and horizontal wet-venting. In both diagrams the water closet connections to the vertical drain are at the same level, and all other wet-vented connections are at or above the water closet level. This is in accordance with Section P3108.4. In Diagram B, each horizontal wet-vented branch must be vented as required by Sections P3108.1 and P3108.2. Sizing of the dry vents must comply with Section P3113.

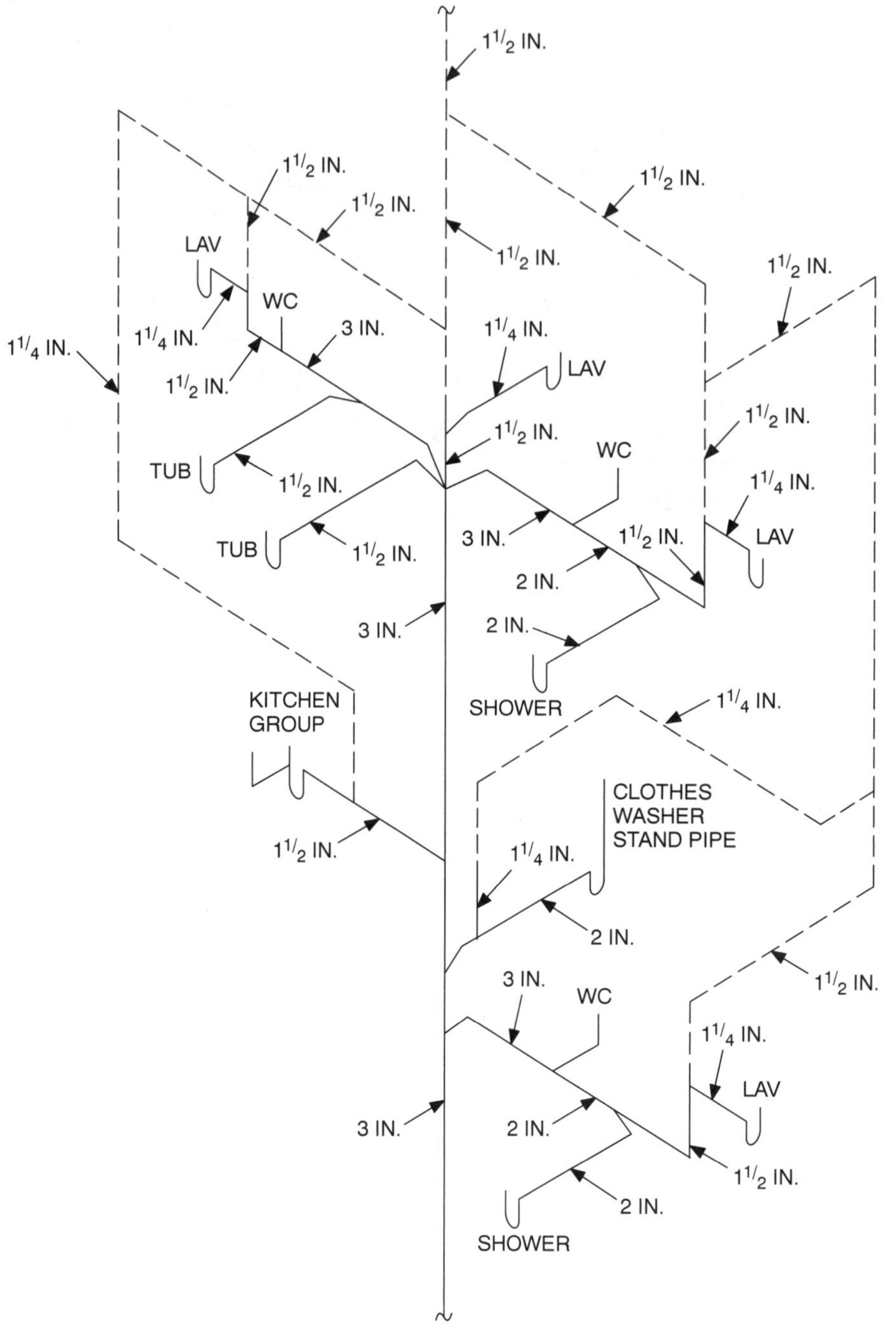

For SI: 1 inch = 25.4 mm.

Note: This figure is for illustration purposes only and the written text shall apply.

FIGURE P3108.1(5)
TYPICAL SINGLE-STACK SYSTEM FOR A TWO-STORY DWELLING

❖ This figure illustrates how a single stack can serve an entire three-bathroom, two-story dwelling. The two bathroom groups on the uppermost floor use a wet-vented installation, combining both horizontal and vertical wet-venting applications. A horizontal wet-vented branch serves the lower single bathroom group. Individual vents serve the kitchen group and the clothes washer.

SECTION P3109
WASTE STACK VENT

P3109.1 Waste stack vent permitted. A waste stack shall be considered a vent for all of the fixtures discharging to the stack where installed in accordance with the requirements of this section.

❖ The principles of waste stack use are based on some of the original research that was performed in plumbing. The system has been identified by a variety of names, including vertical wet venting, single-stack venting, and multi-floor stack venting. Common venting and vertical wet-venting installations are allowed only when the various fixtures connecting into the stack are all on the same floor. Waste stack venting, not to be confused with common venting and wet venting, is allowed to serve fixtures on multiple floors.

P3109.2 Stack installation. The waste stack shall be vertical, and both horizontal and vertical offsets shall be prohibited. Every fixture drain shall connect separately to the waste stack. The stack shall not receive the discharge of water closets or urinals. (See Figure P3109.2 for typical waste stack venting.)

❖ The waste stack vent must be vertical without any vertical or horizontal offsets. Water closets and urinals are not approved for use in a waste stack vent installation. All other fixture drains connecting to the stack must be connected separately.

P3109.3 Stack vent. A stack vent shall be provided for the waste stack. The size of the stack vent shall be equal to the size of the waste stack.

❖ The waste stack vent serves as both a drain and a vent. As with wet venting, air is necessary to accommodate pressure fluctuations within the stack. However, since a waste stack vent configuration can serve multiple floors, research has proven the need for over-

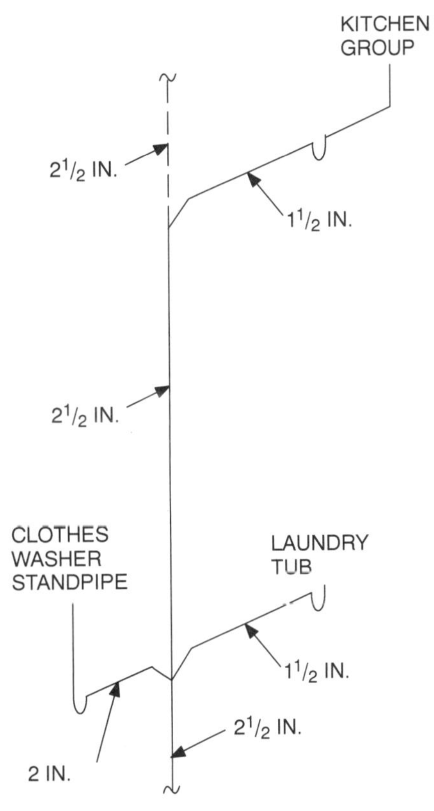

For SI: 1 inch = 25.4 mm.

Note: This figure is for illustration purposes only and the written text shall apply.

FIGURE P3109.2
WASTE STACK VENTING

❖ This illustration shows how fixtures on separate floors might be connected by means of a waste stack vent. The stack is without offsets as specified in Section P3109.2, and the stack vent is sized in accordance with Section 3109.3.

sizing the stack vent as is done with the waste stack. This section requires the vent to be the same diameter as the stack. Installation of the stack vent must conform to Section P3104.

P3109.4 Waste stack size. The waste stack shall be sized based on the total discharge to the stack and the discharge within a branch interval in accordance with Table P3109.4. The waste stack shall be the same size throughout the length of the waste stack.

❖ This section references Table P3109.4, which gives the minimum size limitations for a waste stack vent. The larger the stack, the higher the threshold is for d.f.u. allowances.

In Figure P3109.2, the total d.f.u.'s of all the fixtures could be handled by a 2-inch (51 mm) diameter vertical drain. See Tables P3004.1 and P3005.4.1. However, since the kitchen group is a total of 3 d.f.u.'s, the waste stack vent cannot be less than 2 inches (66 mm), the smallest size of waste stack vent that will accommodate the connecting of a branch interval (or fixture drain) of more than 2 d.f.u.'s. The minimum size of the stack is based on Table P3109.4 and must be maintained for the entire length of the waste stack; this applies also the stack vent as prescribed in Section P3109.3.

TABLE P3109.4
WASTE STACK VENT SIZE

STACK SIZE (inches)	MAXIMUM NUMBER OF FIXTURE UNITS (d.f.u.)	
	Total discharge into one branch interval	Total discharge for stack
$1^1/_2$	1	2
2	2	4
$2^1/_2$	No limit	8
3	No limit	24
4	No limit	50

For SI: 1 inch = 25.4 mm.

❖ This table provides the minimum size requirements for waste stack vents based on the d.f.u. totals for all fixtures and fixture drains served. See the commentary for Section P3109.4.

SECTION P3110
CIRCUIT VENTING

P3110.1 Circuit vent permitted. A maximum of eight fixtures connected to a horizontal branch drain shall be permitted to be circuit vented. Each fixture drain shall connect horizontally to the horizontal branch being circuit vented. The horizontal branch drain shall be classified as a vent from the most downstream fixture drain connection to the most upstream fixture drain connection to the horizontal branch.

❖ Circuit venting is based on many of the same principles as wet venting. However, it will allow for the connection of other types or totals of similar fixtures that may be outside the definition of a bathroom group. The total of fixtures allowed to be circuit vented on one branch is eight. In addition, unlike wet venting, each fixture drain being circuit vented must connect horizontally to the horizontal circuit-vented branch drain.

P3110.2 Vent connection. The circuit vent connection shall be located between the two most upstream fixture drains. The vent shall connect to the horizontal branch and shall be installed in accordance with Section P3104. The circuit vent pipe shall not receive the discharge of any soil or waste.

❖ Circuit venting is unique in its design. The circuit vent itself, the main air supply artery for the entire branch, must connect between the two uppermost fixture drains. It must conform to Section 3104 and cannot serve as a drain; it must be a dry vent.

P3110.3 Slope and size of horizontal branch. The maximum slope of the vent section of the horizontal branch drain shall be one unit vertical in 12 units horizontal (8-percent slope). The entire length of the vent section of the horizontal branch drain shall be sized for the total drainage discharge to the branch in accordance with Table P3005.4.1.

❖ The maximum slope for a circuit-vented horizontal branch drain is one unit vertical in 12 units horizontal (8 percent), or 1 inch (28 mm) per foot (304.8 mm). Unlike wet venting, circuit venting does not depend on a unique sizing table. Rather, the circuit-vented horizontal drain branch is to be sized in accordance with Table P3005.4.1.

P3110.4 Additional fixtures. Fixtures, other than the circuit-vented fixtures, are permitted to discharge to the horizontal branch drain. Such fixtures shall be located on the same floor as the circuit-vented fixtures and shall be either individually or common vented. (See Figure P3110.4 for a typical circuit vent with additional noncircuit vented fixtures.)

❖ While a total of eight fixture drains are allowed to connect horizontally to the circuit vent, other fixtures may connect also, provided they are individually or common vented and are located on the same floor level as the circuit-vented fixtures.

SECTION P3111
COMBINATION WASTE AND VENT SYSTEM

P3111.1 Type of fixtures. A combination waste and vent system shall not serve fixtures other than floor drains, standpipes, sinks, lavatories and drinking fountains. A combination waste and vent system shall not receive the discharge of a food waste grinder.

❖ Combination waste and vent is another type of venting system that allows a drain to serve as a vent. Floor drains, standpipes, sinks, lavatories, and drinking fountains discharge clear- or gray-water waste, so an unlimited number of these fixtures can be added to the system. Because of the low flow velocities and volumes that occur in these systems, any solids introduced would accumulate, causing blockages and reduction of the internal cross-sectional area of the piping. Therefore, sinks equipped with food waste grinders and fixtures that discharge solid wastes are not allowed to be served by a combination drain and vent. Ground food waste can collect in the drain causing blockages. Combination drain and vent systems are oversized drains; therefore, the flow velocity will be low, which reduces drain line carry and scouring action.

P3111.2 Installation. The only vertical pipe of a combination drain and vent system shall be the connection between the fixture drain of a sink, lavatory, standpipe or drinking fountain, and the horizontal combination waste and vent pipe. The maximum vertical distance shall be 8 feet (2438 mm).

❖ The only vertical extension allowed in a combination waste and vent system is one that is a maximum of 8

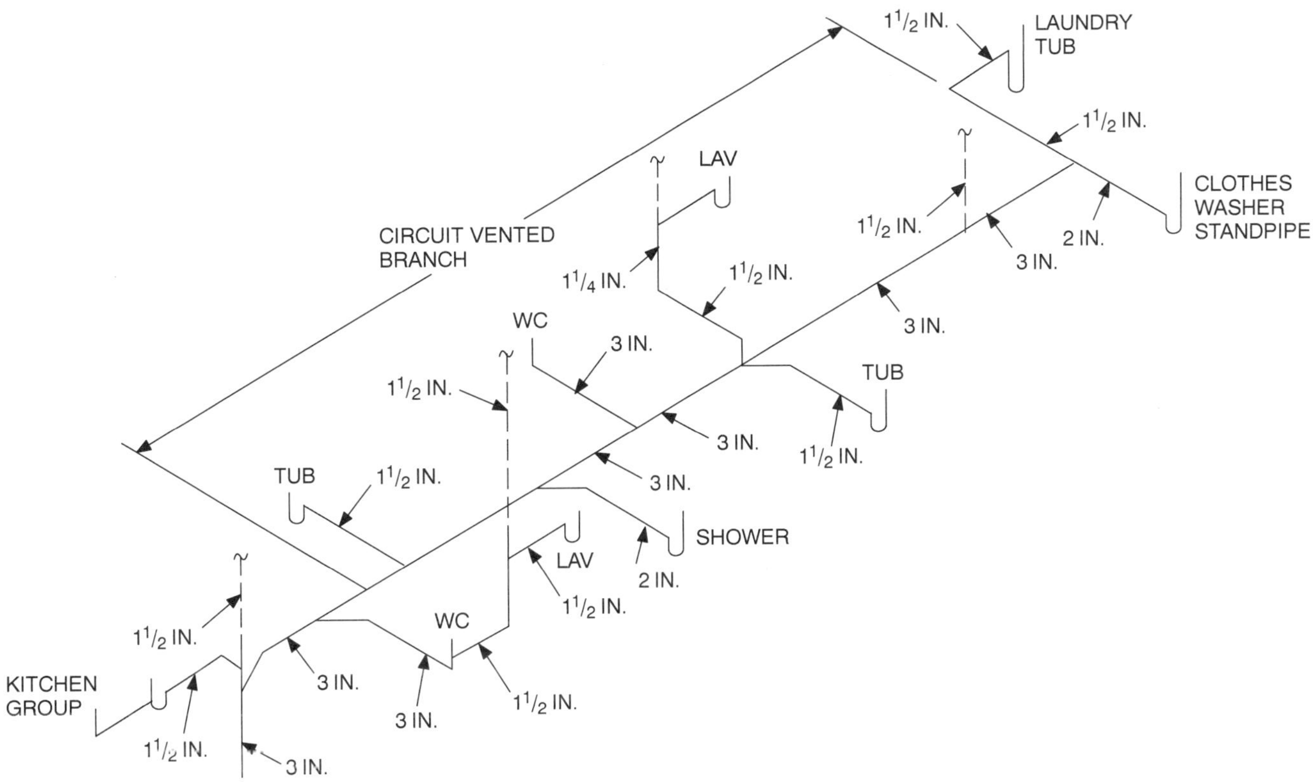

For SI: 1 inch = 25.4 mm.

Note: This figure is for illustration purposes only and the written text shall apply.

FIGURE P3110.4
CIRCUIT VENT WITH ADDITIONAL NONCIRCUIT VENTED BRANCHES

❖ In this diagram, the circuit vent is a dry vent and is installed as required in Section P3110.2. Even though the circuit vent itself is connecting to a 2-inch (51 mm) drain (normally requiring only a $1^1/_4$-inch [32 mm] vent per Section P3113.1) it must be a minimum $1^1/_2$-inch (38 mm) vent because it serves the entire circuit-vented horizontal branch drain. Since the largest size required of the horizontal branch drain is 3 inches (76 mm) per Table P3005.4.1, the circuit vent must be at least one-half the diameter of the drain served as specified in Section P3113.1.

The water closet and lavatory connection prior to the vertical stack and the kitchen group are vented by other venting provisions and are not part of the circuit-vented horizontal branch drain. In addition, the lavatory that connects to the circuit vented horizontal branch drain downstream of the dry vent is also vented independently of the circuit-vented branch; therefore, it is not counted as one of the eight allowable fixtures for a circuit vent. However, additional fixtures connected will count toward the total d.f.u. load to determine the size of the horizontal branch drain.

feet (2438 mm) long. Commentary Figures P3111.2(1), P3111.2(2), and P3111.2.2 show a combination waste and vent system.

P3111.2.1 Slope. The horizontal combination waste and vent pipe shall have a maximum slope of one-half unit vertical in 12 units horizontal (4-percent slope). The minimum slope shall be in accordance with Section P3005.3.

❖ This section specifies the maximum slope to be one-half unit vertical in 12 units horizontal, or a 4-percent slope. Table 3005.4.2 shows the minimum slope. A review of that table shows that the minimum slope for

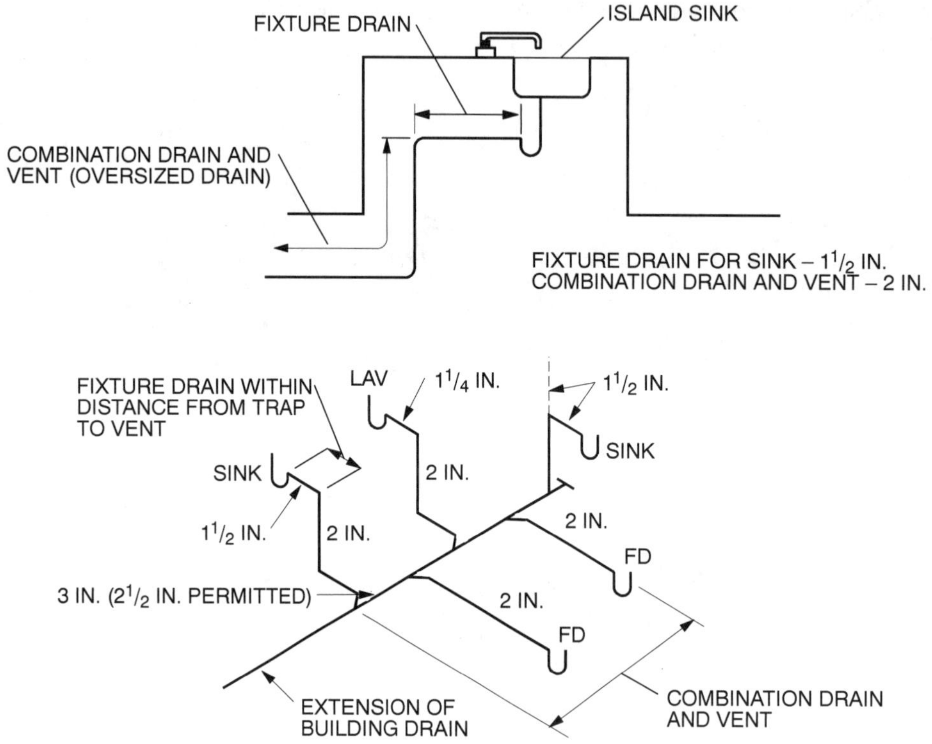

For SI: 1 inch = 25.4 mm.

Figure P3111.2(1)
COMBINATION DRAIN AND VENT

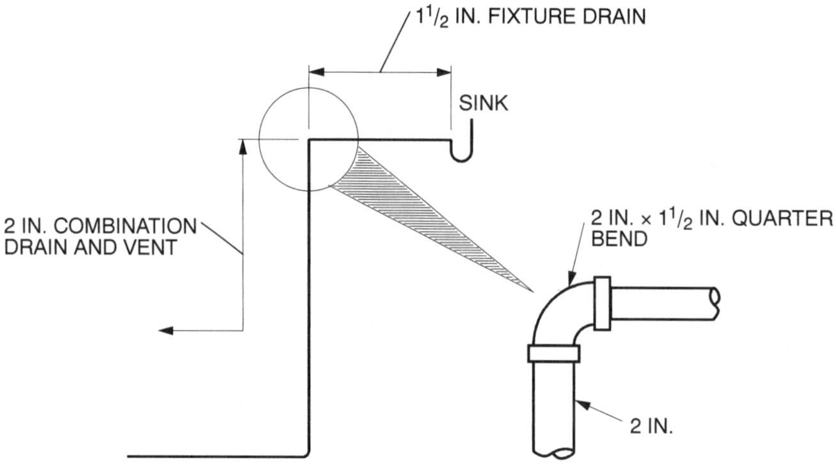

For SI: 1 inch = 25.4 mm.

Figure P3111.2(2)
COMBINATION DRAIN AND VENT

2-inch (51 mm) and 2½-inch (64 mm) diameter pipe is one-quarter unit vertical in 12 units horizontal (2-percent slope), and one-eighth unit vertical in 12 units horizontal (1-percent slope) for 3-inch (76 mm) and 4-inch (102 mm) diameter pipe.

P3111.2.2 Connection. The combination waste and vent pipe shall connect to a horizontal drain that is vented or a vent shall connect to the combination waste and vent. The vent connecting to the combination waste and vent pipe shall extend vertically a minimum of 6 inches (152 mm) above the flood level rim of the highest fixture being vented before offsetting horizontally.

❖ Where a combination waste and vented horizontal branch extends laterally from a vented horizontal drain, an additional dry vent is not required; otherwise, the combination waste and vent system will require a vent as outlined in this section and in section P3111.2.3. See Commentary Figure P3111.2.2.

P3111.2.3 Vent size. The vent shall be sized for the total fixture unit load in accordance with Section P3113.1.

❖ See Section P3113.1 for vent sizing provisions.

P3111.2.4 Fixture branch or drain. The fixture branch or fixture drain shall connect to the combination waste and vent within a distance specified in Table P3105.1. The combination waste and vent pipe shall be considered the vent for the fixture.

❖ The combination waste and vent system is oversized and vented so that that the entire horizontal and vertical portions of the system provide sufficient air to maintain the trap seals for the fixtures being connected. Since the combination waste and vent system is the vent, the fixture drains or branches used for connect-

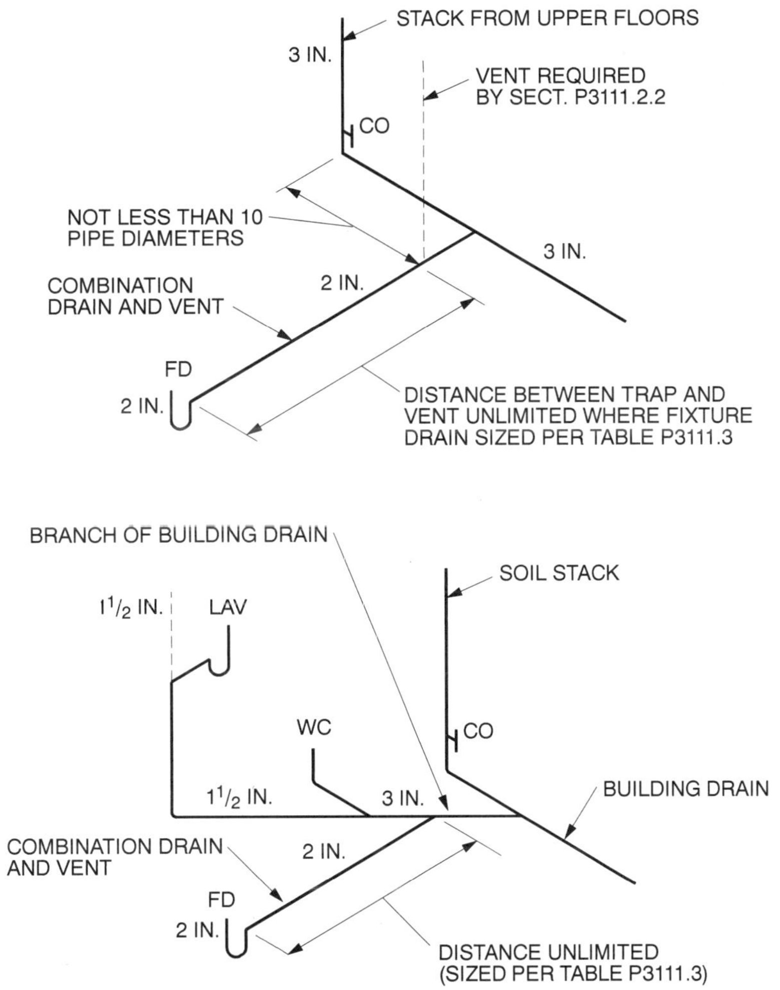

For **SI:** 1 inch = 25.4 mm.

Figure P3111.2.2
COMBINATION DRAIN AND VENT

❖ The vent must rise vertically and cannot offset horizontally or connect to other vents unless it is at least 6 inches (152 mm) above the flood level rim of the highest fixture served by the combination waste and vent system.

ing them to the system need not be oversized, so the size and length limitations of Table P3105.1 apply.

P3111.3 Size. The minimum size of a combination waste and vent pipe shall be in accordance with Table P3111.3.

❖ See Table 3111.3 for the sizing of a combination waste and vent system.

TABLE P3111.3
SIZE OF COMBINATION WASTE AND VENT PIPE

DIAMETER PIPE (inches)	MAXIMUM NUMBER OF FIXTURE UNITS (d.f.u.)	
	Connecting to a horizontal branch or stack	Connecting to a building drain or building subdrain
2	3	4
$2^1/_2$	6	26
3	12	31
4	20	50

For SI: 1 inch = 25.4 mm.

❖ This table outlines the maximum d.f.u.'s that can be loaded on various sizes of piping. The amounts depend on whether the combination waste and vent branch extends from a building drain or sub-drain or from a horizontal branch or stack. See the definitions of "Building Drain" and "Horizontal Branch, Drainage."

SECTION P3112
ISLAND FIXTURE VENTING

P3112.1 Limitation. Island fixture venting shall not be permitted for fixtures other than sinks and lavatories. Kitchen sinks with a dishwasher waste connection, a food waste grinder, or both, in combination with the kitchen sink waste, shall be permitted to be vented in accordance with this section.

❖ Often, with many of the diverse designs in construction, a fixture will be located without the nearby walls or partitions necessary for conventional venting methods to be used. In these cases, island fixture venting, which is limited to sinks and lavatories, can prove useful. However, since a kitchen sink can have a food-waste grinder and a dishwasher connected with it, island fixture venting can serve a kitchen group also; see "Kitchen group" in Table P3004.1.

P3112.2 Vent connection. The island fixture vent shall connect to the fixture drain as required for an individual or common vent. The vent shall rise vertically to above the drainage outlet of the fixture being vented before offsetting horizontally or vertically downward. The vent or branch vent for multiple island fixture vents shall extend to a minimum of 6 inches (152 mm) above the highest island fixture being vented before connecting to the outside vent terminal.

❖ The island fixture vent (sometimes referred to as a loop vent) must rise vertically to a height above the fixture drain before it turns (or loops) down. See Commentary Figure P3112.2. The loop is typically below the counter height. The horizontal vent extending below the floor extends vertically in a wall and connects to other vents. The point of connection must be at least 6 inches (152 mm) above the flood level rim of the island fixture.

P3112.3 Vent installation below the fixture flood level rim. The vent located below the flood level rim of the fixture being vented shall be installed as required for drainage piping in accordance with Chapter 30, except for sizing. The vent shall be sized in accordance with Section P3113.1. The lowest point of the island fixture vent shall connect full size to the drainage system. The connection shall be to a vertical drain pipe or to the top half of a horizontal drain pipe. Cleanouts shall be provided in the island fixture vent to permit rodding of all vent piping located below the flood level rim of the fixtures. Rodding in both directions shall be permitted through a cleanout.

❖ This section contains further provisions for the vent discussed in Section P3112.2. It refers to Section

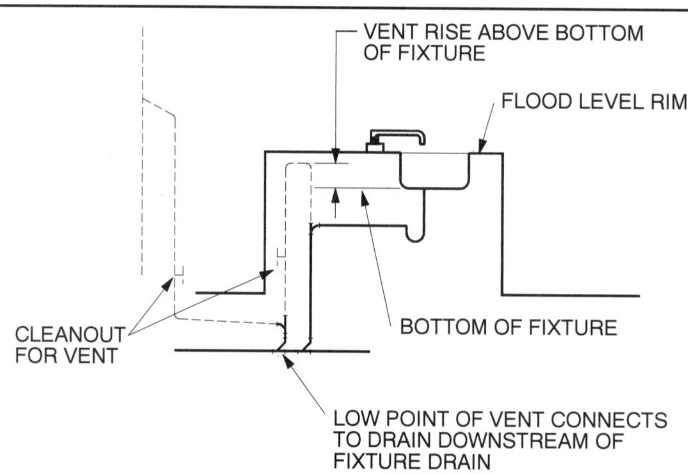

Figure 3112.2
ISLAND FIXTURE VENT

P3113.1 for proper sizing of the vent. In addition, any portion of the vent that is below the flood level rim of the island sink must be installed with drainage fittings and slope in accordance with Chapter 30.

The vent must be connected at its lowest point to the drain, either vertically or horizontally. This will prevent the low portion of the vent from becoming filled with liquid. Since the vent is installed horizontally and could be subject to backflow conditions, a cleanout must be provided for servicing with drain cleaning equipment. See Commentary Figure 3112.2.

SECTION P3113
VENT PIPE SIZING

P3113.1 Size of vents. The minimum required diameter of individual vents, branch vents, circuit vents, vent stacks and stack vents shall be at least one-half the required diameter of the drain served. The required size of the drain shall be determined in accordance with Chapter 30. Vent pipes shall be not less than $1^1/_4$ inch (32 mm) in diameter. Vents exceeding 40 feet (12 192 mm) in developed length shall be increased by one nominal pipe size for the entire developed length of the vent pipe.

❖ This section provides sizing specifications for different types of vents. Sizing is based on the minimum required diameter of the drain served and the total developed length. Any vent must be a minimum of one-half the diameter of the drain that it serves. However, no vent may be less than $1^1/_4$-inch (32 mm) in diameter. This does not exempt the specific provisions for vent sizing outlined in the sections addressing some of the venting concepts, such as those found in Sections P3108.3, P3109.3, P3110.3, and P3111.3.

This section also stipulates that a vent exceeding 40 feet (12 192 mm) in developed length must be increased by one pipe size for the entire length. See Section P3113.2.

P3113.2 Developed length. The developed length of individual, branch, and circuit vents shall be measured from the farthest point of vent connection to the drainage system, to the point of connection to the vent stack, stack vent or termination outside of the building.

❖ The 40-foot (12 192 mm) developed length limitation of Section P3113.1 applies to individual vents, branch vents, and circuit vents, measured from their starting point to their point of termination. Termination is at a vent stack, stack vent, or outside the building. However, termination can be an air admittance valve in accordance with Section P3114. See Commentary Figure P3113.2.

P3113.3 Branch vents. Where branch vents are connected to a common branch vent, the common branch vent shall be sized in accordance with this section, based on the size of the common horizontal drainage branch that is or would be required to serve the total drainage fixture unit (dfu) load being vented.

❖ Multiple branch vents could connect to a common branch vent, thus serving a large number of fixture units. In other words, one can have a number of horizontal branches, each with a vent, all of which connect together to a common branch vent prior to termination. In order to determine the proper size of the vents, add the total sum of d.f.u.'s for each horizontal branch, then determine what size of drain would be required for this total in accordance with Section P3005.4. The vent must then be sized per Section P3113 based on the required drain size for a drain that could serve all of the d.f.u.'s.

P3113.4 Sump vents. Sump vent sizes shall be determined in accordance with Sections P3113.4.1 and P3113.4.2.

❖ The following two sub-sections provide venting requirements for sewage pumps and ejectors.

P3113.4.1 Sewage pumps and sewage ejectors other than pneumatic. Drainage piping below sewer level shall be vented in a similar manner to that of a gravity system. Building sump vent sizes for sumps with sewage pumps or sewage ejectors, other than pneumatic, shall be determined in accordance with Table P3113.4.1.

❖ Sumps receiving sanitary drainage must be vented. Where a sewage pump is used, the vent can connect to the building venting system. Table P3113.4.1 specifies the minimum vent pipe size based on the discharge capacity of the pump and the developed length of vent pipe. Building sub-drain systems, including their venting systems, are designed like gravity drainage systems, including their venting systems, except that they require a sewage pump or ejector. The vents for both gravity and sub-drain systems can be combined provided that pneumatic ejector vents terminate independently in the outside air.

P3113.4.2 Pneumatic sewage ejectors. The air pressure relief pipe from a pneumatic sewage ejector shall be connected to an independent vent stack terminating as required for vent extensions through the roof. The relief pipe shall be sized to relieve air pressure inside the ejector to atmospheric pressure, but shall not be less than $1^1/_4$ inches (32 mm) in size.

❖ Pneumatic sewage ejectors operate at high air pressure, forcing the sewage from the ejector receiver to the gravity drainage system. An independent vent is necessary to relieve the pressure in the receiver at the completion of the ejector cycle. If such a vent were connected to the building venting system, it would create tremendous pressure differentials, disrupting the normal operation of the drainage system.

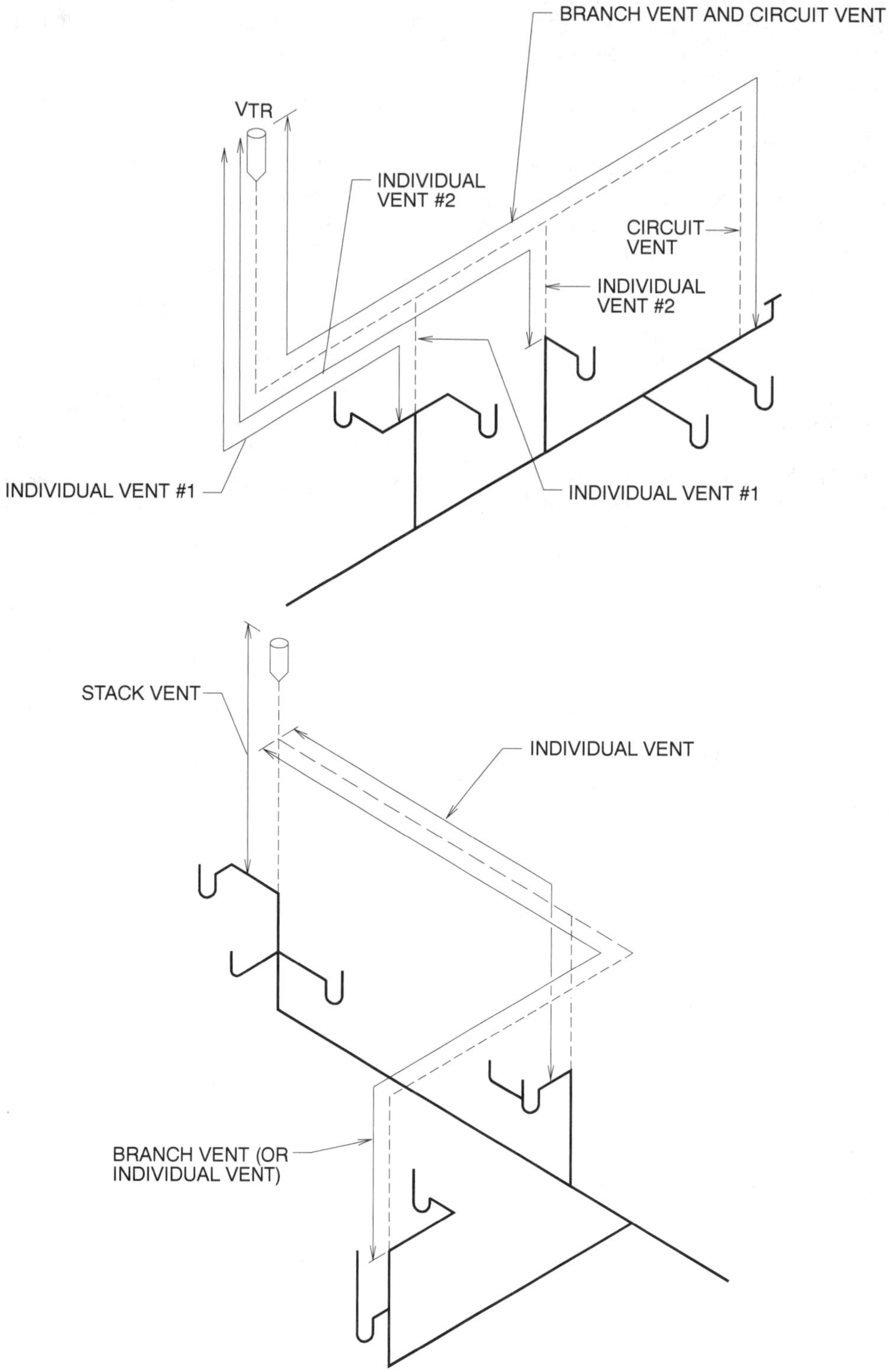

Figure P3113.2
DEVELOPED LENGTH OF VENTS

TABLE P3113.4.1
SIZE AND LENGTH OF SUMP VENTS

DISCHARGE CAPACITY OF PUMP (gpm)	MAXIMUM DEVELOPED LENGTH OF VENT (feet)[a]				
	Diameter of vent (inches)				
	$1^1/_4$	$1^1/_2$	2	$2^1/_2$	3
10	No limit[b]	No limit	No limit	No limit	No limit
20	270	No limit	No limit	No limit	No limit
40	72	160	No limit	No limit	No limit
60	31	75	270	No limit	No limit

For SI: 1 inch = 25.4 mm, 1 foot = 304.8 mm, 1 gallon-per-minute (gpm) = 3.785 L/m.

a. Developed length plus an appropriate allowance for entrance losses and friction due to fittings, changes in direction and diameter. Suggested allowances shall be obtained from NBS Monograph 31 or other approved sources. An allowance of 50 percent of the developed length shall be assumed if a more precise value is not available.

b. Actual values greater than 500 feet.

❖ This table specifies the minimum vent pipe size based on the discharge capacity of the pump and the developed length of the vent.

SECTION P3114
AIR ADMITTANCE VALVES

P3114.1 General. Vent systems utilizing air admittance valves shall comply with this section. Individual- and branch-type air admittance valves shall conform to ASSE 1051. Stack type air admittance valves shall conform to ASSE 1050.

❖ Air admittance valves (AAV) can be used in lieu of other vent termination methods provided for in this chapter. See the definition of "air admittance valve" in Chapter 2.

There are two standards to which AAV's are manufactured. These are ASSE 1051 for individual and branch-type valves and ASSE 1050 for stack-type valves. The standards require that the AAV be marked with the name of the manufacturer, the model number or device description, and the device classification.

P3114.2 Installation. The valves shall be installed in accordance with the requirements of this section and the manufacturer's installation instructions. Air admittance valves shall be installed after the DWV testing required by Section P2503.5.1 or P2503.5.2 has been performed.

❖ Both this code (Section P3114) and the manufacturer's installation instructions apply. Where there is a conflict between this section and the manufacturer's installations, the provisions of Section R102.4 apply.

Section P2503.5.1 provides for a test of the DWV system. That test will produce pressure differentials greater than those that AAV's are designed to accommodate. The test specified in Section P2503.5.2 does not produce pressure differentials beyond those that AAV's can handle. In any case, one of the two DWV tests must be conducted before the installation of the AAV.

P3114.3 Where permitted. Individual vents, branch vents, circuit vents and stack vents shall be permitted to terminate with a connection to an air admittance valve.

❖ This section establishes the specific locations within the vent system where AAV's can be installed. Individual vents, branch vents, and circuit vents can terminate to an AAV conforming to ASSE 1051, and a stack vent or vent stack can terminate to an AAV conforming to ASSE 1050. Sump vents cannot terminate to an AAV due to the higher pressure differentials likely to be experienced. Commentary Figures P3114.3(1) through P3114.3(5) show various AAV applications.

P3114.4 Location. Individual and branch air admittance valves shall be located a minimum of 4 inches (102 mm) above the horizontal branch drain or fixture drain being vented. Stack-

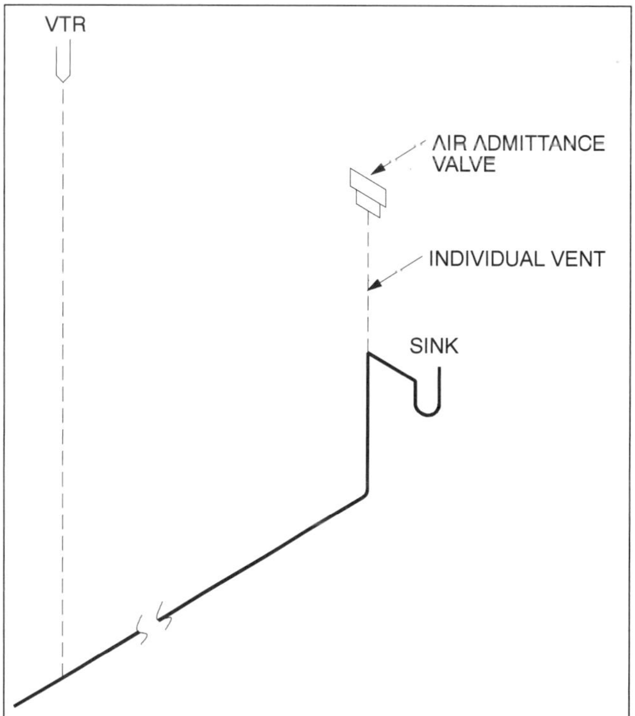

Figure P3114.3(1)
AIR ADMITTANCE VALVES — INDIVIDUAL VENT

type air admittance valves shall be located a minimum of 6 inches (152 mm) above the flood level rim of the highest fixture being vented. The air admittance valve shall be located within the maximum developed length permitted for the vent. The air admittance valve shall be installed a minimum of 6 inches (152 mm) above insulation materials where installed in attics.

❖ This section provides three specific provisions applying to the installation of AAV's. The first refers to the AAV being above the trap level. The height of air opening into an individual or branch type AAV must be at least 4 inches (102 mm) above the fixture drain or branch, and for a stack-type AAV it must be at least 6 inches (152 mm) above the flood level rim of the highest fixture being vented. The second states that since it is the vent for the trap, it must be located within the distance required for the size of trap as provided for in Table 3105.1; see Commentary Figure P3114.4.

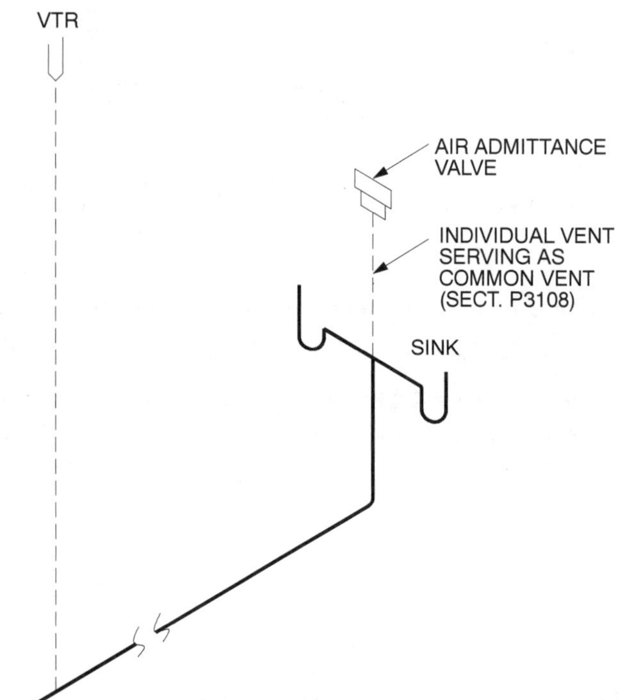

Figure P3114.3(2)
COMMON VENT

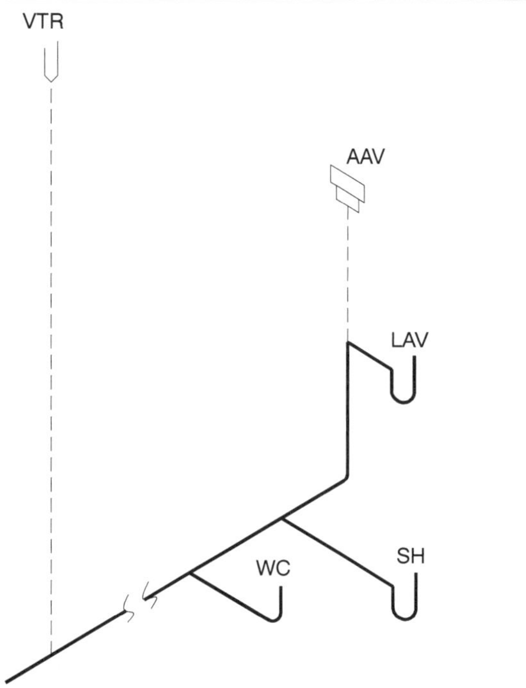

Figure P3114.3(3)
WET VENT

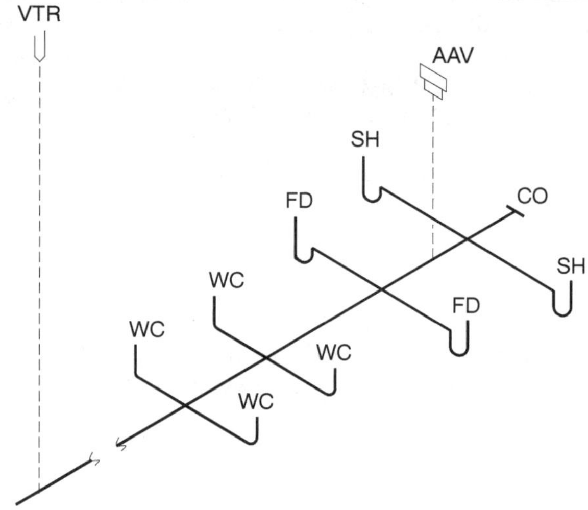

Figure P3114.3(4)
CIRCUIT VENT

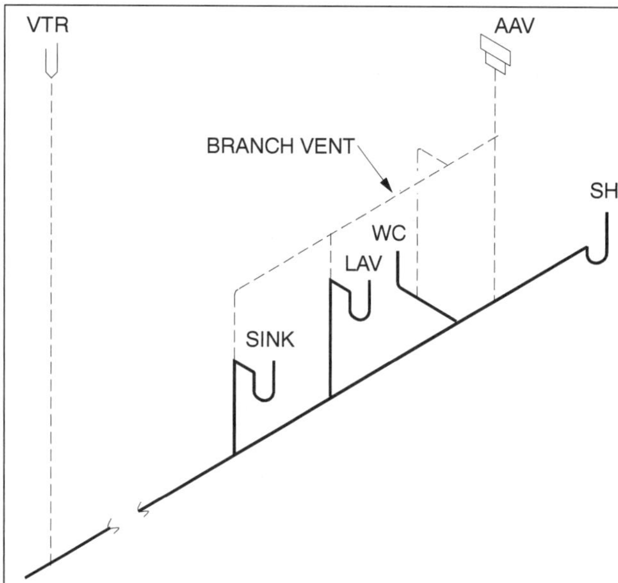

Figure P3114.3(5)
BRANCH VENT

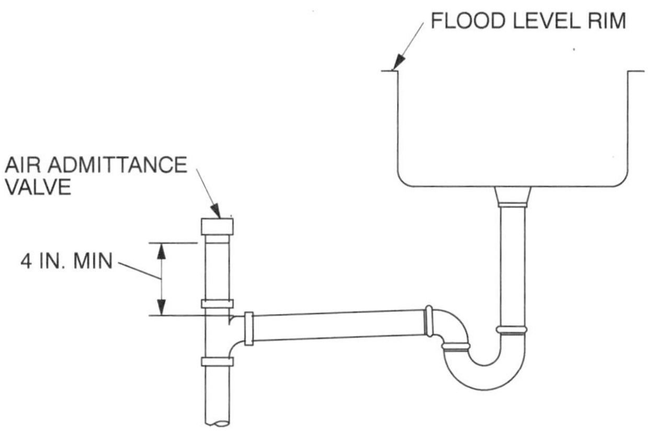

For SI: 1 inch = 25.4 mm.

Figure P3114.4
DISTANCE ABOVE DRAIN

P3114.5 Access and ventilation. Access shall be provided to all air admittance valves. The valve shall be located within a ventilated space that allows air to enter the valve.

❖ Because an air admittance valve has a moving part, it must be accessible for inspection, service, repair or replacement. The code defines accessible as being able to be reached, but first the removal of a panel, door, or similar obstruction might be required. Access to an air admittance valve is not achieved if it is necessary to remove any portion of the structure's permanent finish materials such as drywall, plaster, paneling, built-in furniture or cabinets, or any other similar permanently affixed building component.

Where located in confined spaces, ventilation openings are required in those spaces. Air admittance valves admit air into the drainage system; therefore, an unimpeded air supply must be available to the devices at all times. Additionally, the manufacturer's installation instructions address access and ventilation requirements. The space within a cabinet under a sink counter has an unimpeded air supply and is accessible.

P3114.6 Size. The air admittance valve shall be rated for the size of the vent to which the valve is connected.

❖ The size of the vent pipe is regulated by Section P3113.1 or as provided for in the more specific venting provisions found in Chapter 31. The AAV is sized based on the required size of the vent pipe and the d.f.u. load limitations found in the manufacturer's instructions. The d.f.u. loading is based on testing performed in accordance with ASSE 1051 and ASSE 1050.

P3114.7 Vent required. Within each plumbing system, a minimum of one stack vent or a vent stack shall extend outdoors to the open air.

❖ While AAV's are useful in protecting traps from siphoning, larger pressure surges sometimes occur in a sewer. For this reason, the plumbing system must have at least one vent stack or a stack vent extending to the outdoors. Most likely, this will serve also as the main vent required in Section P3102.1.

Bibliography

The following resource materials are referenced in this chapter or are relevant to the subject matter addressed in this chapter.

ASHRAE *Handbook: Fundamentals.* Atlanta, GA: American Society of Heating, Refrigerating and Air-Conditioning Engineers, 2001.

ASSE 1050—91, *Performance Requirements for Air Admittance Valves for Plumbing Drainage, Waste and Vented Systems-Stack Type Device.* Westlake OH: American Society of Sanitary Engineering, 1991.

ASSE 1051—98, *Performance Requirements for Air Admittance Valves for Plumbing Drainage Systems, Fixture and Branch Devices.* Westlake, OH: American Society of Sanitary Engineering, 1998.

Chapter 32: Traps

General Comments

Chapter 32 covers fixture traps. It provides criteria for the material, design, installation, and minimum size of traps as determined by the fixture served and outlines the approved methods available for maintaining the trap seal. It also specifies types of traps that are prohibited for use in one- and two-family occupancies.

Purpose

Traps prevent sewer gas from escaping from the drainage piping into the building. Water seal traps are one of the simplest means of preventing sewer gas from entering the interior environment.

SECTION P3201 FIXTURE TRAPS

P3201.1 Design of traps. Traps shall be of standard design, shall have smooth uniform internal waterways, shall be self-cleaning and shall not have interior partitions except where integral with the fixture. Traps shall be constructed of lead, cast iron, cast or drawn brass or approved plastic. Tubular brass traps shall be not less than No. 20 gage (0.8 mm) thickness. Solid connections, slip joints and couplings are permitted to be used on the trap inlet, trap outlet, or within the trap seal. Slip joints shall be accessible.

❖ This section outlines the types of material that can be used for the construction of fixture traps. Most traps are pre-manufactured, though a trap can be fabricated with approved pipe and fittings. It is at the discretion of the installer as to whether the trap, when field constructed, will incorporate solid or slip joints. When slip joints are used, they must be accessible as required in Section P2704.1. See Commentary Figure P3201.1(1).

A trap must have a pattern allowing unobstructed flow to the drain. Some fixtures, such as water closets, have an integral trap and thus are not designed to be connected to an additional trap. See Commentary Figure P3201.1(2).

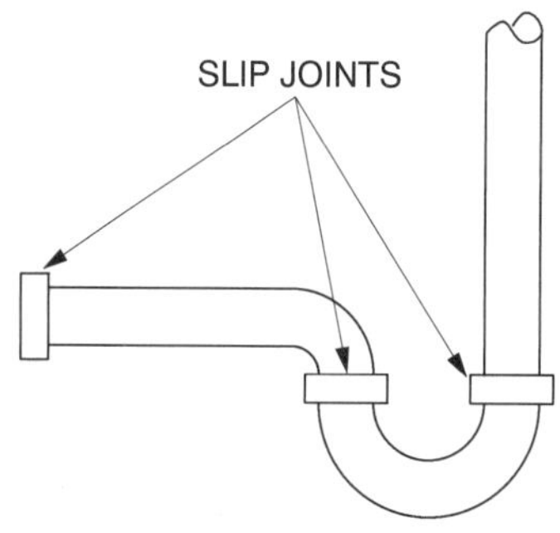

Figure P3201.1(1)
SLIP JOINTS IN TRAPS

P3201.2 Trap seals and trap seal protection. Traps shall have a liquid seal not less than 2 inches (51 mm) and not more than 4 inches (102 mm).

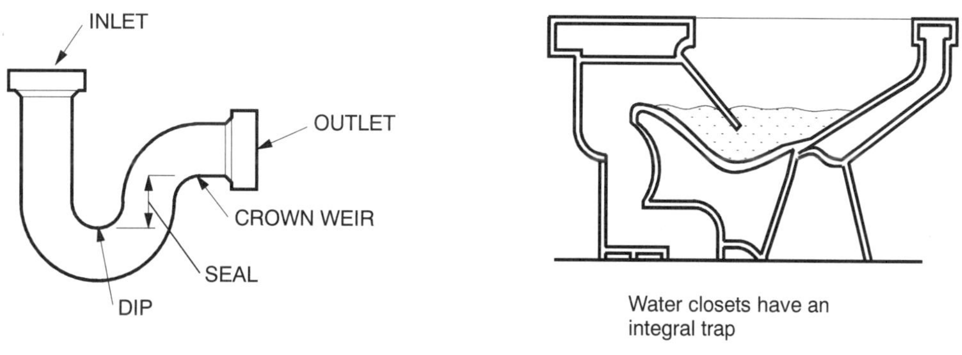

For SI: 1 inch = 25.4 mm.

Figure P3201.1(2)
TRAP REQUIREMENTS

Exception: Traps for floor drains shall be fitted with a trap primer or shall be of the deep seal design.

❖ A trap creates a seal of standing water in the elbow so that sewer gases will not enter the building through the fixture at the connection to the drainage system. The trap must hold a minimum of 2 inches (52 mm) and a maximum of 4 inches (102 mm) of water. See Commentary Figure P3201.1(2). For this reason, a trap seal must be deep enough to resist the pressures that develop in the drainage system but not deep enough to hold solids, which promote the growth of bacteria.

Floor drains usually have trap seals that are subject to evaporation. This is why the code requires a trap for a floor drain to be provided with a means of maintaining its seal. There are two methods allowed by the code: a deep-seal trap and a trap primer. See Commentary Figure P3201.2.

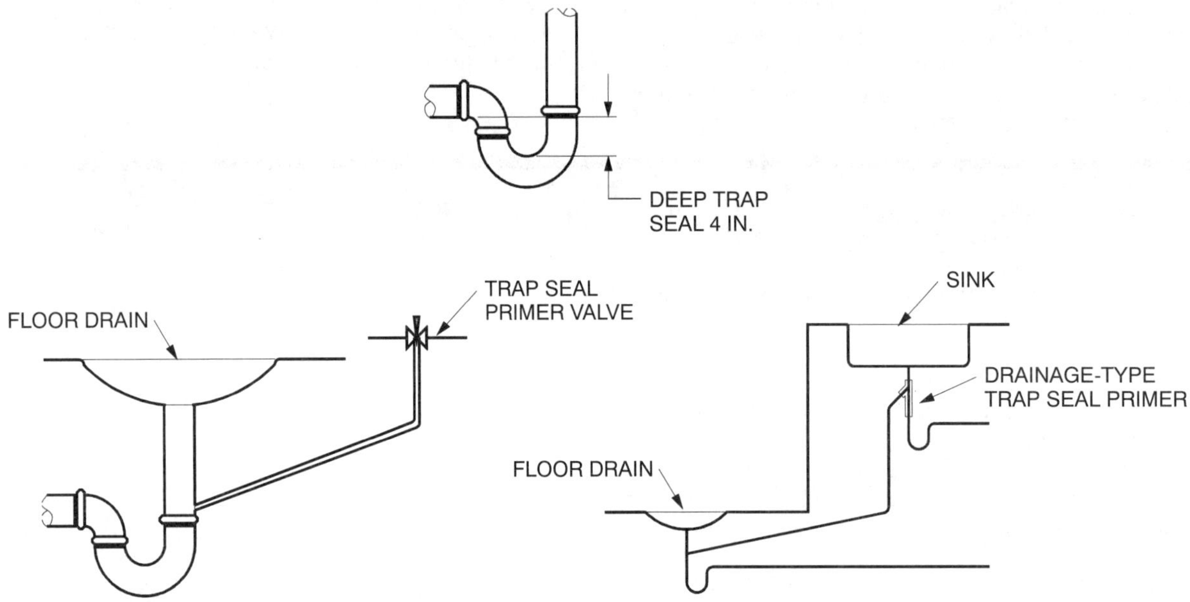

For SI: 1 inch = 25.4 mm.

Figure P3201.2
TRAP SUBJECT TO EVAPORATION

❖ A deep seal does not eliminate evaporation. It merely slows down the evaporation process. Trap primers are more effective. They come in either a water-supply-fed type or a drainage type. They cause a small amount of water to discharge to the drain. The water-supply-fed-type does so automatically; the drainage type does so with each use of the fixture.

P3201.3 Trap setting and protection. Traps shall be set level with respect to their water seals and shall be protected from freezing. Trap seals shall be protected from siphonage, aspiration or back pressure by an approved system of venting (see Section P3101).

❖ Traps must be set level with their seal in order to perform as designed. See Commentary Figure P3201.3.

Because they maintain water at all times, traps must be protected where they are subject to freezing. In colder climates, freezing can occur in unheated areas like garages or workshops and also in cantilevered construction where the trap for a bathtub or shower may be installed below the floor insulation.

When properly installed to code, the venting portion of the drainage system compensates for both negative and positive pressures that can occur in the plumbing system. This helps maintain the water in the traps. The venting provisions of Chapter 31 must be adhered to for a properly vented system. This is addressed in Section P3101.2.

P3201.4 Building traps. Building traps shall not be installed, except in special cases where sewer gases are extremely corrosive or noxious, as directed by the building official.

❖ Originally, building traps were installed on the building drain, before it was connected to the building sewer, as a form of rat control. In major cities, rats would breed in the sewer system and enter a building through the sewer connection. A water trap seal was believed to prevent rats from entering the building.

Building traps may still be necessary in those rare cases where the public sewer exerts a positive back pressure on the connected building sewers. The back pressure can cause the building vent terminations to emit strong sewer gases, which can create a serious odor problem around the building.

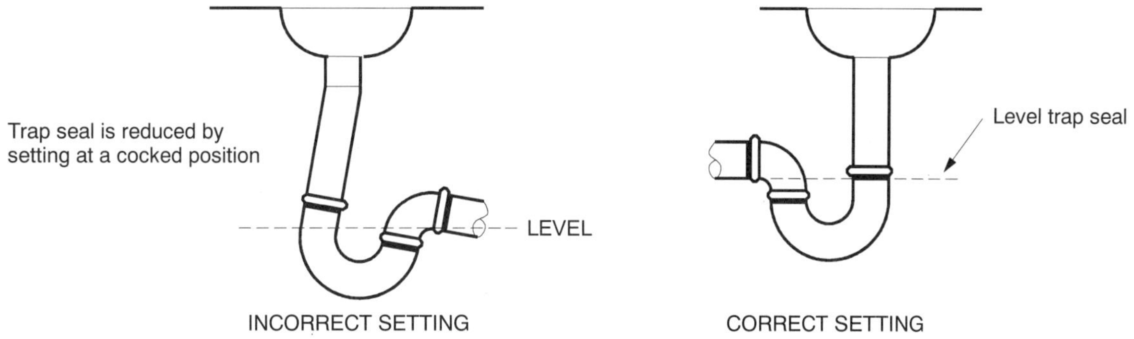

Figure P3201.3
INSTALLATION OF TRAP

Section 1002.6 of the 2003 *International Plumbing Code* provides further guidance for the installation of building traps.

P3201.5 Prohibited trap designs. The following types of traps are prohibited:

1. Bell traps.
2. Separate fixture traps with interior partitions, except those lavatory traps made of plastic, stainless steel or other corrosion-resistant material.
3. "S" traps (except as permitted under Section P3105.3).
4. Drum traps.
5. Trap designs with moving parts.

❖ A trap is a simple U-shaped piping arrangement that offers minimal resistance to flow. Certain traps or arrangements are prohibited.

Item 1 refers to bell traps, which, because of their design, tend to clog with debris. The trap seal has a larger exposed surface area, which accelerates evaporation.

Item 2 refers to separate fixture traps with interior partitions, except those lavatory traps made of plastic, stainless steel, or other corrosion-resistant material. The item refers to independent traps that are not integral with the fixture. For example, water closet traps are partitions, but because they are part of the fixture they are not prohibited. Partition traps are single-wall traps that rely on a single partition to separate the house side of the trap from the sewer side of the trap. Although they may be made of noncorrosive material, they are still subject to corrosion and can structurally fail. Because such failure does not result in leakage, there will be no indication of the lost trap seal.

Item 3 refers to "S" traps (except as permitted under Section P3105.3). "S" configuration traps tend to create a siphon effect, which causes loss of the trap seal. Because of its over-sizing and dimensional requirements, the "S" configuration described in Section P3105.3 is not subject to siphonage when connected to a properly vented drainage system.

Item 4 refers to drum traps, which, because they are not self-cleaning, are subject to clogging. Although in certain cases they may be desired in nonresidential occupancies, there is no special need for them in dwellings; thus, they are prohibited by this code.

Item 5 refers to trap designs with moving parts. These are mechanical traps that employ moving parts such as floats or flappers. Because of corrosion, clogging, and waste deposits that interfere with the operation of moving components and seals, such designs are not dependable.

P3201.6 Number of fixtures per trap. Each plumbing fixture shall be separately trapped by a water seal trap placed as close as possible to the fixture outlet. The vertical distance from the fixture outlet to the trap weir shall not exceed 24 inches (610 mm). The distance of a clothes washer standpipe above a trap shall conform to Section P2706.2. Fixtures shall not be double trapped. Exceptions to the separate trapping requirements are as follows:

1. Fixtures that have integral traps.
2. A single trap shall be permitted to serve two or three like fixtures limited to kitchen sinks, laundry tubs and lavatories. Such fixtures shall be adjacent to each other and located in the same room with a continuous waste arrangement. The trap shall be installed at the center fixture where three such fixtures are installed. Common trapped fixture outlets shall not be more than 30 inches (762 mm) apart.
3. The connection of a laundry tray waste line is permitted to be made into a standpipe for the automatic clothes-washer drain in accordance with Section P2706.2.1.

❖ Double trapping of a fixture will cause obstruction of flow and potential stoppages. This is because air is trapped between the two trap seals.

Except for a clothes washer standpipe, which has a maximum vertical distance of 42 inches (1067 mm), the maximum vertical distance of 24 inches (610 mm) from a fixture outlet to the trap weir is required for all other fixtures for two reasons. First, buildup on the wall of the pipe can breed bacteria and cause odors. It is therefore best to locate the trap as close as possible to the fixture to minimize the amount of drain pipe that occurs on the inlet (house) side of the trap. Second, the velocity of the drainage flow must be controlled. If the vertical distance is too great, a self-siphoning effect could occur.

This section allows a single trap to serve a set of not more than three like fixtures (lavatories, laundry tubs, or kitchen sinks). In most residential applications this could apply to a double lavatory vanity or a dual or triple compartment kitchen sink. Where a single trap is used, fixtures must not be spaced more than 30 inches (762 mm) apart center-to-center and must be located in the same room. Where fixtures are connected to a single trap, the trap must be installed at the center fixture. This minimizes the amount of piping on the inlet (house) side of the trap, thus reducing the likelihood of bacteria growth and subsequent odors.

Reference is made to Section P2706.2.1, where the provision for connecting a laundry tray to a clothes washer standpipe is allowed. Here, a maximum distance of 30 inches (762 mm) between the point of connection at the standpipe and the outlet of the laundry tray applies.

P3201.7 Size of fixture traps. Fixture trap size shall be sufficient to drain the fixture rapidly and not less than the size indicated in Table P3201.7. A trap shall not be larger than the drainage pipe into which the trap discharges.

❖ A properly sized trap allows the fixture to drain rapidly and function as intended. The code does not prescribe a maximum trap size; however, if a trap is oversized, it will not scour (cleanse) itself and will be prone to clogging. A trap that is larger than the drainage pipe into which it discharges is further prone to clogging because the reduced size outlet pipe does not allow a water velocity that is necessary to scour the trap.

TABLE P3201.7
SIZE OF TRAPS AND TRAP ARMS FOR PLUMBING FIXTURES

PLUMBING FIXTURE	TRAP SIZE MINIMUM (inches)
Bathtub (with or without shower head and/or whirlpool attachments)	$1^1/_2$
Bidet	$1^1/_4$
Clothes washer standpipe	2
Dishwasher (on separate trap)	$1^1/_2$
Floor drain	2
Kitchen sink (one or two traps, with or without dishwasher and garbage grinder)	$1^1/_2$
Laundry tub (one or more compartments)	$1^1/_2$
Lavatory	$1^1/_4$
Shower	2
Water closet	Note a

For SI: 1 inch = 25.4 mm.

a. Consult fixture standards for trap dimensions of specific bowls.

❖ This table provides minimum trap sizes for most fixtures. The provisions of Section 3201.7, which allow a set of not more than three similar fixtures (lavatories, laundry trays, or kitchen sink compartments) to be drained by one trap do not increase the size of the trap. In most cases, side-by-side fixtures are either served by one faucet (such as a multiple compartment kitchen sink) or do not experience heavy use simultaneously.

Although a bathtub equipped with a shower riser and a shower enclosure are assigned to the same drainage fixture unit loading in Table P3004.1, a shower stall or receptor requires a minimum 2-inch (52 mm) trap, whereas a bathtub only requires a minimum $1^1/_2$-inch (38 mm) trap. This is primarily due to the lower threshold of a shower base as opposed to a bathtub. An occupant may inadvertently impede the drain of a shower while standing over the inlet. Showers also do not have a built-in overflow drain similar to most bathtub installations.

Bibliography

The following resource materials are referenced in this chapter or are relevant to the subject matter addressed in this chapter.

CSA-B181.1-99, *ABS Drain, Waste, and Vent Pipe and Pipe Fittings.* Toronto, Ontario: Canadian Standards Association, 1999.

CSA-B181.2-99, *PVC Drain, Waste, and Vent Pipe and Pipe Fittings* — with Revisions through December 1993. Toronto, Ontario: Canadian Standards Association, 1999.

ASSE 1018-86, *Performance Requirements for Trap Seal Primer Valves; Water Supply Fed* — with 1986 Revisions. Bay Village, OH: American Society of Sanitary Engineering, 1986.

ASSE 1044-86, *Performance Requirements for Trap Seal Primer Valves; Drainage Type*. Bay Village, OH: American Society of Sanitary Engineering, 1986.

ASTM F 409-99a, *Specification for Thermoplastic Accessible and Replaceable Plastic Tube and Tubular Fittings.* West Conshohocken, PA: ASTM International, 1999.

Part VIII — Electrical

Chapter 33: General Requirements

See Appendix L, the *International Residential Code/National Electrical Code* cross reference.

General Comments

This chapter contains broadly applicable requirements including provisions for the protection of the structural elements of a building, inspection of work, general installation provisions, and conductor identification. Section E3303.3 requires all electrical system components to be listed and labeled by an approved agency.

The portions of the NEC text that are not applicable to dwelling construction have been excluded; therefore, the provisions herein are appropriate for the scope of coverage of the IRC. The NEC text has been reformatted, reorganized and, in some cases, restated for the purpose of making the provisions easier to locate, understand, and apply. The IRC is formatted in a manner consistent with the other ICC codes, is designed to be a "cookbook" for dwelling construction and is meant to be user friendly for both builders and code officials.

The electrical code provisions of the IRC are identical to the intent of the NEC provisions except the IRC requires all electrical system components to be listed and labeled; the NEC does not. The IRC does *not* contain unique electrical requirements. A dwelling built to the IRC will have electrical systems identical to those required by the respective edition of the NEC. The IRC addresses only those electrical systems that are common to dwelling construction, and the NEC is referenced for any subject not addressed in the IRC.

The IRC electrical chapters are produced and maintained by the NFPA in accordance with a contract between the ICC and the NFPA. The contract requires the NFPA to update the IRC to the new edition of the NEC and the IRC and the NEC to remain technically consistent. The ICC technical staff works with the NFPA technical staff to present the updated text in the ICC style and format.

Purpose

This chapter establishes the scope of coverage of this code and provides reference to NFPA 70 where this code lacks coverage of any wiring method or material. The provisions of this chapter intend to protect life, limb and property from the hazards of electricity, attempting to make sure that electrical equipment and conductors are properly located, guarded and identified.

SECTION E3301 GENERAL

E3301.1 Applicability. The provisions of Chapters 33 through 42 shall establish the general scope of the electrical system and equipment requirements of this code. Chapters 33 through 42 cover those wiring methods and materials most commonly encountered in the construction of one- and two-family dwellings and structures regulated by this code. Other wiring methods, materials and subject matter covered in NFPA 70 are also allowed by this code.

❖ These chapters cover the wiring methods, materials, equipment, fixtures, and appliances that are most commonly used in residential buildings. Where special equipment, wiring methods, or materials are used in dwellings or on residential property, the provisions of the *National Electrical Code* (NEC) apply. See Section R101.2.

E3301.2 Scope. Chapters 33 through 42 shall cover the installation of electrical systems, equipment and components indoors and outdoors that are within the scope of this code, including services, power distribution systems, fixtures, appliances, devices and appurtenances. Services within the scope of this code shall be limited to 120/240-volt, 0- to 400-ampere, single-phase systems. These chapters specifically cover the equipment, fixtures, appliances, wiring methods and materials that are most commonly used in the construction or alteration of one- and two-family dwellings and accessory structures regulated by this code. The omission from these chapters of any material or method of construction provided for in the referenced standard NFPA 70 shall not be construed as prohibiting the use of such material or method of construction. Electrical systems, equipment or components not specifically covered in these chapters shall comply with the applicable provisions of NFPA 70.

❖ This section states that any material or method of construction not covered by these chapters must comply with the applicable provisions of the NEC. Section R102.4 states that codes and standards referenced in this code must be considered part of the requirements of this code. This section in the Administration chapter goes on to state that "where differences occur between provisions of this code and referenced codes and standards, the provisions of this code must apply."

The provisions of Chapters 33 through 42 cover the electrical installations of wiring and equipment most commonly used for one- and two-family dwellings and associated structures. In the development of these chapters, it was considered that the vast majority of one- and two-family dwellings have a 120/240-volt single phase service of between 100 and 400 amperes. Therefore, it was deemed adequate for this code to include provisions considered necessary only for one- and two-family dwellings with a service of up to a maximum of 400 amperes. See Section R101.2. For larger services and methods that are less common for one- and two-family dwellings, the provisions of the NEC would apply.

In residential electrical work, many times the electrician or electrical contractor on the job designs the wiring system. The plans may have only the locations of lighting and receptacle outlets, so the electrician or electrical contractor must calculate the load for the dwelling as well as the load for the branch circuits and then design the layout of the wiring job. The electrical chapters of this coded are written in such a way that makes them simple to use while still being technically accurate and complete for the most common methods of installation. They contain minimum provisions that if followed will result in a safe and sufficient installation. However, larger conductor sizes, more branch circuits and more lighting and receptacle outlets than the minimum required by code are sometimes desirable for added convenience.

The electrical provisions in these chapters are based on the 1999 NEC, so the requirements are the same as in that referenced standard. For consistency with the building, mechanical and plumbing provisions, the electrical chapters of this code are written in a somewhat different style and format from those in the NEC. The requirements of the NEC pertain to many different occupancies and varied installations. In the development of the electrical provisions of the IRC, it was desired to develop an organization scheme that pertained specifically to dwellings. One result of this is that fewer exceptions are needed because the code covers only residential buildings. Also, several tables are included to summarize the various code rules. For example, many of the requirements for the different wiring methods are consolidated into tables.

An effort has been made to arrange the chapters in a logical order, beginning with general requirements. The start of the wiring system and electrical supply is the service and feeders, so provisions concerning them became the next chapter in the sequence. Grounding as a general subject was separated into two areas: service and feeder grounding and equipment grounding. Since system grounding is part of the electrical service, the service grounding and bonding provisions including the grounding electrode system are included in Chapter 35. The next chapter contains the branch circuit and feeder requirements. The following chapters are on wiring methods and requirements for power and lighting distribution. These include provisions for outlets and equipment grounding which in turn include rules for sizing equipment grounding conductors. The final three chapters contain rules on appliances, swimming pools, and low-voltage circuits.

E3301.3 Not covered. Chapters 33 through 42 do not cover the following:

1. Installations, including associated lighting, under the exclusive control of communications utilities and electric utilities.
2. Services over 400 amperes.

❖ This code does not directly cover requirements for the service-drop or service-lateral conductors. In most cases, the service-drop and service-lateral conductors are owned and maintained by the utility company, so it is responsible for these conductors. The clearances for these conductors and the depth of cover for laterals that are beyond the service point are not in the code. A dwelling unit with a service rated over 400 amperes may require equipment not covered by this code and is not within the scope of this code.

E3301.4 Additions and alterations. Any addition or alteration to an existing electrical system shall be made in conformity with the provisions of Chapters 33 through 42. Where additions subject portions of existing systems to loads exceeding those permitted herein, such portions shall be made to comply with Chapters 33 through 42.

❖ This section correlates with the administrative provisions of Chapter 1, Section R101.2, which state that the IRC applies to the construction, alteration, movement, enlargement, replacement, repair, removal and demolition of one- and two-family dwellings. It is important to consider the provisions of IRC Chapter 1 with the electrical chapters. For example, Section R105 has the requirement to obtain a permit from the building official before work is started. Electrical work that is exempt from a permit is also listed in Section R105.

Additions, alterations or repairs to any structure must conform to this code, but they do not require that the entire structure be updated or brought into current code compliance. Where the additions, alterations or repairs do not cause the existing electrical system to be unsafe, it is not required to upgrade the old system. For example, a simple bedroom addition may not result in a big difference in the electrical load on the service or feeders. The addition of a circuit for a bedroom light and four or five receptacle outlets would probably not be a reason to replace the electrical service equipment with one of larger ampacity. However, in a very old house with a 60-ampere service, the addition of one circuit could well cause an overload on the service, and the service would need to be upgraded. This is not to imply that the electrical load required for a small addition of lighting, outlets or equipment can be neglected. The load for additions to existing dwellings should always be calculated. If the existing service and/or feeder conductors are found to be inadequate to handle the new load, an upgrade would be necessary. Chapters 35 and 36 contain the load calculation requirements.

It is important to look not only at the electrical chapters when considering electrical requirements. A bedroom addition would require that a smoke alarm be installed in the bedroom. Where smoke alarms are wired and not just battery powered, they are installed as part of the wiring job and are in the domain of the electrician and electrical inspector. Although the requirement to install smoke alarms is in Part III—Building Planning and Construction and not in Part VIII—Electrical, the provisions for the materials and methods for wiring the smoke alarms are included in the electrical chapters. Section R313.1.1 requires that when one or more sleeping rooms are added or created in an existing dwelling, the dwelling unit must be provided with smoke alarms as required for new dwellings. If this dwelling did not already have smoke alarms installed, the code requires that they be installed in all bedrooms and outside each separate sleeping area in the immediate vicinity of the bedrooms and in each additional story of the dwelling including basements and cellars. All of these smoke alarms must be interconnected and hard wired, including the one in the new bedroom. The exception to this rule is that where the wall or ceiling finish is not removed during the construction or remodeling and there is no access to install the wiring, smoke alarms would not be required to be interconnected. However, the smoke alarm in the new bedroom would be required to be hard wired. Whether and how to interconnect additional smoke alarms in the existing part of the dwelling may be decisions reached between the building official and the installer of the electrical wiring. See the Exceptions to Section R313.1.1.

Whenever interior alterations, repairs, or additions are done that require a permit, smoke alarms must be installed in the dwelling in accordance with the smoke alarm requirements for new construction. This provision indicates that in a house with no existing smoke alarms, where the kitchen is remodeled for example, the kitchen wiring must be installed according to current code rules, and smoke alarms must be installed in all the required locations in the existing house, hard wired and interconnected where possible.

SECTION E3302 BUILDING STRUCTURE PROTECTION

E3302.1 Drilling and notching. Wood-framed structural members shall not be drilled, notched or altered in any manner except as provided for in this code.

❖ The installation of wire and cable requires drilling and notching of floor joists, studs and other wood framing members. This provision is another example of the necessity to look in other parts of the code when installing or inspecting the electrical wiring. When drilling and notching, two factors must be remembered: the integrity of the framing member and the protection of the wire or cable. To protect the strength and structural load bearing integrity of wood framed structural members, the provisions of Sections R502.8 and R602.6 must be followed. Requirements in Table E3702.1 are somewhat different and are for the protection of the wire, cable or other wiring method.

E3302.2 Penetrations of fire-resistance-rated assemblies. Electrical installations in hollow spaces, vertical shafts, and ventilation or air-handling ducts shall be made so that the possible spread of fire or products of combustion will not be substantially increased. Electrical penetrations through fire-resistance rated walls, partitions, floors or ceilings shall be protected by approved methods to maintain the fire-resistance rating of the element penetrated. Penetrations of fire-resistance-rated walls shall be limited as specified in Section R317.3.

❖ Penetrations through fire-resistance-rated walls or floors must meet the requirements of Section R317.3. Wall and/or floor assemblies of at least a 1-hour fire-resistance rating separate two-family dwellings from each other. For the installation of electrical wire, cable, conduit or electrical outlet boxes, it is often necessary to make a through penetration or a membrane penetration. A through penetration is an opening that passes through an entire assembly, accommodating an item such as a cable or conduit. Where membrane construction is provided, such as gypsum board applied to both sides of a stud wall, a through penetration would pass entirely through both membranes and the cavity of the wall. An electrical box installed in the wall or ceiling/floor assembly separating a two-family dwelling and a recessed light fixture in the ceiling of a dwelling that has a separate dwelling unit on the floor above are examples of membrane penetrations, which is an opening through only one membrane of a wall, ceiling or floor. Where non-metallic electrical outlet boxes are used in a membrane penetration, they must be installed in accordance with their listings. For steel electrical boxes, the specific requirements of Section R317.3.2 must be followed.

E3302.3 Penetrations of fire blocking and draftstops. Penetrations through fire blocking and draftstopping shall be protected in an approved manner to maintain the integrity of the element penetrated.

❖ See Section R317.3.2 for requirements regarding the penetration of firestop systems. Penetrations through fire blocking and draftstopping by electrical conduit, cables and equipment must be installed in a way that will maintain the integrity of the building element being penetrated. There are many fire blocking and draftstopping materials available, and both the electrician and the builder must be familiar with these products and methods.

SECTION E3303 INSPECTION AND APPROVAL

E3303.1 Approval. Electrical materials, components and equipment shall be approved.

❖ Electrical materials and equipment installed in one- and two-family dwellings must be approved by the building official for the specific use or application.

E3303.2 Inspection required. New electrical work and parts of existing systems affected by new work or alterations shall be inspected by the building official to ensure compliance with the requirements of Chapters 33 through 42.

❖ In small towns and municipalities, the building inspector may perform all of the inspections. However, in many larger jurisdictions, inspectors are hired to work under the direction of the building official who manages the building department from the office and does not spend much time in the field. Where the code requires inspections by the building official, it means that the building official can designate the inspector to perform the actual inspection in the field.

E3303.3 Listing and labeling. Electrical materials, components, devices, fixtures and equipment shall be listed for the application, shall bear the label of an approved agency and shall be installed, and used, or both, in accordance with the manufacturer's installation instructions.

❖ This code requires all electrical system components to be listed. Testing laboratories publish lists of materials that have been tested. Laboratories and the testing methods they use are evaluated by an evaluation service. An evaluation report is available on the listed products. Building officials rely on these lists and evaluation reports to approve materials for use in the installation of electrical systems. Electrical materials must be approved and installed in accordance with the manufacturer's installation instructions. It is necessary to know these listing requirements and not only the code rules.

SECTION E3304 GENERAL EQUIPMENT REQUIREMENTS

E3304.1 Voltages. Throughout Chapters 33 through 42, the voltage considered shall be that at which the circuit operates.

❖ For the code provisions in Chapters 33 through 42, the voltage to be considered must be the voltage at which the circuit operates. Nominal voltage for residential wiring as stated in Section E3502.4 is 120/240-volt, single phase, with a grounded neutral. However, the actual voltage at which the circuit operates can vary from the nominal voltage designation.

E3304.2 Interrupting rating. Equipment intended to interrupt current at fault levels shall have a minimum interrupting rating of 10,000 amperes. Equipment intended to interrupt current at levels other than fault levels shall have an interrupting rating at nominal circuit voltage sufficient for the current that must be interrupted.

❖ See the commentary for Section E3304.3.

E3304.3 Circuit characteristics. The overcurrent protective devices, total impedance, component short-circuit current ratings and other characteristics of the circuit to be protected shall be so selected and coordinated as to permit the circuit protective devices that are used to clear a fault to do so without extensive damage to the electrical components of the circuit. This fault shall be assumed to be either between two or more of the circuit conductors or between any circuit conductor and the grounding conductor or enclosing metal raceway. Listed products applied in accordance with their listing shall be considered to meet the requirements of this section.

❖ It is important that the ratings, intended uses and proper types of materials are selected for the appropriate use. In one- and two-family dwellings, circuit breakers are the most commonly used overcurrent protective device (OPD) for circuits originating at the panelboard. Fuses are often used for protection of wiring and components of air conditioning equipment and other motor circuits. If a short circuit or ground fault occurs, the circuit must be de-energized as soon as possible to prevent fire and/or damage to the equipment. Section E3605.5 mandates that overcurrent protective devices be placed in the circuit where the conductors receive their supply. These devices are designed to open the circuit and interrupt the current under other than normal current flow conditions without damage to the device, or items protected by the device.

<u>Insulation Failure</u>

One of the major causes of trouble in wiring is insulation failure. The results of insulation failure are short circuits and ground faults.

A "short circuit" is a conducting connection between any two conductors of an electrical system; it could be between any two ungrounded conductors or between an ungrounded conductor and the grounded conductor.

A "ground fault" is a conducting connection between a conductor of an electrical system and a grounded object such as a conduit, box or enclosure. Where opposite phase conductors are accidentally connected together and then energized, the connection is called a bolted fault, and there is usually no arcing. But where the ungrounded conductor lightly touches another ungrounded conductor or a metal enclosure around the conductors, a fault occurs that would likely be an arcing ground fault.

Purpose of the Overcurrent Protective Device

Overcurrent protective devices serve two separate functions. They protect against: (1) current overload or current higher than normal operating levels for the circuit conductors and (2) short circuits and ground faults. Accordingly, OPDs have two different interrupting ratings.

Current Overload Interrupting Rating

The standard ampere rating of an OPD is for the maximum current flow under normal operating conditions. In common practice in the field, this rating is not usually referred to as an interrupting rating, although that is what it is. Electricians simply call it the ampere rating of the breaker. For example, they would install a 20-ampere breaker on the circuit supplying the clothes washer. The 20-ampere rating is an interrupting rating in that the breaker will open and interrupt the current flow if the load increases to over 20 amperes. The laundry circuit often has a convenience outlet in the same room in addition to the clothes washer receptacle. If a high current use appliance such as a space heater is plugged into this receptacle, the total load could possibly be over 20 amperes. A current flow of greater than 20 amperes, if not interrupted, could cause such a high temperature that the conductor insulation or the wire itself would be damaged. Section E3605.6 lists the standard ampere rating of overcurrent protective devices, which determines the rating of the branch circuit and is coordinated with the ampacity of the circuit conductors. See Section 3602.2: Branch-circuit ampere rating.

Short-circuit or Ground-fault Interrupting Rating

This section requires that all overcurrent protective devices have a short-circuit and ground-fault interrupting rating of at least 10,000 amperes. If the available short-circuit or ground-fault current is higher than 10,000 amperes, the installed OCP devices must have sufficient interrupting ratings above the 10,000 ampere level. The electrician or electrical contractor must have this information and be able to provide it to the inspector or the building official (the authority having jurisdiction).

The Amount of Current Flow in a Short-circuit or Ground-fault Circuit

A high level short circuit or ground fault could destroy the OCP device as well as lead to the ignition of a fire. To determine the amount of current (let-through current) in a short circuit or ground fault, it is necessary to know the amount of available short-circuit current in amperes from the supply system and take into account the impedance of the short-circuit or ground-fault circuit. The impedance is very low in a short circuit, and therefore the current flow is very high.

In a bolted short circuit, there is high current flow. The current can be almost all of what the system can deliver. In most residential systems, the available short-circuit current is usually under 10,000 amperes but could be more, depending on several factors including the transformer size, how close the transformer is to the dwelling, motor load and the wire size.

E3304.4 Protection of equipment. Equipment identified only as "dry locations," "Type 1," or "indoor use only" shall be protected against permanent damage from the weather during building construction.

❖ During construction, equipment delivered to the job site that is identified for indoor use or for dry locations only must be protected from the weather. The electrical equipment manufacturing industry has designated equipment with different type numbers. A Type 1 enclosure is designated for indoor use only. Equipment should be stored in a dry and clean location until it is installed.

E3304.5 Unused openings. Unused cable or raceway openings in boxes, cabinets, meter socket enclosures, equipment cases or housings shall be effectively closed to afford protection substantially equivalent to the wall of the equipment. Where metallic plugs or plates are used with nonmetallic enclosures they shall be recessed at least $^{1}/_{4}$ inch (6 mm) from the outer surface of the enclosure.

❖ Unused openings include knockouts in device boxes, threaded raceway sockets in conduit bodies and boxes, cabinet and enclosure concentric and eccentric knockouts and circuit breaker knockouts in panelboard covers. Boxes and enclosures must have the integrity to keep out foreign objects, protect the conductors therein and contain any arcing or hot materials resulting from faults and splice failures in the box or enclosure. Metal plugs used with nonmetallic boxes will not be grounded, and recessing them in the outer surface will limit contact.

Knockouts being left open and unused in boxes and other cabinets is a very common problem. A box that has the knockouts removed is sometimes covered with tape, which is not equivalent to the wall of the box. This procedure does not comply with the code. Unused openings must be closed with appropriate materials for the protection of the circuit conductors as well as for protection against shock by someone inserting a finger or tool into the enclosure. Approved knockout plugs are available for this purpose.

E3304.6 Integrity of electrical equipment. Internal parts of electrical equipment, including busbars, wiring terminals, insulators and other surfaces, shall not be damaged or contaminated by foreign materials such as paint, plaster, cleaners or abrasives, and corrosive residues. There shall not be any damaged parts that might adversely affect safe operation or mechanical strength of the equipment such as parts that are broken; bent; cut; deteriorated by corrosion, chemical action, or overheating. Foreign debris shall be removed from equipment.

❖ After installation, electrical materials must be protected. A panelboard installed on the rough wiring phase of the job could result in having the bus bars coated with paint while the walls were being spray painted. Thus, the panel must be covered during construction. Also, the inspector should watch for a router tool being used to cut openings for the electrical boxes after the drywall is hung. In many cases, the openings for the electrical boxes are not pre-measured and cut before the wallboard is installed. First the drywall or wallboard is placed on the stud wall, then a router is used to cut the openings where a switch or receptacle box has been installed behind the wallboard. This results in damage to the conductors in the box and can also damage the box; it is a clear violation of the code. Often the conductors are cut and the insulation is scraped from the conductors. Later, after the walls are finished and painted, it is difficult to repair the wiring within the outlet boxes. It is often thought that code violations in the electrical wiring system are caused only by the electrician as a result of improper installation or improper wiring methods. But the electrical system can be installed in total compliance with code requirements and later be found to violate the code through improper construction practices. Correlation between all construction trades is needed to prevent these types of problems and is required by the code.

E3304.7 Mounting. Electric equipment shall be firmly secured to the surface on which it is mounted. Wooden plugs driven into masonry, concrete, plaster, or similar materials shall not be used.

❖ Equipment must be mounted securely. This provision is an example of a code requirement that is not exactly prescriptive, meaning that it does not spell out such things as the weight of the equipment or the number and size of fasteners. It is more of a performance requirement in that it simply requires that the equipment be mounted securely so that it will not come loose or fall off the wall. For example, on a brick house the service equipment is often installed on the surface of the brick wall. Holes lined up with the mounting holes in the back of the meter base or panelboard are drilled into the brick or mortar. If wooden plugs are used, the panelboard may look secure at the time of installation, but the wooden plugs will deteriorate, and the equipment will eventually sag or fall off the wall. Wooden plugs can decay and shrink, causing fasteners to pull out. Proper masonry anchors and fasteners must be used.

For relatively lightweight boxes and enclosures, plastic or metal plugs with lag screws can work well. The weight of the equipment should be taken into account. In some cases, it may be necessary to install additional supports such as lag screws that attach to the framing members. Where service equipment is supplied by a service lateral, conduit is often used from below the ground up to the bottom of the service equipment. This conduit must be securely attached to the house. In new housing developments where the area next to the foundation has been excavated to place the footing and foundation and then backfilled, the ground may settle over the first several months and can pull the conduit containing the service-lateral conductors out of the service equipment. This can easily happen where, for example, a heavy 2-inch rigid metal conduit with a 90-degree elbow underground is installed into a connector through concentric knockouts at the bottom of the outdoor panelboard. A one-hole strap attached with a plastic anchor may not be enough to hold the conduit from pulling away from the equipment. This code requirement is not only for service equipment but pertains to all electrical equipment. Sections E3904.5 and E4001.6 cover the provisions for mounting boxes for lighting fixtures weighing over 50 pounds and boxes for fans; see Section E3703.10.

E3304.8 Energized parts guarded against accidental contact. Approved enclosures shall guard energized parts that are operating at 50 volts or more against accidental contact.

❖ Any electrical circuit of 50 volts or more must have its components enclosed. A door-bell circuit transformer has the line terminals or leads that connect to a 120-volt circuit enclosed in a box, but the load terminals, which may be 24 volts, are permitted to be in the open.

E3304.9 Prevent physical damage. In locations where electrical equipment is likely to be exposed to physical damage, enclosures or guards shall be so arranged and of such strength as to prevent such damage.

❖ Locations where electrical equipment is likely to be exposed to physical damage are determined by the building official in the enforcement of this code. These locations are not spelled out, but many are obvious, such as on the outside of the dwelling where a vehicle or a lawnmower could possibly hit or damage a box or other wiring or equipment. Guards or enclosures may be required in such locations.

E3304.10 Equipment identification. The manufacturer's name, trademark or other descriptive marking by which the organization responsible for the product can be identified shall be placed on all electric equipment. Other markings shall be provided that indicate voltage, current, wattage or other ratings as specified elsewhere in Chapters 33 through 42. The marking shall have the durability to withstand the environment involved.

❖ In addition to the name of the manufacturer, markings must be provided on electrical equipment regarding voltage, current, wattage, and other information necessary to define the correct use of such equipment. The equipment must be identified so that it can be determined if it is being used in compliance with code provisions and according to its listing requirements.

E3304.11 Identification of disconnecting means. Each disconnecting means shall be legibly marked to indicate its purpose, except where located and arranged so that the purpose is evident. The marking shall have the durability to withstand the environment involved.

❖ The main disconnecting means for the dwelling must be marked "Main Disconnect." In the case of a fire, fire department personnel must be able to determine if the disconnect they are switching off is the main disconnect or not. In outdoor panels especially the marking may fade with time, so the building official may consider a requirement for a durable type of marking. Included in this section is the requirement to label the branch circuit panelboard. See also the commentary for Section E3606.2.

Branch circuits must be legibly marked or identified. In many cases, panelboards are marked with illegible handwriting or may have circuits marked with only a simple description. The code does not spell out how detailed the marking must be; that is, the code does not specify whether the marking must indicate the specific rooms a lighting circuit serves or only that it is a lighting circuit. These would be determined by the building official in the enforcement of the code.

SECTION E3305 EQUIPMENT LOCATION AND CLEARANCES

❖ Enough space should be provided to work on electrical equipment without jeopardizing the safety of the worker. It is not left to the arbitrary decision of the electrician or architect as to how much clearance is needed; the code is very specific with regard to the minimum clearances required for electrical equipment. More clearance is always better.

E3305.1 Working space and clearances. Sufficient access and working space shall be provided and maintained around all electrical equipment to permit ready and safe operation and maintenance of such equipment in accordance with this section and Figure E3305.1.

❖ The code does not simply state that sufficient space shall be provided. Figure E3305.1 provides a very complete picture of the basic requirements for location and clearances for panelboards. This also includes the requirements for electrical equipment located outdoors such as the meter base and main disconnect and air conditioner disconnect.

E3305.2 Working clearances for energized equipment and panelboards. Except as otherwise specified in Chapters 33 through 42, the dimension of the working space in the direction of access to panelboards and live parts likely to require examination, adjustment, servicing or maintenance while energized shall be not less than 36 inches (914 mm) in depth. Distances shall be measured from the energized parts where such parts are exposed or from the enclosure front or opening where such parts are enclosed. In addition to the 36-inch dimension (914 mm), the work space shall not be less than 30 inches (762 mm) wide in front of the electrical equipment and not less than the width of such equipment. The work space shall be clear and shall extend from the floor or platform to a height of 6.5 feet (1981 mm). In all cases, the work space shall allow at least a 90-degree opening of equipment doors or hinged panels. Equipment associated with the electrical installation located above or below the electrical equipment shall be permitted to extend not more than 6 inches (152 mm) beyond the front of the electrical equipment.

❖ When a worker is standing in front of a panelboard or air conditioner disconnect switch, for example, he or she should be able to walk up to it and stand in a clear space with no obstructions without having to lean or climb over another piece of electrical equipment or other obstruction. The last sentence of this section states that "equipment associated with the electrical installation located above or below the electrical equipment shall be permitted to extend not more that 6 inches (152 mm) beyond the front of the electrical equipment." At first reading, this may be difficult to understand if taken out of context. The working space in front of any electrical equipment must be clear from the floor or ground to at least 6.5 feet (1981 mm) above the floor or ground. Equipment that extends more than 6 inches (152 mm) past the front of other equipment is considered to protrude into the required clear working space.

On residential properties, such things as the locations of a lawn sprinkler valve box that sits above the surface of the ground, a telephone connection box, a junction box, a subpanel or a current transformer enclosure should be carefully considered when locating the panelboard and service equipment. If these items extend more than 6 inches (152 mm) further out than the front of the adjacent equipment, they are not permitted in the 36-inch (914 mm) depth or 30-inch (762 mm) width space in front of the panelboard or service equipment.

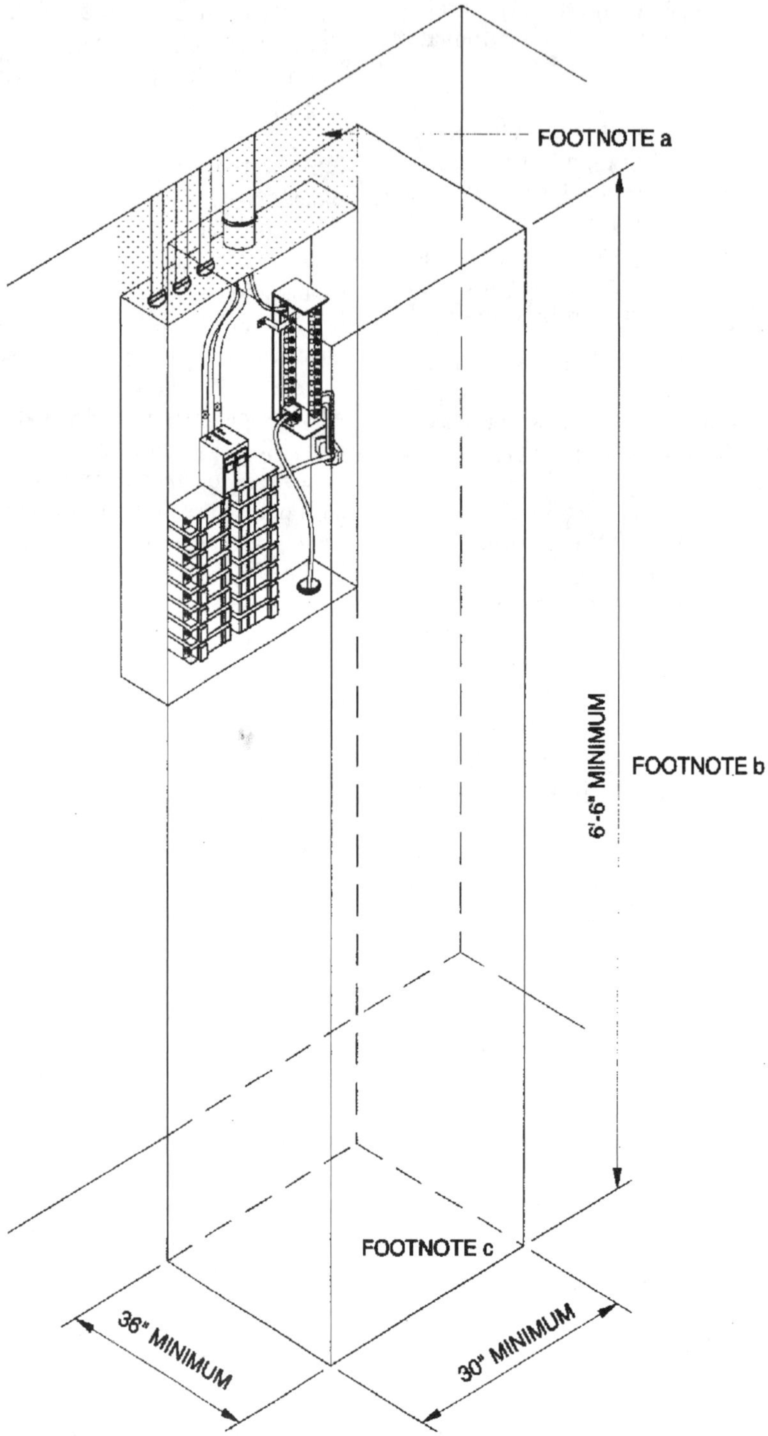

FIGURE E3305.1[a,b,c,d,e]
WORKING SPACE AND CLEARANCES

For SI: 1 inch = 25.4 mm, 1 foot = 304.8 mm.

a. Equipment, piping and ducts shall not be placed in the shaded area extending directly above the top of the panelboard from the panelboard to the ceiling above.

b. The working space shall be clear and unobstructed from the floor to a height of 6.5 feet.

c. The working space shall not be designated for storage.

d. Panelboards, service equipment and similar enclosures shall not be located in bathrooms, toilet rooms and clothes closets.

e. Such work spaces shall be provided with artificial lighting where located indoors.

❖ See the commentary for Section E3305.1.

E3305.3 Clearances over panelboards. A dedicated space directly over a panelboard that extends from the panelboard to the structural ceiling or to a height of 6 feet (1829 mm) above the panelboard, whichever is lower, and has a width and depth equal to the equipment shall be dedicated and kept clear of equipment unrelated to the electrical equipment. Piping, ducts or equipment unrelated to the electrical equipment shall not be installed in such dedicated space.

❖ A dedicated space above the panelboard equal to the depth and width of the panelboard must be maintained clear of all ducts and piping. In basements, the heating and air conditioning ducts are often run against a wall and the ceiling. To locate the panelboard on a wall with a duct installed above and tight to the wall is a code violation, unless the duct is 6 feet or more above the top of the panelboard. That is not likely in most dwellings. Many times, the electrician arrives at the job site first and installs the panelboard and the wiring, and later the HVAC contractor installs the ducts. This may seem to be acceptable if the wiring is installed first; however, no such provision is in the code. It is necessary for the electrician to know the location of the ducts before they are installed or for the installer of the ducts to avoid the required clear space above the panelboard.

E3305.4 Location of working spaces and equipment. Required working space shall not be designated for storage. Panelboards and overcurrent protection devices shall not be located in clothes closets or bathrooms.

❖ This section specifies that the clear working space required in front of electrical equipment not be in a storage area and that overcurrent devices and panelboards must not be located in clothes closets or bathrooms.

Closets are obviously designed for storage, and combustible material in a closet could be stored close to or in contact with the panelboard. To avoid the possibility of easily ignitable material being placed next to a panelboard and to avoid the storage of materials in front of a panelboard, thus blocking access to it, overcurrent protection devices must not be installed in clothes closets. Although the code specifically prohibits the panelboard in a clothes closet, it seems wise to not locate a panelboard in any closet, since no one knows what the occupants of the dwelling will store there.

Overcurrent devices in a bathroom are a problem because of the location of plumbing fixtures and moisture present in the room.

The required clear space cannot be part of an area designated for storage. One common problem occurs where a panelboard or other equipment is installed on the wall of a garage. When the house is being built and there are no shelves along the wall of the garage, it is easy to put the panel on that wall. Later, a workbench is built under the panelboard, or shelves are installed all around it. Although this is not a code violation at the time of the panelboard installation, it can be a major problem later and becomes a code violation, even though the house passed inspection initially. It is difficult to foresee that a home owner may install a cabinet over a panelboard in the garage, cutting out the back of the cabinet to apparently allow access to the panel. Care should be taken by the electrician to select a location for the panelboard that will not likely be used as a storage area later.

E3305.5 Access and entrance to working space. Access shall be provided to the required working space.

❖ The minimum working space is defined, but access dimensions such as the specific dimensions of a door or entrance to the working space are not given. The access to the working space must be sufficient.

E3305.6 Illumination. Artificial illumination shall be provided for all working spaces for service equipment and panelboards installed indoors.

❖ Although the code does not specify how close to the equipment the light must be installed, it must provide enough light to work safely on the equipment.

E3305.7 Headroom. The minimum headroom for working spaces for service equipment and panelboards shall be 6.5 feet (1981mm).

❖ See the commentary for Section E3305.2.

SECTION E3306 ELECTRICAL CONDUCTORS AND CONNECTIONS

E3306.1 General. This section provides general requirements for conductors, connections and splices. These requirements do not apply to conductors that form an integral part of equipment, such as motors, appliances and similar equipment, or to conductors specifically provided for elsewhere in Chapters 33 through 42.

❖ This section deals specifically with residential wiring.

E3306.2 Conductor material. Conductors used to conduct current shall be of copper except as otherwise provided in Chapters 33 through 42. Where the conductor material is not specified, the material and the sizes given in these chapters shall apply to copper conductors. Where other materials are used, the conductor sizes shall be changed accordingly.

❖ Copper conductors are used almost exclusively for branch circuits of 15 and 20 amperes. Nonmetallic sheathed cable with aluminum conductors was used in the 1970s in some regions for these smaller circuits in residential wiring, but now it is not readily available. Aluminum conductors, even in these smaller sizes, are not prohibited by the code. Table E3605.1, which lists the allowable ampacities for conductors, begins with size 12 AWG for aluminum conductors. For repair and remodeling work, it is necessary to have the ampacities included for these smaller aluminum conductors. For larger conductors such as for services and feeders, aluminum is often used.

E3306.3 Minimum size of conductors. The minimum size of conductors for feeders and branch circuits shall be No. 14 copper and No. 12 aluminum. The minimum size of service conductors shall be as specified in Chapter 35. The minimum size of class 2 remote control, signaling and power-limited circuits conductors shall be as specified in Chapter 42.

❖ The absolute minimum size of conductors is listed here, but for the minimum ampacities or allowable ampacities, refer to Table E3605.1. For example, in that table, size 12 AWG aluminum is listed as having an allowable ampacity of 20 amperes but is limited to a 15 ampere circuit per Section E3605.5.3.

E3306.4 Stranded conductors. Where installed in raceways, conductors of size No. 8 and larger shall be stranded. A solid No. 8 conductor shall be permitted to be installed in a raceway only to meet the requirements of Section E4104.

❖ This requirement limits the size of solid conductors installed in raceways to 10 and smaller. Solid conductors of 8 AWG and larger are usually very difficult to pull into conduit or tubing and may be damaged in the process. Of course, Section E4104 requires a minimum 8 AWG solid conductor to be used to bond all the metallic parts of a swimming pool. The bonding conductor normally, where installed in a raceway, should be sized to allow sufficient room to hold this conductor along. This provision should not be confused with the case where a single solid grounding electrode is pulled in a sleeve for physical protection.

Another case where an 8 AWG conductor could be installed in conduit is for the grounding electrode conductor. Table E3503.1, Note b requires No. 8 grounding electrode conductors to be installed in metal or nonmetallic conduit. Section E3306.4 would then require this grounding electrode conductor to be stranded and not solid. However, installing a solid No. 8, 6 or 4 copper conductor in a short length of $^1/_2$-inch (12 mm) or $^3/_4$-inch (19 mm) conduit with no bends is common practice and would not likely damage the conductor, especially considering that it is not required to be insulated.

E3306.5 Individual conductor insulation. Except where otherwise permitted in Sections E3505.1 and E3808.9, and E4203, current-carrying conductors shall be insulated. Insulated conductors shall have insulation types identified as RH, RHW, RHW-2, THHN, THHW,THW, THW-2, THWN, THWN-2, TW, UF, USE, USE-2, XHHW or XHHW-2. Insulation types shall be approved for the application.

❖ The common specific conductor insulation types listed are the conductors included in Table E3605.1. The provisions of this code cover the installations in which these conductor types would be used in residential wiring. Conductor insulation types have different temperature ratings, indicating the temperature that the conductor insulation can withstand before being damaged. Wherever conductors are used in wet locations, such as service entrance conductors that are exposed at the service head, the conductors must have a "W" in the insulation type. The following are letters in the conductor insulation types and the corresponding use or meaning.

R - Rubber
T - Thermoplastic
H - Rated for a 15°C temperature rise over the 60°C rating
HH- Rated for a 30°C temperature rise over the 60°C rating
N - Outer nylon jacket or equivalent
W - Rated for wet location
X - Cross linked synthetic polymer

E3306.6 Conductors in parallel. Circuit conductors that are electrically joined at each end to form a single conductor shall be limited to sizes No. 1/0 and larger. Conductors in parallel shall be of the same length, same conductor material, same circular mil area and same insulation type. Conductors in parallel shall be terminated in the same manner. Where run in separate raceways or cables, the raceway or cables shall have the same physical characteristics.

❖ Parallel conductors are typically installed in conduit or raceways, and for one- and two-family dwellings within the scope of this code (dwellings with a service size of up to 400 amperes), only very large circuits or the service or feeders would likely be run in parallel. Although it is permitted for conductors of 10 AWG and above, paralleling conductors in conduit or tubing such as electrical metallic tubing in residential wiring would normally only be used for larger ampere circuits. The size of the conduit, labor involved and the cost of the conductors are major factors in the decision to use paralleled conductors.

E3306.7 Conductors of the same circuit. All conductors of the same circuit and, where used, the grounded conductor and all equipment grounding conductors shall be contained within the same raceway, cable or cord.

❖ This provision requires all conductors of a circuit to be in the same raceway, cable, or cord to prevent magnetic fields from creating heat in metal enclosures and raceways.

E3306.8 Aluminum and copper connections. Terminals and splicing connectors shall be approved for the material of the conductors joined. Conductors of dissimilar metals shall not be joined in a terminal or splicing connector where physical contact occurs between dissimilar conductors such as copper and aluminum, copper and copper-clad aluminum, or aluminum and copper-clad aluminum, except where the device is listed for the purpose and conditions of application. Materials such as inhibitors and compounds shall be suitable for the application and shall be of a type that will not adversely affect the conductors, installation or equipment.

❖ Copper and aluminum conductors must not be joined unless the device or connector is listed for the purpose. Some wire nut connectors are listed for this purpose. If the connector is marked "AL/CU," it will be approved for aluminum and copper conductors to be connected together in the same pressure connector. It is very important to use an approved oxide-inhibiting compound on aluminum conductors. Where aluminum conductors are subject to moisture or even very high humidity, oxidation can occur. Oxidation of aluminum can cause a thin film or layer, which looks like a powder, on the conductor. It will result in heat build-up by impeding the current flow. Wire connectors that are listed and approved for connection of aluminum to aluminum or for copper to aluminum conductors are available. Most wire nuts are not specifically listed for use with aluminum conductors, and one cannot assume that they can be used. If they are approved, the container will so indicate with a marking such as "AL/CU" and may indicate that an approved oxide inhibitor should be used where aluminum conductors are connected. Many of these approved wire nuts are made with a specially formulated corrosion-resistant compound to help prevent the build up of aluminum corrosion. There is a need for these connectors, for example where an old house wired with aluminum is being remodeled or a room is being added. If any of the old aluminum conductors are spliced, the installer should be careful to use approved connectors. Where connections of larger conductors are made, for example feeder or service conductors, aluminum is commonly used. Again, there may be a need in a retrofit job to make connections between the existing aluminum and copper conductors. Split bolts may be used for this connection. These bolts and some clamp-on type bolts will have a separator between the two barrels or lugs where the wire is inserted. The bolts must have marking on them such as "AL/CU," indicating they may be used for aluminum to copper connections.

E3306.9 Terminals. Connection of conductors to terminal parts shall be made without damaging the conductors and shall be made by means of pressure connectors, including set-screw type, by means of splices to flexible leads, or for conductor sizes of No. 10 and smaller, by means of wire binding screws or studs and nuts having upturned lugs or the equivalent. Terminals for more than one conductor and terminals for connecting aluminum conductors shall be identified for the application.

❖ Connections must be made without damaging the conductors. When stripping the insulation from the ends of conductors, it is easy to nick or cut into the wire. Using improper techniques or tools when stripping the conductors may damage the softer aluminum conductors.

In many installations, two or more conductors are connected under one terminal since it is easy to install them and they will fit. However, terminals for connecting multiple conductors must be listed for this purpose. Most modern equipment such as panelboards has terminal strips with a sufficient number of terminals to connect all necessary circuit conductors. In some cases, it may be necessary to install additional terminal strips in order to terminate each conductor under a separate terminal. If the terminals are approved for the connection of more than one conductor, the approval will be stated in the listing information or supplied with the packaging of the equipment.

E3306.10 Splices. Conductors shall be spliced or joined with splicing devices listed for the purpose. Splices and joints and the free ends of conductors shall be covered with an insulation equivalent to that of the conductors or with an insulating device listed for the purpose. Wire connectors or splicing means installed on conductors for direct burial shall be listed for such use.

❖ Conductors must be spliced with devices listed for the purpose. The points in the electrical wiring system where conductors are spliced or connected are the weakest links. Here problems could occur such as deterioration of the conductor, overheating, arcing, breaking, or opening of the connection. Therefore, it is very important that conductors be properly spliced. Splicing and terminating conductors in high-voltage wiring, for example, is a skill that must be acquired and developed. But even in low-voltage wiring, the splicing of small conductors is a skill and critical to the proper operation of the system. Bad connections are the source of many of the problems in an electrical wiring system, including fires.

One unapproved method of making a connection is to rely only on twisting the stripped ends of the conductors together using pliers and then using tape to cover the connection without any approved device. Wire nuts or pressure connectors of the appropriate size and type must be used. Some installers claim that the stripped ends of conductors should be securely twisted together with pliers before applying the wire nut, while the wire nut manufacturers typically say the stripped wire ends should be held together and then twisted together using only the wire nut. The installer should consult the instructions for how to install the particular type of wire connector to be used. The instructions on the container will state, for example, "Pre-twisting unnecessary. Hold the stripped wires together with the ends even." Most wire nuts or pressure connectors that are sold for use in residential wiring are for connecting copper wires only, and the container will have such a message on it. See the commentary for Section E3306.8 for details about joining aluminum and copper conductors.

Where a split bolt or clamp-on pressure connector is used, the code requires that the splice be covered with an insulating material equivalent to the conductor insulation. Some of these bolts are supplied with an approved plastic or insulating cover that snaps into place around the splice, but many are bare, and the electrician must cover the connection with splicing compounds and tape. If the insulating tape is listed for the purpose, this could be an approved method if the thickness of the covering is equivalent to that of the conductor insulation.

E3306.10.1 Continuity. Conductors in raceways shall be continuous between outlets, devices and junctions and shall be without splices or taps in the raceway.

Exception: Splices shall be permitted within surface-mounted raceways that have a removable cover.

❖ See the commentary for Section E3306.10.2.

E3306.10.2 Device connections. The continuity of a grounded conductor in multiwire branch circuits shall not be dependent on connection to devices such as receptacles and lampholders. The continuity of equipment grounding conductors shall not be dependent on such connections in any type of branch circuit.

❖ The grounded conductor, which is commonly referred to in the field as the neutral conductor, is often connected or spliced only by means of the receptacle in an outlet box. This is commonly done where 2 two-wire or three-wire cables are brought into an outlet box for the installation of a 120-volt duplex receptacle, and the two grounded conductors are connected only to the receptacle and are not tied together before being connected to the receptacle. A multiwire circuit has two ungrounded conductors (usually one red and one black) with a voltage of 240 between them and a grounded (neutral) conductor (white) with a voltage of 120 volts between it and each of the ungrounded conductors. When the receptacle is replaced or one of the wires is disconnected, an open neutral could result. If this is done when the circuit is energized, it could cause higher voltages to be imposed on any appliances or equipment connected to the circuit. Therefore, the code requires that the white grounded conductors be connected together and then connected with a short jumper to the receptacle (pigtailed). This way, the continuity of the grounded conductor is maintained at all times even when the device is removed.

Where multiple equipment grounding conductors are brought into a junction box for the connection of devices, they must be connected together. If a receptacle is installed in the box, a pigtail must be used so that only one conductor connects to the grounding terminal of the receptacle.

E3306.10.3 Length of conductor for splice or termination. Where conductors are to be spliced, terminated or connected to fixtures or devices, a minimum length of 6 inches (152 mm) of free conductor shall be provided at each outlet, junction or switch point. The required length shall be measured from the point in the box where the conductor emerges from its raceway or cable sheath. Where the opening to an outlet, junction, or switch point is less than 8 inches (203 mm) in any dimension, each conductor shall be long enough to extend at least 3 inches (76 mm) outside of such opening.

❖ This code provision is prescriptive and leaves little room for misinterpretation. It is very difficult to make splices and device terminations while working inside a device box on very short wires. In many cases, especially when remodeling on older houses, it is found that existing wires have been cut or installed very short. When an attempt is made to remove the switch or receptacle, it cannot be pulled out and turned so that the electrician might see or work on the connections. These device boxes are often smaller than would be normally used in modern wiring, which adds to the difficulty. The alternative can also be a problem. If too much wire is left at the outlet, it is difficult to put the extra length back into the box when installing the device. Although the requirement is very precise, care must be taken to tuck the conductors into the outlet box when the switch or receptacle is installed. If the box is too small, the wires could be damaged when they are pushed back into the box to make room for the device. Minimum box sizes are specified in Section E3805, but consideration should be given to the size of the wiring device to be installed. Some devices, such as fan control switches, dimmers, and ground-fault circuit interrupter receptacles, take up more room in the box than ordinary switches or receptacles. Therefore, it may be necessary to size the box larger than the minimum required in order to have room to tuck the required conductor length into the box without damaging the conductors. For junction boxes with an opening of at least 8 inches in any of its dimensions, the requirement is only for 6 inches of free conductor to be provided from the point in the box where the conductor emerges from the raceway or cable sheath. It may be easier for electricians to reach into the box and work on conductors in a junction box with openings of at least 8 inches. For all other boxes, the conductor must be long enough to extend at least 3 inches beyond the front edge of the box. This provision does not override the requirement for a 6-inch length of free conductor. See Commentary Figure E3306.10.3.

SECTION E3307 CONDUCTOR AND TERMINAL IDENTIFICATION

E3307.1 Grounded conductors. Insulated grounded conductors of sizes No. 6 and smaller shall be identified by a continuous white or natural gray outer finish or by three continuous white stripes on other than green insulation along the entire length of the conductors. Conductors of sizes larger than No. 6 shall be identified by a continuous white or natural gray outer finish or by three continuous white stripes on other than green insulation along the entire length of the conductor, or shall be identified by a distinctive white marking that is applied at the time of installation and that encircles the conductor at all terminations.

❖ Grounded conductors are commonly referred to as neutral conductors. Sometimes these conductors are referred to as the "ground wire," but the code does not define ground wire, so this term should be avoided. If a conductor is called the ground wire, it could mean the grounded or grounding conductor, and the term would

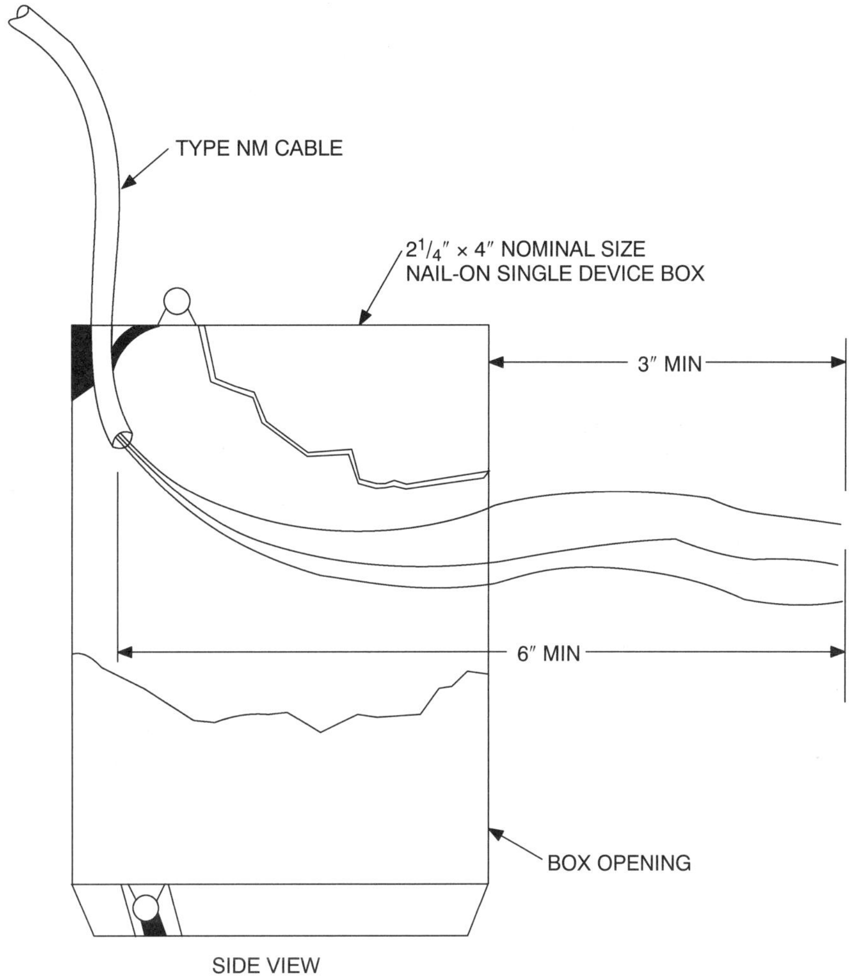

Figure E3306.10.3
MEASURING CONDUCTOR LENGTH AT OUTLET BOX

not be clear. The correct terminology should be used to identify conductors, namely: grounded, grounding and ungrounded. Conductors with green insulation are prohibited from being used as grounded or neutral circuit conductors; white or gray insulation is always used for the grounded conductors. The provision for using a color, other than green, with three white stripes as a grounded conductor is included for cases where equipment, lighting fixtures or flexible cords may contain conductors with this coloring. One of the conductors of flexible cord and fixture wire must be identified as the grounded conductor.

For grounded conductors up to and including size 6 AWG, the white insulation, gray insulation, or three white stripes must be continuous for the length of the grounded conductor. For 4 AWG and larger, any color conductor other than green may be used for the grounded conductor as long as it is identified at the terminations by white marking at the time of installation. The most common method of marking the grounded conductors of colors other than white or gray is the use of white tape, although white paint that would remain permanently on the conductor could be used. Some paints react with the insulation causing damage and should be avoided for marking. At junction boxes, cabinets, and panelboards where grounded conductors are identified with white tape, there is no requirement that the grounded conductor be taped or marked white for the entire length of the exposed conductor. Marking the entire length might be assumed and might help to make the grounded conductor more readily identifiable at these locations. However, the length of the white marking at the termination locations is not defined; the only requirement is that the marking encircle the conductor.

In the wiring of one- and two-family dwellings, the installation of grounded conductors with the wrong color of insulation is usually not a concern since the smaller conductors used in residential wiring are part of a multiconductor cable assembly that is manufactured with the proper color identification. However, for the connections of the conductors within the cable assembly and for conductors installed in conduit or tubing, it is very important that these code provisions be followed. Using the white conductor of a multiconductor cable for an ungrounded conductor is permitted only under the exception to Section E3307.3.

E3307.2 Equipment grounding conductors. Equipment grounding conductors shall be identified by a continuous green color or a continuous green color with one or more yellow stripes on the insulation or covering, except where bare.

❖ Equipment grounding conductors may be bare. In most cases the code will indicate when the insulation is required like for swimming pool grounding. If they are insulated they must be identified by insulation of a continuous green color or continuous green with one or more yellow stripes. An insulated equipment grounding conductor larger than 6 AWG is permitted to be identified at termination points such as junction boxes, cabinets, and panelboards by having the insulation stripped from the entire exposed length or by having the insulation marked with green or taped with green tape. For example, a 4 AWG aluminum equipment grounding conductor used on a circuit with an overcurrent device of 200 amperes could have insulation of any color if it is marked with green tape where it is exposed to view at junction boxes and termination points. Although not specifically stated in Section E3307.2, green taping of conductors larger than 6 AWG used for equipment grounding is permitted based on the fourth sentence of Section E3301.2.

E3307.3 Ungrounded conductors. Insulation on the ungrounded conductors shall be a continuous color other than white, gray or green.

Exception: An insulated conductor that is part of a cable or flexible cord assembly and that has a white or gray finish or a finish marking with three continuous white stripes shall be permitted to be used as an ungrounded conductor where it is permanently reidentified to indicate its use as an ungrounded conductor at all terminations and at each location where the conductor is visible and accessible.

❖ Any color of insulation other than white, gray, or green is permitted for ungrounded conductors. The only exception occurs where the conductor with white or gray insulation or with white stripes is part of a multiconductor cable or cord assembly and is permanently re-identified by marking or taping with a color other than white, gray, or green at all termination points on the circuit and at all locations where the conductor is visible or accessible. One example of this is a switch loop in which conductors supplying power to a light are brought to the lighting outlet. Only one cable with a black, white, and bare conductor is run to the switch point. The white conductor is connected to the ungrounded conductor at the lighting outlet and supplies power to the switch. When the switch is closed, power is carried back to the lighting fixture on the black conductor of the switch-loop cable. At the device box for the switch and in the lighting outlet box, the white conductor must be colored, marked, or taped to indicate it is an ungrounded conductor. After the installation is completed and the wall is covered, the wiring layout will not be visible. If the switch box has only one cable run to it, and the switch terminals have only the white and black conductors connected, a person opening the box could assume that the white conductor is used as an ungrounded conductor. However, the code requirement is for it to be marked as ungrounded in order to avoid such assumptions.

A conductor with white or gray insulation installed in conduit or tubing cannot be used as an ungrounded conductor. Thus, a white conductor to be used as an ungrounded conductor but taped with black tape and installed in conduit or tubing would be in violation of the code.

E3307.4 Identification of terminals. Terminals for attachment to conductors shall be identified in accordance with Sections E3307.4.1 and E3307.4.2.

❖ See the commentary for Sections E3307.4.1 and E3307.4.2.

E3307.4.1 Device terminals. All devices, excluding lighting and appliance branch-circuit panelboards, provided with terminals for the attachment of conductors and intended for connection to more than one side of the circuit shall have terminals properly marked for identification, except where the terminal intended to be connected to the grounded conductor is clearly evident.

Exception: Terminal identification shall not be required for devices that have a normal current rating of over 30 amperes, other than polarized attachment caps and polarized receptacles for attachment caps as required in Section E3307.4.2.

❖ Devices that have provisions for terminating both grounded and ungrounded conductors must have the terminals identified to indicate the connections. This is required for most of the devices that would be used in residential wiring except for certain specified devices with a normal current rating of over 30 amperes.

E3307.4.2 Receptacles, plugs, and connectors. Receptacles, polarized attachment plugs and cord connectors for plugs and polarized plugs shall have the terminal intended for connection to the grounded (white) conductor identified.

Identification shall be by a metal or metal coating substantially white in color or by the word "white" or the letter "W" located adjacent to the identified terminal.

Where the terminal is not visible, the conductor entrance hole for the connection shall be colored white or marked with the word "white or the letter "W".

❖ Most devices used for residential wiring have the grounded conductor terminals identified by a white (silver) metal composition or metal plating that is lighter in color than the terminals for the ungrounded conductors, which are usually brass colored. Some receptacles and connectors may have corrosion-resistant terminals, which are all white or a light color, or they may have all brass-colored terminals. For these devices especially, the terminals for the connection of the grounded or neutral conductor must be marked with the word "white," marked with the letter "W" or have white paint adjacent to the terminal. Many receptacles have both the distinctive colored terminals as well as the word "white" to indicate the grounded conductor connection.

E3307.5 Tag marking. AC cable shall be marked by means of a printed tag attached to the coil, reel or carton.

❖ All cable used in residential wiring must be marked to indicate size, type, voltage rating and the manufacturer's name. The conductor size must be marked on the cable at maximum intervals of 24 inches (610 mm). AC cable must have this information on a printed tag attached to the coil, reel or carton. Code provisions such as this one and the one above regarding identification of terminals are requirements that the manufacturer must satisfy, but they are necessary for the proper installation and function of the wiring system and should be verified by the installer and inspector.

Bibliography

The following resource materials are referenced in this chapter or are relevant to the subject matter addressed in this chapter.

NFPA 70-02, *National Electrical Code.* Quincy, MA: National Fire Protection Association, 2002.

Chapter 34: Electrical Definitions

General Comments

The electrical trade, like other construction trades, has its vernacular. If people do not understand the language of the code text, they will have difficulty achieving a correct interpretation. Many words have a unique meaning in the context of a particular code and may be undefined or defined differently in a dictionary.

Purpose

Because much code text meaning is dependent upon the definitions of the terms used in the text, Chapter 34 lists the definitions of terms that do not have everyday, common, universally accepted, or dictionary-defined meanings. Chapter 34 is provided as the key to understanding the electrical text.

SECTION E3401 GENERAL

E3401.1 Scope. This chapter contains definitions that shall apply only to the electrical requirements of Chapters 33 through 42. Unless otherwise expressly stated, the following terms shall, for the purpose of this code, have the meanings indicated in this chapter. Words used in the present tense include the future; the singular number includes the plural and the plural the singular. Where terms are not defined in this section and are defined in Sections R202, M1202 and P3002 of this code, such terms shall have the meanings ascribed to them in those sections. Where terms are not defined in these sections, they shall have their ordinarily accepted meanings or such as the context implies.

❖ In addition to the definitions in this chapter, the *International Residential Code* provides definitions at the beginning of Chapter 41 on Swimming Pools and the beginning of Chapter 42 on Class 2 Remote-Control, Signaling and Power-Limited Circuits. It is necessary that everyone understand the meaning of the terms used when they are using the code provisions. There are many technical terms in the code that would not be understood without a definition. Also, there are many common terms that have a different meaning or very specific meaning when applied to code rules. An individual may try to use terms such as "ground" or "accessible" according to a commonly understood meaning, but the specific definitions in this chapter are necessary to apply the code rules properly. Any definitions not included in this chapter or in Chapters 41 or 42 are to be understood as the commonly used definition in the English language. In addition, there are some terms used in the code such as "service head," "service mast," "threadless coupling," "raceway seal," "multioutlet branch circuit," "opening" and "yoke" that are neither defined here nor used in their common English sense but are understood with exposure to the electrical industry.

ACCESSIBLE. (As applied to wiring methods.) Capable of being removed or exposed without damaging the building structure or finish, or not permanently closed in by the structure or finish of the building.

❖ Where wiring methods are required to be accessible, they must not be concealed behind finished walls or behind or in a part of the dwelling construction that is permanent. The wiring is accessible even if reaching it requires someone to climb a ladder or remove an access door. For example, access to the wiring connections for a hydromassage bathtub may be through an access door that is secured with several screws. The connections are accessible as long as reaching them does not require the removal of a permanent cabinet or finished wall. It may take some time to remove the screws securing the access door, but the door does provide access. However, if the access door were inside a linen closet on the opposite side of a wall where the bathtub is located, there may not be room to physically get to the wiring and adequately and safely work on it. If the hydromassage bathtub pump motor needed to be replaced, it may be practically impossible to reach the enclosure for the wiring connections. Thus, it may technically be accessible by removal of the access door, but the accessible space must be considered also. If it is too small a space, the wiring would not be accessible. Wiring in a crawl space, unfinished basement, or attic is accessible as long as a person can get to the wiring for inspection, maintenance, repair or replacement. See the commentary for Section 3805.1 for making connections and splices accessible.

ACCESSIBLE. (As applied to equipment.) Admitting close approach; not guarded by locked doors, elevation or other effective means.

❖ Where equipment is required to be accessible, it must not be guarded by such things as locked doors or elevation. This does not mean that the equipment cannot be behind a locked door or installed at a high elevation. The key word of this definition is "guarded"; refer to the definition of "guarded" in this chapter. Equipment is accessible as long as it is not guarded. Also, equipment may be on a roof and still be accessible by someone climbing a permanent or portable ladder. If it is mounted at a high elevation for the purpose of preventing people from getting to it, then it is guarded.

ACCESSIBLE, READILY. Capable of being reached quickly for operation, renewal or inspections, without requiring those to whom ready access is requisite to climb over or remove obstacles or to resort to portable ladders, chairs, etc.

❖ A student of or person experienced in using the *International Building Code* may automatically think of accessible and readily accessible in terms of something being accessible to physically disabled persons. The definition in the IRC as it pertains to equipment has not been a part of building codes in the past but has been peculiar to electrical code requirements and also more recently included in mechanical and plumbing codes. The difference between accessible and readily accessible is very important in applying code rules.

The definition here is also in Section R202 of this code. Where the code states that equipment must be readily accessible, it means that a person can walk up to the equipment or device and work on it or operate it without climbing or removing any obstacles. Disconnect switches and panelboards containing overcurrent devices must be readily accessible. They should be located where it is not likely that someone will store materials nearby that could block access. See the commentary for Section 3305.4.

AMPACITY. The current in amperes that a conductor can carry continuously under the conditions of use without exceeding its temperature rating.

❖ This term may be thought of as the combination of two terms, ampere and capacity. The current that a conductor can carry continuously without exceeding its temperature rating is not set at a certain value but will vary with the conditions of use. The two conditions of use that must be considered in order to determine the allowable ampacity are the ambient temperature and the number of conductors in the cable or raceway. See the commentary for Section E3604.6 for more information on ampacity.

APPLIANCE. Utilization equipment, normally built in standardized sizes or types, that is installed or connected as a unit to perform one or more functions such as clothes washing, air conditioning, food mixing, deep frying, etc.

❖ Unlike devices, appliances consume power. Chapter 40 has the general code provisions covering appliances. Chapter 36 contains the rules covering the load calculations and branch circuit conductors for appliances such as clothes dryers and ranges. The provisions for the installation of ceiling fans are in Chapter 40: Appliance Installation. Many times a ceiling fan is thought of as a lighting fixture, and a person might look in Chapter 39, which covers devices and lighting fixtures. But a ceiling fan is an appliance.

APPROVED. Acceptable to the authority having jurisdiction.

❖ The term "approved" is used to describe a specific material or method of installation. It means that the material or method is acceptable to the code official. This term is also defined in Section R202 of this code. That definition states that "approved" means acceptable to the building official. The term "authority having jurisdiction" is used in the *National Electrical Code*. In the enforcement of the IRC, the authority having jurisdiction is the same as the code official. As stated in Section R202, it is imperative that the building official (or authority having jurisdiction) bases her or his decision of approval on the result of investigations, tests, or accepted principals or practices.

ARC-FAULT CIRCUIT INTERRUPTER. A device intended to provide protection from the effects of arc-faults by recognizing characteristics unique to arcing and by functioning to de-energize the circuit when an arc-fault is detected.

❖ An arc-fault circuit interrupter (AFCI) is designed to detect arcing faults that could cause a fire, and de-energize a circuit. The proper use of AFCI's will help reduce the risk of fire even more than conventional circuit breakers. For example, at a bad (loose) connection at the terminal of a receptacle, arcing can occur under load. Over time, as the arcing persists, a layer of carbon builds even more resistance in the connection and the result is heat, which could eventually build up enough to cause a fire. AFCI's are required for bedroom circuits where flammable materials would likely be close to the electrical outlets, and people could be sleeping when a fire starts. An AFCI is designed to de-energize the circuit before ignition could occur.

Several different AFCI devices have been developed and are available for use at different points in a circuit. A circuit-breaker type of device that mounts in the panelboard at the origin of the branch circuit or feeder provides protection for the entire branch circuit or feeder. An outlet branch circuit AFCI is installed as a receptacle in the branch circuit and provides protection for downstream wiring, including extension cords and power-supply cords plugged into the circuit. Other AFCI devices can be used such as a portable plug-in device with multiple outlets that provides protection for extension cords and power-supply cords plugged into it, and a cord AFCI that has the device built into or attached to the extension cord and is intended to protect only that cord. These devices are designed to clear the arc fault or open the circuit in a specified time of between one second and 200 milliseconds, depending on the current in the arc fault, thus helping to prevent the ignition of a fire.

ATTACHMENT PLUG (PLUG CAP) (PLUG). A device that, by insertion into a receptacle, establishes connection between the conductors of the attached flexible cord and the conductors connected permanently to the receptacle.

❖ Receptacles are often incorrectly referred to as plugs. Attachment plugs have voltage and ampere ratings. Specific configurations are designed so that the attachment plug for an appliance such as a clothes dryer, for example, will not fit the receptacle for a range since they have different ampere ratings.

AUTOMATIC. Self-acting, operating by its own mechanism when actuated by some impersonal influence as, for example, a change in current strength, pressure, temperature or mechanical configuration.

❖ Examples of devices that operate automatically are timing switches, thermostats, and overcurrent devices.

BATHROOM. An area, including a basin, with one or more of the following: a toilet, a tub or a shower.

❖ The definition of bathroom is important in determining where outlets are required and which ones must have GFCI receptacles. See Sections E3501.6.2, E3605.7, E3801.6 and E3802.1.

BONDING. The permanent joining of metallic parts to form an electrically conductive path that will ensure electrical continuity and the capacity to conduct safely any current likely to be imposed.

❖ Bonding is necessary to help ensure continuity between all the metallic parts of the electrical wiring system and will provide an effective path for fault current. If the fault current path is interrupted or open, circuit breakers and fuses will not operate properly or at the appropriate time. Bonding can be provided with a conductor or a device, as long as one of these provides a solid metallic connection between the parts being bonded. See Section E3509 for bonding provisions and commentary.

BONDING JUMPER. A reliable conductor to ensure the required electrical conductivity between metal parts required to be electrically connected.

❖ A bonding jumper is required where the continuity between metallic parts may be broken or opened. An example of this is where a metal conduit is installed containing underground service conductors and is attached to the bottom of an outdoor panelboard with a fitting through an opening made by breaking out the concentric rings of the pre-punched knockout, which is there to adjust the opening to the size of conduit being installed. The electrical conductivity through the remaining two or three tabs holding the remaining concentric rings in place may be negligible, and in the case of ground fault, the fault current would be impeded. Installing the correct size bonding jumper between the conduit and the panelboard provides for a solid fault current path.

BONDING JUMPER (EQUIPMENT). The connection between two or more portions of the equipment grounding conductor.

❖ Where there is a possibility of an incomplete fault current path or a fault current path that would have high impedance, a bonding jumper must be installed to maintain the ampacity of the fault current path.

BONDING JUMPER, MAIN. The connection between the grounded circuit conductor and the equipment grounding conductor at the service.

❖ The main bonding jumper is a connection between the service grounded conductor and the equipment grounding conductor. Since the code uses the term "jumper" as part this term, an individual might think of the main bonding jumper only as a conductor or short piece of wire. Indeed, a wire is used in some cases to make the connection, but the connection is often accomplished with a strap, busbar, or screw.

For example, in the service equipment, at the location of the main disconnect, the grounded service conductor is connected to the same busbar as the equipment grounding conductor and possibly the grounding electrode conductor. Therefore, that busbar functions as a main bonding jumper because it is the connection between the grounded service conductor and the equipment grounding conductor.

The busbar and the equipment grounding conductor must be bonded to the metal panelboard enclosure, which may be accomplished with a screw or bolt. The screw or bolt then becomes a main bonding jumper because it is the connection between the grounded service conductor and the service equipment. Where a screw or bolt is used, it must be listed for the purpose and is usually supplied with the panelboard, or the busbar and screw comes as a kit. It would be a code violation to use an ordinary sheet metal screw, for example, for this purpose.

In some types of service equipment, a short strap or length of wire—a "jumper"—is used to connect or bond the busbar to the metal panelboard enclosure. In this case, the strap or wire is a main bonding jumper because it is the connection between the service grounded conductor and the equipment grounding conductor.

The purpose of the main bonding jumper is to provide a low-impedance fault-current path back to the power supply to facilitate the tripping of the overcurrent device in case of a ground fault.

BRANCH CIRCUIT. The circuit conductors between the final overcurrent device protecting the circuit and the outlet(s).

❖ "Branch circuit" is a general term that includes the conductors run between the final overcurrent device in the panelboard and an outlet such as a receptacle outlet or a lighting fixture. This term helps distinguish a branch circuit from a feeder, which supplies power to the final overcurrent devices.

BRANCH CIRCUIT, APPLIANCE. A branch circuit that supplies energy to one or more outlets to which appliances are to be connected, and that has no permanently connected luminaires that are not a part of an appliance.

❖ This term refers to dedicated branch circuits such as those for laundry appliances and the small-appliance branch circuits required for kitchen and dining areas by Section E3603.2. These are receptacle circuits intended for appliance use only.

BRANCH CIRCUIT, GENERAL PURPOSE. A branch circuit that supplies two or more receptacle outlets or outlets for lighting and appliances.

❖ General purpose branch circuits supply lighting and receptacle outlets. The receptacle and lighting outlets are not required to be supplied by separate circuits as both are permitted on the same circuit. For dwellings, the number of lighting or receptacle outlets permitted on each general-purpose branch circuit is not specified. The minimum number of general purpose branch circuits is determined from the total computed load. See Section 3603.5.

BRANCH CIRCUIT, INDIVIDUAL. A branch circuit that supplies only one utilization equipment.

❖ Individual branch circuits are used for most major appliances and motors and are sized to provide power to specific loads according to the rating of the equipment, appliance, or load served. The circuits supplying power for dishwashers, dryers, ranges, and many motor-driven appliances are examples of individual branch circuits. See Section E3602 for rules on sizing individual branch circuits for motor loads. Where an individual branch circuit supplies a receptacle that the appliance is plugged into, it must be a single receptacle. An appliance such as an instant electric water heater installed under a kitchen sink may require an individual branch circuit because of the high current load. If a duplex receptacle were installed under the sink, a disposal could also be connected to that circuit, and if both appliances were operated at once, the overcurrent device could trip. Therefore, this electric water heater should have a single receptacle installed and should be supplied by a separate circuit—an individual branch circuit. Not all individual branch circuits supply receptacles. For example, the individual branch circuit conductors are run directly to the furnace and not to a receptacle. Section E3603.1 requires that the furnace be supplied by an individual branch circuit regardless of the horsepower rating of the motor.

BRANCH CIRCUIT, MULTIWIRE. A branch circuit consisting of two or more ungrounded conductors having voltage difference between them, and a grounded conductor having equal voltage difference between it and each ungrounded conductor of the circuit, and that is connected to the neutral or grounded conductor of the system.

❖ See the commentaries for "Branch Circuit" and "Branch Circuit, Individual."

CABINET. An enclosure designed either for surface or flush mounting and provided with a frame, mat or trim in which a swinging door or doors are or may be hung.

❖ A cabinet is an enclosure such as a meter base. A panelboard enclosure is a cabinet, which is essentially a metal or plastic box that houses apparatus such as busbars, terminals, and overcurrent devices. In everyday usage in the field, many electricians and electrical contractors involved in residential wiring refer to the cabinet and panelboard together as simply the panel, or they use slang terms such as "breaker box" and do not necessarily distinguish between the cabinet and the panelboard as separate items. Perhaps this is because most panelboards used in one- and two-family dwellings come from the manufacturer already installed in a cabinet. For simplicity, the term "panelboard" is also used in this commentary in some places to refer to the cabinet and panelboard as a unit. However, it is important to know the definition for cabinet to understand code rules that apply specifically to the panelboard enclosure or other enclosures such as the meter base. See Section E3807 for the rules covering cabinets, which include positioning, mounting, and running cables and raceways into the cabinet. Pre-assembled cabinet and panelboard units comply with requirements for sizing and space such as providing sufficient wire bending space.

CIRCUIT BREAKER. A device designed to open and close a circuit by nonautomatic means and to open the circuit automatically on a predetermined overcurrent without damage to itself when properly applied within its rating.

❖ Circuit breakers are the most commonly used form of overcurrent protection in residential wiring. Except for service-entrance conductors and some tap conductors, the code requires that an overcurrent protection device be installed in series with each ungrounded conductor at the point where the conductor receives its supply. Conductors are rated to operate at maximum temperatures, and circuit breakers serve to open the circuit and protect the conductor if the temperature in the conductor rises to a level that would damage the conductor. Circuit breakers also protect conductors from short circuits and ground faults. A ground fault is current flowing to something that is grounded, such as a flow between a hot conductor and a metal enclosure. A short circuit is the flow of current between ungrounded conductors or between an ungrounded and a grounded conductor. The short circuit current is usually quite high and flows for a very short time. Circuit breakers must have an interrupting rating sufficient to stop the flow of current before damage occurs from excessive heating or arcing.

CONCEALED. Rendered inaccessible by the structure or finish of the building. Wires in concealed raceways are considered concealed, even though they may become accessible by withdrawing them. [See "Accessible (As applied to wiring methods)."]

❖ Wiring and equipment that is installed within the finished construction is considered concealed. Access to it would require the removal of finished construction. The wiring in a crawl space or attic is not concealed if one could get to the wiring to work on it. In some cases, the space in the attic and crawl space or some parts of the attic or crawl space of a dwelling is very limited. For example, where a junction box is visible from one part of a crawl space but the area is too small to actually crawl to the box or reach the box for working on it, is the junction box considered concealed? Junction boxes are considered concealed if a person would need to remove some finished construction to work on them.

CONDUCTOR

Bare. A conductor having no covering or electrical insulation whatsoever.

Covered. A conductor encased within material of composition or thickness that is not recognized by this code as electrical insulation.

Insulated. A conductor encased within material of composition and thickness that is recognized by this code as electrical insulation.

❖ An example of a covered conductor is the grounding conductor within the sheathing of Type NM cable. When the sheathing is stripped and the separate conductors are examined, it can be seen that this grounding conductor is covered with paper. This is not an insulated conductor. An improper use of this paper-covered bare conductor would be to use it as an ungrounded or grounded conductor. Insulated conductors are tested to meet standards for temperature and conditions of use and are marked as to the type of insulation. The conductor types most widely used in wiring of one- and two-family dwellings are listed in Section E3306.5. See Commentary Figure E3401(1).

CONDUIT BODY. A separate portion of a conduit or tubing system that provides access through a removable cover(s) to the interior of the system at a junction of two or more sections of the system or at a terminal point of the system. Boxes such as FS and FD or larger cast or sheet metal boxes are not classified as conduit bodies.

❖ A conduit body is a fitting designed to provide access to the wiring through a separate cover. It is used mostly where a 90-degree bend with a full radius would be difficult to install. Many different styles and shapes of conduit bodies are available for both metal and non-metallic conduit systems. Some of the different types are LB, where the opening is on the back; LL and LR, which are intended to lay flat against the surface and have the opening on the left or right side respectively; and T, which is a combination of an LL and an LR. A Type C conduit body is used in a straight run of conduit to provide an opening in the conduit run or to make a splice. A splice can be made in a conduit body if it is marked with the cubic inch capacity.

CONNECTOR, PRESSURE (SOLDERLESS). A device that establishes a connection between two or more conductors or between one or more conductors and a terminal by means of mechanical pressure and without the use of solder.

❖ This term applies to the common bolted-type pressure connector used for terminals in panelboards and other equipment where a screw or bolt binds against the end of a conductor. This term also applies to devices commonly referred to as "wire nuts," which are used connect two or more conductors. Wire nuts are a screw-on type of connector with an insulated cap made of plastic or other non-conductive material. Most wire nuts contain a spiral metallic spring or coil inside to bind the conductors together. Wire nuts are typically color-coded to indicate the maximum size and number of conductors for which the device is rated. These colors are not determined by the code and may be different for different manufacturers. It is necessary to use the proper size connector for the wires to be spliced. See Section 3306.10 for detailed information on making connections.

CONTINUOUS LOAD. A load where the maximum current is expected to continue for 3 hours or more.

❖ Continuous loads could result in higher conductor temperatures. Section E3601.2 requires that conductors for continuous loads be sized at 125 percent of the load.

COOKING UNIT, COUNTER-MOUNTED. A cooking appliance designed for mounting in or on a counter and consisting of one or more heating elements, internal wiring, and built-in or separately mountable controls.

❖ This definition must be considered when calculating the feeder neutral load for cooking appliances as prescribed in Section E3604.4. See also Table E3502.2 and Section E3602.9.

COPPER-CLAD ALUMINUM CONDUCTORS. Conductors drawn from a copper-clad aluminum rod with the copper metallurgically bonded to an aluminum core. The copper forms a minimum of 10 percent of the cross-sectional area of a solid conductor or each strand of a stranded conductor.

❖ Copper-clad aluminum conductors were used in non-metallic sheathed cable in the late 1950s and early 1960s. It is important to know the terminating and splicing provisions for repair and retrofitting work that involves older dwellings that are wired with copper-clad aluminum conductors.

CUTOUT BOX. An enclosure designed for surface mounting and having swinging doors or covers secured directly to and telescoping with the walls of the box proper. (See "Cabinet.")

❖ This is a variation of a cabinet defined above, which is an enclosure for electrical devices and equipment such as fuses and circuit breakers, switches, current transformers, etc. A cutout box may be used for a motor disconnect switch with fuses for overcurrent protection. A junction box has a cover attached with screws or bolts, but a cutout box has a swinging door with a hinge. A cutout box is made to contain devices that may need to be worked on, inspected or maintained; the hinged cover makes it easier to get into than a junction box.

DEAD FRONT. Without live parts exposed to a person on the operating side of the equipment.

❖ The door or cover of a cabinet or cutout box is a "dead front" since the live parts within are not exposed to a person standing in front of a panelboard or equipment. This term is used in measuring the working clearances for electrical equipment. The clearance from the dead front or front of the enclosure of the cabinet for a panelboard is a minimum of 36 inches (914 mm).

DEMAND FACTOR. The ratio of the maximum demand of a system, or part of a system, to the total connected load of a system or the part of the system under consideration.

❖ Demand factors are used in the calculations for various loads such as service and feeder loads. Demand factors are expressed as a percent of the calculated load and are used when there are multiple loads that are not likely to be used at the same time. Feeders are not sized to have the current-carrying capacity to supply the total current for all the loads at once. For example, for certain fastened-in-place appliances in a dwelling, the total nameplate rating load must be calculated at 100 percent if the number of appliances is three or fewer. But if there are four or more, then the load is calculated at 75 percent of the total load of the ap-

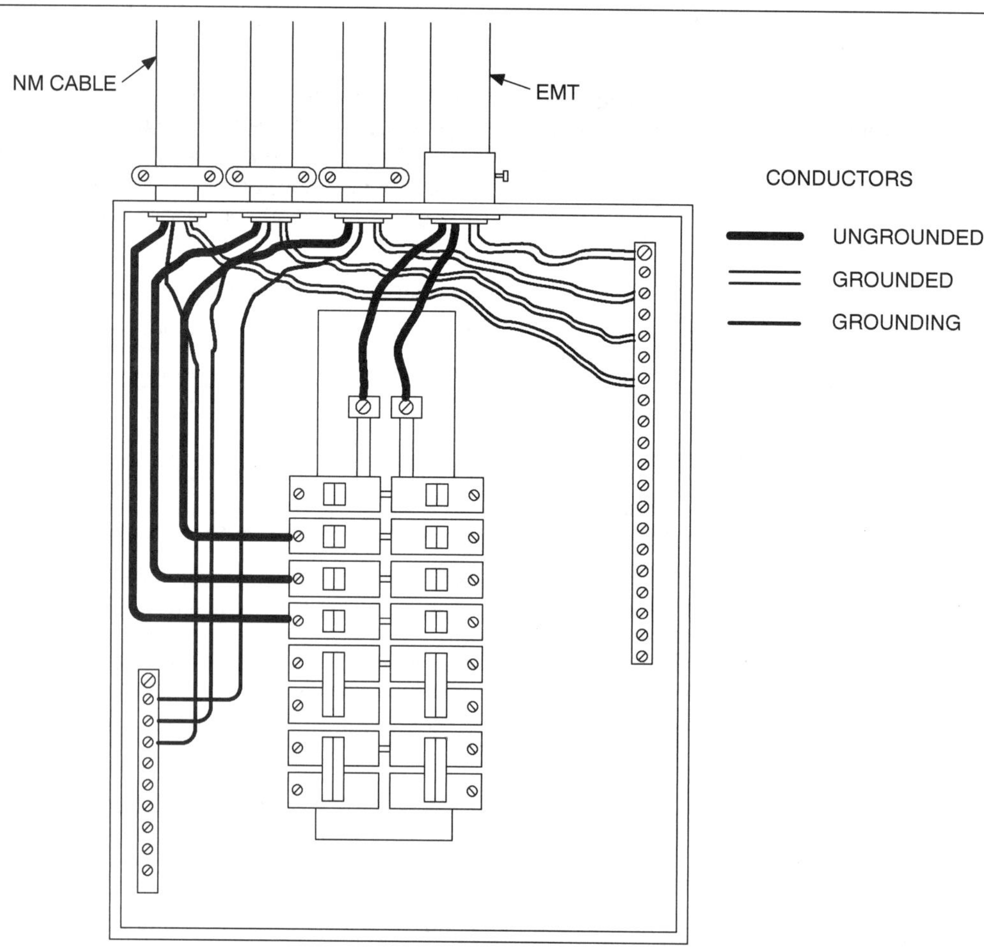

Figure E3401(1)
CONDUCTOR TERMINOLOGY

pliances. It is unlikely that all four of the appliances would be used at full load at the same time. See Tables E3502.2 and E3604.3(1).

DEVICE. A unit of an electrical system that is intended to carry but not utilize electric energy.

❖ A switch, receptacle, circuit breaker, fuse, meter, and thermostat are all examples of a "device" since they do not use or consume electrical energy; they just control or carry the current.

DISCONNECTING MEANS. A device, or group of devices, or other means by which the conductors of a circuit can be disconnected from their source of supply.

❖ This term is usually used in conjunction with, for example, the service, appliances, or motors, rather than for a lighting fixture, even though technically a switch is the disconnecting means for the light. Code provisions for disconnecting means are very specific for such things as identification, location, type, and ratings. The disconnection of power to a circuit is an obvious safety issue.

DWELLING

Dwelling unit. One or more rooms for the use of one or more persons as a housekeeping unit with space for eating, living and sleeping, and permanent provisions for cooking and sanitation.

One-family dwelling. A building consisting solely of one dwelling unit.

Two-family dwelling. A building consisting solely of two dwelling units.

❖ A guest room in a hotel or motel is not a dwelling unit unless it contains permanent provisions for cooking. A portable microwave is not permanent, so even a motel room with a kitchenette is not a dwelling unit unless the cooking equipment is a built-in range or cook top. Since this code covers only one- and two-family dwellings and townhouses, the definition for a multiple-family dwelling is not included. See Section R101.2.

ENCLOSED. Surrounded by a case, housing, fence or walls that will prevent persons from accidentally contacting energized parts.

❖ This is a general term that applies to enclosing things such as devices inside a cabinet. It can also apply to equipment inside a room such as an electrical equipment room.

ENCLOSURE. The case or housing of apparatus, or the fence or walls surrounding an installation, to prevent personnel from accidentally contacting energized parts or to protect the equipment from physical damage.

❖ An enclosure contains electrical equipment. All enclosures, from recessed light fixtures to outdoor panelboards, are evaluated by a testing agency to designate proper use and location. The designation is marked on or comes with the equipment. Examples of types of enclosures are Type 1 for general indoor use and Type 3R for outdoor use. A Type 3R enclosure provides a degree of protection against rain, snow, and sleet.

ENERGIZED. Electrically connected to a source of voltage.

❖ Energized conductors and devices have a voltage impressed upon them. Some jurisdictions require that a house be energized in order for an inspection to be completed. Often, temporary power can be connected to the house for inspection purposes to check such things as proper polarity of receptacles and the proper functioning of the GFCI receptacles.

EQUIPMENT. A general term including material, fittings, devices, appliances, luminaires, apparatus and the like used as a part of, or in connection with, an electrical installation.

❖ This is a general catchall term used to refer to anything from large items such appliances, air conditioning units, cabinets, panelboards, and meter bases to items such as a 4-inch-square junction box or receptacle. Note that equipment is defined differently in Chapter 2.

EXPOSED. (As applied to live parts.) Capable of being inadvertently touched or approached nearer than a safe distance by a person. It is applied to parts not suitably guarded, isolated or insulated.

❖ The term "exposed" live parts refers to the uninsulated portion of conductors, terminals, or devices that are energized and that are not guarded or enclosed. The conductors and overcurrent devices of a panelboard in a cabinet are exposed when the cover is removed.

EXPOSED. (As applied to wiring methods.) On or attached to the surface or behind panels designed to allow access.

❖ Wiring methods that are not concealed in the construction of the dwelling are considered exposed. For example, non-metallic sheathed cable is permitted to be run exposed where not subject to physical damage. The provisions of Section E3702.3 cover the installation of exposed cables.

EXTERNALLY OPERABLE. Capable of being operated without exposing the operator to contact with live parts.

❖ Circuit breakers, for example, are externally operable because a dead front cover is installed over the panelboard containing the breakers so that the operator will not be exposed to live parts.

FEEDER. All circuit conductors between the service equipment, or the source of a separately derived system, or other power supply source and the final branch-circuit overcurrent device.

❖ In residential wiring systems, a feeder is usually a large set of conductors run in cable or in a raceway from the service equipment to the main panel or to a sub-panel, which has the final branch circuit overcurrent devices. A feeder carries power to a distribution center or panel that is the origination point for one or more branch circuits. The arrangement and layout of the service equipment can vary, and some houses may not have feeder conductors. Where the meter base, main disconnect, and distribution panel are combined into one unit or piece of equipment, this unit may be referred to in common usage or slang as a meter-main combination panel. The main disconnect and all the branch circuit overcurrent devices are located in this equipment, usually on the outside of the house. If all the circuit breakers except for the main disconnect are for branch circuits, and there are no sub-panels anywhere else in or on the house, there is no feeder.

Another arrangement may be where a meter base only is on the outside of the house and the service-entrance conductors are run to the meter base and continue on to a panelboard inside the house. This panel inside the house may be called, in common usage (but not in the code), a main breaker panel since the main breaker is installed in or is part of the panelboard along with the final branch circuit overcurrent devices. Again, if there are no sub-panels where additional final overcurrent devices for branch circuits would be installed, there would be no feeder.

Another arrangement of service equipment is where the meter base and main disconnect are located on the outside of the house and the panelboard is separate. The conductors that are run from the main disconnect to the distribution panel are the main power feeder. The main disconnect may consist of up to six switches (which are usually circuit breakers), and the conductors from any of these switches or circuit breakers which feed a distribution panel are feeders. See Section E3606.3. A large house may have two or more distribution panels where final branch circuit overcurrent devices are installed. For example, the conductors from the main service to a panel for a detached garage and to a panel serving swimming pool equipment are feeder conductors. All of the main disconnect switches, which could number up to six, do not necessarily supply feeder conductors. For example, one of the six main disconnects at the service equipment may serve an air conditioning condenser unit near the service equipment. These conductors would be branch circuit conductors, not feeder conductors.

FITTING. An accessory such as a locknut, bushing or other part of a wiring system that is intended primarily to perform a mechanical rather than an electrical function.

❖ Fittings connect cables, raceways, boxes and other equipment together to form a complete wiring system for the wiring method used. Wiring methods for dwellings include cable and raceway systems along with the associated boxes and hardware. Fittings include accessories such as cable and conduit connectors, couplings, adapters, reducers, clamps, bushings, locknuts, nipples, knock-out seals, conduit bodies, capped elbows, pipe and rod ground clamps, brackets, hubs, and service heads.

GROUND. A conducting connection, whether intentional or accidental, between an electrical circuit or equipment and the earth, or to some conducting body that serves in place of the earth.

❖ A "ground" is a conducting connection. Two examples of how this term is used are "ground clamp" and "ground bus," which refer to the conducting connection. This term is not used by the code to describe a wire or conductor. To say the "ground conductor" or the "ground wire" may lead to miscommunication. In talking about a conductor, it is helpful to use either "grounded conductor" or "grounding conductor." In some publications the term "ground wire" is used to indicate the "grounding wire." The terms used to refer to ground and grounding have very specific meanings and should be understood and used properly. "Ground wire" usually means the grounding conductor, but someone could possibly use it to mean the grounded conductor. Many informal words and phrases are used in the electrical trade to describe materials, devices, equipment, and conductors. Although not defined in the code or other publications, these informal terms are understood by all individuals with at least some experience. Some terms, however, should be used according to the code definition to make communication easier.

GROUNDED. Connected to earth or to some conducting body that serves in place of the earth.

❖ Grounded conductors are intentionally connected to the earth through a ground rod, water service pipe, or other grounding electrode.

GROUNDED, EFFECTIVELY. Intentionally connected to earth through a ground connection or connections of sufficiently low impedance and having sufficient current-carrying capacity to prevent the buildup of voltages that may result in undue hazards to connected equipment or to persons.

❖ Grounding an object prevents it from being at a voltage potential different from the earth. An electrical system must be effectively grounded by connecting it to the earth in a manner that will provide a current-carrying path of sufficient capacity to limit the voltage to ground (earth), limit the voltage imposed by a lightning strike or a line surge and provide a reference level for the normal system voltage. See Section E3507 for more information on grounding.

GROUNDED CONDUCTOR. A system or circuit conductor that is intentionally grounded.

❖ The grounded conductor of the service, feeder, or branch circuits is commonly called the "neutral" and is required to be white or gray. Although the word "neutral" is not included in this chapter and is seldom used in the code, in trade jargon it is used interchangeably with "grounded conductor." The term "neutral" is used in Section E3604.4 regarding the load on a feeder. The code requires that one of the three service conductors supplying the dwelling be grounded for the reasons listed under "Effectively Grounded," above. Electrical power is supplied to one- and two-family dwellings from a single-phase, 3-wire transformer, the midpoint of which is grounded at the transformer. The conductor that is connected to the midpoint of the transformer winding is the grounded conductor. The other two service conductors are ungrounded and have a potential difference of 240 volts between them. The grounded conductor has a potential difference of 120 volts between it and each ungrounded conductor. The grounded conductor is the neutral conductor of the system and carries only the unbalanced current between the two phase conductors. According to Section E3507.1, the grounded conductor is grounded at the service disconnect location through its connection to the grounding electrode conductor, which is run to an acceptable grounding electrode. See Section E3508 for the "Grounding Electrode System."

GROUNDING CONDUCTOR. A conductor used to connect equipment or the grounded circuit of a wiring system to a grounding electrode or electrodes.

❖ The grounding conductor is the green or bare wire. For the premises wiring system or the service, it is the grounding electrode conductor, and for feeders and branch circuits it is the equipment grounding conductor. Some electricians and some publications call it the "ground wire." For code terminology, it should be called the "grounding electrode conductor" or the "equipment grounding conductor" as defined below.

GROUNDING CONDUCTOR, EQUIPMENT. The conductor used to connect the noncurrent-carrying metal parts of equipment, raceways and other enclosures to the system grounded conductor, the grounding electrode conductor or both, at the service, at each building or structure where supplied from a common service, or at the source of a separately derived system.

❖ The equipment grounding conductor is not intended to carry current except in case of a ground fault. It must be sized according to the rating of the overcurrent device serving the feeder or branch circuit. Section E3808 covers equipment grounding. All of the noncurrent carrying metallic parts of the electrical wiring system are tied together to form a continuous low-impedance current path for any fault current that could occur. Equipment is grounded so that in case of a failure in the insulation or for any other reason the metal equipment becomes energized, current will flow along the equipment grounding conductor path back to the source and trip the circuit breaker or fuse. In one- and two-family dwellings, equipment is grounded with the equipment grounding conductor, which is the green or bare conductor in the non-metallic sheathed cable run for the branch circuits. Or, where metal conduit or tubing (such as electrical metallic tubing or EMT) is used, the code permits the conduit or tubing itself to be the equipment grounding conductor. There are several conditions in Section E3808.8 for using flexible metal conduit itself as the equipment grounding conductor. In actual practice, it is easier to install the green or bare equipment grounding conductor. The equipment grounding conductor is connected to the grounded conductor only at the service equipment. It must not be tied to the grounded conductor or neutral conductor path at any point downstream from the service equipment. See Section E3507.2.

GROUNDING ELECTRODE CONDUCTOR. The conductor used to connect the grounding electrode(s) to the equipment grounding conductor, to the grounded conductor, or to both, at the service equipment or at the source of a separately derived system.

❖ The grounding electrode conductor is a copper, aluminum, or copper-clad aluminum conductor, and it is not intended to carry current except in case of a ground fault. It can be bare, and the code does not specify a color for the grounding electrode conductor. The grounding electrode conductor is sized based on the size of the largest service-entrance conductor as shown in Table E3503.1.

The grounding electrode conductor is connected to the grounded service conductor at any accessible point between the load end of the service drop or service lateral and the main disconnect. This connection is usually made in the service equipment where the main disconnect is located or in the meter enclosure. The grounding electrode conductor and the grounded service conductor are bonded together and to the metal service enclosure by the main bonding jumper. The equipment grounding conductors are also brought to this point and tied to the grounded service conductor and to the grounding electrode conductor. The other end of the grounding electrode conductor is connected to a grounding electrode such as underground water pipe and ground rod.

GROUND-FAULT CIRCUIT-INTERRUPTER. A device intended for the protection of personnel that functions to de-energize a circuit or portion thereof within an established period of time when a current to ground exceeds the value for a Class A device.

❖ A ground-fault circuit-interrupter (GFCI) is a device that will open a circuit when current flows outside the circuit. A current flowing outside the circuit conductors could occur, for example, from a damaged conductor that could create an outside path for current to flow to a grounded object. The small current flow may not be enough to trip the normal circuit breaker. Or, it could be a low- resistance current path through an individual who touches something like a cord, box, wire, or tool surface that has become "hot" or energized by a ground fault. It has been determined that even a small current flow of 6 milliamperes could be harmful to people under certain circumstances.

The GFCI device contains a coil through which both circuit conductors are routed. If between 4 and 6 milliamperes of current is flowing outside the circuit conductors, this current transformer coil senses the unbalance of current in the two branch circuit conductors and generates a small current that activates the trip mechanism and opens the circuit. A GFCI receptacle provides only ground-fault protection, and the GFCI circuit breaker provides overcurrent protection as well.

GUARDED. Covered, shielded, fenced, enclosed or otherwise protected by means of suitable covers, casings, barriers, rails, screens, mats or platforms to remove the likelihood of approach or contact by persons or objects to a point of danger.

❖ The code requires live parts to be guarded against accidental contact. Where live parts are installed inside approved boxes, cabinets, and other enclosures, they are considered guarded.

IDENTIFIED. (As applied to Equipment.) Recognizable as suitable for the specific purpose, function, use, environment, application, etc., where described in a particular code requirement.

❖ Equipment that is identified for a specific purpose is listed and labeled or marked. In this way, it is recognizable as suitable for the use and for the conditions in which it is installed. For example, a cable that is suitable for direct burial underground is identified for this use by its listing and labeling as Type UF (Underground Feeder) cable. Another example is the disconnect switch installed on the outside of a house for an air conditioner. It must be identified for outdoor use where it could be subject to rain or snow. A Type 3R enclosure for this disconnect switch would survive the usual outdoor weather conditions. An installer and inspector can rely on the identification of equipment in its listing and labeling since a nationally recognized testing laboratory has tested it.

INTERRUPTING RATING. The highest current at rated voltage that a device is intended to interrupt under standard test conditions.

❖ This rating is the highest current that can flow through a circuit breaker or fuse without damage to the device and its enclosure. This is not the overcurrent rating, such as the identification of a fuse as a 15-ampere or 20-ampere fuse. Where a short circuit occurs as two ungrounded conductors touch, or where a ground faults occurs as an ungrounded conductor comes in contact with a grounded conductor, a very high current can flow. This can damage the overcurrent device and the equipment in which it is mounted. Circuit breakers and fuses are designed to trip without being damaged provided the rating is high enough for the current imposed on the device. The amount of short-circuit or ground-fault current available must be calculated and the devices rated accordingly. Most circuit breakers are made with an interrupting rating of 10,000 amperes. See Sections E3304.2 and E3304.3 for more information.

ISOLATED. (As applied to location.) Not readily accessible to persons unless special means for access are used.

❖ This term may be used to refer to some wiring method or equipment that is not readily accessible.

Switches or devices could be considered isolated if they are located inside a locked cabinet. This term is related to location and does not relate to usage such as isolated grounding receptacles.

LABELED. Equipment or materials to which has been attached a label, symbol or other identifying mark of an organization acceptable to the authority having jurisdiction and concerned with product evaluation that maintains periodic inspection of production of labeled equipment or materials and by whose labeling the manufacturer indicates compliance with appropriate standards or performance in a specified manner.

❖ The building official must approve the materials and equipment installed in an electrical wiring system. A label of a nationally recognized testing laboratory (NRTL) identifies a product or material and provides other information that the building official uses to determine compliance with the code. Equipment and products that have been labeled have been tested for conformance to a standard and subject to third-party inspection, which verifies that the minimum level of quality required by the appropriate standard is maintained. The labeling agency performing the third-party inspection must be approved by the building official, and the basis for this approval may include, but is not necessarily limited to, the capacity and capability of the agency to perform the specific testing and inspection.

The label will have the name and/or symbol of the NRTL and will contain the product category and type of the equipment or product, which has its proper use and application described in the listing publication from the NRTL. The applicable reference standard often states the minimum identifying information that must be on a label. Labels or marks are applied to the tested product by stamping, molding, affixing a sticker, affixing a printed decal, or other similar means. If the product is too small or has a shape, texture, or is of a material that makes the attachment of a label difficult, the product container will have the label or mark. For this purpose, the installer should make available the product container for the inspector if requested. The product container might have, in addition to the label, installation instructions for the product.

LIGHTING OUTLET. An outlet intended for the direct connection of a lampholder, a luminaire (lighting fixture) or a pendant cord terminating in a lampholder.

❖ An outlet box designed for the attachment of a lamp holder, lighting fixture, or a pendant cord terminating in a lamp holder is installed at a lighting outlet, which is an opening in the wiring system. The word "opening" is not defined in the code but is commonly used in the trade. It is the point where junction boxes, device boxes, or boxes for the attachment of lighting fixtures are installed. Lighting outlets are required in specific locations in a dwelling as covered in Section E3803.1.

LISTED. Equipment, materials, or services included in a list published by an organization that is acceptable to the authority having jurisdiction and concerned with product evaluation of products or services, that maintains periodic inspection of production of listed equipment or materials or periodic evaluation of services, and whose listing states either that the equipment, material, or services meets identified standards or has been tested and found suitable for a specified purpose.

❖ Nationally recognized testing laboratories (NRTL's) publish lists of electrical materials, components, devices, fixtures and equipment that they have subjected to standardized tests and found suitable for installation and use in a specific manner. The listing information is necessary if there is a question as to the suitability of the product or material for a specific installation. Section E3303.3 requires the listing and labeling of electrical materials, components, devices, fixtures and equipment.

LIVE PARTS. Energized conductive components.

❖ Live parts must be enclosed or guarded to prevent people from receiving electric shock. For example, live parts should not be left exposed during the construction of the dwelling. When the cover (dead front) of a panelboard is removed, live parts will be exposed if the panelboard is energized. Cabinet doors and junction box covers are usually not installed until the finish phase of the project, but means should be taken where necessary to enclose or guard live parts.

LOCATION, DAMP. Location protected from weather and not subject to saturation with water or other liquids but subject to moderate degrees of moisture. Examples of such locations include partially protected locations under canopies, marquees, roofed open porches and like locations, and interior locations subject to moderate degrees of moisture, such as some basements, some barns and some cold-storage warehouses.

LOCATION, DRY. A location not normally subject to dampness or wetness. A location classified as dry may be temporarily subject to dampness or wetness, as in the case of a building under construction.

LOCATION, WET. Installations underground or in concrete slabs or masonry in direct contact with the earth and locations subject to saturation with water or other liquids, such as vehicle-washing areas, and locations exposed to weather.

❖ The designation of certain areas as a type of location is very important in determining what equipment and materials are permitted to be installed at a specific place in the dwelling. The location must be considered when selecting almost any material such as boxes, cabinets, conductors, lighting fixtures, cables and raceways.

It may seem apparent that all locations inside a dwelling would be "dry locations" for the purpose of installing wire and materials. But, as the code states, some basements may be damp locations since moisture could seep through concrete or masonry walls and since water may be present in some basements. In most cases it is evident that an area might be subject to dampness, and the appropriate wiring methods must be used. If the classification of location is questionable, the code official would make the determination.

All conductors used in a wet location should be rated for such and must have a "W" in the marking of the conductor type such as THW and THWN. The "W" indicates that it is approved for use in a wet location. The inside of a conduit where the conduit is installed underground is a wet location because of the chance that water could accumulate there. Conductors installed in underground conduits must have the "W" rating.

LUMINAIRE. A complete lighting unit (lighting fixture) consisting of a lamp or lamps together with parts designed to distribute the light, to position and protect the lamps and ballast, where applicable, and to connect the lamps to the power supply.

❖ The term "luminaire," pronounced lumen-air, has replaced the term "lighting fixture."

MULTIOUTLET ASSEMBLY. A type of surface, or flush, or freestanding raceway; designed to hold conductors and receptacles, assembled in the field or at the factory.

❖ A multioutlet assembly is a raceway type of enclosure that is a long narrow strip with single receptacles spaced at regular intervals such as every 6 inches. An 8-foot-long multioutlet assembly, for example, could be installed on the wall above a workbench where it would be convenient to have several tools plugged in at once. It is designed to be mounted directly on the surface of the wall and is either factory-made or built on the job. It may come as a channel with fittings for couplings, connectors, corner installations, and a snap-on cover that holds the receptacles. It could be direct wired by mounting over a receptacle outlet box. If it were used on a garage wall over a workbench, the receptacles would require GFCI protection. A multi-outlet assembly may be used in a kitchen where several appliances need to be plugged in or at a desk or built-in counter where there is a need to plug in computers and related equipment.

OUTLET. A point on the wiring system at which current is taken to supply utilization equipment.

❖ An "outlet" is not a junction box or any device box such as a switch box. A switch is installed in a device box, and it may be referred to by the electrician as an opening in the wiring system, but it is not an outlet since power is not taken from the distribution system at this point. A box where a receptacle or lighting fixture will be installed is an outlet.

OVEN, WALL-MOUNTED. An oven for cooking purposes and consisting of one or more heating elements, internal wiring, and built-in or separately mountable controls.

❖ The name-plate volt-ampere ratings of ovens and ranges are used for making load calculations. Since electric cooking equipment can be a considerably large load in amperes, this specific definition is included in the code. Certain demand factors can be applied where a number of ovens and/or ranges are used in a dwelling.

OVERCURRENT. Any current in excess of the rated current of equipment or the ampacity of a conductor. Such current might result from overload, short circuit or ground fault.

❖ Conductors have a rated current which is the allowable ampacity. An overcurrent is a current in excess of the rated current. Overcurrent is caused by overloads and short circuits or ground faults. An overcurrent will cause the temperature of the conductor and/or equipment to rise to a level that could cause damage. In some cases, an overcurrent can exist for a certain length of time without causing damage. For installation in these cases, some overcurrent protection devices are built with a time delay so they will not trip instantaneously.

OVERLOAD. Operation of equipment in excess of normal, full-load rating, or of a conductor in excess of rated ampacity that, when it persists for a sufficient length of time, would cause damage or dangerous overheating. A fault, such as a short circuit or ground fault, is not an overload.

❖ Heat is produced as a result of an overload, which is due to a level of current above the rating of the conductors. The severity of the overload is determined by (1) the amount of current, (2) the resistance of the conductor and (3) the amount of time the overload persists. Overloading a conductor results in overcurrent that can cause conductor and device damage, excessive voltage drop, and if severe enough, can cause a fire.

PANELBOARD. A single panel or group of panel units designed for assembly in the form of a single panel, including buses and automatic overcurrent devices, and equipped with or without switches for the control of light, heat or power circuits, designed to be placed in a cabinet or cutout box placed in or against a wall or partition and accessible only from the front.

❖ A panelboard contains the final branch-circuit overcurrent devices, which are the points where the branch circuits originate. The panelboard is an assembly of busbars and mounting hardware designed to be enclosed in a cabinet or cut-out box. In residential wiring, the panelboard is usually not distinguished as a separate piece of equipment from its enclosure since it is normally installed inside the cabinet at the factory. Panelboards intended for flush mounting in the wall have adjusting screws so that the dead front or cover can be set to the right depth.

PLENUM. A compartment or chamber to which one or more air ducts are connected and that forms part of the air distribution system.

❖ A plenum is a ductwork component that serves to connect multiple ducts together and is associated with furnaces and air handlers. See Section E3804.7, which addresses cavities within the building construction.

POWER OUTLET. An enclosed assembly that may include receptacles, circuit breakers, fuseholders, fused switches, buses and watt-hour meter mounting means, intended to supply and control power to mobile homes, recreational vehicles or boats, or to serve as a means for distributing power required to operate mobile or temporarily installed equipment.

❖ A pedestal-mounted power outlet is an assembly supplied by an underground feeder or service lateral. A power outlet can also be mounted on a pole or other surface and supplied by an overhead service drop. It has receptacles and overcurrent devices installed in a weatherproof cabinet or box, and can include a meter. It can also be used to provide temporary power during construction.

PREMISES WIRING (SYSTEM). That interior and exterior wiring, including power, lighting, control and signal circuit wiring together with all of their associated hardware, fittings and wiring devices, both permanently and temporarily installed, that extends from the service point of utility conductors or source of power such as a battery, a solarphotovoltaic system, or a generator, transformer, or converter winding, to the outlet(s). Such wiring does not include wiring internal to appliances, luminaires (fixtures), motors, controllers, and similar equipment.

❖ The premises wiring system includes all of the installed conductors and equipment inside and outside the dwelling. This definition makes it clear that the premises wiring system begins at the service point. For an overhead service, this is at the load end of the service drop or at the service head. For an underground service, it is at the load end of the service lateral. All of the wiring and equipment from that point on is covered by this code except internal wiring of appliances and factory-made equipment.

QUALIFIED PERSON. One who has the skills and knowledge related to the construction and operation of the electrical equipment and installations and has received safety training on the hazards involved.

❖ The code may limit access to, operation of or installation of certain equipment to persons who have the ability to work safely with such equipment. For example, NFPA 70 will relax some safety provisions where it can be demonstrated that only qualified persons will be working on or around certain equipment.

RACEWAY. An enclosed channel of metal or nonmetallic materials designed expressly for holding wires, cables, or busbars, with additional functions as permitted in this code. Raceways include, but are not limited to, rigid metal conduit, rigid nonmetallic conduit, intermediate metal conduit, liquid-tight flexible conduit, flexible metallic tubing, flexible metal conduit, electrical nonmetallic tubing, electrical metallic tubing, underfloor raceways, cellular concrete floor raceways, cellular metal floor raceways, surface raceways, wireways and busways.

❖ Table E3701.2 lists the types of raceways used in residential wiring. Raceways include any channel that encloses conductors including conduit and tubing. The term "conduit" is used in the descriptive name of various types of conduit and is different from "tubing." Electrical metallic tubing (EMT), for example, is not conduit. This distinction is important in relation to the rules of the support of boxes with devices and lighting fixtures. The use of EMT for this support is not permitted, but the use of conduit is. See Sections E3806.8.4 and E3806.8.5.

RAINPROOF. Constructed, protected or treated so as to prevent rain from interfering with the successful operation of the apparatus under specified test conditions.

❖ The definition of "rainproof" is similar to the definition of "weatherproof" in that the box, enclosure, or fitting, is made and installed so that rain will "not interfere with the successful operation" of the contained apparatus. When the enclosure is rainproof, it has been tested to ensure that exposure to a beating rain will not interfere with the operation or result in the live parts and wiring within the enclosure becoming wet. It does not mean that the enclosure is sealed against the entry of water.

RAIN TIGHT. Constructed or protected so that exposure to a beating rain will not result in the entrance of water under specified test conditions.

❖ The definition of "rain tight" is similar to the definition of "water tight," which concerns the entrance of water. The testing of equipment designed to be rain tight simulates exposure to a beating rain, and the result must be that water will not enter the enclosure. For example, the weatherhead or service head must be "rain tight." See Section E3505.9.1.

RECEPTACLE. A receptacle is a contact device installed at the outlet for the connection of an attachment plug. A single receptacle is a single contact device with no other contact device on the same yoke. A multiple receptacle is two or more contact devices on the same yoke.

❖ The distinction between "outlet" and "receptacle" is important. The outlet is the point on the wiring system where a a device box and receptacle are installed. A receptacle is the device installed in the box at the outlet. A receptacle can be single, duplex or multiple. The term "yoke" is not defined in this chapter but is used to describe the strap on which the receptacle is mounted and that is made to attach to the device box. A strap or yoke can hold a single device or multiple devices.

Chapter 38 of this code requires receptacle outlets to be installed at specific locations. Duplex receptacles are typical, however, there are some situations in which a single receptacle is required, for example where a food freezer is to be located in a garage. Receptacles in the garage walls must be GFCI protected, and an exception is made for the freezer since it is not readily moved and occupies the single receptacle provided for the freezer. Where a freezer (or other such appliance) is to be installed in a dedicated space in the garage, a single receptacle is installed that is non-GFCI protected. If a duplex receptacle were installed, an extension cord could be plugged into the other receptacle on that yoke and used to power tools that would require GFCI protection.

RECEPTACLE OUTLET. An outlet where one or more receptacles are installed.

❖ See the commentary for "Receptacle."

SERVICE. The conductors and equipment for delivering energy from the serving utility to the wiring system of the premises served.

❖ Service is a general term used to refer to all of the service conductors and equipment. A key part of this definition is that the service begins at the load end of the service drop or lateral and ends at the enclosure that houses the main disconnecting means.

SERVICE CABLE. Service conductors made up in the form of a cable.

❖ Service conductors for dwellings are installed in a raceway or in a cable assembly called the service-entrance cable, although almost all modern services are separate conductors in a raceway. Cable types SE and USE are service entrance cables.

SERVICE CONDUCTORS. The conductors from the service point or other source of power to the service disconnecting means.

❖ It is clear that service conductors include service-entrance conductors and any other conductors up to the main disconnect. This definition typically excludes the service drop or service lateral conductors, because such conductors are on the supply side of the service point. What is included and what is excluded depends on where the service point is. See the definition of "Service Point."

SERVICE DROP. The overhead service conductors from the last pole or other aerial support to and including the splices, if any, connecting to the service-entrance conductors at the building or other structure.

❖ Service drop conductors run from the overhead utility line or pole and are connected at the weatherhead to the service-entrance conductors. They are usually owned and maintained by the utility company. Where utility owned, the size of the service drop conductors is decided by the utility company and is not dictated by this code.

SERVICE-ENTRANCE CONDUCTORS, OVERHEAD SYSTEM. The service conductors between the terminals of the service equipment and a point usually outside the building, clear of building walls, where joined by tap or splice to the service drop.

❖ Overhead service-entrance conductors run from the weatherhead to the main disconnect. The layout and configuration of the service equipment may be such that only a meter is on the outside of the house with the conductors run to the main disconnect on the inside of the house. The code does not dictate that the main disconnect be on the outside of the house except as might be required by Section E3501.6.2. An electric meter installed between the weatherhead and the main disconnect is not relevant. The service-entrance conductors continue inside the house to the main disconnect. Where the meter is in the same box or enclosure as the main disconnect, the service-entrance conductors would only run from the weatherhead down (in raceway or cable) to the meter-main enclosure.

SERVICE-ENTRANCE CONDUCTORS, UNDERGROUND SYSTEM. The service conductors between the terminals of the service equipment and the point of connection to the service lateral.

❖ For an underground service, the service-entrance conductors connect to the service lateral and run to the terminals of the service equipment. Where the meter-main combination cabinet is on the outside of the house, there are no service-entrance conductors because the service lateral is defined as the conductors from the utility supply to the first terminal box, meter or enclosure at the structure. If the first enclosure is the location of the main disconnect, which could be an enclosure with a meter and the main disconnect together, then the service-lateral conductors connect at the terminals of this equipment, and there are essentially no service-entrance conductors. In this installation, if the panelboard were also in the same enclosure, there would be no feeder to supply the panelboard, and the installer would not need to calculate the size of the service-entrance conductors. If the panelboard were on the inside of the house, then a feeder would be installed to serve the panelboard, and the size of feeder would need to be calculated.

SERVICE EQUIPMENT. The necessary equipment, usually consisting of a circuit breaker(s) or switch(es) and fuse(s), and their accessories, connected to the load end of the service conductors to a building or other structure, or an otherwise designated area, and intended to constitute the main control and cutoff of the supply.

❖ See the commentary for "Service."

SERVICE LATERAL. The underground service conductors between the street main, including any risers at a pole or other structure or from transformers, and the first point of connection to the service-entrance conductors in a terminal box or meter or other enclosure, inside or outside the building wall. Where there is no terminal box, meter or other enclosure with adequate space, the point of connection shall be considered to be the point of entrance of the service conductors into the building.

❖ The utility company may have an underground supply system with transformers on pads at or near the property line. The service lateral runs underground to the building from a transformer or connection box installed at the ground level. Where there is an overhead transmission line, the homeowner may not want a service drop (overhead line) run across the property. In such cases, the conductors can be run down a utility pole in a raceway and then run underground to the house. Both of these installations would be called a service lateral, which is defined as the conductors from the utility main line to the point at which they connect to the service equipment or service-entrance conductors at the house. The service lateral typically terminates at a meter or main disconnect enclosure.

Usually the utility company is responsible for maintaining the conductors of the service lateral.

The electrician or contractor is usually responsible for installing a raceway conductor enclosure from the service equipment down to the required burial depth of the lateral.

SERVICE POINT. Service point is the point of connection between the facilities of the serving utility and the premises wiring.

❖ The service point is determined by the electric utility company. This is the point of connection between the electric-utility-owned conductors and the premises wiring-system conductors. If the overhead service drop is installed and owned by the utility company, then the service point would be at the connection of the service-entrance conductors to the service drop at a point close to the the weatherhead. If the utility company owns the underground service lateral conductors, then the service point would be where the service lateral conductors connect to the terminals of the service equipment.

STRUCTURE. That which is built or constructed.

❖ Structures include buildings, houses, towers, fences, sheds and swimming pools.

SWITCHES

General-use snap switch. A form of general-use switch constructed so that it can be installed in device boxes or on box covers or otherwise used in conjunction with wiring systems recognized by this code.

❖ General-use snap switches come from the factory mounted on a yoke or strap, the same as a receptacle yoke, and are installed in a device box with two screws. They also may be used in other ways such as attached to the raised cover of a four-square metal box. General-use snap switches are the common toggle switches used throughout the dwelling for the majority of the switching needs. They are installed to control basically all of the lighting fixtures and also motor loads such as the garbage disposal and pump motors. The switches are rated in amperes and voltage, and these ratings are stamped or marked on the switch. For example, the yoke of the switch might have stamped on it "15A, 120VAC." It is important to use these switches within their ratings.

AC general-use snap switches are designed and permitted to be used up to a maximum of the full current rating of the switch to control (1) resistive loads such as incandescent lighting and heating loads, (2) inductive loads such as fluorescent lighting and (3) tungsten-filament lamp loads of 120 volts. For example, an AC general-use snap switch rated at 15 amperes can control a resistive load of 15 amperes. For stationary motors of 2 horsepower or less, an AC general-use snap switch is permitted as the disconnect where the full-load current rating of the motor is a maximum of 80 percent of the rating of the switch. For example, if an AC general-use snap switch were used as the disconnect for a 1-horsepower motor with a full-load current rating of 16 amperes, a 20-ampere rated switch would be required. The full-load motor current of 16 amperes times 125 percent is 20 amperes. A 15-ampere rated switch would be permitted to control a motor with a maximum full-load current of 12 amperes or 80 percent of the 15 amperes.

AC/DC general-use snap switches are designed and permitted to be used up to a maximum of the full current rating of the switch to control resistive loads and tungsten-filament lamp loads. These switches are tested for a tungsten-filament lamp load with a rating of 125 volts and are marked with a "T" where intended to be used for this type of lamp, which draws a high current when first turned on. Where an AC/DC general-use snap switch is used for inductive loads, the load cannot exceed 50 percent of the rating of the switch. Some residential lighting circuits could serve relatively large inductive lighting loads such as fluorescent lights, mercury vapor lights, or high-intensity discharge fixtures. If a 20-ampere rated AC/DC general-use snap switch were used for these types of lights, the load could only be a maximum of 10 amperes.

General-use switch. A switch intended for use in general distribution and branch circuits. It is rated in amperes and is capable of interrupting its rated current at its rated voltage.

❖ A general-use switch may be used to control equipment, appliances, or a feeder. These switches are rated in amperes and voltage and are designed and permitted to interrupt up to their rated current at their rated voltage. An example of a general-use switch is the disconnect switch for a condensing unit of an air-conditioner system mounted on the outside of the house. These switches are either fused or non-fused and are often referred to by electricians as safety switches. They are installed in a cabinet or cutout box, which is usually mounted on the surface of the wall and has a hinged door that opens on the side or the top. In many of these installations, the switch handle is on the side and is operated vertically. In some safety switch enclosures, the switch is a pull-out block in the front with a handle. A safety feature of some of these switches is that the door will not open when the switch is in the closed position, and some are built so that when the door is opened, a mechanism moves an insulated cover over the live parts.

Isolating switch. A switch intended for isolating an electric circuit from the source of power. It has no interrupting rating and is intended to be operated only after the circuit has been opened by some other means.

❖ This type of switch is used mainly for large equipment or for entire parts of an electric wiring system. It is not designed to be operated under load conditions. Such switches are uncommon in residential installations. For a dwelling, a meter socket can serve as an isolation switch by pulling the meter from its socket and placing a cover over the meter opening in the enclosure.

Motor-circuit switch. A switch, rated in horsepower that is capable of interrupting the maximum operating overload current of a motor of the same horsepower rating as the switch at the rated voltage.

❖ A motor-circuit switch must have a horsepower rating of at least the horsepower of the motor and must be able to interrupt the maximum overload current of the motor. A horsepower-rated motor-circuit switch may be a toggle switch and will have the rating stamped or marked on it. But not all toggle switches are motor-circuit switches, and not all motors in a dwelling require a motor-circuit switch. The disconnect switch for a motor could be the branch-circuit overcurrent device if the motor is 1/8 horsepower or less. And an AC general-use snap switch could be used for a motor of 2 horsepower or less if the switch has a rating of 125 percent of the motor full-load current.

UTILIZATION EQUIPMENT. Equipment that utilizes electric energy for electronic, electromechanical, chemical, heating, lighting or similar purposes.

❖ This equipment, such as a range, exhaust fan, luminaire, or electric furnace, is differentiated by its use from equipment that does not use energy. Junction boxes, switches and raceways are not utilization equipment.

VENTILATED. Provided with a means to permit circulation of air sufficient to remove an excess of heat, fumes or vapors.

❖ Some types of equipment are ventilated such as motors, transformers, recessed lighting fixtures, and cabinets. This definition for the electrical chapters does not necessarily refer to the ventilation system for the occupants of the building.

VOLTAGE (OF A CIRCUIT). The greatest root-mean-square (rms) (effective) difference of potential between any two conductors of the circuit concerned.

❖ See "Voltage (Nominal)."

VOLTAGE, NOMINAL. A nominal value assigned to a circuit or system for the purpose of conveniently designating its voltage class (e.g., 120/240). The actual voltage at which a circuit operates can vary from the nominal within a range that permits satisfactory operation of equipment.

❖ The voltage of the power supplied by the electric utility varies somewhat by regions and by power demand but typically not enough to cause any harm to the components of the premises wiring system or the utilization equipment. The circuit voltage and the voltage to ground are the actual voltages as measured. The nominal voltage is used as a common reference and for calculation purposes. See Commentary Figure E3401(2).

VOLTAGE TO GROUND. For grounded circuits, the voltage between the given conductor and that point or conductor of the circuit that is grounded. For ungrounded circuits, the greatest voltage between the given conductor and any other conductor of the circuit.

❖ See "Voltage (Nominal)."

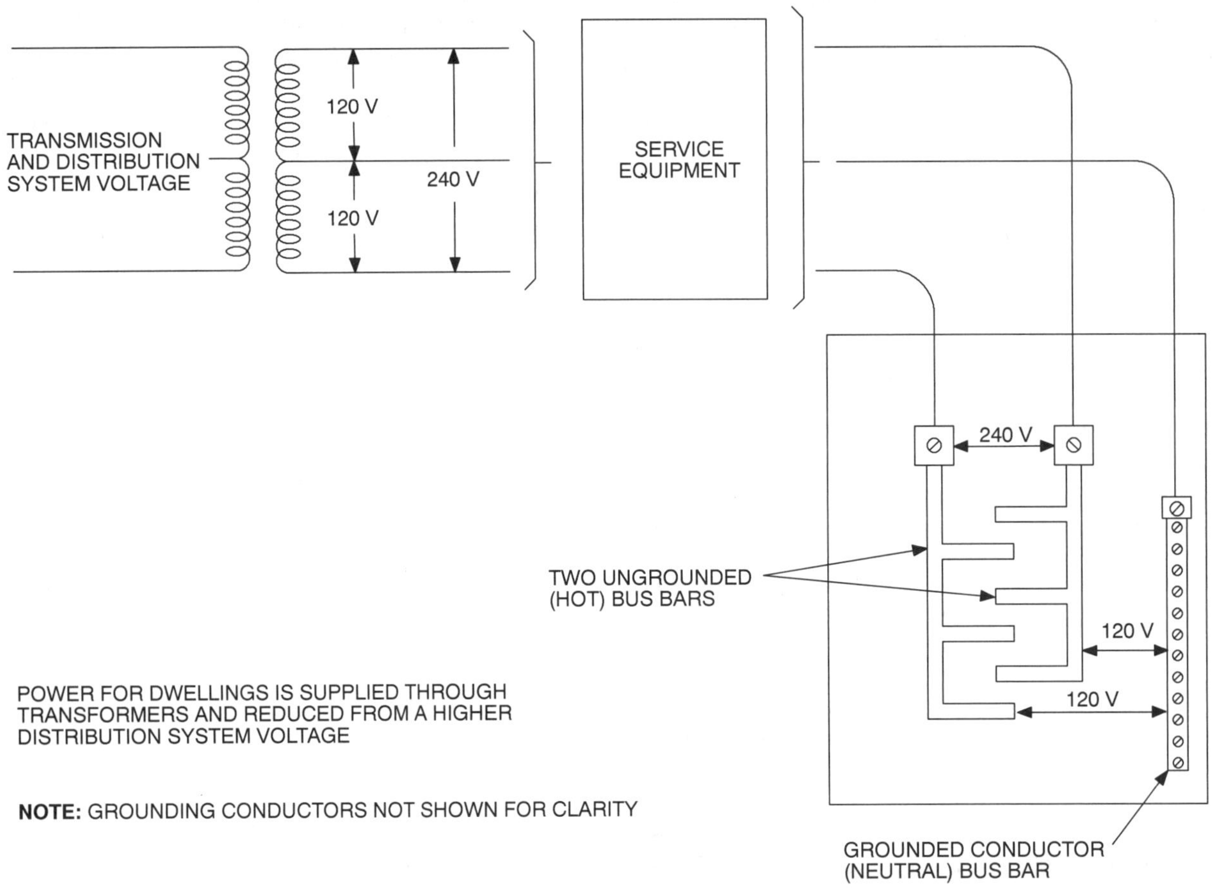

Figure E3401(2)
POWER SUPPLY VOLTAGES

WATER TIGHT. So constructed that moisture will not enter the enclosure under specified test conditions.

❖ See the commentary for "Rain tight."

WEATHERPROOF. So constructed or protected that exposure to the weather will not interfere with successful operation.

❖ See the commentary for "Rainproof."

Bibliography

The following resource materials are referenced in this chapter or are relevant to the subject matter addressed in this chapter.

NFPA 70-02, *National Electrical Code.* Quincy, MA: National Fire Protection Association, 2002.

Chapter 35: Services

General Comments

This chapter is the first of the logical order of chapters that mimics the normal sequence of dwelling construction. The first step in dwelling wiring is typically the sizing, design and installation of the service that is the source of power for the building. This chapter addresses the sizing of services, service conductor sizing and installation, system grounding and bonding, overcurrent protection, disconnecting means and the grounding electrode system.

Purpose

This chapter requires services to be properly sized to serve the load. This is intended to prevent overloading and to provide the utility expected by the occupants. This chapter also intends to protect occupants and the building from fire and protect the occupants from electrical shock hazards associated with service conductors and equipment.

SECTION E3501 GENERAL SERVICES

E3501.1 Scope. This chapter covers service conductors and equipment for the control and protection of services and their installation requirements.

❖ A logical place to begin when planning the wiring for a house is the service. The size and location of the service equipment and main distribution panelboard are usually determined before the branch circuits are laid out. The service delivers energy from the serving utility to the wiring system of the premises and includes all of the equipment and conductors for this purpose. The electric utility provides power to the dwelling in an overhead service drop or in an underground service lateral. This chapter covers these conductors and all wiring and equipment up to the main disconnect, including the system grounding and bonding requirements. Equipment grounding is covered in later chapters, but the rules for the grounding electrode system are related directly to the service.

E3501.2 Number of services. A dwelling unit shall be supplied by only one service.

❖ The code requires that a building be supplied by only one service, that is, one overhead service drop or one underground service lateral run to the one- or two-family dwelling. In some cases, there may be a need for each unit of a two-family dwelling to be considered as a separate building. Generally, where a building is separated into two areas or occupancies, the code requires a 2-hour fire-resistive wall and/or floor assembly between the two areas. For example, this code defines a townhouse as a single-family dwelling and considers the dwelling as a separate building. See the *IRC Commentary*, Volume I, Sections R202 and R317.2. The adjoining walls of each townhouse must have at least a 1-hour fire-resistive rating, or the townhouse separation must have a common wall with a 2-hour fire-resistive rating. Therefore, where the townhouses are joined, the wall would have a combined rating of 2 hours. For each townhouse to be considered a separate building, a 2-hour rating is required.

Since Section R317.1 of the IRC states that dwelling units in two-family dwellings shall be separated from each other by wall and/or floor assemblies of not less than 1-hour fire-resistive rating, the two dwellings in a two-family dwelling would not be considered as separate buildings, and two services would not be permitted. If the need existed for two services, it would be with special permission from the code official, and the fire-resistance rating of the separation between the two dwelling units would be considered in the decision.

Anyone observing existing older dwellings has probably seen a house that was remodeled to create a separate apartment with a separate service drop and separate service equipment. Special written permission of the code official is required for this kind of installation.

E3501.3 One building or other structure not to be supplied through another. Service conductors supplying a building or other structure shall not pass through the interior of another building or other structure.

❖ The service entrance conductors are not permitted to be run through one building to serve another. For example, the service equipment for a townhouse could not pass through the crawl space or attic of another townhouse; see Commentary Figure E3501.3.

E3501.4 Other conductors in raceway or cable. Conductors other than service conductors shall not be installed in the same service raceway or service cable.

Exceptions:

1. Grounding conductors and bonding jumpers.
2. Load management control conductors having overcurrent protection.

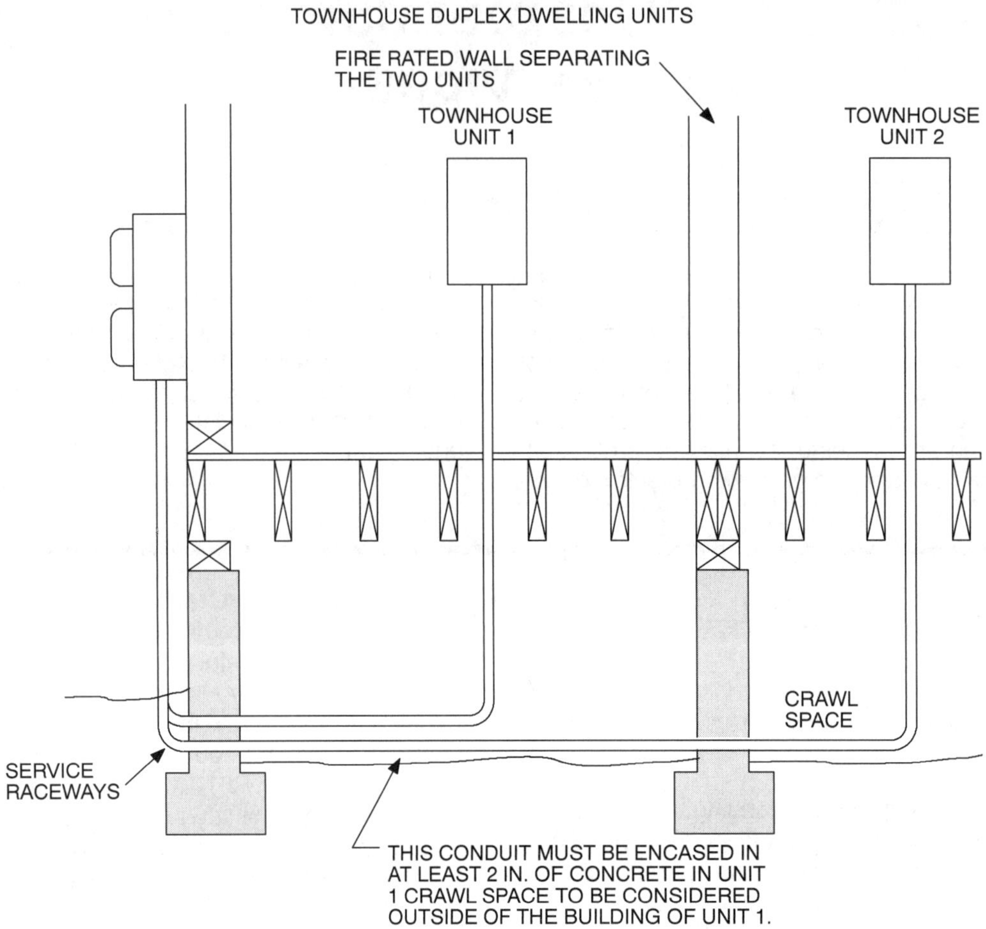

Figure E3501.3
ONE BUILDING NOT TO BE SUPPLIED THROUGH ANOTHER

❖ The set of fuses or a circuit breaker that is the overcurrent protection for the service conductors is an integral part of the service disconnecting means and is located at the load end of the service lateral or service entrance conductors. By definition, service conductors end at the main disconnect, and any conductors after that point on the system are feeders and branch-circuit conductors. Since service conductors are run directly into the service equipment from the utility company's supply lines, they are protected only at the load end. If feeder or branch-circuit conductors were in the same raceway with the service conductors, they could be subject to fault current.

E3501.5 Raceway seal. Where a service raceway enters from an underground distribution system, it shall be sealed in accordance with Section E3703.7.

❖ See the commentary for Section E3703.7.

E3501.6 Service disconnect required. A means shall be provided to disconnect all conductors in a building or other structure from the service entrance conductors.

❖ The code requires that there be a means for disconnecting all conductors from the service entrance conductors. The disconnecting means for the ungrounded conductors is almost always an overcurrent device in the form of a circuit breaker or set of fuses. If the service overcurrent device is not an integral part of the service disconnecting means, then the main disconnect must be adjacent to the overcurrent device. It is rarely done, but a fuseless safety switch could be installed as the main disconnect; however, it must be next to the service overcurrent device.

The main service disconnect switch, circuit breaker, or set of fuses provides a means to disconnect the ungrounded conductors from the supply. No overcurrent device should be placed in series with the grounded (neutral) conductor unless that device will disconnect both ungrounded and grounded conductors. Such devices are not used in residential services.

The code requires that the service equipment have a means to disconnect all conductors from the supply, which includes the grounded service conductor or neutral. The neutral is cut and terminated in a bus bar or lugs provided in the service equipment for this purpose. For example, where the main disconnect is located on the outside of the house and the panelboard is on the inside, it would not be permitted to run the neutral through the service equipment and on to the panelboard without cutting it. Means must be provided to disconnect the neutral or remove it from the terminal at the location of the main disconnect. It is not required to be disconnected with a switch or as part of the overcurrent device. The code rule is satisfied if the neutral can be removed from the terminal or busbar.

E3501.6.1 Marking of service equipment and disconnects. Service disconnects shall be permanently marked as a service disconnect. Service equipment shall be listed for the purpose. Individual meter socket enclosures shall not be considered service equipment.

❖ In a house with an overhead service, it may be obvious that the power is supplied through the service drop and continues down through the service entrance conductors to the enclosure where the main disconnect is located. In an emergency, for example in the case of a fire, the fire department personnel could easily see where the main disconnect is located and turn off the power. But in many houses there may be enclosures, gutters, raceways and cables on the house. These might be part of the service equipment or other boxes on the load side of the service equipment. Therefore, it is important to have the main disconnect clearly marked. On some older dwellings, the only way to disconnect the power was by pulling the meter. An individual meter socket is not considered service equipment because service equipment constitutes the main control and cutoff of the supply. See the definition of "Service Equipment" in Chapter 34. The code does not require a meter, but permits it.

E3501.6.2 Service disconnect location. The service disconnecting means shall be installed at a readily accessible location either outside of a building or inside nearest the point of entrance of the service conductors. Service disconnecting means shall not be installed in bathrooms. Each occupant shall have access to the disconnect serving the dwelling unit in which they reside.

❖ The code permits the service disconnect to be located either on the outside or inside of a building; see Section E3503.3.3. It is not required to be attached to or mounted on the house, although it almost always is. Various reasons from practical to personal have been given for locating it on the outside and inside. Practical reasons may include making it easier for fire service personnel to find it outside the house and turn off the power in case of a fire. In other cases, it is installed inside the house for personal reasons like reducing the likelihood of a burglar turning off the power before attempting to enter.

Wherever the main disconnect is installed, it must be at a readily accessible location. See the definition of "Readily Accessible" in Chapter 34. The maximum height permitted for the grip of the operating handle of the disconnect is 6 feet, 7 inches above the floor or ground. See Section E3901.6. On a house under construction, the final grade level around the house might not be established when the service equipment is installed. After the electrician installs the main disconnect, the contractor might excavate and lower the surrounding grade, making the disconnect out of reach. For an underground service, by grading the yard, the contractor might decrease the burial depth of the service lateral, resulting in a code violation. It is important to consider what the final grade will be before mounting the main disconnect. The same caution also applies to meter placement; although the code does not specify any dimension, the power company usually has a required range of height above the ground.

Where the main disconnect is located on the inside of the house, a question arises of how far inside the dwelling the unprotected service conductors are allowed to be run before being terminated at the main disconnect. The concern is that the service conductors are unprotected from fault-current and short-current up to the point of the main disconnect, which is usually the main overcurrent device. The available short-circuit current or fault current is usually very high. In case of a fault current or short circuit, the current flow could be of a magnitude that would heat up and damage the conductors, which could start a fire. If the decision is made to locate the main disconnect inside the house, the code indicates that it must be at a location "nearest the point of entrance of the service conductors."

Many code provisions are very prescriptive and state an exact dimension, but because of the many variables and configurations of laying out the service equipment, the length of the service conductors inside the dwelling is left to the discretion of the code official. Some jurisdictions establish a length limit for this. For instance, the length of the conductors might be permitted to run a maximum of 15 feet inside the building before being terminated at the load end at the main disconnect and overcurrent device. If the requirement stated that the main disconnect must be placed on the inside wall of the point at which the service conductors enter, the disconnect might be at a location that is either not readily accessible or convenient. Since there are so many different house plans and designs, it is impossible to have detailed prescriptive code provisions that would address every situation. See Commentary Figures E3501.6.2(1) and (2).

E3501.7 Maximum number of disconnects. The service disconnecting means shall consist of not more than six switches or six circuit breakers mounted in a single enclosure or in a group of separate enclosures.

❖ A person should be able to turn off all the power to a house with no more than six motions of the hand while at one location. The main disconnect does not have to

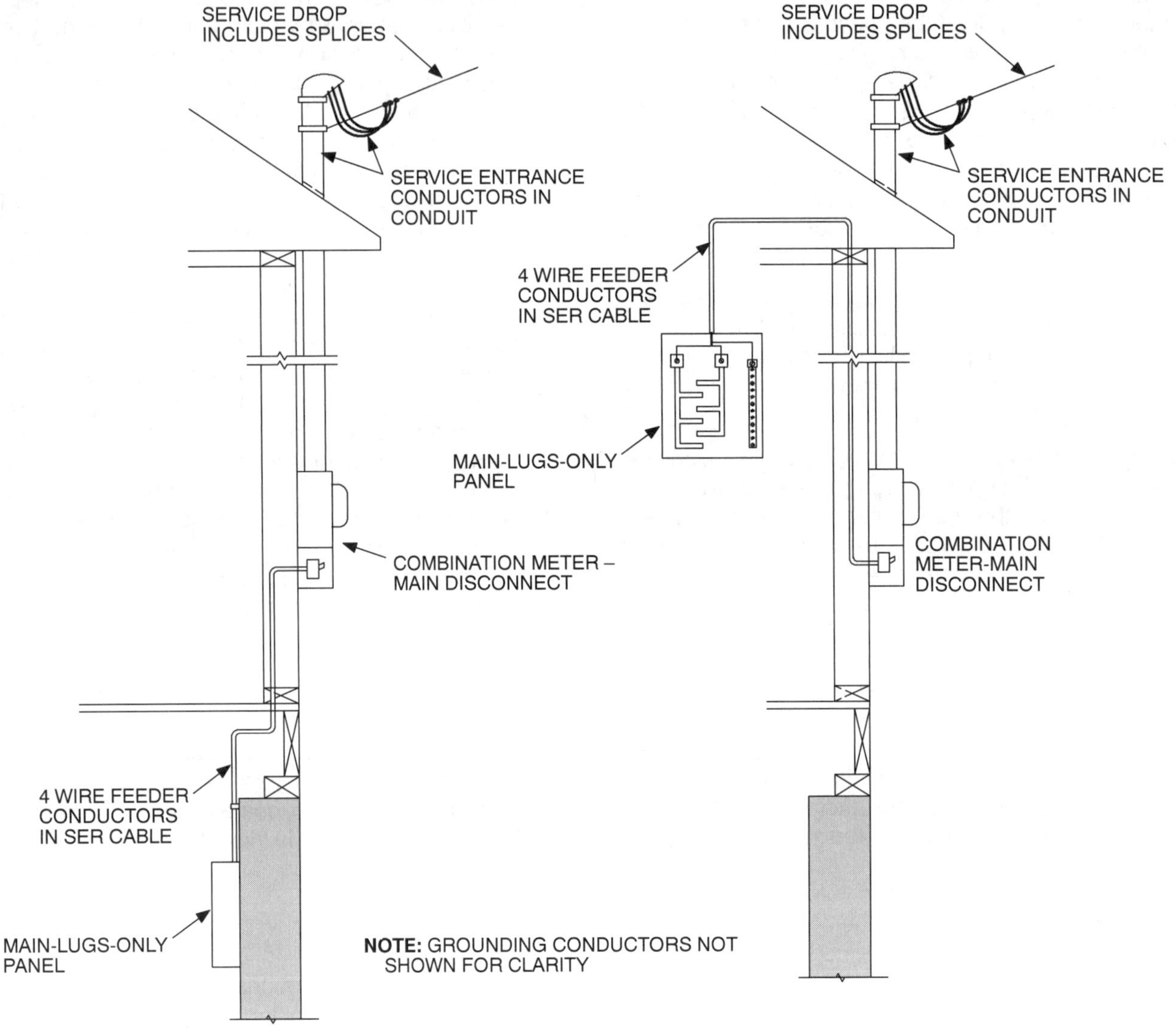

Figure E3501.6.2(1)
METER AND MAIN DISCONNECT ON OUTSIDE OF HOUSE

be one switch, one circuit breaker, or one set of fuses and does not have to be in only one enclosure. For new construction, the service disconnect or disconnects are usually in one enclosure, but on an existing dwelling, for example where a load is being added, an additional main disconnect could be installed in a separate enclosure. However, all the disconnects must be grouped or located together and marked. There may be up to six enclosures that constitute the main disconnecting means for the dwelling, and each one must be clearly marked.

This code provision is usually not a problem in new construction. A house that has only one panelboard with all loads are fed from that panelboard would require only one main disconnect, such as a two-pole circuit breaker. Although it is not a code requirement, installing service equipment that has room for additional circuit breakers may be helpful for future expansion. Service equipment enclosures are available for residential wiring with spaces for up to 6 two-pole circuit breakers. With this design, feeders could be run from the main service equipment instead of from the panelboard for such things as a sub-panel for swimming pool equipment or for an out-building, detached garage or air conditioner.

In an older dwelling, there might be six circuit breakers or fuses that serve the branch circuits with no other main disconnect. If that dwelling is being remodeled and circuits are run for new kitchen appliances or for a room addition, it may be necessary to change the service equipment since the addition of another branch circuit overcurrent device would be over the six maximum permitted for turning off all the power to the house. Obviously, when adding more circuits, the size of the existing service conductors would be a major concern in the decision to install new service equip-

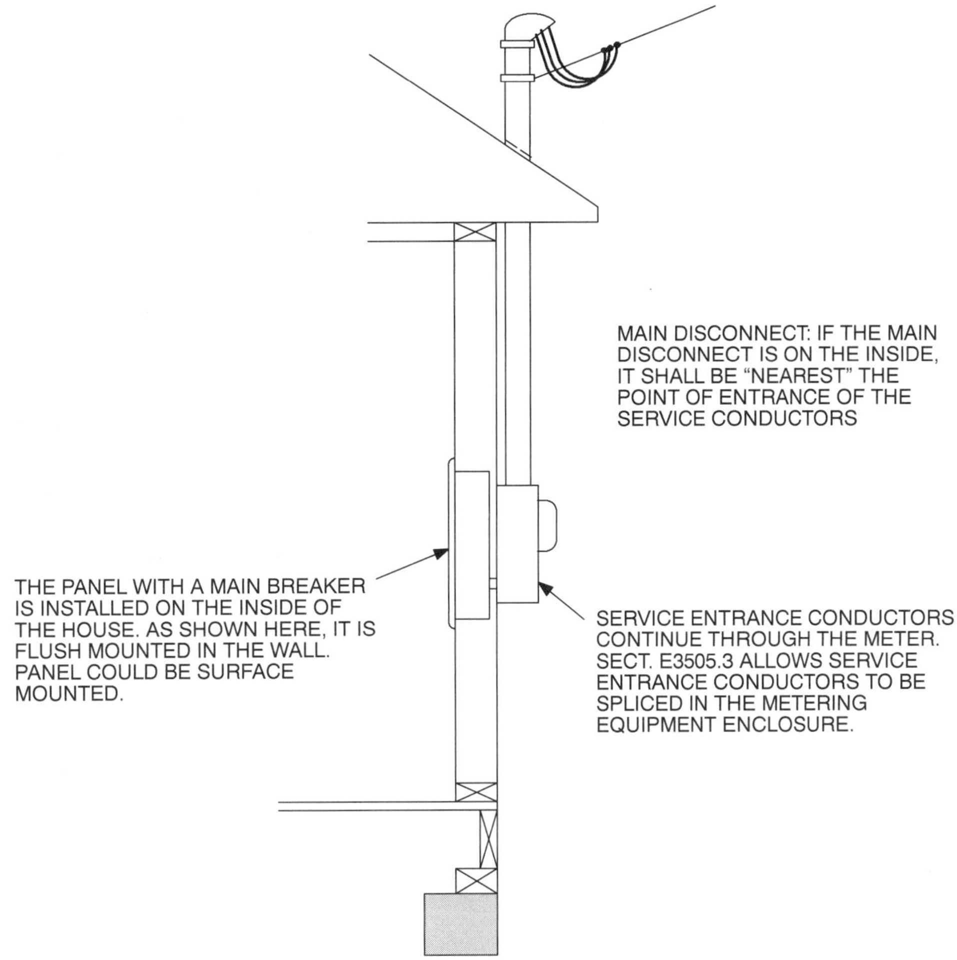

Figure E3501.6.2(2)
METER ON OUTSIDE OF HOUSE; MAIN DISCONNECT ON INSIDE OF HOUSE

ment, but it would not be the only concern. The number of disconnects could be the deciding factor.

See Section E3503.3.1, which provides for the overload protection of ungrounded service conductors, and Commentary Figure E3501.7.

SECTION E3502
SERVICE SIZE AND RATING

E3502.1 Rating of ungrounded conductors. Ungrounded service conductors shall have an ampacity of not less than the load served. For one-family dwellings, the rating of the ungrounded conductors shall be not less than 100 amperes, 3 wire. For all other installations, the rating of the ungrounded conductors shall be not less than 60 amperes:

❖ The minimum service rating for any dwelling unit is 100 amperes, but the load should be calculated to determine if 100 amperes is sufficient. The load depends on many factors, such as use of gas appliances or electric space heating and water heating. The square footage of the house is another factor, but it only one of the elements that should be taken into account. Some electricians may have a guideline that would require a certain size service corresponding to the square footage of the house, such as a minimum 100-ampere service for a house of 1,000 square feet (93 m^2), a 125-ampere rating for a house up to 1,250 square feet (116 m^2), a 150-ampere rating for a house of 1,500 square feet (139 m^2) and so on. This kind of rule is not a code provision, but it may work in some cases. Because of the various loads that may be present in a dwelling, the code official, in enforcing the code, would require a calculation to be done to demonstrate that the rating of the service is sufficient.

The provision in this paragraph for a 60-ampere rated service does not apply to the dwelling but would be for something like a separate service on the residential property such as on a farm where a service might be run to a remote building. Exceptions to the minimum 60-ampere rating are in Section E3503.2.

E3502.2 Service load. The minimum load for ungrounded service conductors and service devices that serve 100 percent of the dwelling unit load shall be computed in accordance

Figure E3501.7
SERVICE DISCONNECTING MEANS CONSISTING OF SIX OR FEWER SWITCHES OR CIRCUIT BREAKERS

with Table E3502.2. Ungrounded service conductors and service devices that serve less than 100 percent of the dwelling unit load shall be computed as required for feeders in accordance with Chapter 36.

❖ Where the service conductors serve 100 percent of the load, Table E3502.2 is used to calculate the size of the service conductors. For example, where an overhead service drop feeds the service entrance conductors, which are run to a service equipment cabinet that includes the main disconnect and the panelboard, the table must be used. This equipment is referred to by electricians in some areas as a meter-main combination, meaning that the enclosure has a meter base and a panelboard with the main disconnect or main breaker in the panel. In other words, it is a main-breaker panel combined with a meter base. In this case, the service entrance conductors terminated at the main breaker serve 100 percent of the load.

E3502.2.1 Services under 100 amperes. Services that are not required to be 100 amperes shall be sized in accordance with Chapter 36.

❖ The load calculation method in Table E3502.2 is only for services of 100 amperes and larger. Therefore, smaller services must be sized by the method in Chapter 36. Where a service supplies a separate load such as a separate workshop or a panelboard at a small building that houses swimming pool equipment and a bath house, the service conductors would be calculated as required for feeders according to Chapter 36. See Section E3502.1.

E3502.3 Rating of service disconnect. The combined rating of all individual service disconnects serving a single dwelling unit shall not be less than the load determined from Table 3502.2 and shall not be less than as specified in Section 3502.1.

❖ The service load is calculated to determine the ampacity rating of the service conductors and service equipment. When designing the service and performing the load calculation, it is commonly assumed that there will be only one switch for the main disconnecting means in the service equipment. However, Section E3501.7 permits the service disconnecting means to consist of up to six switches or circuit breakers, which could all be in the same enclosure or in a group of separate enclosures. The combined rating of the switches or circuit breakers must be at least equal to the calculated load and not less than the minimums dictated by Section E3502.1. If the load calculation for a dwelling is under 100 amperes, then the minimum required ampere rating for the service disconnect would be 100 amperes according to this section. See Section E3503.3.1.

TABLE E3502.2
MINIMUM SERVICE LOAD CALCULATION

LOADS AND PROCEDURE
3 volt-amperes per square foot of floor area for general lighting and convenience receptacle outlets.
Plus
1,500 volt-amperes × total number of 20-ampere-rated small appliance and laundry circuits.
Plus
The nameplate volt-ampere rating of all fastened-in-place, permanently connected or dedicated circuit-supplied appliances such as ranges, ovens, cooking units, clothes dryers and water heaters.
Apply the following demand factors to the above subtotal:
The minimum subtotal for the loads above shall be 100 percent of the first 10,000 volt-amperes of the sum of the above loads plus 40 percent of any portion of the sum that is in excess of 10,000 volt-amperes.
Plus the largest of the following:
Nameplate rating(s) of air-conditioning and cooling equipment including heat pump compressors.
Nameplate rating of the electric thermal storage and other heating systems where the usual load is expected to be continuous at the full nameplate value. Systems qualifying under this selection shall not be figured under any other category in this table.
Sixty-five percent of nameplate rating of central electric space-heating equipment, including integral supplemental heating for heat pump systems.
Sixty-five percent of nameplate rating(s) of electric space-heating units if less than four separately controlled units.
Forty percent of nameplate rating(s) of electric space-heating units of four or more separately controlled units.
The minimum total load in amperes shall be the volt-ampere sum calculated above divided by 240 volts.

❖ The service conductors must be of a sufficient size to carry the load. The procedure set forth in this table is used to calculate the minimum load, in amperes, required for the service conductors. Once the load is determined, the conductor size can be selected from Table E3503.1. For many smaller houses, especially where gas is used for space heating and water heating, the calculation using Table E3502.2 will usually result in a load of 100 amperes or less. Of course, the minimum rating required for ungrounded service conductors is 100 amperes as stated in Section E3502.1, even if the calculated load is less. The code requirement is to calculate the minimum load in order to select the appropriate size conductors that are necessary for safety. However, the minimum conductor size and service rating may not be adequate for additional loads that could be added later on. A good practice is to calculate the minimum load as required and then consider the feasibility and costs of installing a service larger than required to provide for future expansion.

The following is a description of the loads listed in the table.

General Lighting and Convenience Receptacle Loads:

A unit load of not less than 3 volt-amperes per square foot constitutes the minimum lighting load for each square foot of floor area. The floor area must be computed from the outside dimensions of the dwelling and does not include open porches, garages, or unused spaces not adaptable for future use. However, in all cases these unit values are based on minimum load conditions with a 100-percent power factor, and they may not provide sufficient capacity for the installation contemplated.

Included in the lighting load are convenience receptacle outlets of 15- and 20-ampere rating. The bathroom receptacle outlets supplied by the separate branch circuit required by Section E3603.4 are also included in the general lighting load.

Small Appliance Branch Circuit Loads:

A load of 1,500 volt-amperes is included in the calculation for each two-wire small appliance branch circuit for portable appliances supplied by 15- or 20-ampere receptacles on 20-ampere branch circuits in the kitchen, pantry, dining room, and breakfast room. Since the code requires a minimum of two small appliance circuits regardless of the size of the house, the minimum load to be included in the calculation is 3,000 volt-amperes. Where a dwelling has more than two small appliance branch circuits, a load of 1,500 volt-amperes must be included for each circuit. An electrician may choose to supply the refrigerator receptacle on a separate circuit. Since this is not required by the code, an additional 1,500 volt-amperes would not need to be added to the load calculation for this circuit.

Laundry Circuit Load:

A load of 1,500 volt-amperes is included for the circuit supplying the clothes washer and gas dryer. Where the clothes washer is in a laundry room and additional receptacle outlets are provided, such as an outlet for the iron, these receptacles are permitted on the laundry circuit, and only one unit load of 1,500 volt-amperes is included.

Appliance Load:

This is the nameplate rating of all appliances that are fastened in place, permanently connected, or connected to a dedicated circuit including ranges, dryers, and water heaters. For the calculation, an appliance that has the nameplate rating given only in amperes must have its rating converted to volt-amperes by multiplying the amperes rating by the rated voltage.

Air Conditioning and Heating Loads:

It is unlikely that air-conditioning (cooling equipment) and heating loads will be used simultaneously. Therefore, only the largest load is used in the calculation. The bottom half of Table E3502.2 lists cooling equipment loads and various combinations of space- heating loads. The largest of any of these loads present in the dwelling is used for the calculation.

SAMPLE CALCULATION

Description of loads in a dwelling:

Volt-amperes are indicated as VA. Depending upon the power factor, VA is not always equivalent to wattage. However, wattage is considered the same as volt-amperes for purposes of calculating service loads.

Sample Calculation #1: A dwelling with no gas supply for space or water heating has the following:

- Outside dimensions of 30 feet x 65 feet, which does not include the garage, porches, attic, or cellar.
- Two small appliance circuits in the kitchen and related rooms.
- One laundry circuit to serve the clothes washer and outlets in the laundry room.
- Water heater rated at 4,500 watts.
- Clothes dryer rated at 5,000 VA.
- Range rated at 12,000 VA.
- Dishwasher rated at 1,200 VA.
- Disposal rated at 1,000 VA.
- Microwave oven rated at 1,800 VA.
- Air conditioner condensing unit with a nameplate rating of 28 amperes, 240 volts.
- Gas furnace rated at 7 amperes, 120 volts for heating and cooling but on a separate circuit.

Performing the calculation:

1. General lighting and convenience receptacle loads		
30 x 65 = 1,950 sq. ft. x 3 VA /sq. ft. =		5,850 VA
2. Small appliance and laundry circuit loads		
2 small appliance branch circuits at 1,500 VA each	= 3,000 VA	
1 laundry branch circuit at 1,500 VA	= 1,500 VA	
	4,500 VA	4,500 VA
3. Nameplate rating of appliances		
Water heater	4,500 VA	
Clothes dryer	5,000 VA	
Range	12,000 VA	
Dishwasher	1,200 VA	
Disposer	1,000 VA	
Microwave oven	1,800 VA	
Gas furnace: 7 amperes x 120 volts	840 VA	
	26,340 VA	26,340 VA
4. Combine loads from 1, 2, and 3		36,690 VA

Apply demand factor:

		Add VA in this column
Up to first 10,000 at 100%	10,000 VA	
Remainder at 40% = 36,690 - 10,000 = 26,690 x 0.4	10,676 VA	
Total	20,676 VA	20,676 VA
4. Cooling / Heating Loads		
Gas furnace rated a 840 VA		
Cooling equipment load is 28 amperes x 240 volts =	6,720 VA	
The largest of these is the cooling load:		6,720 VA
	Total	27,396 VA

Sizing the ungrounded service conductors:

27,396 VA / 240 volts = 114.15 amperes. The standard ampere ratings of fuses and circuit breakers include 100, 110, and 125. Therefore, the main disconnect would need to be at least 125 amperes since the calculated load is over 110 amperes and below 125 amperes. From Table E3503.1, the minimum size permitted for an ungrounded service conductor would be a size 2 AWG copper conductor or a size 1/0 AWG aluminum with one of the insulation types listed at the top of the table. The main breaker or set of fuses would be 125 amperes.

Sizing the grounded (neutral) service conductor:

The neutral conductor must be of a size to adequately carry the maximum unbalanced load in accordance with Sections E3503.1 and E3604.4. The size of the service ungrounded conductors that are part of the "main power feeder" is dictated by Section E3503.1. Since Section E3503.1 does not directly address the sizing of grounded service conductors, Section E3604.4 provides for the sizing of such conductors by applying the feeder load calculation method. Section E3604.4 provides the definition of "maximum unbalanced load," which is applicable for any feeder neutral and for the grounded service conductor. It states: "The maximum unbalanced load shall be the maximum net computed load between the neutral and any one ungrounded conductor." Accordingly, the maximum unbalanced load for the 120/240-volt, three-wire range and dryer branch circuits is 70 percent of the load calculated for the ungrounded conductors serving these two appliances. Two-wire 240-volt loads, such as electric water heaters, do not add to the neutral load. The cooling equipment is not counted since it is a 240-volt load that does not have a neutral. Where a blower motor is used for air handling in air-conditioning, it is usually a 120-volt motor and must be included in the calculation of the neutral load. Of course, when the circuit breakers are installed in the panelboard, the loads should be balanced as much as possible between the two "hot" legs of the service. The following example assumes that the 120-volt loads are equally distributed between the two ungrounded service conductors.

Calculating the Service Neutral Load Using Table E3604.3(1) and Section E3604.4:

	Add VA in this column
1. General lighting and convenience receptacle loads 3 VA per sq. ft. 30 x 65 = 1950 x 3 = 5,850 VA	
2. Small appliance and laundry circuit loads 2 small appliance branch circuits at 1,500 VA each = 3,000 VA 1 laundry branch circuit at 1,500 VA = 1,500 VA 10,350 VA	
Applying demand factor, first 3,000 at 100% = 3,000 VA	
10,350 - 3,000 = 7,350 at 35% = 2,573 VA	
Total for lighting, receptacle, small appliances, and laundry 5,573 VA	5,573 VA
3. Appliance Loads Water heater (two-wire, 240 volts, no neutral load) 0 VA Dishwasher 1,200 VA Disposer 1,000 VA Largest motor load: Disposer: 1,000 x 25% 250 VA Microwave Oven 1,800 VA Gas furnace, 7 amperes x 120 volts 840 VA Clothes dryer 5,000 VA x 70% 3,500 VA Range* 8,000 VA x 70% 5,600 VA Subtotal 14,190 VA 14,190 VA	14,190 VA
Total	19,763 VA
4. Cooling / Heating: No neutral loads other than furnace.	

*The branch circuit load for the range is permitted to be calculated in accordance with Table E3604.3(2), which lists the demand load for one range with a rating of 12 kVA or less, as 8,000 VA. This demand load is not used for the range load when calculating the load for the service in accordance with Section E3502.2.

19,763 VA / 240 volts = 82.34 amperes.

From Table E3503.1, at 75°C, the minimum size permitted for the grounded service conductor would be a size 4 AWG copper conductor or a size 2 AWG aluminum conductor.

The total load was divided by 240 volts rather than 120 volts because this example uses the common calculation procedure, which assumes that the 120 volt loads are equally distributed between the two ungrounded conductors. The only alternative would be to add all of the 120-volt loads that are actually connected to one ungrounded conductor and divide the total by 120 volts, and this procedure must be repeated for the other ungrounded conductor to determine the greatest load (maximum unbalance). Of course, the lengthy exercise of calculating the neutral conductor size could be avoided by simply making the neutral conductor the same size as the ungrounded conductors. A "full-size" neutral has the same ampacity as the ungrounded conductors and will certainly be able to carry the maximum unbalanced load.

E3502.4 Voltage rating. Systems shall be three-wire, 120/240-volt, single-phase with a grounded neutral.

❖ For purposes of performing load calculations, the voltage used should be 120/240 volts. The services considered under the provisions of this code are three-wire, single phase with a grounded neutral.

SECTION E3503 SERVICE, FEEDER AND GROUNDING ELECTRODE CONDUCTOR SIZING

E3503.1 Grounded and ungrounded service conductor size. Conductors used as ungrounded service entrance conductors, service lateral conductors, and feeder conductors that serve as the main power feeder to a dwelling unit shall be those listed in Table E3503.1. Ungrounded service conductors shall have a minimum size in accordance with Table E3503.1. The grounded conductor size shall not be less than the maximum unbalance of the load and shall not be smaller than the required minimum grounding electrode conductor size specified in Table E3503.1.

❖ Table E3503.1 contains the allowable ampacity for service conductors and feeder conductors that serve as the main power feeder for a dwelling. For the same size conductor, the allowable ampacities for service conductors and main power feeders are higher as compared to Table E3605.1. The loads in a dwelling operate with a great degree of diversity and for short periods of time, hence the reason for the demand factors used in the load calculations. Compare, for example, a size 4 AWG copper conductor with THW insulation under different circumstances. Where used for the service conductor, it has an allowable ampacity of 100 amperes, but where used as an individual branch circuit to supply equipment with 75°C terminals, it has an allowable ampacity of 85 per Table E3605.1.

The grounded service conductor (neutral) is often reduced in size compared to the ungrounded conductors, but it should not be assumed that the neutral can always be smaller than the ungrounded conductors. It must be large enough to carry the maximum unbalance of current between it and either of the ungrounded conductors. Also, it must not be smaller than the minimum grounding conductor electrode size because it must carry any fault current back to the supply transformer. The neutral becomes the low-impedance path back to the power source.

Electricians and inspectors may remember that in the past, the service grounded conductor (neutral) was permitted to be two sizes smaller than the ungrounded conductors, and that became a rule of thumb used by many individuals. Multiconductor cables used as service conductors are called Service Entrance (SE) cable. On many new dwellings, Type SER cable is used, which is basically round and has four conductors: three insulated and one bare, all contained within the outer sheathing. One of the insulated conductors is gray or white. Type SER cable is available with the gray or white conductor two sizes smaller than the other two insulated conductors used for the hot legs. Because of this, many people assume that this ratio is an automatic rule, but the code requires that the neutral size be calculated.

E3503.2 Ungrounded service conductors for accessory buildings and structures. Ungrounded conductors for other than dwelling units shall have a minimum rating of 60 amperes and shall be sized as required for feeders in Chapter 36.

Exceptions:

1. For loads consisting of not more than two two-wire branch circuits, the minimum size shall be No. 8 copper or No. 6 aluminum or copper-clad aluminum.
2. For loads limited by demand or by the source of supply, the minimum size shall be No. 8 copper or No. 6 aluminum or copper-clad aluminum.
3. For limited loads of a single branch circuit, the minimum size shall be No. 12 copper or No. 10 aluminum or copper-clad aluminum, but in no case smaller than the branch-circuit conductors.

❖ Power supplied to accessory buildings and detached structures is usually by a feeder, although in some cases it could be a separate service. This section specifies the size of these conductors. Section E3507.3 contains important rules for system grounding for an accessory building.

E3503.3 Overload protection. Each ungrounded service conductor shall have overload protection.

❖ See the commentary for Section E3503.3.1.

TABLE E3503.1
SERVICE CONDUCTOR AND GROUNDING ELECTRODE CONDUCTOR SIZING

CONDUCTOR TYPES AND SIZES—THHW, THW, THWN, USE, XHHW (Parallel sets of 1/0 and larger conductors are permitted in either a single raceway or in separate raceways)		ALLOWABLE AMPACITY	MINIMUM GROUNDING ELECTRODE CONDUCTOR SIZE[a]	
Copper (AWG)	Aluminum and copper-clad aluminum (AWG)	Maximum load (amps)	Copper (AWG)	Aluminum (AWG)
4	2	100	8[b]	6[c]
3	1	110	8[b]	6[c]
2	1/0	125	8[b]	6[c]
1	2/0	150	6[c]	4
1/0	3/0	175	6[c]	4
2/0	4/0 or two sets of 1/0	200	4[d]	2[d]
3/0	250 kcmil or two sets of 2/0	225	4[d]	2[d]
4/0 or two sets of 1/0	300 kcmil or two sets of 3/0	250	2[d]	1/0[d]
250 kcmil or two sets of 2/0	350 kcmil or two sets of 4/0	300	2[d]	1/0[d]
350 kcmil or two sets of 3/0	500 kcmil or two sets of 250 kcmil	350	2[d]	1/0[d]
400 kcmil or two sets of 4/0	600 kcmil or two sets of 300 kcmil	400	1/0[d]	3/0[d]

For SI: 1 inch = 25.4 mm.

a. Where protected by a metal raceway, grounding electrode conductors shall be electrically bonded to the metal raceway at both ends.

b. No. 8 grounding electrode conductors shall be protected with metal conduit or nonmetallic conduit.

c. Where not protected, No. 6 grounding electrode conductors shall closely follow a structural surface for physical protection. The supports shall be spaced not more than 24 inches on center and shall be within 12 inches of any enclosure or termination.

d. Where the sole grounding electrode system is a ground rod or pipe as covered in Section E3508.2, the grounding electrode conductor shall not be required to be larger than No. 6 copper or No. 4 aluminum. Where the sole grounding electrode system is the footing steel as covered in Section E3508.1.2, the grounding electrode conductor shall not be required to be larger than No. 4 copper conductor.

❖ See the commentary for Section E3503.1.

E3503.3.1 Ungrounded conductor. Overload protection shall be provided by an overcurrent device installed in series with each ungrounded service conductor. The overcurrent device shall have a rating or setting not higher than the allowable ampacity specified in Table E3503.1. A set of fuses shall be considered all the fuses required to protect all of the ungrounded conductors of a circuit. Single pole circuit breakers, grouped in accordance with Section E3501.7, shall be considered as one protective device.

> **Exception:** Two to six circuit breakers or sets of fuses shall be permitted as the overcurrent device to provide the overload protection. The sum of the ratings of the circuit breakers or fuses shall be permitted to exceed the ampacity of the service conductors, provided that the calculated load does not exceed the ampacity of the service conductors.

❖ The overload protection must be located at the service equipment and is typically integral with the main disconnect in the form of a circuit breaker or set of fuses. The rating of the overcurrent protection device must be not more than the allowable ampacity of the service conductors as determined by Table E3503.1, except as allowed by the exception. Section E3605.5 contains the general rule for overload protection of conductors and states that overcurrent protection be installed at the point where the conductors receive their supply, which is the power company's transformer. On the line side of the main disconnect, such conductors are not protected against short-circuit or fault current. This requirement for overload protection of the ungrounded service conductors correlates with Section E3501.7.

The main disconnecting means may consist of any combination of single-pole or two-pole circuit breakers, or single fuses or a set of fuses. A set of fuses used as the main disconnect or as one of the six disconnecting means would be used on a 240 volt circuit and would have a handle tie or means to disconnect both hot legs of the circuit. The set of fuses would count as one of the maximum six disconnects. Two-pole circuit breakers are manufactured with a master handle or a handle tie that connects the handles of each pole together. Thus, a two-pole circuit breaker would be one of the six disconnects. Where two single-pole switches or circuit breakers are tied together with a handle tie, they would count as one of the six disconnecting means. The intent of this code requirement is that all ungrounded service conductors can be disconnected from the source of supply with no more than six operations of the hand. Section E3606.3 requires panelboards to be protected by a maximum of two circuit breakers or sets of fuses.

E3503.3.2 Not in grounded conductor. Overcurrent devices shall not be connected in series with a grounded service conductor except where a circuit breaker is used that simultaneously opens all conductors of the circuit.

❖ The grounded service conductor must not be switched because it also must serve as the low-impedance ground path in case of a fault to facilitate the operation of the overcurrent device.

E3503.3.3 Location. The service overcurrent device shall be an integral part of the service disconnecting means or shall be located immediately adjacent thereto.

❖ Most often, the service overcurrent device and the service disconnecting means are the same (a main circuit breaker). These devices could be separate but grouped together, such as a set of fuses or a circuit breaker and an unfused switch, although this would be rare today. The key to the intent of this provision lies in section E3501.6.2; the code intends to limit the length of service-entrance conductors that are both inside of a building and on the line side of the service overcurrent device. Service-entrance conductors on the line side of the service overcurrent device are not protected against short circuits and ground faults. Conductors on the load side of the service overcurrent device are feeders and are not limited in length; see Commentary Figure E3503.3.3.

E3503.4 Grounding electrode conductor size. The grounding electrode conductors shall be sized based on the size of the service entrance conductors as required in Table E3503.1.

❖ This is a prescriptive rule based on the size of the largest service conductor for the dwelling. The size of the grounding electrode conductor is not based on the rating of the overcurrent device, nor is it based on the calculated load. The service conductor size is determined from the calculation, but if a larger service conductor is used, for example, to provide capacity for future expansion, the grounding electrode conductor would be based on the size installed.

SECTION E3504 OVERHEAD SERVICE-DROP AND SERVICE CONDUCTOR INSTALLATION

E3504.1 Clearance from building openings. Open conductors and multiconductor cables without an overall outer jacket shall have a clearance of not less than 3 feet (914 mm) from the sides of doors, porches, decks, stairs, ladders, fire escapes, and balconies, and from the sides and bottom of windows that open. See Figure E3504.1.

❖ This code requirement pertains to overhead service drops. Where service entrance conductors are installed as Service Entrance (SE) cable or in a raceway, there is no minimum distance specified between openings and doors and the cable or raceway. The main concern here is that an individual standing on a balcony or similar place or reaching out of an openable window should not be able to reach open conductors. See Figure E3504.1 and Commentary Figure E3504.1.

Figure E3503.3.3
SERVICE OVERCURRENT DEVICE INTEGRAL WITH SERVICE DISCONNECTING MEANS

E3504.2 Vertical clearances. The vertical clearances of all service-drop conductors shall be based on a conductor temperature of 60°F (15°C), with no wind and with final unloaded sag in the wire, conductor or cable. Service-drop conductors shall not have ready access and shall comply with Sections E3504.2.1 and E3504.2.2.

❖ The installer is responsible for the required clearance of the service drop. The sag in the overhead span of conductors will change with the temperature and with loading from ice and wind. The conductors will contract in colder temperatures and expand in warmer temperatures. The conductors must be installed such that when they are extended because of the increased sag, required clearances will still be maintained.

E3504.2.1 Above roofs. Conductors shall have a vertical clearance of not less than 8 feet (2438 mm) above the roof surface. The vertical clearance above the roof level shall be maintained for a distance of not less than 3 feet (914 mm) in all directions from the edge of the roof. See Figure E3504.2.1.

Exceptions:

1. Conductors above a roof surface subject to pedestrian traffic shall have a vertical clearance from the roof surface in accordance with Section E3504.2.2.
2. Where the roof has a slope of 4 inches (102 mm) in 12 inches (305 mm), or greater, the minimum clearance shall be 3 feet (914 mm).
3. The minimum clearance above only the overhanging portion of the roof shall not be less than 18 inches (457 mm) where not more than 6 feet (1829 mm) of conduc-

tor length passes over 4 feet (1219 mm) or less of roof surface measured horizontally and such conductors are terminated at a through-the-roof raceway or approved support.

4. The requirement for maintaining the vertical clearance for a distance of 3 feet (914 mm) from the edge of the roof shall not apply to the final conductor span where the service drop is attached to the side of a building.

❖ Most roofs of one- and two-family dwellings are not subject to pedestrian traffic but may be walked on. This code provision considers that where the slope of the roof is less than 4 inches (102 mm) vertical in 12 inches (305 mm) horizontal, is it easier to walk upon than a steeper roof; therefore, the vertical clearance of conductors must be at least 8 feet (2538 mm) above the roof surface to provide for an individual walking on the roof. A roof that has a slope of 4 or more inches (102 mm) in 12 inches (305 mm) horizontal is more difficult to walk on, and the conductor clearance may be reduced to 3 feet (915 mm).

Exception 3 pertains only to the overhanging portion of the roof, that is, where the roof extends past the wall of the dwelling. Where the service equipment is mounted on the wall with the riser extending through the roof, the service drop conductors will usually pass over only the overhanging portion of the roof. The clearance can be reduced to 18 inches (457 mm) above the roof surface if a length of 6 feet (1829 mm) or less of service conductor passes over a maximum of 4 feet (1219 mm) of overhanging roof. The 4 feet (1219 mm) is measured in any direction, not just perpendicular to the wall of the house.

Exception 4 is for an installation where the service drop is not attached to the service mast but to the building, and the service riser does not extend through the roof. The service drop conductors may run near the edge of the roof, and the vertical clearance for a

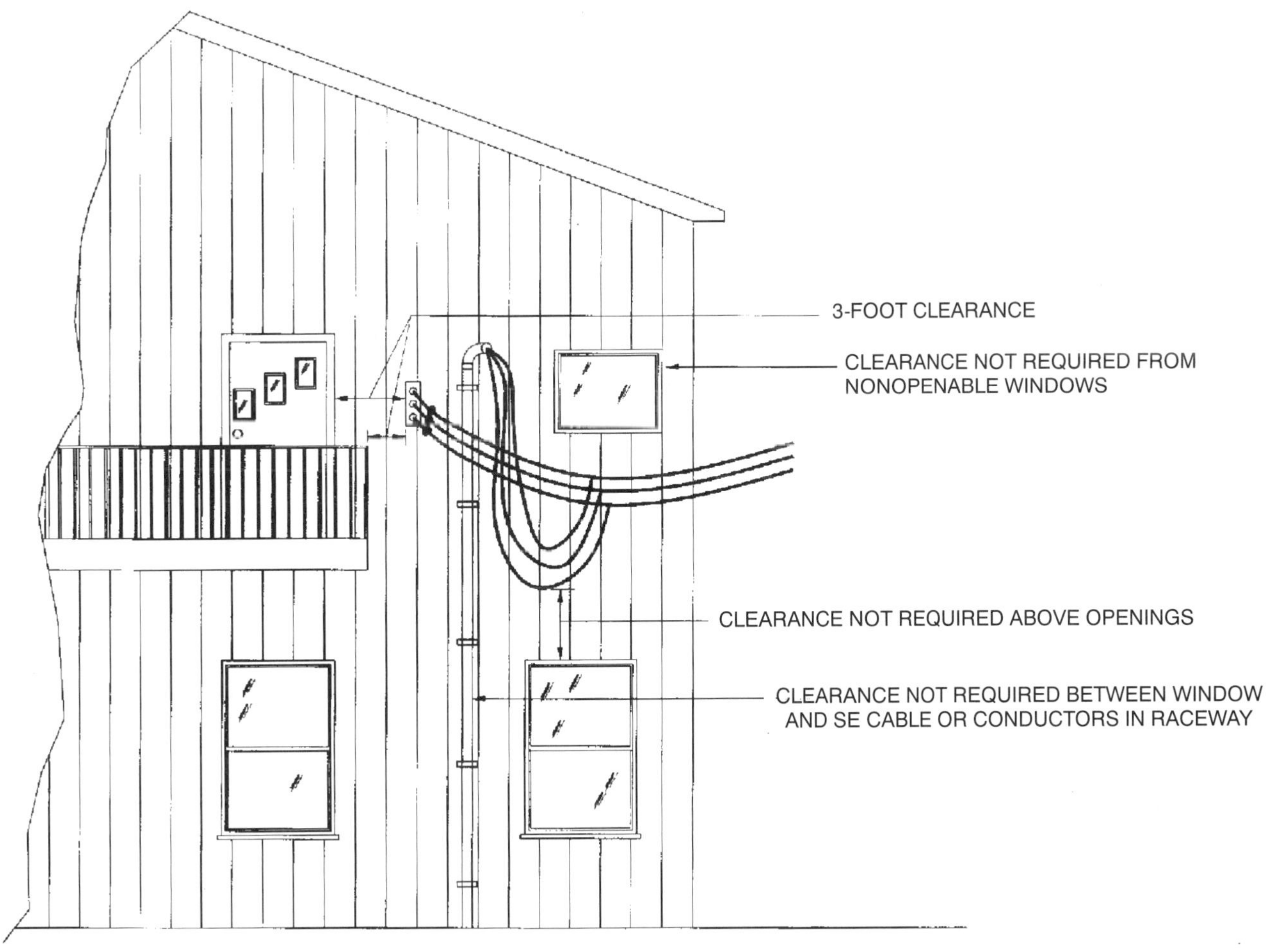

For SI: 1 foot = 304.8 mm.

FIGURE E3504.1
CLEARANCES FROM BUILDING OPENINGS

❖ See the commentary for Section E3504.1.

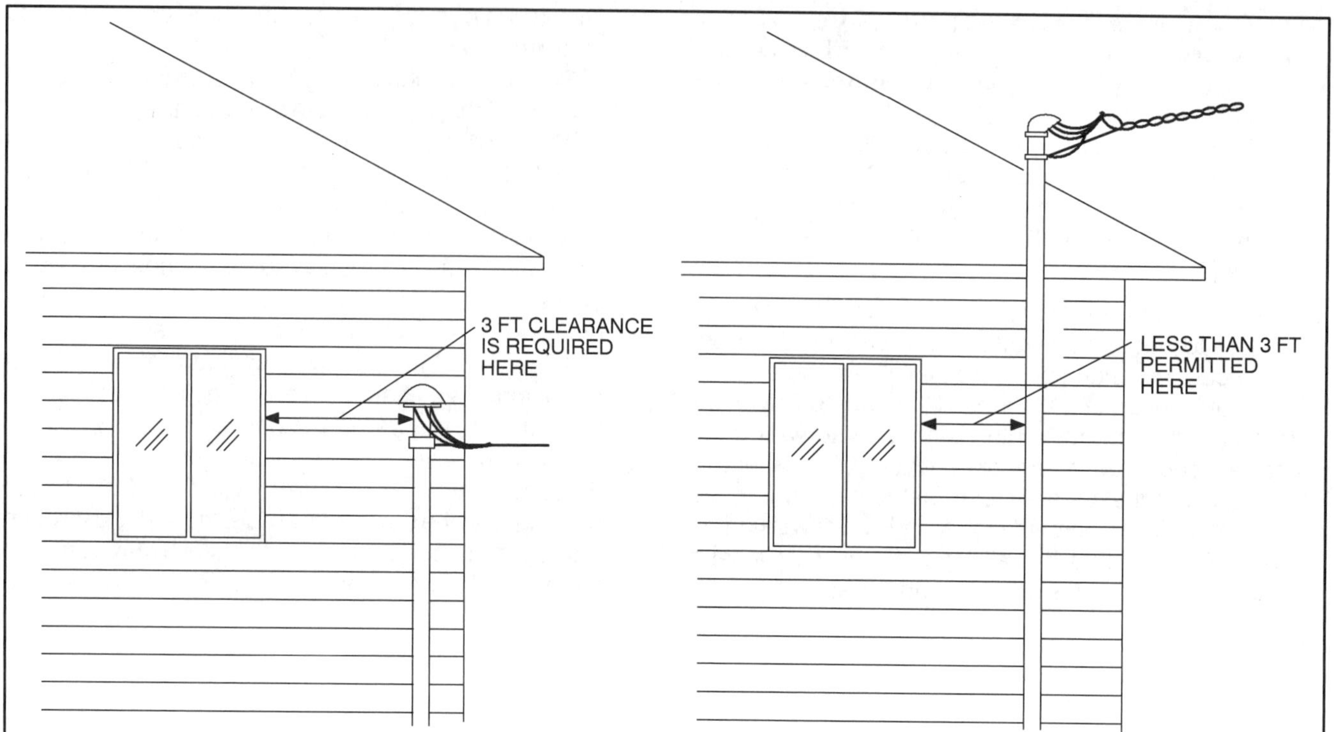

For SI: 1 foot = 304.8 mm.

Figure E3504.1
CLEARANCE FROM BUILDING OPENINGS

distance of 3 feet (914 mm) from the edge of the roof is not required for this installation.

E3504.2.2 Vertical clearance from grade. Service-drop conductors shall have the following minimum clearances from final grade:

1. For service-drop cables supported on and cabled together with a grounded bare messenger wire, the minimum vertical clearance shall be 10 feet (3048 mm) at the electric service entrance to buildings, at the lowest point of the drip loop of the building electric entrance, and above areas or sidewalks accessed by pedestrians only. Such clearance shall be measured from final grade or other accessible surfaces.
2. Twelve feet (3658 mm)—over residential property and driveways.
3. Eighteen feet (5486 mm)—over public streets, alleys, roads or parking areas subject to truck traffic.

❖ The 10-foot (3048 mm) vertical clearance requirement is from final grade to the drip loop and not to the weatherhead or service drop.

E3504.3 Point of attachment. The point of attachment of the service-drop conductors to a building or other structure shall provide the minimum clearances as specified in Sections E3504.1 through E3504.2.2. In no case shall the point of attachment be less than 10 feet (3048 mm) above finished grade.

❖ This code section indicates that the height above grade of the point of attachment governs the clearance of the service drop conductors. Several factors must be considered in determining the height of the point of attachment such as the length of conductor left to form the drip loop; the length of span and sag of the service drop conductors; and the 10-, 12-, or 18-foot (3048, 3658, 5486 mm) clearance rules. The point of attachment could be 10 feet (3048 mm) above grade at the lowest.

E3504.4 Means of attachment. Multiconductor cables used for service drops shall be attached to buildings or other structures by fittings approved for the purpose.

❖ Only fittings that are approved and designed for the purpose can be used for the attachment of the service drops. It may be necessary to install the supports or fittings to the framing members before the outside or inside walls are finished.

E3504.5 Service masts as supports. Where a service mast is used for the support of service-drop conductors, it shall be of adequate strength or be supported by braces or guys to withstand the strain imposed by the service drop. Where raceway-type service masts are used, all equipment shall be approved. Only power service drop conductors shall be permitted to be attached to a service mast.

❖ The code does not contain a prescriptive requirement for the size and type of conduit to be used or for the height above the roof it can extend before being supported by braces or guy wires where overhead service drop conductors are attached to the service mast. In many jurisdictions, the power utility company specifies that if the service riser is installed through the roof, a

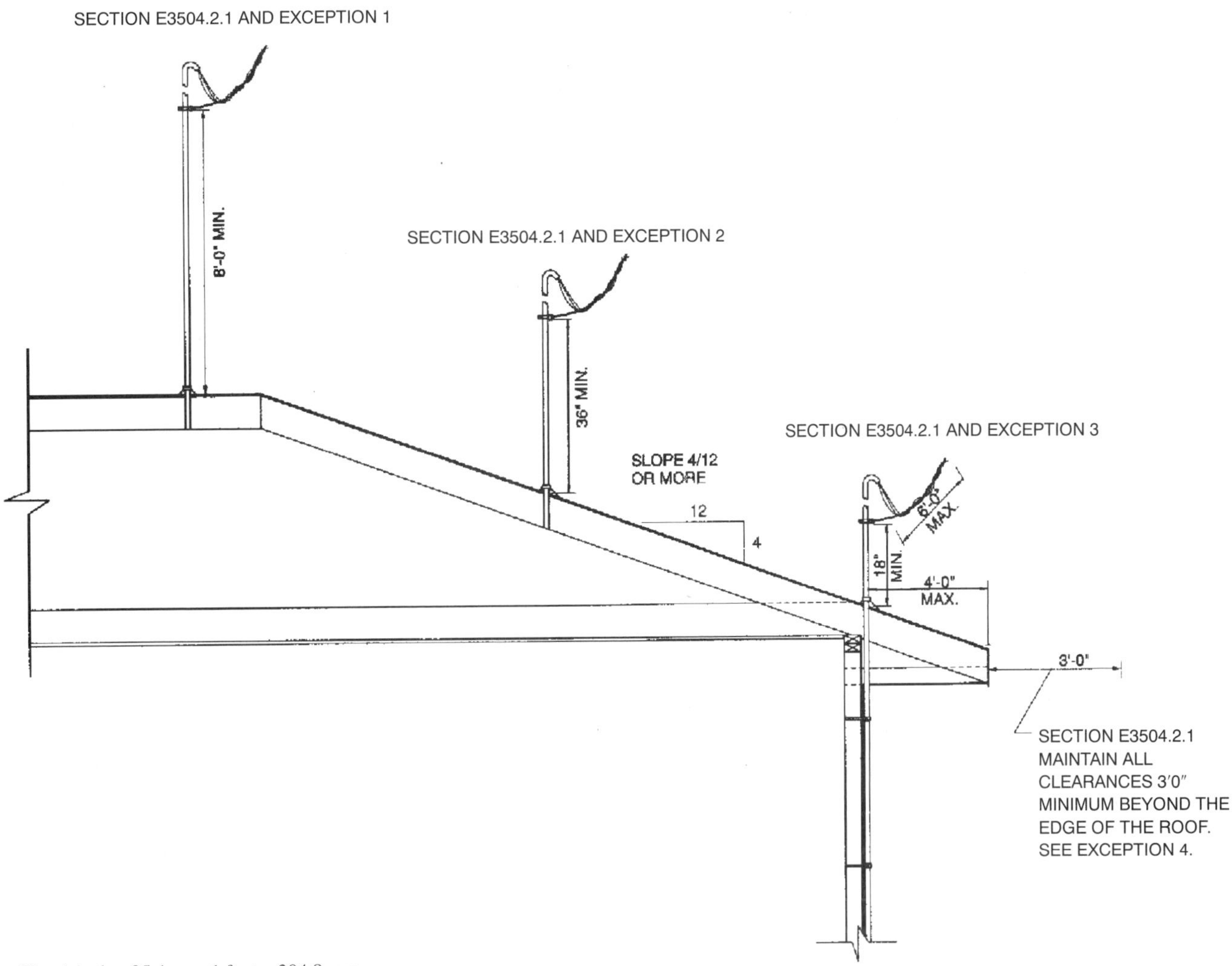

For SI: 1 inch = 25.4 mm, 1 foot = 304.8 mm.

FIGURE E3504.2.1
CLEARANCES FROM ROOFS

❖ See the commentary for Section E3504.2.1.

minimum 2-inch (51 mm) rigid conduit must be installed to provide adequate strength where the conduit connects the overhead service drop. In addition to the height above the roof of the point of attachment of the service drop, other factors should be considered such as the length of the span of the overhead conductors and the conductor weight and loading in case of wind and ice build-up. The overhead service mast must not support such things as cable TV and telephone wires.

E3504.6 Supports over buildings. Service-drop conductors passing over a roof shall be securely supported. Where practicable, such supports shall be independent of the building.

❖ Where it is necessary for a service drop to pass over the roof of a building, the service drop should be supported independently of the building over which it crosses.

SECTION E3505
SERVICE-ENTRANCE CONDUCTORS

E3505.1 Insulation of service-entrance conductors. Service-entrance conductors entering or on the exterior of buildings or other structures shall be insulated in accordance with Section E3306.5.

Exceptions:

1. A copper grounded conductor shall not be required to be insulated where it is:
   - 1.1. In a raceway or part of a service cable assembly,
   - 1.2. Directly buried in soil of suitable condition or
   - 1.3. Part of a cable assembly listed for direct burial without regard to soil conditions.

2. An aluminum or copper-clad aluminum grounded conductor shall not be required to be insulated where part of a cable or where identified for direct burial or utilization in underground raceways.

❖ Service entrance conductors are almost always installed in raceways or they are part of a multiwire cable assembly. The grounded conductor is not required to be insulated if it is in a raceway or cable. Some service entrance (SE) cables are made with a bare conductor that is used as the neutral.

E3505.2 Wiring methods for services. Service-entrance wiring methods shall be installed in accordance with the applicable requirements in Chapter 37.

❖ Table E3701.4 includes eight different types of wiring methods as allowable for service entrance wiring methods.

E3505.3 Spliced conductors. Service-entrance conductors shall permitted to be spliced or tapped by clamped or bolted connections. Splices shall be made in enclosures or, if directly buried, with listed underground splice kits. Conductor splices shall be made in accordance with Chapters 33, 36, 37 and 38.

❖ Service entrance conductors are spliced for the service of a two-family dwelling. On an overhead service, the riser is run down to a gutter where the service entrance conductors are spliced and run to the two separate main disconnects.

E3505.4 Protection against physical damage. Underground service-entrance conductors shall be protected against physical damage in accordance with Chapter 37.

❖ See the commentary for Section E3703 for rules on the protection of underground service entrance conductors.

E3505.5 Protection of service cables against damage. Above-ground service-entrance cables, where subject to physical damage, such as where installed in exposed areas near sidewalks, walkways and driveways or where subject to contact with awnings, shutters or similar objects, shall be protected by one or more of the following: rigid metal conduit, intermediate metal conduit, rigid nonmetallic conduit suitable for the location, electrical metallic tubing or other approved means.

❖ Where above-ground service entrance cables are deemed by the code official to be subject to physical damage, they must be protected by installing them in one of the types of conduit listed, or they must be installed in electrical metallic tubing. For new work, separate service entrance conductors are normally installed in one of the listed wiring methods. If service entrance cable is used, it could be run exposed. It is not required to be installed in one of the raceways on the list except where it could be damaged.

E3505.6 Direct sunlight exposure. Where exposed to direct rays of the sun, insulated conductors and cables shall be of a type listed for sunlight resistance or listed and marked "sunlight resistant."

❖ Ultraviolet radiation from sunlight will attack and degrade many plastics such as those used for conductor insulation and cable jackets (sheaths). Therefore, any conductor exposed to sunlight, such as the exposed service-entrance conductors at the drip loop at the connection to a service drop, must be "sunlight resistant." Cables run on the exterior surfaces of buildings, such as SE cables, must be "sunlight resistant" as well.

E3505.7 Mounting supports. Service cables shall be supported by straps or other approved means within 12 inches (305 mm) of every service head, gooseneck or connection to a raceway or enclosure and at intervals not exceeding 30 inches (762 mm).

❖ Where installed on the surface of a wall, service entrance cable must be supported at least every 30 inches to keep the cable flat against the wall surface and protected from damage.

E3505.8 Raceways to drain. Where exposed to the weather, raceways enclosing service-entrance conductors shall be raintight and arranged to drain. Where embedded in masonry, raceways shall be arranged to drain.

❖ Although a weatherhead installed at the top of an overhead riser is considered rain tight, differences in temperature will cause condensation. The raceway and the associated equipment have drain holes for this purpose, which must not be plugged or covered during installation.

E3505.9 Overhead service locations. Connections at service heads shall be in accordance with Sections E3505.9.1 through E3505.9.7.

❖ See the commentary for Section E3505.9.7.

E3505.9.1 Rain-tight service head. Service raceways shall be equipped with a rain-tight service head at the point of connection to service-drop conductors.

❖ See the commentary for Section E3505.9.7.

E3505.9.2 Service cable, service head or gooseneck. Service cable shall be equipped with a rain-tight service head or shall be formed into a gooseneck in an approved manner.

❖ See the commentary for Section E3505.9.7.

E3505.9.3 Service head location. Service heads, and goosenecks in service-entrance cables, shall be located above the point of attachment of the service-drop conductors to the building or other structure.

> **Exception:** Where it is impracticable to locate the service head above the point of attachment, the service head location shall not be more than 24 inches (610 mm) from the point of attachment.

❖ See the commentary for Section E3505.9.7.

E3505.9.4 Separately bushed openings. Service heads shall have conductors of different potential brought out through separately bushed openings.

❖ See the commentary for Section E3505.9.7.

E3505.9.5 Drip loops. Drip loops shall be formed on individual conductors. To prevent the entrance of moisture, service-entrance conductors shall be connected to the service-drop conductors either below the level of the service head or below the level of the termination of the service-entrance cable sheath.

❖ See the commentary for Section E3505.9.7.

E3505.9.6 Conductor arrangement. Service-drop conductors and service-entrance conductors shall be arranged so that water will not enter service raceways or equipment.

❖ See the commentary for Section E3505.9.7.

E3505.9.7 Secured. Service cables shall be held securely in place.

❖ A weatherhead or service head is installed at the top of the conduit or electrical metallic tubing used as the riser for the service entrance conductors. A separate hole is provided in the weatherhead for each conductor. The weatherhead is rain tight, which means that when it is installed properly, rain water will not enter into the raceway through the weatherhead. Some condensation will occur within the raceway for which drain holes are provided. The point of attachment of the service-drop conductors is below the weatherhead so that rain water running down along the service conductors will not enter the service raceway and get into the service equipment. At the point the service entrance conductors emerge from the weatherhead, they are bent downward to form a drip loop so that water will not run along these conductors and enter the service equipment.

SECTION E3506 SERVICE EQUIPMENT—GENERAL

E3506.1 Service equipment enclosures. Energized parts of service equipment shall be enclosed.

❖ All energized parts are enclosed within the panelboard or service equipment. See the commentary for Section E3304.8.

E3506.2 Working space. In no case shall the working space in the vicinity of service equipment be less than that specified in Chapter 33.

❖ See the commentary for Section E3305.

E3506.3 Available short-circuit current. Service equipment shall be suitable for the maximum fault current available at its supply terminals, but not less than 10,000 amperes.

❖ See the commentary for Section E3304.3.

E3506.4 Marking. Service equipment shall be marked to identify it as being suitable for use as service equipment. Individual meter socket enclosures shall not be considered service equipment.

❖ See the commentary for Section E3501.6.1.

SECTION E3507 SYSTEM GROUNDING

E3507.1 System service ground. The premises wiring system shall be grounded at the service with a grounding electrode conductor connected to a grounding electrode system as required by this code. Grounding electrode conductors shall be sized in accordance with Table E3503.1.

❖ The electrical system must be connected to earth to limit voltage imposed by lightning, line surges, or unintentional contact with higher voltage lines and to stabilize voltage to earth during normal operation. The electrical system is connected to earth or grounded by connecting the neutral, referred to as the grounded service conductor, to a grounding electrode conductor, which is run to the grounding electrode system. This is part of the fault-current path and must have sufficiently low impedance to facilitate the operation of the overcurrent device and clear the fault. An electrically continuous and permanent path must be established that will allow enough current to flow during a fault condition to trip the main circuit breaker or set of fuses. When a fault occurs at some point in the electrical wiring system, the current will return to the source of the power supply, which is the utility company transformer. The grounding electrode conductor is sized according to Table E3503.1, which lists the sizes based on the size of the ungrounded service conductors. The size of the grounding electrode conductor is not based on the rating of the main service disconnect, and it is not based strictly on the calculated load. Although the calculated load determines the minimum acceptable ungrounded conductor size, a larger conductor could be selected and installed. The amount of current that will flow in a ground-fault (line-to-ground) condition is determined by the impedance in the circuit from the source (ungrounded conductor) to the fault and the impedance in the fault-current return path. Obviously, the larger the size of the source or ungrounded conductor, the lower the impedance will be, and the larger the grounding electrode conductor must be.

The grounding electrode conductor sizes in Table E3503.1, listed for the corresponding service conductor and installed in a permanent and continuous manner, are appropriate to carry the fault current likely to be imposed on the grounding electrode conductor.

E3507.2 Location of grounding electrode conductor connection. The grounding electrode conductor shall be connected to the grounded service conductor at any accessible point from the load end of the service drop or service lateral to and including the terminal or bus to which the grounded service conductor is connected at the service disconnecting means. A grounding connection shall not be made to any grounded circuit conductor on the load side of the service disconnecting means, except as provided in Section E3507.3.

❖ Commonly, the grounded conductor is connected to the grounding electrode conductor at a grounding terminal in the service equipment at the location of the main disconnect; see Commentary Figure 3507.2. The code permits this connection to be made at any accessible location between the load end of the service drop or service lateral and the main disconnect and it is often made in a meter cabinet. The meter cabinet is an accessible location for this connection. For an overhead service, the connection could be made at the weatherhead, the meter cabinet, a current transformer cabinet or the main disconnect. If the connection is made in the meter cabinet, it may be considered inaccessible and therefore not permitted by the utility company. The main disconnect is usually the most convenient place to make the connection.

It is extremely important that no connection is made between the grounded conductors and the grounding conductors downstream of the main disconnect. Such a connection on the load side of the main disconnect could create a parallel path for fault-current and normal current. An exception to this rule is for the circuits to existing ranges and dryers. Another exception is where two or more buildings are supplied from a common service and an equipment grounding conductor is not run with the supply conductors to the other building. See the commentary for Section E3507.3.

E3507.3 Two or more buildings or structures supplied from a common service. Where two or more buildings or structures are supplied from a common service, the grounding electrode(s) required in Section E3508 at each building or structure shall be connected in a manner specified in Section E3507.3.1 or E3507.3.2. Where there is no existing grounding electrode, the grounding electrode(s) required in Section E3508 shall be installed.

Exception: A grounding electrode at separate buildings or structures shall not be required where only one branch circuit supplies the building or structure and the branch circuit includes an equipment grounding conductor for grounding the noncurrent-carrying parts of all equipment.

❖ Where two or more buildings are supplied from one service, each building must have a grounding electrode. For example, a dwelling has a service supplied from the power company to service equipment on the outside of the dwelling and has two 2-pole circuit breakers as main service disconnects. One 2-pole circuit breaker serves a feeder run to the panelboard inside the dwelling, and the other serves a feeder run to a panelboard at a separate workshop building 100 feet away from the dwelling. Both the dwelling and the workshop must have a grounding electrode as described in Section E3508. If there is not a grounding electrode at the workshop, one of the grounding electrodes listed in Section E3508 must be installed. A ground rod, concrete encased electrode, or ground ring could be installed. If water is supplied to the separate workshop building through metal pipe, then an underground metal water pipe could serve as the grounding electrode. This section states that the grounding electrode at the separate building must be connected in the building disconnecting means in one of the ways described in the two sub-sections that follow. A separate building must have a main disconnect and not simply a panelboard. A main disconnect could be a main circuit breaker in the panelboard that would disconnect all the power to the panelboard.

A grounding electrode is not required in a separate building if it is served by only one branch circuit. If a separate building supplied by only one branch circuit does not have any grounding electrode, it does not have to have a ground rod installed. If a metal underground water pipe is already installed at the building, it does not have to be bonded to the equipment grounding conductor of the one branch circuit. A building like a small tool shed, for example, may have only a light and receptacle. The equipment grounding conductor run with the branch circuit conductors is sufficient to ground any equipment for this separate building. Of course, a receptacle in a tool storage shed or accessory building at grade level would require ground-fault circuit-interrupter protection. But the grounding conductor would not be required to be connected to a grounding electrode such as a ground rod.

E3507.3.1 Equipment grounding conductor. An equipment grounding conductor as described in Section 3808 shall be run with the supply conductors and connected to the building or structure disconnecting means and to the grounding electrode(s). The equipment grounding conductor shall be used for grounding or bonding of equipment, structures, or frames required to be grounded or bonded. Any installed grounded conductor shall not be connected to the equipment grounding conductor or to the grounding electrode(s).

❖ Where the feeder that supplies the separate building is run with an equipment grounding conductor, the rules of this sub-section apply. Consider the example of the separate workshop building in the commentary to Section E3507.3. The panelboard at the workshop is served by a feeder in a raceway or cable consisting of four conductors: two ungrounded, one grounded (neutral) and one equipment grounding conductor of the appropriate sizes. The equipment grounding conductor run along with the supply conductors will be connected to the separate building disconnecting means and to the separate building grounding electrode(s). The feeder neutral or grounded conductor of the supply must be insulated and is not bonded to the panelboard cabinet. It is kept separate from the grounding conductors and is not bonded to the separate building grounding electrode(s). The panelboard cabinet of the

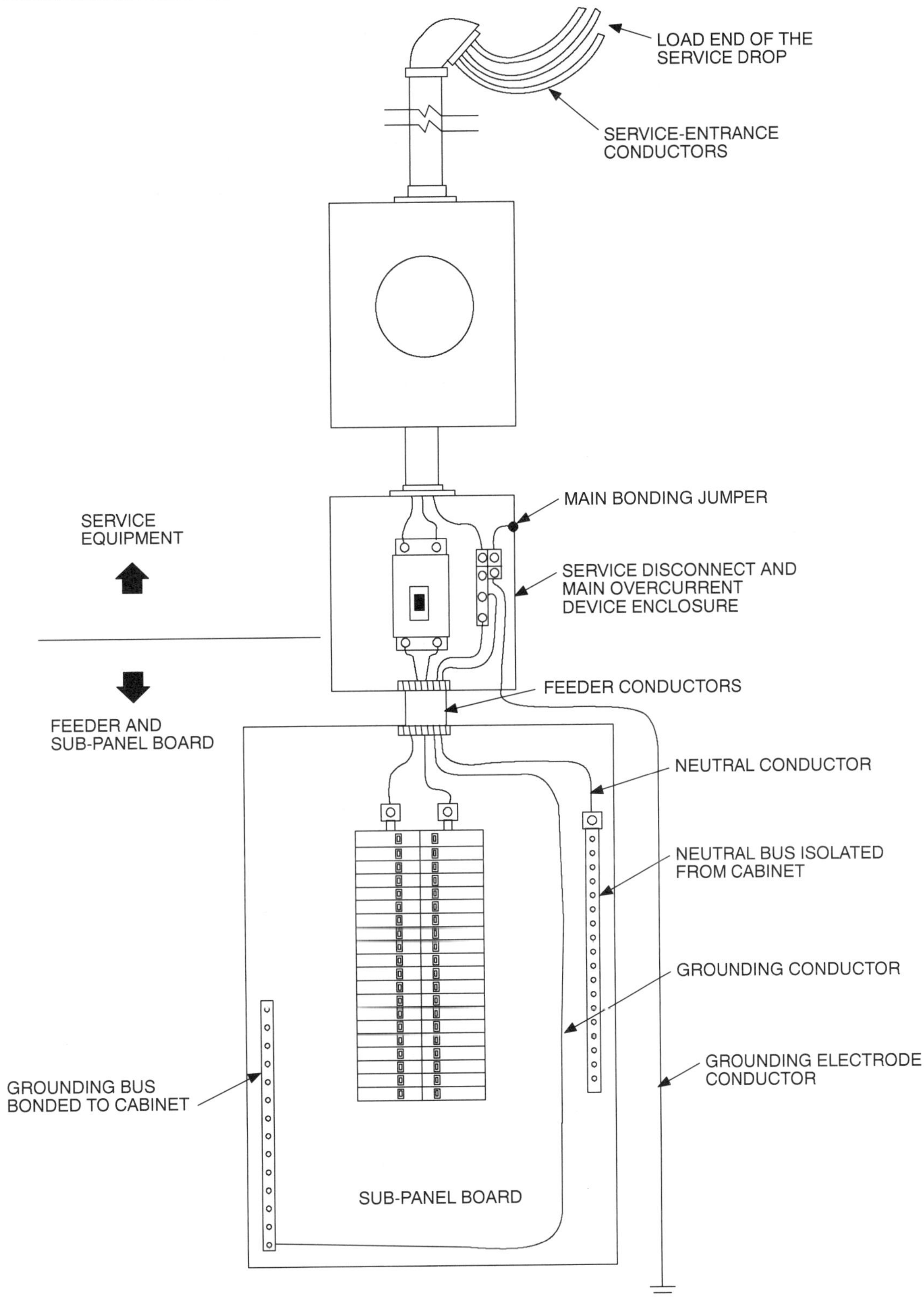

Figure E3507.2

SERVICE EQUIPMENT AND EQUIPMENT ON THE LOAD SIDE OF THE DISCONNECTING MEANS

separate building has one terminal bus for the grounded conductors and another for the grounding conductors. The bonding screw supplied for the purpose of bonding the neutral terminal strip to the metal cabinet must not be installed, and this neutral terminal strip is insulated from the metal cabinet. The grounding electrode is connected to the equipment grounding terminal, which is bonded to the metal cabinet. If the neutral terminal bus and the equipment grounding conductor terminal bus were bonded together, the current intended to return via the neutral could also flow on the equipment grounding conductor. This is a fire and life safety hazard.

The equipment-grounding conductor run with the feeder could be any of the types listed in Section E3808.8 including the conduit or tubing enclosing the circuit conductors where the installation complies with the provisions of Section E3808; see Commentary Figure E3507.3.1.

E3507.3.2 Grounded conductor. Where an equipment grounding conductor is not run with the supply conductors to the building or structure, and there are no continuous metallic paths bonded to the grounding system in both buildings or structures involved, and ground-fault protection of equipment has not been installed on the common ac service, the grounded circuit conductor run with the supply conductors to the building or structure shall be connected to the building or structure disconnecting means and to the grounding electrode(s) and shall be used for grounding or bonding of equipment, structures, or frames required to be grounded or bonded.

❖ Where the feeder that supplies the separate building is run without an equipment grounding conductor, the rules of this sub-section apply. In this case, the panelboard at the workshop is served by a feeder in a raceway or cable consisting of three conductors: two ungrounded and one neutral conductor of the appropriate sizes. The neutral or grounded conductor of the feeder is grounded again at the separate building by a grounding electrode conductor connected to the separate building grounding electrode(s). The panelboard cabinet of the separate building may have one terminal bus for the grounded conductors and another for the grounding conductors; if so, they must be bonded together or bonded to the cabinet by a main bonding jumper. Or there may be only one terminal bus to connect all grounded and grounding conductors. In this case, the bonding screw supplied for the purpose of bonding the neutral terminal strip to the metal cabinet is installed, and this bonding screw (with a green head) is considered a main bonding jumper. The grounding electrode is also bonded to the equipment-grounding terminal, which is bonded to the metal cabinet.

An important rule that applies where only a three-conductor feeder is run to the separate building and the feeder grounded conductor is bonded to the grounding electrode system at the separate building is that there must be no common continuous metallic paths bonded to the grounding electrode system in both buildings. Bonding at both ends of such objects as a metal water pipe or gas pipe that was run from the house to the separate building could result in a parallel path for neutral current as well as for fault current. If the neutral were broken or became disconnected, the return current could be through the metal pipe, which would cause a dangerous potential on non-current

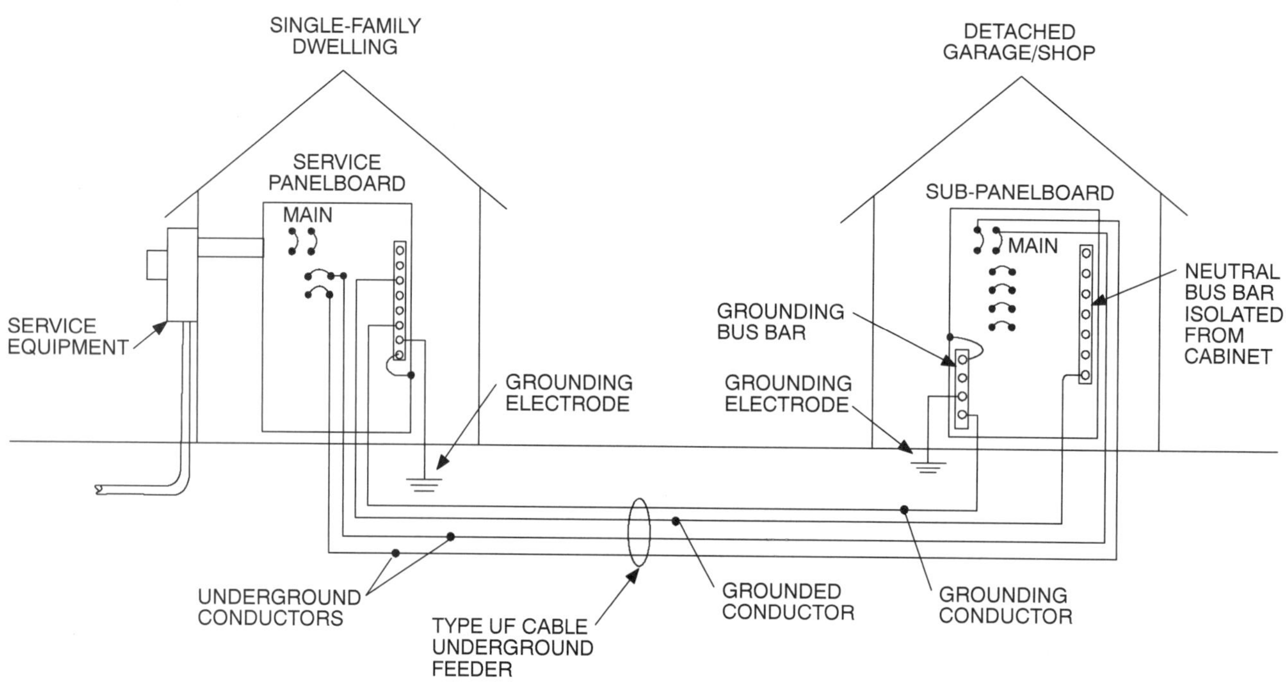

Figure E3507.3.1
TWO BUILDINGS SUPPLIED FROM A COMMON SERVICE

carrying metal surfaces; see Commentary Figure E3507.3.2.

E3507.4 Grounding electrode conductor. A grounding electrode conductor shall be used to connect the equipment grounding conductors, the service equipment enclosures, and the grounded service conductor to the grounding electrode(s).

❖ The grounded service conductor, grounding conductors and the metal service enclosures must be stabilized at a potential voltage to earth of zero volts. The grounding electrode conductor is used for the purpose of connecting these components to the grounding electrode buried in the earth.

E3507.5 Main bonding jumper. An unspliced main bonding jumper shall be used to connect the equipment grounding conductor(s) and the service-disconnect enclosure to the grounded conductor of the system within the enclosure for each service disconnect.

❖ The grounded service conductor, grounding conductors and the metal service enclosures must be connected together by the main bonding jumper. This connection is made at the main disconnect location.

E3507.6 Common grounding electrode. The service grounding electrode in or at a building shall be used to ground conductor enclosures and equipment in or on that building. Two or more grounding electrodes that are effectively bonded together shall be considered as a single grounding electrode system.

❖ This section requires that the grounding electrode used to ground the service neutral conductor must also be used to ground all the metallic enclosures in the electrical system. With one grounding electrode used to ground all of these components, there will not be a different potential; all enclosures will be at the same zero voltage potential as the grounded service conductor. Where two or more grounding electrodes are used at the premises, they must be effectively bonded together and considered as one grounding electrode. This keeps all non-current carrying metal surfaces in the electrical system at earth potential.

SECTION E3508 GROUNDING ELECTRODE SYSTEM

E3508.1 Grounding electrode system. Where available on the premises at each building or structure served, electrodes specified in Sections E3508.1.1, E3508.1.2 and E3508.1.3, and any made electrodes specified in Section E3508.2, shall be bonded together to form the grounding electrode system. Interior metal water piping located more than 5 feet (1524 mm) from the point of entrance to the building shall not be used as part of the grounding electrode system or as a conductor to interconnect the electrodes that are part of the grounding electrode system. The bonding jumper shall be sized and installed in accordance with Section E3509. The unspliced grounding electrode conductor shall run to any convenient grounding electrode available in the grounding electrode system. Such conductor shall be sized based on the largest grounding electrode conductor required for any of the available electrodes.

❖ The code requires the premises wiring system, which includes the wiring for the house, any outbuildings and

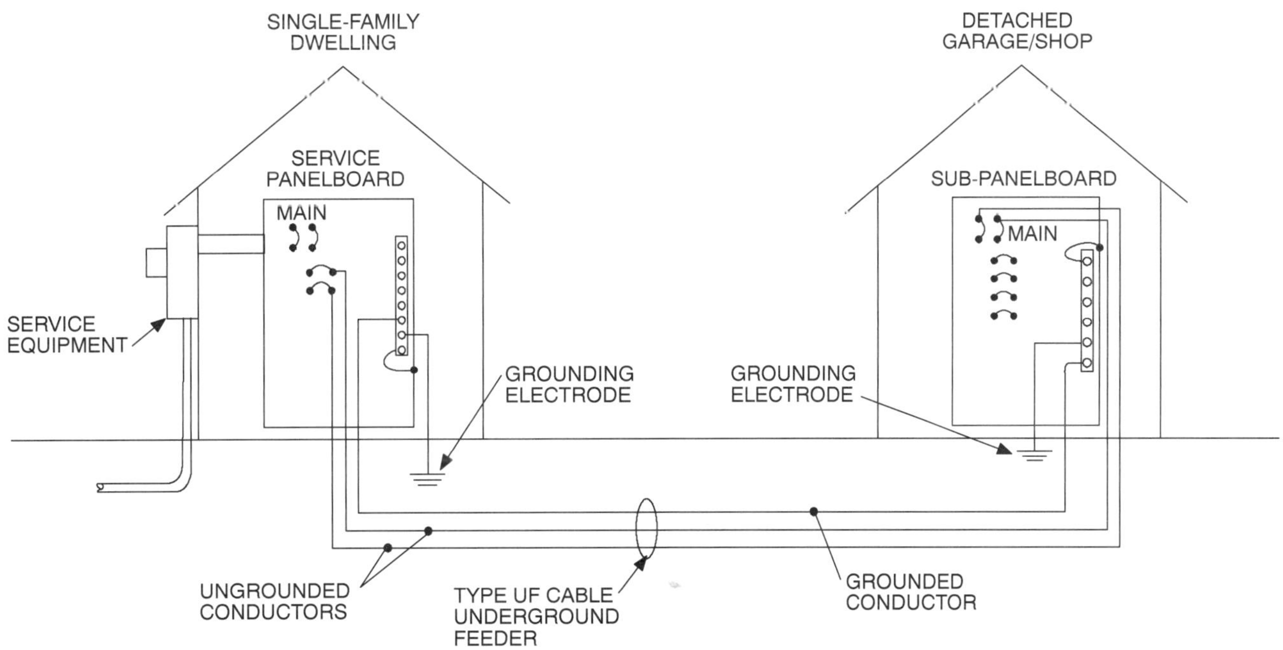

Figure E3507.3.2
TWO BUILDINGS SUPPLIED FROM A COMMON SERVICE

appurtenances, to be connected to the earth or grounded at the service. The term "grounded" as defined in Chapter 34 means connected to the earth. The electrical system must be connected to the earth to limit voltage imposed by lightning, line surges or unintentional contact with higher voltage lines and to stabilize voltage to earth during normal operation.

The electrical system is connected to the earth through the grounding electrode system. An electrode is a conductor or object serving as a conductor placed in an electrolyte or substance through which electricity could enter or leave. In this case, the electrode is placed in the earth, which contains electrolytes such as minerals and salts that are at different potencies depending on the moisture content of the soil. All of the available grounding electrodes at the premises that are identified by the code must be bonded together to form a "grounding electrode system." The code does not specify that these electrodes be bonded together all at one point or location, or at a specific point or location; it just specifies that they be bonded together. The grounded service conductor or neutral is connected to the grounding electrode system with a grounding electrode conductor. This conductor is run from the service equipment, usually at the point of the main disconnect or on the line side of the main disconnect, to any convenient point on the grounding electrode system, convenient meaning at any of the electrodes or at any point along the grounding electrode conductor. The size of the grounding electrode conductor depends on the size required for the largest of the available grounding electrodes.

Since a metal underground water pipe is one of electrodes required to be connected to the grounding electrode system, it would be convenient to run a short grounding electrode conductor to the nearest cold water pipe, but this is prohibited. In many houses, there is usually a cold water pipe just inside the house near the service equipment. The general concept would be to let the water piping system serve as the conductor all the way through the house to where the water pipe enters the house and is buried underground. This type of grounding was a common practice in some areas before the early 1990s when code provisions were changed to prohibit using the water-piping system as a grounding electrode conductor. For various reasons, the water piping system may not always remain continuous or remain the ideal conductor, and the wiring system could be left in an ungrounded condition. Piping could be disconnected during repairs of the plumbing system and a section of nonmetallic water piping could be installed. The code permits only the first five feet of the water piping system from the point where the pipe enters the building to be used to connect the electrodes together. There are usually not any or not very many interconnections of the piping system in the first five feet. In most new houses, if the underground water pipe is metal for 10 feet (5048 mm) to the point where it enters the building, it is an available grounding electrode, and the grounding electrode conductor is run continuously from the service to the point where the water pipe enters the house.

Where any or all of the three electrodes mentioned under Section E3508.1 are available or exist on the premises, they must be bonded together as part of the system. A metal underground water pipe is available if it is metal where it enters the dwelling and can be bonded to the grounding electrode conductor. However, there is no code definition for an available concrete encased grounding electrode. So, going by the common dictionary meaning, this would be an electrode capable of being used, within reach or accessible. The code does not specify that a reinforcing bar be made available or left exposed above the concrete for connection of a grounding electrode conductor, or that 20 feet of size 4 AWG bare copper be encased in the concrete and left exposed to run to the service equipment. It is only required to be connected to the grounding electrode system if it is available, which is left up to the jurisdiction. However, the concrete-encased electrode has been proven to be one of the most effective electrodes in any type of soil. See Commentary Figure E3508.1.

E3508.1.1 Metal underground water pipe. A metal underground water pipe that is in direct contact with the earth for 10 feet (3048 mm) or more, including any well casing effectively bonded to the pipe, and that is electrically continuous, or made electrically continuous by bonding around insulating joints or insulating pipe to the points of connection of the grounding electrode conductor and the bonding conductors, shall be considered as a grounding electrode. Interior metal water pipe located more than 5 feet (1524 mm) from the entrance to the building shall not be used as a part of the grounding electrode system or as a conductor to interconnect electrodes that are part of the grounding electrode system. Continuity of the grounding path or the bonding connection to interior piping shall not rely on water meters or filtering devices and similar equipment. A metal underground water pipe shall be supplemented by an additional electrode of a type specified in this section or in Section E3508.2. The supplemental electrode shall be bonded to the grounding electrode conductor, the grounded service-entrance conductor, the grounded service raceway or any grounded service enclosure.

Where the supplemental electrode is a made electrode in accordance with Section E3508.2, that portion of the bonding jumper that is the sole connection to the supplemental grounding electrode shall not be required to be larger than No. 6 copper wire or No. 4 aluminum wire.

❖ Any metal water pipe in contact with the earth for 10 feet or more is considered a grounding electrode. It is often thought that only the main water service piping fits this description, but any metal water pipe, such as an irrigation pipe or pipe from a water well, if buried in the earth for 10 feet (3048 mm) or more, must be bonded as part of the grounding electrode system.

The pipe could be of any material such as copper or steel since the code does not mention the type of metal. In many houses, a water meter, water pressure reducing valve, or similar equipment is installed in the water supply line. A bonding jumper of the same size as the grounding electrode conductor is installed around such devices because in many of these devices there are non-conductive materials. In many cases, when the device is removed and/or replaced, the grounding electrode is disconnected. A useful practice is to leave enough slack in the bonding jumper around the device that the jumper will not have to be taken off the piping in the event the device is replaced.

The code assumes that the first 5 feet (1524 mm) of water piping, measured from the point that the piping penetrates an outside wall or floor slab on grade, will not be disturbed or altered by plumbing work. Any piping beyond 5 feet (1524 mm) into the building is more likely to be altered such that electrical continuity is lost. This alteration could take the form of the installation of plastic piping or dielectric fittings or the removal of clamps.

A supplemental grounding electrode is always required when a water pipe electrode is used. The most commonly used supplemental grounding electrode is the ground rod. That is why in so many dwellings a grounding electrode conductor is run from the service equipment to the entry point of the water pipe and another grounding electrode conductor is run to a ground rod. The ground rod is quite often driven in the ground close to the service equipment. The grounding electrode conductor from the service equipment to a ground rod is not required to be larger than size 6 AWG. If the supplemental grounding electrode is a concrete encased electrode, the maximum size 6 AWG bonding jumper rule does not apply.

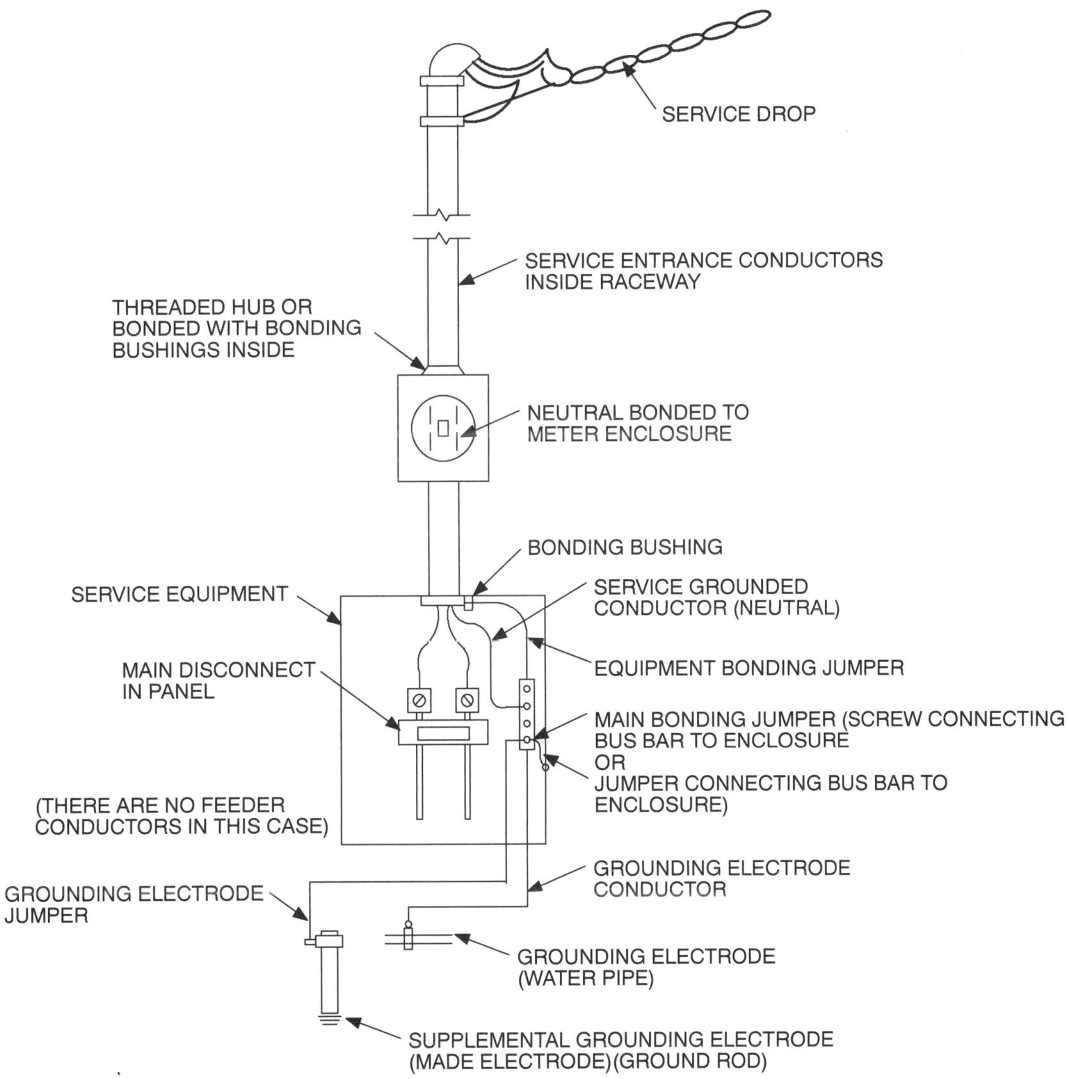

Figure E3508.1
GROUNDING ELECTRODE SYSTEM

E3508.1.2 Concrete-encased electrode. An electrode encased by at least 2 inches (51 mm) of concrete, located within and near the bottom of a concrete foundation or footing that is in direct contact with the earth, consisting of at least 20 feet (6096 mm) of one or more bare or zinc-galvanized or other electrically conductive coated steel reinforcing bars or rods of not less than $^1/_2$ inch (12.7 mm) diameter, or consisting of at least 20 feet (6096 mm) of bare copper conductor not smaller than No. 4 shall be considered as a grounding electrode. Reinforcing bars shall be permitted to be bonded together by the usual tie wires or other effective means.

❖ Encasing an electrode in concrete must be done, of course, before the concrete is placed for the footings. The 20 feet of size 4 AWG bare copper conductor is run in the forms close to the bottom of the footing before the concrete is placed. Enough conductor is left extending out of the footing to be run to the service equipment without a splice. At the point where the copper conductor exits the concrete, it could be broken off by being moved back and forth several times before and during construction. To help protect this conductor, a non-metallic conduit around the conductor where it leaves the footing would be helpful. If enough conductor is not left to extend to the metal service enclosure, the grounding electrode conductor, or the service equipment containing the grounded service conductor, the concrete-encased electrode can be spliced with exothermic welding or approved bolted-type connection. In some installations, a reinforcing bar is left exposed so that a grounding electrode conductor can be connected to it and run to the grounding electrode system.

E3508.1.3 Ground rings. A ground ring encircling the building or structure, in direct contact with the earth at a depth below the earth's surface of not less than 2.5 feet (762 mm), consisting of at least 20 feet (6096 mm) of bare copper conductor not smaller than No. 2 shall be considered as a grounding electrode.

❖ Where a ground ring is installed around the building, it must be bonded as part of the grounding electrode system. A ground ring may be necessary because of the resistivity of the soil, the type of equipment present at the house, or other conditions. It is not considered a made electrode, but it must be bonded to the system.

E3508.2 Made and other electrodes. Where none of the electrodes specified in Section E3508.1 is available, one or more of the electrodes specified in Section E3508.2.1 shall be used. Where practicable, made electrodes shall be embedded below permanent moisture level. Made electrodes shall be free from nonconductive coatings such as paint or enamel. Where more than one electrode is used, each electrode of one grounding system shall be not less than 6 feet (1829 mm) from any other electrode of another grounding system. Two or more grounding electrodes that are effectively bonded together shall be considered as a single grounding electrode system.

❖ Whereas the electrodes discussed above are used if they are available in order to form the grounding electrode system, a made electrode is one that is made and installed specifically for the electrical installation. An electrode is "made" in a case where none of the above electrodes are available. Dwellings have a water piping system, but not all will necessarily be metal pipe and underground.

These code provisions also apply to a separate service or feeder at an outbuilding or detached structure, which may not have water piping. As noted above in Section E3508.1.1, a metal underground water pipe must be supplemented by another electrode. If a rod or pipe is selected as the supplemental electrode, the rules of this section apply. It is important to install the rod or pipe electrode at least 6 feet away from any other electrode. It cannot always be driven in the most convenient place, which is directly below the service equipment. In a case where multiple ground rods are installed, they must be 6 feet apart and bonded together.

E3508.2.1 Rod and pipe electrodes. Rod and pipe electrodes shall be not less than 8 feet (2438 mm) in length and shall consist of the following materials:

1. Electrodes of pipe or conduit shall not be smaller than $^3/_4$-inch trade size (metric designator 21) and, where of iron or steel, shall have the outer surface galvanized or otherwise metal-coated for corrosion protection.
2. Electrodes of rods of iron or steel shall be at least $^5/_8$ inch (15.9 mm) in diameter. Stainless steel rods less than $^5/_8$ inch (15.9 mm) in diameter, nonferrous rods or their equivalent shall be listed and shall be not less than $^1/_2$ inch (12.7 mm) in diameter.

❖ Rod and pipe electrodes must be at least 8 feet long, and must be made of steel, stainless steel, or nonferrous material. If not of stainless steel or nonferrous material, the rod must have a coating for corrosion protection.

E3508.2.2 Installation. The electrode shall be installed such that at least 8 feet (2438 mm) of length is in contact with the soil. It shall be driven to a depth of not less than 8 feet (2438 mm) except that, where rock bottom is encountered, the electrode shall be driven at an oblique angle not to exceed 45 degrees from the vertical or shall be buried in a trench that is at least 2.5 feet (762 mm) deep. The upper end of the electrode shall be flush with or below ground level unless the aboveground end and the grounding electrode conductor attachment are protected against physical damage.

❖ The previous section requires that the ground rod must be at least 8 feet (2438 mm) long. This section requires that at least 8 feet (2438 mm) of the ground rod is in contact with the earth. Therefore, an 8-foot (2438 mm) long ground rod cannot be installed with the top exposed for the connection of the grounding electrode conductor. The ground rod clamp must be suitable for direct burial, or an exothermic welding process must be used. In some cases, the installers leave the upper end of the rod exposed so that the connection of the grounding electrode conductor can be inspected. The only way this would be acceptable is if the rod were longer than 8 feet (2438 mm) , and then the attachment of the grounding electrode conductor must be protected against physical damage. In some cases the rod may be driven away from the house to miss the footing as it is being driven. As a result, it is several inches away from the wall of the house in a flower bed or in a future driveway or patio. Some electricians leave it exposed rather than cover it with concrete for a sidewalk or patio. Where the rod is installed in such a location, the code official would need to inspect the connection and the length of the driven or buried ground rod before the concrete slab was placed.

E3508.2.3 Aluminum electrodes. Aluminum electrodes shall not be permitted.

❖ See the commentary for Section E3510.1.

E3508.3 Resistance of made electrodes. A single electrode consisting of a rod or pipe that does not have a resistance to ground of 25 ohms or less shall be augmented by one additional electrode of any of the types specified in Sections E3508.1 or E3508.2. Where multiple rod or pipe electrodes are installed to meet the requirements of this section, they shall be not less than 6 feet (1829 mm) apart.

❖ The requirement for a maximum of 25 ohms of resistance between the grounding electrode and the earth is only for made electrodes. It is not required that the resistance measurement be taken between an underground water pipe and the earth, for instance, and there is no resistance measurement required for a concrete-encased electrode. If the resistance Is over 25 ohms for a made electrode, the code requires that an additional ground rod be installed at least 6 feet away from the first one. There is no requirement to keep adding electrodes until the resistance is reduced to at least 25 ohms. There are many different types of soil, and the resistivity or conductivity of the soil changes with the amount of moisture in the soil and even with the temperature of the soil. This code rule provides a low-resistance path to ground to limit the voltage imposed on the system by lightning or ground faults. But the earth is not to be used as the sole grounding conductor or ground- fault path because it has such high resistance. The current flow through the earth would not facilitate the operation of the overcurrent device under fault conditions.

This rule may be implemented differently by code officials in different jurisdictions. The code official might permit a single ground rod to be installed per service as the supplemental grounding electrode and not require the resistance measurement to be taken on every house in a tract of houses or in the town since the resistance may have been previously measured and known to be low for the soil in the area.

The code does not state specifically that the resistance measurement must be taken for each house and submitted. In some areas, it may be well known that the soil resistivity is high and there will almost always be over 25 ohms of resistance to ground from the ground rod. The installer may not have the equipment to measure the resistance to ground, and if he or she is wiring only one house, it may be easier and less expensive to go ahead and install two ground rods. This would satisfy the intent of the code requirement. In applying this code provision, the code official might simply take the approach that two ground rods must be installed unless the resistance measurement is taken and shown to be 25 ohms or less.

E3508.4 Metal underground gas piping system. A metal underground gas piping system shall not be used as a grounding electrode.

❖ Where a dwelling is served with a metal underground gas piping system, the gas utility usually installs an insulating fitting in the line at or near the point the gas line enters the house. This interrupts the continuity of the metal pipe to the earth. If there is not an insulating fitting in the line, current could flow in the gas piping, creating a dangerous situation.

SECTION E3509 BONDING

E3509.1 General. Bonding shall be provided where necessary to ensure electrical continuity and the capacity to conduct safely any fault current likely to be imposed.

❖ Bonding is the connection of metallic parts to make one continuous current path to ensure the fault current path has low impedance. Bonding and grounding are often thought of as the same general subject since bonding is related to grounding as far as the fault-current path is concerned.

E3509.2 Bonding of services. The noncurrent-carrying metal parts of the following equipment shall be effectively bonded together:

1. The service raceways or service cable armor.
2. All service enclosures containing service conductors, including meter fittings, and boxes, interposed in the service raceway or armor.
3. Any metallic raceways or armor enclosing a grounding electrode conductor. Bonding shall apply at each end and to all intervening raceways, boxes and enclosures between the service equipment and the grounding electrode.

❖ All of the service equipment, raceways, and metallic surfaces must be bonded together to ensure a low impedance fault-current path of electrical continuity. The enclosures and cabinets such as the service mast, weatherhead, conduit to and from the meter base, meter base, cable armor, boxes and the can for the main disconnecting means all contain conductors that have overcurrent protection only on the load end. The service equipment and conductors may be subject to a high fault current, which would require the utility company transformer switch or overcurrent device to open to clear the fault. Where metal conduit or electrical metallic tubing is used to enclose the grounding electrode conductor, the conduit or tubing must be bonded to the grounding electrode conductor to make one continuous fault-current path.

E3509.3 Bonding to other systems. An accessible means external to enclosures for connecting intersystem bonding and grounding conductors shall be provided at the service by at least one of the following means:

1. Exposed nonflexible metallic service raceways.
2. Exposed grounding electrode conductor.
3. Approved means for the external connection of a copper or other corrosion-resistant bonding or grounding conductor to the service raceway or equipment.

For the purposes of providing an accessible means for intersystem bonding, the disconnecting means at a separate building or structure as permitted in Section E3507.3 shall be considered the service equipment.

❖ Communications systems, cable TV, and similar systems must be grounded. If the enclosures, raceways, and non-current carrying metal surfaces of these "other systems" are not bonded to the premises wiring system, they could operate at a difference in potential. Intersystem bonding is required to reduce the shock hazard and minimize the possible fire danger. Where the service equipment cabinet and meter base are flush with a brick or stucco wall, the raceways are concealed within the wall, the grounding electrode conductor is not accessible, and the ground rod has been buried, there is nothing available to which to bond the metallic jacket of the communications cable. If the installer drives a separate ground rod to bond the cable TV cable shield and boxes to, it would be a code violation if not bonded to the service electrode system. An accessible means external to the service equipment enclosures must be provided. If the underground service lateral is run in metal conduit installed on the surface of the wall from the ground up, or if the overhead service riser is external, which it most often is, or if the grounding electrode conductor is run to a ground rod along the outside wall, these would be accessible means for intersystem bonding. In some cases, especially where the service equipment is flush with the wall, it may be necessary to bond a size 6 AWG bare copper conductor to the enclosure and leave several inches of the conductor outside the equipment so that it is accessible for the connection of bonding jumpers of other systems.

E3509.4 Method of bonding at the service. Electrical continuity at service equipment, service raceways, and service conductor enclosures shall be ensured by one or more of the methods specified in Sections E3509.4.1 through E3509.4.4.

Bonding jumpers meeting the other requirements of this code shall be used around concentric or eccentric knockouts that are punched or otherwise formed so as to impair the electrical connection to ground. Standard locknuts or bushings shall not be the sole means for the bonding required by this section.

❖ The non-current-carrying metal parts of the service equipment must be bonded together. The methods and means to do this are covered in the following four sub-sections. Punching concentric and eccentric knockouts greatly reduces the current carrying capacity or conductivity of the connection. Fault current will flow along the raceway and be impeded at the point where the raceway fitting is attached to the cabinet through one of these knockouts. In some cases, where part of but not the entire knockout is removed, the mechanical connection is so weak that it is necessary to install reducing washers just to strengthen the connection. There certainly is not a solid electrical connection at this point. In addition, standard locknuts are not permitted to provide the only bonding connection of the service equipment. In some instances, a good practice is to use standard locknuts on the outside and inside of the cabinet to connect the raceway or fitting and then use a special bonding bushing for the bonding of the raceway or fitting to the enclosure.

E3509.4.1 Grounded service conductor. Equipment shall be bonded to the grounded service conductor in a manner provided in this code.

❖ Bonding to the grounded service conductor is done by using listed pressure connectors such as a terminal strip in a panelboard cabinet, listed lugs, clamps, or exothermic welding. These conductor splicing and connecting provisions are covered in Section E3511.1 and Section E3306.

E3509.4.2 Threaded connections. Equipment shall be bonded by connections using threaded couplings or threaded bosses on enclosures. Such connections shall be made wrench tight.

❖ Raceways joined by the use of threaded couplings and connectors must be wrench tight. High impedance in the ground-fault current path can occur if the installer only hand tightens threaded couplings. "Wrench tight" is another term that is not defined in Chapter 34, but the meaning is obvious. Tightening the threaded couplings with a wrench is necessary not only for mechanical strength but to ensure that the fault current path has good electrical continuity. For overhead service risers, a typical installation uses a 2-inch (51 mm) rigid conduit. A hub is provided with the service equipment that has pre-drilled holes for the attachment of the conduit. It is important that the conduit is threaded into the hub wrench tight.

E3509.4.3 Threadless couplings and connectors. Equipment shall be bonded by threadless couplings and connectors for metal raceways and metal-clad cables. Such couplings and connectors shall be made wrench tight. Standard locknuts or bushings shall not be used for the bonding required by this section.

❖ Threadless couplings and connectors are permitted for bonding the parts of service equipment together, but these must be rain-tight where installed outdoors. For example, a set-screw connector is permitted to connect the conduit of an underground service lateral to the service equipment cabinet where it is run up to the bottom of the cabinet. Also, where the cabinet is installed in the wall and is flush with the surface, set-screw connectors and couplings are permitted, but they must be made wrench tight.

E3509.4.4 Other devices. Equipment shall be bonded by approved devices, such as bonding-type locknuts and bushings.

❖ To ensure electrical continuity throughout the interconnection of the various components of the service equipment, special bonding bushings are required where raceways and fittings are connected to concentric and eccentric knockouts punched in boxes and cabinets. Bonding bushings have a set screw that positively bonds to the threads of the raceway or fitting and a lug for the connection of a bonding jumper.

E3509.5 Sizing bonding jumper on supply side of service and main bonding jumper. The bonding jumper shall not be smaller than the sizes shown in Table E3503.1 for grounding electrode conductors. Where the service-entrance conductors are paralleled in two or more raceways or cables, the equipment bonding jumper, where routed with the raceways or cables, shall be run in parallel. The size of the bonding jumper for each raceway or cable shall be based on the size of the service-entrance conductors in each raceway or cable. The bonding jumper for a grounding electrode conductor raceway shall be the same size or larger than the required enclosed grounding electrode conductor.

❖ Paralleled service conductors are used in relatively few dwelling unit services. For all except paralleled service conductors, Table E3503.1 provides the minimum size for the bonding jumper. For example, for a 200-ampere service using size 2/0 AWG copper conductors, the table indicates that bonding jumpers would be a minimum of size 4 AWG copper. The table also provides the minimum sizes for grounding electrode conductors, which are the same sizes for the line-side bonding jumpers for paralleled sets of services conductors for services of 200 amperes and over. If the supply side equipment-bonding jumpers are routed in separate raceways with the service conductors, the minimum size is based on the conductors in each raceway.

Where a grounding electrode conductor is installed in a metallic raceway, the raceway must be bonded to the grounding electrode conductor at each end. The bonding jumper must be the same size as the grounding electrode conductor.

E3509.6 Metal water piping bonding. The interior metal water piping system shall be bonded to the service equipment enclosure, the grounded conductor at the service, the grounding electrode conductor where of sufficient size, or to the one or more grounding electrodes used. The bonding jumper shall be sized in accordance with Table E3503.1. The points of attachment of the bonding jumper(s) shall be accessible.

❖ Because the interior metal water piping might become energized, it must be bonded to the grounding electrode system. This is not the same requirement that the metal underground water pipe be used as a grounding electrode. If the water service is a metal pipe for 10 feet (3048 mm) underground and metal where it enters the house, it is used as the grounding electrode. In some dwellings, the water piping system changes to plastic inside the house, and much of the water supply piping is plastic; however, any piping that is metal must be bonded to one of the parts of the grounding electrode system listed in this section. If the entire piping system is metal, it may seem obvious that the grounding electrode conductor being run to the first 5 feet of the point at which the water pipe enters the house would bond the entire water piping system. In many cases there is a dielectric union at the entry of the water piping system, and a jumper is required from the line side of the union to the downstream side of the water piping. It may be necessary to bond between the hot and cold water pipes since these could have nonconductive materials at the water heater. The bonding jumper must be at least the size of the grounding electrode conductor determined for the service from Table E3503.1. The pipe clamp or fitting used to attach the bonding jumper to the water piping system must be accessible. The water heater is a convenient place for this since the attachment point will always be accessible. Also, this may be the best place to bond between the hot and cold water piping if necessary.

E3509.7 Bonding other metal piping. Where installed in or attached to a building or structure, metal piping systems, including gas piping, capable of becoming energized shall be bonded to the service equipment enclosure, the grounded conductor at the service, the grounding electrode conductor where of sufficient size, or to the one or more grounding electrodes used. The bonding jumper shall be sized in accordance with Table E3808.12 using the rating of the circuit capable of energizing the piping. The equipment grounding conductor for the circuit that is capable of energizing the piping shall be permitted to serve as the bonding means. The points of attachment of the bonding jumper(s) shall be accessible.

❖ The aboveground metal gas piping upstream from equipment shutoff valves must be bonded since it could become energized. The rating of the circuit serving the gas appliance or equipment determines the size of the bonding jumper in accordance with Table E3808.12. Under this practice, the size of a bonding conductor for a gas pipe serving a furnace supplied with a 20 ampere circuit would be 12 AWG according to the table. A three-conductor nonmetallic sheathed cable with size 12 AWG conductors (12-2 with ground) that is to run to the furnace is adequate for bonding the metal gas pipe since the grounding conductor of the cable is bonded to the grounding electrode system. The gas piping is connected to the appliance by metal-to-metal joints; therefore, the piping is bonded to the appliance and the appliance grounding conductor. The gas piping is considered as bonded where connected to the equipment-grounding conductor for a circuit capable of energizing the piping. This is specifically stated in the 2002 edition of NFPA 70, Section 250.104(B).

E3509.8 Bonding other metal piping. Interior metal piping capable of becoming energized shall be bonded to the service equipment enclosure, the grounded conductor at the service, the grounding electrode conductor where of sufficient size, or to the one or more grounding electrodes used. The bonding jumper shall be sized in accordance with Table E3808.12 using the rating of the circuit capable of energizing the piping. A piping system shall be considered as bonded where connected to the equipment grounding conductor for the circuit capable of energizing such piping system.

❖ All other interior metal piping is bonded the same as described above for gas piping.

SECTION E3510
GROUNDING ELECTRODE CONDUCTORS

E3510.1 Installation. A grounding electrode conductor or its enclosure shall be securely fastened to the surface on which it is carried. A No. 4 or larger conductor shall be protected where exposed to severe physical damage. A No. 6 grounding conductor that is free from exposure to physical damage shall be permitted to be run along the surface of the building construction without metal covering or protection where it is and securely fastened to the construction; otherwise, it shall be in rigid metal conduit, intermediate metal conduit, rigid nonmetallic conduit, electrical metallic tubing or cable armor. Grounding conductors smaller than No. 6 shall be in rigid metal conduit, intermediate metal conduit, rigid nonmetallic conduit, electrical metallic tubing or cable armor.

Insulated or bare aluminum or copper-clad aluminum grounding conductors shall not be used where in direct contact with masonry or the earth or where subject to corrosive conditions. Where used outside, aluminum or copper-clad aluminum grounding conductors shall not be installed within 18 inches (457 mm) of the earth.

❖ This section provides rules for the physical protection of the grounding electrode conductors. Two terms are used in this section that are not defined in Chapter 34 of this code: "severe physical damage" and "physical damage." It is up to the code official to decide if the grounding electrode conductors as installed are exposed to severe physical damage or physical damage.

Where run on the interior of the dwelling, it is unlikely that these conductors would be exposed to physical damage. Where installed along the outside wall or surface of the house, the possibility of physical damage is often a concern. In most new houses, the grounding electrode conductor is run on the interior, and another grounding electrode conductor is run to a made electrode on the outside of the building. The conductor that is the sole connection to the supplemental grounding electrode or ground rod is considered one of the grounding electrode conductors, and the requirements of this section apply to it as well as to a grounding electrode conductor run to the entry point of the metal underground water pipe.

Quite often a solid bare copper conductor is installed on the outside of the house as the grounding electrode conductor and is run from the service equipment enclosure to grade along the wall of the house. The smallest size grounding conductor permitted to run exposed without any covering or protection is 6 AWG, but only if it is free from exposure to physical damage. Although it can be done, it is difficult to install these conductors in a neat and workmanlike manner where they are run exposed. Many electricians install these surface conductors in one of the raceways listed in this section.

Aluminum conductors are not permitted to be installed within 18 inches of the earth where used outside because of the moisture and potential for corrosion.

E3510.2 Enclosures for grounding electrode conductors. Metal enclosures for grounding electrode conductors shall be electrically continuous from the point of attachment to cabinets or equipment to the grounding electrode, and shall be securely fastened to the ground clamp or fitting. Metal enclosures that are not physically continuous from cabinet or equipment to the grounding electrode shall be made electrically continuous by bonding each end to the grounding conductor. Where a raceway is used as protection for a grounding conductor, the installation shall comply with the requirements of Chapter 37.

❖ Where grounding electrode conductors are installed inside a metal raceway, the metal conduit or tubing must be bonded at each end to the conductor so that it is electrically continuous. If this is not done and a high fault current flows through the grounding electrode conductor, the current could create a magnetic flux field around the raceway enclosing the conductor and, depending on the amount of fault current, this would create a high impedance in the fault current return path. One solution to this is to install non-metallic conduit where permitted for protection of the grounding electrode conductor.

SECTION E3511 GROUNDING ELECTRODE CONDUCTOR CONNECTION TO THE GROUNDING ELECTRODES

E3511.1 Methods of grounding conductor connection to electrodes. The grounding conductor shall be connected to the grounding electrode by listed lugs, listed pressure connectors, listed clamps or other listed means. Connections depending on solder shall not be used. Ground clamps shall be listed for the materials of the grounding electrode and the grounding electrode conductor and, where used on pipe, rod or other buried electrodes, shall also be listed for direct soil burial or concrete encasement. Not more than one conductor shall be connected to the grounding electrode by a single clamp or fitting unless the clamp or fitting is listed for multiple conductors. One of the methods indicated in the following items shall be used:

1. A pipe fitting, pipe plug, or other approved device screwed into a pipe or pipe fitting.
2. A listed bolted clamp of cast bronze or brass, or plain or malleable iron.
3. For indoor telecommunications purposes only, a listed sheet metal strap-type ground clamp having a rigid metal base that seats on the electrode and having a strap of such material and dimensions that it is not likely to stretch during or after installation.
4. Other means. Equally substantial approved means.

❖ This section lists items that can be used to connect the grounding electrode conductor to the grounding electrode.

E3511.2 Accessibility. The connection of the grounding electrode conductor or bonding jumper to the grounding electrodes that are not buried or concrete encased shall be accessible.

❖ This connection is required to be accessible except for buried ground rod clamps. A concrete patio, sidewalk or driveway may be installed where the ground rod is driven. In cases like these, especially where the connection and/or conductor could be damaged, it is not required to be accessible.

E3511.3 Effective grounding path. The connection of the grounding electrode conductor or bonding jumper shall be made in a manner that will ensure a permanent and effective grounding path. Where necessary to ensure effective grounding for a metal piping system used as a grounding electrode, effective bonding shall be provided around insulated joints and sections and around any equipment that is likely to be disconnected for repairs or replacement. Bonding conductors shall be of sufficient length to permit removal of such equipment while retaining the integrity of the bond.

❖ Where a metal piping system is used as the grounding electrode, a permanent and effective ground-fault current path must be maintained. Water meters, pressure reducing devices and the like may be installed in the piping system, and it is necessary to install a bonding jumper around such devices. The bonding jumper must be of sufficient length that the item can be repaired or replaced without disconnecting the bonding jumper.

E3511.4 Protection of ground clamps and fittings. Ground clamps or other fittings shall be approved for applications without protection or shall be protected from physical damage by installing them where they are not likely to be damaged or by enclosing them in metal, wood or equivalent protective coverings.

❖ If not protected from physical damage by their location, ground clamps and fittings must be protected with approved enclosures. The code official must decide if physical damage is likely to occur.

E3511.5 Clean surfaces. Nonconductive coatings (such as paint, enamel, and lacquer) on equipment to be grounded shall be removed from threads and other contact surfaces to ensure good electrical continuity or shall be connected by fittings that make such removal unnecessary.

❖ Nonconductive coatings such as paint must be removed to help ensure the continuity of a grounding circuit that may need to carry high levels of fault current. Fittings that make the removal of paint unnecessary may not be available or may be very hard to find.

Bibliography

The following resource materials are referenced in this chapter or are relevant to the subject matter addressed in this chapter.

NFPA 70-02, *National Electrical Code.* Quincy, MA: National Fire Protection Association, 2002.

Chapter 36: Branch Circuit and Feeder Requirements

General Comments

Chapter 36 addresses the sizing of conductors for feeders and branch circuits, specifies the required branch circuits, provides for overcurrent protection of such conductors, specifies limitations for branch circuit loading, and addresses panel board ratings and protection. After the service is designed or installed, the next stage in the design or installation is the feeder and branch circuit design or installation.

Purpose

The primary purpose of this chapter is to provide the required number and type/rating of branch circuits to serve the anticipated loads and to protect conductors from overload (overcurrent). This translates to protection from fire and property damage.

SECTION E3601 GENERAL

E3601.1 Scope. This chapter covers branch circuits and feeders and specifies the minimum required branch circuits, the allowable loads and the required overcurrent protection for branch circuits and feeders that serve less than 100 percent of the total dwelling unit load. Feeder circuits that serve 100 percent of the dwelling unit load shall be sized in accordance with the procedures in Chapter 35.

❖ Feeders carry power from the service to the branch circuit overcurrent protective devices or the panelboard, which is where the branch circuits begin. The figure shows that the feeder runs from the service equipment to supply power to the panelboard, in accordance with the definition given in Chapter 34. The terms "subfeeder" and "subpanel" are not defined in the code but are terms commonly used in the field. A subfeeder runs from a panelboard to serve another panel or subpanel. In other words, feeders do not serve loads directly but serve a panelboard from which a branch circuit is run to supply the appliances or equipment. Overcurrent protection is provided at the supply end of the feeder or subfeeder, and the panel or subpanel it serves is not required to have a main circuit breaker or set of main fuses. These panelboards have main lugs only.

The conductor size for a feeder that carries 100 percent of the load is calculated under the provisions of Chapter 35. An example of a feeder that carries 100 percent of the load is one where the service consists of only one main disconnect, and the conductors from the service to the panelboard (the feeder conductors) supply the power for all the loads of the dwelling, and there is only one set of feeder conductors.

Where the feeder conductors do not serve 100 percent of the total dwelling unit load, the provisions of this chapter are used to calculate the conductor size.

E3601.2 Branch-circuit and feeder ampacity. Branch-circuit and feeder conductors shall have ampacities not less than the maximum load to be served. Where a branch circuit or a feeder supplies continuous loads or any combination of continuous and noncontinuous loads, the minimum branch-circuit or feeder conductor size, before the application of any adjustment or correction factors, shall have an allowable ampacity equal to or greater than the noncontinuous load plus 125 percent of the continuous load.

❖ This requirement pertains to the size or the allowable ampacity of the conductors and not to the rating of the branch circuit. The branch circuit rating does not always correspond to the conductor size. See Section E3602 for branch circuit ratings. The branch circuit and feeder loads must be calculated to determine the appropriate size conductors and overcurrent protection devices to be used. Demand factors may be used in sizing conductors. For demand factors, see the commentary under Section E3604.3.

In calculating conductor ampacity, the type of load also must be taken into account. See the commentary under Section E3605.1. A load that is in operation for three hours or more is a continuous load. When a load is operated continuously, the heating effect on the conductors is a concern because, without time to dissipate, the heat can build up. Although it may seem that the lights are sometimes on for many hours, there is usually much diversity in the operation of residential lighting. Generally, dwelling unit lighting loads are not considered continuous loads for calculation purposes. Some loads in residential wiring may need to be considered continuous and be calculated at 125 percent of the full load. However, the code does not specify certain loads as continuous. For residential wiring, it may be a judgement call as to how long the load will operate at any one time.

For any load considered continuous, the allowable conductor ampacity must be at least 125 percent of the calculated load. This is the same as saying that the continuous load on a conductor should not be more than 80 percent of its allowable ampacity.

Even for noncontinuous loads, this is a good practice, although not required. For example, a 20-ampere-rated branch circuit serving a noncontinuous load is permitted to be loaded up to 20 amperes per Table E3602.13. A good rule of thumb is to try to estimate a maximum load of 16 amperes for a 20-ampere circuit. However, for general purpose branch circuits in a dwelling supplying receptacles and lighting outlets, it is impossible to know at the time the wiring is being installed what load will be connected. Even for a circuit consisting of only lighting outlets, it is difficult to know what the load will be, as light fixtures and lamps can be changed.

E3601.3 Selection of ampacity. Where more than one calculated or tabulated ampacity could apply for a given circuit length, the lowest value shall be used.

Exception: Where two different ampacities apply to adjacent portions of a circuit, the higher ampacity shall be permitted to be used beyond the point of transition, a distance equal to 10 feet (3048 mm) or 10 percent of the circuit length figured at the higher ampacity, whichever is less.

❖ If the conductors of a circuit are exposed to different ambient temperatures (a cable run from a basement through an attic, for example), they could have different ampacities based on the application of Section E3605.2. The portion of the circuit in the higher ambient temperature location is the weakest link; therefore, the lowest calculated ampacity must apply. The exception relaxes this requirement slightly by allowing the higher ampacity to apply for conductors that extend only a short distance into an area of higher ambient temperature.

E3601.4 Multi-outlet branch circuits. Conductors of multi-outlet branch circuits supplying more than one receptacle for cord- and plug-connected portable loads shall have ampacities of not less than the rating of the branch circuit.

❖ For multi-outlet branch circuits, the rating of the branch circuit overcurrent protection device determines the minimum size conductor that can be used. Conductors must have an ampacity not less than the rating of the branch circuit. For example, the use of a 20-ampere circuit breaker requires that the conductors have an ampacity of at least 20 amperes. This is different for a dedicated circuit to a specific load such as a large motor or appliance. The conductors must have an ampacity sufficient for the load, but this ampacity for the motor or appliance may not correspond to a standard size of overcurrent device. In this case, the next higher size circuit breaker could be used, and therefore the conductor ampacity could be less than the rating of the branch circuit. The rating of the branch circuit is set by the size of the overcurrent protection device. However, on multi-outlet branch circuits that supply receptacles for cord- and plug-connected loads, the conductor ampacity must be matched exactly to the branch circuit rating because it is unknown what load might be plugged into the circuit.

E3601.5 Multiwire branch circuits. All conductors for multiwire branch circuits shall originate from the same panelboard. Where two or more devices on the same yoke or strap are supplied by a multiwire branch circuit, a means shall be provided at the panelboard where the circuit originates to simultaneously disconnect all ungrounded conductors of the multiwire circuit. Except where all ungrounded conductors are opened simultaneously by the branch-circuit overcurrent device, multiwire branch circuits shall supply only line-to-neutral loads or only one appliance.

❖ In residential wiring, a multiwire branch circuit is a three-wire circuit that consists of two ungrounded conductors that have a potential of 240 volts between them and a grounded conductor that has a potential of 120 volts between it and each ungrounded conductor. The circuit that supplies the range or dryer is a multiwire branch circuit.

A multiwire branch circuit could also be two separate circuits that share a neutral conductor. In commercial locations where conduit is used, this type of circuit is used very commonly: three conductors are pulled into the raceway, and the neutral carries only the unbalanced current of the two circuits. However, in residential work where cable is used, it is not as advantageous. Some electricians like to install a three-conductor cable. It actually contains four conductors: a red and black for the ungrounded or hot conductors, a white for the shared neutral or grounded conductor and a bare or green for the grounding conductor. This installation can result in problems, and care must be taken to ensure that all code provisions are complied with. One concern is that in dwellings a homeowner or individual who does not understand the special code provisions sometimes rearranges the circuit breakers in the panelboard by placing both hot conductors of the multiwire circuit on the same leg (phase) of the service, resulting in the neutral conductor carrying the sum of the return current of both circuits instead of only the unbalanced current of the two circuits. On a 20-ampere multiwire circuit, it is conceivable that the neutral could carry up to 40 amperes on a size 12 AWG conductor. This could heat up the neutral conductor and melt or burn the insulation and possibly result in a fire.

One use where this is encountered is for the two kitchen small appliance circuits. Duplex receptacles with break-off tabs are used so that the two separate receptacles on the same yoke or strap can be supplied by the two hot conductors of the multiwire circuit. A person working on the circuit might check one of the "split-wired" receptacles to determine that the power is turned off at the circuit breaker but not turn off the power to both circuits. The worker is then subject to the risk of shock. This is why multiwire circuits must be supplied by a two-pole circuit breaker or two single-pole circuit breakers with an approved handle tie. An approved handle tie is a clip made specifically for and listed for clamping over the handles of the two single-pole circuit breakers. If one of the circuit breakers trips open, it will operate the handle of the other breaker. Some circuit breakers are made with a hole in the handle through which a nail can be inserted. A nail holding the handles of two breakers together should not be approved. Where the handle of the circuit breaker does not already have a hole, attempts have been made to drill a small hole through the handle to insert a nail or wire. This is an obvious code violation. Not tying the handles together is certainly a code violation.

The shared neutral conductor of the multiwire circuit must be pigtailed in each outlet box where the duplex receptacle is installed. If the duplex receptacle was used to splice or connect the neutral conductor of the circuit, the circuit would be broken when the receptacle is removed or worked on. If this is done when the circuit is energized, it would place appliances plugged into the circuit in a 240-volt series circuit. The voltage drop across each appliance would depend on its resistance and could cause abnormal voltages and thus damage to the appliances. See the commentary for Section E3306.10.2.

SECTION E3602 BRANCH CIRCUIT RATINGS

E3602.1 Branch-circuit voltage limitations. The voltage rating of branch circuits that supply luminaires, receptacles for cord- and plug-connected loads of up to 1,440 volt-amperes, and loads of less than 0.25 horsepower shall be limited to a maximum rating of 120 volts, nominal, between conductors.

Branch circuits that supply cord- and plug-connected or permanently connected utilization equipment and appliances rated at over 1,440 volt-amperes or $^1/_4$ horsepower and greater shall be rated at 120 volts or 240 volts, nominal.

❖ The code requires that lighting fixtures in dwelling units be supplied by circuits with a maximum voltage of 120 volts. Lighting fixtures designed for higher voltages are for use in commercial and industrial occupancies. Receptacles supplying cord- and plug-connected loads of up to 1,440 volt-amperes or 12 amperes are also limited to 120 volts. A multiwire circuit with a potential of 240 volts between the two ungrounded conductors that supplies a range or dryer is permitted under the second paragraph because the load is well over 12 amperes. See Section E3601.5. A multiwire circuit connected to one duplex receptacle with the break-off tab removed so that it is a split-wired receptacle is a 240-volt branch circuit, and the cord- and plug-connected load will usually be less than 1,440 volt-amperes; however, this multiwire circuit would be considered as two separate 120-volt circuits. See the commentary for Section E3601.5, which states that a multiwire branch circuit could also be two separate circuits that share a neutral conductor.

E3602.2 Branch-circuit ampere rating. Branch circuits shall be rated in accordance with the maximum allowable ampere rating or setting of the overcurrent protection device. The rating for other than individual branch circuits shall be 15, 20, 30, 40 and 50 amperes. Where conductors of higher ampacity are used, the ampere rating or setting of the specified overcurrent device shall determine the circuit rating.

❖ The rating of the branch circuit is set by the size of the overcurrent protection device.

The rating of the overcurrent device, not the size or ampacity of the branch circuit conductors, determines the rating of the branch circuit. The conductor size is sometimes increased to compensate for voltage drop such as where a size 10 AWG copper conductor is used on a 20-ampere branch circuit where a size 12 AWG would normally be used. It is a 20-ampere branch circuit because a 20-ampere circuit breaker is used.

Most of the loads for residential applications can be supplied by branch circuits of the standard 15, 20, 30, 40, or 50 ampere ratings given in this section. An individual branch circuit (a circuit that supplies only one piece of utilization equipment) is permitted to be of any rating necessary to supply the load it serves.

E3602.3 Fifteen- and 20-ampere branch circuits. A 15- or 20-ampere branch circuit shall be permitted to supply lighting units, or other utilization equipment, or a combination of both. The rating of any one cord- and plug-connected utilization equipment shall not exceed 80 percent of the branch-circuit ampere rating. The total rating of utilization equipment fastened in place, other than lighting fixtures, shall not exceed 50 percent of the branch-circuit ampere rating where lighting units, cord- and plug-connected utilization equipment not fastened in place, or both, are also supplied.

❖ This section covers combination loads for 15- and 20-ampere branch circuits that supply multiple outlets, appliances, and equipment.

In this code section, appliances and equipment are referred to as two types: (1) as cord- and plug-connected and (2) as fastened in place. An appliance or equipment that is fastened in place or connected by permanent wiring methods is sometimes called a fixed appliance or fixed equipment. For this discussion, "fixed" will be used for appliances and equipment fastened in place and connected with permanent wiring methods. "Appliance" will be used for appliances and utilization equipment.

A 15- or 20-ampere branch circuit is permitted to supply a combination of loads except for the two 20-ampere small appliance branch circuits required for the kitchen area receptacles in accordance with

Sections E3603.2 and E3801.3 through E3801.3.2. It is common practice to run a separate branch circuit to appliances such as a dishwasher, refrigerator, disposal, etc. It is also permissible to supply two or more loads of different types on the same circuit. A furnace, considered an appliance, is required to be supplied by a separate individual branch circuit per Section E3603.1.

Where a separate circuit is installed for appliances, the circuit is called an appliance branch circuit or individual branch circuit. Appliances can also be supplied by a general purpose branch circuit, but the rating of the appliance is restricted. The total ampere rating of fixed appliances must not exceed 50 percent of the rating of the branch circuit if other loads are also served on the same branch circuit. A 120-volt, 20-ampere-rated branch circuit that supplies a room air conditioner, for example, could also supply light fixtures or a receptacle for cord- and plug-connected loads, as long as the air conditioner load does not exceed 10 amperes. More than one fixed appliance could be served by the mixed load circuit. For example, consider a 20-ampere general purpose branch circuit that supplies several receptacles, light fixtures, and two fixed attic fans. The rating of this circuit would comply with code as long as the combined rating of the two attic fans was not more than 10 amperes. The number of receptacle outlets on the mixed load circuit is not restricted.

In summary, for a 15- or-20 ampere-rated general purpose branch circuit that serves a combination of loads, the maximum rating of a fixed appliance served by that circuit is 50 percent of the circuit rating. Any cord- and plug-connected appliance not fastened in place and served by this general purpose branch circuit must have an ampere rating not more than 80 percent of the circuit rating.

E3602.4 Thirty-ampere branch circuits. A 30-ampere branch circuit shall be permitted to supply fixed utilization equipment. A rating of any one cord- and plug-connected utilization equipment shall not exceed 80 percent of the branch-circuit ampere rating.

❖ A 30-ampere branch circuit is used for fixed utilization equipment, which is considered equipment fastened in place or connected by permanent wiring methods. Water heaters and condensing units are examples of 30-ampere circuits that would be installed in a dwelling. The load of the appliance should be checked to make sure it does not exceed 80 percent of the branch circuit rating or 5,760 volt-amperes or 24 amperes. In most cases, when a new house is being wired, the exact rating of the appliances is not known. For a clothes dryer, it is common practice to run size 10 AWG copper conductors on a 30-ampere circuit. For calculating a feeder load, a clothes dryer for which the nameplate rating is not known is considered as 5,000 volt-amperes. This would require a 240 volt, 30-ampere rated circuit (5,000 volt-amperes/240 volts = 20.8 amperes, which would require a 30-ampere rated circuit).

E3602.5 Branch circuits serving multiple loads or outlets. General-purpose branch circuits shall supply lighting outlets, appliances, equipment or receptacle outlets, and combinations of such. The rating of a fastened-in-place appliance or equipment, where used in combination on the same branch circuit with light fixtures, receptacles, and/or other appliances or equipment not fastened in place, shall not exceed 50 percent of the branch-circuit rating. Multioutlet branch circuits serving lighting or receptacles shall be limited to a maximum branch-circuit rating of 20 amperes.

❖ The code permits circuits serving general purpose convenience receptacles to also serve lighting outlets. In commercial and industrial buildings, it is common practice to install receptacles on one circuit and lighting loads on another circuit and not to mix the two. But in residential wiring, it is very common to mix the two loads on the same circuit. See the commentary for Section E3602.5. The number of convenience receptacle outlets on a branch circuit is not limited.

E3602.6 Branch circuits serving a single motor. Branch-circuit conductors supplying a single motor shall have an ampacity not less than 125 percent of the motor full-load current rating.

❖ Some examples of motor loads in residential wiring are an attic fan, a sump pump, a circulating pump for hot water space heating, a garage door opener, and a garbage disposal. Such loads are typically served by individual dedicated branch circuits. Circuits that run to motors such as these must have conductors rated for at least 125 percent of the full load rating of the motor to allow for starting current.

E3602.7 Branch circuits serving motor-operated and combination loads. For circuits supplying loads consisting of motor-operated utilization equipment that is fastened in place and that has a motor larger than 0.125 horsepower in combination with other loads, the total computed load shall be based on 125 percent of the largest motor load plus the sum of the other loads.

❖ To correctly size branch circuit conductors, the load must be calculated. Branch circuits that supply fastened-in-place utilization equipment having a motor larger than 1/8 horsepower must have the branch circuit load calculated at 125 percent of the motor load plus the other loads served. If none of the motors on a circuit is larger than 1/8 horsepower, it is not necessary to add the extra 25 percent.

E3602.8 Branch-circuit inductive lighting loads. For circuits supplying luminaires having ballasts, the computed load shall be based on the total ampere ratings of such units and not on the total watts of the lamps.

❖ On circuits serving light fixtures that contain a ballast such as fluorescent, mercury-vapor, high intensity discharge, and high pressure sodium, there is an inductive load. In addition to the wattage or volt-amperes of the lamps, the inductive load must be considered. Therefore, the total ampere rating of the light fixtures must be used for determining the load on the circuit. Such light fixtures are marked with a rating that includes the ballast and lamp load.

E3602.9 Branch-circuit load for ranges and cooking appliances. It shall be permissible to compute the branch-circuit load for one range in accordance with Table E3604.3(2). The branch-circuit load for one wall-mounted oven or one counter-mounted cooking unit shall be the nameplate rating of the appliance. The branch-circuit load for a counter-mounted cooking unit and not more than two wall-mounted ovens all supplied from a single branch circuit and located in the same room shall be computed by adding the nameplate ratings of the individual appliances and treating the total as equivalent to one range.

❖ The oven and all of the heating elements of a range are usually not used all at the same time. Because of this diversity, the demand factors of Table E3604.3(2) can be used. A counter-mounted cooking unit (a stove top) and one or two ovens are treated as one range. If only a separate counter-mounted cooking unit or oven is installed, the nameplate rating must be used as the load for the branch circuit. See the sample feeder load calculations for Section E3604.3.

E3602.9.1 Minimum ranch-circuit for ranges. Ranges with a rating of 8.75 kVA or more shall be supplied by a branch circuit having a minimum rating of 40 amperes.

❖ A minimum rating of 40 amperes is required for any range rated 8.75 kVA or more. Size 8 AWG copper conductors are typically used for this circuit.

E3602.10 Branch circuits serving heating loads. Branch circuit ratings for space-heating and water-heating appliances shall be 125 percent of the nameplate rating of the appliance. Branch circuits supplying two or more outlets for fixed electric space-heating equipment shall be rated 15, 20 or 30 amperes.

❖ The appliance load cannot be greater than 80 percent of the rating of the circuit. A 30-ampere circuit is the maximum permitted for fixed electric space heating such as baseboard resistance heaters.

E3602.11 Branch circuits for air-conditioning and heat pump equipment. The ampacity of the conductors supplying multimotor and combination load equipment shall not be less than the minimum circuit ampacity marked on the equipment. The branch-circuit overcurrent device rating shall be the size and type marked on the appliance and shall be listed for the specific purpose.

❖ For these circuits, it is necessary to know the nameplate information in order to determine the proper size conductors. The equipment label will dictate the required branch circuit ampacity and rating of the overcurrent device.

E3602.12 Branch circuits serving room air conditioners. A room air conditioner shall be considered as a single motor unit in determining its branch-circuit requirements where all the following conditions are met:

1. It is cord- and attachment plug-connected.
2. The rating is not more than 40 amperes and 250 volts; single phase.
3. Total rated-load current is shown on the room air-conditioner nameplate rather than individual motor currents.
4. The rating of the branch-circuit short-circuit and ground-fault protective device does not exceed the ampacity of the branch-circuit conductors, or the rating of the branch-circuit conductors, or the rating of the receptacle, whichever is less.

❖ Dwelling unit air-conditioning is usually provided by whole-house systems, but room air-conditioners, such as the window type or the through-the-wall type used in motels and hotels, usually cool or heat only a small room or area. Room air-conditioners are considered fastened-in-place appliances but are subject to specific rules that are different from those for whole-house systems, mainly because these units are connected with cord and attachment plugs. These appliances can be supplied either by an individual branch circuit or by a branch circuit that also supplies other loads including lighting and/or other appliances. In the planning stages of wiring a dwelling unit, an individual branch circuit is commonly used; however, the room air-conditioner could be installed later and connected to a branch circuit also serving other loads.

A room air-conditioner usually has a compressor motor and a fan motor, but the code rule indicates that for calculating the branch circuit conductor size and overcurrent device rating, it is considered a unit. The total rated-load current must be shown on the nameplate. Where these units are served by a 15-ampere or 20-ampere branch circuit, they are subject to the provisions of Section E3602.3. Those same provisions are reiterated specifically for room air-conditioners in the two sections below.

E3602.12.1 Where no other loads are supplied. The total marked rating of a cord- and attachment plug-connected room air conditioner shall not exceed 80 percent of the rating of a branch circuit where no other appliances are also supplied.

❖ Where supplied by an individual branch circuit, the unit as a whole, which includes both the compressor and the fan motor, must have a rating not more than 80 percent of the branch circuit rating. For a 15-ampere circuit, the rating can be a maximum of 12 amperes, and for a 20-ampere circuit, a maximum of 16 amperes.

E3602.12.2 Where lighting units or other appliances are also supplied. The total marked rating of a cord- and attachment plug-connected room air conditioner shall not exceed 50 percent of the rating of a branch circuit where lighting or other appliances are also supplied. Where the circuitry is interlocked to prevent simultaneous operation of the room air conditioner and energization of other outlets on the same branch circuit, a cord-and-attachment-plug-connected room air conditioner shall not exceed 80 percent of the branch-circuit rating.

❖ Where the branch circuit serves other loads, the total rating of the room air-conditioner, whether it has been planned from the initial stages of the wiring job or installed in a window with the cord being plugged into an existing circuit, must not exceed 50 percent of the branch circuit rating, that is, 50 percent of the size of the overcurrent protection device, not 50 percent of the ampacity of the branch circuit conductors. For example, one room air-conditioner with a total unit rating of up to 10 amperes could be plugged into a receptacle on a 20-ampere circuit. An interlock arrangement that prevents other outlets on the same circuit from being used when the room air conditioner is running would allow the 50-percent limit to increase to 80 percent.

E3602.13 Branch-circuit requirement—summary. The requirements for circuits having two or more outlets, or receptacles, other than the receptacle circuits of Section E3603.2, are summarized in Table E3602.13. Branch circuits in dwelling units shall supply only loads within that dwelling unit or loads associated only with that dwelling unit. Branch circuits required for the purpose of lighting, central alarm, signal, communications, or other needs for public or common areas of a two-family dwelling shall not be supplied from equipment that supplies an individual dwelling unit.

❖ Notice from Table E3602.13 that a 20-ampere rated receptacle is not permitted to be installed on a 15-ampere rated circuit, but a 15-ampere rated receptacle is permitted on a 20-ampere circuit. Equipment may have a 20-ampere rated attachment plug intended for use on a 20-ampere circuit. If a circuit is not really a 20-ampere circuit, it may not provide the required amount of current. Section E3902.1.1 requires that a single receptacle on an individual branch be rated equal to the rating of the branch circuit.

A two-family dwelling unit may have areas common to both residences. For example, yard lighting, carport lighting, or entry lighting could be used by occupants of both dwellings. If this is the case, the power for the common area loads must be supplied from a separate service and panelboard. Even where a two-family dwelling is owned or managed by one occupant, the common power needs cannot be supplied from the panel in the manager's unit, for example. Common area lighting and other power needs are more prevalent in multiple-family dwellings and are not often needed for two-family dwellings.

TABLE E3602.13
BRANCH-CIRCUIT REQUIREMENTS—SUMMARY[a,b]

	CIRCUIT RATING		
	15 amp	20 amp	30 amp
Conductors: Minimum size (AWG) circuit conductors	14	12	10
Maximum overcurrent-protection device rating Ampere rating	15	20	30
Outlet devices: Lampholders permitted Receptacle rating (amperes)	 Any type 15 maximum	 Any type 15 or 20	 N/A 30
Maximum load (amperes)	15	20	30

a. These gages are for copper conductors.

b. N/A means not allowed.

❖ See Section E3602.13.

SECTION E3603 REQUIRED BRANCH CIRCUITS

E3603.1 Branch circuits for heating. Central heating equipment other than fixed electric space heating shall be supplied by an individual branch circuit. Auxiliary equipment, such as pumps, motorized valves, humidifiers and electrostatic air cleaners directly associated with the central heating equipment, shall not be prohibited from connecting to the same branch circuit as the central heating equipment.

❖ For most furnaces, the motor load, even with the extra 25 percent, would not exceed the maximum 80 percent of the branch circuit rating; however, if the furnace was connected to the same circuit serving other loads and an overload occurred, opening the circuit breaker, the heat supply for the house would be out of order. In some climates, this could be a very serious problem. Thus, the heating equipment must be supplied from a separate circuit. Associated auxiliary equipment and controls may be connected to the same branch circuit.

E3603.2 Kitchen and dining area receptacles. A minimum of two 20-ampere-rated branch circuits shall be provided to serve receptacles located in the kitchen, pantry, breakfast area, dining area or similar area of a dwelling. The kitchen countertop receptacles shall be served by a minimum of two 20-ampere-rated branch circuits, either or both of which shall also be permitted to supply other receptacle outlets in the kitchen, pantry, breakfast and dining area including receptacle outlets for refrigeration appliances.

❖ This section requires at least two 20-ampere rated circuits for the kitchen and dining area receptacles. Section E3801 provides rules on location and spacing of the receptacles. The code specifies that there must be at least two circuits but does not specify or limit the number of receptacle outlets. The countertop receptacle branch circuits are permitted to supply receptacles installed for refrigerators.

E3603.3 Laundry circuit. A minimum of one 20-ampere-rated branch circuit shall be provided for receptacles located in the laundry area and shall serve only receptacle outlets located in the laundry area.

❖ A dedicated branch circuit is required for the laundry area. This circuit can serve receptacles that, for example, are used for ironing, clothes washers, and gas dryers.

E3603.4 Bathroom branch circuits. A minimum of one 20-ampere branch circuit shall be provided to supply the bathroom receptacle outlet(s). Such circuits shall have no other outlets.

Exception: Where the 20-ampere circuit supplies a single bathroom, outlets for other equipment within the same bathroom shall be permitted to be supplied in accordance with Section E3602.

❖ This section requires a dedicated branch circuit for the bathroom receptacle outlet(s). Section E3801.6 provides rules for the location of the receptacle(s). Two choices are available for the branch circuit serving the bathroom(s) in the house.

1. One 20-ampere rated circuit may serve the required receptacle(s) in all the bathrooms in the house. In this case, no other outlets may be served by this circuit. The bathroom lighting and any fan or heat lamps would be served by a distinct general purpose branch circuit.
2. A 20-ampere circuit may serve only one bathroom. For additional bathrooms in the house, additional separate circuits would be run. The one circuit for the single bathroom supplies the required receptacle(s) and is also permitted to serve lighting and equipment in the bathroom such as an exhaust fan and/or heat lamp. Because this circuit is subject to the provisions of Section E3602, the load of the other equipment can be a maximum of 50 percent of the rating of the branch circuit.

The bathroom receptacle must be GFCI protected, but lighting and other equipment is not required to be GFCI protected unless otherwise stated as a condition of the listing of a fixture or equipment.

E3603.5 Number of branch circuits. The minimum number of branch circuits shall be determined from the total computed load and the size or rating of the circuits used. The number of circuits shall be sufficient to supply the load served. In no case shall the load on any circuit exceed the maximum specified by Section E3602.

❖ The minimum number of branch circuits required for various specific loads such as the laundry, furnace, bathroom, and small appliance circuits, is clearly prescribed in Section E3603. However, a simple calculation is necessary to find the minimum number of branch circuits required for the general lighting load.

The general lighting load includes all of the general purpose convenience receptacles and lighting outlets and is calculated at 3 volt-amperes per square foot per Section E3604.5. The square footage of the house used for this calculation does not include open porches, garages, or unused or unfinished areas that are not adaptable for future use. Certainly, these areas have receptacle and lighting outlets, and even areas such as an attic or crawl space could have lighting and receptacles. See Section 3803.4 for lighting required in storage spaces. Although these outlets are part of the general lighting load, these rooms and spaces are not part of the floor area used in the general lighting load calculation. In many jurisdictions, new houses are completed with an unfinished basement, which is intended for future habitable space. The square footage of such basements must be included in the general lighting load calculation.

To find the number of general purpose lighting and receptacle circuits needed, multiply the dwelling square footage by 3 volt-amperes. This is the total lighting and receptacle load. Divide this load by the circuit voltage to find the total amperage needed. The total amperage is then divided by the rating of the circuit breaker to find the minimum number of circuits needed.

For a 2,200 square (204 m^2) foot house, the total lighting load is 3VA x 2,200 = 6,600 volt-amperes. To get the total amperage load for the 120-volt branch circuits, divide the volt-amperes by 120 volts, which is 55 amperes. For a 15-ampere-rated circuit, divide the 55 amperes by 15. The result is 3.67 circuits, which must be rounded up to four circuits. For a 20-ampere-rated circuit, divide the 55 amperes by 20. The result is 2.75 circuits, which must be rounded up to three 20-ampere circuits to serve all the general purpose convenience receptacles and lighting load of the house. This is the minimum number of circuits required; however, many more are typically installed as a matter of convenience and good planning.

The 2,200 square (204 m^2) foot house in this example could have three or four bedrooms, a family room and a living room. With the addition of hallway, entry hall and outdoor receptacle outlets, it may easily contain 40 or more general purpose receptacle outlets. (This does not include the laundry, bathroom, kitchen and dining area receptacles on the small appliance circuits.) Add to this all the lighting outlets, (which do include the laundry, kitchen and dining room, and could include the bathroom lights, depending on how the bathroom branch circuits are run). See Section E3603.4 for bathroom branch circuits. The outside porch lights, garage lights, attic storage area lights, etc., are included in the general lighting branch circuits, although these areas are not included in the 3 volt-amperes per square foot of floor area. The total number of lighting outlets could easily be 20 or more.

Say there were 40 general purpose receptacle outlets and 15 lighting outlets for this house. The total of 55 outlets would be supplied by the minimum required number of four 15-ampere or three 20-ampere circuits. This lighting load must be evenly proportioned among the multioutlet branch circuits per Section E3603.6. Therefore, there could be approximately 17 to 18 outlets per 20-ampere circuit or 13 to 14 outlets per 15-ampere circuit for this house; however, the number of outlets per circuit is not limited by the code. Evenly proportioned loading of the required number of branch circuits is required.

This calculation covers the first sentence of the section: "The minimum number of branch circuits shall be determined from the total computed load and the size or rating of the circuits used." However, it is very important not to overlook the rest of the paragraph, which requires that the number of circuits be sufficient to supply the load served and that the load not exceed the limitations of Section E3602.

Where the loads are known, such as the number and wattage of the lighting fixtures to be installed or the cord- and plug-connected appliances or equipment to be used, additional circuits will need to be installed according to the load. For good practice, the wiring system should provide for increased loads and the use of added future electrical appliances. These factors should be taken into account in addition to the mere minimum number of circuits.

Many electricians use a rule of thumb method based on experience and practice for deciding on the number of branch circuits for a dwelling. In most cases, these traditional practices result in the actual number of branch circuits installed exceeding the minimum number required by the code calculation. Because the number and wattage of lamps to be used in light fixtures is not known at the time of deciding the number of circuits, and the load of the plug- and cord-connected appliances cannot be known, one rule of thumb method used in some areas is a maximum of 8 to 10 outlets included on any one general purpose branch circuit. This is not a code rule, but this kind of installation would provide a safety and convenience margin so that circuit breakers are not as likely to open on overload-for example, during a cold season when a homeowner may use portable space heaters in the house. Limiting of outlets per circuit is not a code rule unless the loads are known.

One reason that some electricians limit the number of outlets per circuit may be because of the influence of the method used to determine the number of circuits for commercial buildings. The number of outlets per circuit is limited on commercial buildings. For other than residential wiring, each receptacle outlet is assigned a value of 1.5 amperes, which is 180 volt-amperes divided by 120 volts. Under this method, a 20-ampere circuit could have 13 outlets, and a 15-ampere circuits could have 10 outlets. This is sometimes confused with the requirements for dwellings, and if the commercial method is applied to calculating the number of outlets and branch circuits for dwellings, then the number of circuits would most likely satisfy the residential code requirements. In residential wiring, as long as the minimum number of circuits based on the square footage is provided, there is no limit on the number of outlets per branch circuit. The spacing and location requirements of receptacle outlets for dwellings results in the installation of many outlets that most definitely will not all be used at the same time. Because of the intermittent use and great diversity in the loads and duration of operation of the loads connected to the receptacle outlets in a house, any number of outlets can be installed on each general purpose branch circuit.

Consider a three bedroom, two bathroom, 1,500 square foot (139 m^2) house served with fuel gas for space and water heating and with the following lighting and receptacle outlets. The house has a 500 square foot (46 m^2) garage.

Number of Lighting Outlets	
3 Bedrooms	3
2 Bathrooms	2
Hall	1
Living Room	1
Kitchen/Dining Rooms	2
Laundry Room	1
Outside Doors and Garage	3
Total Lighting Outlets*	13
Number of Receptacle Outlets	
3 Bedrooms, 4 receptacles each	12
Hall	1
Outdoors (front and back)	2
Garage	2
Living Room	6
Total Receptacles*	23

*These numbers are used for this example only. See Chapter 38 for required outlets.

What is the minimum number of branch circuits needed for the total of 36 lighting and receptacle outlets for this house?

To determine the minimum required number of circuits, from Table E3604.3(1) find 3 volt-amperes per square foot for lighting and receptacles.

1,500 square feet x 3 VA = 4,500 VA/120 Volts = 37.5 Amps minimum required for the house.

For 15-amp circuits, divide 15 into
37.5 amperes   37.5/15 = 2.5, round up to 3

For 20-amp circuits, divide 20 into
37.5 amperes   37.5/20 = 1.8, round up to 2

Another method is to take the voltage times the rating of the circuit in amperes to find how many volt-amperes are allowed on the circuit.

A 15-ampere circuit has a capacity for 1,800 VA (120 x 15 = 1,800 VA)

A 20-ampere circuit has a capacity for 2,400 VA (120 x 20 = 2,400 VA)

The minimum lighting load required for this house is 3 x 1,500 or 4,500 VA.

Divide the VA each circuit can handle into 4,500 VA to find the minimum number of general lighting circuits.

Using 15 amp circuits: 4,500/1,800 = 2.5, round up to 3

Using 20 amp circuits: 4,500/2,400 = 1.8, round up to 2

The 36 lighting and receptacle outlets should be evenly proportioned between the two or three general purpose branch circuits. More circuits could be needed, such as a separate circuit to a receptacle in the garage to serve a freezer.

In addition to the general lighting branch circuits, the minimum required circuits for this house are as follows:

- Kitchen and dining area receptacles served by two small appliance branch circuits. See Sections E3603.2 and E3801.3.
- Refrigerator. The refrigerator is permitted to be on one of the kitchen small appliance circuits, or it can be on a separate 15-ampere circuit. See Section E3801.3 and Exception 2.
- Laundry branch circuit. See Sections E3603.3 and E3801.8.
- Bathroom circuit. One 20-ampere circuit could serve the receptacles in both bathrooms. See Sections E3503.4 and E3801.6.
- Central heating or furnace circuit. See Section E3603.1.
- Individual or appliance branch circuits. Section E3602 requires that circuits be provided to supply such loads as a disposal, dishwasher, range and dryer sized according to the rating of the appliance or equipment.

The code does not prescribe or state in words that the dishwasher and disposal should each be supplied by a separate circuit as it does for the furnace, laundry, and bathroom circuits. This must be determined by a simple calculation. Because the dishwasher and disposal are closely associated in proximity and function, some electricians wire both appliances on the same 20-ampere circuit. This might be permitted if all code requirements are satisfied. Because the appliances to be installed are not usually available on the job when the circuits are being installed at the rough wiring stage, and because the appliances could be upgraded at a future time using the same branch circuits, it is best to run a separate 20-ampere circuit to each appliance. Section E3602.3 requires that if two appliances are supplied by the same branch circuit, the rating of any one appliance (utilization equipment), in this case the dishwasher or disposal, not exceed 80 percent of the branch circuit rating. And Section E3602.7, for combination loads, requires that the branch circuit total computed load be based on the 125 percent of the largest motor plus the sum of the other loads. A dishwasher has a pump motor and heating element, but they usually do not operate at the same time. A typical dishwasher nameplate lists the total load as 11.0 amperes and the motor load as 6.5 amperes. This dishwasher could be supplied by a separate 15-ampere circuit. The disposal is rated at 7.3 amperes and could be supplied by a separate 15-ampere circuit because the computed load is 7.3 x 125 percent, or 9.125 amperes. Could these two appliances be supplied on the same 20-ampere branch circuit? For combination loads, take 125 percent of the largest motor load plus the sum of the other loads. The disposal is the largest motor load at 7.3 x 125 percent = 9.125 amperes + the dishwasher at 11.0 amperes = 20.125 amperes. Neither one exceeds 80 percent of the 20-ampere branch circuit rating, but the total computed load is over 20 amperes. Therefore, two separate circuits are needed for these two appliances. Two separate 15-ampere circuits would satisfy code requirements, but good practice would indictate that two 20-ampere circuits be installed to provide sufficient capacity for future upgrades to appliances of higher load ratings.

E3603.6 Branch-circuit load proportioning. Where the branch-circuit load is computed on a volt-amperes/per-square-foot (m^2) basis, the wiring system, up to and including the branch-circuit panelboard(s), shall have the capacity to serve not less than the calculated load. This load shall be evenly proportioned among multioutlet branch circuits within the panelboard(s). Branch-circuit overcurrent devices and circuits shall only be required to be installed to serve the connected load.

❖ What if the actual connected load on the branch circuits was known and was less than the calculated load? The first sentence of this section requires that the wiring system have the capacity to serve at least the calculated load even if the actual load is less.

The branch circuit load calculated for the general lighting and general purpose receptacles must be evenly proportioned among the number of multi-outlet branch circuits. Although distributing the loads over a greater number of circuits than the required minimum would provide for future load increases, the code requires only the number of branch circuits necessary to carry the connected load.

SECTION E3604
FEEDER REQUIREMENTS

E3604.1 Conductor size. Feeder conductors that do not serve 100 percent of the dwelling unit load and branch-circuit conductors shall be of a size sufficient to carry the load as determined by this chapter. Feeder conductors shall not be required to be larger than the service-entrance conductors that supply the dwelling unit. The load for feeder conductors that serve as the main power feeder to a dwelling unit shall be determined as specified in Chapter 35 for services.

❖ Feeder conductors that are run to a subpanel serving a detached garage, swimming pool equipment building, detached shop or upper story are examples of feeder conductors that do not serve 100 percent of the dwelling unit load.

But how are the main power feeder conductors sized? The term "main power feeder" refers to the feeder conductors between the main disconnect at the service and the panelboard serving branch circuits for lighting, receptacles and appliances. The main power feeder can be, but is not always, the only feeder in the wiring system and as such might not serve 100 percent of the dwelling unit load. In the example feeder load calculation for Section E3604.3, the main power feeder is not the only feeder in the system.

Use Chapter 35 for sizing the main power feeder conductors. Section E3503.1 states that feeder conductors that serve as the main power feeder are listed in Table E3503.1. This table provides for higher ampacities than those in Table E3605.1. Because of the diversity in the use of the dwelling unit lighting, receptacle, and appliance loads, the demand on the main power feeder conductors is lower than the demand on the other feeders. In addition to the demand factors allowed in the actual calculation, Table E3503.1 has a built-in demand factor or reduction in the conductor size. Feeders other than the main power feeder can be expected to be more fully loaded and thus are sized by a different, more conservative method.

E3604.2 Minimum feeder conductor size. The size of feeder conductors shall not be less than 10 AWG copper or 8 AWG aluminum where the load supplied consists of any of the following number and types of circuits: (1) two or more two-wire branch circuits supplied by a two-wire feeder; (2) three or more two-wire branch circuits supplied by a three-wire feeder; or (3) two or more three-wire branch circuits supplied by a three-wire feeder.

❖ Feeder conductors are sized to serve the calculated load. However, for a load consisting of only two or three branch circuits, as indicated below, the minimum size feeder must have a minimum ampacity of 30-amperes. This is a prescriptive requirement to ensure that feeders provide a minimum level of capacity. The conductors must be at least size 10 AWG copper or size 8 AWG aluminum.

E3604.3 Feeder loads. The minimum load in volt-amperes shall be computed in accordance with the load calculation procedure prescribed in Table E3604.3(1). The associated table demand factors shall be applied to the actual load to determine the minimum load for feeders.

❖ Table E3604.3(1) is used to calculate the load for a feeder that supplies power to a sub-panelboard serving branch circuits for lighting, receptacles, motors, and appliances. Once the feeder load is determined, the conductor size is selected. For sizing a main power feeder (serves 100 percent of the dwelling unit load), use Table E3503.1.

The code provisions for calculating the feeder loads allow for the fact that in most cases all of the lighting, receptacles, motors, and appliances served by the feeder will not be in use at the same time. Because of this diversity in the operation of the various loads, demand factors can be used. For example, a demand factor of 75 percent can be applied to the total load of four or more appliances served by a feeder. If it is known that the load will be operated 100 percent of the time, the demand factor would be 100 percent.

EXAMPLE OF A FEEDER LOAD CALCULATION

Using Table E3604.3(1), the following is an example of the calculation for a feeder to a sub-panelboard serving branch circuits for lighting, receptacles, motors, and appliances. In this example, the feeder is protected by circuit breaker #4.

A house that measures 30 feet by 60 feet (9144 mm by 18 288 mm) exclusive of the garage and porches has service equipment containing four main disconnects, which are two-pole, 240-volt circuit breakers. The four disconnects serve the following:

Disconnect (1) serves a feeder to a sub-panelboard in a separate bathhouse building used to house swimming pool equipment,

Disconnect (2) serves a branch circuit to a 240-volt, 18,000-volt-ampere heat pump,

Disconnect (3) serves a branch circuit to a 240-volt, 8,500-volt-ampere house air conditioner, and

Disconnect (4) serves a feeder to a sub-panelboard inside the house, which supplies the following loads:

Appliance or Equipment	Rating
240-volt water heater:	6,000 volt-amperes
120-volt dishwasher:	1,600 volt-amperes
(motor rated 6.5 amperes included in 1,600 VA total rating)	
120-volt disposal:	8.3 amperes
120-volt microwave oven:	1,400 volt-amperes
240-volt dryer:	5,500 volt-amperes
240-volt range:	11,000 volt-amperes
120-volt pump motor for a hydromassage bathtub:	10 amperes
120-volt room air-conditioner:	1,400 volt-amperes
120-volt garage door opener:	7.2 amperes
120-volt clothes washer:	1,500 volt-amperes
Three 20-ampere kitchen and dining area receptacle branch circuits:	4,500 volt-amperes

Step 1: Lighting load, kitchen and dining area receptacle branch circuits, and laundry branch circuit

Total floor area 30 feet x 60 feet = 1,800 x 3 VA/sq ft = 5,400 volt-amperes
Three 20-ampere kitchen and dining area receptacle branch circuits: 4,500 volt-amperes*
*(Only two are required, but this example indicates three installed.)
One 20-ampere laundry branch circuit: 1,500 volt-amperes
11,400 volt-amperes
Applying the demand factor, 3,000 at 100 percent: 3,000
35 percent of remainder, 8,400 x .35 = 2,940
Total for Step 1: 5,940 volt-amperes

Step 2: Fastened-in-place appliances

Water heater: 6,000 volt-amperes
Dishwasher: 1,600 volt-amperes
Disposal: 120 volts x 8.3 amperes = 996 volt-amperes
Microwave oven: 1,400 volt-amperes
Hydromassage bathtub pump motor 120 volts x 10 amperes = 1,200 volt-amperes
Room air-conditioner: 1,400 volt-amperes
Garage door opener:
120 volts x 7.2 amperes = 864 volt-amperes
Total rating of appliances: 13,460volt-amperes
Applying the demand factor 13,460 x .75 = 10,095
Total for Step 2: 10,095 volt-amperes

Step 3: Motors

25 percent of largest motor:
Hydromassage bathtub pump motor
120 volts x 10 amperes = 1,200 x 0.25
Total for Step 3: 300 volt-amperes

Step 4: Clothes Dryer

Greater of 5,000 VA or nameplate rating.
Total for Step 4: 5,500 volt-amperes

Step 5: Cooking Appliances

8,000 VA for the range per column A of Table E3604.3(2)
Total for Step 5: 8,000 volt-amperes

Total computed feeder load: 29,835 volt-amperes

The minimum ampacity of the ungrounded feeder conductors is 29,835/240 volts = 124.3 amperes. A two-pole, 125-ampere circuit breaker can be used. Referring to Table E3605.1, copper conductors of size 1 AWG with a 75°C rated insulation type can be used for ungrounded conductors for this feeder.

The heating and cooling loads for the house in this example are not served by the feeder from Disconnect #4, but would be used in calculating the load for the service conductors. If the heating and cooling loads were served by this feeder, it would not be necessary to consider both loads in the calculation. As shown in Table E3604.3(1), only the largest of these two loads would be used, because it is unlikely that the heating and cooling loads will be used simultaneously.

TABLE E3604.3(1)
FEEDER LOAD CALCULATION

LOAD CALCULATION PROCEDURE	APPLIED DEMAND FACTOR
Lighting and receptacles: A unit load of not less than 3 VA per square foot of total floor area shall constitute the lighting and 120-volt, 15- and 20-ampere convenience receptacle load. 1,500 VA shall be added for each 20-ampere branch circuit serving receptacles in the kitchen, dining room, pantry, breakfast area and laundry area.	100 percent of first 3,000 VA or less and 35 percent of that in excess of 3,000 VA.
Plus	
Appliances and motors: The nameplate rating load of all fastened-in-place appliances other than dryers, ranges, air-conditioning and space-heating equipment.	100 percent of load for three or less appliances. 75 percent of load for four or more appliances.
Plus	
Fixed motors: Full-load current of motors plus 25 percent of the full load current of the largest motor.	
Plus	
Electric clothes dryer: The dryer load shall be 5,000 VA for each dryer circuit or the nameplate rating load of each dryer, whichever is greater.	
Plus	
Cooking appliances: The nameplate rating of ranges, wall-mounted ovens, counter-mounted cooking units, and other cooking appliances rated in excess of 1.75 kVA shall be summed.	Demand factors shall be as allowed by Table E3604.3(2).
Plus the largest of either the heating or cooling load	
Largest of the following two selections: 1. 100 percent of the nameplate rating(s) of the air conditioning and cooling, including heat pump compressors. 2. 100 percent of the fixed electric space heating.	

For SI: 1 square foot = 0.0929 m^2.

❖ See the commentary for Section E3604.3.

TABLE E3604.3(2)
DEMAND LOADS FOR ELECTRIC RANGES, WALL-MOUNTED OVENS, COUNTER-MOUNTED COOKING UNITS AND OTHER COOKING APPLIANCES OVER $1^3/_4$ kVA RATING[a,b]

NUMBER OF APPLIANCES	MAXIMUM DEMAND[b,c]	DEMAND FACTORS (percent)[d]	
	Column A maximum 12 kVA rating	Column B less than $3^1/_2$ kVA rating	Column C $3^1/_2$ to $8^3/_4$ kVA rating
1	8 kVA	80	80
2	11 kVA	75	65

a. Column A shall be used in all cases except as provided for in Footnote d.

b. For ranges all having the same rating and individually rated more than 12 kVA but not more than 27 kVA, the maximum demand in Column A shall be increased 5 percent for each additional kVA of rating or major fraction thereof by which the rating of individual ranges exceeds 12 kVA.

c. For ranges of unequal ratings and individually rated more than 8.75 kVA, but none exceeding 27 kVA, an average value of rating shall be computed by adding together the ratings of all ranges to obtain the total connected load (using 12 kVA for any ranges rated less than 12 kVA) and dividing by the total number of ranges; and then the maximum demand in Column A shall be increased 5 percent for each kVA or major fraction thereof by which this average value exceeds 12 kVA.

d. Over 1.75 kVA through 8.75 kVA. As an alternative to the method provided in Column A, the nameplate ratings of all ranges rated more than 1.75 kVA but not more than 8.75 kVA shall be added and the sum shall be multiplied by the demand factor specified in Column B or C for the given number of appliances.

❖ See the commentary for Section E3604.3.

E3604.4 Feeder neutral load. The feeder neutral load shall be the maximum unbalance of the load determined in accordance with this chapter. The maximum unbalanced load shall be the maximum net computed load between the neutral and any one ungrounded conductor. For a feeder or service supplying electric ranges, wall-mounted ovens, counter-mounted cooking units and electric dryers, the maximum unbalanced load shall be considered as 70 percent of the load on the ungrounded conductors.

❖ For a three-wire, 120/240-volt single-phase system, the grounded (neutral) conductor run from the transformer is also grounded at the service equipment. The voltage between the grounded neutral and each ungrounded conductor is 120 volts. A 240-volt branch circuit to a motor or space heater, for example, has no neutral conductor because the load is supplied by the two ungrounded conductors. There is no current in the feeder neutral from the 240-volt load. See Commentary Figure E3401(2).

To calculate the feeder neutral load from the example in Section E3604.3, do not include the water heater load, because it is 240 volts and does not use a neutral conductor. The maximum unbalanced load for the dryer and range is considered 70 percent of the load used for the ungrounded conductor load calculation.

Appliance or Equipment	Rating
120-volt dishwasher:	1,600 volt-amperes
(motor rated 6.5 amperes included in 1,600 VA total rating)	
120-volt disposal:	8.3 amperes
120-volt microwave oven:	1,400 volt-amperes
240-volt dryer	5,500 volt-amperes
240-volt range	11,000 volt-amperes
120-volt pump motor for a hydromassage bathtub	10 amperes
120-volt room air-conditioner	1,400 volt-amperes
120-volt garage door opener:	7.2 amperes
120-volt clothes washer (laundry circuit):	1,500 volt-amperes
Three 20-ampere kitchen and dining area receptacle branch circuits:	4,500 volt-amperes

Step 1: Lighting load, kitchen and dining area receptacle branch circuits and laundry branch circuit

Total floor area 30 feet x 60 feet = 1,800 x 3 VA/sq ft =	5,400 volt-amperes
Three 20-ampere kitchen and dining area receptacle branch circuits:	4,500 volt-amperes*
*(Only two are required, but this example indicates three installed.)	
One 20-ampere laundry branch circuit	1,500 volt-amperes
	11,400 volt-amperes
Applying the demand factor:	
3,000 at 100 percent:	3,000
35 percent of remainder: 8,400 x .35 =	2,940
Total for Step 1:	5,940 volt-amperes

Step 2: Fastened-in-place appliances that use a neutral conductor

Dishwasher:	1,600 volt-amperes
Disposal: 120 volts x 8.3 amperes =	996 volt-amperes
Microwave oven:	1,400 volt-amperes
Hydromassage bathtub pump motor 120 volts x 10 amperes =	1,200 volt-amperes
Room air-conditioner:	1,400 volt-amperes
Garage door opener 120 volts x 7.2 amperes =	864 volt-amperes
Total rating of appliances:	7,460 volt-amperes
Applying the demand factor: 7,460 x .75 =	10,095
Total for Step 2:	5,595 volt-amperes

Step 3: Motors.

25 percent of largest motor:
Hydromassage bathtub pump motor

120 volts x 10 amperes = 1,200 x 25 percent
Total for Step 3: 300 volt-amperes

Step 4: Clothes Dryer.
70 percent of rating: 5,500 VA x
70 percent = 3,850
Total for Step 4: 3,850 volt-amperes

Step 5: Cooking Appliances.
8,000 VA x 70 percent = 5,600
Total for Step 5: 5,600 volt-amperes
Total computed feeder load: 21,285 volt-amperes

The minimum ampacity of the neutral feeder conductor is 21,285/240 volts = 88.68 amperes. Using the 75°C column of Table E3605.1, a size 3 AWG Type THWN copper conductor is the minimum size permitted.

E3604.5 Lighting and convenience receptacle load. A unit load of not less than 3 volt-amperes shall constitute the minimum lighting and convenience receptacle load for each square foot of floor area (33 VA for each square meter of floor area). The floor area for each floor shall be computed from the outside dimensions of the building. The computed floor area shall not include open porches, garages, or unused or unfinished spaces not adaptable for future use.

❖ See the commentary for Section E3604.3.

E3604.6 Ampacity and computed loads. The computed load of a feeder shall be not less than the sum of the loads on the branch circuits supplied, as determined by Section E3604, after any applicable demand factors permitted by Section E3604 have been applied.

Feeder conductors shall have sufficient ampacity to supply the load served. In no case shall the computed load of a feeder be less than the sum of the loads on the branch circuits supplied as determined by this chapter after any permitted demand factors have been applied.

❖ A feeder must have an ampacity at least equal to the calculated feeder load. Feeder load is calculated in accordance with Section E3604.

E3604.7 Feeder and branch-circuit conductors. Feeder and branch-circuit conductors shall be protected by overcurrent-protective devices connected at the point the conductors receive their supply.

❖ See the commentary for Section E3605.5.

SECTION E3605 CONDUCTOR SIZING AND OVERCURRENT PROTECTION

E3605.1 General. Ampacities for conductors shall be determined based in accordance with Table E3605.1 and Sections E3605.2 and E3605.3.

❖ Table E3605.1 covers the allowable ampacities for other than service conductors and main power feeder conductors. Table E3503.1 covers the conductor sizes for services and feeders that serve 100 percent of the dwelling unit load.

The selection of conductors for dwelling unit wiring begins with finding the allowable conductor ampacity. Several variables are involved in determining the ampacity of a conductor. These include the ambient temperature, the number of conductors in close proximity, the temperature rating of the terminations, and the maximum overcurrent protection ratings for specific conductor sizes such as given in Table E3605.5.3.

Table E3605.1 cannot be used to determine conductor ampacity without knowing the "conditions of use" of the conductor. It is important here to review the definition of ampacity from Chapter 34: "The current in amperes a conductor can carry continuously under the conditions of use without exceeding its temperature rating." This definition applies to continuous use. There is not one ampacity for continuous use and another for noncontinuous use. Where a load is operated continuously (for three hours or more), the ampacity of the branch circuit conductors must be 125 percent of the continuous load. If the load is not operated continuously, the branch circuit conductors must have an ampacity at least equal to the load being served. See Section E3601.2.

Because of resistance, current passing through a conductor causes heat to build up in the wire, which must be dissipated into the ambient air. Ampacity is the amount of current that a conductor can carry without exceeding its thermal limit, which is the temperature rating. Conductors are classified by the different types of insulation that are capable of withstanding a certain level of heat before the insulation degrades, melts, or burns. The conductor temperature ratings of 60°C, 75°C, and 90°C at the top of the table indicate the maximum temperature a given size conductor with the type of insulation shown can withstand. The maximum current the conductor can carry without raising the temperature higher than 60°C, 75°C, or 90°C is shown in the table as the ampacity for that size of conductor with the type of insulation shown at the top of the column. Ampacities shown in table E3605.1 are based on an ambient temperature of 86°F (30°C). Look at the first column in Table E3605.2 for ambient temperature. The second row is 26°C - 30°C, and there is no adjustment factor for the ampacity of conductors with this ambient temperature.

For example, a size 8 AWG copper conductor with Type XHHW insulation can carry 50 amperes continuously at 30°C without raising the temperature of the conductor over 167°F (75°C). If this conductor carries a higher current, a temperature higher than 75°C could develop in the conductor, thus causing damage.

E3605.2 Correction factor for ambient temperatures. For ambient temperatures other than 30°C (86°F), multiply the allowable ampacities specified in Table E3605.1 by the appropriate correction factor shown in Table E3605.2.

❖ The ampacity values shown in Table E3605.1 must be reduced according to the factors in Table E3605.2. This table is usually referred to for a reduction in allowable ampacity; however, the ambient temperature correction factors could result in a slight increase in ampacity if the ambient temperature where the conductors are installed never exceeds 77°F (25°C) as indicated in the first row of the table.

It is reasonable to assume that attic temperatures will reach 105°F (41°C) and greater. Where conductors run in attics are exposed to the possibility of such high temperatures, a correction factor for conductor ampacity must be applied. Consider an air-conditioner supplied by a 40-ampere branch circuit with copper size 8 AWG, Type THHN conductors run through an attic where temperatures in summer months reach 125°F (52°C). To find the ampacity of these conductors under this condition of use, find Type THHN conductors in the 90°C column (on the left half of the table, for copper) of Table E3605.1. Reading down to the row for size 8 AWG, these conductors would have an ampacity of 55 amperes if installed in a maximum ambient temperature of 30°C. However, read down the far right hand column to the row that includes 125°F (52°C). In this row, move left to the column for Type THHN conductors. THHN is actually at the top of Table E3605.1. The correction factor is 0.76, which indicates a correction factor of 76 percent of the 55 amperes, or 41.8 amperes. The maximum allowable ampacity for the size 8 AWG conductors is 41.8 amperes.

If these size 8 AWG Type THHN copper conductors supplying the air-conditioner are in nonmetallic sheathed cable, the code simply states that the ampacity shall be that of the 60°C column or 40 amperes. See Section E3605.4.4. So why start with the 90°C column and have to apply the correction factor? If we started with the 60°C column, the correction factor would be 0.41, which would reduce the ampacity of the size 8 AWG conductors to 16.4 amperes. Section E3605.4.4 states that the 90°C rating is permitted to be used for ampacity correction and adjustment purposes.

Most nonmetallic sheathed cable available today has Type THHN insulation on each conductor for this very purpose. The de-rating is done from the ampacity values in the 90°C column. The lowest value, of either the 60°C column or the calculated ampacity using the correction factor, is then used to determine the ampacity. If the ambient temperature was 132°F (55°C), size 8 AWG Type THHN conductors would have an ampacity of 39.05 (55 x 0.71) and could not be used on a 40 ampere circuit. For this installation, size 6 AWG Type THHN conductors could be used. The ampacity would be (75 x 0.71) 53.25 amperes. This would be acceptable for the 40-ampere branch circuit under the conditions of use where the conductors pass through the 132°F (55°C) ambient temperature.

TABLE E3605.1
ALLOWABLE AMPACITIES

CONDUCTOR SIZE	CONDUCTOR TEMPERATURE RATING						CONDUCTOR SIZE
	60°C	75°C	90°C	60°C	75°C	90°C	
AWG kcmil	Types TW, UF	Types RHW, THHW, THW, THWN, USE, XHHW	Types RHW-2, THHN, THHW, THW-2, THWN-2, XHHW, XHHW-2	Types TW, UF	Types RHW, THHW, THW, THWN, USE, XHHW	Types RHW-2, THHN, THHW, THW-2, THWN-2, XHHW, XHHW-2	AWG kcmil
	Copper			Aluminum or copper-clad aluminum			
18	—	—	14	—	—	—	—
16	—	—	18	—	—	—	—
14	20	20	25	—	—	—	—
12	25	25	30	20	20	25	12
10	30	35	40	25	30	35	10
8	40	50	55	30	40	45	8
6	55	65	75	40	50	60	6
4	70	85	95	55	65	75	4
3	85	100	110	65	75	85	3
2	95	115	130	75	90	100	2
1	110	130	150	85	100	115	1
1/0	125	150	170	100	120	135	1/0
2/0	145	175	195	115	135	150	2/0
3/0	165	200	225	130	155	175	3/0
4/0	195	230	260	150	180	205	4/0

For SI: °C = [(°F)-32]/1.8.

❖ See the commentary for Section E3605.1.

TABLE E3605.2
AMBIENT TEMPERATURE CORRECTION FACTORS

AMBIENT TEMP. °C	FOR AMBIENT TEMPERATURES OTHER THAN 30°C (86°F), MULTIPLY THE ALLOWABLE AMPACITIES SPECIFIED IN TABLE E3605.1 BY THE APPROPRIATE FACTOR SHOWN BELOW						AMBIENT TEMP. °F
21-25	1.08	1.05	1.04	1.08	1.05	1.04	70-77
26-30	1.00	1.00	1.00	1.00	1.00	1.00	78-86
31-35	0.91	0.94	0.96	0.91	0.94	0.96	87-95
36-40	0.82	0.88	0.91	0.82	0.88	0.91	96-104
41-45	0.71	0.82	0.87	0.71	0.82	0.87	105-113
46-50	0.58	0.75	0.82	0.58	0.75	0.82	114-122
51-55	0.41	0.67	0.76	0.41	0.67	0.76	123-131
56-60	—	0.58	0.71	—	0.58	0.71	132-140
61-70	—	0.33	0.58	—	0.33	0.58	141-158
71-80	—	—	0.41	—	—	0.41	159-176

For SI: °C = [(°F)-32]/1.8.

❖ See the commentary for Section E3605.2.

E3605.3 Adjustment factor for conductor proximity. Where the number of current-carrying conductors in a raceway or cable exceeds three, or where single conductors or multiconductor cables are stacked or bundled for distances greater than 24 inches (610 mm) without maintaining spacing and are not installed in raceways, the allowable ampacity of each conductor shall be reduced as shown in Table E3605.3.

Exceptions:

1. Adjustment factors shall not apply to conductors in nipples having a length not exceeding 24 inches (610 mm).
2. Adjustment factors shall not apply to underground conductors entering or leaving an outdoor trench if those conductors have physical protection in the form of rigid metal conduit, intermediate metal conduit, or rigid nonmetallic conduit having a length not exceeding 10 feet (3048 mm) and the number of conductors does not exceed four.

❖ The allowable ampacity of branch circuit conductors varies not only according to the ambient temperature surrounding the conductors but also according to the number of current-carrying conductors in a raceway or cable or in close proximity. Where multiconductor cables (such as nonmetallic sheathed cables used in residential wiring) are stacked or bundled together for 24 inches or more without maintaining spacing, the allowable ampacity is reduced according to the factors in Table E3605.3. For example, the allowable ampacity of size 12 AWG copper THHN conductors (found in nonmetallic sheathed cable) used with 90°C terminals is 30 amperes. If several of these cables were bundled together for two feet or more such that seven current-carrying conductors were in the bundle, then the allowable ampacity would be reduced 70 percent to 21 amperes; see Commentary Figure E3605.3.

In some cases where the allowable ampacity is reduced, either a smaller circuit breaker would need to be installed or the conductors would need to be installed with sufficient spacing between them and not bundled. Some judgement may enter into the interpretation and enforcement of this code rule because in residential loads there is so much diversity in the amount of current use in the branch circuit conductors and in the duration of operation of the loads. To drill one hole in the top plate of the framing above the panelboard and run all the branch circuit cables in a bundle for two or three feet into the panel would be a code violation. This is not a good practice, because the heat developed in the current-carrying conductors cannot readily dissipate. Spacing should be maintained or the allowable ampacity of the conductors would be reduced.

The correction factors for ambient temperature and for the number of current-carrying conductors must both be applied. For example, if a 30-ampere electric water heater branch circuit using size 10 AWG copper conductors of Type THHN in nonmetallic sheathed cable was run through an attic where the temperature reached 123°F (51°C), the ampacity would be 30.4 amperes. If this cable was bundled together for over two feet with other cables containing six current carrying conductors, a further reduction would be required. The 30.4 amperes would be reduced by 80 percent (indicated in Table E3605.3) to 24.32 amperes, and the size 10 AWG conductors would be a violation if protected by a 30-ampere circuit breaker.

TABLE E3605.3
CONDUCTOR PROXIMITY ADJUSTMENT FACTORS

NUMBER OF CURRENT-CARRYING CONDUCTORS IN CABLE OR RACEWAY	PERCENT OF VALUES IN TABLE E3605.1
4-6	80
7-9	70
10-20	50
21-30	45
31-40	40
41 and above	35

❖ See the commentary for Section E3605.3.

Figure E3605.3
MULTICONDUCTOR CABLES BUNDLED TOGETHER

E3605.4 Temperature limitations. The temperature rating associated with the ampacity of a conductor shall be so selected and coordinated to not exceed the lowest temperature rating of any connected termination, conductor or device. Conductors with temperature ratings higher than specified for terminations shall be permitted to be used for ampacity adjustment, correction, or both.

❖ The allowable conductor ampacity is not the only consideration in determining what size conductor to use. In addition to the temperature rating of the conductor, the temperature rating of the terminations or terminals the conductor connects to at each end of the conductor is critical. The terminals on devices such as receptacles, switches, circuit breakers, and panelboards all have maximum temperature ratings. Current passing through these terminations can heat up the device just as it does in a conductor. The lowest temperature rating of any of the devices in the circuit determines the allowable ampacity. Many switches and receptacles used in residential work have a temperature rating of 60°C. Circuit breakers might have temperature ratings of 75°C.

E3605.4.1 Conductors rated 60°C. Termination provisions of equipment for circuits rated 100 amperes or less, or marked for 14 AWG through 1 AWG conductors, shall be used only for one of the following, except where the equipment is marked otherwise:

1. Conductors rated 60°C (140°F), or
2. Conductors with higher temperature ratings, provided that the ampacity of such conductors is determined based on the 60°C (140°F) ampacity of the conductor size used, or

3. Conductors with higher temperature ratings where the equipment is listed and identified for use with such conductors, or
4. For motors marked with design letters B, C, D, or E, conductors having an insulation rating of 75°C (167°F) or higher shall be permitted to be used provided that the ampacity of such conductors does not exceed the 75°C (167°F) ampacity.

❖ See the commentary for Section E3605.4.2.

E3605.4.2 Conductors rated 75°C. Termination provisions of equipment for circuits rated over 100 amperes, or marked for conductors larger than 1 AWG, shall be used only for:

1. Conductors rated 75°C (167°F).
2. Conductors with higher temperature ratings provided that the ampacity of such conductors does not exceed the 75°C (167°F) ampacity of the conductor size used, or provided that the equipment is listed and identified for use with such conductors.

❖ Where the temperature rating of the termination on the device is 60°C, the allowable conductor ampacity is found in the 60°C column of Table E3605.1. Conductors with higher temperature rated insulation are allowed to be used, and this is often the case when derating is necessary for ambient temperature or for the number of conductors in close proximity.

An example of this would be THHN copper conductors connected to a 70-ampere 2-pole circuit breaker that is listed and marked at 60/75°C. The conductors feed equipment that has terminals listed and marked at 60°C. Assuming an ambient temperature of 80°F, and not over three current carrying conductors in the raceway, the smallest size THHN copper conductors that may be used in this installation are size 4 AWG, determined from the 60°C column.

E3605.4.3 Separately installed pressure connectors. Separately installed pressure connectors shall be used with conductors at the ampacities not exceeding the ampacity at the listed and identified temperature rating of the connector.

❖ The temperature rating of wire nuts and other pressure connectors is one factor in determining the allowable ampacity of the conductors.

E3605.4.4 Conductors of Type NM cable. Conductors in NM cable assemblies shall be rated at 90°C (194°F). Types NM, NMC, and NMS cable identified by the markings NM-B, NMC-B, and NMS-B meet this requirement. The ampacity of Types NM, NMC, and NMS cable shall be at 60°C (140°F) conductors and shall comply with Section E3605.1 and Table E3605.5.3. The 90°C (194°F) rating shall be permitted to be used for ampacity correction and adjustment purposes provided that the final corrected or adjusted ampacity does not exceed that for a 60°C (140°F) rated conductor.

❖ The code requires that the individual conductors in nonmetallic sheathed cable have an insulation rating of 90°C, but the ampacity is that of the 60°C column. See the commentary for Section E3605.2.

E3605.5 Overcurrent protection required. All ungrounded branch-circuit and feeder conductors shall be protected against overcurrent by an overcurrent device installed at the point where the conductors receive their supply. Overcurrent devices shall not be connected in series with a grounded conductor. Overcurrent protection and allowable loads for branch circuits and feeders that do not serve as the main power feeder to the dwelling unit load shall be in accordance with this chapter.

Branch-circuit conductors and equipment shall be protected by overcurrent protective devices having a rating or setting not exceeding the allowable ampacity specified in Table E3605.1 and Sections E3605.2, E3605.3 and E3605.4 except where otherwise permitted or required in Sections E3605.5.1 through E3605.5.3.

❖ The overcurrent devices consist of circuit breakers or fuses and are installed at the beginning of the circuit where the conductors receive their supply. The overcurrent protection devices prevent the conductors from carrying currents above the allowable ampacity determined from Table E3605.1 and the associated de-rating provisions. The overcurrent protection device will automatically open and protect the conductors if the current is high enough to cause a dangerous temperature.

E3605.5.1 Cords. Cords shall be protected in accordance with Section E3809.2.

❖ See the commentary for Section E3809.2.

E3605.5.2 Overcurrent devices of the next higher rating. The next higher standard overcurrent device rating, above the ampacity of the conductors being protected, shall be permitted to be used, provided that all of the following conditions are met:

1. The conductors being protected are not part of a multi-outlet branch circuit supplying receptacles for cord- and plug-connected portable loads.
2. The ampacity of conductors does not correspond with the standard ampere rating of a fuse or a circuit breaker without overload trip adjustments above its rating (but that shall be permitted to have other trip or rating adjustments).
3. The next higher standard device rating does not exceed 400 amperes.

❖ Where a circuit supplies only a specific load, the maximum ampacity is known. For multi-outlet branch circuits, it is not known what the load will be for cord- and plug-connected equipment and appliances. But where the load is known, and it does not match one of the standard ratings of overcurrent devices, an overcurrent device of the next higher rating can be used. In these cases, the conductor ampacity could be less than the rating of overcurrent device, which is the branch circuit rating.

E3605.5.3 Small conductors. Except as specifically permitted by Section E3605.5.4, the rating of overcurrent protection devices shall not exceed the ratings shown in Table E3605.5.3 for the conductors specified therein.

❖ Referring to Table E3605.1 on conductor ampacity, the allowable ampacity of size 14 AWG is 20 to 25 amperes, depending on the type of insulation and the temperature ratings. Various ampacities are also given for sizes 12 and 10 AWG. However, Table E3605.5.3 provides the overcurrent device rating for these small size conductors regardless of the allowable ampacity given in Table E3605.1.

A 15-ampere circuit breaker is used on circuits with size 14 AWG conductors, a 20-ampere circuit breaker with size 12 AWG conductors, and a 30-ampere circuit breaker with size 10 AWG conductors. The only exception is as allowed by Section E3605.5.4.

TABLE E3605.5.3
OVERCURRENT-PROTECTION RATING

COPPER		ALUMINUM OR COPPER-CLAD ALUMINUM	
Size (AWG)	Maximum overcurrent-protection-device rating[a] (amps)	Size (AWG)	Maximum overcurrent-protection-device rating[a] (amps)
14	15	12	15
12	20	10	25
10	30	8	30

a. The maximum overcurrent-protection-device rating shall not exceed the conductor allowable ampacity determined by the application of the correction and adjustment factors in accordance with Sections E3605.2 and E3605.3.

❖ See the commentary for Section E3605.5.3.

E3605.5.4 Air conditioning and heat pump equipment. Air-conditioning and heat pump equipment circuit conductors shall be permitted to be protected against overcurrent in accordance with Section E3602.11.

❖ The overcurrent protection for air-conditioning and heat-pump circuits is stated on the equipment label.

E3605.6 Fuses and fixed trip circuit breakers. The standard ampere ratings for fuses and inverse time circuit breakers shall be considered 15, 20, 25, 30, 35, 40, 45, 50, 60, 70, 80, 90, 100, 110, 125, 150, 175, 200, 225, 250, 300, 350 and 400 amperes.

❖ Although a standard size circuit breaker is 25 amperes, and the allowable ampacity from Table E3605.1 shows 25 amperes for size 12 AWG Type THW, it would be a code violation to install a 25-ampere circuit breaker on a branch circuit of this size conductor. An electrician might try to solve a problem of a circuit breaker tripping from overload by looking at Table E3605.1 and, seeing that a size 12 THHN conductor has an ampacity of 30 amperes, installing a 20- or 30-ampere circuit breaker. This has been done and is a code violation and potential fire hazard.

E3605.7 Location of overcurrent devices in or on premises. Overcurrent devices shall:

1. Be readily accessible.
2. Not be located where they will be exposed to physical damage.
3. Not be located where they will be in the vicinity of easily ignitible material such as in clothes closets.
4. Not be located in bathrooms.

Exceptions:

1. This section shall not apply to supplementary overcurrent protection that is integral to utilization equipment.
2. Overcurrent devices installed adjacent to the utilization equipment that they supply shall be permitted to be accessible by portable means.

❖ The panelboard (which contains overcurrent protection devices) is not permitted to be installed in a clothes closet or in any closet. It might be determined that a pantry closet off a kitchen would not have "easily ignitable material." Whether the code official would permit a panelboard to be installed in such a closet is a matter of interpretation. Because of the dampness and humidity, and potential shock hazard, overcurrent devices are not permitted in bathrooms. A panelboard would be permitted to be installed in a laundry room or utility room.

E3605.8 Ready access for occupants. Each occupant shall have ready access to all overcurrent devices protecting the conductors supplying that occupancy.

❖ In a two-family dwelling, each occupant must have access to all overcurrent protective devices for their own dwelling. This includes the service disconnecting means. The overcurrent protection devices must not be locked such as to prevent occupant access.

E3605.9 Enclosures for overcurrent devices. Overcurrent devices shall be enclosed in cabinets or cutout boxes except where an overcurrent device is part of an assembly that provides equivalent protection. The operating handle of a circuit breaker shall be permitted to be accessible without opening a door or cover.

❖ Panelboards are installed in a cabinet or cutout box. See the commentary in Chapter 34 under the definitions for these items.

SECTION E3606 PANELBOARDS

E3606.1 Panelboard rating. All panelboards shall have a rating not less than that of the minimum service entrance or feeder capacity required for the computed load.

❖ Panelboards are located in a cabinet or cutout box and hold the circuit breakers for the branch circuits. The busbar to which the circuit breakers are connected must have an ampere rating at least equal to the rating of the feeder serving the panelboard. Some common panelboard ratings are 60, 100, 125, 150, and 200 ampere ratings.

E3606.2 Panelboard circuit identification. All circuits and circuit modifications shall be legibly identified as to purpose or use on a circuit directory located on the face or inside the door of the enclosure.

❖ The panelboard circuit directory must be permanent and legible. In many homes, the door of the panelboard cabinet reveals illegible markings. The outlets a circuit serves are a mystery to an observer because the markings on the panelboard have become faded or are incomplete. Some electricians and installers write only such things as "lights" or "plugs" or "kitchen plugs" to satisfy this section. Of course, acceptance of such markings is according to the judgment of the code official, but if the markings are not legible a code violation exists. Correct terminology must be used; for example, branch circuits can supply receptacles but not "plugs." Obviously, the circuits for appliances such as dishwashers, disposals, furnaces, etc. should be marked. Commentary Figure E3501.7 illustrates panelboard directories on the inside of the doors.

E3606.3 Panelboard overcurrent protection. Panelboards shall be protected on the supply side by not more than two main circuit breakers or two sets of fuses having a combined rating not greater than that of the panelboard.

> **Exception:** Individual protection for a panelboard shall not be required if the panelboard feeder has overcurrent protection not greater than the rating of the panelboard.

❖ The rating of the overcurrent device supplying the panelboard must have a rating not more than the ampere rating of the panelboard busbars. The exception indicates that the panelboard can be protected by the main breaker installed in the panelboard or at the supply end of the feeder supplying the panelboard.

Bibliography

The following resource materials are referenced in this chapter or are relevant to the subject matter addressed in this chapter.

NFPA 70-02, *National Electrical Code.* Quincy, MA: National Fire Protection Association, 2002.

Chapter 37: Wiring Methods

General Comments

This chapter provides installation details for the wiring methods commonly found in dwelling unit construction, and it dictates where and under what conditions specific wiring methods can be used.

Purpose

The purpose of this chapter is to protect life and property by making sure that the wiring in a building is suitable for the environment to which it is exposed and that the wiring is properly supported and protected from damage.

SECTION E3701 GENERAL REQUIREMENTS

E3701.1 Scope. This chapter covers the wiring methods for services, feeders and branch circuits for electrical power and distribution.

❖ Wiring methods consist of the cable, tubing, or conduit in which the individual wiring is installed. Cable comes with the wire already installed. Tubing and conduit are installed first, with the wire inserted after. This chapter includes the provisions for installing cable, tubing, or conduit. Chapter 38 covers the number of conductors permitted in the tubing or conduit, and boxes. This chapter covers how and where the cable, tubing, or conduit may be installed.

E3701.2 Allowable wiring methods. The allowable wiring methods for electrical installations shall be those listed in Table E3701.2. Single conductors shall be used only where part of one of the recognized wiring methods listed in Table E3701.2. As used in this code, abbreviations of the wiring-method types shall be as indicated in Table E3701.2.

❖ Single insulated conductors are not permitted as a wiring method without being part of a cable assembly or installed in tubing or conduit. In many older houses, a wiring method known as "knob-and-tube" was used, consisting of single insulated conductors supported in free air on porcelain insulators (knobs) and, where run through a wood framing member, installed through insulating tubes. Where remodeling, retrofitting, or repair is being done in an old house that has this type of wiring method, it is permitted to connect into or tap into one of these circuits to make an extension of the circuit. But, instead of running new individual conductors, the transition to the new wiring should be to cable. In early years, this type of wiring method was installed in the hollow spaces of walls, ceilings, and attics. It was not installed directly on the wood framing as cable is today, but supported in free air, thus allowing for the dissipation of any heat buildup in the conductors. Care should be taken if insulation is blown in the wall or attic, or if foam insulation is installed around these wires, thereby affecting heat dissipation. In many older houses, the wiring is still sound and in good shape, but when additions or extensions are made, the possibility for arcing, bad connections, heating, and overloading comes into play.

TABLE E3701.2
ALLOWABLE WIRING METHODS

ALLOWABLE WIRING METHOD	DESIGNATED ABBREVIATION
Armored cable	AC
Electrical metallic tubing	EMT
Electrical nonmetallic tubing	ENT
Flexible metal conduit	FMC
Intermediate metal conduit	IMC
Liquidtight flexible conduit	LFC
Metal-clad cable	MC
Nonmetallic sheathed cable	NM
Rigid nonmetallic conduit	RNC
Rigid metallic conduit	RMC
Service entrance cable	SE
Surface raceways	SR
Underground feeder cable	UF
Underground service cable	USE

❖ See the commentary for Section E3701.2.

E3701.3 Circuit conductors. All conductors of a circuit, including equipment grounding conductors, shall be contained in the same raceway, trench, cable or cord.

❖ Current flow in a conductor produces a magnetic flux around the conductor. Through this induction process, heating of the conductor and the surrounding metal enclosure occurs, and impedance increases. In alternating current circuits, the polarity of the magnetic field is constantly reversing, which causes heat. When current in the ungrounded conductor is positive, the current in the grounded conductor is negative. Where both conductors of the circuit are run together in close proximity, the magnetic field of one conductor is canceled or balanced by that of the other.

E3701.4 Wiring method applications. Wiring methods shall be applied in accordance with Table E3701.4.

❖ Table E3701.4 is a summary of the applications or locations permitted for the various wiring methods. The footnotes of the table are provisions of the code and should not be overlooked. The most common wiring method for interior wiring is nonmetallic sheathed cable (NM). Type NM and NMC cables are permitted for one- and two-family dwellings regardless of the number of stories.

TABLE E3701.4
ALLOWABLE APPLICATIONS FOR WIRING METHODS[a,b,c,d,e,f,g,h,i]

ALLOWABLE APPLICATIONS (application allowed where marked with an "A")	AC	EMT	ENT	FMC	IMC RMC RNC	LFC[a]	MC	NM	SR	SE	UF	USE
Services	—	A	A[h]	A[i]	A	A[i]	A	—	—	A	—	A
Feeders	A	A	A	A	A	A	A	A	—	A[b]	A	A[b]
Branch circuits	A	A	A	A	A	A	A	A	A	A[c]	A	—
Inside a building	A	A	A	A	A	A	A	A	A	A	A	—
Wet locations exposed to sunlight	—	A	A[h]	A[d]	A	A	A	—	—	A	A[e]	A[e]
Damp locations	—	A	A	A[d]	A	A	A	—	—	A	A	A
Embedded in noncinder concrete in dry location	—	A	A	—	A	—	—	—	—	—	—	—
In noncinder concrete in contact with grade	—	A[f]	A	—	A[f]	—	—	—	—	—	—	—
Embedded in plaster not exposed to dampness	A	A	A	A	A	A	A	—	—	A	A	—
Embedded in masonry	—	A	A	—	A[f]	A	A	—	—	—	—	—
In masonry voids and cells exposed to dampness or below grade line	—	A[f]	A	A[d]	A[f]	A	A	—	—	A	A	—
Fished in masonry voids	A	—	—	A	—	A	A	A	—	A	A	—
In masonry voids and cells not exposed to dampness	A	A	A	A	A	A	A	A	—	A	A	—
Run exposed	A	A	A	A	A	A	A	A	A	A	A	A
Run exposed and subject to physical damage	—	—	—	—	A[g]	—	—	—	—	—	—	—
For direct burial	—	A[f]	—	—	A[f]	A	A[f]	—	—	—	A	A

For SI: 1 foot = 304.8 mm.

a. Liquid-tight flexible nonmetallic conduit without integral reinforcement within the conduit wall shall not exceed 6 feet in length.

b. The grounded conductor shall be insulated except where used to supply other buildings on the same premises. Type USE cable shall not be used inside buildings.

c. The grounded conductor shall be insulated.

d. Conductors shall be a type approved for wet locations and the installation shall prevent water from entering other raceways.

e. Shall be listed as sunlight resistant.

f. Metal raceways shall be protected from corrosion and approved for the application.

g. RNC shall be Schedule 80.

h. Shall be listed as "Sunlight Resistant" where exposed to the direct rays of the sun.

i. Conduit shall not exceed 6 feet in length.

❖ See the commentary for Section E3701.4.

SECTION E3702
ABOVE-GROUND INSTALLATION REQUIREMENTS

E3702.1 Installation and support requirements. Wiring methods shall be installed and supported in accordance with Table E3702.1.

❖ Table E3702.1 covers the requirements for the support of aboveground wiring methods used in dwellings. Where run parallel to a framing member, cables and nonmetallic wiring methods must be held back at least 1.25 inches from the edge. This is so that when the drywall is applied, a nail or screw that misses the stud of the framing member cannot penetrate and damage the cable or conduit.

Where holes are bored in studs and vertical framing members for cables and nonmetallic wiring methods, the nearest edge of the hole must be at least 1.25 inches (31 mm) from the edge of the framing member. This is to help prevent the nail or screw from penetrating the wiring installed through the holes. Where this is impossible, a steel plate at least $^1/_{16}$ inch (1.5 mm) thick must be installed over the framing member where the wiring passes through.

Many electricians who have studied electrical codes for years know the 1.25 inch (31 mm) rule because this is an electrical code provision for protection of cables and nonmetallic wiring methods. But for drilling and notching of framing members, structural integrity must be considered also. Electrical code provisions cover protection of wiring but not structural framing rules.

Building code provisions require that holes drilled in horizontal framing members must be at least 2 inches back from the edge. This requirement is to help prevent weakening of the structural members such as floor joists. See Section R502.8 and Figure R502.8 of this code for details of drilling and notching provisions for structural floor members. This requirement is included in Table E3702.1.

Detailed provisions for drilling and notching of studs in an exterior wall or bearing partition are in Section R602.6. In addition to following the electrical code provisions for protection of wiring, these provisions must be followed for structural strength. See Commentary Figure E3702.1.

TABLE E3702.1
GENERAL INSTALLATION AND SUPPORT REQUIREMENTS FOR WIRING METHODS[a,b,c,d,e,f,g,h,i]

INSTALLATION REQUIREMENTS (requirement applicable only to wiring methods marked "A")	AC MC	EMT IMC RMC	ENT	FMC LFC	NM UF	RNC	SE	SR[a]	USE
Where run parallel with the framing member, the wiring shall be 1.25 inches from the edge of a framing member such as a joist, rafter or stud or shall be physically protected.	A	—	A	A	A	—	A	—	—
Bored holes in studs and vertical framing members for wiring shall be located 1.25 inches from the edge or shall be protected with a minimum 0.0625-inch steel plate or sleeve or other physical protection.	A	—	A	A	A	—	A	—	—
Where installed in grooves, to be covered by wallboard, siding, paneling, carpeting, or similar finish, wiring methods shall be protected by 0.0625-inch-thick steel plate, sleeve, or equivalent or by not less than 1.25-inch free space for the full length of the groove in which the cable or raceway is installed.	A	—	A	A	A	—	A	A	A
Bored holes in joists, rafters, beams and other horizontal framing members shall be 2 inches from the edge of the structural framing member.	A	A	A	A	A	A	A	—	—
Securely fastened bushings or grommets shall be provided to protect wiring run through openings in metal framing members.	—	—	A	—	A	—	A	—	—
The maximum number of 90-degree bends shall not exceed four between junction boxes.	—	A	A	A	—	A		—	—
Bushings shall be provided where entering a box, fitting or enclosure unless the box or fitting is designed to afford equivalent protection.	A	A	A	A		A	—	A	—
Ends of raceways shall be reamed to remove rough edges.	—	A	A	A	—	A	—	A	—
Maximum allowable on center support spacing for the wiring method in feet.	4.5[b, c]	10	3[b]	4.5[b]	4.5[i]	3[d]	2.5[e]	—	2.5[e]
Maximum support distance in inches from box or other terminations.	12[b, f]	36	36	12[b, g]	12[h,i]	36	12	—	12

For SI: 1 inch = 25.4 mm, 1 foot = 304.8 mm, 1 degree = 0.009 rad.

a. Installed in accordance with listing requirements.

b. Supports not required in accessible ceiling spaces between light fixtures where lengths do not exceed 6 feet.

c. Six feet for MC cable.

d. Five feet for trade sizes greater than 1 inch.

e. Two and one-half feet where used for service or outdoor feeder and 4.5 feet where used for branch circuit or indoor feeder.

f. Twenty-four inches where flexibility is necessary.

g. Thirty-six inches where flexibility is necessary.

h. Within 8 inches of boxes without cable clamps.

i. Flat cables shall not be stapled on edge.

❖ See the commentary for Section E3702.1.

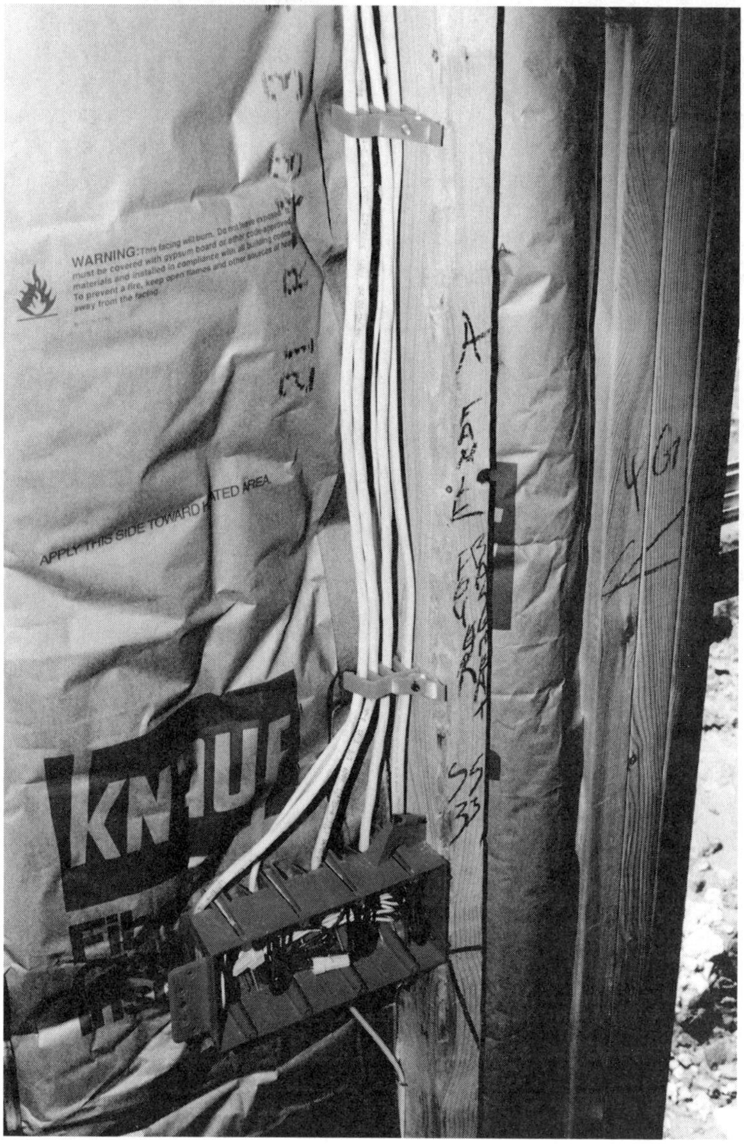

Figure E3702.1
CABLES HELD $1^1/_4$ INCHES BACK FROM EDGE OF FRAMING MEMBER

E3702.2 Cables in accessible attics. Cables in attics or roof spaces provided with access shall be installed as specified in Sections E3702.2.1 and E3702.2.2.

❖ An attic can be totally enclosed or accessible through a small 22 x 30-inch (559 mm by 762 mm) opening or provided with a ladder or stairs. Section R807 of this code requires an attic access opening for attic areas over 30 square feet (3 m^2) and a vertical height of at least 30 inches (762 mm). Where cables are run in an attic that has an access opening, specific rules apply as stated in the two following sections.

E3702.2.1 Across structural members. Where run across the top of floor joists, or run within 7 feet (2134 mm) of floor or floor joists across the face of rafters or studding, in attics and roof spaces that are provided with access, the cable shall be protected by substantial guard strips that are at least as high as the cable. Where such spaces are not provided with access by permanent stairs or ladders, protection shall only be required within 6 feet (1829 mm) of the nearest edge of the attic entrance.

❖ If the attic is provided with access by a ladder or stairs, any exposed cable run must be protected as stated. These code provisions are intended as protection for the cables where it is likely that people will be in the attic. Any cable run exposed along the surface of rafters or studs could easily be pulled on or used to hang things. Cables that are exposed and run across the surface of joists could be walked on or have storage items placed on them, causing damage to the cables. Thus, they are required to be protected or kept at least seven feet above the floor of the attic. The majority of new dwellings do not have an unfinished attic accessible by stairs or a ladder where cables are exposed, but most do have an attic accessible by a scuttle hole (a term used in the industry for the small attic entrance). In attics with only the scuttle hole entrance, cables run across the face of rafters or studs must be protected with guard strips for a distance of six feet from the entrance. This provision is sometimes overlooked during the rough inspection. On the finish inspection a ladder is then necessary to look into the attic, and the wiring is usually covered by insulation.

E3702.2.2 Cable installed parallel to framing members. Where cables are installed parallel to the sides of rafters, studs or floor joists, guard strips and running boards shall not be required, and the installation shall comply with Table E3702.1.

❖ In an attic where cable is run along the sides of rafters, studs, or joists, no extra protection such as guard strips is required, because the cable is not subject to damage by being walked on or having storage items placed on it as would be the case if it was run over the tops of joists. Table E3702.1 requires that cable be kept back at least 1.25 inches (31 mm) from the face of the rafter, stud or joist.

E3702.3 Exposed cable. In exposed work, except as provided for in Sections E3702.2 and E3702.4, cable assemblies shall be installed as specified in Sections E3702.3.1 and E3702.3.2.

❖ See the commentary for Section E3702.3.2 and Commentary Figures E3702.3.1(1) and E3702.3.1(2).

E3702.3.1 Surface installation. Cables shall closely follow the surface of the building finish or running boards.

❖ See the commentary for Section E3702.3.2 and Commentary Figures E3702.3.1(1) and E3702.3.1(2).

E3702.3.2 Protection from physical damage. Where subject to physical damage, cables shall be protected by conduit, electrical metallic tubing, Schedule 80 PVC rigid nonmetallic conduit, pipe, guard strips or other approved means. Where passing through a floor, the cable shall be enclosed in rigid metal conduit, intermediate metal conduit, electrical metallic tubing, Schedule 80 PVC rigid nonmetallic conduit or other metal pipe extending not less than 6 inches (152 mm) above the floor.

❖ The provisions of this section apply to exposed cable in other than attics and joist spaces of unfinished basements. They apply to cables installed in such places as on the wall of an unfinished basement. Although it may not look good to have cable installed on the surface of a wall, it is permitted where the cable is not subject to physical damage. Running it through conduit is one common protection method.

For example, in an unfinished basement, a switch is installed on the concrete or masonry wall at the entrance to the outside stairway. The nonmetallic sheathed cable run to the switch must be protected. One way to do this is to frame a short, two or three stud section of unfinished wall to install a switch box to and secure the cable. Another way is to install a short, three- or four-foot length of conduit or EMT from the ceiling or joist space down to an appropriate switch box attached to the concrete wall. This makes for a neat looking installation along the wall and provides protection for the cable. However, the cable could be subject to damage where it enters the conduit or EMT. The part of the cable in the basement ceiling or joist area could be pulled to gain slack for various reasons, such as when installing heating ducts. At the top of the short run of conduit or EMT, a fitting must be installed to protect the cable where it enters the EMT.

Where cable is installed through a floor, it must be protected such that it cannot be damaged by being walked on or kicked or having objects placed against it; see Commentary Figures E3702.3.1(1) and E3702.3.1(2).

E3702.4 In unfinished basements. Where the cable is run at angles with joists in unfinished basements, cable assemblies containing two or more conductors of sizes 6 AWG and larger and assemblies containing three or more conductors of sizes 8 AWG and larger shall not require additional protection where attached directly to the bottom of the joists. Smaller cables shall be run either through bored holes in joists or on running boards.

❖ Although there are options, in most cases in unfinished basement ceilings, cables are installed through holes in the joists. Cables that are run across the bottom of floor joists are subject to damage resulting from occupants hanging things from or storing things on the cables. If the basement is left unfinished in a new house, the electrician is usually expected to run the cables within the joist area instead of on the lower surface of the joists because it is expected that at a future time a ceiling membrane will be attached directly to the bottom edge of the floor joists to finish the basement ceiling. Even larger cables such as feeders to a panelboard are run through drilled holes. But to accommodate piping and ducts, the ceiling of an unfinished basement may need to be furred down below the surface of the floor joists to prepare for applying the membrane. For whatever reason, if the electrician chooses to run the cables on the surface of the lower edge of the joists, these specific requirements apply.

Only the specified cables can be installed across the bottom of joists without further protection. Cables with conductor sizes 14, 12 and 10 AWG must be either run through drilled holes or installed on running boards. Note that the building portion of this code limits the size and location of bored holes in joists. See Commentary Figure E3702.4.

E3702.5 Bends. Bends shall be made so as not to damage the wiring method or reduce the internal diameter of raceways.

For types NM and SE cable, bends shall be so made, and other handling shall be such that the cable will not be damaged and the radius of the curve of the inner edge of any bend shall be not less than five times the diameter of the cable.

❖ Where installed through drilled holes in studs and joists, it is important to avoid kinks and knots in the cable as it is pulled off the spool or coil. Although a slang term used in some areas for cable is "rope," it should not be installed as if it were rope. Care must be used when pulling through holes at various angles.

SECTION E3703 UNDERGROUND INSTALLATION REQUIREMENTS

E3703.1 Minimum cover requirements. Direct buried cable or raceways shall be installed in accordance with the minimum cover requirements of Table E3703.1.

❖ Underground wiring must be protected by burial at a sufficient depth. For example, a 120-volt, 20-ampere branch circuit with GFCI protection serving a yard light must be buried at least 12 inches (305 mm) deep where it is under the residential driveway. A service lateral installed in a PVC conduit under the backyard lawn must be buried at least 18 inches (457 mm) deep. Many utility companies require the service lateral to be installed in conduit. Where the service lateral is a direct buried cable under the backyard lawn with no protec-

Figure E3702.3.1(1)
PROHIBITED APPLICATION OF TYPE NM CABLE

tion such as a concrete slab, it is required to be buried at least 24 inches (610 mm) deep, according to Table E3703.1.

E3703.2 Warning ribbon. Underground service conductors that are not encased in concrete and that are buried 18 inches (457 mm) or more below grade shall have their location identified by a warning ribbon that is placed in the trench not less than 12 inches (300 mm) above the underground installation.

❖ The warning ribbon, typically red plastic tape with printed warnings, is required for service conductor installations to help prevent excavators from damaging the conductors and risking injury or death. Table E3703.1 dictates the conditions under which service conductors must be buried at least 18 inches (457 mm). The ribbon must be at least 12 inches (305 mm) above the conductors to provide warning well in advance of the conductor depth.

E3703.3 Grounding. All underground installations shall be grounded and bonded in accordance with this code.

❖ An example of this is a rigid metal conduit or intermediate metal conduit used to enclose the service lateral conductors from under the ground to where they enter the service equipment enclosure. This conduit is bonded to the service equipment.

E3703.4 Protection from damage. Direct buried conductors and cables emerging from the ground shall be protected by enclosures or raceways extending from the minimum cover distance below grade required by Section 3703.1 to a point at least 8 feet (2438 mm) above finished grade. In no case shall the protection be required to exceed 18 inches (457 mm) below finished grade. Service laterals that are not encased in concrete and that are buried 18 inches (457 mm) or more below grade shall have their location identified by a warning ribbon that is placed in the trench at least 12 inches (305 mm) above the underground installation. Conductors entering a building shall be protected to the point of entrance. Where the enclosure or raceway is subject to physical damage, the conductors shall be installed in rigid metal conduit, intermediate metal conduit, Schedule 80 rigid nonmetallic conduit or the equivalent.

❖ Where a light installed on a pole is fed by direct buried cable such as a UF cable, the cable must be protected from at least 18 inches (457 mm) below the ground surface to a height of at least 8 feet (2438 mm) up the pole.

Most service laterals to a dwelling would not be encased in concrete (such as a service lateral enclosed in non-metallic conduit that is encased in concrete) unless under special circumstances. In all other cases, whether direct-buried or enclosed in conduit, the warning ribbon must be placed in the trench 12 inches (305 mm) or more above the service lateral to help avoid

Figure E3702.3.1(2)
PROHIBITED APPLICATION OF TYPE NM CABLE

damage or electrocution when digging occurs near the buried conductors. The service lateral is not protected against overcurrent caused by a short-circuit.

E3703.5 Splices and taps. Direct buried conductors or cables shall be permitted to be spliced or tapped without the use of splice boxes. The splices or taps shall be made by approved methods with materials listed for the application.

❖ Splicing kits are available for use on direct-buried conductors. These are used in most cases where the conductors have been damaged and it becomes necessary to repair the conductors rather than replace the entire underground run.

E3703.6 Backfill. Backfill containing large rock, paving materials, cinders, large or sharply angular substances, or corrosive material shall not be placed in an excavation where such materials cause damage to raceways, cables or other substructures or prevent adequate compaction of fill or contribute to corrosion of raceways, cables or other substructures. Where necessary to prevent physical damage to the raceway or cable, protection shall be provided in the form of granular or selected material, suitable boards, suitable sleeves or other approved means.

❖ Unless the soil is suitable to prevent damage to the buried conductors or raceways, sand must be brought in and placed in the trench below and on top of the underground wiring method, or other approved means must be used. In many cases where this is not done, direct buried cable can be damaged upon backfilling by rocks falling into the trench on top of the cable. Where aluminum cable is used, a small nick or cut in the conductor insulation can cause moisture to penetrate the sheath or insulation, which in turn can cause

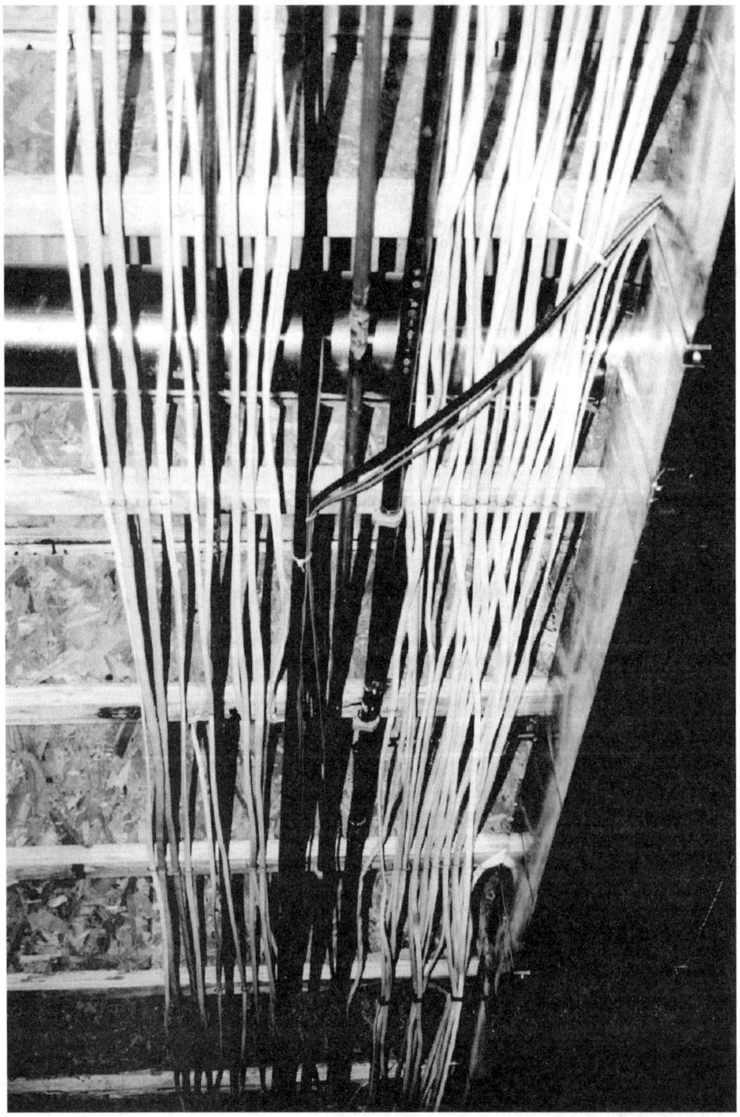

Figure E3702.4
PROHIBITED INSTALLATION OF TYPE NM CABLE ACROSS THE BOTTOM OF FLOOR JOISTS

the aluminum to oxidize. Where this happens, it appears that the aluminum has pulverized at that spot and degenerated into a substance like powder. For underground service laterals, many utility companies have equipment to locate the exact point of the damaged or open conductor. It can then be repaired by digging down to the cable and making a splice allowed by Section E3703.4. This is a very costly and time-consuming process and requires digging. Therefore, in landscaped areas, the utility companies often require that service laterals be run in conduit from the point of supply to the service equipment.

E3703.7 Raceway seals. Conduits or raceways shall be sealed or plugged at either or both ends where moisture will enter and contact energized parts.

❖ Where an underground conduit is run through the basement wall, for example, the temperature difference will cause condensation, and moisture could enter an electrical enclosure such as a box or cabinet. An example is a service lateral that is run directly into service equipment located on the inside wall of a basement. The conduit must be plugged or a sealing compound used to prevent moisture from entering and contacting energized parts.

TABLE E3703.1
MINIMUM COVER REQUIREMENTS, BURIAL IN INCHES[a,b,c,d]

LOCATION OF WIRING METHOD OR CIRCUIT	TYPE OF WIRING METHOD OR CIRCUIT				
	1 Direct burial cables or conductors	2 Rigid metal conduit or intermediate metal conduit	3 Nonmetallic raceways listed for direct burial without concrete encasement or other approved raceways	4 Residential branch circuits rated 120 volts or less with GFCI protection and maximum overcurrent protection of 20 amperes	5 Circuits for control of irrigation and landscape lighting limited to not more than 30 volts and installed with type UF or in other identified cable or raceway
All locations not specified below	24	6	18	12	6
In trench below 2-inch-thick concrete or equivalent	18	6	12	6	6
Under a building	0 (In raceway only)	0	0	0 (In raceway only)	0 (In raceway only)
Under minimum of 4-inch-thick concrete exterior slab with no vehicular traffic and the slab extending not less than 6 inches beyond the underground installation	18	4	4	6 (Direct burial) 4 (In raceway)	6 (Direct burial) 4 (In raceway)
Under streets, highways, roads, alleys, driveways and parking lots	24	24	24	24	24
One- and two-family dwelling driveways and outdoor parking areas, and used only for dwelling-related purposes	18	18	18	12	18
In solid rock where covered by minimum of 2 inches concrete extending down to rock	2 (In raceway only)	2	2	2 (In raceway only)	2 (In raceway only)

For SI: 1 inch = 25.4 mm.

a. Raceways approved for burial only where encased concrete shall require concrete envelope not less than 2 inches thick.

b. Lesser depths shall be permitted where cables and conductors rise for terminations or splices or where access is otherwise required.

c. Where one of the wiring method types listed in columns 1 to 3 is combined with one of the circuit types in columns 4 and 5, the shallower depth of burial shall be permitted.

d. Where solid rock prevents compliance with the cover depths specified in this table, the wiring shall be installed in metal or nonmetallic raceway permitted for direct burial. The raceways shall be covered by a minimum of 2 inches of concrete extending down to the rock.

❖ See the commentary for Section E3702.1 and Section E3703.1.

E3703.8 Bushing. A bushing, or terminal fitting, with an integral bushed opening shall be installed on the end of a conduit or other raceway that terminates underground where the conductors or cables emerge as a direct burial wiring method. A seal incorporating the physical protection characteristics of a bushing shall be considered equivalent to a bushing.

❖ An example of this requirement is where direct buried service lateral conductors are installed in rigid or intermediate metal conduit from a point underground up to the service equipment. At the point where the conductors emerge from the conduit, a bushing must be installed to protect the conductors. The ground around the conductors and the conduit will settle and pull the conductors against the conduit, which, if installed in compliance with code provisions, is fastened solidly to the wall of the house. Where the conduit is not fastened solidly to the wall of the house, the settling of the ground has been observed to pull the conduit connector away from the service equipment enclosure, exposing the conductor under the service equipment cabinet. It is important to leave slack in the direct buried conductors of the service lateral where they enter the riser conduit in the trench. Where the earth has been backfilled around a new house, settlement of the ground will occur that can also cause the conductors to move.

E3703.9 Single conductors. All conductors of the same circuit and, where present, the grounded conductor and all equipment grounding conductors shall be installed in the same raceway or shall be installed in close proximity in the same trench.

Exception: Where conductors are installed in parallel in raceways, each raceway shall contain all conductors of the same circuit including grounding conductors.

❖ Conductors must be grouped together in the same trench to reduce induction heating in raceways.

E3703.10 Ground movement. Where direct buried conductors, raceways, or cables are subject to movement by settlement or frost, direct buried conductors, raceways, or cables shall be arranged to prevent damage to the enclosed conductors or to equipment connected to the raceways.

❖ See the commentary for Section E3703.8.

Bibliography

The following resource materials are referenced in this chapter or are relevant to the subject matter addressed in this chapter.

NFPA 70-02, *National Electrical Code.* Quincy, MA: National Fire Protection Association, 2002.

Chapter 38: Power and Lighting Distribution

General Comments

Chapter 38 addresses the "rough-in" stage of construction in which the wiring system is installed for the purpose of distributing receptacle and lighting outlets throughout the dwelling. This chapter covers receptacle outlet spacing, GFCI and AFCI protection, lighting outlet locations, raceway and box fill limitations, box and panel board installation, equipment grounding, and flexible cords.

Purpose

The intent of Chapter 38 is to protect occupants from fire, electrical shock, and accidents. By eliminating the need for extension cords, by providing adequate lighting, and by requiring equipment grounding and GFCI and AFCI protection, the code substantially increases the level of safety in homes.

SECTION E3801 RECEPTACLE OUTLETS

E3801.1 General. Receptacle outlets shall be provided in accordance with Sections E3801.2 through E3801.11. Receptacle outlets required by this section shall be in addition to any receptacle that is part of a luminaire or appliance, that is located within cabinets or cupboards, or that is located over 5.5 feet (1676 mm) above the floor.

Permanently installed electric baseboard heaters equipped with factory-installed receptacle outlets, or outlets provided as a separate assembly by the baseboard manufacturer shall be permitted as the required outlet or outlets for the wall space utilized by such permanently installed heaters. Such receptacle outlets shall not be connected to the heater circuits.

❖ Section E3801 covers the requirements for the locations of receptacle outlets. A receptacle that is built in or is an integral part of light fixture, appliance, or cabinet is not counted as one of the outlets required by this section. However, a factory-installed receptacle in a baseboard heater is permitted to serve as one of the required outlets, but such a receptacle must not be connected to the circuit supplying the heater.

A baseboard heater must not be installed on the wall below a receptacle outlet unless specifically listed for such installations. Where a cord is plugged into a receptacle, it could drape over the baseboard heater or come into contact with it; thus, the cord would become hot, which could melt or weaken the insulation and create a possible fire or shock hazard. If the plan calls for baseboard heaters, the required receptacles must be either integral with the heater or be positioned so that cords will not contact the heater.

E3801.2 Convenience receptacle distribution. In every kitchen, family room, dining room, living room, parlor, library, den, sun room, bedroom, recreation room, or similar room or area of dwelling units, receptacle outlets shall be installed in accordance with the general provisions specified in Sections E3801.2.1 through E3801.2.3. See Figure E3801.2.

❖ The receptacles in all rooms of a dwelling unit except bathrooms and laundry rooms are referred to as convenience receptacles. This section and the following three subsections cover the location and spacing requirements for convenience receptacles. There is no minimum or maximum number of receptacles for a room. There is no height requirement for convenience receptacles except that any receptacle located over $5^1/_2$ feet above the floor is not counted. There are certain height requirements for accessibility standards; however, they usually do not apply to a private dwelling unit. Although there are specific requirements for countertop receptacles in a kitchen, the location and spacing of all other receptacles in kitchens and dining areas are included in these provisions.

E3801.2.1 Spacing. Receptacles shall be installed so that no point along the floor line in any wall space is more than 6 feet (1829 mm), measured horizontally, from an outlet in that space. Receptacles shall, insofar as practicable, be spaced equal distances apart.

❖ This requirement is intended to eliminate the need for extension cords. In any of the rooms listed (or a similar area of a dwelling), a lamp, appliance, radio, TV, or other electrical appliance can be placed at any point along the floor line, and a receptacle will be available within six feet of that location. It is important to not "cut corners" when measuring. The floor line is measured along the wall all the way into and around corners. Where a door opens against a wall, it may seem reasonable to begin the measurement at the end of the door swing, about 30 to 36 inches (762 to 914 mm) from the hinged side of the door casing, but the code requires that the wall space be counted all the way to the door jamb.

E3801.2.2 Wall space. As used in this section, a wall space shall include the following:

1. Any space that is 2 feet (610 mm) or more in width, (including space measured around corners), and that is un-

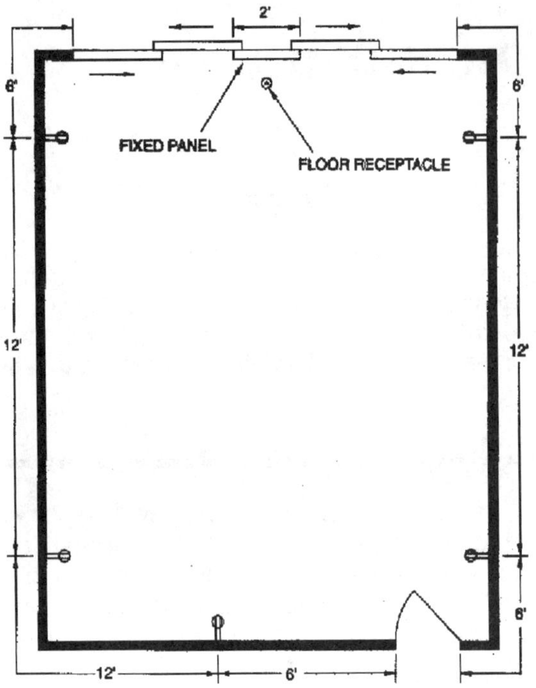

For SI: 1 foot = 304.8 mm.

FIGURE E3801.2
GENERAL USE RECEPTACLE DISTRIBUTION

❖ See the commentary for Section E3801.2.

broken along the floor line by doorways, fireplaces, and similar openings.

2. The space occupied by fixed panels in exterior walls, excluding sliding panels.
3. The space created by fixed room dividers such as railings and freestanding bar-type counters.

❖ An appliance or lamp, for example, could be placed in front of the fixed portion of a sliding door, which must be counted as wall space for determining the spacing requirements of receptacle outlets. An example of a fixed room divider is a railing installed along an open balcony or loft. Appliances could be placed by the homeowner along the railing, and a receptacle must be installed at a point no further than six feet from any point along the railing. A floor outlet may be necessary in this situation.

E3801.2.3 Floor receptacles. Receptacle outlets in floors shall not be counted as part of the required number of receptacle outlets except where located within 18 inches (457 mm) of the wall.

❖ Where the furniture layout is known, floor receptacles are sometimes located in the middle of a room. For example, power may be required for a lamp on an end table at each end of a couch that is placed away from the walls of a living room. Such receptacles are not counted as part of the required number of receptacles as measured along the wall line. In the example under Section E3801.2.2 above, a floor outlet must be installed a maximum of 18 inches (457 mm) from the railing to count as the required receptacle.

E3801.3 Small appliance receptacles. In the kitchen, pantry, breakfast room, dining room, or similar area of a dwelling unit, the two or more 20-ampere small-appliance branch circuits required by Section E3603.2, shall serve all receptacle outlets covered by Sections E3801.2 and E3801.4 and those receptacle outlets provided for refrigeration appliances.

Exceptions:

1. In addition to the required receptacles specified by Sections E3801.1 and E3801.2, switched receptacles supplied from a general-purpose branch circuit as defined in Section E3803.2, Exception 1, shall be permitted.
2. The receptacle outlet for refrigeration appliances shall be permitted to be supplied from an individual branch circuit rated at 15 amperes or greater.

❖ Small appliance receptacles are supplied by at least two 20-ampere branch circuits and include all of the receptacles in the kitchen, pantry, dining, and similar areas of a house. The countertop receptacles as well as all of the other "low" receptacles (usually installed around 12 to 18 inches (305 to 457 mm) above the floor) in these rooms are included on one of the small appliance receptacle branch circuits. In a dining room, a receptacle controlled by a wall switch can serve as the required lighting outlet. Such a receptacle would not be served by one of the small appliance branch circuits but would be served by a general purpose lighting circuit.

The receptacle serving a refrigerator can be included on one of the small appliance circuits or can be supplied by a separate individual branch circuit. Where it is an individual branch circuit, it can be a 15-ampere or 20-ampere rated branch circuit.

E3801.3.1 Other outlets prohibited. The two or more small-appliance branch circuits specified in Section 3801.3 shall serve no other outlets.

Exceptions:

1. A receptacle installed solely for the electrical supply to and support of an electric clock in any of the rooms specified in Section E3801.3.
2. Receptacles installed to provide power for supplemental equipment and lighting on gas-fired ranges, ovens, and counter-mounted cooking units.

❖ Receptacles for specific fixed-in-place appliances such as a disposal, dishwasher, or trash compactor cannot be served by the small-appliance branch circuits. The small-appliance circuits cannot supply anything else, such as the hood fan over the range.

A half century ago, a receptacle for an electric clock installed high on the wall was a standard feature in the kitchen in some areas of the country. The power for a clock is minimal, and although this type of receptacle outlet is not as common now, it has been permitted for many years to be served by one of the small appliance branch circuits. Where a gas range is installed in the kitchen, a 120-volt receptacle outlet is needed to operate such items on the range as the timer, lights, and ignitor. This receptacle can be connected to one of the small appliance branch circuits.

E3801.3.2 Limitations. Receptacles installed in a kitchen to serve countertop surfaces shall be supplied by not less than two small-appliance branch circuits, either or both of which shall also be permitted to supply receptacle outlets in the same kitchen and in other rooms specified in Section E3801.3. Additional small-appliance branch circuits shall be permitted to supply receptacle outlets in the kitchen and other rooms specified in Section E3801.3. A small-appliance branch circuit shall not serve more than one kitchen.

❖ Small appliance branch circuits can serve receptacles in a dining room. In a formal dining room, where no cooking is done and it is not likely that any appliances will be used, it may seem that the receptacles could be served by a general purpose lighting branch circuit, but these receptacles must be included on one of the two or more small-appliance branch circuits.

There is not a limit on the number of outlets served by a small appliance branch circuit. Some electrical contractors have traditional ways of laying out these circuits. For example, one practice is that all countertop receptacles on one side of the kitchen sink are wired on one circuit, and another circuit is used for all receptacles on the other side of the sink. Another method is to alternate countertop receptacles on two different circuits so that no two adjacent receptacles are on the same circuit. Yet another method is to use three-conductor cable and wire the top half of the duplex receptacle on one circuit and the bottom half on another circuit. These methods are not code requirements but only preferences. The code simply requires that all receptacles in the kitchen, pantry, and dining areas be served by two or more small appliance branch circuits.

E3801.4 Countertop receptacles. In kitchens and dining rooms of dwelling units, receptacle outlets for counter spaces shall be installed in accordance with Sections E3801.4.1 through E3801.4.5. (See Figure E3801.4.)

❖ This section introduces five subsections. The first four subsections cover the countertop spaces where receptacles are required, and the fifth covers the location of receptacles in those required spaces.

E3801.4.1 Wall counter space. A receptacle outlet shall be installed at each wall counter space 12 inches (305 mm) or wider. Receptacle outlets shall be installed so that no point along the wall line is more than 24 inches (610 mm), measured horizontally from a receptacle outlet in that space.

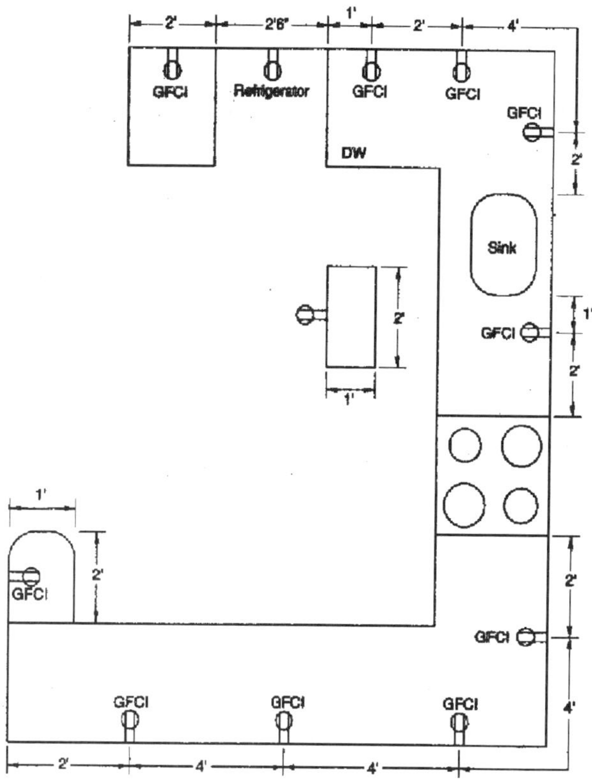

For SI: 1 inch = 25.4 mm, 1 foot = 304.8 mm.

FIGURE E3801.4
COUNTERTOP RECEPTACLES

❖ See the commentary for Section E3801.4.

❖ In many kitchens, there is a short section of countertop between the refrigerator and the range on which it is helpful to place items when getting them in and out of the refrigerator. The literal code requirement is that if this countertop is 12 inches (305 mm) or more wide, a receptacle is required along the wall at this space. Some installers have omitted the receptacle where the countertop space is only 9 or 10 inches. This is a case where good judgment should prevail. For example, a toaster could be placed at such a countertop space, and if there were not a receptacle available, the toaster cord could be placed across the range to be plugged in. This would create an unsafe situation. Providing a receptacle at a 10-inch (254 mm), for example, countertop space is a good idea, although not a code requirement.

As indicated in Figure E3801.4 of the code, receptacle outlets must not be spaced more than 48 inches (1219 mm) apart along the wall line of the countertop. For this measurement, it is not permitted to cut corners. Notice the measurement at the left of the sink in the figure. The edge of the sink is considered the beginning of the countertop, and an outlet must be available within 24 inches (610 mm) of that point. Continuing to the left from that point, the measurement follows the wall line.

Using Figure E3801.4, it is not correct, for example, to draw an imaginary line along the front edge of the countertop in front of the sink all the way to the wall left of the sink, consider that one "countertop," then begin at that point to measure to the refrigerator. The wall line in the corner must be included in the measurement.

E3801.4.2 Island counter spaces. At least one receptacle outlet shall be installed at each island counter space with a long dimension of 24 inches (610 mm) or greater and a short dimension of 12 inches (305 mm) or greater.

❖ At an island counter space, one receptacle is required. More may be installed, for example, at each end. The 24-inch (610 mm) spacing requirement does not apply to the island countertop space, because there is no wall line. An island might be divided into two counter spaces by the presence of a sink or cooktop.

E3801.4.3 Peninsular counter space. At least one receptacle outlet shall be installed at each peninsular counter space with a long dimension of 24 inches (610 mm) or greater and a short dimension of 12 inches (305 mm) or greater. A peninsular countertop is measured from the connecting edge.

❖ The connecting edge of a peninsular countertop is an imaginary line along the front edge of the adjacent countertop. At least one receptacle is required at the peninsular counter space. It is not required to be at the end of the peninsula.

E3801.4.4 Separate spaces. Countertop spaces separated by range tops, refrigerators, or sinks shall be considered as separate countertop spaces in applying the requirements of Sections E3801.4.1, E3801.4.2 and E3801.4.3.

❖ The intent of this code requirement is to prevent cords from being stretched or laid across the range or sink or other separations of the countertop space. The left and right edges of the kitchen sink are points where the measurement begins for the countertop space.

E3801.4.5 Receptacle outlet location. Receptacle outlets shall be located not more than 20 inches (508 mm) above the countertop. Receptacle outlets shall not be installed in a face-up position in the work surfaces or countertops. Receptacle outlets rendered not readily accessible by appliances fastened in place, appliance garages or appliances occupying dedicated space shall not be considered as these required outlets.

Exception: Receptacle outlets shall be permitted to be mounted not more than 12 inches (305 mm) below the countertop in construction designed for the physically impaired and for island and peninsular countertops where the countertop is flat across its entire surface and there are no means to mount a receptacle within 20 inches (508 mm) above the countertop, such as in an overhead cabinet. Receptacles mounted below the countertop in accordance with this exception shall not be located where the countertop extends more than 6 inches (152 mm) beyond its support base.

❖ The intent of the code rule is that receptacles will be available for small appliances used on the countertop. It can be a safety hazard to have appliances plugged in below the countertop surface because the cord would be draped over the edge of the countertop, and the cord could be easily snagged, pulling the appliance off of the countertop. Where an island or peninsular countertop does not overhang the cabinet more than 6 inches (152 mm), it is permitted to install the required receptacle not more than 12 inches (305 mm) below the countertop.

E3801.5 Appliance outlets. Appliance receptacle outlets installed for specific appliances, such as laundry equipment, shall be installed within 6 feet (1829 mm) of the intended location of the appliance.

❖ Where the outlet is no more than 6 feet from the location of an appliance, it is assumed that the appliance cord will reach the outlet. The intent is that extension cords will not have to be used. Common sense should be used here, and the 6-foot (1829 mm) measurement may not be adequate if the measured distance is simply from the edge of the appliance space. For example, if a 120-volt receptacle for a gas clothes dryer is 6 feet from the edge of the dryer space, the receptacle placement may satisfy the code rule, but the dryer cord at the rear of a gas clothes dryer may be at the opposite side of the space closest to the receptacle. Consideration should be given to the location of the appliances such as garage door openers, sump pumps, clothes washers, etc., so that extension cords will not have to be used.

E3801.6 Bathroom. At least one wall receptacle outlet shall be installed in bathrooms and such outlet shall be located within 36 inches (914 mm) of the outside edge of each lavatory basin. The receptacle outlet shall be located on a wall that is adjacent to the lavatory basin location.

Receptacle outlets shall not be installed in a face-up position in the work surfaces or countertops in a bathroom basin location.

❖ Where a bathroom lighting fixture contains a receptacle, it does not satisfy the requirement for the bathroom wall receptacle per Section E3801.1, because required receptacles are in addition to any that are part of a lighting fixture. A receptacle in a lighting fixture must be GFCI protected, and the bathroom lighting circuit is usually not supplied from a GFCI protected circuit.

E3801.7 Outdoor outlets. At least one receptacle outlet accessible at grade level and not more than 6 feet, 6 inches (1981 mm) above grade, shall be installed outdoors at the front and back of each dwelling unit having direct access to grade.

❖ Any dwelling unit that has direct access to grade requires an outdoor receptacle both at the front and back of the dwelling. But a receptacle that is part of a yard light mounted on a post, for example, would not satisfy this requirement. Where a dwelling unit has a balcony or deck accessible only through a door from the inside of the house, an outdoor receptacle at the balcony or deck location may not satisfy this requirement if the receptacle is not accessible at grade level. If it is not over $6^1/_2$ feet from the ground and can be reached from standing on the ground, it counts as the required receptacle.

E3801.8 Laundry areas. At least one receptacle outlet shall be installed to serve laundry appliances.

❖ Section E3603.3 requires at least one 20-amp rated branch circuit for the laundry area. Typically, this circuit supplies a duplex receptacle to serve a clothes washer and a gas-fired clothes dryer. See Section E3801.5 for receptacle location requirements. "Laundry area" is not defined but is interpreted as including an ironing area within the laundry area, meaning that multiple receptacle locations could be supplied by the required laundry branch circuit. Bear in mind that a 20-amp rated branch circuit is nearly at capacity when serving both a washer and dryer.

E3801.9 Basements and garages. At least one receptacle outlet, in addition to any provided for laundry equipment, shall be installed in each basement and in each attached garage, and in each detached garage that is provided with electrical power. Where a portion of the basement is finished into a habitable room(s), the receptacle outlet required by this section shall be installed in the unfinished portion.

❖ The code does not require that a detached garage be supplied with electrical power. But if it has power, then it must have at least one receptacle. Because there are many uses and activities in garages and unfinished basements that require power, an unfinished basement or an unfinished portion of a basement must have at least one receptacle to avoid the overuse or unsafe use of extension cords. Section E3809.1 prohibits the use of flexible cord as a substitute for fixed wiring in the dwelling and indicates that cords must not be run through holes in walls, ceilings, or floors. Where receptacles are installed at appropriate locations, the receptacles will reduce the need for extension cords.

E3801.10 Hallways. Hallways of 10 feet (3048 mm) or more in length shall have at least one receptacle outlet. The hall length shall be considered the length measured along the centerline of the hall without passing through a doorway.

❖ The length of an L-shaped hallway is also measured along the centerline and includes the total L- shaped length. The purpose for this requirement is to mandate supply power for a vacuum cleaner. This results in convenience and safety by helping avoid the use of extension cords.

E3801.11 HVAC outlet. A 125-volt, single-phase, 15 or 20 ampere-rated convenience receptacle outlet shall be installed for the servicing of heating, air-conditioning and refrigeration equipment located in attics and crawl spaces. The receptacle shall be accessible and shall be located on the same level and within 25 feet (7620 mm) of the heating, air-conditioning and refrigeration equipment. The receptacle outlet shall not be connected to the load side of the HVAC equipment disconnecting means and shall be protected in accordance with Section E3802.4.

❖ When servicing or repairing HVAC equipment, a technician usually needs power tools, instruments, or equipment. To avoid the necessity of a technician dragging an extension cord through the attic or crawl space, a receptacle outlet provides the necessary power.

SECTION E3802 GROUND-FAULT AND ARC-FAULT CIRCUIT-INTERRUPTER PROTECTION

E3802.1 Bathroom receptacles. All 125-volt, single-phase, 15- and 20-ampere receptacles installed in bathrooms shall have ground-fault circuit-interrupter protection for personnel.

❖ There are many unique layouts for a dwelling unit bathroom. This and the requirement for bathroom receptacles and ground-fault circuit-interrupter protection are the reasons that there is a code definition of bathroom. A bathroom is not necessarily a room. It is an area that includes a wash basin (lavatory) and also has a water closet, a tub, or a shower. The water closet, tub, or shower could be in a separate room from the wash basin, and the area would still be defined as a bathroom. There is no exception to the requirement for GFCI protection of receptacles in bathrooms. All receptacles would include, for example, a receptacle in a light fixture, a receptacle for a clothes washer, or a receptacle for any other appliance installed in a bathroom; see Commentary Figure E3802.1.

E3802.2 Garage and accessory building receptacles. All 125-volt, single-phase, 15- or 20-ampere receptacles installed in garages and grade-level portions of unfinished accessory buildings used for storage or work areas shall have ground-fault circuit-interrupter protection for personnel.

Exceptions:

1. Receptacles that are not readily accessible.
2. A single receptacle or a duplex receptacle for two appliances located within dedicated space for each appliance that in normal use is not easily moved from one place to another, and that is cord- and plug-connected.

❖ A receptacle in an accessory building, an implement shed, or even a storage shed also requires GFCI

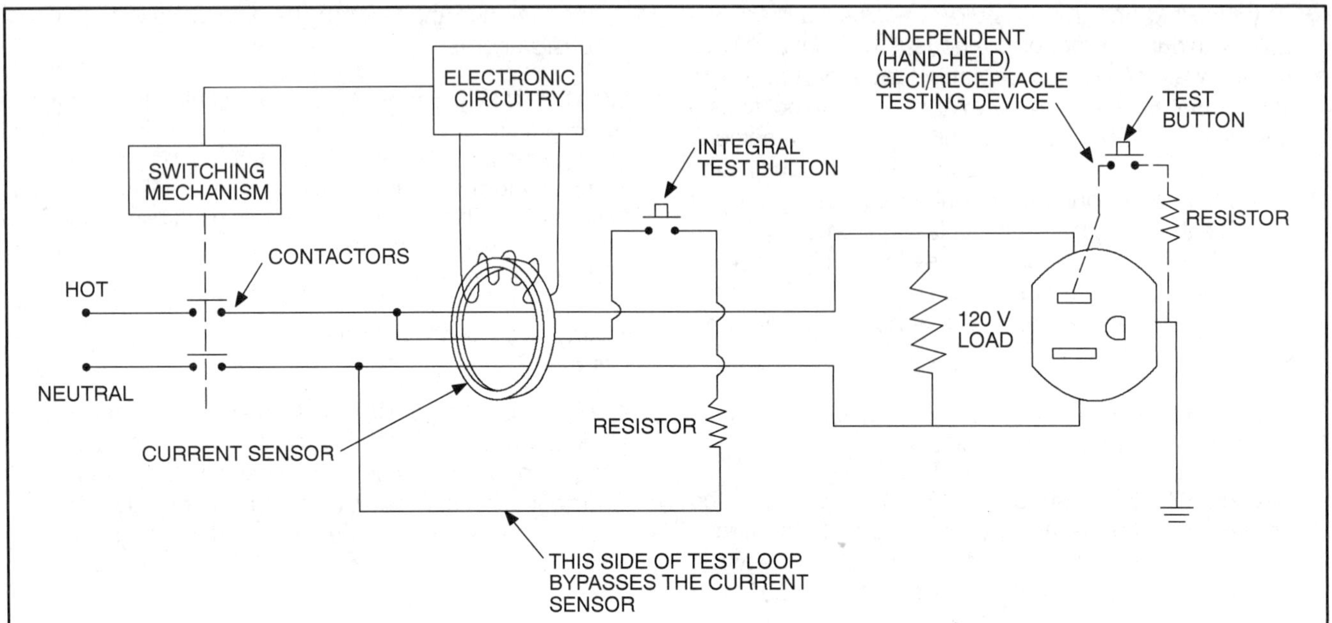

Figure E3802.1
BASIC PRINCIPLES OF OPERATION OF GFCI RECEPTACLE DEVICE AND DEVICE TESTER

protection. Sometimes an appliance such as a food freezer or refrigerator is located in a garage, and where such an appliance is plugged into a GFCI receptacle, nuisance tripping of the GFCI device could cause food to spoil. Therefore, it is common to have some appliances supplied by non-GFCI circuits. Where a space is dedicated for a cord- and plug-connected appliance that is "not easily moved," the receptacle in such space can be non-GFCI protected, but it must be a single receptacle. If a duplex receptacle were used, an extension cord could be plugged in to the vacant receptacle and used to supply power for hand tools, trimmers, etc. A non-GFCI protected duplex receptacle could be used in a garage in a space dedicated for two appliances that are big enough to be considered "not easily moved"; each appliance would be supplied by one of the receptacles on the yoke of the duplex receptacle. Receptacles described by the term "not readily accessible" are garage door opener receptacles. It is not likely that these receptacles would be used to supply power for hand-held tools and other cord- and plug-connected equipment.

E3802.3 Outdoor receptacles. All 125-volt, single-phase, 15- and 20-ampere receptacles installed outdoors shall have ground-fault circuit-interrupter protection for personnel.

Exception: Receptacles as covered in Section E4001.7.

❖ The only outdoor receptacles that are exempt from GFCI protection are those located in the eave of the dwelling for cord- and plug-connected deicing equipment. These receptacle outlets are not readily accessible, are supplied from a dedicated circuit especially for the deicing equipment or cable, and are usually separate from eave receptacles that would be used for decorative holiday lighting.

E3802.4 Crawl space receptacles. Where a crawl space is at or below grade level, all 125-volt, single-phase, 15- and 20-ampere receptacles installed in such spaces shall have ground-fault circuit-interrupter protection for personnel.

❖ This requirement is for GFCI protection of a crawl space receptacle outlet. According to Section E3801.11, a crawl space is not required to have a receptacle outlet unless there is equipment such as HVAC appliances that require servicing located there. Plumbing pipes and electrical cable run in the crawl space area are not considered equipment that would need servicing, so a receptacle would not be required because of the piping or wiring. However, if the choice is made to install a crawl space receptacle, then it must be GFCI protected.

E3802.5 Unfinished basement receptacles. All 125-volt, single-phase, 15- and 20-ampere receptacles installed in unfinished basements shall have ground-fault circuit-interrupter protection for personnel. For purposes of this section, unfinished basements are defined as portions or areas of the basement not intended as habitable rooms and limited to storage areas, work areas, and the like.

Exceptions:

1. Receptacles that are not readily accessible.
2. A single receptacle or duplex receptacle for two appliances located within dedicated space for each appliance that in normal use is not easily moved from one place to another, and that is cord- and plug-connected.
3. A receptacle supplying only a permanently installed fire alarm or burglar alarm system.

❖ Masonry and concrete can provide a conductive path to earth. Bare concrete or masonry walls might be subject to moisture. Where a basement masonry or concrete wall has a framed stud wall in front of it, it should not be considered a damp location, because any cable installed in the wall framing would not be in contact with the bare masonry or concrete. Review the definition of "damp location" and also the allowable applications for wiring methods in Table E3701.4. There is no detailed description of what constitutes "moderate degrees of moisture" from the Chapter 34 definition of damp location.

The exceptions are the same as for receptacle in a garage that are not readily accessible or are dedicated for an appliance or two appliances.

A single receptacle (not a duplex) supplying a plug-in transformer for an alarm system is exempt since the receptacle is occupied, and nuisance tripping could disable the alarm system.

E3802.6 Kitchen receptacles. All 125-volt, single-phase, 15- and 20-ampere receptacles that serve countertop surfaces shall have ground-fault circuit-interrupter protection for personnel.

❖ In a kitchen, all receptacles serving any countertop are required to be GFCI protected. In the past, the only kitchen countertop receptacles that required GFCI protection were the ones close to the kitchen sink. Now, any countertop receptacle, even in an area far away from the sink, is required to have GFCI protection.

E3802.7 Bar sink receptacles. All 125-volt, single-phase, 15- and 20-ampere receptacles that serve a countertop surface, and are located within 6 feet (1829 mm) of the outside edge of a wet bar sink shall have ground-fault circuit-interrupter protection for personnel. Receptacle outlets shall not be installed in a face-up position in the work surfaces or countertops.

❖ This requirement is different from the kitchen countertop receptacles in that only receptacles within 6 feet measured in any direction from the wet bar sink must be GFCI protected.

E3802.8 Boathouse receptacles. All 125-volt, single-phase, 15- or 20-ampere receptacles installed in boathouses used for storage or work areas shall have ground-fault circuit-interrupter protection for personnel.

❖ The potential shock hazards in buildings having unfinished concrete floors on grade and wet surfaces are comparable to shock hazards outdoors and in kitchens, bathrooms, garages, basements and workshops.

E3802.9 Electrically heated floors. Ground-fault circuit-interrupter protection for personnel shall be provided for electrically heated floors in bathrooms, and in hydromassage bathtub, spa and hot tub locations.

❖ Heat-producing cables and grids can be imbedded in concrete floors to warm the floor for occupant comfort. The possibility of current leakage combined with wet surfaces and grounded plumbing fixtures justifies this requirement; the branch circuit overcurrent device serving the heating cables must provide GFCI protection.

E3802.10 Exempt receptacles. Receptacles installed under exceptions to Sections E3802.2 and E3802.5 shall not be considered as meeting the requirements of Section E3801.9.

❖ This section correlates with Section E3801.9, which requires that at least one receptacle be installed in an unfinished basement and at least one in a garage. Garage receptacles and unfinished basement receptacles must have GFCI protection, except for those not readily accessible and those at a dedicated appliance space. Those "exempt receptacles," which are the exceptions to GFCI protection in garages and unfinished basements, are not permitted to be the required garage or basement receptacles.

E3802.11 Bedroom outlets. All branch circuits that supply 125-volt, single-phase, 15- and 20-ampere outlets installed in dwelling unit bedrooms shall be protected by an arc-fault circuit interrupter listed to provide protection of the entire branch circuit.

❖ This section refers to "outlets," which includes both receptacles and other outlets, such as lighting outlets. Arc-fault circuit interrupters (AFCI) combine a standard circuit breaker with electronic circuitry that can detect the signature waveform caused by arcing in an AC circuit; see the commentary for the definition of AFCI. Bedrooms were chosen as the location in which to introduce AFCI protection requirements for a number of reasons, including the type of electrical loads and combustible materials found there. Because arcing can occur in any circuit in a dwelling, it is possible that future editions of this code will require AFCI devices for all 120-volt branch circuits.

Section R313 requires smoke alarm outlets in bedrooms, and although such outlets must also be on an AFCI circuit, the code official should consider whether or not this arrangement was the intent and whether it has a downside. Smoke alarms must be interconnected and would all conveniently be served by the same branch circuit. If such a circuit were an AFCI circuit, and a nuisance trip occurred, all smoke alarms would lose primary power. Smoke alarm circuits powered by the same branch circuit that supplies other bedroom outlets will be monitored, since power loss to the smoke alarms will also mean loss of power to bedroom lights and receptacles. In such cases, the occupant will restore power to the alarms in their effort to restore bedroom lighting.

SECTION E3803
LIGHTING OUTLETS

E3803.1 General. Lighting outlets shall be provided in accordance with Sections E3803.2 through E3803.3.

❖ See the commentary for Sections E3803.2 and E3803.4.

E3803.2 Habitable rooms. At least one wall switch-controlled lighting outlet shall be installed in every habitable room and bathroom.

Exceptions:

1. In other than kitchens and bathrooms, one or more receptacles controlled by a wall switch shall be considered equivalent to the required lighting outlet.
2. Lighting outlets shall be permitted to be controlled by occupancy sensors that are in addition to wall switches, or that are located at a customary wall switch location and equipped with a manual override that will allow the sensor to function as a wall switch.

❖ According to Chapter 2, a space for living, sleeping, eating, or cooking is a habitable space and does not include a bathroom. Therefore, a bathroom is included as an area where a wall-switch-controlled lighting outlet is required. In all rooms other than kitchens and bathrooms, the lighting outlet may be a switched receptacle.

It seems unlikely that the dining room lighting outlet would be a switched receptacle. However, if this is the case, such a receptacle must not be supplied by one of the small appliance branch circuits. This is clear from the exception to Section E3801.3, which allows for a switched receptacle to be supplied from a general purpose branch circuit. Thus, all the receptacles in a dining room must be supplied by a small appliance circuit except the one that is switched and considered to be the lighting outlet. Whether this switched receptacle is permitted to be a duplex receptacle or required to be a single receptacle is not stated; this would be a judgment call by the inspector.

E3803.3 Additional locations. At least one wall switch-controlled lighting outlet shall be installed in hallways, stairways, attached garages, and detached garages with electric power. At least one wall-switch-controlled lighting outlet shall be installed to provide illumination on the exterior side of each outdoor egress door having grade level access. A vehicle door in a garage shall not be considered as an outdoor egress door. Where lighting outlets are installed in interior stairways, there shall be a wall switch at each floor level to control the lighting outlet where the stairway has 6 or more risers.

Exception: In hallways, stairways, and at outdoor egress doors, remote, central, or automatic control of lighting shall be permitted.

❖ Additional areas that are not habitable and would not be included in Section E3803.2 are hallways, stairways, and garages. There is no requirement for a three-way switch at each end of a hallway or, for that matter, in any room. The only requirement for three-way switches (where a lighting outlet is controlled from two separate switches at different locations) is at a stairway that has at least six risers. Installing three-way and four-way switches at separate entrances of a large room does provide for convenience and also an added measure of safety, but it is not required by the code. See Section R303.6.1, which is more stringent than the 6-riser criterion of this section.

E3803.4 Storage or equipment spaces. In attics, under-floor spaces, utility rooms and basements, at least one lighting outlet shall be installed where these spaces are used for storage or contain equipment requiring servicing. Such lighting outlet shall be controlled by a wall switch or shall have an integral switch. At least one point of control shall be at the usual point of entry to these spaces. The lighting outlet shall be provided at or near the equipment requiring servicing.

❖ In attics or crawl spaces that are used for storage or have equipment that requires servicing, a lighting outlet is required with a switch at the point of entrance. If an attic or crawl space is not used for storage or contains no equipment, no lighting outlet is required. A utility room and a basement would have equipment and/or be used for storage and would require a lighting outlet.

SECTION E3804
GENERAL INSTALLATION REQUIREMENTS

E3804.1 Electrical continuity of metal raceways and enclosures. Metal raceways, cable armor and other metal enclosures for conductors shall be mechanically joined together into a continuous electric conductor and shall be connected to all boxes, fittings and cabinets so as to provide effective electrical continuity. Raceways and cable assemblies shall be mechanically secured to boxes, fittings cabinets and other enclosures.

Exceptions:

1. As provided in Section E3805.3.2 for nonmetallic boxes used with Type NM cable.
2. Short sections of raceway used to provide cable assemblies with support or protection against physical damage.

❖ Where armored cable is used, the armor must be continuous to be able to serve as a conductor to carry fault current. In a raceway system, the metal raceway must serve as a conductor to carry fault current. Where the mechanical connection at a box or fitting is not tight, high resistance will occur and prevent the flow of any fault current necessary to trip the overcurrent protective device.

E3804.2 Mechanical continuity—raceways and cables. Metal or nonmetallic raceways, cable armors and cable sheaths shall be continuous between cabinets, boxes, fittings or other enclosures or outlets.

Exception: Short sections of raceway used to provide cable assemblies with support or protection against physical damage.

❖ Cable is permitted to be protected from physical damage by running it through a section of raceway.

E3804.3 Securing and supporting. Raceways, cable assemblies, boxes, cabinets and fittings shall be securely fastened in place.

❖ Raceways and cables are wiring methods covered in Chapter 37. The supporting requirements are covered in Table E3702.1. Boxes and cabinets must be securely supported according to provisions in Section E3806.8. Detailed requirements are included for support of boxes.

E3804.4 Raceways as means of support. Raceways shall be used as a means of support for other raceways, cables or nonelectric equipment only under the following conditions:

1. Where the raceway or means of support is identified for the purpose; or
2. Where the raceway contains power supply conductors for electrically controlled equipment and is used to support Class 2 circuit conductors or cables that are solely for the purpose of connection to the control circuits of the equipment served by such raceway; or
3. Where the raceway is used to support boxes or conduit bodies in accordance with Sections E3806.8.4 and E3806.8.5.

❖ Raceways are permitted to be used to support other raceways, cables, or wires only under the limited circumstances listed. A thermostat wire for the control of heating and air-conditioning may be supported by the raceway supplying power to the equipment. Where such cables as telephone wiring and cable TV wiring are supported by a conduit or electrical metallic tubing, the cables could prevent the dissipation of heat produced inside the raceway.

E3804.5 Raceway installations. Raceways shall be installed complete between outlet, junction or splicing points prior to the installation of conductors.

❖ The proper way to install wiring in raceways is to assemble the complete raceway system first and properly support it. After the entire system is in place and secured, the wire can be inserted without damaging it. Trying to install the wire piecemeal as the conduit is installed could result in damage to the conductor insulation.

E3804.6 Conduit and tubing fill. The maximum number of conductors installed in conduit or tubing shall be in accordance with Tables E3804.6(1) through E3804.6(10).

❖ The code includes ten tables to indicate how many conductors of the various types and sizes are allowed to be installed in the various sizes and types of raceway.

E3804.7 Air handling—stud cavity and joist spaces. Where wiring methods having a nonmetallic covering pass through stud cavities and joist spaces used for air handling, such wiring shall pass through such spaces perpendicular to the long dimension of the spaces.

❖ It is not permitted to run wiring through or within an air-handling duct. Limiting the wiring in an air-handling space will help minimize the production of toxic fumes created during a fire. Such fumes carried by an air handling duct could cause injury and death sooner than the actual fire. But in many situations it is necessary to run nonmetallic sheathed cable through a stud or joist space that is intended for use as a cold air return. The code permits this for dwelling wiring on a limited basis only where it is run perpendicular to the stud or joist space. On a new house under construction that has been framed and is ready for installation of the rough wiring, plumbing, and duct work, it is important to know which stud or joist spaces will be used for a cold-air return duct. If the heating and air-conditioning contractor arrives first and installs the duct work, and the electrician comes along after that, then the wiring can be run correctly. That is, it is allowed to install cable through a cold-air return perpendicular through the stud or joist duct space. But if the electrician arrives first and has no information on which, if any, stud spaces may be used as an air handling duct, wiring could be installed parallel to the stud or joist in the space intended for the duct. The mechanical contractor might then enclose the space with a sheet metal pan and create the air handling space, which would cause the wiring to be in violation of the code. The installers of the wiring and of the heating and air-conditioning ducts must coordinate and not just ignore each other's work.

TABLE E3804.6(1)
MAXIMUM NUMBER OF CONDUCTORS IN ELECTRICAL METALLIC TUBING (EMT)[a]

TYPE LETTERS	CONDUCTOR SIZE AWG/kcmil	TRADE SIZES (inches)					
		1/2	3/4	1	1 1/4	1 1/2	2
RHW, RHW-2	14	4	7	11	20	27	46
	12	3	6	9	17	23	38
	10	2	5	8	13	18	30
	8	1	2	4	7	9	16
	6	1	1	3	5	8	13
	4	1	1	2	4	6	10
	3	1	1	1	4	5	9
	2	1	1	1	3	4	7
	1	0	1	1	1	3	5
	1/0	0	1	1	1	2	4
	2/0	0	1	1	1	2	4
	3/0	0	0	1	1	1	3
	4/0	0	0	1	1	1	3
TW	14	8	15	25	43	58	96
	12	6	11	19	33	45	74
	10	5	8	14	24	33	55
	8	2	5	8	13	18	30
RHW[a], RHW-2[a], THHW, THW, THW-2	14	6	10	16	28	39	64
	12	4	8	13	23	31	51
	10	3	6	10	18	24	40
	8	1	4	6	10	14	24
RHW[a], RHW-2[a], TW, THW, THHW, THW-2	6	1	3	4	8	11	18
	4	1	1	3	6	8	13
	3	1	1	3	5	7	12
	2	1	1	2	4	6	10
	1	1	1	1	3	4	7
	1/0	0	1	1	2	3	6
	2/0	0	1	1	1	3	5
	3/0	0	1	1	1	2	4
	4/0	0	0	1	1	1	3
THHN, THWN, THWN-2	14	12	22	35	61	84	138
	12	9	16	26	45	61	101
	10	5	10	16	28	38	63
	8	3	6	9	16	22	36
	6	2	4	7	12	16	26
	4	1	2	4	7	10	16
	3	1	1	3	6	8	13
	2	1	1	3	5	7	11
	1	1	1	1	4	5	8
	1/0	1	1	1	3	4	7
	2/0	0	1	1	2	3	6
	3/0	0	1	1	1	3	5
	4/0	0	1	1	1	2	4
XHHW, XHHW-2	14	8	15	25	43	58	96
	12	6	11	19	33	45	74
	10	5	8	14	24	33	55
	8	2	5	8	13	18	30
	6	1	3	6	10	14	22
	4	1	2	4	7	10	16
	3	1	1	3	6	8	14
	2	1	1	3	5	7	11
	1	1	1	1	4	5	8
	1/0	1	1	1	3	4	7
	2/0	0	1	1	2	3	6
	3/0	0	1	1	1	3	5
	4/0	0	1	1	1	2	4

For SI: 1 inch = 25.4 mm.

a. Types RHW, and RHW-2 without outer covering.

❖ See the commentary for Section E3804.6.

TABLE E3804.6(2)
MAXIMUM NUMBER OF CONDUCTORS IN ELECTRICAL NONMETALLIC TUBING (ENT)[a]

TYPE LETTERS	CONDUCTOR SIZE AWG/kcmil	TRADE SIZES (inches)					
		1/2	3/4	1	1 1/4	1 1/2	2
RHW, RHW-2	14	3	6	10	19	26	43
	12	2	5	9	16	22	36
	10	1	4	7	13	17	29
	8	1	1	3	6	9	15
	6	1	1	3	5	7	12
	4	1	1	2	4	6	9
	3	1	1	1	3	5	8
	2	0	1	1	3	4	7
	1	0	1	1	1	3	5
	1/0	0	0	1	1	2	4
	2/0	0	0	1	1	1	3
	3/0	0	0	1	1	1	3
	4/0	0	0	1	1	1	2
TW	14	7	13	22	40	55	92
	12	5	10	17	31	42	71
	10	4	7	13	23	32	52
	8	1	4	7	13	17	29
RHW[a], RHW-2[a], THHW, THW, THW-2	14	4	8	15	27	37	61
	12	3	7	12	21	29	49
	10	3	5	9	17	23	38
	8	1	3	5	10	14	23
RHW[a], RHW-2[a], TW, THW, THHW, THW-2	6	1	2	4	7	10	17
	4	1	1	3	5	8	13
	3	1	1	2	5	7	11
	2	1	1	2	4	6	9
	1	0	1	1	3	4	6
	1/0	0	1	1	2	3	5
	2/0	0	1	1	1	3	5
	3/0	0	0	1	1	2	4
	4/0	0	0	1	1	1	3
THHN, THWN, THWN-2	14	10	18	32	58	80	132
	12	7	13	23	42	58	96
	10	4	8	15	26	36	60
	8	2	5	8	15	21	35
	6	1	3	6	11	15	25
	4	1	1	4	7	9	15
	3	1	1	3	5	8	13
	2	1	1	2	5	6	11
	1	1	1	1	3	5	8
	1/0	0	1	1	3	4	7
	2/0	0	1	1	2	3	5
	3/0	0	1	1	1	3	4
	4/0	0	0	1	1	2	4
XHHW, XHHW-2	14	7	13	22	40	55	92
	12	5	10	17	31	42	71
	10	4	7	13	23	32	52
	8	1	4	7	13	17	29
	6	1	3	5	9	13	21
	4	1	1	4	7	9	15
	3	1	1	3	6	8	13
	2	1	1	2	5	6	11
	1	1	1	1	3	5	8
	1/0	0	1	1	3	4	7
	2/0	0	1	1	2	3	6
	3/0	0	1	1	1	3	5
	4/0	0	0	1	1	2	4

For SI: 1 inch = 25.4 mm.

a. Types RHW, and RHW-2 without outer covering.

❖ See the commentary for Section E3804.6.

TABLE E3804.6(3)
MAXIMUM NUMBER OF CONDUCTORS IN FLEXIBLE METALLIC CONDUIT (FMC)[a]

TYPE LETTERS	CONDUCTOR SIZE AWG/kcmil	TRADE SIZES (inches)					
		1/2	3/4	1	1 1/4	1 1/2	2
RHW, RHW-2	14	4	7	11	17	25	44
	12	3	6	9	14	21	37
	10	3	5	7	11	17	30
	8	1	2	4	6	9	15
	6	1	1	3	5	7	12
	4	1	1	2	4	5	10
	3	1	1	1	3	5	7
	2	1	1	1	3	4	7
	1	0	1	1	1	2	5
	1/0	0	1	1	1	2	4
	2/0	0	1	1	1	1	3
	3/0	0	0	1	1	1	3
TW	14	9	15	23	36	53	94
	12	7	11	18	28	41	72
	10	5	8	13	21	30	54
	8	3	5	7	11	17	30
RHW[a], RHW-2[a], THHW, THW, THW-2	14	6	10	15	24	35	62
	12	5	8	12	19	28	50
	10	4	6	10	15	22	39
	8	1	4	6	9	13	23
RHW[a], RHW-2[a], TW, THW, THHW, THW-2	6	1	3	4	7	10	18
	4	1	1	3	5	7	13
	3	1	1	3	4	6	11
	2	1	1	2	4	5	10
	1	1	1	1	2	4	7
	1/0	0	1	1	1	3	6
	2/0	0	1	1	1	3	5
	3/0	0	1	1	1	2	4
	4/0	0	0	1	1	1	3
	4/0	0	0	1	1	1	2
THHN, THWN, THWN-2	14	13	22	33	52	76	134
	12	9	16	24	38	56	98
	10	6	10	15	24	35	62
	8	3	6	9	14	20	35
	6	2	4	6	10	14	25
	4	1	2	4	6	9	16
	3	1	1	3	5	7	13
	2	1	1	3	4	6	11
	1	1	1	1	3	4	8
	1/0	1	1	1	2	4	7
	2/0	0	1	1	1	3	6
	3/0	0	1	1	1	2	5
	4/0	0	1	1	1	1	4
XHHW, XHHW-2	14	9	15	23	36	53	94
	12	7	11	18	28	41	72
	10	5	8	13	21	30	54
	8	3	5	7	11	17	30
	6	1	3	5	8	12	22
	4	1	2	4	6	9	16
	3	1	1	3	5	7	13
	2	1	1	3	4	6	11
	1	1	1	1	3	5	8
	1/0	1	1	1	2	4	7
	2/0	0	1	1	2	3	6
	3/0	0	1	1	1	3	5
	4/0	0	1	1	1	2	4

For SI: 1 inch = 25.4 mm.

a. Types RHW, and RHW-2 without outer covering.

❖ See the commentary for Section E3804.6.

TABLE E3804.6(4)
MAXIMUM NUMBER OF CONDUCTORS IN INTERMEDIATE METALLIC CONDUIT (IMC)[a]

TYPE LETTERS	CONDUCTOR SIZE AWG/kcmil	TRADE SIZES (inches)					
		1/2	3/4	1	1 1/4	1 1/2	2
RHW, RHW-2	14	4	8	13	22	30	49
	12	4	6	11	18	25	41
	10	3	5	8	15	20	33
	8	1	3	4	8	10	17
	6	1	1	3	6	8	14
	4	1	1	3	5	6	11
	3	1	1	2	4	6	9
	2	1	1	1	3	5	8
	1	0	1	1	2	3	5
	1/0	0	1	1	1	3	4
	2/0	0	1	1	1	2	4
	3/0	0	0	1	1	1	3
	4/0	0	0	1	1	1	3
TW	14	10	17	27	47	64	104
	12	7	13	21	36	49	80
	10	5	9	15	27	36	59
	8	3	5	8	15	20	33
RHW[a], RHW-2[a], THHW, THW, THW-2	14	6	11	18	31	42	69
	12	5	9	14	25	34	56
	10	4	7	11	19	26	43
	8	2	4	7	12	16	26
RHW[a], RHW-2[a], TW, THW, THHW, THW-2	6	1	3	5	9	12	20
	4	1	2	4	6	9	15
	3	1	1	3	6	8	13
	2	1	1	3	5	6	11
	1	1	1	1	3	4	7
	1/0	1	1	1	3	4	6
	2/0	0	1	1	2	3	5
	3/0	0	1	1	1	3	4
	4/0	0	1	1	1	2	4
THHN, THWN, THWN-2	14	14	24	39	68	91	149
	12	10	17	29	49	67	109
	10	6	11	18	31	42	68
	8	3	6	10	18	24	39
	6	2	4	7	13	17	28
	4	1	3	4	8	10	17
	3	1	2	4	6	9	15
	2	1	1	3	5	7	12
	1	1	1	2	4	5	9
	1/0	1	1	1	3	4	8
	2/0	1	1	1	3	4	6
	3/0	0	1	1	2	3	5
	4/0	0	1	1	1	2	4
XHHW, XHHW-2	14	10	17	27	47	64	104
	12	7	13	21	36	49	80
	10	5	9	15	27	36	59
	8	3	5	8	15	20	33
	6	1	4	6	11	15	24
	4	1	3	4	8	11	18
	3	1	2	4	7	9	15
	2	1	1	3	5	7	12
	1	1	1	2	4	5	9
	1/0	1	1	1	3	5	8
	2/0	1	1	1	3	4	6
	3/0	0	1	1	2	3	5
	4/0	0	1	1	1	2	4

For SI: 1 inch = 25.4 mm.

a. Types RHW, and RHW-2 without outer covering.

❖ See the commentary for Section E3804.6.

TABLE E3804.6(5)
MAXIMUM NUMBER OF CONDUCTORS IN LIQUID-TIGHT FLEXIBLE NONMETALLIC CONDUIT (FNMC-B)[a]

TYPE LETTERS	CONDUCTOR SIZE AWG/kcmil	TRADE SIZES (inches)						
		3/8	1/2	3/4	1	1 1/4	1 1/2	2
RHW, RHW-2	14	2	4	7	12	21	27	44
	12	1	3	6	10	17	22	36
	10	1	3	5	8	14	18	29
	8	1	1	2	4	7	9	15
	6	1	1	1	3	6	7	12
	4	0	1	1	2	4	6	9
	3	0	1	1	1	4	5	8
	2	0	1	1	1	3	4	7
	1	0	0	1	1	1	3	5
	1/0	0	0	1	1	1	2	4
	2/0	0	0	1	1	1	1	3
	3/0	0	0	0	1	1	1	3
	4/0	0	0	0	1	1	1	2
TW	14	5	9	15	25	44	57	93
	12	4	7	12	19	33	43	71
	10	3	5	9	14	25	32	53
	8	1	3	5	8	14	18	29
RHW[a], RHW-2[a], THHW, THW, THW-2	14	3	6	10	16	29	38	62
	12	3	5	8	13	23	30	50
	10	1	3	6	10	18	23	39
	8	1	1	4	6	11	14	23
RHW[a], RHW-2[a], TW, THW, THHW, THW-2	6	1	1	3	5	8	11	18
	4	1	1	1	3	6	8	13
	3	1	1	1	3	5	7	11
	2	0	1	1	2	4	6	9
	1	0	1	1	1	3	4	7
	1/0	0	0	1	1	2	3	6
	2/0	0	0	1	1	2	3	5
	3/0	0	0	1	1	1	2	4
	4/0	0	0	0	1	1	1	3
THHN, THWN, THWN-2	14	8	13	22	36	63	81	133
	12	5	9	16	26	46	59	97
	10	3	6	10	16	29	37	61
	8	1	3	6	9	16	21	35
	6	1	2	4	7	12	15	25
	4	1	1	2	4	7	9	15
	3	1	1	1	3	6	8	13
	2	1	1	1	3	5	7	11
	1	0	1	1	1	4	5	8
	1/0	0	1	1	1	3	4	7
	2/0	0	0	1	1	2	3	6
	3/0	0	0	1	1	1	3	5
	4/0	0	0	1	1	1	2	4
XHHW, XHHW-2	14	5	9	15	25	44	57	93
	12	4	7	12	19	33	43	71
	10	3	5	9	14	25	32	53
	8	1	3	5	8	14	18	29
	6	1	1	3	6	10	13	22
	4	1	1	2	4	7	9	16
	3	1	1	1	3	6	8	13
	2	1	1	1	3	5	7	11
	1	0	1	1	1	4	5	8
	1/0	0	1	1	1	3	4	7
	2/0	0	0	1	1	2	3	6
	3/0	0	0	1	1	1	3	5
	4/0	0	0	1	1	1	2	4

For SI: 1 inch = 25.4 mm.

a. Types RHW, and RHW-2 without outer covering.

❖ See the commentary for Section E3804.6.

**TABLE E3804.6(6)
MAXIMUM NUMBER OF CONDUCTORS IN LIQUID-TIGHT FLEXIBLE NONMETALLIC CONDUIT (FNMC-A)[a]**

TYPE LETTERS	CONDUCTOR SIZE AWG/kcmil	TRADE SIZES (inches)						
		3/8	1/2	3/4	1	1 1/4	1 1/2	2
RHW, RHW-2	14	2	4	7	11	20	27	45
	12	1	3	6	9	17	23	38
	10	1	3	5	8	13	18	30
	8	1	1	2	4	7	9	16
	6	1	1	1	3	5	7	13
	4	0	1	1	2	4	6	10
	3	0	1	1	1	4	5	8
	2	0	1	1	1	3	4	7
	1	0	0	1	1	1	3	5
	1/0	0	0	1	1	1	2	4
	2/0	0	0	1	1	1	1	4
	3/0	0	0	0	1	1	1	3
	4/0	0	0	0	1	1	1	3
TW	14	5	9	15	24	43	58	96
	12	4	7	12	19	33	44	74
	10	3	5	9	14	24	33	55
	8	1	3	5	8	13	18	30
RHW[a], RHW-2[a], THHW, THW, THW-2	14	3	6	10	16	28	38	64
	12	3	4	8	13	23	31	51
	10	1	3	6	10	18	24	40
	8	1	1	4	6	10	14	24
RHW[a], RHW-2[a], TW, THW, THHW, THW-2	6	1	1	3	4	8	11	18
	4	1	1	1	3	6	8	13
	3	1	1	1	3	5	7	11
	2	0	1	1	2	4	6	10
	1	0	1	1	1	3	4	7
	1/0	0	0	1	1	2	3	6
	2/0	0	0	1	1	1	3	5
	3/0	0	0	1	1	1	2	4
	4/0	0	0	0	1	1	1	3
THHN, THWN, THWN-2	14	8	13	22	35	62	83	137
	12	5	9	16	25	45	60	100
	10	3	6	10	16	28	38	63
	8	1	3	6	9	16	22	36
	6	1	2	4	6	12	16	26
	4	1	1	2	4	7	9	16
	3	1	1	1	3	6	8	13
	2	1	1	1	3	5	7	11
	1	0	1	1	1	4	5	8
	1/0	0	1	1	1	3	4	7
	2/0	0	0	1	1	2	3	6
	3/0	0	0	1	1	1	3	5
	4/0	0	0	1	1	1	2	4
XHHW, XHHW-2	14	5	9	15	24	43	58	96
	12	4	7	12	19	33	44	74
	10	3	5	9	14	24	33	55
	8	1	3	5	8	13	18	30
	6	1	1	3	5	10	13	22
	4	1	1	2	4	7	10	16
	3	1	1	1	3	6	8	14
	2	1	1	1	3	5	7	11
	1	0	1	1	1	4	5	8
	1/0	0	1	1	1	3	4	7
	2/0	0	0	1	1	2	3	6
	3/0	0	0	1	1	1	3	5
	4/0	0	0	1	1	1	2	4

For SI: 1 inch = 25.4 mm.

a. Types RHW, and RHW-2 without outer covering.

❖ See the commentary for Section E3804.6.

TABLE E3804.6(7)
MAXIMUM NUMBER OF CONDUCTORS IN LIQUID-TIGHT FLEXIBLE METAL CONDUIT (LFMC)[a]

TYPE LETTERS	CONDUCTOR SIZE AWG/kcmil	TRADE SIZES (inches)					
		1/2	3/4	1	1 1/4	1 1/2	2
RHW, RHW-2	14	4	7	12	21	27	44
	12	3	6	10	17	22	36
	10	3	5	8	14	18	29
	8	1	2	4	7	9	15
	6	1	1	3	6	7	12
	4	1	1	2	4	6	9
	3	1	1	1	4	5	8
	2	1	1	1	3	4	7
	1	0	1	1	1	3	5
	1/0	0	1	1	1	2	4
	2/0	0	1	1	1	1	3
	3/0	0	0	1	1	1	3
	4/0	0	0	1	1	1	2
TW	14	9	15	25	44	57	93
	12	7	12	19	33	43	71
	10	5	9	14	25	32	53
	8	3	5	8	14	18	29
RHW[a], RHW-2[a], THHW, THW, THW-2	14	6	10	16	29	38	62
	12	5	8	13	23	30	50
	10	3	6	10	18	23	39
	8	1	4	6	11	14	23
RHW[a], RHW-2[a], TW, THW, THHW, THW-2	6	1	3	5	8	11	18
	4	1	1	3	6	8	13
	3	1	1	3	5	7	11
	2	1	1	2	4	6	9
	1	1	1	1	3	4	7
	1/0	0	1	1	2	3	6
	2/0	0	1	1	2	3	5
	3/0	0	1	1	1	2	4
	4/0	0	0	1	1	1	3
THHN, THWN, THWN-2	14	13	22	36	63	81	133
	12	9	16	26	46	59	97
	10	6	10	16	29	37	61
	8	3	6	9	16	21	35
	6	2	4	7	12	15	25
	4	1	2	4	7	9	15
	3	1	1	3	6	8	13
	2	1	1	3	5	7	11
	1	1	1	1	4	5	8
	1/0	1	1	1	3	4	7
	2/0	0	1	1	2	3	6
	3/0	0	1	1	1	3	5
	4/0	0	1	1	1	2	4
XHHW, XHHW-2	14	9	15	25	44	57	93
	12	7	12	19	33	43	71
	10	5	9	14	25	32	53
	8	3	5	8	14	18	29
	6	1	3	6	10	13	22
	4	1	2	4	7	9	16
	3	1	1	3	6	8	13
	2	1	1	3	5	7	11
	1	1	1	1	4	5	8
	1/0	1	1	1	3	4	7
	2/0	0	1	1	2	3	6
	3/0	0	1	1	1	3	5
	4/0	0	1	1	1	2	4

For SI: 1 inch = 25.4 mm.

a. Types RHW, and RHW-2 without outer covering.

❖ See the commentary for Section E3804.6.

TABLE E3804.6(8)
MAXIMUM NUMBER OF CONDUCTORS IN RIGID METAL CONDUIT (RMC)[a]

TYPE LETTERS	CONDUCTOR SIZE AWG/kcmil	TRADE SIZES (inches)					
		1/2	3/4	1	1 1/4	1 1/2	2
RHW, RHW-2	14	4	7	12	21	28	46
	12	3	6	10	17	23	38
	10	3	5	8	14	19	31
	8	1	2	4	7	10	16
	6	1	1	3	6	8	13
	4	1	1	2	4	6	10
	3	1	1	2	4	5	9
	2	1	1	1	3	4	7
	1	0	1	1	1	3	5
	1/0	0	1	1	1	2	4
	2/0	0	1	1	1	2	4
	3/0	0	0	1	1	1	3
	4/0	0	0	1	1	1	3
TW	14	9	15	25	44	59	98
	12	7	12	19	33	45	75
	10	5	9	14	25	34	56
	8	3	5	8	14	19	31
RHW[a], RHW-2[a], THHW, THW, THW-2	14	6	10	17	29	39	65
	12	5	8	13	23	32	52
	10	3	6	10	18	25	41
	8	1	4	6	11	15	24
RHW[a], RHW-2[a], TW, THW, THHW, THW-2	6	1	3	5	8	11	18
	4	1	1	3	6	8	14
	3	1	1	3	5	7	12
	2	1	1	2	4	6	10
	1	1	1	1	3	4	7
	1/0	0	1	1	2	3	6
	2/0	0	1	1	2	3	5
	3/0	0	1	1	1	2	4
	4/0	0	0	1	1	1	3
THHN, THWN, THWN-2	14	13	22	36	63	85	140
	12	9	16	26	46	62	102
	10	6	10	17	29	39	64
	8	3	6	9	16	22	37
	6	2	4	7	12	16	27
	4	1	2	4	7	10	16
	3	1	1	3	6	8	14
	2	1	1	3	5	7	11
	1	1	1	1	4	5	8
	1/0	1	1	1	3	4	7
	2/0	0	1	1	2	3	6
	3/0	0	1	1	1	3	5
	4/0	0	1	1	1	2	4
XHHW, XHHW-2	14	9	15	25	44	59	98
	12	7	12	19	33	45	75
	10	5	9	14	25	34	56
	8	3	5	8	14	19	31
	6	1	3	6	10	14	23
	4	1	2	4	7	10	16
	3	1	1	3	6	8	14
	2	1	1	3	5	7	12
	1	1	1	1	4	5	9
	1/0	1	1	1	3	4	7
	2/0	0	1	1	2	3	6
	3/0	0	1	1	1	3	5
	4/0	0	1	1	1	2	4

For SI: 1 inch = 25.4 mm.

a. Types RHW, and RHW-2 without outer covering.

❖ See the commentary for Section E3804.6.

TABLE E3804.6(9)
MAXIMUM NUMBER OF CONDUCTORS IN RIGID PVC CONDUIT, SCHEDULE 80 (PVC-80)[a]

TYPE LETTERS	CONDUCTOR SIZE AWG/kcmil	TRADE SIZES (inches)					
		1/2	3/4	1	1 1/4	1 1/2	2
RHW, RHW-2	14	3	5	9	17	23	39
	12	2	4	7	14	19	32
	10	1	3	6	11	15	26
	8	1	1	3	6	8	13
	6	1	1	2	4	6	11
	4	1	1	1	3	5	8
	3	0	1	1	3	4	7
	2	0	1	1	3	4	6
	1	0	1	1	1	2	4
	1/0	0	0	1	1	1	3
	2/0	0	0	1	1	1	3
	3/0	0	0	1	1	1	3
	4/0	0	0	0	1	1	2
TW	14	6	11	20	35	49	82
	12	5	9	15	27	38	63
	10	3	6	11	20	28	47
	8	1	3	6	11	15	26
RHW[a], RHW-2[a], THHW, THW, THW-2	14	4	8	13	23	32	55
	12	3	6	10	19	26	44
	10	2	5	8	15	20	34
	8	1	3	5	9	12	20
RHW[a], RHW-2[a], TW, THW, THHW, THW-2	6	1	1	3	7	9	16
	4	1	1	3	5	7	12
	3	1	1	2	4	6	10
	2	1	1	1	3	5	8
	1	0	1	1	2	3	6
	1/0	0	1	1	1	3	5
	2/0	0	1	1	1	2	4
	3/0	0	0	1	1	1	3
	4/0	0	0	1	1	1	3
THHN, THWN, THWN-2	14	9	17	28	51	70	118
	12	6	12	20	37	51	86
	10	4	7	13	23	32	54
	8	2	4	7	13	18	31
	6	1	3	5	9	13	22
	4	1	1	3	6	8	14
	3	1	1	3	5	7	12
	2	1	1	2	4	6	10
	1	0	1	1	3	4	7
	1/0	0	1	1	2	3	6
	2/0	0	1	1	1	3	5
	3/0	0	1	1	1	2	4
	4/0	0	0	1	1	1	3
XHHW, XHHW-2	14	6	11	20	35	49	82
	12	5	9	15	27	38	63
	10	3	6	11	20	28	47
	8	1	3	6	11	15	26
	6	1	2	4	8	11	19
	4	1	1	3	6	8	14
	3	1	1	3	5	7	12
	2	1	1	2	4	6	10
	1	0	1	1	3	4	7
	1/0	0	1	1	2	3	6
	2/0	0	1	1	1	3	5
	3/0	0	1	1	1	2	4
	4/0	0	0	1	1	1	3

For SI: 1 inch = 25.4 mm.

a. Types RHW, and RHW-2 without outer covering.

❖ See the commentary for Section E3804.6.

**TABLE E3804.6(10)
MAXIMUM NUMBER OF CONDUCTORS IN RIGID PVC CONDUIT SCHEDULE 40 (PVC-40)[a]**

TYPE LETTERS	CONDUCTOR SIZE AWG/kcmil	TRADE SIZES (inches)					
		1/2	3/4	1	1 1/4	1 1/2	2
RHW, RHW-2	14	4	7	11	20	27	45
	12	3	5	9	16	22	37
	10	2	4	7	13	18	30
	8	1	2	4	7	9	15
	6	1	1	3	5	7	12
	4	1	1	2	4	6	10
	3	1	1	1	4	5	8
	2	1	1	1	3	4	7
	1	0	1	1	1	3	5
	1/0	0	1	1	1	2	4
	2/0	0	0	1	1	1	3
	3/0	0	0	1	1	1	3
	4/0	0	0	1	1	1	2
TW	14	8	14	24	42	57	94
	12	6	11	18	32	44	72
	10	4	8	13	24	32	54
	8	2	4	7	13	18	30
RHW[a], RHW-2[a], THHW, THW, THW-2	14	5	9	16	28	38	63
	12	4	8	12	22	30	50
	10	3	6	10	17	24	39
	8	1	3	6	10	14	23
RHW[a], RHW-2[a], TW, THW, THHW, THW-2	6	1	2	4	8	11	18
	4	1	1	3	6	8	13
	3	1	1	3	5	7	11
	2	1	1	2	4	6	10
	1	0	1	1	3	4	7
	1/0	0	1	1	2	3	6
	2/0	0	1	1	1	3	5
	3/0	0	1	1	1	2	4
	4/0	0	0	1	1	1	3
THHN, THWN, THWN-2	14	11	21	34	60	82	135
	12	8	15	25	43	59	99
	10	5	9	15	27	37	62
	8	3	5	9	16	21	36
	6	1	4	6	11	15	26
	4	1	2	4	7	9	16
	3	1	1	3	6	8	13
	2	1	1	3	5	7	11
	1	1	1	1	3	5	8
	1/0	1	1	1	3	4	7
	2/0	0	1	1	2	3	6
	3/0	0	1	1	1	3	5
	4/0	0	1	1	1	2	4
XHHW, XHHW-2	14	8	14	24	42	57	94
	12	6	11	18	32	44	72
	10	4	8	13	24	32	54
	8	2	4	7	13	18	30
	6	1	3	5	10	13	22
	4	1	2	4	7	9	16
	3	1	1	3	6	8	13
	2	1	1	3	5	7	11
	1	1	1	1	3	5	8
	1/0	1	1	1	3	4	7
	2/0	0	1	1	2	3	6
	3/0	0	1	1	1	3	5
	4/0	0	1	1	1	2	4

For SI: 1 inch = 25.4 mm.

a. Types RHW, and RHW-2 without outer covering.

❖ See the commentary for Section E3804.6.

SECTION E3805
BOXES, CONDUIT BODIES AND FITTINGS

E3805.1 Box, conduit body or fitting—where required. A box or conduit body shall be installed at each conductor splice point, outlet, switch point, junction point and pull point except as otherwise permitted in Sections E3805.1.1 through E3805.1.7.

Fittings and connectors shall be used only with the specific wiring methods for which they are are designed and listed.

❖ The point in the wiring system that is the most vulnerable for possible problems due to overheating caused by arcing or sparking is a splice or connection. Except for some cases of underground wiring, a box or enclosure must always be used to make a splice or connection. Some individuals may believe that if the splice has plenty of tape, is not against a stud or joist, and is supported in air, it will be fine. A splice or connection made outside of a box is a violation and a serious safety concern. Nonmetallic sheathed cable must be used in continuous lengths from box to box. There must be no splices, and the box must be accessible. See Section E3805.10.

E3805.1.1 Equipment. An integral junction box or wiring compartment that is part of listed equipment shall be permitted to serve as a box or conduit body.

❖ An example of this is on a hood fan unit installed above a range. If a box were installed and splices made, the box would probably be covered with the hood. Because the exact height of the hood is not always known, a sufficient length of the cable is left stapled to a stud on the rough wiring without any box and brought through a hole in the drywall. The hood fan unit has a built-in box where the splices can be made. This is similar for a bathroom exhaust fan, a dishwasher, and many other appliances. It is important to use a proper fitting such as a listed cable connector where the cable enters the appliance junction box. If not protected with a proper connector, the cable could be damaged by the metal edge of the knockout on the appliance. The proper installation is to use a connector and not simply to wrap tape around the cable where it enters the box.

E3805.1.2 Protection. A box or conduit body shall not be required where cables enter or exit from conduit or tubing that is used to provide cable support or protection against physical damage. A fitting shall be provided on the end(s) of the conduit or tubing to protect the cable from abrasion.

❖ A box is not required, because there are no splices. An example of the protection required is where a switch box is attached to a concrete or masonry wall in a basement and a short run of EMT is used to protect the cable along the wall. At the top of the short run of EMT, a connector with a bushing or a cable transition fitting must be installed. See the commentary for Section E3702.3.2.

E3805.1.3 Integral enclosure. A wiring device with integral enclosure identified for the use, having brackets that securely fasten the device to walls or ceilings of conventional on-site frame construction, for use with nonmetallic-sheathed cable, shall be permitted in lieu of a box or conduit body.

❖ An example of an integral enclosure is a receptacle for a range or dryer outlet. It has a base section that is attached to the wall and a cover that is attached by screws after the connections are made to terminals. This type of device is usually for surface mounting.

E3805.1.4 Fitting. A fitting identified for the use shall be permitted in lieu of a box or conduit body where such fitting is accessible after installation and does not contain spliced or terminated conductors.

❖ Where a change is made in the type of wiring method, a fitting is permitted as long as it does not contain splices. Where transition is made from nonmetallic-sheathed cable to EMT, a box is not needed if a proper fitting is provided at the transition.

E3805.1.5 Buried conductors. Splices and taps in buried conductors and cables shall not be required to be enclosed in a box or conduit body where installed in accordance with Section 3703.5.

❖ See the commentary for Section E3703.5.

E3805.1.6 Luminaires. Where a luminaire is listed to be used as a raceway, a box or conduit body shall not be required for wiring installed therein.

❖ In order to make the connection of the circuit wires to the fixtures wire, generally a box is required where lighting fixtures are installed. In some older homes it can be observed that the cable was brought through the plaster ceiling and the connection made without the use of a box. The light fixture was attached directly to the ceiling with long wood screws, and the canopy of the light fixture contained the splices. This is a code violation because a box is required. A common violation in newer homes occurs where a fluorescent lighting fixture is installed over a junction box that contains splices or connections. The box must be accessible but is covered by the back of the fluorescent fixture. In this installation, the opening in the back of the fluorescent fixture must be large enough to permit access into the junction box. If not, then the entire fixture would have to be removed to get into the junction box.

According to this section, however, if the fluorescent fixture is listed to be used like a box to contain the splices, then it would not be a code violation, but only if it is not covering a flush mounted box in the ceiling. If the cable or cables are run through the drywall ceiling and brought into the wiring section of the fluorescent fixture through a small $^{1}/_{2}$-inch (12.7 mm) knockout opening that is usually provided, the installation would be acceptable and would serve as the enclosure for splicing the circuit conductors to the fixture wires.

Where the nonmetallic sheathed cable extends through the finished ceiling wallboard, it is permissible to omit the junction box in the ceiling and use the wiring compartment of the listed fixture as the junction box to make the connections. But if a box is installed in the ceiling with other circuit conductors in the ceiling box, it would be a code violation insofar as it would be necessary to remove the fluorescent lighting fixture in order to get into the junction box.

E3805.1.7 Closed loop. Where a device identified and listed as suitable for installation without a box is used with a closed-loop power-distribution system, a box or conduit body shall not be required.

❖ A closed loop system is the wiring system used by "smart house" design, which uses a cable containing conductors for 120-volt supply as well as low-voltage signaling conductors. The receptacles used on this system are different from the conventional receptacles and will only accept attachment plugs designed for the system. The receptacle outlets are contained in "convenience centers," which do not utilize a conventional device or outlet box.

E3805.2 Metal boxes. All metal boxes shall be grounded.

❖ See Section E3808 on grounding.

E3805.3 Nonmetallic boxes. Nonmetallic boxes shall be used only with nonmetallic-sheathed cable, cabled wiring methods, flexible cords and nonmetallic raceways.

Exceptions:

1. Where internal bonding means are provided between all entries, nonmetallic boxes shall be permitted to be used with metal raceways and metal-armored cables.
2. Where integral bonding means with a provision for attaching an equipment grounding jumper inside the box are provided between all threaded entries in nonmetallic boxes listed for the purpose, nonmetallic boxes shall be permitted to be used with metal raceways and metal-armored cables.

❖ A nonmetallic device box (used for directly attaching switches or receptacles) is designed for nonmetallic-sheathed cable. A nonmetallic four-inch square box can be used, but a plaster ring would be needed for the attachment of switches and receptacles. Nonmetallic 4-inch square boxes come with knockouts for the entry of nonmetallic-sheathed cable and also have standard size knockouts for conduit and tubing fittings. To use a nonmetallic box in a run of metal conduit or tubing requires special attention to the grounding and bonding requirements.

A metal raceway is permitted to serve as a grounding conductor. Where a nonmetallic box is inserted between two metal raceway entries, the continuity of the grounding conductor is interrupted. Therefore, bonding jumpers are required.

E3805.3.1 Nonmetallic-sheathed cable and nonmetallic boxes. Where nonmetallic-sheathed cable is used, the cable assembly, including the sheath, shall extend into the box not less than $^1/_4$ inch (6.4 mm) through a nonmetallic-sheathed cable knockout opening.

❖ The individual conductors should not be exposed outside of the box. Where the cable enters the box, the clamp should not bind against the individual conductors but against the sheathing. Some electricians strip the sheathing off the conductors before installing the cable in the nonmetallic device box, and this sometimes results in the cable sheath being too short to extend into the box. On the rough inspection, it is an indication that the job has been done in a careless manner where the sheathing has been stripped short of entering the box. Wiring should be installed in a neat manner. However, there is no code rule that the cable sheathing should be stripped before or after it is inserted into the box; it is a preference only.

The amount of sheathing left on the conductors inside the box affects the box fill, because it takes more room to tuck the conductors back into the box when there is too much sheathing left. This leaves less room to install the device. The length of sheathing left in the box, the length of the conductors, the size of the box, and the type of device all should be considered, not only the minimum requirements of the code for box size.

E3805.3.2 Securing to box. All permitted wiring methods shall be secured to the boxes.

Exception: Where nonmetallic-sheathed cable is used with boxes not larger than a nominal size of $2^1/_4$ inches by 4 inches (57 mm by 102 mm) mounted in walls or ceilings, and where the cable is fastened within 8 inches (203 mm) of the box measured along the sheath, and where the sheath extends through a cable knockout not less than $^1/_4$ inch (6.4 mm), securing the cable to the box shall not be required.

❖ The basic rule here is that the cable, such as armored cable and nonmetallic-sheathed cable, should be secured to the box. Where metal boxes are used, this is required. All metal device boxes come with clamps, usually inside the box. If the metal box used is not a device box, such as a four-square box with a plaster ring installed to make it possible to attach devices, then the cable is secured to the box by approved cable connectors through the knockouts. This code provision refers to a single-gang box, which is usually not larger than $2^1/_4$ inches by 4 inches (25 mm by 102 mm). For single-gang nonmetallic boxes, it is permitted to not secure the cable to the box itself if the cable is stapled to the stud within 8 inches of cable length outside the box. This provision was not limited to a single-gang box until 1978, when the length of 8 inches was also specified as "measured along the sheath" and not just 8 inches (203 mm) from the box. Before that, a two-gang, three-gang, or larger nonmetallic box did not have to have the cable fastened to the box. Then the manufacturers of nonmetallic boxes started making boxes with internal clamps.

The reason for this requirement is that the electrician may run the cable against the stud and then, at the top of the box, make a bend and run the cable over to a knockout opening furthest away from the stud. When the wires are tucked back into the box or when the device is installed, the cable could be pushed up and the conductors exposed outside the box where the sheathing has been stripped off. Many nonmetallic single-gang boxes have internal clamps that work like a spring, putting pressure against the cable so that it is difficult if not impossible to remove the cable simply by pulling it back out of the box. But the cable sheathing should be long enough so that a nonmetallic internal clamp puts pressure against the sheathing and not against the individual conductors. In other words, in addition to extending into the box at least $^1/_4$ inch (6.3 mm), the sheathing should extend into the box beyond any cable clamp.

It is permitted to run multiple cables through a single knockout opening in the box.

E3805.3.3 Conductor rating. Nonmetallic boxes shall be suitable for the lowest temperature-rated conductor entering the box.

❖ All conductors have a temperature rating according to the insulation type, as indicated in Table E3605.1. Nonmetallic boxes have a temperature rating also, and it should be suitable for the lowest temperature rating of any conductor installed in the box. For example, it is not necessary for a nonmetallic box in a normal temperature environment to have a rating of 90°C only because the conductors in the box have that temperature rating.

E3805.4 Minimum depth of outlet boxes. Boxes shall have an internal depth of not less than 0.5 inch (12.7 mm). Boxes designed to enclose flush devices shall have an internal depth of not less than 0.938 inch (2.4 mm).

❖ An outlet box could be for a lighting fixture or for a device such as a receptacle or switch. A box for a lighting fixture that is an outlet box but does not contain devices could be as shallow as $^1/_2$ inch (12.7 mm). A term used for this type of round shallow box is "pancake box," and it can be used, for example, on the surface of a joist or truss where it is necessary to center the lighting fixture in a room or over a bathroom sink area. Where a device is installed, the box must be at least $^{15}/_{16}$ inch (2.4 mm) deep. This provision for such a shallow box does not supercede the box fill requirements.

E3805.5 Boxes enclosing flush-mounted devices. Boxes enclosing flush-mounted devices shall be of such design that the devices are completely enclosed at the back and all sides and shall provide support for the devices. Screws for supporting the box shall not be used for attachment of the device contained therein.

❖ Many devices require a box of greater depth than the minimum $^{15}/_{16}$ inch (2.4 mm) because of the size of the device. For example, where a GFCI or dimmer is installed, the box must have sufficient depth and capacity to completely enclose the device on the sides and back. If a cover is installed but the sides of the device are exposed, it would be a code violation. The box must be secured independently of the device.

E3805.6 Boxes at luminaire outlets. Boxes for luminaire outlets shall be designed for the purpose. At every outlet used exclusively for lighting, the box shall be designed or installed so that a luminaire may be attached.

> **Exception:** A wall-mounted luminaire weighing not more than 3 kg (6 lb) shall be permitted to be supported on other boxes or plaster rings that are secured to other boxes, provided the luminaire or its supporting yoke is secured to the box with no fewer than two No. 6 or larger screws.

❖ Lighting fixtures are mounted to round or octagonal boxes, which are usually close to 4 inches in diameter. Two sets of mounting holes are provided in many round fixture boxes. The holes, machined to accept 8-32 mounting screws (size No. 8 in diameter with 32 threads per inch), are $2^3/_4$ inches (70 mm) and $3^1/_2$ inches (89 mm) apart. This is a standard measurement. Many nonmetallic round boxes have the same two sets of mounting holes. Metal round or octagonal boxes usually have only one set of mounting holes, which are $3^1/_2$ inches (89 mm) apart. Some lighting fixtures are secured by two 8-32 screws directly through the base of the fixture into the mounting holes of the box, and the holes in the fixture base are exactly $3^1/_2$ inches (89 mm) apart to match the holes on the round box. Porcelain keyless lamp holder and pull chain type fixtures are examples of fixtures having the mounting holes the same as those on the round "fixture boxes," as they are sometimes referred to in the field. Most lighting fixtures come with a bracket designed to be attached to the outlet box. The bracket serves as a means to adjust the alignment of the holes for mounting the fixture. The bracket must be used as indicated in the instructions provided with the lighting fixture.

The code does not specify that fixtures must be mounted only on a round or octagonal box. A device box could be used for a small fixture that weighs not over 6 pounds provided that two 6-32 screws are used. Device boxes are used for switches and receptacles and have mounting holes machined to accept 6-32 screws, which obviously would not provide as much support as the 8-32 screws that are supplied with fixtures. A problem is that the mounting holes on a device box are $3^1/_4$ inches (83 mm) apart, and they do not line up with those on the lighting fixtures. They also might not line up with the mounting holes on many of the brackets supplied with lighting fixtures.

A device box could be installed in the rough wiring stage with the intention that a small fixture weighing less than 6 pounds will be installed at that location on the finish stage. But in the future, the small fixture could be changed for a larger and heavier fixture, thus making the device box not suitable and not in compliance with the code. In some areas, single gang device boxes have been used in a brick wall for mounting outdoor fixtures. On a brick house, it is easy to install a device box in one of the courses of brick without cutting any brick. If a round box is used, the brick mason has to cut or notch one or more bricks to install the round box flush with the wall. If the light fixture weighs 6 pounds or less, it would be acceptable to use a device box, but it would be good practice to install a round or octagonal box.

See Section E3903 for more provisions for light fixture installation.

E3805.7 Maximum luminaire weight. Outlet boxes or fittings installed as required by Section E3804.3 shall be permitted to support luminaires weighing 23 kg (50 lb) or less. A luminaire that weighs more than 23 kg (50 lb) shall be supported independently of the outlet box unless the outlet box is listed for the weight to be supported.

❖ Luminaires (lighting fixtures) that weigh more than 50 pounds must be supported by building structural members, not by an outlet box, except where the outlet box is specifically listed for support of the weight. It is common to find that heavy luminaires, such as chandeliers, have been hung from standard outlet boxes that were not designed to support the weight. Such fixtures could fall, causing injury. It would be wise to rough-in special luminaire support boxes in all locations where homeowners would likely install chandeliers and similar heavy luminaires in the future. See the commentary for Section E4001.6.

E3805.8 Floor boxes. Where outlet boxes for receptacles are installed in the floor, such boxes shall be listed specifically for that application.

❖ It is important to check the listing for the box used. For example, an ordinary device box may not be listed to be installed in a floor and used for a receptacle flush with the floor. Obviously, the support of a floor box should be considered as needing more than just the two nails with which most device boxes are mounted, because the floor could be walked on, and furniture could be placed on it. The finish or front edge of floor boxes is usually adjustable because the thickness of the flooring material is not known at the rough stage of the wiring. Even if a regular device box were approved for a floor outlet, it would be difficult to estimate the thickness of the flooring so that the box could be mounted at the appropriate height.

E3805.9 Boxes at fan outlets. Outlet boxes shall not be depended on as the sole support for ceiling (paddle) fans except where such boxes are listed for that application.

❖ See Section E4001.6.

E3805.10 Conduit bodies and junction, pull and outlet boxes to be accessible. Conduit bodies and junction, pull and outlet boxes shall be installed so that the wiring therein can be accessed without removing any part of the building or, in underground circuits, without excavating sidewalks, paving, earth or other substance used to establish the finished grade.

Exception: Boxes covered by gravel, light aggregate or noncohesive granulated soil shall be listed for the application, and the box locations shall be effectively identified and access shall be provided for excavation.

❖ Boxes must be accessible and have an appropriate cover. A junction box could have a blank cover in a finished wall or ceiling. The homeowner may not want a blank cover showing on a finished wall or ceiling, but it would be a code violation to "bury" or hide the box behind the drywall. In many locations of the country, dwellings are constructed with a basement that is left unfinished when the house is purchased and moved into by the owner. The wiring to the lights in the ceiling of the unfinished basement is part of a circuit that extends to or from the floor above. At a later date, when the basement is finished, the part of the cable exposed in the basement is often too short to extend to the desired location of a new lighting outlet or receptacle for the finished room. If the conductors are spliced in the existing box and then covered with drywall, it would be a code violation insofar as the box would then be concealed behind the ceiling or wall, and the only way to get to the connection would be to cut open or remove the drywall. If the splice is not in a box, that would be a separate code violation.

This code provision requiring a box where a splice is made is not only for the convenience of being able to work on the circuit in the future, but is a safety concern because overheating could occur from arcing or sparking in the splice. In the example of the basement lighting outlet, if it were not possible or feasible to replace the entire length of the cable back to the nearest box, because that may be on the floor above, then the homeowner may simply have to have a junction box in the ceiling of the basement with a blank cover on the finished ceiling. That would make the splice accessible. No matter what choice is made, the cable must be continuous from box to box, and the box must be accessible. It may become necessary to remove some of the existing drywall in some parts of the house. Another example is in the remodeling of a kitchen. The old outlets may not be in the desired or required location for the new kitchen cabinet layout. It would be a code violation to permanently cover or conceal junction boxes or splices.

E3805.11 Damp or wet locations. In damp or wet locations, boxes, conduit bodies and fittings shall be placed or equipped so as to prevent moisture from entering or accumulating within the box, conduit body or fitting. Boxes, conduit bodies and fittings installed in wet locations shall be listed for use in wet locations.

❖ Where an outdoor receptacle is installed and a box is surface mounted, it must be listed for the purpose. Commonly used boxes for this purpose are Type FS or FD. The cover must be installed with a gasket to seal the interior of the box from moisture.

E3805.12 Number of conductors in outlet, device, and junction boxes, and conduit bodies. Boxes and conduit bodies shall be of sufficient size to provide free space for all enclosed conductors. In no case shall the volume of the box, as calculated in Section E3805.12.1, be less than the box fill calculation as calculated in Section E3805.12.2. The minimum volume for conduit bodies shall be as calculated in Section E3805.12.3. The provisions of this section shall not apply to terminal housings supplied with motors.

❖ Selecting the proper size box requires two tasks. One is to determine the cubic inch capacity of the box, and the second is to determine the box fill allowed or needed. This section deals with determining the cubic inch capacity or volume of the box.

E3805.12.1 Box volume calculations. The volume of a wiring enclosure (box) shall be the total volume of the assembled sections, and, where used, the space provided by plaster rings, domed covers, extension rings, etc., that are marked with their volume in cubic inches or are made from boxes the dimensions of which are listed in Table E3805.12.1.

❖ See the commentary for Section E3805.12.1.2.

E3805.12.1.1 Standard boxes. The volumes of standard boxes that are not marked with a cubic-inch capacity shall be as given in Table E3805.12.1.

❖ See the commentary for Section E3805.12.1.2.

TABLE E3805.12.1
MAXIMUM NUMBER OF CONDUCTORS IN METAL BOXES[a]

BOX DIMENSIONS (inches trade size and type)	MAXIMUM CAPACITY (cubic inches)	MAXIMUM NUMBER OF CONDUCTORS[a]						
		No. 18	No. 16	No. 14	No. 12	No. 10	No. 8	No. 6
$4 \times 1^1/_4$ round or octagonal	12.5	8	7	6	5	5	4	2
$4 \times 1^1/_2$ round or octagonal	15.5	10	8	7	6	6	5	3
$4 \times 2^1/_8$ round or octagonal	21.5	14	12	10	9	8	7	4
$4 \times 1^1/_4$ square	18.0	12	10	9	8	7	6	3
$4 \times 1^1/_2$ square	21.0	14	12	10	9	8	7	4
$4 \times 2^1/_8$ square	30.3	20	17	15	13	12	10	6
$4^{11}/_{16} \times 1^1/_4$ square	25.5	17	14	12	11	10	8	5
$4^{11}/_{16} \times 1^1/_2$ square	29.5	19	16	14	13	11	9	5
$4^{11}/_{16} \times 2^1/_8$ square	42.0	28	24	21	18	16	14	8
$3 \times 2 \times 1^1/_2$ device	7.5	5	4	3	3	3	2	1
$3 \times 2 \times 2$ device	10.0	6	5	5	4	4	3	2
$3 \times 2 \times 2^1/_4$ device	10.5	7	6	5	4	4	3	2
$3 \times 2 \times 2^1/_2$ device	12.5	8	7	6	5	5	4	2
$3 \times 2 \times 2^3/_4$ device	14.0	9	8	7	6	5	4	2
$3 \times 2 \times 3^1/_2$ device	18.0	12	10	9	8	7	6	3
$4 \times 2^1/_8 \times 1^1/_2$ device	10.3	6	5	5	4	4	3	2
$4 \times 2^1/_8 \times 1^7/_8$ device	13.0	8	7	6	5	5	4	2
$4 \times 2^1/_8 \times 2^1/_8$ device	14.5	9	8	7	6	5	4	2
$3^3/_4 \times 2 \times 2^1/_2$ masonry box/gang	14.0	9	8	7	6	5	4	2
$3^3/_4 \times 2 \times 3^1/_2$ masonry box/gang	21.0	14	12	10	9	8	7	4

For SI: 1 inch = 25.4 mm, 1 cubic inch = 16.4 cm^3.

a. Where no volume allowances are required by Sections E3805.12.2.2 through E3805.12.2.5.

❖ See the commentary for Section E3805.11.1.2.

E3805.12.1.2 Other boxes. Boxes 100 cubic inches (1640 cm^3) or less, other than those described in Table E3805.12.1, and nonmetallic boxes shall be durably and legibly marked by the manufacturer with their cubic-inch capacity. Boxes described in Table E3805.12.1 that have a larger cubic inch capacity than is designated in the table shall be permitted to have their cubic-inch capacity marked as required by this section.

❖ Section E3805.12.1 and the two subsequent subsections are included in the code to establish the criteria to determine the volume or cubic inch capacity of the box. The code requires the cubic inch capacity to be marked on all nonmetallic boxes and any metal box of over 100 cubic inches. For standard size metal boxes listed in Table E3805.12.1 that do not have the cubic inch capacity marked, the volume or cubic inch capacity used for conductor fill is that listed in Table E3805.12.1. Because most boxes used in residential wiring do have the cubic inch capacity marked, Table E3805.12.1 is not often needed to determine the box volume.

E3805.12.2 Box fill calculations. The volumes in Section E3805.12.2.1 through Section E3805.12.2.5, as applicable, shall be added together. No allowance shall be required for small fittings such as locknuts and bushings.

❖ The various components of box fill are described in the five subsections below (E3805.12.2.1 through E3805.12.2.5). These components are (1) conductors, (2) internal clamps, (3) support fittings, (4) devices and (5) grounding conductors. The volume allowance for each of these components must be added and multiplied by the cubic inch volume in Table E3805.12.2.1 for the size conductor used. In Table E3805.12.2.1, the volume allowances are listed in the second column as free space for each conductor. The volume allowance for all nonconductor components, i.e., clamps, fittings, devices, and grounding conductors, is based on the largest conductor in the box. For devices, two volume allowances are counted.

The Exception to Section E3805.12.2.1 allows for fixtures that have several fixture wires from different lamps on the fixture terminating in the fixture canopy to be omitted from the calculation. Some fixtures may attach to the box with the use of a support fitting like a threaded stud or hickey, which is a stud with openings on the sides. These were used in the past but are not very common in modern wiring.

<u>Example of box fill calculation</u>

An inspector finds that a device box with no internal cable clamps is marked with a volume of 20 cubic inches and contains two size 12 AWG nonmetallic-sheathed cables of 12-3 with ground and a duplex receptacle. (Each cable has two ungrounded conductors, one grounded conductor, and one grounding conductor). Does this box comply with code provisions?

Determine the minimum cubic inch capacity required for each component of the box from Table E3805.12.2.1. The table indicates that the volume allowance for size 12 AWG conductors is 2.25 cubic inches.

Six conductors (ungrounded grounded conductors)	6 x 2.25 =	13.5
All grounding conductors counted as one	1 x 2.25 =	2.25
One receptacle counts as two volume allowances	2 x 2.25 =	<u>4.5</u>
Total cubic inches required		20.25

The use of this box is a code violation because it does not have the volume required.

E3805.12.2.1 Conductor fill. Each conductor that originates outside the box and terminates or is spliced within the box shall be counted once, and each conductor that passes through the box without splice or termination shall be counted once. The conductor fill, in cubic inches, shall be computed using Table E3805.12.2.1. A conductor, no part of which leaves the box, shall not be counted.

Exception: An equipment grounding conductor or not more than four fixture wires smaller than No. 14, or both, shall be permitted to be omitted from the calculations where such conductors enter a box from a domed fixture or similar canopy and terminate within that box.

❖ See the commentary for Section E3805.12.2.

TABLE E3805.12.2.1
VOLUME ALLOWANCE REQUIRED PER CONDUCTOR

SIZE OF CONDUCTOR (AWG)	FREE SPACE WITHIN BOX FOR EACH CONDUCTOR (cubic inches)
No. 18	1.50
No. 16	1.75
No. 14	2.00
No. 12	2.25
No. 10	2.50
No. 8	3.00
No. 6	5.00

For SI: 1 cubic inch = 16.4 cm^3.

❖ See the commentary for Section E3805.12.2.

E3805.12.2.2 Clamp fill. Where one or more internal cable clamps, whether factory or field supplied, are present in the box, a single volume allowance in accordance with Table E3805.12.2.1 shall be made based on the largest conductor present in the box. No allowance shall be required for a cable connector with its clamping mechanism outside the box.

❖ See the commentary for Section E3805.12.2.

E3805.12.2.3 Support fittings fill. Where one or more fixture studs or hickeys are present in the box, a single volume allowance in accordance with Table E3805.12.2.1 shall be made for each type of fitting based on the largest conductor present in the box.

❖ See the commentary for Section E3805.12.2.

E3805.12.2.4 Device or equipment fill. For each yoke or strap containing one or more devices or equipment, a double volume allowance in accordance with Table E3805.12.2.1 shall be made for each yoke or strap based on the largest conductor connected to a device(s) or equipment supported by that yoke or strap.

❖ See the commentary for Section E3805.12.2.

E3805.12.2.5 Equipment grounding conductor fill. Where one or more equipment grounding conductors or equipment bonding jumpers enters a box, a single volume allowance in accordance with Table E3805.12.2.1 shall be made based on the largest equipment grounding conductor or equipment bonding jumper present in the box.

❖ See the commentary for Section E3805.12.2.

E3805.12.3 Conduit bodies. Conduit bodies enclosing No. 6 conductors or smaller, other than short radius conduit bodies, shall have a cross-sectional area not less than twice the cross-sectional area of the largest conduit or tubing to which it is attached. The maximum number of conductors permitted shall be the maximum number permitted by Table E3804.6 for the conduit to which it is attached.

❖ A listed conduit body will have the minimum required cross-sectional area. This is a manufacturing requirement. Following the minimum fill requirements of Table E3804.6 for the type and size of conduit or tubing used will fulfill the requirements for maximum conductor fill of conduit bodies.

E3805.12.3.1 Splices, taps or devices. Only those conduit bodies that are durably and legibly marked by the manufacturer with their cubic inch capacity shall be permitted to contain splices, taps or devices. The maximum number of conductors shall be computed using the same procedure for similar conductors in other than standard boxes.

❖ It is difficult to calculate the cubic inch capacity of a conduit body. If the body is not marked with its capacity, it cannot contain any splices. Conduit bodies used in residential wiring are usually installed to provide access to pull conductors and make a splice.

SECTION E3806 INSTALLATION OF BOXES, CONDUIT BODIES AND FITTINGS

E3806.1 Conductors entering boxes, conduit bodies or fittings. Conductors entering boxes, conduit bodies or fittings shall be protected from abrasion.

❖ See the commentary for Section E3806.1.1.

E3806.1.1 Insulated fittings. Where raceways containing ungrounded conductors No. 4 or larger enter a cabinet, box enclosure, or raceway, the conductors shall be protected by a substantial fitting providing a smoothly rounded insulating surface, unless the conductors are separated from the fitting or raceway by substantial insulating material securely fastened in place.

Exception: Where threaded hubs or bosses that are an integral part of a cabinet, box enclosure, or raceway provide a smoothly rounded or flared entry for conductors.

Conduit bushings constructed wholly of insulating material shall not be used to secure a fitting or raceway. The insulating fitting or insulating material shall have a temperature rating not less than the insulation temperature rating of the installed conductors.

❖ Conduit entering a panelboard cabinet or box is usually attached or connected with a fitting that enters the box. A conduit could be threaded and have one locknut on the outside of the box and one on the inside to secure it in place. Whether this method is used or a fitting is used to attach the raceway to the box or cabinet, it is a good practice to install an insulating bushing on all raceway entries. The code requires this for conductors of size 4 AWG or larger because there is more of chance of damaging the insulation because these conductors are more difficult to bend than the smaller conductors.

E3806.2 Openings. Openings through which conductors enter shall be adequately closed.

❖ The fitting used to secure the cable into the box should close the knockout opening sufficiently. If a small conductor is installed in a large knockout with a connector or fitting too large for the conductor, it will leave an opening through which foreign items could enter the cabinet or box.

E3806.3 Metal boxes, conduit bodies and fittings. Where raceway or cable is installed with metal boxes, or conduit bodies, the raceway or cable shall be secured to such boxes and conduit bodies.

❖ The exception in Section E3805.3.2 pertains to nonmetallic boxes only. For metal boxes, there are no exceptions. The cable must be secured to the box. It is obvious that conduit and tubing installed with metal boxes must be secured to the box.

E3806.4 Unused openings. Unused cable or raceway openings in boxes and conduit bodies shall be effectively closed to afford protection substantially equivalent to that of the wall of the box or conduit body. Metal plugs or plates used with nonmetallic boxes or conduit bodies shall be recessed at least 0.25 inch (6.4 mm) from the outer surface of the box or conduit body.

❖ Where a knockout on a box or cabinet is opened by mistake or not used, it is necessary to install a plug designed for the purpose of closing the opening. Taping over the opening is not an allowable method of closing unused openings.

E3806.5 In wall or ceiling. In walls or ceilings of concrete, tile or other noncombustible material, boxes shall be installed so that the front edge of the box will not be set back from the finished surface more than 0.25 inch (6.4 mm). In walls and ceilings constructed of wood or other combustible material, outlet boxes shall be flush with the finished surface or project therefrom.

❖ Boxes installed in a noncombustible surface are permitted to be recessed a maximum of $^1/_4$ inch (6.4 mm) from the surface. In a wall or ceiling of combustible material, the front edge of the box must be flush with the surface. A problem could occur where the electrician installs the box for $^1/_2$-inch (12.7 mm) drywall and then later it is decided to install wood paneling or trim on top of the drywall. The box then is recessed and must have an extension installed. It may seem sufficient to use longer 6-32 screws than are provided with the devices, but whenever longer screws are needed, the box is probably recessed too far from the surface.

E3806.6 Plaster, gypsum board and plasterboard. Openings in plaster, gypsum board or plasterboard surfaces that accommodate boxes shall be made so that there are no gaps or open spaces greater than 0.125 inch (3.2 mm) around the edge of the box or fitting.

❖ The maximum gap in the wall opening from the edge of the box is $^1/_8$ inch (3.2 mm). Usually the drywall installer will repair all gaps with drywall joint compound.

E3806.7 Exposed surface extensions. Surface extensions from an outlet box of a concealed wiring system shall be made by mounting and mechanically securing a box or extension ring over the concealed box.

Exception: A surface extension shall be permitted to be made from the cover of a concealed box where the cover is designed so it is unlikely to fall off, or be removed if its securing means becomes loose. The wiring method shall be flexible and arranged so that any required grounding continuity is independent of the connection between the box and cover.

❖ A plaster ring or box extension should be secured to the box and not just to the wall or ceiling surface. It is acceptable to attach a flexible conduit to the cover of the box rather than use a box extension. In this case, the grounding conductor must be independent of the conduit.

E3806.8 Supports. Boxes and enclosures shall be supported in accordance with one or more of the provisions in Sections E3806.8.1 through E3806.8.6.

❖ The six subsections of Section E3806.8 cover the code provisions for the support of boxes and enclosures. There is no specific provision for the number of nails or screws to be used to attach boxes and enclosures, and there are no exact provisions pertaining to the weight of the box. How securely the box is supported is a judgment call by the installer and the inspector. Many variables enter into consideration here including the type of surface or structure the box is mounted on. In some cases, special support must be constructed in addition to the building structure or components. This code section is similar to Section E3304.7 in the General Requirements chapter; see the commentary for that section.

E3806.8.1 Surface mounting. An enclosure mounted on a building or other surface shall be rigidly and securely fastened in place. If the surface does not provide rigid and secure support, additional support in accordance with other provisions of Section E3806.8 shall be provided.

❖ See Section E3806.8 and Section E3304.7.

E3806.8.2 Structural mounting. An enclosure supported from a structural member of a building or from grade shall be rigidly supported either directly, or by using a metal, polymeric or wood brace.

❖ See Section E3806.8 and Section E3304.7.

E3806.8.2.1 Nails. Nails, where used as a fastening means, shall be attached by using brackets on the outside of the enclosure, or they shall pass through the interior within 0.25 inch (6.4 mm) of the back or ends of the enclosure.

❖ See Section E3806.8 and Section E3304.7.

E3806.8.2.2 Braces. Metal braces shall be protected against corrosion and formed from metal that is not less than 0.020 inch (.508 mm) thick uncoated. Wood braces shall have a cross section not less than nominal 1 inch by 2 inches (25 mm by 51 mm). Wood braces in wet locations shall be treated for the conditions. Polymeric braces shall be identified as being suitable for the use.

❖ See Section E3806.8 and Section E3304.7.

E3806.8.3 Mounting in finished surfaces. An enclosure mounted in a finished surface shall be rigidly secured thereto by clamps, anchors, or fittings identified for the application.

❖ Where the house is finished and the homeowner wants to add an outlet, the provisions of this section apply. A box, for a switch or receptacle for example, is permitted to be supported solely by the wallboard if clamps, anchors, and/or fittings identified for the purpose are used. Boxes are available specifically designed and approved for this kind of installation. In some areas, these boxes are called an "old work box" from the idea that outlets are being added in an existing house or in an old house. Another term is "cut-in box" from the idea that the wall is being cut into for the installation of the box. A hole is cut in the wall exactly to the shape of the box. The box has a bracket that seats against the face or surface of the drywall and a bracket that tightens against the inside of the drywall with a screw or screws. Another approved anchor is called a "hold it" or "F" strap, which is a thin piece of metal that slips into the wall along each side of the box and is bent into the box. For these types of installations, it is usually necessary that the box be located at least an inch away from a stud. See also Section E3806.8 and Section E3304.7.

E3806.8.4 Raceway supported enclosures without devices or fixtures. An enclosure that does not contain a device(s) or support a luminaire or other equipment, and that is supported by entering raceways shall not exceed 100 cubic inches (1640 cm^3) in size. The enclosure shall have threaded entries or have hubs identified for the purpose. The enclosure shall be supported by two or more conduits threaded wrenchtight into the enclosure or hubs. Each conduit shall be secured within 3 feet (914 mm) of the enclosure, or within 18 inches (457 mm) of the enclosure if all entries are on the same side of the enclosure.

Exception: Rigid metal, intermediate metal, or rigid nonmetallic conduit or electrical metallic tubing shall be permitted to support a conduit body of any size, provided that the conduit body is not larger in trade size than the largest trade size of the supporting conduit or electrical metallic tubing.

❖ Where boxes are used outdoors in landscaping, the provisions of this section apply to the support of the boxes, where outdoor receptacles or decorative lighting are installed around patios or in flower beds, for example. This section pertains to junction boxes only or boxes that do not contain any devices or fixtures. The box must be supported by at least two conduits that are threaded into the box or enclosure. If the two conduits are run into the same side of the box, such as in the installation of boxes for landscape wiring where the two conduits emerge from the ground and are run into the bottom of the box for support, the box must be no more than 18 inches above the ground or from the secure support. These types of threaded conduits are rigid metal conduit and intermediate metal conduit. Electrical metallic tubing (EMT), sometimes referred to as "thin wall," is not considered to be conduit and is not permitted to be used to support a box or enclosure in this manner.

It would be a code violation to install the box on only one conduit that emerges from the ground, even if it were a rigid metal conduit and secured by cement. The code requires two conduits. In some cases, at the end of a circuit, a second conduit is installed just for support of the box and is stubbed into the ground for a short distance but not connected to any other box and has no wire. Nonmetallic conduit is not permitted to support boxes in the situation described above.

The exception allows a conduit body to be supported by rigid metal conduit, intermediate metal conduit, EMT, or rigid nonmetallic conduit. Conduit bodies are usually not supported independently from the conduit but are intended to be part of the conduit run and are similar to a fitting in this manner.

E3806.8.5 Raceway supported enclosures, with devices or luminaire. An enclosure that contains a device(s) or supports a luminaire or other equipment and is supported by entering raceways shall not exceed 100 cubic inches (1640 cm^3) in size. The enclosure shall have threaded entries or have hubs identified for the purpose. The enclosure shall be supported by two or more conduits threaded wrench-tight into the enclosure or hubs. Each conduit shall be secured within 18 inches (457 mm) of the enclosure.

Exceptions:

1. Rigid metal or intermediate metal conduit shall be permitted to support a conduit body of any size, provided that the conduit bodies are not larger in trade size than the largest trade size of the supporting conduit.
2. An unbroken length(s) of rigid or intermediate metal conduit shall be permitted to support a box used for luminaire or lampholder support, or to support a wiring enclosure that is an integral part of a luminaire and used in lieu of a box in accordance with Section E3805.1.1, where all of the following conditions are met:
   - 2.1. The conduit is securely fastened at a point so that the length of conduit beyond the last point of conduit support does not exceed 3 feet (914 mm).
   - 2.2. The unbroken conduit length before the last point of conduit support is 12 inches (305 mm) or greater, and that portion of the conduit is securely fastened at some point not less than 12 inches (305 mm) from its last point of support.
   - 2.3. Where accessible to unqualified persons, the luminaire or lampholder, measured to its lowest point, is not less than 8 feet (2438 mm) above grade or standing area and at least 3 feet (914 mm) measured horizontally to the 8-foot (2438 mm) elevation from windows, doors, porches, fire escapes, or similar locations.
   - 2.4. A luminaire supported by a single conduit does not exceed 12 inches (305 mm) in any direction from the point of conduit entry.
   - 2.5. The weight supported by any single conduit does not exceed 20 pounds (9.1 kg).
   - 2.6. At the luminaire or lampholder end, the conduit(s) is threaded wrenchtight into the box, conduit body, or integral wiring enclosure, or into hubs identified for the purpose. Where a box or conduit body is used for support, the luminaire shall be secured directly to the box or conduit body, or through a threaded conduit nipple not over 3 inches (76 mm) long.

❖ The main difference in this section and Section E3806.8.4 is that if the box contains a switch or receptacle or supports a light fixture, then the conduit must be secured within 18 inches (457 mm) of the enclosure, and the provision for being secured within 36 inches (914 mm) from the box is not allowed. Note that the exceptions omit EMT and rigid nonmetallic conduit.

E3806.8.6 Enclosures in concrete or masonry. An enclosure supported by embedment shall be identified as being suitably protected from corrosion and shall be securely embedded in concrete or masonry.

E3806.9 Covers and canopies. Outlet boxes shall be effectively closed with a cover, faceplate or fixture canopy.

❖ All outlets must have a cover, whether it is a blank cover, the faceplate of a switch or receptacle, or the canopy of a light fixture. A residence should not pass inspection if a junction box is left open with the conductors exposed, even if they are tucked or folded back into the box. Where a light fixture is not yet installed, the box must be covered with a blank cover.

E3806.10 Metal covers and plates. Metal covers and plates shall be grounded.

❖ Metal covers are grounded by the metal attachment screws that screw into the metal yoke of the receptacle. Receptacles have a grounding terminal that is connected to the yoke or strap on which the receptacle is mounted. Therefore, the metal plate is grounded through the mounting screws into the yoke and through the grounding terminal. However, some switches do not have a grounding terminal, and the metal yoke or strap on which the switch handle is mounted is a separate piece of metal not connected to the grounding conductor. If a metal cover plate is used, it might not be grounded. The same is true where metal screws are used to fasten a nonmetallic cover on the switch. If the yoke of the switch becomes energized by coming into contact with an ungrounded conductor, the metal screws or a metal cover plate could become energized. If a metal box is used and the yoke seats against the metal, as in the case of a surface mounted box, then there will be direct metal-to-metal contact between the yoke and the metal box. This is an approved method of grounding the device and any metal screws or metal cover plates. If there is not metal-to-metal contact, as in the case where a nonmetallic box is used, the mounting strap of a switch must have a grounding terminal to ensure that a metal face plate will be grounded.

See the code requirements on grounding terminals of wiring devices in Section E3808.14 and switch faceplate grounding in Section E3901.11.1.

E3806.11 Exposed combustible finish. Combustible wall or ceiling finish exposed between the edge of a fixture canopy or pan and the outlet box shall be covered with noncombustible material.

❖ This code provision applies, for example, where a round 4-inch (102 mm) box for a wall hung light fixture is used on a wall with a wood paneling finish. If installed properly, the front edge of the box will be flush with the finished wall surface. A light fixture is installed having a canopy 6 inches (152 mm) in diameter that seats against the wall, leaving a wall surface area 2 inches (51 mm) wide around the box exposed inside the canopy. This area becomes part of the space where the connections are made between the circuit wires and the fixture wires. If arcing occurs, the wood wall surface around the box could ignite if enough heat builds up. Thus that area of combustible wall must be covered with a noncombustible material.

SECTION E3807 CABINETS AND PANELBOARDS

E3807.1 Enclosures for switches or overcurrent devices. Enclosures for switches or overcurrent devices shall not be used as junction boxes, auxiliary gutters, or raceways for conductors feeding through or tapping off to other switches or overcurrent devices, except where adequate space for this purpose is provided. The conductors shall not fill the wiring space at any cross section to more than 40 percent of the cross-sectional area of the space, and the conductors, splices, and taps shall not fill the wiring space at any cross section to more than 75 percent of the cross-sectional area of that space.

❖ This section allows splices within panelboards and switch and circuit breaker enclosures provided there is ample room to accomplish and contain such splices without crowding the conductors. It is commonly necessary to splice branch circuit conductors within panelboards as a result of relocating circuit breakers in the panel or replacing standard circuit breakers with GFCI or AFCI-type breakers. Splices are also necessary when the panelboard is replaced as part of a service upgrade.

E3807.2 Damp or wet locations. In damp or wet locations, cabinets and panelboards of the surface type shall be placed or equipped so as to prevent moisture or water from entering and accumulating within the cabinet, and shall be mounted to provide an airspace not less than 0.25 inch (6.4 mm) between the enclosure and the wall or other supporting surface. Cabinets installed in wet locations shall be weatherproof.

❖ Cabinets and other boxes and enclosures that are mounted in wet or damp locations must have at least a $^1/_4$-inch (6.4 mm) airspace between the wall, cabinet, or box so that water or moisture is not trapped against the metal. In some areas and locations, corrosion will result from moisture behind the cabinet or box.

E3807.3 Position in wall. In walls of concrete, tile or other noncombustible material, cabinets and panelboards shall be installed so that the front edge of the cabinet will not set back of the finished surface more than 0.25 inch (6.4 mm). In walls constructed of wood or other combustible material, cabinets shall be flush with the finished surface or shall project therefrom.

❖ See Section E3806.5.

E3807.4 Unused openings. Unused openings in cabinets and panelboards shall be effectively closed to afford protection equivalent to that of the wall of the cabinet. Metal plugs or plates used with nonmetallic cabinets shall be recessed at least 0.25 inch (6.4 mm) from the outer surface.

❖ See Section E3806.4.

E3807.5 Conductors entering cabinets. Conductors entering cabinets and panelboards shall be protected from abrasion and shall comply with Section E3806.1.1.

❖ See Section 3806.1.1.

E3807.6 Openings to be closed. Openings through which conductors enter cabinets, panelboards and meter sockets shall be adequately closed.

❖ See Section E3806.2.

E3807.7 Cables. Where cables are used, each cable shall be secured to the cabinet, panelboard, cutout box, or meter socket enclosure.

Exception: Cables with entirely nonmetallic sheaths shall be permitted to enter the top of a surface-mounted enclosure through one or more sections of rigid raceway not less than 18 inches (457 mm) nor more than 10 feet (3048 mm) in length, provided all the following conditions are met:

1. Each cable is fastened within 12 inches (305 mm), measured along the sheath, of the outer end of the raceway.
2. The raceway extends directly above the enclosure and does not penetrate a structural ceiling.
3. A fitting is provided on each end of the raceway to protect the cable(s) from abrasion and the fittings remain accessible after installation.
4. The raceway is sealed or plugged at the outer end using approved means so as to prevent access to the enclosure through the raceway.
5. The cable sheath is continuous through the raceway and extends into the enclosure beyond the fitting not less than $^{1}/_{4}$ inch (6.4 mm).
6. The raceway is fastened at its outer end and at other points in accordance with Section E3702.1, the applicable article.

❖ Nonmetallic cable must be secured with cable connectors to the cabinet. Where a cable enters a cabinet through a knockout, it would be a code violation to run the cable directly through the hole without any type of connector. Where the connector is omitted, the sharp edge of the knockout could damage the sheathing on the cable. Installers have been known to wrap the cable with several layers of tape at the point where it enters the cabinet through the knockout. This method is not an acceptable alternative to cable connectors.

The exception recognizes what has been common practice in some localities, and that is running type NM cables through a conduit riser and into the top of a panelboard cabinet. This practice can overcome a shortage of $^{1}/_{2}$-inch (12.7 mm) knockouts on the top of a cabinet and provides physical protection for the cables. It also saves labor in some cases and makes a neat appearance. Section E3605.3 applies where the raceway is longer than 24 inches (610 mm).

SECTION E3808 GROUNDING

E3808.1 Metal enclosures. Metal enclosures of conductors, devices and equipment shall be grounded.

Exceptions:

1. Short sections of metal enclosures or raceways used to provide cable assemblies with support or protection against physical damage.
2. A metal elbow that is installed in an underground installation of rigid nonmetallic conduit and is isolated from possible contact by a minimum cover of 18 inches (457 mm) to any part of the elbow.

❖ Two general subjects of grounding are system grounding and equipment grounding. This section covers provisions for equipment grounding. In general, all metal enclosures must be grounded.

E3808.2 Equipment fastened in place or connected by permanent wiring methods (fixed). Exposed noncurrent-carrying metal parts of fixed equipment likely to become energized shall be grounded where any of the following conditions apply:

1. Where within 8 feet (2438 mm) vertically or 5 feet (1524 mm) horizontally of earth or grounded metal objects and subject to contact by persons,
2. Where located in a wet or damp location and not isolated, or
3. Where in electrical contact with metal.

❖ Examples of equipment fastened in place include a furnace or a water heater. Permanent wiring methods include cable or conduit where a dishwasher, for example, is permanently connected by cable instead of by cord and plug. Some motors are cord and plug connected, and others, like an evaporative cooler mounted on a roof top, are connected by permanent wiring methods. The equipment may be supplied by a circuit run in liquid-tight flexible metal conduit. There is almost no equipment in a residential wiring system that is exempt from grounding requirements. Section E3808.9 lists the methods of grounding this type of equipment.

E3808.3 Specific equipment fastened in place or connected by permanent wiring methods. Exposed noncurrent-carrying metal parts of the following equipment and enclosures shall be grounded:

1. Luminaires as provided in Chapter 39.
2. Motor-operated water pumps, including submersible types. Where a submersible pump is used in a metal well casing, the well casing shall be bonded to the pump circuit equipment grounding conductor.

❖ See Section 3903.3.

Although a submersible water pump may not be attached to the metal well casing, the code requires that the casing be bonded to the pump circuit equipment grounding conductor.

E3808.4 Performance of fault current path. The fault current path shall be permanent and electrically continuous, shall be capable of safely conducting the maximum fault current likely to be imposed on it, and shall have sufficiently low impedance to facilitate the operation of overcurrent devices under fault conditions.

❖ When a ground-fault occurs at some point in the electrical wiring system, the current will return to the source of the power supply if there is an electrically continuous path. Where an equipment grounding conductor is used, it must be sized according to Table E3808.12.

E3808.5 Earth as a grounding conductor. The earth shall not be used as the sole equipment grounding conductor or fault current path.

❖ Equipment must be grounded by connection to a grounding conductor and not only to the earth. For example, driving a ground rod at the equipment location and grounding only to the ground rod would not create the electrically continuous and permanent path that must be established to allow enough current to flow during a fault condition to trip the circuit breaker or set of fuses. Some current flows through the earth, but not enough to trip the circuit breaker.

E3808.6 Load-side neutral. A grounding connection shall not be made to any grounded circuit conductor on the load side of the service disconnecting means.

Exception: A grounding conductor connection shall be made at each separate building where required by Section E3507.3.

❖ The grounding and grounded conductors are connected together only at the location of the main disconnect. At any point downstream, meaning on the load side of the main disconnect, the grounding conductor and the grounded conductor must be electrically isolated except at the service or feeder to a separate building as provided for in Section E3507.3. A shock hazard and fire hazard is created where grounding conductors, raceways and enclosures are connected to a grounded conductor at any point other than in or ahead of the service equipment. See Commentary Figure E3507.2.

E3808.7 Load-side equipment. A grounded circuit conductor shall not be used for grounding noncurrent-carrying metal parts of equipment on the load side of the service disconnecting means.

Exception: For separate buildings, in accordance with Section E3507.3.

❖ See Commentary Figure E3808.7 and the commentary for Sections E3507.2 and E3808.6. Commentary Figure E3808.7 illustrates what can happen when a grounded circuit conductor is used to ground an appliance or when the grounded circuit conductor and the grounding conductor are not electrically isolated from each other. Because the grounded circuit conductor is carrying current, the entire appliance housing will be energized, creating a potential shock hazard. Also, any metallic raceway will be energized, causing it to become a current carrier in parallel with the grounded conductor and a potential shock hazard as well. A fire hazard is also introduced because of possible arcing at any weak links in the continuity of the raceway. If the grounded conductor were lost for some reason, the raceway would carry the entire load, and the appliance would continue to operate even though it would be hazardous. Voltage drop along the circuit conductors and raceway can cause voltage potentials to exist between the appliance housing or raceway and any other grounded surface.

E3808.8 Types of equipment grounding conductors. The equipment grounding conductor run with or enclosing the circuit conductors shall be one or more or a combination of the following:

1. A copper or other corrosion-resistant conductor. This conductor shall be solid or stranded; insulated, covered or bare; in the form of a wire or a busbar of any shape.
2. Rigid metal conduit.
3. Intermediate metal conduit.
4. Electrical metallic tubing.
5. Flexible metal conduit, where both the conduit and fittings are listed for grounding.
6. Armor of Type AC cable.
7. Surface metal raceway.
8. Metal-clad cable, where both the cable and fittings are listed for grounding.
9. Liquid-tight flexible metal conduit terminated with fittings that are listed for grounding.

❖ Where conduit is used as the grounding conductor, it is very important that the connections be made tight so that there is an electrically continuous path. The conduit system can be considered a grounding conductor, but it is really a conductor that is "manufactured" or assembled on the job. A grounding conductor can be pulled into the conduit system where there is a lack of confidence in the integrity of all the connectors and couplings of the assembled system. If the electrician forgets to tighten one set screw or locknut on a fitting, for example, a high resistance could be created in the grounding return path.

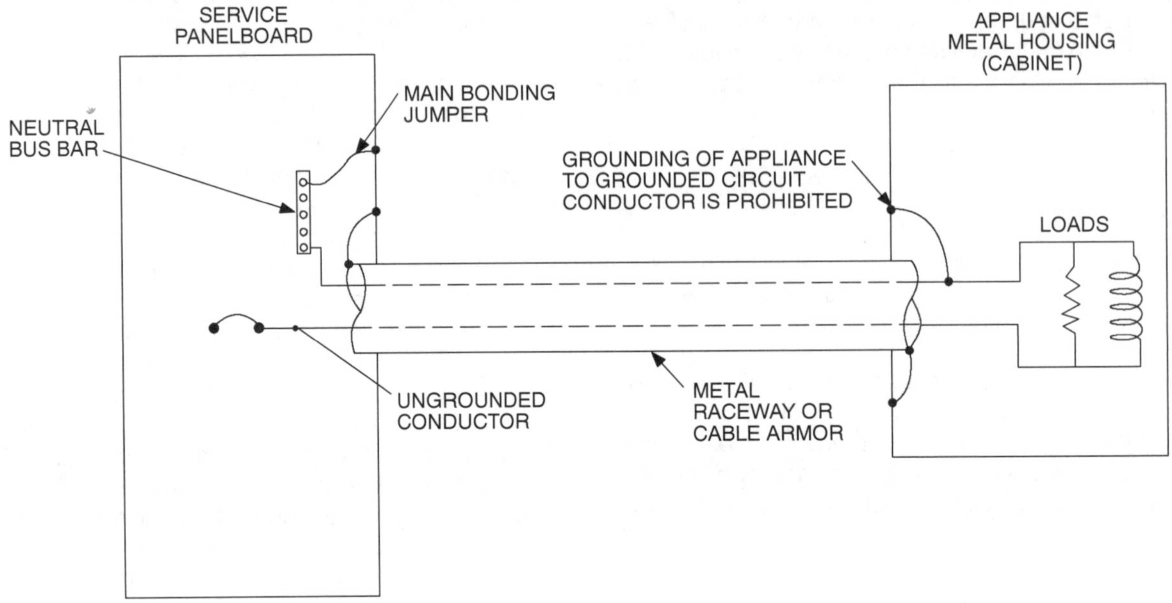

Figure E3808.7
PROHIBITED USE OF GROUNDED CIRCUIT CONDUCTOR

E3808.8.1 Flexible metal conduit. Flexible metal conduit that is not listed for grounding shall not be used as a grounding conductor except where all of the following conditions are met:

1. The conduit is terminated in fittings listed for grounding.
2. The circuit conductors contained in the conduit are protected by overcurrent devices rated at 20 amperes or less.
3. The combined length of flexible metal conduit and flexible metallic tubing and liquid-tight flexible metal conduit in the same ground return path does not exceed 6 feet (1829 mm).
4. The conduit is not installed to provide flexibility.

❖ Flexible metal conduit is suitable as a grounding conductor to a certain extent, but the conditions are very limiting. It can be used only for circuits of 20 amperes or less, only in lengths of 6 feet or less, and only with fittings listed for grounding. All flexible metal conduit runs over 6 feet in length require that a grounding conductor be installed within the conduit. It is usually easier to install a grounding conductor in a run of 6 feet (152 mm) or less than find the listed grounding fittings and comply with all the conditions. Where flexible metal conduit is used to provide flexibility for the equipment being served, such as a motor that may need to be replaced or equipment that may vibrate, the flexible metal conduit cannot be relied on for grounding, and a grounding conductor must be installed within the conduit.

Because flexible conduit is formed as a spiral, fault current would have to travel along the spiral path and not straight along the length of the conduit. The fault current return path is much longer following the long the spiral path, and the impedance could be too high to allow the current flow necessary to trip the overcurrent device.

E3808.8.2 Liquid-tight flexible metal conduit. Liquid-tight flexible metal conduit shall not be used as a grounding conductor except where all of the following conditions are met:

1. The conduit is terminated in fittings listed for grounding.
2. For trade sizes $^3/_8$ inch through $^1/_2$ inch (9.5 mm through 12.7 mm), the circuit conductors contained in the conduit are protected by overcurrent devices rated at 20 amperes or less.
3. For trade sizes $^3/_4$ inch through $1^1/_4$ inches (19.1 mm through 32 mm), the circuit conductors contained in the conduit are protected by overcurrent devices rated at not more than 60 amperes and there is no flexible metal conduit, flexible metallic tubing, or liquid-tight flexible metal conduit in trade sizes $^3/_8$ inch or $^1/_2$ inch (9.5 mm through 12.7 mm) in the grounding path.
4. The combined length of flexible metal conduit and flexible metallic tubing and liquidtight flexible metal conduit in the same ground return path does not exceed 6 feet (1829 mm).
5. The conduit is not installed for flexibility.

❖ The same conditions apply to liquid-tight flexible conduit as to flexible metal conduit mentioned above, with the exception of size. For a circuit up to 60 amperes, the code requires the larger conduit sizes indicated for the liquid-tight flexible metal conduit that is to be used as a grounding conductor. Again, the conditions are so limiting that for residential wiring most electricians choose to run an appropriate size grounding conductor within the conduit. If the conduit is used as a grounding conductor, the conditions must be strictly adhered to.

E3808.8.3 Nonmetallic sheathed cable (Type NM). In addition to the insulated conductors, the cable shall be permitted to have an insulated or bare conductor for equipment grounding purposes only. Equipment grounding conductors shall be sized in accordance with Table E3808.12.

❖ The equipment grounding conductor in a nonmetallic sheathed cable is uninsulated and must be used only as a grounding conductor and not as a grounded conductor.

E3808.9 Equipment fastened in place or connected by permanent wiring methods. Noncurrent-carrying metal parts of equipment, raceways and other enclosures, where required to be grounded, shall be grounded by one of the following methods:

1. By any of the equipment grounding conductors permitted by Sections E3808.8 through E3808.8.3.
2. By an equipment grounding conductor contained within the same raceway, cable or cord, or otherwise run with the circuit conductors. Individually, covered or insulated equipment grounding conductors shall have a continuous outer finish that is either green or green with one or more yellow stripes.

❖ This section simply requires that equipment be grounded by one of the metallic wiring methods in Section E3808.8 or by a separate conductor run in the conduit, tubing, or cable. If the grounding conductor is insulated, it must be green. See Section E3307 for conductor color requirements.

E3808.10 Methods of equipment grounding. Fixtures and equipment shall be considered grounded where mechanically connected to an equipment grounding conductor as specified in Sections E3808.8 through E3808.8.3 and sized in accordance with Section E3808.12.

❖ See Section E3808.8.

E3808.11 Equipment grounding conductor installation. Where an equipment grounding conductor consists of a raceway, cable armor or cable sheath or where such conductor is a wire within a raceway or cable, it shall be installed in accordance with the provisions of this chapter and Chapters 33 and 37 using fittings for joints and terminations approved for installation with the type of raceway or cable used. All connections, joints and fittings shall be made tight using suitable tools.

❖ See Section E3808.8.

E3808.12 Equipment grounding conductor size. Copper, aluminum and copper-clad aluminum equipment grounding conductors shall be not smaller than shown in Table E3808.12, but shall not be required to be larger than the circuit conductors supplying the equipment. Where a raceway or a cable armor or sheath is used as the equipment grounding conductor, as provided in Section E3808.8, it shall comply with Section E3808.4.

❖ The rule that the equipment grounding conductor does not have to be larger than the circuit conductors may apply where the rating of the circuit breaker has been increased to allow for motor starting. In such a case, the minimum size equipment grounding conductor listed in the table may be larger than the circuit conductors. This is not usually a concern in residential wiring but could be in some rare situations.

Where the equipment grounding conductor is an actual conductor and not a conduit system, using the table to size this conductor is simple because it is prescriptive. But where a raceway like a conduit or the sheath of an armored cable is used as the grounding conductor as allowed in Section E3808.8, the code states here that it must comply with Section E3808.4, which is actually a performance requirement. How is it determined that the conduit or cable sheath is of a sufficient size or has a low enough impedance as an equipment grounding conductor? It is usually assumed that the conduit has enough metal to serve as a conductor and will do so as long as the conduit system is assembled properly. This is why the use of flexible conduit as a grounding conductor is limited by strict conditions.

As an example of sizing the equipment conductor, a 240 volt, 70-ampere circuit to an air-conditioner run with two size 4 AWG copper ungrounded conductors and no grounded conductor would be required to have a minimum size 8 AWG copper equipment grounding conductor.

TABLE E3808.12
EQUIPMENT GROUNDING CONDUCTOR SIZING

RATING OR SETTING OF AUTOMATIC OVERCURRENT DEVICE IN CIRCUIT AHEAD OF EQUIPMENT, CONDUIT, ETC., NOT EXCEEDING THE FOLLOWING RATINGS (amperes)	MINIMUM SIZE	
	Copper wire No. (AWG)	Aluminum or copper-clad aluminum wire No. (AWG)
15	14	12
20	12	10
30	10	8
40	10	8
60	10	8
100	8	6
200	6	4
300	4	2
400	3	1

❖ See the commentary for Section E3808.12.

E3808.12.1 Multiple circuits. Where a single equipment grounding conductor is run with multiple circuits in the same raceway or cable, it shall be sized for the largest overcurrent device protecting conductors in the raceway or cable.

❖ It is not necessary to run multiple equipment grounding conductors in the same conduit or cable that supplies multiple equipment. But the equipment grounding conductor must be sized for the largest rated circuit in the conduit or cable. For example, where a 20-ampere circuit and a 30-ampere circuit are run in the same conduit, the equipment grounding conductor must be at least size 10 AWG copper. Another example would be where a three conductor cable is used. A cable marked 12-3 w/ground may have a red, black, white, and a bare conductor. The red and black conductors could supply two separate circuits to two different appliances or equipment. The one size 12 AWG bare conductor serves as the equipment grounding conductor for both circuits.

E3808.13 Continuity and attachment of equipment grounding conductors to boxes. Where more than one equipment grounding conductor enters a box, and is spliced, joined, or terminated to equipment within or supported by a box, such conductors shall be spliced or joined within the box or to the box with devices suitable for the use. Splices shall be made in accordance with Section E3306.10 except that insulation shall not be required. The arrangement of grounding connections shall be such that the disconnection or removal of a receptacle, fixture or other device fed from the box will not interfere with or interrupt the grounding continuity.

❖ See Section E3808.15 and Section E3808.16 for equipment grounding conductors in metal boxes and in nonmetallic boxes.

Where several nonmetallic sheathed cables are installed in a box, the equipment grounding conductors are required to be joined together with a listed device. These conductors must not be twisted together with pliers as the only means of splicing. A listed connector such as a wire nut must be used. A jumper or "pigtail" must be used to connect to the grounding terminal of the device so that when the device is removed it will not interrupt the continuity of the grounding return path. Special pressure connectors (e.g. wire nuts) are available that provide a lead wire (jumper) for connection to devices.

E3808.14 Connecting receptacle grounding terminal to box. An equipment bonding jumper shall be used to connect the grounding terminal of a grounding-type receptacle to a grounded box except where grounded in accordance with one of the following:

1. Surface mounted box. Where the box is mounted on or at the surface, direct metal-to-metal contact between the device yoke and the box shall be permitted to ground the receptacle to the box. This provision shall not apply to cover-mounted receptacles unless the box and cover combination are listed as providing satisfactory ground continuity between the box and the receptacle.
2. Contact devices or yokes. Contact devices or yokes designed and listed for the purpose shall be permitted in conjunction with the supporting screws to establish the grounding circuit between the device yoke and flush-type boxes.
3. Floor boxes. The receptacle is installed in a floor box designed for and listed as providing satisfactory ground continuity between the box and the device.

❖ Where a metal box is used, a receptacle must be grounded by a jumper from the device grounding terminal to the box unless there is metal-to-metal contact between the yoke and the box. A contact device listed for the purpose of equipment grounding could be used. The contact device is a spring type of clip that binds against the 6-32 supporting screw of the device to provide a positive connection between the receptacle yoke and the metal box.

E3808.15 Metal boxes. A connection shall be made between the one or more equipment grounding conductors and a metal box by means of a grounding screw that shall be used for no other purpose, or by means of a listed grounding device. Sheet-metal screws shall not be used to connect grounding conductors to boxes.

❖ Where a metal box is used, the equipment grounding conductor must be bonded or connected to the metal box by means of a listed grounding screw or grounding device. It may be handy to wrap the equipment grounding conductor around the screw used for attaching the box to the structure, but this would be a code violation.

E3808.16 Nonmetallic boxes. One or more equipment grounding conductors brought into a nonmetallic outlet box shall be arranged to allow connection to fittings or devices installed in that box.

❖ See Section E3808.13.

E3808.17 Clean surfaces. Nonconductive coatings such as paint, lacquer and enamel on equipment to be grounded shall be removed from threads and other contact surfaces to ensure electrical continuity or the equipment shall be connected by means of fittings designed so as to make such removal unnecessary.

❖ See Section E3808.18.

E3808.18 Bonding other enclosures. Metal raceways, cable armor, cable sheath, enclosures, frames, fittings and other metal noncurrent-carrying parts that serve as grounding conductors, with or without the use of supplementary equipment grounding conductors, shall be effectively bonded where necessary to ensure electrical continuity and the capacity to conduct safely any fault current likely to be imposed on them. Any nonconductive paint, enamel and similar coating shall be removed at threads, contact points and contact surfaces, or connections shall be made by means of fittings designed so as to make such removal unnecessary.

❖ Paint is an insulator, and a positive connection cannot be made without removing the paint at the point where the equipment grounding conductor is attached. See also Section E3808.9 and Section E3806.10.

E3808.19 Size of equipment bonding jumper on load side of service. The equipment bonding jumper on the load side of the service overcurrent devices shall be sized, as a minimum, in accordance with Table E3808.12, but shall not be required to be larger than the circuit conductors supplying the equipment. An equipment grounding conductor shall be not smaller than No. 14 AWG.

A single common continuous equipment bonding jumper shall be permitted to bond two or more raceways or cables where the bonding jumper is sized in accordance with Table 3808.12 for the largest overcurrent device supplying circuits therein.

❖ Bonding jumpers can be used where two or more raceways enter a junction box to ensure the continuity of the conduit as an equipment grounding conductor. A common bonding jumper can be used as mentioned in the comment for Section E3808.12.1.

E3808.20 Installation—equipment bonding jumper. The equipment bonding jumper shall be permitted to be installed inside or outside of a raceway or enclosure. Where installed on the outside, the length of the equipment bonding jumper shall not exceed 6 feet (1829 mm) and shall be routed with the raceway or enclosure. Where installed inside of a raceway, the equipment bonding jumper shall comply with the requirements of Sections E3808.9, Item 2; E3808.13; E3808.15; and E3808.16.

❖ An equipment bonding jumper is limited to 6 feet if installed on the outside of the raceway, but the length is not limited if installed on the inside and run with the circuit conductors.

SECTION E3809 FLEXIBLE CORDS

E3809.1 Where permitted. Flexible cords shall be used only for the connection of appliances where the fastening means and mechanical connections of such appliances are designed to permit ready removal for maintenance, repair or frequent interchange and the appliance is listed for flexible cord connection. Flexible cords shall not be installed as a substitute for the fixed wiring of a structure; shall not be run through holes in walls, ceilings or floors; shall not be concealed behind building walls, ceilings or floors; and shall not be installed in raceways.

❖ A flexible cord could not be used for the power supply to a furnace or water heater, but could be used for a range. Where a flexible cord is used for an appliance, a receptacle must be placed within reach of the cord in the same area as the appliance so the cord does not have to be run through a wall, ceiling or floor.

E3809.2 Loading and protection. The ampere load and overcurrent protection of flexible cords serving fixed appliances shall be in accordance with Table E3809.2. This table shall be used in conjunction with applicable end use product standards to ensure selection of the proper size and type.

❖ Where a flexible cord is field installed, it is important to follow the ampere load ratings in Table E3809.2. Many cord- and plug-connected appliances and equipment come with a cord already installed, which is sized to carry the load.

TABLE E3809.2
MAXIMUM AMPERE LOAD AND OVERCURRENT-PROTECTION-DEVICE RATING FOR FLEXIBLE CORDS

CORD SIZE (AWG)	CORD TYPES S, SE, SEO, SJ, SJE, SJEO, SJO, SJOO, SJT, SJTO, SJTOO, SO, SOO, SRD, SRDE, SRDT, ST, STD, SV, SVO, SVOO, SVTO, SVTOO — Maximum ampere load: Three current-carrying conductors	Two current-carrying conductors	MAXIMUM AMPERE RATING OF BRANCH-CIRCUIT OVERCURRENT-PROTECTIVE DEVICE[a]
18	7	10	20
16	10	13	30
14	15	18	30
12	20	25	40

a. This column applies only where the cord is used in accordance with the appliance listing.

❖ See the commentary for Section E3809.2.

E3809.3 Splices. Flexible cord shall be used only in continuous lengths without splices or taps.

❖ No provision is made for splicing or making tap connections to a flexible cord. Flexible cord is not be used as a substitute for the fixed wiring of the dwelling in any way.

E3809.4 Attachment plugs. Where used in accordance with Section E3809.1, each flexible cord shall be equipped with an attachment plug and shall be energized from a receptacle outlet.

❖ Attachment plugs are used to receive power, not to deliver power to a circuit.

Bibliography

The following resource materials are referenced in this chapter or are relevant to the subject matter addressed in this chapter.

NFPA 70-02, *National Electrical Code.* Quincy, MA: National Fire Protection Association, 2002.

Chapter 39: Devices and Luminaires

General Comments

This chapter addresses the "trim-out" (final) stage of construction in which devices and fixtures are installed and connected to the installed wiring system.

Chapter 39 covers receptacle ratings and installation, lighting fixture installation, construction and location, and grounding of devices and fixtures.

Purpose

The purpose of Chapter 39 is to prevent fire and electrical shock hazards associated with the exposed devices and fixtures to which occupants and dwelling contents are exposed.

SECTION E3901 SWITCHES

E3901.1 Rating and application of snap switches. General-use snap switches shall be used within their ratings and shall control only the following loads:

1. Resistive and inductive loads, including electric-discharge lamps, not exceeding the ampere rating of the switch at the voltage involved.
2. Tungsten-filament lamp loads not exceeding the ampere rating of the switch at 120 volts.
3. Motor loads not exceeding 80 percent of the ampere rating of the switch at its rated voltage.

❖ "Snap switch" is the term for the common types of switches used in dwelling units for lighting control, small appliances, and motor-driven equipment. Most snap switches installed in dwellings are rated 15 amperes and 20 amperes. For resistive loads and non-motor inductive loads such as fluorescent lighting, the switch must have an ampere rating of at least 100 percent of the load. A routine practice is to install 15-ampere switches because the switch often controls only one light fixture in a room. However, where a switch controls many lights of heavy wattage, it is important to consider the total load and make sure the ampere rating is adequate. A switch controlling a motor must have a rating of 125 percent of the motor load.

E3901.2 CO/ALR snap switches. Snap switches rated 20 amperes or less directly connected to aluminum conductors shall be marked CO/ALR.

❖ Aluminum conductors are not used in modern wiring for 15- and 20-ampere branch circuits. However, this code provision is important because it also covers remodeling and rewiring work in older homes where aluminum conductors may be found. If a snap switch is replaced on aluminum wire, the switch must be marked CO/ALR or otherwise indicate "suitable for copper or aluminum conductors." If a switch is not available marked in this way, one alternative is to connect a copper pigtail to the end of each aluminum conductor to be terminated on the switch. A splicing device listed for the application must be used to connect the pigtail. In some cases, a decision is made to go through the entire house and add copper pigtails and replace all receptacles and switches where there is evidence of terminal overheating. See also Section 3306.8.

E3901.3 Indicating. General-use and motor-circuit switches and circuit breakers shall clearly indicate whether they are in the open OFF or closed ON position. Where single-throw switches or circuit breaker handles are operated vertically rather than rotationally or horizontally, the up position of the handle shall be the ON position.

❖ It is assumed that a vertically operated switch will be "on" in the up position and "off" in the down position, because this is the convention.

E3901.4 Time switches and similar devices. Time switches and similar devices shall be of the enclosed type or shall be mounted in cabinets or boxes or equipment enclosures. A barrier shall be used around energized parts to prevent operator exposure when making manual adjustments or switching.

❖ Some modern time switches are electronic, and the setting buttons and controls are not near the energized parts, but many time switches have the dial or controls near the live terminals. It is important that the insulating barrier not be removed to help prevent shock or electrocution when adjusting the settings.

E3901.5 Grounding of enclosures. Metal enclosures for switches or circuit breakers shall be grounded. Where nonmetallic enclosures are used with metal raceways or metal-armored cables, provisions shall be made for maintaining grounding continuity.

Metal boxes for switches shall be effectively grounded. Nonmetallic boxes for switches shall be installed with a wiring method that provides or includes an equipment grounding conductor.

❖ The exposed metal parts of a cabinet or switch box must be grounded by connection to the equipment grounding conductor by approved devices. Where nonmetallic boxes are used, the equipment grounding conductors, whether they be wires or raceways, must be bonded together to assure continuity of the ground fault return path.

E3901.6 Access. All switches and circuit breakers used as switches shall be located to allow operation from a readily accessible location. Such devices shall be installed so that the center of the grip of the operating handle of the switch or circuit breaker, when in its highest position, will not be more than 6 feet 7 inches (2007 mm) above the floor or working platform.

❖ This provision for access to switches includes access to circuit breakers, so when the contractor is considering the location for the panelboard and the location of the main disconnect, this rule must be followed. For example, if the main disconnect is installed on the outside of the house and the level of the grade is lowered, the height of the main breaker will be raised. Most ordinary switches are installed lower than 6 feet, 7 inches (2007 mm). However, in some cases for special equipment, this rule will come into play.

E3901.7 Wet locations. A switch or circuit breaker located in a wet location or outside of a building shall be enclosed in a weatherproof enclosure or cabinet. Switches shall not be installed within wet locations in tub or shower spaces unless installed as part of a listed tub or shower assembly.

❖ There is no exception to the rule that a common light switch wired to control a fan, lighting outlet, or other load cannot be installed in the bathtub or shower space; however, the code does not provide an exact definition or dimension of tub or shower space. It seems logical, though, that the switch should be out of the reach of someone in the tub or shower.

E3901.8 Grounded conductors. Switches or circuit breakers shall not disconnect the grounded conductor of a circuit except where the switch or circuit breaker simultaneously disconnects all conductors of the circuit.

❖ Switching the grounded conductor of a two-wire circuit will turn the load on and off, but a safety hazard is created. When someone is working on the wiring in a lighting outlet or on a light fixture, the ungrounded or hot conductor must be de-energized when the switch is off.

E3901.9 Switch connections. Three- and four-way switches shall be wired so that all switching occurs only in the ungrounded circuit conductor. Color coding of switch connection conductors shall comply with Section E3307.4. Where in metal raceways or metal-jacketed cables, wiring between switches and outlets shall be in accordance with Section 3306.7.

Exception: Switch loops do not require a grounded conductor.

❖ One standard way to wire three-way switches is to have an ungrounded and grounded conductor feed into one switch box (sometimes called the power end), then run "travelers" along with the grounded conductor to the other switch box (sometimes called the switch leg end), then run a switch leg along with the grounded conductor up to the light. In this circuit layout, it is very simple to keep the grounded conductor from being switched. But there are many different variations on designing a three-way switch circuit. and some of these over the years have involved switching the grounded conductor. Therefore this provision is included in the code to specifically prohibit this.

This section also requires, by reference to Section E3306.7, that all conductors of a circuit be in the same raceway or cable. This prohibits running the travelers without the grounded conductor and picking up the grounded conductor on the switch leg from a close-by outlet box that has a grounded conductor such as from a receptacle.

In a switch loop, the conductors supplying power to a light are brought first to the lighting outlet. Only one cable with a black, white, and bare conductor is run to the switch point. The white conductor is connected to the ungrounded conductor at the lighting outlet and supplies power down to the switch. When the switch is closed, power is carried back to the lighting fixture on the black conductor of the switch-loop cable. This hook-up of a switch loop does not require the grounded or grounded conductor to be run with the "hot" switch conductors. See Section E3307.3

E3901.10 Box mounted. Flush-type snap switches mounted in boxes that are recessed from the wall surfaces as covered in Section 3806.5 shall be installed so that the extension plaster ears are seated against the surface of the wall. Flush-type snap switches mounted in boxes that are flush with the wall surface or project therefrom shall be installed so that the mounting yoke or strap of the switch is seated against the box.

❖ Following the rules of this code provision will result in the switch being secure so that it will not move, become loose, or cause accidental shorting to the metal box. The ears of the switch are designed to seat solidly against the wall surface or the box itself. Where the wall opening at the outlet is too big because the drywall or plaster has been cut too large, the switch might "float" with only the cover-plate screws holding it in place.

E3901.11 Snap switch faceplates. Snap switches mounted in boxes shall have faceplates installed so as to completely cover the opening and seat against the finished surface.

❖ The faceplate is designed to seat against the finished wall surface when the switch is installed properly with its extension plaster ears also seated against the wall surface. If the wall surface around the outlet is not finished properly, the faceplate will not fit properly. An oversized faceplate may cover the opening that has been left too large for the box, but Section E3806.6 requires that the openings in plaster, gypsum board, or plasterboard surfaces around boxes be made so that there is no gap more than $^1/_8$ inch (3.2 mm) from the edge of the box.

E3901.11.1 Faceplate grounding. Snap switches, including dimmer switches, shall be effectively grounded and shall provide a means to ground metal faceplates, whether or not a metal faceplate is installed. Snap switches shall be considered effectively grounded where either of the following conditions are met:

1. The switch is mounted with metal screws to a metal box or to a nonmetallic box with integral means for grounding devices.
2. An equipment grounding conductor or equipment bonding jumper is connected to an equipment grounding termination of the snap switch.

❖ See the commentary for Section E3806.10 and E3808.14.

SECTION E3902
RECEPTACLES

E3902.1 Rating and type. Receptacles installed for the attachment of portable cords shall be rated at not less than 15 amperes, 125 volts, or 15 amperes, 250 volts, and shall not be a lampholder type. Receptacles shall be rated in accordance with this section.

❖ The minimum rating allowed for receptacles in dwelling units is 15 amperes. Receptacles are not permitted to be of a type suitable for use as lampholders.

E3902.1.1 Single receptacle. A single receptacle installed on an individual branch circuit shall have an ampere rating not less than that of the branch circuit.

❖ A receptacle can be single, duplex, or multiple. The term "yoke" describes the strap the receptacle is mounted on; the yoke is used to attach to the device box. A single receptacle is only one device on the strap or yoke. The rating of the circuit breaker or overcurrent device is the rating of branch circuit, and on an individual branch circuit serving only a single receptacle, all of the circuit current flow might be taken through it. Therefore, the single receptacle must be rated the same as the branch circuit overcurrent device.

E3902.1.2 Two or more receptacles. Where connected to a branch circuit supplying two or more receptacles or outlets, receptacles shall conform to the values listed in Table E3902.1.2.

❖ Where two or more receptacles are fed by a 20-ampere branch circuit, Table E3902.1.2 indicates that the receptacles could be rated 15 amperes. This could mean that where a 20-ampere branch circuit feeds only one duplex receptacle, that duplex receptacle could be rated 15 amperes because there is more than a single receptacle on the circuit.

**TABLE E3902.1.2
RECEPTACLE RATINGS FOR VARIOUS SIZE
MULTIOUTLET CIRCUITS**

CIRCUIT RATING (amperes)	RECEPTACLE RATING (amperes)
15	15
20	15 or 20
30	30
40	40 or 50
50	50

E3902.2 Grounding type. Receptacles installed on 15- and 20-ampere-rated branch circuits shall be of the grounding type.

❖ All receptacles must be of the grounding type. That is, they must accept a three-prong grounding plug (cord cap).

E3902.3 CO/ALR receptacles. Receptacles rated at 20 amperes or less and directly connected to aluminum conductors shall be marked CO/ALR.

❖ See Section E3901.2.

E3902.4 Faceplates. Metal faceplates shall be grounded.

❖ See the commentary for Sections E3806.10 and E3808.14.

E3902.5 Position of receptacle faces. After installation, receptacle faces shall be flush with or project from faceplates of insulating material and shall project a minimum of 0.015 inch (0.381 mm) from metal faceplates. Faceplates shall be installed so as to completely cover the opening and seat against the mounting surface.

❖ This section makes a distinction between the face of the actual receptacle and the faceplate or cover plate. The face of the receptacle must project out from the faceplate at least 1/64th of an inch. If the face of the receptacle does not project out from the faceplate where a metal faceplate is used, the blades of the cord attachment plug could touch the metal faceplate and create a ground fault or short. See also Section E3901.11.

E3902.6 Receptacle mounted in boxes. Receptacles mounted in boxes that are set back of the wall surface shall be installed so that the mounting yoke or strap of the receptacle is held rigidly at the surface of the wall. Receptacles mounted in boxes that are flush with the wall surface or project therefrom shall be so installed that the mounting yoke or strap is seated against the box or raised cover.

❖ The way the drywall (gypsum board) is cut out around the box has an effect on compliance with this code provision. If the drywall is cut out too far at the top and bottom of the box so that the extension plaster ears of the receptacle do not seat firmly against the finished wall surface, then the receptacle will "float" between the front edge of the box and the finished wall surface, and the face of the receptacle will be recessed from the faceplate. The one screw in the center of the faceplate will be the only thing holding the receptacle in place and will likely break. If washers or spacers are used to space the receptacle mounting yoke out from the face of the box, the faceplate will then be anchored to the receptacle and seated firmly against the wall surface.

E3902.7 Receptacles mounted on covers. Receptacles mounted to and supported by a cover shall be secured by more than one screw or shall be a device assembly or box cover listed and identified for securing by a single screw.

❖ An example of this is where a 4-inch square metal box is used, and the receptacle is installed in the box raised cover. A raised cover, for example, may have a $^1/_2$-inch (12.7 mm) rise. It is the same size as the four-square box and fits directly on the face of the box. A switch or receptacle is secured to the raised cover and not to the box, and then the cover is secured to the box. A switch installed in the cover would be secured by two 6-32 screws, the same as regular cover-plate screws, but the cover of a receptacle is normally secured by only one screw. The problem is that when attachment plugs are inserted, the receptacle will break away from the cover. This provision was put into the code to require that in these types of installations, two or more screws must be used to support the receptacle in the cover.

E3902.8 Damp locations. A receptacle installed outdoors in a location protected from the weather or in other damp locations shall have an enclosure for the receptacle that is weatherproof when the receptacle cover(s) is closed and an attachment plug cap is not inserted. An installation suitable for wet locations shall also be considered suitable for damp locations. A receptacle shall be considered to be in a location protected from the weather where located under roofed open porches, canopies and similar structures and not subject to rain or water runoff.

❖ In this section and the following, the distinction is made between receptacles in damp locations and in wet locations. The type of receptacle cover used is the main issue here. The conventional type of receptacle faceplate used on the interior of the dwelling is not permitted outdoors or in any damp or wet location. There are several different types of receptacle covers available for use in damp and wet locations. This section requires that the receptacle must have a cover that will make it weatherproof when the cover is closed and the receptacle is not in use. On a receptacle installed on the outside of the house, the receptacle terminals could become corroded from condensation, and the corrosion causes resistance. The result might be that because of corrosion and the development over time of a thin insulating layer on the contacts and terminals, continuity of the circuit might be lost. Damage could result also from the build-up of heat as current is drawn through the poor connection caused by corrosion. The weatherproof cover will help prevent the entrance of moisture on the interior of the receptacle.

The definition of the term "weatherproof" is important to remember: "So constructed or protected that exposure to the weather will not interfere with successful operation." If the receptacle is in a damp location as described here and in Chapter 34, the cover must make the receptacle weatherproof when the cover is closed and no attachment plug is inserted. When an attachment plug is inserted, the plug might help prevent the entrance of moisture in the damp location.

E3902.9 Outdoor wet locations. Where installed outdoors in a wet location, 15- and 20-ampere, 125- and 250-volt receptacles shall have an enclosure that is weatherproof whether or not the attachment plug cap is inserted.

❖ This section applies to receptacles that are both outdoors and in a wet location. The receptacle box and cover must be weatherproof in all cases, thus requiring what is commonly referred to as a "bubble" cover. Such covers provide the space necessary to allow the cover to close fully over an inserted plug and cord. This provision should end the debate over whether or not the item plugged into the receptacle will be attended while in use. See Section E3902.10.

E3902.10 Wet locations other than outdoors. Where a receptacle is installed in a wet location other than outdoors and where the product intended to be plugged into it is not attended while in use, the receptacle shall have an enclosure that is weatherproof both when the attachment plug cap is inserted and when it is removed. Where a receptacle is installed in a wet location and where the product intended to be plugged into it will be attended while in use, the receptacle shall have an enclosure that is weatherproof when the attachment plug cap is removed.

❖ Where a receptacle is installed in a wet location, two conditions or situations are described: (1) where an attachment plug is left connected to the receptacle continually, such as for landscape lighting, and (2) where the attachment plug is connected only when attended; for example, when someone is occasionally using a tool such as a lawn edger or power saw for a temporary time period. In the first situation, the cover used must make the receptacle weatherproof all the time. There are covers that completely cover the receptacle and the attachment plug when it is inserted. If the receptacle will only be used when attended, the cover could be one that makes the receptacle weatherproof only when not in use.

E3902.11 Bathtub and shower space. A receptacle shall not be installed within a bathtub or shower space.

❖ See Section E3901.7.

E3902.12 Flush mounting with faceplate. In damp or wet locations, the enclosure for a receptacle installed in an outlet box flush-mounted on a wall surface shall be made weatherproof by means of a weatherproof faceplate assembly that provides a water-tight connection between the plate and the wall surface.

❖ Weatherproof covers on outdoor receptacles must seat completely on the wall surface. A receptacle installed in a brick wall must have a gasket or sealant between the wall and the receptacle cover. On a house that has horizontal or vertical siding that overlaps, the receptacle cover must be completely in one of the sections of siding so that there is not a gap of airspace where moisture could enter. Siding manufacturers make components of the siding system to be installed at the receptacle location that provides a flat surface for the installation of a receptacle cover.

E3902.13 Outdoor installation. A receptacle outlet installed outdoors shall be located so that water accumulation is not likely to contact the outlet cover or plate.

❖ There is not a specific height provision for outdoor receptacles, but it is important to install them at a location where water will not accumulate. For example, during a heavy rainstorm, the water runoff from a roof or from a yard could flow against a house, and if the receptacle were low enough, the weatherproof cover would not prevent water from entering the receptacle.

SECTION E3903 FIXTURES

E3903.1 Energized parts. Luminaires, lampholders, lamps and receptacles shall not have energized parts normally exposed to contact.

❖ Energized parts of light fixtures must be enclosed so that they will not be exposed to accidental contact. Suitable covers must be installed to prevent accidental contact with energized parts.

E3903.2 Luminaires near combustible material. Luminaires shall be installed so that combustible material will not be subjected to temperatures in excess of 90°C (194°F).

❖ A significant amount of heat is developed in some incandescent fixtures and fixtures containing ballasts. If there is no means to dissipate the heat, it could cause damage to the surface on which it is mounted and also could cause a fire. It may be necessary to hold the light fixture away from a combustible ceiling or wall surface if the light fixture will create heat in excess of 90°C. Compliance with this code provision will help prevent fires in combustible structural elements as well as combustible storage.

E3903.3 Exposed conductive parts. The exposed conductive parts of luminaires and equipment directly wired or attached to outlets shall be grounded. Lamp tie wires, mounting screws, clips and decorative bands on glass spaced at least 1.5 inches (38 mm) from lamp terminals shall not be required to be grounded.

❖ All lighting fixtures must be grounded. In older houses, only lighting fixtures installed in a bathroom or outdoors were grounded. A remodeler can observe that some interior fixtures for a bedroom or living room do not contain a grounding wire, although the wiring method has a grounding conductor. When these fixtures are replaced, the grounding wire must be connected. All new light fixtures are supplied with a grounding wire or grounding means.

E3903.4 Screw-shell type. Lampholders of the screw-shell type shall be installed for use as lampholders only.

❖ It is not permitted to install or use lampholders with a screw-plug adapter. Lampholders must be used only for lamps.

E3903.5 Recessed incandescent luminaries. Recessed incandescent luminaires shall have thermal protection and shall be listed as thermally protected.

Exceptions:

1. Thermal protection shall not be required in recessed luminaires listed for the purpose and installed in poured concrete.
2. Thermal protection shall not be required in recessed luminaires having design, construction, and thermal performance characteristics equivalent to that of thermally protected luminaires, and such luminaires are identified as inherently protected.

❖ The buildup of heat in an incandescent recessed fixture can be a severe problem. A recessed fixture should have a clearance of at least 3 inches from insulation installed around or over it. When the heat builds up to a level that could be dangerous, the required thermal protection device automatically turns off the lamp. After the heat in the fixture has dissipated, the device closes the circuit and the light automatically turns back on. Where this happens repeatedly, the fixture might have a higher wattage lamp than the maximum allowed. The problem could be solved by installing lower-wattage lamps. The recessed fixture that cycles off to allow the heat to dissipate is doing what it is designed to do; however, the cause of the overheating must be eliminated.

E3903.6 Thermal protection. The ballast of a fluorescent luminaires installed indoors shall have integral thermal protection. Replacement ballasts shall also have thermal protection integral with the ballast. A simple reactance ballast in a fluorescent luminaire with straight tubular lamps shall not be required to be thermally protected.

❖ In addition to incandescent recessed fixtures, fluorescent fixtures must have thermal protection. The protection required must be part of the ballast. The thermal trip protection integral with the ballast responds to an abnormally high temperature in the ballast.

E3903.7 High-intensity discharge luminaires. Recessed high-intensity luminaires designed to be installed in wall or ceiling cavities shall have thermal protection and be identified as thermally protected. Thermal protection shall not be required in recessed high-intensity luminaires having design, construction and thermal performance characteristics equivalent to that of thermally protected luminaires, and such luminaires are identified as inherently protected. Thermal protection shall not be required in recessed high-intensity discharge luminaires installed in and identified for use in poured concrete. A recessed remote ballast for a high-intensity discharge luminaire shall have thermal protection that is integral with the ballast and shall be identified as thermally protected.

❖ High-intensity discharge fixtures such as metal-halide, sodium, and mercury-vapor operate at high temperatures and must have thermal protection unless identified as inherently protected.

E3903.8 Wet or damp locations. Luminaires installed in wet or damp locations shall be installed so that water cannot enter or accumulate in wiring compartments, lampholders or other electrical parts. All luminaires installed in wet locations shall be marked SUITABLE FOR WET LOCATIONS. All luminaires installed in damp locations shall be marked SUITABLE FOR WET LOCATIONS or SUITABLE FOR DAMP LOCATIONS.

❖ An indoor fixture is not permitted to be installed outdoors, even if it is under a porch or on a porch ceiling where the roof overhangs several feet. A partially protected location under canopies and roofed open porches is considered a damp location, as stated in Chapter 34. If a homeowner chooses lighting fixtures for a house and the ones for the ceiling of the open porch are the same as the ones for the entryway just inside the door, it would be necessary to check the fixtures to determine if they are marked as suitable for a damp location. If not, it would be a code violation to install the fixture outdoors.

A wall porch light that is exposed to weather and unprotected by the roof overhang is in a wet location, and the fixture must be marked as suitable for wet locations.

A fixture that is suitable only for damp locations is not permitted to be installed in wet locations. But a fixture that is marked as suitable for wet locations could also be installed in damp locations. An indoor light fixture is not permitted to be installed outdoors.

E3903.9 Lampholders in wet or damp locations. Lampholders installed in wet or damp locations shall be of the weatherproof type.

❖ An example of a weatherproof type of lampholder is floodlights installed under the eaves of a house. These weatherproof fixtures have a canopy with a gasket that seats against the surface of the underside of the eave (if that is where it is installed). They have one or more sockets that swivel for adjusting to the desired position. The lamps used are rated for direct exposure to the weather, and the lampholder has a gasket that helps prevent moisture from entering the socket. Fixtures used for this type of application must be weatherproof, and the lampholder must be weatherproof as well.

E3903.10 Bathtub and shower areas. Cord-connected luminaires, hanging luminaires, lighting track, pendants, and ceiling-suspended paddle fans shall not have any parts located within a zone measured 3 feet (914 mm) horizontally and 8 feet (2438 mm) vertically from the top of a bathtub rim or shower stall threshold. This zone is all encompassing and includes the zone directly over the tub or shower.

❖ In most homes, a bathroom ceiling is not high enough to permit any fixture other than a surface mounted or recessed type fixture above a bathtub or shower. Fixtures such as chain hung or pendant types where a cord is exposed above a bathtub or shower could be a safety hazard, and there is more of a danger of electrocution where water is present. The code does not specifically list a bathroom as a damp or wet location; however, the inspector might require a light in the ceiling of a shower to be suitable for a wet location. A recessed light with a gasketed glass cover is quite often the fixture used in the ceiling of a shower. There is not a code definition for "out of reach," but a hanging fixture or paddle fan installed at least 8 feet (2438 mm) above the rim of the bathtub would, in most cases, be out of reach of someone standing in or on the bathtub.

E3903.11 Luminaires in clothes closets. For the purposes of this section, storage space shall be defined as a volume bounded by the sides and back closet walls and planes extending from the closet floor vertically to a height of 6 feet (1829 mm) or the highest clothes-hanging rod and parallel to the walls at a horizontal distance of 24 inches (610 mm) from the sides and back of the closet walls respectively, and continuing vertically to the closet ceiling parallel to the walls at a horizontal distance of 12 inches (305 mm) or the width of the shelf, whichever is greater. For a closet that permits access to both sides of a hanging rod, the storage space shall include the volume below the highest rod extending 12 inches (305 mm) on either side of the rod on a plane horizontal to the floor extending the entire length of the rod (see Figure E3903.11).

The types of luminaires installed in clothes closets shall be limited to surface-mounted or recessed incandescent luminaires with completely enclosed lamps, and surface-mounted or recessed fluorescent luminaires. Incandescent luminaires with open or partially enclosed lamps and pendant luminaires or lamp-holders shall be prohibited. Luminaire installations shall be in accordance with one or more of the following:

1. Surface-mounted incandescent luminaires shall be installed on the wall above the door or on the ceiling, pro-

vided there is a minimum clearance of 12 inches (305 mm) between the fixture and the nearest point of a storage space.

2. Surface-mounted fluorescent luminaires shall be installed on the wall above the door or on the ceiling, provided there is a minimum clearance of 6 inches (152 mm) between the fixture and the nearest point of a storage space.
3. Recessed incandescent luminaires with a completely enclosed lamp shall be installed in the wall or the ceiling provided there is a minimum clearance of 6 inches (152 mm) between the luminaire and the nearest point of a storage space.
4. Recessed fluorescent luminaires shall be installed in the wall or on the ceiling provided there is a minimum clearance of 6 inches (152 mm) between the fixture and the nearest point of a storage space.

❖ In a clothes closet, a fixture with an exposed light bulb is a definite fire hazard and has been strictly prohibited for several years. Some older houses still have a porcelain fixture in closets, with an exposed lamp such as a keyless lampholder or lampholder with a pullchain for the switch. Storage items piled or stacked on closet shelves could easily come into contact with the bare light bulb and cause a fire. This section clearly defines the storage space and provides minimum clearances from the permitted light fixtures to the storage space. For a small closet where this clearance would not be possible, a lighting fixture would be prohibited.

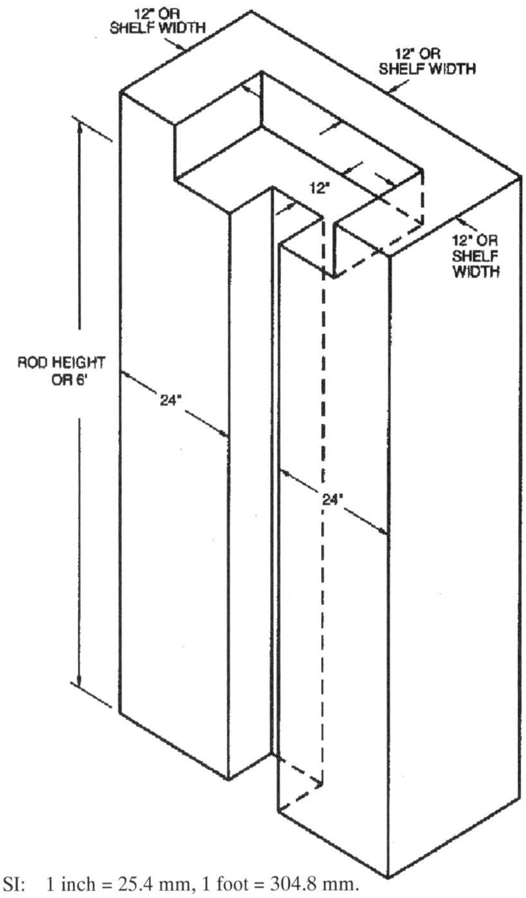

For SI: 1 inch = 25.4 mm, 1 foot = 304.8 mm.

FIGURE E3903.11
CLOSET STORAGE SPACE

❖ See the commentary for Section E3903.11.

E3903.12 Luminaire wiring—general. Wiring within luminaires or luminaire canopies shall not be exposed to physical damage. Conductors shall be arranged so that they are not subjected to temperatures above those for which the conductors are rated.

❖ Excessive heat will damage fixture wires and the wall or ceiling surface and can cause a fire. Most ceiling-mounted incandescent light fixtures for indoor use for general lighting are marked with a statement indicating the maximum wattage of the lamp. Many typical light fixtures indicate a maximum lamp wattage of 60. Where higher wattage lamps are installed, the heat produced will likely damage the fixture wires and the wall finish within the canopy of the fixture. Many fixtures come with thermal insulation inside the canopy, which must not be removed, because it helps protect both the circuit wires and the fixture wires from excessive heat.

E3903.12.1 Polarization of luminaires. Luminaires shall be wired so that the screw shells of lampholders will be connected to the same luminaire or circuit conductor or terminal. The grounded conductor shall be connected to the screw shell.

❖ The grounded conductor of the circuit must be connected to the screw shell of the socket. The "hot" conductor is connected to the pin in the bottom of the socket. When a person is replacing the lamp in a light fixture that is energized and touches the screw shell part of the lamp, the person could be electrocuted if the screw shell of the lampholder was connected to the ungrounded (hot) conductor.

E3903.12.2 Luminaires as raceways. Luminaires shall not be used as raceways for circuit conductors except where such luminaires are listed for the purpose.

❖ This is usually only a concern where several fluorescent fixtures are installed end-to-end and the supply conductors run through one fixture to another. Where two or more fluorescent fixtures are installed end-to-end, the supply conductors may enter at one fixture, and the switched circuit conductors are continued to all other fixtures. This is permitted if the fixture wireway is listed as a raceway and if the conductors are rated for the temperature in the fixture. Heat is created by the ballast, and the circuit wires likely are run next to the ballast. Branch-circuit conductors within three inches of a ballast must have an insulation temperature rating of at least 90°C.

SECTION E3904
LUMINAIRE INSTALLATION

E3904.1 Outlet box covers. In a completed installation, each outlet box shall be provided with a cover except where covered by means of a luminaire canopy, lampholder or device with a faceplate.

❖ See the commentary for Section E3806.9.

E3904.2 Combustible material at outlet boxes. Combustible wall or ceiling finish exposed between the inside edge of a luminaire canopy or pan and the outlet box to which the luminaire connects shall be covered with a noncombustible material.

❖ See the commentary for Section E3806.11.

E3904.3 Access. Luminaires shall be installed so that the connections between the luminaire conductors and the circuit conductors can be accessed without requiring the disconnection of any part of the wiring.

❖ To access connections to circuit wiring, a person might need to move a lighting fixture; therefore, sufficient slack in the fixture and/or circuit wires is necessary. The fixture must be hung or supported independently of the wiring.

E3904.4 Supports. Luminaires and lampholders shall be securely supported. A luminaire that weighs more than 6 pounds (2.72 kg) or exceeds 16 inches (406 mm) in any dimension shall not be supported by the screw shell of a lampholder.

❖ A fixture with a lampholder may be designed to have a decorative part attached and supported directly by the lampholder. The part supported by the lampholder is limited in size and weight. The entire fixture that is supported directly by the outlet box is covered in the next section.

E3904.5 Means of support. Outlet boxes or fittings installed as required by Sections E3805 and E3806 shall be permitted to support luminaires weighing 50 pounds (22.7 kg) or less. A luminaire that weighs more than 50 pounds (22.7 kg) shall be supported independently of an outlet box except where the outlet box is listed for the weight to be supported.

❖ Most light fixtures are designed to be supported directly from the outlet box. Any light fixture weighing over 50 pounds (23 kg) must be supported from the building structure or by a box listed for the application. However, for a large fixture such as a chandelier that may be a little less than 50 pounds (23 kg), it may be well to check how securely the outlet box is supported before assuming that it will be acceptable simply because the code allows it. The ceiling outlet box may be attached to the structure with only two nails or by a bracket between two framing members, which may not provide the support necessary for a heavy fixture.

E3904.6 Exposed ballasts. Luminaires having exposed ballasts or transformers shall be installed so that such ballasts or transformers are not in contact with combustible material.

❖ Ballasts and transformers often operate at high temperatures under normal conditions and even higher temperatures under abnormal conditions. Clearance to combustible materials is a precaution against these devices starting a fire. Because winding failure in such devices can cause extreme temperatures and/or open flames, ballasts are equipped with internal thermal overload protection devices. See Section E3903.6.

E3904.7 Combustible low-density cellulose fiberboard. Where a surface-mounted luminaire containing a ballast is installed on combustible low-density cellulose fiberboard, the luminaire shall be listed for the purpose or it shall be spaced not less than 1.5 inches (38 mm) from the surface of the fiberboard. Where such luminaires are partially or wholly recessed, the provisions of Sections E3904.8 and E3904.9 shall apply.

❖ Where a fluorescent lighting fixture is installed directly on combustible low-density cellulose fiberboard, the fixture must be marked very clearly for this installation. The heat from the ballast could exceed the temperature rating of this type of ceiling finish. Thermal insulation installed above a ceiling of this type will add to the buildup of heat. Combustible low-density cellulose fiberboard includes sheets, panels, and tiles that have a density of 20 pounds per cubic foot (320 kg/m^3) or less and are formed from bonded plant fiber material.

E3904.8 Recessed luminaire clearance. A recessed luminaire that is not identified for contact with insulation shall have all recessed parts spaced at least 0.5 inch (12.7 mm) from combustible materials. The points of support and the finish trim parts at the opening in the ceiling or wall surface shall be permitted to be in contact with combustible materials. A recessed luminaire that is identified for contact with insulation, Type IC, shall be permitted to be in contact with combustible materials at recessed parts, points of support, and portions passing through the building structure and at finish trim parts at the opening in the ceiling or wall.

❖ A recessed fixture that is designed to allow the insulation above the ceiling to be in contact with the fixture is referred to as Type IC.

E3904.9 Recessed luminaire installation. Thermal insulation shall not be installed above a recessed luminaire or within 3 inches (76 mm) of the recessed luminaire's enclosure, wiring compartment or ballast except where such luminaire is identified for contact with insulation, Type IC.

❖ Only Type IC recessed fixtures can have insulation installed above the fixture. There should be coordination between the electrician and the insulation installer to be sure that insulation is not placed above and is kept at least three inches away from the recessed lights, unless Type IC fixtures are used. Heat will be trapped and can cause damage or a fire.

SECTION E3905
TRACK LIGHTING

E3905.1 Installation. Lighting track shall be permanently installed and permanently supplied by a branch circuit having a rating not more than that of the track.

❖ Lighting track cannot be supplied by a cord and plug; rather, it must be hard wired as a permanent fixture. To prevent overloading of the track, the branch circuit serving the track must not have a higher current rating than the track.

E3905.2 Fittings. Fittings identified for use on lighting track shall be designed specifically for the track on which they are to be installed. Fittings shall be securely fastened to the track, shall maintain polarization and grounding, and shall be designed to be suspended directly from the track. Only lighting track fittings shall be installed on lighting track. Lighting track fittings shall not be equipped with general-purpose receptacles.

❖ Track fittings and the track to which they connect must be designed to mate. Ill-fitting, mismatched or improperly installed components could cause poor electrical connections and could cause lighting units to fall. Tracks are intended to function as a component of a lighting system and are not intended to serve as a substitute for convenience receptacles.

E3905.3 Connected load. The connected load on lighting track shall not exceed the rating of the track.

❖ If too many lighting units are connected to a track, the total load could exceed the rating of the track, resulting in a hazard. Section E3905.1 requires that the branch circuit rating not exceed the track rating; thus the overcurrent device will be sized to protect the track.

E3905.4 Prohibited locations. Lighting track shall not be installed in the following locations:

1. Where likely to be subjected to physical damage.
2. In wet or damp locations.
3. Where subject to corrosive vapors.
4. In storage battery rooms.
5. In hazardous (classified) locations.
6. Where concealed.
7. Where extended through walls or partitions.
8. Less than 5 feet (1524 mm) above the finished floor except where protected from physical damage or the track operates at less than 30 volts rms open-circuit voltage.
9. Within the zone measured 3 feet (900 mm) horizontally and 8 feet (2438 mm) vertically from the top of a bathtub rim.

❖ Track lighting must not be located where it is subject to damage from impact or moisture. It must not be concealed by structural or finish elements of a building and must not pass through walls or partitions. Item 9 is consistent with Section E3903.10.

E3905.5 Fastening. Lighting track shall be securely mounted so that each fastening will be suitable for supporting the maximum weight of luminaires that can be installed. Except where identified for supports at greater intervals, a single section 4 feet (1219 mm) or shorter in length shall have two supports and, where installed in a continuous row, each individual section of not more than 4 feet (1219 mm) in length shall have one additional support.

❖ Support for track lighting is very important. The code does not state that the track must be supported directly into a truss or framing member, only that it must be securely mounted. It is usually not sufficient to support the track lighting solely by attaching it to the drywall of the ceiling. When track lighting is first installed, it may well support the type of fixture initially installed. However, the quantity and type of fixtures that will be attached to the track at a future time is unknown.

E3905.6 Grounding. Lighting track shall be grounded in accordance with Chapter 38, and the track sections shall be securely coupled to maintain continuity of the circuitry, polarization and grounding throughout.

❖ Where several sections of lighting track are joined together, it is important that the grounding continuity be secured. The track has the hot or ungrounded conductor on one side and the grounded conductor on the other side. If a section is reversed, the polarity could be reversed.

Bibliography

The following resource materials are referenced in this chapter or are relevant to the subject matter addressed in this chapter.

NFPA 70-02, *National Electrical Code*. Quincy, MA: National Fire Protection Association, 2002.

Chapter 40: Appliance Installation

General Comments

Chapter 40 covers appliance installation, which is typically the final stage of construction after all wiring, devices, and fixtures are installed. This chapter covers flexible cords, overcurrent protection, disconnecting means, and installation provisions.

Purpose

This chapter intends to make appliance installations safe by regulating power cords, overcurrent/overload protection support, and means of disconnecting from the power supply.

SECTION E4001 GENERAL

E4001.1 Scope. This section covers installation requirements for appliances and fixed heating equipment.

❖ Appliances are also referred to as utilization equipment. The classification includes small kitchen appliances like a food mixer and deep fryer that are cord and plug connected as well as larger equipment that is fixed in place such as a dishwasher, water heater, air conditioner, or central heating furnace.

E4001.2 Installation. Appliances and equipment shall be installed in accordance with the manufacturer's installation instructions. Electrically heated appliances and equipment shall be installed with the required clearances to combustible materials.

❖ The code does not have prescriptive provisions for the minimum clearance of appliances to combustible materials. Required clearances for various appliances are given in the manufacturer's installation instructions.

E4001.3 Flexible cords. Cord- and plug-connected appliances shall use cords suitable for the environment and physical conditions likely to be encountered. Flexible cords shall only be used where the appliance is listed to be connected with a flexible cord. Specific appliances have additional requirements as specified in Table E4001.3. (See Section E3809.)

❖ Flexible cords are permitted to be used for appliances to permit easy removal or replacement of the unit, but only if the appliance is listed to be cord- and plug-connected. Kitchen appliances listed in Table E4001.3 are permitted to be connected with flexible cord. A disposal is connected to the sink and to the drain piping, but these connections are designed to permit disconnection and removal of the disposal. In many areas, power to a dishwasher is provided by direct connection of nonmetallic sheathed cable in the junction box or termination box at the base of the dishwasher; however, flexible cord is allowed.

TABLE E4001.3
FLEXIBLE CORD LENGTH

APPLIANCE	MINIMUM CORD LENGTH (inches)	MAXIMUM CORD LENGTH (inches)
Kitchen waste disposal	18	36
Built-in dishwasher	36	48
Trash compactor	36	48

For SI: 1 inch = 25.4 mm.

❖ See the commentary for Section E4001.3.

E4001.4 Overcurrent protection. Each appliance shall be protected against overcurrent in accordance with the rating of the appliance and its listing.

❖ Some appliances come with installation instructions or are marked with the minimum and maximum ratings of the overcurrent device for the branch circuit feeding the appliance.

E4001.4.1 Single nonmotor-operated appliance. The overcurrent protection for a branch circuit that supplies a single nonmotor-operated appliance shall not exceed that marked on the appliance. Where the overcurrent protection rating is not marked and the appliance is rated at over 13.3 amperes, the overcurrent protection shall not exceed 150 percent of the appliance rated current. Where 150 percent of the appliance rating does not correspond to a standard overcurrent device ampere rating, the next higher standard rating shall be permitted. Where the overcurrent protection rating is not marked and the appliance is rated at 13.3 amperes or less, the overcurrent protection shall not exceed 20 amperes.

❖ The overcurrent device protects the conductors, but it also serves to protect the appliance or equipment itself. An appliance marked with the rated ampere load might not provide information on the rating of branch circuit overcurrent protection. In this case and for appliances rated over 13.3 amperes, the provisions of this section limit the size of the overcurrent device on the branch circuit to 150 percent of the appliance rated current.

E4001.5 Disconnecting means. Each appliance shall be provided with a means to disconnect all ungrounded supply conductors. For fixed electric space-heating equipment, means shall be provided to disconnect the heater and any motor controller(s) and supplemental overcurrent-protective devices. Switches and circuit breakers used as a disconnecting means shall be of the indicating type. Disconnecting means shall be as set forth in Table E4001.5.

❖ In Table E4001.5, permanently connected appliances are those with fixed wiring (as opposed to cord- and plug-connected appliances). A circuit breaker in a panelboard could serve as a disconnect if it has a rated load of less than 1/8 horsepower or under 300 VA. If it is rated 1/8 horsepower or more, or 300 VA or more, the circuit breaker is permitted to serve as the disconnecting means if it can be locked in the open position. The second from bottom row in Table E4001.5 indicates that if the appliance has a unit switch with a marked OFF position, it can serve as the disconnect if there is an additional redundant disconnect. In residential wiring, the main service disconnect is permitted to serve as the additional disconnect for this purpose. Air-conditioning condensing units and heat pump units must have a disconnect within sight from the unit as the only allowable means.

E4001.6 Support of ceiling-suspended paddle fans. Ceiling-suspended paddle fans that do not exceed 35 pounds (15.88 kg) in weight, with or without accessories, shall be permitted to be supported by outlet boxes identified for such use and supported in accordance with Sections E3805 and E3806. Ceiling-suspended paddle fans exceeding 35 pounds (15.88 kg) in weight, with or without accessories, shall be supported independently of the outlet box (see Section 3806).

Exception: Listed outlet boxes or outlet box systems that are identified for the purpose shall be permitted to support ceiling-suspended paddle fans, with or without accessories, that weigh 70 pounds (31.76 kg) or less.

❖ The provisions for paddle fans are included in this chapter since they are considered appliances and not lighting fixtures by the code. The outlet boxes permitted by this section for the support of paddle fans of 35 pounds (16 kg) or less are not the standard light fixture boxes that are used for most lighting outlets.

TABLE E4001.5
DISCONNECTING MEANS

DESCRIPTION	ALLOWED DISCONNECTING MEANS
Permanently connected appliance rated at not over 300 volt-amperes or 0.125 horsepower.	Branch-circuit overcurrent device.
Permanently connected appliances rated in excess of 300 volt-amperes or 0.125 horsepower.	Branch circuit breaker or switch located within sight of appliance or such devices in any location that are capable of being locked in the open position.
Appliances listed for cord-and-plug connection.	A separable connector or attachment plug and receptacle provided with access.
Permanently installed heating equipment with motors rated at not over 0.125 horsepower with supplementary overcurrent protection.	Disconnect, on the supply side of fuses, in sight from the supplementary overcurrent device, and in sight of the heating equipment or, in any location, if capable of being locked in the open position.
Heating equipment containing motors rated over 0.125 horsepower with supplementary overcurrent protection.	Disconnect permitted to serve as required disconnect for both the heating equipment and the controller where, on the supply side of fuses, and in sight from the supplementary overcurrent devices, if the disconnecting means is also in sight from the controller, or is capable of being locked off.
Heating equipment containing no motor rated over 0.125 horsepower without supplementary overcurrent protection.	Branch-circuit switch or circuit breaker where within sight from the heating equipment or capable of being locked off.
Heating equipment containing motors rated over 0.125 horsepower without supplementary overcurrent protection.	Disconnecting means in sight from motor controller or as provided for heating equipment with motor rated over 0.125 horsepower with supplementary overcurrent protection.
Air-conditioning condensing units and heat pump units.	A readily accessible disconnect within sight from unit as the only allowable means.[a]
Appliances and fixed heating equipment with unit switches having a marked OFF position.	Unit switch where an additional individual switch or circuit breaker serves as a redundant disconnecting means.
Thermostatically controlled fixed heating equipment.	Thermostats with a marked OFF position that directly open all ungrounded conductors, which when manually placed in the OFF position are designed so that the circuit cannot be energized automatically and that are located within sight of the equipment controlled.

a. The disconnecting means shall be permitted to be installed on or within the unit. It shall not be located on panels designed to allow access to the unit.

❖ See the commentary for Section 4001.5.

E4001.7 Snow-melting and deicing equipment protection. Outdoor receptacles that are not readily accessible and are supplied from a dedicated branch circuit for electric snow-melting or deicing equipment shall be permitted to be installed without ground-fault circuit-interrupter protection for personnel. However, ground-fault protection of equipment shall be provided for fixed outdoor electric deicing and snow-melting equipment.

❖ Receptacles are installed under the eaves for the connection of deicing cables installed along the roof edge. These receptacles are not required to be GFCI protected. However, the deicing equipment must have ground-fault protection. Circuit breakers are available with ground-fault equipment protection (GFEP), which will open the circuit due to leakage current and will help prevent fires caused by arcing and sparking due to low level fault currents.

Bibliography

The following resource materials are referenced in this chapter or are relevant to the subject matter addressed in this chapter.

NFPA 70-02, *National Electrical Code.* Quincy, MA: National Fire Protection Association, 2002.

Chapter 41: Swimming Pools

General Comments

This chapter addresses all aspects of wiring, fixtures, motors, and electrical accessories for swimming pools, wading pools, hot tubs, spas, and hydromassage bathtubs.

Purpose

This chapter focuses on protection of occupants from electrical shock. The dangers of electrical usage around water, wet surfaces, grounded surfaces, and plumbing are well known, and this chapter is intended to abate or eliminate those hazards.

SECTION E4101 GENERAL

E4101.1 Scope. The provisions of this chapter shall apply to the construction and installation of electric wiring and equipment associated with all swimming pools, wading pools, decorative pools, fountains, hot tubs and spas, and hydromassage bathtubs, whether permanently installed or storable, and shall apply to metallic auxiliary equipment, such as pumps, filters and similar equipment. Sections E4102 through E4106 provide general rules for permanent pools, spas and hot tubs. Section E4107 provides specific rules for storable pools. Section E4108 provides specific rules for spas and hot tubs. Section E4109 provides specific rules for hydromassage bathtubs.

❖ Because of a greater potential of electrical shock hazard around pools, a separate chapter is included in the code to provide specific detailed requirements regarding the wiring in and around swimming pools, wading pools, spas and hot tubs, and hydromassage tubs. A conductive path for fault current can easily be established through the water to the earth. When a person in a pool touches a metallic surface that is energized, the fault-current path through the individual can be fatal. Also, a person in the pool not touching anything but the water could become part of a fault-current path. For example, if an energized device such as an electric appliance drops into the pool, an electrical potential could be established in the pool where a voltage gradient could cause the person to be surrounded by different levels of voltage in the water acting like a conductor. Because of these types of hazards, more stringent code provisions apply to wiring around swimming pools.

E4101.2 Definitions.

CORD- AND PLUG-CONNECTED LIGHTING ASSEMBLY. A lighting assembly consisting of a cord and plug-connected transformer and a lighting fixture intended for installation in the wall of a spa, hot tub, or storable pool.

❖ This is a definition for a lighting fixture used in the wall of a storable swimming pool, spa, or hot tub. It is made with a nonmetallic housing and is supplied with power from a cord- and plug-connected transformer.

DRY-NICHE LUMINAIRE. A luminaire intended for installation in the wall of a pool or fountain in a niche that is sealed against the entry of pool water.

❖ A dry-niche lighting fixture is installed behind a clear window below the water level and does not allow pool water to enter the forming shell or the area around the lamp housing. For swimming pool wiring in dwellings, access to the lamp would typically be through a deck box, and relamping is done from behind the waterproof lens or glass window. The dry-niche area where the lamp is installed must have drainage for any water that may enter from the deck area of the pool when the cover is off for servicing. The glass window is sealed against the entry of pool water.

FORMING SHELL. A structure designed to support a wet-niche lighting fixture assembly and intended for mounting in a pool or fountain structure.

❖ A forming shell is built into the wall of the pool and is the supporting part of the wet-niche lighting fixture.

FOUNTAIN. Fountains, ornamental pools, display pools, and reflection pools. The definition does not include drinking fountains.

❖ Decorative pools and fountains fall under the scope of this chapter.

HYDROMASSAGE BATHTUB. A permanently installed bathtub equipped with a recirculating piping system, pump, and associated equipment. It is designed so it can accept, circulate and discharge water upon each use.

❖ A hydromassage bathtub is similar to a normal bathtub in that the water is drained after each use. It is not intended that the hydromassage bathtub stores water or is left filled with water like a spa or hot tub. It has a pump and piping system for jetting water when filled. The pump must be plugged into a GFCI protected receptacle. All receptacles in the same bathroom with the hydromassage bathtub must be GFCI protected.

MAXIMUM WATER LEVEL. The highest level that water can reach before it spills out.

❖ In the plumbing part of this code the term "flood-level rim" is used; the meaning is the same as "maximum water level."

NO-NICHE LUMINAIRE. A luminaire intended for installation above or below the water without a niche.

❖ This is an underwater lighting fixture installed on the surface of the wall of the pool on a mounting bracket. For changing the lamp, the cord should be long enough to detach the fixture from the bracket underwater and bring it out of the water. A no-niche lighting fixture is typically used on an aboveground pool that does not have a forming shell.

PACKAGED SPA OR HOT TUB EQUIPMENT ASSEMBLY. A factory-fabricated unit consisting of water-circulating, heating and control equipment mounted on a common base, intended to operate a spa or hot tub. Equipment may include pumps, air blowers, heaters, luminaires, controls and sanitizer generators.

❖ The packaged unit comes already assembled from the manufacturer minimizing the need to design and install each component in the field. The spa or hot tub vessel comes as a separate component and is installed on the job along with the packaged assembly of equipment.

PERMANENTLY INSTALLED SWIMMING AND WADING POOLS. Those that are constructed in the ground or partially in the ground, and all others capable of holding water with a depth greater than 42 inches (1067 mm), and all pools installed inside of a building, regardless of water depth, whether or not served by electrical circuits of any nature.

❖ Special provisions are included in the code for swimming pools because of the high risk of electric shock due to the conductive path from a human body to the water surrounding a human body through the concrete and reinforcing steel in the wall of the pool to the earth.

POOLCOVER, ELECTRICALLY OPERATED. Motor-driven equipment designed to cover and uncover the water surface of a pool by means of a flexible sheet or rigid frame.

❖ Because of the close proximity of the motor of an electrically-operated pool cover to the edge of the pool, the code has special provisions regarding these pool covers.

SELF-CONTAINED SPA OR HOT TUB. A factory-fabricated unit consisting of a spa or hot tub vessel with all water-circulating, heating and control equipment integral to the unit. Equipment may include pumps, air blowers, heaters, luminaires, controls and sanitizer generators.

❖ The equipment needed to operate the spa or hot tub such as pumps, heaters, controls, etc. is manufactured together with the spa or hot vessel and comes as a complete unit.

SPA OR HOT TUB. A hydromassage pool, or tub for recreational or therapeutic use, not located in health care facilities, designed for immersion of users, and usually having a filter, heater, and motor-driven blower. They are installed indoors or outdoors, on the ground or supporting structure, or in the ground or supporting structure. Generally, a spa or hot tub is not designed or intended to have its contents drained or discharged after each use.

❖ A hot tub is typically built with wood, such as redwood, while a spa is usually made of fiberglass, concrete, or tile. They may have much the same equipment that a swimming pool could have, such as stainless-steel handrails and electric equipment for heating and circulating the water.

STORABLE SWIMMING OR WADING POOL. Those that are constructed on or above the ground and are capable of holding water with a maximum depth of 42 inches (1067 mm), or a pool with nonmetallic, molded polymeric walls or inflatable fabric walls regardless of dimension.

❖ This definition was included in the code to help differentiate between a storable pool and a permanent pool built in the ground. Also, the code provisions and definition for storable pools are included in the code because these structures have most of the same equipment as an in-ground pool could have. They are not built with concrete and reinforcing steel, but similar electrical hazards exist. Although storable pool are intended to be temporary, and many storable pools are disassembled during the cold months of the year, an on-ground pool that is left installed all year could still be defined as a storable pool if it is capable of holding water at least 42 inches (1067 mm) deep. How big the pool is in surface area or how much water the pool can hold are not determining factors in defining a pool as a storable pool.

THROUGH-WALL LIGHTING ASSEMBLY. A lighting assembly intended for installation above grade, on or through the wall of a pool, consisting of two interconnected groups of components separated by the pool wall.

❖ Such assemblies allow the installations of pool lighting in above-ground pools. The components house the luminaire and also serve to seal the opening made in the wall of the pool.

WET-NICHE LUMINAIRE. A luminaire intended for installation in a forming shell mounted in a pool or fountain structure where the luminaire will be completely surrounded by water.

❖ A wet-niche lighting fixture is designed to have the pool water completely surround the fixture within the forming shell. A wet-niche lighting fixture is connected with a cord that extends to a junction box typically at the pool deck. It has enough cord so that when the fixture is relamped, it is brought up out of the water without the cord being disconnected and then the cord is re-coiled and placed behind the fixture in the forming shell.

SECTION E4102 WIRING METHODS FOR POOLS, SPAS, HOT TUBS AND HYDROMASSAGE BATHTUBS

E4102.1 General. Wiring methods used in conjunction with permanently installed swimming pools, spas, hot tubs or hydromassage bathtubs shall be installed in accordance with Table E4102.1 and Chapter 37 except as otherwise stated in this section. Storable swimming pools shall comply with Section E4107.

❖ Table E4102.1 covers the rules for the raceways and cables that can be used in wiring associated with swimming pools. The rules in Chapter 37 for wiring methods apply except as modified by the provisions of this chapter. It is important to consider the footnotes of Table E4102.1. For example, Note c indicates that EMT can only be used on or within buildings.

TABLE E4102.1
ALLOWABLE APPLICATIONS FOR WIRING METHODS[a,b,c,d,e,f,g]

WIRING LOCATION OR PURPOSE	AC, FMC, NM, SR, SE[b]	EMT[c]	ENT[b]	IMC, RMC, RNC	LFMC	LFNMC	UF	MC	Flex Cord[g]
Panelboard(s) that supply pool equipment: from service equipment to panelboard	A[e]	A	—	A	—	A	A[e]	A[e]	—
Wet niche and no niche fixtures: from branch circuit OCPD to deck or junction box	—	A	A	A	—	A	—	—	—
Wet niche and no niche fixtures: from deck or junction box to forming shell	—	—	—	A[d]	—	A	—	—	A
Dry niche: from branch circuit OCPD to fixture	—	A	A	A	—	A	—	—	—
Pool-associated motors: from branch circuit OCPD to motor	A	A	A	A	A[f]	A[f]	A[b]	A	A
Packaged or self-contained outdoor spas and hot tubs with underwater lighting fixture: from branch circuit OCPD to spa or hot tub	—	A	A	A	—	A[f]	—	—	A
Packaged or self-contained outdoor spas and hot tubs without underwater lighting fixture: from branch circuit OCPD to spa or hot tub	A	A	A	A	A[f]	A[f]	A[b]	A	A
Indoor spa and hot tubs, hydromassage bathtubs, and other pool, spa or hot tub associated equipment: from branch circuit OCPD to equipment	A	A	A	A	A	A	A	A	A

For SI: 1 foot = 304.8 mm.

a. For all wiring methods, see Section E4105.2 for equipment grounding conductor requirements.

b. Limited to use within buildings.

c. Limited to use on or within buildings.

d. Metal conduit shall be constructed of brass or other approved corrosion resistant metal

e. Permitted only for existing feeder panelboards where the equipment grounding conductor is insulated or covered.

f. Limited to use in lengths not to exceed 3 feet at pool, spa or hot tub equipment where flexibility is necessary.

g. Flexible cord shall be installed in accordance with Section E4102.2.

❖ See the commentary for Section E4102.1.

E4102.2 Flexible cords. Flexible cords used in conjunction with a pool, spa, hot tub or hydromassage bathtub shall be installed in accordance with the following:

1. For other than underwater luminaires, fixed or stationary equipment, rated at 20 amperes or less shall be permitted to be connected with a flexible cord to facilitate the removal or disconnection for maintenance or repair. For other than storable pools, the flexible cord shall not exceed 3 feet (914 mm) in length. Cords that supply swimming pool equipment, shall have a copper equipment grounding conductor not smaller than 12 AWG and shall be provided with a grounding-type attachment plug.
2. Flexible cord that is supplied as part of a listed under-water swimming pool lighting luminaire shall be permitted to be installed in any of the permitted wiring methods from the luminaire to a deck box or other enclosure. Splices shall not be made within a raceway. The equipment grounding conductor shall be an insulated copper conductor that is not smaller than the supply conductors and not smaller than 16 AWG.
3. A listed packaged spa or hot tub installed outdoors that is GFCI protected, shall be permitted to be cord and plug connected provided that such cord does not exceed 15 feet (4572 mm) in length.
4. A listed packaged spa or hot tub rated at 20 amperes or less and installed indoors shall be permitted to be cord and plug connected to facilitate maintenance and repair.
5. For other than underwater and storable pool lighting luminaire, the requirements of Item 1 shall apply to any cord equipped luminaire that is located within 16 feet (4877 mm) radially from any point on the water surface.

❖ Flexible cords are used to facilitate easy removal or repair of fixed or stationary equipment. For example, a pool cover motor could be connected by a flexible cord not over 3 feet long. Limiting the length of the cord will help prevent excessive cord length from being in the pool water or otherwise exposed. Underwater fixtures such as no-niche and wet-niche fixtures are permitted to have longer cords so that the fixture can be removed and brought up on the deck for servicing. Some equipment on a storable pool could have cords longer than 3 feet since it is not fixed or stationary, being installed such that it can be removed and stored.

It is important to check the size of the equipment grounding conductor within the cord. Some cords have an equipment grounding conductor smaller than the hot conductor(s).

An example of a permitted flexible cord in the area of a swimming pool is one for a lighting fixture within 16 feet of the edge of the water. It is permitted to be cord- and plug-connected if the cord is a maximum of 3 feet long, has a grounding conductor of at least size 12 AWG, and has a grounding-type attachment plug (cap).

SECTION E4103 EQUIPMENT LOCATION AND CLEARANCES

E4103.1 Receptacle outlets. Receptacles outlets shall be installed and located in accordance with Sections E4103.1.1 through E4103.1.5. Distances shall be measured as the shortest path that an appliance supply cord connected to the receptacle would follow without penetrating a floor, wall, ceiling, doorway with hinged or sliding door, window opening, or other effective permanent barrier.

❖ The measurement between the edge of the pool and a receptacle is the shortest unobstructed route of an appliance supply cord plugged into the receptacle. The code considers such things as doors, windows, walls, floors, and ceilings to be "effective permanent barriers" in determining the distance of a receptacle from the edge of the pool. When measuring the distance from the pool water to the receptacle, if the measurement is, for example, a total of 9 feet (2743 mm) from the pool to the receptacle, but the measurement is through a doorway that has a sliding or hinged door installed (not an opening without a door), the door is considered a barrier and the receptacle would not be considered 9 feet from the edge of the pool and would not require GFCI protection because of the proximity of the swimming pool. The receptacle in this example could be required to have GFCI protection for a different reason, such as if it were in a bathroom. See the commentary for Section E3802.1.

E4103.1.1 Location. Receptacles that provide power for water-pump motors or other loads directly related to the circulation and sanitation system shall be permitted to be located between 5 feet and 10 feet (1524 mm and 3048 mm) from the inside walls of pools and outdoor spas and hot tubs, and, where so located, shall be single and of the locking and grounding type and shall be protected by ground-fault circuit interrupters.

Other receptacles on the property shall be located not less than 10 feet (3048 m) from the inside walls of pools and outdoor spas and hot tubs.

❖ Only receptacles for specific equipment are permitted between 5 and 10 feet (1524 mm and 3048 mm) from the inside wall of the pool, and they must be a single receptacle of the locking and grounding type so that a typical radio, for example, could not be plugged into it. The code prohibits receptacles that supply power to appliances from being installed within 10 feet (3048 mm) of the inside wall of the pool.

E4103.1.2 Where required. At least one 125-volt 15- or 20-ampere receptacle supplied by a general-purpose branch circuit shall be located a minimum of 10 feet (3048 mm) from and not more than 20 feet (6096 mm) from the inside wall of pools and outdoor spas and hot tubs. This receptacle shall be located not more than 6 feet, 6 inches (1981 mm) above the floor, platform or grade level serving the pool, spa or hot tub.

❖ At least one receptacle is required in the area between 10 and 20 feet (3048 mm and 6096 mm) from the inside wall of a swimming pool and outdoor spas and hot tubs. This requirement is for indoor and outdoor swimming pools but only for outdoor spas and hot tubs. This provides a power source for appliances to be used near the pool without resorting to the use of extension cords. Since a typical appliance supply cord is about 6 feet (1829 mm), it is less likely that an appliance would be bumped and knocked into the pool where it is plugged into a receptacle located 10 feet from the pool water. This required receptacle could be on a wall above a raised deck, for example, as long as the receptacle is not over 78 inches (1981 mm) above the swimming pool deck.

E4103.1.3 GFCI protection. All 125-volt receptacles located within 20 feet (6096 mm) of the inside walls of pools and outdoor spas and hot tubs shall be protected by a ground-fault circuit-interrupter.

❖ All receptacles within 20 feet (6096 mm) of the pool water must have GFCI protection including any single, locking and grounding type receptacles installed for specific equipment. This applies to storable pools as well as permanently installed indoor and outdoor pools. It may not be common to have an indoor pool with a deck over 20 feet (6096 mm) wide, [20 feet (6096 mm) from the inside wall of the pool to the wall of the building], but where this occurs, any receptacle over 20 feet (6096 mm) from the pool would not have to be GFCI protected. Where a pool is installed outdoors, the 20-foot (6096 mm) rule is irrelevant since all outdoor receptacles at a dwelling require GFCI protection per Section E3802.3.

E4103.1.4 Indoor locations. Receptacles shall be located not less than 5 feet (1524 mm) from the inside walls of indoor spas and hot tubs. A minimum of one 125-volt receptacle shall be located between 5 feet (1524 mm) and 10 feet (3048 mm) from the inside walls of indoor spas or hot tubs.

❖ This code provision is specifically for indoor spas and hot tubs. The rule to have at least one receptacle between 5 feet and 10 feet (1524 mm and 3048 mm) from the inside wall of the spa or hot tub is to help prevent the hazards inherent with the use of extension cords.

E4103.1.5 Indoor GFCI protection. One hundred twenty-five-volt receptacles located within 10 feet (3048 mm) of the inside walls of spas and hot tubs installed indoors shall be protected by ground-fault circuit-interrupters. One hundred twenty-five-volt receptacles located within 5 feet (1524 mm) of the inside walls of hydromassage bathtubs shall be protected by a ground-fault circuit-interrupter.

❖ All receptacles within 10 feet (3048 mm) of the inside wall of the spa or hot tub require GFCI protection. The code provisions for locating receptacles around hydromassage bathtubs are different from those for spas and hot tubs. A receptacle is not prohibited in the area within 5 feet and is not required between 5 feet and 10 feet from a hydromassage bathtub, but any receptacles located within 5 feet of the inside wall of a hydromassage bathtub must be GFCI protected.

E4103.2 Switching devices. Switching devices shall be located not less than 5 feet (1524 mm) horizontally from the inside walls of pools, spas and hot tubs except where separated from the pool, spa or hot tub by a solid fence, wall, or other permanent barrier. Switching devices located in a room or area containing a hydromassage bathtub shall be located in accordance with the general requirements of this code.

❖ This section refers not only to ordinary light switches but also to such devices as a timer, panelboard, etc. By requiring these devices to be at least 5 feet (1524 mm) from the inside walls of the pool, spa, or hot tub, the code intends that they cannot be reached by someone in the pool. People will need to get out of the pool, spa, or hot tub to operate the switching device, thus reducing the possible shock hazard. This specific rule does not pertain to switching devices around hydromassage bathtubs.

E4103.3 Disconnecting means. An accessible disconnecting means to disconnect all ungrounded conductors for all utilization equipment, other than lighting, shall be provided and located within sight from all pools, spas, and hot tub equipment, and shall be located not less than 5 feet (1524 mm) from the inside walls of the pool, spa or hot tub.

❖ The disconnect must be at least 5 feet (1524 mm) away and within sight from the pool, spa or hot tub. This means it must be visible and not more than 50 feet (15 240 mm) away. This allows the power to lighting fixtures and equipment such as motors and heaters to be turned off while someone is working on these units.

E4103.4 Luminaires and ceiling fans. Lighting outlets, luminaires, and ceiling-suspended paddle fans shall be installed and located in accordance with Sections E4103.4.1 through E4103.4.5.

❖ Compliance with these prescriptive code provisions will help reduce the increased hazards of electrocution near water. These requirements are shown in Figure E4103.

E4103.4.1 Outdoor location. In outdoor pool, outdoor spas and outdoor hot tubs areas, luminaires, lighting outlets, and ceiling-suspended paddle fans shall not be installed over the pool or over the area extending 5 feet (1524 mm) horizontally from the inside walls of a pool except where no part of the luminaire or ceiling-suspended paddle fan is less than 12 feet (3658 mm) above the maximum water level.

❖ For new installations of an outdoor pool, spa, or hot tub, light fixtures and paddle fans are not permitted in the area over the water and extending 5 feet horizontally from the inside edge of the pool to a distance of 12 feet above the water level. There are no exceptions to this rule. Any fixtures or paddle fans located over 12 feet above the water level for a distance of at least 5 feet horizontally from the inside wall of the pool do not require GFCI protection. See Section E4103.4.3 below for the code provisions regarding a pool installed in an area where light fixtures have already been installed.

E4103.4.2 Indoor locations. In indoor pool areas, the limitations of Section E4103.4.1 shall apply except where the luminaires, lighting outlets and ceiling-suspended paddle fans comply with all of the following conditions:

1. The luminaires are of a totally enclosed type; and
2. A ground-fault circuit interrupter is installed in the branch circuit supplying the luminaires or ceiling-suspended (paddle) fans; and
3. The distance from the bottom of the luminaire or ceiling-suspended (paddle) fan to the maximum water level is not less than 7 feet, 6 inches (2286 mm).

❖ In indoor pool, spa, or hot tub areas, light fixtures and paddle fans can be installed at least 12 feet above the water level without GFCI protection, the same as for outdoor locations. However, if GFCI protection is provided, such as a GFCI circuit breaker in the panelboard feeding the circuit, fixtures of the totally enclosed type and paddle fans are permitted as close as $7^1/_2$ feet (2286 mm) above the water level. An example of a totally enclosed light fixture is one that has a glass or plastic lens.

E4103.4.3 Existing lighting outlets and luminaires. Existing lighting outlets and luminaires that are located within 5 feet (1524 mm) horizontally from the inside walls of pools and outdoor spas and hot tubs shall be permitted to be located not less than 5 feet (1524 mm) vertically above the maximum water level, provided that such luminaires and outlets are rigidly attached to the existing structure and ground-fault circuit-interrupter protection is provided for the branch circuit that supplies such luminaires and outlets.

❖ This section covers light fixtures that are already installed on the wall of a house, or structure when the construction of the pool begins. This rule on existing lighting outlets and fixtures is for indoor and outdoor pools but only for outdoor spas and hot tubs. If the existing lighting fixtures are rigidly attached to the house or structure and end up being within 5 feet (1524 mm) horizontally from the inside walls, they must be at least 5 feet (1524 mm) above the maximum water level, and the supply circuit must have GFCI protection.

E4103.4.4 Indoor spas and hot tubs.

1. Luminaires, lighting outlets, and ceiling-suspended paddle fans located over the spa or hot tub or within 5 feet (1524 mm) from the inside walls of the spa or hot tub shall be a minimum of 7 feet, 6 inches (2286 mm) above the maximum water level and shall be protected by a ground-fault circuit interrupter.

   Luminaires, lighting outlets, and ceiling-suspended paddle fans that are located 12 feet (3658 mm) or more above the maximum water level shall not require ground-fault circuit interrupter protection.
2. Luminaires protected by a ground-fault circuit interrupter and complying with Item 2.1. or 2.2. shall be permitted to be installed less than 7 feet, 6 inches (2286 mm) over a spa or hot tub.
   - 2.1. Recessed luminaires shall have a glass or plastic lens and nonmetallic or electrically isolated metal trim, and shall be suitable for use in damp locations.
   - 2.2. Surface-mounted luminaires shall have a glass or plastic globe and a nonmetallic body or a metallic body isolated from contact. Such luminaires shall be suitable for use in damp locations.

❖ The general rules for lighting fixtures and paddle fans over indoor spas and hot tubs are the same as for indoor swimming pools found in Section E4103.4.2. However, where lighting is desired above an indoor spa or hot tub, the ceiling is often not high enough to meet the $7^1/_2$ foot (2286 mm) rule. The room containing the spa or hot tub would need to have a high enough ceiling so that the bottom of a light fixture or fan would be at least $7^1/_2$ feet (2286 mm) above the water level. Recessed or surface mounted light fixtures are permitted to be less than $7^1/_2$ feet (2286 mm) above the water level of the spa or hot tub but must have GFCI protection, and the fixtures must be suitable for use in damp locations, must be all nonmetallic or metallic parts must be electrically isolated from the fixture, and must have a suitable lens or cover so that the lamp is not exposed. A cord-hung or pendant fixture less than $7^1/_2$ feet (2286 mm) above the water level is not permitted above an indoor spa or hot tub.

E4103.4.5 GFCI protection. Luminaires and outlets that are installed in the area extending between 5 feet (1524 mm) and 10 feet (3048 mm) from the inside walls of pools and outdoor spas and hot tubs shall be protected by ground-fault circuit-interrupters except where such fixtures and outlets are installed not less than 5 feet (1524 mm) above the maximum water level and are rigidly attached to the structure.

❖ In the area extending horizontally between 5 feet (1524 mm) and 10 feet (3048 mm) from the inside wall of a pool or outdoor spa or hot tub and at a height vertically of up to 5 feet above the water level, light fixtures and lighting outlets (not receptacle outlets) are permitted, but these must have GFCI protection. (This section is not addressing the single locking, grounding type receptacle outlet permitted for specific equipment less than 5 feet [1524 mm] from the pool.) In the same area, but at a height of over 5 feet above the water level, the circuits feeding rigidly attached light fixtures are not required to have GFCI protection.

E4103.5 Overhead conductor clearances. Except where installed with the clearances specified in Table E4103.5, the following parts of pools and outdoor spas and hot tubs shall not be placed under existing service-drop conductors or any other open overhead wiring; nor shall such wiring be installed above the following:

1. Pools and the areas extending 10 feet (3048 mm) horizontally from the inside of the walls of the pool;
2. Diving structures; or

3. Observation stands, towers, and platforms.

Utility-owned, -operated and -maintained communications conductors, community antenna system coaxial cables and the supporting messengers shall be permitted at a height of not less than 10 feet (3048 mm) above swimming and wading pools, diving structures, and observation stands, towers, and platforms.

❖ Service drop conductors are not permitted above a pool or within 10 feet (3048 mm) horizontally from the edge of the pool unless they are at least 22 feet (6706 mm) above the water in any direction and 14 feet (4267 mm) above a diving platform. Telephone and cable TV lines are required to be at least 10 feet (3048 mm) above a pool or diving platform. Typically, swimming pools are not installed under the service drop lines since this minimum height is difficult to maintain.

E4103.6 Underground wiring. Underground wiring shall not be installed under or within the area extending 5 feet (1524 mm) horizontally from the inside walls of pools and outdoor hot tubs and spas except where the wiring is installed to supply pool, spa or hot tub equipment or where space limitations prevent wiring from being routed 5 feet (1524 mm) or more horizontally from the inside walls. Where installed within 5 feet (1524 mm) of the inside walls, the wiring method shall be rigid metal conduit, intermediate metal conduit or a nonmetallic raceway system. Metal conduit shall be corrosion resistant and suitable for the location. The minimum raceway burial depth shall be in accordance with Table E4103.6.

❖ Where underground wiring is installed, corrosion of the raceway is possible that could result in a hazardous situation under ground fault conditions. Because of dampness and the presence of chemicals such as chlorine in the area, corrosion of metallic parts is more likely in a pool area and in the ground around a pool. For this reason, where it is necessary to install wiring underground within 5 feet of the pool, it must be in rigid metal conduit, intermediate metal conduit that is corrosion resistant, or nonmetallic raceway. A typical corrosion resistant rigid metal conduit has an outer nonmetallic covering. The other option that is commonly used is to run the wiring in a nonmetallic raceway.

TABLE E4103.5
OVERHEAD CONDUCTOR CLEARANCES

	INSULATED SUPPLY OR SERVICE DROP CABLES, 0-750 VOLTS TO GROUND, SUPPORTED ON AND CABLED TOGETHER WITH AN EFFECTIVELY GROUNDED BARE MESSENGER OR EFFECTIVELY GROUNDED NEUTRAL CONDUCTOR (feet)	ALL OTHER SUPPLY OR SERVICE DROP CONDUCTORS (feet)	
		Voltage to ground	
		0-15 kV	Greater than 15 to 50 kV
A. Clearance in any direction to the water level, edge of water surface, base of diving platform, or permanently anchored raft	22	25	27
B. Clearance in any direction to the diving platform	14	17	18
C. Horizontal limit of clearance measured from inside wall of the pool	This limit shall extend to the outer edge of the structures listed in Rows (A) and (B) above but not less than 10 feet.		

For SI: 1 foot = 304.8 mm.

❖ See the commentary for Section E4103.5.

TABLE E4103.6
MINIMUM BURIAL DEPTHS

WIRING METHOD	MINIMUM BURIAL DEPTH (inches)
Rigid metal conduit	6
Intermediate metal conduit	6
Nonmetallic raceways listed for direct burial without concrete encasement	18
Other approved raceways[a]	18

For SI: 1 inch = 25.4 mm.
a. Raceways approved for burial only where concrete-encased shall require a concrete envelope not less than 2 inches in thickness.

❖ See the commentary for Section E4103.6.

SECTION E4104 BONDING

E4104.1 Bonded parts. The following parts shall be bonded together:

1. All metallic parts of pool, spa and hot tub structures, including the reinforcing metal of pool, spa and hot tub shells, coping stones, and decks. The usual steel tie wires shall be considered suitable for bonding the reinforcing steel together, and welding or special clamping shall not be required. Such tie wires shall be made tight. Where reinforcing steel is effectively insulated by a listed encapsulating nonconductive compound, at the time of manufacture and installation, it shall not be required to be bonded. Where reinforcing steel is encapsulated with a nonconductive compound, provisions shall be made for an alternate means to eliminate voltage gradients that would otherwise be provided by unencapsulated bonded reinforcing steel.
2. All forming shells and mounting brackets of no-niche luminaires except where a listed low-voltage lighting system is used that does not require bonding.
3. All metal fittings within or attached to pool, spa and hot tub structures. Isolated parts that are not over 4 inches (102 mm) in any dimension and do not penetrate into the pool structure more than 1 inch (25.4 mm) shall not require bonding. The metal bands or hoops used to secure wooden staves for a hot tub or spa shall not be required to be bonded.
4. Metal parts of electrical equipment associated with pool, spa and hot tub water circulating systems, including pump motors and metal parts of equipment associated with pool covers, including electric motors. Metal parts of listed equipment incorporating an approved system of double insulation and providing a means for grounding internal nonaccessible, noncurrent-carrying metal parts shall not be bonded. Where a double-insulated water-pump motor is installed under the provisions of this section, a solid 8 AWG copper conductor that is of sufficient length to make a bonding connection to a replacement motor shall be extended from the bonding grid to an accessible point in the motor vicinity. Where there is no connection between the swimming pool bonding grid and the equipment grounding system for the premises, this bonding conductor shall be connected to the equipment grounding conductor of the motor circuit.
5. Metal-sheathed cables and raceways, metal piping and all fixed metal parts that are within 5 feet (1524 mm) horizontally of the inside walls of the pool, spa or hot tub and that are within 12 feet (3658 mm) above the maximum water level of the pool or any observation stands, towers or platforms, or from any diving structures, and that are not separated from the pool by a permanent barrier.

For pool water heaters rated at more than 50 amperes and having specific instructions regarding bonding and grounding, only those parts designated to bonded shall be bonded and only those parts designated to be grounded shall be grounded.

❖ Bonding is the joining of metallic parts to form an electrically conductive path that will result in electrical continuity between each component of a common grid to ensure that the electrical potential will be the same throughout. Keeping the electrical potential at the same level will reduce the shock hazard created by stray currents in the pool or in the ground around the pool. The risk of electric shock is great in damp locations, and it is increased where a person is in a pool surrounded by water. An electrical potential may occur in a pool due to such things as lightning, ground faults, or the operation of electrical equipment in close proximity. Bonding together of all the metallic items in and around a pool will help eliminate the voltage gradients or differences in electrical potential from one part of the pool to another or from metallic equipment to the pool water. The various items enumerated in the five paragraphs above provide a clear description of what parts must be bonded together and what parts are exempt from bonding.

Many of the bonding connections, such as the tie wires connecting the reinforcing steel together, must be inspected prior to placing the concrete or gunite.

E4104.2 Parts not required to be bonded. Small conductive surfaces not likely to become energized, such as towel bars, mirror frames, and air and water jets and drain fittings that are not connected to metallic piping, and similar equipment installed on or within indoor spas and hot tubs shall not be required to be bonded.

❖ Metallic objects that are not likely to be energized are exempt from bonding. The items in the list given as examples in this section are obviously not likely to become energized.

E4104.3 Methods of bonding. It shall not be the intent to require that the 8 AWG or larger solid copper bonding conductor be extended or attached to any remote panelboard, service equipment, or any electrode, but only that it shall be employed to eliminate voltage gradients in the pool area as prescribed. Bonding shall be accomplished by one or more of the following methods:

1. Common Bonding Grid. The parts specified in Section E4104.1 above shall be connected to a common bonding grid with a solid copper conductor, insulated, covered, or bare, not smaller than 8 AWG. Connection shall be made by exothermic welding or by pressure connectors or clamps that are labeled as being suitable for the purpose and that are made of stainless steel, brass, copper or copper alloy.

   The common bonding grid shall be permitted to be any of the following:

1.1. The structural reinforcing steel of a concrete pool where the reinforcing rods are bonded together by the usual steel tie wires made up tight or the equivalent; or

1.2. The wall of a bolted or welded metal pool; or

1.3. A solid copper conductor, insulated, covered, or bare, not smaller than 8 AWG.

2. For hot tubs and spas, metal to metal mounting on a common frame or base
3. The interconnection of threaded metal piping and fittings.

❖ The purpose of bonding all metal parts together is to ensure that all the metal components are at the same potential to eliminate the shock hazard from stray currents. The size 8 AWG pool bonding conductor does not have to be run back to the panelboard or service equipment. The code does not exactly require the bonding grid to be grounded, but it is tied back to the service grounding system through the equipment grounding conductor of the circuit that serves the underwater lighting fixtures and/or the pool circulating pump motor.

SECTION E4105 GROUNDING

E4105.1 Equipment to be grounded. The following equipment shall be grounded:

1. Wet-niche, dry-niche and no-niche underwater luminaires other than those low-voltage systems listed for the application without a grounding conductor.
2. All electrical equipment located within 5 feet (1524 mm) of the inside wall of the pool, spa or hut tub.
3. All electrical equipment associated with the recirculating system of the pool, spa or hot tub.
4. Junction boxes.
5. Transformer enclosures.
6. Ground-fault circuit-interrupters.
7. Panelboards that are not part of the service equipment and that supply any electrical equipment associated with the pool, spa or hot tub.

❖ Grounding and bonding are required for different reasons. Bonding is required for all metal parts of the electrical equipment and for non-electrical metal parts of the structure to establish an equipotential grid. Bonding conductors in many cases are permitted to be externally clamped or connected to the noncurrent-carrying metal parts of equipment. The attachment is not required to be accessible. Bonding of metal parts of electrical equipment is required to ensure a low-impedance path for fault current back to the source of the circuit to facilitate the operation of the overcurrent device. For equipment grounding, a separate insulated copper grounding conductor must be run to the equipment grounding terminal in the main service panel. For wiring of swimming pool equipment, metal raceways are not used as the equipment grounding conductor.

E4105.2 Luminaires and related equipment. Wet-niche, dry-niche, or no-niche luminaires shall be connected to an insulated copper equipment grounding conductor sized in accordance with Table E3808.12 but not smaller than 12 AWG. The equipment grounding conductor between the wiring chamber of the secondary winding of a transformer and a junction box shall be sized in accordance with the overcurrent device in such circuit. The junction box, transformer enclosure, or other enclosure in the supply circuit to a wet-niche or no-niche luminaire and the field-wiring chamber of a dry-niche luminaire shall be grounded to the equipment grounding terminal of the panelboard. The equipment grounding terminal shall be directly connected to the panelboard enclosure. The equipment grounding conductor shall be installed without joint or splice.

Exceptions:

1. Where more than one underwater luminaire is supplied by the same branch circuit, the equipment grounding conductor, installed between the junction boxes, transformer enclosures, or other enclosures in the supply circuit to wet-niche luminaires, or between the field-wiring compartments of dry-niche luminaires, shall be permitted to be terminated on grounding terminals.
2. Where an underwater luminaire is supplied from a transformer, ground-fault circuit-interrupter, clock-operated switch, or a manual snap switch that is located between the panelboard and a junction box connected to the conduit that extends directly to the underwater luminaire, the equipment grounding conductor shall be permitted to terminate on grounding terminals on the transformer, ground-fault circuit-interrupter, clock-operated switch enclosure, or an outlet box used to enclose a snap switch.

❖ Of particular note here is the requirement that the equipment grounding conductor must be insulated and sized in accordance with Table E3808.12. An insulated conductor run within the metal raceway is used since the conduit could corrode and open the conductive grounding path. The junction box or enclosure in the supply circuit to an underwater light fixture or related equipment must be grounded to the equipment grounding terminal of the panelboard. This is done by running a minimum size 12 AWG insulated equipment grounding conductor directly between the terminals of the junction box or enclosure and the panelboard. This conductor must not be confused with the size 8 AWG bonding conductor used to connect the forming shell to the bonding grid.

E4105.3 Nonmetallic conduit. Where a nonmetallic conduit is installed between a wet-niche luminaire and a junction box, transformer enclosure, or other enclosure, a 8 AWG insulated copper conductor shall be installed in this conduit with provisions for terminating in the forming shell, junction box or transformer enclosure, or ground-fault circuit-interrupter enclosure. The termination of the 8 AWG conductor in the forming shell shall be covered with, or encapsulated in, a listed potting compound to protect such connection from the possible deteriorating effect of pool water.

❖ In this section, the size 8 AWG conductor required to be run within the nonmetallic-conduit is not the same bonding conductor used to connect the forming shell to the common bonding grid. The size 8 AWG or larger bonding conductor connects to the outside of the forming shell and is a solid copper conductor permitted to be bare or insulated. The size 8 AWG conductor run within the nonmetallic-conduit provides electrical continuity between the forming shell and the junction box or transformer enclosure. It must be insulated (and green according to Section E3307.2) and is permitted to be a stranded conductor since the code does not specify solid. Corrosion can occur when the connection is exposed to pool water, and thus the connection must be encapsulated in a listed potting compound.

E4105.4 Flexible cords. Wet-niche or no-niche luminaires that are supplied by a flexible cord or cable shall have all exposed noncurrent-carrying metal parts grounded by an insulated copper equipment grounding conductor that is an integral part of the cord or cable. This grounding conductor shall be connected to a grounding terminal in the supply junction box, transformer enclosure, or other enclosure. The grounding conductor shall not be smaller than the supply conductors and not smaller than 16 AWG.

❖ The equipment grounding conductor for the branch circuit serving underwater lighting fixtures is the insulated equipment grounding conductor in the flexible cord that is run from a terminal in the forming shell to a terminal in the supply junction box. Except for the two situations given at the end of Section E4105.2, an equipment grounding conductor is run without splice from the terminal in the junction box back to the panelboard. The end of the flexible cord jacket must be sealed, and the terminations within the fixture must be covered with a suitable potting compound to prevent corrosion.

E4105.5 Motors. Pool-associated motors shall be connected to a copper equipment grounding conductor sized in accordance with Table E3808.12, but not smaller than 12 AWG.

❖ Motors used with pool equipment must be grounded with a minimum size 12 AWG equipment grounding conductor. It must be insulated or within an overall sheath such as nonmetallic-sheathed cable.

E4105.6 Panelboards. A panelboard that is not part of the service equipment, or source of a separately derived system shall have an equipment grounding conductor installed between its grounding terminal and the grounding terminal of the applicable service equipment or source of a separately derived system. The equipment grounding conductor shall be insulated, shall be sized in accordance with Table E3808.12, and shall be not smaller than 12 AWG.

❖ Where a separate panelboard that supplies swimming pool equipment is fed from service equipment, it must have an insulated equipment grounding conductor not smaller than size 12 AWG run with the feeders from the service equipment.

E4105.7 Cord-connected equipment. Where fixed or stationary equipment is connected with a flexible cord to facilitate removal or disconnection for maintenance, repair, or storage, as provided in Section E4102.2, the equipment grounding conductors shall be connected to a fixed metal part of the assembly. The removable part shall be mounted on or bonded to the fixed metal part.

❖ The equipment grounding conductor is required in order to provide a path of low impedance that helps limit the voltage to ground and serves to carry enough current to trip the overcurrent protective device. Flexible cord is used in some cases to supply power to equipment that may need to be removed for replacement or repair, and the equipment grounding conductor must be connected to a part of the equipment that is not removable.

E4105.8 Other equipment. Other electrical equipment shall be grounded in accordance with Section E3808.

❖ Equipment not addressed specifically in this section must comply with the grounding provisions of Section E3808.

SECTION E4106
EQUIPMENT INSTALLATION

E4106.1 Transformers. Transformers used for the supply of underwater luminaires, together with the transformer enclosure, shall be listed for the purpose. Such transformers shall be of an isolated winding type having a grounded metal barrier between the primary and secondary windings.

❖ Transformers used to supply power for underwater swimming pool and spa light fixtures typically are of the independent two-winding type and have a primary rating of 120 volts and a maximum secondary rating of 15 volts. They are of the isolated winding type, meaning they have a shield or metal barrier between the primary and secondary windings. The metal barrier is grounded by an equipment grounding conductor connected to the equipment grounding terminal in the transformer housing. Without a metal barrier or shield, deterioration of the insulation of the primary and secondary windings could cause the two windings to come into contact with each other, creating a higher voltage on the secondary windings.

E4106.2 Ground-fault circuit-interrupters. Ground-fault circuit-interrupters shall be self-contained units, circuit-breaker types, receptacle types or other approved types.

❖ A ground-fault circuit interrupter (GFCI) is a device designed to interrupt or open the circuit when a fault current to ground exceeds a certain value that is less than that required to trip the circuit breaker supplying the branch circuit. A Class A GFCI trips when the ground-fault current is from 4 to 6 milliamperes. Class A GFCIs are suitable for use in branch circuits feeding swimming pool equipment. However, existing branch circuits in older swimming pool installations might have sufficient leakage current to cause a Class A GFCI to trip.

E4106.3 Wiring on load side of ground-fault circuit-interrupters and transformers. For other than grounding conductors, conductors installed on the load side of a ground-fault circuit-interrupter or transformer used to comply with the provisions of Section E4106.4, shall not occupy raceways, boxes, or enclosures containing other conductors except where the other conductors are protected by ground-fault circuit interrupters or are grounding conductors. Supply conductors to a feed-through type ground-fault circuit interrupter shall be permitted in the same enclosure. Ground-fault circuit interrupters shall be permitted in a panelboard that contains circuits protected by other than ground-fault circuit interrupters.

❖ Except within panelboard cabinets, the load side conductors of a GFCI circuit are not permitted to share the same wiring space with other conductors that are not GFCI protected other than for equipment grounding conductors. The magnetic inductance caused by current flow in adjacent conductors could cause current to flow in circuit conductors on the load side of a GFCI simulating a ground fault. Also, insulation failure could energize GFCI protected conductors.

E4106.4 Underwater luminaires. The design of an underwater luminaire supplied from a branch circuit either directly or by way of a transformer meeting the requirements of Section E4106.1, shall be such that, where the fixture is properly installed without a ground-fault circuit-interrupter, there is no shock hazard with any likely combination of fault conditions during normal use (not relamping). In addition, a ground-fault circuit-interrupter shall be installed in the branch circuit supplying luminaires operating at more than 15 volts, so that there is no shock hazard during relamping. The installation of the ground-fault circuit-interrupter shall be such that there is no shock hazard with any likely fault-condition combination that involves a person in a conductive path from any ungrounded part of the branch circuit or the luminaire to ground. Compliance with this requirement shall be obtained by the use of a listed underwater luminaire and by installation of a listed ground-fault circuit-interrupter in the branch circuit. Luminaires that depend on submersion for safe operation shall be inherently protected against the hazards of overheating when not submerged.

❖ An underwater light fixture must be of such a design and listed to assure freedom from electric shock hazard during operation of the fixture without a GFCI device in the circuit. Any light fixture operating at over 15 volts must have GFCI protection. This will provide safety from electric shock during relamping. Wet-niche and no-niche fixtures are detached and brought above the water level for relamping.

E4106.4.1 Maximum voltage. Luminaires shall not be installed for operation on supply circuits over 150 volts between conductors.

❖ Where light fixtures are designed to operate at I20 volts, the circuit must have GFCI protection.

E4106.4.2 Luminaire location. Luminaires mounted in walls shall be installed with the top of the fixture lens not less than 18 inches (457 mm) below the normal water level of the pool, except where the luminaire is listed and identified for use at a depth of not less than 4 inches (102 mm) below the normal water level of the pool. A luminaire facing upward shall have the lens adequately guarded to prevent contact by any person.

❖ If a person is hanging onto the edge in the deep end of a pool, directly in front of a wet-niche light fixture with a broken bulb or damaged lens, fault current could leak into the pool directly to the person. Where the light fixture is at least 18 inches (457 mm) below the normal water level, the person's chest, the most sensitive area for shock hazard, would most likely be above and not be directly in line with the fixture.

E4106.5 Wet-niche luminaires. Forming shells shall be installed for the mounting of all wet-niche underwater luminaires and shall be equipped with provisions for conduit entries. Conduit shall extend from the forming shell to a suitable junction box or other enclosure located as provided in Section E4106.8. Metal parts of the luminaire and forming shell in contact with the pool water shall be of brass or other approved corrosion-resistant metal.

The end of flexible-cord jackets and flexible-cord conductor terminations within a luminaire shall be covered with, or encapsulated in, a suitable potting compound to prevent the entry of water into the luminaire through the cord or its conductors. In addition, the grounding connection within a luminaire shall be similarly treated to protect such connection from the deteriorating effect of pool water in the event of water entry into the luminaire.

Luminaires shall be bonded to and secured to the forming shell by a positive locking device that ensures a low-resistance contact and requires a tool to remove the luminaire from the forming shell.

❖ These fixtures are typically installed in the wall of a pool or field-fabricated spa. A forming shell is installed before concrete or gunite is placed. The forming shell is not sealed from pool water, and the fixture is designed to be surrounded by water. Wet-niche fixtures that are UL listed have a permanently attached, factory installed flexible cord at least 12 feet (3658 mm) long. With the sufficient length of flexible cord, the fixture can be removed from the housing or forming shell and brought up to the pool deck for relamping without lowering the water level and without disconnecting the cord from the junction box above the pool deck.

E4106.6 Dry-niche luminaires. Dry-niche luminaires shall be provided with provisions for drainage of water and means for accommodating one equipment grounding conductor for each conduit entry. Junction boxes shall not be required but, if used, shall not be required to be elevated or located as specified in Section E4106.8 if the luminaire is specifically identified for the purpose.

❖ These underwater lighting fixtures are designed to be serviced or relamped from the rear in a passage-way behind the pool wall. This is not very common in residential pools. A no-niche lighting fixture could be installed in the wall of a pool or spa that is partially above ground level where an elevated deck is built above ground to match the top edge of the pool. Some dry-niche fixtures are installed with a box flush with the deck (where listed for use with the dry-niche fixture, it is not required to be raised above the deck) from which the lamp can be replaced.

E4106.7 No-niche luminaires. No-niche luminaires shall be listed for the purpose and shall be installed in accordance with the requirements of Section E4106.5. Where connection to a forming shell is specified, the connection shall be to the mounting bracket.

❖ See Section E4106.5.

E4106.8 Junction boxes and enclosures for transformers or ground-fault circuit interrupters. Junction boxes for underwater luminaires and enclosures for transformers and ground-fault circuit-interrupters that supply underwater luminaires shall comply with the following:

❖ The branch circuit wiring for underwater swimming pool fixtures is brought to a junction box, transformer, or enclosure installed a relatively short distance from the pool or spa. A flexible cord from the fixture is installed in a raceway and connected in the junction box or transformer enclosure, which must be listed specifically for use with underwater light fixtures. It must be labeled as a swimming pool junction box.

E4106.8.1 Junction boxes. A junction box connected to a conduit that extends directly to a forming shell or mounting bracket of a no-niche luminaire shall be:

1. Listed and labled for the purpose; and
2. Equipped with threaded entries or hubs or a nonmetallic hub listed for the purpose; and
3. Constructed of copper, brass, suitable plastic, or other approved corrosion-resistant material; and
4. Provided with electrical continuity between every connected metal conduit and the grounding terminals by means of copper, brass, or other approved corrosion-resistant metal that is integral with the box; and
5. Located not less than 4 inches (102 mm), measured from the inside of the bottom of the box, above the ground level, or pool deck, or not less than 8 inches (203 mm) above the maximum pool water level, whichever provides the greatest elevation, and shall be located not less than 4 feet (1219 mm) from the inside wall of the pool, unless separated from the pool by a solid fence, wall or other permanent barrier. Where used on a lighting system operating at 15 volts or less, a flush deck box shall be permitted provided that an approved potting compound is used to fill the box to prevent the entrance of moisture; and the flush deck box is located not less than 4 feet (1219 mm) from the inside wall of the pool.

❖ The junction box must be sized according the provisions of Section 3805. The junction box must be large enough to accommodate the raceways connected to it. A nonmetallic conduit between a wet-niche light fixture and the deck junction box must be sized large enough for both a size 8 AWG insulated copper bonding conductor and the flexible cord that supplies the fixture. This conduit would need to be at least $^3/_4$ inch. Terminals must be provided within the box for connection of the bonding conductors. A box that is listed under swimming pool junction boxes will have the correct number of integral grounding and bonding terminals for the number of conduit entries it has. The box must have at least one grounding terminal for each conduit entry plus one. A grounding terminal or grounding bar is not permitted to be field installed. See Section E4106.8.4.

E4106.8.2 Other enclosures. An enclosure for a transformer, ground-fault circuit-interrupter or a similar device connected to a conduit that extends directly to a forming shell or mounting bracket of a no-niche luminaire shall be:

1. Listed and labeled for the purpose, comprised of copper, brass, suitable plastic, or other approved corrosion-resistant material; and
2. Equipped with threaded entries or hubs or a nonmetallic hub listed for the purpose; and
3. Provided with an approved seal, such as duct seal at the conduit connection, that prevents circulation of air between the conduit and the enclosures; and
4. Provided with electrical continuity between every connected metal conduit and the grounding terminals by means of copper, brass or other approved corrosion-resistant metal that is integral with the enclosures; and
5. Located not less than 4 inches (102 mm), measured from the inside bottom of the enclosure, above the ground level or pool deck, or not less than 8 inches (203 mm) above the maximum pool water level, whichever provides the greater elevation, and shall be located not less than 4 feet (1219 mm) from the inside wall of the pool, except where separated from the pool by a solid fence, wall or other permanent barrier.

❖ A transformer enclosure that has conduits that extend directly to the underwater lighting fixture must have the conduit entries sealed against air circulation. Corrosion of internal components of the enclosure could occur at a faster rate if not sealed.

E4106.8.3 Protection of junction boxes and enclosures. Junction boxes and enclosures mounted above the grade of the finished walkway around the pool shall not be located in the walkway unless afforded additional protection, such as by location under diving boards or adjacent to fixed structures.

❖ Junction boxes must not be installed in the open at walkways and similar locations where subject to damage from people walking, carts, patio furniture, etc.

E4106.8.4 Grounding terminals. Junction boxes, transformer enclosures, and ground-fault circuit-interrupter enclosures connected to a conduit that extends directly to a forming shell or mounting bracket of a no-niche luminaire shall be provided with grounding terminals in a quantity not less than the number of conduit entries plus one.

❖ See the commentary for Section E4106.8.1.

E4106.8.5 Strain relief. The termination of a flexible cord of an underwater luminaire within a junction box, transformer enclosure, ground-fault circuit-interrupter, or other enclosure shall be provided with a strain relief.

❖ A flexible cord installed from an underwater lighting fixture within a raceway must be connected within the box with a strain relief device. When the fixture is removed from the housing or forming shell for relamping, the cord is pulled out of the forming shell and it could be pulled against the connections in the junction box resulting in bad connections or damage to the cord.

E4106.9 Underwater audio equipment. Underwater audio equipment shall be identified for the purpose.

❖ Underwater audio equipment is installed under similar provisions as underwater lighting fixtures. See Section E4106.4.

E4106.9.1 Speakers. Each speaker shall be mounted in an approved metal forming shell, the front of which is enclosed by a captive metal screen, or equivalent, that is bonded to and secured to the forming shell by a positive locking device that ensures a low-resistance contact and requires a tool to open for installation or servicing of the speaker. The forming shell shall be installed in a recess in the wall or floor of the pool.

❖ Underwater speakers are installed in a listed forming shell that is recessed in the wall of the pool. A metal screen must be installed over the forming shell to prevent contact with the speaker. The screen must require a tool to open. Typically, the screen is attached with screws so that it is not easily removed.

E4106.9.2 Wiring methods. Rigid metal conduit or intermediate metal conduit of brass or other identified corrosion-resistant metal, rigid nonmetallic conduit, or liquidtight flexible nonmetallic conduit (LFNC-B) shall extend from the forming shell to a suitable junction box or other enclosure as provided in Section E4106.8. Where rigid nonmetallic conduit or liquidtight flexible nonmetallic conduit is used, a 8 AWG solid or stranded insulated copper equipment grounding conductor shall be installed in this conduit with provisions for terminating in the forming shell and the junction box. The termination of the 8 AWG conductor in the forming shell shall be covered with, or encapsulated in, a suitable potting compound to protect such connection from the possible deteriorating effect of pool water.

❖ The wiring method of extending a conduit from the speaker forming shell to a deck junction box is the same as for a wet- niche lighting fixture. See the commentary for Section E4106.4.

E4106.9.3 Forming shell and metal screen. The forming shell and metal screen shall be of brass or other approved corrosion-resistant metal. All forming shells shall include provisions for terminating an 8 AWG copper conductor.

❖ Listed equipment is available for installation of underwater audio equipment.

E4106.10 Electrically operated pool covers. The electric motors, controllers, and wiring for pool covers shall be located not less than 5 feet (1524 mm) from the inside wall of the pool except where separated from the pool by a wall, cover, or other permanent barrier. Electric motors installed below grade level shall be of the totally enclosed type. The electric motor and controller shall be connected to a circuit protected by a ground-fault circuit-interrupter. The device that controls the operation of the motor for an electrically operated pool cover shall be located so that the operator has full view of the pool.

❖ Pool cover motors and the controllers are typically located at the side of the pool in a box or recessed with a cover flush with the pool deck. Obviously the box or recessed enclosure must be provided with means to drain water that will accumulate, and the motor must be totally enclosed and served by a GFCI protected circuit.

The operating control for an electrically operated pool cover must be located so as to allow the person operating the device to see if people are clear of moving parts. This is a common-sense safety approach for the operation of all machinery.

E4106.11 Electric pool water heaters. All electric pool water heaters shall have the heating elements subdivided into loads not exceeding 48 amperes and protected at not more than 60 amperes. The ampacity of the branch-circuit conductors and the rating or setting of overcurrent protective devices shall be not less than 125 percent of the total nameplate load rating.

❖ See Sections E4001.4.1 and E3602.10.

E4106.12 Pool area heating. The provisions of Sections E4106.12.1 through E4106.12.3 shall apply to all pool deck areas, including a covered pool, where electrically operated comfort heating units are installed within 20 feet (6096 mm) of the inside wall of the pool.

❖ Where electric heating is installed in an area within 20 feet (6096 mm) of the inside wall of the pool, specific code rules apply to the wiring for heating units.

E4106.12.1 Unit heaters. Unit heaters shall be rigidly mounted to the structure and shall be of the totally enclosed or guarded types. Unit heaters shall not be mounted over the pool or within the area extending 5 feet (1524 mm) horizontally from the inside walls of a pool.

❖ Where unit heaters are used in an indoor pool, spa, or hot tub area, they must be kept at least 5 feet from the inside wall of the pool. They must be guarded or enclosed. These two terms have clear definitions, given in Chapter 34.

E4106.12.2 Permanently wired radiant heaters. Electric radiant heaters shall be suitably guarded and securely fastened to their mounting devices. Heaters shall not be installed over a pool or within the area extending 5 feet (1524 mm) horizontally from the inside walls of the pool and shall be mounted not less than 12 feet (3658 mm) vertically above the pool deck.

❖ Radiant heaters in an indoor pool area must be at least 12 feet (3058 mm) above the pool deck. Although these heating units are required to be securely fastened, there is no exception permitting installation over the pool. They must be at least 5 feet (1524 mm) from the inside wall of the pool.

E4106.12.3 Radiant heating cables prohibited. Radiant heating cables embedded in or below the deck shall be prohibited.

❖ Radiant heating cables are prohibited from being installed in the deck of the pool area because of the possible breakdown of the insulation on the cable leading to ground fault or current leakage.

E4106.13 Double insulated pool pumps. A permanently installed pool shall be permitted to be provided with listed cord- and plug-connected pool pumps incorporating an approved system of double insulation that provides a means for grounding only the internal and nonaccessible, noncurrent-carrying metal parts of the pump.

❖ See Section E4107.1.

SECTION E4107 STORABLE SWIMMING POOLS

E4107.1 Pumps. A cord-connected pool filter pump for use with storable pools shall incorporate an approved system of double insulation or its equivalent and shall be provided with means for grounding only the internal and nonaccessible non-current-carrying metal parts of the appliance.

The means for grounding shall be an equipment grounding conductor run with the power-supply conductors in a flexible cord that is properly terminated in a grounding-type attachment plug having a fixed grounding contact.

❖ Filter pumps must be protected by a system of double insulation or the equivalent. The means for grounding the internal non-current carrying metal parts must be an equipment grounding conductor run with the power-supply conductors in the flexible cord terminated in a three-prong or grounding-type attachment plug.

E4107.2 Ground-fault circuit-interrupters required. Electrical equipment, including power-supply cords, used with storable pools shall be protected by ground-fault circuit-interrupters.

❖ See the commentary for Section E4103.4.5.

E4107.3 Luminaires. Luminaires for storable pools shall not have exposed metal parts and shall be listed for the purpose as an assembly. In addition, luminaires for storable pools shall comply with the requirements of Section E4107.3.1 or E4107.3.2.

❖ Lighting fixtures for storable pools are intended for temporary installation on or in the wall of the pool, generally between 8 and 10 inches (203 and 254 mm) below the top of the pool wall. These light fixtures are manufactured as an assembly with flexible cords of 25 feet (762 mm) or more so that they can be routed away from the pool to an enclosure for a transformer or GFCI enclosure. The transformer or GFCI assembly typically has a 3 to 6-foot-long (914 to 1829 mm) power supply cord, and the unit is designed to be temporarily installed on a structure near a receptacle.

E4107.3.1 Fifteen (15) volts or less. A luminaire installed in or on the wall of a storable pool shall be part of a cord- and plug-connected lighting assembly. The assembly shall:

1. Have a luminaire lamp that operates at 15 volts or less; and
2. Have an impact-resistant polymeric lens, luminaire body, and transformer enclosure; and
3. Have a transformer meeting the requirements of section E4106.1 with a primary rating not over 150 volts.

❖ For use on a storable pool, a cord- and plug-connected low-voltage light fixture is a listed assembly that receives its power through a flexible cord plugged into a transformer. The primary of the transformer is supplied from a 120-volt branch circuit.

E4107.3.2 Not over 150 volts. A lighting assembly without a transformer, and with the luminaire lamp(s) operating at not over 150 volts, shall be permitted to be cord- and plug-connected where the assembly is listed as an assembly for the purpose and complies with all of the following:

1. It has an impact-resistant polymeric lens and luminaire body.
2. A ground-fault circuit interrupter with open neutral protection is provided as an integral part of the assembly.
3. The luminaire lamp is permanently connected to the ground-fault circuit interrupter with open-neutral protection.
4. It complies with the requirements of Section E4106.4.
5. It has no exposed metal parts.

❖ These fixtures are connected through flexible cords plugged into GFCI devices that have open-neutral protection. A typical ground-fault circuit-interrupter device monitors the current in the ungrounded and grounded conductors, and when it is not balanced, it will trip. A GFCI with open-neutral protection will trip if the neutral is opened or lost.

SECTION E4108 SPAS AND HOT TUBS

E4108.1 Ground-fault circuit-interrupters. The outlet(s) that supplies a self-contained spa or hot tub, or a packaged spa or hot tub equipment assembly, or a field-assembled spa or hot tub with a heater load of 50 amperes or less, shall be protected by a ground-fault circuit-interrupter.

A listed self-contained unit or listed packaged equipment assembly marked to indicate that integral ground-fault circuit-interrupter protection is provided for all electrical parts within the unit or assembly, including pumps, air blowers, heaters, luminaires, controls, sanitizer generators and wiring, shall not require that the outlet supply be protected by a ground-fault circuit interrupter.

A combination pool/hot tub or spa assembly commonly bonded need not be protected by a ground-fault circuit interrupter.

❖ The definitions provided at the beginning of this chapter for "packaged spa or hot tub equipment assembly" and "self-contained spa or hot tub" are useful in understanding and applying the rules on GFCI protection. Where these units are listed and provided with integral GFCI protection, the flexible cord is not required to be plugged into a GFCI receptacle.

E4108.2 Electric water heaters. Electric spa and hot tub water heaters shall be listed and shall have the heating elements subdivided into loads not exceeding 48 amperes and protected at not more than 60 amperes. The ampacity of the branch-circuit conductors, and the rating or setting of overcurrent protective devices, shall be not less than 125 percent of the total nameplate load rating.

❖ The ampere load on the circuits supplying water heaters for spas and hot tubs is limited to 80 percent of the branch circuit rating, and the maximum load is 48 amperes. Therefore, 60 amperes is the maximum rating for these water heater circuits. If the water heater load consists of more than 48 amperes, it must be broken up into groups, and the power supply to the heating elements must be divided so that any one circuit is supplying a load not over 48 amperes.

E4108.3 Underwater audio equipment. Underwater audio equipment used with spas and hot tubs shall comply with the provisions of Section E4106.9.

❖ See Section E4106.9.

SECTION E4109 HYDROMASSAGE BATHTUBS

E4109.1 Ground-fault circuit-interrupters. Hydromassage bathtubs and their associated electrical components shall be protected in accordance with Section E4108. All 125-volt, single-phase receptacles not exceeding 30 amperes and located within 5 feet (1524 mm) measured horizontally of the inside walls of a hydromassage tub shall be protected by a ground-fault circuit interrupter(s).

❖ Hydromassage bathtubs typically have a cord- and plug-connected motor and a heater that come together with the unit as a package. A flexible cord, usually from 18 inches to 3 feet (457 mm to 914 mm) long, is provided for the power supply to the motor and/or heater. The GFCI protected receptacle is usually located in the accessible area near this equipment.

Where located farther than 5 feet (1524 mm) from the inside wall of a bathtub, the receptacles are not required to have GFCI protection because such receptacles are not easily accessed by or are not within the immediate reach of bathers. Receptacles located such that the occupants would be able to use electrical appliances and electronic products in close proximity to the bathtub should be avoided; however, if they are installed with 5 feet, they must be GFCI protected. Section E3802.1 requires all receptacles in a bathroom to be GFCI protected; therefore, if a hydromassage bathtub is located in a room defined as a bathroom, the distance between the receptacles and the bathtub is irrelevant.

E4109.2 Other electric equipment. Luminaires, switches, receptacles, and other electrical equipment located in the same room, and not directly associated with a hydromassage bathtub, shall be installed in accordance with the requirements of this code relative to the installation of electrical equipment in bathrooms.

❖ Receptacles and lighting fixtures in the area are subject to the code provisions for bathrooms. There are no specific rules for locating receptacles at a certain distance from the hydromassage bathtub as there are for pool, spas or hot tubs.

E4109.3 Accessibility. Hydromassage bathtub electrical equipment shall be accessible without damaging the building structure or building finish.

❖ The equipment, usually at the base of the hydromassage bathtub, must be accessible. This access is usually provided through an access panel. The cover or door must not be permanently sealed by tile, etc.

E4109.4 Bonding. All metal piping systems, metal parts of electrical equipment, and pump motors associated with the hydromassage tub shall be bonded together using a copper bonding jumper, insulated, covered, or bare, not smaller than 8 AWG solid.

Metal parts of listed equipment incorporating an approved system of double insulation and providing a means for grounding internal nonaccessible, noncurrent-carrying metal parts shall not be bonded.

❖ Metal parts of the hydromassage bathtub must be bonded together with at least a size 8 AWG copper conductor. This is usually done at the base of the unit where the equipment is located.

Bibliography

The following resource materials are referenced in this chapter or are relevant to the subject matter addressed in this chapter.

NFPA 70-02, *National Electrical Code.* Quincy, MA: National Fire Protection Association, 2002.

Chapter 42: Class 2 Remote-control, Signaling and Power-limited Circuits

General Comments

This chapter covers low-voltage power-limited circuits such as alarm, door bell, remote control, and signaling circuits.

Purpose

By regulating installation methods, conductor types and locations, and power supplies, this chapter intends to mitigate the potential fire and shock hazards associated with low-voltage power-limited circuits.

SECTION E4201 GENERAL

E4201.1 Scope. This chapter contains requirements for power supplies and wiring methods associated with Class 2 remote-control, signaling, and power-limited circuits that are not an integral part of a device or appliance. Other classes of remote-control, signaling and power-limited conductors shall comply with Article 725 of the *National Electrical Code®*.

❖ Wiring for heating and cooling controls, security systems, intercom systems, and door bell chimes are examples of Class 2 wiring in one- and two-family dwellings. Low-voltage circuits that are part of an appliance or equipment are not covered in this chapter. For example, the low-voltage wiring that is integral with a furnace is not covered but the thermostat wiring is, since it is run in the dwelling exterior to the furnace. For a garage door opener, the wiring to the safety sensors at the base of the door and the push button control wiring is subject to the provisions of this chapter.

E4201.2 Definitions.

CLASS 2 CIRCUIT. That portion of the wiring system between the load side of a Class 2 power source and the connected equipment. Due to its power limitations, a Class 2 circuit considers safety from a fire initiation standpoint and provides acceptable protection from electric shock.

❖ There is little or no shock or fire hazard from the operation of devices on power-limited low-voltage circuits. Class 2 circuits typically operate at 30 volts or less and use power supplies that limit current flow to reasonably safe levels.

REMOTE-CONTROL CIRCUIT. Any electrical circuit that controls any other circuit through a relay or an equivalent device.

❖ A low-voltage remote-control circuit is used to open or close a switch to control current flow in another circuit. Current is supplied by a low-voltage power source to a relay, which is used to control the energy supply to equipment, appliances, or light fixtures. A thermostat is part of a remote-control circuit for the HVAC system that it controls.

SIGNALING CIRCUIT. Any electrical circuit that energizes signaling equipment.

❖ Examples of signaling circuits in one- and two-family dwellings include wiring for security systems, fire alarm systems, chimes, bells, and buzzers. For the hearing impaired, a doorbell button is used to energize a relay to turn on a dedicated lamp or to initiate a sequence of strobe lights. Also for the hearing impaired, signaling devices are available that cause the normal lighting fixtures to flash on and off rapidly for a short period. The signaling devices are usually controlled by low-voltage circuits.

SECTION E4202 POWER SOURCES

E4202.1 Power sources for Class 2 circuits. The power source for a Class 2 circuit shall be one of the following:

1. A listed Class 2 transformer.
2. A listed Class 2 power supply.
3. Other listed equipment marked to identify the Class 2 power source.
4. Listed computer equipment limited power circuits.
5. A dry cell battery provided that the voltage is 30 volts or less and the capacity is equal to or less than that available from series connected No. 6 carbon zinc cells.

❖ Class 2 low-voltage circuits in residential wiring usually have the voltage and current inherently limited by a transformer. These transformers have inherent protection against overcurrent. Low-voltage transformers are designed to limit current flow to safe values even when a short circuit occurs. Low-voltage transformers are rated in volt-amperes, and a typical step-down transformer for a 24-volt system is rated at 40 volt-amperes or less. The power rating of Class 2 low-voltage circuits in typical residential applications is less than 100 volt-amperes.

E4202.2 Interconnection of power sources. A Class 2 power source shall not have its output connections paralleled or otherwise interconnected with another Class 2 power source except where listed for such interconnection.

❖ Power sources must not be interconnected so as to allow excessive current flow in Class II circuits. The power supply for a remote-control circuit is usually a step-down transformer, from 120 volts to 24 volts for example. The transformer can supply the power for an entire system. The devices on a low-voltage system do not all operate at the same time; therefore, power supplied by one transformer is sufficient.

SECTION E4203 WIRING METHODS

E4203.1 Wiring methods on supply side of Class 2 power source. Conductors and equipment on the supply side of the power source shall be installed in accordance with the appropriate requirements of Chapters 33 through 40. Transformers or other devices supplied from electric light or power circuits shall be protected by an overcurrent device rated at not over 20 amperes. The input leads of a transformer or other power source supplying Class 2 circuits shall be permitted to be smaller than No. 14, if not over 12 inches (305 mm) long and if the conductor insulation is rated at not less than 600 volts. In no case shall such leads be smaller than No. 18.

❖ The maximum rating of a branch circuit used to supply power to Class 2 low-voltage circuit is 20 amperes. The input leads to a transformer are required to be at least size 18 AWG. All of the code provisions covering branch circuits must be followed for the wiring up to and including the primary side of the transformer or power supply. The primary leads of a transformer are installed in the same box or enclosure with the branch circuit conductors and must have at least the same voltage rating as that of the branch circuit wires.

E4203.2 Wiring methods and materials on load side of the Class 2 power source. Class 2 cables installed as wiring within buildings shall be listed as being resistant to the spread of fire and listed as meeting the criteria specified in Sections E4203.2.1 through E4203.2.3. Cables shall be marked in accordance with Section E4203.2.4. Cable substitutions as described in Table E4203.2 and wiring methods covered in Chapter 37 shall also be permitted.

❖ Class 2 low-voltage wiring on the load side of the power source is limited to the cable types listed in Table E4302.2 and wiring methods covered in Chapter 37. The conductors for Class 2 low-voltage circuits are listed based on their flame spread characteristics. They are typically rated as conductors having low smoke-producing characteristics during standardized tests for fire and smoke. The cable types in Table E4203.2 are listed in descending order of reistance to spread of fire.

TABLE E4203.2
CABLE USES AND PERMITTED SUBSTITUTIONS

CABLE TYPE	USE	PERMITTED SUBSTITUTIONS[a]
CL2P	Class 2 Plenum Cable	MPP, CMP, FPLP, CL3P
CL2	Class 2 Cable	MPP, CMP, FPLP, CL3P, CL2P, MPR, CMR, FPLR, CL3R, CL2R MPG, MP, CMG, CM, FPL, CL3
CL2X	Class 2 Cable, Limited Use	MPP, CMP, FPLP, CL3P CL2P, MPR, CMR, FPLR, CL3R, CL2R, MPG, MP, CMG, CM, FPL, CL3, CL2, CMX, CL3X

a. For identification of cables other than Class 2 cables, see the *National Electrical Code®*.

E4203.2.1 Type CL2P cables. Cables installed in ducts, plenums and other spaces used to convey environmental air shall be Type CL2P cables listed as being suitable for the use and listed as having adequate fire-resistant and low smoke-producing characteristics.

❖ Under standardized test methods, Class 2 plenum cable (CL2P) is determined to have adequate fire-resistant and low smoke-producing characteristics. The Underwriters Laboratories (UL) *Marking Guide for Wire and Cable* (1993) lists the testing criteria for power-limited circuit cable. Cable Type CL2P exhibits a maximum peak optical density of 0.5, a maximum optical density of 0.15, and maximum flame-spread distance of 5 feet (1524 mm).

E4203.2.2 Type CL2 cables. Cables for general-purpose use, shall be listed as being resistant to the spread of fire and listed for the use.

❖ The technical criterion for listing cable Type CL2 as being resistant to the spread of fire is based on a vertical tray flame test where the cable "does not spread flame to the top of the tray" as stated in the (UL) *Marking Guide for Wire and Cable* (1993).

E4203.2.3 Type CL2X cables. Type CL2X limited-use cable shall be listed as being suitable for use in dwellings and for the use and in raceways and shall also be listed as being flame retardant. Cables with a diameter of 0.25 inches (6.4 mm) or less shall be permitted to be installed as open cable without a raceway.

❖ According to the (UL) *Marking Guide for Wire and Cable* (1993), cable Type CL2X is intended for use in Class 2 circuits in one- and two-family dwellings or multifamily dwellings when the cable diameter is less than 0.25 inches. This cable complies with "vertical wire" flame test requirements in UL 1581.

E4203.2.4 Marking. Cables shall be marked in accordance with Table E4203.2. Voltage ratings shall not be marked on cables.

❖ The voltage rating is not marked on cables listed for use in Class 2 low-voltage circuits because a voltage rating could be misinterpreted to mean that the cable is suitable for use in Class 1 circuits.

SECTION E4204 INSTALLATION REQUIREMENTS

E4204.1 Separation from other conductors. In cables, compartments, enclosures, outlet boxes, device boxes, and raceways, conductors of Class 2 circuits shall not be placed in any cable, compartment, enclosure, outlet box, device box, raceway, or similar fitting with conductors of electric light, power, Class 1 and nonpower-limited fire alarm circuits.

Exceptions:

1. Where the conductors of the electric light, power, Class 1 and nonpower-limited fire alarm circuits are separated by a barrier from the Class 2 circuits. In enclosures, Class 2 circuits shall be permitted to be installed in a raceway within the enclosure to separate them from Class 1, electric light, power and nonpower-limited fire alarm circuits
2. Class 2 conductors in compartments, enclosures, device boxes, outlet boxes and similar fittings where electric light, power, Class 1 or nonpower-limited fire alarm circuit conductors are introduced solely to connect to the equipment connected to the Class 2 circuits. The electric light, power, Class 1 and nonpower-limited fire alarm circuit conductors shall be routed to maintain a minimum of 0.25 inch (6.4 mm) separation from the conductors and cables of the Class 2 circuits; or the electric light power, Class 1 and nonpower-limited fire alarm circuit conductors operate at 150 volts or less to ground and the Class 2 circuits are installed using Types CL3, CL3R, or CL3P or permitted substitute cables, and provided that these Class 3 cable conductors extending beyond their jacket are separated by a minimum of 0.25 inch (6.4 mm) or by a nonconductive sleeve or nonconductive barrier from all other conductors.

❖ Class 2 low-voltage wiring is not permitted to be run in the same raceway or installed in the same box or enclosure with 120-volt branch-circuit wiring. For example, the 120-volt branch circuit conductors for a furnace are run in conduit to the furnace junction box. The Class 2 thermostat wiring controlling the furnace cannot be run in the same raceway.

Class 2 low-voltage conductors could be installed in an enclosure with higher voltage power conductor if a barrier or partition is installed to separate the wiring of each system. This does not apply to conductors such as the leads to the primary side of the transformer as indicated in Section E4203.1.

E4204.2 Other applications. Conductors of Class 2 circuits shall be separated by not less than 2 inches (51 mm) from conductors of any electric light, power, Class 1 or nonpower-limited fire alarm circuits except where one of the following conditions is met:

1. All of the electric light, power, Class 1 and nonpower-limited fire alarm circuit conductors are in raceways or in metal-sheathed, metal-clad, nonmetallic-sheathed or Type UF cables.
2. All of the Class 2 circuit conductors are in raceways or in metal-sheathed, metal-clad, nonmetallic-sheathed or Type UF cables.

❖ Class 2 circuit conductors must be separated by at least 2 inches (51 mm) from any power and lighting branch circuit conductors except where either the Class 2 or power and lighting conductors are enclosed in raceways, sheathed cable, or metal-clad cable. Low-voltage door bell wiring, for example, can be run through the same bored holes as used for the NM cable.

E4204.3 Class 2 circuits with communications circuits. Where Class 2 circuit conductors are in the same cable as communications circuits, the Class 2 circuits shall be classified as communications circuits and shall meet the requirements of Article 800 of the *National Electrical Code*®. The cables shall be listed as communications cables or multipurpose cables.

Exception: Cables constructed of individually listed Class 2 and communications cables under a common jacket shall not be required to be classified as communications cables. The fire-resistance rating of the composite cable shall be determined by the performance of the composite cable.

❖ In some cases, a composite cable is used that has telephone wire and Class 2 signaling wire. For these installations, the code provisions for communication circuits apply.

E4204.4 Class 2 cables with other circuit cables. Jacketed cables of Class 2 circuits shall be permitted in the same enclosure or raceway with jacketed cables of any of the following:

1. Power-limited fire alarm systems in compliance with Article 760 of the *National Electrical Code*®.
2. Nonconductive and conductive optical fiber cables in compliance with Article 770 of the *National Electrical Code*®.
3. Communications circuits in compliance with Article 800 of the *National Electrical Code*®.
4. Community antenna television and radio distribution systems in compliance with Article 820 of the *National Electrical Code*®.

❖ Jacketed or sheathed Class 2 cables are permitted in the same raceway or enclosure with the cables listed since they are protected from faults by the jackets on the cables, and the cables listed in this code section operate at low voltage.

E4204.5 Installation of conductors and cables. Cables shall be supported by the building structure in such a manner that the cable will not be damaged by normal building use. The installation of wires and cables shall not prevent access to equipment nor prevent removal of panels, including suspended ceiling panels. Raceways shall not be used as a means of support for

Class 2 circuit conductors, except where the supporting raceway contains conductors supplying power to the functionally associated equipment controlled by the Class 2 conductors.

❖ While Class 2 low-voltage cables can be installed in the same bored holes as NM cable, they cannot be taped, strapped, or attached by any means to a raceway. The exception to this is where a low-voltage thermostat cable is run along with and strapped to the raceway to a furnace or condensing unit. In this case, the Class 2 cable must be part of the control circuitry for the equipment supplied by the raceway wiring method. Low voltage wiring is sometimes run in a careless manner since there is not as much concern for shock or fire hazard. However, low-voltage wiring should be supported from the building structure and run in a manner that will protect it from damage.

Bibliography

The following resource materials are referenced in this chapter or are relevant to the subject matter addressed in this chapter.

NFPA 70-02, *National Electrical Code.* Quincy, MA: National Fire Protection Association, 2002.

Wire and Cable Marking Guide. Northbrook, IL: Underwriters Laboratories, 1993.

Part IX — Referenced Standards

Chapter 43: Referenced Standards

General Comments

Chapter 43 contains a comprehensive list of standards that are referenced in the code. It is organized to make locating specific document references easy.

It is important to understand that not every document related to residential design, installation, and construction is qualified to be a "referenced standard." The International Code Council (ICC) has adopted a criterion that standards referenced in the International Codes and standards intended for adoption into the International Codes must meet to qualify as a referenced standard. The policy is summarized as follows:

- Code references: The scope and application of the standard must be clearly identified in the code text.
- Standard content: The standard must be written in mandatory language and be appropriate for the subject covered. The standard cannot have the effect of requiring proprietary materials or pre-scribing a proprietary testing agency.
- Standard promulgation: The standard must be readily available and developed and maintained in a consensus process such as those used by ASTM or ANSI.

The ICC Code Development Procedures, of which the standards policy is a part, are updated periodically. A copy of the latest version can be obtained from the ICC offices.

Once a standard is incorporated into the code through the code development process, it becomes an enforceable part of the code. When the code is adopted by a jurisdiction, the standard also is part of that jurisdiction's adopted code. It is for this reason that the criteria were developed. Compliance with this policy means that documents incorporated into the code are developed through the use of a consensus process, are written in mandatory language, and do not mandate the use of proprietary materials or agencies. The requirement for a standard to be developed through a consensus process means that the standard is representative of the most current body of available knowledge on the subject as determined by a broad range of interested or affected parties without dominance by any single interest group. A true consensus process has many attributes, including but not limited to:

- An open process that has formal (published) procedures that allow for the consideration of all viewpoints;
- A definitive review period that allows for the standard to be updated and/or revised;
- A process of notification to all interested parties; and
- An appeals process.

Many available documents related to mechanical system design, installation, and construction, though useful, are not "standards" and are not appropriate for reference in the code. Often, these documents are developed or written with the intention of being used for regulatory purposes and are unsuitable for use as a standard because of extensive use of recommendations, advisory comments, and nonmandatory terms. Typical examples include installation instructions, guidelines, and practices.

The objective of ICC's standards policy is to provide regulations that are clear, concise, and enforceable; thus, the requirement for standards to be written in mandatory language. This requirement is not intended to mean that a standard cannot contain informational or explanatory material that will aid the user of the standard in its application. When the standard's promulgating agency wants such material to be included, however, the information must appear in a nonmandatory location, such as an annex or appendix, and be clearly identified as not being part of the standard.

Overall, standards referenced by the code must be authoritative, relevant, up to date, and, most important, reasonable and enforceable. Standards that comply with ICC's standards policy fulfill these expectations.

Purpose

This code contains numerous references to documents that are used to regulate materials and methods of construction. The references to these documents within the code text consist of the promulgating agency's acronym and its publication designation (for example, AWS A5.8) and a further indication that the document being referenced is the one that is listed in Chapter 43. Chapter 43 contains all of the information that is necessary to identify the specific referenced document. Included is the following information on a document's promulgating agency:

- The promulgating agency (the agency's title);
- The promulgating agency's acronym; and
- The promulgating agency's address.

For example, a reference to an ASME standard within the code indicates that the document is promulgated by the American Society of Mechanical Engineers (ASME), which is located in New York, New York. Chapter 43 lists the standards agencies alphabetically for ease of identification.

Chapter 43 also includes the following information on the referenced document itself:

- The document's publication designation;

- The document's edition year;
- The document's title;
- Any addenda or revisions to the document known at the time of the code's publication; and
- Every section of the code in which the document is referenced.

For example, a reference to ASME A112.1.3 indicates that this document can be found in Chapter 43 under the heading ASME. The specific standard's designation is A112.1.3. For convenience, these designations are listed in alphanumeric order. Chapter 43 identifies that ASME A112.1.3 is titled *Air Gap Fittings for Use with Plumbing Fixtures, Appliances, and Appurtenances*; the applicable edition (that is, its year of publication) is 2000; and it is referenced in one specifically identified section of the code. See Commentary Figure 43.

Chapter 43 also indicates when a document has been discontinued or replaced by its promulgating agency. When a document is replaced by a different one, a note appears to tell the user the designation and title of the new document.

The key aspect of the manner in which standards are referenced by the code is that a specific edition of a specific standard is clearly identified. The requirements necessary for compliance can be readily determined. The basis for code compliance is, therefore, established and available to the code official, the mechanical contractor, the designer, and the owner.

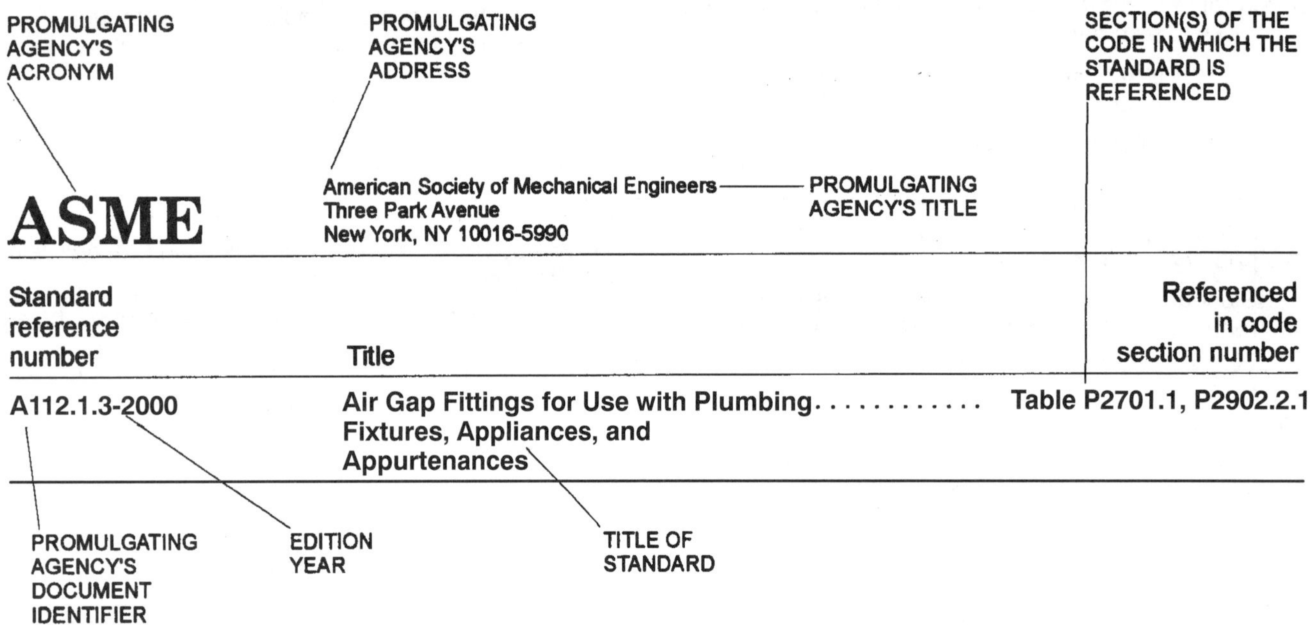

FIGURE 43
REFERENCED STANDARDS

AAMA

American Architectural Manufacturers Association
1827 Walden Office Square, Suite 104
Schaumburg, IL 60173

Standard reference number	Title	Referenced in code section number
101/I.S2—97	Voluntary Specifications for Aluminum, Vinyl (PVC) and Wood Windows and Glass Doors	R613.3, N1101.3.2.2
101/I.S2/NAFS—02	Voluntary Performance Specification for Windows, Skylights and Glass Doors	R308.6.9, R613.3, N1101.3.2.2

ACI

American Concrete Institute
38800 Country Club Drive
Farmington Hills, MI 48331

Standard reference number	Title	Referenced in code section number
318—02	Building Code Requirements for Structural Concrete	R402.2, R404.1, R404.4, R404.4.6.1, Table R404.4(1), Table R404.4(2), Table R404.4(3), Table R404.4(4), Table R404.4(5), R611.1, Table R611.3(1), Table R611.7(1), Table R611.7(2), Table R611.7(3), Table R611.7(4), Table R611.7(5), Table R611.7(6), Table R611.7(7), Table R611.7(9), Table R611.7(10), R611.7.1.1, Table R611.7.4, R612.1
530—02	Building Code Requirements for Masonry Structures	R404.1, R606.1, R606.1.1, R606.11.1, R606.11.2.2.1, R606.11.2.2.2, R606.11.3.1
530.1—02	Specifications for Masonry Structures	R404.1, R606.1, R606.1.1, R606.11.1, R606.11.2.2.1, R606.11.2.2.2, R606.11.3.1

ACCA

Air Conditioning Contractors of America
1228 17th Street, NW
Washington, DC 20046

Standard reference number	Title	Referenced in code section number
Manual D—95	Residential Duct Systems	M1601.1, M1602.2
Manual J—87	Residential Load Calculation—Seventh Edition	M1401.3

AFPA

American Forest and Paper Association
111 19th Street, NW, #800
Washington, DC 20036

Standard reference number	Title	Referenced in code section number
NDS—01	National Design Specification (NDS) for Wood Construction—with 2001 Supplement	R404.2.2, R502.2, Table R503.1, R602.3, R802.2
WFCM—2001	Wood Frame Construction Manual for One- and Two-family Dwellings	R301.2.1.1
AFPA—93	Span Tables for Joists and Rafters	R502.3, R802.4, R802.5
T.R. No. 7—87	Basic Requirements for Permanent Wood Foundation System	R401.1

AHA

American Hardboard Association
1210 West Northwest Highway
Palatine, IL 60067

Standard reference number	Title	Referenced in code section number
A135.4—95	Basic Hardboard	Table R602.3(1)
A135.5—95	Prefinished Hardboard Paneling	R702.5
A135.6—98	Hardboard Siding	Table R703.4
A194.1—85	Cellulosic Fiber Board	Table R602.3(1)

AISI

American Iron and Steel Institute
1140 Connecticut Ave, Suite 705
Washington, DC 20036

Standard reference number	Title	Referenced in code section number
Header—2001	Standard for Cold-formed Steel Framing-Header Design, 2001	R603.6
PM—2001	Standard for Cold-formed Steel Framing-Prescriptive Method for One- and Two-family Dwellings	R301.1.1, R301.2.1.1(4), R301.2.2.4.1, R301.2.2.4.5
Truss—2001	Standard for Cold-formed Steel Framing-Truss Design	R804.1.3

AITC

American Institute of Timber Construction
7012 S. Revere Parkway, Suite 140
Englewood, CO 80112

Standard reference number	Title	Referenced in code section number
AITC A 190.1—92	Structural Glued Laminated Timber	R502.1.5, R602.1.2, R802.1.4

ANSI

American National Standards Institute
25 West 43rd Street, Fourth Floor
New York, NY 10036

Standard reference number	Title	Referenced in code section number
A108.1A—99	Installation of Ceramic Tile in the Wet-set Method, with Portland Cement Mortar	R702.4.1
A108.1B—99	Installation of Ceramic Tile, Quarry Tile on a Cured Portland Cement Mortar Setting Bed with Dry-set or Latex-Portland Mortar	R702.4.1
A108.4—99	Installation of Ceramic Tile with Organic Adhesives or Water Cleanable Tile-setting Epoxy Adhesive	R702.4.1
A108.5—99	Installation of Ceramic Tile with Dry-set Portland Cement Mortar or Latex-Portland Cement Mortar	R702.4.1
A108.6—99	Installation of Ceramic Tile with Chemical Resistant, Water Cleanable Tile-setting and -grouting Epoxy	R702.4.1
A108.11—99	Interior Installation of Cementitious Backer Units	R702.4.1
A118.1—99	American National Standard Specifications for Dry-set Portland Cement Mortar	R702.4.1
A118.3—99	American National Standard Specifications for Chemical Resistant, Water Cleanable Tile-setting and Grouting Epoxy and Water Cleanable Tile-setting Epoxy Adhesive	R702.4.1
A136.1—99	American National Standard Specifications for Organic Adhesives for Installation of Ceramic Tile	R702.4.1
A137.1—88	American National Standard Specifications for Ceramic Tile	R702.4.1
A208.1—99	Particleboard	R503.3.1, R605.1
LC1—97	Interior Fuel Gas Piping Systems Using Corrugated Stainless Steel Tubing	G2414.5.3
Z21.1—00	Household Cooking Gas Appliances	G2447.1
Z21.5.1—99	Gas Clothes Dryers—Volume I—Type I Clothes Dryers	G2438.1
Z21.8—94	Installation of Domestic Gas Conversion Burners	G2443.1
Z21.10.1—00	Gas Water Heaters—Volume I—Storage, Water Heaters with Input Ratings of 75,000 Btu per hour or Less	G2448.1
Z21.10.3—98	Gas Water Heaters—Volume III—Storage, Water Heaters with Input Ratings above 75,000 Btu per hour, Circulating and Instantaneous Water Heaters—with Z21.10.3a-99 Addendum	G2448.1
Z21.11.1—91	Gas-fired Room Heaters—Volume I—Vented Room Heaters—with 1993 Addendum (Replaced by Z21.86-98/CSA 2.32 - M98, Vented Gas-fired Space Heating Appliances)	G2446.1
Z21.11.2—96	Gas-fired Room Heaters—Volume II—Unvented Room Heaters—with Addendum A-97 and Addendum B-98	G2445.1
Z21.13—99	Gas-fired Low Pressure Steam and Hot Water Boilers	G2452.1
Z21.15—97	Manually Operated Gas Valves for Appliances, Appliance Connector Valves and Hose End Valves	G2420.1.1
Z21.22—99	Relief Valves for Hot Water Supply Systems	P2803.2
Z21.40.1—96	Gas-fired Heat Activated Air Conditioning and Heat Pump Appliances—with Z21.40.1a-98 Addendum	G2449.1
Z21.40.2—96	Gas-fired Work Activated Air Conditioning and Heat Pump Appliances (Internal Combustion)—with Z21.40.2a-97 Addendum	G2449.1
Z21.42—93	Gas-fired Illuminating Appliances	G2450.1
Z21.47—00	Gas-fired Central Furnaces—with Addendum Z21.47a-00	R617.1, G2442.1
Z21.48—92	Gas-fired Gravity and Fan-type Floor Furnace—with Addendum A-1993 (Replaced by Z21.86—98/CSA 2.32—M98, Vented Gas-fired Space-heating Appliances)	G2437.1
Z21.49—92	Gas-fired Gravity and Fan-type Vented Wall Furnaces—with Addendum B-1994 (Replaced by Z21.86—98/CSA 2.32—M98, Vented Gas-fired Space-heating Appliances)	G2436.1

ANSI—continued

Z21.50—2000	Vented Gas Fireplaces—with Addendum A-1999 and Addendum B-1999	G2434.1
Z21.56—98	Gas-fired Pool Heaters—with Addendum Z21.56a-99	R616.1, G2441.1
Z21.58—95	Outdoor Cooking Gas Appliances—with Addendum Z21.58a-1998	R622.1, G2447.1
Z21.60—2000	Decorative Gas Appliances for Installation in Vented Fireplaces	G2432.1
Z21.84—99	Manually-listed, Natural Gas Decorative Gas Appliances for Installation in Solid Fuel Burning Fireplaces	G2432.1, G2432.2
Z21.86—98/CSA 2.32 M98	Gas-fired Vented Space Heating Appliances	G2436.1, G2437.1, G2446.1
Z21.88—99	Vented Gas Fireplace Heaters	G2435.1
Z83.6—90 (R 1998)	Gas-fired Infrared Heaters	G2442.1, G2449.1, G2451.1
Z83.8—96	Gas-fired Unit Heaters—with Addendum Z83.8a-1997	G2444.1
Z97.1—84 (R1994)	Safety Glazing Materials Used in Buildings—Safety Performance Specifications and Methods of Test (Reaffirmed 1994)	R308.3
Z124.1—95	Plastic Bathtub Units	Table P2701.1
Z124.2—95	Plastic Shower Receptors and Shower Stalls	Table P2701.1
Z124.3—95	Plastic Lavatories	Table P2701.1, P2711.1, P2711.2
Z124.4—96	Plastic Water Closet Bowls and Tanks	Table P2701.1, P2712.1
Z124.6—97	Plastic Sinks	Table P2701.1

APA

APA-The Engineered Wood Association
P. O. Box 11700
Tacoma, WA 98411-0700

Standard reference number	Title	Referenced in code section number
APA E30—01	Engineered Wood Construction Guide	R803.2.3

ASCE

American Society of Civil Engineers
1801 Alexander Bell Drive
Reston, VA 20191

Standard reference number	Title	Referenced in code section number
5—02	Building Code Requirements for Masonry Structures	R404.1, R606.1, R606.1.1, R606.11.1, R606.11.2.2.1, R606.11.2.2.2, R606.11.3.1
6—02	Specifications for Masonry Structures	R404.1, R606.1, R606.1.1, R606.11.1, R606.11.2.2.1, R606.11.2.2.2, R606.11.3.1
7—02	Minimum Design Loads for Buildings and Other Structures	R301.2.1.1
32—01	Design and Construction of Frost Protected Shallow Foundations	R403.1.4.1

ASHRAE

American Society of Heating, Refrigerating
and Air-Conditioning Engineers, Inc.
1791 Tullie Circle, NE
Atlanta, GA 30329

Standard reference number	Title	Referenced in code section number
34—2001	Designation and Safety Classification of Refrigerants	M1411.1
ASHRAE—2001	ASHRAE Fundamentals Handbook—2001	P3001.2, P3002.3, P3101.4, P3103.2

ASME

American Society of Mechanical Engineers
Three Park Avenue
New York, NY 10016-5990

Standard reference number	Title	Referenced in code section number
A112.1.2—1991(R1998)	Air Gaps in Plumbing Systems	Table P2902.2, P2902.2.1
A112.1.3—2000	Air Gap Fittings for Use with Plumbing Fixtures, Appliances, and Appurtenances	Table P2701.1, P2902.2.1
A112.3.4—2000	Macerating Toilet Systems and Related Components	Table P2701.1, P3007.2.1
A112.4.1—1993(R1998)	Water Heater Relief Valve Drain Tubes	P2803.6.2
A112.4.3—1999	Plastic Fittings for Connecting Water Closets to the Sanitary Drainage System	P3003.4.5
A112.6.1M—1997	Floor Affixed Supports for Off-the-floor Plumbing Fixtures for Public Use	P2702.4
A112.6.2—2000	Framing-Affixed Supports for Off-the-floor Water Closets with Concealed Tanks	Table P2701.1, P2702.4
A112.6.3—2001	Floor and Trench Drains	Table P2701.1
A112.18.1—2000	Plumbing Fixture Fittings	Table P2701.1, P2722.1
A112.18.6—1999	Flexible Water Connectors	P2904.7
A112.19.1M—1994 (R1999)	Enameled Cast Iron Plumbing Fixtures	Table P2701.1, P2711.1
A112.19.2M—1998	Vitreous China Plumbing Fixtures	Table P2701.1, P2711.1 P2712.2
A112.19.3M—1987 (R1996)	Stainless Steel Plumbing Fixtures (Designed for Residential Use)	Table P2701.1, P2711.1
A112.19.4M—1994 (R1999)	Porcelain Enameled Formed Steel Plumbing Fixtures	Table P2701.1, P2711.1
A112.19.5—1999	Trim for Water-closet Bowls, Tanks, and Urinals	Table P2701.1
A112.19.6—1995	Hydraulic Performance Requirements for Water Closets and Urinals	Table P2701.1, P2712.1, P2712.2
A112.19.7M—1995	Whirlpool Bathtub Appliances	Table P2701.1
A112.19.8M—1987 (R1996)	Suction Fittings for Use in Swimming Pools, Wading Pools, Spas, Hot Tubs, and Whirlpool Bathtub Appliances	Table P2701.1
A112.19.9M—1998	Non-vitreous Ceramic Plumbing Fixtures	Table P2701.1, P2712.1
A112.19.12—2000	Wall Mounted and Pedestal Mounted, Adjustable and Pivoting Lavatory and Sink Carrier Systems	Table P2701.1, P2711.4, P2714.2
A112.19.13—2001	Electrohydraulic Water Closets	P2712.9
A112.19.15—2001	Bathtub/Whirlpool Bathtubs with Pressure Sealed Doors	Table P2701.1, P2713.2
B1.20.1—1983(R1999)	Pipe Threads, General Purpose (Inch)	G2414.9
B16.3—1999	Malleable Iron Threaded Fittings Classes 150 and 300	Table P2904.6
B16.4—1998	Gray-iron Threaded Fittings Classes 125 and 250	Table P2904.6
B16.9—1993	Factory-made Wrought Steel Buttwelding Fittings	Table P2904.6
B16.11—1996	Forged Fittings, Socket-welding and Threaded	Table P2904.6
B16.12—1998	Cast Iron Threaded Drainage Fittings	Table P2904.6
B16.15—1985(R1994)	Cast Bronze Threaded Fittings	Table P2904.6
B16.18—1984(R1994)	Cast Copper Alloy Solder Joint Pressure Fittings	Table P2904.6
B16.22—1995	Wrought Copper and Copper Alloy Solder Joint Pressure Fittings	Table P2904.6
B16.23—1992	Cast Copper Alloy Solder Joint Drainage Fittings (DWV)	Table P2904.6
B16.26—1988	Cast Copper Alloy Fittings for Flared Copper Tubes	Table P2904.6
B16.28—1994	Wrought Steel Buttwelding Short Radius Elbows and Returns	Table P2904.6
B16.29—1994	Wrought Copper and Wrought Copper Alloy Solder Joint Drainage Fittings-DWV	Table P2904.6
B16.33—1990	Manually Operated Metallic Gas Valves for Use in Gas Piping Systems up to 125 psig (Sizes $^1/_2$ through 2)	G2420.1
B36.10M—2000	Welded and Seamless Wrought-steel Pipe	G2414.4.2
BPVC—1998	Boiler and Pressure Vessel Code (Sections I, II, IV & VI)	M2001.1.1, G2452.1
CSD-1—1996	Controls and Safety Devices for Automatically Fired Boilers	M2001.1.1, G2452.1

ASSE

American Society of Sanitary Engineering
28901 Clemens Road, Suite A
Westlake, OH 44145

Standard reference number	Title	Referenced in code section number
1001—90	Performance Requirements for Pipe Applied Atmospheric Type Vacuum Breakers	Table P2902.2, P2902.2.2
1002—99	Performance Requirements for Water Closet Flush Tank Fill Valves	Table P2701.1, Table P2902.2, P2902.3.1
1003—95	Performance Requirements for Water Pressure Reducing Valves	P2903.3.1
1006—89	Performance Requirements for Residential Use (Household) Dishwashers	Table P2701.1
1007—92	Performance Requirements for Home Laundry Equipment	Table P2701.1
1008—89	Performance Requirements for Household Food Waste Disposer Units	Table P2701.1

ASSE—continued

1010—98	Performance Requirements for Water Hammer Arresters	P2903.5
1011—95	Performance Requirements for Hose Connection Vacuum Breakers	Table P2902.2, P2902.2.2, P2902.3.3
1012—95	Performance Requirements for Backflow Preventers with Intermediate Atmospheric Vent	Table P2902.2, P2902.2.3, P2902.3.3, P2902.4.1, P2902.4.5
1013—99	Performance Requirements for Reduced Pressure Principle Backflow Preventers and Reduced Pressure Fire Protection Principle Backflow Preventers	Table P2902.2, P2902.2.5, P2902.4.1, P2902.4.5
1014—90	Performance Requirements for Handheld Shower	Table P2701.1
1015—99	Performance Requirements For Double Check Backflow Prevention Assemblies and Double Check Fire Protection Backflow Prevention Assemblies	Table P2902.2, P2902.2.6
1016—96	Performance Requirements for Individual Thermostatic, Pressure Balancing and Combination Control Valves for Bathing Facilities	Table P2701.1, P2708.3, P2722.2
1017—99	Performance Requirements for Temperature Actuated Mixing Valves for Hot Water Distribution Systems	P2802.2
1019—97	Performance Requirements for Wall Hydrants, Freezeless, Automatic Draining, Anti-backflow Types	Table P2701.1, Table P2902.2, P2902.2.2, P2902.3.3
1020—98	Performance Requirements for Pressure Vacuum Breaker Assembly	Table P2902.2, P2902.2.4, P2902.3.3
1023—79	Performance Requirements for Hot Water Dispensers Household Storage Type Electrical	Table P2701.1
1024—98	Performance Requirements for Dual Check Valve Type Backflow Preventers (For Residential Supply Service or Individual Outlets)	Table P2902.2
1025—78	Performance Requirements for Diverters for Plumbing Faucets with Hose Spray, Anti-siphon Type, Residential Applications	Table P2701.1
1035—95	Performance Requirements for Laboratory Faucet Backflow Preventers	Table P2902.2, P2902.2.2
1037—90	Performance Requirements for Pressurized Flushing Devices for Plumbing Fixtures	Table P2701.1
1047—99	Performance Requirements for Reduced Pressure Detector Fire Protection Backflow Prevention Assemblies	P2902.2.5, Table P2902.2
1048—99	Performance Requirements for Double Check Detector Fire Protection Backflow Prevention Assemblies	P2902.2.6, Table P2902.2
1050—91	Performance Requirements for Air Admittance Valves for Plumbing Drainage, Waste and Vented Systems—Stack Type Device	P3114.1
1051—98	Performance Requirements for Air Admittance Valves for Plumbing Drainage Systems—Fixture and Branch Devices	P3114.1
1052—94	Performance Requirements for Hose Connection Backflow Preventers	Table P2701.1, Table P2902.2, P2902.2.2, P2902.3.3
1056—95	Performance Requirements for Back Siphonage Vacuum Breaker	Table P2902.2, P2902.2.4, P2902.3.3
1062—97	Performance Requirements for Temperature Actuated, Flow Reduction Valves to Individual Fixture Fittings	Table P2701.1, P2724.1
1066—97	Performance Requirements for Individual Pressure Balancing In-Line Valves for Individual Fixture Fittings	P2722.3

ASTM

ASTM International
100 Barr Harbor Drive
West Conshohocken, PA 19428

Standard reference number	Title	Referenced in code section number
A 36/A 36M—01	Specification for Carbon Structural Steel	R606.14
A 53/A53M—01	Specification for Pipe, Steel, Black and Hot-dipped, Zinc-coated Welded and Seamless	Table M2101.1, Table 2904.4.1, Table P2904.5, Table P3002.1, G2414.4.2
A 74—98	Specification for Cast Iron Soil Pipe and Fittings	Table P3002.1,Table P3002.2
A 106—99e	Specification for Seamless Carbon Steel Pipe for High Temperature Service	Table M2101.1, G2414.4.2
A 126—01	Specification for Gray Iron Castings for Valves, Flanges and Pipe Fittings	Table P3002.1
A 153—01a	Specification for Zinc Coating (Hot Dip) on Iron and Steel Hardware	Table R606.14.1
A 167—99	Specification for Stainless and Heat-resisting Chromium-nickel Steel Plate, Sheet, and Strip	R606.14, Table R606.14.1
A 197/A197M—00	Specification for Cupola Malleable Iron	Table P3002.1
A 254—97	Specification for Copper Brazed Steel Tubing	Table M2101.1, G2414.5.1
A 312/A 312M—01	Specification for Seamless and Welded Austenitic Stainless Steel Pipes	Table P2904.4.1, Table P2904.5, Table P2904.6, P2904.10.2
A 377—99	Specification for Ductile Iron Pressure Pipe	Table P2904.4.1
A 510M—00	Specification for General Requirements for Wire Rods and Coarse Round Wire, Carbon Steel Metric	R606.14
A 525-87	Specification for Steel Sheet Zinc-coated (Galvanized) Steel Wire	Table R606.14.1, M1601.1.1
A 539—99	Specification for Electronic-resistance-welded Coiled Steel Tubing for Gas and Fuel Oil Lines	M2202.1, G2414.5.1
A 615/A 0615M—01b	Specification for Deformed and Plain Billet-steel Bars for Concrete Reinforcement	R404.4.6.1, R611.6.2
A 641/A 0641M—98	Specification for Zinc-coated (Galvanized) Carbon Steel Wire	Table R606.14.1
A 653/A 0653M—00	Specification for Steel Sheet, Zinc-coated (Galvanized) or Zinc-iron Alloy-coated (Galvanized) by the Hot-dip Process	R505.2.1, R505.2.3, R603.2.1, R603.2.3, R804.2.1, R804.2.3, Table R905.10.3

ASTM—continued

Standard	Title	Referenced in code section number
A 706M—00	Specification for Low-alloy Steel Deformed and Plain Bars for Concrete Reinforcement	R404.4.6.1, R611.6.2
A 755M—99	Specification for Steel Sheet, Metallic Coated by the Hot-dip Process and Prepainted by the Coil-coating Process for Exterior Exposed Building Products	Table R905.10.3
A 778—01	Specification for Welded Unannealed Austenitic Stainless Steel Tubular Products	Table P2904.4.1, Table P2904.5, Table P2904.6
A 792M—99	Specification for Steel Sheet, 55% Aluminum-zinc Alloy-coated by the Hot-dip Process	R505.2.1, R603.2.1, R603.2.3, R804.2.1, R804.2.3
A 875/0875M—01a	Specification for Steel Sheet, Zinc-5%, Aluminum Alloy-coated by the Hot-dip Process	R505.2.1, R505.2.3, R603.2.1, R603.2.3, R804.2.1, R804.2.3
A 888—98e1	Specification for Hubless Cast Iron Soil Pipe and Fittings for Sanitary and Storm Drain, Waste, and Vent Piping Application	Table P3002.1, Table P3002.2
A 951—00	Specification for Masonry Joint Reinforcement	R606.14
A 996/A 996M-00	Specifications for Rail-steel and Axel-steel Deformed Bars Reinforcement for Concrete	R404.4.6.1, R611.6.2
B 32—00	Specification for Solder Metal	P3003.3.4
B 42—98	Specification for Seamless Copper Pipe, Standard Sizes	Table M2101.1, Table P2904.5, Table P3002.1
B 43—98	Specification for Seamless Red Brass Pipe, Standard Sizes	Table M2101.1, G2413.5.2, Table P3002.1, Table P2904.4.1, Table P2904.5
B 75—99	Specification for Seamless Copper Tube	Table M2101.1, Table P2904.4.1, Table P2904.5, Table P3002.1, Table P3002.2
B 88—99e1	Specification for Seamless Copper Water Tube	Table M2101.1, G2414.5.2, Table P2904.4.1, Table P2904.5, Table P3002.1, Table P3002.2
B 101—96	Specification for Lead-coated Copper Sheet and Strip for Building Construction	Table R905.10.3
B 135—00	Specification for Seamless Brass Tube (Metric)	Table M2101.1
B 209—96	Specification for Aluminum and Aluminum-alloy Sheet and Plate	Table 905.10.3
B 227—98	Specification for Hard-drawn Copper-clad Steel Wire	R606.14
B 251—97	Specification for General Requirements for Wrought Seamless Copper and Copper-alloy Tube	Table M2101.1, Table P2904.4.1, Table P2904.5
B 280—99	Specification for Seamless Copper Tube for Air Conditioning and Refrigeration Field Service	G2414.5.2
B 302—00	Specification for Threadless Copper Pipe, Standard Sizes	Table M2101.1, Table P2904.5, P2904.4
B 306—99	Specification for Copper Drainage Tube (DWV)	Table M2101.1, Table P3002.1, Table P3002.2
B 370—98	Specification for Copper Sheet and Strip for Building Construction	Table R905.10.3, Table P2701.1
B 447—00	Specification for Welded Copper Tube	Table P2904.4.1, Table P2904.5
B 633—98e01	Specification for Electrodeposited Coatings of Zinc on Iron and Steel	R505.2.4, R603.2.4, R804.2.4
B 813—00e01	Specification for Liquid and Paste Fluxes for Soldering Applications of Copper and Copper Alloy Tube	Table M2101.1, P2904.11, P3003.3.4
B 828—00	Practice for Making Capillary Joints by Soldering of Copper and Copper Alloy Tube and Fittings	P2904.11
C 5—79(97)	Specification for Quicklime for Structural Purposes	R702.2
C 14—99	Specification for Concrete Sewer, Storm Drain, and Culvert Pipe	Table P3002.2
C 27—98	Classification of Fireclay and High-alumina Refractory Brick	R1003.5, R1003.8
C 28/C28M—00	Specification for Gypsum Plasters	R702.2
C 34—96(2001)	Specification for Structural Clay Load-bearing Wall Tile	Table R301.2(1)
C 35—95(2001)	Specification for Inorganic Aggregates for Use in Gypsum Plaster	R702.2
C 36/C 0036M—01	Specification for Gypsum Wallboard	R702.3.1
C 37/C 0037M—01	Specification for Gypsum Lath	R702.2
C 55—01	Specification for Concrete Brick	R202, Table R301.2(1)
C 59/C 0059M—00	Specification for Gypsum Casting and Molding Plaster	R702.2
C 61/C 0061M—00	Specification for Gypsum Keene's Cement	R702.2
C 62—01	Specification for Building Brick (Solid Masonry Units Made from Clay or Shale)	R202, Table R301.2(1)
C 73—99a	Specification for Calcium Silicate Face Brick (Sand Lime Brick)	R202, Table R301.2(1)
C 79—01	Specification for Treated Core and Nontreated Core Gypsum Sheathing Board	Table R602.3(1), R702.3.1
C 90—01	Specification for Load-bearing Concrete Masonry Units	Table R301.2(1)
C 129—01	Specification for Nonload-bearing Concrete Masonry Units and Related Units	Table R301.2(1)
C 143/C 0143M—00	Test Method for Slump or Hydraulic Cement Concrete	R404.4.5, R611.6.1
C 145—85	Specification for Solid Load-bearing Concrete Masonry Units	R202, Table R301.2(1)
C 199—84(2000)	Test Method for Pier Test for Refractory Mortar	R1001.9, R1003.5, R1003.8
C 207—97	Specification for Hydrated Lime for Masonry Purposes	Table R607.1
C 208—95(2001)	Specification for Cellulosic Fiber Insulating Board	Table R602.3(1)
C 216—01a	Specification for Facing Brick (Solid Masonry Units Made from Clay or Shale)	R202, Table R301.2(1)
C 236—89 (1993)e1	Test Method for Steady-State Thermal Performance of Building Assemblies by Means of a Guarded Hot Box	N1102.1.1.1
C 270—01a	Specification for Mortar for Unit Masonry	R607.1
C 296—00	Specification for Asbestos Cement Pressure Pipe	Table P2904.4.1
C 315—00	Specification for Clay Flue Linings	Table R1001.11(1), Table R1001.11(2), R1001.8.1, G2425.12
C 406—00	Specifications for Roofing Slate	R905.6.4

ASTM—continued

Standard	Title	Referenced in code section number
C 411—97	Test Method for Hot-surface Performance of High-temperature Thermal Insulation	M1601.2.1
C 425—01	Specification for Compression Joints for Vitrified Clay Pipe and Fittings	Table P3002.2, P3003.3.5
C 443—01	Specification for Joints for Concrete Pipe and Manholes, Using Rubber Gaskets	P3003.3.5
C 475—01	Specification for Joint Compound and Joint Tape for Finishing Gypsum Wallboard	R702.3.1
C 476—01	Specification for Grout for Masonry	R609.1.1
C 514—01	Specification for Nails for the Application of Gypsum Wallboard	R702.3.1
C 557—99	Specification for Adhesives for Fastening Gypsum Wallboard to Wood Framing	R702.3.1
C 578—01	Specification for Rigid, Cellular Polystyrene Thermal Insulation	R403.3
C 587—97	Specification for Gypsum Veneer Plaster	R702.2
C 588/C0588M—99	Specification for Gypsum Base for Veneer Plasters	R702.2
C 630/0630M—01	Specification for Water-resistant Gypsum Backing Board	R702.3.1, R702.4.2
C 631—(2000)	Specification for Bonding Compounds for Interior Gypsum Plastering	R702.2
C 645—00	Specification for Nonstructural Steel Framing Members	R702.3.3
C 652—01a	Specification for Hollow Brick (Hollow Masonry Units Made from Clay or Shale)	R202, Table R301.2(1)
C 700—00	Specification for Vitrified Clay Pipe, Extra Strength, Standard Strength, and Perforated	Table P3002.2
C 836—00	Specification for High Solids Content, Cold Liquid-applied Elastomeric Waterproofing Membrane for Use with Separate Wearing Course	R905.15.2
C 843—99	Specification for Application of Gypsum Veneer Plaster	R702.2
C 844—99	Specification for Application of Gypsum Base to Receive Gypsum Veneer Plaster	R702.2
C 847—(2000)	Specification for Metal Lath	R702.2
C 887—79(96)e[1]	Specification for Packaged, Dry, Combined Materials for Surface Bonding Mortar	R406.1
C 897—00	Specification for Aggregate for Job-mixed Portland Cement-based Plasters	R702.2
C 931—97	Specification for Exterior Gypsum Soffit Board	R702.3.1
C 933—96a(2001)	Specification for Welded Wire Lath	R702.2
C 954—00	Specification for Steel Drill Screws for the Application of Gypsum Panel Products or Metal Plaster Bases to Steel Studs from 0.033 in. (0.84 mm) to 0.112 in. (2.84 mm) in Thickness	R505.2.4, R603.2.4, R702.3.6, R804.2.4
C955—01	Specification for Load-Bearing Transverse and Axial Steel Studs, Runners Tracks, and Bracing or Bridging for Screw Application of Gypsum Panel Products and Metal Plaster Bases	R702.3.3
C 957—93(1998)	Specification for High-solids Content, Cold Liquid—Applied Elastomeric Waterproofing Membrane for Use with Integral Wearing Surface	R905.15.2
C 960—01	Specification for Predecorated Gypsum Board	R702.3.1
C 976—90(1996)e1	Test Method for Thermal Performance of Building Assemblies by Means of a Calibrated Hot Box	N1102.1.1.1
C 1002—01	Specification for Steel Drill Screws for the Application of Gypsum Panel Products or Metal Plaster Bases	R702.3.1, R702.3.6, Table R702.3.4
C 1029—96	Specification for Spray-applied Rigid Cellular Polyurethane Thermal Insulation	R905.14.2
C 1032—96	Specification for Woven Wire Plaster Base	R702.2
C 1047—99	Specification for Accessories for Gypsum Wallboard and Gypsum Veneer Base	R702.2, R702.3.1
C 1063—99	Specification for Installation of Lathing and Furring to Receive Interior and Exterior Portland Cement based Plaster	R702.2
C 1157—00a	Performance Specification for Hydraulic Cements	R402.2
C 1167—96	Specification for Clay Roof Tiles	R905.3, R905.3.4
C 1173—97	Specification for Flexible Transition Couplings for Underground Piping Systems	P3003.3.5
C 1177/C 1177M—01	Specification for Glass Mat Gypsum Substrate for Use as Sheathing	R702.3.1
C 1178/C 1178M—01	Specification for Glass Mat Water-resistant Gypsum Backing Panel	R702.3.1, R702.4.2
C 1186—99	Specification for Flat Nonasbestos Fiber Cement Sheets	R703.4
C 1261—98	Specification for Firebox Brick for Residential Fireplaces	R1003.5, R1003.8
C 1277—97	Specification for Shielded Couplings Joining Hubless Cast Iron Soil Pipe and Fittings	Table P3002.1, Table P3002.2
C 1278/C 1278M—01	Specification for Fiber-reinforced Gypsum Panels	R702.3.1
C 1283—00e01	Practice for Installing Clay Flue Lining	R1001.9
C 1395/C 1395M—01	Specification for Gypsum Ceiling Board	R702.3.1
C 1396—99	Specification for Gypsum Board	R702.3.1
C 1460—00	Specification for Shielded Transition Couplings for Use with Dissimilar DWV Pipe and Fittings Above Ground	Table P3002.1, Table P3002.2
C 1461—00	Specification for Mechanical Couplings Using Thermoplastic Elastomeric (TPE) Gaskets for Joining Drain, Waste, and Vent (DWV) Sewer, Sanitary and Storm Plumbing Systems for Above and Below Ground Use	Table P3002.1, Table P3002.2
C 1492—01	Specification for Concrete Roof Tile	R905.3.5
D 41—e01	Specification for Asphalt Primer Used in Roofing, Dampproofing, and Waterproofing	Table R905.9.2, Table R905.11.2
D 43—00	Specification for Coal Tar Primer Used in Roofing, Dampproofing and Waterproofing	Table R905.9.2
D 224—89(1996)	Specification for Smooth-surfaced Asphalt Roll Roofing (Organic Felt)	R905.2.8.2
D 225—01	Specification for Asphalt Shingles (Organic Felt) Surfaced with Mineral Granules	R905.2.4
D 226—97a	Specification for Asphalt-saturated (Organic Felt) Used in Roofing and Waterproofing	R703.2, R703.9.1, Table R905.2.3, R905.4.3, R905.5.3, R905.5.4, R905.6.3, R905.7.3, R905.8.3, R905.8.4, Table 905.9.2

ASTM—continued

Standard	Title	Referenced in code section number
D 227—97a	Specification for Coal Tar Saturated (Organic Felt) Used in Roofing and Waterproofing	Table R905.9.2
D 249—89(96)	Specification for Asphalt Roll Roofing (Organic Felt) Surfaced with Mineral Granules	Table R905.9.2, R905.2.8.2, R905.3.3, R905.5.4
D 0312—00	Specification for Asphalt Used in Roofing	R905.9.2
D 371—89(1996)	Specification for Asphalt Roll Roofing (Organic Felt) Surfaced with Mineral Granules: Wide-selvage	R905.5.4
D 422—63(1998)	Test Method for Particle-size Analysis of Soils	R403.1.7.5.1
D 449—89(1999)e01	Specification for Asphalt Used in Dampproofing and Waterproofing	R406.2
D 450—(00)e[1]	Specification for Coal-tar Pitch Used in Roofing , Dampproofing and Waterproofing	Table R905.9.2
D 1227—00	Specification for Emulsified Asphalt Used as a Protective Coating for Roofing	Table R905.9.2, Table R905.11.2, R905.15.2
D 1248—00a	Specification for Polyethylene Plastics Extrusion Materials for Wire and Cable	M1601.1.2
D 1527—99	Specification for Acrylonite-Butadiene-Styrene (ABS) Plastic Pipe, Schedules 40 and 80	Table P2904.4.1
D 1693—01	Test Method for Environmental Stress-cracking of Ethylene Plastics	Table M2101.1
D 1784—99a	Standard Specification for Rigid Poly (Vinyl Chloride) (PVC) Compounds and Chlorinated Poly(Vinyl Chloride) (CPVC) Compounds	M1601.1.2
D 1785—99	Specification for Poly (Vinyl Chloride) (PVC) Plastic Pipe, Schedules 40, 80 and 120	Table P2904.4.1
D 1863—93(2000)	Specification for Mineral Aggregate Used in Built-up Roofs	Table R905.9.2, Table R906.3.2
D 1869—95(2000)	Specification for Rubber Rings for Asbestos-cement Pipe	P2904.15
D 1970—01	Specification for Self-adhering Polymer Modified Bitumen Sheet Materials Used as Steep Roofing Underlayment for Ice Dam Protection	R905.2.3, R905.2.8.3
D 2104—01	Specification for Polyethylene (PE) Plastic Pipe, Schedule 40	Table P2904.4.1
D 2178—97a	Specification for Asphalt Glass Felt Used in Roofing and Waterproofing	Table R905.9.2
D 2235—01	Specification for Solvent Cement for Acrylonitrile-Butadiene-Styrene (ABS) Plastic Pipe and Fittings	P2904.9.1.1, Table P3002.1, Table P3002.2
D 2239—01	Specification for Polyethylene (PE) Plastic Pipe (SIDR-PR) Based on Controlled Inside Diameter	Table P2904.4.1
D 2241—00	Specification for Poly (Vinyl Chloride) (PVC) Pressure-rated Pipe (SDR-Series)	Table P2904.4.1
D 2282—99	Specification for Acrylonitrile-Butadiene-Styrene (ABS) Plastic Pipe (SDR-PR)	Table P2904.4.1
D 2412—96a	Test Method for Determination of External Loading Characteristics of Plastic Pipe by Parallel-plate Loading	M1601.1.2
D 2447—01	Specification for Polyethylene (PE) Plastic Pipe Schedules 40 and 80, Based on Outside Diameter	Table M2101
D 2464—99	Specification for Threaded Poly (Vinly Chloride) (PVC) Plastic Pipe Fittings, Schedule 80	Table P2904.6
D 2466—01	Specification for Poly(Vinyl Chloride) (PVC) Plastic Pipe Fittings, Schedule 40	Table P2904.6
D 2467—99	Specification for Poly (Vinyl Chloride) (PVC) Plastic Pipe Fittings, Schedule 80	Table P2904.6
D 2468—96a	Specification for Acrylonitrile-Butadiene-Styrene (ABS) Plastic Pipe Fittings, Schedule 40	Table P2904.6
D 2513—01a	Specification for Thermoplastic Gas Pressure Pipe, Tubing, and Fittings	Table M2101.1, M2104.2.1.3, G2414.6, G2414.6.1, G2414.11, G2415.14.3
D 2564—96a	Specification for Solvent Cements for Poly (Vinyl Chloride) (PVC) Plastic Piping Systems	P2904.9.1.3, Table P3002.1, Table P3002.2
D 2609—00	Specification for Plastic Insert Fittings for Polyethylene (PE) Plastic Pipe	Table P2904.6
D 2626—97b	Specification for Asphalt-saturated and Coated Organic Felt Base Sheet Used in Roofing	R905.3.3, Table R905.9.2
D 2661—01	Specification for Acrylonitrile-Butadiene-Styrene (ABS) Schedule 40 Plastic Drain, Waste, and Vent Pipe and Fittings	Table P3002.1, Table P3002.2
D 2662—96a	Specification for Polybutylene (PB) Plastic Pipe (SDR-PR) Based on Controlled Inside Diameter	Table P2904.4.1
D 2665—01	Specification for Poly (Vinyl Chloride) (PVC) Plastic Drain, Waste, and Vent Pipe and Fittings	Table P3002.1, Table P3002.2
D 2666—96a	Specification for Polybutylene (PB) Plastic Tubing	Table P2904.4.1
D 2672—96a	Specification for Joints for IPS PVC Pipe Using Solvent Cement	Table P3002.1, Table, P3002.2, Table P2904.4.1
D 2683—98	Specification for Socket-type Polyethylene Fittings for Outside Diameter-controlled Polyethylene Pipe and Tubing	Table M2101.1, M2104.2.1.1
D 2737—01	Specification for Polyethylene (PE) Plastic Tubing	Table P2904.4.1
D 2751—96a	Specification for Acrylonitrile-Butadiene-Styrene (ABS) Sewer Pipe and Fittings	Table P3002.2
D 2822—91(1997)e[1]	Specification for Asphalt Roof Cement	Table R905.9.2, Table R905.11.2
D 2823—90(1997)e[1]	Specification for Asphalt Roof Coatings	Table R905.9.2, Table R905.11.2
D 2824—94	Specification for Aluminum-pigmented Asphalt Roof Coatings, Non-fibered, Asbestos Fibered, and Fibered without Asbestos	Table R905.9.2, Table R905.11.2
D 2837—98a	Test Method for Obtaining Hydrostatic Design Basis for Thermoplastic Pipe Materials	Table M2101.1
D 2846/D 2846M—99	Specification for Chlorinated Poly (Vinyl Chloride) (CPVC) Plastic Hot- and Cold-water Distribution Systems	Table P2904.4.1, Table P2904.5, P2904.9.1.2, M2101.1
D 2898—94(1999)	Test Methods for Accelerated Weathering of Fire-retardant-treated Wood for Fire Testing	R802.1.3.3
D 2949—00a	Specification for 3.25-in. Outside Diameter Poly (Vinyl Chloride) (PVC) Plastic Drain, Waste, and Vent Pipe and Fittings	Table P3002.1, Table P3002.2
D 3019—e01	Specification for Lap Cement Used with Asphalt Roll Roofing, Non-fibered, Asbestos Fibered, and Non-asbestos Fibered	Table R905.9.2, Table R905.11.2
D 3034—00	Specification for Type PSM Poly (Vinyl Chloride) (PVC) Sewer Pipe and Fittings	Table P3002.2
D 3035—01	Specification for Polyethylene (PE) Plastic Pipe (DR-PR) Based On Controlled Outside Diameter	Table M2101.1
D 3161—99a	Test Method for Wind Resistance of Asphalt Shingles (Fan Induced Method)	R905.2.6

ASTM—continued

Standard	Title	Referenced in code section number
D 3201—94(1998)el	Test Method for Hygroscopic Properties of Fire-retardant Wood and Wood-base Products	R802.1.3.4
D 3212—96a	Specification for Joints for Drain and Sewer Plastic Pipes Using Flexible Elastomeric Seals	Table P3002.2
D 3309—96a	Specification for Polybutylene (PB) Plastic Hot- and Cold-water Distribution Systems	Table P2904.4.1, Table P2904.5, Table M2101.1
D 3350—01	Specification for Polyethylene Plastic Pipe and Fitting Materials	Table M2101.1
D 3462—01e01	Specification for Asphalt Shingles Made From Glass Felt and Surfaced with Mineral Granules	R905.2.4
D 3468—99	Specification for Liquid-applied Neoprene and Chlorosulfanated Polyethylene Used in Roofing and Waterproofing	Table R905.15.2
D 3679—01c	Specification for Rigid Poly (Vinyl Chloride) (PVC) Siding	Table R703.4
D 3737—99	Practice for Establishing Allowable Properties for Structural Glued Laminated Timber (Glulam)	R502.1.5, R602.1.2, R802.1.4
D 3747—e01	Specification for Emulsified Asphalt Adhesive for Adhering Roof Insulation	Table R905.9.2, Table R905.11.2
D 3909—97b	Specification for Asphalt Roll Roofing (Glass Felt) Surfaced with Mineral Granules	R905.5.4, Table R905.9.2, Table R906.3.2
D 4022—94(00)e[1]	Specification for Coal Tar Roof Cement, Asbestos Containing	Table R905.9.2
D 4068—01	Specification for Chlorinated Polyethylene (CPE) Sheeting for Concealed Water Containment Membrane	P2709.2
D 4318—00	Test Methods for Liquid Limit, Plastic Limit and Plasticity Index of Soils	R403.1.7.5.1
D 4434—96	Specification for Poly (Vinyl Chloride) Sheet Roofing	R905.13.2
D 4479—00	Specification for Asphalt Roof Coatings-Asbestos-free	Table R905.9.2
D 4586—00	Specification for Asphalt Roof Cement-Asbestos-free	Table R905.9.2
D 4601—98	Specification for Asphalt-coated Glass Fiber Base Sheet Used in Roofing	Table R905.9.2
D 4637—96	Specification for EPDM Sheet Used in Single-ply Roof Membrane	R905.12.2
D 4829—95	Test Method for Expansion Index of Soils	R403.1.8.1
D 4869—88(1993)e[1]	Specification for Asphalt-saturated (Organic Felt) Underlayment Used in Steep Slope Roofing	R905.2.3
D 4897—01	Specification for Asphalt Coated Glass-fiber Venting Base Sheet Used in Roofing	Table R905.9.2
D 4990—97a	Specification for Coal Tar Glass Felt Used in Roofing and Waterproofing	Table R905.9.2
D 5019-96	Specification for Reinforced Non-vulcanized Polymeric Sheet Used in Roofing Membrane	R905.12.2
D 5055—00	Specification for Establishing and Monitoring Structural Capacities of Prefabricated Wood I-Joists	R502.1.4
D 5516—99a	Test Method for Evaluating the Flexural Properties of Fire-retardant-treated Softwood Plywood Exposed to the Elevated Temperatures	R802.1.3.2.1
D 5643—94(00)e[1]	Specification for Coal Tar Roof Cement Asbestos-free	Table R905.9.2
D 5664—01	Test Methods For Evaluating the Effects of Fire-retardant Treatments and Elevated Temperatures on Strength Properties of Fire-retardant-treated Lumber	R802.1.3.2.2
D 5665—99a	Specification for Thermoplastic Fabrics Used in Cold-applied Roofing and Waterproofing	Table R905.9.2
D 5726—98	Specification for Thermoplastic Fabrics Used in Hot-applied Roofing and Waterproofing	Table R905.9.2
D 6083—97a	Specification for Liquid Applied Acrylic Coating Used in Roofing	Table R905.9.2, Table R905.11.2, Table R905.15.2
D 6162—00a	Specification for Styrene Butadiene Styrene (SBS) Modified Bituminous Sheet Materials Using a Combination of Polyester and Glass Fiber Reinforcements	Table R905.11.2
D 6163—00	Specification for Styrene Butadiene Styrene (SBS) Modified Bituminous Sheet Materials Using Glass Fiber Reinforcements	Table R905.11.2
D 6164—00	Specification for Styrene Butadiene Styrene (SBS) Modified Bituminous Sheet Materials Using Polyester Reinforcements	Table R905.11.2
D 6221—00	Specification for Reinforced Bituminous Flashing Sheets for Roofing and Waterproofing	Table R905.11.2
D 6222—00e01	Specification for Atactic Polypropelene (APP) Modified Bituminous Sheet Materials Using Polyester Reinforcement	Table R905.11.2
D 6223—00e01	Specification for Atactic Polypropelene (APP) Modified Bituminous Sheet Materials Using a Combination of Polyester and Glass Fiber Reinforcement	Table R905.11.2
D 6298—98	Specification for Fiberglass Reinforced Styrene-Butadiene-Styrene (SBS) Modified Bituminous Sheets with a Factory Applied Metal Surface	Table R905.11.2
D 6305—98e1	Practice for Calculating Bending Strength Design Adjustment Factors for Fire-Retardant-Treated Plywood Roof Sheathing	R802.1.3.2.1
E 84—01	Test Method for Surface Burning Characteristics of Building Materials	R202, R3141.1, R3142.6, R314.3, R315.3, R316.1, R316.2, R802.1.3, M1601.2.1, M1601.4.2
E 96—00	Test Method for Water Vapor Transmission of Materials	R202, M1411.4, M1601.3.4, N1102.1.7
E 108—99	Test Methods for Fire Tests of Roof Coverings	R902.1, R902.2
E 119—00a	Test Methods for Fire Tests of Building Construction and Materials	R314.1.2, R317.1, R317.3.1
E 136—99	Test Method for Behavior of Materials in a Vertical Tube Furnace at 750°C	R202
E 152—95	Methods of Fire Tests of Door Assemblies	R314.3
E 330—97e[1]	Test Method for Structural Performance of Exterior Windows, Curtain Walls, and Doors by Uniform Static Air Pressure Difference	R613.3
E 814—00	Test Method for Fire Tests of Through-penetration Firestops	R317.3.1.2
E 970—00	Test Method for Critical Radiant Flux of Exposed Attic Floor Insulation Using a Radiant Heat Energy Source	R316.5
E 1602—94	Guide for Construction of Solid Fuel Burning Masonry Heaters	R1006.2
E 1886—97	Test Method for Performance of Exterior Windows, Curtain Walls, Doors and Storm Shutters Impacted by Missles and Exposed to Cyclic Pressure Differentials	R301.2.1.2

ASTM—continued

E 1996—01	Specification for Performance of Exterior Windows, Curtain Walls, Doors and Storm Shutters Impacted by Windborne Debris in Hurricanes	R301.2.1.2
F 409—99a	Specification for Thermoplastic Accessible and Replaceable Plastic Tube and Tubular Fittings	Table P2701.1, P2702.3
F 437—99	Specification for Threaded Chlorinated Poly (Vinyl Chloride) (CPVC) Plastic Pipe Fittings, Schedule 80	Table P2904.6
F 438—01	Specification for Socket-type Chlorinated Poly (Vinyl Chloride) (CPVC) Plastic Pipe Fittings, Schedule 40	Table P2904.6
F 439—01	Specification for Socket-type Chlorinated Poly (Vinyl Chloride) (CPVC) Plastic Pipe Fittings, Schedule 80	Table P2904.6
F 441/F 441M—99	Specification for Chlorinated Poly (Vinyl Chloride) (CPVC) Plastic Pipe, Schedules 40 and 80	Table P2904.4.1, Table P2904.5
F 442/F 442M—99	Specification for Chlorinated Poly (Vinyl Chloride) (CPVC) Plastic Pipe (SDR-PR)	Table P2904.4.1, Table P2904.5
F 477—99	Specification for Elastomeric Seals (Gaskets) for joining Plastic Pipe	P2904.15
F 493—97	Specification for Solvent Cements for Chlorinated Poly (Vinyl Chloride) (CPVC) Plastic Pipe and Fittings	P2904.9.1.2
F 628—01	Specification for Acrylonitrile-Butadiene-Styrene (ABS) Schedule 40 Plastic Drain, Waste, and Vent Pipe with a Cellular Core	Table P3002.1, Table P3002.2
F 656—96a	Specification for Primers for Use in Solvent Cement Joints of Poly (Vinyl Chloride) (PVC) Plastic Pipe and Fittings	P2904.9.1.3, Table P3002.1, Table P3002.2
F 714—00	Specification for Polyethylene (PE) Plastic Pipe (SDR-PR) Based on Outside Diameter	Table P3002.2
F 789—95a	Specification for Type PS-46 and Type PS-115 Poly (Vinyl Chloride) (PVC) Plastic Gravity Flow Sewer Pipe and Fittings	Table P3002.2
F 876—00	Specification for Cross-linked Polyethylene (PEX) Tubing	Table M2101.1, Table P2904.4.1
F 877—00	Specification for Cross-linked Polyethylene (PEX) Plastic Hot- and Cold-water Distribution Systems	Table M2101.1, Table P2904.4.1, Table P2904.5
F 891—00	Specification for Coextruded Poly (Vinyl Chloride) (PVC) Plastic Pipe with a Cellular Core	Table P3002.1, Table P3002.2
F 1055—98	Specification for Electrofusion Type Polyethylene Fittings for Outside Diameter Controlled Polyethylene Pipe and Fittings	Table M2101.1, M2104.2.1.2
F 1281—e01	Specification for Crosslinked Polyethylene/Aluminum/Crosslinked Polyethylene (PEX-AL-PEX) Pressure Pipe	Table P2904.4.1, Table P2904.5, Table M2101.1
F 1282—01a	Specification for Polyethylene/Aluminum/Polyethylene (PE-AL-PE) Composite Pressure Pipe	Table P2904.4.1, Table P2904.5
F 1488—00	Specification for Coextruded Composite Pipe	Table P3002.1, P3002.2
F 1667—01a	Specification for Driven Fasteners, Nails, Spikes, and Staples	R905.2.5
F 1807—99	Specification for Metal Insert Fittings Utilizing a Copper Crimp Ring for SDR9 Cross-linked Polyethylene (PEX) Tubing	Table P2904.6, P2904.9.1.4.2, Table M2101.1
F 1960—99	Specification for Cold Expansion Fittings with PEX Reinforcing Rings for Use with Cross-linked Polyethylene (PEX) Tubing	Table P2904.6
F 1974—00	Specification for Metal Insert Fittings for Polyethylene/Aluminum/Polyethylene and Crosslinked Polyethylene/Aluminum/Crosslinked Polyethylene Composite Pressure Pipe	Table P2904.6
F 2080—01	Specification for Cold-expansion Fittings with Metal Compression-Sleeves for Crosslinked Polyethylene (PEX) Pipe	P2904.6, P2904.9.1.4.2

AWPA

American Wood-Preservers' Association
P.O. Box 5690
Granbury, Texas 76049

Standard reference number	Title	Referenced in code section number
C1—00	All Timber Products — Preservative Treatment by Pressure Processes	R319.1, R323.1.7, R902.2.1
C2—01	Lumber, Timbers, Bridge Ties and Mine Ties — Preservative Treatment by Pressure Processes	R319.1, R323.1.7, Table R905.8.5
C3—99	Piles — Preservative Treatment by Pressure Processes	R319.1, R323.1.7
C4—99	Poles — Preservative Treatment by Pressure Processes	R319.1, R323.1.7
C9—00	Plywood — Preservative Treatment by Pressure Processes	R319.1, R323.1.7
C15—00	Wood for Commercial-Residential Construction Preservative Treatment by Pressure Process	R319.1, R323.1.7
C18—99	Standard for Pressure Treated Material in Marine Construction	R319.1, R323.1.7
C20—99	Structural Lumber—Fire-retardant Treatment by Pressure Processes	R802.1.3.2
C22—96	Lumber and Plywood for Permanent Wood Foundations—Preservative Treatment by Pressure Processes	R319.1, R323.1.7, R402.1.2, R504.3
C23—00	Round Poles and Posts Used in Building Construction—Preservative Treatment by Pressure Processes	R319.1, R323.1.7
C24—96	Sawn Timber Used to Support Residential and Commercial Structures	R319.1, R323.1.7
C27—88	Plywood—Fire-retardant Treatment by Pressure Process	R802.1.3.2
C28—99	Standard for Preservative Treatment by Pressure Process of Structural Glued Laminated Members and Laminations Before Gluing	R319.1, R323.1.7
C31—01	Lumber Used Out of Contact With the Ground and Continuously Protected From Liquid Water—Treatment by Pressure Processes	R319.1

AWPA—continued

C33—00	Standard for Preservative Treatment of Structural Composite Lumber by Pressure Processes	R323.1
M4—01	Standard for the Care of Preservative-treated Wood Products	R320.3.1
P1/P13—01	Standard for Creosote Preservative	R319.1, R323.1.7
P2—01	Standard for Creosote Solutions	R319.1, R323.1.7
P3—01	Standard for Creosote-petroleum Solutions	R319.1, R323.1.7

AWWA

American Water Works Association
6666 West Quincy Avenue
Denver, CO 80235

Standard reference number	Title	Referenced in code section number
C110—98	Standard for Ductile-iron and Gray-iron Fittings, 3 Inches through 48 Inches, for Water	Table P2904.6
C115—99	Standard for Flanged Ductile-iron Pipe with Ductile-iron or Gray-iron Threaded Flanges	Table P2904.4.1
C151—96	Standard for Ductile-iron Pipe, Centrifugally Cast, for Water	Table P2904.4.1
C153—00	Standard for Ductile-iron Compact Fittings for Water Service	Table P2904.6
C510—00	Double Check Valve Backflow Prevention Assembly	Table P2902.2, P2902.2.6
C511—00	Reduced-Pressure Principle Backflow Prevention Assembly	Table P2902.2, P2902.2.5, P2902.4.1

CDA

The Copper Development Association, Inc.
260 Madison Avenue
New York, NY 10016

Standard reference number	Title	Referenced in code section number
4050	Copper in Architecture—Design Handbook	Table R905.10.3

CGSB

Canadian General Standards Board
222 Queens Street
14th Floor, Suite 1402
Ottawa, Ontario, Canada KIA 1G6

Standard reference number	Title	Referenced in code section number
37-GP—52M—(1984)	Roofing and Waterproofing Membrane, Sheet Applied, Elastomeric	R905.12.2
37-GP—56M—(1980)	Membrane, Modified Bituminous, Prefabricated and Reinforced for Roofing—with December 1985 Amendment	Table R905.11.2
CAN/CGSB-37.54—95	Polyvinyl Chloride Roofing and Waterproofing Membrane	R905.13.2

CISPI

Cast Iron Soil Pipe Institute
5959 Shallowford Road, Suite 419
Chattanooga, TN 37421

Standard reference number	Title	Referenced in code section number
301—97	Standard Standard Specification for Hubless Cast Iron Soil Pipe and Fittings for Sanitary and Storm Drain, Waste and Vent Piping Applications	Table P3002.1, Table P3002.2
310—97	Standard Specification for Coupling for Use in Connection with Hubless Cast Iron Soil Pipe and Fittings for Sanitary and Storm Drain, Waste, and Vent Piping Applications	Table P3002.1, Table P3002.2

CPSC

Consumer Product Safety Commission
4330 East West Highway
Bethesda, MD 20814-4408

Standard reference number	Title	Referenced in code section number
16 CFR Part 1201—(1977)	Safety Standard for Architectural Glazing	R308.1.1, R308.3
16 CFR Part 1209—(1979)	Interim Safety Standard for Cellulose Insulation	R316.3
16 CFR Part 1404—(1979)	Cellulose Insulation	R316.3

CSA

Canadian Standards Association
178 Rexdale Blvd.
Rexdale (Toronto), Ontario, Canada M9W 1R3

Standard reference number	Title	Referenced in code section number
8—93	Requirements for Gas Fired Log Lighters for Wood Burning Fireplaces—with Revisions through January 1999	G2433.1
0325.0—92	Construction Sheathing (Reaffirmed 1998)	R503.2.1
0437-Series—93	Standards on OSB and Waferboard (Reaffirmed 2001)	R503.2.1, R803.2.1
B45.1—99	Ceramic Plumbing Fixtures	P2711.1, P2712.1
B45.2—99	Enameled Cast Iron Plumbing Fixtures	P2711.1
B45.3—99	Porcelain Enameled Steel Plumbing Fixtures	P2711.1
B45.4—99	Stainless Steel Plumbing Fixtures	P2711.1, P2712.1
B45.5—99	Plastic Plumbing Fixtures	P2711.2, P2712.1
B45.9—99	Macerating Systems and Related Components	P3007.1
B64.7—94	Vacuum Breakers, Laboratory Faucet Type (LFVB)	Table P2902.2, P2902.2.2
B137.1—99	Polyethylene Pipe, Tubing and Fittings for Cold Water Pressure Services	Table P2904.4.1
B137.2—99	PVC Injection-moulded Gasketed Fittings for Pressure Applications	Table P2904.6
B137.3—99	Rigid Poly (Vinyl Chloride) (PVC) Pipe for Pressure Applications	Table P2904.4.1
B137.5—99	Cross-linked Polyethylene (PEX) Tubing Systems for Pressure Applications—with Revisions through September 1992	Table P2904.4.1, Table P2904.5
B137.6—99	CPVC Pipe, Tubing and Fittings For Hot and Cold Water Distribution Systems—with Revisions through May 1986	Table P2904.4.1, Table P2904.5
LC3—00	Appliance Stands and Drain Pans	P2801.5
CAN3-B137.8M—99	Polybutylene (PB) Piping for Pressure Applications—with Revisions through July 1992	Table P2904.4.1, Table P2904.5
CAN/CSA A257.3M—92	Joints for Circular Concrete Sewer and Culvert Pipe, Manhole Sections, and Fittings Using Rubber Gaskets	P3003.3.5
CAN/CSA B64.1.1—01	Vacuum Breakers, Atmospheric Type (AVB)	Table P2902.2, P2902.2.2
CAN/CSA B64.2—01	Vacuum Breakers, Hose Connection Type (HCVP)	Table P2902.2, P2902.2.2
CAN/CSA B64.2.2—01	Vacuum Breakers, Hose Connection Type (HCVP) with Automatic Draining Feature	Table P2902.2, P2902.2.2
CAN/CSA B64.3—01	Backflow Preventers, Dual Check Valve Type with Atmospheric Port (DCAP)	Table P2902.2, P2902.2.3, P2902.4.1
CAN/CSA B64.4—01	Backflow Preventers, Reduced Pressure Principle Type (RP)	Table P2902.2, P2902.2.5
CAN/CSA B137.9—99	Polyethylene/Aluminum/Polyethylene Composite Pressure Pipe Systems	Table P2904.4.1
CAN/CSA B137.10M—99	Crosslinked Polethylene/Aluminum/Polyethylene Composite Pressure Pipe Systems	Table P2904.4.1, Table P2904.5, Table M2101.1
CAN/CSA B602—99	Mechanical Couplings for Drain, Waste, and Vent Pipe and Sewer Pipe	Table P3002.1, Table P3002.2

CSSB

Cedar Shake & Shingle Bureau
515 116th Avenue, NE, #275
Bellevue, WA 98004-5294

Standard reference number	Title	Referenced in code section number
CSSB—97	Grading and Packing Rules for Western Red Cedar Shakes and Western Red Shingles of the Cedar Shake and Shingle Bureau	Table R905.7.4, R702.6, R703.5
CSSB—97	Grading and Packing Rules for Western Red Cedar Shakes and Western Red Shingles of the Cedar Shake and Shingle Bureau	Table R905.7.4, R703.5

DOC

United States Department of Commerce
National Institute of Standards and Technology
Gaithersburg, MD 20899

Standard reference number	Title	Referenced in code section number
PS 1—95	Construction and Industrial Plywood	R404.2.1, Table R404.2.3, R503.2.1, R604.1, R803.2.1
PS 2—92	Performance Standard for Wood-based Structural-use Panels	R404.2.1, Table R404.2.3, R503.2.1, R604.1, R803.2.1
PS 20—99	American Softwood Lumber Standard	R404.2.1, R502.1, R602.1, R802.1

DOTn

Department of Transportation
400 Seventh St. S.W.
Washington, DC 20590

Standard reference number	Title	Referenced in code section number
49 CFR, Parts 192.281(e) & 192.283 (b)	Transportation of Natural and Other Gas by Pipeline: Minimum Federal Safety Standards	G2414.6.1

FEMA

Federal Emergency Management Agency
500 C Street, SW
Washington, DC 20472

Standard reference number	Title	Referenced in code section number
TB-2—93	Flood-resistant Materials Requirements	R323.1.7
TB-11—01	Crawlspace Construction for Buildings Located in Special Flood Hazard Area	R408.6

FM

Factory Mutual
Standards Laboratories Department
1151 Boston Providence Turnpike
Norwood, MA 02062

Standard reference number	Title	Referenced in code section number
4450—(1989)	Approval Standard for Class 1 Insulated Steel Deck Roofs—with Supplements through July 1992	R906.1
4880—(2001)	American National Standard for Evaluating Insulated Wall or Wall and Roof/Ceiling Assemblies, Plastic Interior Finish Materials, Plastic Exterior Building Panels,Wall/Ceiling Coating Systems, Interior or Exterior Finish Systems	R314.3

GA

Gypsum Association
810 First Street, Northeast, Suite 510
Washington, DC 20002-4268

Standard reference number	Title	Referenced in code section number
GA-253—99	Recommended Standard Specification for the Application of Gypsum Sheathing	Table R602.3(1)

HPVA

Hardwood Plywood & Veneer Association
1825 Michael Faraday Drive
Reston, Virginia 20190-5350

Standard reference number	Title	Referenced in code section number
HP-1—2000	The American National Standard for Hardwood and Decorative Plywood	R702.5

ICC

International Code Council
5203 Leesburg Pike, Suite 600
Falls Church, VA 22041

Standard reference number	Title	Referenced in code section number
IBC—03	International Building Code®	R110.2, R322.1, R323.1, R323.1.5, R403.1.8, R1001.8.2, G2402.3
ICC EC—03	ICC Electrical Code®—Administrative Provisions	R107.3, G2402.3
IEBC—03	International Existing Building Code®	R101.2, G2401.1
IECC—03	International Energy Conservation Code®	R104.11
IFC—03	International Fire Code®	R102.7, G2402.3, G2412.2, G2423.1
IFGC—03	International Fuel Gas Code®	R104.11, G2401.1
IMC—03	International Mechanical Code®	R104.11, M2106.1, G2402.3
IPC—03	International Plumbing Code®	R104.11, G2402.3
IPSDC—03	International Private Sewage Disposal Code®	R323.1.6
IPMC—03	International Property Maintenance Code®	R102.7
SBCCI SSTD 10—99	Standard for Hurricane Resistant Construction	R301.2.1.1

MSS

Manufacturers Standardization Society of the Valve and Fittings Industry
127 Park Street, Northeast
Vienna, VA 22180

Standard reference number	Title	Referenced in code section number
SP-58—93	Pipe Hangers and Supports—Materials, Design and Manufacture	G2418.2

NAIMA

North American Insulation Manufacturers Association
44 Canal Center Plaza, Suite 310
Alexandria, VA 22314

Standard reference number	Title	Referenced in code section number
AH 116 06—02	Fibrous Glass Duct Construction Standards	M1601.1.1

NCMA

National Concrete Masonry Association
2302 Horse Pen Road
Herndon, VA 20171-3499

Standard reference number	Title	Referenced in code section number
TR 68-A—75	Design and Construction of Plain and Reinforced Concrete Masonry and Basement and Foundation Walls	R404.1

NFPA

National Fire Protection Association
Batterymarch Park
Quincy, MA 02269

Standard reference number	Title	Referenced in code section number
13—99	Installation of Sprinkler Systems	R317.1
13D—99	Installation of Sprinkler Systems in One- and Two-family Dwellings and Manufactured Homes	R317.1
13R—99	Installation of Sprinkler Systems in Residential Occupancies Up to and Including Four Stories in Height	R317.1
31—01	Installation of Oil-burning Equipment	M1801.3.1, M1805.3
58—01	Liquefied Petroleum Gas Code	G2412.2, G2414.6.2
70—02	National Electrical Code	E3301.1, E3301.2, E4201.1, Table E4203.2, E4204.3, E4204.4
72—99	National Fire Alarm Code	R313.1

NFPA—continued

85—01	Boiler and Construction Sysems Hazards Code	G2452.1
259—98	Standard Test Method for Potential Heat of Building Materials	R314.2.5
286—00	Standard Methods of Fire Tests for Evaluating Contribution of Wall and Ceiling Interior Finish to Room Fire Growth	R314.3, R315.4
501—99	Standard on Manufactured Housing	R202
853—00	Standard for the Installation of Stationary Fuel Cell Power Plants	M1903.1

NFRC

National Fenestration Rating Council, Inc.
1300 Spring Park, Suite 120
Silver Spring, MD 20910

Standard reference number	Title	Referenced in code section number
100—2001	Procedure for Determining Fenestration Product U-factors	N1101.3.2, N1101.3.2.1
200—2001	Procedure for Determining Fenestration product Solar Heat Gain Coefficients and Visible Transmittance at Normal Incidence	N1101.3.2, N1101.3.2.1
400—2001	Procedure for Determining Fenestration Product Air Leakage	N1101.3.2.2

NSF

NSF International
789 N. Dixboro Road
Ann Arbor, MI 48105

Standard reference number	Title	Referenced in code section number
14—1999	Plastic Piping System Components and Related Materials	P2608.3, P2907.3
42—2000	Drinking Water Treatment Units—Anesthetic Effects	P2907.1, P2907.3
44—2001	Residential Cation Exchange Water Softeners	P2907.1, P2907.3
53—2001	Drinking Water Treatment Units—Health Effects	P2907.1, P2907.3
58—2001	Reverse Osmosis Drinking Water Treatment Systems	P2907.2, P2907.3
61—2001	Drinking Water System Components—Health Effects	P2608.5, P2722.1, P2904.4, P2904.5, P2907.3

RMA

Rubber Manufacturer's Association
1400 K Street, NW
Washington, DC 20005

Standard reference number	Title	Referenced in code section number
RP1—90	Minimum Requirements for Non-Reinforced Black EPDM Rubber Sheets	R905.12.2
RP2—90	Minimum Requirements for Fabric-Reinforced Black EPDM Rubber Sheets	R905.12.2
RP3—85	Minimum Requirements for Fabric-Reinforced Black Polychloroprene Rubber Sheets	R905.12.2

SAE

Society of Automotive Engineers
4700 West Lake Avenue
Glenview, IL 60025

Standard reference number	Title	Referenced in code section number
J 78—(1998)	Steel Self-drilling Tapping Screws	R505.2.4, R603.2.4, R804.2.4

SMACNA

Sheet Metal & Air Conditioning Contractors National Assoc., Inc.
4021 Lafayette Center Road
Chantilly, VA 22021

Standard reference number	Title	Referenced in code section number
SMACNA—92	Fibrous Glass Duct Construction Standards	M1601.1.1

TMS

The Masonry Society
3970 Broadway, Suite 201-D
Boulder, CO 80304

Standard reference number	Title	Referenced in code section number
402—02	Building Code Requirements for Masonry Structures	R404.1, R606.1, R606.1.1, R606.11.1, R606.11.2.2.1, R606.11.2.2.2, R606.11.3.1
602—02	Specification for Masonry Structures	R404.1, R606.1, R606.1.1, R606.11.1, R606.11.2.2.1, R606.11.2.2.2, R606.11.3.1

TPI

Truss Plate Institute
583 D'Onofrio Drive, Suite 200
Madison, WI 53719

Standard reference number	Title	Referenced in code section number
TPI 1—2002	National Design Standard for Metal-plate-connected Wood Truss Construction	R502.11.1, R502.11.2, R802.10.2, R802.10.3, R802.11.1

UL

Underwriters Laboratories, Inc.
333 Pfingsten Road
Northbrook, IL 60062

Standard reference number	Title	Referenced in code section number
17—94	Vent or Chimney Connector Dampers for Oil-fired Appliances—with Revisions through September 1999	M1802.2.2
58—96	Steel Underground Tanks for Flammable and Combustible Liquids—with Revisions through July 1998	M2201.1
80—96	Steel Inside Tanks for Oil-burner Fuel	M2201.1
103—98	Factory-built Chimneys for Residential Type and Building Heating Appliances—with Revision through March 1999	R202, R1002.3, G2430.1
127—99	Factory-built Fireplaces—with Revisions through November 1999	R1002.4, R1003.12, R1004.1, R1004.4, G2445.7
174—98	Household Electric Storage Tank Water Heaters—with Revisions through October 1999	M2005.1
181—96	Factory-made Air Ducts and Air Connectors—with Revisions through December 1998	M1601.2, M1601.3.1
181A—95	Closure Systems for Use with Rigid Air Ducts and Air Connectors—with Revisions through December 1998	M1601.2, M1601.3.1
181B—98	Closure Systems for Use with Flexible Air Ducts and Air Connectors—with Revisions through December 1998	M1601.2, M1601.3.1
325—95	Door, Drapery, Louver and Window Operations and Systems—with Revisions through June 2001	R309.6
343—97	Pumps for Oil-Burning Appliances—with Revisions through December 22, 1999	M2204.1
441—98	Gas Vents	G2426.1
536—97	Flexible Metallic Hose—with Revisions through October 2000	M2202.3
641—95	Type L, Low-temperature Venting Systems—with Revisions through April 1999	R202, R1001.8.5, M1804.2.4, G2426.1
726—98	Oil-fired Boiler Assemblies—with Revisions through January 1999	M2001.1.1, M2006.1, G2425.1
727—98	Oil-fired Central Furnaces—with Revisions through January 1999	M1402.1
729—98	Oil-fired Floor Furnaces—with Revisions through January 1999	M1408.1
730—98	Oil-fired Wall Furnaces—with Revisions through January 1999	M1409.1
732—95	Oil-fired Storage Tank Water Heaters—with Revisions through January 1999	M2005.1
737—96	Fireplaces Stoves—with Revisions through May 1998	R1002.3, M1414.1
790—98	Tests for Fire Resistance of Roof Covering Materials—with Revisions through July 1998	R902.1
795—99	Commercial-Industrial Gas Heating Equipment	G2442.1, G2452.1
834—98	Heating, Water Supply, and Power-Electric—with Revisions through November 1998	M2001.1.1
896—93	Oil-burning Stoves—with Revisions through November 1999	M1410.1
959—01	Medium Heat Appliance Factory-built Chimneys	R1002.6
923—95	Microwave Cooking Appliances—with Revisions through July 20, 1998	M1503.1
1040—96	Fire Test of Insulated Wall Construction—with Revisions through April 2001	R314.3
1256—98	Fire Test of Roof Deck Construction—with Revisions through March 2000	R906.1
1261—96	Electric Water Heaters for Pools and Tubs—with Revisions through November 25, 1998	M2006.1
1453—95	Electronic Booster and Commercial Storage Tank Water Heaters—with Revisions through September 1998	M2005.1

UL—continued

1479—94	Fire Tests of Through-Penetration Firestops	R317.3.1.2
1482—98	Solid-fuel Type Room Heaters—with Revisions through January 2000	M1410.1
1715—97	Fire Test of Interior Finish Material	R314.3
1738—00	Venting Systems for Gas-burning Appliances, Categories II, III and IV—with Revisions through December 2000	G2426.1
1777—98	Chimney Liners—with Revisions through July 1998	R1001.8.1, R1001.15, M1801.3.4, G2425.12, G2425.15.4
1995—98	Heating and Cooling Equipment—with Revisions through August 1999	M1402.1, M1403.1, M1407.1
2158A—96	Outline of Investigation for Clothes Dryer Transition Duct	M1501.1

ULC

Underwriters' Laboratories of Canada
7 Crouse Road
Scarborough, Ontario, Canada M1R 3A9

Standard reference number	Title	Referenced in code section number
S 102—M88	Surface Burning Characteristics of Building Materials and Assemblies	R316.2

WDMA

Window & Door Manufacturers Association
1400 East Touhy Avenue, Suite 470
Des Plaines, IL 60018

Standard reference number	Title	Referenced in code section number
101/I.S. 2—97	Voluntary Specifications for Aluminum, Vinyl (PVC) and Wood Windows and Glass Doors	R613.3
101/I.S.2/NAFS—02	Voluntary Performance Specification for Windows, Skylights and Glass Doors	R308.6.9, R613.3, N1101.3.2.2

This appendix is a direct excerpt from the 2003 *International Fuel Gas Code*, and no attempt has been made to coordinate with Chapter 24 of the *International Residential Code.*

APPENDIX A (IFGS)

SIZING AND CAPACITIES OF GAS PIPING

(This appendix is informative and is not part of the code.)

A.1 General. To determine the size of piping used in a gas piping system, the following factors must be considered:

(1) Allowable loss in pressure from point of delivery to equipment

(2) Maximum gas demand

(3) Length of piping and number of fittings

(4) Specific gravity of the gas

(5) Diversity factor

For any gas piping system, or special gas utilization equipment, or for conditions other than those covered by the tables provided in this code, such as longer runs, greater gas demands, or greater pressure drops, the size of each gas piping system should be determined by standard engineering practices acceptable to the code official.

A.2 Description of tables

A.2.1 General. The quantity of gas to be provided at each outlet should be determined, whenever possible, directly from the manufacturer's British thermal unit (Btu) input rating of the equipment that will be installed. In case the ratings of the equipment to be installed are not known, Table 402.2 shows the approximate consumption (in Btu per hour) of certain types of typical household appliances.

To obtain the cubic feet per hour of gas required, divide the total Btu input of all equipment by the average Btu heating value per cubic feet of the gas. The average Btu per cubic feet of the gas in the area of the installation can be obtained from the serving gas supplier.

A.2.2 Low pressure natural gas tables. Capacities for gas at low pressure [0.5 psig (3.5 kPa gauge) or less] in cubic feet per hour of 0.60 specific gravity gas for different sizes and lengths are shown in Tables 402.4(1) and 402.4(2) for iron pipe or equivalent rigid pipe, in Tables 402.4(7) through 402.4(9) for smooth wall semi-rigid tubing, and in Tables 402.4(14) through 402.4(16) for corrugated stainless steel tubing. Tables 402.4(1) and 402.4(7) are based upon a pressure drop of 0.3-inch water column (w.c.) (75 Pa), whereas Tables 402.4(2), 402.4(8) and 402.4(14) are based upon a pressure drop of 0.5-inch w.c. (125 Pa). Tables 402.4(9), 402.4(15) and 402.4(16) are special low-pressure applications based upon pressure drops greater than 0.5-inch w.c. (125 Pa). In using these tables, an allowance (in equivalent length of pipe) should be considered for any piping run with four or more fittings [see Table A.2.2].

A.2.3 Undiluted liquefied petroleum tables. Capacities in thousands of Btu per hour of undiluted liquefied petroleum gases based on a pressure drop of 0.5-inch w.c. (125 Pa) for different sizes and lengths are shown in Table 402.4(24) for iron pipe or equivalent rigid pipe, in Table 402.4(26) for smooth wall semi-rigid tubing, in Table 402.4(28) for corrugated stainless steel tubing, and in Tables 402.4(31) and 402.4(33) for polyethylene plastic pipe and tubing. Tables 402.4(29) and 402.4(30) for corrugated stainless steel tubing and Table 402.4(32) for polyethylene plastic pipe are based on operating pressures greater than 0.5 pounds per square inch (psi) (3.5 kPa) and pressure drops greater than 0.5-inch w.c. (125 Pa). In using these tables, an allowance (in equivalent length of pipe) should be considered for any piping run with four or more fittings [see Table A.2.2].

A.2.4 Natural gas specific gravity. Gas piping systems that are to be supplied with gas of a specific gravity of 0.70 or less can be sized directly from the tables provided in this code, unless the code official specifies that a gravity factor be applied. Where the specific gravity of the gas is greater than 0.70, the gravity factor should be applied.

Application of the gravity factor converts the figures given in the tables provided in this code to capacities for another gas of different specific gravity. Such application is accomplished by multiplying the capacities given in the tables by the multipliers shown in Table A.2.4. In case the exact specific gravity does not appear in the table, choose the next higher value specific gravity shown.

A.2.5 Higher pressure natural gas tables. Capacities for gas at pressures greater than 0.5 psig (3.5 kPa gauge) in cubic feet per hour of 0.60 specific gravity gas for different sizes and lengths are shown in Tables 402.4(3) through 402.4(6) for iron pipe or equivalent rigid pipe, Tables 402.4(10) to 402.4(13) for semi-rigid tubing, Tables 402.4(17) and 402.4(18) for corrugated stainless steel tubing, and Tables 402.4(19) through 402.4(21) for polyethylene plastic pipe.

TABLE A.2.2
EQUIVALENT LENGTHS OF PIPE FITTINGS AND VALVES

		SCREWED FITTINGS[2]				90° WELDING ELBOWS AND SMOOTH BENDS[3]					
		45°/Ell	90°/Ell	180° close return bends	Tee	$R/d = 1$	$R/d = 1^1/_3$	$R/d = 2$	$R/d = 4$	$R/d = 6$	$R/d = 8$
k factor =		0.42	0.90	2.00	1.80	0.48	0.36	0.27	0.21	0.27	0.36
L/d ′ratio[4] *n* =		14	30	67	60	16	12	9	7	9	12
Nominal pipe size, inches	Inside diameter *d*, inches, Schedule 40[6]	*L* = Equivalent Length in Feet of Schedule 40 (Standard-Weight) Straight Pipe[6]									
$^1/_2$	0.622	0.73	1.55	3.47	3.10	0.83	0.62	0.47	0.36	0.47	0.62
$^3/_4$	0.824	0.96	2.06	4.60	4.12	1.10	0.82	0.62	0.48	0.62	0.82
1	1.049	1.22	2.62	5.82	5.24	1.40	1.05	0.79	0.61	0.79	1.05
$1^1/_4$	1.380	1.61	3.45	7.66	6.90	1.84	1.38	1.03	0.81	1.03	1.38
$1^1/_2$	1.610	1.88	4.02	8.95	8.04	2.14	1.61	1.21	0.94	1.21	1.61
2	2.067	2.41	5.17	11.5	10.3	2.76	2.07	1.55	1.21	1.55	2.07
$2^1/_2$	2.469	2.88	6.16	13.7	12.3	3.29	2.47	1.85	1.44	1.85	2.47
3	3.068	3.58	7.67	17.1	15.3	4.09	3.07	2.30	1.79	2.30	3.07
4	4.026	4.70	10.1	22.4	20.2	5.37	4.03	3.02	2.35	3.02	4.03
5	5.047	5.88	12.6	28.0	25.2	6.72	5.05	3.78	2.94	3.78	5.05
6	6.065	7.07	15.2	33.8	30.4	8.09	6.07	4.55	3.54	4.55	6.07
8	7.981	9.31	20.0	44.6	40.0	10.6	7.98	5.98	4.65	5.98	7.98
10	10.02	11.7	25.0	55.7	50.0	13.3	10.0	7.51	5.85	7.51	10.0
12	11.94	13.9	29.8	66.3	59.6	15.9	11.9	8.95	6.96	8.95	11.9
14	13.13	15.3	32.8	73.0	65.6	17.5	13.1	9.85	7.65	9.85	13.1
16	15.00	17.5	37.5	83.5	75.0	20.0	15.0	11.2	8.75	11.2	15.0
18	16.88	19.7	42.1	93.8	84.2	22.5	16.9	12.7	9.85	12.7	16.9
20	18.81	22.0	47.0	105.0	94.0	25.1	18.8	14.1	11.0	14.1	18.8
24	22.63	26.4	56.6	126.0	113.0	30.2	22.6	17.0	13.2	17.0	22.6

continued

TABLE A.2.2—continued
EQUIVALENT LENGTHS OF PIPE FITTINGS AND VALVES

		MITER ELBOWS[3] (No. of miters)					WELDING TEES		VALVES (screwed, flanged, or welded)			
		1-45°	1-60°	1-90°	2-90°[5]	3-90°[5]	Forged	Miter[3]	Gate	Globe	Angle	Swing Check
k factor =		0.45	0.90	1.80	0.60	0.45	1.35	1.80	0.21	10	5.0	2.5
L/d 'ratio[4] *n* =		15	30	60	20	15	45	60	7	333	167	83
Nominal pipe size, inches	Inside diameter *d*, inches, Schedule 40[6]	*L* = Equivalent Length in Feet of Schedule 40 (Standard-Weight) Straight Pipe[6]										
1/2	0.622	0.78	1.55	3.10	1.04	0.78	2.33	3.10	0.36	17.3	8.65	4.32
3/4	0.824	1.03	2.06	4.12	1.37	1.03	3.09	4.12	0.48	22.9	11.4	5.72
1	1.049	1.31	2.62	5.24	1.75	1.31	3.93	5.24	0.61	29.1	14.6	7.27
1 1/4	1.380	1.72	3.45	6.90	2.30	1.72	5.17	6.90	0.81	38.3	19.1	9.58
1 1/2	1.610	2.01	4.02	8.04	2.68	2.01	6.04	8.04	0.94	44.7	22.4	11.2
2	2.067	2.58	5.17	10.3	3.45	2.58	7.75	10.3	1.21	57.4	28.7	14.4
2 1/2	2.469	3.08	6.16	12.3	4.11	3.08	9.25	12.3	1.44	68.5	34.3	17.1
3	3.068	3.84	7.67	15.3	5.11	3.84	11.5	15.3	1.79	85.2	42.6	21.3
4	4.026	5.04	10.1	20.2	6.71	5.04	15.1	20.2	2.35	112.0	56.0	28.0
5	5.047	6.30	12.6	25.2	8.40	6.30	18.9	25.2	2.94	140.0	70.0	35.0
6	6.065	7.58	15.2	30.4	10.1	7.58	22.8	30.4	3.54	168.0	84.1	42.1
8	7.981	9.97	20.0	40.0	13.3	9.97	29.9	40.0	4.65	222.0	111.0	55.5
10	10.02	12.5	25.0	50.0	16.7	12.5	37.6	50.0	5.85	278.0	139.0	69.5
12	11.94	14.9	29.8	59.6	19.9	14.9	44.8	59.6	6.96	332.0	166.0	83.0
14	13.13	16.4	32.8	65.6	21.9	16.4	49.2	65.6	7.65	364.0	182.0	91.0
16	15.00	18.8	37.5	75.0	25.0	18.8	56.2	75.0	8.75	417.0	208.0	104.0
18	16.88	21.1	42.1	84.2	28.1	21.1	63.2	84.2	9.85	469.0	234.0	117.0
20	18.81	23.5	47.0	94.0	31.4	23.5	70.6	94.0	11.0	522.0	261.0	131.0
24	22.63	28.3	56.6	113.0	37.8	28.3	85.0	113.0	13.2	629.0	314.0	157.0

For SI: 1 foot = 305 mm, 1 degree = 0.01745 rad.

Note: Values for welded fittings are for conditions where bore is not obstructed by weld spatter or backing rings. If appreciably obstructed, use values for "Screwed Fittings."

1. Flanged fittings have three-fourths the resistance of screwed elbows and tees.
2. Tabular figures give the extra resistance due to curvature alone to which should be added the full length of travel.
3. Small size socket-welding fittings are equivalent to miter elbows and miter tees.
4. Equivalent resistance in number of diameters of straight pipe computed for a value of $(f - 0.0075)$ from the relation $(n - k/4f)$.
5. For condition of minimum resistance where the centerline length of each miter is between d and $2^1/_2d$.
6. For pipe having other inside diameters, the equivalent resistance may be computed from the above *n* values.

Source: Crocker, S. *Piping Handbook*, 4th ed., Table XIV, pp. 100-101. Copyright 1945 by McGraw-Hill, Inc. Used by permission of McGraw-Hill Book Company.

TABLE A.2.4
MULTIPLIERS TO BE USED WITH TABLES 402.4(1) THROUGH 402.4(21) WHERE THE SPECIFIC GRAVITY OF THE GAS IS OTHER THAN 0.60

SPECIFIC GRAVITY	MULTIPLIER	SPECIFIC GRAVITY	MULTIPLIER
.35	1.31	1.00	.78
.40	1.23	1.10	.74
.45	1.16	1.20	.71
.50	1.10	1.30	.68
.55	1.04	1.40	.66
.60	1.00	1.50	.63
.65	.96	1.60	.61
.70	.93	1.70	.59
.75	.90	1.80	.58
.80	.87	1.90	.56
.85	.84	2.00	.55
.90	.82	2.10	.54

A.3 Use of capacity tables

A.3.1 Longest length method. This sizing method is conservative in its approach by applying the maximum operating conditions in the system as the norm for the system and by setting the length of pipe used to size any given part of the piping system to the maximum value.

To determine the size of each section of gas piping in a system within the range of the capacity tables, proceed as follows. (also see sample calculations included in this Appendix).

(1) Divide the piping system into appropriate segments consistent with the presence of tees, branch lines and main runs. For each segment, determine the gas load (assuming all appliances operate simultaneously) and its overall length. An allowance (in equivalent length of pipe) as determined from Table A.2.2 shall be considered for piping segments that include four or more fittings.

(2) Determine the gas demand of each appliance to be attached to the piping system. Where Tables 402.4(1) through 402.4(23) are to be used to select the piping size, calculate the gas demand in terms of cubic feet per hour for each piping system outlet. Where Tables 402.4(24) through 402.4(33) are to be used to select the piping size, calculate the gas demand in terms of thousands of Btu per hour for each piping system outlet.

(3) Where the piping system is for use with other than undiluted liquefied petroleum gases, determine the design system pressure, the allowable loss in pressure (pressure drop), and specific gravity of the gas to be used in the piping system.

(4) Determine the length of piping from the point of delivery to the most remote outlet in the building/piping system.

(5) In the appropriate capacity table, select the row showing the measured length or the next longer length if the table does not give the exact length. This is the only length used in determining the size of any section of gas piping. If the gravity factor is to be applied, the values in the selected row of the table are multiplied by the appropriate multiplier from Table A.2.4.

(6) Use this horizontal row to locate ALL gas demand figures for this particular system of piping.

(7) Starting at the most remote outlet, find the gas demand for that outlet in the horizontal row just selected. If the exact figure of demand is not shown, choose the next larger figure left in the row.

(8) Opposite this demand figure, in the first row at the top, the correct size of gas piping will be found.

(9) Proceed in a similar manner for each outlet and each section of gas piping. For each section of piping, determine the total gas demand supplied by that section.

When a large number of piping components (such as elbows, tees and valves) are installed in a pipe run, additional pressure loss can be accounted for by the use of equivalent lengths. Pressure loss across any piping component can be equated to the pressure drop through a length of pipe. The equivalent length of a combination of only four elbows/tees can result in a jump to the next larger length row, resulting in a significant reduction in capacity. The equivalent lengths in feet shown in Table A.2.2 have been computed on a basis that the inside diameter corresponds to that of Schedule 40 (standard-weight) steel pipe, which is close enough for most purposes involving other schedules of pipe. Where a more specific solution for equivalent length is desired, this may be made by multiplying the actual inside diameter of the pipe in inches by n/12, or the actual inside diameter in feet by n (n can be read from the table heading). The equivalent length values can be used with reasonable accuracy for copper or brass fittings and bends although the resistance per foot of copper or brass pipe is less than that of steel. For copper or brass valves, however, the equivalent length of pipe should be taken as 45 percent longer than the values in the table, which are for steel pipe.

A.3.2 Branch length method. This sizing method reduces the amount of conservatism built into the traditional Longest Length Method. The longest length as measured from the meter to the furthest remote appliance is only used to size the initial parts of the overall piping system. The Branch Length Method is applied in the following manner:

(1) Determine the gas load for each of the connected appliances.

(2) Starting from the meter, divide the piping system into a number of connected segments, and determine the length and amount of gas that each segment would carry assuming that all appliances were operated simultaneously. An allowance (in equivalent length of pipe) as determined from Table A.2.2 should be considered for piping segments that include four or more fittings.

(3) Determine the distance from the outlet of the gas meter to the appliance furthest removed from the meter.

(4) Using the longest distance (found in Step 3), size each piping segment from the meter to the most remote appliance outlet.

(5) For each of these piping segments, use the longest length and the calculated gas load for all of the connected appliances for the segment and begin the sizing process in Steps 6 through 8.

(6) Referring to the appropriate sizing table (based on operating conditions and piping material), find the longest length distance in the first column or the next larger distance if the exact distance is not listed. The use of alternative operating pressures and/or pressure drops will require the use of a different sizing table, but will not alter the sizing methodology. In many cases, the use of alternative operating pressures and/or pressure drops will require the approval of both the code official and the local gas serving utility.

(7) Trace across this row until the gas load is found or the closest larger capacity if the exact capacity is not listed.

(8) Read up the table column and select the appropriate pipe size in the top row. Repeat Steps 6, 7 and 8 for each pipe segment in the longest run.

(9) Size each remaining section of branch piping not previously sized by measuring the distance from the gas meter location to the most remote outlet in that branch, using the gas load of attached appliances and following the procedures of Steps 2 through 8.

A.3.3 Hybrid pressure method. The sizing of a 2 psi (13.8 kPa) gas piping system is performed using the traditional Longest Length Method but with modifications. The 2 psi (13.8 kPa) system consists of two independent pressure zones, and each zone is sized separately. The Hybrid Pressure Method is applied as follows.

The sizing of the 2 psi (13.8 kPa) section (from the meter to the line regulator) is as follows:

(1) Calculate the gas load (by adding up the name plate ratings) from all connected appliances. (In certain circumstances the installed gas load may be increased up to 50 percent to accommodate future addition of appliances.) Ensure that the line regulator capacity is adequate for the calculated gas load and that the required pressure drop (across the regulator) for that capacity does not exceed $^3/_4$ psi (5.2 kPa) for a 2 psi (13.8 kPa) system. If the pres-

sure drop across the regulator is too high (for the connected gas load), select a larger regulator.

(2) Measure the distance from the meter to the line regulator located inside the building.

(3) If there are multiple line regulators, measure the distance from the meter to the regulator furthest removed from the meter.

(4) The maximum allowable pressure drop for the 2 psi (13.8 kPa) section is 1 psi (6.9 kPa).

(5) Referring to the appropriate sizing table (based on piping material) for 2 psi (13.8 kPa) systems with a 1 psi (6.9 kPa) pressure drop, find this distance in the first column, or the closest larger distance if the exact distance is not listed.

(6) Trace across this row until the gas load is found or the closest larger capacity if the exact capacity is not listed.

(7) Read up the table column to the top row and select the appropriate pipe size.

(8) If there are multiple regulators in this portion of the piping system, each line segment must be sized for its actual gas load, but using the longest length previously determined above.

The low pressure section (all piping downstream of the line regulator) is sized as follows:

(1) Determine the gas load for each of the connected appliances.

(2) Starting from the line regulator, divide the piping system into a number of connected segments and/or independent parallel piping segments, and determine the amount of gas that each segment would carry assuming that all appliances were operated simultaneously. An allowance (in equivalent length of pipe) as determined from Table A.2.2 should be considered for piping segments that include four or more fittings.

(3) For each piping segment, use the actual length or longest length (if there are sub-branchlines) and the calculated gas load for that segment and begin the sizing process as follows

(a) Referring to the appropriate sizing table (based on operating pressure and piping material), find the longest length distance in the first column or the closest larger distance if the exact distance is not listed. The use of alternative operating pressures and/or pressure drops will require the use of a different sizing table, but will not alter the sizing methodology. In many cases, the use of alternative operating pressures and/or pressure drops may require the approval of the code official.

(b) Trace across this row until the appliance gas load is found or the closest larger capacity if the exact capacity is not listed.

(c) Read up the table column to the top row and select the appropriate pipe size.

(d) Repeat this process for each segment of the piping system.

A.4 Use of sizing equations. Capacities of smooth wall pipe or tubing can also be determined by using the following formulae:

(1) High Pressure [1.5 psi (10.3 kPa) and above]:

$$Q = 181.6\sqrt{\frac{D^5 \cdot (P_1^2 - P_2^2) \cdot Y}{C_r \cdot fba \cdot L}}$$

$$= 2237\, D^{2.623}\left[\frac{(P_1^2 - P_2^2) \cdot Y}{C_r \cdot L}\right]^{0.541}$$

(2) Low Pressure [Less than 1.5 psi (10.3 kPa)]:

$$Q = 187.3\sqrt{\frac{D^5 \cdot \Delta H}{C_r \cdot fba \cdot L}}$$

$$= 2313\, D^{2.623}\left(\frac{\Delta H}{C_r L}\right)^{0.541}$$

where:

Q = Rate, cubic feet per hour at 60°F and 30-inch mercury column

D = Inside diameter of pipe, in.

P_1 = Upstream pressure, psia

P_2 = Downstream pressure, psia

Y = Superexpansibility factor = 1/supercompressibility factor

C_r = Factor for viscosity, density and temperature*

$= 0.00354\, ST\left(\frac{Z}{S}\right)^{0.152}$

Note: See Table 402.4 for Y and C_r for natural gas and propane.

S = Specific gravity of gas at 60°F and 30-inch mercury column (0.60 for natural gas, 1.50 for propane), or = 1488μ

T = Absolute temperature, °F or = t + 460

t = Temperature, °F

Z = Viscosity of gas, centipoise (0.012 for natural gas, 0.008 for propane), or = 1488μ

fba = Base friction factor for air at 60°F (CF=1)

L = Length of pipe, ft

ΔH = Pressure drop, in. w.c. (27.7 in. H_2O = 1 psi) (For SI, see Section 402.4)

A.5 Pipe and tube diameters. Where the internal diameter is determined by the formulas in Section 402.4, Tables A.5.1 and A.5.2 can be used to select the nominal or standard pipe size based on the calculated internal diameter.

TABLE A.5.1
SCHEDULE 40 STEEL PIPE STANDARD SIZES

NOMINAL SIZE (in.)	INTERNAL DIAMETER (in.)	NOMINAL SIZE (in.)	INTERNAL DIAMETER (in.)
1/4	0.364	1 1/2	1.610
3/8	0.493	2	2.067
1/2	0.622	2 1/2	2.469
3/4	0.824	3	3.068
1	1.049	3 1/2	3.548
1 1/4	1.380	4	4.026

TABLE A.5.2
COPPER TUBE STANDARD SIZES

TUBE TYPE	NOMINAL OR STANDARD SIZE inches	INTERNAL DIAMETER inches
K	$^1/_4$	0.305
L	$^1/_4$	0.315
ACR (D)	$^3/_8$	0.315
ACR (A)	$^3/_8$	0.311
K	$^3/_8$	0.402
L	$^3/_8$	0.430
ACR (D)	$^1/_2$	0.430
ACR (A)	$^1/_2$	0.436
K	$^1/_2$	0.527
L	$^1/_2$	0.545
ACR (D)	$^5/_8$	0.545
ACR (A)	$^5/_8$	0.555
K	$^5/_8$	0.652
L	$^5/_8$	0.666
ACR (D)	$^3/_4$	0.666
ACR (A)	$^3/_4$	0.680
K	$^3/_4$	0.745
L	$^3/_4$	0.785
ACR	$^7/_8$	0.785
K	1	0.995
L	1	1.025
ACR	$1^1/_8$	1.025
K	$1^1/_4$	1.245
L	$1^1/_4$	1.265
ACR	$1^3/_8$	1.265
K	$1^1/_2$	1.481
L	$1^1/_2$	1.505
ACR	$1^5/_8$	1.505
K	2	1.959
L	2	1.985
ACR	$2^1/_8$	1.985
K	$2^1/_2$	2.435
L	$2^1/_2$	2.465
ACR	$2^5/_8$	2.465
K	3	2.907
L	3	2.945
ACR	$3^1/_8$	2.945

A.6 Use of sizing charts. A third method of sizing gas piping is detailed below as an option that is useful when large quantities of piping are involved in a job (e.g., an apartment house) and material costs are of concern. If the user is not completely familiar with this method, the resulting pipe sizing should be checked by a knowledgeable gas engineer. The sizing charts are applied as follows.

(1) With the layout developed according to Section 106.3.1 of the code, indicate in each section the design gas flow under maximum operation conditions. For many layouts, the maximum design flow will be the sum of all connected loads. However, in some cases, certain combinations of utilization equipment will not occur simultaneously (e.g., gas heating and air conditioning). For these cases, the design flow is the greatest gas flow that can occur at any one time.

(2) Determine the inlet gas pressure for the system being designed. In most cases, the point of inlet will be the gas meter or service regulator, but in the case of a system addition, it could be the point of connection to the existing system.

(3) Determine the minimum pressure required at the inlet to the critical utilization equipment. Usually, the critical item will be the piece of equipment with the highest required pressure for satisfactory operation. If several items have the same required pressure, it will be the one with the greatest length of piping from the system inlet.

(4) The difference between the inlet pressure and critical item pressure is the allowable system pressure drop. Figures A.6(a) and A.6(b) show the relationship between gas flow, pipe size and pipe length for natural gas with 0.60 specific gravity.

(5) To use Figure A.6(a) (low pressure applications), calculate the piping length from the inlet to the critical utilization equipment. Increase this length by 50 percent to allow for fittings. Divide the allowable pressure drop by the equivalent length (in hundreds of feet) to determine the allowable pressure drop per hundred feet. Select the pipe size from Figure A.6(a) for the required volume of flow.

(6) To use Figure A.6(b) (high pressure applications), calculate the equivalent length as above. Calculate the index number for Figure A.6(b) by dividing the difference between the squares of the absolute values of inlet and outlet pressures by the equivalent length (in hundreds of feet). Select the pipe size from Figure A.6(b) for the gas volume required.

A.7 Examples of piping system design and sizing

A.7.1 Example 1: Longest length method. Determine the required pipe size of each section and outlet of the piping system shown in Figure A.7.1, with a designated pressure drop of 0.5-inch w.c. (125 Pa) using the Longest Length Method. The gas to be used has 0.60 specific gravity and a heating value of 1,000 Btu/ft^3 (37.5 MJ/m^3).

Solution:

(1) Maximum gas demand for Outlet A:

$$\frac{\text{Consumption (rating plate input, or Table 402.2 if necessary)}}{\text{Btu of gas}} =$$

$$\frac{35,000 \text{ Btu per hour rating}}{1,000 \text{ Btu per cubic foot}} = 35 \text{ cubic feet per hour} = 35 \text{ cfh}$$

Maximum gas demand for Outlet B:

$$\frac{\text{Consumption}}{\text{Btu of gas}} = \frac{75,000}{1,000} = 75 \text{ cfh}$$

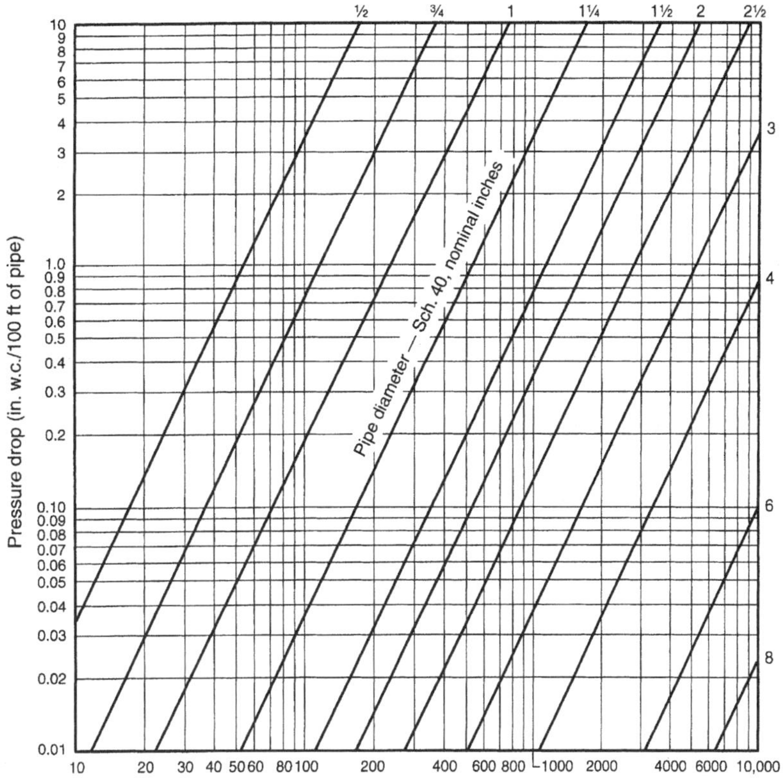

FIGURE A.6(a)
CAPACITY OF NATURAL GAS PIPING, LOW PRESSURE (0.60 WC)

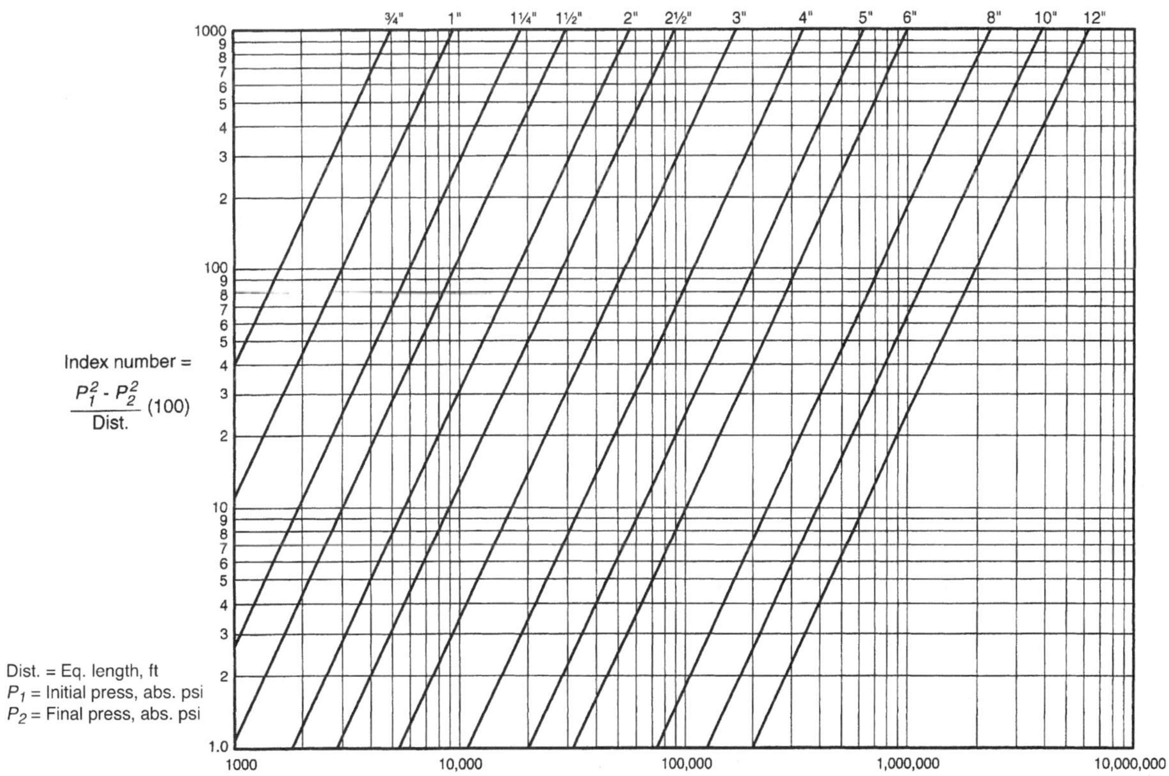

FIGURE A.6(b)
CAPACITY OF NATURAL GAS PIPING, HIGH PRESSURE (1.5 psi and above)

Maximum gas demand for Outlet C:

$$\frac{\text{Consumption}}{\text{Btu of gas}} = \frac{35,000}{1,000} = 35 \text{ cfh}$$

Maximum gas demand for Outlet D:

$$\frac{\text{Consumption}}{\text{Btu of gas}} = \frac{100,000}{1,000} = 100 \text{ cfh}$$

(2) The length of pipe from the point of delivery to the most remote outlet (A) is 60 feet (18 288 mm). This is the only distance used.

(3) Using the row marked 60 feet (18 288 mm) in Table 402.4(2):

(a) Outlet A, supplying 35 cfh (0.99 m^3/hr), requires $^3/_8$-inch pipe.

(b) Outlet B, supplying 75 cfh (2.12 m^3/hr), requires $^3/_4$-inch pipe.

(c) Section 1, supplying Outlets A and B, or 110 cfh (3.11 m^3/hr), requires $^3/_4$-inch pipe.

(d) Section 2, supplying Outlets C and D, or 135 cfh (3.82 m^3/hr), requires $^3/_4$-inch pipe.

(e) Section 3, supplying Outlets A, B, C and D, or 245 cfh (6.94 m^3/hr), requires 1-inch pipe.

(4) If a different gravity factor is applied to this example, the values in the row marked 60 feet (18 288 mm) of Table 402.4(2) would be multiplied by the appropriate multiplier from Table A.2.4 and the resulting cubic feet per hour values would be used to size the piping.

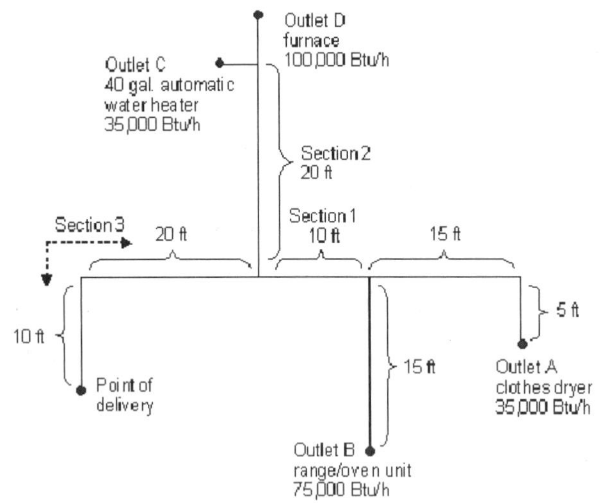

FIGURE A.7.1
PIPING PLAN SHOWING A STEEL PIPING SYSTEM

A.7.2 Example 2: Hybrid or dual pressure systems. Determine the required CSST size of each section of the piping system shown in Figure A.7.2, with a designated pressure drop of 1 psi (6.9 kPa) for the 2 psi (13.8 kPa) section and 3-inch w.c. (0.75 kPa) pressure drop for the 13-inch w.c. (2.49 kPa) section. The gas to be used has 0.60 specific gravity and a heating value of 1,000 Btu/ft^3 (37.5 MJ/ m^3).

Solution

(1) Size 2 psi (13.8 kPa) line using Table 402.4(17).

(2) Size 10-inch w.c. (2.5 kPa) lines using Table 402.4(15).

(3) Using the following, determine if sizing tables can be used.

(a) Total gas load shown in Figure A.7.2 equals 110 cfh (3.11 m^3/hr).

(b) Determine pressure drop across regulator [see notes in Table 402.4 (17)].

(c) If pressure drop across regulator exceeds $^3/_4$ psig (5.2 kPa), Table 402.4 (17) cannot be used. Note: If pressure drop exceeds $^3/_4$ psi (5.2 kPa), then a larger regulator must be selected or an alternative sizing method must be used.

(d) Pressure drop across the line regulator [for 110 cfh (3.11 m^3/hr)] is 4-inch w.c. (0.99 kPa) based on manufacturer's performance data.

(e) Assume the CSST manufacturer has tubing sizes or EHDs of 13, 18, 23 and 30.

(4) Section A [2 psi (13.8 kPa) zone]

(a) Distance from meter to regulator = 100 feet (30 480 mm).

(b) Total load supplied by A = 110 cfh (3.11 m^3/hr) (furnace + water heater + dryer).

(c) Table 402.4 (17) shows that EHD size 18 should be used.

Note: It is not unusual to oversize the supply line by 25 to 50 percent of the as-installed load. EHD size 18 has a capacity of 189 cfh (5.35 m^3/hr).

(5) Section B (low pressure zone)

(a) Distance from regulator to furnace is 15 feet (4572 mm).

(b) Load is 60 cfh (1.70 m^3/hr).

(c) Table 402.4 (15) shows that EHD size 13 should be used.

(6) Section C (low pressure zone)

(a) Distance from regulator to water heater is 10 feet (3048 mm).

(b) Load is 30 cfh (0.85 m^3/hr).

(c) Table 402.4 (15) shows that EHD size 13 should be used.

(7) Section D (low pressure zone)

(a) Distance from regulator to dryer is 25 feet (7620 mm).

(b) Load is 20 cfh (0.57 m^3/hr).

(c) Table 402.4(15) shows that EHD size 13 should be used.

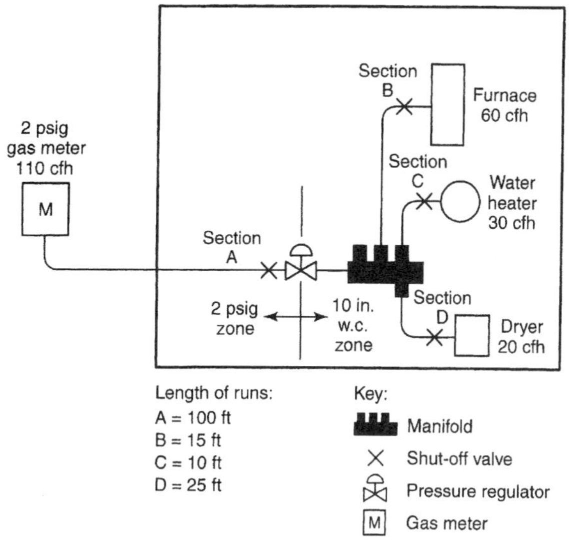

FIGURE A.7.2
PIPING PLAN SHOWING A CSST SYSTEM

A.7.3 Example 3: Branch length method. Determine the required semi-rigid copper tubing size of each section of the piping system shown in Figure A.7.3, with a designated pressure drop of 1-inch w.c. (250 Pa) (using the Branch Length Method). The gas to be used has 0.60 specific gravity and a heating value of 1,000 Btu/ft^3 (37.5 MJ/m^3).

Solution

(1) Section A

(a) The length of tubing from the point of delivery to the most remote appliance is 50 feet (15 240 mm), A + C.

(b) Use this longest length to size Sections A and C.

(c) Using the row marked 50 feet (15 240 mm) in Table 402.4(9), Section A, supplying 220 cfh (6.2 m^3/hr) for four appliances requires 1-inch tubing.

(2) Section B

(a) The length of tubing from the point of delivery to the range/oven at the end of Section B is 30 feet (9144 mm), A + B.

(b) Use this branch length to size Section B only.

(c) Using the row marked 30 feet (9144 mm) in Table 402.4(9), Section B, supplying 75 cfh (2.12 m^3/hr) for the range/oven requires $^1/_2$-inch tubing.

(3) Section C

(a) The length of tubing from the point of delivery to the dryer at the end of Section C is 50 feet (15 240 mm), A + C.

(b) Use this branch length (which is also the longest length) to size Section C.

(c) Using the row marked 50 feet (15 240 mm) in Table 402.4(9), Section C, supplying 30 cfh (0.85 m^3/hr) for the dryer requires $^3/_8$-inch tubing.

(4) Section D

(a) The length of tubing from the point of delivery to the water heater at the end of Section D is 30 feet (9144 mm), A + D.

(b) Use this branch length to size Section D only.

(c) Using the row marked 30 feet (9144 mm) in Table 402.4(9), Section D, supplying 35 cfh (0.99 m^3/hr) for the water heater requires $^3/_8$-inch tubing.

(5) Section E

(a) The length of tubing from the point of delivery to the furnace at the end of Section E is 30 feet (9144 mm), A + E.

(b) Use this branch length to size Section E only.

(c) Using the row marked 30 feet (9144 mm) in Table 402.4(9), Section E, supplying 80 cfh (2.26 m^3/hr) for the furnace requires $^1/_2$-inch tubing.

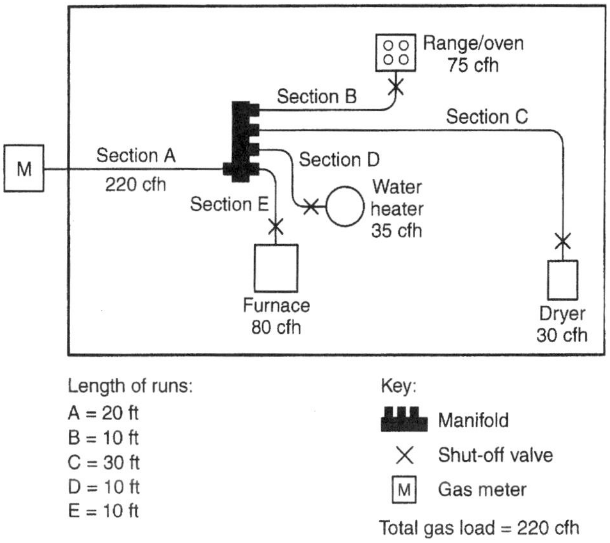

FIGURE A.7.3
PIPING PLAN SHOWING A COPPER TUBING SYSTEM

A.7.4 Example 4: Modification to existing piping system. Determine the required CSST size for Section G (retrofit application) of the piping system shown in Figure A.7.4, with a designated pressure drop of 0.5-inch w.c. (125 Pa) using the branch length method. The gas to be used has 0.60 specific gravity and a heating value of 1,000 Btu/ft^3 (37.5 MJ/m^3).

Solution

(1) The length of pipe and CSST from the point of delivery to the retrofit appliance (barbecue) at the end of Section G is 40 feet (12 192 mm), A + B + G.

(2) Use this branch length to size Section G.

(3) Assume the CSST manufacturer has tubing sizes or EHDs of 13, 18, 23 and 30.

(4) Using the row marked 40 feet (12 192 mm) in Table 402.4(14), Section G, supplying 40 cfh (1.13 m^3/hr) for the barbecue requires EHD 18 CSST.

(5) The sizing of Sections A, B, F and E must be checked to ensure adequate gas carrying capacity since an appliance has been added to the piping system (see A.7.1 for details).

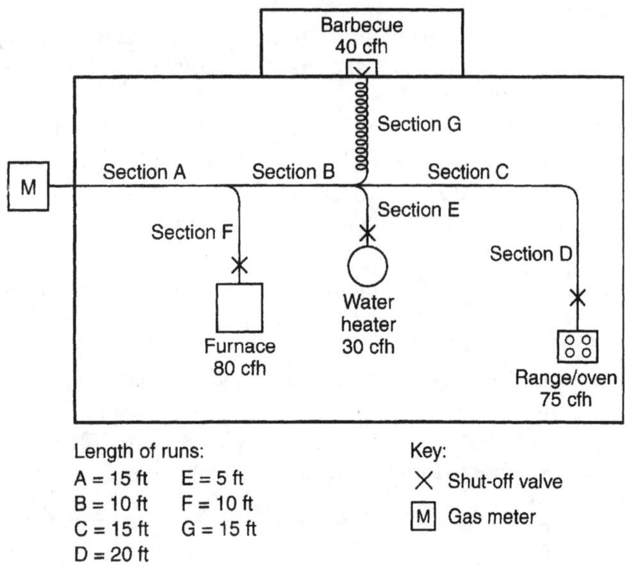

FIGURE A.7.4
PIPING PLAN SHOWING A MODIFICATION TO EXISTING PIPING SYSTEM

This appendix is a direct excerpt from the 2003 *International Fuel Gas Code*, and no attempt has been made to coordinate with Chapter 24 of the *International Residential Code.*

APPENDIX B (IFGS)

SIZING OF VENTING SYSTEMS SERVING APPLIANCES EQUIPPED WITH DRAFT HOODS, CATEGORY I APPLIANCES, AND APPLIANCES LISTED FOR USE WITH TYPE B VENTS

(This appendix is informative and is not part of the code.)

EXAMPLES USING SINGLE APPLIANCE VENTING TABLES

Example 1: Single draft-hood-equipped appliance

An installer has a 120,000 British thermal unit (Btu) per hour input appliance with a 5-inch-diameter draft hood outlet that needs to be vented into a 10-foot-high Type B vent system. What size vent should be used assuming (a) a 5-foot lateral single-wall metal vent connector is used with two 90-degree elbows, or (b) a 5-foot lateral single-wall metal vent connector is used with three 90-degree elbows in the vent system?

Solution:

Table 504.2(2) should be used to solve this problem, because single-wall metal vent connectors are being used with a Type B vent.

(a) Read down the first column in Table 504.2(2) until the row associated with a 10-foot height and 5-foot lateral is found. Read across this row until a vent capacity greater than 120,000 Btu per hour is located in the shaded columns labeled "NAT Max" for draft-hood-equipped appliances. In this case, a 5-inch-diameter vent has a capacity of 122,000 Btu per hour and may be used for this application.

(b) If three 90-degree elbows are used in the vent system, then the maximum vent capacity listed in the tables must be reduced by 10 percent (see Section 504.2.3 for single appliance vents). This implies that the 5-inch-diameter vent has an adjusted capacity of only 110,000 Btu per hour. In this case, the vent system must be increased to 6 inches in diameter (see calculations below).

122,000 (.90) = 110,000 for 5-inch vent

From Table 504.2(2), Select 6-inch vent

186,000 (.90) = 167,000; This is greater than the required 120,000. Therefore, use a 6-inch vent and connector where three elbows are used.

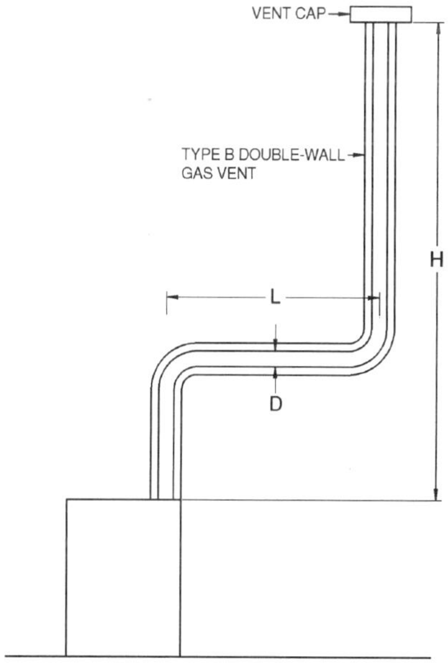

For SI: 1 foot = 304.8 mm, 1 British thermal unit per hour = 0.2931 W.
Table 504.2(1) is used when sizing Type B double-wall gas vent connected directly to the appliance.
Note: The appliance may be either Category I draft hood equipped or fan-assisted type.

FIGURE B-1
TYPE B DOUBLE-WALL VENT SYSTEM SERVING A SINGLE APPLIANCE WITH A TYPE B DOUBLE-WALL VENT

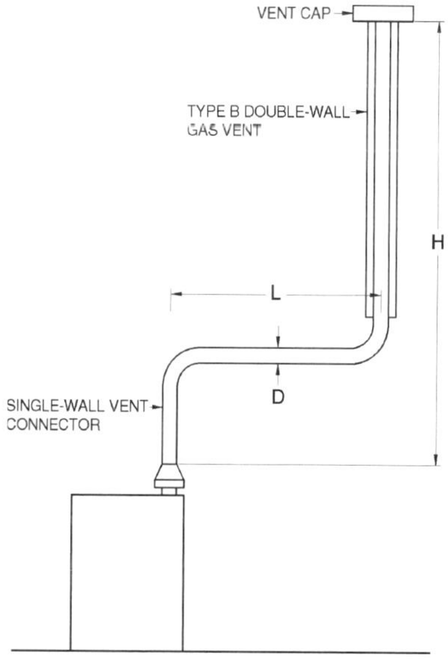

For SI: 1 foot = 304.8 mm, 1 British thermal unit per hour = 0.2931 W.
Table 504.2(2) is used when sizing a single-wall metal vent connector attached to a Type B double-wall gas vent.
Note: The appliance may be either Category I draft hood equipped or fan-assisted type.

FIGURE B-2
TYPE B DOUBLE-WALL VENT SYSTEM SERVING A SINGLE APPLIANCE WITH A SINGLE-WALL METAL VENT CONNECTOR

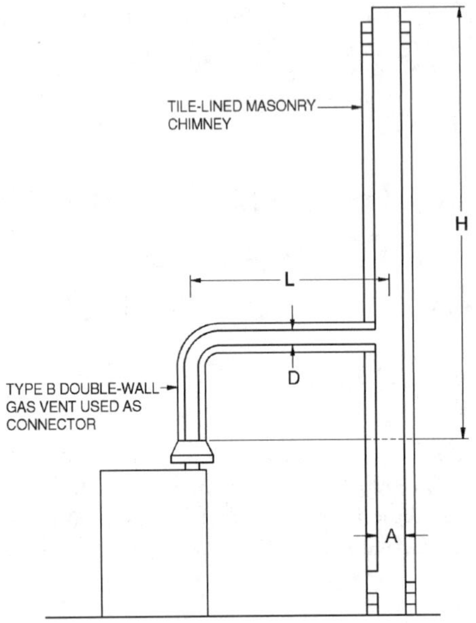

Table 504.2(3) is used when sizing a Type B double-wall gas vent connector attached to a tile-lined masonry chimney.
Note: "A" is the equivalent cross-sectional area of the tile liner.
Note: The appliance may be either Category I draft hood equipped or fan-assisted type.

FIGURE B-3
VENT SYSTEM SERVING A SINGLE APPLIANCE WITH A MASONRY CHIMNEY OF TYPE B DOUBLE-WALL VENT CONNECTOR

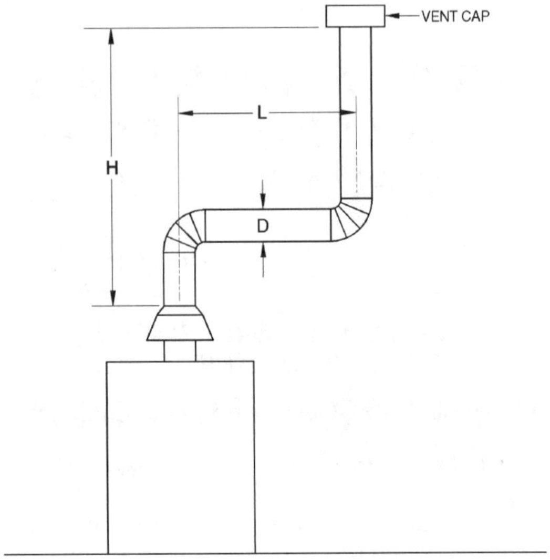

Asbestos cement Type B or single-wall metal vent serving a single draft-hood-equipped appliance [see Table 504.2(5)].

FIGURE B-5
ASBESTOS CEMENT TYPE B OR SINGLE-WALL METAL VENT SYSTEM SERVING A SINGLE DRAFT-HOOD-EQUIPPED APPLIANCE

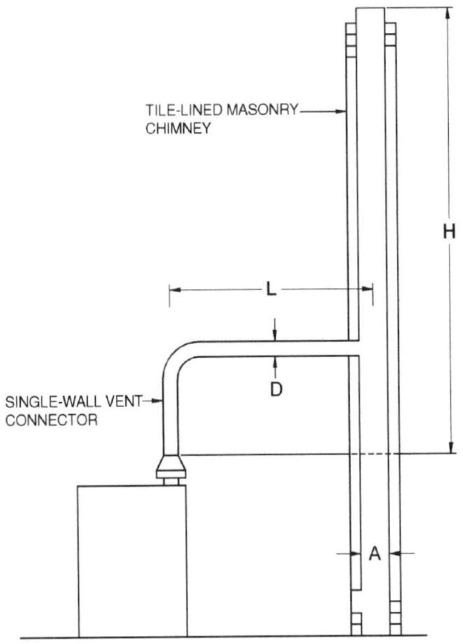

Table 504.2(4) is used when sizing a single-wall gas vent connector attached to a tile-lined masonry chimney.
Note: "A" is the equivalent cross-sectional area of the tile liner.
Note: The appliance may be either Category I draft hood equipped or fan-assisted type.

FIGURE B-4
VENT SYSTEM SERVING A SINGLE APPLIANCE USING A MASONRY CHIMNEY AND A SINGLE-WALL METAL VENT CONNECTOR

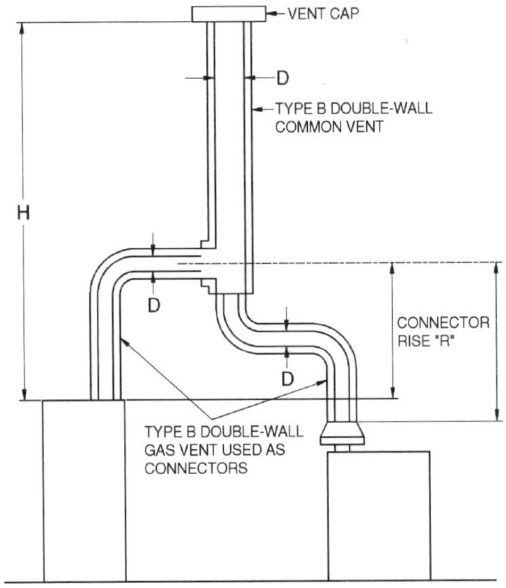

Table 504.3(1) is used when sizing Type B double-wall vent connectors attached to a Type B double-wall common vent.
Note: Each appliance may be either Category I draft hood equipped or fan-assisted type.

FIGURE B-6
VENT SYSTEM SERVING TWO OR MORE APPLIANCES WITH TYPE B DOUBLE-WALL VENT AND TYPE B DOUBLE-WALL VENT CONNECTOR

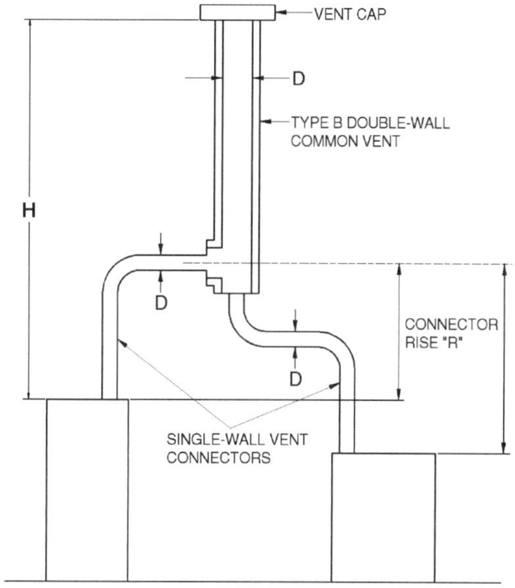

Table 504.3(2) is used when sizing single-wall vent connectors attached to a Type B double-wall common vent.
Note: Each appliance may be either Category I draft hood equipped or fan-assisted type.

FIGURE B-7
VENT SYSTEM SERVING TWO OR MORE APPLIANCES WITH TYPE B DOUBLE-WALL VENT AND SINGLE-WALL METAL VENT CONNECTORS

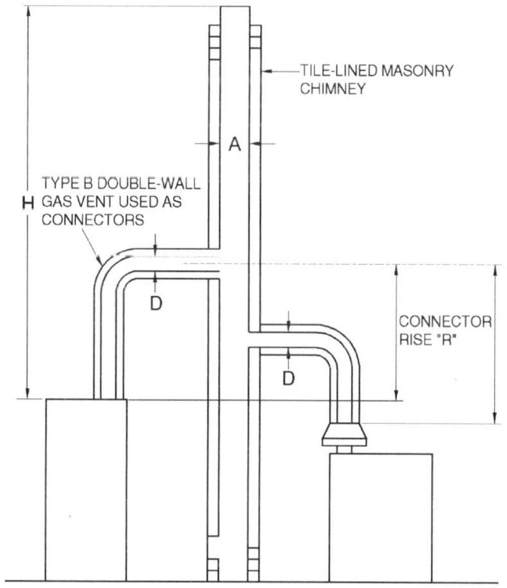

Table 504.3(3) is used when sizing Type B double-wall vent connectors attached to a tile-lined masonry chimney.
Note: "A" is the equivalent cross-sectional area of the tile liner.
Note: Each appliance may be either Category I draft hood equipped or fan-assisted type.

FIGURE B-8
MASONRY CHIMNEY SERVING TWO OR MORE APPLIANCES WITH TYPE B DOUBLE-WALL VENT CONNECTOR

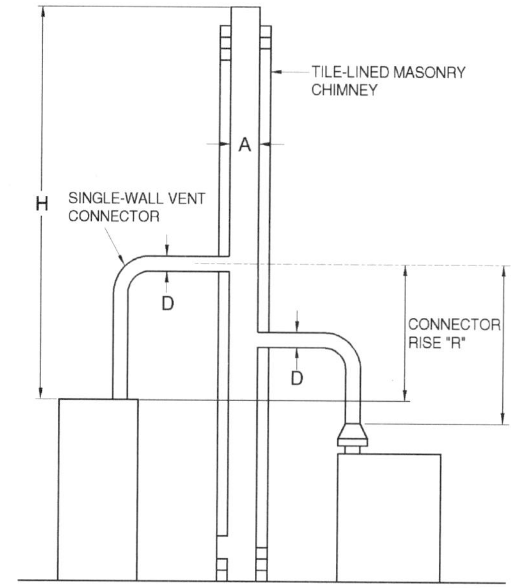

Table 504.3(4) is used when sizing single-wall metal vent connectors attached to a tile-lined masonry chimney.
Note: "A" is the equivalent cross-sectional area of the tile liner.
Note: Each appliance may be either Category I draft hood equipped or fan-assisted type.

FIGURE B-9
MASONRY CHIMNEY SERVING TWO OR MORE APPLIANCES WITH SINGLE-WALL METAL VENT CONNECTORS

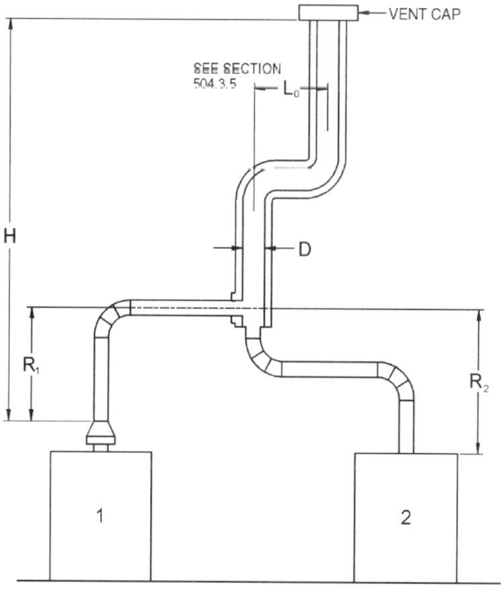

Asbestos cement Type B or single-wall metal pipe vent serving two or more draft-hood-equipped appliances [see Table 504.3(5)].

FIGURE B-10
ASBESTOS CEMENT TYPE B OR SINGLE-WALL METAL VENT SYSTEM SERVING TWO OR MORE DRAFT-HOOD-EQUIPPED APPLIANCES

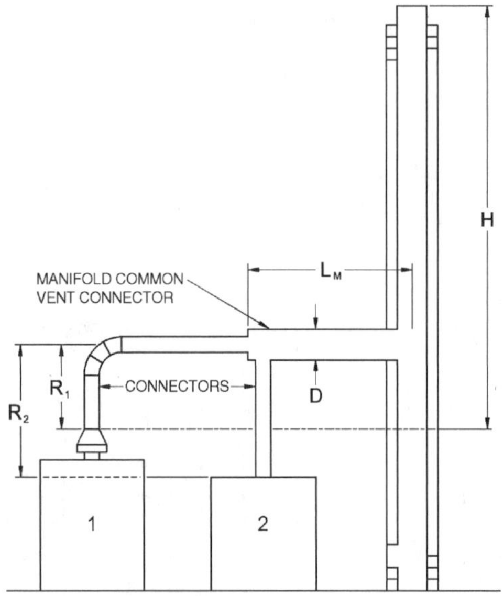

Example: Manifolded Common Vent Connector L_M shall be no greater than 18 times the common vent connector manifold inside diameter; i.e., a 4-inch (102 mm) inside diameter common vent connector manifold shall not exceed 72 inches (1829 mm) in length (see Section 504.3.4).

Note: This is an illustration of a typical manifolded vent connector. Different appliance, vent connector, or common vent types are possible. Consult Section 502.3.

FIGURE B-11
USE OF MANIFOLD COMMON VENT CONNECTOR

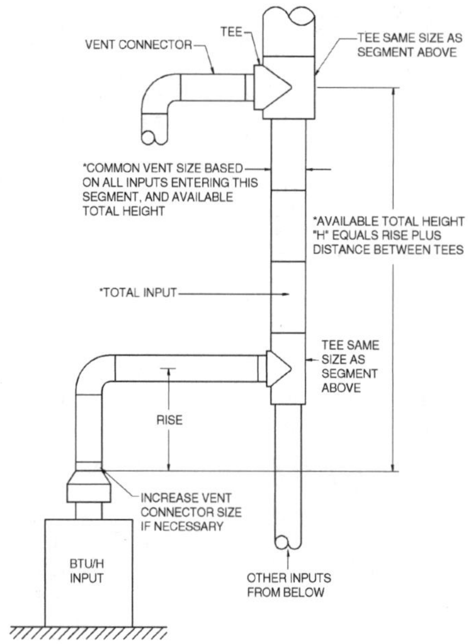

Vent connector size depends on:
- Input
- Rise
- Available total height *"H"*
- Table 504.3(1) connectors

Common vent size depends on
- Combined inputs
- Available total height *"H"*
- Table 504.3(1) common vent

FIGURE B-13
MULTISTORY GAS VENT DESIGN PROCEDURE FOR EACH SEGMENT OF SYSTEM

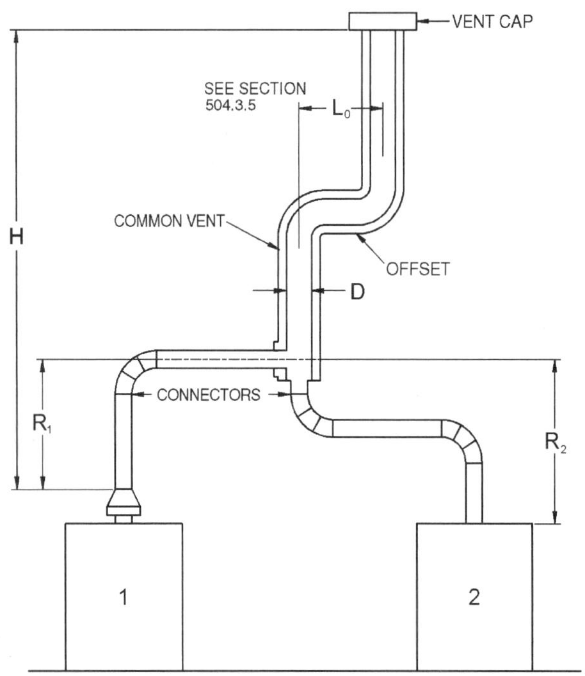

Example: Offset Common Vent

Note: This is an illustration of a typical offset vent. Different appliance, vent connector, or vent types are possible. Consult Sections 504.2 and 504.3.

FIGURE B-12
USE OF OFFSET COMMON VENT

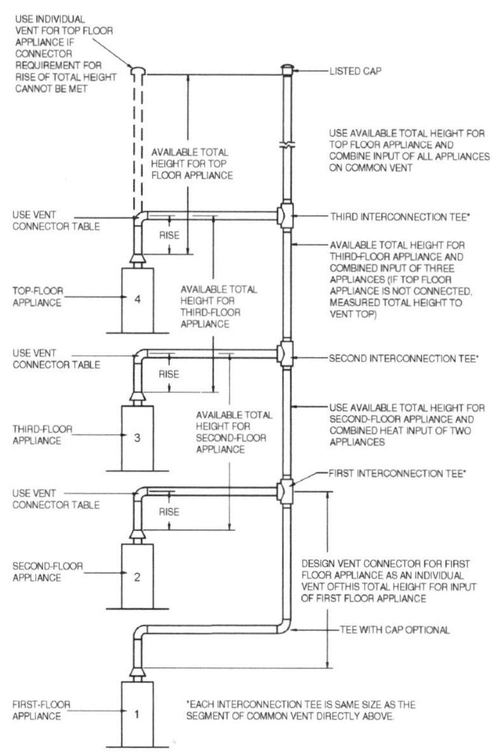

Principles of design of multistory vents using vent connector and common vent design tables (see Sections 504.3.11 through 504.3.17).

FIGURE B-14
MULTISTORY VENT SYSTEMS

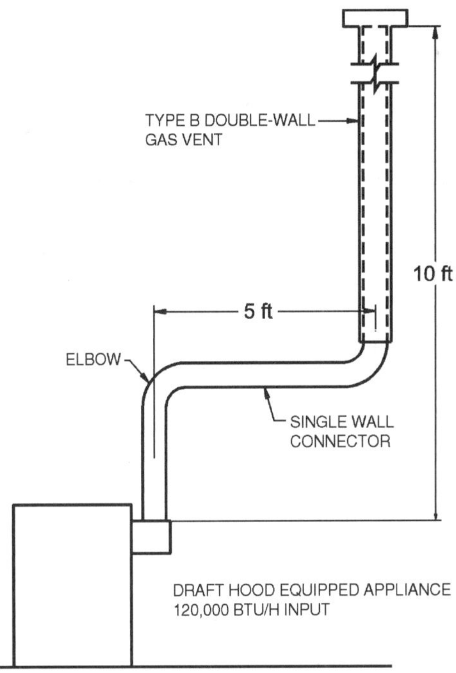

For SI: 1 foot = 304.8 mm, 1 British thermal unit per hour = 0.2931 W.

**FIGURE B-15 (EXAMPLE 1)
SINGLE DRAFT-HOOD-EQUIPPED APPLIANCE**

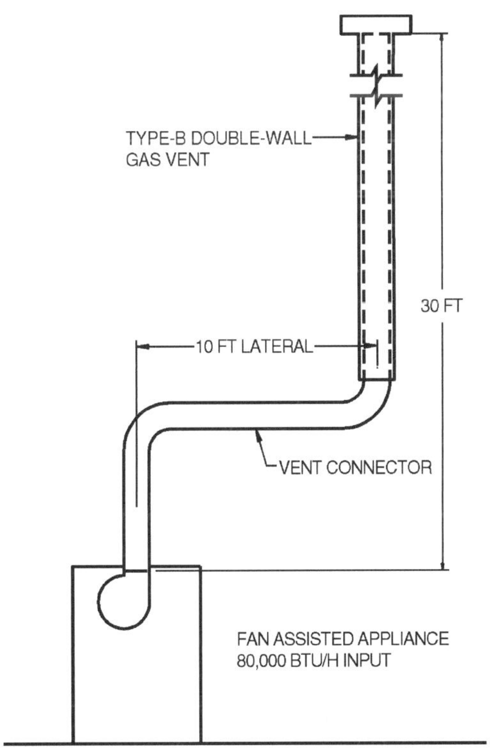

For SI: 1 foot = 304.8 mm, 1 British thermal unit per hour = 0.2931 W.

**FIGURE B-16 (EXAMPLE 2)
SINGLE FAN-ASSISTED APPLIANCE**

Example 2: Single fan-assisted appliance

An installer has an 80,000 Btu per hour input fan-assisted appliance that must be installed using 10 feet of lateral connector attached to a 30-foot-high Type B vent. Two 90-degree elbows are needed for the installation. Can a single-wall metal vent connector be used for this application?

Solution:

Table 504.2(2) refers to the use of single-wall metal vent connectors with Type B vent. In the first column find the row associated with a 30-foot height and a 10-foot lateral. Read across this row, looking at the FAN Min and FAN Max columns, to find that a 3-inch-diameter single-wall metal vent connector is not recommended. Moving to the next larger size single wall connector (4 inches), note that a 4-inch-diameter single-wall metal connector has a recommended minimum vent capacity of 91,000 Btu per hour and a recommended maximum vent capacity of 144,000 Btu per hour. The 80,000 Btu per hour fan-assisted appliance is outside this range, so the conclusion is that a single-wall metal vent connector cannot be used to vent this appliance using 10 feet of lateral for the connector.

However, if the 80,000 Btu per hour input appliance could be moved to within 5 feet of the vertical vent, then a 4-inch single-wall metal connector could be used to vent the appliance. Table 504.2(2) shows the acceptable range of vent capacities for a 4-inch vent with 5 feet of lateral to be between 72,000 Btu per hour and 157,000 Btu per hour.

If the appliance cannot be moved closer to the vertical vent, then Type B vent could be used as the connector material. In this case, Table 504.2(1) shows that for a 30-foot-high vent with 10 feet of lateral, the acceptable range of vent capacities for a 4-inch-diameter vent attached to a fan-assisted appliance is between 37,000 Btu per hour and 150,000 Btu per hour.

Example 3: Interpolating between table values

An installer has an 80,000 Btu per hour input appliance with a 4-inch-diameter draft hood outlet that needs to be vented into a 12-foot-high Type B vent. The vent connector has a 5-foot lateral length and is also Type B. Can this appliance be vented using a 4-inch-diameter vent?

Solution:

Table 504.2(1) is used in the case of an all Type B vent system. However, since there is no entry in Table 504.2(1) for a height of 12 feet, interpolation must be used. Read down the 4-inch diameter NAT Max column to the row associated with 10-foot height and 5-foot lateral to find the capacity value of 77,000 Btu per hour. Read further down to the 15-foot height, 5-foot lateral row to find the capacity value of 87,000 Btu per hour. The difference between the 15-foot height capacity value and the 10-foot height capacity value is 10,000 Btu per hour. The capacity for a vent system with a 12-foot height is equal to the capacity for a 10-foot height plus $^2/_5$ of the difference between the 10-foot and 15-foot height values, or 77,000 + $^2/_5$ (10,000) = 81,000 Btu per hour. Therefore, a 4-inch-diameter vent may be used in the installation.

EXAMPLES USING COMMON VENTING TABLES

Example 4: Common venting two draft-hood-equipped appliances

A 35,000 Btu per hour water heater is to be common vented with a 150,000 Btu per hour furnace using a common vent with a total height of 30 feet. The connector rise is 2 feet for the water heater with a horizontal length of 4 feet. The connector rise for the furnace is 3 feet with a horizontal length of 8 feet. Assume single-wall metal connectors will be used with Type B vent. What size connectors and combined vent should be used in this installation?

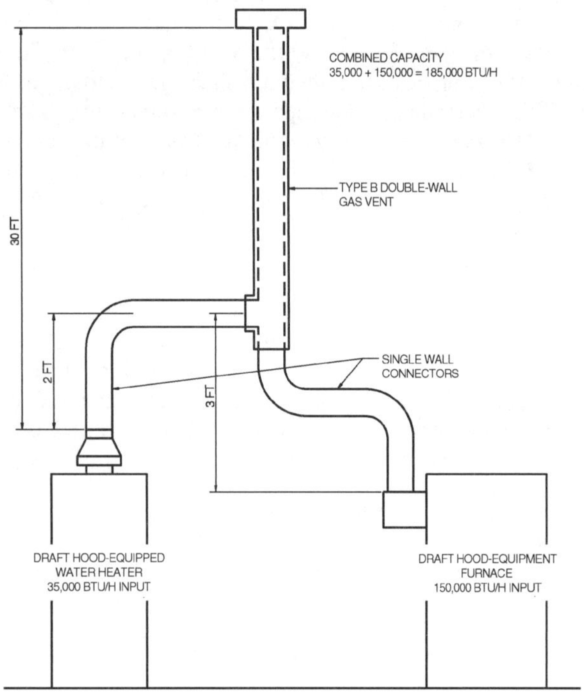

**FIGURE B-17 (EXAMPLE 4)
COMMON VENTING TWO DRAFT-HOOD-EQUIPPED APPLIANCES**

Solution:

Table 504.3(2) should be used to size single-wall metal vent connectors attached to Type B vertical vents. In the vent connector capacity portion of Table 504.3(2), find the row associated with a 30-foot vent height. For a 2-foot rise on the vent connector for the water heater, read the shaded columns for draft-hood-equipped appliances to find that a 3-inch-diameter vent connector has a capacity of 37,000 Btu per hour. Therefore, a 3-inch single-wall metal vent connector may be used with the water heater. For a draft-hood-equipped furnace with a 3-foot rise, read across the appropriate row to find that a 5-inch-diameter vent connector has a maximum capacity of 120,000 Btu per hour (which is too small for the furnace) and a 6-inch-diameter vent connector has a maximum vent capacity of 172,000 Btu per hour. Therefore, a 6-inch-diameter vent connector should be used with the 150,000 Btu per hour furnace. Since both vent connector horizontal lengths are less than the maximum lengths listed in Section 504.3.2, the table values may be used without adjustments.

In the common vent capacity portion of Table 504.3(2), find the row associated with a 30-foot vent height and read over to the NAT + NAT portion of the 6-inch-diameter column to find a maximum combined capacity of 257,000 Btu per hour. Since the two appliances total only 185,000 Btu per hour, a 6-inch common vent may be used.

Example 5a: Common venting a draft-hood-equipped water heater with a fan-assisted furnace into a Type B vent

In this case, a 35,000 Btu per hour input draft-hood-equipped water heater with a 4-inch-diameter draft hood outlet, 2 feet of connector rise, and 4 feet of horizontal length is to be common vented with a 100,000 Btu per hour fan-assisted furnace with a 4-inch-diameter flue collar, 3 feet of connector rise, and 6 feet of horizontal length. The common vent consists of a 30-foot height of Type B vent. What are the recommended vent diameters for each connector and the common vent? The installer would like to use a single-wall metal vent connector.

Solution: - [Table 504.3(2)]

Water Heater Vent Connector Diameter. Since the water heater vent connector horizontal length of 4 feet is less than the maximum value listed in Section 504.3.2, the venting table values may be used without adjustments. Using the Vent Connector Capacity portion of Table 504.3(2), read down the Total Vent Height (*H*) column to 30 feet and read across the 2-foot Connector Rise (*R*) row to the first Btu per hour rating in the NAT Max column that is equal to or greater than the water heater input rating. The table shows that a 3-inch vent connector has a maximum input rating of 37,000 Btu per hour. Although this is greater than the water heater input rating, a 3-inch vent connector is prohibited by Section 504.3.21. A 4-inch vent connector has a maximum input rating of 67,000 Btu per hour and is equal to the draft hood outlet diameter. A 4-inch vent connector is selected. Since the water heater is equipped with a draft hood, there are no minimum input rating restrictions.

Furnace Vent Connector Diameter. Using the Vent Connector Capacity portion of Table 504.3(2), read down the Total Vent Height (*H*) column to 30 feet and across the 3-foot Connector Rise (*R*) row. Since the furnace has a fan-assisted combustion system, find the first FAN Max column with a Btu per hour rating greater than the furnace input rating. The 4-inch vent connector has a maximum input rating of 119,000 Btu per hour and a minimum input rating of 85,000 Btu per hour. The 100,000 Btu per hour furnace in this example falls within this range, so a 4-inch connector is adequate. Since the furnace vent connector horizontal length of 6 feet does not exceed the maximum value listed in Section 504.3.2, the venting table values may be used without adjustment. If the furnace had an input rating of 80,000 Btu per hour, then a Type B vent connector [see Table 504.3(1)] would be needed in order to meet the minimum capacity limit.

Common Vent Diameter. The total input to the common vent is 135,000 Btu per hour. Using the Common Vent Capacity portion of Table 504.3(2), read down the Total Vent Height (*H*) column to 30 feet and across this row to find the smallest vent diameter in the FAN + NAT column that has a Btu per hour rating equal to or greater than 135,000 Btu per hour. The 4-inch common vent has a capacity of 132,000 Btu per hour and the 5-inch common vent has a capacity of 202,000 Btu per hour. Therefore, the 5-inch common vent should be used in this example.

Summary. In this example, the installer may use a 4-inch-diameter, single-wall metal vent connector for the water heater and a 4-inch-diameter, single-wall metal vent connector for the furnace. The common vent should be a 5-inch-diameter Type B vent.

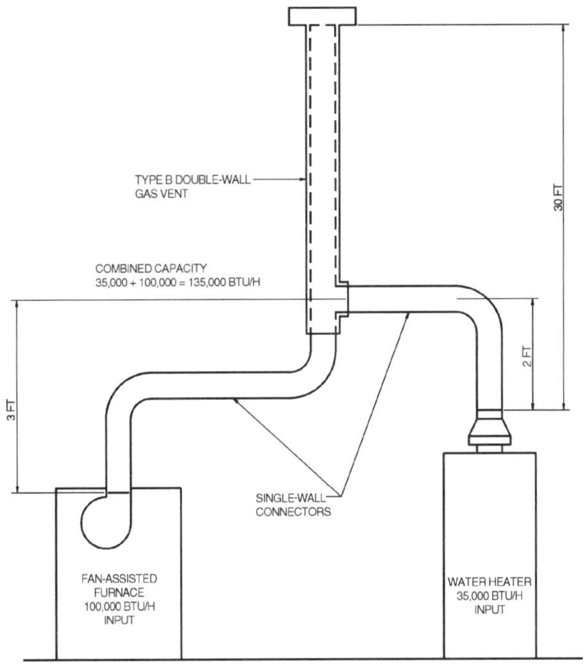

FIGURE B-18 (EXAMPLE 5A)
COMMON VENTING A DRAFT HOOD WITH A FAN-ASSISTED FURNACE INTO A TYPE B DOUBLE-WALL COMMON VENT

Example 5b: Common venting into a masonry chimney

In this case, the water heater and fan-assisted furnace of Example 5a are to be common vented into a clay tile-lined masonry chimney with a 30-foot height. The chimney is not exposed to the outdoors below the roof line. The internal dimensions of the clay tile liner are nominally 8 inches by 12 inches. Assuming the same vent connector heights, laterals, and materials found in Example 5a, what are the recommended vent connector diameters, and is this an acceptable installation?

Solution:

Table 504.3(4) is used to size common venting installations involving single-wall connectors into masonry chimneys.

Water Heater Vent Connector Diameter. Using Table 504.3(4), Vent Connector Capacity, read down the Total Vent Height (*H*) column to 30 feet, and read across the 2-foot Connector Rise (*R*) row to the first Btu per hour rating in the NAT Max column that is equal to or greater than the water heater input rating. The table shows that a 3-inch vent connector has a maximum input of only 31,000 Btu per hour while a 4-inch vent connector has a maximum input of 57,000 Btu per hour. A 4-inch vent connector must therefore be used.

Furnace Vent Connector Diameter. Using the Vent Connector Capacity portion of Table 504.3(4), read down the Total Vent Height (*H*) column to 30 feet and across the 3-foot Connector Rise (*R*) row. Since the furnace has a fan-assisted combustion system, find the first FAN Max column with a Btu per hour rating greater than the furnace input rating. The 4-inch vent connector has a maximum input rating of 127,000 Btu per hour and a minimum input rating of 95,000 Btu per hour. The 100,000 Btu per hour furnace in this example falls within this range, so a 4-inch connector is adequate.

Masonry Chimney. From Table B-1, the equivalent area for a nominal liner size of 8 inches by 12 inches is 63.6 square inches. Using Table 504.3(4), Common Vent Capacity, read down the FAN + NAT column under the Minimum Internal Area of Chimney value of 63 to the row for 30-foot height to find a capacity value of 739,000 Btu per hour. The combined input rating of the furnace and water heater, 135,000 Btu per hour, is less than the table value, so this is an acceptable installation.

Section 504.3.17 requires the common vent area to be no greater than seven times the smallest listed appliance categorized vent area, flue collar area, or draft hood outlet area. Both appliances in this installation have 4-inch-diameter outlets. From Table B-1, the equivalent area for an inside diameter of 4 inches is 12.2 square inches. Seven times 12.2 equals 85.4, which is greater than 63.6, so this configuration is acceptable.

Example 5c: Common venting into an exterior masonry chimney

In this case, the water heater and fan-assisted furnace of Examples 5a and 5b are to be common vented into an exterior masonry chimney. The chimney height, clay tile liner dimensions, and vent connector heights and laterals are the same as in Example 5b. This system is being installed in Charlotte, North Carolina. Does this exterior masonry chimney need to be relined? If so, what corrugated metallic liner size is recommended? What vent connector diameters are recommended?

Solution:

According to Section 504.3.20, Type B vent connectors are required to be used with exterior masonry chimneys. Use Table 504.3(8) to size FAN+NAT common venting installations involving Type-B double wall connectors into exterior masonry chimneys.

The local 99-percent winter design temperature needed to use Table 504.3(8) can be found in the ASHRAE *Handbook of Fundamentals*. For Charlotte, North Carolina, this design temperature is 19°F.

Chimney Liner Requirement. As in Example 5b, use the 63 square inch Internal Area columns for this size clay tile liner. Read down the 63 square inch column of Table 504.3(8a) to the 30-foot height row to find that the combined appliance maximum input is 747,000 Btu per hour. The combined input rating of the appliances in this installation, 135,000 Btu per hour, is less than the maximum value, so this criterion is satisfied. Table 504.3(8b), at a 19°F design temperature, and at the same vent height and internal area used above, shows that the minimum allowable input rating of a space-heating appliance is 470,000 Btu per hour. The furnace input rating of 100,000 Btu per hour is less than this minimum value. So this criterion is not satisfied, and an alternative venting design needs to be used, such as a Type B vent shown in Example 5a or a listed chimney liner system shown in the remainder of the example.

According to Section 504.3.19, Table 504.3(1) or 504.3(2) is used for sizing corrugated metallic liners in masonry chimneys, with the maximum common vent capacities reduced by 20 percent. This example will be continued assuming Type B vent connectors.

Water Heater Vent Connector Diameter. Using Table 504.3(1), Vent Connector Capacity, read down the Total Vent Height (*H*) column to 30 feet, and read across the 2-foot Connector Rise (*R*) row to the first Btu/h rating in the NAT Max column that is equal to or greater than the water heater input rating. The table shows that a 3-inch vent connector has a maximum capacity of 39,000 Btu/h. Although this rating is greater than the water heater input rating, a 3-inch vent connector is prohibited by Section 504.3.21. A 4-inch vent connector has a maximum input rating of 70,000 Btu/h and is equal to the draft hood outlet diameter. A 4-inch vent connector is selected.

Furnace Vent Connector Diameter. Using Table 504.3(1), Vent Connector Capacity, read down the Vent Height (*H*) column to 30 feet, and read across the 3-foot Connector Rise (*R*) row to the first Btu per hour rating in the FAN Max column that is equal to or greater than the furnace input rating. The 100,000 Btu per hour furnace in this example falls within this range, so a 4-inch connector is adequate. Chimney Liner Diameter. The total input to the common vent is 135,000 Btu per hour. Using the Common Vent Capacity Portion of Table 504.3(1), read down the Vent Height (*H*) column to 30 feet and across this row to find the smallest vent diameter in the FAN+NAT column that has a Btu per hour rating greater than 135,000 Btu per hour. The 4-inch common vent has a capacity of 138,000 Btu per hour. Reducing the maximum capacity by 20 percent (Section 504.3.17) results in a maximum capacity for a 4-inch corrugated liner of 110,000 Btu per hour, less than the total input of 135,000 Btu per hour. So a larger liner is needed. The 5-inch common vent capacity listed in Table 504.3(1) is 210,000 Btu per hour, and after reducing by 20 percent is 168,000 Btu per hour. Therefore, a 5-inch corrugated metal liner should be used in this example.

Single-Wall Connectors. Once it has been established that relining the chimney is necessary, Type B double-wall vent connectors are not specifically required. This example could be redone using Table 504.3(2) for single-wall vent connectors. For this case, the vent connector and liner diameters would be the same as found above with Type B double-wall connectors.

TABLE B-1
MASONRY CHIMNEY LINER DIMENSIONS WITH CIRCULAR EQUIVALENTS[a]

NOMINAL LINER SIZE (inches)	INSIDE DIMENSIONS OF LINER (inches)	INSIDE DIAMETER OR EQUIVALENT DIAMETER (inches)	EQUIVALENT AREA (square inches)
4 × 8	$2^1/_2 \times 6^1/_2$	4	12.2
		5	19.6
		6	28.3
		7	38.3
8 × 8	$6^3/_4 \times 6^3/_4$	7.4	42.7
		8	50.3
8 × 12	$6^1/_2 \times 10^1/_2$	9	63.6
		10	78.5
12 × 12	$9^3/_4 \times 9^3/_4$	10.4	83.3
		11	95
12 × 16	$9^1/_2 \times 13^1/_2$	11.8	107.5
		12	113.0
		14	153.9
16 × 16	$13^1/_4 \times 13^1/_4$	14.5	162.9
		15	176.7
16 × 20	13 × 17	16.2	206.1
		18	254.4
20 × 20	$16^3/_4 \times 16^3/_4$	18.2	260.2
		20	314.1
20 × 24	$16^1/_2 \times 20^1/_2$	20.1	314.2
		22	380.1
24 × 24	$20^1/_4 \times 20^1/_4$	22.1	380.1
		24	452.3
24 × 28	$20^1/_4 \times 24^1/_4$	24.1	456.2
28 × 28	$24^1/_4 \times 24^1/_4$	26.4	543.3
		27	572.5
30 × 30	$25^1/_2 \times 25^1/_2$	27.9	607
		30	706.8
30 × 36	$25^1/_2 \times 31^1/_2$	30.9	749.9
		33	855.3
36 × 36	$31^1/_2 \times 31^1/_2$	34.4	929.4
		36	1017.9

For SI: 1 inch = 25.4 mm, 1 square inch = 645.16 m^2.

a. Where liner sizes differ dimensionally from those shown in Table B-1, equivalent diameters may be determined from published tables for square and rectangular ducts of equivalent carrying capacity or by other engineering methods.

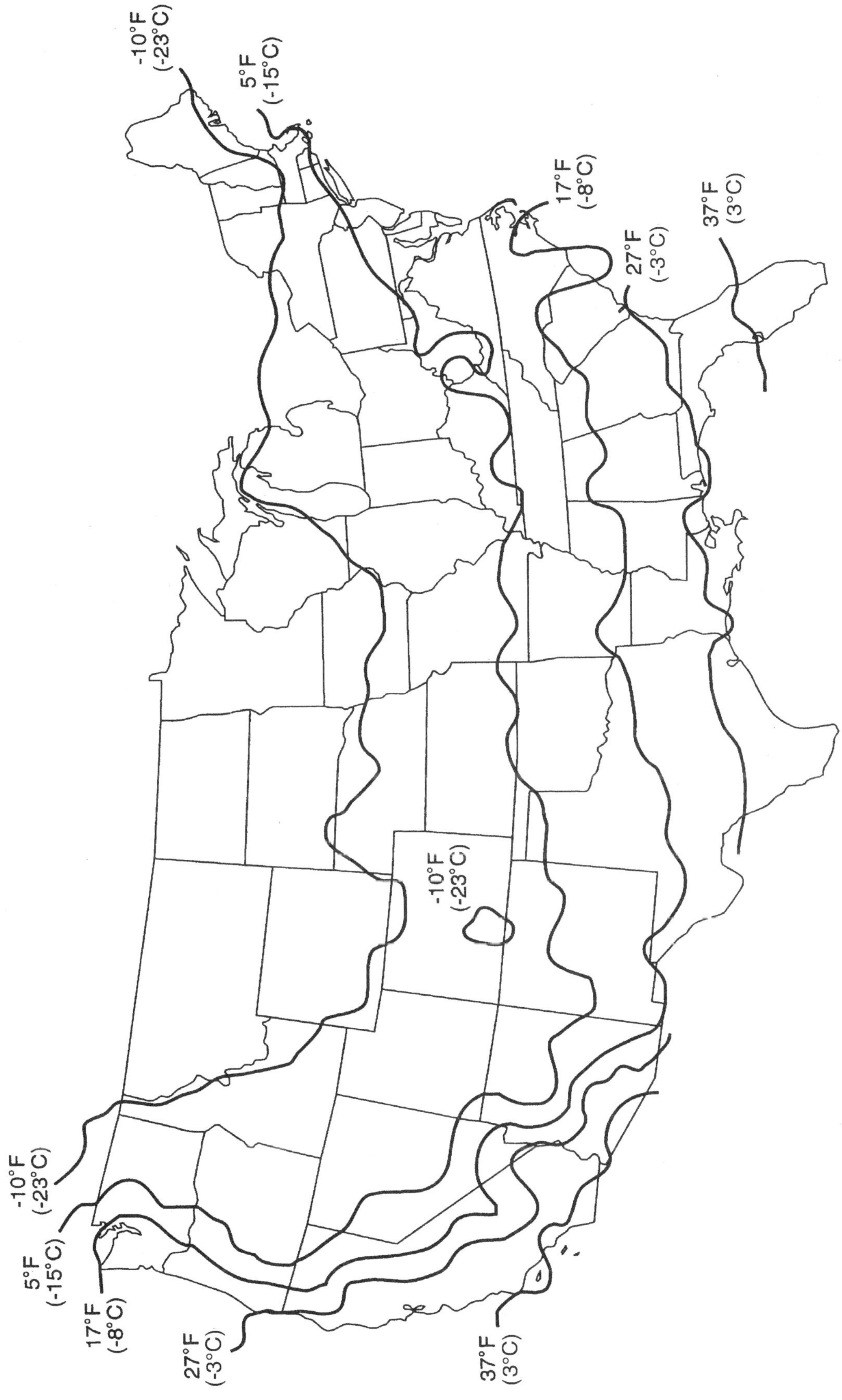
-10°F
(-23°C)
5°F
(-15°C)
17°F
(-8°C)
27°F
(-3°C)
37°F
(3°C)
-10°F
(-23°C)
-10°F
(-23°C)
5°F
(-15°C)
17°F
(-8°C)
27°F
(-3°C)
37°F
(3°C)

FIGURE B-19

This appendix is a direct excerpt from the 2003 *International Fuel Gas Code*, and no attempt has been made to coordinate with Chapter 24 of the *International Residential Code*.

APPENDIX C (IFGS)

EXIT TERMINALS OF MECHANICAL DRAFT AND DIRECT-VENT VENTING SYSTEMS

(This appendix is informative and is not part of the code.)

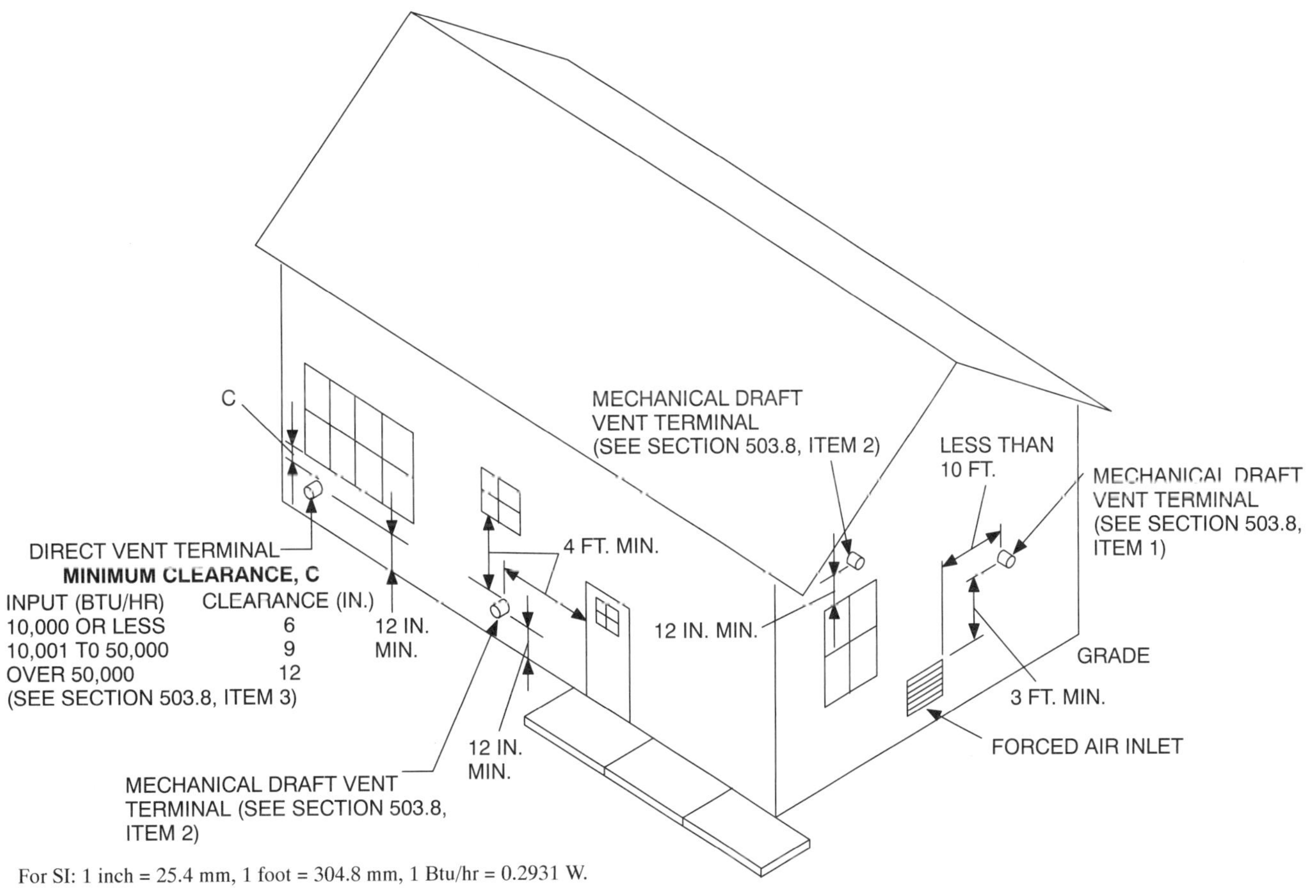

For SI: 1 inch = 25.4 mm, 1 foot = 304.8 mm, 1 Btu/hr = 0.2931 W.

APPENDIX C
EXIT TERMINALS OF MECHANICAL DRAFT AND DIRECT-VENT SYSTEMS

This appendix is a direct excerpt from the 2003 *International Fuel Gas Code*, and no attempt has been made to coordinate with Chapter 24 of the *International Residential Code*.

APPENDIX D (IFGS)

RECOMMENDED PROCEDURE FOR SAFETY INSPECTION OF AN EXISTING APPLIANCE INSTALLATION

(This appendix is informative and is not part of the code.)

The following procedure is intended as a guide to aid in determining that an appliance is properly installed and is in a safe condition for continuing use.

This procedure is predicated on central furnace and boiler installations, and it should be recognized that generalized procedures cannot anticipate all situations. Accordingly, in some cases, deviation from this procedure is necessary to determine safe operation of the equipment.

(a) This procedure should be performed prior to any attempt at modification of the appliance or of the installation.

(b) If it is determined there is a condition that could result in unsafe operation, the appliance should be shut off and the owner advised of the unsafe condition. The following steps should be followed in making the safety inspection:

1. Conduct a test for gas leakage. (See Section 406.6)
2. Visually inspect the venting system for proper size and horizontal pitch and determine there is no blockage or restriction, leakage, corrosion, and other deficiencies that could cause an unsafe condition.
3. Shut off all gas to the appliance and shut off any other fuel-gas-burning appliance within the same room. **Use the shutoff valve in the supply line to each appliance.**
4. Inspect burners and crossovers for blockage and corrosion.
5. **Applicable only to furnaces.** Inspect the heat exchanger for cracks, openings, or excessive corrosion.
6. **Applicable only to boilers.** Inspect for evidence of water or combustion product leaks.
7. Insofar as is practical, close all building doors and windows and all doors between the space in which the appliance is located and other spaces of the building. Turn on clothes dryers. Turn on any exhaust fans, such as range hoods and bathroom exhausts, so they will operate at maximum speed. Do not operate a summer exhaust fan. Close fireplace dampers. If, after completing Steps 8 through 13, it is believed sufficient combustion air is not available, refer to Section 304 of this code for guidance.
8. Place the appliance being inspected in operation. **Follow the lighting instructions.** Adjust the thermostat so appliance will operate continuously.
9. Determine that the pilot(s), where provided, is burning properly and that the main burner ignition is satisfactory by interrupting and reestablishing the electrical supply to the appliance in any convenient manner. If the appliance is equipped with a continuous pilot(s), test the pilot safety device(s) to determine if it is operating properly by extinguishing the pilot(s) when the main burner(s) is off and determining, after 3 minutes, that the main burner gas does not flow upon a call for heat. If the appliance is not provided with a pilot(s), test for proper operation of the ignition system in accordance with the appliance manufacturer's lighting and operating instructions.
10. Visually determine that the main burner gas is burning properly (i.e., no floating, lifting, or flashback). Adjust the primary air shutter(s) as required. If the appliance is equipped with high and low flame controlling or flame modulation, check for proper main burner operation at low flame.
11. Test for spillage at the draft hood relief opening after 5 minutes of main burner operation. Use a flame of a match or candle or smoke.
12. Turn on all other fuel-gas-burning appliances within the same room so they will operate at their full inputs. **Follow lighting instructions for each appliance.**
13. Repeat Steps 10 and 11 on the appliance being inspected.
14. Return doors, windows, exhaust fans, fireplace dampers, and any other fuel-gas-burning appliance to their previous conditions of use.
15. **Applicable only to furnaces.** Check both the limit control and the fan control for proper operation. Limit control operation can be checked by blocking the circulating air inlet or temporarily disconnecting the electrical supply to the blower motor and determining that the limit control acts to shut off the main burner gas.
16. **Applicable only to boilers.** Determine that the water pumps are in operating condition. Test low water cutoffs, automatic feed controls, pressure and temperature limit controls, and relief valves in accordance with the manufacturer's recommendations to determine that they are in operating condition.

Appendix E: Manufactured Housing Used as Dwellings

General Comments

Manufactured housing in the United States represents a major portion of housing construction. The criteria for the construction of the structure are governed by the National Manufactured Housing Construction and Safety Act. This act, in effect for quite some time, provides a minimum baseline for such structures no matter where they are installed; manufactured housing is constructed in a factory and not subject to inspection by the local jurisdiction when installed.

While this act may seem to cover the bulk of the construction of manufactured housing, it does not cover those areas related to the placement of the housing on the property. This appendix chapter works in conjunction with the act in providing a complete set of regulations that can assure the owner and the jurisdiction alike that the structure is safe.

Section AE101 is the scope of the appendix. Section AE102 addresses the application of these criteria to existing manufactured housing. Section AE201 contains the definitions necessary for the proper interpretation of the regulations. Section AE301 discusses the type of work that requires permits for such installations. Section AE302 identifies the information needed on such permits. Section AE303 addresses permit issuance. Section AE304 contains the criteria for the establishment of fees for such permits. Section AE305 is the material on required permits. Section AE306 addresses when special inspections are necessary for the work regulated by this appendix, and Section AE307 discusses how to handle the connection to the utilities.

Section AE401 addresses the issues related to the occupancy of this type of housing. Section AE402 contains the provisions for locating the structure on the property. Section AE501 discusses the design of the foundation systems for manufactured housing. Section AE502 contains the specific criteria for the foundation system chosen. Section AE503 addresses the enclosure of the perimeter of the underfloor area. Section AE504 discusses the structural concerns of adding to such housing. Section AE505 identifies which set of regulations apply to building service equipment. Section AE506 is concerned with the path of egress that occupants will need to use in the event of an emergency. Section AE507 provides a cross-reference to the HUD Safety Standards. Section AE601 provides more specific criteria for the use of various foundation systems. Section AE602 addresses the use of piers as foundation material. Section AE603 discusses the allowable height of such piers. Section AE604 refers the reader to the manufacturer's installation instructions for the use of anchors for the foundation system, and Section AE605 addresses the use of ties.

Purpose

This appendix chapter provides a complete set of regulations in conjunction with federal law for the installation of manufactured housing. Application of the law and the provisions together provides safe housing.

SECTION AE101 SCOPE

AE101.1 General. These provisions shall be applicable only to a manufactured home used as a single dwelling unit installed on privately owned (nonrental) lots and shall apply to the following:

1. Construction, alteration and repair of any foundation system which is necessary to provide for the installation of a manufactured home unit.
2. Construction, installation, addition, alteration, repair or maintenance of the building service equipment which is necessary for connecting manufactured homes to water, fuel, or power supplies and sewage systems.
3. Alterations, additions or repairs to existing manufactured homes. The construction, alteration, moving, demolition, repair and use of accessory buildings and structures and their building service equipment shall comply with the requirements of the codes adopted by this jurisdiction.

These provisions shall not be applicable to the design and construction of manufactured homes and shall not be deemed to authorize either modifications or additions to manufactured homes where otherwise prohibited.

> **Exception:** In addition to these provisions, new and replacement manufactured homes to be located in flood hazard areas as established in Table R301.2(1) of the *International Residential Code* shall meet the applicable requirements of Sections R323 of the *International Residential Code.*

❖ This appendix chapter discusses the use of the IRC, which is to be used in conjunction with federal laws that govern the construction of manufactured housing. Since the federal laws only cover the design and construction of the housing itself, other provisions must be used to address the design and construction of the foundation system, the connection to utilities, and the construction of accessory buildings. These issues have typically fallen under the adopted codes of the locality.

Item 3 indicates that alterations, additions, and repairs must comply with the IRC. This seems to indicate that this code governs over such activities. While that might be true in some cases, in most it is not. In fact, the National Manufactured Housing Construction and Safety Act is referred to in Section AE102.2 for these alterations, additions, and repairs.

Flood plain construction is referred to and the criteria are found in Section R327. See Sections R327.1.8 and R327.2 for flood requirements for manufactured housing.

SECTION AE102 APPLICATION TO EXISTING MANUFACTURED HOMES AND BUILDING SERVICE EQUIPMENT

AE102.1 General. Manufactured homes and their building service equipment to which additions, alterations or repairs are made shall comply with all the requirements of these provisions for new facilities, except as specifically provided in this section.

❖ This section discusses the use of this appendix for existing manufactured housing. As a general rule, this section requires that the activity must comply with the criteria for new construction; however, the following subsections contain many variances to that rule.

AE102.2 Additions, alterations or repairs. Additions made to a manufactured home shall conform to one of the following:

1. Be certified under the National Manufactured Housing Construction and Safety Standards Act of 1974 (42 U.S.C. Section 5401, et seq.).
2. Be designed and constructed to conform with the applicable provisions of the National Manufactured Housing Construction and Safety Standards Act of 1974 (42 U.S.C. Section 5401, et seq.).
3. Be designed and constructed in conformance with the code adopted by this jurisdiction.

Additions shall be structurally separated from the manufactured home.

Exception: A structural separation need not be provided when structural calculations are provided to justify the omission of such separation.

Alterations or repairs may be made to any manufactured home or to its building service equipment without requiring the existing manufactured home or its building service equipment to comply with all the requirements of these provisions, provided the alteration or repair conforms to that required for new construction, and provided further that no hazard to life, health or safety will be created by such additions, alterations or repairs.

Alterations or repairs to an existing manufactured home which are nonstructural and do not adversely affect any structural member or any part of the building or structure having required fire protection may be made with materials equivalent to those of which the manufactured home structure is constructed, subject to approval by the building official.

Exception: The installation or replacement of glass shall be required for new installations.

Minor additions, alterations and repairs to existing building service equipment installations may be made in accordance with the codes in effect at the time the original installation was made subject to approval of the building official, and provided such additions, alterations and repairs will not cause the existing building service equipment to become unsafe, insanitary or overloaded.

❖ This section discusses alterations, additions, and repairs to manufactured homes and is important because of confusion over the application of federal regulations to these topics. In many areas of the country, manufactured housing is ordered and installed, and a patio cover or a carport is added later. The jurisdiction then has to make a determination of which set of criteria applies. This section allows the use of any one of the three specified alternative methods of compliance for additions, making it clear that the National Manufactured Housing Construction and Safety Standards Act can apply, or the additions can be designed using the code that is adopted by the jurisdiction. See Section AE507.1 for alterations made to a manufactured home.

AE102.3 Existing installations. Building service equipment lawfully in existence at the time of the adoption of the applicable codes may have their use, maintenance or repair continued if the use, maintenance or repair is in accordance with the original design and no hazard to life, health or property has been created by such building service equipment.

❖ Existing building service equipment is generally "grandfathered" as an allowable condition when a newer edition of a code is adopted, provided that a minimum level of safety is maintained. In most cases, that level is the set of criteria or the code under which the original building service equipment was installed. Unfortunately, there are a number of areas where codes have only recently been adopted. In those cases, the code official must apply those criteria that can be reasonably applied to an existing installation.

AE102.4 Existing occupancy. Manufactured homes which are in existence at the time of the adoption of these provisions may have their existing use or occupancy continued if such use or occupancy was legal at the time of the adoption of these provisions, provided such continued use is not dangerous to life, health and safety.

The use or occupancy of any existing manufactured home shall not be changed unless evidence satisfactory to the building official is provided to show compliance with all applicable provisions of the codes adopted by this jurisdiction. Upon any change in use or occupancy, the manufactured home shall cease to be classified as such within the intent of these provisions.

❖ Very much like the previous section, this section's provisions describe a minimum level of safety for the occupants of an existing structure. Also, if the occupancy is changed, evidence must be provided indicating that all applicable codes and standards are complied with by the new use.

AE102.5 Maintenance. All manufactured homes and their building service equipment, existing and new, and all parts thereof shall be maintained in a safe and sanitary condition. All device or safeguards which are required by applicable codes or by the Manufactured Home Standards shall be maintained in conformance with the code or standard under which it was installed. The owner or the owner's designated agent shall be responsible for the maintenance of manufactured homes, accessory buildings, structures and their building service equipment. To determine compliance with this subsection, the building official may cause any manufactured home, accessory building or structure to be reinspected.

❖ As with all structures, building service equipment, and parts, proper maintenance is vital to maintaining a safe and sanitary environment for the occupants. This section specifically designates the owner or her or his representative as being responsible for such maintenance. In order to determine if the structure and associated equipment are properly maintained, the code official can have an inspection done.

AE102.6 Relocation. Manufactured homes which are to be relocated within this jurisdiction shall comply with these provisions.

❖ This section is a clarification that even relocated manufactured housing must comply with these provisions.

SECTION AE201 DEFINITIONS

AE201.1 General. For the purpose of these provisions, certain abbreviations, terms, phrases, words and their derivatives shall be construed as defined or specified herein.

❖ These terms are necessary for the proper design and enforcement of this appendix chapter. Any terms not listed herein will be found in Chapter 2 of this code, in the other ICC codes, or will use their normally accepted meanings.

ACCESSORY BUILDING. Any building or structure, or portion thereto, located on the same property as a manufactured home which does not qualify as a manufactured home as defined herein.

❖ This definition addresses those structures that are not manufactured housing. An example would be a patio cover or shed.

BUILDING SERVICE EQUIPMENT. Refers to the plumbing, mechanical and electrical equipment including piping, wiring, fixtures and other accessories which provide sanitation, lighting, heating ventilation, cooling, fire protection and facilities essential for the habitable occupancy of a manufactured home or accessory building or structure for its designated use and occupancy.

❖ This term covers the plumbing, mechanical, and electrical equipment serving the manufactured home.

MANUFACTURED HOME. A structure transportable in one or more sections which, in the traveling mode, is 8 body feet (2438 body mm) or more in width or 40 body feet (12 192 body mm) or more in length or, when erected on site, is 320 or more square feet (30 m^2), and which is built on a permanent chassis and designed to be used as a dwelling with or without a permanent foundation when connected to the required utilities, and includes the plumbing, heating, air-conditioning and electrical systems contained therein; except that such term shall include any structure which meets all the requirements of this paragraph except the size requirements and with respect to which the manufacturer voluntarily files a certification required by the secretary (HUD) and complies with the standards established under this title.

For mobile homes built prior to June 15, 1976, a label certifying compliance to the Standard for Mobile Homes, NFPA 501, ANSI 119.1, in effect at the time of manufacture is required. For the purpose of these provisions, a mobile home shall be considered a manufactured home.

❖ The term "Manufactured Home" in the IRC is consistent with the HUD definition. An existing manufactured home requires a label certifying compliance with the Standard for Mobile Homes, NFPA 501, ANSI 119.1 if built prior to June 15, 1976. Included in this definition is everything that is built into the home.

MANUFACTURED HOME INSTALLATION. Construction which is required for the installation of a manufactured home, including the construction of the foundation system, required structural connections thereto and the installation of on-site water, gas, electrical and sewer systems and connections thereto which are necessary for the normal operation of the manufactured home.

❖ The components covered are those that have to be installed on the site prior to the arrival of the manufactured home. Included are the foundation systems, connections to the utilities, and anything else necessary for the operation of the home.

MANUFACTURED HOME STANDARDS. The Manufactured Home Construction and Safety Standards as promulgated by the United States Department of Housing and Urban Development.

❖ This definition gives the title of the federal standard that is applicable for the construction of the manufactured home.

PRIVATELY OWNED (NONRENTAL) LOT. A parcel of real estate outside of a manufactured home rental community (park) where the land and the manufactured home to be installed thereon are held in common ownership.

❖ This in one of the few instances that an ICC code addresses ownership. This definition makes it clear that both the land and manufactured home are to be held by the same owner or owners.

SECTION AE301 PERMITS

AE301.1 Initial installation. A manufactured home shall not be installed on a foundation system reinstalled or altered without first obtaining a permit from the building official. A separate permit shall be required for each manufactured home installation. When approved by the building official, such permit may include accessory buildings and structures and their building service equipment when the accessory buildings or structures will be constructed in conjunction with the manufactured home installation.

❖ This section specifies that a permit is needed for the installation, reinstallation, or alteration of each manufactured home. See Section AE302.1 for the information that is required to apply for a permit. The permit process gives the jurisdiction the opportunity to verify that the manufactured home installation complies with the specified requirements in this appendix chapter.

AE301.2 Additions, alterations and repairs to a manufactured home. A permit shall be obtained to alter, remodel, repair or add accessory buildings or structures to a manufactured home subsequent to its initial installation. Permit issuance and fees therefor shall be in conformance with the codes applicable to the type of work involved.

An addition made to a manufactured home as defined in these provisions shall comply with these provisions.

❖ Any alteration, remodeling, repair, or addition to a manufactured home must have a permit. This section also coordinates with Section AE102.2, which discusses the application of the appendix provisions.

AE301.3 Accessory buildings. Except as provided in Section AE301.1, permits shall be required for all accessory buildings and structures and their building service equipment. Permit issuance and fees therefor shall be in conformance with the codes applicable to the types of work involved.

❖ This section requires a separate permit for the construction/installation of all accessory structures. Any fee that is charged is intended to cover those services provided by the jurisdiction for the enforcement of the code.

AE301.4 Exempted work. A permit shall not be required for the types of work specifically exempted by the applicable codes. Exemption from the permit requirements of any of said codes shall not be deemed to grant authorization for any work to be done in violation of the provisions of said codes or any other laws or ordinances of this jurisdiction.

❖ Certain types of work are exempted from permit in most codes. See Section R105.2 for items that do not require a permit. This does not mean, however, that the exempted work does not have to meet the code. Such work must still comply with all applicable codes and standards.

SECTION AE302 APPLICATION FOR PERMIT

AE302.1 Application. To obtain a manufactured home installation permit, the applicant shall first file an application inwriting on a form furnished by the building official for that purpose. At the option of the building official, every such application shall:

1. Identify and describe the work to be covered by the permit for which application is made.
2. Describe the land on which the proposed work is to be done by legal description, street address or similar description that will readily identify and definitely locate the proposed building or work.
3. Indicate the use or occupancy for which the proposed work is intended.
4. Be accompanied by plans, diagrams, computations and specifications and other data as required in Section AE302.2.
5. Be accompanied by a soil investigation when required by Section AE502.2.
6. State the valuation of any new building or structure or any addition, remodeling or alteration to an existing building.
7. Be signed by permittee, or permittee's authorized agent, who may be required to submit evidence to indicate such authority.
8. Give such other data and information as may be required by the building official.

❖ This section lists the items that must be found on an application for a permit. The purpose of requiring this information is so that the installation can be verified to comply with this appendix chapter.

AE302.2 Plans and specifications. Plans, engineering calculations, diagrams and other data as required by the building official shall be submitted in not less than two sets with each application for a permit. The building official may require plans, computations and specifications to be prepared and designed by an engineer or architect licensed by the state to practice as such.

Where no unusual site conditions exist, the building official may accept approved standard foundation plans and details in conjunction with the manufacturer's approved installation instructions without requiring the submittal of engineering calculations.

❖ The submittal of plans and other supporting information is necessary to verify that the manufactured home installation complies with all necessary codes and standards. Under certain circumstances the code official has the discretion to waive the submittal of engineering calculations.

AE302.3 Information on plans and specifications. Plans and specifications shall be drawn to scale on substantial paper or cloth and shall be of sufficient clarity to indicate the location, nature and extent of the work proposed and shown in detail that it will conform to the provisions of these provisions and all relevant laws, ordinances, rules and regulations. The building official shall determine what information is required on plans and specifications to ensure compliance.

❖ This section requires that the information necessary to verify the compliance of the installation of the manufactured home be provided on the submittal documents.

SECTION AE303 PERMITS ISSUANCE

AE303.1 Issuance. The application, plans and specifications and other data filed by an applicant for permit shall be reviewed by the building official. Such plans may be reviewed by other departments of this jurisdiction to verify compliance with any applicable laws under their jurisdiction. If the building official finds that the work described in an application for a permit and the plans, specifications and other data filed therewith conform to the requirements of these provisions and other data filed therewith conform to the requirements of these provisions and other pertinent codes, laws and ordinances, and that the fees specified in Section AE304 have been paid, the building official shall issue a permit therefor to the applicant.

When the building official issues the permit where plans are required, the building official shall endorse in writing or stamp the plans and specifications APPROVED. Such approved plans and specifications shall not be changed, modified or altered without authorization from the building official, and all work shall be done in accordance with the approved plans.

❖ This section requires that the code official issue a permit when the information submitted has been shown to comply with the applicable codes and standards. These plans are the approved set and there must be no alteration to them; otherwise, authorization from the code official is necessary.

AE303.2 Retention of plans. One set of approved plans and specifications shall be returned to the applicant and shall be kept on the site of the building or work at all times during which the work authorized thereby is in progress. One set of approved plans, specification and computations shall be retained by the building official until final approval of the work.

❖ A set of the approved plans and specifications must be kept by the jurisdiction for a period of time that is typically set out in state law. These retained documents are very useful in resolving disputes that can arise during the construction process and beyond.

AE303.3 Validity of permit. The issuance of a permit or approval of plans and specifications shall shall not be construed to be a permit for, or an approval of, any violation of any of the provisions of these provisions or other pertinent codes of any other ordinance of the jurisdiction. No permit presuming to give authority to violate or cancel these provisions shall be valid.

The issuance of a permit based on plans, specifications and other data shall not prevent the building official from thereafter requiring the correction of errors in said plans, specifications and other data, or from preventing building operations being carried on thereunder when in violation of these provisions or of any other ordinances of this jurisdiction.

❖ This section states the fundamental premise that the permit is only a license to proceed with the intended work. It is not a license to violate, cancel, or set aside any provisions of this code or any other law or regulation. This places the burden of complying with all codes and regulations on the permit applicant and not on the building official.

AE303.4 Expiration. Every permit issued by the building official under these provisions shall expire by limitation and become null and void if the work authorized by such permit is not commenced within 180 days from the date of such permit, or if the work authorized by such permit is suspended or abandoned at any time after the work is commenced for a period of 180 days. Before such work can be recommenced, a new permit shall be first obtained, and the fee therefor shall be one-half the amount required for a new permit for such work, provided no changes have been made or will be made in the original plans and specifications for such work, and provided further that such suspension or abandonment has not exceeded one year. In order to renew action on a permit after expiration, the permittee shall pay a new full permit fee.

Any permittee holding an unexpired permit may apply for an extension of the time within which work may commence under that permit when the permittee is unable to commence work within the time required by this section for good and satisfactory reasons. The building official may extend the time for action by the permittee for a period not exceeding 180 days upon written request by the permittee showing that circumstances beyond the control of the permittee have prevented action from being taken. No permit shall be extended more than once.

❖ A permit becomes invalid under two separate conditions. The first is when work is not commenced within 180 days; the second is when such work has been suspended for 180 days or more. To reinstate the permit, a new permit must first be obtained, and one-half of the fees paid. If the builders know that they can run into a problem with having work delayed for 180 days or longer, then they can submit a request for an extension to the building official if the reason is a good one. The building official can extend the permit for up to an additional 180 day period. This can only happen once; after that a new permit must be obtained.

AE303.5 Suspension or revocation. The building official may, in writing, suspend or revoke a permit issued under these provisions whenever the permit is issued in error or on the basis of incorrect information supplied, or in violation of any ordinance or regulation or any of these provisions.

❖ As with any license to do work, a permit can be revoked or suspended if it has been shown that the permit has been issued in error or that erroneous information has been provided. A permit to do work cannot violate any codes, laws, or regulations imposed on such a project.

SECTION AE304 FEES

AE304.1 Permit fees. The fee for each manufactured home installation permit shall be established by the building official.

When permit fees are to be based on the value or valuation of the work to be performed, the determination of value or valuation under these provisions shall be made by the building official. The value to be used shall be the total value of all work required for the manufactured home installation plus the total value of all work required for the construction of accessory buildings and structures for which the permit is issued as well as all finish work, painting, roofing, electrical, plumbing, heating, air conditioning, elevators, fire-extinguishing systems and any other permanent equipment which is a part of the accessory building or structure. The value of the manufactured home itself shall not be included.

❖ Each jurisdiction must establish a set of fees for permit issuance. The fees collected cover only the services provided by the city in regard to the enforcement of the code and other laws in relation to the work anticipated. There is additional guidance for the building officials in establishing the valuation of a project found in this section.

AE304.2 Plan review fees. When a plan or other data are required to be submitted by Section AE302.2, a plan review fee shall be paid at the time of submitting plans and specifications for review. Said plan review fee shall be as established by the building official. Where plans are incomplete or changed so as to require additional plan review, an additional plan review fee shall be charged at a rate as established by the building official.

❖ If the review of a set of construction documents is required to verify that the project is in compliance with the code, then an additional fee is imposed to cover the service of that review. If the plans are incomplete, then an additional fee is imposed to cover the additional plan review of the project. Each jurisdiction must establish a set of fees to be applied where plans are incomplete or changed.

AE304.3 Other provisions.

AE304.3.1 Expiration of plan review. Applications for which no permit is issued within 180 days following the date of application shall expire by limitation, and plans and other data submitted for review may thereafter be returned to the applicant or destroyed by the building official. The building official may extend the time for action by the applicant for a period not exceeding 180 days upon request by the applicant showing that circumstances beyond the control of the applicant have prevented action from being taken. No application shall be extended more than once. In order to renew action on an application after expiration, the applicant shall resubmit plans and pay a new plan review fee.

❖ This section is similar to the description of the expiration of permit in Section AE303.4. The applicant is given a period of time to complete the plan review process and obtain a permit for the work. There is an opportunity to extend this time period since it is understood that problems can be encountered while obtaining a permit.

AE304.3.2 Investigation fees: work without a permit.

AE304.3.2.1 Investigation. Whenever any work for which a permit is required by these provisions has been commenced without first obtaining said permit, a special investigation shall be made before a permit may be issued for such work.

❖ This fee is imposed to cover the services of reviewing a project after it has been commenced without a permit. Oftentimes work is covered or concealed by the time the inspection takes place, and the fees involved may have to reflect the difficulty of investigations to verify compliance with the codes and other regulations.

AE304.3.2.2 Fee. An investigation fee, in addition to the permit fee, shall be collected whether or not a permit is then or subsequently issued. The investigation fee shall be equal to the amount of the permit fee required. The minimum investigation fee shall be the same as the minimum fee established by the building official. The payment of such investigation fee shall not exempt any person from compliance with all other provisions of either these provisions or other pertinent codes or from any penalty prescribed by law.

❖ This section indicates that the building official must determine the fees involved for such investigations.

E304.3.3 Fee refunds.

AE304.3.3.1 Permit fee erroneously paid or collected. The building official may authorize the refunding of any fee paid hereunder which was erroneously paid or collected.

❖ Any fees that were collected in error can be refunded to the applicants.

AE304.3.3.2 Permit fee paid when no work done. The building official may authorize the refunding of not more than 80 percent of the permit fee paid when no work has been done under a permit issued in accordance with these provisions.

❖ This section addresses the situation where a permit was issued and fees paid and the work was not begun. Eighty percent of such fees can be refunded to applicants by the building official. Twenty percent of the permit fee is retained to cover the administrative work that has already been done.

AE304.3.3.3 Plan review fee. The building official may authorize the refunding of not more than 80 percent of the plan review fee paid when an application for a permit for which a plan review fee has been paid is withdrawn or canceled before any plan reviewing is done.

The building official shall not authorize the refunding of any fee paid except upon written application by the original permittee not later than 180 days after the date of the fee payment.

❖ If a plan review is not done and the permit withdrawn or canceled, then the building official can authorize the reimbursement of most of the fees paid. A refund request must be in writing.

SECTION AE305 INSPECTIONS

AE305.1 General. All construction or work for which a manufactured home installation permit is required shall be subject to inspection by the building official, and certain types of construction shall have continuous inspection by special inspectors as specified in Section AE306. A survey of the lot may be required by the building official to verify that the structure is located in accordance with the approved plans.

It shall be the duty of the permit applicant to cause the work to be accessible and exposed for inspection purposes. Neither the building official nor this jurisdiction shall be liable for expense entailed in the removal or replacement of any material required to allow inspection.

❖ This is one of the most important functions of a building department. This section authorizes the building official to have a project inspected for compliance with the approved documents and codes. The work must be exposed and accessible to the inspectors. If the work to be inspected is not exposed, the jurisdiction is not responsible for any expenses incurred for making the work exposed and accessible.

AE305.2 Inspection requests. It shall be the duty of the person doing the work authorized by a manufactured home installation permit to notify the building official that such work is ready for inspection. The building official may require that every request for inspection be filed at least one working day before such inspection is desired. Such request may be in writing or by telephone at the option of the building official.

It shall be the duty of the person requesting any inspections required either by these provisions or other applicable codes to provide access to and means for proper inspection of such work.

❖ It is the responsibility of the people doing the work to ask for inspections at the proper intervals. Also, the people doing the work need to provide the means for making such inspections. For example, this can be in the form of a ladder or other such device.

AE305.3 Inspection record card. Work requiring a manufactured home installation permit shall not be commenced until the permit holder or the permit holder's agent shall have posted an inspection record card in a conspicuous place on the premises and in such position as to allow the building official conveniently to make the required entries thereon regarding inspection of the work. This card shall be maintained in such position by the permit holder until final approval has been issued by the building official.

❖ The purpose of keeping the permit in a conspicuous location on the job site is to make it available for the inspector to make notes and keep track of the previous inspections.

AE305.4 Approval required. Work shall not be done on any part of the manufactured home installation beyond the point indicated in each successive inspection without first obtaining the approval of the building official. Such approval shall be given only after an inspection has been made of each successive step in the construction as indicated by each of the inspections required in Section AE305.5. There shall be a final inspection and approval of the manufactured home installation, including connections to its building service equipment, when completed and ready for occupancy or use.

❖ No work on the project can progress beyond a required inspection without having approval from the building official.

AE305.5 Required inspections.

AE305.5.1 Structural inspections for the manufactured home installation. Reinforcing steel or structural framework of any part of any manufactured home foundation system shall not be covered or concealed without first obtaining the approval of the building official. The building official, upon notification from the permit holder or the permit holder's agent, shall make the following inspections and shall either approve that portion of the construction as completed or shall notify the permit holder or the permit holder's agent wherein the same fails to comply with these provisions or other applicable codes:

1. Foundation inspection: To be made after excavations for footings are completed and any required reinforcing steel is in place. For concrete foundations, any required forms shall be in place prior to inspection. All materials for the foundation shall be on the job, except where concrete from a central mixing plant (commonly termed "transit mixed") is to be used, the concrete materials need not be on the job. Where the foundation is to be

constructed of approved treated wood, additional framing inspections as required by the building official may be required.

2. Concrete slab or under-floor inspection: To be made after all in-slab or underfloor building service equipment, conduit, piping accessories and other ancillary equipment items are in place but before any concrete is poured or the manufactured home is installed.
3. Anchorage inspection: To be made after the manufactured home has been installed and permanently anchored.

❖ The inspections specified in Section AE305.5 are the minimum thought necessary to verify the compliance of the installation. This section lists the structural inspections for the foundation, underfloor equipment, concrete slab, and anchorage of the manufactured home.

AE305.5.2 Structural inspections for accessory building and structures. Inspections for accessory buildings and structures shall be made as set forth in this code.

❖ This section suggests that the jurisdiction have the same inspections for accessory structures that it would have for any building.

AE305.5.3 Building service equipment inspections. All building service equipment which is required as a part of a manufactured home installation, including accessory buildings and structures authorized by the same permit, shall be inspected by the building official. Building service equipment shall be inspected and tested as required by the applicable codes. Such inspections and testing shall be limited to site construction and shall not include building service equipment which is a part of the manufactured home itself. No portion of any building service equipment intended to be concealed by any permanent portion of the construction shall be concealed until inspected and approved. Building service equipment shall not be connected to the water, fuel or power supply or sewer system until authorized by the building official.

❖ All equipment serving the manufactured home must be inspected. This section also requires that no connection be made to the utilities without the building official's approval.

AE305.5.4 Final inspection. When finish grading and the manufactured home installation, including the installation of all required building service equipment, is completed and the manufactured home is ready for occupancy, a final inspection shall be made.

❖ When all work is complete, this section requires one final inspection to verify that the approved plans have been followed. Once this inspection is approved, then the home is ready for occupancy.

AE305.6 Other inspections. In addition to the called inspections specified above, the building official may make or require other inspections of any construction work to ascertain compliance with these provisions or other codes and laws which are enforced by the code enforcement agency.

❖ This section allows a building official to impose other inspections as may be necessary to verify compliance with this set of criteria or other laws and regulations.

SECTION AE306 SPECIAL INSPECTIONS

AE306.1 General. In addition to the inspections required by Section AE305, the building official may require the owner to employ a special inspector during construction of specific types of work as described in this code.

❖ A special inspector may be needed to verify compliance with the code. This could be due to the use of high strength concrete, special soil conditions or anchorage, or even flood plain installations.

SECTION AE307 UTILITY SERVICE

AE307.1 General. Utility service shall not be provided to any building service equipment which is regulated by these provisions or other applicable codes and for which a manufactured home installation permit is required by these provisions until approved by the building official.

❖ The building official must approve the final inspection before connection can be made to any utilities that serve the manufactured home.

SECTION AE401 OCCUPANCY CLASSIFICATION

AE401.1 Manufactured homes. A manufactured home shall be limited in use to use as a single dwelling unit.

❖ There can be instances of applications to change the use of the structure to something other than a dwelling. This chapter was not developed with that in mind, and in those cases the *International Building Code* should be used.

AE401.2 Accessory buildings. Accessory buildings shall be classified as to occupancy by the building official as set forth in this code.

❖ The use of accessory structures is indeed accessory to the home, so the structures allowed would be those normally attached to such dwellings. Examples would be tool sheds, patio covers, carports, and gazebos.

SECTION AE402
LOCATION ON PROPERTY

AE402.1 General. Manufactured homes and accessory buildings shall be located on the property in accordance with applicable codes and ordinances of this jurisdiction.

❖ This section refers to other codes and laws that may be in effect for locating the manufactured home on the site. Section R302 of the IRC also applies, as well as any zoning ordinance of the jurisdiction.

SECTION AE501
DESIGN

AE501.1 General. A manufactured home shall be installed on a foundation system which is designed and constructed to sustain within the stress limitations specified in this code and all loads specified in this code.

Exception: When specifically authorized by the building official, foundation and anchorage systems which are constructed in accordance with the methods specified in Section AE600 of these provisions, or in the United States Department of Housing and Urban Development Handbook, *Permanent Foundations for Manufactured Housing,* 1984 Edition, Draft, shall be deemed to meet the requirements of this Appendix E.

❖ This section specifies that an appropriate foundation system be used for manufactured housing. The provided exception allows the use of special foundations when approved by the building official. These systems are called out in Section AE600 of this appendix chapter and in the United States Department of Housing and Urban Development Handbook: *Permanent Foundations for Manufactured Housing*. While the code specifies the 1984 edition of the HUD *Permanent Foundations* document, the 1996 edition is the most current and should be used.

AE501.2 Manufacturer's installation instructions. The installation instructions as provided by the manufacturer of the manufactured home shall be used to determine permissible points of support for vertical loads and points of attachment for anchorage systems used to resist horizontal and uplift forces.

❖ The installation instructions from the manufacturer of the housing must be followed. The code requirements of Section AE501.1 always apply, and where those foundations requirements are more stringent than the manufacturer's requirements, the code overrides the manufacturer's installation instructions. For example, the HUD *Permanent Foundations* document would override the manufacturer's installation instructions.

AE501.3 Rationality. Any system or method of construction to be used shall admit to a rational analysis in accordance with well-established principles of mechanics.

❖ Any structural system must comply with standard engineering practice.

SECTION AE502
FOUNDATION SYSTEMS

AE502.1 General. Foundation systems designed and constructed in accordance with this section may be considered as a permanent installation.

❖ Foundation systems designed in accordance with this section are permanent installations. That is, they are not subject to any time limitations that may be imposed on temporary uses.

AE502.2 Soil classification. The classification of the soil at each manufactured home site shall be determined when required by the building official. The building official may require that the determination be made by an engineer or architect licensed by the state to conduct soil investigations.

The classification shall be based on observation and any necessary tests of the materials disclosed by borings or excavations made in appropriate locations. Additional studies may be necessary to evaluate soil strength, the effect of moisture variation on soil-bearing capacity, compressibility and expansiveness.

When required by the building official, the soil classification design bearing capacity and lateral pressure shall be shown on the plans.

❖ The determination of the classification of soil at the manufactured-home site is the same as for buildings and structures that are constructed under any building code. It is left to the discretion of the local building official to determine if an investigation is necessary.

AE502.3 Footings and foundations. Footings and foundations, unless otherwise specifically provided, shall be constructed of materials specified by this code for the intended use and in all cases shall extend below the frost line. Footings of concrete and masonry shall be of solid material. Foundations supporting untreated wood shall extend at least 8 inches (203 mm) above the adjacent finish grade. Footings shall have a minimum depth below finished grade of 12 inches (305 mm) unless a greater depth is recommended by a foundation investigation.

Piers and bearing walls shall be supported on masonry or concrete foundations or piles, or other approved foundation systems which shall be of sufficient capacity to support all loads.

❖ This section requires that footings and foundations supporting manufactured housing extend both above and below ground a specified amount. These numbers are consistent with those found in IRC Sections R323.1 and R403.1.4. A footing that is not below the frost line would result in uneven frost heave movement and possible damage to the manufactured home.

AE502.4 Foundation design. When a design is provided, the foundation system shall be designed in accordance with the applicable structural provisions of this code and shall be designed to minimize differential settlement. Where a design is not provided, the minimum foundation requirements shall be as set forth in this code.

❖ This section states that a foundation design must follow the structural provisions of the IRC. If a design is not provided, then the prescriptive criteria in the appendix must be followed.

AE502.5 Drainage. Provisions shall be made for the control and drainage of surface water away from the manufactured home.

❖ Manufactured housing must meet the same criteria for drainage of surface water away from other buildings and structures as other structures in the IRC. Such water directed toward a structure would cause the decay of that structure.

AE502.6 Under-floor clearances—ventilation and access. A minimum clearance of 12 inches (305 mm) shall be maintained beneath the lowest member of the floor support framing system. Clearances from the bottom of wood floor joists or perimeter joists shall be as specified in this code.

Under-floor spaces shall be ventilated with openings as specified in this code. If combustion air for one or more heat-producing appliances is taken from within the under-floor spaces, ventilation shall be adequate for proper appliance operation.

Under-floor access openings shall be provided. Such openings shall be not less than 18 inches (457 mm) in any dimension and not less than 3 square feet (0.279 m^2) in area and shall be located so that any water supply and sewer drain connections located under the manufactured home are accessible.

❖ This section requires a nominal amount of space for access to the underfloor area and for its proper ventilation. This section also specifies that all underfloor connections of the water supply system and the drainage system be accessible.

SECTION AE503 SKIRTING AND PERIMETER ENCLOSURES

AE503.1 Skirting and permanent perimeter enclosures. Skirting and permanent perimeter enclosures shall be installed only where specifically required by other laws or ordinances. Skirting, when installed, shall be of material suitable for exterior exposure and contact with the ground. Permanent perimeter enclosures shall be constructed of materials as required by this code for regular foundation construction.

Skirting shall be installed in accordance with the skirting manufacturer's installation instructions. Skirting shall be adequately secured to assure stability, to minimize vibration and susceptibility to wind damage, and to compensate for possible frost heave.

❖ Skirting is not required by this appendix chapter. However, the skirting of the perimeter of the underfloor area of manufactured housing is required by other laws or ordinances in many areas of the United States. Such skirting is intended to hide unsightly areas and to prevent the intrusion of rodents. The manufacturer's installation instructions must be followed, and the materials themselves must be able to resist the effects of being installed outdoors.

AE503.2 Retaining walls. Where retaining walls are used as a permanent perimeter enclosure, they shall resist the lateral displacements of soil or other materials and shall conform to this code as specified for foundation walls. Retaining walls and foundation walls shall be constructed of approved treated wood, concrete, masonry or other approved materials or combination of materials as for foundations as specified in this code. Siding materials shall extend below the top of the exterior of the retaining or foundation wall or the joint between siding and enclosure wall shall be flashed in accordance with this code.

❖ This section applies where a continuous foundation wall is used, and the elevation of the exterior grade is different from the grade within the crawl space.

Also, this section requires that when skirting is used as a retaining wall, the material must be designed to withstand the loads imposed and the exposure to the earth itself.

SECTION AE504 STRUCTURAL ADDITIONS

AE504.1 General. Accessory buildings shall not be structurally supported by or attached to a manufactured home unless engineering calculations are submitted to substantiate any proposed structural connection.

Exception: The building official may waive the submission of engineering calculations if it is found that the nature of the work applied for is such that engineering calculations are not necessary to show conformance to these provisions.

❖ Additions to manufactured homes must be structurally independent of the housing itself. If not, an engineered design consistent with that of the manufacturer of the dwelling is acceptable. An exception is provided to allow for simple structures such as carports and some sheds to be added to the housing without engineering calculations.

SECTION AE505 BUILDING SERVICE EQUIPMENT

AE505.1 General. The installation, alteration, repair, replacement, addition to or maintenance of the building service equipment within the manufactured home shall conform to regulations set forth in the Manufactured Home Standards. Such work which is located outside the manufactured home shall comply with the applicable codes adopted by this jurisdiction.

❖ This section specifies which set of regulations are to be used for the installation, alteration, repair, replacement, addition to, or maintenance of any building service equipment that serves the home. The HUD Manufactured Home Construction and Safety Standards govern when the equipment is within the home.

SECTION AE506 EXITS

AE506.1 Site development. Exterior stairways and ramps which provide egress to the public way shall comply with applicable provisions of this code.

❖ The path of egress from the home must comply with the means of egress criteria in the body of the IRC. See IRC Sections R311 through R316 for stairway and ramp requirements. Any means of egress element within the structure of the manufactured housing itself is covered by the HUD safety standards.

AE506.2 Accessory buildings. Every accessory building or portion thereof shall be provided with exits as required by this code.

❖ Accessory buildings must have adequate means of egress for the occupants.

SECTION AE507 OCCUPANCY, FIRE SAFETY AND ENERGY CONSERVATION STANDARDS

AE507.1 General. Alterations made to a manufactured home subsequent to its initial installation shall conform to the occupancy, fire-safety and energy conservation requirements set forth in the Manufactured Home Standards.

❖ This is a reference to the HUD safety standards for any alteration to manufactured housing.

SECTION AE600 SPECIAL REQUIREMENTS FOR FOUNDATION SYSTEMS

AE600.1 General. Section AE600 is applicable only when specifically authorized by the building official.

❖ This gives the building official the discretion to use the prescriptive criteria found in the following sections for the foundation system of manufactured housing. See Section AE501.1 for alternative HUD criteria that are commonly used.

SECTION AE601 FOOTINGS AND FOUNDATIONS

AE601.1 General. The capacity of individual load-bearing piers and their footings shall be sufficient to sustain all loads specified in this code within the stress limitations specified in this code. Footings, unless otherwise approved by the building official, shall be placed level on firm, undisturbed soil or an engineered fill which is free of organic material, such as weeds and grasses. Where used, an engineered fill shall provide a minimum load-bearing capacity of not less than 1,000 psf (48 kN/m^2). Continuous footings shall conform to the requirements of this code. Section AE502 of these provisions shall apply to footings and foundations constructed under the provisions of this section.

❖ These are the general criteria for the design of foundation systems, similar to those found in Chapter 4. If fill is to be used for foundation support, the load-bearing capacity should be verified by field testing prior to foundation construction. Section AE502.3 requires the foundation to extend below the frost line.

SECTION AE602 PIER CONSTRUCTION

AE602.1 General. Piers shall be designed and constructed to distribute loads evenly. Multiple section homes may have concentrated roof loads which will require special consideration. Load-bearing piers may be constructed utilizing one of the methods listed below. Such piers shall be considered to resist only vertical forces acting in a downward direction. They shall not be considered as providing any resistance to horizontal loads induced by wind or earthquake forces.

1. A prefabricated load-bearing device that is listed and labeled for the intended use.
2. Mortar shall comply with ASTM C 270 Type M, S or N; this may consist of one part portland cement, one-half part hydrated lime and four parts sand by volume. Lime shall not be used with plastic or waterproof cement.
3. A cast-in-place concrete pier with concrete having specified compressive strength at 28 days of 2,500 psi (17 225 kPa).

Alternate materials and methods of construction may be used for piers which have been designed by an engineer or architect licensed by the state to practice as such.

Caps and leveling spacers may be used for leveling of the manufactured home. Spacing of piers shall be as specified in the manufacturer's installation instructions, if available, or by an approved designer.

❖ This section allows the use of devices that have been developed to provide the foundation for manufactured housing. These devices and other such products can only resist gravity loads. As always, an engineered foundation or a foundation that meets the requirements of the HUD *Permanent Foundations* document referenced in Section AE501.1 is acceptable.

SECTION AE603
HEIGHT OF PIERS

AE603.1 General. Piers constructed as indicated in Section AE602 may have heights as follows:

1. Except for corner piers, piers 36 inches (914 mm) or less in height may be constructed of masonry units, placed with cores or cells vertically. Piers shall be installed with their long dimension at right angles to the main frame member they support and shall have a minimum cross-sectional area of 128 square inches (82 560 mm^2). Piers shall be capped with minimum 4-inch (102 mm) solid masonry units or equivalent.
2. Piers between 36 and 80 inches (914 mm and 2032 mm) in height and all corner piers over 24 inches (610 mm) in height shall be at least 16 inches by 16 inches (406 mm by 406 mm) consisting of interlocking masonry units and shall be fully capped with minimum 4-inch (102 mm) solid masonry units or equivalent.
3. Piers over 80 inches (2032 mm) in height may be constructed in accordance with the provisions of Item 2 above, provided the piers shall be filled solid with grout and reinforced with four continuous No. 5 bars. One bar shall be placed in each corner cell of hollow masonry unit piers or in each corner of the grouted space of piers constructed of solid masonry units.
4. Cast-in-place concrete piers meeting the same size and height limitations of Items 1, 2 and 3 above may be substituted for piers constructed of masonry units.

❖ Masonry or concrete piers have specific height limitations for stability. Minimum cross-sectional areas are also specified for adequate load-bearing capacity.

SECTION AE604
ANCHORAGE INSTALLATIONS

AE604.1 Ground anchors. Ground anchors shall be designed and installed to transfer the anchoring loads to the ground. The load-carrying portion of the ground anchors shall be installed to the full depth called for by the manufacturer's installation directions and shall extend below the established frost line into undisturbed soil.

Manufactured ground anchors shall be listed and installed in accordance with the terms of their listing and the anchor manufacturer's instructions and shall include means of attachment of ties meeting the requirements of Section AE605. Ground anchor manufacturer's installation instructions shall include the amount of preload required and load capacity in various types of soil. These instructions shall include tensioning adjustments which may be needed to prevent damage to the manufactured home, particularly damage that can be caused by frost heave. Each ground anchor shall be marked with the manufacturer's identification and listed model identification number which shall be visible after installation. Instructions shall accompany each listed ground anchor specifying the types of soil for which the anchor is suitable under the requirements of this section.

Each approved ground anchor, when installed, shall be capable of resisting an allowable working load at least equal to 3,150 pounds (14 kN) in the direction of the tie plus a 50 percent overload [4,725 pounds (21 kN) total] without failure. Failure shall be considered to have occurred when the anchor moves more than 2 inches (51 mm) at a load of 4,725 pounds (21 kN) in the direction of the tie installation. Those ground anchors which are designed to be installed so that loads on the anchor are other than direct withdrawal shall be designed and installed to resist an applied design load of 3,150 pounds (14 kN) at 40 to 50 degrees from vertical or within the angle limitations specified by the home manufacturer without displacing the tie end of the anchor more than 4 inches (102 mm) horizontally. Anchors designed for connection of multiple ties shall be capable of resisting the combined working load and overload consistent with the intent expressed herein.

When it is proposed to use ground anchors and the building official has reason to believe that the soil characteristics at a given site are such as to render the use of ground anchors i advisable, or when there is doubt regarding the ability of the ground anchors to obtain their listed capacity, the building official may require that a representative field installation be made at the site in question and tested to demonstrate ground anchor capacity. The building official shall approve the test procedures.

❖ Approved devices for anchoring manufactured housing is allowed by this section. The manufacturer's installation instructions must be followed. This section describes the conditions by which the anchors will be tested and designed. Building officials can either verify the information themselves or ask for a third party evaluation of the product to verify its compliance with these criteria.

The site soil condition must be suitable for the use of ground anchors. The capacity of a ground anchor is significantly affected by the soil moisture content. Pull-out and lateral load capacity should be verified by tests.

AE604.2 Anchoring equipment. Anchoring equipment, when installed as a permanent installation, shall be capable of resisting all loads as specified within these provisions. When the stabilizing system is designed by an engineer or architect licensed by the state to practice as such, alternative designs may be used, providing the anchoring equipment to be used is capable of withstanding a load equal to 1.5 times the calculated load. All anchoring equipment shall be listed and labeled as being capable of meeting the requirements of these provisions. Anchors as specified in this code may be attached to the main frame of the manufactured home by an approved $^{3}/_{16}$-inch-thick (4.76 mm) slotted steel plate anchoring device. Other anchoring devices or methods meeting the requirements of these provisions may be permitted when approved by the building official.

Anchoring systems shall be so installed as to be permanent. Anchoring equipment shall be so designed to prevent self-disconnection with no hook ends used.

❖ This section applies to the devices that hold the manufactured home to its ground anchors. This set of equipment should also be addressed in the approved manufacturer's installation instructions.

AE604.3 Resistance to weather deterioration. All anchoring equipment, tension devices and ties shall have a resistance to deterioration as required by this code.

❖ All anchorage devices and equipment must be able to withstand exposure to weather without deterioration.

AE604.4 Tensioning devices. Tensioning devices, such as turnbuckles or yoke-type fasteners, shall be ended with clevis or welded eyes.

❖ This section requires that the ends of tensioning devices for the anchorage system must have either clevises or welded eyes. Such eyes constitute a substantial connection and are required to resist a substantial tension load.

SECTION AE605
TIES, MATERIALS AND INSTALLATION

AE605.1 General. Steel strapping, cable, chain or other approved materials shall be used for ties. All ties shall be fastened to ground anchors and drawn tight with turnbuckles or other adjustable tensioning devices or devices supplied with the ground anchor. Tie materials shall be capable of resisting an allowable working load of 3,150 pounds (14 kN) with no more than 2 percent elongation and shall withstand a 50 percent overload [4,750 pounds (21 kN)]. Ties shall comply with the weathering requirements of Section AE604.3. Ties shall connect the ground anchor and the main structural frame. Ties shall not connect to steel outrigger beams which fasten to and intersect the main structural frame unless specifically stated in the manufacturer's installation instructions. Connection of cable ties to main frame members shall be $^5/_8$-inch (15.9 mm) closed-eye bolts affixed to the frame member in an approved manner. Cable ends shall be secured with at least two U-bolt cable clamps with the "U" portion of the clamp installed on the short (dead) end of the cable to assure strength equal to that required by this section.

Wood floor support systems shall be fixed to perimeter foundation walls in accordance with provisions of this code. The minimum number of ties required per side shall be sufficient to resist the wind load stated in this code. Ties shall be evenly spaced as practicable along the length of the manufactured home with the distance from each end of the home and the tie nearest that end not exceeding 8 feet (2438 mm). When continuous straps are provided as vertical ties, such ties shall be positioned at rafters and studs. Where a vertical tie and diagonal tie are located at the same place, both ties may be connected to a single anchor, provided the anchor used is capable of carrying both loadings. Multisection manufactured homes require diagonal ties only. Diagonal ties shall be installed on the exterior main frame and slope to the exterior at an angle of 40 to 50 degrees from the vertical or within the angle limitations specified by the home manufacturer. Vertical ties which are not continuous over the top of the manufactured home shall be attached to the main frame.

❖ All tie material used must meet the minimum criteria found in this section. The fact that it does meet the criteria should be a part of the information submitted by the proponent of an approved evaluation by a third party. Tie systems are commonly pre-tested systems that are expressly manufactured for this purpose.

Bibliography

ASTM C270-99, *Standard Specification for Mortar for Unit Masonry.* West Conshohocken, PA: ASTM International, 1999.

National Manufactured Housing Construction and Safety Standards Act of 1974. (42 United States Code, Section 5401, et seq.).

NFPA 501A-92, *Standard on Manufactured Housing.* Quincy, MA: National Fire Protection Association, 1992.

United States Department of Housing and Urban Development. *Permanent Foundations for Manufactured Housing.* Draft. Washington, D.C.: United States Department of Housing and Urban Development, 1984.

Appendix F: Radon Control Methods

General Comments

Radon is a radioactive gas that has been identified as a cancer-causing agent. According to the Environmental Protection Agency (EPA), it is estimated to cause many thousands of deaths each year and increases the potential for lung cancer. Radon comes from the natural (radioactive) breakdown of uranium in soil, rock, and water and finds its way into the air. The primary concern of this appendix chapter is the transfer of radon gases from the soil into the dwelling through openings in the floor system.

The provisions of this chapter regulate the design and construction of radon-resistant measures intended to reduce the entry of radon gases into the living space of residential buildings.

Section AF101 establishes the scope to the appendix chapter, Section AF102 defines the specific terms related to the chapter, and Section AF103 discusses the construction techniques for radon-resistant construction.

Purpose

In the case of residential construction, radon is created in the soil beneath the house. Varying from one area of the United States to another, even from one house to another, the amount of radon gas in the soil is based on the soil chemistry. Since the movement of radon from the soil into the living area of a residence is enhanced as the house warms, the areas of high radon potential are typically found in portions of the United States with colder climates. The construction of an effective and efficient radon mitigation system is necessary where the radon potential reaches a point considered unacceptable. This appendix chapter establishes prescriptive provisions to reduce the amount of radon entering a dwelling unit from the soil beneath the residence.

SECTION AF101 SCOPE

AF101.1 General. This appendix contains requirements for new construction in jurisdictions where radon-resistant construction is required.

Inclusion of this appendix by jurisdictions shall be determined through the use of locally available data or determination of Zone 1 designation in Figure AF101.

❖ When adopted by the jurisdiction, the provisions of this appendix chapter provide regulations for radon-resistant construction. The jurisdiction may choose to adopt this chapter based on available data, or, alternatively, through designation as a Zone 1 structure based on Figure AF101. Zone 1 areas have a relatively high potential for radon contamination, deemed to measure at more than 4 pCi/L.

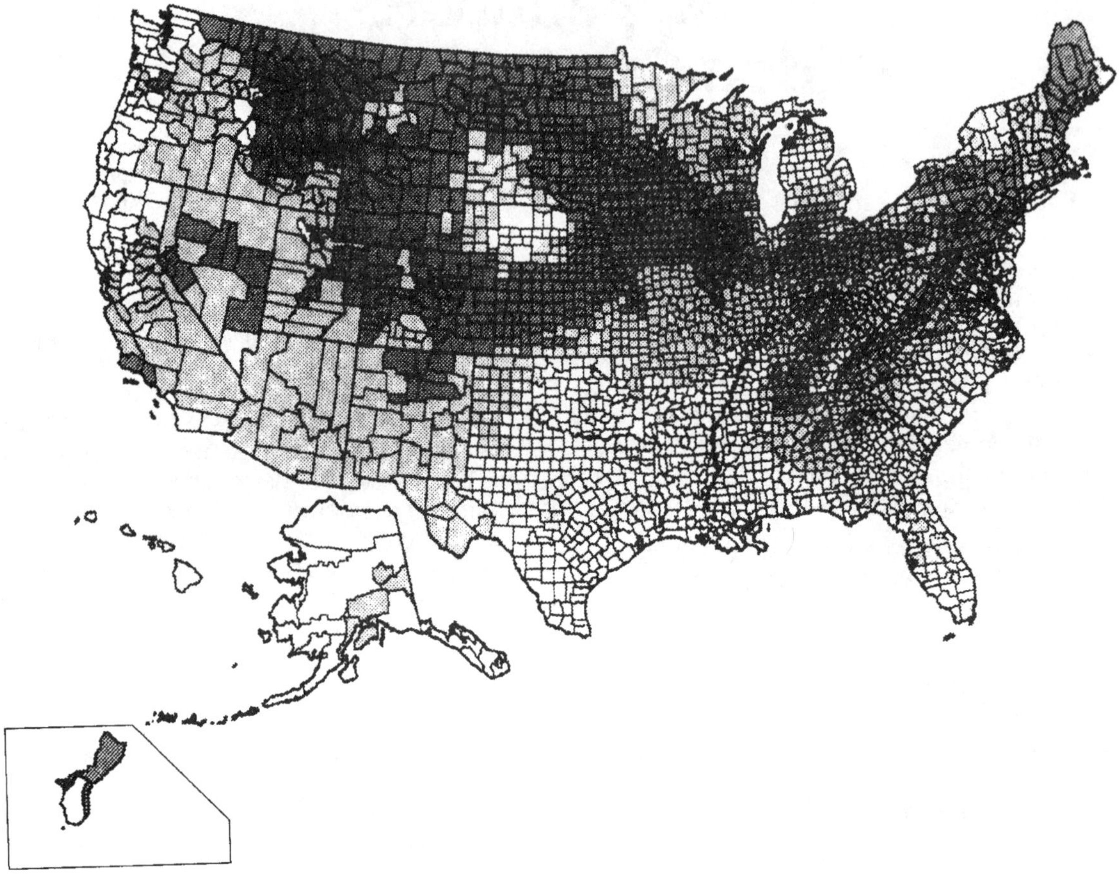

LEGEND

■	ZONE 1 HIGH POTENTIAL (GREATER THAN 4 pCi/L[a])
▦	ZONE 2 MODERATE POTENTIAL (FROM 2 TO 4 pCi/L)
□	ZONE 3 LOW POTENTIAL (LESS THAN 2 pCi/L)

a. pCi/L standard for picocuries per liter of radon gas. EPA recommends that all homes that measure 4 pCi/L and greater be mitigated.

The United States Environmental Protection Agency and the United States Geological Survey have evaluated the radon potential in the United States and have developed a map of radon zones designed to assist building building officials in deciding whether radon-resistant features are applicable in new construction.

The map assigns each of the 3,141 counties in the United States to one of three zones based on radon potential. Each zone designation reflects the average short-term radon measurement that can be expected to be measured in a building without the implementation of radon control methods. The radon zone designation of highest priority is Zone 1. Table 1 of this appendix lists the Zone 1 counties illustrated on the map. More detailed information can be obtained from state-specific booklets (EPA-402-R-93-021 through 070) available through State Radon Offices or from U.S. EPA Regional Offices.

FIGURE AF101
EPA MAP OF RADON ZONES

❖ The map assigns each of the 3,141 counties in the United States to one of three zones (Zone 1, Zone 2 or Zone 3), based on radon potential. Each zone designation reflects the expected level of radon in a building not protected by radon control methods. The short-term measurement method provides the number of picocuries per liter (pCi/L) of radon gas. Zone 3 counties have a low potential (less than 2 pCi/L), and Zone 2 counties have a moderate potential (2 pCi/L to 4 pCi/L), while Zone 1 has the highest priority based on more than 4 pCi/L.

TABLE AF101(1)
HIGH RADON POTENTIAL (ZONE 1) COUNTIES[a]

ALABAMA
Calhoun
Clay
Cleburne
Colbert
Coosa
Franklin
Jackson
Lauderdale
Lawrence
Limestone
Madison
Morgan
Talladega

CALIFORNIA
Santa Barbara
Ventura

COLORADO
Adams
Arapahoe
Baca
Bent
Boulder
Chaffee
Cheyenne
Clear Creek
Crowley
Custer
Delta
Denver
Dolores
Douglas
El Paso
Elbert
Fremont
Garfield
Gilpin
Grand
Gunnison
Huerfano
Jackson
Jefferson
Kiowa
Kit Carson
Lake
Larimer
Las Animas
Lincoln
Logan
Mesa
Moffat
Montezuma
Montrose
Morgan
Otero
Ouray
Park
Phillips
Pitkin
Prowers
Pueblo
Rio Blanco
San Miguel
Summit
Teller
Washington
Weld
Yuma

CONNECTICUT
Fairfield
Middlesex
New Haven
New London

GEORGIA
Cobb
De Kalb
Fulton
Gwinnett

IDAHO
Benewah
Blaine
Boise
Bonner
Boundary
Butte
Camas
Clark
Clearwater
Custer
Elmore
Fremont
Gooding
Idaho
Kootenai
Latah
Lemhi
Shoshone
Valley

ILLINOIS
Adams
Boone
Brown
Bureau
Calhoun
Carroll
Cass
Champaign
Coles
De Kalb
De Witt
Douglas
Edgar
Ford
Fulton
Greene
Grundy
Hancock
Henderson
Henry
Iroquois
Jersey
Jo Daviess
Kane
Kendall
Knox
La Salle
Lee
Livingston
Logan
Macon
Marshall
Mason
McDonough
McLean
Menard
Mercer
Morgan
Moultrie
Ogle
Peoria
Piatt
Pike
Putnam
Rock Island
Sangamon
Schuyler
Scott
Stark
Stephenson
Tazewell
Vermilion
Warren
Whiteside
Winnebago
Woodford

INDIANA
Adams
Allen
Bartholomew
Benton
Blackford
Boone
Carroll
Cass
Clark
Clinton
De Kalb
Decatur
Delaware
Elkhart
Fayette
Fountain
Fulton
Grant
Hamilton
Hancock
Harrison
Hendricks
Henry
Howard
Huntington
Jay
Jennings
Johnson
Kosciusko
Lagrange
Lawrence
Madison
Marion
Marshall
Miami
Monroe
Montgomery
Noble
Orange
Putnam
Randolph
Rush
Scott
Shelby
Steuben
St. Joseph
Tippecanoe
Tipton
Union
Vermillion
Wabash
Warren
Washington
Wayne
Wells
White
Whitley

IOWA
All Counties

KANSAS
Atchison
Barton
Brown
Cheyenne
Clay
Cloud
Decatur
Dickinson
Douglas
Ellis
Ellsworth
Finney
Ford
Geary
Gove
Graham
Grant
Gray
Greeley
Hamilton
Haskell
Hodgeman
Jackson
Jewell
Johnson
Kearny
Kingman
Kiowa
Lane
Leavenworth
Lincoln
Logan
Marion
Marshall
McPherson
Meade
Mitchell
Nemaha
Ness
Norton
Osborne
Ottawa
Pawnee
Phillips
Pottawatomie
Pratt
Rawlins
Republic
Rice
Riley
Rooks
Rush
Russell
Saline
Scott
Sheridan
Sherman
Smith
Stanton
Thomas
Trego
Wallace
Washington
Wichita
Wyandotte

KENTUCKY
Adair
Allen
Barren
Bourbon
Boyle
Bullitt
Casey
Clark
Cumberland
Fayette
Franklin
Green
Harrison
Hart
Jefferson
Jessamine
Lincoln
Marion
Mercer
Metcalfe
Monroe
Nelson
Pendleton
Pulaski
Robertson
Russell
Scott
Taylor
Warren
Woodford

MAINE
Androscoggin
Aroostook
Cumberland
Franklin
Hancock
Kennebec
Lincoln
Oxford
Penobscot
Piscataquis
Somerset
York

MARYLAND
Baltimore
Calvert
Carroll
Frederick
Harford
Howard
Montgomery
Washington

MASS.
Essex
Middlesex
Worcester

MICHIGAN
Branch
Calhoun
Cass
Hillsdale
Jackson
Kalamazoo
Lenawee
St. Joseph
Washtenaw

MINNESOTA
Becker
Big Stone
Blue Earth
Brown
Carver
Chippewa
Clay
Cottonwood
Dakota
Dodge
Douglas
Faribault
Fillmore
Freeborn
Goodhue
Grant
Hennepin
Houston
Hubbard
Jackson
Kanabec
Kandiyohi
Kittson
Lac Qui Parle
Le Sueur
Lincoln
Lyon
Mahnomen
Marshall
Martin
McLeod
Meeker
Mower
Murray
Nicollet
Nobles
Norman
Olmsted
Otter Tail
Pennington
Pipestone
Polk
Pope
Ramsey
Red Lake
Redwood
Renville
Rice
Rock
Roseau
Scott
Sherburne
Sibley
Stearns
Steele
Stevens
Swift
Todd
Traverse
Wabasha
Wadena
Waseca
Washington
Watonwan
Wilkin
Winona
Wright
Yellow Medicine

MISSOURI
Andrew
Atchison
Buchanan
Cass
Clay
Clinton
Holt
Iron
Jackson
Nodaway
Platte

MONTANA
Beaverhead
Big Horn
Blaine
Broadwater
Carbon
Carter
Cascade
Chouteau
Custer
Daniels
Dawson
Deer Lodge
Fallon
Fergus
Flathead
Gallatin
Garfield
Glacier
Granite
Hill
Jefferson
Judith Basin
Lake
Lewis and Clark
Liberty
Lincoln
Madison
McCone
Meagher
Mineral
Missoula
Park
Phillips
Pondera
Powder River
Powell
Prairie
Ravalli
Richland
Roosevelt
Rosebud
Sanders
Sheridan
Silver Bow
Stillwater
Teton
Toole
Valley
Wibaux
Yellowstone
National Park

NEBRASKA
Adams
Boone
Boyd
Burt
Butler
Cass
Cedar
Clay
Colfax
Cuming
Dakota
Dixon
Dodge
Douglas
Fillmore
Franklin
Frontier
Furnas
Gage
Gosper
Greeley
Hamilton
Harlan
Hayes
Hitchcock

a. EPA recommends that this county listing be supplemented with other available State and local data to further understand the radon potential of Zone 1 areas.

(continued)

TABLE AF101(1)—continued
HIGH RADON POTENTIAL (ZONE 1) COUNTIES[a]

Hurston
Jefferson
Johnson
Kearney
Knox
Lancaster
Madison
Nance
Nemaha
Nuckolls
Otoe
Pawnee
Phelps
Pierce
Platte
Polk
Red Willow
Richardson
Saline
Sarpy
Saunders
Seward
Stanton
Thayer
Washington
Wayne
Webster
York

NEW HAMPSHIRE
Carroll

NEW JERSEY
Hunterdon
Mercer
Monmouth
Morris
Somerset
Sussex
Warren

NEW MEXICO
Bernalillo
Colfax
Mora
Rio Arriba
San Miguel
Santa Fe
Taos

NEVADA
Carson City
Douglas
Eureka
Lander
Lincoln
Lyon
Mineral
Pershing
White Pine

NEW YORK
Albany
Allegheny
Broome
Cattaraugus
Cayuga
Chautauqua
Chemung
Chenango
Columbia
Cortland
Delaware
Dutchess
Erie
Genesee
Greene
Livingston
Madison
Onondaga
Ontario
Orange
Otsego
Putnam
Rensselaer
Schoharie
Schuyler
Seneca
Steuben
Sullivan
Tioga
Tompkins
Ulster
Washington
Wyoming
Yates

N. CAROLINA
Allegheny
Buncombe
Cherokee
Henderson
Mitchell
Rockingham
Transylvania
Watauga

N. DAKOTA
All Counties

OHIO
Adams
Allen
Ashland
Auglaize
Belmont
Butler
Carroll
Champaign
Clark
Clinton
Columbiana
Coshocton
Crawford
Darke
Delaware
Fairfield
Fayette
Franklin
Greene
Guernsey
Hamilton
Hancock
Hardin
Harrison
Holmes
Huron
Jefferson
Knox
Licking
Logan
Madison
Marion
Mercer
Miami
Montgomery
Morrow
Muskingum
Perry
Pickaway
Pike
Preble
Richland
Ross
Seneca
Shelby
Stark
Summit
Tuscarawas
Union
Van Wert
Warren
Wayne
Wyandot

PENNSYLVANIA
Adams
Allegheny
Armstrong
Beaver
Bedford
Berks
Blair
Bradford
Bucks
Butler
Cameron
Carbon
Centre
Chester
Clarion
Clearfield
Clinton
Columbia
Cumberland
Dauphin
Delaware
Franklin
Fulton
Huntingdon
Indiana
Juniata
Lackawanna
Lancaster
Lebanon
Lehigh
Luzerne
Lycoming
Mifflin
Monroe
Montgomery
Montour
Northampton
Northumberland
Perry
Schuylkill
Snyder
Sullivan
Susquehanna
Tioga
Union
Venango
Westmoreland
Wyoming
York

RHODE ISLAND
Kent
Washington

S. CAROLINA
Greenville

S. DAKOTA
Aurora
Beadle
Bon Homme
Brookings
Brown
Brule
Buffalo
Campbell
Charles Mix
Clark
Clay
Codington
Corson
Davison
Day
Deuel
Douglas
Edmunds
Faulk
Grant
Hamlin
Hand
Hanson
Hughes
Hutchinson
Hyde
Jerauld
Kingsbury
Lake
Lincoln
Lyman
Marshall
McCook
McPherson
Miner
Minnehaha
Moody
Perkins
Potter
Roberts
Sanborn
Spink
Stanley
Sully
Turner
Union
Walworth
Yankton

TENNESEE
Anderson
Bedford
Blount
Bradley
Claiborne
Davidson
Giles
Grainger
Greene
Hamblen
Hancock
Hawkins
Hickman
Humphreys
Jackson
Jefferson
Knox
Lawrence
Lewis
Lincoln
Loudon
Marshall
Maury
McMinn
Meigs
Monroe
Moore
Perry
Roane
Rutherford
Smith
Sullivan
Trousdale
Union
Washington
Wayne
Williamson
Wilson

UTAH
Carbon
Duchesne
Grand
Piute
Sanpete
Sevier
Uintah

VIRGINIA
Allegheny
Amelia
Appomattox
Augusta
Bath
Bland
Botetourt
Bristol
Brunswick
Buckingham
Buena Vista
Campbell
Chesterfield
Clarke
Clifton Forge
Covington
Craig
Cumberland
Danville
Dinwiddie
Fairfax
Falls Church
Fluvanna
Frederick
Fredericksburg
Giles
Goochland
Harrisonburg
Henry
Highland
Lee
Lexington
Louisa
Martinsville
Montgomery
Nottoway
Orange
Page
Patrick
Pittsylvania
Powhatan
Pulaski
Radford
Roanoke
Rockbridge
Rockingham
Russell
Salem
Scott
Shenandoah
Smyth
Spotsylvania
Stafford
Staunton
Tazewell
Warren
Washington
Waynesboro
Winchester
Wythe

WASHINGTON
Clark
Ferry
Okanogan
Pend Oreille
Skamania
Spokane
Stevens

W. VIRGINIA
Berkeley
Brooke
Grant
Greenbrier
Hampshire
Hancock
Hardy
Jefferson
Marshall
Mercer
Mineral
Monongalia
Monroe
Morgan
Ohio
Pendleton
Pocahontas
Preston
Summers
Wetzel

WISCONSIN
Buffalo
Crawford
Dane
Dodge
Door
Fond du Lac
Grant
Green
Green Lake
Iowa
Jefferson
Lafayette
Langlade
Marathon
Menominee
Pepin
Pierce
Portage
Richland
Rock
Shawano
St. Croix
Vernon
Walworth
Washington
Waukesha
Waupaca
Wood

WYOMING
Albany
Big Horn
Campbell
Carbon
Converse
Crook
Fremont
Goshen
Hot Springs
Johnson
Laramie
Lincoln
Natrona
Niobrara
Park
Sheridan
Sublette
Sweetwater
Teton
Uinta
Washakie

❖ This table identifies those counties within the United States that have been designated as Zone 1 radon zones. These counties have enough radon potential that mitigation measures are deemed necessary. In addition to those counties listed on the table, other jurisdictions may desire to implement radon control measures based upon local or state data.

a. EPA recommends that this county listing be supplemented with other available State and local data to further understand the radon potential of Zone 1 areas.

SECTION AF102
DEFINITIONS

AF102.1 General. For the purpose of these requirements, the terms used shall be defined as follows:

❖ This section clarifies the terminology used in this appendix chapter. The terms take on unique and specific meanings, with many of the terms used solely in the context of radon-resistant construction.

SUB-SLAB DEPRESSURIZATION SYSTEM (Passive). A system designed to achieve lower sub-slab air pressure relative to indoor air pressure by use of a vent pipe routed through the conditioned space of a building and connecting the sub-slab area with outdoor air, thereby relying on the convective flow of air upward in the vent to draw air from beneath the slab.

❖ One of several methods available for mitigating radon entry into a dwelling unit, this passive system uses the convective movement of air to remove radon from the area below the slab. Illustrated in Figure AF102, this method uses a vertical vent pipe between the sub-slab area and the exterior of the building to draw air from the sub-slab area to the outside.

SUB-SLAB DEPRESSURIZATION SYSTEM (Active). A system designed to achieve lower sub-slab air pressure relative to indoor air pressure by use of a fan-powered vent drawing air from beneath the slab.

❖ This method is similar to the passive sub-slab depressurization system. However, in this case a fan-powered vent is used to pull the air from beneath the slab and to the exterior. As it is with the passive system, the intent is to reduce the air pressure beneath the slab to a lower level than the indoor air pressure.

DRAIN TILE LOOP. A continuous length of drain tile o perforated pipe extending around all or part of the internal or external perimeter of a basement or crawl space footing.

❖ Much like a drainage system for moving water away from a foundation or basement wall, a drain tile loop can consist of either drain tile or perforated pipe. It can be located on the interior or exterior side of a crawl space or basement footing.

RADON GAS. A naturally-occurring, chemically inert, radioactive gas that is not detectable by human senses. As a gas, it can move readily through particles of soil and rock and can accumulate under the slabs and foundations of homes where it can easily enter into the living space through construction cracks and openings.

❖ A radioactive gas, radon occurs naturally. It is not detectable by sight, smell, or other human senses and can enter a building's interior by passing from the soil through small cracks and openings in the foundation and slab construction.

SOIL-GAS-RETARDER. A continuous membrane of 6-mil (0.15 mm) polyethylene or other equivalent material used to retard the flow of soil gases into a building.

❖ 6-mil (0.15 mm) polyethylene is specifically identified as an acceptable material for retarding the flow of gas from the soil into a structure, if the polyethylene is applied in a continuous manner. Any equivalent membrane material is also a soil-gas retarder provided it accomplishes an equal or better result.

SUB-MEMBRANE DEPRESSURIZATION SYSTEM. A system designed to achieve lower-sub-membrane air pressure relative to crawl space air pressure by use of a vent drawing air from beneath the soil-gas-retarder membrane.

❖ Where a crawl space is present rather than a slab on grade, this system can be used in much the same manner a sub-slab depressurization system. Shown in Figure AF102, this method draws air from beneath the soil-gas retarder membrane and vents it to the exterior of the building.

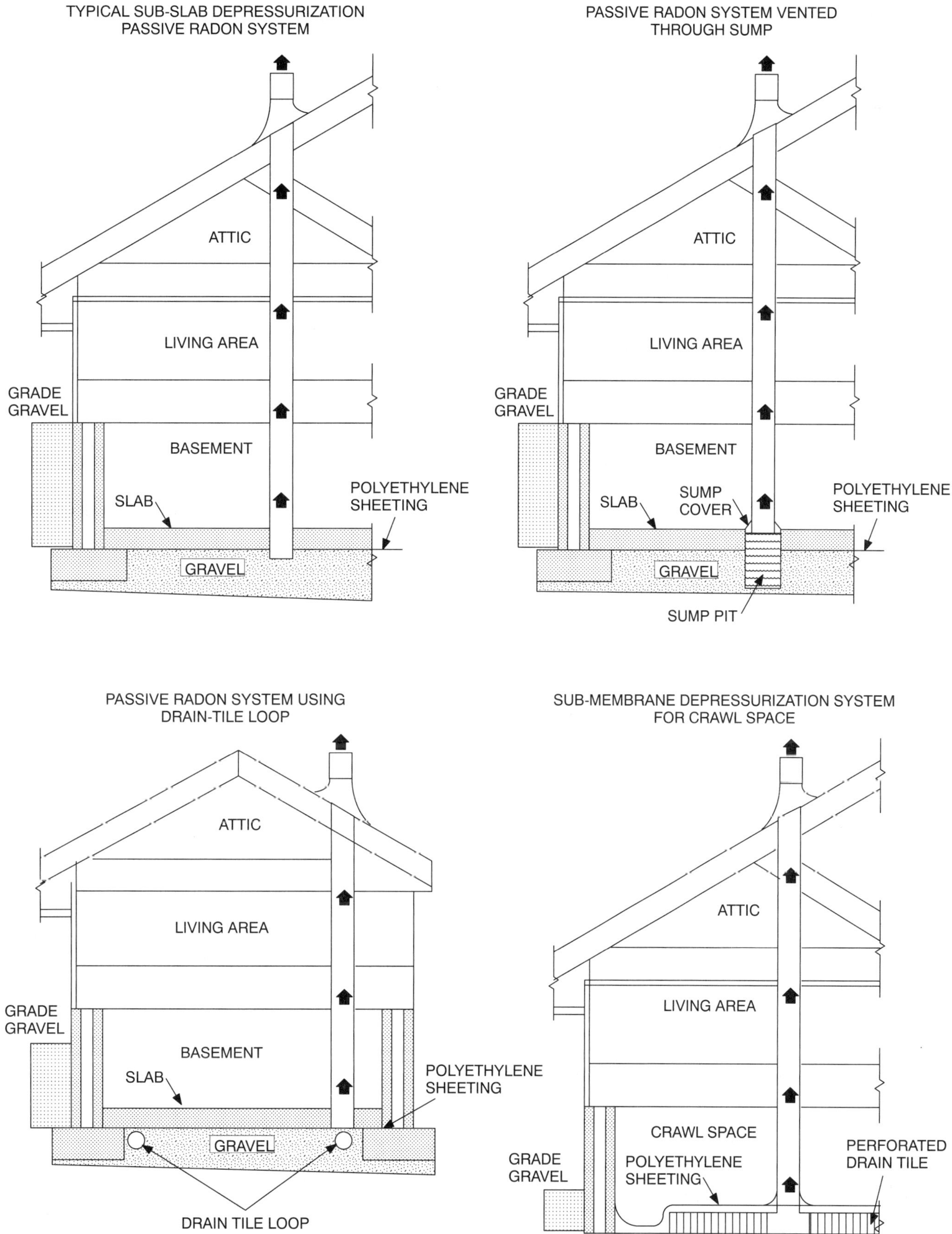

FIGURE AF102
RADON-RESISTANT CONSTRUCTION DETAILS FOR FOUR FOUNDATION TYPES

❖ The illustrations in this figure depict various methods of radon removal based on the type of foundation under consideration. The different conditions illustrated include (1) a basement slab scenario with a passive system vented through a sump, (2) a crawl space design with the use of perforated drain tile, (3) a basement slab condition with a passive sub-slab pressurization system, and (4) a basement slab floor with a drain tile loop. Section AF103 contains more specific details of construction.

SECTION AF103 REQUIREMENTS

AF103.1 General. The following construction techniques are intended to resist radon entry and prepare the building for post-construction radon mitigation, if necessary (see Figure AF102). These techniques are required in areas where designated by the jurisdiction.

❖ This section sets forth a number of construction methods designed to reduce the potential for radon movement from the soil to the interior of the building. It is up to the jurisdiction to determine which methods are necessary based upon the specific potential for radon. Four drawings are referenced in Figure AF102 that illustrate several basic construction methods for radon mitigation.

AF103.2 Subfloor preparation. A layer of gas-permeable material shall be placed under all concrete slabs and other floor systems that directly contact the ground and are within the walls of the living spaces of the building, to facilitate future installation of a sub-slab depressurization system, if needed. The gas-permeable layer shall consist of one of the following:

1. A uniform layer of clean aggregate, a minimum of 4 inches (102 mm) thick. The aggregate shall consist of material that will pass through a 2-inch (51 mm) sieve and be retained by a $^{1}/_{4}$-inch (6.4 mm) sieve.
2. A uniform layer of sand (native or fill), a minimum of 4 inches (102 mm) thick, overlain by a layer or strips of geotextile drainage matting designed to allow the lateral flow of soil gases.
3. Other materials, systems or floor designs with demonstrated capability to permit depressurization across the entire sub-floor area.

❖ Where a sub-slab depressurization system is to be used, it is necessary to correctly prepare the base prior to installation of a soil-gas-retarder. The use of aggregate, sand overlaid with drainage matting, or any other approved gas-permeable material or system is required beneath all concrete slabs or similar floor systems in direct contact with the ground. The layer of material must be located so that it protects all interior floor areas of the building. Where aggregate or sand are used, they must be placed in a uniform layer at least 4 inches (102 mm) thick. The sand must also be overlaid with a drainage matting material that will assist in causing the soil gases to travel in a lateral direction.

AF103.3 Soil-gas-retarder. A minimum 6-mil (0.15 mm) [or 3-mil (0.075 mm) cross-laminated] polyethylene or equivalent flexible sheeting material shall be placed on top of the gas-permeable layer prior to casting the slab or placing the floor assembly to serve as a soil-gas-retarder by bridging any cracks that develop in the slab or floor assembly and to prevent concrete from entering the void spaces in the aggregate base material. The sheeting shall cover the entire floor area with separate sections of sheeting lapped at least 12 inches (305 mm). The sheeting shall fit closely around any pipe, wire or other penetrations of the material. All punctures or tears in the material shall be sealed or covered with additional sheeting.

❖ An acceptable sheeting material must be installed on top of the gas-permeable base layer. The material is to be a minimum 6-mil (0.15 mm) polyethylene membrane or any other flexible sheeting that provides equivalent protection. The soil-gas-retarder resists the vertical flow of radon gas into the slab or other type of floor assembly. Therefore, the membrane must cover the entire floor area of the building, with joints adequately lapped and penetrations tightly sealed. Any tears, rips, or punctures are to be adequately repaired with additional sheeting material.

AF103.4 Entry routes. Potential radon entry routes shall be closed in accordance with Sections AF103.4.1 through AF103.4.10.

❖ This section identifies the various points at which radon may enter the building and specifies the appropriate methods for sealing or otherwise protecting the potential entry routes.

AF103.4.1 Floor openings. Openings around bathtubs, showers, water closets, pipes, wires or other objects that penetrate concrete slabs or other floor assemblies shall be filled with a polyurethane caulk or equivalent sealant applied in accordance with the manufacturer's recommendations.

❖ It is typical for a floor slab or other type of floor assembly to be penetrated by underslab or underfloor plumbing, mechanical, and electrical components. Polyurethane caulk or an equivalent sealant material must be installed at all penetrations created by the passage of piping, vents, conduit, cable, or other items penetrating the floor. The sealant is to be installed in accordance with the recommendations of the manufacturer.

AF103.4.2 Concrete joints. All control joints, isolation joints, construction joints and any other joints in concrete slabs or between slabs and foundation walls shall be sealed with a caulk or sealant. Gaps and joints shall be cleared of loose material and filled with polyurethane caulk or other elastomeric sealant applied in accordance with the manufacturer's recommendations.

❖ Where joints occur within a concrete slab floor or at the point where the floor intersects the foundation walls, they must be filled with an appropriate caulk or sealant material. All loose material must be removed prior to installation of the sealant material. Polyurethane caulk or any other type of elastomeric sealant will allow for slight movement of the penetrating items without reducing the integrity of the seal. The manufacturer's installation instructions should be consulted for application procedures.

AF103.4.3 Condensate drains. Condensate drains shall be trapped or routed through nonperforated pipe to daylight.

❖ Condensate drains that pass through the floor system are to be run to the exterior using non-perforated pipe or must be provided with an approved trap.

AF103.4.4 Sumps. Sump pits open to soil or serving as the termination point for sub-slab or exterior drain tile loops shall be covered with a gasketed or otherwise sealed lid. Sumps used as the suction point in a sub-slab depressurization system shall have a lid designed to accommodate the vent pipe. Sumps used as a floor drain shall have a lid equipped with a trapped inlet.

❖ A gasketed or sealed lid must be provided on any sump pit that serves as the end point for a sub-slab or exterior drain tile loop system. Such a lid is also required if the sump pit is open to the soil. The sump lid must be designed to accommodate the vent pipe where the sump is used as the suction point in a sub-slab decompression system. Where used as a floor drain, the sump pit lid is to be equipped with a trapped inlet.

AF103.4.5 Foundation walls. Hollow block masonry foundation walls shall be constructed with either a continuous course of solid masonry, one course of masonry grouted solid, or a solid concrete beam at or above finished ground surface to prevent passage of air from the interior of the wall into the living space. Where a brick veneer or other masonry ledge is installed, the course immediately below that ledge shall be sealed. Joints, cracks or other openings around all penetrations of both exterior and interior surfaces of masonry block or wood foundation walls below the ground surface shall be filled with polyurethane caulk or equivalent sealant. Penetrations of concrete walls shall be filled.

❖ Where the foundation is made up of hollow masonry units, it is necessary to provide a means to prohibit the flow of air and potential soil gas within the cavities of the block masonry. Several methods are identified, including the use of solid masonry or solid-grouted masonry for a minimum of one course. The solid barrier must be located at or directly above the finished ground surface to prevent gases that enter the wall cavity from traveling up and into the living spaces. In those situations where a ledge for brick, stone, or other masonry material is provided, the barrier must be located directly below the ledge.

All penetrations, joints, cracks, and other openings that occur below ground level in the masonry or wood foundation walls are to be filled with polyurethane caulk or a similar type of flexible sealant. Penetrations in concrete foundation walls must also be appropriately filled. The required penetration and opening protection must be provided on both the interior and exterior sides of the foundation walls.

AF103.4.6 Dampproofing. The exterior surfaces of portions of concrete and masonry block walls below the ground surface shall be dampproofed in accordance with Section R406 of this code.

❖ Dampproofing of the exterior surfaces of concrete and masonry block walls located below ground level must be done in accordance with the provisions of Section R406.1. A variety of methods are established for dampproofing concrete and masonry foundations.

AF103.4.7 Air-handling units. Air-handling units in crawl spaces shall be sealed to prevent air from being drawn into the unit.

Exception: Units with gasketed seams or units that are otherwise sealed by the manufacturer to prevent leakage.

❖ Unless sealed by the manufacturer or provided with gasketed seams to prevent leakage, air-handling units located in crawl spaces must be field-sealed to eliminate the potential for air and gas to be drawn into the unit and distributed throughout the building.

AF103.4.8 Ducts. Ductwork passing through or beneath a slab shall be of seamless material unless the air-handling system is designed to maintain continuous positive pressure within such ducting. Joints in such ductwork shall be sealed to prevent air leakage.

Ductwork located in crawl spaces shall have all seams and joints sealed by closure systems in accordance with Section M1601.3.1.

❖ Where ductwork passes through or is installed beneath a concrete floor slab, the ducts must be free of seams that may allow air and gas to enter the duct system. Seams are only permitted where it can be demonstrated that the air-handling equipment will maintain continuous positive pressure within the ducting. It such situations, the seams must be sealed to eliminate any air leakage.

The provision allows ductwork passing through a crawl space to have seams and joints provided they are sealed by one of the methods prescribed in Section M1601.3.1. This will allow the use of fibrous glass and seamed metal ducts and field fabricated ductwork.

AF103.4.9 Crawl space floors. Openings around all penetrations through floors above crawl spaces shall be caulked or otherwise filled to prevent air leakage.

❖ Air leakage is to be prevented between a crawl space and the floor above by the sealing of all penetrations and openings. Caulking or other approved materials are acceptable for providing the necessary airtight separation.

AF103.4.10 Crawl space access. Access doors and other openings or penetrations between basements and adjoining crawl spaces shall be closed, gasketed or otherwise filled to prevent air leakage.

❖ The provisions of Section R408.3 mandate a minimum of one 18-inch by 24-inch (457 mm by 610 mm) opening to access a crawl space. Section M1305.1.4 addresses access to under-floor mechanical equipment. Where such openings or any other access points to the crawl space are provided, the doors or panels must be closed and gasketed to create an airtight separation.

AF103.5 Passive sub-membrane depressurization system. In buildings with crawl space foundations, the following components of a passive sub-membrane depressurization system shall be installed during construction.

Exception: Buildings in which an approved mechanical crawl space ventilation system or other equivalent system is installed.

❖ In those buildings having a floor system that includes a crawl space, the installation of a passive sub-membrane depressurization system requires a number of components, which are identified and regulated by this section.

AF103.5.1 Ventilation. Crawl spaces shall be provided with vents to the exterior of the building. The minimum net area of ventilation openings shall comply with Section R408.1 of this code.

❖ Crawl spaces are to be naturally ventilated directly to the exterior of the building. The provisions of Section R408.1 for under-floor ventilation are applicable only to the extent of determining the minimum required net ventilating area.

AF103.5.2 Soil-gas-retarder. The soil in crawl spaces shall be covered with a continuous layer of minimum 6-mil (0.15 mm) polyethylene soil-gas-retarder. The ground cover shall be lapped a minimum of 12 inches (305 mm) at joints and shall extend to all foundation walls enclosing the crawl space area.

❖ An acceptable sheeting material must be installed on top of a gas-permeable base layer. The material is to be a minimum 6-mil (0.15 mm) polyethylene membrane or any other flexible sheeting that provides equivalent protection. The soil-gas-retarder resists the vertical flow of radon gas into the slab or other type of floor assembly. Therefore, the membrane must cover the entire floor area of the building, with joints adequately lapped and penetrations tightly sealed. Any tears, rips, or punctures are to be adequately repaired with additional sheeting material.

AF103.5.3 Vent pipe. A plumbing tee or other approved connection shall be inserted horizontally beneath the sheeting and connected to a 3- or 4-inch-diameter (76 mm or 102 mm) fitting with a vertical vent pipe installed through the sheeting. The vent pipe shall be extended up through the building floors, terminate at least 12 inches (305 mm) above the roof in a location at least10 feet (3048 mm) away from any window or other opening into the conditioned spaces of the building that is less than 2 feet (610 mm) below the exhaust point, and 10 feet (3048 mm) from any window or other opening in adjoining or adjacent buildings.

❖ Prior to placement of the concrete floor slab, at least one radon vent pipe is to be embedded vertically into the aggregate or other approved permeable material. Each pipe must be at least 3 inches (76 mm) in diameter and constructed of a gas-tight material such as PVC or ABS. An acceptable method of permanency is necessary so that the vent pipe is not moved from its location within the sub-slab material. This can be accomplished by providing a tee fitting or any other approved equivalent method, or by insertion of the pipe directly into an interior perimeter drain tile loop. It is also acceptable to attach the vent pipe directly to a sealed sump provided the sump is exposed to the sub-slab aggregate or connected to the aggregate area by a drainage system.

The vent pipe must extend vertically through the building until it terminates at least 12 inches (305 mm) above the roof. In all cases, the termination point must be at least 10 feet (3048 mm) from any building opening that is located less than 2 feet (610 mm) below the exhaust point. The exhaust point also cannot be located within 10 feet (3048 mm) of any openings in nearby buildings.

AF103.6 Passive sub-slab depressurization system. In basement or slab-on-grade buildings, the following components of a passive sub-slab depressurization system shall be installed during construction.

❖ In those buildings having a slab-on-grade floor system or a basement, the installation of a passive sub-slab depressurization system requires a number of components, which are identified and regulated by this section.

AF103.6.1 Vent pipe. A minimum 3-inch-diameter (76 mm) ABS, PVC or equivalent gas-tight pipe shall be embedded vertically into the sub-slab aggregate or other permeable material before the slab is cast. A "T" fitting or equivalent method shall be used to ensure that the pipe opening remains within the sub-slab permeable material. Alternatively, the 3-inch (76 mm) pipe shall be inserted directly into an interior perimeter drain tile loop or through a sealed sump cover where the sump is exposed to the sub-slab aggregate or connected to it through a drainage system.

The pipe shall be extended up through the building floors, terminate at least 12 inches (305 mm) above the surface of the roof in a location at least 10 feet (3048 mm) away from any window or other opening into the conditioned spaces of the building that is less than 2 feet (610 mm) below the exhaust point, and 10 feet (3048 mm) from any window or other opening in adjoining or adjacent buildings.

❖ Prior to placement of the concrete floor slab, at least one radon vent pipe is to be embedded vertically into the aggregate or other approved permeable material. Each pipe must be at least 3 inches (76 mm) in diameter and constructed of a gas-tight material such as PVC or ABS. An acceptable method of permanency is necessary so that the vent pipe is not moved from its intended location within the sub-slab material. This can be accomplished by providing a tee fitting or any other approved equivalent method, or by insertion of the pipe directly into an interior perimeter drain tile loop. It is also acceptable to attach the vent pipe directly to a sealed sump provided the sump is exposed to the sub-slab aggregate or connected to the aggregate area by a drainage system.

The vent pipe must extend vertically through the building until it terminates at least 12 inches (305 mm) above the roof. In all cases, the termination point must be at least 10 feet (3048 mm) from any building opening that is located less than 2 feet (610 mm) below the exhaust point. The exhaust point also cannot be located within 10 feet (3048 mm) of any openings in nearby buildings.

AF103.6.2 Multiple vent pipes. In buildings where interior footings or other barriers separate the sub-slab aggregate or other gas-permeable material, each area shall be fitted with an individual vent pipe. Vent pipes shall connect to a single vent that terminates above the roof or each individual vent pipe shall terminate separately above the roof.

❖ An individual vent pipe is required for each unique under-slab area, such as those spaces separated by interior footings. The vent pipes may terminate individually above the roof or may be connect to a single vent.

AF103.7 Vent pipe drainage. All components of the radon vent pipe system shall be installed to provide positive drainage to the ground beneath the slab or soil-gas-retarder.

❖ The manner of installation of a radon vent pipe system must be such that positive drainage is created to the ground beneath the floor slab or soil-gas-retarder.

AF103.8 Vent pipe accessibility. Radon vent pipes shall be accessible for future fan installation through an attic or other area outside the habitable space.

Exception: The radon vent pipe need not be accessible in an attic space where an approved roof-top electrical supply is provided for future use.

❖ In order to accommodate the future installation of equipment to provide for a fan-powered vent system, access to radon vent pipes must be provided through the attic or other nonhabitable space. If an approved rooftop electrical supply is provided, the access from the building's interior is not necessary.

AF103.9 Vent pipe identification. All exposed and visible interior radon vent pipes shall be identified with at least one label on each floor and in accessible attics. The label shall read: "Radon Reduction System."

❖ Interior vent pipes installed as a portion of the radon venting system must be adequately identified to reduce the potential for improper use or modification of the venting system. The identification is required for every floor level and in all accessible attics where the radon vents are exposed and visible. At a minimum, the identification label must state: "Radon Reduction System."

AF103.10 Combination foundations. Combination basement/crawl space or slab-on-grade/crawl space foundations shall have separate radon vent pipes installed in each type of foundation area. Each radon vent pipe shall terminate above the roof or shall be connected to a single vent that terminates above the roof.

❖ Where the design of the structure combines a basement with a crawl space foundation, or a slab-on-grade floor with a crawl space foundation, separate radon vent pipes are to be provided for each individual type of foundation system. The vent piping must extend above the roof, either as individual vent terminations or as a single vent termination connected to the multiple vents.

AF103.11 Building depressurization. Joints in air ducts and plenums in unconditioned spaces shall meet the requirements of Section M1601. Thermal envelope air infiltration requirements shall comply with the energy conservation provisions in Chapter 11. Firestopping shall meet the requirements contained in Section R602.8.

❖ In those situations where other provisions of the code apply to the installation of a radon reduction system, they are applicable to the extent prescribed by the code. Specific reference is made to Section M1601 for joints in air ducts and plenums in unconditioned spaces, Section R602.8 for firestopping in concealed spaces, and Chapter 11 for energy conservation features.

AF103.12 Power source. To provide for future installation of an active sub-membrane or sub-slab depressurization system, an electrical circuit terminated in an approved box shall be installed during construction in the attic or other anticipated location of vent pipe fans. An electrical supply shall also be accessible in anticipated locations of system failure alarms.

❖ It is possible that a passive depressurization system will be converted to an active system at some future time. In anticipation of such an occurrence, an electrical circuit must be provided to an approved box. The box should be located in the attic or other location that provides access to the vent pipe fans. An additional electrical supply must be provided at the anticipated future locations of system failure alarms.

Appendix G: Swimming Pools, Spas and Hot Tubs

General Comments

Drowning is the second leading cause of accidental death in the home for children under five years of age. It has been the number one cause of accidental deaths in the home for that age group in a number of states, including Arizona, California, Florida, and Texas. The use of effective residential swimming pool barriers is the best way to reduce these tragic losses.

This appendix chapter sets forth the regulations for swimming pools, hot tubs, and spas. The primary focus of the provisions is the need for an effective barrier surrounding the water area to reduce the potential for young children to gain uncontrolled access.

Section AG101 establishes the scope of the chapter. Section AG102 defines those terms specific to this appendix chapter. Section AG103 identifies specification standards for the design and construction of swimming pools. Section AG104 identifies specification standards for the design and construction of spas and hot tubs. Section AG105 discusses barrier requirements for swimming pools, hot tubs, and spas. Section AG106 contains provisions for entrapment protection for suction outlets. Section AG107 indicates the abbreviations for three standards-writing organizations, and Section AG108 specifies the various standards used in this appendix chapter.

Purpose

According to the Consumer Product Safety Commission (CPSC), approximately 350 children under 5 years of age drown each year in residential swimming pools, spas, and hot tubs. A CPSC study, *Child Drowning Study: A Report on the Epidemiology of Drownings in Residential Pools of Children Under Age Five*, found that the majority of the victims lived in or were visiting the residence where the accident happened. Less than 2 percent of the drowning incidents occurred when a child trespassed on the property. For these reasons, this appendix chapter states that all swimming pools, spas, and hot tubs must be enclosed to prevent young children from gaining unsupervised access to pool areas. This chapter provides prescriptive details for the construction of enclosures around swimming pools, spas, and hot tubs to make it more difficult for children, particularly those 5 years old and younger, to enter such areas unsupervised.

SECTION AG101 GENERAL

AG101.1 General. The provisions of this appendix shall control the design and construction of swimming pools, spas and hot tubs installed in or on the lot of a one- and two-family dwelling.

❖ This section provides the scope of the appendix chapter on swimming pools, spas, and hot tubs. It regulates the design and construction of such facilities where they are located inside a dwelling unit or on the lot of a one- or two-family dwelling.

SECTION AG102 DEFINITIONS

AG102.1 General. For the purposes of these requirements, the terms used shall be defined as follows and as set forth in Chapter 2.

❖ This section clarifies the terminology used in this appendix chapter. The terms take on specific meanings, often different from the way they are typically used.

ABOVE-GROUND/ON-GROUND POOL. See "Swimming pool."

❖ These two terms have essentially the same meaning. If a side of a swimming pool projects above the adjacent ground level, the pool is referred to as an above-ground pool. If the bottom of the pool rests on the ground with no portion recessed except for leveling purposes, it is referred to as an on-ground pool. The important factor in both situations is that access to the pool surface is elevated and requires a vertical ascent (from at least one side) to gain access to the water.

A swimming pool situated on the ground or located above the ground is in the same category as other similar facilities such as spas, hot tubs, and in-ground pools. All such facilities are simply regulated as swimming pools.

BARRIER. A fence, wall, building wall or combination thereof which completely surrounds the swimming pool and obstructs access to the swimming pool.

❖ Any system of components that encloses a swimming pool to the degree that access is obstructed is a barrier. Enclosure components include the exterior wall of the dwelling unit, a fence, and any doors or gates included as a portion of the enclosure. Any construction or natural element that does not surround the pool will allow access at some point. The vast majority of provisions in this appendix chapter relate to the installation of a complying barrier to restrict access to swimming pools, spas, and hot tubs. Left unprotected, these facilities present the potential for drownings and near-drownings.

HOT TUB. See "Swimming pool."

❖ Typically regarded as a small soaking tub, a hot tub is defined as a swimming pool and is regulated in the same manner as spas and the various types of swimming pools. Hot tubs often are equipped to introduce bubbles or jets of water into the tub.

IN-GROUND POOL. See "Swimming pool."

❖ An in-ground pool is a swimming pool in which the top of the pool structure is roughly at the same elevation as the adjoining surface surrounding the pool. Unlike an above-ground or on-ground pool, the pool construction itself does not limit access to the pool.

A swimming pool constructed in the ground is in the same category as similar facilities such as spas, hot tubs, above-ground pools, and on-ground pools. All such facilities are simply regulated as swimming pools.

RESIDENTIAL. That which is situated on the premises of a detached one- or two-family dwelling or a one-family townhouse not more than three stories in height.

❖ Where a pool is located on the property of a single-family dwelling, two-family dwelling, or one-family townhouse, it is "residential." The scope of the provisions in this appendix chapter coincides with this definition.

SPA, NONPORTABLE. See "Swimming pool."

❖ Typically regarded as a whirlpool tub, a spa is defined as a swimming pool and is regulated in the same manner as hot tubs and the various types of swimming pools.

SPA, PORTABLE. A nonpermanent structure intended for recreational bathing, in which all controls, water-heating and water-circulating equipment are an integral part of the product.

❖ A nonpermanent structure, a portable spa is self-contained, with all of the controls and equipment integrated.

SWIMMING POOL. Any structure intended for swimming or recreational bathing that contains water over 24 inches (610 mm) deep. This includes in-ground, aboveground and on-ground swimming pools, hot tubs and spas.

❖ In order to be considered a swimming pool for the provisions of this appendix chapter, the structure used for swimming or recreational bathing must be more than 24 inches (610 mm) deep. Hot tubs, spas, in-ground pools, on-ground pools, and above-ground pools are included in this definition, provided they are the minimum depth prescribed.

SWIMMING POOL, INDOOR. A swimming pool which is totally contained within a structure and surrounded on all four sides by walls of said structure.

❖ Where a swimming pool is located in a enclosed structure, fully surrounded by walls, it is an indoor pool. Of critical concern is the easy access afforded to children by an indoor pool.

SWIMMING POOL, OUTDOOR. Any swimming pool which is not an indoor pool.

❖ Where a swimming pool is not fully enclosed, as is required in the definition of an indoor pool, it is an outdoor swimming pool. A pool that may be partially inside and partially outside is defined as an outdoor pool because it is not completely surrounded by a structure.

SECTION AG103 SWIMMING POOLS

AG103.1 In-ground pools. In-ground pools shall be designed and constructed in conformance with ANSI/NSPI-5 as listed in Section AG108.

❖ The requirements of ANSI/NSPI-5 regulating residential in-ground swimming pools are applicable to all in-ground pools regulated by this appendix chapter.

AG103.2 Above-ground and on-ground pools. Above-ground and on-ground pools shall be designed and constructed in conformance with ANSI/NSPI-4 as listed in Section AG108.

❖ The requirements of ANSI/NSPI-4 regulating residential above-ground and on-ground swimming pools are applicable to all such pools regulated by this appendix chapter.

SECTION AG104 SPAS AND HOT TUBS

AG104.1 Permanently installed spas and hot tubs. Permanently installed spas and hot tubs shall be designed and constructed in conformance with ANSI/NSPI-3 as listed in Section AG108.

❖ The requirements of ANSI/NSPI-3 regulating permanently installed residential spas are applicable to all nonportable spas and hot tubs.

AG104.2 Portable spas and hot tubs. Portable spas and hot tubs shall be designed and constructed in conformance with ANSI/NSPI-6 as listed in Section AG108.

❖ The requirements of ANSI/NSPI-6 regulating residential portable spas are applicable to all such spas.

SECTION AG105
BARRIER REQUIREMENTS

AG105.1 Application. The provisions of this chapter shall control the design of barriers for residential swimming pools, spas and hot tubs. These design controls are intended to provide protection against potential drownings and near-drownings by restricting access to swimming pools, spas and hot tubs.

❖ This section describes the provisions for barriers around residential swimming pools, hot tubs, and spas. A swimming pool or similar facility creates an attractive temptation to children, including very young children and infants who do not know how to swim. The installation of an effective barrier can help reduce the number of children who die or are injured as the result of open access to a swimming pool, spa, or hot tub.

AG105.2 Outdoor swimming pool. An outdoor swimming pool, including an in-ground, aboveground or on-ground pool, hot tub or spa shall be provided with a barrier which shall comply with the following:

1. The top of the barrier shall be at least 48 inches (1219 mm) above grade measured on the side of the barrier which faces away from the swimming pool. The maximum vertical clearance between grade and the bottom of the barrier shall be 2 inches (51 mm) measured on the side of the barrier which faces away from the swimming pool. Where the top of the pool structure is above grade, such as an aboveground pool, the barrier may be at ground level, such as the pool structure, or mounted on top of the pool structure. Where the barrier is mounted on top of the pool structure, the maximum vertical clearance between the top of the pool structure and the bottom of the barrier shall be 4 inches (102 mm).
2. Openings in the barrier shall not allow passage of a 4-inch diameter (102 mm) sphere.
3. Solid barriers which do not have openings, such as a masonry or stone wall, shall not contain indentations or protrusions except for normal construction tolerances and tooled masonry joints.
4. Where the barrier is composed of horizontal and vertical members and the distance between the tops of the horizontal members is less than 45 inches (1143 mm), the horizontal members shall be located on the swimming pool side of the fence. Spacing between vertical members shall not exceed 1.75 inches (44 mm) in width. Where there are decorative cutouts within vertical members, spacing within the cutouts shall not exceed 1.75 inches (44 mm) in width.
5. Where the barrier is composed of horizontal and vertical members and the distance between the tops of the horizontal members is 45 inches (1143 mm) or more, spacing between vertical members shall not exceed 4 inches (102 mm). Where there are decorative cutouts within vertical members, spacing within the cutouts shall not exceed 1.75 inches (44 mm) in width.
6. Maximum mesh size for chain link fences shall be a 2.25-inch (57 mm) square unless the fence is provided with slats fastened at the top or the bottom which reduce the openings to not more than 1.75 inches (44 mm).
7. Where the barrier is composed of diagonal members,such as a lattice fence, the maximum opening formed by the diagonal members shall not be more than 1.75 inches (44 mm).
8. Access gates shall comply with the requirements of Section AG105.2, Items 1 through 7, and shall be equipped to accommodate a locking device. Pedestrian access gates shall open outward away from the pool and shall be self-closing and have a self-latching device. Gates other than pedestrian access gates shall have a self-latching device. Where the release mechanism of the self-latching device is located less than 54 inches (1372 mm) from the bottom of the gate, the release mechanism and openings shall comply with the following:
   - 8.1. The release mechanism shall be located on the pool side of the gate at least 3 inches (76 mm) below the top of the gate, and
   - 8.2. The gate and barrier shall have no opening greater than 0.5 inch (12.7 mm) within 18 inches (457 mm) of the release mechanism.
9. Where a wall of a dwelling serves as part of the barrier one of the following conditions shall be met:
   - 9.1. The pool shall be equipped with a powered safety cover in compliance with ASTM F1346; or
   - 9.2. All doors with direct access to the pool through that wall shall be equipped with an alarm which produces an audible warning when the door and its screen, if present, are opened. The alarm shall sound continuously for a minimum of 30 seconds immediately after the door is opened and be capable of being heard throughout the house during normal household activities. The alarm shall automatically reset under all conditions. The alarm system shall be equipped with a manual means, such as touchpad or switch, to temporarily deactivate the alarm for a single opening. Such deactivation shall last for not more than 15 seconds. The deactivation switch(es) shall be located at least 54 inches (1372 mm) above the threshold of the door; or
   - 9.3. Other means of protection, such as self-closing doors with self-latching devices, which are approved by the governing body, shall be acceptable so long as the degree of protection afforded is not less than the protection afforded by Item 9.1 or 9.2 described above.
10. Where an aboveground pool structure is used as a barrier or where the barrier is mounted on top of the pool structure, and the means of access is a ladder or steps, then:
    - 10.1. The ladder or steps shall be capable of being secured, locked or removed to prevent access, or
    - 10.2. The ladder or steps shall be surrounded by a barrier which meets the requirements of Section AG105.2, Items 1 through 9. When the ladder or steps are secured, locked or removed, any opening created shall not allow the passage of a 4-inch-diameter (102 mm) sphere.

❖ This section provides prescriptive requirements for the construction of the swimming pool barrier.

1. The barrier height requirement of 48 inches (1219 mm) above the ground is based on reports that document the ability of children under the age of 5 to climb over barriers that are less than 48 inches (1219 mm) in height. The basis for the 4-inch (102 mm) criterion for an opening between the barrier and the top of the pool frame is the same as for guard construction as addressed in Section R312. Refer to Commentary Figure AG105.2(1).
2. The general provision is applicable only when one of the conditions addressed in Items 4, 5, 6, and 7 is not present. For example, a chain-link fence would be regulated by the requirements of Item 6, which reduces the general opening criterion of 4 inches (102 mm) to 2.25 inches (57 mm). The basis for the 4-inch (102 mm) criterion is the same as for guard construction per Section R312. It is based on studies of the body measurements of children 13 to 18 months old.
3. This provision reduces the potential for gaining a foothold and climbing the barrier.
4. The more stringent 1.75-inch (44 mm) provision for spacing between vertical members applies when the spacing between horizontal members is less than 45 inches (1143 mm). It acknowledges the potential for a child to gain both a handhold and a foothold on closely spaced horizontal members and reduces the potential for a child to gain a foothold by limiting the space between the vertical members on the same barrier. If the horizontal members are spaced less than 45 inches (1143 mm) apart, they must also be located on the swimming pool side of the fence as shown in Commentary Figure AG105.2(2) so that they are not available to be used to climb the barriers.
5. This requirement is the counterpart to Item 4 in that it permits the opening in the barrier to be 4 inches (102 mm) provided the vertical spacing of the horizontal members equals or exceeds 45 inches (1143 mm) as illustrated in Commentary Figure AG105.2(2). It is consistent with Item 2, which limits openings in the barrier to a 4-inch (102 mm) diameter. The spacing of horizontal members 45 inches (1143 mm) apart precludes them from being used by small children to climb the barrier.
6. The $2^1/_4$-inch (57 mm) dimension is intended to reduce the potential for a child to gain a foothold [see Figure AG105.2(3)]. The mesh size is permitted to be larger than $2^1/_4$-inches (57 mm) square if slats are used to reduce the mesh opening to $1^3/_4$ inches (44 mm) in order to decrease the potential for a child to obtain a foothold or handhold.
7. A slightly larger opening is permitted for barriers composed of diagonal members other than chain link fences, on the basis that such barriers would be more difficult to gain a foothold and handhold on than a chain link fence. The 1.75-inch (44 mm) dimension is consistent with Items 4, 5 and 6.
8. A gate represents the same potential hazard relative to climbing as do the other portions of the barrier; therefore, it must be constructed in accordance with applicable Items 1 through 7. Additionally, since the gate also represents a potential breech of the barrier because the gate can be opened, the code provides prescriptive details for the construction and operation of the gate. A self-closing pedestrian gate must open away from the pool because if the latch fails to operate, a child pushing on the gate will not gain immediate access to the pool. Pushing on the gate may also engage the latch. Large, nonpedestrian gates are not required to be self-closing because of prohibitive cost and maintenance concerns coupled with the fact that these gates are typically operated by persons other than small children. The 54-inch (1372 mm) latch height requirement limits the potential for small children to reach and activate the latch. If the latch is located lower than 54 inches (1372 mm), the code's prescriptive location requirements preclude the latch from being activated by small children who are not on the pool side of the gate.
9. Many residential settings with backyard pools use the dwelling as a portion of the barrier required around the pool, such as where the fence bounding the property terminates at the dwelling. This limits access to the pool by unsupervised children around the perimeter of the fence, but there is still a potential for children to access the pool from within the dwelling. Indeed, almost half the children involved in drowning or near-drowning accidents gained access to the pool from the dwelling.

   The provisions of this section restrict such access by small children and are applicable to all doors in walls that form a portion of the barrier required around swimming pools.

   Protection of such door openings to pool areas can be achieved in any one of the methods described in Items 9.1 through 9.3. The first alternative does not require protection of the exterior door itself but limits access to the pool by means of a power safety cover. The performance criteria specified when this option is selected assures that the power safety cover is an adequate and reliable barrier to the pool. In Item 9.2, the alarm is configured to allow adults who are accessing the house to open the door, enter the house, and deactivate the system to prevent a false alarm. The touchpad permitted to deactivate the system is required to be mounted 54 inches (1372 mm) above the floor, which is presumed to be beyond the reach of small children.

Item 9.3 permits doors to pool areas to be protected by devices that render the door self-closing and self-latching. Any other requirements would be performance based, as the code only requires equivalency with Items 9.1 or 9.2. One possible criterion could require the release mechanism for the latching device to be located a minimum of 54 inches (1372 mm) above the floor, which is presumed to be beyond the reach of small children. In addition, doors protected by the method specified in Item 9.3 should probably open away from the pool area. This is so that if the door failed to latch, a child outside the pool area pushing against the door would cause it to close and not swing to an open position.

10. The code permits the wall of the pool itself to serve as the barrier to the pool, provided that the wall extends at least 48 inches (1219 mm) above the finished ground level around the perimeter of the pool. Unless capable of being secured, locked, or removed, the ladder must be surrounded by a complying barrier to limit access to the ladder.

AG105.3 Indoor swimming pool. All walls surrounding an indoor swimming pool shall comply with Section AG105.2, Item 9.

❖ Indoor pools represent the same hazards as outdoor pools. For this reason, the walls and doors surrounding an indoor swimming pool are regulated in the same manner as an exterior wall of a dwelling where the wall is used as a barrier for an outdoor pool. The provisions of Section AG105.2, Item 9 apply in their entirety.

AG105.4 Prohibited locations. Barriers shall be located so as to prohibit permanent structures, equipment or similar objects from being used to climb the barriers.

❖ The purpose of a swimming pool barrier would be defeated if children could climb on benches, planters, pumps, and similar permanent features adjacent to the barrier and gain access to the pool area. Therefore, the area adjacent to the barrier must be carefully designed and constructed to avoid such a condition. This provision is performance in character and must be reviewed on a case-by-case basis.

AG105.5 Barrier exceptions. Spas or hot tubs with a safety cover which complies with ASTM F 1346, as listed in Section AG108, shall be exempt from the provisions of this appendix.

❖ The provisions of this appendix chapter are not applicable to spas and hot tubs where an approved safety cover serves as the protective barrier. The requirements of ASTM F 1346 contain a number of criteria so that the safety cover can provide an equivalent level of

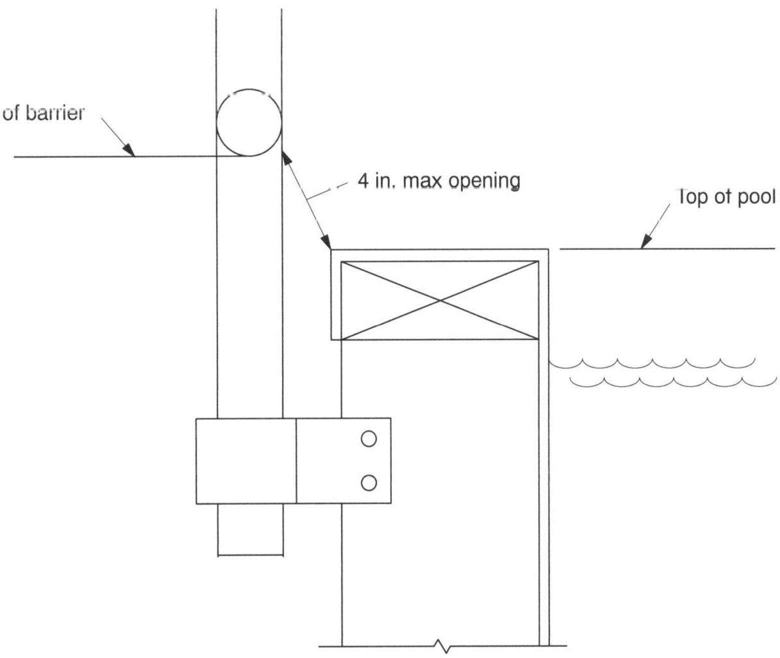

For SI: 1 inch = 25.4 mm.

Figure AG105.2(1)
OPENING LIMITATIONS

protection as that provided by a swimming pool enclosure barrier. The following requirements are representative of several of the specifications found in the standard:

1. There should be a means of fastening the safety cover to the hot tub or spa, such as key locks, combination locks, special tools or similar devices that will prohibit children from removing or operating the cover. The fastening mechanism, design, and location are vital components that help prevent a child's entry to the water.
2. The safety cover should have a label that provides a warning and message regarding the risk of drowning. The label is also very important for

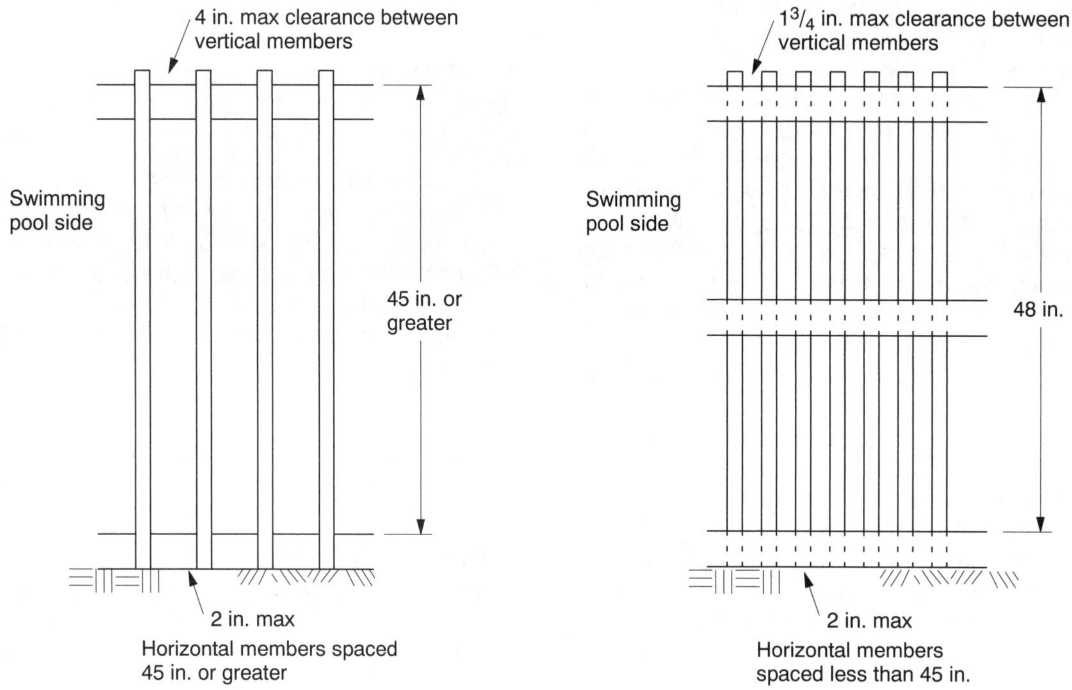

For SI: 1 inch = 25.4 mm.

Figure AG105.2(2)
PRIVATE SWIMMING POOL BARRIER CONSTRUCTION

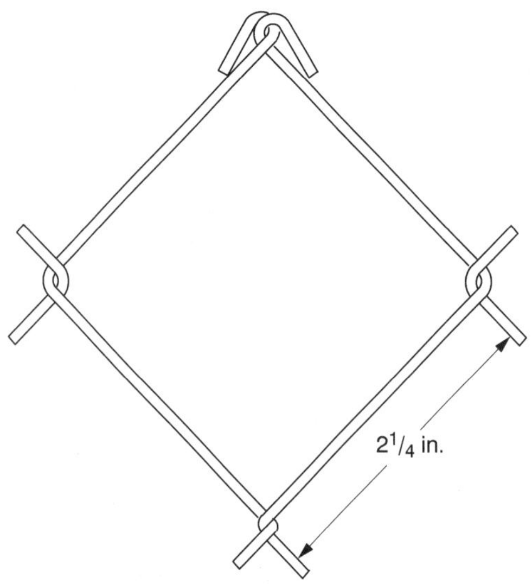

For SI: 1 inch = 25.4 mm.

Figure AG105.2(3)
CHAIN-LINK FENCE MESH FOR PRIVATE SWIMMING POOLS

the transfer of information to second owners and temporary users.

3. The cover should have been tested to demonstrate that it is capable of supporting the weight of one child [50 pounds (23 kg)] and one adult [225 pounds (102 kg)] so an adult and a child can be supported during a rescue operation.
4. There should be no openings in the cover itself or at any point where the cover joins the surface of the hot tub or spa that would allow a child's head to pass through. The 4-inch (102 mm) spacing for guards in Section R312 and openings in pool enclosures of Section AG105.2 are also applicable.
5. Safety covers are to be installed in accordance with the manufacturer's instructions.

SECTION AG106 ENTRAPMENT PROTECTION FOR SWIMMING POOL AND SPA SUCTION OUTLETS

AG106.1 General. Suction outlets shall be designed to produce circulation throughout the pool or spa. Single outlet systems, such as automatic vacuum cleaner systems, or other such multiple suction outlets whether isolated by valves or otherwise shall be protected against user entrapment.

❖ Vacuum devices for suction inlet systems in pool water circulation are a safety hazard. Body entrapment or hair entrapment can cause drowning and evisceration. Therefore it is important that protection be provided against possible entrapment at the pool entrances to suction inlets and that vacuum relief be provided for the vacuum system. Sections AG106.2 through AG106.5 contain requirements for the various types of safety devices.

AG106.2 Suction fittings. All Pool and Spa suction outlets shall be provided with a cover that conforms with ANSI/ASME A112.19.8M, or a 12″ × 12″ drain grate or larger, or an approved channel drain system.

Exception: Surface skimmers

❖ ANSI/ASME A112.19.8M requires cover material for pool and spa suction outlets to be tested for structural integrity and for entrapment/entanglement potential. It also requires the cover to be marked with the maximum flow rate for which the cover has been tested. Exceeding the maximum flow rate will increase the potential for a child or small adult being entrapped due to the increased suction. The code also allows 12-inch by 12-inch (305 mm by 305 mm) drain grates or approved channel drain systems as alternative protection methods. Both of these will provide larger surface areas to maintain the desired flow and will minimize the entrapment hazard because it will be difficult to seal off the entire one-foot-square area.

AG106.3 Atmospheric vacuum relief system required. All pool and spa single or multiple outlet circulation systems shall be equipped with atmospheric vacuum relief should grate covers located therein become missing or broken. Such vacuum relief systems shall include at least one approved or engineered method of the type specified herein, as follows:

1. Safety vacuum release system conforming to ASME A112.19.17, or
2. An approved gravity drainage system

❖ Safety Vacuum Relief Systems (SRVS) are required for all pool and spa circulating systems to serve as a backup safety system in case the suction outlet cover or grate is removed or becomes broken. If the outlet becomes sealed off, the SRVS will activate and eliminate the high vacuum forces at the drain, avoiding body entrapment. An SRVS is not required if the pool or spa has a gravity drain system instead of a pumped circulation system.

AG106.4 Dual drain separation. Single or multiple pump circulation systems shall be provided with a minimum of two (2) suction outlets of the approved type. A minimum horizontal or vertical distance of three (3) feet shall separate such outlets. These suction outlets shall be piped so that water is drawn through them simultaneously through a vacuum relief-protected line to the pump or pumps.

❖ The principle for installing dual or multiple drain systems is to prevent a single drain opening from being the sole inlet to the suction side of the pump. The installation of additional drains effectively divides the suction between the drains, provided the interconnecting piping configuration produces hydraulic balance.

AG106.5 Pool cleaner fittings. Where provided, vacuum or pressure cleaner fitting(s) shall be located in an accessible position(s) at least (6) inches and not greater than twelve (12) inches below the minimum operational water level or as an attachment to the skimmer(s).

❖ Pool cleaner fittings are not required for all pools and spas, but where they are provided, they need to be located below the normal water surface but not more than 12 inches (305 mm) below the surface. This location provides ease of access to the fittings for cleaning and prevents them from contributing to an entrapment situation at the bottom of the pool.

SECTION AG107 ABBREVIATIONS

AG107.1 General.

ANSI—American National Standards Institute
11 West 42nd Street, New York, NY 10036

ASTM—ASTM International
100 Bar Harbor Drive, West Conshohocken, PA 19428

NSPI—National Spa and Pool Institute
2111 Eisenhower Avenue, Alexandria, VA 22314

❖ This section sets forth the full names and addresses of organizations that develop standards referenced in this appendix chapter. The abbreviations for the names of the organizations are used throughout the code text.

SECTION AG108 STANDARDS

AG108.1 General.

ANSI/NSPI

ANSI/NSPI-3 Standard for Permanently Installed Residential Spas AG104.1

ANSI/NSPI-4 Standard for Above-ground/On-ground Residential Swimming Pools AG103.2

ANSI/NSPI-5 Standard for Residential In-ground Swimming Pools AG103.1

ANSI/NSPI-6 Standard for Residential Portable Spas AG104.2

ANSI/ASME A112.19.8M-1987 Suction Fittings for Use in Swimming Pools, Wading Pools, Spas, Hot Tubs and Whirlpool Bathing Appliances AG106.2

ASTM

ASTM F 1346-91 Standard Performance Specifications for Safety Covers and Labeling Requirements for All Covers for Swimming Pools, Spas and Hot Tubs AG105.2, AG105.5

ASME

ASME A112.19.17 Manufacturers Safety Vacuum Release Systems (SVRS) for Residential and Commercial Swimming Pool, Spa, Hot Tub and Wading Pool AG106.3

❖ The seven referenced standards found in this appendix chapter are listed in this section.

Bibliography

The following resource materials are referenced in this chapter or are relevant to the subject matter addressed in this chapter.

ANSI/NSPI-3—1999, *Standard for Permanently Installed Residential Spas.* Alexandria, VA: National Spa and Pool Institute, 1999.

ANSI/NSPI-4—1999, *Standard for Aboveground/On-ground Residential Swimming Pools.* Alexandria, VA: National Spa and Pool Institute, 1999.

ANSI/NSPI-5—1999, *Standard for Residential Inground Swimming Pools.* Alexandria, VA: National Spa and Pool Institute, 1999.

ANSI/NSPI-6—1999, *Standard for Residential Portable Spas.* Alexandria, VA: National Spa and Pool Institute, 1999.

ASME A112.19.8M-1987 (1996), *Suction Fittings for Use in Swimming Pools, Wading Pools, Spas, Hot Tubs and Whirlpool Bathtub Appliances*. New York: American Society of Mechanical Engineers, 1996.

ASME A112.19.17-2002, *Manufactured Safety Vacuum Release Systems (SVRS) for Residential and Commercial Swimming Pool, Spa, Hot Tub and Wading Pool Suction System*. New York: American Society of Mechanical Engineers, 2002.

ASTM F 1346-91 (1996), *Performance Specification for Safety Covers and Labeling Requirements for All Covers for Swimming Pools, Spas and Hot Tubs*. West Conshohocken, PA: ASTM International, 1991.

Appendix H: Patio Covers

General Comments

Patio covers and similar structures are regulated under the general provisions of the code unless this appendix chapter is adopted to address their design and construction. The provisions of this chapter address several specific areas where the general criteria of the code would establish more stringent requirements. The floor area below the patio cover is viewed independently from the remainder of the dwelling unit, with only a limited number of provisions.

This appendix chapter sets forth the regulations and limitations for patio covers.

Section AH101 establishes the scope of the appendix chapter. Section AH102 describes the limitations for structures designated as patio covers. Section AH103 sets forth the uses permitted for patio covers. Section AH104 establishes the design load requirements for patio structures. Section AH105 addresses the affect a patio cover has on light and ventilation, emergency escape and rescue, and egress requirements. Section AH106 addresses the construction of patio covers without footings. Section AH107 contains the requirements for aluminum screen enclosures in hurricane-prone regions.

Purpose

In many climates, the interior living space of a dwelling unit is extended to create an outdoor living area. At times, a separate structure is constructed to create an outdoor area that is protected from the environment. Such structures are patio covers, with a limited number of requirements as set forth by this appendix chapter. The provisions address those uses permitted in patio cover structures, the minimum design loads to be assigned for structural purposes, and the effect of the patio cover on egress and emergency escape or rescue from sleeping rooms.

SECTION AH101 GENERAL

AH101.1 Scope. Patio covers shall conform to the requirements of this appendix chapter.

❖ This appendix chapter regulates the design and construction of patio covers accessory to residential dwelling units.

SECTION AH102 DEFINITION

Patio covers. One-story structures not exceeding 12 feet (3657 mm) in height. Enclosure walls shall be permitted to be of any configuration, provided the open or glazed area of the longer wall and one additional wall is equal to at least 65 percent of the area below a minimum of 6 feet 8 inches (2032 mm) of each wall, measured from the floor. Openings shall be permitted to be enclosed with (1) insect screening, (2) approved translucent or transparent plastic not more than 0.125 inch (3.2 mm) in thickness, (3) glass conforming to the provisions of Section R308, or (4) any combination of the foregoing.

❖ Limited in height to a single story and 12 feet (3657 mm), patio covers are open structures. However, it is common for a portion of the covered area to be enclosed by solid wall construction, with the remainder of the enclosure provided by glazing or screening. This section indicates that the minimum opening length of 65 percent only need be provided at a height of 80 inches (2032 mm) and below. Insect screening, approved minimum $^{1}/_{8}$-inch (3.2 mm) plastic panels, and glazing materials in compliance with Section R308 are permitted as enclosure materials.

SECTION AH103 PERMITTED USES

AH103.1 General. Patio covers shall be permitted to be detached from or attached to dwelling units. Patio covers shall be used only for recreational, outdoor living purposes and not as carports, garages, storage rooms or habitable rooms.

❖ A patio cover may be constructed as a freestanding structure or may be attached directly to a dwelling unit. In either case, the acceptable uses of the floor area beneath the patio cover are limited. This section allows a patio cover to be used for recreational use only, as an outdoor living area. A further provision specifically prohibits the area under a patio cover from use for vehicle parking, general storage, or habitable space.

SECTION AH104 DESIGN LOADS

AH104.1 General. Patio covers shall be designed and constructed to sustain, within the stress limits of this code, all dead loads plus a minimum vertical live load of 10 pounds per square foot (0.48 kN/m^2) except that snow loads shall be used where such snow loads exceed this minimum. Such covers shall be designed to resist the minimum wind loads set forth in Table R301.2(1).

❖ In addition to the dead loads imposed on the structure, a minimum design live load of 10 pounds per square foot (0.48 kN/m^2) is to be used for vertical loading. If the snow load provides for a higher design load, then it must be used as the design criteria. Any wind loads must also be considered in the design and construction of a patio cover.

SECTION AH105 LIGHT AND VENTILATION/EMERGENCY EGRESS

AH105.1 General. Exterior openings required for light and ventilation shall be permitted to open into a patio structure conforming to Section AH101, provided that the patio structure shall be unenclosed if such openings are serving as emergency egress or rescue openings from sleeping rooms. Where such exterior openings serve as an exit from the dwelling unit, the patio structure, unless unenclosed, shall be provided with exits conforming to the provisions of Section R311.4 of this code.

❖ Due to the general openness required of patio structures, it is permissible to have the exterior openings required for light and ventilation purposes by Section R303 open into a patio cover structure. If the exterior openings are also required for emergency escape or rescue by the provisions of Section R310, the patio cover structure cannot be enclosed. Therefore, the allowances for full enclosure by insect screening, light-transmitting plastics, or glazing are not applicable. In addition, where an exit passes through the patio cover structure, the structure must be either unenclosed or provided with at least one complying exit.

SECTION AH106 FOOTINGS

AH106.1 General. In areas with a frostline depth of zero as specified in Table R301.2(1), a patio cover shall be permitted to be supported on a slab on grade without footings, provided the slab conforms to the provisions of Section R506 of this code, is not less than 3.5 inches (89 mm) thick and the columns do not support live and dead loads in excess of 750 pounds (3.34 kN) per column.

❖ Footings supporting a patio cover must extend below the frostline depth established by the jurisdiction. Where the frostline does not extend below ground level and is established as zero, the patio cover structure need only be supported by a slab on grade. Use of a concrete slab without footings is only permitted if the slab is at least 3.5 inches (89 mm) thick and in compliance with Section R506. In addition, the maximum loading supported by each column is limited to 750 pounds (3.34 kN), including both live loads and dead loads.

SECTION AH107 SPECIAL PROVISIONS FOR ALUMINUM SCREEN ENCLOSURES IN HURRICANE-PRONE REGIONS

AH107.1 General. Screen enclosures in hurricane-prone regions shall be in accordance with the provisions of this Section.

❖ Because screened enclosures are open structures and many of them have solid roofs, they must be designed to withstand the wind loads in hurricane-prone regions of the country. Sections AH107.1.1 through AH107.5 provide guidance and minimum requirements for the design of these enclosures.

AH107.1.1 Habitable spaces. Screen enclosures shall not be considered habitable spaces.

❖ Screened enclosures are classified as non-habitable spaces to avoid having to comply with some of the requirements for habitable spaces such as space heating. Obviously, if the owner encloses the space at a later date, compliance with all applicable code provisions for habitable spaces would be required.

AH107.1.2 Minimum ceiling height. Screen enclosures shall have a ceiling height of not less than 7 feet (2134 mm).

❖ The minimum ceiling height of 7 feet (2134 mm) is necessary because these enclosures are typically an addition to a house, and tying an enclosure into an existing roof while maintaining an adequate slope for runoff would make a 7$^1/_2$-foot (2286 mm) ceiling difficult to achieve. The code already permits other spaces such as bathrooms and kitchens to have 7-foot (2134 mm) ceilings, so applying this exception to screened enclosures, which are classified as nonhabitable, is reasonable.

AH107.2 Definitions.

SCREEN ENCLOSURE. A building or part thereof, in whole or in part self-supporting, and having walls of insect screening and a roof of insect screening, plastic, aluminum, or similar lightweight material.

❖ The definition of screen enclosures is provided to clarify that they do not have to be completely self-supporting and that the roof may be of screen or a lightweight, solid material such as aluminum.

AH107.3 Screen enclosures.

AH107.3.1 Thickness. Actual wall thickness of extruded aluminum members shall be not less than 0.040 inches (1.02 mm).

❖ This section specifies a minimum acceptable wall thickness for extruded aluminum members to provide the structural integrity of the screened enclosure during high windload periods.

AH107.3.2 Density. Screen density shall be a maximum of 20 threads per inch by 20 threads per inch mesh.

❖ This section specifies a maximum screen mesh density of 20 threads per inch by 20 threads per inch. A more dense screen mesh could develop more resistance to wind and result in failure of the structure due to increased loads. The minimum screen thread count is 16 per inch for window openings in the *International Property Maintenance Code*.

AH107.4 Design.

AH107.4.1 Wind load. Structural members supporting screen enclosures shall be designed to support minimum wind loads given in Table AH107.4(1) and AH107.4(2). Where any value is less than 10 psf (0.479 kN/m^2) use 10 psf (0.479 kN/m^2).

❖ The basic wind speed used for design is determined from Table R301.2(1) or Figure R301.2(4). Wind exposure category is established on a site-specific basis in accordance with Section R301.2.1.4. Once the wind speed and exposure category is determined, the wind load from Table AH107.4(1), adjusted for height and exposure from Table AH107.4(2), is used to design the structural members of the screen enclosure.

AH107.4.2 Deflection limit. For members supporting screen surfaces only, the total load deflection shall not exceed *l/60*. Screen surfaces shall be permitted to include a maximum of 25 percent solid flexible finishes.

❖ The deflection limit of *l/60* is from the Aluminum Association, Standard ADM - 1, *Aluminum Design Manual*. The limit on solid flexible finishes that is permitted to be included in screen surfaces addresses elements such as kick plates.

AH107.4.3 Importance factor. The wind factor for screen enclosures shall be 0.77 in accordance with Table 1604.5. of the *International Building Code*.

❖ The importance factor for screened enclosures is 0.77, which corresponds to Category I structures in hurricane-prone regions in Table 1604.5 of the *International Building Code.*

AH107.4.4 Roof live load. The minimum roof live load shall be 10 psf (0.479 kN/m^2).

❖ The 10 psf minimum roof live load is the same as the roof live load for patio covers as specified in the IBC.

AH107.5 Footings. In areas with a frost line is zero, a screen enclosure shall be permitted to be supported on a concrete slab on grade without footings, provided the slab conforms to the provisions of Section R506, is not less than $3^1/_2$ inches (89 mm) thick, and the columns do not support loads in excess of 750 pounds (3.36 kN) per column.

❖ This allowance, which permits the enclosures to be constructed without foundations, is due to the minor value of these structures and the lack of any substantive safety issues involved. This provision addresses the exemption of footings in regard to frost penetration; it does not address allowable load-bearing capacity of the soil.

TABLE AH107.4(1)
DESIGN WIND PRESSURES FOR ALUMINUM SCREEN ENCLOSURE FRAMING WITH AN IMPORTANCE FACTOR OF 0.77[a, b, c]

Load Case	WALL	Basic Wind Speed (mph)											
		100		110		120		130		140		150	
		Exposure Category Design Pressure (psf)											
		C	B	C	B	C	B	C	B	C	B	C	B
A[d]	Windward and leeward walls (flow thru) and windward wall (non-flow thru) L/W = 0-1	12	8	14	10	17	12	19	14	23	16	26	18
A[d]	Windward and leeward walls (flow thru) and windward wall (non-flow thru) L/W = 2	13	9	16	11	19	14	22	16	26	18	30	21
B[e]	Windward: Non-gable roof	16	12	20	14	24	17	28	20	32	23	37	26
B[e]	Windward: Gable roof	22	16	27	19	32	23	38	27	44	31	50	36
	ROOF												
All[f]	Roof-screen	4	3	5	4	6	4	7	5	8	6	9	7
All[f]	Roof-solid	12	9	15	11	18	13	21	15	24	17	28	20

For SI: 1 mile per hour = 0.44 m/s, 1 pound per square foot = 0.0479 kPa, 1 foot = 304.8 mm.

a. Values have been reduced for 0.77 Importance Factor in accordance with Table 1604.5 of the *International Building Code.*

b. Minimum design pressure shall be 10 psf in accordance with Section 1609.1.2 of the *International Building Code.*

c. Loads are applicable to screen enclosures with a mean roof height of 30 feet or less. For screen enclosures of different heights the pressures given shall be adjusted by multiplying the table pressure by the adjustment factor given in Table AH107.4(2).

d. For Load Case A flow thru condition the pressure given shall be applied simultaneously to both the upwind and downwind screen walls acting in the same direction as the wind. The structure shall also be analyzed for wind coming from the opposite direction. For the non-flow thru condition the screen enclosure wall shall be analyzed for the load applied acting toward the interior of the enclosure.

e. For Load Case B the table pressure multiplied by the projected frontal area of the screen enclosure is the total drag force, including drag on screen surfaces parallel to the wind, which must be transmitted to the ground. Use Load Case A for members directly supporting the screen surface perpendicular to the wind. Load Case B loads shall be applied only to structural members which carry wind loads from more than one surface.

f. The roof structure shall be analyzed for the pressure given occurring both upward and downward.

❖ This table lists the design wind pressures to be applied to the aluminum screen enclosure framing. These pressures were developed using rational analysis based on the wind loads in the IBC, ASCE 7 and test data. The test data was obtained from wind tunnel investigations of wind loads on screen enclosures. Full scale testing was done to determine drag coefficients for common screen meshes and typical framing members. Scale model testing of the entire screen enclosure was done to determine drag and lift force coefficients. The scale models included freestanding and attached enclosures with solid and screen monoslope, gable and hip roofs. The wind tunnel tests were conducted at the Virginia Polytechnic Institute Wind Stability Facility and Clemson University's Wind Load Test Facility. For additional information refer to *Wind Loads on Screen Enclosures* by Timothy Reinhold et. al.

TABLE AH107.4(2)
HEIGHT ADJUSTMENT FACTORS

Mean	Exposure	
Roof Height (ft)	B	C
15	1	0.86
20	1	0.92
25	1	0.96
30	1	1.00
35	1.05	1.03
40	1.09	1.06
45	1.12	1.09
50	1.16	1.11
55	1.19	1.14
60	1.22	1.16

For SI: 1 foot = 304.8 mm.

❖ This table provides height adjustment factors for the wind pressures in Table AH107.4(1). Table AH107.4(1) is based on a mean roof height of 30 feet (9144 mm) or less. For screen enclosures of different height, the pressures from Table AH107.4(1) are to be multiplied by the height adjustment factor obtained from Table AH107.4(2).

Bibliography

The following resource materials are referenced in this chapter or are relevant to the subject matter addressed in this chapter.

IBC-2003, *International Building Code.* Falls Church, VA: International Code Council, 2003.

IPMC-2003, *International Property Maintenance Code.* Falls Church, VA: International Code Council, 2003.

ASCE 7-02, *Minimum Design Loads for Buildings and Other Structures.* Reston, VA: American Society of Civil Engineers, 2002.

Reinhold, T. A., J. D. Belcher, C. O. Everly and David Miller. *Wind Loads on Screen Enclosures.* Clemson, SC: Department of Civil Engineering, Clemson University, 2000.

APPENDIX I

PRIVATE SEWAGE DISPOSAL

SECTION AI101 GENERAL

AI101.1 Scope. Private sewage disposal systems shall conform to the *International Private Sewage Disposal Code.*

Appendix J: Existing Buildings and Structures

General Comments

Appendix Chapter J contains the provisions for the repair, renovation, alteration, and reconstruction of existing buildings and structures that are within the scope of the *International Residential Code*. The appendix chapter, when adopted, includes or refers to all the code requirements for existing buildings. The format of the chapter readily identifies the alternative methods of code compliance, developed to provide consistency with the purpose of the *International Residential Code*.

Section AJ101 sets forth the intent and purpose of the chapter. Section AJ102 contains methods to obtain compliance with the appendix chapter. Section AJ103 establishes the use of a preliminary meeting to discuss the proposed work. Section AJ104 sets forth the potential means for investigating and evaluating an existing building. Section AJ105 requires the identification of a work area on the permit. Section AJ201 establishes definitions for specific terms used in the appendix chapter. Section AJ301 regulates repairs of existing buildings and structures. Section AJ401 addresses renovations of existing buildings and structures. Section AJ501 regulates alterations to existing buildings and structures, and Section AJ601 establishes requirements for reconstruction work in existing buildings and structures.

Purpose

Many existing residential buildings do not meet the code requirements that currently apply to new construction. Although these buildings are potentially salvageable, rehabilitation is often cost prohibitive, as the building may not be able to comply with all aspects of new construction requirements. At the same time, it is necessary to regulate construction activity in existing buildings that undergo repair, renovation, alteration, or reconstruction. Such activity represents an opportunity for the jurisdiction to determine both that new work conforms to the code (or the intent of the code) and that existing conditions either remain at their current level of compliance or are improved. To accomplish this objective and to make the rehabilitation process more available, this chapter allows for a controlled departure from full code compliance, without compromising minimum life safety, fire safety, structural, and environmental features of the rehabilitated building.

SECTION AJ101 PURPOSE AND INTENT

AJ101.1 General. The purpose of these provisions is to encourage the continued use or reuse of legally existing buildings and structures. These provisions are intended to permit work in existing buildings that is consistent with the purpose of the *International Residential Code*. Compliance with these provisions shall be deemed to meet the requirements of the *International Residential Code*.

❖ This section allows for residential structures to continue their existing use while permitting repair, renovation, alteration, or reconstruction work that may not be in full compliance with the general provisions of the code. While strict compliance with the general requirements in the code may be modified by the provisions of this chapter, the resultant work will be consistent with the intent and purpose criteria set forth in Chapter 1. As a rule, the general requirements of the code are applicable unless modified by the provisions of this chapter. Upon compliance with the provisions of this appendix chapter, the work meets the requirements of the *International Residential Code*.

AJ101.2 Classification of work. For purposes of this appendix, all work in existing buildings shall be classified into the categories of repair, renovation, alteration and reconstruction. Specific requirements are established for each category of work in these provisions.

❖ This appendix chapter addresses four unique and specific types of work that may occur on an existing residential structure. Each type is regulated independently based on the provisions of the chapter. Repairs, where work is performed to restore a building element or system back to working order, are regulated by Section AJ301. Renovation, the act of restoring a building element to its original condition, is addressed in Section AJ401. Section AJ501 deals with alterations, where elements, components, or systems of the building are modified through the use of new materials and methods, and Section AJ601 addresses reconstruction of existing building elements. While each of these areas seems to overlap in their scope, each has unique characteristics that are identified and regulated by this chapter.

AJ101.3 Multiple categories of work. Work of more than one category may be part of a single work project. All related work permitted within a 12-month period shall be considered a single work project. Where a project includes one category of work in one building area and another category of work in a separate and unrelated area of the building, each project area shall comply with the requirements of the respective category of work. Where a project with more than one category of work is performed in the same area or in related areas of the building, the project shall comply with the requirements of the more stringent category of work.

❖ It is not uncommon for two or more types of construction activity to be taking place at the same time, all a part of a single work project. Under such conditions, there are two situations that may occur. 1) Where the categories of work are isolated in different areas of the building and unrelated to each other, each project area is regulated by the specific requirements of the work being performed. 2) Where the multiple categories of work occur in the same project area, the more restrictive provisions of the categories will apply.

SECTION AJ102 COMPLIANCE

AJ102.1 General. Regardless of the category of work being performed, the work shall not cause the structure to become unsafe or adversely affect the performance of the building; shall not cause an existing mechanical or plumbing system to become unsafe, hazardous, insanitary or overloaded; and unless expressly permitted by these provisions, shall not make the building any less conforming to this code or to any previously approved alternative arrangements than it was before the work was undertaken.

❖ The general limitation on any work done under the provisions of this chapter is that the level of safety, health, and public welfare of the existing building must not be reduced by any work being performed. In the case of structural stability, the existing degree of structural strength must be maintained or increased. In general terms, the structure is not to be made unsafe. This requirement can be broadly interpreted, as its application can vary on a case-by-case situation. As far as existing mechanical and plumbing systems are involved, there must not be a reduced level of protection or sanitation.

Where this appendix chapter does not specifically modify the general provisions of the code, such provisions are applicable and conformance is required. Only in those areas where this chapter addresses specific materials or methods for specific elements, components, or systems do the general code requirements not apply.

AJ102.2 Requirements by category of work. Repairs shall conform to the requirements of Section AJ301. Renovations shall conform to the requirements of Section AJ401. Alterations shall conform to the requirements of Section AJ501 and the requirements for renovations. Reconstructions shall conform to the requirements of Section AJ601 and the requirements for alterations and renovations.

❖ Depending on the category of work being performed, conformance with one or more groups of requirements is necessary. If the work is simply a repair job or a renovation, only the provisions of Sections AJ301 or AJ401, respectively, are applicable. If the building is to undergo an alteration, the requirements of both Sections AJ501 and AJ401 must be applied. Section AJ601 must be used for reconstruction work.

AJ102.3 Smoke detectors. Regardless of the category of work, smoke detectors shall be provided where required by Section R313.1.1.

❖ The benefit of smoke detectors in residential buildings is unquestioned. Occupants of such buildings spend about one-third of the day asleep. The potential for a fire getting out of control before the occupants are awakened is quite probable, as reflected in statistics that indicate residential fire deaths far exceed those of any other building classification. As a result, the general provisions of Section R317.1.1 for smoke alarms are applicable in all cases and for all categories of work. This includes specific requirements where a dwelling unit undergoes interior alterations or repairs.

AJ102.4 Replacement windows. Regardless of the category of work, when an entire existing window, including frame, sash and glazed portion is replaced, the replacement window shall comply with the requirements of Section N1102.4.

❖ Under all circumstances, a replacement window must comply with the energy efficiency provisions of Section N1102.4. The provision is applicable only where the entire window is changed out, including the frame, sash, and glazing. Where such conditions exist, the new window is to have a maximum fenestration *U*-factor in compliance with Table 502.2.5 of the 2000 *International Energy Conservation Code*. If the replacement window is in a hazardous location, safety glazing complying with Section R308 must be installed.

AJ102.5 Flood Hazard areas. Work performed in existing buildings located in a flood hazard area as established by Table R301.2(1) shall be subject to the provisions of R105.3.1.1.

❖ Long-term reduction in exposure to flood hazards is one of the reasons floodplain development is regulated. If alterations or repairs of an existing building, including restoration of damage from any cause, constitute substantial improvement, the existing building is to be brought into compliance as required in R105.3.1.1. See the commentary for that section. Improvements are deemed to be "substantial" if the cost, including the value of labor as well as donated labor and materials, equals or exceeds 50 percent of the market value of the building before the improvements are made. For additional guidance, refer to FEMA 259, *Engineering Principles and Practices for Retrofitting Flood Prone Residential Buildings* and FEMA 213, *Answers to Questions About Substantially Damaged Buildings.*

The "substantial" improvement provisions specify that the cost of correcting existing health, sanitary, or safety code violations that have been identified by the code official are not included in the determination of whether the cost of an improvement or repair project is a substantial improvement, provided such cited corrections are the minimum necessary to assure safe living conditions. The provision also specifies that alteration of a historic structure is not a substantial improvement provided the structure will continue to be designated as a historic structure. The code official may require applicants to consult with the appropriate historic preservation authority to determine if a proposed alteration will jeopardize a structure's historic designation. The exception for historic structures does not apply to structures located within designated historic districts, unless such structures are individually listed as historic structures.

AJ102.6 Equivalent alternatives. These provisions are not intended to prevent the use of any alternate material, alternate design or alternate method of construction not specifically prescribed herein, provided any alternate has been deemed to be equivalent and its use authorized by the building official.

❖ A comprehensive regulatory document such as a construction code cannot envision and then address all future innovations in the industry. As a result, the code must be applicable to, and provide a basis for, the approval of a number of newly developed, innovative materials, systems, and methods for which no code text or referenced standards yet exist. The building official is expected to exercise sound technical judgment in accepting materials, systems, or methods that, while not anticipated by the code text, can be demonstrated to offer equivalent performance. By virtue of this section, the code regulates new and innovative construction practices while contributing to the safety of building occupants.

AJ102.7 Other alternatives. Where compliance with these provisions or with this code as required by these provisions is technically infeasible or would impose disproportionate costs because of structural, construction or dimensional difficulties, other alternatives may be accepted by the building official. These alternatives may include materials, design features and/or operational features.

❖ The building official may accept alternative solutions involving materials, design features, or operational features where compliance with the code provisions creates practical difficulties. A practical difficulty, in this context, means it is technically infeasible to meet the code requirements, or the costs involved in providing code compliance are highly disproportionate to the overall costs of the work. It is up to the building official to evaluate the legitimacy of these two exemptions from the general requirements. This section identifies structural, construction, or dimensional difficulties as a basis for using these modifications.

AJ102.8 More restrictive requirements. Buildings or systems in compliance with the requirements of this code for new construction shall not be required to comply with any more restrictive requirement of these provisions.

❖ Where the existing building is in compliance with the requirements of the code for new construction, it is not necessary to comply with any higher level requirements as set forth in this appendix chapter. If the work to be performed is in compliance with the general provisions of the code, as would be required of a newly-constructed building, there is no reason to require a higher level of performance for existing structures.

AJ102.9 Features exceeding International Residential Code requirements. Elements, components and systems of existing buildings with features that exceed the requirements of this code for new construction, and are not otherwise required as part of approved alternative arrangements or deemed by the building official to be required to balance other building elements not complying with this code for new construction, shall not be prevented by these provisions from being modified as long as they remain in compliance with the applicable requirements for new construction.

❖ Unless used as a portion of an alternative solution as an equivalency, modifications to existing elements, components, or systems need only meet the requirements of the code for new construction. While the modification may actually reduce the previous level of compliance, the modification is acceptable provided it does not reduce compliance to a level below that accepted for new work.

SECTION AJ103 PRELIMINARY MEETING

AJ103.1 General. If a building permit is required at the request of the prospective permit applicant, the building official or his designee shall meet with the prospective applicant to discuss plans for any proposed work under these provisions prior to the application for the permit. The purpose of this preliminary meeting is for the building official to gain an understanding of the prospective applicant's intentions for the proposed work, and to determine, together with the prospective applicant, the specific applicability of these provisions.

❖ Where the provisions of the code are performance provisions, such as those set forth in this appendix chapter for existing buildings, the party or parties involved with the anticipated work must meet with the building official at the onset of the project to discuss any concerns or questions. Those issues that may be interpreted or applied in various ways must be addressed so that an understanding is reached as to the position of the jurisdiction.

This section indicates that the issues must be discussed prior to application for a building permit, with an emphasis on the scope of the work and the applicability of the code provisions for existing buildings. The effort should be cooperative, with input and discussion by both the building official and the potential permit applicant.

SECTION AJ104
EVALUATION OF AN EXISTING BUILDING

AJ104.1 General. The building official may require an existing building to be investigated and evaluated by a registered design professional in the case of proposed reconstruction of any portion of a building. The evaluation shall determine the existence of any potential nonconformities with these provisions, and shall provide a basis for determining the impact of the proposed changes on the performance of the building. The evaluation shall utilize the following sources of information, as applicable:

1. Available documentation of the existing building.
   1.1. Field surveys.
   1.2. Tests (nondestructive and destructive).
   1.3. Laboratory analysis.

Exception: Detached one- or two-family dwellings that are not irregular buildings under Section R301.2.2.6 and are not undergoing an extensive reconstruction shall not be required to be evaluated.

❖ It is possible that the nature of the reconstruction work is such that an evaluation of the existing building by a registered design professional should be performed. Based on the investigation and evaluation of the existing conditions that occur in the building or a portion of the building, the building official is able to analyze the necessary work that must take place to comply with the provisions of the code. The need for an evaluation is determined by the building official based on the individual circumstances of the situation. In addition, the code recognizes that such an evaluation is unnecessary for single-family or two-family dwelling units of conventional construction unless the building is undergoing a major reconstruction.

The evaluation should focus on those existing conditions that are not in conformance with the code and its intended purpose, as well as determining how the proposed modifications will address building performance. All resources that may assist in the investigation and evaluation of the existing conditions must be considered along with the three general sources identified in the code.

SECTION AJ105
PERMIT

AJ105.1 Identification of work area. The work area shall be clearly identified on all permits issued under these provisions.

❖ Once the limits of one or more work areas have been established based on the scope of the work involved, the work areas must be properly identified and clearly described. The provisions of this appendix chapter are typically limited to those portions of the building characterized as the work areas. The work areas must be described in satisfactory detail on any permits that are issued so that all parties have a consistent understanding of the area limits.

SECTION AJ201
DEFINITIONS

AJ201.1 General. For purposes of this appendix, the terms used shall be defined as follows.

❖ This section clarifies the terminology used in this appendix chapter. The terms take on very specific meanings that are often different from the way they are typically used.

ALTERATION. The reconfiguration of any space, the addition or elimination of any door or window, the reconfiguration or extension of any system, or the installation of any additional equipment.

❖ Where the work is extensive enough to cause a change in the layout of a building or a portion of a building, the work is an alteration. This activity could include taking an existing space and changing the configuration of the floor plan by adding or removing walls or partitions. Additional rooms may be created or multiple rooms could be transformed into a single room or area. As a result of such an alteration, the exiting system may be modified or the occupant load revised.

Where an additional door or window is installed, or where an existing door or window is removed, the work is an alteration. An alteration is also an increase to any system of the building, including the expansion of any electrical wiring, plumbing piping, or mechanical ducts. If additional equipment, such as another electrical panel or HVAC unit, is installed in an existing building, the installation must be regulated as an alteration.

CATEGORIES OF WORK. The nature and extent of construction work undertaken in an existing building. The categories of work covered in this Appendix, listed in increasing order of stringency of requirements, are repair, renovation, alteration and reconstruction.

❖ This term describes the four types of construction activity that are regulated by this appendix chapter. The code sets forth specific criteria for each of the defined types of work, with each category placed in a hierarchical position based on the mandated level of requirements. The least restrictive category of work is defined as repair, with an increased degree of regulation for renovations and an even higher level for alterations. The most stringent requirements are applicable for work categorized as reconstruction.

DANGEROUS. Where the stresses in any member; the condition of the building, or any of its components or elements or attachments; or other condition that results in an overload exceeding 150 percent of the stress allowed for the member or material in this code.

❖ While this term is recognized in general terms as a situation that is unsafe, perilous, or likely to cause injury or death, for the purposes of this chapter it is much more narrow in scope. It is limited to a structural condition where the structural member or material is subjected to a load significantly higher than that which it is designed to support.

EQUIPMENT OR FIXTURE. Any plumbing, heating, electrical, ventilating, air conditioning, refrigerating and fire protection equipment, and elevators, dumb waiters, boilers, pressure vessels, and other mechanical facilities or installations that are related to building services.

❖ A multitude of building service components are included in the definition of equipment or fixtures. These include elements of the plumbing, mechanical, and electrical systems, as well as fire protective features that are installed in the building.

LOAD-BEARING ELEMENT. Any column, girder, beam, joist, truss, rafter, wall, floor or roof sheathing that supports any vertical load in addition to its own weight, and/or any lateral load.

❖ The structural components of a building that carry loads other than their own are load-bearing elements. These include members that carry gravity loads, lateral loads, or both. The definition lists the various elements that are load-bearing where they support more than their own weight.

MATERIALS AND METHODS REQUIREMENTS. Those requirements in this code that specify material standards, details of installation and connection, joints, penetrations, and continuity of any element, component or system in the building. The required quantity, fire resistance, flame spread, acoustic or thermal performance, or other performance attribute is specifically excluded from materials and methods requirements.

❖ Materials standards, installation details, and similar specific requirements make up the materials and methods requirements. On the other hand, performance-related requirements tend to be excluded by definition.

RECONSTRUCTION. The reconfiguration of a space that affects an exit, a renovation and/or alteration when the work area is not permitted to be occupied because existing means of egress and fire protection systems, or their equivalent, are not in place or continuously maintained; and/or there are extensive alterations as defined in Section AJ501.3.

❖ The most stringent requirements for the four categories of work are assigned to reconstruction activities. Reconstruction causes a high level of concern because of its impact on fire safety and life safety. Where an existing exit is affected, where the existing level of egress or fire protection is not acceptable, or where extensive alterations are to be undertaken, the work is considered as reconstruction.

REHABILITATION. Any repair, renovation, alteration or reconstruction work undertaken in an existing building.

❖ This is the general term for describing any or all of the four categories of work involved on an existing building.

RENOVATION. The change, strengthening or addition of load-bearing elements; and/or the refinishing, replacement, bracing, strengthening, upgrading or extensive repair of existing materials, elements, components, equipment and/or fixtures. Renovation involves no reconfiguration of spaces. Interior and exterior painting are not considered refinishing for purposes of this definition, and are not renovation.

❖ A renovation can occur for either structural or nonstructural work. The definition includes structural members that are replaced, modified, or added to a building, as well as nonstructural materials that are changed in some form. The act of renovation focuses on specific elements, rather than the reconfiguration of floor area.

REPAIR. The patching, restoration and/or minor replacement of materials, elements, components, equipment and/or fixtures for the purposes of maintaining such materials, elements, components, equipment and/or fixtures in good or sound condition.

❖ The least stringent requirements for the four categories of work on existing building regulate repairs. Repair work maintains elements and systems of buildings in sound condition. Repair activities do not change the configuration of the space, nor do they address new construction or equipment.

WORK AREA. That portion of a building affected by any renovation, alteration or reconstruction work as initially intended by the owner and indicated as such in the permit. Work area excludes other portions of the building where incidental work entailed by the intended work must be performed, and portions of the building where work not initially intended by the owner is specifically required by these provisions for a renovation, alteration or reconstruction.

❖ The work area must be correctly identified in order to define the extent of the rehabilitation activities. Most of the requirements in this appendix chapter are only applicable to the work area under consideration and do not apply to other portions of the building. The work area, determined at the time of permitting, is limited to that part of the building affected by the rehabilitation work.

SECTION AJ301 REPAIRS

AJ301.1 Materials. Except as otherwise required herein, work shall be done using like materials or materials permitted by this code for new construction.

❖ There are two possible options for materials used in repair work on an existing building. Unless prohibited by other provisions of the section, it is acceptable to use materials consistent with those that are already present. This allowance follows the general concept that the repair work is making the building no more unsafe or hazardous than it was prior to the work being done. In lieu of using the same type of materials, the code permits the use of any materials presently allowed by the code.

AJ301.1.1 Hazardous materials. Hazardous materials no longer permitted, such as asbestos and lead-based paint, shall not be used.

❖ It is generally possible to repair a structure, its components, and its systems with materials consistent with those materials that were used previously. However, where materials that are now deemed hazardous are involved in the repair work, they may no longer be used. For example, the code identifies asbestos and lead-based paint as two common hazardous materials that cannot be used in the repair process. Certain materials previously considered acceptable for building construction are now a threat to the health of the occupants.

AJ301.1.2 Plumbing materials and supplies. The following plumbing materials and supplies shall not be used:

1. All-purpose solvent cement, unless listed for the specific application;
2. Flexible traps and tailpieces, unless listed for the specific application; and
3. Solder having more than 0.2 percent lead in the repair of potable water systems.

❖ Specific methods and materials of plumbing installations are identified as no longer acceptable due to their negative impact on public health and safety. Where such existing materials and supplies are a part of the repair of a plumbing element or system, alternate materials must be used. The use of all-purpose solvent cement is only permitted for plumbing repair work if listed for the specific application. The same is true for flexible traps and flexible tailpieces.

AJ301.2 Water closets. When any water closet is replaced with a newly manufactured water closet, the replacement water closet shall comply with the requirements of Section P2903.2.

❖ Where a new water closet replaces an existing water closet, the new fixture must be designed for a maximum water consumption of 1.6 gallons for each flushing cycle. Addressing environmental concerns, this limitation assists in reducing the amount of water consumed during the ongoing use of the building.

AJ301.3 Safety glazing. Replacement glazing in hazardous locations shall comply with the safety glazing requirements of Section R308.1.

❖ Due to the obvious hazards associated with glazing located in an area subject to human impact, glazing located in such a hazardous location must be replaced with safety glazing. The provisions of Section R308 must be referenced to fully comply with these requirements. Section R308.3 identifies the test requirements and criteria for glazing installed in hazardous locations. Typical materials meeting these requirements include tempered glass and laminated glass. Section R308.4 identifies hazardous locations where it is possible that an individual could impact the glazing, including glazing in and adjacent to doors.

AJ301.4 Electrical. Existing electrical wiring and equipment undergoing repair shall be permitted to be repaired or replaced with like material.

Exceptions:

1. Replacement of electrical receptacles shall comply with the requirements of Chapters 33 through 42.
2. Plug fuses of the Edison-base type shall be used for replacements only where there is no evidence of over-fusing or tampering per the applicable requirements of Chapters 33 through 42.
3. For replacement of nongrounding-type receptacles with grounding-type receptacles and for branch circuits that do not have an equipment grounding conductor in the branch circuitry, the grounding conductor of a grounding type receptacle outlet shall be permitted to be grounded to any accessible point on the grounding electrode system, or to any accessible point on the grounding electrode conductor, as allowed and described in Chapters 33 through 42.

❖ Under most conditions, it is acceptable to repair existing electrical installations with the same types of wiring materials and electrical equipment as were used previously. However, this section identifies three conditions where additional criteria must be considered. Electrical receptacles must comply as for new construction as described in Chapters 33 through 42. Edison-base-type plug fuses can only be replaced with like fuses if it can be shown that no tampering or overfusing has occurred. Alternate methods of grounding are also set forth when nongrounding-type receptacles are replaced.

SECTION AJ401
RENOVATIONS

AJ401.1 Materials and methods. The work shall comply with the materials and methods requirements of this code.

❖ The general provisions of the code are to be used for renovation work. There are several modifications to these provisions for door and window dimensions, interior finish materials, and the parapets of unreinforced masonry buildings assigned to a high seismic design category.

AJ401.2 Door and window dimensions. Minor reductions in the clear opening dimensions of replacement doors and windows that result from the use of different materials shall be allowed, whether or not they are permitted by this code.

❖ During many renovation projects, it is common for existing doors and windows to be removed and replaced. Quite often the new doors and windows are of different materials and do not provide the same clear opening dimensions as the originals. Reductions in opening dimensions beyond those allowed by the code are permitted provided they are minor in nature. While not specifically defined, a minor reduction would remain consistent with the intent and purpose of the provisions.

AJ401.3 Interior finish. Wood paneling and textile wall coverings used as an interior finish shall comply with the flame spread requirements of Section R315.

❖ Where wood paneling or textile wall covering materials are being replaced in a renovation project, the new materials must be in compliance with the provisions of Section R319. The new interior finishes are to be regulated for flame spread and smoke development as they are for new construction.

AJ401.4 Structural. Unreinforced masonry buildings located in Seismic Design Categories D_2 or E shall have parapet bracing and wall anchors installed at the roofline whenever a reroofing permit is issued. Such parapet bracing and wall anchors shall be of an approved design.

❖ When reroofing work requiring a building permit takes place on an unreinforced masonry building located in Seismic Design Categories D_2 or E, it may be necessary to strengthen the parapet by providing bracing in an approved manner. In addition, wall anchors must be provided at the roof line. An evaluation of the existing parapet and wall anchors should be done to determine the extent of the structural strengthening that needs to be accomplished. The scope of the structural work and the methods of compliance are to be approved by the building official.

SECTION AJ501 ALTERATIONS

AJ501.1 Newly constructed elements. All newly constructed elements, components and systems shall comply with the requirements of this code.

Exceptions:

1. Openable windows may be added without requiring compliance with the light and ventilation requirements of Section R303.
2. Newly installed electrical equipment shall comply with the requirements of Section AJ501.5.

❖ Where the alteration of any portion of a building includes new construction, the work must be accomplished in accordance with the requirements of the code. An exception permits the installation of openable windows as additional features without requiring adherence to the light and ventilation provisions of Section R303. In addition, new electrical components and equipment need only comply with the requirements of Section AJ501.5.

AJ501.2 Nonconformities. The work shall not increase the extent of noncompliance with the requirements of Section AJ601, or create nonconformity with those requirements which did not previously exist.

❖ The extent of noncompliance is limited in regard to stairways, handrails, guards, interior finish materials, and dwelling separation walls as addressed in Section AJ601. In addition, nonconformity must not be created regarding any requirements that did not previously exist.

AJ501.3 Extensive alterations. When the total area of all the work areas included in an alteration exceeds 50 percent of the area of the dwelling unit, the work shall be considered as a reconstruction and shall comply with the requirements of these provisions for reconstruction work.

Exception: Work areas in which the alteration work is exclusively plumbing, mechanical or electrical shall not be included in the computation of total area of all work areas.

❖ If the amount of construction activity in a dwelling unit involves more than 50 percent of the unit's floor area, the stringency of the requirements is increased. The category of work is reconstruction, with the requirements for both alterations and reconstruction to be followed. This increase to a higher level category is not required where the alteration being performed consists only of plumbing, mechanical, or electrical work.

AJ501.4 Structural. The minimum design loads for the structure shall be the loads applicable at the time the building was constructed, provided that no dangerous condition is created. Structural elements that are uncovered during the course of the alteration and that are found to be unsound or dangerous shall be made to comply with the applicable requirements of this code.

❖ As building codes have progressed over the years, structural design values have been reviewed and modified. Unless a dangerous condition will be created, it is permissible to use the minimum structural design loads in place at the time the building was constructed.

As the alteration of the building is progressing, there may be occasion where structural elements are exposed and found to be damaged, unsound, or otherwise dangerous. In such situations, it is mandatory that the structural integrity of the building components be achieved. All necessary steps must be taken to ensure that the applicable structural requirements of the code are met.

AJ501.5 Electrical equipment and wiring.

❖ This section identifies the specific provisions applicable to electrical equipment and systems in buildings undergoing an alteration.

AJ501.5.1 Materials and methods. All newly installed electrical equipment and wiring relating to work done in any work area shall comply with the materials and methods requirements of Chapters 33 through 42.

Exception: Electrical equipment and wiring in newly installed partitions and ceilings shall comply with all applicable requirements of Chapters 33 through 42.

❖ In any work area, electrical equipment and wiring that is installed must comply with the requirements of Chapters 33 through 42. Such requirements are also applicable in the construction of new walls, partitions, and ceiling systems.

AJ501.5.2 Electrical service. Service to the dwelling unit shall be a minimum of 100 ampere, three-wire capacity and service equipment shall be dead front having no live parts exposed whereby accidental contact could be made. Type "S" fuses shall be installed when fused equipment is used.

Exception: Existing service of 60 ampere, three-wire capacity, and feeders of 30 ampere or larger two- or three-wire capacity shall be accepted if adequate for the electrical load being served.

❖ It is typical that minimum 100-ampere service be provided to each dwelling unit that is undergoing alteration of electrical service. However, where it can be determined that the service loading does not exceed 60 amperes, an existing 60-ampere service is acceptable.

AJ501.5.3 Additional electrical requirements. When the work area includes any of the following areas within a dwelling unit, the requirements of Sections AJ501.5.3.1 through AJ501.5.3.5 shall apply.

❖ This section sets forth additional requirements for specified enclosed spaces, kitchens, laundry rooms, ground-fault circuit interruption, and lighting outlets in conjunction with work areas of dwelling units.

AJ501.5.3.1 Enclosed areas. All enclosed areas other than closets, kitchens, basements, garages, hallways, laundry areas and bathrooms shall have a minimum of two duplex receptacle outlets, or one duplex receptacle outlet and one ceiling or wall type lighting outlet.

❖ Those areas and rooms typically viewed as habitable spaces must be provided with at least two duplex receptacle outlets or one duplex receptacle outlet and a ceiling- or wall-type lighting outlet. A minimum number of receptacles must be available in order to reduce the potential for dangerous electrical conditions.

AJ501.5.3.2 Kitchen and laundry areas. Kitchen areas shall have a minimum of two duplex receptacle outlets. Laundry areas shall have a minimum of one duplex receptacle outlet located near the laundry equipment and installed on an independent circuit.

❖ To reduce the possibility that dangerous conditions are created in kitchen and laundry areas, a minimum number of receptacle outlets is mandated. In the kitchen, at least two duplex receptacle outlets are required, while a minimum of one duplex receptacle outlet is mandated near laundry equipment. The receptacle outlet in the laundry area must be located on an independent circuit.

AJ501.5.3.3 Ground-fault circuit-interruption. Ground fault circuit interruption shall be provided on newly installed receptacle outlets if required by Chapters 33 through 42.

❖ The installation of ground-fault circuit interruption for all new receptacle outlets is covered by the provisions of Chapters 33 through 42. Those locations in new construction identified by the code to be protected by ground-fault circuit interruption are the same in alterations to existing buildings, but interruption is required only in those locations where new receptacle outlets are installed.

AJ501.5.3.4 Lighting outlets. At least one lighting outlet shall be provided in every bathroom, hallway, stairway, attached garage and detached garage with electric power to illuminate outdoor entrances and exits, and in utility rooms and basements where these spaces are used for storage or contain equipment requiring service.

❖ Lighting outlets are to be installed in a variety of specific locations identified by this section. Areas where at least one lighting outlet is required include bathrooms, hallways, stairways, and attached garages. Detached garages provided with electrical power must have facilities for illuminating exterior entrances and exits. Basements and utility rooms used as storage areas or to house equipment that must be serviced must also be provided with at least one lighting outlet.

AJ501.5.3.5 Clearance. Clearance for electrical service equipment shall be provided in accordance with Chapters 33 through 42.

❖ All electrical equipment is to be provided with the necessary working space and other clearances as set forth in Chapters 33 through 42.

AJ501.6 Ventilation. All reconfigured spaces intended for occupancy and all spaces converted to habitable or occupiable space in any work area shall be provided with ventilation in accordance with Section R303.

❖ Adequate ventilation, provided by either natural or mechanical means, must be provided in occupiable spaces located within a work area. The requirement is applicable to spaces that are altered in shape or size, as well as those areas that are converted to habitable space. The ventilation requirements of Section R303 provide the minimum ventilation criteria that are to be used.

SECTION AJ601
RECONSTRUCTION

AJ601.1 Stairways, handrails and guards.

AJ601.1.1 Stairways. Stairways within the work area shall be provided with illumination in accordance with Section R303.6.

❖ Stairways located in the work area of a building undergoing reconstruction activities must be provided with illumination in accordance with Section R303.4. The provisions set forth two options for locating the light sources. In addition, the required locations of the control switches for the lighting are specified.

AJ601.1.2 Handrails. Every required exit stairway that has three or more risers; is part of the means of egress for any work area; and is not provided with at least one handrail, or in which the existing handrails are judged to be in danger of collapsing, shall be provided with handrails designed and installed in accordance with Section RNO TAG for the full length of the run of steps on at least one side.

❖ The handrail requirements of Section R315 apply to reconstruction work only in certain circumstances. The stairway under consideration must be a required exit element for a work area of the building and consist of at least three risers. Either there are no handrails provided for the stairway, or the existing rail or rails are unsafe due to the possibility of collapse. If all of these conditions exist, a complying handrail is to be installed on at least one side of the stairway.

Handrail height, continuity, termination, clearance, and gripping surface must all be reviewed for conformance with the provisions of Section R315.

AJ601.1.3 Guards. Every open portion of a stair, landing or balcony that is more than 30 inches (762 mm) above the floor or grade below; is part of the egress path for any work area; and is not provided with guards or in which the existing guards are judged to be in danger of collapsing, shall be provided with guards designed and installed in accordance with Section R312.

❖ The guard requirements of Section R316 apply to reconstruction work only in certain circumstances. The guard under consideration must be serving a walking surface at least 30 inches (762 mm) above the floor below and located along an exit path for a work area of the building. Either there is no guard provided, or the existing guard creates an unsafe condition due to the potential for collapse. If all of these conditions exist, a complying guard is to be installed as protection at the elevation change.

Guard height and opening limitations must be reviewed for conformance with the provisions of Section R316.

AJ601.2 Wall and ceiling finish. The interior finish of walls and ceilings in any work area shall comply with the requirements of Section R315. All existing interior finish materials that do not comply with those requirements shall be removed or shall be treated with an approved fire retardant coating in accordance with the manufacturer's instructions to secure compliance with the requirements of this section.

❖ In a reconstruction work area, all wall and ceiling finish materials must be in compliance with the flame-spread and smoke-development limitations of Section R319. The flame-spread classification is limited to 200, with a maximum smoke-developed index of 450. Where the existing interior finish materials do not comply with these requirements, the materials are to be removed or treated with an approved fire retardant coating. Where the treatment method is used to obtain compliance, the manufacturer's instructions must be followed.

AJ601.3 Separation walls. Where the work area is in an attached dwelling unit, all walls separating dwelling units that are not continuous from the foundation to the underside of the roof sheathing shall be constructed to provide a continuous fire separation using construction materials consistent with the existing wall or complying with the requirements for new structures. Work shall only be required to be performed on the side of the wall of the dwelling unit that is part of the work area.

❖ Where reconstruction work takes place in a dwelling unit that is attached to one or more additional dwelling units, the wall or walls separating the units are to be continuous from the foundation to the underside of the roof deck. If such conditions do not exist, the wall must be extended to the roof sheathing to maintain the necessary separation of the units. The materials used to provide the wall construction must be at least equivalent to those of the existing wall, or alternatively, as required for new construction. There is no requirement for work to be done outside the work area; therefore, the wall construction need be provided only on the side of the dwelling where the actual work area occurs.

BIBLIOGRAPHY

The following resource materials are referenced in this chapter or are relevant to the subject matter addressed in this chapter.

FEMA 213, *Answers to Questions About Substantially Damaged Buildings.* Washington, DC: Federal Emergency Management Agency, 1991.

FEMA 259, *Engineering Principles and Practices for Retrofitting Flood Prone Residential Buildings*. Washington, DC: Federal Emergency Management Agency, 1995.

Appendix K: Sound Transmission

General Comments

Regulation of sound transmission bears directly on the psychological well-being of a building's occupants and the occupants' long-term physical well-being. By implementing the minimum sound transmission criteria set forth in this appendix chapter, a jurisdiction can assist in providing a more livable environment for the occupants.

Section AK101 provides a general overview of the requirements for limiting sound transmission. Section AK102 establishes the specific provisions for airborne sound insulation in wall and floor-ceiling assemblies. Section AK103 establishes the specific provisions for structural-borne sound transfer in floor-ceiling assemblies.

Purpose

In an effort to appropriately regulate the environmental and physiological aspects of habitable spaces within dwelling units, the code addresses a variety of conditions. In addition to the fundamental requirements for adequate lighting, ventilation, and sanitation, regulations are often desirable to control the transmission of sound and noise between individual dwelling units within a multiple-family residential building. The provisions of this appendix chapter set forth a minimum Sound Transmission Class (STC) rating for common walls and floor-ceiling assemblies between dwelling units. In addition, a minimum Impact Insulation Class (IIC) rating is also established to limit structure-borne sound through common floor-ceiling assemblies separating dwelling units.

SECTION AK101 GENERAL

AK101.1 General. Wall and floor-ceiling assemblies separating dwelling units shall provide airborne sound insulation for walls, and both airborne and impact sound insulation for floor-ceiling assemblies.

❖ Since noise transmission can be quantified, and noise affects the quality of life, the code incorporates regulations that address noise transmission in multiple-family residential construction. In order to provide for an acceptable degree of livability in structures housing two or more dwelling units, the code mandates a minimum level of sound insulation between units. In such residential structures, the occupants of a dwelling unit typically have little control over noise generated in other units. Where the dwelling units are separated by a common floor-ceiling assembly, both airborne sound and structural-borne sound are present. Where a common wall system separates the units, only airborne sound is regulated. The components of construction regulated are those through which noise is primarily transmitted.

SECTION AK102 AIRBORNE SOUND

AK102.1 General. Airborne sound insulation for wall and floor-ceiling assemblies shall meet a Sound Transmission Class (STC) rating of 45 when tested in accordance with ASTM E 90.

❖ The code requires common walls and floor-ceiling assemblies between dwelling units to have a minimum Sound Transmission Class (STC) rating of 45. The STC is a measure of an assembly's ability to resist sound transmission. The higher the number (rating), the higher the resistance (less sound transmission). Standard architectural wall and floor-ceiling construction assemblies have been tested for sound transmission ratings, and reference to the construction specifications will yield such information. Airborne sound, such as a voice or music, is transferred through the air.

As a rule, vertical assemblies meeting the requirements of this section consist of double walls or walls containing insulation similar to exterior walls. Horizontal assemblies typically contain some type of insulating materials within the assembly or resilient furring channels to isolate the ceiling membrane from the structural members of the floor construction.

SECTION AK103 STRUCTURAL-BORNE SOUND

AK103.1 General. Floor/ceiling assemblies between dwelling units or between a dwelling unit and a public or service area within a structure shall have an Impact Insulation Class (IIC) rating of not less than 45 when tested in accordance with ASTM E 492.

❖ The Impact Insulation Class (IIC) rating is a measure of an assembly's ability to resist sound transmission. The higher the number (rating), the higher the resistance (less sound transmission). Floors between dwelling units are required to have a minimum IIC rating of 45.

Usually, carpeted floor assemblies meet the minimum requirement for an IIC rating of 45. Other areas with hard-surfaced finishes may require additional treatment or insulation to comply with these requirements.

Various resource documents containing Sound Transmission Class (STC) ratings and IIC ratings are listed in the Bibliography. These sources, or any other similar sources, can be submitted as a basis for approval once it is demonstrated to the building official that the data are based on ASTM E90 and ASTM E492.

Bibliography

The following resource materials are referenced in this chapter or are relevant to the subject matter addressed in this chapter.

ASTM E90—97, *Standard Test Method for Laboratory Measurement of Airborne Sound Transmission Loss of Building Partitions and Elements.* West Conshohocken, PA: ASTM International, 1997.

ASTM E492—90 (1996), *Standard Specification for Laboratory Measurement of Impact Sound Transmission through Floor-Ceiling Assemblies using the Tapping Machine.* West Conshohocken, PA: ASTM International, 1996.

BIA TN 5A—83, *Sound Insulation—Clay Masonry Walls.* Reston, VA: Brick Institute of America, 1983.

Gypsum Association. GA 600—97: *Fire-resistance Design Manual.* 15th Edition. Washington, D.C.: Gypsum Association, 1997.

National Concrete Masonry Association. NCMA TEK 69A—78: *New Data on Sound Reduction with Concrete Masonry Walls.* Herndon, VA: National Concrete Masonry Association, 1978.

APPENDIX L

ICC INTERNATIONAL RESIDENTIAL ELECTRICAL PROVISIONS/ NATIONAL ELECTRICAL CODE CROSS REFERENCE

This table is a cross-reference of the ICC *International Residential Code,* Chapters 33 through 42, and the 2002 *National Electrical Code* (NFPA 70-2002).

International Residential Code		National Electrical Code
CHAPTER 33	**GENERAL REQUIREMENTS**	
SECTION E3301	**GENERAL**	
E3301.1	Applicability	None
E3301.2	Scope	Sections 90.1 and 90.2
E3301.3	Not covered	Section 90.2
E3301.4	Additions and alterations	None
SECTION E3302	**BUILDING STRUCTURE PROTECTION**	
E3302.1	Drilling and notching	None
E3302.2	Penetrations of fire-resistance-rated assemblies	Section 300.21
E3302.3	Penetrations of firestops and draftstops	Section 300.21
SECTION E3303	**INSPECTION AND APPROVAL**	
E3303.1	Approval	Section 110.2
E3303.2	Inspection required	None
E3303.3	Listing and labeling	Section 110.3
SECTION E3304	**GENERAL EQUIPMENT REQUIREMENTS**	
E3304.1	Voltages	Section 110.4
E3304.2	Interrupting rating	Section 110.9
E3304.3	Circuit characteristics	Section 110.10
E3304.4	Protection of equipment	Section 110.11
E3304.5	Unused openings	Section 110.12(a)
E3304.6	Integrity of electrical equipment	Section 110.12(c)
E3304.7	Mounting	Section 110.13(a)
E3304.8	Energized parts guarded against accidental contact	Section 110.27(a)
E3304.9	Prevent physical damage	Section 110.27(b)
E3304.10	Equipment identification	Section 110.21
E3304.11	Identification of disconnecting means	Section 110.22
SECTION E3305	**EQUIPMENT LOCATION AND CLEARANCES**	
E3305.1	Working space and clearances	Section 110.26
Figure E3305.1	Working space and clearances	Sections 110.26
	Footnote a.	Section 110.26(f)
	Footnote b.	Section 110.26(e)
	Footnote c.	Section 110.26(b)
	Footnote d.	Sections 230.70(a), 240.24(d) and 240.24(e)
	Footnote e.	Section 110.26(d)
E3305.2	Working clearances for energized equipment and panelboards	Sections 110.26(a), 110.26(d), and 110.26(f)
E3305.3	Clearances over panelboards	Section 110.26(f)(1)(a)
E3305.4	Location of clear spaces	Sections 110.26(b), 230.70(a) and 240.24
E3305.5	Access and entrance to working space	Section 110.26(c)
E3305.6	Illumination	Section 110.26(d)
E3305.7	Headroom	Section 110.26(e)

International Residential Code		National Electrical Code
SECTION E3306	**ELECTRICAL CONDUCTORS AND CONNECTIONS**	
E3306.1	General	Article 110 and Section 310.1
E3306.2	Conductor material	Section 110.5
E3306.3	Minimum size of conductors	Section 310.5
E3306.4	Stranded conductors	Section 310.3
E3306.5	Individual conductor insulation	Section 310.2(a) and 310.8
E3306.6	Conductors in parallel	Section 310.4
E3306.7	Conductors of the same circuit	Section 300.3(b)
E3306.8	Aluminum and copper connections	Section 110.14
E3306.9	Terminals	Section 110.14(a)
E3306.10	Splices	Section 110.14(b)
E3306.10.1	Continuity	Section 300.13(a)
	Exception	Section 300.13(a)
E3306.10.2	Device connections	Sections 250.148 and 300.13(b)
E3306.10.3	Length of conductor for splice or termination	Section 300.14
SECTION E3307	**CONDUCTOR AND TERMINAL IDENTIFICATION**	
E3307.1	Grounded conductors	Sections 200.6(a), 200.6(b) and 310.12(a)
E3307.2	Equipment grounding conductors	Sections 250.119 and 310.12(b)
E3307.3	Ungrounded conductors	Section 310.12(c)
	Exception	Section 200.7(c)(2)
E3307.4	Identification of terminals	Section 200.10
E3307.4.1	Device terminals	Section 200.10(a)
E3307.4.2	Receptacles, plugs, and connectors	Section 200.10(b)
E3307.5	Tag marking	Section 310.11(b)(3)
CHAPTER 34	**ELECTRICAL DEFINITIONS**	
SECTION E3401	**GENERAL**	Article 100, Definitions
CHAPTER 35	**SERVICES**	
SECTION 3501	**GENERAL SERVICES**	
E3501.1	Scope	Section 230.1
E3501.2	Number of services	Section 230.2
E3501.3	One building or other structure not to be supplied through another	Section 230.3
E3501.4	Other conductors in raceway or cable	Section 230.7
E3501.5	Raceway seal	Section 230.8
E3501.6	Service disconnect required	Section 230.70
E3501.6.1	Marking of service equipment and disconnects	Sections 230.66 and 230.70(b)
E3501.6.2	Service disconnect location	Sections 230.70(a) and 230.72(c)
E3501.7	Maximum number of disconnects	Section 230.71(a)
SECTION E3502	**SERVICE SIZE AND RATING**	
E3502.1	Rating of ungrounded conductors	Section 230.79(c) and (d)
E3502.2	Service load	Section 220.30
Table E3502.2	Minimum service load calculation	Table 220.30
E3502.2.1	Services under 100 amperes	None
E3502.3	Rating of service disconnect	Section 230.79
E3502.4	Voltage rating	Section 220.30(a)
SECTION E3503	**SERVICE, FEEDER AND GROUNDING ELECTRODE CONDUCTOR SIZING**	
E3503.1	Grounded and ungrounded service conductor size	310.15(b)(6)
Table E3503.1	Service conductor and grounding electrode conductor sizing	Table 310.16 and Table 250.66
	Footnote a.	Section 250.64(e)
	Footnotes b. and c.	Section 250.64(b)
	Footnote d.	Section 250.66(a) and (b)
E3503.2	Ungrounded service conductors for accessory buildings and structures	Section 230.42(b)

International Residential Code		National Electrical Code
	Exceptions 1, 2, and 3	Section 230.42(b) and 230.79
E3503.3	Overload protection	Section 230.90
E3503.3.1	Ungrounded conductor	Section 230.90(a)
	Exception	Section 230.90(a), Exception No. 3
E3503.3.2	Not in grounded conductor	Section 230.90(b)
E3503.3.3	Location	Section 230.91
E3503.4	Grounding electrode conductor size	Section 250.66
SECTION E3504	**OVERHEAD SERVICE DROP AND SERVICE CONDUCTOR INSTALLATION**	
E3504.1	Clearance from building openings	Section 230.9
Figure E3504.1	Clearances from building openings	Section 230.9
E3504.2	Vertical clearances	Section 230.24
E3504.2.1	Above roofs	Section 230.24(a)
	Exception 1	Section 230.24(a), Exception No. 1
	Exception 2	Section 230.24(a), Exception No. 2
	Exception 3	Section 230.24(a), Exception No. 3
	Exception 4	Section 230.24(a), Exception No. 4
Figure E3504.2.1	Clearances from roofs	Section 230.24
E3504.2.2	Vertical clearance from grade	Section 230.24(b)
	Items 1, 2, and 3	Section 230.24(b)
E3504.3	Point of attachment	Section 230.26
E3504.4	Means of attachment	Section 230.27
E3504.5	Service masts as supports	Section 230.28
E3504.6	Supports over buildings	Section 230.29
SECTION E3505	**SERVICE-ENTRANCE CONDUCTORS**	
E3505.1	Insulation of service-entrance conductors	Section 230.41
	Exceptions 1 and 2	Section 230.41, Exception
E3505.2	Wiring methods for services	Section 230.43
E3505.3	Spliced conductors	Section 230.46
E3505.4	Protection against physical damage	Section 230.49
E3505.5	Protection of service cables against damage	Section 230.50(a)
E3505.6	Direct sunlight exposure	Section 310.8(D)
E3505.7	Mounting supports	Section 230.51
E3505.8	Raceways to drain	Section 230.53
E3505.9	Overhead service locations	Section 230.54
E3505.9.1	Raintight service head	Section 230.54(a)
E3505.9.2	Service cable, service head or gooseneck	Section 230.54(b)
E3505.9.3	Service head location	Section 230.54(c)
	Exception	Section 230.54(c), Exception
E3505.9.4	Separately bushed openings	Section 230.54(e)
E3505.9.5	Drip loops	Section 230.54(f)
E3505.9.6	Conductor arrangement	Section 230.54(g)
E3505.9.7	Secured	Section 230.54(d)
SECTION E3506	**SERVICE EQUIPMENT—GENERAL**	
E3506.1	Service equipment enclosures	Section 230.62
E3506.2	Working space	Section 110.26
E3506.3	Available short-circuit current	None
E3506.4	Marking	Section 230.66
SECTION E3507	**SYSTEM GROUNDING**	
E3507.1	System service ground	Sections 250.20(b)(1) and 250.24(a)
E3507.2	Location of grounding electrode conductor connection	Section 250.24(a)(1) and (a)(5)
E3507.3	Two or more buildings or structures supplied from a common service	250.32(a)
	Exception	Section 250.32(a), Exception
E3507.3.1	Equipment grounding conductor	Section 250.32(b)(1) and Table 250.122

International Residential Code		National Electrical Code
E3507.3.2	Grounded conductor.	Section 250.32(b)(2)
E3507.4	Grounding electrode conductor	Section 250.24(c)
E3507.5	Main bonding jumper	Section 250.28
E3507.6	Common grounding electrode	Section 250.58
SECTION E3508	**GROUNDING ELECTRODE SYSTEM**	
E3508.1	Grounding electrode system	Section 250.50
E3508.1.1	Metal underground water pipe	Section 250.50(a), 250.53(D)(1), 250.53(E)
E3508.1.2	Concrete-encased electrode	Section 250.50(c)
E3508.2	Made and other electrodes	Section 250.52
E3508.2.1	Rod and pipe electrodes	Section 250.52(c)(1) and (2)
E3508.2.2	Installation	Section 250.52(c)(2)
E3508.2.3	Aluminum electrodes	Section 250.52(b)(1)
E3508.3	Resistance of made electrodes	Section 250.56
E3508.4	Metal underground gas piping system	Section 250.52(b)(2)
SECTION E3509	**BONDING**	
E3509.1	General	Section 250.90
E3509.2	Bonding of service equipment	Section 250.92(a)
E3509.3	Bonding to other systems	Section 250.94
E3509.4	Method of bonding at the service	Section 250.92(b)
E3509.4.1	Grounded service conductor	Section 250.94(1)
E3509.4.2	Threaded connections	Section 250.94(2)
E3509.4.3	Threadless couplings and connectors	Section 250.94(3)
E3509.4.4	Other devices	Section 250.94(4)
E3509.5	Sizing bonding jumper on supply side of service and main bonding jumper	Section 250.102(c)
E3509.6	Metal water piping bonding	Section 250.104(a)
E3509.7	Bonding other metal piping	Section 250.104(b)
SECTION E3510	**GROUNDING ELECTRODE CONDUCTORS**	
E3510.1	Installation	Section 250.64(a) and (b)
E3510.2	Enclosures for grounding electrode conductors	Section 250.64(e)
SECTION E3511	**GROUNDING ELECTRODE CONDUCTOR CONNECTION TO GROUNDING ELECTRODES**	
E3511.1	Methods of grounding connection to electrodes	Section 250.70
E3511.2	Accessibility	Section 250.68
E3511.3	Effective grounding path	Section 250.68
E3511.4	Protection of ground clamps and fittings	Section 250.10
E3511.5	Clean surfaces	Section 250.12
CHAPTER 36	**BRANCH CIRCUIT AND FEEDER REQUIREMENTS**	
SECTION E3601	**GENERAL**	
E3601.1	Scope	None
E3601.2	Branch-circuit and feeder ampacity	Sections 210.19(A) and 215.2(A)(1)
E3601.3	Selection of ampacity	Section 310.15(A)(2)
E3601.4	Multioutlet branch circuits	Section 210.19(A)(2)
E3601.5	Multiwire branch circuits	Section 210.4
SECTION E3602	**BRANCH CIRCUIT RATINGS**	
E3602.1	Branch-circuit voltage limitations	Sections 210.6(A) and (B)
E3602.2	Branch-circuit ampere rating	Section 210.3
E3602.3	Fifteen- and 20-ampere branch circuits	Section 210.23(A)
E3602.4	Thirty-ampere branch circuits	Section 210.23(b)
E3602.5	Branch circuits serving multiple loads or outlets	Section 210.23(A)
E3602.6	Branch circuits serving a single motor	Section 430.22(A)
E3602.7	Branch circuits serving motor-operated and combination loads	Section 220.4(A)

International Residential Code		National Electrical Code
E3602.8	Branch-circuit inductive lighting loads	Section 220.4(B)
E3602.9	Branch-circuit load for ranges and cooking appliances	Table 220.19, Note 4
E3602.9.1	Minimum branch circuit for ranges	Section 210.19(A)(3)
E3602.10	Branch circuits serving heating loads	Sections 422.13 and 424.3(A)
E3602.11	Branch circuits for air-conditioning and heat pump equipment	Sections 440.35, 440.4(B), 440.52(A)
E3602.12	Branch circuits serving room air conditioners	Section 440.62(A)
E3602.12.1	Where no other loads are supplied	Section 440.62(B)
E3602.12.2	Where lighting units or other appliances are also supplied	Section 440.62(C)
E3602.13	Branch-circuit requirement—summary	Section 210.24 and 210.25
Table E3602.13	Branch-circuit requirement—summary	Table 210.24
SECTION E3603	**REQUIRED BRANCH CIRCUITS**	
E3603.1	Branch circuits for heating	Section 422.12 Exception
E3603.2	Kitchen and dining area receptacles	Section 210.52(B)(1)
E3603.3	Laundry circuit	Sections 210.23(A), Exception and 210.11(C)(2)
E3603.4	Bathroom branch circuits	Section 210.11(C)(3)
E3603.5	Number of branch circuits	Section 210.11(A)
E3603.6	Branch-circuit load proportioning	Section 210.11(B)
SECTION E3604	**FEEDER REQUIREMENTS**	
E3604.1	Conductor size	Table 310.15(B)(6), Sections 215.2(A)(3), (A)(4), and 220.10(A)
E3604.2	Minimum feeder conductor size	Section 215.2(A)(2)
E3604.3	Feeder loads	Table 220.11
Table E3604.3(1)	Feeder load calculation	Table 220.11, Table 220.30(C), Sections 220.14, 220.15, 220.17, 220.18 and 220.19
Table E3604.3(2)	Demand loads for electric ranges, wall-mounted ovens, counter-mounted cooking units and other cooking appliances over $1^3/_4$ kVA rating	Table 220.19
E3604.4	Feeder neutral load	Section 220.22
E3604.5	Lighting and convenience receptacle load	Section 220.3(A), Table 220.3(A)
E3604.6	Ampacity and computed loads	Section 220.10
E3604.7	Feeder and branch-circuit conductors	Section 240.21(A)
SECTION E3605	**CONDUCTOR SIZING AND OVERCURRENT PROTECTION**	
E3605.1	General	310.15(a)
Table E3605.1	Allowable ampacities	Table 310.16 and 240.4(D)
E3605.2	Correction factors for ambient temperatures	Table 310.16
Table E3605.2	Ambient temperature correction factors	Table 310.16, Correction factors
E3605.3	Adjustment for conductor proximity	Section 310.15(B)(2)(a)
Table E3605.3	Conductor proximity adjustment factors	Table 310.15(B)(2)(a)
E3605.4	Temperature limitations	Section 110.14(C)
E3605.4.1	Conductors rated 60°C.	Section 110.14(C)(1)(a)
E3605.4.2	Conductors rated 75°C.	Section 110.14(C)(b)
E3605.4.3	Separately installed pressure connectors	Section 110.14(C)(2)
E3605.4.4	Conductors of type NM cable	Section 334.8(D) and 334.112
E3605.5	Overcurrent protection required	Table 310.15(B)(6), Sections 240.4, 240.5, 240.21
E3605.5.1	Cords	Section 240.5
E3605.5.2	Overcurrent devices of the next higher size	Section 240.4(B)
E3605.5.3	Small conductors	Section 240.4(D)
E3605.5.4	Air conditioning and heat pump equipment	Section 240.4(G)
E3605.6	Fuses and fixed trip circuit breakers	Section 240.6
E3605.7	Location of overcurrent devices in or on premises	Section 240.24(A)(C)(D)(E)
E3605.8	Ready access for occupants	Section 240.24(B)
E3605.9	Enclosures for overcurrent devices	Section 240.30(A)(B)
SECTION E3606	**PANELBOARDS**	
E3606.1	Panelboard rating	Section 408.13
E3606.2	Panelboard circuit identification	Section 408.4
E3606.3	Panelboard overcurrent protection	Section 408.16(A)

International Residential Code		National Electrical Code
CHAPTER 37	**WIRING METHODS**	
SECTION E3701	**GENERAL REQUIREMENTS**	
E3701.1	Scope	Section 300.1
E3701.2	Allowable wiring methods	Sections 110.8 and 300.3
Table E3701.2	Allowable wiring methods	None
E3701.3	Circuit conductors	Section 300.3(B)
E3701.4	Wiring method applications	Chapter 3 and Section 300.2
Table E3701.4	Allowable applications for wiring methods	Chapter 3 and Section 300.2
SECTION E3702	**ABOVEGROUND INSTALLATION REQUIREMENTS**	
E3702.1	Installation and support requirements	Chapter 3
Table E3702.1	General installation and support requirements for wiring methods	Chapter 3
E3702.2	Cables in accessible attics	Sections 320.23 and 334.23
E3702.2.1	Across structural members	Section 320.23(A) and 334.23
E3702.2.2	Cable installed parallel to framing members	Section 320.17, 320.23(A), 334.17 and 334.23
E3702.3	Exposed cable	Sections 320.15 and 334.15
E3702.3.1	Surface installation	Section 334.15
E3702.3.2	Protection from physical damage	Section 334.15(B)
E3702.3.3.3	Location exposed to direct sunlight	Section 310.8(D)
E3702.4	In unfinished basements	Section 334.15(C)
E3702.5	Bends	Sections 320.24 and 334.24
SECTION E3703	**UNDERGROUND INSTALLATION REQUIREMENTS**	
E3703.1	Minimum cover requirements	Section 300.5(A)
Table E3703.1	Minimum cover requirements, burial in inches	Table 300.5
E3703.2	Warning ribbon	Section 300.5(D)(3)
E3703.3	Grounding	Section 300.5(B)
E3703.4	Protection from damage	Section 300.5(D)
E3703.5	Splices and taps	Section 300.5(E)
E3703.6	Backfill	Section 300.5(F)
E3703.7	Raceway seals	Section 300.5(G)
E3703.8	Bushing	Section 300.5(H)
E3703.9	Single conductors	Section 300.5(I)
E3703.10	Ground movement	Section 300.5(J)
CHAPTER 38	**POWER AND LIGHTING DISTRIBUTION**	
SECTION E3801	**RECEPTACLE OUTLETS**	
E3801.1	General	Section 210.50 and 210.52
E3801.2	Convenience receptacle distribution	Section 210.52(A)
E3801.2.1	Spacing	Section 210.52(A)(1)
E3801.2.2	Wall space	Section 210.52(A)(2)
E3801.2.3	Floor receptacles	Section 210.52(A)(3)
E3801.3	Small appliance receptacles	Section 210.52(B)
E3801.3.1	Other outlets prohibited	Section 210.52(B)(2)
E3801.3.2	Limitations	Section 210.52(B)(3)
E3801.4	Countertop receptacles	Section 210.52(C)
E3801.4.1	Wall counter space	Section 210.52(C)(1)
E3801.4.2	Island counter spaces	Section 210.52(C)(2)
E3801.4.3	Peninsular counter space	Section 210.52(C)(3)
E3801.4.4	Separate spaces	Section 210.52(C)(4)
E3801.4.5	Receptacle outlet location	Section 210.52(C)(5)
E3801.5	Appliance outlets	Section 210.50(C)
E3801.6	Bathroom and toilet room	Section 210.52(D)
E3801.7	Outdoor outlets	Section 210.52(E)
E3801.8	Laundry areas	Section 210.52(F)

International Residential Code		National Electrical Code
E3801.9	Basements and garages	Section 210.52(G)
E3801.10	Hallways	Section 210.52(H)
E3801.11	HVAC outlet	Section 210.63
SECTION E3802	**GROUND-FAULT AND ARC-FAULT CIRCUIT-INTERRUPTER PROTECTION**	
E3802.1	Bathroom receptacles	Section 210.8(A)(1)
E3802.2	Garage receptacles	Section 210.8(A)(2)
E3802.3	Outdoor receptacles	Section 210.8(A)(3)
E3802.4	Crawl space receptacles	Section 210.8(A)(4)
E3802.5	Unfinished basement receptacles	Section 210.8(A)(5)
E3802.6	Kitchen receptacles	Section 210.8(A)(6)
E3802.7	Bar sink receptacles	Section 210.8(A)(7)
E3802.8	Boathouse receptacles	Section 210.8(A)(8)
E3802.9	Electrically heated floors	Section 424.44(G)
E3802.10	Exempt receptacles	Section 210.8(A)(2)
E3802.11	Bedroom outlets	Section 210.12(B)
SECTION E3803	**LIGHTING OUTLETS**	
E3803.1	General	Section 210.70(A)
E3803.2	Habitable rooms	Section 210.70(A)(1)
E3803.3	Additional locations	Section 210.70(A)(2)
E3803.4	Storage or equipment spaces	Section 210.70(A)(3)
SECTION E3804	**GENERAL INSTALLATION REQUIREMENTS**	
E3804.1	Electrical continuity of metal raceways and enclosures	Section 300.10
E3804.2	Mechanical continuity—raceways and cables	Section 300.12
E3804.3	Securing and supporting	Section 300.11(A)
E3804.4	Raceways as means of support	Section 300.11(B)
E3804.5	Raceway installations	Section 300.18
E3804.6	Conduit and tubing fill	Section 300.17 and Chapter 9, Table 1
Tables E3804.6(1)-E3804.6(10)	Maximum number of conductors in conduit or tubing	Section 300.17 and Chapter 9, Table 1, Note 1
E3804.7	Air handling—stud cavity and joist spaces	Section 300.22(C), Exception
CHAPTER E3805	**BOXES, CONDUIT BODIES, AND FITTINGS**	
E3805.1	Box, conduit body, or fitting—where required	Section 300.15
E3805.1.1	Equipment	Section 300.15(B)
E3805.1.2	Protection	Section 300.15(C)
E3805.1.3	Integral enclosure	Section 300.15(E)
E3805.1.4	Fitting	Section 300.15(F)
E3805.1.5	Buried conductors	Section 300.15(G)
E3805.1.6	Fixtures	Section 300.15(J)
E3805.1.7	Closed loop	Section 300.15(M)
E3805.2	Metal boxes	Section 314.4
E3805.3	Nonmetallic boxes	Section 314.3
E3805.3.1	Nonmetallic–sheathed cable and nonmetallic boxes	Section 314.17(C)
E3805.3.2	Securing to box	Section 314.17(C)
E3805.3.3	Conductor rating	Section 314.17(C)
E3805.4	Minimum depth of outlet boxes	Section 314.24
E3805.5	Boxes enclosing flush–mounted devices	Section 314.19
E3805.6	Boxes at luminaire (lighting fixture) outlets	Section 314.27(A)
E3805.7	Maximum luminaire (fixture) weight	Section 314.27(B)
E3805.8	Floor boxes	Section 314.27(C)
E3805.9	Boxes at fan outlets	Section 314.27(C)
E3805.10	Conduit bodies, junction, pull and outlet boxes to be accessible	Section 314.29
E3805.11	Damp or wet locations	Section 314.15(A)
E3805.12	Number of conductors in outlet, device, and junction boxes, and conduit boxes	Section 314.16

International Residential Code		National Electrical Code
E3805.12.1	Box volume calculations	Section 314.16(A)
Table E3805.12.1	Maximum number of conductors in metal boxes	Table 314.16
E3805.12.1.1	Standard boxes	Section 314.16(A)
E3805.12.1.2	Other boxes	Section 314.16(A)(2)
E3805.12.2	Box fill calculations	Section 14.16(B)
E3805.12.2.1	Conductor fill	Section 314.16(B)(1), Table 314.16
Table E3805.12.2.1	Volume allowance required per conductor	Table 314.16
E3805.12.2.2	Clamp fill	Section 314.16(B)(2)
E3805.12.2.3	Support fittings fill	Section 314.16(B)(3)
E3805.12.2.4	Device or equipment fill	Section 314.16(B)(4)
E3805.12.2.5	Equipment grounding conductor fill	Section 314.16(B)(5)
E3805.12.3	Conduit bodies	Section 314.16(C)
E3805.12.3.1	Splices, taps or devices	Section 314.16(C)(2)
SECTION E3806	**INSTALLATION OF BOXES, CONDUIT BODIES AND FITTINGS**	
E3806.1	Conductors entering boxes, conduit bodies or fittings	Section 314.17
E3806.1.1	Insulated fittings	Section 300.4(F)
E3806.2	Openings	Section 314.17(A), 110.12(A)
E3806.3	Metal boxes, conduit bodies and fittings	Section 314.17, 314.17(B)
E3806.4	Unused openings	Section 314.17, 110.12(A)
E3806.5	In wall or ceiling	Section 314.20
E3806.6	Plaster, gypsum board and plasterboard	Section 314.21
E3806.7	Exposed surface extensions	Section 314.22
E3806.8	Supports	Section 314.23
E3806.8.1	Surface mounting	Section 314.23(A)
E3806.8.2	Structural mounting	Section 314.23(B)
E3806.8.2.1	Nails	Section 314.23(B)(1)
E3806.8.2.2	Braces	Section 314.23(B)(2)
E3806.8.3	Mounting in finished surfaces	Section 314.23(C)
E3806.8.4	Raceway supported enclosures without devices or fixtures	Section 314.23(E)
E3806.8.5	Raceway supported enclosures, with devices or fixtures	Section 314.23(F)
E3806.8.6	Enclosures in concrete or masonry	Section 314.23(G)
E3806.9	Covers and canopies	Section 314.25
E3806.10	Metal covers and plates	Section 314.25(A)
E3806.11	Exposed combustible finish	Section 314.25(B)
SECTION E3807	**CABINETS AND PANELBOARDS**	
E3807.1	Enclosures for switches or overcurrent devices	Section 312.8
E3807.2	Damp or wet locations	Section 312.2(A)
E3807.3	Position in wall	Section 312.3
E3807.4	Unused openings	Section 312.5(A), 110.12(A)
E3807.5	Conductors entering cabinets	Sections 300.4(F) and 312.5(B)
E3807.6	Openings to be closed	Section 312.5(A)
E3807.7	Cables	Section 312.5(B)
SECTION E3808	**GROUNDING**	
E3808.1	Metal enclosures	Section 250.86, 250.110(1)
E3808.2	Equipment fastened in place or connected by permanent wiring methods (fixed)	Section 250.110(I)
E3808.3	Specific equipment fastened in place or connected by permanent wiring methods	Section 250.112(J), (L)
E3808.4	Performance of fault current path	Section 250.4(A)(5)
E3808.5	Earth as a grounding conductor	Section 250.4(A)(5)
E3808.6	Load-side neutral	Section 250.142(B)
E3808.7	Load-side equipment	Section 250.142(B)
E3808.8	Types of equipment grounding conductors	Section 250.118
E3808.8.1	Flexible metal conduit	Section 250.118(6)
E3808.8.2	Liquidtight flexible metal conduit	Section 250.118(7)

International Residential Code		National Electrical Code
E3808.8.3	Nonmetallic sheathed cable (Type NM)	Section 334.10(B)
E3808.9	Equipment fastened in place or connected by permanent wiring methods	Section 250.134
E3808.10	Methods of equipment grounding	Section 250 Part (VII)
E3808.11	Equipment grounding conductor installation	Section 250.120
E3808.12	Equipment grounding conductor size	Section 250.122
Table E3808.12	Equipment grounding conductor sizing	Table 250.122
E3808.12.1	Multiple circuits	Section 250.122(C)
E3808.13	Continuity and attachment of branch-circuit equipment grounding conductors to boxes	Section 250.148
E3808.14	Connecting receptacle grounding terminal to box	Section 250.146
E3808.15	Metal boxes	Section 250.148(A)
E3808.16	Nonmetallic boxes	Section 250.148(B)
E3808.17	Clean surfaces	Section 250.12
E3808.18	Bonding other enclosures	Section 250.96(A)
E3808.19	Size of equipment bonding jumper on load side of service	Section 250.102(D)
E3808.20	Installation—equipment bonding jumper	Section 250.102(E)
SECTION E3809	**FLEXIBLE CORDS**	
E3809.1	Where permitted	Section 400.7 and 400.8
E3809.2	Loading and protection	Section 400.5
Table E3809.2	Maximum ampere load and overcurrent-protection-device rating for flexible cords	Section 240.4, Table 400.5(A) and 240.5(B)(1)
E3809.3	Splices	Section 400.9
E3809.4	Attachment plugs	Section 400.7(B)
CHAPTER 39	**DEVICES AND LIGHTING FIXTURES**	
SECTION E3901	**SWITCHES**	
E3901.1	Rating and application of snap switches	Section 404.14(A)
E3901.2	CO/ALR snap switches	Section 404.14(C)
E3901.3	Indicating	Section 404.7
E3901.4	Time switches and similar devices	Section 404.5
E3901.5	Grounding of enclosures	Section 404.12
E3901.6	Access	Section 404.4
E3901.7	Wet locations	Section 404.8
E3901.8	Grounded conductors	Section 404.2(B)
E3901.9	Switch connections	Section 404.2(A)
E3901.10	Box mounted	Section 404.10(B)
E3901.11	Snap switch faceplates	Section 404.9(A)
E3901.11.1	Faceplate grounding	Section 404.9(B)
SECTION E3902	**RECEPTACLES**	
E3902.1	Rating and type	Section 406.2(A), (B)
E3902.1.1	Single receptacle	Section 210.21(B)(1)
E3902.1.2	Two or more receptacles	Section 210.21(B)(3)
Table E3902.1.2	Receptacle ratings for various size multioutlet circuits	Section 210.21
E3902.2	Grounding type	Section 406.3(A)
E3902.3	CO/ALR receptacles	Section 406.2(C)
E3902.4	Faceplates	Section 406.5(B)
E3902.5	Position of receptacle faces	Section 406.4(D)
E3902.6	Receptacles mounted in boxes	Section 406.4, 406.4(A), 406.5
E3902.7	Receptacles mounted on covers	Section 406.4(C)
E3902.8	Damp locations	Section 406.8(A)
E3902.9	Wet locations	Section 406.8(B)
E3902.10	Bathtub and shower space	Section 406.8(C)
E3902.11	Flush mounting with faceplate	Section 406.8(E)
E3902.12	Outdoor installation	Section 406.8

International Residential Code		National Electrical Code
SECTION E3903	**FIXTURES**	
E3903.1	Energized parts	Section 410.3
E3903.2	Fixtures near combustible material	Section 410.5
E3903.3	Exposed conductive parts	Section 410.18(A)
E3903.4	Screw-shell type	Section 410.47
E3903.5	Recessed incandescent fixtures	Section 410.65(C)
E3903.6	Thermal protection	Section 410.73(E)
E3903.7	High-intensity discharge fixtures	Section 410.73(F)
E3903.8	Wet or damp locations	Section 410.4(A)
E3903.9	Lampholders in wet or damp locations	Section 410.49
E3903.10	Pendants	Section 410.4(D)
E3903.11	Fixtures in clothes closets	Section 410.8(A)
Figure E3903.11	Closet storage space	Figure 410.8
E3903.12	Fixture wiring—general	Section 410.22
E3903.12.1	Polarization of fixtures	Section 410.23
E3903.12.2	Fixtures as raceways	Section 410.31
SECTION E3904	**FIXTURE INSTALLATION**	
E3904.1	Outlet box covers	Section 410.12
E3904.2	Combustible material at outlet boxes	Section 410.13
E3904.3	Access	Section 410.16(B)
E3904.4	Supports	Section 410.15(A)
E3904.5	Means of support	Section 410.16, 314.27(B), 410.15(A)
E3904.6	Exposed ballasts	Section 410.76(A)
E3904.7	Combustible low-density cellulose fiberboard	Section 410.76(B)
E3904.8	Recessed fixture clearance	Section 410.66(A)(1), (A)(2)
E3904.9	Recessed fixture installation	Section 410.66(B)
SECTION E3905	**TRACK LIGHTING**	
E3905.1	Installation	Section 410.101(A)
E3905.2	Connected load	Section 410.101(B)
E3905.3	Locations not permitted	Section 410.101(C)
E3905.4	Support	Section 410.101(D)
E3905.5	Fastening	Section 410.104
E3905.6	Grounding	Section 410.105(B)
CHAPTER 40	**APPLIANCE INSTALLATION**	
SECTION E4001	**GENERAL**	
E4001.1	Scope	Sections 422.1 and 424.1
E4001.2	Installation	Sections 110.3(B) and 422.17
E4001.3	Flexible cords	Section 422.16
Table E4001.3	Flexible cord length	Sections 422.16(B)(1) and (2)
E4001.4	Overcurrent protection	Section 422.11
E4001.4.1	Single motor–operated appliance	Section 422.11(E)
E4001.5	Disconnecting means	Sections 422.30, 422.35 and 424.19
Table E4001.5	Disconnecting means	Sections 422.31(A), (B), 422.35, 424.19, 424.20, and 440.14
E4001.6	Ceiling fans	Section 422.18
E4001.7	Snow–melting and deicing equipment protection	Section 210.8(a)(3) Exception and 426.28
CHAPTER 41	**SWIMMING POOLS**	
SECTION E4101	**GENERAL**	
E4101.1	Scope	Section 680.1
E4101.2	Definitions	Section 680.2

International Residential Code		National Electrical Code
SECTION E4102	**WIRING METHODS FOR POOLS, SPAS, HOT TUBS AND HYDROMASSAGE BATHTUBS**	
E4102.1	General	Sections 608.7, 680.23(B), 680.21(C), 680.23(F), 680.21(A), 680.40, 680.42 and 680.43
E4102.2	Flexible cords	Sections 680.22(B)(5), 680.7(A), (B), 680.42 and 680.21(F)(1)
SECTION E4103	**EQUIPMENT LOCATION AND CLEARANCES**	
E4103.1	Receptacle outlets	Section 680.22(A)(6)
E4103.1.1	Location	Section 680.22(A)(1)
E4103.1.2	Where required	Section 680.22(A)(3)
E4103.1.3	GFCI protection	Section 680.22(A)(5)
E4103.1.4	Indoor locations	Section 680.43(A) and (A)(1)
E4103.1.5	Indoor GFCI protection	Sections 680.43(A)(2), 680.71
E4103.2	Switching devices	Sections 680.22(C), 680.43(C), 680.72
E4103.3	Disconnecting means	Section 680.12
E4103.4	Luminaires and ceiling fans	Section 680.22(B)
E4103.4.1	Outdoor location	Section 680.22(B)
E4103.4.2	Indoor location	Section 680.22(B)(2)
E4103.4.3	Existing lighting outlets and luminaires	680.22(B)(3)
E4103.4.4	Indoor spas and hot tubs	Section 680.43(B)
E4103.4.5	GFCI protection	Sections 680.22(B)(4)
E4103.5	Overhead conductor clearances	Section 680.8
Table E4103.5	Overhead conductor clearances	Table 680.8
Figure E4103.5	Overhead conductor clearances	Figure 680.8
E4103.6	Underground wiring	Section 680.10
Table E4103.6	Minimum burial depths	Section 680.10
SECTION E4104	**BONDING**	
E4104.1	Bonded Parts	Sections 680.26(B), 680.42(B) and 680.43(D)(4)
E4104.2	Parts not required to be bonded	Sections 680.42(B) and 680.43(D)(4) Exception
E4104.3	Methods of bonding	Sections 680.26(B)(4), 680.26(C), 680.42(B) and 680.43(E)
SECTION E4105	**GROUNDING**	
E4105.2	Luminaires	Section 680.23(F)(1), (2)
E4105.3	Nonmetallic conduit	Section 680.23(B)(1), 680.23(B)(2)(b)
E4105.4	Flexible cords	Section 680.23(B)(3)
E4105.5	Motors	Section 680.21(A)(1)
E4105.6	Panelboards	Section 680.25(B), (B)(1)
E4105.7	Cord-connected equipment	Section 680.7(E)
E4105.8	Other equipment	Section 250.1
SECTION E4106	**EQUIPMENT INSTALLATION**	
E4106.1	Transformers	Section 680.23(A)(2)
E4106.2	Ground-fault circuit-interrupters	Section 680.5
E4106.3	Wiring on load side of ground-fault circuit-interrupters and transformers	Section 680.23(F)(3)
E4106.4	Underwater luminaires	Section 680.23(A)(1), (A)(3), (A)(7)
E4106.4.1	Maximum voltage	Section 680.23(A)(4)
E4106.4.2	Luminaire location	Section 680.23(A)(5), (A)(6)
E4106.5	Wet-niche luminaire	Section 680.23(B)(1), (B)(2), 680.23(B)(3), (B)(4), (B)(5)
E4106.6	Dry-niche luminaire	Section 680.23(C)(1), (C)(2)
E4106.7	No-niche luminaire	Section 680.23(D)
E4106.8	Junction boxes and enclosures for transformers or ground-fault circuit interrupters	Section 680.24(A)–(E)
E4106.8.1	Junction boxes	Section 680.24(A)
E4106.8.2	Other enclosures	Section 680.24(B)
E4106.8.3	Protection of junction boxes and enclosures	Section 680.24(C)
E4106.8.4	Grounding terminals	Section 680.24(D)
E4106.8.5	Strain relief	Section 680.24(E)
E4106.9	Underwater audio equipment	Section 680.27(A)
E4106.9.1	Speakers	Section 680.27(A)(1)
E4106.9.2	Wiring methods	Section 680.27(A)(2)

International Residential Code		National Electrical Code
E4106.9.3	Forming shell and metal screen	Section 680.27(A)(3)
E4106.10	Electrically operated pool covers	Section 680.27(B)
E4106.11	Electric pool water heaters	Section 680.9
E4106.12	Pool area heating	Section 680.27(C)
E4106.12.1	Unit heaters	Section 680.27(C)(1)
E4106.12.2	Permanently wired radiant heaters	Section 680.27(C)(2)
E4106.12.3	Radiant heating cables prohibited	Section 680.27(C)(3)
E4106.13	Double insulated pool pumps	Section 680.21
SECTION E4107	**STORABLE SWIMMING POOLS**	
E4107.1	Pumps	Section 680.31
E4107.2	Ground–fault circuit-interrupters required	Section 680.32
E4107.3	Luminaires	Section 680.33
E4107.3.1	15 volts or less	680.33(A)
E4107.3.2	Not over 150 volts	680.33(B)
SECTION E4108	**SPAS AND HOT TUBS**	
E4108.1	Ground-fault circuit-interrupters	Section 680.44
E4108.2	Electric water heaters	Section 680.9
E4108.3	Underwater audio equipment	Section 680.43(G)
SECTION E4109	**HYDROMASSAGE BATHTUBS**	
E4109.1	Ground-fault circuit-interrupters	Section 680.71
E4109.2	Other electric equipment	Section 680.72
E4109.3	Accessibility	Section 680.73
E4109.4	Bonding	Section 680.74
CHAPTER 42	**CLASS 2 REMOTE CONTROL, SIGNALING, AND POWER-LIMITED CIRCUITS**	
SECTION E4201	**GENERAL**	
E4201.1	Scope	Section 725.1
E4201.2	Definitions	Article 100 and Section 725.2
SECTION E4202	**POWER SOURCES**	
E4202.1	Power sources for Class 2 circuits	Section 725.41(A)
E4202.2	Interconnection of power sources	Section 725.41(B)
SECTION E4203	**WIRING METHODS**	
E4203.1	Wiring methods on supply side of Class 2 power source	Section 725.51
E4203.2	Wiring methods and materials on load side of the Class 2 power source	Sections 725.52, 725.61, 725.71, Tablc 760.61
E4203.2.1	Type CL2P cables	Section 725.71(A)
E4203.2.2	Type CL2 cables	Section 725.71(C)
E4203.2.3	Type CL2X cables	Sections 725.61(E) and 725.71(D)
E4203.2.4	Marking	Section 725.71(H)
SECTION E4204	**INSTALLATION REQUIREMENTS**	
E4204.1	Separation from other conductors	Section 725.55
E4204.2	Other applications	Section 725.55(J)
E4204.3	Class 2 circuits with communication circuits	Section 725.56(D)
E4204.4	Class 2 cables with other circuit cables	Section 725.56(E)
E4204.5	Installation of conductors and cables	Sections 725.5, 725.6, 725.58, 300.11(B)(2)

INDEX

A

B

C

D

E

F

G

M

T

Y